Spatiotemporal Random Fields

Spatiotemporal Random Fields

Theory and Applications

Second Edition

George Christakos

Department of Geography, San Diego State University, San Diego, California, USA;
Institute of Islands and Coastal Ecosystems, Ocean College, Zhejiang University, Zhoushan, Zhejiang, China

Elsevier
Radarweg 29, PO Box 211, 1000 AE Amsterdam, Netherlands
The Boulevard, Langford Lane, Kidlington, Oxford OX5 1GB, United Kingdom
50 Hampshire Street, 5th Floor, Cambridge, MA 02139, United States

Library of Congress Cataloging-in-Publication Data

A catalog record for this book is available from the Library of Congress

British Library Cataloguing-in-Publication Data

A catalogue record for this book is available from the British Library

ISBN: 978-0-12-803012-7

For Information on all Elsevier publications visit our website at https://www.elsevier.com/books-and-journals

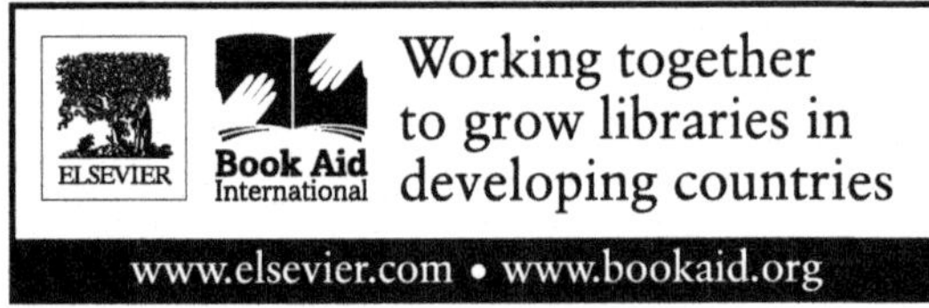

Publisher: Candice Janco
Acquisition Editor: Marisa LaFleur
Editorial Project Manager: Marisa LaFleur
Production Project Manager: Paul Prasad Chandramohan
Designer: Greg Harris

Typeset by TNQ Books and Journals

Dedicated

to

Yongmei

Contents

Preface

The study of real-world phenomena relies on theories of natural (physical, biological, etc.) sciences that generally involve mathematical models. These models are usually defined by a set of equations and supplemented by a set of logical reasoning rules for rigorously translating the quantitative analysis results into meaningful statements about the phenomenon of interest. Additionally, and very importantly, the real-world study of a phenomenon is faced with various sources of uncertainty, ontic, and epistemic (including phenomenal, technical, conceptual, and computational sources related to quantitative modeling, data selection, and processing). As a result, exact deterministic model solutions in terms of well-known analytic functions often turn out to be unrealistic and lack any visible means of meaningful interpretation.

In light of the above considerations, in most real-world applications the mathematical models that we currently use to describe the attributes of a natural phenomenon are stochastic in nature, i.e., these attributes and the associated boundary/initial conditions are represented by random fields with arguments in a composite space–time domain. In this domain, space represents the order of coexistence, and time represents the order of successive existence of the attribute. Randomness manifests itself as an ensemble of possible realizations regarding the attribute distribution, where the likelihood that each one of these possible realizations occurs is expressed by the corresponding probability law. Thus, spatiotemporal random model solutions are considerably more flexible and realistic than the deterministic single-valued solutions. Attribute distributions are well represented by theoretical probability laws, and this permits us to calculate various space–time properties of these distributions with reasonable accuracy.

The above considerations are the primary reasons for devoting this book to the spatiotemporal random field theory and its potential applications in natural sciences. In this context, for any such theory there is first the mathematical problem of analyzing, as far as possible, the stochastic model governing the relevant attributes together with the available data sets (hard or exact and soft or uncertain, in general), and of finding as realistic and complete a solution as possible to the problem of interest that maintains good contact with the real-world phenomenon in conditions of in situ uncertainty. Next comes the interpretation (mathematical and physical) of the conclusions thus obtained, and their utilization to make informative predictions. It should be pointed out that certain exact models and equations have played very important roles in the study of natural phenomena. It should also be noticed that because many models and equations describing real-world phenomena are necessarily complicated (multiparametered, highly nonlinear, and heterogeneous, whereas potentially critical features of the phenomena remain unspecified), it is very useful to understand what qualitative features these models and equations might possess, since they have been proven to offer an invaluable guide about the phenomenon of interest.

Certainly, there are several important issues related to the distinction between theory and implementation. Concerning the in situ implementation of random field theory and techniques, one should be aware that, although the fact is not always appreciated, the real-world complexities of the phenomena mentioned above also mean that approximate techniques used as part of implementation could run into hidden complications that have a tendency to distract attention from more useful issues. The situation may also be partially the fault of those of us working in this discipline, when we occasionally propose abstract theories for the sake of greater "generality." Yet this is not an excuse for the fact that, as real-world experience shows, in the vast majority of cases the "ineffectiveness" is not a feature of the theory or the modeling technique used but rather of the practitioners often attempting to use them in a "black-box" manner. In this framework, exact models and equations that can be compared with approximate or numerical results are very useful in checking the validity of approximation techniques used in an application.

In addition to the above reasons for devoting this book to the theory of spatiotemporal random fields, it should be noted that although much work has been done concerning the subject, it is often not generally known because of the plethora of disciplines, journals, and mathematical terminologies and notations in which it has appeared. It is hoped that one beneficial effect of the present effort will be to save the interested readers from spending their time rediscovering already known results. And I hope the present attempt to characterize the known results invariantly will help readers to identify any new findings that may emerge. Throughout the book, for the basic issue regarding fundamental concepts of probability, statistics, and random variables, I refer the reader to other texts, when necessary.

Naturally, I begin by introducing the basic notions of the space−time continuum (points, lags, metrics, and geometrical transformations), mathematical field and related functions, probability, uncertainty, and randomness (Chapter 1). Two chapters (Chapters 2 and 4) are devoted to the presentation of standard results of the ordinary spatiotemporal random field theory, including much of the terminology used later in the book. Among these two chapters I have interpolated one chapter (Chapter 3) on space−time metrics. Its position is due to the fact that the space−time metric properties can be used to elucidate the validity of certain random field issues introduced in the preceding two chapters. This chapter also discusses the classification of space−time metrics for scalar and vector random fields, and a physical law−based metric determination technique is outlined and applied in several cases. Intuitively, a natural attribute represented as a spatiotemporal random field is projected on the physical (real) domain. Yet, there may be constructed other domains on which an attribute could be projected. Such domains are the spectral domain, the reduced dimensionality, and the traveling domains, which provide equivalent representations of the attribute defined on it. And while one's intuition may be better adapted to the physical domain, in certain cases it may be more convenient to work in the alternative domains. So in Chapter 5 we discuss important concepts and methods associated with these alternative domains. Chapter 6 focuses on spatiotemporal random field geometry (continuity, differentiability, and integrability). This is one of the subjects that would warrant a book of its own and, thus, I had to be very selective in the choice and manner of the material presented. Because of its special physical and mathematical interest, the topic of homostationarity (space homogeneous/time stationary) was given a chapter of its own (Chapter 7). Similar reasons are valid for devoting Chapter 8 to isostationary (space isotropic/time stationary) random fields. In this chapter a large number of spatiotemporal variability functions (covariance, variogram, and structure functions of high order) are presented. Chapter 9 deals with multivariate and vectorial random fields varying in the space−time domain, including their main mathematical features and differences as regards their interpretation (mathematical and physical). In Chapter 10, I discuss a selected group of spatiotemporal random fields with special properties of particular interest to applications (this group includes the frozen random field and its variations, the plane-wave, the lognormal, the spherical, and the Lagrangian random fields). Chapter 11 focuses on techniques for constructing multivariate probability density functions that offer a complete characterization of the spatiotemporal random field in stochastic terms. Due to their fundamental role in the study of space–time heterogeneous random fields, an entire chapter (Chapter 12) is devoted to the theory of spatiotemporal random functionals. Indeed, the functional description of randomness naturally involves more complex mathematics, but it has its rewards on both theoretical and application grounds (e.g., many real-world phenomena and their measurements need to be expressed in terms of random functionals). Chapter 13 provides a rather detailed account of the theory of space–time heterogeneous (generalized) random fields that is useful in the case of natural attributes characterized by complex variations and patterns (varying trends, fluctuations of varying magnitude, coarse-grained measurements, etc.). Interestingly, since the first edition (1992) of the present book, only certain limited aspects of this theory have been thoroughly discussed in the literature. Chapter 14 emphasizes the importance of accounting for physical laws, scientific models, and empirical relationships in the development of a spatiotemporal random field theory. This valuable core knowledge concerning a phenomenon is usually quantitatively expressed in terms of stochastic partial differential equations, several of which are reviewed in this chapter. Admittedly, I only tangentially deal with the solution of these equations and relevant topics. The strongest reason for excluding the omitted topics is that each would fill another book (I do, of course, give references to the relevant literature). Chapter 15 presents a series of permissibility criteria for space–time covariance functions (ordinary and generalized) that are widely used in applied stochastics. Certain of these criteria are necessary and sufficient, whereas some others are only sufficient, but they have the advantage that they refer directly to the covariance function itself. Further, some important practical implications of permissibility in different kinds of applications are discussed. Chapter 16 presents a rather large number of techniques for constructing space–time covariance models, which can be used in a variety of scientific applications. Formal and substantive model-building techniques are examined, each of which has its own merits and limitations. There are many covariance model construction techniques in use and they could not all be discussed in full: my choice of what to present in detail and what to mention only as a reference simply reflects my personal taste and experience.

The book has benefited by the contributions in the field of my colleagues, collaborators, and students during the last few decades. The second edition of the book was written mainly during my leave of absence year at the Ocean College of Zhejiang University (China). I am grateful for the support of Zhejiang University and of the CNSF (Grant no. 41671399). I am also grateful to my colleagues at the Ocean College, particularly Professor Jiaping Wu, who did everything possible to create the right environment for writing such a book. Last but not least, this work would not have been completed without Yongmei's infinite patience during the long process of writing the book, which is why to Yongmei this book is dedicated.

George Christakos

CHAPTER

SPACE, TIME, SPACE–TIME, RANDOMNESS, AND PROBABILITY

I

CHAPTER OUTLINE

1. INTRODUCTION

Due to its importance in almost any scientific discipline, *random field* theory is an active area of ongoing research. Significant work has been done, indeed, in the theory of *spatial* random field, but much less so in the theory and applications of *spatiotemporal* random field, where many important topics still need to be studied and notions to be advanced. On the other hand, many practitioners argue that random field remains a tough theory to work with, due to the difficulty of the "nondeterministic" mathematics involved. This kind of mathematics is also known as *stochastics*, a term that generally refers to the mathematical representation of phenomena that vary jointly in space and time under conditions of in situ *uncertainty*. In a formal sense, deterministic mathematics can be viewed as a special case of stochastics under the limiting and rare conditions that the phenomenon under study is known with certainty. To phrase it in more words, stochastics deals with any topic covered by the deterministic theory of functions, and, in addition, the presence of uncertainty (technically, sometimes characterized as randomness) makes stochastics a much larger, considerably more complex and surely more challenging subject than the deterministic theory of functions. Historically, the development of stochastics can be traced back in the works of some of the world's greatest scientists, such as Maxwell (1860), Boltzmann (1868), Gibbs (1902), Einstein (1905), Langevin (1908), Wiener (1930), Heisenberg (1930) Khinchin (1934), Kolmogorov (1941), Chandrasekhar (1943), Lévy (1948), Ito (1954), Gel'fand (1955), von Neumann (1955), Yaglom (1962), and Bohr (1963), among many others.

Spatiotemporal Random Fields. http://dx.doi.org/10.1016/B978-0-12-803012-7.00001-5

It should be pointed out that random field *modeling* is at the heart of many theoretical advances in stochastics. It has led to the development of new mathematical concepts and techniques, and, also, it has raised several interesting theoretical questions worthy of investigation. *Computational* random field modeling, on the other hand, deals with computational and numerical aspects of the systematic implementation of random field theory in the study of complex real-world phenomena, which covers almost every scientific and engineering discipline. The term "computational" used here should not create any confusion with computational mathematics and statistics: while computational mathematics and statistics are concerned with numbers, computational random fields are concerned with physical quantities.[1]

In applied sciences, random field modeling deals with spatiotemporal *natural attributes*, that is, real-world attributes that develop simultaneously in space and time, and they are measurable or observable. These natural attributes occur in nearly all the areas of applied sciences, such as ecology and environment (e.g., concentrations of pollutants in environmental media—water/air/soil/biota), climate predictions and meteorology (e.g., variations of atmospheric temperature, density, moisture content, and velocity), hydrology (e.g., water vapor concentrations, soil moisture content, and precipitation data consisting of long time series at various locations in space), oil reservoir engineering (e.g., porosities, permeabilities, and fluid saturations during the production phase), environmental health (e.g., human exposure indicators and dose–effect associations), and epidemiology (e.g., breast cancer incidence, and Plague mortality). In all these cases, a central issue of random field modeling is factual accuracy in the informational statements that describe what was observed and experienced.

For sure, the application of random fields in the study of real-world phenomena is not an unconstrained theoretical exercise. It rather follows certain methodological criteria that involve the identification of the bounds of the specific application, the evaluation of the context in making sense of empirical data, a focus on probative evidence from diverse sources, an openness to inductive insights, and an in-depth analysis justified by the generation of interpretable results. Induction, interpretation, and abstraction are not competing objectives in this approach, but mutually reinforcing operations.

Random field modeling is concerned, although to varying extends, about both its internal and external validity. *Internal validity* relates to whether the findings or results of the random field modeling relate to and are caused by the phenomena under investigation, and not by other unaccounted for influences. On the other hand, *external validity* is assessed by the extent to which these findings or results can be generalized, and thus applied to other real-world situations. While internal validity is the primary concern of random field modeling, external validity is also a very important goal.

In this Chapter, I present the fundamentals regarding the conceptual and quantitative characterization of *space–time* (or *space/time*, or *spacetime*) within which random fields will be defined in subsequent chapters. Arguably, there are many issues surrounding the use and nature of the notion of "space–time" in scientific modeling, and some of them are even controversial. Yet, "space" and "time" are fundamental concepts that were invented by humans in their effort to describe Nature, but the map is not the territory. The formulation of space–time introduced in this chapter has the considerable merit of maintaining close contact between mathematical description and physical reality. Among the central goals of this formulation are to direct us toward a correct interpretation of space–time, and, to the extend possible, to help us avoid asking the wrong questions and focus on the insignificant issues.

---

[1]Surely, physical quantities are represented by numbers, but they also possess physical meaning and content (which are ignored by computational mathematics and statistics), and they are also associated with space–time arguments.

In this book, random quantities like the random variable, the random field, and the vector random field will be studied in both the physical (real) and the frequency domains. Notationally, a random variable is represented by lowercase Latin letters (x, y, etc.), a random field by uppercase Latin letters (X, Y, etc.), and a vector random field by uppercase bold Latin letters ($\boldsymbol{X}$, $\boldsymbol{Y}$, etc.). Lowercase Greek letters (χ, ψ, etc.) denote random variable or random field values (realizations), and lowercase bold Greek letters ($\boldsymbol{\chi}$, $\boldsymbol{\psi}$, etc.) denote vector random field values. The frequency domain counterparts of the above random quantities will be denoted by inserting the symbol $\sim$ above them (e.g., $\widetilde{X}$). The N denotes the set of natural numbers ($\{0, 1, 2, \ldots\}$ or $\{1, 2, \ldots\}$, depending on the context). The R^1 (or R) and C denote, respectively, the spaces of real and complex numbers. In the latter case, $\zeta = \chi + i\psi \in C$, where $i = \sqrt{-1}$, and the $\chi \equiv \mathrm{Re}(\zeta) \in R^1$ and $\psi \equiv \mathrm{Im}(\zeta) \in R^1$ denote, respectively, the real and imaginary parts of ζ. The complex conjugate of ζ is denoted by $\zeta^* = \chi - i\psi \in C$; also, $|\zeta| = |\zeta^*| = \sqrt{\chi^2 + \psi^2} \in R^1_{+,0}$ (i.e., the positive part of the real line including zero) is the modulus of ζ, ζ^*. The symbol T is sometimes used to represent the time domain as a subset of R^1 ($T \subseteq R^1$), which is in agreement with the physical irreversibility of most real-world phenomena. On the real line R^1, I use the convention for closed, open, and half-open intervals written as $[\chi, \psi]$, (χ, ψ), $[\chi, \psi)$, and $(\chi, \psi]$. Also, R^n is the Euclidean space of dimension $n \geq 1$.

Scalar, vector, and matrix notation will be used, noticing that scalars can be seen as tensors of rank zero, vectors have rank one, and matrices have rank two. A vector in R^n will be denoted as $\boldsymbol{s} = (s_1, \ldots, s_n)$ or $\boldsymbol{s} = \sum_{i=1}^n s_i \boldsymbol{\varepsilon}_i$, where $\boldsymbol{\varepsilon}_i$, $i = 1, \ldots, n$, are base vectors along the coordinate directions. The simplest choice of an (orthonormal) basis is, of course, the set of unit length vectors $\boldsymbol{\varepsilon}_i$, where the ith component is 1 and all others 0. For any two vectors $\boldsymbol{s}$ and $\boldsymbol{s}' = (s_1', \ldots, s_n')$, their scalar product is defined as $\boldsymbol{s} \cdot \boldsymbol{s}' = \sum_{i=1}^n s_i s_i'$. The length of the vector $\boldsymbol{s}$ is $|\boldsymbol{s}| = (\boldsymbol{s} \cdot \boldsymbol{s})^{\frac{1}{2}}$, and $|\boldsymbol{s} - \boldsymbol{s}'| = ((\boldsymbol{s} - \boldsymbol{s}') \cdot (\boldsymbol{s} - \boldsymbol{s}'))^{\frac{1}{2}}$ is the distance between $\boldsymbol{s}$ and $\boldsymbol{s}'$ in R^n. The space–time domain is denoted as R^{n+1}, or as $R^{n,1}$, if we want to explicitly distinguish space from time (for the same reason, we may also denote the space–time domain as the Cartesian product $R^n \times R^1$ or $R^n \times T$). I.e., in the case of space–time, the domain dimensionality increases to $n + 1$ by including the additional term s_0 or t representing time. Table 1.1 lists some commonly used symbols. Table 1.2 gives a list of special

Table 1.1 Commonly Used Symbols

Symbol	Mathematical Expression
Dirac delta	$\int ds\delta(\boldsymbol{s}) = 1$, $\delta(\boldsymbol{s} \neq \boldsymbol{0}) = 0$.
Kronecker delta	$\delta_{ij} = \begin{cases} 1 & \text{if } i = j, \\ 0 & \text{otherwise.} \end{cases}$
Levi-Chivita	$\varepsilon_{ijk} = \begin{cases} +1 & \text{if } (i,j,k) = (1,2,3) \text{ or } (2,3,1) \text{ or } (3,1,2), \\ -1 & \text{if } (i,j,k) = (3,2,1) \text{ or } (1,3,2) \text{ or } (2,1,3), \\ 0 & \text{if } i = j \text{ or } j = k \text{ or } k = i. \end{cases}$
Kronecker/Levi-Chivita relationship	$\varepsilon_{ijk}\varepsilon_{ipq} = \delta_{jp}\delta_{kq} - \delta_{jq}\delta_{kp}$

Table 1.2 Special Functions and Polynomials

Name	Notation
Gamma function.	$\Gamma(n)$
Bessel function of 1st kind and νth order.	J_ν
Modified Bessel function of the 1st kind.	I_n
Modified Bessel function of 2nd kind and νth order.	K_ν
Gegenbauer polynomials of degrees l, q.	$G_{l,q}$

functions and polynomials that will be used in the mathematical expression of several results throughout the book. For the readers' convenience, the mathematical definitions and basic properties of these special functions and polynomials are briefly reviewed in the book's Appendix.

Although many of the theoretical results presented in each chapter of the book (in the form of propositions and corollaries) will be repeatedly used in subsequent chapters, most proofs and other details will not, so they will not be discussed. Instead, examples illustrating the most important application-related aspects of these proofs will be presented when appropriate. A consistent effort is made throughout the book to keep a balance between abstract mathematical rigor and real-world science. In many cases, this means that by suppressing certain strict mathematical conditions, a more realistic representation of the observed phenomenon is achieved, and, also, a richness of new material is produced (this is, e.g., the case with space–time metric). The remaining of this chapter presents a review of the basic concepts and principles (regarding space, time, field, uncertainty, and probability) around which the *spatiotemporal random field* theory will be developed in the following chapters of the book.

2. SPACE–TIME CONTINUUM AND KOLMOGOROV PROBABILITY SPACE

This section presents an overview of the fundamental notions pertaining to the description of the *space–time* domain shared by the natural phenomena and the mathematical constructions that represent them. A general point to be stressed is that although on formal mathematics grounds many aspects of the analysis in the $n+1$-dimensional (space–time) domain are similar to those in the n-dimensional (spatial) domain, essential differences could exist on physical grounds. On the same grounds, crucial links may emerge between space and time, in which case the deconstruction of the concept of space–time into separate types of space and time may be an unnecessary conceptualization often leading to unsatisfactory conclusions.

Having said that, I start with the introduction of the different geometrical notions and arguments that play a central role in the spatiochronological specification of a phenomenon, an attribute or an event occurring in physical space–time, highlighting similarities and differences as they emerge.

2.1 SPACE–TIME ARGUMENTS: POINTS, LAGS, SEPARATIONS, AND METRICS

Generally, a *continuum* $\mathcal{E}$ of space–time points is associated with the physical space–time domain, and as such, it is assumed to be equipped with the corresponding metric that specifies "distances" between points in space–time. Operationally, a point in $\mathcal{E}$ can be defined in different ways, depending on the situation and the objectives of the analysis.

Definition 2.1

A *point* in $\mathcal{E}$ is denoted by a vector $\boldsymbol{p}$, which can be defined, either as an element of the $n+1$-dimensional domain,

$$\boldsymbol{p} = (s_1, \ldots, s_n, s_0) \in R^{n+1}, \tag{2.1a}$$

where, in notational terms, $s_1, \ldots, s_n$ denote space coordinates and time is considered jointly with space using the convention $s_0 = t$; or, as a pair of elements

$$\boldsymbol{p} = (\boldsymbol{s}, t) \in R^{n,1}, \tag{2.1b}$$

where $\boldsymbol{s} = (s_1, \ldots, s_n) \in R^n$ is the spatial location vector considered separately than the time instant $t \in R^1$. As noted earlier, instead of $R^{n,1}$ some authors use the Cartesian domain notation $R^n \times T$, i.e., the time axis $T\ (\subseteq R^1)$ is indicated separately from space R^n.

The point vector $\boldsymbol{p}$ plays a key role in physical sciences, not only because it uniquely specifies a point in the space–time domain of interest, but also because of the numerous functions of $\boldsymbol{p}$ (also known as *fields*, see Section 2.4) encountered in physical sciences. In Definition 2.1, either the entire real line may be considered as the time axis (referring, e.g., to situations in which the attribute of interest is reversible), or, due to physical requirements (i.e., the attribute is irreversible in the real-world), the domain $R^1_{+,\{0\}}$ (restricted to the positive part of the real line including zero) may be more appropriate to represent the time axis. To phrase it in more words, time may be seen as a coordinate (s_0) of the vector $\boldsymbol{p}$ in the R^{n+1} domain, as in Eq. (2.1a). Intuitively, a space–time point in R^{n+1} is considered as a fusion of a space point and a time point (e.g., the "here and now" exists as a unity not specifying the "here" and the "now" separately; similar is the interpretation of the "there and then"). Alternatively, time may enter the analysis as a distinct variable via the vector–scalar pair ($\boldsymbol{s}$, t) of the $R^{n,1}$ domain, as in Eq. (2.1b). Both approaches have their merits and uses, which will be discussed in various parts of the book. Yet, one should be aware of certain noticeable consequences of the two approaches in applications, as illustrated in the examples below.

Example 2.1

From a modeling viewpoint, in many cases we may treat space and time on essentially the same formal footing. By setting, e.g., $s_0 = iat$, the wave operator[2] can be written as

$$a^{-2}\frac{\partial^2}{\partial t^2} - \sum_{i=1}^{n} \frac{\partial^2}{\partial s_i^2}[\cdot] \mapsto -\sum_{i=1}^{n,0} \frac{\partial^2}{\partial s_i^2}[\cdot], \tag{2.2}$$

and we may view the solution of Poisson's equation in two dimensions and that of the wave equation in one dimension as analogous problems. The physical differences between space and time need to be carefully taken into consideration, though, since certain formal analogies between space and time may be deceptive. Indeed, the boundary conditions and the initial conditions may enter the problem in different ways, even though the governing equation may look symmetric (as in the wave operator above). Also, when we study the space–time variation of an air pollutant, the way pollution concentration changes across space (the distribution of spatial locations in which pollution exceeds a critical threshold) can be essentially different than concentration changes as a function of time (frequency of threshold exceedances at each location).

[2]This operator expresses certain fundamental physical laws like the electromagnetic laws.

Our discussion so far of the issues surrounding the essence and use of the term "space–time" lead to the first postulate.

Postulate 2.1

The vast majority of real-world data are interrelated both in space and time. This space–time connection is ingrained through physical relations and is welcomed in scientific modeling because it allows the representation of the space–time variation of a natural attribute from the limited number of data usually available.

This postulate is supported by reality, including the fact that space–time coupling is known to remove possibly unphysical divergences from the moments of the corresponding transport processes (e.g., Shlesinger et al., 1993). Unfortunately, these crucial facts are often ignored in purely technical treatments of space–time phenomena. Indeed, in statistical inferences (i.e., the inductive process of inferring from a limited sample valid conclusions about the underlying yet unknown population) the proper assessment of space–time correlations is often problematic, since most standard statistics tools of data analysis and processing have been developed based on the key premise of independent (physical relation-free) experiments.

Example 2.2

Working in the classical Newtonian conceptual framework, many practitioners find it tempting to completely separate the space component $\boldsymbol{s}$ from the time component t. Although convenient, indeed, this approach is often inadequate in real-world studies. A common example is the model-fitting procedure in which a valid covariance function of space is fitted at any fixed time (or a valid covariance function of time is fitted at any fixed distance). However, the resulting model is not necessarily a valid spatiotemporal covariance model, as has been discussed in Ma (2003b). Furthermore, in ocean studies involving underwater acoustics propagation, the travel time t is related to the horizontal (s_1, s_2) and vertical (s_3) coordinates in $R^{3,1}$ by means of

$$\left(s_1^2 + s_2^2\right)^{\frac{1}{2}} \cos\theta + s_3 \sin\theta - vt = 0, \tag{2.3}$$

where θ is the angle of a ray element in a refracting medium with sound speed v (Lurton, 2010). Hence, the acoustics of the phenomenon imply that space $\boldsymbol{s}$ and time t are closely linked through the physical relation of Eq. (2.5).

The perspective suggested by Eq. (2.1b) enables the introduction of alternative expressions of space–time point determination, while still accounting for the space–time connection posited in Postulate 2.1. These expressions are called *conditional*, and the reason for this will become obvious below. Consider a space–time domain represented by the nodes i ($i = 1, \ldots, m$) of a grid or lattice. An obvious expression of space–time at each grid node i is

$$(\boldsymbol{s}_i, t_i), \quad i = 1, \ldots m, \tag{2.4a}$$

denoting location $\boldsymbol{s}_i = (s_{i1}, \ldots, s_{in})$ and time t_i. Eq. (2.4a) assigns the same subscript to space and time, and, hence, it allows the consideration of a unique time instant at each node i. Sometimes, the physics of the situation may require that more than one time instants need to be considered at each node, in which case it is convenient to represent space–time at each grid node by

$$\left(\boldsymbol{s}_i, t_{j_i}\right), \quad i = 1, \ldots, m, \text{ and } j = 1, \ldots, m', \tag{2.4b}$$

where now j_i denotes the time instant considered given that we are at spatial position $\boldsymbol{s}_i$ (e.g., at a given node 9, $j_i = 1_9$ means time instant 1 at node 9). Similarly, when more than one nodes need to be considered during each instant, we can write,

$$\left(\boldsymbol{s}_{i_j}, t_j\right), \quad i = 1, \ldots, m, \text{ and } j = 1, \ldots, m', \tag{2.4c}$$

where now i_j denotes the spatial position given that we are at time instant t_j (e.g., at a given instant 9, $i_j = 1_9$ means spatial node 1 at instant 9). In a sense, Eqs. (2.4b) and (2.4c) introduce conditional expressions of the space–time point (i.e., time conditioned to space and space conditioned to time, respectively).

In addition to the physical space–time domain one may refer, equivalently, to the location-instant of a physical attribute occurring in the *frequency* space–time domain $\mathcal{F}$.

Definition 2.2

A *point* in $\mathcal{F}$ is denoted by a vector $\boldsymbol{w}$, which can be defined, either as an element of the $n+1$-dimensional domain,

$$\boldsymbol{w} = \left(w_1, \ldots, w_n, w_0\right) \in R^{n+1}; \tag{2.5a}$$

or, as a pair of elements of the (n, 1)-dimensional domain,

$$\boldsymbol{w} = (\boldsymbol{k}, \omega) \in R^{n,1}, \tag{2.5b}$$

where $\boldsymbol{k} = (k_1, \ldots, k_n)$ denotes spatial frequency (wavevector) and ω denotes temporal (wave scalar) frequency.

In view of Definition 2.2, as was the case with the representations of Eqs. (2.1a) and (2.1b) being formally equivalent, the same is valid for the representations of Eqs. (2.5a) and (2.5b), but they may be both affected by physical context too. The attributes of interest in a real-world study satisfy certain laws of change that may impose some intrinsic links between the location vector $\boldsymbol{s}$ and the time instant t or, equivalently, between the wavevector $\boldsymbol{k}$ and the wave frequency ω. Specifically, the matter of space–time links is discussed in Section 3 of Chapter VII, in which it is shown that physical laws representing wave phenomena impose a link between $\boldsymbol{k}$ and ω, which are not independent but closely linked by means of a physical *dispersion relation*.

Example 2.3

The acoustic pressure $X(\boldsymbol{s}, t)$ of a wave propagating in space–time is governed by the classical law

$$\left[\upsilon^{-2}\frac{\partial^2}{\partial t^2} - \nabla^2\right]X(\boldsymbol{s}, t) = 0, \tag{2.6a}$$

where υ is the local sound speed. The dispersion relation of this underwater acoustics law in $R^{3,1}$ that links $\boldsymbol{k}$ and ω is

$$\boldsymbol{k}^2 - \upsilon^{-2}\omega^2 = 0, \tag{2.6b}$$

where $\boldsymbol{k}^2 = \sum_{i=1}^{3} k_i^2$.[3] Similarly to Eq. (2.3), underwater acoustics laws representing wave phenomena impose a strong link between $\boldsymbol{k}$ and ω, which are not independent anymore, but they are closely connected physically by the dispersion relation of Eq. (2.6b). As a result of this dependence, the space–time frequency domain is restricted, thus affecting the form of the space–time covariance function (more details in Chapter VII).

A real (or complex) function of the space–time vector $\boldsymbol{p}$ (including the random fields to be considered, Chapter II) is termed continuous or discrete parameter according to whether its argument $\boldsymbol{p}$ takes continuous or discrete values. Unless stated otherwise, in the following continuous-parameter functions will be considered, which is most often the case in applied stochastics. Space–time variation analysis requires the simultaneous consideration of pairs of points defined by the vectors $\boldsymbol{p}$ and $\boldsymbol{p}'$, in which case the notion of space–time lag emerges naturally.

Definition 2.3

The *space–time lag*, $\Delta\boldsymbol{p} = \boldsymbol{p}' - \boldsymbol{p}$, between a pair of points $\boldsymbol{p}$ and $\boldsymbol{p}'$ is defined, either as the single vector

$$\Delta\boldsymbol{p} = (\Delta s_1, \ldots, \Delta s_n, \Delta s_0) \in R^{n+1}, \tag{2.7a}$$

where $\Delta s_i = s_i' - s_i$ $(i = 1, \ldots, n, 0)$ with the convention that the vector components with $i = 1, \ldots, n$ denote space lags and the component with $i = 0$ denotes time separation; or, as a pair of vector–scalar components

$$\Delta\boldsymbol{p} = (\boldsymbol{h}, \tau) \in R^{n,1}, \tag{2.7b}$$

where the spatial vector lag $\boldsymbol{h} = (h_1, \ldots, h_n) \in R^n$, with $h_i = s_i' - s_i$ $(i = 1, \ldots, n)$, and the time scalar lag, $\tau = t' - t$, are represented separately.[4]

A direct correspondence can be established between the representations of Eqs. (2.7a) and (2.7b) by observing that

$$\begin{aligned} \Delta s_i &= h_i & (i = 1, \ldots, n); \\ \Delta s_0 &= h_0 = \tau & (i = 0). \end{aligned}$$

For simplicity, in the following the notations Δs_i and h_i $(i = 1, \ldots, n, 0)$ will be used interchangeably. As in Definition 2.1, Eq. (2.7a) allows an intrinsic mixing of space and time lags, whereas Eq. (2.7b) considers them explicitly. As we will see in various parts of the book, this distinction may have considerable consequences in the study of certain aspects of random field theory and its physical applications.

Remark 2.1

At this point, it is worth noticing that the above space–time notation is sometimes termed the *Eulerian* space–time coordinate representation, where $\boldsymbol{s}$ and t are allowed to vary independently. Another

[3] Here, the dispersion relation results from the requirement that $X(\boldsymbol{s}, t) = e^{i(\boldsymbol{k} \cdot \boldsymbol{s} - \omega t)}$ is a solution to the physical law.
[4] The lag sign, i.e., the order of the space–time points $\boldsymbol{p}, \boldsymbol{p}'$ in the lag can affect mathematical manipulations and also can be consequential in certain applications (Chapter VII).

possibility is the *Lagrangian* space–time coordinate representation, in which case s and t do not vary independently, but, instead, s is considered a function of t, which is expressed as $s(t)$. The Eulerian notation has been used, e.g., in turbulence studies to represent correlations between physical attributes (pressure or flow velocity) where $\boldsymbol{s} = (s_1, \ldots, s_n)$ specifies a measurement location, $\boldsymbol{s} + \boldsymbol{h}$ denotes locations at varying distances $\boldsymbol{h}$ from $\boldsymbol{s}$, and τ is the time increment between times t and $t + \tau$. The Lagrangian setting has been used to determine correlations for properties of fluid particles passing through locations $\boldsymbol{s}$ at times t_0 (different for each particle), traveling along certain trajectories and arriving at locations $\boldsymbol{s} + \boldsymbol{h}(t_0 + \tau)$ at times $t_0 + \tau$. The displacement vector $\boldsymbol{h}(t_0 + \tau)$ is a random variable describing the locations at times $t + \tau$ of the particles in the ensemble averaging with respect to the initial locations $\boldsymbol{s}$ at times t_0. In this book, Eulerian space–time coordinates will be predominantly considered, whereas the Lagrangian notation will be used only in special cases.

Space–time points and lags viewed as vectors, standard vector operations can be applied on them. A partial list of such operations that are useful in the following is given in Table 2.1 (see also Appendix). The space–time point vector $\boldsymbol{p}$ can be expressed in terms of the unit vectors along the (orthogonal) coordinate directions (base vectors) $\boldsymbol{\varepsilon}_i$, $i = 1, \ldots, n$. The notation for integer powers involving space–time coordinates, $\boldsymbol{p}^{\boldsymbol{\lambda}}$, where $\boldsymbol{\lambda} = (\lambda_1, \ldots, \lambda_n, \lambda_0)$ is a multi-index of nonnegative

Table 2.1 Standard Vector Operations on Space–Time Points and Lags

$$\boldsymbol{p} = (s_1, \ldots, s_n, s_0) = \sum_{i=1}^{n,0} s_i \boldsymbol{\varepsilon}_i$$

$$\boldsymbol{p}^{\boldsymbol{\lambda}} = \prod_{i=1}^{n,0} s_i^{\lambda_i} = \prod_{i=1}^{n} s_i^{\lambda_i} t^{\lambda_0}, \quad \boldsymbol{\lambda} = (\lambda_1, \ldots, \lambda_n, \lambda_0)$$

$$(\boldsymbol{s} + \boldsymbol{h})^{\boldsymbol{\lambda}} = \sum_{k=0}^{\lambda} C_{\lambda!}^{k!} \boldsymbol{s}^{\boldsymbol{k}} \boldsymbol{h}^{\boldsymbol{\lambda}-\boldsymbol{k}}, \quad k = |\boldsymbol{k}|, \ \lambda = |\boldsymbol{\lambda}|, \ \boldsymbol{k} \leq \boldsymbol{\lambda}$$

$$(t + \tau)^{\lambda} = \sum_{m=0}^{\lambda} C_{\lambda!}^{m!} t^m \tau^{\lambda - m} \tag{2.8a–g}$$

$$|\boldsymbol{s}|^k = \left| \sum_{i=1}^{n} s_i^2 \right|^{\frac{k}{2}} \quad (k = \pm 1, \pm 2, \ldots)$$

$$\frac{\partial |\boldsymbol{s}|}{\partial s_i} = s_i |\boldsymbol{s}|^{-1} = \zeta_i$$

$$\frac{\partial \zeta_k}{\partial s_i} = (\delta_{ik} - \zeta_i \zeta_k) |\boldsymbol{s}|^{-1}$$

$$\frac{\partial \boldsymbol{p}}{\partial s_i} = \boldsymbol{\varepsilon}_i$$

$$\frac{\partial (\boldsymbol{p} \cdot \boldsymbol{p}')}{\partial s_i} = \frac{\partial \boldsymbol{p}}{\partial s_i} \cdot \boldsymbol{p}' + \boldsymbol{p} \cdot \frac{\partial \boldsymbol{p}'}{\partial s_i} \tag{2.9a–c}$$

$$\frac{\partial (\boldsymbol{p} \times \boldsymbol{p}')}{\partial s_i} = \frac{\partial \boldsymbol{p}}{\partial s_i} \times \boldsymbol{p}' + \boldsymbol{p} \times \frac{\partial \boldsymbol{p}'}{\partial s_i}$$

Continued

Table 2.1 Standard Vector Operations on Space–Time Points and Lags—cont'd

$$\Delta \boldsymbol{p} = \boldsymbol{p} - \boldsymbol{p}' = \sum_{i=1}^{n,0} (s_i - s'_i)\boldsymbol{\varepsilon}_i = \sum_{i=1}^{n,0} h_i \boldsymbol{\varepsilon}_i$$

$$\frac{\partial \Delta \boldsymbol{p}}{\partial s_i} = \boldsymbol{\varepsilon}_i \quad (i = 1, \ldots, n, 0) \tag{2.10a–c}$$

$$\frac{\partial \Delta \boldsymbol{p}}{\partial s'_i} = -\boldsymbol{\varepsilon}_i$$

$$\nabla \cdot \boldsymbol{p} = \nabla \cdot \Delta \boldsymbol{p} = n + 1$$

$$\nabla \times \boldsymbol{p} = \nabla \times \Delta \boldsymbol{p} = 0 \tag{2.11a–b}$$

integers such that $|\boldsymbol{\lambda}| = \sum_{i=1}^{n,0} \lambda_i$ and $\boldsymbol{\lambda}! = \Pi_i \lambda_i$ (Table 2.1), is essential in the study of random fields representing the space–time heterogeneous variation pattern of a physical attribute, and can be combined with other functions, such as, e.g., $\boldsymbol{p}^{\lambda} e^{\boldsymbol{\gamma} \cdot \boldsymbol{p}}$ (Chapter XIII).[5] Also, space–time vector differentiation operators in terms of $\boldsymbol{p}$ and $\Delta \boldsymbol{p}$ are useful in the study of derivative random fields, their covariance, variogram, and structure functions (Chapters VI and VII), as well as the physical laws they obey (usually expressed in terms of partial differential equations, PDE, Chapter XIV). In such and similar cases, the proper representation of space–time geometry (in terms of coordinates, distances, lags etc.) allows an efficient approach to random field modeling, which is the subject of this book. More specifically, if the random field represents a space–time heterogeneous attribute distribution (i.e., one exhibiting a space–nonhomogeneous/time–nonstationary, or, simply, space–time heterogeneous pattern),[6] the following space/time polynomial functions appear in the analysis (Chapter XIII)

$$\eta_{\nu/\mu}(\boldsymbol{s}, t) = \sum_{\zeta=0}^{\mu} \sum_{|\boldsymbol{\rho}|=0}^{\nu} c_{\boldsymbol{\rho},\zeta} s_1^{\rho_1} \ldots s_n^{\rho_n} t^{\zeta}, \tag{2.12a}$$

where ν and μ are integers representing, respectively, the spatial and temporal orders of the phenomenon heterogeneity, and $c_{\boldsymbol{\rho},\zeta}$ are known coefficients. The notation used in Eq. (2.12a) shows explicitly the degrees of all spatial and temporal monomials, but it is cumbersome, since it requires keeping track of all the power exponents (ρ_i, $i = 1, \ldots, n$, and ζ). An alternative notation, which turns out to be more efficient, is based on the space–time monomial functions $\boldsymbol{p}^{\lambda}$ introduced earlier. In this setting, a space–time polynomial of space–time degrees ν/μ can be expressed as (Christakos and Hristopulos, 1998)

$$\eta_{\nu/\mu}(\boldsymbol{s}, t) = \sum_{\alpha=1}^{N_n(\nu/\mu)} c_{\alpha} \eta_{\alpha}(\boldsymbol{s}, t), \tag{2.12b}$$

[5] In these applications, $\boldsymbol{\gamma} = (\boldsymbol{\alpha}, \beta)$, $\boldsymbol{\lambda} = (\boldsymbol{\rho}, \zeta)$, and $\boldsymbol{\gamma} \cdot \boldsymbol{p} = \sum_{i=1}^{n} \alpha_i s_i + \beta s_0$.

[6] Etymologically, the words homogeneous and heterogeneous come from Ancient Greek. The Greek words ὁμογενής (*homogenēs*) and ἑτερογενής (*heterogenēs*) come from ὁμός (*homos*, "same") and ἕτερος (*heteros*, "other, different") respectively, followed by γένος (*genos*, "kind"). The "-ous" is an adjectival suffix.

Table 2.2 Number of Monomials for n = 2, 3, and Different Combinations of Continuity Orders ν/μ

ν	μ	$N_2\ (\nu/\mu)$	$N_3\ (\nu/\mu)$
0	0	1	1
1	0	3	4
0	1	2	2
1	1	6	8
2	0	6	10
2	1	12	20
1	2	9	12
2	2	18	30

where $\eta_\alpha(\boldsymbol{s},t) = p^{(\boldsymbol{\rho},\zeta)} = \boldsymbol{s}^{\boldsymbol{\rho}} t^{\zeta} = \prod_{i=1}^{n} s_i^{\rho_i} t^{\zeta}$, and $N_n(\nu/\mu)$ is the number of monomials that depends on the spatial dimension n and the orders ν, μ. This notation is surely more compact than the one presented in Eq. (2.12a), and it involves only a single index α instead of the $n+1$ space–time indices $(\rho_1, ..., \rho_n, \zeta)$. The $N_n(\nu/\mu)$ for any n and ν/μ can be then determined by

$$N_n(\nu/\mu) = (\mu+1)\sum_{|\boldsymbol{\rho}|=0}^{\nu} \frac{(|\boldsymbol{\rho}|+n-1)!}{|\boldsymbol{\rho}|!(n-1)!}, \tag{2.12c}$$

i.e., for $|\boldsymbol{\rho}| \leq \nu$ number of monomials it is equal to the permutations of n integers from $\{0, 1, ..., \nu\}$ that add up to $|\boldsymbol{\rho}|$. Table 2.2 displays the monomials for the ν/μ combinations that are most commonly used in applications.

Example 2.4

For illustration, in this example, I chose to focus on the $R^{2,1}$ domain, where (s_1, s_2) denote the Cartesian coordinates associated with the locational vector $\boldsymbol{s}$, and t is the time scalar. Also, I assume that $\nu/\mu = 2/1$ are the space/time orders of attribute nonhomogeneity/nonstationarity. Under these conditions, the corresponding monomials (12 in number) are listed in Table 2.3.

In light of Definition 2.3, a relevant notion of space–time modeling is the space–time "distance" or metric. In applied sciences, metrics are mathematical expressions of the concept of distance in the space–time continuum $\mathcal{E}$ (an in depth treatment of the metric notion is presented in Chapter III). In general, a space–time metric would be defined as a function of the pair $(\boldsymbol{p},\boldsymbol{p}')$. In many cases, however, the value of the metric function is invariant for each given lag $\Delta\boldsymbol{p} = \boldsymbol{p}' - \boldsymbol{p}$, i.e., the metric is defined as a function of the space–time lag (which is more restrictive, yet satisfactory in many physical situations). These metrics are sometimes referred to as "lag-based metrics" and are denoted by the absolute value of $\Delta\boldsymbol{p}$, i.e., $|\Delta\boldsymbol{p}|$.[7] Using the correct metric plays a crucial role in the study of

[7] In a more conventional mathematical sense, $|\Delta\boldsymbol{p}|$ is a real-valued function on R^1 satisfying three basic conditions: (1) $|\Delta\boldsymbol{p}| \geq 0$ for all $\Delta\boldsymbol{p}$ ($|\Delta\boldsymbol{p}| = 0$ only if $\Delta\boldsymbol{p} = \boldsymbol{0}$); (2) $|a\Delta\boldsymbol{p}| = |a||\Delta\boldsymbol{p}|$ for all $\Delta\boldsymbol{p}$, $a \in R^1$; and (3) $|\Delta\boldsymbol{p}_1 + \Delta\boldsymbol{p}_2| \leq |\Delta\boldsymbol{p}_1| + |\Delta\boldsymbol{p}_2|$ for all $|\Delta\boldsymbol{p}_1|$, $|\Delta\boldsymbol{p}_2|$.

Table 2.3 Monomials in $R^{2,1}$ With $\nu/\mu = 2/1$

	ρ	ζ	Monomials
$g_1(s, t)$	0	0	$s_1^0 s_2^0 t^0$
$g_2(s, t)$	1	0	$s_1^1 s_2^0 t^0$
$g_3(s, t)$	1	0	$s_1^0 s_2^1 t^0$
$g_4(s, t)$	0	1	$s_1^0 s_2^0 t^1$
$g_5(s, t)$	1	1	$s_1^1 s_2^0 t^1$
$g_6(s, t)$	1	1	$s_1^0 s_2^1 t^1$
$g_7(s, t)$	2	0	$s_1^2 s_2^0 t^0$
$g_8(s, t)$	2	0	$s_1^0 s_2^2 t^0$
$g_9(s, t)$	2	0	$s_1^1 s_2^1 t^0$
$g_{10}(s, t)$	2	1	$s_1^1 s_2^1 t^1$
$g_{11}(s, t)$	2	1	$s_1^1 s_2^0 t^1$
$g_{12}(s, t)$	2	1	$s_1^0 s_2^2 t^1$

spatiotemporal phenomena, including the spatiotemporal variability analysis and the rigorous assessment of space–time dependencies and correlations. In applied sciences, the quantitative notions of space–time distances come from physical experience (laws, data, empirical support, etc.), can be made definite only by reference to physical experience, and are subject to change if a reconsideration of experience seems to warrant change. Hence, keeping the physical requirements of applied sciences in mind, which go beyond mere abstract mathematical considerations, the space–time metric may be defined as follows.

Definition 2.4

A *space–time metric* between pairs of points $\boldsymbol{p}$ and $\boldsymbol{p}'$ is defined, either in R^{n+1} as a *composite* metric on space–time lags

$$\mathcal{m}(\boldsymbol{p},\boldsymbol{p}') = |\Delta\boldsymbol{p}| = g(\Delta\boldsymbol{p}), \tag{2.13a}$$

i.e., a scalar $|\Delta\boldsymbol{p}|$[8] related to $\Delta\boldsymbol{p}$ via the function g (the shape of which depends on the phenomenon of interest); or, in $R^{n,1}$ as a pair of *separate* space and time metrics

$$\mathcal{m}(\boldsymbol{p},\boldsymbol{p}') = (|\Delta\boldsymbol{s}|, |\Delta t|) = (g_1(\boldsymbol{h}), g_2(\tau)), \tag{2.13b}$$

where $\Delta\boldsymbol{s} = \boldsymbol{h} = (h_1, \ldots, h_n)$ and $\Delta t = \tau = t' - t$ are as in Eq. (2.7b).[9]

The metric $\mathcal{m}(\boldsymbol{p},\boldsymbol{p}')$ in Eq. (2.13a) is expressed in terms of a single function g that accounts for the physics of the composite space–time, whereas in Eq. (2.13b) the $\mathcal{m}(\boldsymbol{p},\boldsymbol{p}')$ is expressed in terms of two different functions g_1 and g_2 that account separately for the specifics of space and time. Otherwise said, in the composite metric of Eq. (2.13a) space and time mix together, which is not the case of the separate metric of Eq. (2.13b), where space and time are treated as two distinct entities (in which case, τ may be viewed as distance toward time dimension). For future reference, the class of composite (single) space–time metrics, Eq. (2.13a), will be denoted by $\mathcal{M}_c$, and the class of separate

[8] Also, $|\Delta\boldsymbol{p}|$ is called the magnitude of $\Delta\boldsymbol{p}$.
[9] For convenience, in the following we will usually assume that $t' > t$, i.e., $|\tau| = \tau$.

(pair) space–time metrics, Eq. (2.13b), by $\mathcal{M}_s$. Class $\mathcal{M}_s$ turns out to be useful in applications of practical interest in which we consider natural variations within a domain defined by the Cartesian space × time product. The generalization features of the $\mathcal{M}_c$ class may be not necessary in these applications. The readers are also reminded that in applied sciences the metrics $m(\boldsymbol{p},\boldsymbol{p}')$ are mathematical expressions that define substantively the notion of space–time distance in real-world continua rather than abstract, purely mathematical constructs (Chapter III studies the subject of space–time metrics in considerable detail; a discussion of technical and physical aspects of space–time geometry can be also found in Christakos, 2000).

Remark 2.2

A notational comment may be appropriate here. An alternative way to denote the separate space–time metric of Eq. (2.13b), which is sometimes favored in applied stochastics, is as follows,

$$m((\boldsymbol{s},t),(\boldsymbol{s}',t')) = (d_{R^n}(\boldsymbol{s},\boldsymbol{s}'), d_{R^1}(t,t')),$$

where $d_{R^n}(\boldsymbol{s},\boldsymbol{s}')$ denotes spatial distance, say, the standard Euclidean distance between locations $\boldsymbol{s}$ and $\boldsymbol{s}'$, and $d_{R^1}(t,t')$ denotes the time separation between t and t'.

Definition 2.5

A standard formal way to define a metric is as the following inner or dot vector product (in R^{n+1})

$$|\Delta\boldsymbol{p}| = (\Delta\boldsymbol{p}\cdot\Delta\boldsymbol{p})^{\frac{1}{2}} = \left(\sum_{i,j=1}^{n,0}\varepsilon_{ij}h_i h_j\right)^{\frac{1}{2}}, \tag{2.14}$$

where $\varepsilon_{ij} = \boldsymbol{\varepsilon}_i\cdot\boldsymbol{\varepsilon}_j$ ($\boldsymbol{\varepsilon}_i$ are suitable base vectors). The composite space–time metric of Eq. (2.14) is a very general metric of the $\mathcal{M}_c$ class, which is useful in a large number of physical situations, where it is known as the *Riemann* metric, sometimes denoted as r_R.

The metric coefficients ε_{ij} in Eq. (2.14) are themselves space–time dependent, in general, and determined by the physics of the situation. The above metric includes two celebrated space–time metrics: The *Pythagorean* space–time metric

$$|\Delta\boldsymbol{p}| = \left(\sum_{i=1}^{n,0}\varepsilon_{ii}h_i^2\right)^{\frac{1}{2}}, \tag{2.15a}$$

denoted as r_p, and the *Minkowski* or *Einstein* space–time metric

$$|\Delta\boldsymbol{p}| = \left(c^2h_0^2 - \sum_{i=1}^{n}h_i^2\right)^{\frac{1}{2}}, \tag{2.15b}$$

sometimes denoted as r_{Mi} or r_{Ei}, where $\varepsilon_{00} = c^2$ (c is a physical constant), $\varepsilon_{ii} = -1$, and $\varepsilon_{ij} = 0$ ($i \neq j$). Moreover, using transformation $h_0 \mapsto h_0$, $h_j \mapsto ih_j$ ($j = 1,\ldots,n$; i is here the imaginary unit) on the metric of Eq. (2.15b) we get the special case of r_P with $\varepsilon_{00} = c^2$ and $\varepsilon_{ii} = -1$ ($i = 1,\ldots,n$). And, because imaginary numbers are involved in this transformation to a Euclidean domain some authors talk about the "pseudo-Euclidean" domain.

The space–time metrics of Eqs. (2.15a) and (2.15b) belong to the $\mathcal{M}_c$ class too. A metric of the quadratic form of Eq. (2.15a) may emerge even in rather simple space–time situations, as in the following example.

Example 2.5

Consider an attribute (e.g., a pollutant distributed by atmospheric processes) that covers distance $\Delta \boldsymbol{s} = |\Delta \boldsymbol{s}| = \left(\sum_{i=1}^{n} h_i^2\right)^{\frac{1}{2}}$ within time $h_0 = \tau$, so that $\Delta \boldsymbol{s} = \upsilon h_0$, where υ is the speed with which the attribute moves. We can then write $\Delta \boldsymbol{p} = (h_1, \ldots, h_n, h_0) = \left(h_1, \ldots, h_n, \frac{\Delta s}{\upsilon}\right)$, with unit base vectors ($\varepsilon_{ii} = 1$ and $\varepsilon_{ij} = 0$, $i \neq j$, in suitable units), and the corresponding metric will be

$$\Delta \boldsymbol{p}^2 = \Delta \boldsymbol{p} \cdot \Delta \boldsymbol{p} = \upsilon^{-2} \Delta s^2 + \sum_{i=1}^{n} h_i^2,$$

which is a quadratic form.[10]

Eq. (2.15b) is a physical metric used in applied sciences, e.g., geodesy, radiophysics, optics. The different signs of the time and space components denote that in this metric the coordinate time is not a dimension just like the space dimension. Also, it is worth noticing that in physics the symbol $\Delta \boldsymbol{p}^2$ itself is usually taken as a fundamental quantity, and not as the square of some other quantity $\Delta \boldsymbol{p}$ (Carroll, 2004).

Remark 2.3

One could comment that the numerical difference under the square root of Eq. (2.15b) might take a negative value. Yet, in many physical situations to which this metric refers to the numerical difference is always positive.[11] Furthermore, it is not physically possible that $h_0 = t' - t = 0$ and $h_i \neq 0$ ($i \neq 0$), since this will signify that an object moves in space while time stands still. For operational reasons, the $\Delta \boldsymbol{p}$ of Eq. (2.15b) is often squared so that we get an additive quantity along space–time.

The analysis above further emphasizes the point made earlier concerning the difference between strictly mathematical metrics versus physical space–time metrics. In addition, although Eqs. (2.14)–(2.15a–b) introduce symmetric metrics, i.e., $|\Delta \boldsymbol{p}| = |-\Delta \boldsymbol{p}|$, asymmetric metrics may be also considered in applied stochastics when physically justified. Some examples are given next.

Example 2.6

In many situations, the metric $\Delta \boldsymbol{p}$ is asymmetric because the reverse process, $\boldsymbol{p}'$ to $\boldsymbol{p}$, may be realistically impossible (e.g., the evolution and radioactive decay processes are irreversible). Moreover, an asymmetric metric is defined by

$$|\Delta \boldsymbol{p}| = \begin{cases} |\boldsymbol{p} - \boldsymbol{p}'| & \text{if } \boldsymbol{p} \geq \boldsymbol{p}', \\ 1 + 10^{|\boldsymbol{p}' - \boldsymbol{p}|} & \text{otherwise.} \end{cases} \tag{2.16}$$

Also, the *Manhattan* metric (also known as absolute or city-block metric; see Eq. (2.18b), and Chapter III) in the real-world context of one-way streets may not satisfy the symmetry condition $|\Delta \boldsymbol{p}| = |-\Delta \boldsymbol{p}|$, since a path from point $\boldsymbol{p}$ to point $\boldsymbol{p}'$ may include a different set of streets than a path from $\boldsymbol{p}'$ to $\boldsymbol{p}$.

---

[10]We notice that, due to physical considerations, in this metric, $\varepsilon_{ii} = 1$, whereas in that of Eq. (2.15b), $\varepsilon_{ii} = -1$. Also, in Eq. (2.15b) it is formally possible that $|\Delta \boldsymbol{p}| = 0$ (null case).

[11]For example, in many physical situations in geodynamics, radiophysics etc., the c denotes the speed of light and in the real-world nothing can travel faster than c.

Remark 2.4

The examples above stress the point that a distance cannot always be defined unambiguously in space–time, i.e., it may not be possible to decide on purely formal grounds, without additional information, the appropriate form of the g functions in Definition 2.4. Moreover, we notice that real-life metrics are combinations of physical distances and times that may not always satisfy all conditions of the term "metric" considered in a strict mathematical sense (e.g., the symmetry condition may be unnatural in real-world applications, where the best way to go from point $\boldsymbol{p}$ to point $\boldsymbol{p}'$, generally, may not coincide with the best way to go from point $\boldsymbol{p}'$ to point $\boldsymbol{p}$). In sum, the definition of an appropriate metric in the real-world does not depend solely on purely mathematical considerations, but on both the intrinsic links of space and time as well as on physical constraints (natural laws of change, empirical support, boundary, and initial conditions).

In view of the above considerations, an efficient approach favored in applied sciences is to consider some general expression of the space–time metric, and then test if it fits the physical requirements of the case of interest, i.e., if it is, indeed, a substantive metric on physical grounds. To phrase it in more words, I introduce the next postulate.

Postulate 2.2

The definition of an adequate space–time metric that expresses the intrinsic links of space and time should be made on the basis of physical laws and empirical support and not decided on purely formal grounds or merely on computational convenience.

According to the postulate, e.g., it may be impossible to decide using only mathematical considerations how the separation of two points $\boldsymbol{p}_1$ and $\boldsymbol{p}_2$ in a porous medium with a Euclidean spatial lag of 3 km and a time lag of 2 days compares with the separation of two points $\boldsymbol{p}_1'$ and $\boldsymbol{p}_2'$ with a 2-km spatial lag and a 3-day time lag. In fact, as we will see in Example 2.18, the decision regarding which of the two pairs of points has the larger separation needs to consider the underlying natural mechanisms. The following examples further illustrate the perspective suggested by Postulate 2.2 and the interesting issues it raises.

Example 2.7

Consider a space–time varying attribute $X(\boldsymbol{p})$ that satisfies the physical law (in $R^{n,1}$),

$$\left[\frac{\partial}{\partial t}+\boldsymbol{v}\cdot\nabla\right]X(\boldsymbol{s},t)=0, \tag{2.17a}$$

where $\boldsymbol{v}$ is a velocity vector. The solution is of the form $X(\boldsymbol{s}, t) = X(\boldsymbol{s} - \boldsymbol{v}t)$, in which case the associated space–time metric is given by

$$|\Delta\boldsymbol{p}|=|\boldsymbol{h}-\boldsymbol{v}\tau|=\left[\sum_{i=1}^{n}(h_i-v_i\tau)^2\right]^{\frac{1}{2}}=\left[\sum_{i=1}^{n}\left(h_i^2-2v_ih_i\tau+v_i^2\tau^2\right)\right]^{\frac{1}{2}}, \tag{2.17b}$$

($h_0=\tau$); i.e., the metric is of the general form (Eq. 2.14) with coefficients determined by the physical law (Eq. 2.17a) as $\varepsilon_{00}=\sum_{i=1}^{n}v_i^2$, $\varepsilon_{ii}=1$, and $\varepsilon_{0i}=-2v_i$. This is a metric of the $\mathcal{M}_c$ class.

Returning to the formal definitions of space–time metrics of the $\mathcal{M}_s$ class, we notice that the metric of Eq. (2.13b) allows the decomposition of the space–time metric into a spatial distance $|\boldsymbol{h}|$ and a time lag τ, which can be defined separately(always keeping in mind the link between space and time specified by the physics of the phenomenon). Space–time separate metrics of this form include the case in which the time interval is commonly defined as $\tau = t' - t > 0$, whereas the spatial distance may have a variety of forms. The following example presents two well-known metrics worth of our attention.

Example 2.8

Commonly used space–time metrics of the $\mathcal{M}_s$ class $(|\boldsymbol{h}|, \tau)$ are the

$$\left(\left(\sum_{i=1}^{n} \varepsilon_{ii} h_i^2\right)^{\frac{1}{2}}, \tau\right), \tag{2.18a}$$

where the spatial component is the *Euclidean* distance $|\boldsymbol{h}| = r_E$; and the[12]

$$\left(\sum_{i=1}^{n} \varepsilon_{ii} |h_i|, \tau\right), \tag{2.18b}$$

where the spatial component is the *Manhattan* (or *absolute*) distance $|\boldsymbol{h}| = r_M$. Both these metric forms are symmetric. Also, particularly useful is the metric

$$\left(2\rho \sin^{-1}\left(\frac{r_E}{2\rho}\right), \tau\right), \tag{2.18c}$$

where the spatial component is the *arc length* $|\boldsymbol{h}| = r_S$ determining the spatial distance between two points on the surface of the earth (viewed as a sphere), r_E is the (Euclidean) distance between the two points, and ρ is the earth radius.

Although the above examples also stress the point that the space–time metric expressed by Eq. (2.13a) is more general than that of Eq. (2.13b), yet, on occasion, the decomposition assumed in Eq. (2.13b) may have some interesting implications in mathematical analysis that need to be kept in mind.

Example 2.9

To orient the readers, it is reminded that many applications (e.g., space–time statistics, and geostatistics) often assume a space–time metric defined as a pair of positive real numbers, i.e.,

$$(|\boldsymbol{h}|, \tau) = \left(r_E|_{\varepsilon_{ii}=1}, \tau\right) = \left(\left(\sum_{i=1}^{n} h_i^2\right)^{\frac{1}{2}}, \tau\right), \tag{2.19a}$$

i.e., r_E is as in Eq. (2.18a) with $\varepsilon_{ii} = 1$ $(i = 1, \ldots, n)$. This separate space–time metric of the class $\mathcal{M}_s$ should be distinguished from the composite space–time metric of the class $\mathcal{M}_c$ defined as the single positive real number

$$|\Delta \boldsymbol{p}| = r_P|_{\varepsilon_{ii}=1} = \left(\sum_{i=1}^{n,0} h_i^2\right)^{\frac{1}{2}}, \tag{2.19b}$$

[12]The contextual introduction of these metrics is made in Chapter III.

i.e., r_P as in Eq. (2.15a) with $\varepsilon_{ii} = 1$ $(i = 1, \ldots, n, 0)$. Eq. (2.19b) is apparently different than the space–time metric of Eq. (2.15b) that introduces a special physical partition of space and time. This operational difference between the two space–time metrics has considerable consequences from a physical interpretation viewpoint (e.g., as we will see below, Eq. (2.15b) is invariant under the Lorentzian transformation, whereas Eq. (2.19b) is not, in general).

The Riemann space–time metric, Eq. (2.14), and the Pythagorean space–time metric, Eq. (2.15a), are often assumed in formal analysis in the R^{n+1} domain. The special case of the Pythagorean space–time metric, Eq. (2.19b), plays an essential role in certain space–time *geometry* operations (like random field continuity and differentiability, introduced in Chapter VI). Additional insight is gained if the difference between the metrics of Eqs. (2.19a) and (2.19b) is viewed in the context of spatiotemporal random field variability characterization. Specifically, as we shall see in Chapter VII, composite space–time isotropy implies that the covariance function of the natural attribute depends on the metric of Eq. (2.19b), whereas separate space isotropy/time stationarity (isostationarity) means that the covariance is a function of the metric of Eq. (2.19a). Note that these metrics can be extended in the context of the so-called geometrical anisotropy of space–time attribute variation (also, Chapter VII).

Remark 2.5

For future reference, it may be convenient to stress the point that we should distinguish between three kinds of separability, namely, metric separability (as introduced above), and covariance separability and sample path separability (that will be introduced in later chapters).

2.2 TRANSFORMATIONS AND INVARIANCE IN SPACE–TIME

In physical modeling, we often find it useful to apply some transformation of the original space–time coordinates.[13] In the context of group theory (Helgason, 1984), this leads to a so-called *transformation group* involving a space S on which a group $\mathcal{G}$ acts transitively, i.e., for $\boldsymbol{p}, \breve{\boldsymbol{p}} \in S$, there exists a $g \in \mathcal{G}$ such that $g\boldsymbol{p} = \breve{\boldsymbol{p}} \in S$. Not surprisingly, different hypotheses may be associated with different combinations of S and $\mathcal{G}$. Commonly used groups of transformations that are useful in applied stochastics are as follows.

Definition 2.6

The group of *translation* transformations $g = U_\delta$ for $\boldsymbol{p} \in S$ acts on $S = R^{n+1}$ or $R^{n,1}$ so that

$$\begin{aligned} &\boldsymbol{p} \mapsto U_{\delta p}\, \boldsymbol{p} = \boldsymbol{p} + \delta \boldsymbol{p} = \breve{\boldsymbol{p}}, \\ &(\boldsymbol{s}, t) \mapsto U_{\boldsymbol{h},\tau}(\boldsymbol{s}, t) = (\boldsymbol{s} + \boldsymbol{h}, t + \tau) = \left(\breve{\boldsymbol{s}}, \breve{t}\right), \end{aligned} \tag{2.20a–b}$$

i.e., the translation of a vector $\boldsymbol{p}$ by $\delta \boldsymbol{p}$ or a vector–scalar pair $(\boldsymbol{s}, t)$ by $(\boldsymbol{h}, \tau)$, respectively. The group of *orthogonal* transformations $g = \Lambda_\perp$ for $\boldsymbol{p} \in S$ acts on $S = R^{n+1}$ or $R^{n,1}$ so that

[13]In modeling we commonly consider transformations applied on different entities. Here we consider transformations of space–time coordinates (also, in Chapter V we study some interesting modeling implications of transformations of space–time attributes).

$$|\Lambda_{\perp}(\boldsymbol{p})| = |\boldsymbol{p}|,$$
$$|\Lambda_{\perp}(\boldsymbol{s},t)| = (|\boldsymbol{s}|,t), \tag{2.21a–b}$$

i.e., $\Lambda_{\perp}$ is a linear transformation that preserves the length of a vector $\boldsymbol{p}$ or $\boldsymbol{s}$, respectively.

In the stochastics context, the U_δ and $\Lambda_{\perp}$ transformations play a significant role in the determination of vector or multivariate random field isostationarity (Chapter IX), which is why I introduce some basic transformation results here. To start with, attractive features of the U_δ transformation include the possibility of (1) a succession of translations, i.e., $U_{\delta p}U_{\delta p'}\ldots\boldsymbol{p} = U_{\delta p+\delta p'+\ldots}\boldsymbol{p} = \boldsymbol{p}+\delta\boldsymbol{p}+\delta\boldsymbol{p}'+\ldots = \breve{\boldsymbol{p}}$, and (2) unequal translations with equal distances, i.e., if $U_{\delta p}\,\boldsymbol{p}_1 = \breve{\boldsymbol{p}}_1$ and $U_{\delta p'}\boldsymbol{p}_2 = \breve{\boldsymbol{p}}_2$, then $|\boldsymbol{p}_1-\boldsymbol{p}_2| = |\breve{\boldsymbol{p}}_1-\breve{\boldsymbol{p}}_2|$ if $\delta\boldsymbol{p} = \delta\boldsymbol{p}'$.

Example 2.10

A U_δ transformation of coordinates is defined by

$$U_{\delta p}\boldsymbol{p} = (s_1-h, s_2, s_3, s_0+h_0) = \breve{\boldsymbol{p}},$$
$$U_{h,\tau}(\boldsymbol{s},t) = (s_1-h, s_2, s_3, t+\tau) = \left(\breve{\boldsymbol{s}},\breve{t}\right),$$

where h and h_0 (or τ) are specified space and time lags, respectively.

Among the useful features of the $\Lambda_{\perp}$ transformation are that (1) the inverse of $\Lambda_{\perp}$ is orthogonal and (2) the composition of $\Lambda_{\perp}$ is orthogonal too. The most important cases of $\Lambda_{\perp}$ are *rotations*, $\mathcal{R}\boldsymbol{p} = \breve{\boldsymbol{p}}$, about a fixed point, a fixed axis etc., as well as *reflections*, $\mathcal{P}\boldsymbol{p} = \breve{\boldsymbol{p}}$, across any subspace $V \subset R^n$. In the transformation context, when appropriate, I use the notation $R^{n,1}$ to emphasize the rather obvious fact that, naturally, while U_δ transformations apply in both the R^n and the T domains, certain $\Lambda_{\perp}$ transformations (like $\mathcal{R}$ and $\mathcal{P}$) refer to the R^n domain only. Specifically, in $R^{2,1}$ and in $R^{3,1}$ the $\mathcal{R}$ can be specified using, respectively, polar and spherical coordinates. Combinations of $\mathcal{R}$ and $\mathcal{P}$ are also possible. Relevant to the $\mathcal{P}$ transformation is that of symmetry. *Symmetries*, i.e., transformations that preserve certain quantities, form a group in the sense that a composition of two symmetries is also a symmetry, and the inverse transformation to a symmetry is also a symmetry. There exist intimate relationships between random field geometry and space–time coordinate symmetry transformations. The natural conservation laws are closely linked to the notion of the space–time symmetry, whereas the relativity theory essentially elaborates the inherent symmetry of the space–time continuum.

Example 2.11

If the rotation is specified through an angle θ (say, $\boldsymbol{p}$ is rotated in $R^{2,1}$ about the origin by θ), it is usually denoted as $\mathcal{R}_\theta\boldsymbol{p} = \breve{\boldsymbol{p}}$, in which case the components of $\breve{\boldsymbol{p}}$ are functions of the components of $\boldsymbol{p}$ and the angle θ. A similar notational convention is valid for the $\mathcal{P}$ transformation. In $R^{2,1}$, the $\mathcal{R}_\theta$ transformation that rotates points $\boldsymbol{p}$ around the origin by θ (counterclockwise) is given by $\mathcal{R}_\theta\boldsymbol{p} = \breve{\boldsymbol{p}}$, where

$$\mathcal{R}_\theta = \begin{bmatrix} \cos\theta & -\sin\theta & 0 \\ \sin\theta & \cos\theta & 0 \\ 0 & 0 & 1 \end{bmatrix}$$

The $\mathcal{P}$ transformation that maps points $\boldsymbol{p}$ on to their reflected images about a line through the origin that makes an angle θ with the s_1-axis is given by $\mathcal{P}_\theta \boldsymbol{p} = \breve{\boldsymbol{p}}$, where

$$\mathcal{P}_\theta = \begin{bmatrix} \cos 2\theta & \sin 2\theta & 0 \\ \sin 2\theta & -\cos 2\theta & 0 \\ 0 & 0 & 1 \end{bmatrix}.$$

In higher dimensions, more than one angles may be involved, say θ, φ, and ψ, in which case we write $\mathcal{R}_{\theta\varphi\psi}\boldsymbol{p} = \breve{\boldsymbol{p}}$ or $\mathcal{P}_{\theta\varphi\psi}\boldsymbol{p} = \breve{\boldsymbol{p}}$.

Invariance is one of the most important notions in mathematical and physical sciences. It can be a conceptual tool that offers a deeper understanding of a physical phenomenon, or a computational tool that solves complex systems of equations representing the phenomenon. Generally, the invariance of an entity (e.g., a physical concept, a natural law, or an empirical model) under a certain transformation means that the entity does not change by the transformation. For present purposes, invariance can be described as follows.

Definition 2.7

If prior to a transformation $\mathcal{g}$, the attribute of interest had the value $\Phi(\boldsymbol{p})$ for each $\boldsymbol{p}$, and after the transformation $\mathcal{g}\boldsymbol{p}$ the entity has the value $\Phi(\mathcal{g}\boldsymbol{p})$, *invariance* means that the value of the attribute does not change after $\mathcal{g}$, i.e., $\Phi(\mathcal{g}\boldsymbol{p}) = \Phi(\boldsymbol{p})$ for all $\boldsymbol{p}$.[14]

A natural system is *translationally invariant* (in space and/or time) with respect to a certain attribute Φ if translation by certain characteristic spatial vectors and/or time lags, $\mathcal{g}\boldsymbol{p} = U_{\delta\boldsymbol{p}}\boldsymbol{p}$, does not change Φ. For example, periodic systems such as crystals, planetary orbits, and clocks belong to this category. Also, if Φ denotes energy, invariance means that the energy remains the same under the corresponding transformation (e.g., rotation/shifting of electric charge configurations).[15] It is often desirable that the invariance definition does not depend on the choice of scale (e.g., units) or on the choice of a coordinate system. Thus, *coordinate invariance* in physical sciences means that the entities of interest (e.g., laws of physics or theoretical models) should be independent of the coordinate system used. In the same context, if Φ denotes a space–time metric, an important metric property is formally defined as follows.

Definition 2.8

A space–time metric $|\Delta\boldsymbol{p}|$ is called *invariant* under a certain transformation $\mathcal{g}$ of coordinates, i.e.,

---

[14]Otherwise said, invariance implies that the composition of Φ and $\mathcal{g}$ coincides with Φ.

[15]On the basis of invariance, other useful properties may be derived. For example, as we shall see later, a system is called *homogeneous* with respect to a specific attribute if this attribute has a uniform value in space.

$$g\colon |\Delta\boldsymbol{p}| \mapsto g|\Delta\boldsymbol{p}|, \tag{2.22}$$

if it remains unchangeable when we move from one coordinate system to another using this transformation.

For illustration, the first example is concerned with the most commonly assumed translation (U_δ) invariance.

Example 2.12

Consider the general space–time metric definition $|\Delta\boldsymbol{p}| = |\boldsymbol{p}' - \boldsymbol{p}|$. If we apply a U_δ transformation to the coordinate system, the space–time metric $|\Delta\boldsymbol{p}|$ becomes $\left|\Delta\breve{\boldsymbol{p}}\right| = \left|\breve{\boldsymbol{p}}' - \breve{\boldsymbol{p}}\right| = U_{\delta p}|\Delta\boldsymbol{p}| = |(\boldsymbol{p}' + \delta\boldsymbol{p}) - (\boldsymbol{p} + \delta\boldsymbol{p})| = |\boldsymbol{p}' - \boldsymbol{p}| = |\Delta\boldsymbol{p}|$, i.e., the $|\Delta\boldsymbol{p}|$ remained unchanged by the U_δ transformation. The point reflection (or central inversion) of coordinates defined as

$$\boldsymbol{p} \mapsto \breve{\boldsymbol{p}} = \mathcal{R}_{\delta p}\,\boldsymbol{p} = -\boldsymbol{p} \tag{2.23}$$

preserves the metric $|\Delta\boldsymbol{p}|$, i.e., $\left|\Delta\breve{\boldsymbol{p}}\right| = \left|\breve{\boldsymbol{p}}' - \breve{\boldsymbol{p}}\right| = \left|\mathcal{R}_{\delta p}\,\boldsymbol{p}' - \mathcal{R}_{\delta p}\,\boldsymbol{p}\right| = |-\boldsymbol{p}' + \boldsymbol{p}| = |\Delta\boldsymbol{p}|$.

The following examples present two transformations of principal importance in physical sciences, namely, the Galilean and the Lorentz transformations.

Example 2.13

I start by considering the space–time metric of Eq. (2.19a) in the light of the *Galilean* transformation of coordinates in $R^{n,1}$ defined as[16]

$$(\boldsymbol{s}, t) \mapsto \left(\breve{\boldsymbol{s}}, \breve{t}\right) = (\mathcal{G}_{\delta s}\boldsymbol{s}, \mathcal{G}_{\delta t}t) = (\boldsymbol{s} - vt\boldsymbol{\varepsilon}_1, t) = (s_1 - vt, s_2, \ldots, s_n, t), \tag{2.24}$$

where v is a physical constant, and $\boldsymbol{\varepsilon}_1$ is a unit vector along the direction s_1. Let $\boldsymbol{s}$ and $\boldsymbol{s}'$ be the spatial locations of two points at any given time t, and $\breve{t} = t$, $\breve{t}' = t'$ be two different times associated with $\boldsymbol{s}$ and $\boldsymbol{s}'$. By applying $\mathcal{G}_\delta$ to $\boldsymbol{s}$ and $\boldsymbol{s}'$ we can define the spatial lag

$$\left|\breve{\boldsymbol{h}}\right| = \left|\breve{\boldsymbol{s}}' - \breve{\boldsymbol{s}}\right| = |\mathcal{G}_{\delta s}\boldsymbol{s}' - \mathcal{G}_{\delta s}\boldsymbol{s}| = |(\boldsymbol{s}' - vt\boldsymbol{\varepsilon}_1) - (\boldsymbol{s} - vt\boldsymbol{\varepsilon}_1)| = |\boldsymbol{s}' - \boldsymbol{s}| = |\boldsymbol{h}|.$$

Next, let t, t' be two different times associated with $\boldsymbol{s}$ and $\boldsymbol{s}'$. By applying $\mathcal{G}_\delta$ on t, t' we get $\breve{t} = t$, $\breve{t}' = t'$, and the time lag is given by $\breve{\tau} = \breve{t}' - \breve{t} = t' - t = \tau$. Hence, $\left(\left|\breve{\boldsymbol{h}}\right|, \breve{\tau}\right) = (|\boldsymbol{h}|, \tau)$ i.e., the space–time metric of Eq. (2.19a) remained unchanged by the $(\mathcal{G}_{\delta s}, \mathcal{G}_{\delta t})$ transformation. For comparison purposes, consider the *Lorentz* transformation of coordinates defined in R^{n+1} as

$$\boldsymbol{p} \mapsto \breve{\boldsymbol{p}} = \mathcal{L}_{\delta p}\boldsymbol{p} = \left(c^2\,\frac{s_0 - vc^{-2}s_1}{\beta}, -\frac{s_1 - vs_0}{\beta}, -s_2, \ldots, -s_n\right), \tag{2.25}$$

[16]Notice that the transformation is applied separately on the spatial and the time coordinates.

where υ, and c are constant coefficients, and $\beta = \left(1 - \upsilon^2 c^{-2}\right)^{\frac{1}{2}}$.[17] By applying $\mathcal{L}_\delta$ on the Minkowski metric of Eq. (2.15b), we find that

$$\begin{aligned}
\left|\Delta \breve{p}\right| &= \left|\mathcal{L}_{\delta p} p' - \mathcal{L}_{\delta p} p\right| \\
&= \left[c^2\left(\frac{s_0' - \upsilon c^{-2} s_1'}{\beta} - \frac{s_0 - \upsilon c^{-2} s_1}{\beta}\right)^2 - \left(\frac{s_1' - \upsilon s_0'}{\beta} - \frac{s_1 - \upsilon s_0}{\beta}\right)^2 - \sum_{i=2}^{n}\left(s_i' - s_i\right)^2\right]^{\frac{1}{2}} \\
&= \left(c^2 h_0^2 - \sum_{i=1}^{n} h_i^2\right)^{\frac{1}{2}} = |\Delta p| = \mathcal{L}_{\delta p}|\Delta p|,
\end{aligned}$$

i.e., the $|\Delta p|$ remained unchanged by $\mathcal{L}_\delta$. In fact, the Riemann metric of Eq. (2.14) is also invariant under the $\mathcal{L}_\delta$ transformation.

Next, I will discuss cases in which the metric is not *invariant* under the specified transformation.

Example 2.14

Straightforward algebraic manipulations show that the special Pythagorean metric of Eq. (2.19b) of the $\mathcal{M}_c$ class is not invariant under the $\mathcal{L}_\delta$ transformation (Exercise I.5). Thus, operationally, the physical space–time metric of Eq. (2.15b), although it also belongs to the $\mathcal{M}_c$ class, yet it differs from the metric of Eq. (2.19b) in that the former is invariant under the Loretzian transformation, whereas the latter is not, in general. On interpretational grounds, the invariance of Eq. (2.15b) means that although space and time are relative, this metric introduces a certain mixture of space and time that is absolute. Next, consider the space–time metric of Eq. (2.19b) under the *anisotropic dilation* transformation

$$\breve{p} = \mathcal{D}_\lambda p = (\lambda_1 s_1, \ldots, \lambda_n s_n, \lambda_0 s_0). \tag{2.26}$$

Under the $\mathcal{D}_\lambda$ transformation, the metric $|\Delta p|$ becomes

$$\left|\Delta \breve{p}\right| = \left|\breve{p}' - \breve{p}\right| = |\mathcal{D}_\lambda p' - \mathcal{D}_\lambda p| = \left(\sum_{i=1}^{n,0} \lambda_i^2 \left(s_i' - s_i\right)^2\right)^{\frac{1}{2}} = \left(\sum_{i=1}^{n,0} \lambda_i^2 h_i^2\right)^{\frac{1}{2}} \neq |\Delta p|,$$

i.e., the $|\Delta p|$ of Eq. (2.19b) changed by the $\mathcal{D}_\lambda$ transformation.

Additional insight regarding metric invariance is gained if the difference between the two metric groups discussed in Example 2.9 is viewed in the context of the so-called *geometric anisotropy.*

Example 2.15

The spatial geometric anisotropy transformation of coordinates, $\mathcal{A}_{\alpha,0}$ is properly defined as

$$\left(\breve{h}, \breve{\tau}\right) = \mathcal{A}_{\alpha,0}(s, \tau) = (\alpha \cdot h, \tau) = (\alpha_1 h_1 + \ldots + \alpha_n h_n, \tau) \tag{2.27}$$

[17] The $\mathcal{L}_\delta$ transformation, in a sense, may be interpreted as rotation in R^{n+1}.

($\alpha_i > 0,\ i = 1, \ldots, n$), in which case the separate space–time metric of the form of Eq. (2.19a) is given by

$$\left(\left|\breve{\boldsymbol{h}}\right|, \breve{\tau}\right) = \left(\left(\sum_{i=1}^{n} \alpha_i^2 h_i^2\right)^{\frac{1}{2}}, \tau\right). \tag{2.28}$$

On the other hand, the spatiotemporal geometric anisotropy transformation of coordinates, $\mathcal{A}_{\boldsymbol{\alpha},\alpha_0}$, is such that

$$\Delta\breve{\boldsymbol{p}} = \mathcal{A}_{\boldsymbol{\alpha},\alpha_0}\Delta\boldsymbol{p} = (\boldsymbol{\alpha}, \alpha_0)\cdot\Delta\boldsymbol{p} = (\alpha_1 h_1 + \ldots + \alpha_n h_n + \alpha_0 h_0) \tag{2.29}$$

($\alpha_0 > 0$), in which case for the corresponding composite space–time metric of the form of Eq. (2.19b) it is valid that

$$\left|\Delta\breve{\boldsymbol{p}}\right| = \left(\sum_{i=1}^{n,0} \alpha_i^2 h_i^2\right)^{\frac{1}{2}}. \tag{2.30}$$

Clearly, none of the metrics of Example 2.9 is invariant to the geometric transformations of the coordinates. On the other hand, as we shall see in Chapter VII, certain classes of attribute covariances (i.e., functions assessing space–time attribute variability) may be justifiably defined on physical grounds as functions of the metrics of Eqs. (2.28) and (2.30).

Galilean invariance is one of the most important properties of several fundamental physical equations, like the Navier–Stokes equations of fluid dynamics, because without this invariance it will not be possible to compare fluid experiments performed in different parts of the world (i.e., if these equations were not Galilean invariant, they could not describe fluid motion adequately). In line with the Definition 2.7, when applied to a physical law Galilean invariance implies that if prior to the Galilean transformation the physical equation had the form $\mathcal{D}[X(\boldsymbol{p})]$ for each $\boldsymbol{p}$, and after the transformation

$$\mathcal{G}_{\delta p,\delta X}: \boldsymbol{s}, t, X \mapsto \begin{cases} \boldsymbol{s}' = \mathcal{G}_{\delta s}\boldsymbol{s} = \boldsymbol{s} - \boldsymbol{v}t, \\ t' = \mathcal{G}_{\delta t}t = t, \\ X' = \mathcal{G}_{\delta X}X = X + \upsilon, \end{cases} \tag{2.31}$$

the equation has the form $\mathcal{D}\left[\mathcal{G}_{\delta p,\delta X}X(\boldsymbol{p})\right]$, invariance means that $\mathcal{D}\left[\mathcal{G}_{\delta p,\delta X}X(\boldsymbol{p})\right] = \mathcal{D}[X(\boldsymbol{p})]$.

Example 2.16

To demonstrate the significance of Eq. (2.31), I consider Burgers' equation in $R^{1,1}$,

$$\left[\frac{\partial}{\partial t} + X(s,t)\frac{\partial}{\partial s} - \nu\frac{\partial^2}{\partial s^2}\right]X(s,t) = 0, \tag{2.32}$$

where ν is a diffusion coefficient (viscosity). Burgers' equation is a PDE governing a random field $X(s, t)$ occurring in various applied stochastics areas (gas dynamics, fluid mechanics, acoustics, etc.). By applying the Galilean[18] transformation $\mathcal{G}_{\delta p,\delta X}$ of Eq. (2.31) in $R^{1,1}$, which is defined as

[18]Generally, Galilean invariance is used to denote that measured physics must be the same in any nonaccelerating frame of reference.

$$\mathcal{G}_{\delta s, \delta t, v}: s, t, X \mapsto \begin{cases} s' = s - vt, \\ t' = t, \\ X' = X + v, \end{cases}$$

Eq. (2.32) leads to

$$\left[\frac{\partial}{\partial t'} + (X(s',t') + v)\frac{\partial}{\partial s'} - \nu\frac{\partial^2}{\partial s'^2}\right](X(s',t') + v) = 0. \tag{2.33}$$

Taking into account that v is a constant, the following derivative expressions hold,

$$X(s',t') = X(s'(s,t),t'(s,t)) = X(s(s',t'),t(s',t')),\ \frac{\partial}{\partial t'}X(s',t') = -v\frac{\partial X(s,t)}{\partial s} + \frac{\partial X(s,t)}{\partial t},\ \frac{\partial}{\partial s'}X(s',t')$$
$$= \frac{\partial X(s,t)}{\partial s},\ \frac{\partial^2 X(s',t')}{\partial s'^2} = \frac{\partial^2 X(s,t)}{\partial s^2},\ \text{and}\ \frac{\partial^2 X(s',t')}{\partial t'^2} = \frac{\partial^2 X(s,t)}{\partial t^2},$$

so that Eq. (2.33) becomes Eq. (2.32). That is, Burgers' equation is invariant under the Galilean transformation.

For completeness of presentation of coordinate transformations, I should notice that other commonly used transformations are the standard *geometrical* coordinate transformations, where the standard Cartesian set of coordinates is transformed to a new set in terms of polar, cylindrical, or spherical coordinates. For illustration, I present here the *spherical* coordinate transformation, which is particularly useful in the evaluation of $n+1$-dimensional integrals. This coordinate transformation is defined as

$$(s_1, \ldots, s_i, \ldots, s_n, t) \mapsto \left(s \cos\theta_1, \ldots, s\cos\theta_i \prod_{j=1}^{i-1}\sin\theta_j, \ldots, s\prod_{j=1}^{n-1}\sin\theta_j, t\right), \tag{2.34}$$

where $i = 2, 3, \ldots, n-1$, $s = |\boldsymbol{s}| \geq 0$, $\theta_i \in [0, \pi]$ $(i = 1, \ldots, n-2)$, and $\theta_{n-1} \in [0, 2\pi]$. The detailed presentation of these standard coordinate transformations can be found in many textbooks (also, Appendix).

Remark 2.6

Before leaving this section, I would like to reiterate the following general points: (1) On physical grounds, in certain cases modeling may be more straightforward and rigorous in the R^{n+1} domain, whereas in some others the distinction of space and time introduced in the $R^{n,1}$ analysis may be interpretationally more appropriate. (2) On logical grounds, it is often more meaningful and even simpler to consider that a phenomenon occurs in a composite space–time continuum rather than to divide our experience artificially according to unrelated spatial and temporal aspects (especially, since it is not always clear how this distinction of the spatial and temporal aspects should be made).

2.3 SPACE–TIME INTERPRETATIONS

As a geometrical structure, space–time consists of a continuum $\mathcal{E}$ of space–time points $\boldsymbol{p}$ together with a metric $m(\boldsymbol{p},\boldsymbol{p}') = |\Delta\boldsymbol{p}|$ in the space–time composite domain R^{n+1}, or with a metric $m(\boldsymbol{p},\boldsymbol{p}') = (|\Delta\boldsymbol{s}|, \tau)$ in the space–time Cartesian domain $R^{n,1}$. I.e., the space–time structure is formally characterized by the triplet $(\mathcal{E}, \boldsymbol{p}, m)$. The vector $\Delta\boldsymbol{p}$ or the vector–scalar pair $(\Delta\boldsymbol{s}, \tau)$ would count the embeddings of natural attributes and their dynamic relations in space–time, whereas the form of the function g (expressed, e.g., by the values of the coefficients ε_{ij} at every $\boldsymbol{p}$ in the case of the general metric of Eq. (2.14)) enables us to determine the intrinsic geometrical properties of $\mathcal{E}$.

Space–time point representations, Definition 2.1 or 2.2, may be affected by physical context when the impacts of the laws of nature or of the composite space–time distances are considered.

Example 2.17

The traditional system of space–time coordinates $\boldsymbol{p} \in R^{n+1}$ used by continuum theories has the following properties:

(a) The space–time coordinates $\boldsymbol{p}$ of a continuum $\mathcal{E}$ cannot be established in an absolute way, but always relatively to some *origin* (there is no absolute scale in nature).

(b) The *axes* of the coordinate system must be defined (they may be rectilinear or, more generally, curvilinear). In continuum theories, the coordinate system is assumed to be differentiable.

(c) Length and time intervals are *relative* quantities, in the sense that numerical values are assigned to them only by comparison to some elementary intervals called *units.* This implies the existence of a scale relativity that has significant ramifications for physical laws.

(d) Space and time coordinate measurements are always made with some finite *resolution* corresponding to the minimal unit used to assign numbers to the length or time intervals. The minimal unit may depend on the precision of the instrument or on some physical limitation.[19]

Hence, a key aspect of space–time analysis is that sciences are based on measurements of the natural attributes represented as space–time fields. Their laws do not apply to natural phenomena by themselves, but rather to the numerical results of measurements obtained regarding these phenomena. Insofar as numerical measurement adequacy issues are concerned, two common strands emerge: first, whether the means of measurement are accurate, and, second, whether they are actually measuring what they are intended to measure.

The operational definition of a system of coordinates (such as in Definition 2.1) should include all the relevant information that is necessary to describe these numerical results and to relate them in terms of physical laws. The metric, in particular, is the central mathematical structure identified or associated with space–time. Accordingly, some interesting observations are worth making about the interpretation of a space–time metric. Ontologically, the metric may be interpreted:

(a) in a *substantivist* sense, as representing substantive space (i.e., it may be seen as the real representor of space–time) or

(b) in a *relationist* sense, as a structural quality of space–time (i.e., space–time does not claim existence on its own, but only in relation to the physical situation).

Both interpretations share two key features: the aforementioned mathematical structure $(\mathcal{E}, \boldsymbol{p}, m)$ equated with the nature of space–time is identical in both cases, and if the metric is removed, space–time is removed too. Otherwise said, the substantivist and relationist interpretations of space–time

---

[19]If, e.g., the resolution of a rod is 1 mm, it would have no physical meaning to express a measurement in μm, or to measure the distance from Earth to Venus with a precision of 1 cm; and, according to Heisenberg's principles, in microphysics the results of measurements depend on the resolution of the instrument.

embody the same formal metric, but from the standpoint of diverse ontological perspectives of the same underlying reality. In many real-world cases, what we need to know about space–time is its metric structure (i.e., we focus on the crucial role of the metric used in the real-world case) and not necessarily its interpretation. How mathematical space–time structures apply to reality, or are exemplified in the real-world, is identical to the issue of how mathematical structures are exemplified in the real-world. However, in most cases there is an important reciprocal relationship between the metric and the natural attribute (Chapter III). Indeed, the relevance of empirical evidence in assessing space–time structure should be emphasized, since the possible metric constructions are being judged not solely from a mathematical perspective, but from a scientific and empirical standpoint as well.

In view of the above considerations, the following point should be stressed concerning the space–time structure, $(\mathcal{E}, \boldsymbol{p}, \mathcal{m})$: the distance cannot always be defined unambiguously in space–time, i.e., as was argued earlier, it may not be possible to compare on purely mathematical grounds (1) the distance d_{12} between the pair of points $\boldsymbol{p}_1$ and $\boldsymbol{p}_2$ with a 3-km spatial lag and a 2-days time lag versus (2) the distance d'_{12} between the pair of points $\boldsymbol{p}'_1$ and $\boldsymbol{p}'_2$ with a 2-km spatial lag and a 3-day time lag. As a matter of fact to decide which one of the two cases, (1) or (2), has the larger separation distance, we may need to consider the outcome of a natural process.

Example 2.18

Consider an experiment during which a tracer is released inside a porous medium at points $\boldsymbol{p}_1$ and $\boldsymbol{p}'_1$. If the tracer is detected at point $\boldsymbol{p}_2$ but not at point $\boldsymbol{p}'_2$, then we can claim that $d_{12} < d'_{12}$ with respect to the particular experiment. This approach provides a way of ordering distances between space–time points, but without further refinements a specific quantitative notion of space–time distances cannot be obtained. Also, note that the distance defined above is not purely a geometric property of space–time, but it also depends on the medium's properties, and the particular phenomenon that we decide to use in the measurement. That is, measuring distance by means of fluid tracer dispersion can lead to very different results than measuring distance by means of electromagnetic propagation.

Experience and observation in the space–time domain $R^{n,1}$, in particular, ascribes certain fundamental properties to the concepts of space and time related to the concept of invariance (Definition 2.7). *Space homogeneity* means that one point of space is equivalent to any other point, in the sense that a natural attribute will occur the same way under identical conditions, independent of the place in which it occurs. Space homogeneity implies translational invariance of the properties of space in which case the origin of a coordinate system can be arbitrarily chosen. *Time homogeneity* means that one time instant is equivalent to any other instant, in the sense that an experimental setup will generate the same result, whether performed at one time instant or another. Time homogeneity implies that when we observe a physical phenomenon we can choose the zero of time at any time instant.[20] *Space isotropy* means that one direction in space is equivalent to any other direction. Rotational invariance of the properties of space is assumed, and a specific experimental setup will yield the same result whether the setup faces East or North. Space isotropy implies that we can arbitrarily orient the coordinate axes in space. The case of *time isotropy* is a rather peculiar one. Intuitively, the time direction is forward,

[20]This kind of space or time homogeneity should be distinguished from random field homogeneity or stationarity. As we will see in Chapter VII, random field realizations do not exhibit such convenient invariance properties. Yet, it is often true that ensemble moments (mean, covariance, variogram etc.) possess certain invariance properties, i.e., a random field is said to exhibit stochastic translation invariance if its ensemble moments are translationally invariant.

from present to future, i.e., time has only one direction. Yet, in an analogous manner to space isotropy, time isotropy would denote equivalence of time directions, forward and backward direction (from present to past). In this sense, the question whether time is isotropic or not means whether time is reversible or not. So, although the (Newtonian) laws of motion are time reversible, at least operationally, most natural processes in the real-world are time irreversible (described by the laws of evolution, the second law of thermodynamics, etc.).

In sum, mathematical space–time structures $(\mathcal{E}, \boldsymbol{p}, \mathcal{m})$ may vary along with the postulated scientific theories, a fact that explains why one may encounter a number of competing theories, and thus different mathematical structures, compatible with the same evidence. The application of the mathematical space–time structure in such cases requires coordination with the physical hypotheses and measurement/observation conventions (the choice of a hypothesis or a convention is normally based on an assessment of its overall viability and consistency with scientific method). So, we should not sanction otherwise undetectable attribute correlations and relations simply for the sake of using a cherished metric (e.g., the Euclidean one). In the same context, a transformation may be useful that captures invariance features of the space–time structure that we believe to be crucial in the physical context of the phenomenon of interest.

2.4 FUNCTIONS OF SPACE–TIME ARGUMENTS

The point vector $\boldsymbol{p}$ plays a key role in physical sciences, not only because it uniquely specifies a point in the space–time domain of interest, but also because of the numerous functions of $\boldsymbol{p}$ encountered in physical sciences. Keeping the above considerations in mind, it is appropriate to continue with a definition that brings us to the next level of stochastic modeling.

Definition 2.9

A *field* is generally a function of the space–time arguments (points, coordinates, lags, metrics) that, generally, presupposes a continuum of space–time points and represents values of the physical attribute at these points. A set of values at all points in the space–time continuum specifies a realization of the field.

A field may associate mathematical entities (such as a scalar, a vector, or a tensor) with specified space–time points. For the field to obtain a physical meaning these entities must represent values of some natural attribute (e.g., soil moisture, fluid velocity, cloud density, contaminant concentration, or lung cancer mortality rate). Space–time plots of such fields render their behavior in the entire space–time domain, in contrast with standard practice (e.g., varying space and keeping fixed time, or vice versa). This being the case, the composite space–time domain (with varying both space and time) can provide a deeper understanding of the field's space–time variation than simple space or time field plots allow. In this respect, of particular interest are certain major function spaces: the space C of all real and continuous functions in space–time with compact support; the space C^{∞} of all real, continuous, and infinitely differentiable functions with compact support; and the space C_0^{∞} of all real, continuous, and infinitely differentiable functions which, together with their derivatives of all orders, approach zero rapidly at infinity (e.g., faster than $|\boldsymbol{s}|^{-1}$ and t^{-1} as $|\boldsymbol{s}| \to \infty$ and $t \to \infty$).[21]

[21]Here, $C_0^{\infty} \supset C^{\infty}$, as all functions in C^{∞} vanish identically outside a finite support, whereas those in C_0^{∞} merely decrease rapidly at infinity.

In the particular case of a *random* field, which is the subject of this book, the uncertainty that characterizes many natural attributes represented by random fields manifests itself as an ensemble of possible field realizations. Accordingly, the random field model offers a general framework for analyzing data distributed in space–time under conditions of uncertainty. This modeling framework involves operations that enable us to investigate real-world problems in a mathematically effective way, to improve our insight into the relevant physical mechanisms and, thus, to enhance predictive capabilities (a detailed treatment of the general random field model begins in Chapter II).

Certain functions of the space–time arguments are of prime interest in this book. Functions in both the R^{n+1} and $R^{n,1}$ domains will be used interchangeably, depending on the situation. Specifically, I will consider functions of the following arguments:

(i) $\boldsymbol{p}$, (ii) $\boldsymbol{s}$, t, (iii) $\boldsymbol{p}, \boldsymbol{p}'$, (iv) $\Delta\boldsymbol{p}$, (v) $\boldsymbol{h}$, τ, (vi) $|\Delta\boldsymbol{p}|$, and (vii) $|\boldsymbol{h}|$, τ.

Functions of these arguments are mainly real-valued functions (although, in some cases, complex-valued functions are considered too). As we will see in following sections and chapters, functions with arguments of the forms (i) and (ii) may represent random fields, $X(\boldsymbol{p})$, $X(\boldsymbol{s}, t)$. On the other hand, functions of the forms (iii)–(vii) usually represent the various *spatiotemporal variability* models that measure the space–time dependence or correlation structure of the natural attribute represented by the random field, such as

(a) *covariance* functions, $c_X(\Delta\boldsymbol{p})$, $c_X(\boldsymbol{h}, \tau)$, $c_X(|\Delta\boldsymbol{p}|)$, $c_X(|\boldsymbol{h}|, \tau)$,
(b) *variogram* functions, $\gamma_X(\Delta\boldsymbol{p})$, $\gamma_X(\boldsymbol{h}, \tau)$, $\gamma_X(|\Delta\boldsymbol{p}|)$, $\gamma_X(|\boldsymbol{h}|, \tau)$, and
(c) *structure* functions $\xi_X(\Delta\boldsymbol{p})$, $\xi_X(\boldsymbol{h}, \tau)$, $\xi_X(|\Delta\boldsymbol{p}|)$, $\xi_X(|\boldsymbol{h}|, \tau)$.

Moreover, functions with arguments of the forms (i) and (ii) may be formally equivalent, i.e., we may generally refer to $X(\boldsymbol{p})$ as a function in the $n + 1$-dimensional domain, although in some cases it may be operationally more convenient to use $X(\boldsymbol{s}, t)$ as a separately time-dependent field in an n-dimensional spatial domain. Similarly, the functions $c_X(\Delta\boldsymbol{p})$ and $c_X(\boldsymbol{h}, \tau)$ may be operationally equivalent in many applied stochastics situations. However, when, under certain conditions, these become functions solely of the space–time metric, fundamental differences may exist between the mathematical forms and substantive interpretations of $c_X(|\Delta\boldsymbol{p}|)$ and $c_X(|\boldsymbol{h}|, \tau)$.

Example 2.19

The shape of the covariance function

$$c_X(|\Delta\boldsymbol{p}|) = c_X\left(\left(\boldsymbol{h}^2 + (\upsilon\tau)^2\right)^{\frac{1}{2}}\right)$$

is apparently different than that of the covariance function

$$c_X(|\boldsymbol{h}|, \tau) = c_X\left(\left(\boldsymbol{h}^2\right)^{\frac{1}{2}} + \upsilon\tau\right),$$

in the sense that the two covariances involve different functions of their space and time arguments. In the former case the argument is of the r_p form (belonging to the $\mathcal{M}_c$ class of metrics), whereas in the latter case the argument is of the $r_E + \upsilon\tau$ form (belonging to the $\mathcal{M}_s$ class).

In some cases, formulation of our analysis in terms of $c_X(|\boldsymbol{h}|, \tau)$ may be a more convenient way to represent space–time variability than formulation in terms of $c_X(|\Delta\boldsymbol{p}|)$, whereas in some other cases the opposite is valid. Similarly, there are situations in which we can define the spatial isotropy and the time stationarity of an attribute distribution in a more straightforward and intuitive manner in terms of $c_X(|\boldsymbol{h}|, \tau)$, i.e., by assuming that the covariance function depends on $\boldsymbol{s}$ and $\boldsymbol{s}'$ only through the spatial metric $|\boldsymbol{s}-\boldsymbol{s}'| = |\boldsymbol{h}|$ and separately on t and t' only through the time difference $\tau = |t-t'|$. Generally, the choice to work with $X(\boldsymbol{p})$ and $c_X(|\Delta\boldsymbol{p}|)$ versus $X(\boldsymbol{s}, t)$ and $c_X(|\boldsymbol{h}|, \tau)$ is often closely linked to the modeling choice of using the $|\Delta\boldsymbol{p}|$ versus the $(|\boldsymbol{h}|, \tau)$ metric.

By viewing the coordinates $\boldsymbol{p} \in R^{n+1}$ as auxiliary variables for the geometrical description of the space–time distribution of a natural attribute, it may be assumed that the attribute is projected on the physical $\boldsymbol{p}$-domain. Since this domain is an auxiliary element, there may be constructed other domains on which attributes could be projected. Such a domain is, e.g., the *spectral* (*Fourier* or *frequency*) $\boldsymbol{w}$-domain (Chapter V). The $\boldsymbol{p}$- and $\boldsymbol{w}$-domains provide mathematically equivalent representations of the natural attribute defined on them.

In the $R^{n,1}$ domain, on the other hand, it is assumed that the attribute is projected on the physical $(\boldsymbol{s}, t)$ domain, whereas other domains may be also constructed on which attributes could be projected, such as the *spectral* $(\boldsymbol{k}, \omega)$ domain, the *space transformation* $(\boldsymbol{s}\cdot\boldsymbol{\theta}, t)$ domain, or the *traveling transformation* $(\boldsymbol{s}\text{-}\boldsymbol{v}t)$ domain.[22] The $(\boldsymbol{s}, t)$, $(\boldsymbol{k}, \omega)$, $(\boldsymbol{s}\cdot\boldsymbol{\theta}, t)$, and $(\boldsymbol{s}\text{-}\boldsymbol{v}t)$ domains enable equivalent representations of the natural attribute defined on them. And, while a scientist's intuition may be better adopted to the physical $(\boldsymbol{s}, t)$ domain, in certain studies it could be more convenient to work in one of the other three domains.

As in ordinary mathematics, an important part of stochastic mathematics is concerned with the study of functional *geometry*, especially the study of those function properties (like differentiation and integration) that characterize random field regularity across space–time. The derivatives of a function with vector arguments, say, of the space–time covariance function $c_X(\Delta\boldsymbol{p})$, can be defined using some of the fundamental mathematical formulas of Table 2.4. Similar formulas are also valid in terms of the space–time structure function $\xi_X(\Delta\boldsymbol{p})$ and of the space–time variogram function $\gamma_X(\Delta\boldsymbol{p})$. These differentiation formulas involve certain geometric requirements imposed by the idea that space and time can be jointly represented by a specified vector argument. The differentiation of random fields and associated space–time variation functions will be studied more thoroughly, starting with Chapter VI.

Often the question emerges whether certain results that have been derived in a purely spatial domain R^n remain valid in the R^{n+1} or the $R^{n,1}$ domain. This is an important question, especially when these results concern spatial variation functions (covariance, structure, or variogram), which is why the extension of spatial variation results to spatiotemporal variation functions is often considered. I will deal with the matter in due detail in various parts of the book. At the moment, a general postulate will be introduced and some basic observations will be made concerning the matter.

Postulate 2.3

Many established results in the spatial domain can be formally extended to the space–time domain, i.e., their derivations can carry through as before, as long as the time argument does not change during the derivations.

[22]The unit vector $\boldsymbol{\theta}$ specifies the orientation of a hyperplane $H^{n-1} \subset R^n$ with dimension $n-1$, $\boldsymbol{s}\cdot\boldsymbol{\theta}$ is the inner product that defines its distance from the origin, and the velocity vector $\boldsymbol{v}$ denotes traveling direction (Chapter V).

Table 2.4 Fundamental Differentiation Formulas for $c_X(\Delta\boldsymbol{p})$, $\Delta\boldsymbol{p} = \boldsymbol{p} - \boldsymbol{p}'$, in $R^{n,1}$

$\nabla c_X = \sum_{i=1}^{n,0} \frac{\partial c_X}{\partial h_i}\boldsymbol{\varepsilon}_i$	$\nabla c_X \cdot \boldsymbol{p} = \sum_{i=1}^{n,0} \frac{\partial c_X}{\partial h_i} s_i$
$\frac{\partial^{\nu}}{\partial \boldsymbol{s}^{\nu}} c_X(\Delta\boldsymbol{p}) = \frac{\partial^{\nu} c_X(\Delta\boldsymbol{p})}{\partial \boldsymbol{h}^{\nu}}$	$\frac{\partial}{\partial h_i}\nabla c_X = \sum_{k=1}^{n,0} \frac{\partial^2 c_X}{\partial h_i \partial h_k}\boldsymbol{\varepsilon}_k$
$\frac{\partial^{\nu}}{\partial \boldsymbol{s}'^{\nu}} c_X(\Delta\boldsymbol{p}) = (-1)^{\nu} \frac{\partial^{\nu} c_X(\Delta\boldsymbol{p})}{\partial \boldsymbol{h}^{\nu}}$	$\frac{\partial}{\partial s_i}\nabla c_X = \frac{\partial}{\partial h_i}\nabla c_X$
$\boldsymbol{\nu} = (\nu_1, \ldots, \nu_n, \nu_0), \nu = \lvert\boldsymbol{\nu}\rvert = \sum_{i=1}^{n,0} \nu_i$	$\frac{\partial}{\partial s_i'}\nabla c_X = -\frac{\partial}{\partial h_i}\nabla c_X$
$\partial \boldsymbol{s}^{\nu} = \prod_{i=1}^{n,0} \partial s_i^{\nu_i}, \ \partial \boldsymbol{h}^{\nu} = \prod_{i=1}^{n,0} \partial h_i^{\nu_i}$	
$\nabla c_X \cdot d\boldsymbol{p} = \sum_{k=1}^{n,0} \frac{\partial c_X(\Delta\boldsymbol{p})}{\partial h_k} dp_k = dc_X$	$\nabla \times \nabla c_X = \nabla \cdot (\nabla \times \nabla c_X) = 0$

Specifically, Postulate 2.3 may imply that the derivation of the results involve only spatial operations (e.g., spatial derivatives) and, hence, they carry through as before if the time argument does not change (e.g., quantities that are also functions of time may be considered, but, as long as time is held constant, the derivations are unchanged). I continue with a few simple but instructive examples. The first of them involves covariance functions.

Example 2.20

Let us assume that the spatial covariance function $c_X(\boldsymbol{h})$ in R^n satisfies an equation of the form

$$\Phi_S[c_X(\boldsymbol{h})] = 0, \tag{2.35}$$

where Φ_S is a known function (e.g., an empirical model or a theoretical restriction associated with the phenomenon of interest). Under some general conditions, the direct extension of Eq. (2.35) in the space–time $R^{n,1}$ domain,

$$\Phi_{ST}[c_X(\boldsymbol{h}, \tau)] = 0 \tag{2.36}$$

may be formally correct, where Φ_{ST} is the derived function. One such condition is that the corresponding derivation of Eq. (2.35) does not involve time manipulations (e.g., in terms of time derivatives of the covariance), in which case Eq. (2.35) can be applied to time-dependent fields as well, by making the replacement $(\boldsymbol{h}) \mapsto (\boldsymbol{h}, \tau)$, thus leading to Eq. (2.36). Also, Eq. (2.36) holds in space–time domains where space and time are considered to vary independently, see Eq. (2.7b) or (2.13b), and, consequently, algebraic or differential operations can be performed separately. However, in some other domains space and time are linked to each other, see Eq. (2.7a) or (2.13a), in which case Eq. (2.36) may carry a relation with the space–time structure. This is valid, e.g., in a space–time domain with metrics of the $\mathcal{M}_c$ class, where space and time are explicitly linked.

The situation is somehow different when two (or more) fields are involved, as is illustrated in the example below.

Example 2.21

Assume that the spatial fields $X_1(\boldsymbol{s})$ and $X_2(\boldsymbol{s}')$ in R^n are related by means of an expression of the form

$$\Phi_S[X_1(\boldsymbol{s}), X_2(\boldsymbol{s}')] = 0, \tag{2.37}$$

where Φ_S is a known function (e.g., a physical law or a scientific model). Again, there are cases in which the extension from Eq. (2.37) to the space–time domain is formally correct, although the situation may be physically more involved. Indeed, as was the case with Eq. (2.35), if the derivation of Eq. (2.37) does not involve time manipulations, then Eq. (2.37) can be formally applied to time-dependent fields as well by making the replacement $(\boldsymbol{s}) \mapsto (\boldsymbol{s}, t)$ and $(\boldsymbol{s}') \mapsto (\boldsymbol{s}', t)$, i.e., in $R^{n,1}$ it holds that

$$\Phi_{ST}[X_1(\boldsymbol{s}, t), X_2(\boldsymbol{s}', t)] = 0, \tag{2.38}$$

where Φ_{ST} is the derived function. Because it involves $X_1(\boldsymbol{s}, t)$ and $X_2(\boldsymbol{s}', t)$ evaluated at the same time instant t, in some cases Eq. (2.38) may be termed the *time simultaneous* form of Eq. (2.37). Eq. (2.38) is formally correct and can lead to valid results regarding the fields involved. Yet, under certain circumstances, it may exhibit some conceptual and practical inconveniences. For instance, suppose that Eq. (2.38) is a physical law relating the source (cause) $X_1(\boldsymbol{s}, t)$ and the effect $X_2(\boldsymbol{s}', t)$. The time simultaneous connection between $X_1(\boldsymbol{s}, t)$ and $X_2(\boldsymbol{s}', t)$ may disagree with experimentally verified causality principles according to which the occurrence of the cause $X_1(\boldsymbol{s}, t)$ should precede in time the occurrence of the effect $X_2(\boldsymbol{s}', t)$. Nevertheless, this inconvenience may be avoided if the space–time domain is characterized by a meaningful relation of simultaneity.

The approach implied by Postulate 2.3 becomes more explicit when we are dealing with the special case of the so-called space–time separable function,

$$\phi(\boldsymbol{h}, \tau) = \phi_1(\boldsymbol{h})\phi_2(\tau), \tag{2.39}$$

where $\phi_1(\boldsymbol{h})$ is the function in the spatial domain, and $\phi_2(\tau)$ is the function in the temporal domain.

Example 2.22

Let us assume that the spatial domain covariance function satisfies the relationship

$$ac_{X(1)}(\boldsymbol{0}) - bc_{X(1)}(\boldsymbol{h}) = g(\boldsymbol{h}), \tag{2.40}$$

where a and b are known parameters, and g is a known function of $\boldsymbol{h}$. Assuming that the covariance is a separable function, Eq. (2.40) is formally extended in the space–time domain as

$$ac_X(\boldsymbol{0}, \tau) - bc_X(\boldsymbol{h}, \tau) = g_\tau(\boldsymbol{h}), \tag{2.41}$$

where $c_X(\boldsymbol{0}, \tau) = c_{X(1)}(\boldsymbol{0})c_{X(2)}(\tau)$, $c_X(\boldsymbol{h}, \tau) = c_{X(1)}(\boldsymbol{h})c_{X(2)}(\tau)$, $c_{X(2)}(\tau) \geq 0$, and $g_\tau(\boldsymbol{h}) = c_{X(2)}(\tau)g(\boldsymbol{h})$. Obviously, matters are particularly straightforward in the case of spatial differentiation. Generally, the derivative with respect to $c_X(\boldsymbol{h}, \tau)$ is equal to the derivative with respect to $c_{X(1)}(\boldsymbol{h})$ times $c_{X(2)}(\tau)$. Assume that in the spatial domain the covariance function $c_{X(1)}(\boldsymbol{h})$ satisfies the differential equation

$$\mathcal{L}_s c_{X(1)}(\boldsymbol{h}) = g(\boldsymbol{h}), \tag{2.42}$$

where g is a known function of $\boldsymbol{h}$, $\mathcal{L}_s[\cdot] = \sum\limits_{|\boldsymbol{\rho}|=\nu} \alpha_{\boldsymbol{\rho}} \frac{\partial^{\boldsymbol{\rho}}}{\partial \boldsymbol{h}^{\boldsymbol{\rho}}}[\cdot]$ is linear differential operator with known coefficients $\alpha_{\boldsymbol{\rho}}$, and $\boldsymbol{\rho}$ is the set of integers $(\rho_1, \ldots, \rho_n)$ such that $|\boldsymbol{\rho}| = \sum_{i=1}^{n} \rho_i = \nu$ $\left(\rho_i \text{ denotes the order of the partial derivative with respect to } h_i, \text{ and } \partial \boldsymbol{h}^{\boldsymbol{\rho}} = \prod_{i=1}^{n} \partial h_i^{\rho_i}\right)$. The extension of Eq. (2.42) in the space–time domain is straightforward, i.e., it gives

$$\mathcal{L}_s c_X(\boldsymbol{h}, \tau) = g_\tau(\boldsymbol{h}), \tag{2.43}$$

where, as before, $g_\tau(\boldsymbol{h}) = c_{X(2)}(\tau) g(\boldsymbol{h})$.

As we will see in the following chapters, several valid results in R^n concerning certain properties of a random field or a variation function (regularity, permissibility, etc.) can be properly extended to useful results in $R^{n,1}$ (e.g., Chapters VI, VII, and XV).

3. RANDOM VARIABLES IN SPACE–TIME

The study of most natural attributes and phenomena in real-world continua in conditions of *uncertainty* (e.g., in statistical continuum applications) assumes that the set of space–time points $\{\boldsymbol{p}_1, \boldsymbol{p}_2, \ldots\}$ is infinite. In epistemic terms, the "uncertainty" is an honest and valid account of an investigator's experience of the "realities" that exist within the data collected or the events recorded. To rigorously deal with this kind of phenomena, we need to consider Kolmogorov's probability theory (Kolmogorov, 1933).

3.1 KOLMOGOROV'S PROBABILITY THEORY

The formal description of the basic stochastic notions in *Kolmogorov's* theory is based on set-theoretic notions, which are of fundamental importance in developing the random field theory (Chapter II). In this section, I present some results of the set and measure theories, not aiming at completeness, instead, assuming that the readers are familiar with these theories (for a more detailed treatment of such matters, see Yaglom, 1962; Gihman and Skorokhod, 1974a,b,c; Renyi, 2007). Methodologically, the next step is to translate set-theoretic concepts into probabilistic ones.

Definition 3.1

Assume that Ω is a sample space with generic element u, $\mathcal{F}$ is a σ-field of subsets of Ω, A_i are sets of the field $\mathcal{F}$ (called events assigned to locations-instants $\boldsymbol{p}_i$), and P a probability measure on the measurable space $(\Omega, \mathcal{F})$ satisfying Kolmogorov's axioms

(a) $P[\Omega] = 1$

(b) $0 \le P[A_i] \le 1$ for all sets $A_i \in \mathcal{F}$, and

$$\text{(c) if } A_i \cap A_j = \varnothing \;\; (i \neq j), \text{ then } P\left[\bigcup_{i=1}^{\infty} A_i\right] = \sum_{i=1}^{\infty} P[A_i], \tag{3.1a–c}$$

where Ø is the empty set. The triplet $(\Omega, \mathcal{F}, P)$ is called a *probability space* (also known as Kolmogorov's probability space).

The probability space serves as the basic model on which all stochastic calculations are performed. A parenthetical remark may be appropriate at this point. In the context of stochastics theory, axiomatic probability is considered a formal method of manipulating probabilities using Kolmogorov's celebrated axioms. To apply the theory, the probability space Ω must be defined and the probability measure P assigned. These are a priori probabilities that, at the moment, suffice to be considered purely mathematical notions lacking any objective or subjective meaning.

Example 3.1

Recall the air pollution situation in Example 2.1. Let χ_i be the air pollution concentration at a specified location-instant $\boldsymbol{p}_i$ and c_i be the permissible pollution threshold determined on the basis of environmental and ecological considerations. Also, consider the sample space

$$\Omega = \{(\chi_i, c_i) : \chi_i, c_i \in R^1\}. \tag{3.2a}$$

The event A_i, "the permissible level c_i has been exceeded at location-time $\boldsymbol{p}_i$," is defined as the subset

$$A_i = \{(\chi_i, c_i) : \chi_i > c_i\} \tag{3.2b}$$

of Ω, to which one can assign a probability measure $P[A_i]$ satisfying Kolmogorov's axioms of Definition 3.1. This rather simple setup constitutes a powerful methodological construction for our future investigations.

A key concept in applied stochastics is that of a random variable assigned to a space–time point.

Definition 3.2

Let $(R^1, \Im^1)$ be a measurable space, where $\Im^1$ is a σ-field of Borel sets on the real line R^1. A real-valued *random variable* (RV) $x(u)$, where $u \in \Omega$ are elementary events and $\boldsymbol{p} \in R^{n+1}$ denotes a spatiotemporal point, is a measurable mapping x from $(\Omega, \mathcal{F})$ into $(R^1, \Im^1)$, so that for any $B \in \Im^1$,

$$x^{-1}(B) = \{u \in \Omega : x(u) \in B\} \in \mathcal{F}. \tag{3.3}$$

The RV may be denoted as $x(u, \boldsymbol{p})$, if we wish to emphasize the specific point $\boldsymbol{p}$ linked to the RV. Also, the terms "probability measure" and "random variable" never occur isolated from each other. Indeed, on the strength of Definition 3.2, $x(u)$, or simply x, is said to be an $\mathcal{F}$ – measurable (real-valued) RV, where measurability induces a probability measure μ_X on $(R^1, \Im^1)$ such as for any $B \in \Im^1$,

$$\mu_X(B) = P[x^{-1}(B)] = P[x = x(u) \in B]. \tag{3.4}$$

Naturally, the study of an RV x can be accomplished by studying the probability measure μ_X on $(R^1, \Im^1)$. For a real-valued RV, it is convenient to introduce the distribution function of the measure μ_X.

Definition 3.3

Let us define the set

$$I_\alpha = \{\chi : \chi \leq \alpha, \ \alpha \in R^1\}. \tag{3.5}$$

Then, the function

$$F_X(\chi) = \mu_X(I_\chi) = P[x \leq \chi] \tag{3.6}$$

is called the *cumulative distribution function* (*CDF*) of the RV x (or the distribution function of the measure μ_X).

An important property of the RV notion is as follows. Let $g(\chi)$ be a Borel-measurable function of the real variable χ. If x is an RV, then

$$y = g(x) \tag{3.7a}$$

is an RV too with CDF

$$F_Y(\psi) = F_X(g^{-1}(\psi)) = P[x \leq g^{-1}(\psi)] \tag{3.7b}$$

(Exercise I.15).

An RV x is called degenerate at χ if $P[x = \chi] = 1$. In ordinary mathematics, if $\chi = \psi$, the two deterministic variables are considered indistinguishable, whereas in stochastics theory, two RVs are considered indistinguishable if $P[x = y] = 1$. In this case, the two RVs are called *equivalent*. In most applications of practical interest the CDF of Eq. (3.6) can be replaced by another function, which, in many cases, is analytically more tractable and convenient to use, as follows.

Definition 3.4

If the probability measure of Eq. (3.6) is absolutely continuous and λ is the Lebesgue measure (R^1), the function f_X defined on R^1 so that

$$\begin{aligned} d\chi f_X(\chi) &= P[\chi \leq x \leq \chi + d\chi], \\ \int_B d\lambda(\chi) f_X(\chi) &= \mu_X(B) \end{aligned} \tag{3.8a–b}$$

for each Borel set B, is called the *probability density function* (PDF) of the RV x.

Under certain conditions usually applying in practice, the PDF can be conveniently expressed in terms of the corresponding CDF, i.e., Eq. (3.8b) is replaced by

$$\int d\chi f_X(\chi) = F_X(\chi), \tag{3.8c}$$

where the integration is carried out over the entire χ-range, unless other limits are indicated.[23] The PDF expresses the probability of finding a particular value of the RV over its range of definition. We recall that a study of a real-valued RV can be made by studying the probability measure of Eq. (3.4) or, equivalently, the PDF of Eqs. (3.8a–b). The f_X in Eqs. (3.8a–b) is also called the *univariate* or *marginal* PDF.

Example 3.2

Among the most commonly used PDFs are the *Gaussian* (*Normal*), the *Student*, the *Exponential*, the *Lognormal*, the *Elliptical*, the *Cauchy*, the *Beta*, the *Gamma*, the *Logistic*, the *Liouville*, and the *Pareto*

[23]Usually, there is no need to specify integration limits, since if certain χ-values are impossible, the PDF will be zero.

PDF models (explicit mathematical expressions of these PDF models together with their most significant mathematical properties can be found, e.g., in Kotz et al., 2000).

Next, I introduce an RV space that exhibits certain properties that are useful for applied stochastics purposes.

Definition 3.5

An $L_p(\Omega, \mathcal{F}, P)$ space (or, simply, an L_p-space), $p \geq 1$, is a linear normed space of the RV x on $(\Omega, \mathcal{F}, P)$ that satisfies the Lebesgue integral condition

$$\overline{|x|^p} = \int_{\Omega} dP(u)|x(u)|^p < \infty, \tag{3.9a}$$

where the bar denotes stochastic expectation. Because of measurability, Eq. (3.9a) is equivalent to the Stieltjes integral

$$\overline{|x|^p} = \int_{R^1} dP(\chi)|\chi|^p < \infty. \tag{3.9b}$$

The corresponding norm is defined by the usual formula

$$\| x \| = \left[\overline{|x|^p}\right]^{\frac{1}{p}}, \tag{3.9c}$$

and L_p is a complete space.

Some more terminology is in order: If a real-valued RV x satisfies the Definition 3.5 for $p = 1$, then this RV is called integrable. If the RV satisfies Definition 3.5 for $p = 2$, then x is called square integrable. These RVs are also called 2nd-order RVs. The stochastic expectations $\bar{x}$ and $\sigma_X^2 = \overline{x^2} - \bar{x}^2$ are called, respectively, the *mean* and the (centered) *variance* of x. We notice that an L_2 space equipped with the scalar product

$$\langle x_1, x_2 \rangle = \overline{x_1 x_2} = \int_{\Omega} dP(u) x_1(u) x_2(u), \tag{3.10}$$

where x_1 and x_2 are RVs, is called a Hilbert space (denoted as $\mathcal{H}_2$). The expectation of Eq. (3.10) is called the (noncentered) *covariance* of x_1 and x_2. In the following sections, real-valued RVs are mainly considered.

Lastly, another interesting concept associated with the notion of an RV, which has useful applications, is defined below.

Definition 3.6

The *characteristic function* (*CF*)[24] of the RV set $\boldsymbol{x} = (x_1, \ldots, x_m)$ is defined by

$$\phi_X(\boldsymbol{u}) = \overline{e^{i\boldsymbol{u}\cdot\boldsymbol{x}}} = \int_{R^m} d\boldsymbol{\chi} e^{i\boldsymbol{u}\cdot\boldsymbol{\chi}} f_X(\boldsymbol{\chi}), \tag{3.11}$$

[24] To be distinguished from the characteristic functional to be introduced in Chapter XII.

where $\boldsymbol{u} = (u_1, \ldots, u_m)$ are conjugate (to $x_1, \ldots, x_m$) variables, $\boldsymbol{u} \cdot \boldsymbol{x} = \sum_{k=1}^{m} u_k x_k$, $\boldsymbol{\chi} = (\chi_1, \ldots, \chi_m)$ are RV realizations, and $\boldsymbol{u} \cdot \chi = \sum_{k=1}^{m} u_k \chi_k$.

A key property of the CF above is that it uniquely determines the PDF $f_X(\chi)$. Other interesting properties of the CF include the following:

$$\begin{aligned} &|\phi_X(u_1, \ldots, u_m)| \leq 1, \\ &\phi_X(0, \ldots, 0) = 1, \\ &\phi_X(u_1, \ldots, u_{m-k}) = \phi_X(u_1, \ldots, u_{m-k}, u_{m-k+1} = 0, \ldots, u_m = 0). \end{aligned} \tag{3.12a–c}$$

Moreover, since the CF is properly interpreted as the mean value of the exponential function of the RV in Eq. (3.11) makes it possible to directly obtain the statistical moments of the RV by simply differentiating the CF $\phi_X(\boldsymbol{u})$ with respect to $u_1, \ldots, u_m$, i.e., the conjugate (to $x_1, \ldots, x_m$) variables.

3.2 USEFUL INEQUALITIES

To continue with our discussion, Table 3.1 lists a series of inequalities that are of significant usefulness, in both theoretical and practical investigations. In particular, Holder's inequality involves two RVs x_1 and x_2, where the special case $\kappa = \lambda = 2$ is called the Cauchy–Schwartz inequality, which is widely applicable. On the other hand, the extended Holder's inequality involves a series of RVs x_i $(i = 1, 2, \ldots, m)$. The Minkovski and Lyapunov inequalities also involve two RVs x_1 and x_2. Jensen's inequality assumes that g is a convex (or a concave) function.[25] Markov's inequality is valid for any $\varepsilon > 0$ and any $\kappa > 0$. Again, in the special case, $\kappa = 2$, Markov's inequality properly reduces to Chebyshev's inequality, which provides both upper and lower bounds. In the related Chebyshev's inequality, the $g(x)$ is a positive nondecreasing function of x, where x is an RV such that the $g(x)$ is integrable. Lastly, the association inequalities involve combinations of nondecreasing and nonincreasing functions of the RV x.

Example 3.3

Using Jensen's inequality the following interesting results can be proven. Let x be an RV. Then,

$$\begin{aligned} &\overline{e^x} \geq e^{\bar{x}}, \\ &\log\bar{x} \geq \overline{\log x}. \end{aligned} \tag{3.23a–b}$$

Also, let x be an RV taking a series of values χ_i, $i = 1, \ldots, m$, with equal probability m^{-1}. In this case, the following inequalities hold,

$$\frac{1}{m}\sum_{i=1}^{m} \chi_i \geq \left(\prod_{i=1}^{m} \chi_i\right)^{\frac{1}{m}} \geq m\left(\sum_{i=1}^{m} \chi_i^{-1}\right)^{-1}, \tag{3.24}$$

[25]A function g is convex if $g(\lambda\chi_1 + (1-\lambda)\chi_2) \leq \lambda g(\chi_1) + (1-\lambda)g(\chi_2)$, $\lambda \in [0, 1]$, $\chi_1, \chi_2 \in I \subset R^1$. A function g is concave if $-g$ is convex.

Table 3.1 A List of Useful Inequalities

Inequality and Assumptions	Mathematical Expression	
Holder: $\kappa, \lambda > 1, \kappa^{-1} + \lambda^{-1} = 1$	$\lvert\overline{x_1x_2}\rvert \le \overline{\lvert x_1x_2\rvert} \le \left(\overline{\lvert x_1\rvert^{\kappa}}\right)^{\frac{1}{\kappa}}\left(\overline{\lvert x_2\rvert^{\lambda}}\right)^{\frac{1}{\lambda}}$	(3.13)
Cauchy–Schwartz: Holder with $\kappa = \lambda = 2$	$\lvert\overline{x_1x_2}\rvert \le \overline{\lvert x_1x_2\rvert} \le \left(\overline{\lvert x_1\rvert^{2}}\right)^{\frac{1}{2}}\left(\overline{\lvert x_2\rvert^{2}}\right)^{\frac{1}{2}}$	(3.14)
Extended Holder:	$\left\lvert\overline{\prod_{i=1}^{m} x_i}\right\rvert \le \prod_{i=1}^{m}\left(\overline{\lvert x_i\rvert^{m}}\right)^{\frac{1}{m}}$	(3.15)
Minkovski: $\overline{\lvert x_1\rvert^{\kappa}}, \overline{\lvert x_2\rvert^{\kappa}} < \infty, \kappa \ge 1$	$\left(\overline{\lvert x_1 + x_2\rvert^{\kappa}}\right)^{\frac{1}{\kappa}} \le \left(\overline{\lvert x_1\rvert^{\kappa}}\right)^{\frac{1}{\kappa}} + \left(\overline{\lvert x_2\rvert^{\kappa}}\right)^{\frac{1}{\kappa}}$	(3.16)
Jensen: $P[x \in I] = 1,\ \bar{x} < \infty$	$g(\bar{x}) \le \overline{g(x)} < \infty$ if g convex on $I \subset R^1$ $\overline{g(x)} \le g(\bar{x}) < \infty$ if g concave on $I \subset R^1$	(3.17a–b)
Markov: $\varepsilon, \kappa > 0, \overline{\lvert x\rvert^{\kappa}} < \infty$	$P[\lvert x\rvert > \varepsilon] < \varepsilon^{-\kappa}\overline{\lvert x\rvert^{\kappa}}$	(3.18)
Chebyshev: Markov with $\kappa = 2$	$P[\lvert x - \bar{x}\rvert > \varepsilon\sigma_X] \le \varepsilon^{-2}$ $P\left[\lvert x - \varepsilon\bar{x}\rvert \ge \frac{(1-\varepsilon)^2\bar{x}^2}{\sigma_X^2}\right]$	(3.19a–b)
Related Chebyshev: $g(x) > 0$ nondecreasing, integrable	$P[\lvert x\rvert \ge \chi] \le \frac{\overline{g(x)}}{g(\chi)}$	(3.20)
Lyapunov: $0 < \kappa < \lambda$	$\left(\overline{\lvert x_1\rvert^{\kappa}}\right)^{\frac{1}{\kappa}} \le \left(\overline{\lvert x_2\rvert^{\lambda}}\right)^{\frac{1}{\lambda}}$	(3.21)
Association: $\overline{\lvert g_1(x)\rvert}, \overline{\lvert g_2(x)\rvert}, \overline{\lvert g_1(x)g_2(x)\rvert} < \infty$	$\overline{g_1(x)g_2(x)} \ge \overline{g_1(x)}\ \overline{g_2(x)}$ If both g_1, g_2 nondecreasing (or nonincreasing); $\overline{g_1(x)g_2(x)} \le \overline{g_1(x)}\ \overline{g_2(x)}$ If g_1 nondecreasing, g_2 nonincreasing (or *vice versa*)	(3.22a) (3.22b)

or, in more words, the arithmetic mean $\geq$ the geometric mean $\geq$ the harmonic mean. Lastly, using the Cauchy–Schwartz inequality it is found that

$$P[x - \overline{x} > \varepsilon] \leq \frac{\sigma_X^2}{\sigma_X^2 + \varepsilon^2}, \tag{3.25}$$

where $\varepsilon > 0$ and $\overline{x^2} < \infty$. Eq. (3.25) expresses an inequality with many applications in the theory of RVs.

The RV notion we acquaint ourselves with in the next section happens to be of principal value in the development of random field theory and its applications discussed in the following book chapters.

3.3 CONVERGENCE OF RANDOM VARIABLE SEQUENCES

In stochastic calculus, the study of continuity, differentiability, and integrability involves the notion of the *limit* of a sequence of RVs. This means that in stochastics we often deal with an infinite sequence of RVs

$$\{x_m(u)\} \quad \text{as } m \to \infty, \tag{3.26}$$

at points $\boldsymbol{p}_1, \ldots, \boldsymbol{p}_m, \ldots$, and, therefore, we must define the notion of the limit of such sequences. Before we start our discussion of the main types of convergence of RV sequences, we recall that for every $u \in \Omega$ the RV sequence $\{x_m(u)\}$, simply written as $\{x_m\}$, $m = 1, 2\ldots$, is a set of real numbers.

Definition 3.7

Let $\{x_m\}$ be a sequence of RVs of $L_2(\Omega, \mathcal{F}, P)$. The $\{x_m\}$ is said to converge to the RV x:

(a) *Almost surely* (*a.s.*),

$$x_m \xrightarrow{a.s.} x,$$

if

$$P\Big[\lim_{m \to \infty} x_m = x\Big] = 1. \tag{3.27a}$$

Or, there exist a $\mathcal{V} \subseteq \Omega$ satisfying $P[\mathcal{V}] = 0$, and

$$\lim_{m \to \infty} |x_m(u) - x(u)| = 0 \tag{3.27b}$$

for all $u \notin \mathcal{V}$.

(b) With *probability one* (*p*.1.),

$$x_m \xrightarrow{p.1.} x,$$

if

$$P\Big[\lim_{m \to \infty} x_m = x\Big] = 1. \tag{3.28}$$

(c) *In probability* (*P*),

$x_m \xrightarrow{P} x$, or $\text{l.i.p.}_{m \to \infty} x_m = x$, if for all $\varepsilon > 0$

$$\lim_{m \to \infty} P[|x_m - x| > \varepsilon] = 0. \tag{3.29}$$

(d) *Weakly* or *in distribution* (*F*),

$$x_m \xrightarrow{F} x,$$

if

$$\lim_{m \to \infty} F_{x_m}(\chi) = F_X(\chi) \tag{3.30}$$

on the continuity set of F_X.

(e) *Mean squarely* (*m.s.*),

$$x_m \xrightarrow{m.s.} x, \text{ or l.i.m.}_{m \to \infty} x_m = x,$$

if

$$\lim_{m \to \infty} \overline{|x_m - x|^2} = 0. \tag{3.31}$$

The earlier definitions basically reduce stochastic convergence to ordinary convergence.[26] Also, the main types of stochastic convergence considered above are related to each other. Particularly,

$$\left. \begin{array}{l} x_m \xrightarrow{m.s.} x \\ x_m \xrightarrow{a.s.} x \end{array} \right\} \Rightarrow x_m \xrightarrow{P} x \Rightarrow x_m \xrightarrow{F} x.$$

(Exercise I.17).

Given the sequence of RVs $x_m \in L_2(\Omega, \mathcal{F}, P)$, $m = 1, 2\ldots$, it turns out that the L_2-convergence is convergence in the m.s. sense. Then, if $x_m \xrightarrow{m.s.} x$ and $y_{m'} \xrightarrow{m.s.} y$ (as $m, m' \to \infty$), it is valid that

$$\text{if } x_m \xrightarrow{m.s.} x \text{ and } y_{m'} \xrightarrow{m.s.} y, \text{ then } \left\{ \begin{array}{l} \overline{x_m} \to \overline{x} \\ \overline{x_m y_{m'}} \to \overline{xy} \end{array} \right. \tag{3.32a–b}$$

(Exercise I.18).

Two interesting consequences of Eqs. (3.32a–b) are as follows. First, the operators "l.i.m." and "$\overline{[\bullet]}$" commute, and the x_m converges to a constant in the m.s. sense if and only if the $\overline{x_m x_{m'}}$ converges in the ordinary sense to a constant as $m, m' \to \infty$. Second, the following assertion holds concerning the a.s. convergence of RVs (Sobczyk, 1991):

$$\text{If for some } k > 0, \sum_{m=1}^{\infty} \overline{|x_m - x|^k} < \infty, \text{ then } x_m \xrightarrow{a.s.} x. \tag{3.33}$$

I conclude this chapter by noticing that the present section forms the basis for the study of the random field geometry in Chapter VI. Indeed, in accordance with the analysis above, the random field $X(\boldsymbol{p})$ is said to converge to the field $X(\boldsymbol{p}_0)$ in the sense of one of the aforementioned types a–e as $\boldsymbol{p} \to \boldsymbol{p}_0$ if the corresponding sequence of RVs $\{x_m = x(\boldsymbol{p})\}$ assigned to locations $\boldsymbol{p} = \boldsymbol{p}_1, \ldots, \boldsymbol{p}_m, \ldots$ tends to the RV $x_0 = x(\boldsymbol{p}_0)$ as $m \to \infty$. More random field convergence issues will be addressed in the next chapters.

[26]We note that Eq. (3.29) follows from Eq. (3.31) using Chebyshev's inequality, $\lim_{m \to \infty} P[|x_m - x| > \varepsilon] \leq \frac{\overline{|x_m - x|^2}}{\varepsilon^2}$.

CHAPTER

SPATIOTEMPORAL RANDOM FIELDS

CHAPTER OUTLINE

Spatiotemporal Random Fields. http://dx.doi.org/10.1016/B978-0-12-803012-7.00002-7

1. INTRODUCTION

Making the connection with the previous chapter, I argue in this chapter that the *spatiotemporal random field* (*S/TRF*) model plays an important role in sciences that aim at studying the uncertain features of a phenomenon in Nature and connecting them through causal relations and space–time patterns. Before I present a rigorous mathematical definition of the S/TRF model, let us first look at it heuristically.

An S/TRF is a model that generally associates mathematical quantities (such as a scalar, a vector, or a tensor) with the points of a space–time continuum in conditions of uncertainty. In this setting, an interesting interpretation of the notion of "uncertainty" is that it may offer an account of the investigator's experience of the reality that exists within the phenomenon studied or the data recorded. For the S/TRF model to obtain physical meaning, these quantities must represent values of some real-world *natural attribute* (e.g., ocean waves, water table elevation, atmospheric pollutant concentration, refractive index, crime incidences, Plague mortality rates). The representation of an attribute as an S/TRF model can have significant improvements in our understanding of the underlying physical mechanisms and lead to a more realistic representation of the attribute, which is one of the reasons that scientists increasingly champion random field notions and methods.

Example 1.1

The velocity of a random-walk process describing a diffusion phenomenon can be modeled as a temporal random sequence, where the corresponding temporal covariance function is of principal importance (Klafter and Sokolov, 2011). It has been found that the velocity covariance of a random walk cannot last longer than the time between two consecutive reorientation events. However, by unfolding the velocity into the composite space–time domain where it is modeled as an S/TRF, more extended correlations can be detected, thus revealing previously unnoticed random-walk properties and uncovering long-lived velocity correlations that extend beyond the horizon dictated by the standard temporal correlation function. Moreover, the S/TRF modeling of the phenomenon allows for a meaningful description of velocity correlations when the temporal covariance function does not exist.

Example 1.2

The spatial pattern of disease incident in a geographical region is not uniformly distributed throughout each year. Instead, disease incidents are often clustered together during short time periods. Similarly, at certain time periods a rash of incidents is observed in specific areas where the public health agencies need to respond quickly. These observations firmly indicate that the phenomenon is characterized by strong spatial and temporal links that should be better understood in the composite space–time framework of S/TRF modeling than by an ad hoc deconstruction into separate types of space and time models.

This chapter presents an introduction to the basic notions of the S/TRF theory in a broad sense, i.e., I will keep the presentation as general as possible, basically without making any restrictive assumptions about the variability characteristics of the random fields, the shape of the space–time dependence functions used, etc. Yet, just as an assumptionless science is not possible, there is no assumptionless random field modeling too. Hence, a detailed study of the modeling assumptions that underpin certain important random field classes (homostationary, isostationary, heterogeneous, spherical, etc.) will be presented in following chapters, with the important notice that these assumptions should be used only when they are physically justified on the basis of the available evidence regarding the phenomenon of interest.

In the broad sense, an S/TRF is viewed as a real (or complex)-valued random function[1] defined on a space–time domain (i.e., a domain generally marked by the relationship between space and time). Thus, "random function" and "space–time domain" are the two main components of the conceptual characterization of the S/TRF. Let us consider these two components in more detail.

1.1 THE SPACE–TIME COMPONENT

As we saw in Chapter I, the space–time domain of an S/TRF $X(\boldsymbol{p})$ is denoted by the argument $\boldsymbol{p}$ that may be seen as an $n+1$-dimensional vector

$$\boldsymbol{p} = (s_1, \ldots, s_n, s_0) \in R^{n+1}, \tag{1.1a}$$

where, by convention, $s_1, \ldots, s_n$ denote space coordinates and s_0 denotes time in the composite space–time domain R^{n+1}. Alternatively, if for modeling purposes we need to explicitly indicate the space and time arguments, the $\boldsymbol{p}$ can be viewed as an "n-dimensional vector-scalar" pair

$$\boldsymbol{p} = (\boldsymbol{s}, t) \in R^{n,1}, \tag{1.1b}$$

where $R^{n,1}$ denotes the space–time Cartesian product $R^n \times T$, with $\boldsymbol{s} \in R^n$ and $t \in T \subseteq R^1$. Eq. (1.1b) is formally equivalent to Eq. (1.1a), i.e., Eq. (1.1b) can be obtained form Eq. (1.1a), and vice versa.[2]

Generally, modern scientific theories are formulated in the R^{n+1} domain (e.g., in terms of a four-dimensional space–time), whereas classical scientific theories are formulated mainly in the $R^{n,1}$ domain (e.g., in terms of a three-dimensional space and an one-dimensional time axis with distinct physical properties). In either the R^{n+1} or the $R^{n,1}$ domain, the space and time coordinates are formally treated in a way that seeks to describe and quantify the particularities of space and time observed in the real-world. In the $R^{n,1}$ domain, especially, there is a way to identify some regions of space–time as points of space and some others as instants of time, which cannot be always done in the R^{n+1} domain. This property of the $R^{n,1}$ domain is due to its special geometrical features. Indeed, there is a unique and geometrically preferred way to divide up this $n+1$-dimensional space–time domain into a sequence of n-dimensional subdomains (R^n), each of which is well suited to play the role we want instants of time to play: events that occur on the same subdomain occur simultaneously. Otherwise said, an R^n subdomain may be viewed as a "time instant," whereas a "spatial point" may be seen as a line in $R^{n,1}$ perpendicular to each time (events located on the same line, simultaneous or not, occur in the same place).

[1]To be distinguished from the notion of an ordinary (deterministic) function.

[2]In particular, a vector $\boldsymbol{v}$ can be written as $\boldsymbol{v} = (\boldsymbol{v}_1, \boldsymbol{v}_2)$, where $\boldsymbol{v}_1 \in R^{n_1}$ and $\boldsymbol{v}_2 \in R^{n_2}$. In this sense, the vector $\boldsymbol{p} \in R^{n+1}$ of Eq. (1.1a) can be formally written in the same way with $\boldsymbol{v} = \boldsymbol{p}$, $\boldsymbol{v}_1 = \boldsymbol{s}$, $\boldsymbol{v}_2 = s_0 = t$, $n_1 = n$, and $n_2 = 1$, i.e., $\boldsymbol{p} = (\boldsymbol{s}, t) \in R^n \times R^1 = R^{n,1}$, which is Eq. (1.1b).

1.2 THE RANDOMNESS COMPONENT

The second essential component of an S/TRF model is the "random function" notion that needs to be rigorously specified before proceeding further with more technical details. This notion involves two elements, as follows:

(a) *Connected whole*: When we study a natural attribute $X(\boldsymbol{p})$ that is observable at a set of points within the R^{n+1} domain, we can account for real-world uncertainty by considering the attribute value at each individual point $\boldsymbol{p}_i$ $(i = 1,\ldots, m)$ as a realization of the RV assigned to $\boldsymbol{p}_i$ with a probability measure μ_{X_i} as in Eq. (3.4) of Chapter I. Yet, such a study may lose track of the attribute law of spatiotemporal change. Indeed, if we consider the set of RVs x_i $(i = 1,\ldots, m)$ at points $\boldsymbol{p}_i$, the corresponding univariate probability measures μ_{X_i} $(i = 1,\ldots, m)$ are not by themselves sufficient to express all the important features of x_i and of their correlations. Thus, it seems reasonable to consider the RV set as a connected whole $X = (x_1,\ldots, x_m)$ and define a suitable new measure by extending the definition of Eq. (3.4) of Chapter I.

(b) *Infinite family*: In addition, in most applications (such as the study of statistical continuum problems) the set of space–time points $\{\boldsymbol{p}_1, \boldsymbol{p}_2,\ldots\}$ is infinite. The complete characterization of a turbulent velocity field, e.g., requires the joint probability distribution over the infinite family of RVs $\{x_1, x_2,\ldots\}$.

The added value of the combination of elements (a) and (b) above lies in simultaneously enabling a complete study of the set of RVs and of their spatiotemporal relations on the basis of the measure of the random function X.

Example 1.3

We can imagine a space–time domain represented by the nodes $N_1,\ldots, N_m$ of a grid or lattice in R^{n+1} or $R^{n,1}$. Observable attributes (e.g., pressures, temperatures, or disease rates) are represented by numerical values at these nodes. The set of values on the entire grid defines a realization (sample path) of an S/TRF. Since there is a random aspect in the observable, different realizations of the field are possible corresponding to potential states of the observable attribute, so that the ensemble of all possible realizations constitutes the S/TRF.

2. CHARACTERIZATION OF SCALAR SPATIOTEMPORAL RANDOM FIELDS

I need to reemphasize that an S/TRF generally presupposes a continuum of space–time points (Section 2.1 of Chapter I) and properly assigns natural attribute values at these points. A *physically* linked set of values at all points in the space–time continuum specifies a realization of the random field. *Randomness* manifests itself as an ensemble of possible realizations for the attribute under consideration. Hence, to represent an attribute in terms of an S/TRF model, we associate with it a random aspect and an equally important structural aspect. The random field theory (Christakos, 1992, 2000) offers a general framework for analyzing data distributed in space–time, which includes operations that enable us to investigate natural phenomena in a mathematically constructive way that improves our insight into the underlying mechanisms and, thus, it enhances our predictive capabilities.

Generally, the mathematical characterization of an S/TRF model representing the spatiotemporal variation of a natural attribute can be made in terms of:

(i) its *probability laws* (low or high order, depending on the phenomenon of interest),
(ii) its *statistical moments* (e.g., in terms of space–time variation functions such as the mean, the covariance, or the variogram), or
(iii) its *generating processes* (i.e., in terms of the description of the way the random fields are generated).

All three options (i)–(iii) will be discussed in the book. Concerning option (i), in particular, the synthesis of the space–time domain and random function notions discussed in the previous section lead to the following probabilistic characterization of an S/TRF model.

2.1 PROBABILISTIC STRUCTURE

There exist various definitions of an S/TRF. From an applied stochastics perspective, among the most common definitions are a definition that is analogous to that of a set of RVs (i.e., it essentially follows along the lines described in Section 1.2 above), and another definition in terms of linear normed spaces. I now consider the RV-based definition first.[3]

Definition 2.1

An *ordinary S/TRF* model $X(\boldsymbol{p})$ is a family of related RVs $x_1,\ldots, x_m$ at points $\boldsymbol{p}_1,\ldots, \boldsymbol{p}_m$, i.e.,

$$X(\boldsymbol{p}, u) = (x_1, \ldots, x_m), \tag{2.1a}$$

where $X(\boldsymbol{p}, u)$ is a measurable mapping from $(\Omega, \mathcal{F})$ into $(R^m, \Im^m)$. In this context, each individual RV is defined on $(\Omega, \mathcal{F}, P)$ and takes values in $(R^1, \Im^1)$.[4]

In Definition 2.1, the notions of randomness and physical space–time are specified quantitatively by the model's dependence on elementary events $u \in \Omega$ and vectors $\boldsymbol{p} \in R^{n+1}$, respectively.[5] A scalar S/TRF is termed continuous parameter or discrete parameter according to whether the argument $\boldsymbol{p}$ takes discrete or continuous values, respectively (see, also, Section 7). The adjective "ordinary" is used here to distinguish the S/TRF model considered in this and the following Chapters III–X from the S/TRF model in the "generalized" sense to be introduced in Chapters XII and XIII. Having made this clarification, for simplicity the adjective "ordinary" will be dropped when I refer to S/TRFs in this and in the following chapters. The space of all continuous S/TRFs is sometimes denoted by the symbol $\mathcal{K}$. Let us now look at an alternative S/TRF definition, which is usually favored in the applied sciences literature.

---

[3]As was stated in Chapter I, $(\Omega, \mathcal{F}, P)$ is a probability space, with sample space Ω including all possible field realizations, $\mathcal{F}$ is σ-field of subsets of Ω, and $P \in [0,1]$ is the probability associated with each realization. If $\Im^m$ is a suitably chosen σ-field of subsets of R^m, the $(R^m, \Im^m)$ is a measurable space in the standard sense.

[4]Definition 2.1 concerns scalar random fields, but it can be easily extended to vector random fields (Chapter IX).

[5]With a few noticeable exceptions, the randomness argument u will be usually suppressed in the following.

Definition 2.2

An S/TRF model $X(\boldsymbol{p})$ is defined as the mapping

$$X(\boldsymbol{p}): R^{n+1} \rightarrow L_2(\Omega, \mathcal{F}, P), \tag{2.1b}$$

i.e., as a mapping on the R^{n+1} domain with values in the space $L_2(\Omega, \mathcal{F}, P)$.[6]

According to Definition 2.2, for a fixed event $u \in \Omega$ and a coordinate vector $\boldsymbol{p}$ varying in R^{n+1}, the $\chi(u, \boldsymbol{p})$ is called a *realization* or a *sample path* of the S/TRF (that is, for each u, the χ is a deterministic function of $\boldsymbol{p}$). Epistemically speaking, the smaller the number of possible realizations, the more structure the random field model of the natural attribute of interest contains (i.e., a better understanding of the attribute is possible), whereas the larger the number of possible realizations the less structure the random field model contains (i.e., an increasingly limited knowledge is available) and the closer to a highly uncertain state it gets.

I generally refer to $X(\boldsymbol{p})$ as a random field in the $n + 1$-dimensional domain, i.e., $\boldsymbol{p} \in R^{n+1}$. Yet, according to our discussion in Section 2 of Chapter I, in certain cases it may be convenient to view $X(\boldsymbol{p})$ as a time-dependent field in an n-dimensional domain, i.e., $(\boldsymbol{s}, t) \in R^{n,1}$ (working in $R^{n,1}$ may be the obvious choice in certain special cases, e.g., when the random field is space−time separable). This operational equivalence allows us to derive many mathematical results in the R^{n+1} domain, and then consider their counterparts in the $R^{n,1}$ domain when applying these results in site-specific situations.

As a quantitative notion, random field modeling limits itself to what can be measured or quantified. Also, physical insight and intuition are valuable in the (direct or indirect) resolution of random field *inference* issues.

Example 2.1

The surface displacement of ocean water can be seen as an S/TRF varying in space−time (it is often convenient in ocean studies to assume what is called an "unlimited ocean"). Similarly, air temperature, refractive index, and wind velocity exhibit irregular patterns in a turbulent atmosphere that can be all regarded as S/TRFs, i.e., with specified probabilities each of these patterns coincide with one of the possible random field realizations. And in a similar fashion, material parameters of geological structure (density, shear strength, *P*- and *S*-wave velocities or slownesses, etc.) are regarded as S/TRF realizations with experimentally specified statistics.

Remark 2.1

The emphasis on space−time coordinates ($\boldsymbol{p} = (\boldsymbol{s}, t)$) in the above definitions is necessary, because most natural phenomena vary both in space and time, and it is not sufficient to regard them as varying in space alone or in time alone. Yet, as noted earlier, for modeling reasons and to study theoretically certain aspects of the phenomenon, it may be useful to consider separately the two cases (purely spatial vs. purely temporal variations) by occasionally suppressing one of the S/TRF arguments. Therefore, when using an S/TRF model $X(\boldsymbol{p}) = X(\boldsymbol{s}, t)$: (a) we may consider time t fixed, if the focus of our study is on stationary phenomena (e.g., steady flow with an irregular flow pattern); or, (b) we may regard space $\boldsymbol{s}$ fixed, if our focus is on the time irregularities of the phenomenon (e.g., turbulent flow).

---

[6]That is, the Hilbert space of all continuous-parameter RV $x_1,\ldots, x_m$ defined at $\boldsymbol{p}_1,\ldots, \boldsymbol{p}_m$ (L_p with $p \geq 1$ can be also considered).

Certain physically motivated classes of S/TRF may be explicitly defined by combining analytical functions of the space–time argument $\boldsymbol{p}$ and a collection of RVs as in the following example.

Example 2.2

In ocean sciences, the majority of water waves are wind-generated, whereas other wave-generating mechanisms include earthquakes and planetary forces. Such wave random fields may be expressed in $R^{2,1}$ as

$$X(\boldsymbol{p},u) = X(\boldsymbol{s},t,u) = \sum_{i=1}^{m} z_i(u)\cos(\boldsymbol{k}\cdot\boldsymbol{s} - \omega t), \tag{2.2a}$$

where $z_i(u)$ $(i = 1,\ldots, m)$ are RVs, and $\boldsymbol{k}$ and ω are, respectively, the wavevector and the wave frequency. Furthermore, a solution of the Euler equations describing sea level $X(\boldsymbol{p}, u)$ is given by

$$X(\boldsymbol{p},u) = z(u)\cos(\lambda_0 t + \lambda_1 s_1 + \lambda_2 s_2 + \theta(u)), \tag{2.2b}$$

where $z(u)$ and $\theta(u)$ are RVs, and λ_i $(i = 1, 2, 3)$ are pulsations satisfying the Airy relation, $\lambda_0^2 = g(\lambda_1^2 + \lambda_2^2)$, where g denotes the gravity acceleration. If $z(u)$ and $\theta(u)$ are assumed to be independent RVs with $z(u) \sim$ Rayleigh probability law and $\theta(u) \sim$ Uniform$[0, \pi]$ law, the sea level is a Gaussian[7] random field, also known as the sine–cosine process,

$$X(\boldsymbol{p},u) = z_1(u)\sin(\lambda_0 t + \lambda_1 s_1 + \lambda_2 s_2) + z_2(u)\cos(\lambda_0 t + \lambda_1 s_1 + \lambda_2 s_2), \tag{2.2c}$$

where $z_1(u)$ and $z_2(u)$ are independent Gaussian RVs.

In the light of the foregoing considerations, and since the S/TRF is actually a function of both the elementary event $u \in \Omega$ and the space–time point $\boldsymbol{p} \in R^{n+1}$, i.e., $X(\boldsymbol{p}) = X(\boldsymbol{p}, u)$, to a family of RVs $(x_1,\ldots, x_m)$ we can associate a family of probability measures of the multivariate form

$$\mu_X(B) = \mu_{\boldsymbol{p}_1,\ldots,\boldsymbol{p}_m}(B) = P\left[X^{-1}(B)\right] = P[(x_1,\ldots,x_m) \in B] \tag{2.3}$$

for every $B \in \Im^m$.[8] Then, according to Kolmogorov (1933), a necessary and sufficient condition for the existence of the S/TRF $X(\boldsymbol{p})$ is that the probability measure of Eq. (2.3) satisfies the following conditions:

(a) *Symmetry* condition: Let

$$\mu_{\boldsymbol{p}_1,\ldots,\boldsymbol{p}_m}(B) = P[(x_{i_1},\ldots,x_{i_m}) \in B]$$

where $i_1,\ldots, i_m$ is a permutation of the indices $1,\ldots, m$. Symmetry requires that

$$\mu_{\boldsymbol{p}_1,\ldots,\boldsymbol{p}_m}(B) = \mu_{\boldsymbol{p}_{i_1},\ldots,\boldsymbol{p}_{i_m}}(B) \tag{2.4a}$$

for any permutation.

(b) *Consistency* condition: It holds that

$$\mu_{\boldsymbol{p}_1,\ldots,\boldsymbol{p}_{m+k}}\left(B \times R^k\right) = \mu_{\boldsymbol{p}_1,\ldots,\boldsymbol{p}_m}(B) \tag{2.4b}$$

for any $m, k \geq 1$ and $B \in \Im^m$.

---

[7]The class of Gaussian random fields plays a major role in theoretical and applied random field modeling, and, hence, its properties will be studied in various parts of the book (see, e.g., Section 5 of Chapter IV).

[8]The notation $\mu_{\boldsymbol{p}_1,\ldots\boldsymbol{p}_m}(B)$ emphasizes the dependency of the probability measure on the spatiotemporal points.

The above mathematical conditions lead to the following fundamental result, also known as Kolmogorov's existence theorem.

Theorem 2.1

If Eqs. (2.4a) and (2.4b) are satisfied, there exists an S/TRF $X(\boldsymbol{p})$ on a probability space $(\Omega, \mathcal{F}, P)$ with $\mu_{\boldsymbol{p}_1,\ldots,\boldsymbol{p}_m}(B)$ as its finite-dimensional probability measure.[9]

It is convenient to work in terms of the finite-dimensional CDF (or m-CDF) of the measure of Eq. (2.3) defined as

$$F_{\boldsymbol{p}_1,\ldots,\boldsymbol{p}_m}(\chi_1, \ldots, \chi_m) = P[x_1 \leq \chi_1, \ldots, x_m \leq \chi_m] \tag{2.5}$$

for any m. Eq. (2.5) is the probability that an S/TRF realization assumes at m points values less than or equal to $\chi_1,\ldots, \chi_m$. If $A^{(i)}_{\chi_i}$ denotes the level set of all $u \in \Omega$ such that $x_i = x_i(u) < \chi_i$ $(i = 1, 2,\ldots, m)$, then $A^{(i)}_{\chi} \in \mathcal{F}$ and $\prod_{i=1}^{m} A^{(i)}_{\chi_i} \in \mathcal{F}$. In this case, Eq. (2.5) reduces to

$$F_{\boldsymbol{p}_1,\ldots,\boldsymbol{p}_m}(\chi_1, \ldots, \chi_m) = P\left[\prod_{i=1}^{m} A^{(i)}_{\chi_i}\right]. \tag{2.6}$$

To indicate the space–time dependence of m-CDF, the conditions of Eqs. (2.4a) and (2.4b) can be written as, respectively,

$$F_{\boldsymbol{p}_{i_1},\ldots,\boldsymbol{p}_{i_m}}(x_{i_1}, \ldots, x_{i_m}) = F_{\boldsymbol{p}_1,\ldots,\boldsymbol{p}_m}(\chi_1, \ldots, \chi_m) \tag{2.7a}$$

(symmetry condition), and

$$F_{\boldsymbol{p}_1,\ldots,\boldsymbol{p}_m,\boldsymbol{p}_{m+1},\ldots,\boldsymbol{p}_m+k}(\chi_1, \ldots, \chi_m, \infty, \ldots, \infty) = F_{\boldsymbol{p}_1,\ldots,\boldsymbol{p}_m}(\chi_1, \ldots, \chi_m) \tag{2.7b}$$

(consistency condition).

According to Theorem 2.1, if $F_{\boldsymbol{p}_1,\ldots,\boldsymbol{p}_m}$ satisfies Eqs. (2.7a) and (2.7b) for any set $\{\boldsymbol{p}_1,\ldots, \boldsymbol{p}_m\}$, then there exists a Kolmogorov space $(\Omega, \mathcal{F}, P)$ and an S/TRF $X(\boldsymbol{p})$ having this $F_{\boldsymbol{p}_1,\ldots,\boldsymbol{p}_m}$ as its m-CDF.

Remark 2.2

In light of Theorem 2.1, one may notice that the m-CDF determines most S/TRF properties, which seems to make probability space $(\Omega, \mathcal{F}, P)$ rather irrelevant. Yet, important exceptions exist, which constitute limitations of Theorem 2.1. Specifically, several sets of functions, such as the set

$$C = \{u\colon \chi(\boldsymbol{p}, u) \text{ is a continuous function}\} \tag{2.8}$$

do not belong to the field $\mathcal{F}$, and, hence, the probability $P[C]$ is not defined. This may be due to the structure of the space $(\Omega, \mathcal{F})$ and the complexity of the realizations (sample paths) $\chi(\boldsymbol{p})$ for different $u \in \Omega$. Moreover, even if P is defined, it may not be uniquely determined by the finite-dimensional CDF characterizing the $X(\boldsymbol{p})$. There are certain approaches to overcome these limitations including (a) the direct construction of the S/TRF on a different probability space $(\Omega, \mathcal{F}, P)$ with $C \in \mathcal{F}$ so that $P[C]$ is defined; and (b) the construction of a modification of the S/TRF on the Kolmogorov space

[9]We notice that certain random field geometry features (e.g., realization or sample path differentiability, Chapter VI) may not be determined by finite-dimensional probability measures.

$(\Omega, \mathcal{F}, P)$ that has the required features (e.g., sample path continuity; Chapter VI). The matter is of special significance in stochastics and will be studied in more detail in subsequent sections on the continuity and differentiability of S/TRF realizations.

For the sets of RVs considered in the definition of an S/TRF, the corresponding multivariate CDF has certain properties, which are not all direct extensions of the univariate case:

(a) $F_{\boldsymbol{p}_1,\ldots,\boldsymbol{p}_m}$ is a nondecreasing function for every χ_i ($i = 1,\ldots, m$).
(b) $F_{\boldsymbol{p}_1,\ldots,\boldsymbol{p}_m}$ is a left-continuous function in every χ_i ($i = 1,\ldots, m$).
(c) $F_{\boldsymbol{p}_1,\ldots,\boldsymbol{p}_m} = 0$, if at least one χ_i ($i = 1,\ldots, m$) is equal to $-\infty$.
(d) $F_{\boldsymbol{p}_1,\ldots,\boldsymbol{p}_m} = 1$, if all χ_i ($i = 1,\ldots, m$) are equal to ∞.

(e) $U_{r_1}^{(1)}\ldots U_{r_m}^{(m)} F_{\boldsymbol{p}_1,\ldots,\boldsymbol{p}_m}(\chi_1, \ldots, \chi_m) \geq 0$ for all $r_i \geq 0$ and any $\chi_i \in R^1$ ($i = 1,\ldots, m$), where $U_{r_i}^{(i)} F_{\boldsymbol{p}_1,\ldots,\boldsymbol{p}_m}(\chi_1, \ldots, \chi_m) = F_{\boldsymbol{p}_1,\ldots,\boldsymbol{p}_m}(\chi_1, \ldots, \chi_i + r_i, \ldots, \chi_m)$.

In fact, if Property (e) is valid for $r_i = r > 0$ ($i = 1,\ldots, m$), then it holds in general. It can be shown that a function $F_{\boldsymbol{p}_1,\ldots,\boldsymbol{p}_m}$ that satisfies Properties (a)–(e) may be considered as an m-CDF of the corresponding S/TRF (Exercise II.1). Note that Property (e) follows from the standard Properties (a)–(d) in the case $m = 1$, but not in the case $m > 1$. Property (e) is nonconsequential in the univariate case, but it is critical in deciding whether a function can be considered as a CDF in the multivariate case.

Example 2.3

Interestingly, sometimes in practice one uses the standard properties (a)–(d) when deciding about the validity of a CDF, ignoring property (e). The function

$$F_{\boldsymbol{p}_1,\boldsymbol{p}_2}(\chi_1, \chi_2) = \begin{cases} 1 & \text{if } \chi_1 + \chi_2 > 0 \\ 0 & \text{otherwise} \end{cases} \tag{2.9}$$

obviously satisfies properties (a)–(d) but not property (e), and, hence, it is not a bivariate CDF.

Furthermore, the multivariate probability density function (or, m-PDF) of an S/TRF $X(\boldsymbol{p})$ at a set of points $\boldsymbol{p}_1,\ldots, \boldsymbol{p}_m$ is defined by

$$\begin{aligned} f_X(\chi_1, \ldots, \chi_m)d\chi_1\ldots d\chi_m &= f_{\boldsymbol{p}_1,\ldots,\boldsymbol{p}_m}(\chi_1, \ldots, \chi_m)d\chi_1\ldots d\chi_m \\ &= P_X[\chi_1 \leq x_1 < \chi_1 + d\chi_1, \ldots, \chi_m \leq x_m < \chi_m + d\chi_m] \\ f_X(\chi_1, \ldots, \chi_m) = f_{\boldsymbol{p}_1,\ldots,\boldsymbol{p}_m}(\chi_1, \ldots, \chi_m) &= \frac{\partial^m}{\partial\chi_1\ldots\partial\chi_m} F_{\boldsymbol{p}_1,\ldots,\boldsymbol{p}_m}(\chi_1, \ldots, \chi_m), \end{aligned} \tag{2.10a–b}$$

where Eq. (2.10b) is valid if $F_{\boldsymbol{p}_1,\ldots,\boldsymbol{p}_m}$ is absolutely continuous. Eq. (2.10a) is the probability that an S/TRF realization $\chi_1,\ldots, \chi_m$ occurs. Stochastically, the $X(\boldsymbol{p})$ is described completely if all the m-PDFs in Eq. (2.10a) are known. The conditions of symmetry and consistency must be also satisfied. In the following, both symbols f_X (resp. F_X) and $f_{\boldsymbol{p}}$ (resp. $F_{\boldsymbol{p}}$) will be used to denote a PDF (resp. CDF) of an S/TRF. Also, on occasion, the more concise notations,

$$f_X(\boldsymbol{\chi}), \ \boldsymbol{\chi} = (\chi_1, \ldots, \chi_m), \text{ and } d\boldsymbol{\chi} = d\chi_1\ldots d\chi_m = \prod_{i=1}^{m} d\chi_i,$$

will be used. In practice, valuable information about the physics of the phenomenon of interest is often obtained from an approximate description of the S/TRF in terms of only a few low-order PDF.

By means of Eq. (2.10a), the m-PDF $f_X(\boldsymbol{\chi})$ assigns numerical probabilities to the $X(\boldsymbol{p})$ realizations that evolve between multiple space–time points. In fact, the m-PDF describes the comparative likelihoods of the various realizations and not the certain occurrence of a specific realization. Accordingly, the $f_X(\boldsymbol{\chi})$ unit is a probability per realization unit. Technically, by combining $f_X(\boldsymbol{\chi})$ with an efficient Monte Carlo simulator, we can generate numerous S/TRF realizations, and then look at their prevalent features, thus gaining additional insight to that obtained by studying the analytical expression of $f_X(\boldsymbol{\chi})$. In applied stochastics, each $X(\boldsymbol{p})$ realization is allowed only if it is consistent with physical knowledge concerning the attribute and with logical reasoning. Apparently, not all S/TRF realizations are equally probable. Depending on the underlying physical mechanisms of the attribute, some realizations are more probable than others, and this is reflected in the $f_X(\boldsymbol{\chi})$ model of $X(\boldsymbol{p})$.

Noticeably, other useful probability functions can be derived directly from the m-PDF. One of them is the *conditional* PDF of $X(\boldsymbol{p})$ defined by

$$f_X(\chi_{k+1},\ldots,\chi_m|\chi_1,\ldots,\chi_k)=f_{\boldsymbol{p}_1,\ldots,\boldsymbol{p}_m}(\chi_{k+1},\ldots,\chi_m|\chi_1,\ldots,\chi_k)=\frac{f_{\boldsymbol{p}_1,\ldots,\boldsymbol{p}_m}(\chi_1,\ldots,\chi_m)}{f_{\boldsymbol{p}_1,\ldots,\boldsymbol{p}_k}(\chi_1,\ldots,\chi_k)}. \tag{2.11a}$$

A consequence of Eq. (2.11a) is the expression

$$f_X(\chi_1,\ldots,\chi_m)=f_X(\chi_1)f_X(\chi_2|\chi_1)\ldots f_X(\chi_m|\chi_1,\ldots,\chi_{m-1}). \tag{2.11b}$$

By virtue of Eq. (2.11b), the form of the m-PDF can be obtained as the product of the univariate PDF and a set of conditional PDFs. The usefulness of Eq. (2.11b) lies in the fact that in practice it is often easier to calculate conditional PDFs. This is true in a variety of random field applications such as, e.g., the spatiotemporal simulation of natural attributes.

It is important to point out that the RV property of Eq. (3.7a) of Chapter I remains valid in the case of an S/TRF. That is, if $g(\chi)$ is a Borel-measurable function of the real variable χ and $X(\boldsymbol{p})$ is an S/TRF, then the

$$Y(\boldsymbol{p})=g[X(\boldsymbol{p})]$$

is an S/TRF too (Exercise II.2).

Among the standard PDF classes used in random field characterization are those already mentioned in Chapter I: Gaussian, exponential, lognormal, elliptical, Cauchy, gamma, logistic, and Pareto PDFs. Other important PDF classes also emerge in random field studies that do not have the above general forms, but they are rather case specific and very useful in applications. One such class is that of *generalized* PDFs, which are densities involving delta functions in space–time. Some cases of this PDF class that possess a definite physical meaning are examined in the next example.

Example 2.4

A generalized PDF, which has been used in the kinetic description of real dynamical processes, has the form

$$f_X(\boldsymbol{s},t)=f_S(\boldsymbol{s})\delta\left(t-\frac{|\boldsymbol{s}|}{\upsilon(\boldsymbol{s})}\right), \tag{2.12a}$$

where $f_X(\boldsymbol{s}, t)$ is the probability for a displacement $\boldsymbol{s}$ in time t, $f_S(\boldsymbol{s})$ is the probability to make a displacement $\boldsymbol{s}$, and $\upsilon(\boldsymbol{s})$ denotes the process speed. In Eq. (2.12a), the space–time coupling of $f_X(\boldsymbol{s}, t)$ is achieved through the δ-function. This δ-function expresses the fact that a spatial displacement $\boldsymbol{s}$ takes time $t = \frac{|\boldsymbol{s}|}{\upsilon(\boldsymbol{s})}$ to be performed (i.e., $\delta\left(t - \frac{|\boldsymbol{s}|}{\upsilon(\boldsymbol{s})}\right)$ is the conditional probability for the time spent in displacement $\boldsymbol{s}$ to equal t, given that the displacement length is $|\boldsymbol{s}|$). If physically justified, it is assumed that $f_S(\boldsymbol{s})$ is isotropic, $f_S(\boldsymbol{s}) = f_S(|\boldsymbol{s}|)$, and such that $f_S(|\boldsymbol{s}|) \sim |\boldsymbol{s}|^{-\nu}$ as $|\boldsymbol{s}| \rightarrow \infty$, where the value of $\nu > 0$ depends on the physics of the particular process.[10] An illustration of $f_S(|\boldsymbol{s}|)$ function in $R^{3,1}$ is

$$f_S(|\boldsymbol{s}|) = \frac{f_s}{4\pi|\boldsymbol{s}|^2}, \tag{2.12b}$$

where f_S is the probability of displacement $\boldsymbol{s}$ in any direction. Formally, f_S is the marginal probability of $f_S(\boldsymbol{s})$, integrated over all directions so that (spherical coordinates), $f_S = \int d\sigma f_S(\boldsymbol{s})$, $d\sigma = |\boldsymbol{s}|^2 \sin\theta dr d\theta d\phi$, which yields Eq. (2.12b). Another generalized PDF model with many scientific applications is given by

$$f_X(\boldsymbol{s}, t) = \delta(\chi(\boldsymbol{s}, t) - w), \tag{2.12c}$$

where $X(\boldsymbol{s}, t)$ and w may denote, respectively, the velocity in stationary and decaying homogeneous isotropic turbulence and the corresponding sample space variable in $R^{3,1}$. Higher-order PDFs can be defined in a straightforward manner. So, the bivariate PDF corresponding to Eq. (2.12c) is given by

$$f_X(\boldsymbol{s}, \boldsymbol{s}', t, t') = \delta(\chi(\boldsymbol{s}, t) - w)\delta(\chi(\boldsymbol{s}', t') - w').$$

An added value of the generalized PDFs is that they can serve as the starting point for generating new space–time variability models (Chapter XVI).

Example 2.5

Another interesting generalized PDF that has many applications in physical sciences is given by

$$f_X(\boldsymbol{p}_1, \ldots, \boldsymbol{p}_m) = \prod_{l=1}^{m} f_X(\boldsymbol{p}_l) = \prod_{l=1}^{m} \sum_{i=1}^{m} f_i \delta\left(\chi_{\boldsymbol{p}_l} - \chi_{\boldsymbol{p}_i}\right), \tag{2.13}$$

where f_i are real numbers so that $\sum_{i=1}^{m} f_i = 1$ (a simple case is $f_i = \frac{1}{m}$). In weather studies, the distribution of lightning events in $R^{2,1}$, which are natural indicators of thunderstorms (e.g., Finke, 1999), can be described by a univariate PDF written as the space–time-dependent generalized function

$$f_X(\boldsymbol{s}, t) = c \sum_{i=1}^{m} \delta(\boldsymbol{p} - \boldsymbol{p}_i) = c \sum_{i=1}^{m} \delta(\boldsymbol{s} - \boldsymbol{s}_i)\, \delta(t - t_i),$$

where $\boldsymbol{p}_i = (\boldsymbol{s}_i, t_i)$ denote the space–time coordinates of lightning events observed at a set of positions $\boldsymbol{s}_i$ and time moments t_i ($i = 1, \ldots, m$), and c is a normalization constant.

---

[10]For example, the case of a constant υ and $\nu \in (2, 3)$ corresponds to the physical process (Lévy walk) for which the particle moves with a constant speed and the waiting time depends on the displacement.

2.2 THE CHARACTERISTIC FUNCTION

Sometimes, it is convenient to transform the study of the S/TRF $X(\boldsymbol{p})$ from the physical to the spectral (frequency) domain, where a mathematically equivalent to the PDF notion can be introduced. For this purpose, let $X(\boldsymbol{p})$ be determined in terms of the RV set $\boldsymbol{X} = (x_1,\ldots, x_m)$ at points $(\boldsymbol{p}_1,\ldots, \boldsymbol{p}_m)$.[11]

Definition 2.3

The m-dimensional FT of the PDF of Eqs. (2.10a–b) defines the *characteristic function* (*CF*) of the random field $X(\boldsymbol{p})$ as

$$\begin{aligned}\phi_X(\boldsymbol{u}) &= \overline{e^{i\,\boldsymbol{u}\cdot\boldsymbol{X}}},\\ &= \int d\boldsymbol{\chi} e^{i\,\boldsymbol{u}\cdot\boldsymbol{\chi}} f_X(\boldsymbol{\chi}),\end{aligned} \tag{2.14a–b}$$

where $\boldsymbol{u} = (u_1,\ldots, u_m)$ are conjugate (to $x_1,\ldots, x_m$) variables, $\boldsymbol{u}\cdot\boldsymbol{X} = \sum_{k=1}^{m} u_k x_k$, $\boldsymbol{\chi} = (\chi_1,\ldots, \chi_m)$ are random field realizations, and $\boldsymbol{u}\cdot\boldsymbol{\chi} = \sum_{k=1}^{m} u_k \chi_k$.

Eq. (2.14b) is a remarkable result, because it uniquely determines the PDF $f_X(\boldsymbol{\chi})$ of $X(\boldsymbol{p})$.[12] Put differently, since we can define the inverse FT of Eq. (2.14b), there obviously exists a unique relationship between PDF and CF. For instance, the fact that the PDF and the CF constitute an FT pair implies that the narrower the shape of the former, the wider that of the latter. Among the interpretable implications of this result is the celebrated uncertainty principle and its various applications in sciences (for a more detailed presentation of this principle, see Section 6 of Chapter VIII). To clarify the presentation of the theoretical discussion above, some examples are considered below.

Example 2.6

By differentiating Eq. (2.14a) once with respect to any u_k ($k = 1,\ldots, m$), it is immediately found that

$$\frac{\partial \phi_X(\boldsymbol{u})}{\partial u_k} = \overline{i x_k e^{i\,\boldsymbol{u}\cdot\boldsymbol{X}}},$$

and by further letting $\boldsymbol{u} = \boldsymbol{0}$, the first moment of x_k is found as

$$\frac{\partial \varphi_X(\boldsymbol{u})}{\partial u_k}\Big|_{\boldsymbol{u}=\boldsymbol{0}} = \frac{\partial}{\partial u_k}\varphi_X(\boldsymbol{0}) = i\,\overline{x_k}.$$

In a similar fashion, the ρth moment of x_k is obtained as

$$\frac{\partial^\rho \varphi_X(\boldsymbol{u})}{\partial u_k^\rho}\Big|_{\boldsymbol{u}=\boldsymbol{0}} = \frac{\partial^\rho}{\partial u_k^\rho}\phi_X(\boldsymbol{0}) = i^\rho \overline{x_k^\rho}.$$

Thus, any space–time statistical moment can be calculated from the corresponding CF without using the PDF. As should be expected, this analytical feature of CF has far-reaching consequences in S/TRF modeling.

---

[11]Throughout the chapter, integrations are generally carried out over the whole range of a variable or an argument, unless other limits are indicated.

[12]Given that $\phi_X(\boldsymbol{u})$ is the FT of $f_X(\boldsymbol{\chi})$, it is sometimes written as $\widetilde{f}_X(\boldsymbol{u})$.

Let us take stock. A remarkable conclusion of the random field theory so far is that the S/TRF $X(\boldsymbol{p})$ is specified completely by means of all finite-dimensional probability measures of Eq. (2.3), CDF of Eq. (2.6), PDF of Eq. (2.10a–b), or CF of Eq. (2.14a–b) of orders $m = 1, 2, \ldots$ Therefore, ultimately we would prefer a complete characterization of the S/TRF model in terms of its CDF or PDF for any number of points in the space–time domain. Yet, with the notable exception of some special types of random fields, such as the Gaussian and the Poisson fields, this information is generally unavailable in real-world situations, in which case we need to settle for a partial or incomplete yet very useful characterization of the S/TRF model. This is the focus of the next section.

2.3 SPATIOTEMPORAL VARIABILITY FUNCTIONS: COMPLETE (OR FULL) AND PARTIAL

The S/TRF model is equipped with the analytical tools that allow it to describe adequately the manner in which the underlying natural attribute develops over space–time in conditions of uncertainty. This development is a reflection of a certain pattern of combined spatiotemporal correlations between values of the attribute. In the sequel, I will consider second-order S/TRF, that is, the analysis will be based on up to second-order statistical or ensemble moments (averages)[13] assumed to be continuous and finite.

In second-order moment terms, the $X(\boldsymbol{p})$ can be characterized in terms of its *spatiotemporal variability functions* (S/TVFs) in the physical domain, i.e., functions that adequately assess or measure the spatiotemporal variation of $X(\boldsymbol{p})$. A detailed discussion of the random field variability structure and related inferences can be found in Chapter IV. Here, only a brief introduction is given for completeness of presentation purposes.

Definition 2.4

A (second-order) characterization of the spatiotemporal variability of $X(\boldsymbol{p})$ includes the following S/TVFs (ensemble averages):[14]

(a) The *spatiotemporal mean* function

$$\overline{X(\boldsymbol{p})} = \int d\chi_p \chi_p f_X(\chi_p). \tag{2.15}$$

(b) The *complete* or *full spatiotemporal covariance* function

$$\begin{aligned} \overline{[X(\boldsymbol{p}) - \overline{X(\boldsymbol{p})}][X(\boldsymbol{p}') - \overline{X(\boldsymbol{p}')}]} &= \iint d\chi_p d\chi_{p'} \left(\chi_p - \overline{X}\right)\left(\chi_{p'} - \overline{X}'\right) f_X(\chi_p, \chi_{p'}), \\ \overline{X(\boldsymbol{p})X(\boldsymbol{p}')} &= \iint d\chi_p d\chi_{p'} \chi_p \chi_{p'} f_X(\chi_p, \chi_{p'}), \end{aligned} \tag{2.16a–b}$$

where the difference $X'(\boldsymbol{p}) = X(\boldsymbol{p}) - \overline{X(\boldsymbol{p})}$ denotes random field fluctuation. Specifically, Eqs. (2.16a–b) are called the *centered* and the *noncentered* covariance functions, respectively. The centered covariance of Eq. (2.16a) is commonly denoted as $c_X(\boldsymbol{p}, \boldsymbol{p}')$. Yet, many practitioners

---

[13] To be distinguished from *sample averages* calculated on the basis of a single realization.

[14] Generally, there is no need to specify limits of integration in Eqs. (2.15)–(2.17a–b), since if certain values of χ_p, $\chi_{p'}$ are impossible, the PDF $f_X(\chi_p, \chi_{p'})$ will be zero.

find it convenient to use the same notation for a noncenter covariance when the meaning is obvious from the context.

(c) The *complete* or *full* spatiotemporal *variogram* and *structure* functions, respectively,

$$\gamma_X(\boldsymbol{p},\boldsymbol{p}') = \frac{1}{2}\overline{[X(\boldsymbol{p}) - X(\boldsymbol{p}')]^2} = \frac{1}{2}\iint d\chi_{\boldsymbol{p}}\, d\chi_{\boldsymbol{p}'}(\chi_{\boldsymbol{p}} - \chi_{\boldsymbol{p}'})^2 f_X(\chi_{\boldsymbol{p}}, \chi_{\boldsymbol{p}'}),$$
$$\xi_X(\boldsymbol{p},\boldsymbol{p}') = 2\gamma_X(\boldsymbol{p},\boldsymbol{p}'). \tag{2.17a–b}$$

Remark 2.3

At this point, I would like to emphasize the multiple terminologies used in the literature of different scientific fields to describe the S/TVFs (covariance, variogram, and structure functions):

- The terms "complete" or "full" are used when one needs to distinguish the above S/TVFs from special but interpretatively important cases of *partial* S/TVFs obtained by fixing one or more of the space–time arguments, see below (these terms are dropped when the S/TVF form is obvious from the context).
- The S/TVFs are also termed *two-point statistics* in the applied sciences literature, to indicate that they express physics-based correlations between pairs of points in space–time.

The mean function expresses trends and systematic structures in space–time $\boldsymbol{p} = (\boldsymbol{s}, t)$, but it does not tell us anything about the random field fluctuation, which is though of great importance in real-world situations (e.g., random field fluctuation representing the violence of turbulent motion with respect to the mean velocity is just what we need to know in many oceanographic studies). In contrast, S/TVFs, such as the covariance, variogram, and structure functions, express correlations and dependencies between the points $\boldsymbol{p} = (\boldsymbol{s}, t)$ and $\boldsymbol{p}' = (\boldsymbol{s}', t')$. Understanding, e.g., the relationship between space and time in the covariance function $c_X(\boldsymbol{p},\boldsymbol{p}')$ of the velocity random field $X(\boldsymbol{p})$ is a fundamental issue in statistical theories of geophysics and fluid dynamics.

Example 2.5 (cont.)

Some elementary S/TVF expressions can be derived in the case of Example 2.5. The mean, covariance, and variogram functions are given by

$$\overline{X(\boldsymbol{p})} = \int d\chi_{\boldsymbol{p}}\chi_{\boldsymbol{p}} f_X(\chi_{\boldsymbol{p}}) = \int d\chi_{\boldsymbol{p}}\chi_{\boldsymbol{p}} \sum_{i=1}^{m} f_i\delta(\chi_{\boldsymbol{p}} - \chi_{\boldsymbol{p}_i}) = \sum_{i=1}^{m} f_i\chi_{\boldsymbol{p}_i},$$

$$\begin{aligned} c_X(\boldsymbol{p},\boldsymbol{p}') &= \iint d\chi_{\boldsymbol{p}}d\chi_{\boldsymbol{p}'}\chi_{\boldsymbol{p}}\chi_{\boldsymbol{p}'}f_X(\chi_{\boldsymbol{p}}, \chi_{\boldsymbol{p}'}) - \overline{X(\boldsymbol{p})}^2 \\ &= \iint d\chi_{\boldsymbol{p}}d\chi_{\boldsymbol{p}'}\chi_{\boldsymbol{p}}\chi_{\boldsymbol{p}'} \sum_{i,j=1}^{m} f_i f_j\delta(\chi_{\boldsymbol{p}} - \chi_{\boldsymbol{p}_i})\delta\left(\chi_{\boldsymbol{p}'} - \chi_{\boldsymbol{p}_j}\right) - \overline{X(\boldsymbol{p})}^2 \\ &= \sum_{i,j=1}^{m} f_i f_j \chi_{\boldsymbol{p}_i}\chi_{\boldsymbol{p}_j} - \overline{X(\boldsymbol{p})}^2, \end{aligned}$$

$$\gamma_X(\boldsymbol{p},\boldsymbol{p}') = \frac{1}{2}\iint d\chi_{\boldsymbol{p}}d\chi_{\boldsymbol{p}'}(\chi_{\boldsymbol{p}} - \chi_{\boldsymbol{p}'})^2 f_X(\chi_{\boldsymbol{p}}, \chi_{\boldsymbol{p}}) = \frac{1}{2}\sum_{i,j=1}^{m} f_i f_j\left(\chi_{\boldsymbol{p}_i} - \chi_{\boldsymbol{p}_j}\right)^2,$$

where f_i are real numbers so that $\sum_{i=1}^{m} f_i = 1$.

In the light of the discussion in Section 2 of Chapter I, the vector argument $\boldsymbol{p} \in R^{n+1}$ in the above equations can be replaced by the "vector–scalar" arguments $(\boldsymbol{s}, t) \in R^{n,1}$ and $(\boldsymbol{h}, \tau) \in R^{n,1}$ (also denoted as $(\boldsymbol{s}, t)$, $(\boldsymbol{h}, \tau) \in R^n \times T$). So, it is often convenient to write the complete covariance function $c_X(\boldsymbol{p},\boldsymbol{p}')$ as

$$c_X(\boldsymbol{s},\boldsymbol{s}',t,t') \text{ or } c_X(\boldsymbol{s},\boldsymbol{h},t,\tau),$$

where $\boldsymbol{s}' = \boldsymbol{s} + \boldsymbol{h}$ and $t' = t + \tau$. Herein, both types of space–time arguments will be used interchangeably as the situation requires. For example, to explain a certain covariance categorization in the following remark, it is convenient to use the "vector–scalar" argument. This remark points out an important categorization of complete and partial covariance functions based on physical considerations.

Remark 2.4

Focusing on the covariance function (a similar analysis holds for the other S/TVFs), and depending on the physical context, we may occasionally need to become more specific and distinguish in $R^{n,1}$ between the following (the covariance functions considered below are noncentered, but the same distinction obviously applies in the case of centered covariance functions, as well):

- The *simultaneous two-point* (or, *two-point/one-time*) spatiotemporal covariance function

$$c_X(\boldsymbol{s},\boldsymbol{s}',t) = \overline{X(\boldsymbol{s},t)X(\boldsymbol{s}',t)}, \tag{2.18a}$$

which is obtained from the complete covariance function by letting $\tau = 0$. That is, $c_X(\boldsymbol{s},\boldsymbol{s}',t)$ is a partial covariance function that involves simultaneous time arguments and evaluates spatial correlations between pairs of distinct locations in space.

- The *nonsimultaneous one-point* (or, *one-point/two-time*) spatiotemporal covariance

$$c_X(\boldsymbol{s},t,t') = \overline{X(\boldsymbol{s},t)X(\boldsymbol{s},t')}, \tag{2.18b}$$

which is obtained from the complete covariance function by letting $\boldsymbol{h} = \boldsymbol{0}$. That is, $c_X(\boldsymbol{s},t,t')$ offers an assessment of the "memory" of the field, that is, the time period over which the field is correlated with itself.

- The *nonsimultaneous two-point* (or, *two-point/two-time*) spatiotemporal covariance

$$c_X(\boldsymbol{s},t,\boldsymbol{s}',t') = \overline{X(\boldsymbol{s},t)X(\boldsymbol{s}',t')}, \tag{2.18c}$$

which, of course, is the complete covariance function that provides an unrestricted space–time characterization (no fixed space or time) of an attribute distribution.

- The *simultaneous one-point* (or, *one-point/one-time*) spatiotemporal covariance,

$$c_X(\boldsymbol{s},t) = \overline{X(\boldsymbol{s},t)^2}, \tag{2.18d}$$

which is, obviously, the corresponding *variance*.

The earlier distinction between complete and partial S/TVFs is interpretatively appropriate, because in many studies of physical phenomena there exist sound reasons to specifically focus on one of the expressions of Eqs. (2.18a)–(2.18d) above. So, in some cases we may wish to isolate the dynamical character of the phenomenon, whereas in some others we may want to switch focus on spatial physical features. Below I describe some physical situations associated with the above covariance expressions, in the ocean, in the atmospheric, and in the health sciences.

Example 2.6

In many oceanographic applications, it is very difficult or even impossible to measure the complete space–time covariance function, $c_X(\boldsymbol{s},\boldsymbol{s}',t,t') = c_X(\boldsymbol{s},\boldsymbol{h},t,\tau)$, of the surface wave displacement in $R^{2,1}$ (or, its frequency domain counterpart, termed the complete spectral function or spectrum, see Section 2.4). In this case, we let $\tau = 0$ in the covariance function, i.e., our focus is the simultaneous two-point covariance $c_X(\boldsymbol{s},\boldsymbol{s}',t) = c_X(\boldsymbol{s},\boldsymbol{h},t)$ of Eq. (2.18a). This means that we consider only instantaneous pictures of the sea surface, and we concern ourselves with the simultaneous observation of the surface displacement at different spatial locations (as it turns out this partial covariance or the corresponding partial spectrum can be measured in ocean practice). Also, in atmospheric sciences the covariance function of Eq. (2.18a) may describe the spatial pattern of ozone concentration at a fixed time t. The $c_X(\boldsymbol{s},t,t')$ of Eq. (2.18b) may focus on the temporal correlation structure of free surface pressure as observed at a fixed location $\boldsymbol{s}$. In health sciences, the $c_X(\boldsymbol{s},t,\boldsymbol{s}',t')$ of Eq. (2.18c) may give information about the composite space–time correlations of flu incidences in a region and during a period of time.

Therefore, all cases of Remark 2.4 and their physical significance will be studied throughout the book. Also, the readers may notice that the covariances of Eqs. (2.18a) and (2.18b) could be associated with the random field cases (a) and (b) of Remark 2.1.

Since both the covariance and the variogram functions are generally two-point functions, they are directly interrelated. This interrelation is manifested in a variety of ways, two of which I discuss here:

(a) In the case of a constant random field mean, the c_X and γ_X are simply related by

$$\gamma_X(\boldsymbol{p},\boldsymbol{p}') = \frac{1}{2}\left[c_X(\boldsymbol{p},\boldsymbol{p}) + c_X(\boldsymbol{p}',\boldsymbol{p}')\right] - c_X(\boldsymbol{p},\boldsymbol{p}'). \tag{2.19a}$$

Obviously, the covariance and variogram functions are closely linked via the relationship above, and they are even more so in the case where some additional assumptions are imposed concerning space–time variability, such as space–time homostationarity, STHS, and space–time isostationarity, STIS (some introductory definitions of STHS and STIS are given in Section 7.4; a detailed discussion of these important assumptions is given in Chapters VII and VIII).

(b) Properly selected functions of a valid variogram are valid covariances, and vice versa. A few examples of such functions are discussed next.

Example 2.7

The function $\gamma_X(\boldsymbol{p},\boldsymbol{p}')$, with $\gamma_X(\boldsymbol{p},\boldsymbol{p}) = 0$, is a space–time variogram if and only if the exponential function

$$c_X(\boldsymbol{p},\boldsymbol{p}') = e^{-\alpha\gamma_X(\boldsymbol{p},\boldsymbol{p}')} \tag{2.19b}$$

is a space–time covariance function for all nonnegative α (see, also, Example 4.1 of Chapter X). Likewise, the linear function of the variogram

$$c_X(\boldsymbol{p},\boldsymbol{p}') = \gamma_X(\boldsymbol{s},\boldsymbol{0},t,0) + \gamma_X(\boldsymbol{s}',\boldsymbol{0},t',0) - \gamma_X(\boldsymbol{s},\boldsymbol{s}',t,t') \tag{2.19c}$$

is a covariance function (Ma, 2003a).

Depending on the application and the type of space–time heterogeneity characterizing the phenomenon, one of these functions may be preferred over the other. In theoretical studies of STHS fields (Section 7.4 in this chapter, and Chapter VII), the covariance is commonly used. On the other hand, the variogram is often easier to estimate from experimental data. Moreover, in studies involving random fields with STHS increments (Chapter XIII), the structure function is consistently preferred.

The following covariance expressions, which represent different aspects of the function, will be useful in subsequent developments:

$$\begin{aligned} c_X(\boldsymbol{p},\boldsymbol{p}') &= c_X(\boldsymbol{p}',\boldsymbol{p}) \\ |c_X(\boldsymbol{p},\boldsymbol{p}')| &\leq (c_X(\boldsymbol{p})c_X(\boldsymbol{p}'))^{\frac{1}{2}} \\ \overline{X(\boldsymbol{p})X(\boldsymbol{p}')} &= c_X(\boldsymbol{p},\boldsymbol{p}') + \overline{X(\boldsymbol{p})}\ \overline{X(\boldsymbol{p}')}. \end{aligned} \tag{2.20a–c}$$

Eq. (2.20a) indicates a space–time symmetric function. Eq. (2.20b) is known as the *Cauchy–Schwartz* inequality. Eq. (2.20c) simply relates the centered with the noncentered covariance function. A continuous function such as above is the covariance of an S/TRF if and only if it satisfies a *nonnegative-definiteness* (NND) condition (the mathematical expression of NND is presented in Chapter IV; see, also, Chapter XV). For the variogram (or the structure) function, the symmetry condition

$$\gamma_X(\boldsymbol{p},\boldsymbol{p}') = \gamma_X(\boldsymbol{p}',\boldsymbol{p}) \tag{2.21}$$

can hold too. Moreover, for a function $\gamma_X(\boldsymbol{p},\boldsymbol{p}')$ to be a valid variogram needs to satisfy the *conditionally negative-definiteness* (CND) condition (the mathematical expression of CND is presented Section 2.5 of Chapter IV).

Remark 2.5

Spatiotemporal variability analysis involves expressions for stochastic expectations in terms of PDF integrals in continuous space. If the PDF is a discrete function, the stochastic expectations involve summations over the discrete possibilities. Then, the integral expressions of the earlier equations are still valid if the PDF is replaced by

$$f_X(\chi) = \sum_j f_j \delta(\chi - \chi_j), \tag{2.22a}$$

where $f_j = P_X[\chi = \chi_j]$ are the probabilities of the discrete events. For illustration, consider Eq. (2.15), which is valid in continuous space. By using the above substitution it is found that

$$\overline{X(\boldsymbol{p})} = \sum_j f_j \int d\chi_{\boldsymbol{p}} \chi_{\boldsymbol{p}} \delta(\chi_{\boldsymbol{p}} - \chi_j) = \sum_j f_j \chi_j, \tag{2.22b}$$

which is the correct result in the discrete domain.

As stated at the beginning of this section, the S/TVFs (covariance, variogram, and structure functions) provide the means for sound variability inferences. These functions are important tools of random field modeling, where they are used to represent the space–time-dependent structure of natural attributes in conditions of uncertainty (Dee, 1991; Etherton and Bishop, 2004; Zhang, 2002; Purser et al., 2003; Bongajum et al., 2013; He et al., 2014). S/TVFs offer quantitative assessments of interrelations between attribute values at different spatial locations and time instances. These location−instant interrelations may be associated to different kinds of attribute distributions. When the distribution of an attribute is studied in space−time, the observed interrelations need to be quantified for various purposes, including:

(i) testing for the presence of significant correlations among attribute values within the space−time domain of interest,
(ii) assessing the space−time attribute dependence range (i.e., the spatial and temporal lags beyond which attribute observations are spatiotemporally independent);
(iii) selecting theoretical S/TVF models or laws of change to summarize the observed spatiotemporal attribute structure; and
(iv) inferring about the underlying natural mechanism, such as spread velocity components, dispersal distances, and directional differences.

2.4 ANALYSIS IN THE SPECTRAL DOMAIN

In agreement with the transformation theories to be presented in Chapter V, the other mode of second-order variability analysis is developed in the *spectral* or *frequency domain.* Being able to study a problem in both the physical (real) and the spectral domains often reveals important structural features of a problem that is not obvious in just one domain. Indeed, there are differences between physical and spectral domain analysis in what they search for in an attribute's distribution.

Actually, it is customary in applied sciences to first obtain the results in the physical domain and then reproduce them in the spectral domain, where they obtain a life of their own (e.g., the famous "2/3 law" of small-scale turbulence was initially derived in the physical domain, and then reproduced in the spectral domain, where it became known as the "5/3 law"). Furthermore, it is not uncommon that the derivation of many important results turns out to be easier, analytically or computationally, in the spectral domain (e.g., signal processing techniques in the physical domain involve some kind of convolution, which is reduced to simple multiplication in the spectral domain).

The formal treatment of the spectral representation of a random field in the space−time context presents no major technical problems. In particular, the following definition is useful in this respect.

Definition 2.5

An S/TRF can be expressed as the Fourier integral representation

$$X(\boldsymbol{p}) = \int d\boldsymbol{w} e^{i\boldsymbol{w}\cdot\boldsymbol{p}} \widetilde{X}(\boldsymbol{w}) \tag{2.23}$$

where $\boldsymbol{w} = (w_1, \ldots, w_n, w_0) \in R^{n+1}$ is the spectral vector argument, and $d\boldsymbol{w} = \prod_{i=1}^{n,0} dw_i$. The $\widetilde{X}(\boldsymbol{w})$ is termed the *spectral amplitude* of $X(\boldsymbol{p})$. The corresponding full spectrum or *spectral density function* (SDF) $\widetilde{c}_X(\boldsymbol{w}, \boldsymbol{w}')$ is defined by

$$c_X(\boldsymbol{p}, \boldsymbol{p}') = \iint d\boldsymbol{w}\, d\boldsymbol{w}' e^{i(\boldsymbol{w}\cdot\boldsymbol{p} - \boldsymbol{w}'\cdot\boldsymbol{p}')} \widetilde{c}_X(\boldsymbol{w}, \boldsymbol{w}'), \tag{2.24}$$

where $\widetilde{c}_X(\boldsymbol{w}, \boldsymbol{w}')$ is a positive summable function.

An obvious implication of Definition 2.5 is that the SDF forms an FT pair[15] with the spatiotemporal covariance. In the following chapters, I will revisit these topics in considerable detail.

When it is physically appropriate or analytically convenient to explicitly indicate the space and time arguments, we work in the $R^{n,1}$ domain, in which case Eq. (2.23) is replaced by

$$X(\boldsymbol{s},t) = \iint dkd\omega \; e^{i(\boldsymbol{k}\cdot\boldsymbol{s}\pm\omega t)}\widetilde{X}(\boldsymbol{k},\omega), \tag{2.25}$$

where the spatial frequency (wave)vector $\boldsymbol{k} = (k_1,\ldots, k_n)$ and the temporal frequency ω replace the spectral vector $\boldsymbol{w}$ of Eq. (2.23), in which case $d\boldsymbol{k} = \prod_{i=1}^{n} dk_i$. The notation $e^{i(\boldsymbol{k}\cdot\boldsymbol{s}+\omega\tau)}$ is used in the statistics literature, whereas the notation $e^{i(\boldsymbol{k}\cdot\boldsymbol{s}-\omega\tau)}$ is used in the physics literature because of its space−time metric associations. Actually, in Chapter III, I present a metric-based explanation of the process that leads from the spectral representation of Eq. (2.23) to that of Eq. (2.25). Furthermore, in many physical situations the following partial spectral representations of an S/TRF are used,

$$\begin{aligned} X(\boldsymbol{s},t) &= \int d\boldsymbol{k} e^{i\boldsymbol{k}\cdot\boldsymbol{s}}\widetilde{X}(\boldsymbol{k},t), \\ X(\boldsymbol{s},t) &= \int d\omega e^{-i\omega t}\widetilde{X}(\boldsymbol{s},\omega), \end{aligned} \tag{2.26a–b}$$

where $\widetilde{X}(\boldsymbol{k},t)$ and $\widetilde{X}(\boldsymbol{s},\omega)$ are, respectively, the "wavevector−temporal" and the "space−frequency" spectra. The use of the spatial harmonic factor $e^{i\boldsymbol{k}\cdot\boldsymbol{s}}$ is more convenient in certain physical situations (e.g., when studying wave phenomena), whereas the temporal harmonic factor $e^{-i\omega t}$ takes into account distinct temporal dependencies.

2.5 DATA-INDEPENDENT SPATIOTEMPORAL VARIABILITY FUNCTION

I now examine spatiotemporal variability measures that lie outside the framework of mainstream S/TVFs. Indeed, another interesting tool of measuring composite space−time variation (correlation or dependency) is introduced next.

Definition 2.6

The *space−time sysketogram function* is defined as

$$\beta_X(\boldsymbol{p},\boldsymbol{p}') = \overline{\log\left[f_X(\boldsymbol{p},\boldsymbol{p}')f_X^{-1}(\boldsymbol{p})f_X^{-1}(\boldsymbol{p}')\right]} \geq 0 \tag{2.27}$$

for all $\boldsymbol{p},\boldsymbol{p}' \in R^{n+1}$.

The sysketogram function is an information-theoretic S/TVF. It provides a measure of spatiotemporal correlation information in the sense of the amount of information on $X(\boldsymbol{p})$ that is contained in $X(\boldsymbol{p}')$ and vice versa. The sysketogram is definitely the tool of choice when the regime of interest is one in which there are not enough data to accurately calculate the covariance or the variogram function of the natural attribute, but there exists sufficient information about the PDF of the attribute. Another information-theoretic and data-independent measure of space−time correlation is defined next.

[15]The Fourier transformation (FT) is discussed in detail in Chapter V.

Definition 2.7

The space–time *contingogram* function is defined as

$$\psi_X(\boldsymbol{p},\boldsymbol{p}') = \overline{f_X(\boldsymbol{p},\boldsymbol{p}')f_X^{-1}(\boldsymbol{p})f_X^{-1}(\boldsymbol{p}')} - 1. \tag{2.28}$$

The readers may notice certain similarities between the definitions of β_X and ψ_X. Both β_X and ψ_X offer measures of the degree of $X(\boldsymbol{p})$'s departure from stochastic independence. As it turns out, β_X and ψ_X can be also expressed in terms of copulas, ς_X (to be defined in Chapter XI), i.e.,

$$\begin{aligned} \beta_X(\boldsymbol{p},\boldsymbol{p}') &= \overline{\log \varsigma_X(\boldsymbol{p},\boldsymbol{p}')}, \\ \psi_X(\boldsymbol{p},\boldsymbol{p}') &= \overline{\varsigma_X(\boldsymbol{p},\boldsymbol{p}')} - 1. \end{aligned} \tag{2.29a–b}$$

By using series expansions it is found that for small ς_X values, $\beta_X(\boldsymbol{p},\boldsymbol{p}') \approx \psi_X(\boldsymbol{p},\boldsymbol{p}')$. And, using Schwartz's inequality, the following inequality is valid between the correlation and the contingogram functions,

$$\rho_X^2(\boldsymbol{p},\boldsymbol{p}') \leq \psi_X(\boldsymbol{p},\boldsymbol{p}'). \tag{2.29c}$$

The above results motivate a direct comparison between mainstream and information-theoretic S/TVFs. The β_X and ψ_X have properties that sometimes may favor their use in place of the covariance c_X:

- **(i)** β_X and ψ_X measure the degree of departure from stochastic independence, whereas c_X measures the degree of departure from noncorrelation;
- **(ii)** while $\beta_X = \psi_X = 0$ only in the case of probabilistic independence, $c_X = 0$ when only space–time noncorrelation holds;
- **(iii)** β_X and ψ_X depend only on the PDF, whereas the c_X depends on both the PDF and the $X(\boldsymbol{p})$ values.
- **(iv)** β_X and ψ_X are not affected if $X(\boldsymbol{p})$ is replaced by an one-to-one function $\phi(X(\boldsymbol{p}))$, which is not the case with c_X.

Property (i) suggests that there are differences between mainstream and information-theoretic S/TVFs in what they attempt to measure. Property (ii) implies that β_X and ψ_X contain more information about space–time variation than c_X does or that β_X and ψ_X may uncover dependencies that c_X fails to uncover. Property (iii) is remarkable, since it implies that attribute data does not appear to be equally important to mainstream and information-theoretic S/TVFs. Property (iv) is useful in applications in which the concepts of "scale of measurement" and "instrument window" play an important role. Similarly, when the natural attribute has random space–time coordinates (e.g., distribution of aerosol particles), β_X and ψ_X are independent of the coordinate system chosen.[16]

An important point to be stressed is that whatever the differences in definition and classification are among the above functions (mainstream and information-theoretic), it is the accurate measurement of spatiotemporal variability that appears to underpin their objectives and means.

---

[16]This property brings to mind a basic result of modern physics, according to which only absolute quantities (independent of the space–time coordinate system) can be used as essential components of a valid physical law (in which case the term "covariant" is used).

Both the mainstream and the information-theoretic S/TVFs can be, in principle, extended to characterize the joint space–time variation of the random field values at more than two points. In particular, the mainstream two-point S/TVFs can be extended to m-point functions or moments as[17]

$$\overline{\Phi[X(\boldsymbol{p}_1), \ldots X(\boldsymbol{p}_m)]} = \int d\chi_{p_1} \cdots \int d\chi_{p_m} \Phi[\chi_{p_1}, \ldots, \chi_{p_m}] f_X(\chi_{p_1}, \ldots, \chi_{p_m}), \tag{2.30}$$

where the Φ is of the form of a polynomial in $\chi_{p_1}, \ldots, \chi_{p_m}$. Of course, it is not possible to represent all PDF models in terms of m-point moments. Lévy processes, e.g., have infinite moments of order $m > 2$ (Shlesinger et al., 1993). In such cases, the complete information about an S/TRF is included in the PDF itself. Indeed, if a PDF has a complicated shape, the only way to convey the information it contains regarding the S/TRF is to look at the complete picture as provided by Eq. (2.10a).

Similarly, in the multipoint case the sysketogram β_X and contingogram ψ_X functions can be also extended using copulas ς_X, as follows,

$$\begin{aligned} \beta_X(\boldsymbol{p}_1, \ldots \boldsymbol{p}_m) &= \overline{\log \varsigma_X(\boldsymbol{p}_1, \ldots \boldsymbol{p}_m)}, \\ \psi_X(\boldsymbol{p}_1, \ldots \boldsymbol{p}_m) &= \overline{\varsigma_X(\boldsymbol{p}_1, \ldots \boldsymbol{p}_m)} - 1, \end{aligned} \tag{2.31a–b}$$

respectively. Hence, as soon as the copula density $\varsigma_X(\boldsymbol{p}_1, \ldots \boldsymbol{p}_m)$ is calculated using standard techniques, the multipoint sysketogram $\beta_X(\boldsymbol{p}_1, \ldots \boldsymbol{p}_m)$ and contingogram $\psi_X(\boldsymbol{p}_1, \ldots \boldsymbol{p}_m)$ can be calculated too. A more systematic exposition of information-theoretic S/TVF issues can be found in Christakos (1991a,b,c, 1992, 2010).

2.6 SOME NOTICEABLE SPECIAL CASES OF THE SPATIOTEMPORAL RANDOM FIELD THEORY

It does not take much effort to realize that, formally, special cases of the S/TRF model $X(\boldsymbol{s}, t)$ are as follows:

the *spatial* random field (SRF), $X(\boldsymbol{s})$, $\boldsymbol{s} \in R^n$; and
the random *process* (RP) or *unidimensional* random field, $X(t)$, $t \in T$.

This essentially comes down to the issue of attribute independency on one or the other of its arguments. Specifically, natural attributes that are independent of the time argument may be modeled using SRF (e.g., random fields that represent the steady state of natural processes). On the other hand, phenomena that have only a temporal component are modeled as RP (e.g., global variables that represent spatial averages or processes localized in space, such as the level in a water reservoir or the concentration of a toxic substance in a specific organ).

The difference between an RP and an RF (spatial or spatiotemporal) may appear more significant from the viewpoint of fundamental physical theories. In particular, the evolution of an RP $X(t)$ is described by ordinary differential equations (ODEs). The $X(t)$ may obey, e.g., ODEs that represent mathematical expressions of fundamental physical laws governing the movement of particles as a function of time. The RFs $X(\boldsymbol{s})$ and $X(\boldsymbol{p})$, on the other hand, represent natural attributes the evolution of which is expressed by means of partial differential equations (PDEs). Particularly, the fundamental

[17]The calculation of this integral is tractable especially in cases in which the PDF form is relatively simple.

physical laws that these random fields obey are expressed by PDE with three (spatial) or four (spatiotemporal) independent variables.[18] In general, the solution of an ODE is a much simpler operation, both mathematically and computationally, than the solution of a PDE. In any case, in the stochastic and statistical literature the theories of SRF and RP are important topics in their own right. In the following sections, therefore, certain results concerning separately SRF and RP will be also presented.

2.7 SPACE–TIME SEPARABILITY

Because there is no assumptionless modeling, to prepare the readers for the inescapable inevitability of the assumptions to be introduced in following chapters, I consider below a rather simple one that is the result of the fact that the exact structure of the S/TVFs is in many cases not fully known. Due to this indeterminacy, certain constraints imposed on the form of these functions become a modeling decision. When physically justified, a convenient modeling decision is to use the assumption of separable space–time models.

Definition 2.8

A space–time *separable* covariance model has the following representations in the physical and the spectral domain, respectively,

$$\begin{aligned} c_X(\boldsymbol{s},t,\boldsymbol{s}',t') &= c_{X(1)}(\boldsymbol{s},\boldsymbol{s}')c_{X(2)}(t,t'), \\ \widetilde{c}_X(\boldsymbol{k},\omega,\boldsymbol{k}',\omega') &= \widetilde{c}_{X(1)}(\boldsymbol{k},\boldsymbol{k}')\widetilde{c}_{X(2)}(\omega,\omega'), \end{aligned} \tag{2.32a–b}$$

where $c_{X(1)}(\boldsymbol{s},\boldsymbol{s}') = c_X(\boldsymbol{s},0,\boldsymbol{s}',0)$ is a purely spatial function and $c_{X(2)}(t,t') = c_X(\boldsymbol{0},t,\boldsymbol{0},t')$ is a purely temporal covariance function; $\widetilde{c}_{X(1)}(\boldsymbol{k},\boldsymbol{k}')$ and $\widetilde{c}_{X(2)}(\omega,\omega')$ are the separable SDFs associated with the original SDF $\widetilde{c}_X(\boldsymbol{k},\omega,\boldsymbol{k}',\omega')$ of $c_X(\boldsymbol{s},t,\boldsymbol{s}',t')$.

Eq. (2.32b) is obtained from Eq. (2.32a) using the fact that the FT of a space–time separable function is separable itself. If in Eq. (2.32a) the $c_{X(1)}$ is functionally equal to $c_{X(2)}$, the separable covariance function is given by $c_X(\boldsymbol{s},t,\boldsymbol{s}',t') = c_{X(1)}(\boldsymbol{s},\boldsymbol{s}')c_{X(1)}(t,t')$ and is a symmetric function. A trivial case is the exponential covariance $c_X(\boldsymbol{s},t,\boldsymbol{s}',t') = e^{-u(\boldsymbol{s},\boldsymbol{s}')}e^{-u(t,t')}$, where u is a given function.

Separability is a modeling assumption that essentially implies that if we take snapshots of the space–time attribute variation at different times the covariances will be similar, i.e., they will differ only by a constant factor of proportionality throughout the spatial domain. The same is true for the covariance of the time series obtained at two different locations in space. The decoupling of space and time coordinates also simplifies considerably the mathematical analysis of separable models.

Indeed, the assumption of separability is very convenient on both analytical and computational grounds. For example, it simplifies the evaluation of complex integrals involving covariance functions and the estimation of generalized covariance representing attribute heterogeneities (Chapter XIII). Moreover, a large variety of known purely spatial and purely temporal covariance models can be combined in a straightforward manner to produce valid space–time models (Chapter XVI). Also, the composite space–time integral of Bochner's theorem of covariance permissibility reduces to the product of two separate integrals, one in space and one in time (Chapter XV).

[18]Keeping in mind, of course, that in certain applications ODE and PDE constitute approximations rather than direct expressions of fundamental physical laws.

Remark 2.6

Eq. (2.32a) actually introduces a *multiplicatively* separable model. Another well-known separable model is the *additively* separable covariance model of the form

$$c_X(\boldsymbol{h}, \tau) = c_{X(1)}(\boldsymbol{h}) + c_{X(2)}(\tau), \tag{2.33}$$

which is, perhaps, less popular than the multiplicative separable model, which is why our discussion focuses on the separability assumption of Eq. (2.32a).

Since many simplified but valuable results are obtained in the case of space−time separability, I will revisit this assumption and its implications in other parts of the book.

3. PHYSICAL INSIGHT BEHIND THE RANDOM FIELD CONCEPT

Perhaps, the best introduction to any discussion of random field insights is *Albert Einstein*'s famous statement:

> As far as the laws of mathematics refer to reality, they are not certain; and so far as they are certain they do not refer to reality.

To understand random field modeling we need to add to our language as well as to our imagination. This implies that, occasionally, certain aspects of our present thinking about physical reality may need to be reconsidered to take into account the new thinking required for the development and implementation of the random field model.

3.1 RANDOM FIELD REALIZATIONS

In methodological terms, an S/TRF model was defined above as an infinite number of realizations (possibilities) regarding the natural attribute under consideration, whereas the observed or measured attribute values, due to the conditions of in situ uncertainty, constitute only one such realization. The apparent statistical stability is achieved because of random factors rather than despite them. Yet, in random field modeling, physical reasoning, not just statistics, must retain a central role in data interpretation. In this context:

Simplicity follows truth and not the other way around.

Simplicity does not automatically bring truth, i.e., purely statistical ideas of simplicity are insufficient in natural sciences, since statistical simplicity does not necessarily accord with physical simplicity.

In more technical terms, the S/TRF $X(\boldsymbol{p})$ is viewed as the collection of all possible space−time realizations, $\chi_{\boldsymbol{p}}^{(j)} = \left(\chi_{\boldsymbol{p}_1}^{(j)}, \ldots, \chi_{\boldsymbol{p}_m}^{(j)}\right)$, $j = 1, \ldots, \mathcal{R}$ ($\mathcal{R}$ is the number of realizations considered) at the space−time points $\boldsymbol{p} = (\boldsymbol{p}_1, \ldots, \boldsymbol{p}_m)$ of the attribute domain. That is, the S/TRF $X(\boldsymbol{p})$ may be formally represented as the collection of realizations

$$X(\boldsymbol{p}) = \left\{\chi_{\boldsymbol{p}}^{(1)}, \ldots, \chi_{\boldsymbol{p}}^{(\mathcal{R})}\right\}, \tag{3.1}$$

where $\chi_{\boldsymbol{p}}^{(1)} = \left(\chi_{\boldsymbol{p}_1}^{(1)}, \ldots, \chi_{\boldsymbol{p}_m}^{(1)}\right), \ldots, \chi_{\boldsymbol{p}}^{(\mathcal{R})} = \left(\chi_{\boldsymbol{p}_1}^{(\mathcal{R})}, \ldots, \chi_{\boldsymbol{p}_m}^{(\mathcal{R})}\right)$.

The modeling perspective introduced by the S/TRF theory assigns to the possible attribute realizations $\chi_{\boldsymbol{p}}^{(j)}$ some worth noticing features (Christakos, 2010):

(i) The realizations are *consistent* with the physical properties, uncertainty sources, and space–time variations characterizing the attribute distribution (i.e., the multiplicity of realizations makes it possible to account for uncertainty sources and, at the same time, to adequately represent the spatiotemporal variation of a real-world attribute).

(ii) The realizations have the *epistemic quality* of corresponding to ways that are consistent with the known attribute features rather than to all possible ways an attribute could be represented in terms of formal logic.

(iii) The realizations have different *chances of occurrence*, in general; each realization has a distinct probability to occur that depends on the epistemic condition of the investigator and the underlying mechanisms of the in situ phenomenon.

An implication of the above considerations is that Eq. (3.1) could be rewritten in a more informative way as follows:

$$X(\boldsymbol{p}) = \left\{\chi_{\boldsymbol{p}}^{(1)} \text{ with probability } P^{(1)}, \ldots, \chi_{\boldsymbol{p}}^{(\mathcal{R})} \text{ with probability } P^{(\mathcal{R})}\right\} \tag{3.2}$$

where

$$P^{(j)} = P^{(j)}\left[\chi_{\boldsymbol{p}_i}^{(j)} \leq X(\boldsymbol{p}_i) \leq \chi_{\boldsymbol{p}_i}^{(j)} + d\chi_{\boldsymbol{p}}^{(j)};\ i = 1, \ldots, m\right] = f_X^{(j)}\left(\chi_{\boldsymbol{p}}^{(j)}\right) d\chi_{\boldsymbol{p}}^{(j)}, \tag{3.3}$$

where $j = 1, \ldots, \mathcal{R}$, $d\chi_{\boldsymbol{p}}^{(j)} = \prod_{i=1}^{m} d\chi_{\boldsymbol{p}_i}^{(j)}$. The probability model above is constructed in an integrated manner involving the available physical knowledge as well as principles of mathematical rigor and substantive reasoning.

The following example considers a metaphor that may offer additional insight about the notion of possible realizations.

Example 3.1

Consider the analogy between the random field realizations and mapmaking. First, suppose that the mapmaker in constructing his/her map of city *A* considers the motorist. The map should be such that it satisfies the needs of a motorist (highways, one-way streets, gas stations, parking places). Second, suppose that the mapmaker of city *A* considers another type of user, say, a bicyclist or a walker. Such a user will need a different map to follow a proper course (a map that shows sidewalks, dirt trails, small paths through houses and shops, etc.). In fact, the user will see an alternate reality of city *A* depending on how he/she moves through the territory. Each map will, accordingly, reflect that change in viewpoint. Thus, the user must have an alternate map for each possible use. And yet if we look at all of the maps together, we see an obvious pattern. In a similar manner, a proper random field–based mapping must take into account all of the ways that a user will engage in investigating, say, the porous medium. In one sense, there is an infinity-to-one correspondence between that we predict and what exists (but is unknown). Hence, it seems that we need an infinite number of maps (random field realizations) to guide us through the porous medium, and that should be formidable. Fortunately though, these infinities are within the grasp of human intelligence through the invention of mathematics. Yet, random field maps (realizations) are not ordinary maps. These maps are transparent. We can lay one atop the other to see patterns of sensible knowledge that we can use to make predictions with. In one sense, infinity of maps is reduced to one equation. A new map of the space of our mathematical imagination is conceived containing infinity as a mere point.

Remark 3.1

The fact that the S/TRF model consists of infinite realizations leads to the conceptualization of the random field as a *superspace*, which is a space that "contains" all possible realizations. Yet when all the differences between all these model realizations are taken into account, there appears to be an order arising. This order makes many realizations to appear with just the right values to make the model a reality. In other words, the mind of any sentient modeler that is capable of perceiving a reality is capable of reaching into alternate realizations and performing the task of deriving conclusions about that reality.

From a certain point of view, I suggest that the situation is as follows: Due to lack of knowledge regarding the underlying natural attributes, deterministic theories are inadequate to describe the real-world. Stochastic theories, while they face the same problem of lack of knowledge, they shift their focus from a single system to an ensemble of possible systems (sometimes called, the random field superspace). Then, instead of studying directly single system properties (which in many applications is impossible, anyway), one studies random field properties. Certain conclusions about the former can be subsequently derived on the basis of the latter.

3.2 PROBABLE VERSUS ACTUAL

According to the above perspective, the user of the S/TRF model is in control of possibilities (random field realizations), but not actualities. In other words, the control involves the user's mind in a way that the user could predict what was likely to occur, but not what will actually occur. That which does not exist in one realization but exists in some other realization shares certain important characteristics with that which actually exists. The S/TRF model describes the probable structure of a physical attribute but not its actual structure. Although S/TRF models are used in many applications, the concept of an ensemble of possible states is more appropriate in certain systems than in others. This fact is illustrated in the following examples.

Example 3.2

At the microscopic level, quantum effects and the thermal motion of molecules allow particles to explore various states from a large ensemble. The state of individual particles cannot be determined, but the behavior of the entire system is well described by ensemble averages. In the case of porous media and other environmental systems, on the other hand, we are interested in processes that do not have the dynamic freedom to explore states in the ensemble. In addition, it is possible to measure the properties of the medium at individual points in space–time. However, under ergodic conditions, ensemble averages still provide useful estimates of spatially averaged (coarse-grained) properties of the medium. In the case of strong fluctuations or long-range correlations, ensemble averages are not accurate estimators of coarse-grained behavior. In this case, the PDF of the S/TRF provides a useful measure of the space–time variability that can be expected based on the available knowledge.

Remark 3.2

Mutatis mutandis, the S/TRF concept is similar to the *parallel universe*[19] concept of modern physics, which posits the existence of worlds (realizations) within our technologically extended senses that must connect or relate with our own. The development of the random field and the parallel universe models follows similar paths. Both the random field and the parallel universe models introduce a new and apparently paradoxical way of thinking. Nothing in our previous thinking about the physical world can provide definite answers to questions regarding complex weather heterogeneities and variations, or the unexpected findings of new physics and general relativity. Without these models, the gaps of knowledge brought into light by the discoveries of earth and atmospheric sciences on the one side, and the new physics on the other side would remain unbridgeable—incapable of being solved by previous thinking. The random field model conjectures that, despite the apparent disorder presented by the many realizations, there has to be some order in the natural phenomenon they represent. This order should not be viewed in the physical sense. Instead, it should be considered in the sense that this order involves our minds (viz., our model) in a way that we could predict what was likely to occur but not what will actually occur. Quantum physics indicates the existence of the so-called quantum jumps, as well as the effect that an observer has on a physical system, and these phenomena cannot be objectively understood without the help of the parallel universe model. Similarly, in subsurface flow studies the random field model advocates that, due to our incomplete knowledge, we should not focus our study on a single porous medium, but rather on various possible media (parallel universes) each of which provides partial yet complementary information about the phenomenon. Interestingly, in all such cases the parallel universe model shows that the future can influence the present just as much as the past.

3.3 PROBABILITY AND THE OBSERVATION EFFECT

The idea of *probability* enters random field modeling as some kind of an epistemic idea (just as it enters statistical mechanics). Otherwise said, if the job of the natural laws is to describe physical causation at the level of ontology, the job of a probability theory is to describe human inferences at the level of epistemology. Being a *field of possibility*, not a real field, an S/TRF model describes the probable structure of a natural attribute in space–time but not its actual structure. At this point, the readers may recall that the concept of possibility or "tendency" for an event to occur plays a decisive role in Aristotle's philosophy. In sciences this concept is formulated as probability and subjected to laws of Nature.

In this sense, the S/TRF model allows for the existence of the so-called *observation effect*: Since this model consists of various realizations with various probabilities assigned to them, when a specific observation takes place, the probability of the corresponding realization changes from less than certain to certain. Loosely speaking, the random field model requires the consideration of that which (probably) does not exist to explain that which actually exists (but is unknown to us). Furthermore, that which does not exist in one realization but exists in some other realization shares certain important characteristics with that which actually exists. So in a sense:

That which probably isn't affects what is.

---

[19]Since the discoveries of the new physics (including relativity and quantum physics), the question of the existence of parallel universes–worlds that exist side-by-side along with our own has taken on renewed interest well beyond mere speculation. Remarkably, both quantum physics and relativity theory predict the existence of parallel universe.

This is very similar to what is happening with the parallel universe model of the new physics, where things that *are not* play a role in the world of everything that *is*. In other words, in the new physics a thing, such as an atomic electron, is represented by all of its possibilities, even those that may be remote. And what is true for the tiny electron is true for all the objects within the material universe, since all things are made up of electron-like things. An example may simplify the presentation of the theoretical discussion above.

Relevant to the observation effect is that the random field PDF itself must honor *constraints* imposed by the data. Constraining of the PDF is often referred to as conditioning to the data. Since an S/TRF consists of various realizations that are assigned specific probabilities based on prior knowledge, after a specific observation takes place, the PDF must be updated to account for the new constraint. More specifically, if an experimental value $X(\boldsymbol{p}_1) = \chi_1$ is observed at the point $\boldsymbol{p}_1$, the simplest conditioning approach is the collapsing of that PDF, i.e., requiring that $f_X(\chi|\chi_1) = \delta(\chi - \chi_1)$. In other words, at the observation points the probability of the corresponding realization changes from less than certain to certain.

3.4 SELF-CONSISTENCY AND PHYSICAL FIDELITY

A key insight into the development of the random field way of thinking is *self-consistency.* In other words, we base physical behavior on the principle that whatever takes place must be consistent with itself. The game of physical modeling, in general, is to invent principles that are self-consistent. And the rules governing S/TRF models are self-consistent too. The field realizations are the confluence of agreements concerning what is physically and logically consistent with the phenomenon itself. That which is not consistent with a phenomenon (given, say, the available evidence and our understanding of it) is not included in a random field model of the phenomenon. An example of such a process is the methodology underlying the *simulation* techniques that are widely used in physical sciences: The behavior of the simulation must be consistent with the actual behavior of the simulated attribute. Furthermore, the degree to which a random field simulation looks like the simulated attribute, i.e., the degree to which they have the same behavior, defines what is called the *physical fidelity* of the model.

In Section 1 of this chapter it was pointed out that beyond its probability laws and statistical moments, a characterization of an S/TRF can be made in terms of its generating processes (i.e., by means of the description of the way the random fields are generated). Naturally, this generating process involves physical theories and scientific models. In fact, the essence of an S/TRF model appears more significant from the perspective of fundamental physical theories. Next, I present a few examples of physical theories generating S/TRFs in a variety of applied sciences (the matter is addressed in considerable detail in Chapter XIV).

Example 3.3

The *Korteweg-de Vries* (*KdV*) equation is a stochastic PDE that models dispersive waves in a nonlinear (one-dimensional) space–time medium ($R^{1,1}$),

$$\frac{\partial}{\partial t}X(s,t) - 6X(s,t)\frac{\partial}{\partial s}X(s,t) + \frac{\partial^3}{\partial s^3}X(s,t) = Y(t), \tag{3.4}$$

where the S/TRF $X(s, t)$ represents the wave field and the $Y(t)$ is a zero-mean stationary Gaussian field. Given suitable initial conditions, the solution of Eq. (3.4) is (Orlowski and Sobczyk, 1989)

$$X(s,t) = V(t) - 2k^2 \sec h^2\left[ks - 4k^3 t + kw(t)\right], \tag{3.5}$$

where k is a function of known wave parameters, $V(t) = \int_0^t Y(u)du$ and $w(t) = 6\int_0^t V(u)du$. The $Y(t)$ may be assumed to be a white-noise process with covariance $c_Y(t,t') = 2\delta(\tau),\ \ \tau = t' - t$. The model of Eq. (3.4) emerges in the context of nonlinear wave phenomena, which play a very important role in hydrodynamics, meteorology, and geophysics.

Example 3.4

In climate modeling, the S/TRF $X(s, t)$, $(s, t) \in R^{n,1}$, may denote the space–time distribution of earth's surface temperature. Correlation analysis can provide valuable information regarding the patterns followed by temperature. If $V(s)$ denotes the space sampled around location s during the time period $T(t)$ (say, $t \pm \delta t$), the $V(s)T(t)$ will denote the space–time support of the sample, in which case the average temperature within $V(s)T(t)$ will be given by

$$\overline{X}(s,t) = \frac{1}{V(s)T(t)} \int_{V(s)} \int_{T(t)} dudv X(u+s, v+t), \tag{3.6}$$

$du = \prod_{i=1}^n du_i$. On the basis of $\overline{X}(s,t)$, maps of the averaged surface temperature over space and time can be constructed.

Example 3.5

In hydrogeology, the S/TRF $X(s, t)$, $(s, t) \in R^{n,1}$, may represent the precipitation intensity at location s and time t. Then, the total streamflow for an area V over the time period T is written as

$$\overline{X_f} = \int_V \int_T dsdt\, f(s,t)X(s,t), \tag{3.7}$$

where $ds = \prod_{i=1}^n ds_i$, and $f(s, t)$ is a weighting function. Similarly, if $X(s, t)$ is the rainfall rate at location s and time t, the average rainfall over an area V during the time period T is given by

$$\overline{X_{VT}} = \frac{1}{VT} \int_V \int_T dsdt\ X(s,t), \tag{3.8}$$

(e.g., Bell, 1987).

Example 3.6

In meteorological applications, $X(s, t)$ could represent the wind velocity at site s on day t. Then knowledge of the space–time correlation structure of the S/TRF $X(s, t)$ can be translated into quite precise knowledge of the average available kinetic energy in the wind at a new site (e.g., Haslett and Raftery, 1989). Moreover, weather processes such as dynamic, thermodynamic, and cloud microphysical processes operate over a variety of spatiotemporal scales, which are strongly related to each other, but they also interact with other variables such as soil moisture, surface topography, and roughness discontinuities. The spatiotemporal structure of these processes must be explored, if we hope to understand the physical interactions between hydrology, weather, and climate, and to construct reliable predictions.

Example 3.7

If $X(s, t)$, $(s, t) \in R^{n,1}$, is an S/TRF representing the concentration of aerosol substance and β_i are coefficients that account for the aerosol fraction that gets into the soil, then the quantity

$$\overline{X} = \sum_{i=1}^{k} \int_{D_i} \int_{T_i} ds dt \left(\frac{1}{T_i} + \beta_i \right) X(s, t), \tag{3.9}$$

provides a global measure of the environmental pollution over the ecologically important (nonoverlapping) zones D_i during the time periods T_i $(i = 1, \ldots, k)$. The $\overline{X}$ accounts for the total amount of aerosol that settles on the surface of the earth and is of considerable importance in making sound judgments regarding the pollution of soil and water. The effect of the latter on the environmental ecology can be significant within the framework of biocenosis.

Next, I briefly present a real-world case study, which may offer some early insight regarding the implementation of random fields and S/TVFs in practice (it is to be expected that when S/TVFs and other random field tools are implemented in practice, the results may indicate some differences between the theoretical and the empirically calculated S/TVFs).

Case Study 3.1

Fig. 3.1 shows the Shandong province, which is located in the east of China and covers a territory of about 157,900 km^2 (Christakos et al., 2017c). The $PM_{2.5}$ concentration data (in μg/m^3) that were used in this study have been obtained at 96 national air quality monitoring sites during the period January 1–31, 2014 (NAQM, 2014). Specifically, hourly data were recorded at each site, and the daily $PM_{2.5}$ concentrations were calculated by averaging the hourly concentration data. A summary of the descriptive statistics of the $PM_{2.5}$ data is presented in Table 3.1. The measured pollutant concentrations

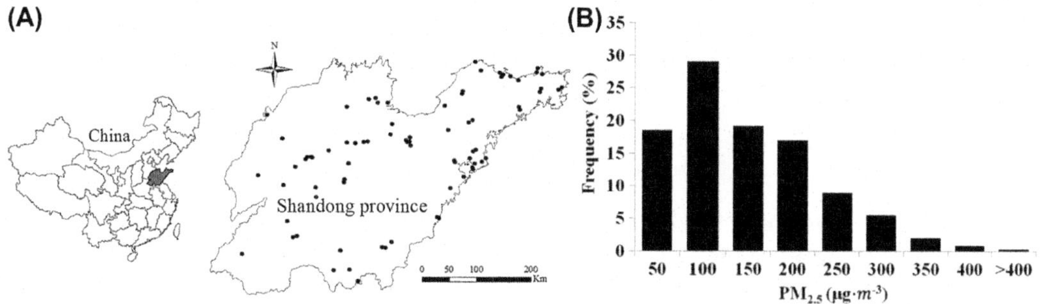

FIGURE 3.1

(A) Locations of $PM_{2.5}$ samples (in μg/m^3) in the Shandong province (China) during the time period January 1–31, 2014 and (B) the corresponding $PM_{2.5}$ data histogram.

Table 3.1 Summary Statistics of the Shandong $PM_{2.5}$ Data Set

Number of Data	Minimum (μg/m^3)	Maximum (μg/m^3)	Mean (μg/m^3)	SD (μg/m^3)	CV (%)
2163	4	666	126	81.02	64.3

Note: SD *and* CV *denote standard deviation and coefficient of variation, respectively.*

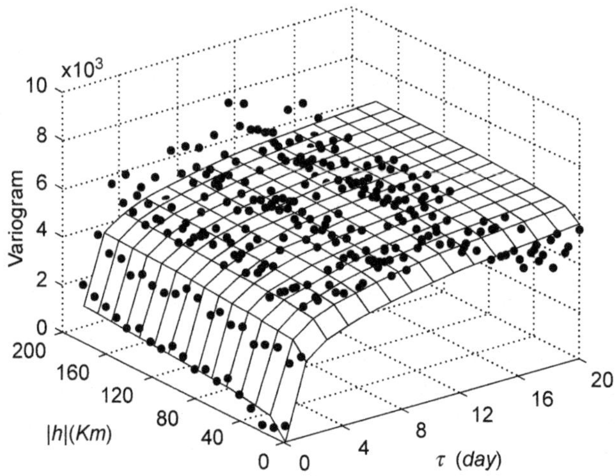

FIGURE 3.2

Empirical variogram [denoted with *black dots*, in $(\mu g/m^3)^2$] and theoretical variogram (*continuous line*) of the original space–time $PM_{2.5}$ distribution $\left(|\boldsymbol{h}| = \left(h_1^2 + h_2^2\right)^{\frac{1}{2}}\right)$.

ranged from 4 to 666 $\mu g/m^3$, and the mean concentration was 126 $\mu g/m^3$. The data coefficient of variation (CV) was 64.3%, which indicates that the $PM_{2.5}$ data exhibited a medium variability ($0.1 < CV < 1$). In this study, the space–time variation of $PM_{2.5}$ concentrations was assessed in terms of the variogram function defined in Eq. (2.17a). Particularly, the empirical space–time variogram of the $PM_{2.5}$ distribution in the Shandong region is plotted in Fig. 3.2. This variogram was calculated using the available $PM_{2.5}$ data at the 96 locations during the period January 1–31, 2014. To this empirical variogram the following theoretical variogram model was fitted (Fig. 3.2).

$$\gamma_X(|\boldsymbol{s}' - \boldsymbol{s}|, t' - t) = \sigma^2 \left[1 - \left(b(t'-t)^\alpha + 1\right)^{-\frac{\beta d}{2}} e^{\frac{-\kappa|\boldsymbol{s}-\boldsymbol{s}'|^{2r}}{(b(t'-t)^\alpha+1)^{\beta r}}} \right], \tag{3.10}$$

where $\sigma = 192.4$ ($\mu g/m^3$), $\alpha = 0.8276$, $\beta = 0.4986$, $b = 8.0215$, $d = 0.1422$, $\kappa = 0.0019$, and $r = 0.1281$. The variogram of Fig. 3.2 measures the strength of the dependence between $PM_{2.5}$ values across space–time (i.e., the variogram interpretation is that the stronger the spatiotemporal dependence, the smaller the variance between nearby pairs of $PM_{2.5}$ observations). The variogram above accounts for different components of $PM_{2.5}$ spread, and, due to its shape, it represents adequately the space–time variation of the spatially homogeneous/temporally stationary $PM_{2.5}$ pattern. The variogram magnitude increases rather fast, implying that strong $PM_{2.5}$ correlations appear at shorter space and time lags, whereas they are weaker at longer lags. At $\tau = 0$ the spatial correlation range is 200 km, whereas at $|\boldsymbol{h}| = 0$ the temporal correlation range is 20 days.

It is worth noticing that the arguments of the variogram of Eq. (3.10) and Fig. 3.2 are not the individual locations $\boldsymbol{s}'$, $\boldsymbol{s}$ and times t, t', as in the definition of Eq. (2.17a), but rather the space lag $|\boldsymbol{s}' - \boldsymbol{s}| = \left(\left(s_1' - s_1\right)^2 + \left(s_2' - s_2\right)^2\right)^{\frac{1}{2}} = |\boldsymbol{h}|$ (in km), and time separation $\tau = t' - t > 0$ (in days).

This means that the variogram of Eq. (3.10) is a limited variogram model that depends only on the distance between two locations and the separation between two time instants, which prepares the readers for the special class of variogram functions to be formally introduced in Chapter VII, which is based on the assumption of space–time homostationarity.

Before leaving this section, it is worth noticing that to study the S/TVFs (ensemble moments or expectations) of a natural attribute rather than specific random field realizations is in many cases a quite legitimate way of confronting scientific theory with physical measurements and making informative predictions. Yet, S/TVFs make sense only when there is some understanding or working hypothesis about the phenomenon. This observation emphasizes the importance of incorporating physical laws and other kinds of data and information in the study of an uncertain situation. Physical laws should be incorporated in the stochastic analysis by means of models, and the models should respect the available information.

4. GEOMETRY OF SPATIOTEMPORAL RANDOM FIELDS

The geometry of an S/TRF concerns its *regularity* properties (continuity, differentiability) and is studied in considerable detail in Chapter VI. Yet, for completeness of presentation in this chapter, I make some initial and rather general comments concerning the matter.

Just as for ordinary (deterministic) functions, the space–time regularity characteristics of a random field $X(\boldsymbol{p})$ will be considered with respect to its varying space–time coordinates $\boldsymbol{p} = (s_1,\ldots, s_n, s_0) = (\boldsymbol{s}, t)$. In particular, we may consider regularity with respect to

(i) varying both space $\boldsymbol{s}$ and time $t = s_0$,
(ii) varying space $\boldsymbol{s}$ and fixed time $t = s_0$ (i.e., simultaneous space variation), or
(iii) varying time $t = s_0$ and fixed space $\boldsymbol{s}$.

In many applications this distinction makes sense not only on formal but on physical grounds too, i.e., the existing physical conditions may give priority to one of these three possibilities over the others (e.g., the physical law focuses on attribute change with space, whereas the temporal dynamics are occasionally suppressed).

More specifically, concerning Possibility (i) above, spatial and temporal regularity properties can be considered in the same formal setting; that is, with respect to each coordinate of the space–time point vector $\boldsymbol{p}$. In physical situations, on the other hand, a natural attribute may depend on the space argument in a different way than it depends on the time argument. Concerning Possibility (ii), many regularity results that have been derived for spatial random fields can be formally extended to composite space–time fields (i.e., the study of the regularity properties carries through as before), as long as the time argument does not change, or by assuming that the space–time metric has a specific form (see, Section 2 of Chapter I and Chapter III). Physically, Possibility (ii) makes sense when the particular study is interested about the spatial regularity features of an otherwise spatiotemporal phenomenon, in which case time may be suppressed. Lastly, Possibility (iii) is usually seen as a special (unidimensional) case of Possibility (ii), where now location is fixed and time varies.

As we saw in Section 2, we can view an S/TRF $X(\boldsymbol{p})$ as producing a sample path or realization $\chi(\boldsymbol{p})$ for each random event $u \in \Omega$ that goes through all space–time points $\boldsymbol{p}$. And, just as for ordinary functions where questions about the regularity properties of $\chi(\boldsymbol{p})$ are transferred to questions about the convergence of the ordinary series $\{\chi(\boldsymbol{p}_m)\}, m = 1, 2\ldots$, regularity questions concerning the S/TRF $X(\boldsymbol{p})$ boil down to the notion of stochastic convergence of the corresponding sequence of RVs $\{x_m(u)\}$ studied in Section 3 of Chapter I, with the important feature that these RVs are here assumed to be continuously $\boldsymbol{p}$ dependent. That is, the sequence $\{x_m(u)\}$ in Eq. (3.26) of Chapter I is written as the infinite sequence

$$\{x_m(u, \boldsymbol{p}_m)\} \text{ as } m \to \infty. \tag{4.1}$$

Accordingly, the various modes of RV convergence correspond to the various modes of S/TRF regularity. In particular, the $X(\boldsymbol{p})$ (or $X(\boldsymbol{p},u)$):

(a) is continuous at $\boldsymbol{p}^*$ in R^{n+1} if

$$\lim_{\boldsymbol{p}\to\boldsymbol{p}*} X(\boldsymbol{p}) = X(\boldsymbol{p}^*), \tag{4.2}$$

where the limit is considered in the sense of the corresponding RV sequence; and

(b) is differentiable of orders ν in space and μ in time at point $\boldsymbol{p}^*$ in R^{n+1} if

$$\lim_{\Delta\boldsymbol{p}\to 0} \frac{\Delta^{\nu+\mu}_{\Delta\boldsymbol{p}} X(\boldsymbol{p}^*)}{\prod_{i=1}^{\nu} \Delta s_i \Delta s_0^{\mu}} = \lim_{h_1,\ldots,h_\nu,h_0\to 0} \frac{\Delta^{\nu+\mu}_{\boldsymbol{h},h_0} X(\boldsymbol{p}^*)}{\prod_{i=1}^{\nu} h_i h_0^{\mu}} = X^{(\nu,\mu)}(\boldsymbol{p}^*), \tag{4.3}$$

where the limit is also considered in the earlier sense (Section 3.3 of Chapter I), $X^{(\nu,\mu)}(\boldsymbol{p})$ denotes the partial random field derivative of orders ν in space and μ in time, and $\Delta^{\nu+\mu}_{\boldsymbol{h},h_0} X(\boldsymbol{p})$ is a space−time difference operator (see detailed expressions in Section 4 of Chapter VI). Often, the differentiation of $X(\boldsymbol{p})$ is along the coordinate directions in the $R^{n,1}$ domain with $\nu = n$.

As it turns out, two key aspects characterize space−time random field geometry: (i) several important features of the S/TRF regularity definitions of Eqs. (4.2) and (4.3) are linked to the corresponding regularity properties of the S/TVFs,[20] and (ii) the quantitative representations of these properties are particularly tractable analytically in the case of certain restrictive but very important classes of random fields, such as the space−time homostationary, the space−time isostationary, and the Gaussian classes of random fields. In view of the aspects (i) and (ii), the more detailed study of random field regularity will have to be postponed until Chapters VI and VII, necessarily following a more detailed and adequate presentation of the most important results regarding these special classes of random fields.

5. VECTOR SPATIOTEMPORAL RANDOM FIELDS

The foregoing theory deals with a single S/TRF, which is why it has been termed a scalar S/TRF. However, the theory can be extended to include a set of S/TRFs correlated to one another.

Example 5.1

For illustration, in geophysics we often need to use a set of correlated random fields to represent mathematically key material parameters, such as density, stiffness matrix components, and P- and S-wave velocities or slownesses.

[20]This comment further supports the essential role of S/TVFs in random field modeling.

Definition 5.1

A *vector S/TRF* (VS/TRF), $\boldsymbol{X}(\boldsymbol{p})$, is a set of random fields described by the vector notation

$$\boldsymbol{X}(\boldsymbol{p}) = [X_1(\boldsymbol{p})\ldots X_k(\boldsymbol{p})]^T, \tag{5.1}$$

where the individual S/TRFs $X_1(\boldsymbol{p}),\ldots, X_k(\boldsymbol{p})$ are called the component random fields of the VS/TRF.

A note should be made here that, as we shall see in Chapter IX, for physical interpretation purposes it may be appropriate to use the term "vector S/TRF" for the specific case $k=n$, and the term "multivariate S/TRF" for the case $k \neq n$, in general. At the moment, the terminology of Definition 5.1 is kept, for the sake of generality. The specification of a VS/TRF $\boldsymbol{X}(\boldsymbol{p})$ is made by analogy with that of a scalar S/TRF $X(\boldsymbol{p})$. That is, all the multivariate PDFs of the component S/TRF $X_1(\boldsymbol{p}),\ldots, X_k(\boldsymbol{p})$ should be determined. By extending the scalar PDF definition of Eq. (2.10a) above, the *km*-PDF of a VS/TRF $\boldsymbol{X}(\boldsymbol{p})$ with k components at a set of points $\boldsymbol{p}_1,\ldots, \boldsymbol{p}_m$ is defined by

$$f_X\left(\chi_1^{(1)},\ldots,\chi_m^{(k)}\right)d\chi_1^{(1)}\ldots d\chi_m^{(k)} = P_X\left[\chi_j^{(i)} \leq x_j^{(i)} < \chi_j^{(i)} + d\chi_j^{(k)};\ i = 1,\ldots,k,\ j = 1,\ldots,m\right] \tag{5.2}$$

A VS/TRF $X(\boldsymbol{p})$ is described completely if all *km*-PDFs in Eq. (5.2) are known. Of course, the *km*-PDF must be such that the usual conditions, like PDF symmetry and compatibility, are satisfied.

Some other earlier results can be extended in the case of the VS/TRF. An important property of the RV notion is Eq. (3.7) of Chapter I, which can be extended in the present setting, as follows. Let $g(\chi_1,\ldots, \chi_k)$ be a Borel-measurable function of the real variables $\chi_1,\ldots, \chi_k$. Under these conditions, if $X_1(\boldsymbol{p}),\ldots, X_k(\boldsymbol{p})$ are S/TRFs, then the

$$Y(\boldsymbol{p}) = g[X_1(\boldsymbol{p}),\ldots,X_k(\boldsymbol{p})] \tag{5.3}$$

is an S/TRF too. The space of g-functions includes continuous functions, summations, and products.

Example 5.2

Let $X_1(\boldsymbol{p}),\ldots, X_k(\boldsymbol{p})$ be S/TRFs in R^{n+1}. Then, the

$$Y_1(\boldsymbol{p}) = \sum_{i=1}^{k} a_i X_i(\boldsymbol{p}), \tag{5.4}$$

where $a_i \in R^1$, and the

$$Y_2(\boldsymbol{p}) = \prod_{i=1}^{k} X_i(\boldsymbol{p}) \tag{5.5}$$

are valid S/TRF too.

The S/TVFs (second-order statistics) of a VS/TRF $\boldsymbol{X}(\boldsymbol{p})$ can be characterized in a manner analogous to that used in the scalar case. Specifically, the stochastic characterization of the *vector* S/TVF includes the following:

(a) The $k \times 1$ *spatiotemporal mean* functions of the elements of the vector $\boldsymbol{X}(\boldsymbol{p})$ of Eq. (5.1) are given by

$$\overline{X_l(\boldsymbol{p})} = \int d\chi_{l,\boldsymbol{p}}\chi_{l,\boldsymbol{p}} f_X(\chi_{l,\boldsymbol{p}}),\quad (l = 1,\ldots,k). \tag{5.6}$$

(b) The $k \times k$ *centered spatiotemporal covariance matrix* $\boldsymbol{c}_X(\boldsymbol{p},\boldsymbol{p}')$ is a square $k \times k$ matrix

$$\boldsymbol{c}_X(\boldsymbol{p},\boldsymbol{p}') = \left[c_{X_l X_{l'}}(\boldsymbol{p},\boldsymbol{p}')\right],\quad (l,l' = 1,2,\ldots,k), \tag{5.7a}$$

with elements

$$\begin{aligned} c_{X_l X_{l'}}(\boldsymbol{p},\boldsymbol{p}') &= \overline{[X_l(\boldsymbol{p}) - \overline{X_l(\boldsymbol{p})}]\,[X_{l'}(\boldsymbol{p}') - \overline{X_{l'}(\boldsymbol{p}')}]} \\ &= \iint d\chi_{l,\boldsymbol{p}} d\chi_{l',\boldsymbol{p}'} \left(\chi_{l,\boldsymbol{p}} - \overline{X}_l\right)\left(\chi_{l',\boldsymbol{p}'} - \overline{X}'_{l'}\right) f_X\left(\chi_{l,\boldsymbol{p}}, \chi_{l',\boldsymbol{p}'}\right). \end{aligned} \quad (5.7b\text{–}c)$$

Specifically, the diagonal elements of $\boldsymbol{c}_X(\boldsymbol{p},\boldsymbol{p}')$ are single-field covariances, whereas the off-diagonal elements are called *cross-covariances*, i.e., they involve a random field pair, and are such that

$$\left|c_{X_l X_{l'}}(\boldsymbol{p},\boldsymbol{p}')\right| \leq \left[\sigma^2_{X_l}(\boldsymbol{p})\sigma^2_{X_{l'}}(\boldsymbol{p}')\right]^{\frac{1}{2}}. \quad (5.8)$$

The matrix $\boldsymbol{c}_X(\boldsymbol{p},\boldsymbol{p}')$ is not necessarily symmetric, since the cross-covariances $c_{X_l X_{l'}}(\boldsymbol{p},\boldsymbol{p}')$ and $c_{X_{l'} X_l}(\boldsymbol{p},\boldsymbol{p}')$ do not necessarily coincide when $l \neq l'$. Yet, the $\boldsymbol{c}_X(\boldsymbol{p},\boldsymbol{p}')$ is symmetric in the sense that $\boldsymbol{c}^T_X(\boldsymbol{p},\boldsymbol{p}') = \boldsymbol{c}_X(\boldsymbol{p}',\boldsymbol{p})$, since it holds that $c_{X_l X_{l'}}(\boldsymbol{p},\boldsymbol{p}') = c_{X_{l'} X_l}(\boldsymbol{p}',\boldsymbol{p})$. The matrix $\boldsymbol{c}_X(\boldsymbol{p},\boldsymbol{p}')$ may not be NND, yet the $\boldsymbol{c}_X(\boldsymbol{p}, \boldsymbol{p})$ is symmetric and NND. The spatiotemporal variance matrix is given by $\boldsymbol{c}_X(\boldsymbol{p},\boldsymbol{p}) = \boldsymbol{\sigma}^2_X(\boldsymbol{p})$.

(c) The $k \times k$ *spatiotemporal variogram matrix* $\boldsymbol{\gamma}_X(\boldsymbol{p},\boldsymbol{p}')$ with elements

$$\gamma_{X_l X_{l'}}(\boldsymbol{p},\boldsymbol{p}') = \frac{1}{2}\overline{[X_l(\boldsymbol{p}) - X_l(\boldsymbol{p}')][X_{l'}(\boldsymbol{p}) - X_{l'}(\boldsymbol{p}')]}. \quad (5.9)$$

The diagonal elements of the matrix $\boldsymbol{\gamma}_X(\boldsymbol{p},\boldsymbol{p}')$ are single-field variograms, whereas the off-diagonal elements are called *cross-variograms*. The variogram matrix is CND and symmetric,

$$\begin{aligned} \boldsymbol{\gamma}_X(\boldsymbol{p},\boldsymbol{p}') &= \boldsymbol{\gamma}_X(\boldsymbol{p}',\boldsymbol{p}) \\ \boldsymbol{\gamma}_X(\boldsymbol{p},\boldsymbol{p}') &= \boldsymbol{\gamma}^T_X(\boldsymbol{p}',\boldsymbol{p}), \end{aligned} \quad (5.10a\text{–}b)$$

and, also, $\boldsymbol{\gamma}_X(\boldsymbol{p}, \boldsymbol{p}) = \boldsymbol{0}$, in principle. The *spectral density matrices* $\widetilde{\boldsymbol{c}}_X(\boldsymbol{w},\boldsymbol{w}')$ and $\widetilde{\boldsymbol{\gamma}}_X(\boldsymbol{w},\boldsymbol{w}')$ of the vector random field $\widetilde{\boldsymbol{X}}(\boldsymbol{w})$ in the spectral domain can be defined in an analogous manner as those of $\widetilde{X}(\boldsymbol{w})$ in the scalar case.

Example 5.3

Let us recall the geophysical material parameters described in Example 5.1. Microscale heterogeneities (e.g., at scales shorter than the wavelength) of material parameters $\boldsymbol{X}(\boldsymbol{p})$ can affect the propagation of seismic waves at certain frequencies, so that they correspond to frequency-dependent effective material parameters $\widetilde{\boldsymbol{X}}(\boldsymbol{s},\omega)$ that are "locally homogeneous." As regards the study of macroscale heterogeneities (e.g., seismic travel-time tomography), the material parameters may also be replaced by the effective material parameters (Klimes, 2002a,b).

Remark 5.1

As has been emphasized on several occasions throughout the book, on theoretical modeling grounds any set of random fields can be considered as a VS/TRF $\boldsymbol{X}(\boldsymbol{p})$. Yet, on physical grounds some distinctions certainly apply. Consider, e.g., the modeling of the attributes, temperature $T(\boldsymbol{p})$, pressure $P(\boldsymbol{p})$, density $D(\boldsymbol{p})$, and velocity $\boldsymbol{V}(\boldsymbol{p})$, in a three-dimensional space–time physical situation. Although mathematically these attributes jointly form a six-component vector random field

$$\boldsymbol{X}(\boldsymbol{p}) = [T(\boldsymbol{p})\ P(\boldsymbol{p})\ D(\boldsymbol{p})\ V_1(\boldsymbol{p})\ V_2(\boldsymbol{p})\ V_3(\boldsymbol{p})]^T,$$

physically it may be more appropriate to refer to three scalar fields, $T(\boldsymbol{p})$, $P(\boldsymbol{p})$, $D(\boldsymbol{p})$, and a vector field $\boldsymbol{V}(\boldsymbol{p}) = [V_1(\boldsymbol{p})\ V_2(\boldsymbol{p})\ V_3(\boldsymbol{p})]^T$.

6. COMPLEX SPATIOTEMPORAL RANDOM FIELDS

In addition to the *real*-valued random fields discussed so far, one may also consider *complex*-valued random fields, scalar and vector. Among the reasons for this consideration is that in some cases complex-valued random fields are mathematically more convenient or they better represent the actual phenomenon (in other words, a complex S/TRF model is adequate and useful if it represents accurately those features of the actual phenomena that it is intended to describe, explain, or theorize). I will continue with the standard definition of a complex random field, which is a rather simple and straightforward combination of real random fields.

Definition 6.1

A *complex S/TRF* is defined as the decomposition

$$X(\boldsymbol{p}) = X_R(\boldsymbol{p}) + i\,X_I(\boldsymbol{p}), \tag{6.1}$$

where

$$\begin{aligned} X_R(\boldsymbol{p}) &\equiv \mathrm{Re}\{X(\boldsymbol{p})\}, \\ X_I(\boldsymbol{p}) &\equiv \mathrm{Im}\{X(\boldsymbol{p})\}, \end{aligned} \tag{6.2a–b}$$

are real random fields in R^{n+1}.

In a way analogous to Eqs. (2.10a–b), a complex random field $X(\boldsymbol{p})$ is stochastically described completely by the corresponding PDFs

$$\begin{aligned} f_{X_R}(\chi_{R,1},\ldots,\chi_{R,m})d\chi_{R,1}\ldots d\chi_{R,m} &= P\left[\chi_{R,i} \leq x_{R,i} < \chi_{R,i} + d\chi_{R,i};\ i=1,\ldots,m\right], \\ f_{X_I}(\chi_{I,1},\ldots,\chi_{I,m})d\chi_{I,1}\ldots d\chi_{I,m} &= P\left[\chi_{I,i} \leq x_{I,i} < \chi_{I,i} + d\chi_{I,i};\ i=1,\ldots,m\right]. \end{aligned} \tag{6.3a–b}$$

In some other cases, the joint probabilities of $X_R(\boldsymbol{p})$ and $X_I(\boldsymbol{p})$ may be needed. Multipoint PDFs of complex random fields are obtained in a similar manner. Matters simplify in special cases, such as the following. A complex random field $X(\boldsymbol{p})$ is said to be Gaussian if $\sum_{i=1}^{m} a_i x(\boldsymbol{p}_i)$ is a complex Gaussian RV for all m and (complex) numbers a_i. I will revisit complex random fields in Section 6 of Chapter IV. It must be noted, however, that in the majority of applications, real random fields are encountered.

7. CLASSIFICATIONS OF THE SPATIOTEMPORAL RANDOM FIELD MODEL

The S/TRF model may be classified in at least five distinct ways. Brief descriptions of the classifications are given here (extensive discussions of them are given in later chapters, where more detailed assessments of their potential uses are also presented).

7.1 FIRST CLASSIFICATION: DISCRETE VERSUS CONTINUOUS ARGUMENTS

The first way of classifying random fields depends on whether the spatiotemporal argument, $\boldsymbol{p}$ and the realizations of the S/TRF $X(\boldsymbol{p})$ are discrete or continuous. In particular, the following combinations are possible:

(i) discrete $\boldsymbol{p}$—discrete $X(\boldsymbol{p})$;
(ii) discrete $\boldsymbol{p}$—continuous $X(\boldsymbol{p})$;
(iii) continuous $\boldsymbol{p}$—discrete $X(\boldsymbol{p})$; and
(iv) continuous $\boldsymbol{p}$—continuous $X(\boldsymbol{p})$.

The cases (i) and (ii) can be derived from cases (iii) and (iv), respectively, by means of direct discretization.

Example 7.1

Real-world illustrations of the above random field classification include the "number of disease cases" at a finite number of geographical locations during specific days (case i), the "monthly maximum wind speed" at specific monitoring locations of the sea shore (case ii), the "soil type indicated by a numerical value" versus depth during a specified time period (case iii), and the "contaminant concentration distribution" in a geographical region as a function of time (case iv).

7.2 SECOND CLASSIFICATION: SCALAR VERSUS VECTOR RANDOM FIELDS AND ARGUMENTS

The second way of classifying S/TRFs depends on whether the argument and the random field are scalar or vector. More specifically, four possibilities can be distinguished here:

(i) scalar argument and scalar random field, viz., RP $X(s)$, $s \in R^1$, or $X(t)$, $t \in T$;
(ii) scalar argument and several random fields, viz., vector RP $\boldsymbol{X}(s)$, $s \in R^1$, or $\boldsymbol{X}(t)$, $t \in T$;
(iii) vector argument and scalar random field, viz., S/TRF $X(\boldsymbol{p})$; and
(iv) vector argument and vector random field, viz., vector S/TRF $\boldsymbol{X}(\boldsymbol{p})$.

Example 7.2

Numerous illustrations of the above classification can be found in the real-world, including the temporal change of ocean wave heights (case i), the combined variation of atmospheric pressure, temperature, and humidity with height (case ii), the space–time pattern of contaminant concentration (case iii), and the distributions of the three fluid velocity components as functions of space–time (case iv).

7.3 THIRD CLASSIFICATION: PROBABILITY LAW SHAPES

Another way of classifying S/TRFs is by means of the form of the corresponding probability law. Broadly speaking, according to this classification one can distinguish between Gaussian and non-Gaussian random fields.

The S/TRF $X(\boldsymbol{p})$ is termed *Gaussian* (or *Normal*) if all its finite-dimensional PDFs are multivariate Gaussian, i.e., they are given by

$$f_{\boldsymbol{p}_1,\ldots,\boldsymbol{p}_m}(\chi_1,\ldots,\chi_m) = c^{-1} e^{-\frac{1}{2}\sum_{i,j=1}^{m} c_{ij}(\chi_i - a_i)(\chi_j - a_j)}, \tag{7.1}$$

Table 7.1 Relationships Between the Statistics of the Normal and Lognormal Random Fields

$\overline{Y} = e^{\overline{X}+\frac{1}{2}\sigma_X^2}$	$\overline{X} = \ln\frac{\overline{Y}^2}{\sqrt{\overline{Y}^2+\sigma_Y^2}}$
$\overline{Y^k} = e^{k\overline{X}+\frac{1}{2}k^2\sigma_X^2},\ k \geq 1$	
$\sigma_Y^2 = e^{2\overline{X}+\sigma_X^2}\left[e^{\sigma_X^2} - 1\right]$	$\sigma_X^2 = \ln\left[\frac{\sigma_Y^2}{\overline{Y}^2} + 1\right]$
$c_Y(\boldsymbol{p},\boldsymbol{p}') = \overline{Y}^2\left[e^{c_X(\boldsymbol{p},\boldsymbol{p}')} - 1\right]$	$c_X(\boldsymbol{p},\boldsymbol{p}') = \ln\left[\frac{c_Y(\boldsymbol{p},\boldsymbol{p}')}{\overline{Y}^2} + 1\right]$

where c, a_i, a_j, and c_{ij} $(i, j = 1,\ldots, m)$ are coefficients determined in terms of the corresponding means $(\overline{X})$, variances (σ_X^2), and covariance functions $(c_X(\boldsymbol{p},\boldsymbol{p}'))$, see Christakos (1992). This special kind of random field is also called the *m-variate* Gaussian (or Normal) S/TRF and is denoted as

$$\begin{aligned} X(\boldsymbol{p}) &\sim G^m\left(\overline{X}_i, \sigma_{X_i}^2,\quad i = 1, \ldots, m\right), \\ &\text{or } N^m\left(\overline{X}_i, \sigma_{X_i}^2,\quad i = 1, \ldots, m\right). \end{aligned} \tag{7.2a–b}$$

This sort of random field is frequently encountered in applications (e.g., classical statistical modeling of ocean waves are based on the assumption that the waves follow a Gaussian process), and it has several important properties, including the following: (a) a Gaussian random field is completely characterized by its mean and covariance, (b) if a random field is Gaussian, any linear transformation of the field is a Gaussian field too, and (c) many useful analytical expressions are obtained strictly under this assumption.

Example 7.3

Among the most notable non-Gaussian, but Gaussian-related, S/TRFs is the m-variate *lognormal* (L^m) random field $Y(\boldsymbol{p})$ defined by

$$\begin{aligned} &Y(\boldsymbol{p}) \sim L^m\left(\overline{Y}_i, \sigma_{Y_i}^2,\quad i = 1, \ldots, m\right), \\ &\text{s.t. } \ln Y(\boldsymbol{p}) = X(\boldsymbol{p}) \sim G^m\left(\overline{X}_i, \sigma_{X_i}^2,\quad i = 1, \ldots, m\right). \end{aligned} \tag{7.3a–b}$$

The relationships between the statistics of the lognormal random field $Y(\boldsymbol{p})$ and those of the associated Normal random field $X(\boldsymbol{p})$ can be formulated analytically, and they are given in Table 7.1.

Additional results about the Gaussian and the lognormal classes of S/TRFs will be presented in the following chapters. Also, the fundamental random field classification that is based on the shape of the underlying probability law includes the class of the so-called space–time *factorable* random fields (to be discussed in Chapter XI).

7.4 FOURTH CLASSIFICATION: SPACE–TIME VARIABILITY

The multidimensionality of the space–time domain opens new possibilities as regards the random field's behavior. Specifically, an S/TRF $X(\boldsymbol{p})$ can be classified according to its spatiotemporal variability pattern over $\boldsymbol{p} \in R^{n+1}$ (or $R^{n,1}$). An observed $X(\boldsymbol{p})$ may exhibit a space–time *heterogeneous* (i.e., space nonhomogeneous/time nonstationary) variation or a space–time *homostationary* (i.e., space homogeneous/time stationary) variation, briefly STHS.

Mathematically, these types of variation can be expressed in terms of PDFs (*strict sense, s.s.*) or in terms of S/TVFs (*wide sense, w.s.*). Generally, s.s. STHS variation refers to a random field with probabilistic properties that are independent of the space–time origin, whereas in the case of w.s. STHS variation only the S/TVFs of the random field are independent of the space–time origin. There is a substantial difference in the severity of the restrictions that these two types of assumptions impose on random field variation. It immediately becomes apparent that w.s. STHS is a weaker assumption than s.s. STHS. The two assumptions are not mutually exclusive. In fact, s.s. STHS implies w.s. STHS, but not the other way around.

In more technical terms, the random field $X(\boldsymbol{p})$ will be called s.s. STHS in $\boldsymbol{p}$ if all PDFs are invariant under the translation $U_{\delta p}\boldsymbol{p} = \boldsymbol{p} + \delta\boldsymbol{p}$,[21] i.e.,

$$f_{\boldsymbol{p}_1,\ldots,\boldsymbol{p}_m}(\chi_1,\ldots,\chi_m) = f_{U_{\delta p}\boldsymbol{p}_1,\ldots,U_{\delta p}\boldsymbol{p}_m}(\chi_1,\ldots,\chi_m). \tag{7.4}$$

(also, Chapter VII). A vector S/TRF $X(\boldsymbol{p}) = [X_1(\boldsymbol{p})\ldots X_k(\boldsymbol{p})]^T$ is called s.s. STHS if all km-PDFs of Eq. (5.2) are invariant under U_δ, in which case the component fields $X_l(\boldsymbol{p})$ $(l = 1,\ldots, k)$ are called STHS and homostationarily connected. On the other hand, the $X(\boldsymbol{p})$ is termed a w.s. STHS random field if a similar expression is valid in terms of its up to second-order space–time statistics, viz.,

$$\begin{aligned} \overline{X(\boldsymbol{p})} &= \overline{X(U_{\delta p}\boldsymbol{p})}, \\ c_X(\boldsymbol{p},\boldsymbol{p}') &= c_X(U_{\delta p}\boldsymbol{p}, U_{\delta p}\boldsymbol{p}') = c_X(\boldsymbol{p}-\boldsymbol{p}'), \\ \gamma_X(\boldsymbol{p},\boldsymbol{p}') &= \gamma_X(U_{\delta p}\boldsymbol{p}, U_{\delta p}\boldsymbol{p}') = \gamma_X(\boldsymbol{p}-\boldsymbol{p}') \end{aligned} \tag{7.5a–c}$$

for all U_δ (again, more details are given in Chapter VII).

Matters are technically more complicated for random fields that are assumed to be space–time *isostationary* (STIS). In the scalar case, w.s. STIS of $X(\boldsymbol{p})$ is uniquely defined and it implies that the mean is constant and the covariance (or variogram) function is rotation ($\mathcal{R}$) invariant, i.e.,

$$\begin{aligned} \overline{X(\boldsymbol{p})} &= \overline{X}, \\ c_X(\boldsymbol{p},\boldsymbol{p}') &= c_X(\mathcal{R}_\theta\boldsymbol{p}, \mathcal{R}_\theta\boldsymbol{p}') = c_X(|\boldsymbol{p}-\boldsymbol{p}'|), \\ \gamma_X(\boldsymbol{p},\boldsymbol{p}') &= \gamma_X(\mathcal{R}_\theta\boldsymbol{p}, \mathcal{R}_\theta\boldsymbol{p}') = \gamma_X(|\boldsymbol{p}-\boldsymbol{p}'|) \end{aligned} \tag{7.6a–c}$$

for all rotations $\mathcal{R}_\theta$ (θ denotes the angle of rotation for the spatial coordinates).[22] In the case of vector random fields the notion of isotropy has more than one interpretation (the matter will be discussed in Chapter VIII). I also bring to the readers' attention that the different kinds of spatiotemporal variability may be interrelated.

A space–time heterogeneous random field does not satisfy any of the assumptions above concerning the field's spatial and temporal variation, which can be generally space nonhomogeneous and time nonstationary. Yet, a middle ground is reached by assuming some specific form of space–time heterogeneity, like the class of S/TRFs with *STHS increments* (or, more precisely, random fields with increments of order ν in space and μ in time). This assumption means that, although the S/TRF $X(\boldsymbol{p})$ itself is space–time heterogeneous, there exists a linear transformation $\mathcal{T}$ such that the

$$Y(\boldsymbol{p}) = \mathcal{T}[X(\boldsymbol{p})] \tag{7.7}$$

is an STHS or an STSI random field, see Chapter XIII.

---

[21] An introduction to the translation transformation was given in Section 2.2 of Chapter I.

[22] An introduction to the rotation transformation was given in Section 2.2 of Chapter I.

Table 7.2 Sea States and Their Corresponding Space and Time Variability Features

Sea State	Spatial Variation	Temporal Variation
Swell	Homogeneous	Stationary
Fully aroused	Homogeneous	Stationary
Duration limited	Nonhomogeneous	Stationary
Fetch limited	Homogeneous	Nonstationary

Example 7.4

As an illustration of the real-world applicability of the space–time variation assumptions considered above, Table 7.2 lists a few cases of sea states with their corresponding physical space and time variability characteristics.

A point to be stressed is that working in the $R^{n,1}$ domain opens certain additional possibilities concerning the modeling of spatiotemporal variability: in addition to STHS and space–time heterogeneous fields in R^{n+1}, a random field may be space homogeneous/time nonstationary, or space nonhomogeneous/time stationary in $R^{n,1}$. Accordingly, an S/TRF $X(\boldsymbol{p})$ can be classified according to its spatiotemporal variability over $\boldsymbol{p} \in R^{n+1}$ (vs. $R^{n,1}$).

Remark 7.1

That there exist S/TRFs that are only homogeneous in space or only stationary in time is a fact that further supports the viewpoint that in many cases space–time analysis is much more than simply an extension of spatial techniques.

7.5 FIFTH CLASSIFICATION: SPATIOTEMPORAL RANDOM FIELD MEMORY VERSUS INDEPENDENCE

This sort of classification is based on the *memory* of the random field model considered. Let us review the key features of some very important special random field cases.

First, a random field model with zero space–time memory (i.e., an *independent* random field) is completely specified by means of its univariate (or marginal) PDFs, since in this case the following simple factorization holds

$$f_{\boldsymbol{p}_1,\ldots,\boldsymbol{p}_m}(\chi_1,\ldots,\chi_m) = \prod_{i=1}^{m} f_{\boldsymbol{p}_i}(\chi_i) \tag{7.8}$$

for all m. Eq. (7.8) is sometime said to represent complete chaos.

Second, a random field with a useful memory structure is the celebrated *Markov* random process. According to the Markov assumption, knowledge of only the present determines the future. Considering a fixed $\boldsymbol{s}$, and assuming that $t_1 < t_2 < \ldots < t_m$, it holds that

$$f_{\boldsymbol{s},t_1,\ldots,t_m}(\chi_1,\ldots,\chi_m) = f_{\boldsymbol{s},t_1}(\chi_1) f_{\boldsymbol{s},t_2|t_1}(\chi_2|\chi_1)\ldots\ f_{\boldsymbol{s},t_m|t_{m-1}}(\chi_m|\chi_{m-1}), \tag{7.9}$$

that is, the m-variate PDF can be calculated for all m given the corresponding bivariate conditional PDFs (Eq. (7.9) is a special case of Eq. (2.11b) under the Markov assumption). As we saw in Eq. (2.11a) the conditional PDF is defined as

$$f_{s,t_k|t_{k-1}}(\chi_k|\chi_{k-1}) = \frac{f_{s,t_k,t_{k-1}}(\chi_{k-1},\chi_k)}{f_{s,t_{k-1}}(\chi_{k-1})}. \tag{7.10}$$

A consequence of Eqs. (7.9) and (7.10) is the expression

$$f_{s,t_1,\ldots,t_m}(\chi_1,\ldots,\chi_m) = \frac{f_{s,t_1,t_2}(\chi_1,\chi_2)\cdot\cdot f_{s,t_{m-1},t_m}(\chi_{m-1},\chi_m)}{f_{s,t_2}(\chi_2)f_{s,t_3}(\chi_3)\cdot\cdot f_{s,t_{m-1}}(\chi_{m-1})}, \tag{7.11}$$

i.e., the m-variate PDF of a Markov random field is specified for all m given the bivariate PDFs. When justified, this assumption enables stochastic reasoning, analysis, and computation with the random field model that would otherwise be intractable.

In the light of the two special cases of Eqs. (7.8) and (7.9), a generalization in multiple dimensions may be considered, although such a generalization should be approached with some caution. So, a Markov random field may be visualized as a collection of RVs distributed in the space–time domain with the PDF of each RV depending on its neighboring RVs. That is, the joint distribution for any set of RVs in the space–time domain can be computed as the product of a form similar to that of Eq. (7.9) involving the conditionals that contain these RVs. Since in the random field context, each RV x_i $(i = 1,\ldots, m)$ is assigned a space–time point $\boldsymbol{p}_i$, Eq. (7.9) may involve RVs that are considered along different directions in space and at prespecified "closest" distances along these directions.

Lastly, it must be noticed that each of the above classifications is more or less independent of the others. Any possible combinations to be considered involve a degree of selection and choice. For example, a Gaussian S/TRF can be STHS or space–time heterogeneous, Markov or non-Markov; an STHS random field can be assumed to be either Gaussian or non-Gaussian, etc. In the following chapters, I will provide more details about the most important of the above classifications and their consequences in random field modeling.

8. CLOSING COMMENTS

I would like to conclude this chapter with a few comments about the methodology that underpins random field modeling. These comments also offer the links with the developments that will be presented in the following chapters.

8.1 THE METHODOLOGICAL IMPORTANCE OF SPACE–TIME

All the events, processes, and phenomena in the world occur in space and time, i.e., spatiochronologically. Accordingly, the fundamental laws that govern composite space–time connections are the most general and hold for all forms of matter and energy. Space–time (or spacetime, or space/time) models combine space and time into a single continuum. Space–time is usually interpretated with space being three dimensional and time playing the role of a fourth dimension that is of a different sort from the spatial dimensions.[23] Because space and time are indissolubly related to each other to form a single whole, any modeling perspective should account for the fact that the occurrence of a phenomenon in space must be necessarily linked to its temporal evolution. This methodological observation indicates that space and time become direct participants of natural processes, although physically the space and time coordinates may not be treated in the same way.

Yet, in other real-world applications the existing physical conditions may require that at some modeling stage one should distinguish between the spatial and the temporal properties of a phenomenon, and identify certain regions of space−time as points of space and other regions as instants of time. There are many reasons for this methodological distinction, such as:

- it follows from the particular physical considerations that either the time or the space coordinates must play a special role in the description of the phenomenon;
- certain properties of the phenomenon (e.g., anisotropy, multidirectionality) refer to space represented as a vector, whereas other properties (e.g., one time instant unambiguously defines the next, consecutive order in which one event succeeds another[24]) are strongly linked with time viewed as a scalar;
- multiple attribute variation degrees along different spatial directions need to be quantified and compared to the single dynamic (temporal) variation degree;
- substantive differences exist between spatial and temporal attribute regularity features (smoothness, heterogeneity);
- spatial and temporal dependency scales may provide distinct measures of long-distance correlations versus short-time dynamic interrelationships; and
- different kinds of information are transmitted spatially and different kinds through temporal channels.

Example 8.1

In social studies, several possible relationships between space and time have been observed. (a) A spatial pattern often occurs independent of time (e.g., certain communities are prone to a particular kind of crime, say, car thefts or robberies are concentrated in particular areas). There is no space−time interaction, since the spatial pattern is observed all the time. (b) A spatial pattern is observed during particular time periods (e.g., usually due to road congestion, car crashes occur with much higher frequency in late afternoon and early evening hours). (c) A number of events are observed within a short time period and in a concentrated area, and the cluster moves from one location to another (e.g., this situation is very common with car thefts occurring within a limited period in one neighborhood and then moving to another one). This is a case of space−time clustering, where there is an interaction between space and time (spatial hot spots appear at particular times, but are temporary). (d) A more involved relationship between space and time is observed,

---

[23]Obvious yet somehow pedantic differences between space and time are that there are a few space coordinates but only one time coordinate; one can point one's finger toward the east but one cannot point it toward the future; or, one can move to the right but one cannot move toward the past. Otherwise said, in human consciousness there exist clear qualitative differences between space and time. Consciousness perceives time as something that runs continuously. The space−time cross section of "now" is continuously experienced and sequentially recognized, whereas space allows us to travel freely through it and to experience its entirety at a time.

[24]Also known as the dynamical or causality principle.

i.e., the space–time interaction follows a more complex pattern (e.g., car thefts shift toward seaside communities during summer months when the number of vacationers increases, or there could be a diffusion of drug sales from a central location to a more dispersed region and this diffusion occurs at different times of the year).

If "space" and "time" are concepts that were invented by humans to describe Nature, yet it should be kept in mind that the map is not the territory. Hence, the methodological distinction between the spatial and the temporal aspects of a natural attribute can be materialized in different ways, e.g., by formally suppressing time in our modeling approach and focusing on space, by fixing space and studying dynamic (temporal) variations, or by making some kind of space–time separability assumption. All the above possibilities are considered in random field modeling.

8.2 A CONCEPTUAL MEETING POINT FOR MODELERS AND EXPERIMENTALISTS

No doubt, the quality of the data and their analysis are decisive for the validity of the conclusions we draw on the basis of random field modeling. Yet, much more than data analysis is often needed to make physical modeling compelling. Among the well-known cases where relying on data alone can be a disastrous perspective are the studies of extreme events that can turn out to be catastrophic, high-throughput screening concerned with the anticipation of new scientific findings, and space–time extrapolation beyond data range. In all these cases, as Bouleau (1991) has noticed, "The approach attributing a precise numerical value for the probability of a rare phenomenon is suspect, unless the laws of Nature governing the phenomenon are explicitly and exhaustively known."

In view of the above considerations, I suggest the notion of *interpretation* as a conceptual meeting point for theoretical modelers and experimentalists. Interpretation should maintain a close contact between the mathematical description and the physical phenomenon described. For this to happen, a random field model should be treated so that both its rigorousness and its interpretative value are maintained. Interpretation can also affect the ways in which attribute observations are reported and evaluated. As such, interpretation creates some serious challenges for both modelers and experimentalists:

(a) For our study results and findings to be physically meaningful and interpretable, they should be based on sound theory and high-quality empirical evidence. The decision over which random field models are the most useful for interpretative purposes can influence how experiments are set up and performed.

(b) The determination of the accuracy of the observation and the type of observational interpretation that emerges can influence the effectiveness of random field modeling in the real-world. A central issue here is factual accuracy in the informational statements that describe what was observed and experienced.

(c) Some data analyses are arbitrary and ad hoc, showing utter disregard for mathematical and physical constraints, thus leading to misapplication of interpretations.

As with all empirical results, we may have to consider multiple interpretations. There are both dominant and alternative theoretical interpretations of a phenomenon and its evidential support. Thought experiments allowed by random field modeling are more likely to alert us to problems of interpretational indeterminacy than data processing.

8.3 THERE IS NO ASSUMPTIONLESS MODELING

It was suggested earlier that, just as an assumptionless science is not a real possibility, so it is that there is no assumptionless modeling. As a consequence, in the following chapters several conditions will be considered that, in certain respects, limit the generality of the random field theory presented in this chapter.

The reason that makes the consideration of modeling assumptions an inescapable inevitability is twofold:

(a) in many cases the modeling assumptions are fully justified or even necessitated by the phenomenon under study (in which case our ability to suggest these assumptions is directly related to the depth of our understanding of the phenomenon), and

(b) the random field modeling of real-world phenomena, like any kind of modeling, should account for the fact that reality is often much more complex than any assumption-free model can handle (it goes without saying that in this case too, any assumption should be justified by the available information about the phenomenon).

To understand we must simplify, which is why in the following parts of the book certain random field modeling assumptions will be introduced in the form of simplifications, but, at the same time, we must be aware that every simplification could lead us away from reality. To eliminate this possibility, we rely on the physically meaningful and valid interpretation of the results an assumption leads to. In this respect, interpretation is not only a very important element of sound random field modeling, but also it is typically viewed as an inextricable element of data collection and assessment. Assumptions are at the center of quantitative modeling, in general, and random field modeling, in particular. When an assumption is introduced, it may limit the generalizability or the external validity of a random field model but it is, nevertheless, deemed necessary to assure the internal validity of the model or to improve its physical fidelity. An assumption will be characterized as reasonable when it can be justified by the core knowledge of the phenomenon or by sound evidence present at one or more stages of the data.

For sure, there is a certain degree of selection and choice involved in the process of assumption making, but the central issue is the factual accuracy of the random field modeling assumptions that purport to describe what was observed and experienced and to contribute to the process of gaining interpretative scientific insights. Surely, critiques of random field modeling assumptions are part of the process followed in this book, and they often take the form of assessing observable quantitative implications of these assumptions.

With all this being said, the random field models that will be the concern of the following chapters are primarily theoretical constructs based on adequate assumptions whose realistic objective is not necessarily to perfectly match the actual phenomenon (something that is most of the time not practically possible given Nature's complexity), but to derive physically meaningful and logically sound *representations* of it, to increase the richness of interpretation, and to draw useful conclusions about the phenomenon.

CHAPTER

SPACE–TIME METRICS

CHAPTER OUTLINE

1. BASIC NOTIONS

In empirical terms, the *space–time distance* (or the *physical space–time metric*) is a numerical assessment of how far apart points in the space–time domain lie. It may be positive (between any two different points), or zero (from a point to itself), and it does not contain information about direction (i.e., it is a scalar quantity that does not depend on the orientation of the coordinate system). As we saw in Chapter I, a basic categorization of space–time metrics $m(\boldsymbol{p},\boldsymbol{p}')$ between two points $\boldsymbol{p}$ and $\boldsymbol{p}'$ distinguishes between

(a) the class $\mathcal{M}_s$ of *separate* spatial and temporal metrics in $R^{n,1}$, i.e., there is a definite way to split space distance from time lag,[1] and

(b) the class $\mathcal{M}_c$ of *composite* spatiotemporal metrics in R^{n+1}, i.e., the space distance and the time lag are considered as a unified whole.

The class $\mathcal{M}_s$ assumes that the space–time domain has a pair of distinct spatial and temporal metrics, whereas the class $\mathcal{M}_c$ assumes that the domain has a single spatiotemporal metric in which space and time are connected via a case-dependent function. Subcategories of the categories (a) and (b) may be also encountered in practice. For example, a space–time metric belonging to $\mathcal{M}_s$ may include a Euclidean or a non-Euclidean spatial distance. The form of a $\mathcal{M}_c$ metric, on the other hand, may not be specifiable

[1]Separate space–time metrics do not necessarily imply separable space–time covariances (Remark 2 of Chapter I).

Spatiotemporal Random Fields. http://dx.doi.org/10.1016/B978-0-12-803012-7.00003-9

a priori by the investigator, but only after careful consideration of the physical constraints governing the phenomenon of interest. Accordingly, the above metric categorization can have significant ramifications in the context of the spatiotemporal random field (S/TRF) theory and its applications.

Surely, assessing a space—time metric is often a considerably more complicated affair than assessing a purely spatial metric. *Inter alia*, this is because the physical space—time metric suggests a certain concept of distance that blends space and time to make space—time, but, simultaneously, it views time as a dissimilar quantity. Adequate and meaningful *space—time metric determination* is a crucial component of spatiotemporal variability assessment and prediction. All quantitative tools assessing the strength of space—time dependence and correlation [like the spatiotemporal variability functions (S/TVFs)] are functions of the metric. This means that these tools can be valid for one kind of metric but invalid for some others (Chapter XV).

Undoubtedly, in current practice the most commonly used spatiotemporal variability models are assumed to have metrics belonging to the $\mathcal{M}_s$ class. In addition, these variability models are often valid only when the $\mathcal{M}_s$ metrics involve a Euclidean spatial distance. The vast majority of software libraries in current use assume such metrics. In applied sciences, however, many metrics emerging from physical observation belong to the $\mathcal{M}_c$ class and the associated spatial coordinates are non-Euclidean (e.g., many data sets are available in geodetic coordinates). These facts underscore the importance of introducing space—time metric classes that are wider than the classes of separate metrics with Euclidean distances.

1.1 FORMAL AND PHYSICAL ASPECTS OF SPACE—TIME METRIC DETERMINATION

Real-world phenomena vary in a composite space—time domain in a way that exhibits the strong links between the physical characteristics of the phenomenon and the geometrical features of the domain. Accordingly, the notion of a metric used in physical sciences may not always coincide with its purely mathematical notion, i.e., some of the conditions of the mathematical metric may be relaxed in the physical metric (Deza and Deza, 2014). Indeed, the definition of an appropriate metric in the latter case depends not on purely mathematical considerations alone, but on both the local properties of space and time (the intrinsic links of space and time) and the physical constraints imposed by the specific phenomenon [natural law of change, empirical support, boundary and initial conditions (BICs)].

This being the case in the real-world, a spatiotemporal metric may be defined explicitly or implicitly. Explicit expressions of the space—time metric are generally obtained on the basis of combined mathematical—physical considerations (involving vector analysis, invariance principles, and physical laws). If such expressions are not available, it may be still possible to implicitly consider the space—time metric in the S/TRFs obtained for specific natural attributes from numerical simulations or experimental observations.[2]

As we saw in Section 2.1 of Chapter I, a space—time lag in R^{n+1} can be mathematically expressed in terms of a vector base $\boldsymbol{\varepsilon}_i$ $(i = 1,\ldots, n)$, i.e.,[3]

$$\Delta \boldsymbol{p} = (h_1, \ldots, h_n, h_0) = \sum_{i=1}^{n,0} h_i \boldsymbol{\varepsilon}_i, \tag{1.1}$$

[2]Attributes that occur in fractal spaces are an example of the latter (Christakos et al., 2000).

[3]Herein, the simpler symbol Δp rather than $m(\boldsymbol{p},\boldsymbol{p}')$ will be mostly used to denote a metric in R^{n+1}.

with the corresponding metric defined in terms of the inner product

$$m(\boldsymbol{p},\boldsymbol{p}')^2 = \Delta\boldsymbol{p}^2 = \Delta\boldsymbol{p}\cdot\Delta\boldsymbol{p} = \sum_{i,j=1}^{n,0} h_i h_j (\boldsymbol{\varepsilon}_i\cdot\boldsymbol{\varepsilon}_j), \tag{1.2}$$

where $|\Delta\boldsymbol{p}| = (\Delta\boldsymbol{p}\cdot\Delta\boldsymbol{p})^{\frac{1}{2}}$ is the magnitude of the vector. If $\boldsymbol{\varepsilon}_i$ are unit vectors along the (orthogonal) coordinate directions, i.e., $\boldsymbol{\varepsilon}_i = (0,\ldots,\varepsilon_i,\ldots,0)$, then $\boldsymbol{\varepsilon}_i\cdot\boldsymbol{\varepsilon}_j = \delta_{ij}\varepsilon_i\varepsilon_j$, and Eq. (1.2) gives

$$\Delta\boldsymbol{p}^2 = \sum_{i=1}^{n,0} \varepsilon_{ii} h_i^2, \tag{1.3}$$

where $\varepsilon_{ii} = \varepsilon_i^2$. If a different vector base is used, say $\boldsymbol{\varepsilon}_i = (\varepsilon_{i1},\ldots,\varepsilon_{in},\varepsilon_{i0})$, then $\boldsymbol{\varepsilon}_i\cdot\boldsymbol{\varepsilon}_j = \sum_{k=1}^{n,0}\varepsilon_{ik}\varepsilon_{jk}$, and Eq. (1.2) gives the more general space–time metric form

$$\Delta\boldsymbol{p}^2 = \sum_{i,j=1}^{n,0} \varepsilon_{ij} h_i h_j, \tag{1.4}$$

where $\varepsilon_{ij} = \boldsymbol{\varepsilon}_i\cdot\boldsymbol{\varepsilon}_j$. While in the case of the metric of Eq. (1.3) the coefficients ε_{ii} form a diagonal matrix, in the case of Eq. (1.4) the matrix of $\{\varepsilon_{ij}\}$ is symmetric but not diagonal. When positive-definite metrics are used, it is assumed that for any real and nonzero h_i ($i = 1,\ldots, n, 0$), the value of the quadratic form in Eq. (1.4) is $\Delta\boldsymbol{p}^2 > 0$.[4] In mathematical analysis, a metric is usually denoted as $|\Delta\boldsymbol{p}|$. However, for notational simplicity, in the following the absolute value ($|\cdot|$) symbol will be often dropped, and the notation Δp will be used, when it is obvious that the value of the metric is positive, and the symbol $\Delta\boldsymbol{p}^2$ will be viewed as a fundamental quantity.

Example 1.1

As we saw in Section 2 of Chapter I, the definition of an appropriate space–time metric that expresses the intrinsic links of space and time often needs to be made on the basis of physical laws and empirical support. For illustration, let the physical law be of the form (in $R^{2,1}$)

$$\left[\frac{\partial}{\partial t} - \frac{t}{2\alpha_0^2}\sum_{i=1}^{2}\frac{\alpha_i^2\partial}{s_i\partial s_i}\right] X(\boldsymbol{s},t) = 0, \tag{1.5}$$

where α_i ($i = 0, 1, 2$) are empirical coefficients. The solution of the law of Eq. (1.5) is of the form $X(\boldsymbol{s},t) = e^{-\left(\frac{s_1^2}{\alpha_1^2}+\frac{s_2^2}{\alpha_2^2}+\frac{t^2}{\alpha_0^2}\right)}$, in which case an internally consistent space–time metric is

$$\Delta\boldsymbol{p}^2 = \frac{h_1^2}{\alpha_1^2} + \frac{h_2^2}{\alpha_2^2} + \frac{h_0^2}{\alpha_0^2} \tag{1.6}$$

[4]For relativistic space–time metrics, however, it is possible that $\Delta\boldsymbol{p}^2 = 0$, but this is beyond the main scope of the current analysis.

($h_0 = \tau$), i.e., the metric is of the form of Eq. (1.3) with coefficients determined by the physical law of Eq. (1.5) as $\varepsilon_{00} = \alpha_0^{-2}$, $\varepsilon_{11} = \alpha_1^{-2}$, $\varepsilon_{22} = \alpha_2^{-2}$. This is a space–time metric of the Pythagorean form.

Example 1.2

Let the coordinates h_1 and h_2 in the metric of Eq. (1.6) correspond to the longitude and the latitude of a point on a spherical surface with unit radius. Then, $\varepsilon_{00} = \alpha_0^{-2}$, $\varepsilon_{11} = \alpha_1^{-2}\cos^2(h_2)$, $\varepsilon_{22} = \alpha_2^{-2}$, and Eq. (1.3) gives

$$\Delta \boldsymbol{p}^2 = \alpha_1^{-2}\cos^2(h_2)h_1^2 + \alpha_2^{-2}h_2^2 + \alpha_0^{-2}h_0^2 \tag{1.7}$$

($h_0 = \tau$), where the metric coefficient ε_{11} is a function of latitude.

Example 1.3

Sometimes, physical requirements make it necessary to calculate the space–time metric in different coordinates. For illustration, consider the metric Eq. (1.3) in the space–time domain $R^{3,1}$ with $\varepsilon_{ii} = 1$ ($i = 0,\ 1,\ 2,\ 3$), and apply the coordinate transformation, $h_0' = h_0$, $h_1' = h_1 + 2h_2$, $h_2' = h_1 - h_2$, and $h_3' = h_3$.[5] Then, the metric becomes

$$\Delta \boldsymbol{p}^2 = h_1^2 + h_2^2 + h_3^2 + h_0^2 = \frac{2}{9}h_1'^2 + \frac{5}{9}h_2'^2 + \frac{2}{9}h_1'h_2' + h_3'^2 + h_0'^2, \tag{1.8}$$

i.e., the transformed metric is now of a different form, namely, that of Eq. (1.4) with $\varepsilon_{00} = \varepsilon_{33} = 1$, $\varepsilon_{11} = \frac{2}{9}$, $\varepsilon_{22} = \frac{5}{9}$, $\varepsilon_{12} = \varepsilon_{21} = \frac{1}{9}$, and all other ε's are zero.

The objective of the following example is to illustrate the fact that more than one vector base can be examined in the metric definition by means of the inner product. This is a process that can reveal certain properties of the space–time metric, such as *metric invariance* under vector base change.

Example 1.4

In terms of the standard unit vectors, $\boldsymbol{\varepsilon}_1 = (\varepsilon_{11}, \varepsilon_{12}, \varepsilon_{10}) = (1, 0, 0)$, $\boldsymbol{\varepsilon}_2 = (\varepsilon_{21}, \varepsilon_{22}, \varepsilon_{20}) = (0, 1, 0)$, and $\boldsymbol{\varepsilon}_0 = (\varepsilon_{01}, \varepsilon_{02}, \varepsilon_{00}) = (0, 0, 1)$, the space–time lag in the $R^{2,1}$ domain can be written as

$$\Delta \boldsymbol{p} = (h_1, h_2, h_0) = h_1\boldsymbol{\varepsilon}_1 + h_2\boldsymbol{\varepsilon}_2 + h_0\boldsymbol{\varepsilon}_0 = h_1(1,0,0) + h_2(0,1,0) + h_0(0,0,1).$$

The corresponding metric is[6]

$$\Delta \boldsymbol{p}^2 = h_1^2\boldsymbol{\varepsilon}_1^2 + h_2^2\boldsymbol{\varepsilon}_2^2 + h_0^2\boldsymbol{\varepsilon}_0^2 = h_1^2 + h_2^2 + h_0^2. \tag{1.9}$$

On the other hand, in terms of the new vector base, $\boldsymbol{\varepsilon}_1 = (\varepsilon_{11}, \varepsilon_{12}, \varepsilon_{10}) = (1,\ 1,\ 0)$, $\boldsymbol{\varepsilon}_2 = (\varepsilon_{21}, \varepsilon_{22}, \varepsilon_{20}) = (0,\ 1,\ 0)$, and $\boldsymbol{\varepsilon}_0 = (\varepsilon_{01}, \varepsilon_{02}, \varepsilon_{00}) = (0,\ 0,\ 1)$, the composite space–time lag is written as

$$\Delta \boldsymbol{p} = (h_1, h_2, h_0) = h_1\boldsymbol{\varepsilon}_1 + (h_2 - h_1)\boldsymbol{\varepsilon}_2 + h_0\boldsymbol{\varepsilon}_0 = h_1(1,1,0) + (h_2 - h_1)(0,1,0) + h_0(0,0,1).$$

---

[5]With inverse transformation, $h_1 = \frac{1}{3}h_1' + \frac{2}{3}h_2'$, $h_2 = \frac{1}{3}h_1' - \frac{1}{3}h_2'$, $h_3 = h_3'$, and $h_0 = h_0'$.

[6]Since $\boldsymbol{\varepsilon}_i \cdot \boldsymbol{\varepsilon}_j = 0$ ($i,j = 0,1,2; i \neq j$)

The corresponding space–time metric is[7]

$$\Delta p^2 = h_1^2 \varepsilon_1^2 + (h_2 - h_1)^2 \varepsilon_2^2 + h_0^2 \varepsilon_0^2 + 2h_1(h_2 - h_1)\varepsilon_1 \cdot \varepsilon_2 = h_1^2 + h_2^2 + h_0^2, \tag{1.10}$$

i.e., of the same form as in Eq. (1.9). Therefore, the metric is invariant under the vector base change above.

Let us take stock. In view of the above examples, the following substantive points should be stressed:

(a) Although one may usually choose a space–time coordinate basis so that the inner product that defines the metric is diagonal and normalized (Euclidean or flat[8] space), in certain real-world cases it may be reasonable, if not necessary, to choose a different coordinate basis.

(b) As far as real-world science is concerned, it is important to determine physical metrics that actually represent or stem from Nature and from experimental evidence rather than simply imposed by pure mathematics often involving convenient conditions not necessarily based on real-world experience. In this spirit, the definition of a space–time metric is not limited to the vector dot product of Eq. (1.2), but it may assume more involved forms, depending on the real-world situation.

(c) Although in mathematical analysis the notation $|\Delta \boldsymbol{p}|$ is typically used to denote the metric, in physical investigations the symbol $\Delta \boldsymbol{p}^2$ is often taken as a fundamental quantity, and not merely the square of some other quantity $\Delta \boldsymbol{p}$. In this sense, the symbol $|\cdot|$ may be dropped, for simplicity, in light of the expression $\Delta p = \left(\Delta \boldsymbol{p}^2\right)^{\frac{1}{2}}$.

1.2 SPACE–TIME METRIC FORMS

As regards purely spatial metrics or distances, both Euclidean and non-Euclidean metric forms have been considered in the applied sciences literature (see, e.g., Turcotte, 1997; Billings et al., 2002; Curriero, 2006; Lloyd, 2010; Lin et al., 2015). As regards spatiotemporal metrics or "distances," certain metric forms were introduced by Christakos and Hristopulos (1998), Christakos et al. (2000), and Christakos (2000) in the context of modern spatiotemporal geostatistics, see also Section 2 of Chapter I in this book. For future reference, a list of space–time metric forms is given in the right column of Table 1.1. Interestingly, as is shown in the left column of Table 1.1, all these space–time metrics are special cases of the following general space–time metric, which can describe the intrinsic geometry of many space–time continua (Christakos et al., 2017b).

Definition 1.1

In a continuum $\mathcal{E}$ of the $n+1$-dimensional space R^{n+1} (or, of the n-dimensional space $\times$ time, $R^{n,1}$), a physical metric may be defined as

$$\Delta \boldsymbol{p}^2 = \left[\boldsymbol{h}^T, 1\right] \breve{\boldsymbol{E}} \left[\boldsymbol{h}^T, 1\right]^T, \tag{1.12}$$

---

[7] Since $\varepsilon_0 \cdot \varepsilon_i = 0$ $(i = 1, 2)$.

[8] Flat space–time is not curved, i.e., objects obey Newton's laws, and a geodesic line is straight.

Table 1.1 Space—Time Metrics in R^{n+1} ($h_0 = \tau > 0$)

Coefficients	Space—Time Metric Form Δp^2	
$\boldsymbol{\varepsilon} = \mathbf{0},\ \boldsymbol{e}_0 = [\varepsilon_{00}\varepsilon_{11}\ldots\varepsilon_{nn}]^T,\ \boldsymbol{E} = \left(\boldsymbol{e}_0\boldsymbol{e}_0^T\right)^{\frac{1}{2}}$	*Manhattan* $\left(r_{Mt}^2\right)$ $\left(\sum_{i=1}^{n} \varepsilon_i \lvert h_i\rvert + \varepsilon_0\tau\right)^2$	(1.11a)
$\boldsymbol{\varepsilon} = \mathbf{0},\ \boldsymbol{E} = diag\ \boldsymbol{e}_0 = \begin{bmatrix} \varepsilon_{00} & 0 & \ldots & 0 \\ 0 & \varepsilon_{11} & \ldots & 0 \\ \vdots & & & \\ 0 & 0 & \ldots & \varepsilon_{nn} \end{bmatrix}$	*Pythagorean* $\left(r_P^2\right)$ $\sum_{i=1}^{n} \varepsilon_{ii}h_i^2 + \varepsilon_{00}\tau^2$	(1.11b)
$\boldsymbol{E} = \mathbf{0}$	*SR Manhattan* $\left(r_{SRM}^2\right)$ $\sum_{i=1}^{n} \varepsilon_i\lvert h_i\rvert + \varepsilon_0\tau$	(1.11c)
$\boldsymbol{\varepsilon} = \mathbf{0},\ \boldsymbol{E} = \begin{bmatrix} \varepsilon_{00} & \varepsilon_{01} & \ldots & \varepsilon_{0n} \\ \varepsilon_{01} & \varepsilon_{11} & \ldots & 0 \\ \vdots & & & \\ \varepsilon_{0n} & 0 & \ldots & \varepsilon_{nn} \end{bmatrix}$	*Riemann* $\left(r_R^2\right)$ $\sum_{i=1}^{n}\left(\varepsilon_{ii}h_i^2 + 2\varepsilon_{00}^{\frac{1}{2}}\varepsilon_{ii}^{\frac{1}{2}}\lvert h_i\rvert\tau\right) + \varepsilon_{00}\tau^2$	(1.11d)
$\boldsymbol{\varepsilon} = \mathbf{0},\ \boldsymbol{E} = \begin{bmatrix} \varepsilon_{00} & 0 & \ldots & 0 \\ 0 & \varepsilon_{11} & \ldots & \varepsilon_{1n} \\ \vdots & & & \\ 0 & \varepsilon_{1n} & \ldots & \varepsilon_{nn} \end{bmatrix}$	*Mixed* $\sum_{i=1}^{n} \varepsilon_{ii}^{\frac{1}{2}}\lvert h_i\rvert + \varepsilon_{00}\tau^2$	(1.11e)
$\boldsymbol{\varepsilon} = \mathbf{0},\ \boldsymbol{E} = \begin{bmatrix} \varepsilon_c & 0 & \cdots & 0 \\ 0 & -\varepsilon & \cdots & 0 \\ \vdots & & & \\ 0 & 0 & \cdots & -\varepsilon \end{bmatrix}$	*Minkowski* $\left(r_{Mi}^2\right)$ $-\varepsilon\sum_{i=1}^{n} h_i^2 + \varepsilon_c\tau^2 = -\varepsilon\boldsymbol{h}^2 + \varepsilon_c\tau^2$	(1.11f)
$\boldsymbol{\varepsilon} = \mathbf{0},\ \boldsymbol{E} = \begin{bmatrix} \varepsilon_{00} & \varepsilon_{01} & \ldots & \varepsilon_{0n} \\ \varepsilon_{10} & \varepsilon_{11} & \ldots & 0 \\ \vdots & & & \\ \varepsilon_{n0} & 0 & \ldots & \varepsilon_{nn} \end{bmatrix}$	*Traveling* $\left(r_T^2\right)$ $\sum_{i=1}^{n} \left(\lvert h_i\rvert + \varepsilon_{0i}\tau\right)^2$	(1.11g)
$\boldsymbol{\varepsilon} = \mathbf{0},\ \boldsymbol{E} = \begin{bmatrix} \varepsilon_{00} & \varepsilon_{01} & \ldots & \varepsilon_{0n} \\ \varepsilon_{01} & \varepsilon_{01}^2\varepsilon_{00}^{-1} & \ldots & \varepsilon_{1n} \\ \vdots & & & \\ \varepsilon_{0n} & \varepsilon_{1n} & \ldots & \varepsilon_{0n}^2\varepsilon_{00}^{-1} \end{bmatrix}$	*Plane wave* $\left(r_W^2\right)$ $\left(\varepsilon_{00}^{\frac{1}{2}}\tau + \varepsilon_{00}^{-\frac{1}{2}}\sum_{i=1}^{n} \varepsilon_{0i}\lvert h_i\rvert\right)^2$	(1.11h)

where in the vector $[\boldsymbol{h}^T, 1]^T = [h_0 \quad h_1 \quad \ldots \quad h_n \quad 1]^T$, the $h_1 \quad \ldots \quad h_n$ denote the coordinates of the space lag and the $h_0 = \tau$ denotes time lag,

$$\breve{\boldsymbol{E}} = \begin{bmatrix} \varepsilon_{00} & \varepsilon_{01} & \ldots & \varepsilon_{0n} & \frac{1}{2}\varepsilon_0 \\ & \vdots & & & \\ \varepsilon_{n0} & \varepsilon_{n1} & \ldots & \varepsilon_{nn} & \frac{1}{2}\varepsilon_n \\ \frac{1}{2}\varepsilon_0 & \frac{1}{2}\varepsilon_1 & \ldots & \frac{1}{2}\varepsilon_n & 0 \end{bmatrix},$$

and the metric coefficients in $\breve{\boldsymbol{E}}$ may be themselves functions of the space–time coordinates, or they may be functions of physical quantities as required by the real-world situation.

The metric of Eq. (1.12) specifies the connection between any two points in the space–time domain. For subsequent analytical manipulations, it may be more convenient to rewrite Eq. (1.12) as

$$\Delta p^2 = \boldsymbol{h}^T(\boldsymbol{E}\boldsymbol{h} + \boldsymbol{\varepsilon}), \tag{1.13}$$

where $\boldsymbol{h} = [h_0 \quad h_1 \quad \ldots \quad h_n]^T$, and

$$\boldsymbol{E} = \begin{bmatrix} \varepsilon_{00} & \varepsilon_{01} & \ldots & \varepsilon_{0n} \\ \varepsilon_{10} & \varepsilon_{11} & \ldots & \varepsilon_{1n} \\ & \vdots & & \\ \varepsilon_{n0} & \varepsilon_{n1} & \ldots & \varepsilon_{nn} \end{bmatrix},$$

$$\boldsymbol{\varepsilon} = [\varepsilon_0 \quad \varepsilon_1 \quad \ldots \quad \varepsilon_n]^T.$$

Other cases of space–time-dependent metric coefficients of practical interest can be similarly analyzed.

Example 1.5

In some special cases a metric may be defined for convenience in terms of the absolute coordinate distances, $|\boldsymbol{h}|$, instead of the signed lags, $\boldsymbol{h}$. Indeed, if $\boldsymbol{E}$ and $\boldsymbol{\varepsilon}$ in Eq. (1.13) are replaced by $\boldsymbol{E}_\Lambda = \Lambda(\boldsymbol{h})\boldsymbol{E}\Lambda(\boldsymbol{h})$ and $\boldsymbol{\varepsilon}_\Lambda = \Lambda(\boldsymbol{h})\boldsymbol{\varepsilon}$, where $\Lambda(\boldsymbol{h})$ is an $(n+1) \times (n+1)$ diagonal matrix with elements

$sign(h_i) = \pm 1$ or 0 ($i = 0,1,\ldots, n$) depending on whether h_i is positive, negative, or null, Eq. (1.13) becomes the metric expression[9]

$$\begin{aligned} \Delta p^2 &= |\boldsymbol{h}|^T(\boldsymbol{E}|\boldsymbol{h}| + \boldsymbol{\varepsilon}), \\ \Delta p^2 &= \boldsymbol{h}^T\boldsymbol{E}\boldsymbol{h} + |\boldsymbol{h}|^T\boldsymbol{\varepsilon}, \end{aligned} \tag{1.14a–b}$$

here the notation $|\boldsymbol{h}|^T = [|h_0|\ |h_1|\ \ldots\ |h_n|]$ is used. Metric symmetry is readily satisfied for space- and time-independent $\boldsymbol{E}$ and $\boldsymbol{\varepsilon}$, and also for space- and time-dependent $\boldsymbol{E}$ and $\boldsymbol{\varepsilon}$ assuming that $\boldsymbol{E}(\boldsymbol{h}) = \boldsymbol{E}(-\boldsymbol{h})$ and $\boldsymbol{\varepsilon}(\boldsymbol{h}) = \boldsymbol{\varepsilon}(-\boldsymbol{h})$. The special case of Eq. (1.14b) consists of metrics that are nonnegative and equal to zero if and only if $\boldsymbol{h} = \boldsymbol{0}$. Eq. (1.13) includes the metrics defined in terms of $|\boldsymbol{h}|$ (like the Manhattan metric in Table 1.1), as far as the matrix $\boldsymbol{E}$ and vector $\boldsymbol{\varepsilon}$ implicitly involve, in these cases, the change to positive sign in the negative coordinates of $\boldsymbol{h}$ (by means of $\Lambda(\boldsymbol{h})$).

Knowing the space–time metric in closed form, the elucidation of its physical properties is based on the features of the phenomenon under consideration, e.g., the interpretation of the metric coefficients and their relationships relies on the corresponding physical law (Christakos and Hristopulos, 1998; Hadsell and Hansen, 1999; Carroll 2004). This implies that even in a classical Newtonian framework the simplified case of a space–time metric consisting of two separate components—a purely spatial (Euclidean) component with constant coefficients and a time component—may be convenient but, nevertheless, inadequate. After all, one should keep in mind that almost all data in applied sciences are closely interrelated both in space and time, and that it is this space–time interrelation that, both, it is at the heart of the physical laws of space–time change and it also allows the representation of the space–time variation of a natural phenomenon from a limited number of observations. Furthermore, a sound metric may be constrained by the realistic invariance transformations it must satisfy to be meaningful (see also discussion in Section 2 of Chapter I).

Remark 1.1

In the following, the notation h_0 and τ will be used interchangeably, and the use will become obvious based on the context of the analysis.

Eq. (1.13) is a general metric form that includes other space–time metrics used in applied sciences as its special cases.

Example 1.6

The left column of Table 1.1 shows the corresponding conditions on $\boldsymbol{E}$ and $\boldsymbol{\varepsilon}$ leading to the space–time metrics of the right column. As noted earlier, the physical meaning of the metric coefficients depends on the corresponding empirical evidence, physical law, associated BIC, and sources of randomness. Eq. (1.11a) corresponds to the space–time *Manhattan* (or *city block* or *taxicab* or *absolute*) metric[10] obtained from the general metric of Eq. (1.13) by letting

$$\begin{aligned} &\varepsilon_i = 0, \\ &\varepsilon_{ij} = (\varepsilon_{ii}\varepsilon_{jj})^{\frac{1}{2}}\ (i,j = 0, 1, \ldots, n); \end{aligned}$$

[9]Notice that $\boldsymbol{h}^T\boldsymbol{E}_{\Lambda(\boldsymbol{h})}\boldsymbol{h} = \boldsymbol{h}^T\Lambda(\boldsymbol{h})\boldsymbol{E}\Lambda(\boldsymbol{h})\boldsymbol{h} = |\boldsymbol{h}|^T\boldsymbol{E}|\boldsymbol{h}|$, and $\boldsymbol{h}^T\boldsymbol{\varepsilon}_{\Lambda(\boldsymbol{h})} = \boldsymbol{h}^T\Lambda(\boldsymbol{h})\boldsymbol{\varepsilon} = |\boldsymbol{h}|^T\boldsymbol{\varepsilon}$.

[10]The Manhattan metric can be also written as $\boldsymbol{h}^T\boldsymbol{\varepsilon}(\boldsymbol{h}) = \boldsymbol{h}^T\Lambda(\boldsymbol{h})\boldsymbol{\varepsilon} = |\boldsymbol{h}|^T\boldsymbol{\varepsilon}$, where $|\boldsymbol{h}|^T = (|h_1|, \ldots, |h_n|, |h_0|)$.

Note that the subscript t is used in the space-time Manhattan metric r_{Mt} in R^{n+1}, Eq. (1.11a), in order to distinguish it from the spatial Manhattan-time metric (r_M, τ) in $R^{n,1}$, see Eq. (2.18b) of Chapter I. Next, Eq. (1.11c) corresponds to the *SR (square root) Manhattan* metric, where

$$\varepsilon_{ij} = 0 \ \ (i,j = 0, 1, \ldots, n);$$

and Eq. (1.11b) corresponds to the *Pythagorean* metric in space–time, where

$$\varepsilon_i = \varepsilon_{ij} = 0 \ \ (i,j = 0, 1, \ldots, n; \ i \neq j)$$

(in all cases, with a rescaling of the dimensions according to the coefficients). The *Riemann* metric of Eq. (1.11d), where

$$\varepsilon_{0i} = (\varepsilon_{00}\varepsilon_{ii})^{\frac{1}{2}},$$

$$\varepsilon_0 = \varepsilon_i = \varepsilon_{ij} = 0 \ \ (i,j = 1, \ldots, n; \ i \neq j),$$

may be seen as an interaction between the space–time Pythagorean and the Manhattan metrics. Eq. (1.11e) is a combination of an absolute space and a squared time metrics, where

$$\varepsilon_0 = \varepsilon_i = \varepsilon_{0i} = 0,$$

$$\varepsilon_{ij} = (\varepsilon_{ii}\varepsilon_{jj})^{\frac{1}{2}} \ \ (i,j = 1, \ldots, n; \ i \neq j).$$

Eq. (1.11f) is a *Minkowski* space–time metric, where

$$\varepsilon_i = \varepsilon_{ij} = 0 \ \ (i,j = 0, 1, \ldots, n; i \neq j),$$

$$\varepsilon_{00} = \varepsilon_c,$$

$$\varepsilon_{ii} = \varepsilon \ \ (i = 1, \ldots, n).$$

This is sometimes called a "pseudo-Euclidean" space–time metric due to the minus sign preceding the spatial coordinates. Eq. (1.11g) is the so-called *traveling* metric, where

$$\varepsilon_0 = \varepsilon_i = \varepsilon_{ij} = 0 \ \ (i,j = 1, \ldots, n; \ i \neq j),$$

$$\varepsilon_{00} = \sum_{i=1}^{n} \varepsilon_{0i}^2,$$

$$\varepsilon_{ii} = 1 \ \ (i = 1, \ldots, n).$$

This metric is of particular interest in statistical turbulence studies. Lastly, Eq. (1.11h), where

$$\varepsilon_0 = \varepsilon_i = 0,$$

$$\varepsilon_{0i} = \varepsilon_{i0},$$

$$\varepsilon_{ii} = \varepsilon_{0i}^2\varepsilon_{00}^{-1},$$

$$\varepsilon_{ij} = \varepsilon_{0i}\varepsilon_{0j}\varepsilon_{00}^{-1} \ \ (i,j = 1, \ldots, n)$$

is the *plane wave* metric used in seismic studies and elsewhere. Although the metric coefficients in Table 1.1 are generally space dependent and/or time dependent, in many cases of practical interest independency may apply.

Under certain circumstances, using complex notation could lead to simpler expressions of the space–time metric.

Example 1.7

Consider the space–time metric in R^{1+1},

$$\Delta p = |\Delta \boldsymbol{p}| = \left(\sum_{i,j=0}^{1} \varepsilon_{ij} h_i h_j\right)^{\frac{1}{2}}.$$

Using complex notation and letting $\hbar = h_1 + ih_2$, $\hbar^* = h_1 - ih_2$, the above metric can be written as

$$\Delta p = |\Delta \boldsymbol{p}| = \theta_1 |\hbar + \theta_2 \hbar^*|,$$

where

$$\theta_1 = 0.5\left(\varepsilon_{00} + \varepsilon_{11} + 2(\varepsilon_{00}\varepsilon_{11} - \varepsilon_{01})^{\frac{1}{2}}\right)^{\frac{1}{2}},$$

$$\theta_2 = (\varepsilon_{00} - \varepsilon_{11} + 2i\varepsilon_{01})\left(\varepsilon_{00} + \varepsilon_{11} + 2(\varepsilon_{00}\varepsilon_{11} - \varepsilon_{01})^{\frac{1}{2}}\right)^{-\frac{1}{2}}.$$

This kind of simpler formulation can be used in other kind of metrics, as well.

Remark 1.2

As noted earlier, many applications are concerned about real-life metrics, i.e., physical distances that may not always satisfy all conditions of the term "metric" considered in a strict mathematical sense. For example, the symmetry condition, $|\Delta \boldsymbol{p}| = |-\Delta \boldsymbol{p}|$, may not be always satisfied in real-world applications. Indeed, the symmetry requirement is unnatural if there is no reason in the real-world that the best way to get from $\boldsymbol{p}$ to point $\boldsymbol{p}'$ is like the best way to get from point $\boldsymbol{p}'$ to point $\boldsymbol{p}$. So, the reverse process from point $\boldsymbol{p}'$ to point $\boldsymbol{p}$ may be physically impossible (e.g., the evolution and radioactive decay processes are irreversible). Also, in the realistic context of one-way streets, the Manhattan metric may not formally satisfy the symmetry condition, since a path from point $\boldsymbol{p}$ to point $\boldsymbol{p}'$ may include a different set of streets than a path from point $\boldsymbol{p}'$ to point $\boldsymbol{p}$. Similar may be the case of flow in a porous medium due to physical forces. In fact, as has been argued (Wilson, 1931), the result of suppressing the condition of symmetry from the definition of a metric may disappoint some mathematicians because it diminishes the number of theorems easily deducible, but it can also lead to a richness of material.

In many cases in applied sciences, it may also be a reasonable choice to work in the $R^{n,1}$ domain and explicitly distinguish between space and time. This is the situation in seismology as discussed next.

Example 1.8

In seismology it is often more convenient to explicitly distinguish R^n from T, and consider the vector–scalar pair $(\boldsymbol{h}, \tau)$ in the $R^{n,1}$ domain, in which the space–time metric (also termed the *link distance*) between two earthquakes at $\boldsymbol{p}'$ and $\boldsymbol{p}$ can be defined as

$$\Delta \boldsymbol{p}^2 = \boldsymbol{h}^2 + \upsilon \tau^2,$$

where $|\boldsymbol{h}|$ is the spatial distance between their epicenters, $\tau = t' - t > 0$ is the time lag, and υ is a scaling coefficient that links spatial distance and time lag.

1.3 DERIVED SPACE–TIME METRICS

Interesting space–time metrics can be derived (*a*) by generalizing the existing metrics, usually in terms of some additional parameters with specified meaning; (*b*) by combining existing space–time metrics; or (c) by suitably transforming the existing metrics. Depending on the circumstances, these new metrics may produce additional insight concerning the space–time structure of the phenomenon.

Example 1.9

An illustration of case (a) above is the λ- and $\boldsymbol{\upsilon}$-parametered generalization of the Minkowski metric r_{Mi} of Eq. (1.11f) leading to the space–time metric

$$\Delta p^2 = \left(\tau^2 - \boldsymbol{h}^2\right)\left[\frac{(\tau - \boldsymbol{\upsilon}\cdot\boldsymbol{h})^2}{\tau^2 - \boldsymbol{h}^2}\right]^{\lambda},$$

where λ and $\boldsymbol{\upsilon}$ are scalar and vector parameters, respectively. In this formulation, the coefficient λ determines the degree of deviation from the space–time Minkowski metric (the r_{Mi} metric is obtained for $\lambda = 0$, and the plane wave metric r_W is obtained for $\lambda = 1$), and the $\boldsymbol{\upsilon}$ may indicate locally dominant spatial directions. For an illustration of case (b), the following space–time metric

$$\Delta p^2 = r_R^2\left[\frac{r_M^2}{r_R^2}\right]^{\lambda},$$

is obtained by combining the Riemannian metric r_R and the Manhattan metric r_M. The λ is here a function of space–time that represents the magnitude of local anisotropies, whereas the metric coefficients may indicate the locally favored directions. Lastly, for an illustration of case (c), one can start from the space–time Pythagorean metric r_p^2 (Eq. 1.11b of Table 1.1) in the R^{n+1} domain, and let $\varepsilon_{ii} = 1$ ($i = 0, 1, \ldots, n$), $\upsilon^2 = \sum_{i=1}^{n} \frac{h_i^2}{h_0^2}$, to get

$$\Delta p^2 = \left(1 + \upsilon^2\right)h_0^2,$$

where h_0 is the time separation, and the $\upsilon = \frac{|\boldsymbol{h}|}{h_0}$ may be interpreted as the magnitude of the velocity $\boldsymbol{\upsilon}$. One can also consider the Pythagorean metric r_p^2 in $R^{n,1}$, Eq. (1.11b), and specify the coordinate transformation $\breve{h}_i = \varphi(r_p)h_i$, $i = 1, \ldots, n$, and $\breve{\tau} = \varphi(r_p)\tau$, where $\varphi(r_p)$ is a function of r_p. The new metric will be

$$\breve{r}_p^2 = \varphi^2(r_p)r_p^2,$$

which is different than r_p^2. A simple way to investigate some consequences of the new metrics is in terms of the space–time covariance structure, see Example 1.14.

1.4 SPACE–TIME METRIC DIFFERENTIALS

At this point, and based on the general space–time metric formulation of Eq. (1.13), I introduce certain expressions of space–time metric derivatives that are important in subsequent developments of physical metric determination.

Proposition 1.1

In light of the space—time metric form of Eq. (1.13), the following *metric differential formulas* (*MDFs*) are valid for metrics with space- and/or time-independent coefficients,

$$\begin{aligned} \frac{\partial \Delta p}{\partial \boldsymbol{h}} &= \left(\boldsymbol{h}^T \boldsymbol{E} + \frac{1}{2}\boldsymbol{\varepsilon}^T\right)\Delta p^{-1} \\ \frac{\partial \Delta p^{-1}}{\partial \boldsymbol{h}} &= -\frac{\partial \Delta p}{\partial \boldsymbol{h}}\Delta \boldsymbol{p}^{-2} \\ \frac{\partial^2 \Delta p}{\partial \boldsymbol{h}^T \partial \boldsymbol{h}} &= \left(\boldsymbol{E} - \frac{\partial \Delta p}{\partial \boldsymbol{h}^T}\frac{\partial \Delta p}{\partial \boldsymbol{h}}\right)\Delta p^{-1}, \end{aligned} \tag{1.15a–c}$$

where the notation is used, $\partial \boldsymbol{h} = [\partial h_0 \quad \partial h_1 \quad \ldots \quad \partial h_n]^T$, $\partial h_0 = \partial \tau$, $\Delta \boldsymbol{p}^2 = \boldsymbol{h}^T(\boldsymbol{E}\boldsymbol{h} + \boldsymbol{\varepsilon})$ and $\Delta p = |\Delta \boldsymbol{p}| = (\Delta \boldsymbol{p}^2)^{\frac{1}{2}}$.

For illustration, an MDF example is presented in Table 1.2. Some more special cases of Eqs. (1.15a–c), in their particular analytical forms, are investigated in the Chapter "Exercises" Exercise III.3.

Example 1.10

As noted earlier, the space—time metric analysis is valid in the general case of metric coefficients that are space dependent and/or time dependent.[11] The metric coefficients would be expressed in terms of standard analytic functions (e.g., polynomials) of the space—time coordinates, whereas in certain cases a separability or symmetry condition may be physically imposed on the metric. In Table 1.3, the MDFs of the general space—time metric of Eq. (1.13) are derived for metric coefficients that are themselves functions of space and/or time, in which case, Eq. (1.13) yields

$$\Delta \boldsymbol{p}^2 = \sum_{i,j=1}^{n}\left(\alpha_{ij;0} + \alpha_{0i;j} + \alpha_{0j;i}\right)h_i h_j \tau + \sum_{i=1}^{n}\left[\left(\alpha_{i;0} + \alpha_{0;i}\right) + \left(2\alpha_{0i;0} + \alpha_{00;i}\right)\tau\right]h_i \tau. \tag{1.16}$$

The space—time metric analysis remains valid with the space—time-dependent metric coefficients ε_{ij}, ε_{0i}, ε_{00}, ε_0, being expressed in terms of the coefficients $\alpha_{ij;0}$, $\alpha_{0i;j}$, $\alpha_{i;0}$, $\alpha_{0i;0}$, $\alpha_{00;i}$, and the space and time lags, as is shown in Table 1.3.

Lastly, certain characteristics of the relationship between space and time may be revealed by the random field geometrical properties. We can relate temporal and spatial derivatives, e.g., of random fields with a traveling space—time metric, Eq. (11.1g), by means of the operator

$$\frac{\partial}{\partial t} = -\boldsymbol{\varepsilon}_0 \cdot \nabla,$$

where $\nabla = \left[\frac{\partial}{\partial s_1} \cdots \frac{\partial}{\partial s_n}\right]^T$ and $\boldsymbol{\varepsilon}_0 = [\varepsilon_{01} \quad \varepsilon_{01} \quad \ldots \quad \varepsilon_{0n}]^T$. This observation suggests a link with the following Sections 1.5 and 2.

[11]Mathematical manipulations may become more tedious than for space- and/or time-independent coefficients, depending on the form of the dependency.

Table 1.2 Metric Differential Formula (MDF) Example With Space- and Time-Independent Metric Coefficients

SR Manhattan Metric	MDF	
$R^{2,1}$ $\Delta p^2 = \boldsymbol{h}^T \boldsymbol{\varepsilon} = \sum_{i=1}^{2} \varepsilon_i h_i + \varepsilon_0 \tau$ $\boldsymbol{h} = [h_1\ h_2\ h_0 = \tau]^T$ $\boldsymbol{E} = \boldsymbol{0}\ (\varepsilon_{00} = \varepsilon_{0i} = \varepsilon_{ij} = 0;\ i,j = 1,2)$	$\frac{\partial \Delta p}{\partial h_i} = \frac{1}{2}\varepsilon_i \Delta p^{-1}$ $\frac{\partial \Delta p}{\partial \tau} = \frac{1}{2}\varepsilon_0 \Delta p^{-1}$ $\frac{\partial^2 \Delta p}{\partial h_i^2} = -\frac{1}{4}\varepsilon_i^2 \boldsymbol{\Delta p}^{-3}$ $\frac{\partial^2 \Delta p}{\partial h_i \partial \tau} = -\frac{1}{4}\varepsilon_0 \varepsilon_i \boldsymbol{\Delta p}^{-3}$	(1.17a–d)

1.5 SPECIFYING SPACE–TIME RELATIONSHIPS IN THE COVARIANCE FUNCTION

As already noted, space–time metrics are the arguments of the S/TVFs (such as the covariance, the variogram, and the structure functions), a fact that makes the adequate specification of these metrics a crucial aspect of a spatiotemporal variability analysis with significant consequences in applied stochastics.

Example 1.11

Illustrations of the close associations between a covariance function and the space–time metric are presented in Table 1.4.[12] The first group of covariance models in Table 1.4 is associated with the separate space–time metric in $R^{n,1}$, where r_E is the R^n-Euclidean distance and τ is the R^1-time separation. This is also the case of the majority of covariance models used in geostatistics and mainstream space–time statistics applications. In the case of the second covariance model, the $R^{n,1}$ metric consists of the Manhattan metric in R^n, r_M, and the R^1-time separation τ. The third group of covariance models is associated with the traveling metric r_T in R^{n+1} (Eq. 1.11g of Table 1.1). The fourth covariance model is associated with the Pythagorean metric r_P in R^{n+1} (Eq. 1.11b). Similar is the case of the fifth covariance model of Table 1.4. The metric of the sixth covariance model is the Minkowski metric r_{Mi} in R^{n+1} (Eq. 1.11f). The seventh covariance model is the von Kármán/Matern model, which is also associated with the Pythagorean metric r_P in R^{n+1} (Eq. 1.11b). Lastly, the eighth covariance model is associated with the mixed traveling-plane wave metric (r_{TW}).

As another illustration, when the phenomenon occurs on the surface of the earth, viewed as a spherical domain

$$S^{n-1} = \{s \in R^n : |s| = \rho\} \subset R^n,$$

[12]Given the essential role of metric in almost every aspect of space–time analysis, most of the equations in this example refer to developments in following chapters of the book.

Table 1.3 Metric Differential Formula (MDF) Example of Metric of Eq. (1.13) With Space- and Time-Dependent Metric Coefficients

Space–Time Metric	MDF
$R^{n,1}$ $\varepsilon_{ij}(\tau) = \alpha_{ij;0}\tau \quad (i,j\neq 0)$ $\varepsilon_{0i}(\boldsymbol{h},\tau) = \sum_{j=1}^{n} \alpha_{0i;j}h_j + \alpha_{0i;0}\tau \quad (i\neq 0)$ $\varepsilon_{00}(\boldsymbol{h}) = \sum_{i=1}^{n} \alpha_{00;i}h_i$ $\varepsilon_i(\tau) = \alpha_{i;0}\tau \quad (i\neq 0)$ $\varepsilon_0(\boldsymbol{h}) = \sum_{i=1}^{n} \alpha_{0;i}h_i$ $(\alpha_{ij;0} = \alpha_{ji;0}; \quad i,j = 0,\ldots,n)$	$\frac{\partial \Delta p}{\partial h_i} = \frac{1}{2}\left[\tau\left(2\sum_{j=1}^{n}(\alpha_{ij;0}+\alpha_{0i;j}+\alpha_{0j;i})h_j + (\alpha_{i;0}+\alpha_{0;i})\right) + \tau^2(2\alpha_{0i;0}+\alpha_{00;i})\right]\Delta p^{-1}$ $\frac{\partial \Delta p^{-1}}{\partial h_i} = -\frac{\partial \Delta p}{\partial h_i}\boldsymbol{\Delta p}^{-2}$ $\frac{\partial \Delta p}{\partial \tau} = \frac{1}{2}\left[\sum_{i,j=1}^{n}(\alpha_{ij;0}+\alpha_{0i;j}+\alpha_{0j;i})h_i h_j + \sum_{i=1}^{n}(\alpha_{i;0}+\alpha_{0;i})h_i + 2\tau\sum_{i=1}^{n}(2\alpha_{0i;0}+\alpha_{00;i})h_i\right]\Delta p^{-1}$ $\frac{\partial^2 \Delta p}{\partial h_i^2} = \left[\alpha_{ii}\tau - \left(\frac{\partial \Delta p}{\partial h_i}\right)^2\right]\Delta p^{-1}$ $\frac{\partial^2 \Delta p}{\partial h_i \partial \tau} = \left[\sum_{j=1}^{n}(\alpha_{ij;0}+\alpha_{0i;j}+\alpha_{0j;i})h_j + \frac{1}{2}(\alpha_{i;0}+\alpha_{0;i}) + \tau(2\alpha_{0i;0}+\alpha_{00;i}) - \frac{\partial \Delta p}{\partial h_i}\frac{\partial \Delta p}{\partial \tau}\right]\Delta p^{-1}$ $\frac{\partial^2 \Delta p}{\partial h_i \partial h_j} = \left[(\alpha_{ij;0}+\alpha_{0i;j}+\alpha_{0j;i})\tau - \frac{\partial \Delta p}{\partial h_i}\frac{\partial \Delta p}{\partial h_j}\right]\Delta p^{-1}$ (1.18a–f)

Table 1.4 Examples of Space–Time Covariance Models and Associated Metrics

Space–Time Covariance Model	Space–Time Metric of Eq. (1.13)
Nos. 1–8 (Table 5.1 of Chapter VIII)	Euclidean-time $(r_E, \tau) \in R^{n,1}$ (Eq. 2.19a of Chapter I)
$[1 + r(a + b + c\tau)]e^{-r-\tau}$	Manhattan-time $(r_M, \tau) \in R^{n,1}$ (Eq. 2.18b of Chapter I)
Nos. 9–11, 32 (Table 5.1 of Chapter VIII)	Traveling $r_T \in R^{n+1}$ (Table 1.1, $\varepsilon_{ii} = 1$, $\varepsilon_{0i} = \upsilon_i$, $i = 1,\ldots, n$)
Nos. 31 (Table 5.1 of Chapter VIII)	Pythagorean $r_P \in R^{n+1}$ (Table 1.1, $\varepsilon_{ii} = \beta_1^2$, $i = 1,\ldots, n$, $\varepsilon_{00} = \beta_2^2$)
Eq. (2.8a) of Chapter VII	Pythagorean $r_P \in R^{n+1}$ (Table 1.1, $\varepsilon_{ii} = \frac{1}{2}a_1^{-2}$, $i = 1,\ldots, n$, $\varepsilon_{00} = \frac{1}{2}a_0^{-2}$)
Eq. (2.8b) of Chapter VII	Minkowski $r_{Mi} \in R^{n+1}$ (Table 1.1, $\varepsilon_{ij} = \frac{1}{2}b_{ij}$, $i = 1,\ldots, n$)
Eq. (2.17b) of Chapter XVI	Pythagorean $r_P \in R^{n+1}$ (Table 1.1, $\varepsilon_{ii} = 1$, $i = 1,\ldots, n$, $\varepsilon_{00} = \frac{a_1}{a_2}$)
Eq. (3.5) of Chapter X	Mixed traveling-plane wave $r_{TW} = \left(r_T^2 + \upsilon^2 r_W^2\right)^{\frac{1}{2}} \in R^{n+1}$ (Table 1.1, $\varepsilon_{ii} = 1$, $\varepsilon_{0i} = -\upsilon_i$ (r_T); $\varepsilon_{00} = 1$, $\varepsilon_{0i} = -\frac{\eta_i}{\upsilon}$ (r_W), $i = 1,\ldots, n$)

where ρ is the Earth radius, see Chapter X, the space–time metric used is of the form (r_s, τ) in $S^{n-1} \times T$, where

$$|\boldsymbol{h}| = r_s = \rho\theta = 2\rho \sin^{-1}\left(\frac{r_E}{2\rho}\right)$$

is the arc length between any two points on the surface of the earth defined by Eq. (2.18c) of Chapter I, θ is the angle of the arc, r_E is the Euclidean metric in R^n, and $\tau \in T \subseteq R^1$ is the time lag. If a covariance in $R^{n,1}$ is a function of r_E and τ, a covariance on the spherical domain $S^{n-1} \times T$ is a function of the arc length r_s and τ. As an approximation, in some cases the covariance on $S^{n-1} \times T$ may be considered a function of the chordal distance $r_c = r_E = 2\rho \sin\left(\frac{\theta}{2}\right)$ and τ.

The form of the space–time metric can also affect the spectral representation of the space–time covariance function.

Example 1.12

The fundamental space–time covariance representation of Eq. (1.5a) of Chapter VIII, which it is written here as (Yadrenko, 1983; Ma, 2003b)

$$c_X(|\boldsymbol{h}|, \tau) = \frac{2\pi^{\frac{n}{2}}}{\Gamma\left(\frac{n}{2}\right)} \int_0^\infty \int_{-\infty}^\infty dk\, d\omega k^{n-1} A_{n,2}(|\boldsymbol{h}|k)\cos(\omega\tau)\widetilde{c}_X(k, \omega), \tag{1.19}$$

with

$$A_{n,2}(u) = 2^{\frac{n}{2}-1}\Gamma\left(\frac{n}{2}\right)u^{1-\frac{n}{2}}J_{\frac{n}{2}-1}(u) \tag{1.20}$$

is valid for a Euclidean spatial metric, $|\boldsymbol{h}| = r_E$. On the other hand, the more involved space–time covariance spectral representation

$$c_X(|\boldsymbol{h}|, \tau) = \frac{2\pi^{\frac{n}{2}}}{\Gamma\left(\frac{n}{2}\right)} \int_0^{\infty} \int_{-\infty}^{\infty} dk\, d\omega\, k^{n-1} A_{n,1}(|\boldsymbol{h}|k)\cos(\omega\tau)\widetilde{c}_X(k, \omega), \tag{1.21}$$

with

$$A_{n,1}(u) = \frac{\Gamma\left(\frac{n}{2}\right)}{\pi^{\frac{1}{2}}\Gamma\left(\frac{n-1}{2}\right)} \int_1^{\infty} d\rho A_{n,2}\left(\rho^{\frac{1}{2}} u\right)\rho^{-\frac{n}{2}}(\rho - 1)^{\frac{n-3}{2}} \tag{1.22}$$

is valid for a Manhattan metric, $|\boldsymbol{h}| = r_M$. Other examples of metrics and the associated covariance models will be discussed in the following.

Remark 1.3

The space–time metric is explicitly involved in the process that leads from the inner product $\boldsymbol{w}\cdot\boldsymbol{p}$ of the composite spectral covariance representation

$$c_X(\boldsymbol{p}) = \int d\boldsymbol{w}\; e^{i\boldsymbol{w}\cdot\boldsymbol{p}}\widetilde{c}_X(\boldsymbol{w}) \tag{1.23a}$$

to the inner product $\boldsymbol{k}\cdot\boldsymbol{h} - \omega\tau$ of the separate spectral covariance representation

$$c_X(\boldsymbol{h}, \tau) = \iint d\boldsymbol{k}d\omega\; e^{i(\boldsymbol{k}\cdot\boldsymbol{h}-\omega\tau)}\widetilde{c}_X(\boldsymbol{k}, \omega). \tag{1.23b}$$

As we saw above, a space–time metric in R^{n+1} is generally defined as the inner product of two space–time vectors with metric coefficients ε_{ij} ($i, j = 0, 1, \ldots, n$). This means that the $\boldsymbol{w}\cdot\boldsymbol{p}$ can be written as

$$\boldsymbol{w}\cdot\boldsymbol{p} = \sum_{i,j=1}^{n,0} \varepsilon_{ij} w_i h_j, \tag{1.24}$$

where in the case of Eq. (1.23a) I let $\varepsilon_{ij} = 0$ ($i \neq j$), $\varepsilon_{00} = -1$, and $\varepsilon_{ii} = 1$ ($i \neq 0$). Then, the inner product in Eq. (1.24) can be expanded as

$$\boldsymbol{w}\cdot\boldsymbol{p} = -w_0 h_0 + \sum_{i=1}^{n} w_i h_i, \tag{1.25}$$

and by using the $R^{n,1}$ notation, $h_0 = \tau$, $w_0 = \omega$, and $w_i = k_i$ ($i \neq 0$), it is finally obtained that

$$\boldsymbol{w}\cdot\boldsymbol{p} = -\omega\tau + \sum_{i=1}^{n} k_i h_i = \boldsymbol{k}\cdot\boldsymbol{h} - \omega\tau, \tag{1.26}$$

which is the notation used in the spectral representation of Eq. (1.23b).

Usually, the covariance is a function with a single peak at the space–time origin that subsequently declines, tending to a zero value at large space–time lag values. Then, given $c_X(|\Delta\boldsymbol{p}|)$, $|\Delta\boldsymbol{p}| = (r, \tau)$, the time lag τ_{max} of the maximum covariance value for varying r can be defined by setting

$$\frac{\partial}{\partial\tau}|\Delta\boldsymbol{p}| = 0. \tag{1.27}$$

Example 1.13

Let the space–time metric be of the form $|\Delta\boldsymbol{p}| = \left[(r - \upsilon\tau)^2 + (\upsilon'\tau)^2\right]^{\frac{1}{2}}$. In this case, the condition of Eq. (1.27) gives

$$\tau_{max} = \frac{\upsilon}{\upsilon^2 + \upsilon'^2} r. \tag{1.28}$$

Working along similar lines, one also finds that

$$r_{max} = \upsilon\tau, \tag{1.29}$$

which is the space lag r_{max} of the maximum covariance value for varying τ.

The contribution of a space–time metric in the characterization of the spatiotemporal variation of a phenomenon can be assessed in terms of the space–time covariance function.

Example 1.14

A simple way to investigate the consequence of the new metric, $\breve{r}_p^2 = \varphi^2(r_p)r_p^2$, obtained in Example 1.9 is to consider the space–time delta covariance $c_X(r_p) = \delta(r_p)$. By applying the coordinate transformation to the delta covariance, the space–time covariance $c_X(r_p) = |\varphi(r_p)|^{-1}\delta(r_p)$ is obtained.

Moreover, determining the connection between space and time through the physical covariance function is an important matter in real-world situations. Understanding, e.g., the relationship between space and time in the fluid turbulence velocity covariance is a fundamental issue in stochastic geophysics theories. Then, as a result of the laws of Nature governing the physical attribute, the attribute covariance functions themselves must be governed by the corresponding physical equations obtained by these laws, which are usually expressed in terms of differential operators $\mathcal{L}[\cdot]$ (several cases of such operators and the associated covariance equations will be examined in Chapter XIV). Proceeding with this train of thought, I consider the following example.

Example 1.15

Under certain circumstances, a fluid dynamics law with attributes represented as frozen random fields (Section 2 of Chapter X) can lead to the following partial differential equation (PDE) governing the space–time change of the attribute covariance function

$$\mathcal{L}[c_X(\Delta\boldsymbol{p})] = \left(\frac{\partial}{\partial\tau} + \boldsymbol{\varepsilon}_0 \cdot \nabla\right) c_X(\Delta\boldsymbol{p}) = 0. \tag{1.30}$$

As the readers may recall, theoretical analysis links the covariance of Eq. (1.30) with the metric $\Delta\boldsymbol{p}$ of the traveling form of Eq. (1.11g).

An appropriate generalization of the observation made in Example 1.15 could suggest the specification of a space–time metric by means of the implication

$$\mathcal{L}[c_X(\Delta\boldsymbol{p})] = 0 \Rightarrow \Delta\boldsymbol{p}, \tag{1.31}$$

i.e., the metric $\Delta\boldsymbol{p}$ may be specified directly from the physical equation governing the space–time change of the attribute covariance function. Motivated by the analysis in Example 1.15 and the methodological implication of Eq. (1.31), the following postulate seems justified.

Postulate 1.1

Since the argument of a covariance function $c_X(\Delta\boldsymbol{p})$ is the space–time metric $\Delta p = |\Delta\boldsymbol{p}|$, the physical equations that are derived from the underlying natural law and describe the change of the covariance function between pairs of space–time points contain information about the structure of this metric (i.e., about its geometrical and physical properties). The covariance equations, then, could be solved with respect to the metric, thus revealing its specific form.

I need to reemphasize that metric selection should not be based on misapplication of common interpretations and utter disregard for mathematical and physical constraints. At its core, the methodological suggestion of Postulate 1.1 is, in fact, the idea behind the metric determination approach to be discussed in more detail in Section 3. But before describing this approach in technical terms, it is appropriate to present some covariance differentiation formulas by means of the space–time metric that will be useful in this description.

2. COVARIANCE DIFFERENTIAL FORMULAS

Space–time covariance functions are important tools of random field modeling, which are used in applied sciences to represent the space–time-dependent structure of natural attributes in conditions of uncertainty (these are key S/TVFs that were formally introduced in Chapter II, and their various properties and applications are studied throughout the book).

I now present certain formulas that express the derivatives of the covariance with respect to the metric, which play a key role in space–time metric determination calculations. For mathematical convenience, for any subset of (possibly repeated) indexes $J = \{j_1, \ldots, j_\nu\}$, I define

$$\partial_J^\nu = \frac{\partial^\nu}{\partial h_{j_1} \ldots \partial h_{j_\nu}} = \frac{\partial^\nu}{\partial \boldsymbol{h}_j^\nu},$$

and introduce the class $\mathcal{P}_{m;k}$ of partitions of $I = \{i_1, \ldots, i_m\}$ into k subsets $\{J_1, \ldots, J_k\} = J \in \mathcal{P}_{m;k}$, so that

$$\bigcup_{p=1}^{k} J_p \equiv I,\ J_p \cap J_{p'} = \varnothing \text{ if } p \neq p',\ |J_p| = \ell_p \left(\sum_{p=1}^{k} \ell_p = m \right).$$

Definition 2.1

The general *covariance differential formula* (*CDF*) is defined by

$$\partial_J^{m+m'} c_X(\Delta\boldsymbol{p}) = \frac{\partial^{m+m'}}{\partial \boldsymbol{h}_i^m \partial \boldsymbol{h}_j^{m'}} c_X(\Delta\boldsymbol{p}) = \sum_{k=1}^{m} \frac{\partial^{k+m'} c_X}{\partial \Delta p^k \partial \boldsymbol{h}_j^{m'}} \left[\sum_{J \in \mathcal{P}_{m;k}} \left(\prod_{p=1}^{k} \partial_{J_p}^{\ell_p} \Delta p \right) \right]. \tag{2.1a}$$

In the special case of $m' = 0$, Eq. (2.1a) reduces to the following CDF expression

$$\partial_J^m c_X(\Delta \boldsymbol{p}) = \frac{\partial^m}{\partial \boldsymbol{h}_j^m} c_X(\Delta \boldsymbol{p}) = \sum_{k=1}^{m} \frac{\partial^k c_X}{\partial \Delta \boldsymbol{p}^k} \left[\sum_{J \in \mathcal{P}_{m;k}} \left(\prod_{p=1}^{k} \partial_{J_p}^{\ell_p} \Delta \boldsymbol{p} \right) \right], \tag{2.1b}$$

which is particularly useful in practice. It is noted that the above formulas express the derivatives of the space–time covariance c_X with respect to the metric and those of the metric with respect to the lags $h_{j_1}...h_{j_m}$. It is also observed that for each $k = 1,\ldots, m$, the number of elements in the class $\mathcal{P}_{m;k}$ is the Stirling number of the second kind, $S(m,k) = \left\{ \begin{matrix} m \\ k \end{matrix} \right\} = \frac{1}{k!} \sum_{j=0}^{k} (-1)^{k-j} \binom{k}{j} j^m$.

Example 2.1

Some worth-noticing special cases of Eqs. (2.1a) and (2.1b), to be used later in the context of the "geometrical metric–physical covariance" analysis, are presented in a matrix–vector format by

$$\begin{aligned} \frac{\partial}{\partial \boldsymbol{h}} c_X(\Delta \boldsymbol{p}) &= \frac{\partial \Delta p}{\partial \boldsymbol{h}} \frac{\partial c_X}{\partial \Delta \boldsymbol{p}}, \\ \frac{\partial^2}{\partial \boldsymbol{h}^T \partial \boldsymbol{h}} c_X(\Delta \boldsymbol{p}) &= \frac{\partial^2 \Delta p}{\partial \boldsymbol{h}^T \partial \boldsymbol{h}} \frac{\partial c_X}{\partial \Delta \boldsymbol{p}} + \left(\frac{\partial \Delta p}{\partial \boldsymbol{h}} \right)^T \left(\frac{\partial \Delta p}{\partial \boldsymbol{h}} \right) \frac{\partial^2 c_X}{\partial \Delta \boldsymbol{p}^2}, \end{aligned} \tag{2.2a–b}$$

where, $\Delta p = \left(\Delta \boldsymbol{p}^2 \right)^{\frac{1}{2}}$ as usual. At this point, I would like to reiterate the significant point that an apparent feature of the above CDF, which is useful for physical metric determination purposes (discussed in Section 3), is that they decompose the ordinary covariance derivatives with respect to space lags $(h_1 \ldots h_n)$ and time lag $(h_0 = \tau)$, rather commonly encountered in physical covariance laws, in terms of the space–time physical metric and the corresponding covariance derivatives with respect to this metric.

By combining MDF with CDF, some interesting expressions are found that explicitly contain the space- and time-independent metric coefficients. This is shown in the following proposition.

Proposition 2.1

In light of the space–time metric of Eq. (1.13), the following expressions involving MDF and CDF are valid,

$$\begin{aligned} \frac{\partial c_X}{\partial \boldsymbol{h}} &= \frac{\boldsymbol{h}^T \boldsymbol{E} + \frac{1}{2} \boldsymbol{\varepsilon}^T}{\Delta p} \frac{\partial c_X}{\partial \Delta \boldsymbol{p}}, \\ \frac{\partial^2 c_X}{\partial \boldsymbol{h}^T \partial \boldsymbol{h}} &= \left[\frac{\boldsymbol{E}}{\Delta p} - \frac{\left(\boldsymbol{E}\boldsymbol{h} + \frac{1}{2} \boldsymbol{\varepsilon} \right) \left(\boldsymbol{h}^T \boldsymbol{E} + \frac{1}{2} \boldsymbol{\varepsilon}^T \right)}{\Delta \boldsymbol{p}^3} \right] \frac{\partial c_X}{\partial \Delta p} + \frac{\left(\boldsymbol{E}\boldsymbol{h} + \frac{1}{2} \boldsymbol{\varepsilon} \right) \left(\boldsymbol{h}^T \boldsymbol{E} + \frac{1}{2} \boldsymbol{\varepsilon}^T \right)}{\Delta \boldsymbol{p}^2} \frac{\partial^2 c_X}{\partial \Delta \boldsymbol{p}^2}. \end{aligned} \tag{2.3a–b}$$

Example 2.2

For illustration purposes, some CDF examples linked to different space–time metrics with space- and time-independent metric coefficients are displayed in Table 2.1. Obviously, as was stressed earlier, analogous formulations are obtained in the case of space- and time-dependent metric coefficients.

The next stage in our discussion is to combine the earlier notions to introduce the following definition that establishes an explicit link between the space–time covariance function and the associated metric.

Table 2.1 Covariance Differential Formula (CDF) Examples

Space–Time Metric	CDF Forms	
$R^{2,1}$ $\Delta p^2 = \boldsymbol{h}^T \boldsymbol{\varepsilon} = \sum_{i=1}^{2} \varepsilon_i h_i + \varepsilon_0 \tau$ $\boldsymbol{E} = \boldsymbol{0}$ $\varepsilon_{ij} = \varepsilon_{0i} = \varepsilon_{00} = 0 \ (i, j = 1, 2)$	$\frac{\partial c_X}{\partial \boldsymbol{h}} = \frac{\boldsymbol{\varepsilon}^T}{2\Delta p} \frac{\partial c_X}{\partial \Delta p},$ $\frac{\partial^2 c_X}{\partial \boldsymbol{h}^T \partial \boldsymbol{h}} = \frac{\boldsymbol{\varepsilon}\boldsymbol{\varepsilon}^T}{4\Delta p^2} \left(\frac{\partial^2 c_X}{\partial \Delta p^2} - \frac{1}{\Delta p} \frac{\partial c_X}{\partial \Delta p} \right).$	(2.4a–b)
$R^{1,1}$ $\Delta p^2 = \boldsymbol{h}^T \boldsymbol{E} \boldsymbol{h} = \varepsilon_{11} h^2 + \varepsilon_{00} \tau^2$ $\boldsymbol{E} = diag\ \boldsymbol{e}_0 = \begin{bmatrix} \varepsilon_{00} & 0 \\ 0 & \varepsilon_{11} \end{bmatrix}$ $\varepsilon_{01} = \varepsilon_0 = \varepsilon_1 = 0$	$\frac{\partial c_X}{\partial h} = \frac{\varepsilon_{11} h}{\Delta p} \frac{\partial c_X}{\partial \Delta p},$ $\frac{\partial^2 c_X}{\partial h^2} = \frac{\varepsilon_{11} - \varepsilon_{11}^2 \frac{h^2}{\Delta p^2}}{\Delta p} \frac{\partial c_X}{\partial \Delta p} + \frac{\varepsilon_{11}^2 h^2}{\Delta p^2} \frac{\partial^2 c_X}{\partial \Delta p^2}.$ $\frac{\partial c_X}{\partial \tau} = \frac{\varepsilon_{00} \tau}{\Delta p} \frac{\partial c_X}{\partial \Delta p},$ $\frac{\partial^2 c_X}{\partial h \partial \tau} = \frac{\varepsilon_{00} \varepsilon_{11} h \tau}{\Delta p^2} \left[\frac{\partial^2 c_X}{\partial \Delta p^2} - \frac{1}{\Delta p} \frac{\partial c_X}{\partial \Delta p} \right].$	(2.4c–f)

Definition 2.2

The κ/κ'^{th}-order *covariance metric ratio* (*CMR*) of the space–time covariance c_X is defined by

$$\partial_{J/J'}^{(\kappa/\kappa')} = \partial_{j_0,\ldots,j_\ell/j'_0,\ldots,j'_m}^{(\kappa_0,\ldots,\kappa_\ell/\kappa'_0,\ldots,\kappa'_m)} = \frac{\dfrac{\partial^\kappa}{\partial h_{j_0}^{\kappa_0} \ldots \partial h_{j_\ell}^{\kappa_\ell}} c_X}{\dfrac{\partial^{\kappa'}}{\partial h_{j'_0}^{\kappa'_0} \ldots \partial h_{j'_m}^{\kappa'_m}} c_X}, \tag{2.5}$$

where $\kappa_p, \kappa'_q \ (p = 0, \ldots, \ell;\ q = 0, \ldots, m)$ are positive integers representing the orders of differentiation and are such that $\kappa = \sum_{p=0}^{\ell} \kappa_p$ and $\kappa' = \sum_{q=0}^{m} \kappa'_q$.

Table 2.2 Some Interesting Covariance Metric Ratio (CMR) Cases

Order of Differentiation	CMR	
$\ell = m = 1$ $\kappa_i = 1\ (i \neq 0),\ \kappa'_0 = 1$	$\partial^{(1/1)}_{i/0} = \frac{\frac{\partial}{\partial h_i} c_X}{\frac{\partial}{\partial \tau} c_X} = \frac{\sum_{j=1}^{n} \varepsilon_{ij} h_j + \varepsilon_{0i}\tau + \frac{1}{2}\varepsilon_i}{\sum_{j=1}^{n} \varepsilon_{0j} h_j + \varepsilon_{00}\tau + \frac{1}{2}\varepsilon_0}$	(2.6a)
$\ell = m = 1$ $\kappa_i = 1\ (i \neq 0),\ \kappa'_j = 1\ (j \neq 0)$	$\partial^{(1/1)}_{i/j} = \frac{\frac{\partial}{\partial h_i} c_X}{\frac{\partial}{\partial h_j} c_X} = \frac{\sum_{j=1}^{n} \varepsilon_{ij} h_j + \varepsilon_{0i}\tau + \frac{1}{2}\varepsilon_i}{\sum_{i=1}^{n} \varepsilon_{ij} h_i + \varepsilon_{0j}\tau + \frac{1}{2}\varepsilon_j}$	(2.6b)
$\ell = 2,\ m = 1$ $\kappa_i = 2\ (i \neq 0),\ \kappa'_0 = 1$	$\partial^{(2/1)}_{i/0} = \frac{\frac{\partial^2}{\partial h_i^2} c_X}{\frac{\partial}{\partial \tau} c_X} = \frac{\partial}{\partial h_i}\left(\partial^{(1/1)}_{i/0}\right) + \zeta_i \partial^{(1/1)}_{i/0}$	(2.6c)
$\ell = 2,\ m = 1$ $\kappa_i = 1\ (i \neq 0),\ \kappa_j = 1\ (j \neq 0, j \neq i),\ \kappa'_0 = 1$	$\partial^{(1,1/1)}_{i,j/0} = \frac{\frac{\partial^2}{\partial h_i \partial h_j} c_X}{\frac{\partial}{\partial \tau} c_X} = \frac{\partial}{\partial h_j}\left(\partial^{(1/1)}_{i/0}\right) + \zeta_j \partial^{(1/1)}_{i/0}$	(2.6d)

Being defined as the ratios of CDFs in Eq. (2.5), it seems natural that the CMRs themselves should be expressed in a geometrical form by means of the corresponding space−time metric. Particularly, the CMRs can be specified in terms of the ratios of the corresponding metric derivatives (i.e., the space−time metric coefficients) through a direct application of the chain rule. This process will become technically clearer in the following discussion.

Below, I focus on up to second-order space−time CMRs (yet, the analysis remains valid for higher orders). For illustration, Table 2.2 displays some special cases of the general CMR formulation of Eq. (2.5) associated with various differentiation orders. As it can be seen, the first-order CMRs are expressed in terms of the space−time metric coefficients ε_i and ε_{ij} $(i, j = 0, 1, \ldots, n)$, whereas the second-order CMRs are functions of the first-order CMRs and the derived metric coefficients ζ_i $(i = 1, \ldots, n)$.[13] Since, together with the ε_i and ε_{ij} coefficients, the ζ_i coefficients are considered the unknowns to be determined in the physical metric determination context, analytically or computationally, a more detailed description of their features and explicit formulation is postponed until Section 5 of this chapter. Two examples are considered next to illustrate some of the CMR features.

[13]As it turns out, see analysis in Section 5 later, the coefficients ζ_i may be also expressed in terms of the ε_i, ε_{ij} coefficients.

Table 2.3 First-Order Covariance Metric Ratio (CMR) Examples of Commonly Used Space–Time Metrics

Space–Time Metric	CMR	
Eq. (1.11a)	$\partial_{i/0}^{(1/1)} = \left(\frac{\varepsilon_{ii}}{\varepsilon_{00}}\right)^{\frac{1}{2}}$ $\partial_{i/j}^{(1/1)} = \left(\frac{\varepsilon_{ii}}{\varepsilon_{jj}}\right)^{\frac{1}{2}}$	(2.7a–b)
Eq. (1.11b)	$\partial_{i/0}^{(1/1)} = \frac{\varepsilon_{ii}h_i}{\varepsilon_{00}\tau}$ $\partial_{i/j}^{(1/1)} = \frac{\varepsilon_{ii}h_i}{\varepsilon_{jj}h_j}$	(2.8a–b)
Eq. (1.11c)	$\partial_{i/0}^{(1/1)} = \frac{\varepsilon_i}{\varepsilon_0}$ $\partial_{i/j}^{(1/1)} = \frac{\varepsilon_i}{\varepsilon_j}$	(2.9a–b)

Example 2.3

For further illustration purposes, some additional CMR expressions associated with commonly encountered space–time metrics of the form of Eq. (1.13) are listed in Table 2.3, assuming space- and time-independent metric coefficients. These CMR expressions play a substantial role in the determination of valid space–time metrics for random fields of arbitrary physical nature.

Example 2.4

Table 2.4, on the other hand, displays the CMR forms in the case of space- and time-dependent coefficients of the metrics described earlier in Table 1.3. In other words, the CMR ratios considered above are determined in terms of the corresponding space–time metric derivative ratios. Obviously, its ability to also consider space- and time-dependent metric coefficients can further enlarge the applicability range of the general metric expression of Eq. (1.13).

Joint "metric–covariance" PDEs are obtained from the CMRs for space- and time-dependent metric coefficients, in general. A unique feature of these PDE is that they explicitly blend physics with geometry.

Table 2.4 First- and Second-Order Covariance Metric Ratio (CMR) for the Space–Time Metric of Table 1.3

Space–Time Metric	CMR
$R^{n,1}$ $\varepsilon_{ij}(\tau) = \alpha_{ij;0}\tau \ \ (i,j\neq 0)$ $\varepsilon_{0i}(\boldsymbol{h},\tau) = \sum_{j=1}^{n} \alpha_{0i;j}h_j + \alpha_{0i;0}\tau \ \ (i\neq 0)$ $\varepsilon_{00}(\boldsymbol{h}) = \sum_{i=1}^{n} \alpha_{00;i}h_i$ $\varepsilon_i(\tau) = \alpha_{i;0}\tau \ \ (i\neq 0)$ $\varepsilon_0(\boldsymbol{h}) = \sum_{i=1}^{n} \alpha_{0;i}h_i$ $\alpha_{ij;0} = \alpha_{ji;0}$ $(i,j = 0,\ldots,n)$	$\partial_{i/0}^{(1/1)} = \frac{2\sum_{j=1}^{n}(\alpha_{ij;0}+\alpha_{0i;j}+\alpha_{0j;i})h_j\tau + (2\alpha_{0i;0}+\alpha_{00;i})\tau^2 + (\alpha_{i;0}+\alpha_{0;i})\tau}{\sum_{k,j=1}^{n}(\alpha_{kj;0}+\alpha_{0k;j}+\alpha_{0j;k})h_k h_j + \sum_{j=1}^{n}(\alpha_{j;0}+\alpha_{0;j})h_j + 2\sum_{j=1}^{n}(2\alpha_{0j;0}+\alpha_{00;j})h_j\tau}$ $\partial_{i/j}^{(1/1)} = \frac{2\sum_{k=1}^{n}(\alpha_{ik;0}+\alpha_{0i;k}+\alpha_{0k;i})h_k + (2\alpha_{0i;0}+\alpha_{00;i})\tau + \alpha_{i;0}+\alpha_{0;i}}{2\sum_{\ell=1}^{n}\alpha_{j\ell}(\alpha_{j\ell;0}+\alpha_{0j;\ell}+\alpha_{0\ell;j})h_\ell + (2\alpha_{0j;0}+\alpha_{00;j})\tau + \alpha_{j;0}+\alpha_{0;j}}$ $\partial_{i/0}^{(2/1)} = \frac{\partial}{\partial h_i}\left(\partial_{i/0}^{(1/1)}\right) + \zeta_i \partial_{i/0}^{(1/1)}$ $\partial_{i,j/0}^{(1,1/1)} = \frac{\partial}{\partial h_j}\left(\partial_{i/0}^{(1/1)}\right) + \zeta_j \partial_{i/0}^{(1/1)}$ (2.10a–d)

Example 2.5

Direct results of Eqs. (2.3a–b) are the PDEs

$$\begin{aligned} \nabla\Delta p \frac{\partial c_X}{\partial \tau} - \frac{\partial \Delta p}{\partial \tau} \nabla c_X &= \mathbf{0}, \\ \nabla\Delta p (\nabla c_X)^T - \nabla c_X (\nabla\Delta p)^T &= \mathbf{0}, \end{aligned} \tag{2.11a–b}$$

respectively, where

$$\begin{aligned} \nabla\Delta \boldsymbol{p} &= \left[\frac{\partial \Delta \boldsymbol{p}}{\partial h_1} \cdots \frac{\partial \Delta \boldsymbol{p}}{\partial h_n}\right]^T, \\ \nabla c_X &= \left[\frac{\partial c_X}{\partial h_1} \cdots \frac{\partial c_X}{\partial h_n}\right]^T. \end{aligned} \tag{2.12a–b}$$

A key feature of Eqs. (2.11a–b) is that they explicitly link physical (covariance) changes with geometrical (metric) changes in space and time. A covariance with the general space–time metric of Eq. (1.13) must satisfy Eqs. (2.11a–b), which is why these equations can play a central role in the determination of a physically meaningful space–time metric.

Because of its special properties, particularly interesting is the case of the traveling metric of Eq. (1.11g) with the coefficients ε_{0i} being themselves functions of space–time and admitting a certain physical interpretation, depending on the situation. This is best illustrated with the help of an example.

Example 2.6

In the case of the traveling metric, the PDE representations of the space–time covariance to be presented in Section 2.3 of Chapter X are obtained from the CMR of Eqs. (2.6a) and (2.6d) with $\varepsilon_{0i} = \upsilon_i$ (denoting velocity components of the attribute spread). As we shall see in Chapter X, the PDE representations are directly linked to certain important features of the frozen random field models. In addition, the readers are referred to the Section 4 in Chapter V.

3. SPACE–TIME METRIC DETERMINATION FROM PHYSICAL CONSIDERATIONS

The above thoughts based on Postulate 1.1 lead to the following three-step approach of Table 3.1, which determines a realistic space–time metric based on physical considerations about the phenomenon or attribute. The regime of interest here is the one in which there exists sufficient information about the attribute in terms of a physical law, a scientific relationship, or an empirical formula. So, if the laws governing a phenomenon have the form of differential equations, these equations contain information on the structure of space and time (apparently, they should contain information not only about the geometrical but also about the physical properties of space–time). An important point to be stressed is that, as laborious as metric determination can be in some cases, it would surely improve on the current state of affairs regarding random field modeling of natural attributes.

Table 3.1 The Three-Step Approach of Space–Time Metric Determination

Step	Description
1	Since the covariance c_X is a function of the metric to be determined, the starting point of the approach is the equation governing c_X, $$\mathcal{L}_c[c_X, \boldsymbol{a}, \lvert\Delta\boldsymbol{p}\rvert] = 0, \quad (3.1)$$ where $\mathcal{L}_c$ denotes the "law of change of c_X," $\lvert\Delta\boldsymbol{p}\rvert = \Delta p$ is the metric of Eq. (1.13), and $\boldsymbol{a} = \{a_k\}$ ($k = 1,\ldots,\eta$) is the set of coefficients determined by the stochastic expectation process leading from the physical law of $X(\boldsymbol{p})$ to the $\mathcal{L}_c$-Eq. (3.1).
2	Using the covariance metric ratio expressions introduced in the previous sections, the $\mathcal{L}_c$-Eq. (3.1) yields $$\mathcal{L}_G\left[\partial^{(\kappa/\kappa')}_{J/J'}, \boldsymbol{a}\right] = 0, \quad (3.2)$$ where $\partial^{(\kappa/\kappa')}_{J/J'}$ were defined earlier in terms of the metric coefficients $(\boldsymbol{E}, \boldsymbol{\varepsilon})$. That is, the physical ($\mathcal{L}_c$) Eq. (3.1) has been replaced by the geometrical ($\mathcal{L}_G$) Eq. (3.2) that the space–time metric coefficients must satisfy.
3	Solve Eq. (3.2) with respect to $(\boldsymbol{E}, \boldsymbol{\varepsilon})$, which should be expressed in terms of the physical law parameters $\boldsymbol{a}$, viz., $$(\boldsymbol{E}, \boldsymbol{\varepsilon}) = (\varepsilon_{00}(a_k), \varepsilon_{0i}(a_k), \varepsilon_{ij}(a_k), \varepsilon_i(a_k), \varepsilon_0(a_k)) \quad (3.3)$$ $(i,j = 1,\ldots,n;\ k = 1,\ldots,\eta)$.

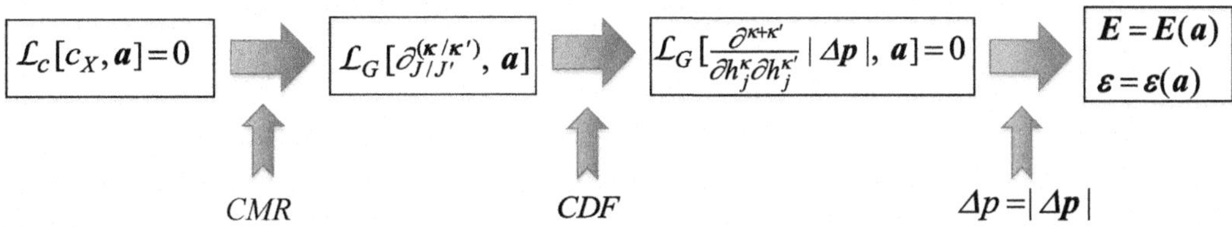

FIGURE 3.1

A graphical outline of the space–time metric determination approach.

Since the attribute covariance is a function of the space–time metric to be determined, the starting point of the proposed approach is the equation governing the covariance obtained from the corresponding physical law, scientific relationship, or empirical formula. As usual, the attribute is modeled as an S/TRF $X(\boldsymbol{p})$ in the $R^{n,1}$ domain. A more detailed outline of the space–time metric determination approach of Table 3.1 is given in Fig. 3.1. It is worth noticing that the approach of Table 3.1 essentially remains valid if, instead of Eq. (1.13), we assume another form of the space–time metric. So, there may exist more than one solution of Eq. (3.2), leading to different metrics that reveal different space–time dependence structures. We may also find that certain solutions recur in various disguises. Surely, the difficulty of deriving metric solutions of the $\mathcal{L}_G$-equation should depend on the complexity of the equation.

In a nutshell, instead of arbitrary selecting a space–time metric for the phenomenon of interest (as is commonly done in many mainstream geostatistical and statistical applications), the approach of Table 3.1 and Fig. 3.1 suggests allowing physical considerations to reveal the metric form. A few more comments can be made concerning the above metric determination approach. A discussion of the existing stochastic methods for deriving the covariance Eq. (3.1) from the corresponding natural laws can be found in Chapter XIV and references therein (e.g. Srinivasan and Vasudevan, 1971; Soong, 1973; Dobrovolski, 2010; Klyatskin, 2015). When BIC equations are considered, they should be included in the $\mathcal{L}_G$ formulation. One cannot specify independently the spatiotemporal metric and the characteristics of the phenomenon (uncertainty sources, BIC, physical law parameters), since they are connected via the geometrical equation $\mathcal{L}_G$. In principle, any metric solution is formally acceptable if it satisfies the $\mathcal{L}_G$-equation. In this sense, two main types of solutions of the $\mathcal{L}_G$-equations can be derived:

- type *a* solutions, i.e., exact metric solutions based on plausible assumptions (which may impose symmetry conditions on the metric, restrict the mathematical structure of the covariance function etc.), and
- type *b* solutions, i.e., approximate analytical or numerical metric solutions able to explore different physical situations.

Certain solutions can be obtained by setting to zero each coordinate term of the $\mathcal{L}_G$-equation, which would lead to a new set of dependent equations (in the sense that a metric coefficient usually appears in more than one equation). Let us denote this set of equations as the set of $\mathcal{L}_G^*$-equations. Using judicious selection (and, occasionally, with the help of computational software) we could simultaneously solve the $\mathcal{L}_G^*$-equations. Moreover, having a general metric form such as Eq. (1.13), we can make assumptions on those components of the physical law for which the corresponding terms in the $\mathcal{L}_G^*$-equations are not automatically determined, and then solve the remaining equations.

Remark 3.1

Concerning their usefulness, type *a* solutions, though often obtained by imposing simplifying assumptions, they are valuable in certain ways: they allow us to get a deeper understanding of the methodological underpinnings of the metric determination approach (some solutions can be of particular importance, physically or mathematically), all quantities are expressed by elementary functions or well-known special functions, and they can complement the type *b* solutions by providing the background on which approximations for real physical situations can be built and by enabling checks of numerical accuracy (some examples are discussed in the following section).

4. EXAMPLES

I start with some type *a* examples to clarify the presentation of the theoretical discussion above. In these examples, the metric solutions are obtained on the basis of certain assumptions: the physical law governing the attribute under consideration is expressed in terms of stochastic PDE (SPDE) with uncertainty sources in the form of random BIC or random law coefficients, the corresponding attribute covariance $\mathcal{L}_c$-equation has been derived from the physical law by means of stochastic expectation or statistical averaging techniques (see Chapter XIV and references therein),[14] and more than one space–time metric solution of the geometrical $\mathcal{L}_G$-equation may be possible, admitting different physical interpretations.

Example 4.1

Eq. (4.1) of Table 4.1 describes the combined space–time change of a covariance function $c_X(\boldsymbol{h}, \tau)$ in the $R^{3,1}$ domain. This kind of $\mathcal{L}_c$-equation may be derived from a transport law under certain turbulent flow conditions (e.g., Monin and Yaglom, 1971) with attributes represented as frozen random fields (Section 2 of Chapter X). Eq. (4.1) can be rewritten in the $\mathcal{L}_G$-form of Eq. (4.2), where the CMR $\partial_{i/0}^{(1/1)}$ is given by Eq. (2.6a). While $\mathcal{L}_c$ outlines the physical correlation behavior of the attribute, $\mathcal{L}_G$ describes the geometrical behavior of space and time within which the physical attribute varies. In this sense, the metric itself may be seen as the geometrical "attribute" of the $\mathcal{L}_G$-equation. Next, by inserting Eq. (2.6a) into Eq. (4.2), the $\mathcal{L}_G$ becomes Eq. (4.3). Eq. (4.3) is an analytical geometrical equation that replaces the physical covariance Eq. (4.1), which means that the metric coefficients must be chosen so that Eq. (4.3) is satisfied. A solution is given in Eqs. (4.4a–d). The space–time metric consistent with the $\mathcal{L}_G$-equation is that expressed by Eq. (4.5). Accordingly, the covariance model representing the traveling random field is[15]

$$c_X(|\Delta \boldsymbol{p}|) = e^{-\left[\sum_{i=1}^{3}(h_i - a\tau)\right]^2}. \tag{4.6}$$

A metric solution implies a certain relationship between the metric coefficients and the physical parameter a that admits a certain interpretation. Specifically, the metric Eq. (4.5) is characterized by the positive effect of a on the correlation strength (range) between points, an effect that decreases with increasing time interval.[16] Another point is worth-noticing. The readers may recall that, in theory Eq. (4.1) can be also associated with the traveling metric of Eq. (1.11g), which is, indeed, confirmed using the metric determination approach of Table 3.1: it is found that the metric of Eq. (1.11g) holds with $\varepsilon_{0i} = -a\ (i = 1,2,3)$. Generally, a derived metric will be meaningful if the corresponding attribute exists physically. This is, indeed, the case here, since the derived metrics are consistent with the frozen attribute field of the underlying physics.

Example 4.2

In the Example 4.2 of Table 4.1 the covariance function satisfies the $\mathcal{L}_c$-Eq. (4.7) in the $R^{3,1}$ domain, where a is a physical constant (Daley, 1999). Eq. (4.7) yields the $\mathcal{L}_G$-Eq. (4.8), and, subsequently, Eq. (4.9). This $\mathcal{L}_G$-equation represents an interaction of geometrical requirements combined with physical parameters, in which the metric coefficients must be chosen so that the $\mathcal{L}_G$-equation is satisfied. Since Eq. (4.9) has more than one metric solution, it is interesting to understand what

[14] As it was pointed out earlier, these SPDEs are assumed to be available (e.g., they have been previously derived or they can be found in the relevant scientific literature) before the metric determination approach is implemented.

[15] This covariance model, indeed, satisfies Eq. (4.2).

[16] This result is obtained by observing the variation of the covariance as a function of the a-dependent metric $|\Delta \boldsymbol{p}|$.

Table 4.1 Space–Time Metric Determination Examples

	Example 4.1 ($R^{3,1}$)		**Example 4.2 ($R^{3,1}$)**	
$\mathcal{L}_c$	$\left(\frac{\partial}{\partial\tau}+a\sum_{i=1}^{3}\frac{\partial}{\partial h_i}\right)c_X=0$	(4.1)	$\sum_{i=1}^{3}\left(h_i\frac{\partial}{\partial\tau}-a^2\tau\frac{\partial}{\partial h_i}\right)c_X=0$	(4.7)
$\mathcal{L}_G$	$\frac{1}{a}+\sum_{i=1}^{3}\partial_{i/0}^{(1/1)}=0$	(4.2)	$\sum_{i=1}^{3}\left(\partial_{i/0}^{(1/1)}-\frac{1}{a^2\tau}h_i\right)=0$	(4.8)
	$\sum_{i=1}^{3}\left(\sum_{j=1}^{3}\varepsilon_{ji}h_j+\varepsilon_{0i}\tau+\frac{1}{2}\varepsilon_i\right)+\frac{1}{a}\left(\sum_{i=1}^{3}\varepsilon_{0i}h_i+\varepsilon_{00}\tau+\frac{1}{2}\varepsilon_0\right)=0$	(4.3)	$a^2\tau\sum_{i=1}^{3}\left(\sum_{j=1}^{3}\varepsilon_{ji}h_j+\varepsilon_{0i}\tau+\frac{1}{2}\varepsilon_i\right)-\left(\sum_{i=1}^{3}\varepsilon_{0i}h_i+\varepsilon_{00}\tau+\frac{1}{2}\varepsilon_0\right)\sum_{i=1}^{3}h_i=0$	(4.9)
$\boldsymbol{\varepsilon}$	$\varepsilon_0=\varepsilon_i=0$ $\varepsilon_{ji}=1$ $\varepsilon_{0j}=-3a$ $\varepsilon_{00}=9a^2$	(4.4a–d)	$\varepsilon_0=\varepsilon_i=\varepsilon_{0i}=\varepsilon_{ji}=0$ $\varepsilon_{ii}=1$ $\varepsilon_{00}=a^2$	(4.10a–c)
			$\varepsilon_0=\varepsilon_i=\varepsilon_{0i}=0$ $\varepsilon_{1j}=\varepsilon_{2j}=\varepsilon_{3j}=1$ $\varepsilon_{00}=a^2\sum_{j=1}^{3}\varepsilon_{1j}=3a^2$	(4.10d–f)
Δp^2	$\left(\sum_{i=1}^{3}h_i-3a\tau\right)^2$	(4.5)	$\sum_{i=1}^{3}h_i^2+a^2\tau^2$ $\sum_{i,j=1}^{3}h_ih_j+3a^2\tau^2$	(4.11a–b)
	Example 4.3 ($R^{2,1}$)		**Example 4.4 ($R^{1,1}$)**	
$\mathcal{L}_c$	$\left[\xi\frac{\partial}{\partial\tau}-\boldsymbol{k}\cdot\nabla-\kappa\nabla^2\right]c_X=0$	(4.15)	$\left[S\frac{\partial}{\partial\tau}-K\frac{\partial^2}{\partial h^2}\right]c_X=0$	(4.24)
			$c_X(h,0)=c_0e^{-\frac{3h}{a_r}}$ $c_X(0,\tau)=c_0e^{\frac{3\tau}{a_\tau}}$ $c_X(\ell,\tau)=0$	(4.25a–c)

Table 4.1 Space–Time Metric Determination Examples—cont'd

	Example 4.3 ($R^{2,1}$)	Example 4.4 ($R^{1,1}$)
$\mathcal{L}_G$	$\xi - \sum_{i=1}^{2}\left(k_i \partial_{i/0}^{(1/1)} - \kappa \partial_{i/0}^{(2/1)}\right) = 0$ (4.16) $\xi - \sum_{i=1}^{2}\left\{ k_i \frac{\sum_{j=1}^{2}\varepsilon_{ji}h_j+\varepsilon_{0i}\tau+\frac{1}{2}\varepsilon_i}{\sum_{j=1}^{2}\varepsilon_{0j}h_j+\varepsilon_{00}\tau+\frac{1}{2}\varepsilon_0} - \kappa \left[\frac{1}{\sum_{j=1}^{2}\varepsilon_{0j}h_j+\varepsilon_{00}\tau+\frac{1}{2}\varepsilon_0}\left(\varepsilon_{ii}-\varepsilon_{0i}\frac{\sum_{j=1}^{2}\varepsilon_{ji}h_j+\varepsilon_{0i}\tau+\frac{1}{2}\varepsilon_i}{\sum_{j=1}^{2}\varepsilon_{0j}h_j+\varepsilon_{00}\tau+\frac{1}{2}\varepsilon_0}\right) + \zeta_i \frac{\sum_{j=1}^{2}\varepsilon_{ji}h_j+\varepsilon_{0i}\tau+\frac{1}{2}\varepsilon_i}{\sum_{j=1}^{2}\varepsilon_{0j}h_j+\varepsilon_{00}\tau+\frac{1}{2}\varepsilon_0}\right]\right\} = 0$ (4.17)	$\frac{S}{K} - \partial_{1/0}^{(2/1)} = 0$ (4.26) $\frac{S}{K} - \frac{1}{\varepsilon_{01}h+\varepsilon_{00}\tau+\frac{1}{2}\varepsilon_0}\left(\varepsilon_{11} - \frac{\varepsilon_{11}h+\varepsilon_{01}\tau+\frac{1}{2}\varepsilon_1}{\varepsilon_{01}h+\varepsilon_{00}\tau+\frac{1}{2}\varepsilon_0}\varepsilon_{01}\right) - \zeta_1 \frac{\varepsilon_{11}h+\varepsilon_{01}\tau+\frac{1}{2}\varepsilon_1}{\varepsilon_{01}h+\varepsilon_{00}\tau+\frac{1}{2}\varepsilon_0} = 0$ (4.27)
$\boldsymbol{\varepsilon}$	$\varepsilon_{ij} = \varepsilon_{00} = \varepsilon_{0i} = 0$ $\varepsilon_0 = \xi^{-1}\sum_{i=1}^{2}\left(k_i\varepsilon_i - \kappa\varepsilon_i^2\right)$ $\zeta_i = -\varepsilon_i$ (4.18a–c)	$\varepsilon_{11} = \varepsilon_{00} = \varepsilon_{01} = 0$ $\varepsilon_0 = -\frac{K}{S}\left(\frac{3}{a_r}\right)^2$ $\zeta_1 = -\varepsilon_1 = -\frac{3}{a_r}$ (4.28a–c)
	$\varepsilon_{ij} = \varepsilon_{00} = \varepsilon_{0i} = 0$ $\varepsilon_i = -\zeta_i = 1$ $\varepsilon_0 = \xi^{-1}\left(\sum_{i=1}^{2}k_i - 2\kappa\right)$ (4.18d–f)	$\varepsilon_0 = \varepsilon_1 = 0$ $\varepsilon_{01} = -\frac{K}{S}$ $\varepsilon_{00} = \varepsilon_{01}^2$ $\varepsilon_{11} = \zeta_1 = -1$ (4.28d–g)
$\Delta \boldsymbol{p}^2$	$\sum_{i=1}^{2}\left[\varepsilon_i h_i + \xi^{-1}\left(k_i\varepsilon_i - \kappa\varepsilon_i^2\right)\tau\right]$ $\sum_{i=1}^{2}h_i + \xi^{-1}\left(\sum_{i=1}^{2}k_i - 2\kappa\right)\tau$ (4.19a–b)	$\frac{3}{a_r}h - \left(\frac{3}{a_r}\right)^2\frac{K}{S}\tau$ $\left(h - \frac{K}{S}\tau\right)^2$ (4.29a–b)

qualitative features each solution may possess. First, an obvious solution of Eq. (4.9) is shown in Eqs. (4.10a–c), with the metric of Eq. (4.11a), and the covariance function

$$c_X(|\Delta \boldsymbol{p}|) = e^{-\left(\sum_{i=1}^{3} h_i^2 + a^2\tau^2\right)}. \tag{4.12}$$

Another solution of Eq. (4.9) is given in Eqs. (4.10d–f), where the space–time metric is now Eq. (4.11b).[17]

So far, all metric coefficients were expressed in terms of elementary functions. It is also observed that

$$\Delta \boldsymbol{p}^2_{(4.11b)} = \Delta \boldsymbol{p}^2_{(4.11a)} + \sum_{i \neq j=1}^{3} h_i h_j + 2a^2\tau^2.$$

Different metrics may rest on different physical bases (e.g., different variability or dependence structures across the space–time domain). Here, $|\Delta \boldsymbol{p}|_{(4.11a)}$ allows a rather smooth attribute variation in space–time, whereas $|\Delta \boldsymbol{p}|_{(4.11b)}$ implies a rougher variation and a higher negative effect of the physical parameter a on the correlation strength between points (this negative effect increases with increasing separation time). Similar to the case of Example 4.1, this result is obtained by observing the variation of the corresponding covariance shapes as functions of $|\Delta \boldsymbol{p}|_{(4.11a)}$ and $|\Delta \boldsymbol{p}|_{(4.11b)}$ (the covariance with metric $|\Delta \boldsymbol{p}|_{(4.11a)}$ has a smaller slope at origin than that with $|\Delta \boldsymbol{p}|_{(4.11b)}$, implying a smoother variation of the attribute distribution). By inserting the Gaussian model

$$c_X(|\Delta \boldsymbol{p}|) = e^{-\left(\sum_{i=1}^{3} b_i h_i^2 + c\tau^2\right)}$$

into $\mathcal{L}_c$-Eq. (4.7), one finds $b_i = \frac{c}{a^2} = b$, i.e., the model coefficients b_i and c may be chosen so that this relationship holds, in which case, the space–time Gaussian covariance is compatible with $\mathcal{L}_c$. From the above relationship the Gaussian model yields

$$c_X(|\Delta \boldsymbol{p}|) = \left[e^{-|\Delta \boldsymbol{p}|_{(4.11a)}}\right]^b, \tag{4.13}$$

i.e., the covariance of Eq. (4.13) is a generalization of that of Eq. (4.12), which, in addition to the physical law parameter a, it includes the Gaussian model parameter b.

Example 4.3

This example (Table 4.1) considers a geohydrologic situation in which the variation of the hydraulic head $X(\boldsymbol{p})$ in the $R^{2,1}$ domain is governed by the subsurface flow law (Dagan, 1989) with uncertainty sources the hydraulic storativity and conductivity parameters. In this case, one can start with the general metric of Eq. (1.13) in $R^{2,1}$

$$\Delta \boldsymbol{p}^2 = \sum_{i=1}^{2} \varepsilon_{ii} h_i^2 + 2\tau \sum_{i=1}^{2} \varepsilon_{0i} h_i + \sum_{i=1}^{2} \varepsilon_i h_i + (\varepsilon_{12} + \varepsilon_{21}) h_1 h_2 + \varepsilon_{00}\tau^2 + \varepsilon_0 \tau, \tag{4.14}$$

[17]The fact that more than one space–time metrics satisfy the $\mathcal{L}_G$-equation is a situation similar to that in which more than one metric has been found to satisfy the field equations of gravitation (e.g., Minkowski, Schwarzschild, Friedmann, and Walker metric solutions; Stephani et al., 2003; Frei, 2014).

and the covariance function characterizing the hydraulic head's space−time variation as is expressed by the $\mathcal{L}_c$-Eq. (4.15), where the deterministic coefficients ξ, κ, and $\boldsymbol{k} = (k_1 \quad k_2)^T$ are expected functions of the above random law parameters. Following the metric determination approach (Table 3.1), the $\mathcal{L}_G$-equation corresponding to $\mathcal{L}_c$-Eq. (4.15) is given by Eq. (4.16), which, in light of Eq. (4.14), is written in the analytical form of Eq. (4.17).[18] As before, Eq. (4.17) must be solved with respect to the metric coefficients ε_{00}, ε_{0i}, ε_{ij}, ε_i, ε_0 and ζ_i $(i, j = 1, 2)$. A solution set is given in Eqs. (4.18a−c), in which case the metric Eq. (4.14) reduces to Eq. (4.19a) with the $\mathcal{L}_c$-Eq. (4.15) producing the covariance solution

$$c_X(|\Delta\boldsymbol{p}|) = e^{-\sum_{i=1}^{2}\varepsilon_i h_i - \varepsilon_0\tau}, \tag{4.20}$$

where ε_0 is given by Eq. (4.18b) in Table 4.1. Particularly, one could chose a unit value for ε_i and then define the corresponding ε_0 value—see Eqs. (4.18d−f) in which the metric coefficients are now functions only of the parameters ξ, κ, k_i $(i = 1, 2)$ of the geohydrologic covariance law. Then, the metric of Eq. (4.19a) reduces to that of Eq. (4.19b).[19]

Space−time plays a dual role in the $\mathcal{L}_G$-equations governing the metric, because it constitutes both the object and the context within which space−time dependence is defined. This kind of self-reference gives the $\mathcal{L}_G$-equations certain characteristics that are different than those of the equations governing the attribute. For example, normally we can formulate the BIC of a physical law by specifying the attribute values at a given space boundary and time instant, and then use the physical law to determine the space−time evolution of the attribute. In contrast, due to the inherent self-referential aspect of the metric, one is not free to specify arbitrary BIC, but only conditions that already satisfy the self-consistency requirements imposed by the $\mathcal{L}_G$-equations themselves.

Example 4.4

This is a special case of the geohydrologic law of the previous Example 4.3 representing transient groundwater flow in the $R^{1,1}$ domain with the random source now being the flow BIC. Specifically, in $R^{1,1}$ the general metric of Eq. (1.13) reduces to

$$\Delta p^2 = \varepsilon_{11}h^2 + \varepsilon_{00}\tau^2 + 2\varepsilon_{01}h\tau + \varepsilon_1 h + \varepsilon_0\tau, \tag{4.21}$$

and the governing space−time physical law is (Oliver and Christakos, 1996)

$$\left[\frac{S}{K}\frac{\partial}{\partial t} - \frac{\partial^2}{\partial s^2}\right]X(s,t) = 0, \tag{4.22}$$

where the hydraulic head $X(s, t)$ is decomposed as $X(s,t) = \overline{X(s,t)} + X'(s,t)$, $\overline{X(s,t)}$ and $X'(s,t)$ denote the head mean and head perturbation, respectively; S is the storage coefficient; and $f(s) = \log K(s)$,

[18] We observe that the $\mathcal{L}_G$-equation has been introduced in terms of a specific coordinate basis, yet a coordinate transformation is a possibility if this makes $\mathcal{L}_G$ more tractable and easier to solve.

[19] On the other hand, one can find covariance models with metrics different than that of Eq. (4.19a), which do not satisfy the covariance law (Eq. 4.15).

with the saturated hydraulic conductivity K considered constant across the study domain. The random BICs are

$$\begin{aligned} X(s, t=0) &= H\left(1-\frac{s}{\ell}\right) + X'(s, t=0) \\ X(s=0, t) &= H + X'(s=0, t) \\ X(s=\ell, t) &= X'(s=\ell, t). \end{aligned} \tag{4.23a–c}$$

where H denotes the hydraulic head at the boundary. Under these conditions, the hydraulic head covariance ($\mathcal{L}_c$) equation associated with the groundwater flow Eq. (4.22) is given by Eq. (4.24), where $h = s - s'$, $\tau = t - t'$, subject to the head covariance BIC of Eqs. (4.25a–c), where the study domain is much larger than the correlation length of head perturbations, i.e., $\ell >> a_r$. The $\mathcal{L}_G$-equation corresponding to Eq. (4.24) is expressed by Eq. (4.26), which, in light of the metric (Eq. 4.21), takes the analytical form of Eq. (4.27). As before, Eq. (4.27) must be solved with respect to the metric coefficients ε_{00}, ε_{01}, ε_{11}, ε_1, ε_0 and ζ_1. A solution set is given by Eqs. (4.28a–c), in which case the space–time metric is Eq. (4.29a).

It is interesting to consider the numerical solution $\widehat{c}_X$ of the hydraulic head covariance $\mathcal{L}_c$-Eq. (4.24), subject to the BIC of Eqs. (4.25a–c) and investigate if the numerical covariance $\widehat{c}_X$ indeed admits a metric of the form of Eq. (4.29a), i.e., it is of the theoretical covariance form

$$c_X(h, \tau) = c_0 e^{-\frac{3}{a_r}h + \left(\frac{3}{a_r}\right)^2 \frac{K}{S}\tau}. \tag{4.30}$$

This is the goal of the next example.

Example 4.5

Assume the following numerical values for the coefficients of the groundwater flow equation above: $S = 10^{-5}$ m, $K(s) = 0.135$ m/day, $a_r = 10$ m, $a_\tau = 0.2$ day, $\ell = 200$ m, $T = 2$ days, and $c_0 = 10$ m^2. The derived numerical $\widehat{c}_X$ solution of the $\mathcal{L}_c$-Eq. (4.24) is plotted in Fig. 4.1A together with the theoretical c_X solution. The perfect fit between c_X and $\widehat{c}_X$ shown in Fig. 4.1B implies that the numerical $\widehat{c}_X$ can be expressed mathematically by the theoretical space–time exponential c_X form of Eq. (4.30) with a space–time metric of the form of Eq. (4.29a). Therefore, for the groundwater flow situation of Eq. (4.22) with the random BIC of Eqs. (4.23a–c), the space–time metric of the form of Eq. (4.29a) is an appropriate choice, indeed. Some interesting findings of this numerical flow example are worth mentioning. The coefficients ε_i of the space–time metric are not determined arbitrarily. In the present numerical illustration it is found that these coefficients are functions of the spatial correlation length a_r of the $\mathcal{L}_c$-BIC. Remarkably, the BIC of Eq. (4.25b) shows that the temporal head covariance can increase exponentially with time. This is the case in practice, e.g., when the flow uncertainty is due to random hydraulic conductivity coefficients.[20] Nevertheless, in

[20] In a following Example 4.6, the numerical solution of the $\mathcal{L}_c$-equation is plotted in Fig. 4.2A.

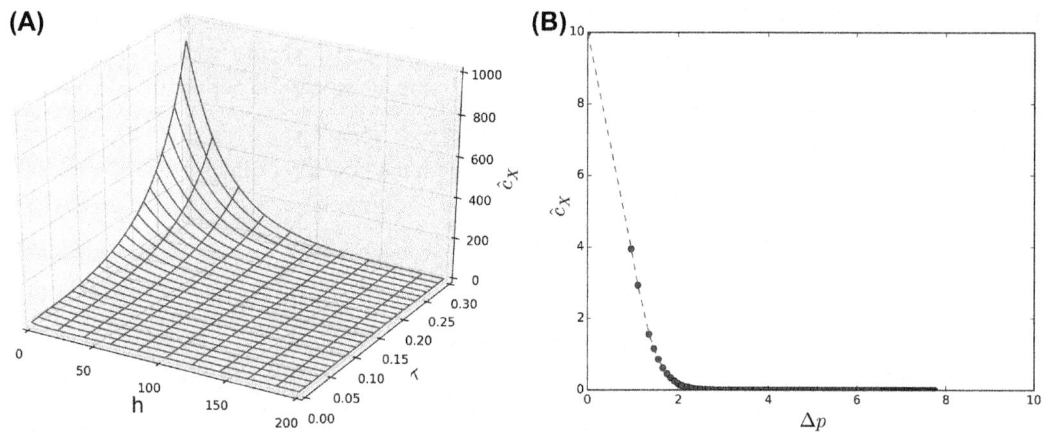

FIGURE 4.1

Plots of (A) the numerical head covariance solution $\widehat{c}_X$ of Eq. (4.24) and (B) head covariance versus the space–time metric of Eq. (4.29a): numerical covariance solution $\widehat{c}_X$ of Eq. (4.24) (*circled line*) and theoretical covariance model c_X of Eq. (4.30) (*dashed line*).

the present flow example the covariance c_X of Eq. (4.30) does not address the regions of extreme high covariance values (i.e., regions with large τ and small h), because the calculation of c_X requires that

$$h - \frac{K}{S}\tau \geq 0. \tag{4.31}$$

This inequality has the physical meaning that hydraulic head variations can only be affected by the noise of the upstream area due to either advection or diffusion mechanisms. Another solution set of $\mathcal{L}_G$-Eq. (4.27) is given by Eqs. (4.28d–g), which corresponds to the metric of Eq. (4.29b).

Lastly, it is instructive to consider a subsurface flow situation that is clearly inconsistent with the space–time metric of Eq. (4.29a).

Example 4.6

For a nonconstant hydraulic conductivity K, the $R^{1,1}$ transient groundwater flow law governing the hydraulic head $X(s, t)$ is given by (Oliver and Christakos, 1996)

$$\left[\frac{S}{K(s)}\frac{\partial}{\partial t} - \frac{\partial f(s)}{\partial s}\frac{\partial}{\partial s} - \frac{\partial^2}{\partial s^2}\right]X(s,t) = 0, \tag{4.32}$$

with BIC

$$\begin{aligned} &X(s,t=0) = H\left(1 - \frac{s}{\ell}\right), \\ &X(s=0,t) = H, \\ &X(s=\ell,t) = 0. \end{aligned} \qquad (4.33a\text{–}c)$$

Unlike Eq. (4.22) earlier, in this case the uncertainty source is the log-hydraulic conductivity $f(s)$, which is considered random and is decomposed as $f(s)=\overline{f(s)}+f'(s)$, where $\overline{f(s)}=\log K(s)$ is the mean and $f'(s)$ is the random fluctuation with variance σ_f^2. Using the stochastic perturbative expansion method (Zhang, 2002), assuming $\frac{\partial \overline{f}}{\partial s}=0$, and a fixed location $(s',t')=(0,0)$ so that $h=s-s'=s$, $\tau=t-t'=t$, the $\mathcal{L}_c$-equation associated with the flow law of Eq. (4.32) is Eq. (4.24), but with new BIC

$$\begin{aligned} c_X(h,0)&=0,\\ c_X(0,\tau)&=0,\\ c_X(\ell,\tau)&=0. \end{aligned} \tag{4.34a–c}$$

Eqs. (4.24) and (4.34a–c) establish the space–time relationships between hydraulic head and conductivity with respect to the first-order hydraulic head or conductivity variation at a fixed location $(s',t')=(0,0)$. To obtain the numerical covariance solution of $\mathcal{L}_c$-Eq. (4.24) subject to Eqs. (4.34a–c), let us assume that the covariance of the log-hydraulic conductivity is given by

$$c_f(h)=e^{-\frac{3h}{a_r}}, \tag{4.35}$$

$H=50$ m, $S=10^{-5}$ m, $K(s)=0.135$ m/day $\left(\text{i.e., } \overline{f}=-2,\ \frac{\partial \overline{f}}{\partial s}=0\right)$, $c_0=0.23$, and $a_r=54$ m. The space and time domains are $\ell=200$ m and $T=2$ days, respectively, and the discretized space and time intervals are $\Delta s=4$ m and $\Delta t=0.0025$ day, respectively. The resulting numerical head covariance $\widehat{c}_X$ is plotted in Fig. 4.2A. One observes that the temporal head covariance component of the plot in Fig. 4.2A can increase exponentially with time (which, as noted earlier, is consistent with Eq. 4.25b). Moreover, in Fig. 4.2B the covariance $\widehat{c}_X$ versus the metric $|\Delta \boldsymbol{p}|$ of Eq. (4.29a) is plotted with $\varepsilon_1=1$ ($a_r=3$). The circles in this figure indicate that for a specified $\Delta p=|\Delta \boldsymbol{p}|$ value there can be assigned several $\widehat{c}_X$ values, corresponding to different combinations of h and τ values in Eq. (4.29a). This means that the covariance of the hydraulic head attribute obeying Eq. (4.32) and subject to the BIC of Eqs. (4.33a–c) cannot have a metric of the form of Eq. (4.29a).

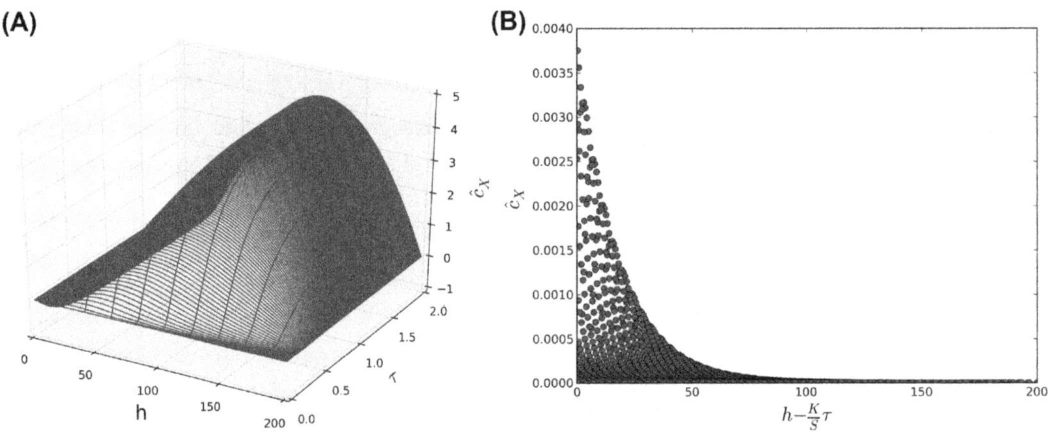

FIGURE 4.2

Plots of the hydraulic head covariance $\widehat{c}_X$ (first-order head) versus (A) (h,τ); and (B) $\Delta p=h-\frac{K}{S}\tau$.

Remark 4.1

In view of the above results the point should be stressed that the source of randomness (physical law BIC vs. law coefficients) can decisively affect the form of the space–time metric. Particularly, the metric of Eq. (4.29a) is appropriate when the randomness in the physical law of Eq. (4.22) is due to its BIC, but it is not appropriate when the randomness source is due to the physical law coefficients.

5. CONCERNING THE ZETA COEFFICIENTS

As was suggested earlier, any space–time metric is formally acceptable if it satisfies the $\mathcal{L}_G$-equation, which means that, in a sense, this equation provides a definition of the metric coefficients. A further interpretation of the zeta coefficients may be useful in this setting. Under general conditions, these coefficients can be expressed as

$$\zeta_{ij} = \frac{\partial}{\partial h_i} \log \frac{\partial}{\partial h_j} c_X(|\Delta \boldsymbol{p}|), \tag{5.1}$$

$(i,j = 0, \ldots, n)$, which is an expression that depends on the geometrical characteristics of the covariance function. In the special space-time setting of interest in the earlier analysis ($j = 0,\ h_0 = \tau$), one gets

$$\zeta_{i0} = \zeta_i = \frac{\partial}{\partial h_i} \log \frac{\partial}{\partial \tau} c_X(|\Delta \boldsymbol{p}|). \tag{5.2}$$

If information about the covariance c_X shape is available, the ζ_{ij} or ζ_i can be calculated directly from Eqs. (5.1)–(5.2), in which case the zeta coefficient is expressed directly in terms of the metric $|\Delta \boldsymbol{p}|$. As a matter of fact, the ζ_i can be calculated and tabulated for several classes of covariance functions.

Example 5.1

Let us consider covariance functions belonging to the exponential-power class

$$\frac{\partial}{\partial \tau} c_X(|\Delta \boldsymbol{p}|) = b \begin{cases} e^{\frac{\theta}{\rho+1}|\Delta \boldsymbol{p}|^{\rho+1}} & (\rho \neq -1) \\ |\Delta \boldsymbol{p}|^{\eta+\theta} & (\rho = -1) \end{cases} \tag{5.3a–b}$$

$(b,\ \theta,\ \rho \in R)$. In this case, one finds that

$$\zeta_i = \theta \frac{\partial \Delta p}{\partial h_i} |\Delta \boldsymbol{p}|^{\rho}, \tag{5.4}$$

i.e., the ζ_i is expressed in terms of the metric $|\Delta \boldsymbol{p}|$, as stated earlier. Then, the CMR Eq. (2.6c) of Table 2.2 becomes

$$\partial^{(2/1)}_{i/0} = \partial^{(2/1)}_{i/\tau} = \frac{\partial}{\partial h_i}\left(\partial^{(1/1)}_{i/\tau}\right) + \theta \frac{\partial \Delta \boldsymbol{p}}{\partial h_i} |\Delta \boldsymbol{p}|^{\rho} \partial^{(1/1)}_{i/\tau}, \tag{5.5}$$

which links the space–time metric with the covariance parameters θ and ρ. Similar expressions can be derived for higher-order CMR.

Example 5.2

As was noted in Example 4.3 the exact shape of model c_X needed not be known in advance to determine the metric, Eq. (4.19a). Rather, coefficients ε_i, ε_{ij}, and ζ_i needed to be chosen so that they satisfy Eqs. (4.18a–c). In light of Eqs. (5.1)–(5.4), for the c_X solution of Eq. (4.15) to be consistent with the metric solution of Eq. (4.19a) it is sufficient that c_X satisfies

$$\frac{\partial}{\partial h_i}\log\frac{\partial}{\partial \tau}c_X(|\Delta \boldsymbol{p}|) = -\varepsilon_i.$$

Indeed, for the covariance of Eq. (4.20), Eq. (5.2) directly yields Eq. (4.18c). The same result is obtained from Eqs. (5.4) and (4.14) by setting $\rho = 1$, $\theta = -2$. Moreover, Eq. (4.20) gives

$$\frac{\partial}{\partial \tau}c_X(|\Delta \boldsymbol{p}|) = -\varepsilon_0 c_X(|\Delta \boldsymbol{p}|),$$

which is the same result as that obtained from Eq. (5.3a) for $b = -\varepsilon_0$, $\rho = 1$, $\theta = -2$.

6. CLOSING COMMENTS

Central among the quantitative features of a physical geometry is its metric structure, i.e., a set of mathematical expressions that define spatiotemporal distance. These expressions cannot always be defined unambiguously, but depend on two entirely different factors: a "relative" factor, the particular coordinate system, and an "absolute" factor, the nature of the space–time continuum. The latter depends on local properties (intrinsic links) of space and time in which a given natural attribute occurs (i.e., whether it is a plane, a sphere, or an ellipsoid), as well as on constraints imposed by the phenomenon itself (physical law governing the attribute of interest, BIC, and randomness sources). If a natural attribute takes place inside a three-dimensional medium with complicated internal structure, the appropriate metric for the space–time correlations is significantly influenced by the structure of the medium. Also, the source of randomness (law coefficients vs. BIC) can affect the form of the space–time metric.

The choice of the spatiotemporal metric can have major consequences in the scientific modeling of natural phenomena. One consequence is related to the choice of mathematical entities that establish linkages between spatiotemporally distributed data, such as covariance functions. These functions need to satisfy certain permissibility criteria that depend on the metric (Chapter XV). The commonly used Gaussian function, e.g., is permissible in the case of a Euclidean metric but not in the case of a Manhattan metric. The exponential function, on the other hand, is permissible for both metrics. In general, the permissibility of a covariance function with respect to one metric form does not guarantee its permissibility for another metric form. Space–time estimation, simulation, and mapping of natural attributes depend on the metric assumed, since the covariance functions are used as inputs in mapping and simulation techniques (e.g., Kriging, Wiener, and Kalman filtering). Hence, the same data set can lead to different attribute maps if estimation or simulation is performed using different space–time metrics. This is, perhaps one of the most dramatic effects of metric choice, as illustrate in the following example.

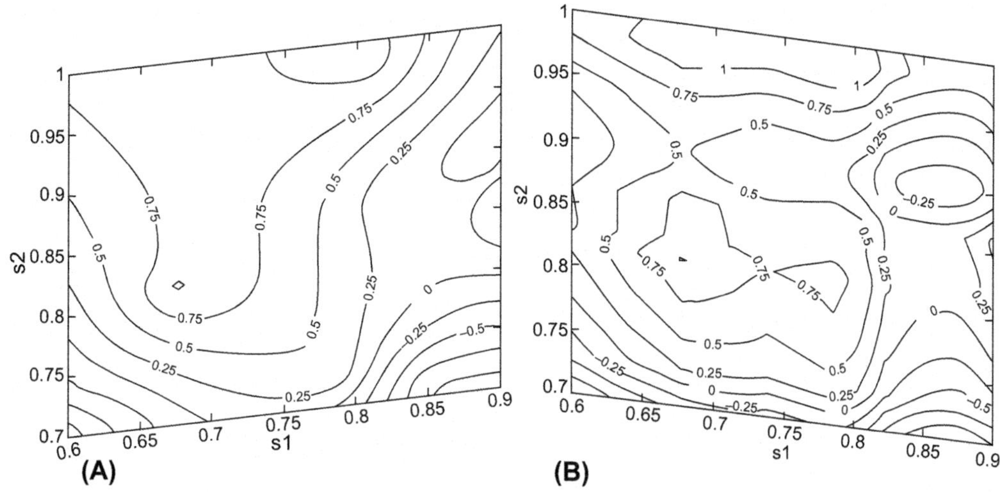

FIGURE 6.1

Maps using the same data but different metrics, (A) r_E and (B) r_M.

Example 6.1

Consider the exponential covariance model

$$c_Y(\boldsymbol{h}) = e^{-0.3|\boldsymbol{h}|}, \tag{6.1}$$

where the metric $|\boldsymbol{h}|$ may be either of the Euclidean (r_E) or the Manhattan (r_M) forms. Using the same data set and the covariance (Eq. 6.1) with r_E and r_M, the two maps of Fig. 6.1A and B, respectively, are obtained (Christakos, 2000). Clearly, the two maps represent very different attribute variations.

There are a number of directions for future work concerning metric determination, including the following:

(i) Different kinds of space–time dependency of the metric coefficients may be studied, considering in particular those that correspond to physically meaningful space–time links. A classification of space–time-dependent metrics would be interesting too. Metric coefficients could be expressed in terms of standard analytic functions (polynomials, trigonometric, and hyperbolic) of space–time coordinates. Symmetry or separability conditions may be imposed on the metric, restricting the mathematical structure of the covariance function, and the effects of BIC and random sources may be accounted for. One possibility is to assume space–time separable metric coefficients with specified analytic functions.

(ii) Knowing the space–time metric in a closed form, elucidation of its physical properties is based on the features of the phenomenon. Certain metric solutions may not necessarily have a unique interpretation, which is an issue in need of further investigation.

(iii) If a metric form shows some resemblance to a known one, it should be investigated if the former can be reduced to the latter by means of a coordinate transformation, and what this means as regards the mathematical and physical interpretation of the $\mathcal{L}_G$-equation.

(iv) Another issue of interest is space–time coordinate transformation, i.e., while the $\mathcal{L}_G$-equations are introduced in terms of a certain coordinate basis, coordinate transformation may be a real possibility if it makes the $\mathcal{L}_G$-equations more tractable and easier to solve.

CHAPTER

SPACE–TIME CORRELATION THEORY

CHAPTER OUTLINE

1. FOCUSING ON SPACE–TIME VARIABILITY FUNCTIONS

In applied sciences, real-world conditions usually make it impossible to obtain a complete characterization of a spatiotemporal random field (S/TRF) on the basis of its probability functions [probability density function (PDF) or cumulative distribution function (CDF)]. This is primarily due to the small number of available attribute realizations, since in most cases we have only one sequence of measurements. Therefore, the central modeling assumption of this chapter is that we are limited to a lesser characterization of the random field in terms of its space–time variability functions, S/TVFs, expressed in terms of statistical or ensemble moments, mainly of order up to two. Accordingly, the

Spatiotemporal Random Fields. http://dx.doi.org/10.1016/B978-0-12-803012-7.00004-0

S/TRF models considered in this chapter are primarily second-order random fields (scalar or vector). The part of the S/TRF theory that studies only the random field properties determined by their second-order statistical moments is traditionally called *space–time correlation theory.* To phrase it in more words, the focus on correlation theory is a consequence of the modeling assumption of this chapter that an adequate S/TRF characterization is possible in terms of S/TVFs. Then, questions about the physical meaning generated by random field modeling come down to issues concerning the interpretive potential and dynamics of the S/TVFs.

1.1 BASICS OF SPACE–TIME CORRELATION THEORY

The mathematical underpinnings of space–time correlation theory are the respective theories of probability and random variables (RVs) rigorously extended in the space–time domain. As usual, $\mathcal{H}_2 = L_2(\Omega, \mathcal{F}, P)$ is the Hilbert space of all continuous-parameter RVs $x_1, \ldots, x_m$ defined at $\boldsymbol{p}_1, \ldots, \boldsymbol{p}_m$ and endowed with the scalar product

$$(x_i, x_j) = \overline{x_i x_j} = \iint d\mathcal{F}_x(\chi_i, \chi_j)\chi_i\chi_j, \tag{1.1}$$

where $i, j = 1, \ldots, m$, and $F_x(\chi_i, \chi_j)$ denotes the bivariate CDF of x_i and x_j indicating the probabilistic character of the representation introduced by Eq. (1.1). The corresponding norm is given by

$$\|x\|^2 = \overline{|x|^2} = \int d\mathcal{F}_x(\chi)\chi^2 < \infty, \tag{1.2}$$

where $F_x(\chi)$ denotes the univariate CDF of the RV x. Usually $F_x(\chi)$ and $F_x(\chi_i, \chi_j)$ are assumed to be differentiable so that they can be conveniently replaced by their PDFs, $f_x(\chi)$ and $f_x(\chi_i, \chi_j)$, respectively.

The above stochastic quantities are typically assumed to be real-valued continuous functions in the R^{n+1} (or, $R^{n,1}$) domain. Also, we will always assume that all the S/TRFs under consideration have finite means and variances. The added value of this assumption is that there is usually no need to further mention the relevant CDFs or PDFs. The S/TRF will be assumed continuous in the mean square (m.s.) sense, that is (see, Section 3.3 of Chapter I),

$$\lim_{m\to\infty} \overline{|x_m - x|^2} = 0,$$

or

$$\lim_{\boldsymbol{p}'\to\boldsymbol{p}} \overline{|X(\boldsymbol{p}') - X(\boldsymbol{p})|^2} = 0, \tag{1.3a–b}$$

where $\boldsymbol{p}' \to \boldsymbol{p}$ means that $\boldsymbol{s}' \to \boldsymbol{s}$, $t' \to t$.

As in previous chapters, in this chapter too the following notation of Eulerian space–time coordinates will be considered,

$$\begin{aligned} \boldsymbol{p} &= (s_1, \ldots, s_n, s_0) \in R^{n+1}, && \text{with } s_0 = t, \\ \boldsymbol{p} &= (\boldsymbol{s}, t) \in R^{n,1}, && \text{with } \boldsymbol{s} = (s_1, \ldots, s_n). \end{aligned} \tag{1.4a–b}$$

Eqs. (1.4a–b) are used interchangeably in this book to represent a space–time point, as appropriate. The term "appropriate" means that, as was pointed out earlier, the mathematical treatment of random fields often favors working in the R^{n+1} domain, whereas in many applications the existing physical conditions may require to explicitly indicate the space and time arguments in the $R^{n,1}$ domain.

Put differently, modeling a natural attribute in the $R^{n,1}$ domain may have a higher potential for meaningful interpretation than in the R^{n+1} domain. Similarly, the notation

$$m(\boldsymbol{p},\boldsymbol{p}') = \begin{cases} |\Delta\boldsymbol{p}| = \Delta\boldsymbol{p}, \\ (|\boldsymbol{h}|, |\tau|), \end{cases} \tag{1.5a–b}$$

has been used in this book (Section 2 of Chapter I) to denote "distances" in space–time, where $\boldsymbol{h} = (h_1,\ldots, h_n)$, $\Delta\boldsymbol{p} = (\Delta s_i = h_i,\ i = 1,\ldots,\ n,\ \Delta s_0 = h_0 = \tau)$. In many cases the notation choice is a purely formal matter, whereas in certain others due care is required in deciding the space–time domain of analysis. As was mentioned in Chapter I (Remark 2.1), in addition to Eulerian space–time coordinates, Lagrangian space–time coordinates are sometimes used in physical applications, where $\boldsymbol{s}$ and t do not vary independently (as they are allowed to do in the Eulerian setting), but, instead, $\boldsymbol{s}$ is considered a function of t, which is expressed as $\boldsymbol{s}(t)$ (Tennekes, 1975).

1.2 PHYSICAL INVESTIGATIONS BASED ON SPACE–TIME CORRELATION THEORY

For several decades, space–time correlation notions and tools, in an Eulerian or a Lagrangian setting, constitute a fundamental part of both stochastic theories of real-world phenomena, and of data analysis and processing techniques investigating these phenomena in situ. A systematic exposition of the numerous applications of the correlation theory in the various branches of sciences can be found in the relevant literature (e.g., a thorough review of the major applications of the space–time correlation theory in turbulence studies is presented in Wallace, 2014). Naturally, the numerous applications of space–time correlation theory in physical sciences cannot be all described here. Instead, my review is very selective and, of necessity, very limited (although, references to specific works are given in various parts of the book).

Correlation theory is widely applicable to many aspects of turbulence research. One of its earliest uses was in the study of the dispersion of fluid particles (Taylor, 1938), where an integral relationship was developed linking the single particle correlation and the mean square distance traveled by an ensemble of particles from a specified flow location and along a specific direction. As is well known, the use of correlation theory in isotropic turbulence led to major early developments by Kolmogorov (1941), Heisenberg (1948), and Batchelor (1951). Subsequently, essential space–time correlations were principally determined in turbulence studies by means of the root–mean–square of the turbulent kinetic energy (Kraichnan, 1958). Borgman (1969) used space–time correlation theory and its frequency domain counterpart to simulate ocean waves in conditions of uncertainty. Space–time correlation techniques are at the center of major developments in statistical fluid mechanics (Monin and Yaglom, 1971). Daley (1999) has reviewed the extensive use of similar techniques in the analysis of spatiotemporal atmospheric data from experimental measurements. Correlation theory has been also successfully implemented in the study of the propagation of optical waves (Strohbehn, 1978) and in the analysis of space–time rainfall patterns (Cox and Isham, 1988). Furthermore, Dagan (1989), and Gelhar (1993) studied subsurface flow and transport based on correlations between flow attributes (hydraulic conductivity, flow head, etc.). Space–time correlations of velocity component fluctuations applied to shear flows were investigated by Phillips (2000) based on the assumption that space–time correlations can be expressed as spatial correlations and their reduction with time. Christakos and Hristopulos (1998) and Christakos (1991a,b,c, 1992, 2000), used space–time correlation theory and techniques in the study of a variety of attributes in earth, atmospheric and health sciences. Sobczyk and Kirkner (2001) used similar techniques in the study of material media microstructures. A model of space–time correlations for flows with mean shear using a Taylor series expansion was proposed by He and Zhang (2006), which relates correlations with spatial separation as the only independent variable to space–time correlations by using two characteristic velocities. Spatiotemporal correlation models of climate dynamics have generated extended-range forecasts in the tropics (Zhu et al., 2015). Lastly, recent solar wind studies are based on the calculation of critical space–time plasma correlations (Matthaeus et al., 2016).

2. SPACE–TIME VARIABILITY FUNCTIONS IN TERMS OF SCALAR SPACE–TIME STATISTICS

As we saw in Section 2 of Chapter II, S/TVFs are functions that are used to assess quantitatively the spatiotemporal variation of an S/TRF representing a natural attribute. To phrase it in more words, S/TVFs quantify how S/TRF variations at one location and one instant covary with those at another location and another instant, and thus describe the behavior of the attribute across spatial and temporal scales. I will revisit below some basic S/TVFs, already introduced in Section 2 of Chapter II. Here, however, I will look at them in the context of correlation theory, which may offer an improved appreciation of their properties, increase the richness of their interpretation, and also enable their extension to higher-order variability functions. This perspective creates challenges that must be addressed, focusing on both the promise and the pitfalls of the S/TVF-based characterization of attribute variation.

In correlation theory, the S/TVFs of a scalar random field are generally expressed in terms of its statistical (or ensemble) moments. The most common among them are the first- and the second-order moments, although in some cases, higher-order moments may be considered, as well. Analysis in terms of the second-order statistical moments is based on the assumption that

$$X(\boldsymbol{p}) \in \mathcal{H}_2$$

($\overline{X(\boldsymbol{p})^2} < \infty$). Continuous CDF are assumed below so that the corresponding integrals are considered to be absolutely convergent, and integration is carried out over the entire space–time domain, unless specific limits are indicated.

In applied stochastics, another terminology is also used. In this terminology, an S/TVF is termed *local* (i.e., considered at a single point), or *nonlocal* (considered between two or more points) in the space–time domain R^{n+1} (or $R^{n,1}$). When two-point S/TVFs are considered, the direction of the variation is of principal interest, in which case a distinction is made between the *omnidirectional* (i.e., they involve all directions) and the *direction-specific* S/TVFs.

Remark 2.1

There is an obvious correspondence between the correlation theory and the stochastics terminology:

First-order statistical moments ↔ Local S/TVF.
Second-order statistical moments ↔ Nonlocal S/TVF.

In this book both terminologies are used, depending on the context. So, when the focus of our analysis is on the statistical features of an attribute, the correlation theory terminology is used, whereas when the emphasis is on its space–time arguments and physical interpretation, the stochastics terminology is used.

The readers will surely notice that some of the basic one- and two-points S/TVFs have already been defined in Chapter II. Below we repeat these early definitions to make the connection with the multipoint and higher-order S/TVFs to be introduced in this chapter, and, also, for completeness of presentation purposes.

2.1 LOCALITY: ONE-POINT SPACE–TIME VARIABILITY FUNCTIONS

We start with a brief review of the specification of the most important one-point S/TVFs (i.e., associated with first-order statistical moments). The nature and scope of these S/TVFs is, by definition, local and context bound.

As has already been stated in Definition 2.4 of Chapter II, the *mean* function of an S/TRF $X(\boldsymbol{p})$ is defined by

$$\overline{X(\boldsymbol{p})} = \int d\chi_{\boldsymbol{p}} \chi_{\boldsymbol{p}} f_X(\chi_{\boldsymbol{p}}), \tag{2.1}$$

where according to standard practice $\chi_{\boldsymbol{p}}$ denotes the corresponding space–time random field realization. The attribute mean, denoted by $\overline{X(\boldsymbol{p})}$, may be assumed constant or it may be any function of $\boldsymbol{p}$. In the latter case the mean is interpreted as describing systematic variations of the attribute represented by the random field $X(\boldsymbol{p})$ and is also known as attribute's space–time *trend* or *drift*.

I suggest that the trend representation of $\overline{X(\boldsymbol{p})}$ is preferable to assumed constant values that usually have much less interpretive value. The random field *fluctuation* is defined by subtracting the mean from the original random field, i.e.,

$$X'(\boldsymbol{p}) = X(\boldsymbol{p}) - \overline{X(\boldsymbol{p})},$$

which is interpreted as the attribute's degree of departure from the trend as a function of space–time. The fluctuation $X'(\boldsymbol{p})$ may validly estimate the space–time effect of the mean in the observed attribute variation, i.e., the attribute variation due to the presence of the mean.

In certain applications, to accommodate the physics of the phenomenon we may need to specifically assume that the mean is a function only of space (in which case the time argument is suppressed) or only of time (in which case the space argument is suppressed).

Example 2.1

A mean function that depends on time only may be assumed in the case of meteorological attributes occurring on a region that is much larger than that covered by the wireless sensor network used to observe them and, also, the attributes have been observed to be homogenously distributed throughout the network area.

The mean function is a first-order statistical (ensemble) moment or a local (single-point) S/TVF. Higher-order single-point moments can be also defined as a straightforward generalization of Eq. (2.1). This is what I do next.

Definition 2.1

The ρth-*order noncentered* single-point moment is defined at point $\boldsymbol{p}$ by

$$\overline{X(\boldsymbol{p})^{\rho}} = \int d\chi_p \chi_p^{\rho} f_X(\chi_p), \tag{2.2a}$$

and the ρth-*order centered* single-point moment is defined by

$$\overline{X'(\boldsymbol{p})^{\rho}} = \overline{\left[X(\boldsymbol{p}) - \overline{X(\boldsymbol{p})}\right]^{\rho}} = \int d\chi_p \left(\chi_p - \overline{X(\boldsymbol{p})}\right)^{\rho} f_X(\chi_p) \tag{2.2b}$$

at point $\boldsymbol{p}$.

It is not difficult to see that for $\rho = 1$ it is valid that $\overline{X'(\boldsymbol{p})} = 0$, whereas for $\rho = 2$ it leads to the familiar *variance* of $X(\boldsymbol{p})$ at point $\boldsymbol{p}$,

$$\sigma_X^2 = \overline{X'(\boldsymbol{p})^2} = \int d\chi_p \left(\chi_p - \overline{X(\boldsymbol{p})}\right)^2 f_X(\chi_p). \tag{2.3}$$

The ρth-order noncentered and the ρth-order centered moments at any point $\boldsymbol{p}$ are related by

$$\overline{X'(\boldsymbol{p})^{\rho}} = \sum_{i=0}^{\rho} (-1)^{\rho - i} C_{\rho}^{i} \overline{X(\boldsymbol{p})^{i}}\, \overline{X(\boldsymbol{p})}^{\rho - i}, \tag{2.4}$$

where $C_{\rho}^{i} = \frac{\rho!}{i!(\rho-i)!}$ (Exercise IV.1). A few examples are given next to shed light on particular cases of the general expressions above.

Example 2.2

Using Eq. (2.4) for $\rho = 2$, 3, and 4, we can find the following useful relationships,

$$\begin{aligned}
\overline{X'(\boldsymbol{p})^2} &= \sigma_X^2 = \overline{X(\boldsymbol{p})^2} - \overline{X(\boldsymbol{p})}^2, \\
\overline{X'(\boldsymbol{p})^3} &= \overline{X(\boldsymbol{p})^3} - 3\overline{X(\boldsymbol{p})}\, \overline{X(\boldsymbol{p})^2} + 2\overline{X(\boldsymbol{p})}^3, \\
\overline{X'(\boldsymbol{p})^4} &= \overline{X(\boldsymbol{p})^4} - 4\overline{X(\boldsymbol{p})}\, \overline{X(\boldsymbol{p})^3} + 6\overline{X(\boldsymbol{p})}^2\, \overline{X(\boldsymbol{p})^2} - 3\overline{X(\boldsymbol{p})}^4.
\end{aligned} \tag{2.5a–c}$$

The third-order moment $\overline{X'(\boldsymbol{p})^3}$ is known as the *skewness* function, and, basically, it quantifies how symmetrical a PDF is. It is particularly useful in applications (e.g., analysis of historical hydrologic series). Skewness expresses the lack of symmetry of the probability distribution. Usually, the skewness function is divided by $\sigma_X^3(\boldsymbol{p})$, leading to the *coefficient of skewness*

$$\varpi_X(\boldsymbol{p}) = \frac{\overline{X'(\boldsymbol{p})^3}}{\sigma_X^3(\boldsymbol{p})} = \frac{\overline{\left[X(\boldsymbol{p}) - \overline{X(\boldsymbol{p})}\right]^3}}{\left(\overline{\left[X(\boldsymbol{p}) - \overline{X(\boldsymbol{p})}\right]^2}\right)^{\frac{3}{2}}}. \tag{2.5d}$$

The fourth-order moment $\overline{X'(\boldsymbol{p})^4}$ is the *kurtosis* function, and it admits a few interpretations (it quantifies how close the PDF shape is to the Gaussian law, it measures the "tailedness" of a PDF etc.). The coefficient of kurtosis is defined as

$$\begin{aligned}
\Bbbk_X(\boldsymbol{p}) &= \frac{\overline{X'(\boldsymbol{p})^4}}{\sigma_X^4(\boldsymbol{p})} = \frac{\overline{\left[X(\boldsymbol{p}) - \overline{X(\boldsymbol{p})}\right]^4}}{\left(\overline{\left[X(\boldsymbol{p}) - \overline{X(\boldsymbol{p})}\right]^2}\right)^{2}}, \\
\Bbbk_X(\boldsymbol{p}) &\geq \varpi_X(\boldsymbol{p})^2 + 1.
\end{aligned} \tag{2.5e–f}$$

(i.e., the kurtosis is bounded below by the squared skewness plus 1). Similar relationships can be derived between higher-order moments, if needed.

The local moments above assume an attractive form in the case of a Gaussian (normal) probability distribution.

Example 2.3

Consider an S/TRF $X(\boldsymbol{p})$ with a Gaussian PDF at a single space–time point $\boldsymbol{p}$,

$$f_X(\chi_p) = \frac{1}{\sqrt{2\pi}\sigma_X} e^{-\frac{1}{2\sigma_X^2(\boldsymbol{p})}\left(\chi_p - \overline{X(\boldsymbol{p})}\right)^2} \tag{2.6a}$$

(this expression is the univariate case of Eq. 7.1 of Chapter II for $m = 1$). After some calculations involving Poisson integrals, we can find that

$$\begin{aligned} \overline{X'(\boldsymbol{p})^{2\rho}} &= (2\rho - 1)!!\sigma_X^{2\rho}(\boldsymbol{p}), \\ \overline{X'(\boldsymbol{p})^{2\rho-1}} &= 0, \end{aligned} \tag{2.6b–c}$$

where $(2\rho - 1)!! = (2\rho - 1)(2\rho - 3)\ldots 3 \times 1$ (Exercise IV.2), i.e., a field $X(\boldsymbol{p})$ with a Gaussian PDF at point $\boldsymbol{p}$ is completely defined locally if the mean $\overline{X(\boldsymbol{p})}$ and variance $\sigma_X^2(\boldsymbol{p})$ are known at $\boldsymbol{p}$.

Moreover, single-point spatiotemporal (ensemble) moments of any order can be derived by differentiating the general characteristic function (CF), see Eqs. (2.14a–b) of Chapter II for $m = 1$, and setting $u = 0$:

$$\begin{aligned} \overline{X^\rho(\boldsymbol{p})} &= i^{-\rho}\left[\frac{\partial^\rho}{\partial u^\rho}\phi_X(u)\right]_{u=0}, \\ \overline{X'^\rho(\boldsymbol{p})} &= i^{-\rho}\left[\frac{\partial^\rho}{\partial u^\rho}e^{-iu\overline{X(\boldsymbol{p})}}\phi_X(u)\right]_{u=0}, \end{aligned} \tag{2.7a–b}$$

where $\phi_X(u) = \mathcal{FT}[f_X(\chi)] = \widetilde{f}_X(u)$—see also, Example 2.6 of Chapter II.

2.2 NONLOCALITY: OMNIDIRECTIONAL TWO-POINT SPACE–TIME VARIABILITY FUNCTIONS

There is much value in extending statistical moments beyond their local context to attain a higher-level understanding of the space–time attribute variability. This is the goal of nonlocal S/TRFs, which are primarily statistical moments defined for two (or more) points in the space–time domain. The two-point S/TVFs describe statistically the similarity or dissimilarity between random field values at different spatial and temporal separations and at different scales of physical interest. Put differently, the aim of the S/TVFs is to obtain averaged information about attribute changes with respect to shifts in space and time.

As has also been stated in Definition 2.4 of Chapter II, the (complete or full) centered spatiotemporal covariance function $c_X(\boldsymbol{p},\boldsymbol{p}')$, also known as the two-point centered spatiotemporal moment, of an S/TRF $X(\boldsymbol{p})$ is defined by

$$c_X(\boldsymbol{p},\boldsymbol{p}') = \overline{X'(\boldsymbol{p})X'(\boldsymbol{p}')} = \iint d\chi'_p d\chi'_{p'} \chi'_p \chi'_{p'} f_X\left(\chi'_p, \chi'_{p'}\right). \tag{2.8a}$$

The class of all centered space–time covariances is denoted by C_{n+1}. And, the (complete) noncentered covariance function, also known as the two-point noncentered spatiotemporal moment, of $X(\boldsymbol{p})$ is defined by

$$\overline{X(\boldsymbol{p})X(\boldsymbol{p}')} = \iint d\chi_p d\chi_{p'} \chi_p \chi_{p'} f_X(\chi_p, \chi_{p'}), \tag{2.8b}$$

where $\boldsymbol{p} = (\boldsymbol{s}, t)$ and $\boldsymbol{p}' = (\boldsymbol{s}', t')$.

Two-point space–time correlations and dependencies are formed in Eqs. (2.8a–b) by letting $\boldsymbol{s}$ and $\boldsymbol{s}'$ be different locations in space and by considering different times t and t' at $\boldsymbol{s}$ and $\boldsymbol{s}'$. We immediately observe that the centered and noncentered space–time covariance functions are related as

$$c_X(\boldsymbol{p},\boldsymbol{p}') = \overline{X(\boldsymbol{p})X(\boldsymbol{p}')} - \overline{X(\boldsymbol{p})}\,\overline{X(\boldsymbol{p}')}. \tag{2.9}$$

Clearly, for S/TRFs with zero mean, the centered and the noncentered covariance functions coincide. Also, the units of c_X are those of X^2. Hence, in the case that the random field is dimensionless, so is the covariance c_X.

Remark 2.1

An interesting result is obtained if we notice that a second-order (zero mean) random field satisfies $\overline{X(\boldsymbol{p})^2} < \infty$, and in light of the Cauchy–Schwartz inequality, we find

$$\overline{X(\boldsymbol{p})X(\boldsymbol{p}')} \leq \left(\overline{X(\boldsymbol{p})^2}\right)^{\frac{1}{2}} \left(\overline{X(\boldsymbol{p}')^2}\right)^{\frac{1}{2}} < \infty,$$

i.e., the covariance exists and is finite. The converse is also valid, i.e., the last equation implies $\overline{X(\boldsymbol{p})^2} = \overline{X(\boldsymbol{p})X(\boldsymbol{p}'=\boldsymbol{p})} < \infty$. Perhaps, a more common version of the Cauchy–Schwartz inequality is

$$|c_X(\boldsymbol{p},\boldsymbol{p}')| \leq \sigma_X(\boldsymbol{p})\sigma_X(\boldsymbol{p}'), \tag{2.10}$$

which provides a means to assess the strength of the covariance measure of space–time attribute dependency by comparing its magnitude with the largest possible value. The covariance attains the limits in Eq. (2.10) when the $X(\boldsymbol{p})$ and $X(\boldsymbol{p}')$ attribute values are proportional. The interpretation of this proportionality is that, in this case, the $c_X(\boldsymbol{p},\boldsymbol{p}')$ measures the degree of linear relationship between the attribute values at the space–time points $\boldsymbol{p}$ and $\boldsymbol{p}'$.

Example 2.4

The space–time covariance functions have considerable physical significance in many scientific fields. In geophysical sciences, covariance functions describing the variation structure of random media play a major role in the understanding of the properties of seismic waves propagating in geological formations. Particular cases include the key role of the space–time covariance function in the interpretation of refraction travel-time tomographic inversion, assessing the accuracy and relation of the seismic model to the geological structure, estimating scattering attenuation, and upscaling material parameter distributions (Wu, 1982; Klimes, 2002b; Müller and Shapiro, 2001).

In addition to the above considerations, the space–time covariance function is a useful quantity operational wise, because it allows us to calculate the variance or mean square value of any linear function of the original attribute represented by the random field.

Example 2.5

Assume that given the random field $X(\boldsymbol{p})$ with covariance $c_X(\boldsymbol{p},\boldsymbol{p}')$, the quantity of interest is the temporal field defined as a linear function of $X(\boldsymbol{p})$, i.e.,

$$X(t) = \frac{1}{A}\int_A ds X(\boldsymbol{s},t)$$

where $A \subset R^n$. The mean square value of $X(t)$ is given by

$$\overline{X(t)^2} = \frac{1}{A^2}\int_A\int_A ds ds' c_X(\boldsymbol{s}-\boldsymbol{s}',t)$$

i.e., it is determined by the space–time covariance.

The following covariance expressions are distinct from one another in the general (space–time heterogeneous) case

$$\begin{aligned} c_X(\boldsymbol{p}+\Delta\boldsymbol{p},\boldsymbol{p}) &= \overline{X'(\boldsymbol{p}+\Delta\boldsymbol{p})X'(\boldsymbol{p})}, \\ c_X(\boldsymbol{p},\boldsymbol{p}+\Delta\boldsymbol{p}) &= \overline{X'(\boldsymbol{p})X'(\boldsymbol{p}+\Delta\boldsymbol{p})}, \\ c_X(\boldsymbol{p},\boldsymbol{p}-\Delta\boldsymbol{p}) &= \overline{X'(\boldsymbol{p})X'(\boldsymbol{p}-\Delta\boldsymbol{p})}, \end{aligned} \tag{2.11a–c}$$

where $\boldsymbol{p},\boldsymbol{p}',\Delta\boldsymbol{p} \in R^{n+1}$. The situation is different in cases where certain restrictive hypotheses are made concerning the space–time dependency structure of $X(\boldsymbol{p})$ (e.g., when the random fields are space–time homostationary, STHS[1]), where, under the specified conditions, the expressions in Eqs. (2.11a–c) can be equivalent. Moreover, the analysis above leads to an interesting proposition.

Proposition 2.1

The S/TRF $X(\boldsymbol{p}) \in \mathcal{H}_2$ if and only if the covariance $c_X(\boldsymbol{p},\boldsymbol{p}') \in C_{n+1}$ exists and is finite.

Once the covariance c_X is known, we can define three more useful statistical moments directly from c_X, as follows:

(a) The spatiotemporal *variance*

$$\sigma_X^2(\boldsymbol{p}) = \overline{[X(\boldsymbol{p}) - \overline{X(\boldsymbol{p})}]^2} = c_X(\boldsymbol{p},\boldsymbol{p}), \tag{2.12}$$

see also, Eq. (2.3), which describes local random variations.

(b) The spatiotemporal *coefficient of variation* (CV),

$$CV_X(\boldsymbol{p}) = \frac{\sigma_X(\boldsymbol{p})}{\overline{X(\boldsymbol{p})}} = \frac{\sqrt{c_X(\boldsymbol{p},\boldsymbol{p})}}{\overline{X(\boldsymbol{p})}}, \tag{2.13}$$

which is a useful measure of the relative strength of fluctuations.

Example 2.6

Random field modeling of environmental attributes in heterogeneous media is usually based on perturbation expansions of the fluctuations around the mean. The CV emerges in such expansions as a

[1]STHS random fields were briefly introduced in Section 7.4 of Chapter II; a detailed presentation is given in Chapter VII.

dimensionless perturbation parameter. Then, first-order perturbation includes only terms linear in the CV. Convergence of the perturbation series can be guaranteed only if the CV is everywhere less than one.

(c) The spatiotemporal *correlation function*

$$\rho_X(\boldsymbol{p},\boldsymbol{p}') = \frac{c_X(\boldsymbol{p},\boldsymbol{p}')}{\sigma_X(\boldsymbol{p})\sigma_X(\boldsymbol{p}')}, \tag{2.14}$$

which expresses space–time variation links. The class of all correlation functions in R^{n+1} is sometimes denoted by Λ_{n+1}. As regards its interpretation, the correlation function provides information about the spatiotemporal distribution of an attribute's fluctuation correlations but not about their strength.

Remark 2.2

A reason to consider using ρ_X in real-world applications is its *scale-independency* property, which has considerable interpretive value. First, bear in mind that this property is not shared by the covariance function. Elementary calculations show that a simple field rescaling (associated, e.g., to units of measurement) leads to

$$c_{aX}(\boldsymbol{p},\boldsymbol{p}') = a^2 c_X(\boldsymbol{p},\boldsymbol{p}') \tag{2.15a}$$

(a is a scaling coefficient), i.e., the covariance is affected by a simple rescaling. Yet, it is physically counter intuitive that the space–time dependence between attribute values (i.e., how much an attribute's value at $\boldsymbol{p}$ depends on its value at $\boldsymbol{p}'$) is affected by, say, a change of measurement units. On the other hand, the spatiotemporal correlation function ρ_X is independent of such a rescaling, i.e.,

$$\rho_{aX}(\boldsymbol{p},\boldsymbol{p}') = \rho_X(\boldsymbol{p},\boldsymbol{p}'). \tag{2.15b}$$

The independence property is worthy of further investigation.

Another interesting situation, at least theoretically, is described as follows: The $X(\boldsymbol{p})$ is termed an *uncorrelated* S/TRF if

$$\rho_X(\boldsymbol{p},\boldsymbol{p}') = \begin{cases} 1 & \text{if } \boldsymbol{p} = \boldsymbol{p}' \\ 0 & \text{otherwise.} \end{cases} \tag{2.16}$$

Such an S/TRF is also called a *white-noise* random field. In general, it is not true that an uncorrelated field is probabilistically independent, too. That is, Eq. (2.16) does not necessarily imply that $f_X(\boldsymbol{p},\boldsymbol{p}') = f_X(\boldsymbol{p})f_X(\boldsymbol{p}')$.

The apparent interpretative limitation of Eq. (2.14) then is that, although it offers a measure about the extent of the relationship between an attribute's values at $\boldsymbol{p}$ and $\boldsymbol{p}'$, it does not represent the entire relationship between them as established by $f_X(\boldsymbol{p},\boldsymbol{p}')$. A way out of this interpretive limitation of the correlation function above is suggested by the information-theoretic S/TVF functions, i.e., the sysketogram and the contingogram functions already defined in Section 2.5 of Chapter II. Indeed, probability independence is a key property of these two functions, which points out the potential advantage of using the sysketogram or contingogram functions as a data-free measure of spatiotemporal variation over the mainstream covariance function. It is clear, however, that independency always implies lack of correlation.

This is a good point to bring to the readers' attention another potentially significant interpretation issue: to a specific covariance function may correspond more than one S/TRF, often quite dissimilar to each other. To illustrate the matter, we consider a simple example.

Example 2.7

Let $X_2(\boldsymbol{p}) = \upsilon X_1(\boldsymbol{p})$, where υ is a zero-mean RV independent of $X_1(\boldsymbol{p})$ for any $\boldsymbol{p}$, and $\overline{\upsilon^2} = 1$. It is easily seen that both fields, $X_1(\boldsymbol{p})$ and $X_2(\boldsymbol{p})$, have the same covariance function although their realizations may be quite different (Exercise IV.3).

In any case, and despite the issues highlighted above, the statistical mean and covariance are two S/TVFs that are very useful tools of the correlation theory for at least five reasons, whose significance has been amply demonstrated in applied stochastics:

(a) A number of S/TRF features can be specified completely in terms of the mean and the covariance.
(b) Under certain circumstances that often apply in practice, the mean and covariance can be calculated accurately and with much less effort than other probabilistic characteristics.
(c) The description of the way the S/TRFs are generated from physical theories and scientific models is commonly made in terms of the mean and covariance functions.
(d) The appropriate S/TVFs of optimal statistical prediction are the mean and the covariance.
(e) The very important class of *Gaussian* S/TRF is completely characterized by the corresponding mean and covariance.

Due to the above reasons, these two statistical moments play a fundamental role in physical applications. But this is not the end of the story. There are other S/TVFs, beyond the mean and the covariance functions, that can turn out to be at least equally important as the above two in applications with particular physical characteristics. Indeed, the statistical theory of locally isotropic turbulence has been based on Kolmogorov's introduction of the powerful mathematical notion of the *structure* function (Kolmogorov, 1941), which still remains the basic tool in the study of many turbulence phenomena in various scientific fields (oceanography, communications, climatology etc.). Due to its importance in random field modeling, I found it appropriate to introduce the spatiotemporal structure function in Chapter II together with the associated *variogram* function. As the readers recall, the (complete or full) spatiotemporal variogram and the structure functions were defined as

$$\begin{aligned} \gamma_X(\boldsymbol{p},\boldsymbol{p}') &= \frac{1}{2}\overline{[X(\boldsymbol{p}) - X(\boldsymbol{p}')]^2}, \\ \xi_X(\boldsymbol{p},\boldsymbol{p}') &= 2\gamma_X(\boldsymbol{p},\boldsymbol{p}'), \end{aligned} \tag{2.17a–b}$$

respectively. As was the case with the covariance, the units of γ_X and ξ_X are those of X^2 (meaning that if X is dimensionless, so are the γ_X and ξ_X). The covariance and variogram functions are directly related, see Eq. (2.19a) in Chapter II.

Remark 2.3

In principle, covariance, variogram, and structure functions are all S/TVFs that describe statistically the random field variations and dependencies in the space–time domain and at different scales of Nature. Also, they are all two-point variation functions, and, thus, they are interrelated. Depending on the application and the type of space–time heterogeneity condition, one of these functions may be preferred over the other. In theoretical studies of STHS random fields the covariance is commonly used. On the other hand, the variogram is often easier to estimate from experimental data. Finally, in studies involving heterogeneous fields but with STHS increments (e.g., turbulence), the structure function is usually the tool of choice.

Multipoint and *higher-order* spatiotemporal moments can be defined in an analogous fashion as the two-point moments.

Definition 2.2

The *m-point* spatiotemporal (ensemble) moment of the ρth-*order* (ρ is a positive integer) of the S/TRF $X(\boldsymbol{p})$ at points $\boldsymbol{p}_1,\ldots, \boldsymbol{p}_m$ is defined by

$$c_{X[m]}^{[\rho]} = \overline{X'^{\rho_1}(\boldsymbol{p}_1)\cdots X'^{\rho_m}(\boldsymbol{p}_m)} = \int\cdots\int d\chi'_1\ldots d\chi'_m \chi'^{\rho_1}_1\cdots\chi'^{\rho_m}_m f_X(\chi'_1,\ldots,\chi'_m). \tag{2.18}$$

Note that the number of points $\boldsymbol{p}_i$ in which $\rho_i \neq 0$ ($i = 1,\ldots, m$) determines the multiplicity m of the moment, whereas the sum, $\sum_{i=1}^{m} \rho_i = \rho$, its order. Clearly, $m \leq \rho$.

Since it is usually assumed that $X(\boldsymbol{p})$ is m.s. continuous, the statistical moment of Eq. (2.18) is a continuous function of $\boldsymbol{p}_i$ ($i = 1,\ldots, m$).

Example 2.8

Special cases of Eq. (2.18) are the m-point spatiotemporal moment of the *same order* (i.e., $m = \rho$),

$$c_{X[m]}^{[m]} = \overline{X'(\boldsymbol{p}_1)\cdots X'(\boldsymbol{p}_m)}, \tag{2.19a}$$

and the ρth-order covariance function,

$$c_X^{[\rho]}(\boldsymbol{p},\boldsymbol{p}') = \overline{[X(\boldsymbol{p}) - \overline{X(\boldsymbol{p})}]^{\rho_1}[X(\boldsymbol{p}') - \overline{X(\boldsymbol{p}')}]^{\rho_2}}, \tag{2.19b}$$

where $m = 2$ and $\rho = \rho_1 + \rho_2$. It seems rather difficult to derive any practically useful formulation of the covariance for $\rho > 2$. Multipoint and higher-order expressions for the variogram (structure) functions are also possible. Of particular interest is the following higher-order S/TVF.

Definition 2.3

The ρth-*order structure* function ($m = 2$) is defined as

$$\xi_X^{[\rho]}(\boldsymbol{p},\boldsymbol{p}') = \overline{[X(\boldsymbol{p}') - X(\boldsymbol{p})]^{\rho}}, \tag{2.20}$$

where $\rho > 2$.

As it turns out, the $\xi_X^{[\rho]}$ is often easier to formulate in practice than the covariance function of Eq. (2.19b). For this reason, $\xi_X^{[\rho]}$ will be studied in sufficient detail in a later section of the book.

To conclude, multipoint spatiotemporal moments of any order are derived by differentiating the multivariate CF $\phi_X(u_1,\ldots, u_m)$ and setting $\boldsymbol{u} = \boldsymbol{0}$, that is,

$$\overline{X^{\rho_1}(\boldsymbol{p}_1)\cdots X^{\rho_m}(\boldsymbol{p}_m)} = i^{-\rho}\left[\frac{\partial^{\rho}}{\partial u_1^{\rho_1}\ldots\partial u_m^{\rho_m}}\phi_X(\boldsymbol{u})\right]_{\boldsymbol{u}=0}, \tag{2.21}$$

where $\phi_X(\boldsymbol{u}) = FT[f_X(\boldsymbol{\chi})]$ and $\sum_{i=1}^{m} \rho_i = \rho$. A similar expression can be easily obtained for multipoint centered covariance functions by replacing $X^{\rho_i}(\boldsymbol{p}_i)$ ($i = 1,\ldots, m$) in Eq. (2.21) with $\left(X(\boldsymbol{p}_i) - \overline{X}(\boldsymbol{p}_i)\right)^{\rho_i}$ (compare with Eq. (2.7b)).

2.3 NONLOCALITY: DIRECTION-SPECIFIC SPACE–TIME VARIABILITY FUNCTIONS

In addition to the S/TVFs above considered along all coordinate directions, expressions for direction-specific S/TVFs are obtained by focusing on one or more particular directions in space. As the ones in the previous section, these S/TVFs too are nonlocal but context bound.

Definition 2.4

The S/TVFs (covariance, variogram, structure functions) along the s_j direction and at distance Δs are obtained by setting $\boldsymbol{p}' = (\boldsymbol{s}', t') = \left(\boldsymbol{s} + \Delta s\boldsymbol{\varepsilon}_j, t'\right)$ in Eqs. (2.8a–b) and (2.17a–b).

The direction-specific covariance and variogram (structure) functions measure, respectively, the averaged centered dependence and the averaged difference between S/TRF values at pairs of points along a line in the s_j direction at specified times. These expressions usually turn out to be more useful in practice when considered in conjunction with physically motivated variability conditions, like space–time homostationarity and isostationarity (Chapter VII).

Example 2.10

Direction-specific covariances are limited along specified directions in space, i.e., they focus on correlations along these directions. The following direction-specific covariance is obtained by measuring the correlation between S/TRF values at a pair of points along a line in the s_j direction,

$$c_X\left(\boldsymbol{s}, \boldsymbol{h}\cdot\boldsymbol{\varepsilon}_j, t, t'\right) = \overline{X(\boldsymbol{s}, t)X\left(\boldsymbol{s} + h\boldsymbol{\varepsilon}_j, t'\right)} = c_X\left(\boldsymbol{s}, s_j + h_j, t, t'\right), \tag{2.22}$$

where $\boldsymbol{\varepsilon}_j$ specifies the unit vector along the direction s_j. Eq. (2.22) is a simultaneous two-point covariance function.

Remark 2.4

To conclude, I would like to stress the point that when deciding which S/TVF to use, it might be helpful if our decision process takes into consideration key modeling questions, like the following:

- Are the necessary conditions for applying an S/TVF to the study of the attribute of interest met in practice?
- Which of the S/TVF parameters can be linked to important features of the attribute pattern and how?
- Are there any important attribute features that can be linked to the parameters of one S/TVF more rigorously than to the parameters of another S/TVF?
- Are there any specific attribute features that can be estimated with one S/TVF but cannot be estimated with another S/TVF?
- What are the measurable consequences of inadequate S/TVF selection in the description of the actual variation of the phenomenon, in uncertainty assessment, or in space–time estimation and prediction?

2.4 PHYSICAL CONSIDERATIONS AND ASSUMPTIONS OF SPACE–TIME VARIABILITY FUNCTIONS

I need to reemphasize that the practical value of random field modeling relies, to a considerable extend, on the adequate *interpretation* of the available data and on the meaningfulness/usefulness of the modeling results obtained. For sure, the data is there but our interpretation expresses what we can perceive from the data (e.g., by confronting the attribute links hypothesized by random field modeling with empirical evidence). In addition, beyond the physical constraints imposed by the data, creative interpretations should be able to extend beyond the physically observable.

As was suggested in Section 2.3 of Chapter II, it is often interpretatively appropriate to distinguish between the complete and the partial S/TVFs used in random field modeling. To put it in more words, depending on the application we may find it necessary to use the *complete* (or *full*) S/TVFs

$$c_X(\boldsymbol{s},t,\boldsymbol{s}',t'), \gamma_X(\boldsymbol{s},t,\boldsymbol{s}',t'), \text{ and } \xi_X(\boldsymbol{s},t,\boldsymbol{s}',t');$$

and in some other applications to use the *partial* S/TVFs

$$c_X(\boldsymbol{s},\boldsymbol{s}',t), c_X(\boldsymbol{s},t,t'), \gamma_X(\boldsymbol{s},\boldsymbol{s}',t), \gamma_X(\boldsymbol{s},t,t'), \xi_X(\boldsymbol{s},\boldsymbol{s}',t), \text{ and } \xi_X(\boldsymbol{s},t,t').$$

To be sure, partial S/TVFs are necessary in studies of physical phenomena in which we want either to switch focus on spatial physical features (i.e., to suppress the time argument, $t = t'$ or $\tau = 0$), or to isolate the dynamical character of the phenomenon (i.e., $\boldsymbol{s} = \boldsymbol{s}'$ or $\boldsymbol{h} = \boldsymbol{0}$). Both situations are considered in the example below.

Example 2.11

If the study focuses only on instantaneous pictures of a sea surface, we need to use the partial S/TVFs, $c_X(\boldsymbol{s},\boldsymbol{s}',t)$, $\gamma_X(\boldsymbol{s},\boldsymbol{s}',t)$, or $\xi_X(\boldsymbol{s},\boldsymbol{s}',t)$, where the varying arguments are the vector locations $(\boldsymbol{s},\boldsymbol{s}')$, and time t is considered a suppressed parameter. On the other hand, if the dynamic (temporal) pattern of a seismic signal is the main interest of the study, then the partial S/TVFs, $c_X(\boldsymbol{s},t,t')$, $\gamma_X(\boldsymbol{s},t,t')$, or $\xi_X(\boldsymbol{s},t,t')$ should be employed, where the varying arguments are the time instants (t,t') considered, and the suppressed parameter is the vector location $\boldsymbol{s}$.

No doubt, the adequate implementation of a complete or a partial S/TVF creates challenges relating to the ways in which attribute observations and their space–time patterns are reported and evaluated. Yet, the provision of sufficient quantitative evidence is only part of the solution to the challenges. The right questions must be first asked and the right assumptions must be formulated (normality, independence, homostationarity, isostationarity, separability, etc.). Then, data gathering and analysis can be used to test these hypotheses and make a choice between a complete and a partial S/TVF. The parameters of these S/TVFs acquire a physical meaning due to their association with key characteristics of the space–time behavior of the attribute represented by the S/TVFs.

Example 2.12

If we consider the exponential space–time covariance model of a natural attribute in $R^{3,1}$,

$$c_X(\Delta\boldsymbol{p}) = c_0 e^{-\left(\sum_{i=1}^{3}\frac{h_i}{a_i} + \frac{\tau}{a_\tau}\right)},$$

its parameters a_i ($i = 1, 2, 3$) and a_τ offer a measure of the physical correlation ranges of the attribute. Particularly, the spatial correlation ranges (along the directions s_i, $i = 1, 2, 3$) are given by $\varepsilon_i \approx 3a_i$, and the temporal correlation range by $\varepsilon_\tau \approx 3a_\tau$ (these are the distances and time interval beyond which, practically, there is no correlation between the attribute values).

It has been stated in various parts of the book that quite frequently in applications certain *assumptions* are made that can simplify the analytical or numerical calculations and, also, make certain quantitative modeling variables measurable with relative ease and accuracy. However, when making these assumptions we must be careful that they do not lead to inconsistencies and they do not violate any physical conditions regarding the phenomenon of interest. In the two following examples, I will examine two cases: the first example presents an assumption that does not lead to inconsistencies, and improves the interpretability of the results obtained, and, it also simplifies calculations; and the second example introduces an assumption, which, although it simplifies calculations considerable, it leads to inconsistent results, and, hence, it is incorrect.

Example 2.13

Let us consider the random fields specified by a two-point average and a two-point difference, i.e.,

$$\begin{aligned} A(\boldsymbol{p},\boldsymbol{p}') &= \frac{1}{2}[X(\boldsymbol{p}) + X(\boldsymbol{p}')], \\ D(\boldsymbol{p},\boldsymbol{p}') &= X(\boldsymbol{p}) - X(\boldsymbol{p}'), \end{aligned} \tag{2.23a–b}$$

where $X^2(\boldsymbol{p}) - X^2(\boldsymbol{p}') = 2A(\boldsymbol{p},\boldsymbol{p}')D(\boldsymbol{p},\boldsymbol{p}')$. An assumption that can simplify considerably the calculation of higher-order statistical moments by expressing them in terms of second-order moments is that the following independence holds,

$$\overline{A^2(\boldsymbol{p},\boldsymbol{p}')D^2(\boldsymbol{p},\boldsymbol{p}')} = \overline{A^2(\boldsymbol{p},\boldsymbol{p}')}\;\overline{D^2(\boldsymbol{p},\boldsymbol{p}')}. \tag{2.24}$$

Let us check whether this assumption holds. For simplicity, let $X(\boldsymbol{p})$ be a zero mean random field, i.e., $\overline{X(\boldsymbol{p})} = 0$ and $\sigma_X^2(\boldsymbol{p}) = \sigma_X^2 = \overline{X^2}$. It is valid that

$$\begin{aligned} \overline{A(\boldsymbol{p},\boldsymbol{p}')} &= \overline{D(\boldsymbol{p},\boldsymbol{p}')} = 0, \\ \overline{D^2(\boldsymbol{p},\boldsymbol{p}')} &= 2\gamma_X(\boldsymbol{p}-\boldsymbol{p}'), \\ \overline{A^2(\boldsymbol{p},\boldsymbol{p}')} &= \sigma_X^2 - \frac{1}{2}\gamma_X(\boldsymbol{p}-\boldsymbol{p}'), \end{aligned} \tag{2.25a–c}$$

where $\frac{1}{2}\left[\overline{X^2(\boldsymbol{p})} + \overline{X^2(\boldsymbol{p}')}\right] = \overline{X^2} = \sigma_X^2$. Then, the cross-covariance of $A(\boldsymbol{p},\boldsymbol{p}')$ and $D(\boldsymbol{p},\boldsymbol{p}')$ is given by

$$c_{AD}(\boldsymbol{0}) = \overline{A(\boldsymbol{p},\boldsymbol{p}')D(\boldsymbol{p},\boldsymbol{p}')} = \frac{1}{2}\overline{[X(\boldsymbol{p})-X(\boldsymbol{p}')][X(\boldsymbol{p})+X(\boldsymbol{p}')]} = \frac{1}{2}\overline{[X^2(\boldsymbol{p})-X^2(\boldsymbol{p}')]} = 0, \tag{2.26a–b}$$

i.e., $A(\boldsymbol{p},\boldsymbol{p}')$ and $D(\boldsymbol{p},\boldsymbol{p}')$ are statistically independent, which means that the assumption of Eq. (2.24) holds, since, according to probability theory (Shreve, 2004), if two random variables are independent, then any measurable functions of them are independent too.[2] The analysis above leads to some

[2]A heuristic sort of proof would be by logical contradiction. First, statistical independence of two RVs y and z means that knowing the value of z offers zero information about the value of y. So, if y and z are independent but $g_1(y)$ and $g_2(z)$ were dependent (g_1 and g_2 are known functions), then knowing the value of z (and, hence, of $g_2(z)$) would imply that inference about $g_1(y)$ (and, hence, about y) can be made, which contradicts the independence of y and z.

interesting, theoretically and computationally, expressions derived in terms of the standard variogram and the mean squared of $X(\boldsymbol{p})$, like

$$\overline{A^2(\boldsymbol{p},\boldsymbol{p}')D^2(\boldsymbol{p},\boldsymbol{p}')} = \overline{A^2(\boldsymbol{p},\boldsymbol{p}')}\,\overline{D^2(\boldsymbol{p},\boldsymbol{p}')} = \gamma_X(\boldsymbol{p}-\boldsymbol{p}')\left[2\sigma_X^2 - \gamma_X(\boldsymbol{p}-\boldsymbol{p}')\right], \tag{2.27}$$

where $\gamma_X(\boldsymbol{p}-\boldsymbol{p}') = \frac{1}{2}\overline{[X(\boldsymbol{p})-X(\boldsymbol{p}')]^2} = \sigma_X^2 - c_X(\boldsymbol{p}-\boldsymbol{p}')$. Moreover, since

$$\overline{A^2(\boldsymbol{p},\boldsymbol{p}')D^2(\boldsymbol{p},\boldsymbol{p}')} = \frac{1}{4}\overline{[X(\boldsymbol{p})-X(\boldsymbol{p}')]^2[X(\boldsymbol{p})+X(\boldsymbol{p}')]^2} = \frac{1}{4}\overline{[X^2(\boldsymbol{p})-X^2(\boldsymbol{p}')]^2}, \tag{2.28}$$

we get the following expression,

$$\overline{[X^2(\boldsymbol{p})-X^2(\boldsymbol{p}')]^2} = 4\overline{A^2(\boldsymbol{p},\boldsymbol{p}')D^2(\boldsymbol{p},\boldsymbol{p}')} = 4\gamma_X(\boldsymbol{p}-\boldsymbol{p}')\left[2\sigma_X^2 - \gamma_X(\boldsymbol{p}-\boldsymbol{p}')\right]. \tag{2.29}$$

Eq. (2.29) involves fourth-order moments expressed in terms of the (second-order) variogram function and the variance of the random field. Other useful expressions of higher-order S/TVFs can be also derived based on the assumption of Eq. (2.24).

Example 2.14

Consider, again, a zero mean random field $X(\boldsymbol{p})$ representing a natural attribute at space—time point $\boldsymbol{p}$, and the two-point difference $D(\boldsymbol{p},\boldsymbol{p}')$ defined in Eq. (2.23b) with $\boldsymbol{p}\neq\boldsymbol{p}'$. Another assumption that could simplify calculations considerably is that the random fields $X(\boldsymbol{p})$ and $D(\boldsymbol{p},\boldsymbol{p}')$ are statistically independent, i.e.,

$$\overline{X(\boldsymbol{p})D(\boldsymbol{p},\boldsymbol{p}')} = \overline{X(\boldsymbol{p})}\;\overline{D(\boldsymbol{p},\boldsymbol{p}')} = 0. \tag{2.30}$$

Now consider the same random field at space—time point $\boldsymbol{p}'$, i.e., $X(\boldsymbol{p}')$. Obviously, $X(\boldsymbol{p}) = X(\boldsymbol{p}') + D(\boldsymbol{p},\boldsymbol{p}')$, which, if inserted into Eq. (2.30) implies that

$$\overline{X(\boldsymbol{p}')D(\boldsymbol{p},\boldsymbol{p}')} = -\overline{D^2(\boldsymbol{p},\boldsymbol{p}')} = -2\gamma_X(\boldsymbol{p},\boldsymbol{p}')\neq 0, \tag{2.31}$$

since $\boldsymbol{p}\neq\boldsymbol{p}'$. Therefore, $X(\boldsymbol{p}')$ and $D(\boldsymbol{p},\boldsymbol{p}')$ are statistically dependent. This makes no sense, since there is neither a logical nor a physical reason that supports the validity of both Eqs. (2.30) and (2.31), i.e., there is no reason to support the choice of $X(\boldsymbol{p})$ over $X(\boldsymbol{p}')$ as statistically independent of $D(\boldsymbol{p},\boldsymbol{p}')$. So, the assumption that $X(\boldsymbol{p})$ and $D(\boldsymbol{p},\boldsymbol{p}')$ are statistically independent leads to inconsistent results, and, therefore, it is physically unjustified and mathematically incorrect.

In the last part of this section, I turn to physical laws and their importance in S/TVF characterization. As was also suggested in Section 3.4 of Chapter II, the S/TRF model of a natural attribute must obey the stochastic partial differential equations (PDE) representing the physical law governing the attribute's distribution in the real-world. Likewise, a deterministic PDE can be derived that describes the lawful change of the corresponding S/TVF characterizing the variability features of the attribute distribution in the space—time domain. Due to its considerable significance, this book devotes an entire chapter on the subject (see, Chapter XIV). In the present section, we are limited to the discussion of a rather simple example, for completeness of presentation.

Example 2.15

S/TRF models can be used to simulate air-pollution processes from routine monitoring data (e.g., Omatu and Seinfeld, 1981). Consider data for sulfur dioxide (SO_2) and assume that the study domain is one-dimensional along the east–west coordinate. The SO_2 concentrations, $X(s, t)$, are described by the atmospheric diffusion equation ($R^{1,1}$),

$$\frac{\partial}{\partial t}X(s,t) + V\frac{\partial}{\partial s}X(s,t) - \mu\frac{\partial^2}{\partial s^2}X(s,t) = Y(s,t), \tag{2.32}$$

where V is the wind velocity, μ is a diffusion coefficient, and $Y(s, t)$ is the rate of emission of SO_2. The PDE governing the (noncentered) SO_2 concentration covariance is obtained directly from the stochastic PDE of Eq. (2.32), i.e.,

$$\left[\frac{\partial}{\partial t} + V\frac{\partial}{\partial s} - \mu\frac{\partial^2}{\partial s^2}\right]c_X(s,t,s',t') = c_{YX}(s,t,s',t'), \tag{2.33}$$

The solution of Eq. (2.33) for c_X requires the consideration of the boundary and initial conditions of the situation, and an estimation of the cross-correlation between SO_2 concentration and SO_2 emission rate. How this can be generally done is discussed in Section 5 of Chapter XVI.

The covariance formulation in terms of a lawful PDE has the great merit of maintaining close contact between the formal definition of the covariance and physical considerations about the phenomenon of interest. The same, of course, is valid for other S/TVFs, like the variogram and structure functions: important physical considerations should be encapsulated in the S/TVF of a natural attribute (e.g., variogram and structure functions with theoretically interesting features can be constructed in terms of sums of cosine functions, yet, often they do not have any practical relevance). This comment is the link with the topic of the following section.

2.5 FORMAL AND PHYSICAL COVARIANCE PERMISSIBILITY

As was noticed in Section 2 of Chapter II, there are certain rather strict conditions a function must satisfy to be a formally permissible (or valid) covariance function. These conditions are clearly stated in the following proposition.

Proposition 2.2

A function $c_X(\boldsymbol{p}_i, \boldsymbol{p}_j) = c_X(\boldsymbol{p}_j, \boldsymbol{p}_i)$ is a covariance function of a multi-Gaussian S/TRF $X(\boldsymbol{p})$ if and only if it is of the *nonnegative-definite* (*NND*) type, that is,

$$\sum_{i=1}^{m}\sum_{j=1}^{m} q_i q_j c_X(\boldsymbol{p}_i, \boldsymbol{p}_j) \geq 0 \tag{2.34}$$

for all nonnegative integers m, all points $\boldsymbol{p}_i$, $\boldsymbol{p}_j \in R^{n+1}$ and all real (or complex) numbers q_i, q_j ($i, j = 1, \ldots, m$).

It is worthwhile to make a few comments related to Proposition 2.2: First, a function $c_X(\boldsymbol{p}_i, \boldsymbol{p}_j)$ in R^{n+1} is a permissible covariance of a multi-Gaussian[3] field $X(\boldsymbol{p})$ if and only if it is an NND function (i.e., some NND function can be the covariance function of some S/TRF). Second, the space C_{n+1} of

[3] Say, m-variate Gaussian, G^m (see also, Definition 5.1).

covariance functions is essential that of NND functions. Third, sometimes an alternative formulation of Proposition 2.2 is used, where Eq. (2.34) is replaced by the condition

$$\sum_{i=1}^{m}\sum_{j=1_i}^{k_i}\sum_{i'=1}^{m}\sum_{j'=1_{i'}}^{k_{i'}} q_{ij}q_{i'j'}c_X(\boldsymbol{s}_i,t_j;\boldsymbol{s}_{i'},t_{j'}) \geq 0 \tag{2.35}$$

for all $m, k_i, k_{i'}(=1,2,\ldots)$, all locations-instants $(\boldsymbol{s}_i, t_j) \in R^n \times T$, and all real (or complex) numbers q_{ij}, $q_{i'j'}$.[4] Fourth, as is noticed in Abrahamsen (1997), for a correlation function, $\rho_X(\boldsymbol{p},\boldsymbol{p}')$, permissibility also requires that $\sigma_X(\boldsymbol{p}) \neq 0$ for all $\boldsymbol{p}$ (this restriction is not necessary for covariance functions).

A cautionary note is due here: NND in the above sense does not guarantee that there is always a non-G^m random field that admits as a covariance an NND function. As a matter of fact, this is not valid, in general (for a more detailed treatment of the subject see Chapter XV, and references therein).

Example 2.16

The *spherical* and the *cosine* space-time covariance models (see explicit mathematical expressions in Chapter VII) are not permissible covariances in the case of a lognormal S/TRF. We will revisit this important topic in Chapter XV.

Furthermore, covariance permissibility depends on the space–time metric considered, i.e., a covariance that is valid for one metric is not necessarily so for another (again, more details can be found in Chapter XV).

Example 2.17

It has been shown (Christakos and Papanicolaou, 2000) that the exponential covariance model is permissible in $\Delta\boldsymbol{p} \in R^{n+1}$ for both the Euclidean r_E and the Manhattan or absolute r_M metrics (Chapter III). However, the Gaussian covariance model is permissible for r_E but not for r_M.[5]

An analogous set of conditions must be satisfied by a function to be a permissible (or valid) variogram function, as stated in the last proposition of this section.

Proposition 2.3

A function $\gamma_X(\boldsymbol{p},\boldsymbol{p}') = \gamma_X(\boldsymbol{p}',\boldsymbol{p})$ and $\gamma_X(\boldsymbol{p},\boldsymbol{p}) = 0$ is a variogram function of a multi-Gaussian S/TRF $X(\boldsymbol{p})$ if and only if it is of the *conditionally negative-definite* (*CND*) type, that is,

$$\sum_{i=1}^{m}\sum_{j=1}^{m} q_i q_j \gamma_X(\boldsymbol{p}_i,\boldsymbol{p}_j) \leq 0 \tag{2.36}$$

for all nonnegative integers m, all points $\boldsymbol{p}_i, \boldsymbol{p}_j \in R^{n+1}$ and all numbers q_i, q_j $(i, j = 1,\ldots, m)$ such that $\sum_{i=1}^{m} q_i = 0$.

Regardless of their mathematical sophistication, it is appropriate that theoretical random field developments always keep an eye on the physical nature of the real-world attributes they are assumed to represent. This is true for the covariance and variogram functions too: It is one thing for a covariance

[4]We recall that k_i denotes the number of time instants $t_j, j = 1_i, 2_i,\ldots k_i$, used, given that we are at the spatial position s_i (Eqs. 2.4b–c in Section 2 of Chapter I).

[5]For the explicit mathematical expressions of the exponential and the Gaussian covariance functions, see Chapter VII.

or variogram function to be *formally* permissible (i.e., to satisfy the NND or CND condition, respectively), and another thing to be *physically* permissible. Therefore, we conclude this section with the following postulate.

Postulate 2.1

For a covariance or variogram model to be an adequate representation of a real-world phenomenon, it is not enough that it is formally permissible and fits reasonably well the data-set available about the phenomenon. It also needs to be physically permissible, i.e., its form needs to be consistent with physical considerations about the phenomenon they represent and to be interpretable.

Postulate 2.1 makes sense in scientific applications, since it is not unusual that covariance or variogram models physically inadequate to describe a real-world phenomenon are often suggested in the literature, simply relying on the false premise of data fitting that the above postulate aims to discredit (see also, comments in later chapters).

3. BASIC PROPERTIES OF COVARIANCE FUNCTIONS

Within the framework of spatiotemporal correlation theory, certain important properties of the covariance function in the general case of S/TRFs are discussed next.

Property 3.1

For $\boldsymbol{p} \neq \boldsymbol{p}' \in R^{n+1}$, the space−time covariances

$$c_X(\boldsymbol{p},\boldsymbol{p}'), c_X(\boldsymbol{p}',\boldsymbol{p}) \in C_{n+1} \tag{3.1a}$$

may be different functions, in general. In the special case of STHS random fields,

$$c_X(\boldsymbol{p},\boldsymbol{p}+\Delta\boldsymbol{p}) = c_X(\boldsymbol{p}+\Delta\boldsymbol{p},\boldsymbol{p}), \tag{3.2}$$

$\Delta\boldsymbol{p} \in R^{n+1}$, discussed later.[6]

Property 3.2

It is valid that

$$\lim_{|\boldsymbol{p}-\boldsymbol{p}'|\to\infty} c_X(\boldsymbol{p},\boldsymbol{p}') = 0, \tag{3.3}$$

which expresses the empirical observation of diminishing attribute correlations at large space−time lags.

Property 3.3

The class C_{n+1} of covariance functions is closed under addition, multiplication, and passages to the limit; in other words, given the sequence of covariances $\{c_{X,k}(\boldsymbol{p},\boldsymbol{p}')\} \in C_{n+1}$ $(k = 1, 2,\ldots)$,

[6] In general, $c_X(\Delta s_1,\ldots, \Delta s_n, \Delta s_0) \neq c_X(\Delta s_1,\ldots, \Delta s_n, -\Delta s_0)$ with $\Delta s_1 \neq 0,\ldots, \Delta s_n \neq 0$, and $c_X(\Delta s_1,\ldots, \Delta s_n, \Delta s_0) \neq c_X(-\Delta s_1,\ldots, -\Delta s_n, \Delta s_0)$ with $\Delta s_0 = \tau \neq 0$.

$$\sum_{\rho \leq k} c_{X,\rho}(\boldsymbol{p},\boldsymbol{p}') \in C_{n+1}, \tag{3.4}$$

and

$$\lim_{k \to \infty} c_{X,k}(\boldsymbol{p},\boldsymbol{p}') = c_X(\boldsymbol{p},\boldsymbol{p}') \in C_{n+1}, \tag{3.5}$$

assuming that the above limit exists for all pairs $(\boldsymbol{p},\boldsymbol{p}')$.

Property 3.4

If $\mu(\mathbf{u})$ is a measure on a space U and $c_{\boldsymbol{u}}(\boldsymbol{p},\boldsymbol{p}' \cdot \boldsymbol{u})$ is a function integrable on a subspace $V$of U for each pair $(\boldsymbol{p},\boldsymbol{p}')$, then

$$\int_V \mu(d\boldsymbol{u}) c_{\boldsymbol{u}}(\boldsymbol{p},\boldsymbol{p}' \cdot \boldsymbol{u}) = c_X(\boldsymbol{p},\boldsymbol{p}') \in C_{n+1}. \tag{3.6}$$

Property 3.5

Every covariance $c_X(\boldsymbol{p},\boldsymbol{p}') \in C_{n+1}$ is also the covariance of a Gaussian S/TRF.

Let us comment a little further on the above properties. Property 3.1 refers to the symmetry structure of the covariance that may vary with time and space directions. Property 3.2 means that when the distance between $\boldsymbol{p}$ and $\boldsymbol{p}'$ tends to infinity the correlation between $X(\boldsymbol{p})$ and $X(\boldsymbol{p}')$ tends to zero. Properties 3.3 and 3.4 provide interesting means of constructing covariance models. Let us discuss an example (more cases are properly discussed in Chapter XVI).

Example 3.1

If the $c_{X,k}(\boldsymbol{p},\boldsymbol{p}')$, $k = 1, 2, \ldots$, are covariance models, the quantities

$$c_X(\boldsymbol{p},\boldsymbol{p}') = \sum_{\rho \leq k} a_\rho \left[c_{X,\rho}(\boldsymbol{p},\boldsymbol{p}')\right]^\rho, \tag{3.7}$$

and

$$c_X(\boldsymbol{p},\boldsymbol{p}') = \lim_{k \to \infty} \sum_{\rho \leq k} a_\rho \left[c_{X,\rho}(\boldsymbol{p},\boldsymbol{p}')\right]^\rho, \tag{3.8}$$

where $a_\rho \geq 0$, are covariance models too.

Property 3.5 is of particular interest for several reasons: First, it is easily seen that a similar property holds for every mean value $\overline{X(\boldsymbol{p})}$, as well. Second, an important implication is that the study of any S/TRF in terms of its mean and covariance functions is equivalent to the study of a Gaussian S/TRF that has the same mean and covariance. Third, the consequences of this fact are significant in various problems of the S/TRF theory, such as random field estimation and simulation on the basis of their mean and covariance functions (Christakos, 1992).

Remark 3.1

In certain cases, the construction of multivariate non-Gaussian probability distributions on the basis of the mean value and the covariance function could lead to formally inconsistent results. However, as already mentioned, in the context of the correlation theory this fact does not create any real problems, because in most applications there is no need to deal with the form of the multivariate probability distribution.

4. CROSS–SPACE–TIME VARIABILITY FUNCTIONS

The circumstances encountered in many real-world phenomena make it necessary to study two or more random fields that are related to each other. We then need to define additional S/TVFs in terms of the corresponding statistical moments. The formal extension of the statistical analysis from a single random field to two or more correlated random fields presents no major technical problems.

Definition 4.1

Let $X_l(\boldsymbol{p})$ and $X_{l'}(\boldsymbol{p}')$ be two S/TRFs. The spatiotemporal *cross-covariance function* is defined as

$$\begin{aligned} c_{X_l X_{l'}}(\boldsymbol{p},\boldsymbol{p}') &= \overline{\left[X_l(\boldsymbol{p}) - \overline{X_l(\boldsymbol{p})}\right]\left[X_{l'}(\boldsymbol{p}') - \overline{X_{l'}(\boldsymbol{p}')}\right]} \\ &= \iint d\chi_{l,\boldsymbol{p}} d\chi_{l',\boldsymbol{p}'} \left[\chi_{l,\boldsymbol{p}} - \overline{X_l(\boldsymbol{p})}\right]\left[\chi_{l',\boldsymbol{p}'} - \overline{X_{l'}(\boldsymbol{p}')}\right] f_{X_l X_{l'}}(\chi_{l,\boldsymbol{p}}, \chi_{l',\boldsymbol{p}'}), \end{aligned} \tag{4.1a–b}$$

where $f_{X_l X_l}(\chi_{l,\boldsymbol{p}}, \chi_{l',\boldsymbol{p}'})$ is now the joint PDF of $X_l(\boldsymbol{p})$ and $X_{l'}(\boldsymbol{p}')$. The *coefficient of spatiotemporal cross-correlation* is defined as

$$\rho_{X_l,X_{l'}}(\boldsymbol{p},\boldsymbol{p}') = \frac{c_{X_l,X_{l'}}(\boldsymbol{p},\boldsymbol{p}')}{\sigma_{X_l}(\boldsymbol{p})\sigma_{X_{l'}}(\boldsymbol{p}')}, \tag{4.2}$$

where $\sigma_{X_l}(\boldsymbol{p})$ and $\sigma_{X_{l'}}(\boldsymbol{p}')$ are the std deviations of $X_l(\boldsymbol{p})$ and $X_{l'}(\boldsymbol{p}')$, respectively.

Remark 4.1

The scale-independency property of the correlation function of a scalar S/TRF (Remark 2.2), also applies in the case of the cross-correlation function of a vector S/TRF, i.e., although a simple field rescaling by the scaling coefficient a causes the rescaling of the cross-covariance to the second power, i.e.,

$$c_{aX_l,aX_{l'}}(\boldsymbol{p},\boldsymbol{p}') = a^2 c_{X_l,X_{l'}}(\boldsymbol{p},\boldsymbol{p}'), \tag{4.3a}$$

the cross-correlation function is independent of such a rescaling, i.e.,

$$\rho_{aX_l,aX_{l'}}(\boldsymbol{p},\boldsymbol{p}') = \rho_{X_l,X_{l'}}(\boldsymbol{p},\boldsymbol{p}'). \tag{4.3b}$$

Moreover, some other standard results derived for a single random field remain valid for two random fields. So, the random fields $X_l(\boldsymbol{p})$ and $X_{l'}(\boldsymbol{p}')$ are considered *uncorrelated* if the cross-correlation function is

$$\rho_{X_l,X_{l'}}(\boldsymbol{p},\boldsymbol{p}') = 0 \tag{4.4}$$

for $\boldsymbol{p} \neq \boldsymbol{p}'$. The space–time noncorrelation between the two S/TRFs expressed by Eq. (4.4) does not necessarily imply probabilistic independence among them, $f_{X_l,X_{l'}}(\boldsymbol{p},\boldsymbol{p}') = f_{X_l}(\boldsymbol{p}) f_{X_{l'}}(\boldsymbol{p}')$. Independency between the two S/TRFs, however, always implies lack of correlation.

Example 4.1

The random fields specified by the relationship (Exercise IV.4)

$$X_{l'}(\boldsymbol{p}') = X_l^2(\boldsymbol{p}) \tag{4.5}$$

exhibit a very strong probabilistic dependency, but the corresponding cross-covariance is $c_{X_l,X_{l'}}(\boldsymbol{p},\boldsymbol{p}') = 0$.

The comprehensive Cauchy–Schwarz inequality can be expressed in terms of the pair of random fields $X_l(\boldsymbol{p})$ and $X_{l'}(\boldsymbol{p}')$, i.e.,

$$\left|c_{X_l,X_{l'}}(\boldsymbol{p},\boldsymbol{p}')\right| \leq \sigma_{X_l}(\boldsymbol{p})\sigma_{X_{l'}}(\boldsymbol{p}'). \tag{4.6}$$

Eq. (4.6) provides a means to assess the strength of the covariance measure of space–time dependency between $X_l(\boldsymbol{p})$ and $X_{l'}(\boldsymbol{p}')$ by comparing the covariance's magnitude versus the largest possible value. Put differently, Eq. (4.6) provides an upper bound for the covariance between any two points in space–time in terms of the local variances. The covariance attains the limits in Eq. (4.6) when $X(\boldsymbol{p})$ and $X(\boldsymbol{p}')$ are proportional, and, in this sense, $c_{X_l,X_{l'}}(\boldsymbol{p},\boldsymbol{p}')$ measures the degree of linear relationship between the two fields. Otherwise said, it is commonly admitted that covariance rather measures a particular kind of dependence, the degree of linear relationship between $X_l(\boldsymbol{p})$ and $X_{l'}(\boldsymbol{p}')$. As a matter of fact, this is why we get zero correlation in the case $X_{l'}(\boldsymbol{p}') = X_l^2(\boldsymbol{p})$ (Example 4.1), where the relationship between $X_l(\boldsymbol{p})$ and $X_{l'}(\boldsymbol{p}')$ is strongly nonlinear (i.e., a purely quadratic relationship).

The definitions above can be easily extended to more than two random fields. Consider the vector S/TRF

$$\boldsymbol{X}(\boldsymbol{p}) = [X_1(\boldsymbol{p})\ldots X_k(\boldsymbol{p})]^T.$$

The corresponding *matrix of cross-covariances* between the component scalar S/TRFs is given by

$$c_X = \left[c_{X_l,X_{l'}}(\boldsymbol{p},\boldsymbol{p}')\right], \tag{4.7}$$

where $l, l' = 1,\ldots,k$. Some interesting properties of the cross-covariance matrix (Eq. 4.7) are summarized below:

Property 4.1

The matrix $\boldsymbol{c}_X$ is of the NND type, that is,

$$\boldsymbol{\lambda}^T \boldsymbol{c}_X\, \boldsymbol{\lambda} \geq 0 \tag{4.8}$$

for all vectors $\boldsymbol{\lambda} = [\lambda_1\ldots\lambda_m]^T$. This is because it must hold that $Var[\boldsymbol{\lambda}^T\boldsymbol{X}(p)] \geq 0$.

Property 4.2

The $\boldsymbol{c}_X$ is a symmetric matrix, in the sense that by definition

$$c_{X_lX_{l'}}(\boldsymbol{p},\boldsymbol{p}') = c_{X_{l'}X_l}(\boldsymbol{p}',\boldsymbol{p}).$$

However there is no symmetry, in general, with respect to $\boldsymbol{p}$ and $\boldsymbol{p}'$.

Property 4.3

Eq. (4.6) is valid for all $l, l' = 1,\ldots,k$ and $\boldsymbol{p},\boldsymbol{p}'$.

To clarify some aspects of the matrix algebra involved in the analysis above, we consider the following example.

Example 4.2

Let $k = 2$ so that

$$c_X = \begin{bmatrix} c_{X_1}(\boldsymbol{p},\boldsymbol{p}') & c_{X_1X_2}(\boldsymbol{p},\boldsymbol{p}') \\ c_{X_2X_1}(\boldsymbol{p}',\boldsymbol{p}) & c_{X_1}(\boldsymbol{p},\boldsymbol{p}') \end{bmatrix}.$$

If we define the S/TRF $Y(\boldsymbol{p}) = \sum_{l=1}^{2} \lambda_l X_l(\boldsymbol{p})$, where $\lambda_l \in R^1$ $(l = 1, 2)$, the corresponding covariance is given by

$$c_Y(\boldsymbol{p},\boldsymbol{p}') = \sum_{l=1}^{2}\sum_{l'=1}^{2} q_l q_{l'} c_{X_l,X_{l'}}(\boldsymbol{p},\boldsymbol{p}'),$$

which must be an NND function.

We chose to conclude this section by introducing the useful class of conditional S/TVFs, as follows.

Definition 4.2

Let $\boldsymbol{X} = [x_1 \ldots x_m]^T$ and $\boldsymbol{Y} = [y_1 \ldots y_m]^T$ be vectors of RVs from the random fields $X(\boldsymbol{p})$ and $Y(\boldsymbol{p})$, respectively. The *conditional* mean vector of $\boldsymbol{X}$ given $\boldsymbol{Y} = \boldsymbol{\Psi} = [\psi_1,\ldots,\psi_m]^T$ is the deterministic vector

$$\overline{\boldsymbol{X}|\boldsymbol{\Psi}} = \int d\boldsymbol{\chi}\boldsymbol{\chi} f_{X|Y}(\boldsymbol{\chi}|\boldsymbol{\Psi}). \tag{4.9}$$

The *conditional* covariance matrix of $\boldsymbol{X}$ given $\boldsymbol{\Psi}$ is the deterministic matrix

$$\boldsymbol{c}_{X|\Psi} = \overline{\left[\boldsymbol{X} - \overline{\boldsymbol{X}|\boldsymbol{\Psi}}\right]\left[\boldsymbol{X} - \overline{\boldsymbol{X}|\boldsymbol{\Psi}}\right]^T \Big|\boldsymbol{\Psi}} = \int d\boldsymbol{\chi}\left(\boldsymbol{\chi} - \overline{\boldsymbol{X}|\boldsymbol{\Psi}}\right)\left(\boldsymbol{\chi} - \overline{\boldsymbol{X}|\boldsymbol{\Psi}}\right)^T f_{X|Y}(\boldsymbol{\chi}|\boldsymbol{\Psi}). \tag{4.10}$$

If $\boldsymbol{\Psi}$ is replaced by the random vector $\boldsymbol{Y}$, the conditional mean vector and covariance matrix become random quantities as well, namely

$$\overline{\boldsymbol{X}|\boldsymbol{Y}} = \int d\boldsymbol{\chi}\boldsymbol{\chi} f_{X|Y}(\boldsymbol{\chi}|\boldsymbol{Y}), \tag{4.11}$$

and

$$c_{X|Y} = \overline{\left[\boldsymbol{X} - \overline{\boldsymbol{X}|\boldsymbol{Y}}\right]\left[\boldsymbol{X} - \overline{\boldsymbol{X}|\boldsymbol{Y}}\right]^T \Big|\boldsymbol{Y}} = \int d\boldsymbol{\chi}\left[\boldsymbol{\chi} - \overline{\boldsymbol{X}|\boldsymbol{Y}}\right]\left[\boldsymbol{\chi} - \overline{\boldsymbol{X}|\boldsymbol{Y}}\right]^T f_{X|Y}(\boldsymbol{\chi}|\boldsymbol{Y}). \tag{4.12}$$

The conditional S/TVFs above play a key role in the space–time estimation of natural attributes (Kolmogorov, Wiener, and Kalman filters) and in statistical decision-making.

Remark 4.2

A particularly useful relationship is provided by the chain rule for conditional random field expectations

$$\overline{\left[\overline{[X(\boldsymbol{p})|Y(\boldsymbol{p})]}_X\right]}_Y = \overline{X(\boldsymbol{p})}, \tag{4.13}$$

where the subscripts X and Y denote expectation with respect to the random field $X(\boldsymbol{p})$, followed by expectation with respect to $Y(\boldsymbol{p})$ (Exercise IV.5).

5. CORRELATION OF GAUSSIAN AND RELATED SPATIOTEMPORAL RANDOM FIELDS

5.1 GENERAL CONSIDERATIONS

Based on our discussion so far, it has already become obvious that the class of Gaussian S/TRFs is a very important one, theoretically and practically, for a variety of reasons, such as the following:

(i) Members of this family have convenient mathematical properties that greatly simplify calculations involving linear transformations; in fact, many results can be worked out only for Gaussian random fields.
(ii) Gaussian is the limit approached by the superposition of a large number of other non-Gaussian random fields; this is a famous result of the well-known central limit theorem (Feller, 1966).
(iii) It has been found that the distribution of many natural attributes can be approximated satisfactorily by means of Gaussian random fields.
(iv) The study of the Gaussian random field is essentially the study of its first- and second-order moments. The class of covariance functions, in particular, coincides with that of NND functions, which have powerful spectral representation features.
(v) For any random field with finite first- and second-order moments, it is always possible to construct a Gaussian random field with the same mean and covariance. This is not true for non-Gaussian random fields. In fact, the construction of multivariate non-Gaussian PDF on the basis of first- and second-order moments can lead to inconsistent results.
(vi) The Gaussian random field maximizes the probabilistic entropy function subject to the constraints associated with the first- and second-order moments.

Definition 5.1

A multivariate Gaussian S/TRF (say, m-variate Gaussian, G^m) is an S/TRF $X(\boldsymbol{p})$ for which the multivariate PDF of the vector RV $\boldsymbol{X} = [x_1 \ldots x_m]^T$ for all m is given by

$$f_X(\chi) = \frac{1}{(2\pi)^{\frac{m}{2}} |\boldsymbol{c}_X|^{\frac{1}{2}}} e^{-\frac{1}{2}[\chi - \overline{X}]^T \boldsymbol{c}_X^{-1} [\chi - \overline{X}]}, \tag{5.1}$$

where $\chi = [\chi_1 \ldots \chi_m]^T$ and $\boldsymbol{c}_X = [c_X(\boldsymbol{p}_i, \boldsymbol{p}_j),\ i, j = 1, \ldots, m]$.

Eq. (5.1) offers a vector-based expression of the G^m PDF that is an alternative to the one introduced by Eq. (7.1) of Chapter II. Two common cases are, for $m = 2$, the $\boldsymbol{X} = [x_1, x_2]^T$ is called *bivariate* (or *two-dimensional*) Gaussian random field and, for $m = 3$, the $\boldsymbol{X} = [x_1, x_2, x_3]^T$ is called a *trivariate* (or *three-dimensional*) Gaussian random field.

5.2 GAUSSIAN PROPERTIES

A multivariate Gaussian random field (G^m) has some additional properties that turn out to be particularly useful in applied stochastics.

Property 5.1

A multivariate Gaussian random field $\boldsymbol{X}(\boldsymbol{p})$ is completely characterized by its first- and second-order moments (mean, covariance, variogram, or structure functions).

Property 5.2

The *conditional* probability density for any two random vectors $\boldsymbol{X}$ and $\boldsymbol{Y}$ from a multivariate Gaussian random field is also multivariate Gaussian and is given by the well-known expression

$$f_{X|Y}(\boldsymbol{\chi}|\boldsymbol{\Psi}) = \frac{1}{(2\pi)^{\frac{m}{2}}\left|\boldsymbol{c}_{X|\Psi}\right|^{\frac{1}{2}}} e^{-\frac{1}{2}\left[\boldsymbol{\chi}-\overline{\boldsymbol{X}|\boldsymbol{\Psi}}\right]^T \boldsymbol{c}_{X|\Psi}^{-1}\left[\boldsymbol{\chi}-\overline{\boldsymbol{X}|\boldsymbol{\Psi}}\right]}, \tag{5.2}$$

where

$$\overline{\boldsymbol{X}|\boldsymbol{\Psi}} = \boldsymbol{c}_{XY}\boldsymbol{c}_Y^{-1}\left[\boldsymbol{\Psi} - \overline{\boldsymbol{Y}}\right] + \overline{\boldsymbol{X}},$$

and

$$\boldsymbol{c}_{X|\Psi} = \boldsymbol{c}_X - \boldsymbol{c}_{XY}\boldsymbol{c}_Y^{-1}\boldsymbol{c}_{XY}^T.$$

Property 5.3

A multivariate Gaussian vector random field $\boldsymbol{X}(\boldsymbol{p})$ is an independent multivariate random field if and only if it is an uncorrelated field.

Property 5.4

As regards the multipoint centered covariance of $\boldsymbol{X}(\boldsymbol{p})$ it is valid that

$$\begin{aligned}&\overline{\left[X(\boldsymbol{p}_1) - \overline{X(\boldsymbol{p}_1)}\right]\left[X(\boldsymbol{p}_2) - \overline{X(\boldsymbol{p}_2)}\right]\cdots\left[X(\boldsymbol{p}_m) - \overline{X(\boldsymbol{p}_m)}\right]}\\ &= \begin{cases} 0 & \text{if } m = \text{odd} \\ \sum\limits_{i_1,\ldots,i_m} c_X(\boldsymbol{p}_{i_1},\boldsymbol{p}_{i_2})c_X(\boldsymbol{p}_{i_3},\boldsymbol{p}_{i_4})\ldots c_X(\boldsymbol{p}_{i_{m-1}},\boldsymbol{p}_{i_m}) & \text{if } m = \text{even}\end{cases}\end{aligned} \tag{5.3}$$

for all distinct pairs of subscripts ($i_1,\ldots, i_m$) that are permutations of ($1,\ldots, m$). This is a remarkable result since it allows the computation of the multipoint covariance function of any order in terms of only the second-order covariance functions.

Property 5.5

If $X(\boldsymbol{p})$ is a Gaussian random field, then so are its derivatives in the m.s. sense. (These derivatives are defined as limits of linear combinations of Gaussian RVs.) In addition, the joint distribution of $X(\boldsymbol{p})$ and its derivative random fields are multivariate Gaussian fields, as well.

Other random fields of considerable importance in real-world applications are derived from Gaussian S/TRFs, like the *lognormal* random field defined as follows. Consider a function of the exponential form

$$Y(\boldsymbol{p}) = e^{X(\boldsymbol{p})}. \tag{5.4}$$

Its inverse is

$$X(\boldsymbol{p}) = \log Y(\boldsymbol{p}), \tag{5.5}$$

where $Y(\boldsymbol{p}) > 0$. Let $X(\boldsymbol{p})$ be a Gaussian random field with univariate PDF (obtained from Eq. 5.1 for $m = 1$)

$$f_X(\chi, \boldsymbol{p}) = \frac{1}{\sqrt{2\pi}\sigma_X(\boldsymbol{p})} e^{-\frac{(\chi - \overline{X(\boldsymbol{p})})^2}{2\sigma_X^2(\boldsymbol{p})}}.$$

Then, it can be shown that the $Y(\boldsymbol{p})$ is a lognormal random field with a univariate PDF of the form

$$f_Y(\psi, \boldsymbol{p}) = \frac{1}{\sqrt{2\pi}\sigma_X(\boldsymbol{p})\psi} e^{-\frac{(\log \psi - \overline{X(\boldsymbol{p})})^2}{2\sigma_X^2(\boldsymbol{p})}}. \tag{5.6}$$

Certain useful relationships between the corresponding means and covariances hold true for the "Gaussian–lognormal" random field pair above, as presented in Table 7.1 of Chapter II (see also, Exercise IV.6), together with the expression

$$\gamma_Y(\boldsymbol{p}, \boldsymbol{p}') = \left[1 + \sigma_Y^2(\boldsymbol{p})\right]\left[1 - e^{-\gamma_X(\boldsymbol{p}, \boldsymbol{p}')}\right], \tag{5.7}$$

relating the corresponding variograms. Lognormal S/TRFs occur in several scientific disciplines, including hydrogeology, mining, climatology, biology, geography, ecology, public health, and epidemiology.

6. CORRELATION THEORY OF COMPLEX SPATIOTEMPORAL RANDOM FIELDS

As we saw in Chapter II, in addition to the real S/TRFs (that are the vast majority of cases encountered in practice), in certain applications we may also consider complex S/TRFs. There are valid reasons for such a consideration. In spectral (Fourier) analysis (Chapter V), e.g., it is often more convenient to work with complex-valued S/TRFs rather than with real-valued S/TRFs. Before proceeding, it is worth noticing that while considerable work on the topic of complex random fields has been done separately in the purely spatial and the purely temporal domains (e.g., Yaglom, 1962; Neeser and Massey, 1991; Lajaunie and Bejaoui, 1991; De Iaco et al., 2003), not surprisingly, much less is available as regards the study of complex random fields in the composite space–time domain.

6.1 BASIC NOTIONS

I start the presentation of complex S/TRF by reminding the readers of the standard definition of such a random field in the space–time domain, already introduced in Chapter II.

Definition 6.1

A *complex* S/TRF $X(\boldsymbol{p})$ is a random field that can be decomposed as

$$X(\boldsymbol{p}) = X_R(\boldsymbol{p}) + i\, X_I(\boldsymbol{p}), \tag{6.1}$$

where the real component, $X_R(\boldsymbol{p}) \equiv \text{Re}\{X(\boldsymbol{p})\}$, and the imaginary component, $X_I(\boldsymbol{p}) \equiv \text{Im}\{X(\boldsymbol{p})\}$, are real-valued S/TRFs such that

$$\overline{|X(\boldsymbol{p})|^2} = \overline{|X_R^2(\boldsymbol{p}) + X_I^2(\boldsymbol{p})|} < \infty. \tag{6.2}$$

There are several applications where the representation of Eq. (6.1) is physically meaningful and interpretable, as is discussed in the following example.

Example 6.1

Consider electromagnetic wavefronts perturbed by atmospheric optical turbulence. The total perturbation can be represented by Eq. (6.1), where $X_R(\boldsymbol{p})$ is the log-amplitude perturbation and $X_I(\boldsymbol{p})$ is the phase perturbation.

Typically, the $X(\boldsymbol{p})$ should be specified by means of the $2m$-variate probability distribution of the real RVs $x_R(\boldsymbol{p}_1),\ldots, x_R(\boldsymbol{p}_m), x_I(\boldsymbol{p}_1),\ldots, x_I(\boldsymbol{p}_m)$ for all values of m and $\boldsymbol{p}_1,\ldots, \boldsymbol{p}_m$. Notice that, by definition, the PDF of $X(\boldsymbol{p})$ is expressed in terms of that of $X_R(\boldsymbol{p})$ and $X_I(\boldsymbol{p})$, i.e.,

$$f_X(\chi_R + i\chi_I) = f_{X_R X_I}(\chi_R, \chi_I).$$

Yet, we already know that random field specification by means of the $2m$-variate probability distribution may be an impractical approach, in which case a partial yet practically adequate specification of the complex S/TRF is obtained in terms of its statistical moments, i.e., by means of correlation theory, which is discussed next.

By virtue of Eq. (6.1), the mean value of the complex S/TRF can be defined in a straightforward manner as the linear transformation of its $X_R(\boldsymbol{p})$ and $X_I(\boldsymbol{p})$ components, in particular,

$$\begin{aligned}\overline{X(\boldsymbol{p})} &= \iint d\chi_R d\chi_I(\chi_R + i\chi_I) f_X(\chi_R, \chi_I) \\ &= \overline{X_R(\boldsymbol{p})} + i\overline{X_I(\boldsymbol{p})}.\end{aligned} \tag{6.3a–b}$$

Thus, the mean of a complex S/TRF is a complex function such that its real and imaginary parts are equal to the means of the real and imaginary parts, respectively, of the complex random field. Similarly, the variance of the complex random field $X(\boldsymbol{p})$ is

$$\begin{aligned}\sigma_X^2 &= \overline{|(X_R(\boldsymbol{p}) + i\,X_I(\boldsymbol{p})) - (\overline{X_R(\boldsymbol{p})} + i\overline{X_I(\boldsymbol{p})})|^2} = \overline{|X_R(\boldsymbol{p}) - \overline{X_R(\boldsymbol{p})}|^2} + \overline{|X_I(\boldsymbol{p}) - \overline{X_I(\boldsymbol{p})}|^2} \\ &= \sigma_{X_R}^2 + \sigma_{X_I}^2.\end{aligned} \tag{6.4}$$

Interestingly, the variance is a real function, whereas the mean Eq. (6.3b) is a complex function. In fact, the covariance is also a complex function (see discussion below).

Definition 6.2

Let $X^*(\boldsymbol{p})$ be the complex conjugate of $X(\boldsymbol{p})$.[7] The space–time (centered) *covariance* function of the complex S/TRF $X(\boldsymbol{p})$ is given by

$$\begin{aligned}c_X(\boldsymbol{p},\boldsymbol{p}') &= \overline{X'(\boldsymbol{p})X'^*(\boldsymbol{p}')} \\ &= \overline{X(\boldsymbol{p})X^*(\boldsymbol{p}')} - \overline{X(\boldsymbol{p})}\ \overline{X^*(\boldsymbol{p}')},\end{aligned} \tag{6.5a–b}$$

where $X'(\boldsymbol{p}) = X(\boldsymbol{p}) - \overline{X(\boldsymbol{p})}$ denotes random field fluctuation as usual.

In view of Definition 6.2, the following relationships between the covariance functions are valid,

$$\begin{aligned}c_X(\boldsymbol{p},\boldsymbol{p}') &= c_X^*(\boldsymbol{p}',\boldsymbol{p}) \\ c_X(\boldsymbol{p},\boldsymbol{p}') &\neq c_X(\boldsymbol{p}',\boldsymbol{p}),\end{aligned} \tag{6.6a–b}$$

which imply that the covariance function of a complex random field is Hermitian but not symmetric.[8]

[7]Complex conjugation is redundant for real-valued S/TRF.

[8]Symmetry is valid if the random field is real valued, $X^*(\boldsymbol{p}) = X(\boldsymbol{p})$.

In defining the variance and the covariance of a complex S/TRF by Eqs. (6.4) and (6.5a–b), respectively, we retained the basic features that these statistics possessed in the case of a real S/TRF, namely, the variance is a nonnegative quantity and it is the covariance of the random field with itself. On the other hand, unlike the real case, the covariance of a complex random field depends on the order of the arguments $\boldsymbol{p}$ and $\boldsymbol{p}'$, i.e., as is shown in Eqs. (6.6a–b) the covariance between points $\boldsymbol{p}$ and $\boldsymbol{p}'$ is equal to the complex conjugate of the covariance between $\boldsymbol{p}'$ and $\boldsymbol{p}$. Moreover, the covariance of a complex S/TRF is an NND function, i.e., it holds that

$$\sum_{i=1}^{m}\sum_{j=1}^{m} q_i q_j^* c_X(\boldsymbol{p}_i,\boldsymbol{p}_j) \geq 0 \tag{6.7}$$

for all nonnegative integers m, points $\boldsymbol{p}_i, \boldsymbol{p}_j \in R^{n+1}$ and complex numbers q_i. Eq. (6.7) is the variance $Var(Y)$ of an arbitrary linear combination $Y = \sum_{i=1}^{m} q_i X(\boldsymbol{p}_i)$.[9] Moreover, the classical Cauchy–Schwartz inequality relating variances with covariance functions holds in the case of complex S/TRFs too.

6.2 OTHER TYPES OF COMPLEX COVARIANCE FUNCTIONS

Interestingly, when we work with complex S/TRFs we can define more than one type of covariance functions. Indeed, it is sometimes convenient to consider a definition of the covariance function of the complex S/TRF that does not use the complex conjugate of the random field $X(\boldsymbol{p})$.

Definition 6.3

The space–time covariance function (sometimes termed the *pseudo*covariance function)[10] of a complex S/TRF is defined as

$$\begin{aligned} c_X^{\circ}(\boldsymbol{p},\boldsymbol{p}') &= \overline{X'(\boldsymbol{p})X'(\boldsymbol{p}')} = \overline{\left[X_R'(\boldsymbol{p}) + i\,X_I'(\boldsymbol{p})\right]\left[X_R'(\boldsymbol{p}') + i\,X_I'(\boldsymbol{p}')\right]} \\ &= \overline{X_R'(\boldsymbol{p})X_R'(\boldsymbol{p}')} + \overline{X_I'(\boldsymbol{p})X_I'(\boldsymbol{p}')} + i\left[\overline{X_I'(\boldsymbol{p})X_R'(\boldsymbol{p}')} + \overline{X_I'(\boldsymbol{p}')X_R'(\boldsymbol{p})}\right] \\ &= c_X^{\circ}(\boldsymbol{p}',\boldsymbol{p}). \end{aligned} \tag{6.7a–c}$$

Unlike $c_X(\boldsymbol{p},\boldsymbol{p}')$ of Eqs. (6.5a–b), the $c_X^{\circ}(\boldsymbol{p},\boldsymbol{p}')$ of Eqs. (6.7a–c) is symmetric with respect to $\boldsymbol{p}$ and $\boldsymbol{p}'$.[11] Keeping in mind that $c_{X_R}(\boldsymbol{p},\boldsymbol{p}')$, $c_{X_I}(\boldsymbol{p},\boldsymbol{p}')$, and $c_{X_I X_R}(\boldsymbol{p},\boldsymbol{p}')$ are the covariances and the cross-covariance of the real fields $X_R(\boldsymbol{p})$, $X_I(\boldsymbol{p})$, whereas $c_X(\boldsymbol{p},\boldsymbol{p}')$ is the covariance of the complex field $X(\boldsymbol{p})$, the covariances between the real and imaginary parts of the complex random field $X(\boldsymbol{p})$ can be also defined as follows,

$$\begin{aligned} c_X(\boldsymbol{p},\boldsymbol{p}') &= c_{X_R}(\boldsymbol{p},\boldsymbol{p}') + c_{X_I}(\boldsymbol{p},\boldsymbol{p}') - i[c_{X_R X_I}(\boldsymbol{p},\boldsymbol{p}') - c_{X_I X_R}(\boldsymbol{p},\boldsymbol{p}')], \\ c_X^{\circ}(\boldsymbol{p},\boldsymbol{p}') &= c_{X_R}(\boldsymbol{p},\boldsymbol{p}') - c_{X_I}(\boldsymbol{p},\boldsymbol{p}') + i[c_{X_R X_I}(\boldsymbol{p},\boldsymbol{p}') + c_{X_I X_R}(\boldsymbol{p},\boldsymbol{p}')]. \end{aligned} \tag{6.8a–b}$$

i.e., the covariance of a complex S/TRF can be expressed in terms of the covariances and cross-covariances of its real and imaginary parts. Conversely, in light of Eqs. (6.8a–b), some more useful

[9] In the case of an uncorrelated random field, $Var(Y) = \sum_{i=1}^{m} |q_i|^2 \sigma_X^2(\boldsymbol{p}_i)$.
[10] This term is used, e.g., in communication sciences.
[11] If the S/TRF is real valued, then $c_X^{\circ}(\boldsymbol{p},\boldsymbol{p}') = c_X(\boldsymbol{p},\boldsymbol{p}')$.

relationships between the covariances of the real fields $X_R(\boldsymbol{p})$, $X_I(\boldsymbol{p})$ and the covariance of the complex field $X(\boldsymbol{p})$ are derived,

$$\begin{aligned} c_{X_R}(\boldsymbol{p},\boldsymbol{p}') &= \frac{1}{2}\mathrm{Re}\{c_X(\boldsymbol{p},\boldsymbol{p}') + \mathring{c}_X(\boldsymbol{p},\boldsymbol{p}')\}, \\ c_{X_I}(\boldsymbol{p},\boldsymbol{p}') &= \frac{1}{2}\mathrm{Re}\{c_X(\boldsymbol{p},\boldsymbol{p}') - \mathring{c}_X(\boldsymbol{p},\boldsymbol{p}')\}, \\ c_{X_R X_I}(\boldsymbol{p},\boldsymbol{p}') &= \frac{1}{2}\mathrm{Im}\{\mathring{c}_X(\boldsymbol{p},\boldsymbol{p}') - c_X(\boldsymbol{p},\boldsymbol{p}')\}, \\ c_{X_I X_R}(\boldsymbol{p},\boldsymbol{p}') &= \frac{1}{2}\mathrm{Im}\{\mathring{c}_X(\boldsymbol{p},\boldsymbol{p}') + c_X(\boldsymbol{p},\boldsymbol{p}')\} \end{aligned} \tag{6.9a–d}$$

(Exercise IV.8). The special case of an uncorrelated complex S/TRF would require that the four covariance functions in the right side of Eqs. (6.8a–b) vanish. Interestingly, in view of Eq. (6.9a–d) uncorrelatedness requires that both $c_X(\boldsymbol{p},\boldsymbol{p}') = \mathring{c}_X(\boldsymbol{p},\boldsymbol{p}') = 0$.

Some noteworthy covariance types emerge in the case of wide sense (w.s.) STHS random fields. Below we use the vector convention $\Delta\boldsymbol{p} = (h_1,\ldots, h_n, h_0)$, where $(h_1,\ldots, h_n)$ denote space lags and $h_0 = \tau$ denotes time separation.

Example 6.1

For an STHS complex random field $X(\boldsymbol{p})$, the mean is constant and the covariance depends only on the vector $\Delta\boldsymbol{p}$, i.e.,

$$\overline{X(\boldsymbol{p})} = \overline{X}, \tag{6.10}$$

and

$$\begin{aligned} c_X(\boldsymbol{p},\boldsymbol{p}') &= c_X(\Delta\boldsymbol{p}) = \overline{X'(\boldsymbol{p})X'^{*}(\boldsymbol{p}-\Delta\boldsymbol{p})}, \\ \mathring{c}_X(\boldsymbol{p},\boldsymbol{p}') &= \mathring{c}_X(\Delta\boldsymbol{p}) = \overline{X'(\boldsymbol{p})X'(\boldsymbol{p}-\Delta\boldsymbol{p})}, \end{aligned} \tag{6.11a–b}$$

which are useful expressions in certain respect. Interestingly, the definition of w.s. STHS in the case of a complex S/TRF entails both covariance functions $c_X(\Delta\boldsymbol{p})$ and $\mathring{c}_X(\Delta\boldsymbol{p})$.

Some more properties of the above covariance functions are described next.

Proposition 6.1

The covariances of an STHS complex S/TRF $X(\boldsymbol{p})$ satisfy the following relationships,

$$\begin{aligned} & c_X(\Delta\boldsymbol{p}) = c_X^{*}(-\Delta\boldsymbol{p}), \\ & |c_X(\Delta\boldsymbol{p})| \leq \sigma_X^2, \\ & \mathrm{Re}\{c_X(-\Delta\boldsymbol{p})\} = \mathrm{Re}\{c_X(\Delta\boldsymbol{p})\}, \\ & \mathrm{Im}\{c_X(-\Delta\boldsymbol{p})\} = -\mathrm{Im}\{c_X(\Delta\boldsymbol{p})\}, \\ & \mathring{c}_X(\Delta\boldsymbol{p}) = \mathring{c}_X(-\Delta\boldsymbol{p}). \end{aligned} \tag{6.12a–e}$$

The properties described in Eqs. (6.12a–e) are similar in form to the respective properties of real-valued S/TRFs. In more words, the real part of $c_X(\Delta\boldsymbol{p})$ is even in $\Delta\boldsymbol{p}$, and the imaginary part is odd in $\Delta\boldsymbol{p}$. The $c_X^{\circ}(\Delta\boldsymbol{p})$ is even in $\Delta\boldsymbol{p}$, but not the $c_X(\Delta\boldsymbol{p})$, in general. In the case that $c_X(\Delta\boldsymbol{p})$ is even,

$$\begin{aligned} c_{X_R}(\Delta\boldsymbol{p}) &= \frac{1}{2}\left[c_X(\Delta\boldsymbol{p}) + \mathrm{Re}\{c_X^{\circ}(\Delta\boldsymbol{p})\}\right], \\ c_{X_I}(\Delta\boldsymbol{p}) &= \frac{1}{2}\left[c_X(\Delta\boldsymbol{p}) - \mathrm{Re}\{c_X^{\circ}(\Delta\boldsymbol{p})\}\right], \\ c_{X_R X_I}(\Delta\boldsymbol{p}) &= c_{X_I X_R}(\Delta\boldsymbol{p}) = \frac{1}{2}\mathrm{Im}\{c_X^{\circ}(\Delta\boldsymbol{p})\}, \end{aligned} \tag{6.13a–c}$$

since if $c_X(\Delta\boldsymbol{p})$ is even, then it is real too, i.e., $c_X(\Delta\boldsymbol{p}) = \mathrm{Re}\{c_X(\Delta\boldsymbol{p})\}$ and $\mathrm{Im}\{c_X(\Delta\boldsymbol{p})\} = 0$ (Exercise IV.9). Otherwise said, the $X_R(\boldsymbol{p})$ and $X_I(\boldsymbol{p})$ of the STHS complex $X(\boldsymbol{p})$ are themselves STHS and cross-STHS.

Example 6.2

In the special case of a space–time isostationary (STIS)[12] random field $X(\boldsymbol{p})$, its covariance is even, and, hence, real; and, from Eqs. (6.13a–c), we find that the $X_R(\boldsymbol{p})$, $X_I(\boldsymbol{p})$ are STIS too, and so is their cross-correlation.

Similar result can be obtained in terms of the space–time variogram function, which in the case of complex S/TRFs can be written as

$$\gamma_X(\boldsymbol{p},\boldsymbol{p}') = \frac{1}{2}\left[c_X(\boldsymbol{p},\boldsymbol{p}) + c_X(\boldsymbol{p}',\boldsymbol{p}') - c_X(\boldsymbol{p},\boldsymbol{p}') - c_X^{*}(\boldsymbol{p},\boldsymbol{p}')\right], \tag{6.14}$$

where $c_X(\boldsymbol{p},\boldsymbol{p}')$ is given by Eqs. (6.5a–b), and $c_X^{*}(\boldsymbol{p},\boldsymbol{p}')$ is its complex conjugate.

The preceding analysis can be extended without much technical difficulty to include the case of *vector* complex S/TRFs. Specifically, if $X_l(\boldsymbol{p})$ and $X_{l'}(\boldsymbol{p})$ are two complex S/TRFs, their cross-covariance function is given by

$$c_{X_l X_{l'}}(\boldsymbol{p},\boldsymbol{p}') = \overline{X_l'(\boldsymbol{p})X_{l'}'^{*}(\boldsymbol{p}')} = \overline{X_l(\boldsymbol{p})X_{l'}^{*}(\boldsymbol{p}')} - \overline{X_l(\boldsymbol{p})}\;\overline{X_{l'}^{*}(\boldsymbol{p}')}, \tag{6.15}$$

which does not exceed the product of the corresponding variances, i.e.,

$$|c_{X_l X_{l'}}(\boldsymbol{p},\boldsymbol{p}')|^2 \le \sigma_{X_l}^2(\boldsymbol{p})\sigma_{X_{l'}}^2(\boldsymbol{p}').$$

It is noteworthy that the $c_{X_l X_{l'}}(\boldsymbol{p},\boldsymbol{p}')$ is neither Hermitian nor NND. Instead, it is such that the property $c_{X_l X_{l'}}^{*}(\boldsymbol{p},\boldsymbol{p}') = c_{X_{l'} X_l}(\boldsymbol{p}',\boldsymbol{p})$ holds.

In the same context, along with statistical moments of the first- and second-order, we can also consider higher-order moments consisting of several space–time points. Specifically, the multipoint and noncentered covariance of the complex random field $X(\boldsymbol{p})$ is generally given by

$$c_{X(m,k)}(\boldsymbol{p}_1,\ldots,\boldsymbol{p}_{m+k}) = \overline{X(\boldsymbol{p}_1)\cdots X(\boldsymbol{p}_m)X^{*}(\boldsymbol{p}_{m+1})\cdots X^{*}(\boldsymbol{p}_{m+k})}, \tag{6.16}$$

where the $X(\boldsymbol{p})$ appears m times, and its complex conjugate $X^{*}(\boldsymbol{p})$ appears k times.

---

[12]STIS random fields were briefly introduced in Section 7.4 of Chapter II. A detailed presentation is given in Chapter VIII.

Remark 6.1

By virtue of Eq. (6.16) several notable properties of $c_{X(m,k)}$ can be proven, including that (1) the $c_{X(m,k)}$ does not change for any permutations within the group of indexes $\{1,\ldots, m\}$ and $\{m+1,\ldots, m+k\}$, (2) the complex conjugate of $c_{X(m,k)}$ is such that $c^*_{X(m,k)}(\boldsymbol{p}_1,\ldots,\boldsymbol{p}_{m+k}) = c_{X(k,m)}(\boldsymbol{p}_{m+k},\ldots,\boldsymbol{p}_1)$, and (3) the $c_{X(k,k)}(\boldsymbol{p}_1,\ldots, \boldsymbol{p}_k; \boldsymbol{p}_k,\ldots, \boldsymbol{p}_1)$, in particular, is a nonnegative function.

6.3 GAUSSIAN COMPLEX SPATIOTEMPORAL RANDOM FIELDS

In this section, I shall focus on the case of a complex Gaussian random field $X(\boldsymbol{p})$. In this case, the summation

$$\sum_{i=1}^{m} a_i X(\boldsymbol{p}_i) \tag{6.17}$$

is also a complex Gaussian random field for all m and (complex) numbers a_i. Accordingly, $X(\boldsymbol{p})$ is determined by its up to second-order statistics, $\overline{X_{R,i}(\boldsymbol{p})}$, $\overline{X_{I,i}(\boldsymbol{p})}$, $\sigma^2_{X_{R,i}}(\boldsymbol{p})$, $\sigma^2_{X_{I,i}}(\boldsymbol{p})$, and $c_{X_{R,i},X_{I,i}}(\boldsymbol{p},\boldsymbol{p}')$, $i = 1,\ldots, m$.

Example 6.3

Consider the complex S/TRF defined as the integral

$$X(\boldsymbol{p}) = \int d\boldsymbol{u}\ e^{2\pi i \boldsymbol{p}\cdot\boldsymbol{u}} \eta(\boldsymbol{u}), \tag{6.18}$$

where η is a zero-mean *Gaussian white noise* (GWN) with covariance $c_\eta(\boldsymbol{p},\boldsymbol{p}') = \delta(\boldsymbol{p}-\boldsymbol{p}')$. The $X(\boldsymbol{p})$ is a Gaussian complex random field with covariance $c_X(\boldsymbol{p},\boldsymbol{p}')$ given by

$$c_X(\boldsymbol{p},\boldsymbol{p}') = \overline{X(\boldsymbol{p})X^*(\boldsymbol{p}')} = \delta(\boldsymbol{p}-\boldsymbol{p}'), \tag{6.19a}$$

and pseudocovariance $c^{\circ}_X(\boldsymbol{p},\boldsymbol{p}')$ given by

$$c^{\circ}_X(\boldsymbol{p},\boldsymbol{p}') = \overline{X(\boldsymbol{p})X(\boldsymbol{p}')} = \delta(\boldsymbol{p}+\boldsymbol{p}'). \tag{6.19b}$$

The two covariance functions are clearly different. In fact, the $c_X(\boldsymbol{p},\boldsymbol{p}')$ is here an STHS covariance, whereas $c^{\circ}_X(\boldsymbol{p},\boldsymbol{p}')$ is not.

The analysis in Example 6.3 can reach an interesting conclusion by taking advantage of the properties of a GWN random field.

Example 6.4

Let us consider a complex random field $Y(\boldsymbol{p})$ with real part

$$Y_R(\boldsymbol{p}) = \frac{1}{2}[\eta(\boldsymbol{p}) + \eta(-\boldsymbol{p})],$$

and imaginary part

$$Y_I(\boldsymbol{p}) = \frac{1}{2}[\eta(\boldsymbol{p}) - \eta(-\boldsymbol{p})],$$

where η is a GWN as in the Example 6.3. Then, the random field $Y(\boldsymbol{p})$ has the same covariance functions as the random field $X(\boldsymbol{p})$ of Example 6.3 (also, Exercise IV.10).

Let me conclude with a note on the relationship between strict sense (s.s.) and w.s. homostationarity. While in the case of a real Gaussian $X(\boldsymbol{p})$, s.s. STHS implies w.s. STHS, in the case of complex $X(\boldsymbol{p})$ it is also required that

$$c_X^{\circ}(\boldsymbol{p},\boldsymbol{p}') = c_X^{\circ}(\boldsymbol{p}-\boldsymbol{p}'). \tag{6.20}$$

This is a useful result, since it implies that when the $X(\boldsymbol{p})$ is a real S/TRF, then it is valid that $c_X^{\circ}(\boldsymbol{p},\boldsymbol{p}') = c_X(\boldsymbol{p},\boldsymbol{p}')$, and, also, the identity holds, w.s. STHS $\equiv$ s.s. STHS, which considerably simplifies the analysis.

6.4 COMPLEX-VALUED VERSUS REAL-VALUED COVARIANCE FUNCTIONS OF SPACE–TIME HOMOSTATIONARY RANDOM FIELDS

The following point should be stressed: in the case of complex STHS random fields, certain of the commonly used definitions of the space–time covariance function may differ from each other. For illustration, consider the following result.

Corollary 6.1

For a complex random field, $X(\boldsymbol{p})$, that is STHS the relationships below hold between the covariance functions,

$$c_X(\Delta\boldsymbol{p})\begin{cases} = c_X(\boldsymbol{p},\boldsymbol{p}-\Delta\boldsymbol{p}) = c_X(\boldsymbol{p}+\Delta\boldsymbol{p},\boldsymbol{p}) = c_X^{*}(\boldsymbol{p},\boldsymbol{p}+\Delta\boldsymbol{p}), \\ \neq c_X(\boldsymbol{p},\boldsymbol{p}+\Delta\boldsymbol{p}), \end{cases} \tag{6.21a–b}$$

where $c_X(\boldsymbol{p}+\Delta\boldsymbol{p},\boldsymbol{p}) = \overline{X'(\boldsymbol{p}+\Delta\boldsymbol{p})X'^{*}(\boldsymbol{p})}$, $c_X(\boldsymbol{p},\boldsymbol{p}-\Delta\boldsymbol{p}) = \overline{X'(\boldsymbol{p})X'^{*}(\boldsymbol{p}-\Delta\boldsymbol{p})}$, and $c_X^{*}(\boldsymbol{p},\boldsymbol{p}+\Delta\boldsymbol{p}) = \left[\overline{X'(\boldsymbol{p})X'^{*}(\boldsymbol{p}+\Delta\boldsymbol{p})}\right]^{*}$.

Eqs. (6.21a–b) are easily proven with the help of standard results regarding the cross-covariances of real S/TRFs, namely,

$$\begin{aligned} c_{X_RX_I}(\Delta\boldsymbol{p}_{ij}) &= c_{X_IX_R}(\Delta\boldsymbol{p}_{ji}) = c_{X_IX_R}(-\Delta\boldsymbol{p}_{ij}) \\ &\neq c_{X_IX_R}(\Delta\boldsymbol{p}_{ij}), \\ c_{X_RX_I}(\Delta\boldsymbol{p}_{ij}) &\neq c_{X_RX_I}(\Delta\boldsymbol{p}_{ji}). \end{aligned} \tag{6.22a–c}$$

Further, in this case we also find that for a complex S/TRF the following relationships hold among the corresponding covariance functions,

$$\begin{aligned} c_X(\boldsymbol{p},\boldsymbol{p}+\Delta\boldsymbol{p}) &= c_{X_R}(\Delta\boldsymbol{p}) + c_{X_I}(\Delta\boldsymbol{p}) - i[c_{X_RX_I}(\Delta\boldsymbol{p}) - c_{X_IX_R}(\Delta\boldsymbol{p})], \\ c_X(\boldsymbol{p}+\Delta\boldsymbol{p},\boldsymbol{p}) &= c_{X_R}(\Delta\boldsymbol{p}) + c_{X_I}(\Delta\boldsymbol{p}) + i[c_{X_RX_I}(\Delta\boldsymbol{p}) - c_{X_IX_R}(\Delta\boldsymbol{p})] \\ &\neq c_X(\boldsymbol{p},\boldsymbol{p}+\Delta\boldsymbol{p}), \\ c_X(\boldsymbol{p},\boldsymbol{p}-\Delta\boldsymbol{p}) &= c_X(\boldsymbol{p}+\Delta\boldsymbol{p},\boldsymbol{p}) \neq c_X(\boldsymbol{p},\boldsymbol{p}+\Delta\boldsymbol{p}). \end{aligned} \tag{6.23a–d}$$

In the above cases, the covariance function c_X is a function of the component covariance functions c_{X_R} and c_{X_I}, as well as the cross-covariance functions $c_{X_RX_I}$ and $c_{X_IX_R}$ (Exercise IV.11).

6.5 SOME METHODOLOGICAL CONSIDERATIONS

Section 6 was devoted to the presentation of complex S/TRFs. The reasons are methodological and practical: complex-valued rather than real-valued S/TRFs sometimes may offer a preferable way to study space–time phenomena, because in these cases it turns out that complex-valued random fields are (1) either mathematically more convenient or (2) they better represent the physical knowledge about the actual phenomena. While the justification of possibility (1) is mostly technical, that of possibility (2) often requires that we delve deeply into data.

In practice, one may need to study the two component random fields $X_R(\boldsymbol{p})$ and $X_I(\boldsymbol{p})$ individually, and then make inferences about the complex random field $X(\boldsymbol{p})$ of Eq. (6.1). As in previous cases, each individual study focuses on the joint space–time analysis of $X_R(\boldsymbol{p})$ and $X_I(\boldsymbol{p})$ in the R^{n+1} or in the $R^{n,1}$ domain. Alternatively, a so-called separate conditional approach may be considered if it fits adequately to the available empirical evidence. This approach includes the combination of spatial analyses considered separately for each time instant (i.e., time-conditional spatial analyses), or the combination of temporal analyses considered separately for each space location (i.e., space-conditional temporal analyses). It is then appropriate to distinguish between the two conditional models depending on whether we are interested in the time-dependent pattern of the phenomenon allowing for spatial correlation between time series or in the space-dependent pattern borrowing information from spatial processes that are close in time. Yet, although the choice of a joint versus a separate conditional space–time analysis will depend on the particular phenomenon, in most cases the former is a more realistic approach.

Which one of the covariance functions introduced above $\left(c_X, c_X^*, c_X^\circ\right)$ we should use will obviously depend on the particular situation we encounter in applications. In some applications, e.g., it may be more realistic to use covariance functions that include an even covariance component c_{X_R} and an odd component c_{X_I}. Some authors (Lajaunie and Bejaoui, 1991) have used an approach based on the Radon–Nikodym theorem that derives the imaginary covariance component of a complex random field from the real component. Spectral representations of complex S/TRFs can be also derived under certain restrictive assumptions (e.g., space–time homostationarity) that allow the direct decomposition of the covariance function in to a real and an imaginary part. In the case of a spatial isotropic $X(\boldsymbol{p})$, the $X_R(\boldsymbol{p})$ and $X_I(\boldsymbol{p})$ are both isotropic and isotropically interconnected. Lastly, spectral representations of random fields will be considered in subsequent chapters.

CHAPTER

TRANSFORMATIONS OF SPATIOTEMPORAL RANDOM FIELDS

CHAPTER OUTLINE

1. INTRODUCTION

Generally, the aim of mathematical *transformations* is to convey the study of a phenomenon or the solution of a problem from the original domain to a new domain that either offers a deeper understanding of the physical phenomenon or makes the solution of complex problems easier, or both. Such problems include, but are not limited to, the modeling of spatiotemporally distributed earth and atmospheric data, the mapping of human exposure and epidemic spread, and the simulation of environmental and ecological processes in multiple scales.

When the notion of transformation as described above is applied in the context of the spatiotemporal random field (S/TRF) theory, the distinction between the following broad transformation categories is methodologically appropriate:

Category I: *Fourier* transformations (*FTs*).
Category II: *Space* transformations (*STs*).
Category III: *Traveling* transformations (*TTs*).

Spatiotemporal Random Fields. http://dx.doi.org/10.1016/B978-0-12-803012-7.00005-2

Two points should be stressed at the outset concerning the transformation categories above. First, these categories share the same basic line of thought:

*T*ransform−*S*olve−*I*nvert (*TSI*).

The TSI line of thought basically reduces the problem into a simpler one, i.e., the reduction transforms the hardness out of the original problem; the problem is solved in its transformed form; and the solution is inverted (back-transformed) to a solution of the original problem. Second, the criterion on which the above categorization relies is the distinct goal of its transformation, namely, (*a*) decomposition of the original attribute into its building blocks (FT), (*b*) reduction of the original attribute space into a unidimensional one (ST), and (*c*) conversion of the original attribute space−time domain into a lower-dimensionality domain in which time is mixed with space (TT).

More specifically, the goal of the FT is to introduce a change of perspective from "what one sees" to "how was it made." The basic idea is that a better understanding of a natural attribute varying in space−time is gained by decomposing it into its building blocks (these building blocks can be, e.g., sinusoidal functions that are special in the sense that they possess an explicit physical meaning in terms of frequencies[1]), which are more accessible for analysis and processing. A set of filters is accordingly introduced that decomposes the attribute into the corresponding frequencies that make it up, so that any space−time varying attribute can be transformed into a different domain called the *spectral* or *frequency* domain (Papoulis, 1962). FT collects information in different ways and may allow measurements that would be very difficult in the original (physical) domain. Moreover, the FT makes it easy to go forward and backward from the physical domain to the frequency domain.

While in the case of FT the variability in space and time is mapped into the spectral domain, the goal of the ST mapping is to involve the projections of the n-dimensional function onto planes orthogonal to transform lines determined by space−direction vectors. Thus, the intuitive idea behind ST is to simplify the study of a natural attribute that takes place in several spatial dimensions by conveying its study to a suitable one-dimensional space. This conveyance has both substance and depth and is mathematically established in terms of the fundamental *Radon* operator (Radon, 1917; Helgason, 1980)[2] that acts on the random field representing the attribute of interest. As we pass from the original $R^{n,1}$ domain into the $R^{1,1}$ domain via the ST, we find that the mathematics simplify considerably but without loss of information. The usefulness of ST in the study of natural phenomena and systems represented by random field models has been discussed extensively in the relevant literature (e.g., Christakos, 1984a, Christakos and Panagopoulos, 1992, Christakos and Hristopulos, 1994; Kolovos et al., 2004).

Lastly, the goal of the mapping introduced by the TT is to project the study of an attribute or the solution of a problem from the original space−time domain onto a lower-dimensionality pseudospatial traveling domain in which space and time are intrinsically dependent and specified in a self-consistent manner (Christakos et al., 2017a). As it turns out, this blending of space and time has certain attractive features, including reduced dimensionality, improved representation of attribute variation in terms of the spatiotemporal variability functions (S/TVFs), higher attribute estimation accuracy at unsampled space−time points, and lower computational cost. The ease of use of TT and its intuitive description

[1]Although in the composite space−time domain it is more precise to talk about time frequencies and space wavevectors, for simplicity reasons sometimes scientists refer to frequencies in both time and space.

[2]Interestingly, in geostatistics and spatial statistics the Radon operation was rediscovered 50 years after Radon's original work and was termed "montée" (Matheron, 1965; Gneiting, 1999).

make it an efficient transformation technique. The simplest case of TT uses a linear combination of space and time, although more involved relationships can be considered, if needed.

Although the main conceptualization frameworks and several important properties of the above transformation categories I–III are presented in this chapter, due to their significance and wide applicability, case-specific investigations of each transformation in the appropriate physical setting are presented in other parts of the book. One of the reasons for this presentation approach is that the specific mathematical formulation of a transformation may depend on the features of the random field model on which the transformation is applied. Space–time homostationary (STHS)[3] random fields, e.g., do not admit an FT based on the standard Riemann integral representation, which is, though, valid for other types of space–time heterogeneous random fields (for STHS, instead, the generalized FT is used).

2. FOURIER TRANSFORMATION

In the applied stochastics context, several interesting notions can be introduced and valuable results can be derived in the spectral domain by applying the space–time FT. This includes a series of spectral random field representations and characteristics that are discussed in this chapter. Unless stated otherwise, the S/TRFs to be considered are assumed to be space–time heterogeneous, in general. FT analysis may be considered in the R^{n+1} or in the $R^{n,1}$ domain. Analysis in the $R^{n,1}$ domain considers space and time as formally separate but physically linked components of a Cartesian product. Although analysis in the R^{n+1} domain formally treats time as an additional dimension, it does not neglect the fact that, as required by the phenomenon of interest, time is not merely an additional dimension (coordinate), but it needs to be physically interpreted. As usual, integrations are generally carried out over the whole range of a variable, unless other limits are indicated. Lastly, although certain of the continuity conditions involve notions to be formally defined in following chapters, yet these notions are instrumental in the description of some of the material of the present sections, which is why I briefly introduce them here, when necessary.

2.1 CHARACTERISTIC FUNCTIONS

Many basic concepts of stochastic analysis are defined in terms of FT. Let me start with such a concept, namely, the probability density function (PDF) $f_X(\chi)$ of an S/TRF $X(\boldsymbol{p})$. As I already noticed in Section 2 of Chapter II, a spectral characteristic of the S/TRF is the so-called *characteristic function (CF)* defined by Eqs. (2.14a–b) of Chapter II as follows,

$$\phi_X(\boldsymbol{u}) = FT[f_X(\boldsymbol{\chi})] = \widetilde{f}_X(\boldsymbol{u}), \tag{2.1}$$

where $\boldsymbol{u} = (u_1,\ldots, u_m)$, $\boldsymbol{X} = (x_1,\ldots, x_m)$, that is, the CF $\phi_X(\boldsymbol{u})$ is formally the FT of $f_X(\boldsymbol{\chi})$ so that $\varphi_X(\boldsymbol{u}) = \overline{e^{i\,\boldsymbol{u}\cdot\boldsymbol{X}}}$. The symmetry and consistency conditions of the probability measure, see Eqs. (2.4a–b) of Chapter II, are satisfied if for all $m \geq 2$, $\boldsymbol{u}$ and $\boldsymbol{X}$, the corresponding CFs satisfy the conditions

$$\phi_X(u_1, \ldots, u_m) = \phi_X(u_{i_1}, \ldots, u_{i_m}), \tag{2.2a}$$

[3]STHS random fields were briefly introduced in Section 7.4 of Chapter II. A detailed presentation is given in Chapter VII. The readers may recall that in the case of STHS random fields, the mean is constant and the covariance depends only on the spatial lag and temporal separation, i.e., $c_X(\boldsymbol{p},\boldsymbol{p}') = c_X(\boldsymbol{p} - \boldsymbol{p}')$, or $c_X(\boldsymbol{s}, \boldsymbol{s}',t,t') = c_X(\boldsymbol{s} - \boldsymbol{s}',t - t')$.

where $i_1, \ldots, i_m$ is a permutation of the indices $1, \ldots, m$, and

$$\phi_X(u_1, \ldots, u_{m-1}) = \phi_X(u_1, \ldots, u_{m-1}, 0) \tag{2.2b}$$

for any permutation.

Useful single-point and multipoint moments of any order can be derived from the CF, such as, respectively[4]

$$\overline{X^{\rho}(\boldsymbol{p})} = i^{-\rho}\left[\frac{\partial^{\rho}}{\partial u^{\rho}}\phi_X(u)\right]_{u=0}, \tag{2.3}$$

and

$$\overline{X^{\rho_1}(\boldsymbol{p}_1)\cdots X^{\rho_m}(\boldsymbol{p}_m)} = i^{-\rho}\left[\frac{\partial^{\rho}}{\partial \boldsymbol{u}^{\rho}}\phi_X(\boldsymbol{u})\right]_{\boldsymbol{u}=0}, \tag{2.4}$$

where $\rho = \sum_{k=1}^{m} \rho_k$, and $\partial \boldsymbol{u}^{\rho} = \prod_{k=1}^{m} \partial u_k^{\rho_k}$. Additional useful results can be derived by applying FT directly on the S/TRF $X(\boldsymbol{p})$ and its statistical properties. This is the subject of the next section.

2.2 HARMONIZABLE RANDOM FIELDS AND COVARIANCE FUNCTIONS

To start with, a space–time heterogeneous, in general, S/TRF can be defined in the spectral domain as follows (without loss of generality, I usually assume zero-mean random fields).

Definition 2.1

An S/TRF $X(\boldsymbol{p})$ is said to be *harmonizable* if there exists a random field $\widetilde{X}(\boldsymbol{w})$, $\boldsymbol{w} = (w_1, \ldots, w_n, w_0) \in R^{n+1}$, such that

$$X(\boldsymbol{p}) = \int d\boldsymbol{w}\, e^{i\boldsymbol{w}\cdot\boldsymbol{p}}\widetilde{X}(\boldsymbol{w}), \tag{2.5}$$

where Eq. (2.5) is a Riemann integral representation of $X(\boldsymbol{p})$.

By applying some basic results of the theory of stochastic integration, it can be shown that the spectral representation

$$\widetilde{X}(\boldsymbol{w}) = \frac{1}{(2\pi)^{\frac{n+1}{2}}}\int d\boldsymbol{p}\, e^{-i\boldsymbol{w}\cdot\boldsymbol{p}}X(\boldsymbol{p}) \tag{2.6}$$

of $X(\boldsymbol{p})$ exists in the mean square (m.s.) sense if and only if

$$\left|\overline{\widetilde{X}(\boldsymbol{w})\widetilde{X}^*(\boldsymbol{w}')}\right| = \left|\frac{1}{(2\pi)^{n+1}}\iint d\boldsymbol{p}\, d\boldsymbol{p}' e^{-i(\boldsymbol{w}\cdot\boldsymbol{p}-\boldsymbol{w}'\cdot\boldsymbol{p}')}c_X(\boldsymbol{p},\boldsymbol{p}')\right| < \infty \tag{2.7}$$

for all $\boldsymbol{p}, \boldsymbol{p}' \in R^{n+1}$, where $\widetilde{X}^*$ denotes the complex conjugate of the random field $\widetilde{X}$. Since most natural attributes studied by S/TRFs are real functions of real arguments, when it is mathematically convenient, I use complex conjugates.

[4]See earlier, Eqs. (2.7a) and (2.21) of Chapter IV.

On the basis of the foregoing, besides the characterization of S/TRFs on the basis of statistical space−time moments of order up to 2, an equivalent characterization can be made in terms of spectral moments of order up to 2. These are the $n+1$-fold FT of the statistical moments. As happened with the statistical moments, the spectral moments are assumed to be continuous functions in R^{n+1}.

Definition 2.2

The *spectral density function* (*SDF*) of an S/TRF $X(\boldsymbol{p})$ in R^{n+1} is defined by

$$\widetilde{c}_X(\boldsymbol{w},\boldsymbol{w}') = \overline{\widetilde{X}(\boldsymbol{w})\widetilde{X}^*(\boldsymbol{w}')} = (2\pi)^{-(n+1)} \iint d\boldsymbol{p}\, d\boldsymbol{p}' e^{-i(\boldsymbol{w}\cdot\boldsymbol{p}-\boldsymbol{w}'\cdot\boldsymbol{p}')} c_X(\boldsymbol{p},\boldsymbol{p}'), \tag{2.8}$$

assuming that the integral exists.

The SDF turns out to be a very useful tool in the investigation of certain covariance properties. In the FT context, the $\widetilde{c}_X$ in Eq. (2.8) may be viewed as the covariance function of the random field $\widetilde{X}(\boldsymbol{w})$ and is such that

$$\int_S \int_S d\boldsymbol{w}\, d\boldsymbol{w}' \widetilde{c}_X(\boldsymbol{w},\boldsymbol{w}') \geq 0$$

for all $S \in R^{n+1}$. By virtue of the above definition, c_X can be also defined as the inverse FT of the SDF, that is,

$$c_X(\boldsymbol{p},\boldsymbol{p}') = \iint d\boldsymbol{w}\, d\boldsymbol{w}' e^{i(\boldsymbol{w}\cdot\boldsymbol{p}-\boldsymbol{w}'\cdot\boldsymbol{p}')} \widetilde{c}_X(\boldsymbol{w},\boldsymbol{w}'). \tag{2.9}$$

Definition 2.3

When a space−time covariance function can take the form of Eq. (2.9), it is said to be a *harmonizable* covariance function.

The $X(\boldsymbol{p})$, $\widetilde{X}(\boldsymbol{w})$ form an FT pair, and as such they have a number of interesting properties listed in Table 2.1. These properties demonstrate that the FT can simplify mathematical calculations significantly (e.g., differentiation and convolution operations in the physical domain turned into mere algebraic operations in the frequency domain).

Table 2.1 Basic Properties of Random Field Fourier Transform (FT) Pairs (R^{n+1} or $R^{n,1}$)

Property	$X(\boldsymbol{p}) \overset{FT}{\longleftrightarrow} \widetilde{X}(\boldsymbol{w})$
Superposition	$\sum_{i=1}^{k} a_i X_i(\boldsymbol{p}) \overset{FT}{\longleftrightarrow} \sum_{i=1}^{k} a_i \widetilde{X}_i(\boldsymbol{w})$
Shifting	$X(\boldsymbol{p} \pm \Delta\boldsymbol{p}) \overset{FT}{\longleftrightarrow} e^{\pm i\boldsymbol{w}\cdot\Delta\boldsymbol{p}} \widetilde{X}(\boldsymbol{w})$
Stretching or scaling	$X(A\boldsymbol{p}) \overset{FT}{\longleftrightarrow} \frac{1}{\lvert \det A \rvert} \widetilde{X}(A^{-T}\boldsymbol{w})$
Convolution	$(X_1 * X_2)(\boldsymbol{p}) \overset{FT}{\longleftrightarrow} \widetilde{X}_1(\boldsymbol{w})\widetilde{X}_2(\boldsymbol{w})$
Differentiation	$X^{(\nu,\mu)}(\boldsymbol{p}) \overset{FT}{\longleftrightarrow} (-1)^{\mu} i^{\nu+\mu} \prod_{i=1}^{\nu} w_i^{\nu_i} w_0^{\mu} \widetilde{X}(\boldsymbol{w})$

As was noticed in Section 2 of Chapter II, in applications it is often more convenient to work in the $R^{n,1}$ domain, explicitly distinguishing between the physical characteristics of space and those of time, in which case Eqs. (2.5) and (2.6) are replaced by

$$\begin{aligned} X(s,t) &= \iint dkd\omega \; e^{i(k\cdot s\pm\omega t)}\widetilde{X}(k,\omega), \\ \widetilde{X}(k,\omega) &= \frac{1}{(2\pi)^{\frac{n+1}{2}}}\iint dsdt \; e^{-i(k\cdot s\pm\omega t)}X(s,t), \end{aligned} \tag{2.10a–b}$$

where the combined spatial frequency (wave) vector $\boldsymbol{k} = (k_1,\ldots, k_n)$ and temporal frequency scalar ω replace the spectral vector $\boldsymbol{w}$ of Eqs. (2.5) and (2.6). In the notation $e^{i(k\cdot s\pm\omega\tau)}$ the "+" sign is mainly used in the statistics literature, whereas the "−" is used in the physics literature, *inter alia*, due to its space−time metric associations (see, discussion in Remark 1.3 of Chapter III). Therefore, herein I will primarily use the second convention. The foregoing developments are linked mathematically to the theory of random functionals to be presented in Chapter XII. In physical applications, noticeably in ocean sciences, for the spectral representation of $X(\boldsymbol{s},t)$ to exist, it is appropriate to express it in terms of the generalized FT (Chapters XII and XIII).

Certain important special classes of S/TRFs, such as the class of STHS random fields, do not possess an FT in the earlier sense, and therefore, they do not admit the Riemann integral representation of Eqs. (2.5) or (2.10a). This difficulty has been addressed in certain applied sciences by assuming that for the spectral representation of STHS random fields to exist in the conventional sense, it is assumed to be STHS in a large finite domain but to vanish quickly outside it, and the transition to the infinite domain is considered following stochastic expectation. Another way is to assume that the spectral random field representations of Eqs. (2.10a−b) exist in the m.s. sense.

More generally, and perhaps mathematically more rigorously, it can be shown that all classes of random fields, including the STHS ones, admit a *Fourier−Stieltjes* representation such as

$$X(\boldsymbol{p}) = \int d\mathcal{M}_X(\boldsymbol{w})\, e^{i\boldsymbol{w}\cdot\boldsymbol{p}}, \tag{2.11}$$

where $\mathcal{M}_X(\boldsymbol{w})$ is a random field that is not necessarily differentiable.[5] The covariance function of the S/TRF of Eq. (2.11) can be written as

$$\begin{aligned} c_X(\boldsymbol{p},\boldsymbol{p}') &= \overline{\int d\mathcal{M}_X(\boldsymbol{w})e^{i\boldsymbol{w}\cdot\boldsymbol{p}} \int d\mathcal{M}^*_X(\boldsymbol{w}')e^{-i\boldsymbol{w}'\cdot\boldsymbol{p}'}} \\ &= \iint \overline{d\mathcal{M}_X(\boldsymbol{w})d\mathcal{M}^*_X(\boldsymbol{w}')}\; e^{i(\boldsymbol{w}\cdot\boldsymbol{p}-\boldsymbol{w}'\cdot\boldsymbol{p}')} \\ &= \iint dQ_X(\boldsymbol{w},\boldsymbol{w}')\; e^{i(\boldsymbol{w}\cdot\boldsymbol{p}-\boldsymbol{w}'\cdot\boldsymbol{p}')}, \end{aligned} \tag{2.12a–c}$$

where Q_X is the so-called *spectral cumulative function* (SCF) of $X(\boldsymbol{p})$ that is not necessarily differentiable.[6] By comparing Eqs. (2.9) and (2.12a), I get

$$\widetilde{c}_X(\boldsymbol{w},\boldsymbol{w}')dw\, dw' = \overline{d\mathcal{M}_X(\boldsymbol{w})d\mathcal{M}^*_X(\boldsymbol{w}')},$$

[5]If $\mathcal{M}_X$ is differentiable, the Fourier−Stieltjes integral reduces to the Riemann integral of Eq. (2.5). Yet, the $\mathcal{M}_X(\boldsymbol{w})$ linked to STHS random fields do not satisfy this property (Chapter VII).

[6]$Q_X(\boldsymbol{w}, \boldsymbol{w}')$ can be also viewed as the covariance of the random field $\mathcal{M}_X(\boldsymbol{w})$.

and in the case that Q_X is indeed differentiable, I can write

$$\widetilde{c}_X(\boldsymbol{w},\boldsymbol{w}') = \frac{\partial^{2(n+1)}}{\partial \boldsymbol{w}^{n+1}\partial \boldsymbol{w}'^{\,n+1}} Q_X(\boldsymbol{w},\boldsymbol{w}'),$$

where $\partial \boldsymbol{w}^{n+1} = \prod_{i=1}^{n,0} \partial w_i$, $\partial \boldsymbol{w}'^{n+1} = \prod_{i=1}^{n,0} \partial w_i'$, and then Eq. (2.12c) coincides with Eq. (2.9). Actually Proposition 2.1 later in this section establishes an important link between an S/TRF and its covariance (Loeve, 1953; Exercise V.1).

In the $R^{n,1}$ domain the above spectral representation of the random field, Eq. (2.11), is replaced by the equivalent expressions

$$X(\boldsymbol{s},t) = \int d\mathcal{M}_X(\boldsymbol{k},\omega)\, e^{i(\boldsymbol{k}\cdot\boldsymbol{s}-\omega t)}. \tag{2.13}$$

And, similarly for the spectral representations of the covariance function, Eq. (2.12b), I can write

$$\begin{aligned} c_X(\boldsymbol{s},\boldsymbol{s}',t,t') &= \iint\iint d\boldsymbol{k}d\boldsymbol{k}'d\omega\, d\omega'\, e^{i(\boldsymbol{k}\cdot\boldsymbol{s}+\boldsymbol{k}'\cdot\boldsymbol{s}'-\omega t-\omega' t')}\widetilde{c}_X(\boldsymbol{k},\boldsymbol{k}',\omega,\omega'), \\ \widetilde{c}_X(\boldsymbol{k},\boldsymbol{k}',\omega,\omega') &= \frac{1}{(2\pi)^{n+1}}\iint\iint d\boldsymbol{s}d\boldsymbol{s}'dt\, dt'\, e^{-i(\boldsymbol{k}\cdot\boldsymbol{s}+\boldsymbol{k}'\cdot\boldsymbol{s}'-\omega t-\omega' t')}c_X(\boldsymbol{s},\boldsymbol{s}',t,t'), \end{aligned} \tag{2.14a–b}$$

which, though, are of limited practical use, unless some physically justified hypotheses are imposed on the random field representation, like STHS or space–time isostationarity (STIS).[7]

Remark 2.1

More comprehensive expressions of the above spectral representation are obtained when the physical conditions justify suppressing some of the arguments. Specifically, suppressing the space argument may be useful when seeking to emphasize the unique role of time variation in the physical application of interest. A similar argument (i.e., focusing on the role of space variation) may be the reason for suppressing the time argument. For illustration, first I consider Eq. (2.13) with $\boldsymbol{s}' = \boldsymbol{s}+\boldsymbol{h}$ and $t' = t+\tau$ so that

$$X(\boldsymbol{s}+\boldsymbol{h},t+\tau) = \int d\mathcal{M}_X(\boldsymbol{k},\omega)\, e^{i(\boldsymbol{k}\cdot(\boldsymbol{s}+\boldsymbol{h})-\omega(t+\tau))},$$

then multiply the last equation with Eq. (2.13), and finally apply the expectation (mean) operator to the product to find

$$c_X(\boldsymbol{s},\boldsymbol{h},t,\tau) = \iint\iint e^{i(\boldsymbol{k}-\boldsymbol{k}'')\cdot\boldsymbol{s}+i(\boldsymbol{k}\cdot\boldsymbol{h}-\omega\tau)-i(\omega-\omega'')t}\,\overline{d\mathcal{M}_X(\boldsymbol{k},\omega)d\mathcal{M}_X^*(\boldsymbol{k}'',\omega'')}.$$

I can further let $\boldsymbol{k}' = \boldsymbol{k}-\boldsymbol{k}''$ and $\omega' = \omega-\omega''$, so that

$$c_X(\boldsymbol{s},\boldsymbol{h},t,\tau) = \iint\iint e^{i(\boldsymbol{k}'\cdot\boldsymbol{s}-\omega' t)+i(\boldsymbol{k}\cdot\boldsymbol{h}-\omega\tau)}\,\overline{d\mathcal{M}_X(\boldsymbol{k},\omega)d\mathcal{M}_X^*(\boldsymbol{k}',\omega')}.$$

[7]This is another random field class, also briefly introduced in Section 7.4 of Chapter II and presented in detail in Chapter VII.

My next step is to define

$$\widetilde{c}_X(\boldsymbol{k},\omega,\boldsymbol{s},t)d\boldsymbol{k}d\omega=\iint e^{i(\boldsymbol{k}'\cdot\boldsymbol{s}-\omega' t)}\,\overline{d\mathcal{M}_X(\boldsymbol{k},\omega)d\mathcal{M}^*_X(\boldsymbol{k}-\boldsymbol{k}',\omega-\omega')},$$

in which case, I end up with a simpler spectral representation pair

$$c_X(\boldsymbol{s},\boldsymbol{h},t,\tau)=\iint d\boldsymbol{k}d\omega\; e^{i(\boldsymbol{k}\cdot\boldsymbol{h}-\omega\tau)}\widetilde{c}_X(\boldsymbol{k},\omega,\boldsymbol{s},t),$$

$$\widetilde{c}_X(\boldsymbol{k},\omega,\boldsymbol{s},t)=\frac{1}{(2\pi)^n}\iint d\boldsymbol{h}d\tau\; e^{-i(\boldsymbol{k}\cdot\boldsymbol{h}-\omega\tau)}c_X(\boldsymbol{s},\boldsymbol{h},t,\tau).$$

The last pair of equations can replace Eqs. (2.14a–b).

As regards $\widetilde{c}_X(\boldsymbol{k},\omega,\boldsymbol{s},t)$, it is an unrestricted spectrum used to model, e.g., surface displacement of ocean waves, where it has six arguments: two for wavevector, one for frequency, two for spatial position, and one for time. Generally, there will be a different spectrum for each position and each time instant. However, in most physical applications there are no measurements available that can be meaningfully interpreted to represent the above spectrum. Again, if some additional assumptions can be made regarding space–time variability (Chapter VII), the above representations can be simplified and physical spectra measurements can be made.

Proposition 2.1

An S/TRF is harmonizable if and only if its space–time covariance function is harmonizable.

According to Proposition 2.1 the stochastic integral representations (2.5) of an S/TRF exist if and only if the deterministic integral representation of the corresponding covariance, see Eq. (2.9), exists.

Remark 2.2

From standard mathematical analysis it is found that there are certain conditions with respect to $c_X(\boldsymbol{p},\boldsymbol{p}')$ and $\widetilde{c}_X(\boldsymbol{w},\boldsymbol{w}')$ for convergence of the deterministic FT (Eq. 2.8) and the inverse FT (Eq. 2.9), respectively. These conditions must be checked carefully when the above formulas are used.

Example 2.1

As a simple illustration of how a covariance can be represented in the FT form of Eq. (2.12c), consider in $R^{1,1}$ the field

$$X(s,t)=\upsilon\, e^{i(s+ct)a},$$

where c is a constant, υ is a zero-mean RV with $E[\upsilon^2]=c^2$, and a is an RV independent of υ with cumulative distribution function (CDF) $F_a(\alpha)=P[a\leq\alpha]$. Then

$$\overline{X(s,t)}=\overline{\upsilon}\overline{e^{i(s+ct)a}}=0,$$

and

$$c_X(h,\tau)=\overline{\upsilon^2}\;\overline{e^{i(h+c\tau)a}}=c^2\int dF_a(\alpha)\,e^{i(h+c\tau)\alpha}=\int dQ_X(\alpha)\,e^{i(h+c\tau)\alpha},$$

where $Q_X(\alpha) = c^2 F_a(\alpha)$ is an arbitrary nondecreasing bounded function such that $Q_X(\alpha) \to 0$ when $\alpha \to -\infty$. Clearly, this is an STHS random field.

In many physical situations in the $R^{n,1}$ domain, it may be appropriate to only transform over the spatial coordinates and use the S/TRF representation pair,

$$X(\boldsymbol{s},t) = \int d\boldsymbol{k} e^{i\boldsymbol{k}\cdot\boldsymbol{s}} \widetilde{X}(\boldsymbol{k},t),$$
$$\widetilde{X}(\boldsymbol{k},t) = \frac{1}{(2\pi)^{\frac{n}{2}}} \int d\boldsymbol{s} e^{-i\boldsymbol{k}\cdot\boldsymbol{s}} X(\boldsymbol{s},t), \tag{2.15a–b}$$

where $\widetilde{X}(\boldsymbol{k},t) = \int d\omega\, e^{-i\omega t} \widetilde{X}(\boldsymbol{k},\omega)$ is the so-called *wavevector–temporal* field representation. Similarly, in some other cases it may be appropriate to only transform over the temporal coordinate and use the random field representation pair,

$$X(\boldsymbol{s},t) = \int d\omega e^{-i\omega t} \widetilde{X}(\boldsymbol{s},\omega),$$
$$\widetilde{X}(\boldsymbol{s},\omega) = \frac{1}{(2\pi)^{\frac{1}{2}}} \int dt e^{i\omega t} X(\boldsymbol{s},t), \tag{2.15c–d}$$

where $\widetilde{X}(\boldsymbol{s},\omega) = \int d\boldsymbol{k}\, e^{i\boldsymbol{k}\cdot\boldsymbol{s}} \widetilde{X}(\boldsymbol{k},\omega)$ is the so-called *spatial–frequency* field representation. Eqs. (2.15a–d) are obviously partial FT representations.

The simultaneous two-point covariance function and the corresponding partial spectral function of the S/TRF representations of Eqs. (2.15a–b) are related by

$$c_X(\boldsymbol{s},\boldsymbol{s}',t) = \iint d\boldsymbol{k} d\boldsymbol{k}'\, e^{i(\boldsymbol{k}'\cdot\boldsymbol{s}'+\boldsymbol{k}\cdot\boldsymbol{s})} \widetilde{c}_X(\boldsymbol{k},\boldsymbol{k}',t),$$
$$\widetilde{c}_X(\boldsymbol{k},\boldsymbol{k}',t) = \frac{1}{(2\pi)^n} \iint d\boldsymbol{s} d\boldsymbol{s}'\, e^{-i(\boldsymbol{k}'\cdot\boldsymbol{s}'+\boldsymbol{k}\cdot\boldsymbol{s})} c_X(\boldsymbol{s},\boldsymbol{s}',t), \tag{2.16a–b}$$

where the $\widetilde{c}_X$ in Eqs. (2.16a–b) is termed the *spectral wavevector–temporal density function* or simply *wavevector–temporal spectrum* to be distinguished from the usual SDF of Eq. (2.8). Moreover, in view of the STHS assumption, the space–time covariance and the wavevector–temporal spectrum representations of Eqs. (2.16a–b) become

$$c_X(\boldsymbol{h},t) = \int d\boldsymbol{k}\, e^{i\boldsymbol{k}\cdot\boldsymbol{h}} \widetilde{c}_X(\boldsymbol{k},t),$$
$$\widetilde{c}_X(\boldsymbol{k},t) = \frac{1}{(2\pi)^n} \int d\boldsymbol{h}\, e^{-i\boldsymbol{k}\cdot\boldsymbol{h}} c_X(\boldsymbol{h},t), \tag{2.17a–b}$$

where $\widetilde{c}_X(\boldsymbol{k},t)\delta(\boldsymbol{k}'-\boldsymbol{k}) = \overline{\widetilde{X}(\boldsymbol{k}',t)\widetilde{X}^*(\boldsymbol{k},t)}$. A more detailed discussion of these STHS representations can be found in Chapter VII.

In certain cases, it is considerably more convenient to work with spectral functions rather than with covariance functions, as is illustrated in the following example.

Example 2.2

Consider in the $R^{3,1}$ domain the u-smoothed S/TRF

$$X_u(\boldsymbol{s},t) = \int d\boldsymbol{s}' u(\boldsymbol{s}')X(\boldsymbol{s}-\boldsymbol{s}',t),$$

where u denotes a suitable smoothing function (kernel). The time-dependent variance of $X_u(\boldsymbol{s},t)$ is

$$\overline{X_u(t)^2} = \iint d\boldsymbol{s}' d\boldsymbol{s}'' u(\boldsymbol{s}')u(\boldsymbol{s}'')c_X(\boldsymbol{s}'-\boldsymbol{s}'',t).$$

On the other hand, in terms of the spectral domain analysis above, and using the convolution theorem of Table 2.1, the spectral representation of the integral $X_u(\boldsymbol{s},t)$ expression above reduces to the simple multiplication

$$\widetilde{X}_u(\boldsymbol{k},t) = \widetilde{X}(\boldsymbol{k},t)\widetilde{u}(\boldsymbol{k}),$$

with wavevector–temporal spectrum $\widetilde{c}_{X_u}(\boldsymbol{k},t) = \left|\widetilde{u}(\boldsymbol{k})\right|^2 \widetilde{c}_X(\boldsymbol{k},t)$. Then, the time-dependent variance of $X_u(\boldsymbol{s},t)$ can be expressed by

$$\overline{X_u(t)^2} = \frac{1}{(2\pi)^3}\int d\boldsymbol{k}|\widetilde{u}(\boldsymbol{k})|^2 \widetilde{c}_X(\boldsymbol{k},t),$$

which is a single rather than a double integral.

In the case of real-valued covariance and spectral functions, the exponentials in the above equations can be replaced by cosine functions. Nevertheless, from a mathematical point of view, it is usually convenient to use exponential forms even in the real case. Lastly, it is worth mentioning that in certain situations in applied stochastics (e.g., statistical continuum media), the FT of higher-order moments may need to be defined, such as

$$FT\left[\overline{X(\boldsymbol{p}_1)\cdots X(\boldsymbol{p}_m)}\right] = \overline{\widetilde{X}(\boldsymbol{w}_1)\cdots\widetilde{X}(\boldsymbol{w}_m)}$$

for any integer m, and $\boldsymbol{p}_i$, $\boldsymbol{w}_i \in R^{n+1}$ ($i = 1,\ldots, m$). Particularly, the second- and higher-order spectral moments (frequency domain) can be defined in terms of the second-order spatiotemporal ensemble moments (physical domain).

Definition 2.4

The *joint* (λ_i, λ_j)*-order spectral moment* of an S/TRF $X(\boldsymbol{p})$ in R^{n+1} is conveniently defined by

$$\begin{aligned} \beta_X^{(\lambda_i\lambda_j)} &= \iint dQ_X(\boldsymbol{w},\boldsymbol{w}')w_i^{\lambda_i}w_j'^{\lambda_j} \\ &= \iint d\boldsymbol{w}\, d\boldsymbol{w}'\; w_i^{\lambda_i}w_j'^{\lambda_j}\; \widetilde{c}_X(\boldsymbol{w},\boldsymbol{w}'), \end{aligned} \tag{2.18a–b}$$

where $\widetilde{c}_X(\boldsymbol{w},\boldsymbol{w}')$ is the FT of the covariance function $c_X(\boldsymbol{p},\boldsymbol{p}')$.

Example 2.3

Taking advantage of the FT properties, the spectral moments are directly related to the covariance functions. For illustration, it can be easily proven that,

$$\beta_X^{(1_i 1_j)} = -\frac{\partial^2}{\partial s_i \partial s'_j} c_X(\boldsymbol{p},\boldsymbol{p}')|_{\boldsymbol{p}=\boldsymbol{p}'=0} \tag{2.19}$$

(the sign in the right hand side of the equation depends on the way the FT pair is defined, Exercise V.2).

More results on spectral moments are presented in the following chapters in a variety of modeling settings and real-world situations. I continue with the examination of some quantities derived from the basic spectral functions that are particularly useful in the study of random field properties in a variety of applications.

2.3 TRANSFER FUNCTION AND EVOLUTIONARY MEAN POWER

As it turns out, additional application-specific quantities can be defined in the spectral domain with useful practical features. Let us start by introducing the notion of the linear operation $\Im$ such that if $X_i(\boldsymbol{p})$, $i = 1, 2,\ldots, m$, are S/TRFs, then

$$\Im\left[\sum_{i=1}^{m} X_i(\boldsymbol{p})\right] = \sum_{i=1}^{m} \Im[X_i(\boldsymbol{p})] \tag{2.20}$$

for all nonnegative integers m. The $\Im$ is translation invariant if it is valid that

$$\Im[U_{\boldsymbol{p}'}X(\boldsymbol{p})] = U_{\boldsymbol{p}'}\Im[X(\boldsymbol{p})] \tag{2.21}$$

for all $\boldsymbol{p}' \in R^{n+1}$, where $U_{\boldsymbol{p}'}X(\boldsymbol{p}) = X(\boldsymbol{p}+\boldsymbol{p}')$ is the familiar translation (shift) operator.

Example 2.4

Common $\Im$ operations in the above sense are the space–time *differentiation* of the (ν,μ) order (Section 4 of Chapter VI)

$$\Im[\cdot] = \frac{\partial^{\nu+\mu}}{\partial \boldsymbol{s}^{\nu}\partial s_0^{\mu}}[\cdot], \tag{2.22}$$

where $\partial \boldsymbol{s}^{\nu} = \prod_{i=1}^{n} \partial s_i^{\nu_i}$, $\sum_{i=1}^{n} \nu_i = \nu$; and the space–time *integration*

$$\Im[\cdot] = \int d\boldsymbol{p}' U_{-\boldsymbol{p}'}\eta(\boldsymbol{p})[\cdot], \tag{2.23}$$

where η is a suitable function in R^{n+1}.

Consider next the S/TRF $Y(\boldsymbol{p})$ defined by applying $\Im$ on the S/TRF $X(\boldsymbol{p})$, i.e.,

$$Y(\boldsymbol{p}) = \Im[X(\boldsymbol{p})]. \tag{2.24}$$

Combining Eqs. (2.23) and (2.24), I find

$$Y(\boldsymbol{p}) = \int d\boldsymbol{p}' X(\boldsymbol{p}') U_{-\boldsymbol{p}'}\eta(\boldsymbol{p}) = \int d\boldsymbol{p}' \, \eta(\boldsymbol{p}') U_{-\boldsymbol{p}'} X(\boldsymbol{p}). \tag{2.25}$$

In view of Eq. (2.25), the means and covariances of the two STHS random fields are related by the simple formulas,

$$\overline{Y} = \overline{X}\tilde{\eta}(\mathbf{0}),$$
$$c_Y(\Delta\boldsymbol{p}) = \iint d\boldsymbol{u}d\boldsymbol{u}'\ \eta(\boldsymbol{u})\eta(\boldsymbol{u}')c_X(\Delta\boldsymbol{p}+\boldsymbol{u}-\boldsymbol{u}'). \tag{2.26a–b}$$

Example 2.5

Let $X(\boldsymbol{p}) = \delta(\boldsymbol{p})$. Then the integral transformation of Eq. (2.25) yields,

$$Y(\boldsymbol{p}) = \int d\boldsymbol{p}'\delta(\boldsymbol{p}')U_{-\boldsymbol{p}'}\eta(\boldsymbol{p}) = \int d\boldsymbol{p}'\ \eta(\boldsymbol{p}')U_{-\boldsymbol{p}'}\delta(\boldsymbol{p}) = \eta(\boldsymbol{p}),$$

which is sometimes called the *unit impulse response* function. Similarly, in the case that $X(\boldsymbol{p})$ is a white random field with $c_X(\Delta\boldsymbol{p}) = c_0\delta(\Delta\boldsymbol{p})$, Eq. (2.26b) reduces to $c_Y(\Delta\boldsymbol{p}) = c_0\int d\boldsymbol{u}\ \eta(\boldsymbol{u})\eta(\boldsymbol{u}+\Delta\boldsymbol{p})$.

The integral representation of Eq. (2.25) with $X(\boldsymbol{p}) = e^{i\boldsymbol{w}\cdot\boldsymbol{p}}$ gives,

$$Y(\boldsymbol{p}) = \Im\left[e^{i\boldsymbol{w}\cdot\boldsymbol{p}}\right] = \int d\boldsymbol{p}'e^{i\boldsymbol{w}\cdot\boldsymbol{p}'}U_{-\boldsymbol{p}'}\eta(\boldsymbol{p}) = \int d\boldsymbol{p}'\eta(\boldsymbol{p}')U_{-\boldsymbol{p}'}\ e^{i\boldsymbol{w}\cdot\boldsymbol{p}} = \tilde{\eta}(\boldsymbol{w})e^{i\boldsymbol{w}\cdot\boldsymbol{p}},$$

which leads to the following central definition of this section.

Definition 2.5

The transfer function (*TF*) of $\Im$ is defined by

$$\tilde{\eta}(\boldsymbol{w}) = \frac{\Im\left[e^{i\boldsymbol{w}\cdot\boldsymbol{p}}\right]}{e^{i\boldsymbol{w}\cdot\boldsymbol{p}}}. \tag{2.27}$$

Assume now that the S/TRF $X(\boldsymbol{p})$ is harmonizable in the sense of Eq. (2.11). Then, I can write

$$Y(\boldsymbol{p}) = \Im[X(\boldsymbol{p})] = \Im\left[\int d\mathcal{M}_X(\boldsymbol{w})e^{i\boldsymbol{w}\cdot\boldsymbol{p}}\right] = \int d\mathcal{M}_X(\boldsymbol{w})\Im\left[e^{i\boldsymbol{w}\cdot\boldsymbol{p}}\right].$$

By comparing the last equation with Eq. (2.27), I obtain

$$Y(\boldsymbol{p}) = \int d\mathcal{M}_X(\boldsymbol{w})e^{i\boldsymbol{w}\cdot\boldsymbol{p}}\tilde{\eta}(\boldsymbol{w})$$
$$= \int d\mathcal{M}_Y(\boldsymbol{w})e^{i\boldsymbol{w}\cdot\boldsymbol{p}}, \tag{2.28a–b}$$

where $d\mathcal{M}_Y(\boldsymbol{w}) = \tilde{\eta}(\boldsymbol{w})d\mathcal{M}_X(\boldsymbol{w})$. Moreover, it is easy to show that the SDFs of the $X(\boldsymbol{p})$ and $Y(\boldsymbol{p})$ (if they exist) are simply related in terms of the TF as

$$\tilde{c}_Y(\boldsymbol{w},\boldsymbol{w}') = \tilde{\eta}(\boldsymbol{w})\tilde{\eta}^*(\boldsymbol{w}')\,\tilde{c}_X(\boldsymbol{w},\boldsymbol{w}'), \tag{2.29}$$

where $\tilde{c}_Y(\boldsymbol{w},\boldsymbol{w}')$ is the FT of the covariance function $c_Y(\boldsymbol{p},\boldsymbol{p}')$ (Exercise V.3).

Some more interesting spectral functions are obtained if I generalize the spectral representation of Eq. (2.11) of the S/TRF $X(\boldsymbol{p})$ as

$$X(\boldsymbol{p}) = \int d\mathcal{M}_X(\boldsymbol{w})\Theta(\boldsymbol{p},\boldsymbol{w})e^{i\boldsymbol{w}\cdot\boldsymbol{p}}, \tag{2.30}$$

where $\Theta(\boldsymbol{p},\boldsymbol{w})$ is an amplitude-modulating function that varies slowly with $\boldsymbol{p}$, and, as usual, $\mathcal{M}_X(\boldsymbol{w})$ is a random field with uncorrelated increments and variance $Q_X(\boldsymbol{w})$. On the basis of the above spectral representations, I can define another useful concept in the spectral analysis of heterogeneous, in general, S/TRFs, as follows.

Definition 2.6

If the S/TRF $X(\boldsymbol{p})$ is represented as in Eq. (2.30), and the $Q_X(\boldsymbol{w})$ is differentiable so that

$$\widetilde{c}_X(\boldsymbol{w}) = \frac{\partial^{n+1}}{\partial \boldsymbol{w}^{n+1}} Q_X(\boldsymbol{w}), \tag{2.31}$$

the *evolutionary mean power* SDF is defined by

$$\widetilde{c}_X(\boldsymbol{p},\boldsymbol{w}) = |\Theta(\boldsymbol{p},\boldsymbol{w})|^2 \widetilde{c}_X(\boldsymbol{w}). \tag{2.32}$$

Note that in this case the evolutionary mean power SDF is not the FT of the covariance function $c_X(\boldsymbol{p},\boldsymbol{p}')$. The TF and the evolutionary mean power SDF are useful tools in the FT analysis of random field models in structural engineering and elsewhere.

2.4 FOURIER TRANSFORM OF VECTOR SPATIOTEMPORAL RANDOM FIELDS

We saw earlier that when one deals with two stochastically correlated S/TRFs, their cross-covariance function should be involved in the analysis. The analysis can be considered in the frequency domain, in which case the corresponding cross-spectral function of the two random fields considered jointly is defined in terms of the FT of the cross-covariance.

Definition 2.7

Let $X_l(\boldsymbol{p})$ and $X_{l'}(\boldsymbol{p}')$, $\boldsymbol{p},\boldsymbol{p}' \in R^{n+1}$ be two correlated S/TRF. The *cross*-SDF is given by

$$\widetilde{c}_{X_l X_{l'}}(\boldsymbol{w},\boldsymbol{w}') = \frac{1}{(2\pi)^{n+1}} \iint d\boldsymbol{p}\, d\boldsymbol{p}' e^{-i(\boldsymbol{w}\cdot\boldsymbol{p}-\boldsymbol{w}'\cdot\boldsymbol{p}')} c_{X_l X_{l'}}(\boldsymbol{p},\boldsymbol{p}'). \tag{2.33}$$

If $\boldsymbol{c_X}$ is the covariance matrix of the vector S/TRF $\boldsymbol{X}(\boldsymbol{p}) = [X_1(\boldsymbol{p})\ldots X_k(\boldsymbol{p})]^T$, the following definition is straightforward.

Definition 2.8

The *cross-spectral density matrix* (*cross-SDM*) of a vector S/TRF $\boldsymbol{X}(\boldsymbol{p})$ is defined as

$$\widetilde{\boldsymbol{c}}_{\boldsymbol{X}} = \left[\widetilde{c}_{X_l X_{l'}}(\boldsymbol{w},\boldsymbol{w}')\right], \tag{2.34}$$

where l, $l' = 1,\ldots, k$, and the component SDFs $\widetilde{c}_{X_l X_{l'}}(\boldsymbol{w},\boldsymbol{w}')$ are given by Eq. (2.33), assuming that the integral exists.

If we need to move our analysis in the $R^{n,1}$ domain, the usual change of FT coordinates

$$(\boldsymbol{w},\boldsymbol{w}') \mapsto (\boldsymbol{k},\boldsymbol{k}',\omega,\omega')$$

applies, and the cross-SDF of Eq. (2.33) can be expressed as $\widetilde{c}_{X_l X_{l'}}(\boldsymbol{k},\boldsymbol{k}',\omega,\omega')$, $l,l' = 1,\ldots, k$. In this case, the cross-SDM becomes

$$\widetilde{c}_X = \left[\widetilde{c}_{X_l X_{l'}}(\boldsymbol{k}, \boldsymbol{k}', \omega, \omega')\right].$$

A more detailed study of vector S/TRFs, including the important case of STHS vector S/TRFs, is presented in Chapter IX.

3. SPACE TRANSFORMATION

It is conceptually attractive and often very useful in practice to transform the representation of a phenomenon into lower-dimensionality domains by means of STs. Naturally, one can find numerous applications of this powerful concept in the real-world.

Example 3.1

The spectrum of atmospheric index fluctuations from $R^{3,1}$ is transformed into the corresponding wavefront in $R^{2,1}$ to physically estimate the observed spectrum in $R^{2,1}$ and then back-transform the observed spectrum into the $R^{3,1}$ domain. Other examples concerning the application of the ST concept are listed in Table 3.1.

STs are mathematical operations that have elegant and comprehensive representations both in the space–time (physical) and the spectral (frequency) domains. Also, STs preserve the second-order space–time correlation structure of a random field in the sense that an STHS random field in $R^{n,1}$ is transferred to an STHS random field in $R^{n-k,1}$.[8] Lastly, using ST a random field in the $R^{n,1}$ domain can be represented by a linear combination of statistically uncorrelated random processes in the $R^{1,1}$ domain (expressions relating the corresponding S/TVFs of the two domains can be established).

3.1 BASIC NOTIONS

Let us begin by introducing some notation. By $\boldsymbol{\theta} = (\theta_1,\ldots, \theta_n)$, I denote here a unit vector, and by $S_n = \frac{2\pi^{\frac{n}{2}}}{\Gamma\left(\frac{n}{2}\right)}$ as usual the surface area of the unit n-dimensional sphere S^{n-1}.[9] As before, integrations are carried

Table 3.1 Examples of Space Transformation (ST) Applications

Much that seemed problematic about the application of stochastic partial differential equation theory in the study of multidimensional physical systems may be resolved by means of ST (Section 3.5).
Development of comprehensive and analytically tractable ST-based expressions of the most important criteria of permissibility (Chapter XV).
Generation of new spatiotemporal variability function (S/TVF) models in $R^{n,1}$ from the corresponding functions in $R^{1,1}$ (this versatile approach of S/TVF model generation is outlined in Chapter XVI).
As a particularly attractive instrument in the simulation context (e.g., ST is the basic notion of geostatistical turning bands simulation; Section 3.4).

[8]Generally, $k < n$, although in most cases in practice, $k = n - 1$.

[9]For simplicity, the same symbol S_n is sometimes used in the literature to denote both the unit n-dimensional sphere and its surface area.

out over the whole range of a variable, unless other limits are indicated. Given a function $\mathcal{X}_n(\boldsymbol{s},t)$ on $R^{n,1}$ (for ST analysis purposes, the subscript n of the function indicates that its domain is $R^{n,1}$), consider:

(i) The mapping

$$\mathcal{R}: \mathcal{X}_n(\boldsymbol{s},t) \mapsto \int d\boldsymbol{s}\, \mathcal{X}_n(\boldsymbol{s},t)\, \delta(\sigma - \boldsymbol{s}\cdot\boldsymbol{\theta}), \tag{3.1}$$

where $\sigma \in R^1$ and δ is the delta function, as usual, and the $\mathcal{X}_n(\boldsymbol{s},t)$ is such that the integral in Eq. (3.1) exists (e.g., it must decay fast at infinity or have a compact support).[10] Eq. (3.1) is known as the *Radon operator* (Radon, 1917) or as the plane wave integral (John, 1955).

(ii) The operator

$$\Omega = \frac{(-1)^{m-1}}{2(2\pi)^{2m-1}} \begin{cases} -\dfrac{S_{2m+1}}{2\pi} \dfrac{\partial^{2m}}{\partial\sigma^{2m}}[\cdot], & \text{if } n = 2m+1 \\ S_{2m}\mathcal{H}\left\{ \dfrac{\partial^{2m-1}}{\partial\sigma^{2m-1}}[\cdot] \right\}, & \text{if } n = 2m, \end{cases} \tag{3.2}$$

where $\mathcal{H}$ denotes the Hilbert transform.

(iii) The set of hyperplanes passing through a given point in R^n,

$$\{H_{n-1}; \boldsymbol{\theta}, \boldsymbol{s}\cdot\boldsymbol{\theta}\}, \tag{3.3}$$

where the subscript $n-1$ denotes the dimension of H_{n-1}. The unit vector $\boldsymbol{\theta}$ defines the orientation of H_{n-1}, and $\boldsymbol{s}\cdot\boldsymbol{\theta}$ is the inner product that defines its distance from the origin so that

$$\begin{aligned} &\textstyle\int d\boldsymbol{s}\delta(\sigma - \boldsymbol{s}\cdot\boldsymbol{\theta}) = 1, \\ &\delta(\sigma - \boldsymbol{s}\cdot\boldsymbol{\theta}) = 0 \quad \text{if} \quad \sigma \neq \boldsymbol{s}\cdot\boldsymbol{\theta}. \end{aligned} \tag{3.4}$$

In general, STs are mathematical operations that reduce a multidimensional space to a unidimensional space. In the random field modeling context, the function $\mathcal{X}_n(\boldsymbol{s},t)$ may be a random one (representing, say, an S/TRF) or a deterministic one (denoting, e.g., an S/TVF).

Definition 3.1

The ST operators T_n^1 and Ψ_n^1, which reduce $\mathcal{X}_n(\boldsymbol{s},t)$ to a unidimensional space–time function by means of the mapping of Eq. (3.1) and the operator of Eq. (3.2), are defined as

$$T_n^1[\mathcal{X}_n](\sigma,\boldsymbol{\theta},t) = \mathcal{R}[\mathcal{X}_n](\sigma,\boldsymbol{\theta},t) = \widehat{\mathcal{X}}_{1,\boldsymbol{\theta}}(\sigma,t), \tag{3.5}$$

and

$$\Psi_n^1[\mathcal{X}_n](\sigma,\boldsymbol{\theta},t) = \Omega T_n^1[\mathcal{X}_n](\sigma,\boldsymbol{\theta},t) = \mathcal{X}_{1,\boldsymbol{\theta}}(\sigma,t). \tag{3.6}$$

The T_n^1 and Ψ_n^1 are considered as completely defined if the integrals expressions in Eqs. (3.5) and (3.6) are known for all $\boldsymbol{\theta}$. In fact, the ST operators of Eqs. (3.5) and (3.6), which represent the projection of $\mathcal{X}_n(\boldsymbol{s},t)$ on hyperplanes that are perpendicular to a specific direction, can be inverted as follows.

---

[10] An interesting class of $\mathcal{X}_n(\boldsymbol{s},t)$ belongs to the Schwartz space (Christakos, 1992).

Definition 3.2

The ST operators T_1^n and Ψ_1^n, which raise $\widehat{X}_{1,\theta}$ and $X_{1,\theta}$ to an n-dimensional space × time function, are defined as

$$\Psi_1^n[X_{1,\theta}](s,t) = S_n^{-1}\int_{S_n} d\theta\, X_{1,\theta}(s\cdot\theta,t) = X_n(s,t); \tag{3.7}$$

and

$$T_1^n\left[\widehat{X}_{1,\theta}\right](s,t) = \Psi_1^n\Omega\left[\widehat{X}_{1,\theta}\right](s,t) = X_n(s,t). \tag{3.8}$$

Some interesting features of the ST are worth noticing. While in the FT context, variability in space and time is mapped into the frequency domain, in ST this mapping involves the projections of an n-dimensional space–time function onto planes orthogonal to transform lines determined by direction vectors in $R^{n,1}$. Specifically, the projections introduced by T_n^1 and Ψ_n^1 involve the integrals of the function over hyperplanes (planes, if $n=3$) defined by means of the equation $\Pi_\theta(s) = \sigma - s\cdot\theta = 0$ (σ denotes the ordinate of the projection onto the transform line). The projections along a single transform line can be viewed as functions of a single variable (i.e., the projection length). On the other hand, in terms of T_1^n and Ψ_1^n the n-dimensional space–time function can be reconstructed exactly from the one-dimensional space–time functions in all directions.[11]

Operationally, the T_n^1 consists of infinitely many integral transforms with the delta functions δ as their kernels (δ defines the transform plane). The T_n^1 of a function may be also considered as a homogeneous function of degree -1 on the projective space. The Ψ_1^n, on the other hand, operates on the set of functions $X_{1,\theta}(\sigma,t)$ defined on a set of hyperplanes $\{H_{n-1};\theta,s\cdot\theta\}$ passing through a given point in R^n.

It is worth noticing that the T_n^1 operator satisfies certain basic properties that are operationally very significant, as follows:

(a) The *scaling* property under dilation D_λ of the spatial vector s, i.e.,

$$T_n^1[D_\lambda X_n(s,t)] = T_n^1[X_n(\lambda s,t)] = \lambda^{1-n}\widehat{X}_{1,\theta}(\lambda\sigma,t). \tag{3.9}$$

(b) The *shifting* property represents the change caused by a translation U_{-a} of the position vector s in space

$$T_n^1[U_{-a}X_n(s,t)] = T_n^1[X_n(s-a,t)] = \widehat{X}_{1,\theta}(\sigma-\theta\cdot a,t). \tag{3.10}$$

(c) The ST includes *redundant* information, a fact that is reflected in the property

$$\widehat{X}_{1,\theta}(\sigma,t) = \widehat{X}_{1,-\theta}(-\sigma,t); \tag{3.11}$$

this property expresses the fact that the mapping from the configuration space $R^{n,1}$ onto the space $S^{n-1}\times R^{n,1}$ leads to a double covering of the physical space if both the projection length and the transform line direction vectors are unrestricted. The redundancy is lifted by restricting the ST to

[11]However, in many practical cases the function can be accurately reconstructed by means of projections along a finite set of lines, which effectively reduces the dimensionality of the reconstructed function.

positive projection lengths $\sigma > 0$ or to direction vectors contained within one hemisphere of the unit hypersphere in S^{n-1}.

Some generalizations of the ST expressions above are possible. Specifically, the Ψ_1^n can be extended so that

$$\Psi_1^n\left[X_{1,\theta}\right](s,t) = \int_{\Theta_n} d\theta\, u(s,\theta) X_{1,\theta}(s\cdot\theta,t) = X_n(s,t), \tag{3.12}$$

where $u(s,\theta)$ is a weight function and the integration is carried out over the closed surface Θ_n in R^n. Clearly, Eq. (3.7) is a special case of Eq. (3.12). Also, ST may be considered in the case of hyperplanes H_{n-k} of dimension $n-k$. Eq. (3.5), e.g., may be written as

$$T_n^{n-k}[X_n](s_1,s_2,\ldots,s_{n-k},t) = \int ds_{n-k+1}\ldots ds_n\, X_n(s,t) = \widehat{X}_{n-k}(s_1,s_2,\ldots,s_{n-k},t). \tag{3.13}$$

Remark 3.1

The STs are related to each other, like

$$\Psi_1^n \Omega T_n^1 = I, \tag{3.14}$$

where I is the identity operator.

Example 3.2

Let $u(s,\theta) = S_n^{-1}$, that is, uniform weight. In this case, Eq. (3.12) reduces to

$$\Psi_1^n\left[X_{1,\theta}\right](s,t) = S_n^{-1}\int_{S_n} d\theta\, X_{1,\theta}(s\cdot\theta,t) = X_n(s,t) \tag{3.15}$$

with weight function S_n^{-1}. Another interesting special case of T_n^1 emerges if I set $\theta = (1, 0,\ldots, 0)$, in which case

$$T_n^1[X_n](\sigma,\theta,t) = \widehat{X}_{1,\theta}(s_1,t) = \int ds\, X_n(s,t)\delta(s_1 - s\cdot\theta) = \int_U ds_2\ldots ds_n X_n(s,t), \tag{3.16}$$

where $U \subset R^n$. That is, in this case the T_n^1 represents the projection of the n-dimensional function on hyperplanes that are perpendicular to the direction $\theta = (1, 0,\ldots, 0)$.

In the spectral domain, the STs turn out to have particularly simple algebraic forms (Christakos, 1992).

Proposition 3.1

In the spectral domain, the T_n^1 can be written as

$$T_n^1\left[\widetilde{X}_n\right](k,\theta,t) = \widetilde{X}_n(\boldsymbol{k},t) = \widetilde{\widehat{X}}_{1,\theta}(k,t) \tag{3.17}$$

along the wavevector (spatial frequency) $\boldsymbol{k} = k\theta$. As regards T_1^n, it is valid that

$$T_1^n\left[\widetilde{\widehat{X}}_{1,\theta}\right](\boldsymbol{k},t) = \widetilde{\widehat{X}}_{1,\theta}(k,t) = \widetilde{X}_n(\boldsymbol{k},t). \tag{3.18}$$

Similarly, for the Ψ_n^1 and Ψ_1^n it can be shown that,

$$\Psi_n^1\left[\widetilde{X}_n\right](k,t) = \frac{S_n k^{n-1}}{(2\pi)^{n-1}} \widetilde{X}_n(\boldsymbol{k},t) = \widetilde{X}_{1,\boldsymbol{\theta}}(k,t), \tag{3.19}$$

and

$$\Psi_1^n\left[\widetilde{X}_{1,\boldsymbol{\theta}}\right](\boldsymbol{k},t) = \frac{(2\pi)^{n-1}}{S_n k^{n-1}} \widetilde{X}_{1,\boldsymbol{\theta}}(k,t) = \widetilde{X}_n(\boldsymbol{k},t), \tag{3.20}$$

respectively.[12]

In the case of a spatially isotropic function $X_n(\boldsymbol{s},t) = X_n(s = |\boldsymbol{s}|,t)$ with infinite support,[13] the T_n^1 is also isotropic, namely the function $\widehat{X_{1,\boldsymbol{\theta}}}(\sigma, t)$ is independent of the orientation of the vector $\boldsymbol{\theta}$. In this case, the following proposition holds (Christakos, 1986).

Proposition 3.2

When $X_n(s,t)$ is a spatially isotropic function, the following expressions are obtained,

$$X_{n+1}(s,t) = T_n^{n+1}[X_n](s,t) = -\frac{1}{\pi}\int_s^\infty \frac{1}{\sqrt{u^2-s^2}} \frac{\partial X_n(u,t)}{\partial u} du, \tag{3.21}$$

$$X_{n+2}(s,t) = T_n^{n+2}[X_n](s,t) = -\frac{1}{2\pi s}\frac{\partial X_n(s,t)}{\partial s}; \tag{3.22}$$

and

$$\begin{aligned} X_n(s,t) = \Psi_1^n[X_n](s,t) &= \frac{2\Gamma\left(\frac{n}{2}\right)s^{2-n}}{\Gamma\left(\frac{1}{2}\right)\Gamma\left(\frac{n-1}{2}\right)} \int_0^s du\,(s^2-u^2)^{\frac{n-3}{2}} X_1(u,t) \\ &= \frac{2\Gamma\left(\frac{n}{2}\right)}{\Gamma\left(\frac{1}{2}\right)\Gamma\left(\frac{n-1}{2}\right)} \int_0^1 dv\,(1-v^2)^{\frac{n-3}{2}} X_1(vs,t), \end{aligned} \tag{3.23}$$

$$X_{n-1}(s,t) = \Psi_n^{n-1}[X_n](s,t) = \frac{2\Gamma\left(\frac{n+1}{2}\right)s^{2-n}}{(n-1)\Gamma\left(\frac{1}{2}\right)\Gamma\left(\frac{n}{2}\right)} \frac{\partial}{\partial s}\left[\int_0^s du \frac{u^{n-1}}{u^2-s^2} X_n(u,t)\right], \tag{3.24}$$

$$X_{n-2}(s,t) = \Psi_n^{n-2}[X_n](s,t) = \frac{s^{3-n}}{(n-2)}\frac{\partial}{\partial s}\left[s^{n-2}X_n(s,t)\right]. \tag{3.25}$$

Interestingly, in the above cases the coefficients of the Ψ operators explicitly depend on the dimensionality n, whereas those of the T operators do not.

---

[12]Note that depending on the definition of the n-fold FT used, these formulas may be multiplied by a constant.

[13]Such functions are, e.g., certain classes of covariance functions.

Example 3.3

For a simple illustration, some special cases of Proposition 3.2 are shown in Table 3.2. The readers may notice the simpler formulas obtained in the frequency domain. In the case of $X_3(s) = e^{-\frac{s^2}{a^2}-\frac{t^2}{b^2}}$ we have a function that is isotropic in space and time but anisotropic in space–time. The T_3^1 transformation is obtained by direct inversion of the last equation in Table 3.2, i.e.,

$$T_3^1[X_3](\sigma) = \pi a^2 e^{-\frac{\sigma^2}{a^2}-\frac{t^2}{b^2}} = \pi a^2 e^{-\frac{s^2}{a^2}-\frac{t^2}{b^2}} = X_1(s) \tag{3.26}$$

($\sigma = s$). The ST isotropy breaks down in the case of finite support that lacks spherical symmetry (e.g., rectangular support). ST depends on the orientation of the transform line and the shape of the support. Finite supports can be used as spatial filters that permit to define the ST even for nonintegrable functions. The trade-off is that compact supports introduce boundary effects.

In applied sciences, the specific conditions of the problem under consideration may require that STs be defined in terms of S/TRF realizations, S/TVFs (covariance, variogram, structure functions, ordinary and generalized), or spectral densities (full and partial spectra). These are the topics of the following sections.

3.2 SPACE TRANSFORMATION OF SPATIOTEMPORAL RANDOM FIELDS

Consider an S/TRF that, for the purposes of the present analysis, will be denoted as $X_n(\boldsymbol{s},t)$ to indicate that its domain is $R^{n,1}$. Definitions 3.1 and 3.2 and other results of the previous section are valid in terms of the S/TRF $X_n(\boldsymbol{s},t)$.

Most physical attributes occur in three spatial dimensions, in which case the STs of random field gradients are particularly useful in the study of stochastic partial differential equations (SPDEs) that govern three-dimensional attributes. Thus, below I focus on the $R^{3,1}$ domain. The ST of a partial derivative can be expressed in terms of the direction cosines of the transform line and the derivative of the ST with respect to the projection length. This is a very useful property, because it allows transforming SPDE into SODE (stochastic ordinary differential equation). In particular, the following proposition holds (Christakos and Hristopulos, 1997).

Table 3.2 Examples of Ψ- and T-operator Forms in $R^{n,1}$ (n = 1, 2, 3) for Isotropic Functions

Space Transformation	Physical Domain	Frequency Domain
Ψ_1^2	$X_2(s,t) = \frac{2}{\pi}\int_0^r \frac{du}{\sqrt{s^2-u^2}} X_1(u,t)$	$\widetilde{X}_2(k,t) = k^{-1}\widetilde{X}_1(k,t)$
Ψ_1^3	$X_3(s,t) = s^{-1}\int_0^s du X_1(s,t)$	$\widetilde{X}_3(k,t) = \pi k^{-2}\widetilde{X}_1(k,t)$
Ψ_2^1	$X_1(s,t) = \frac{\partial}{\partial s}\left[\int_0^s du \frac{u}{u^2-s^2} X_2(u,t)\right]$	$\widetilde{X}_1(k,t) = k\widetilde{X}_2(k,t)$
Ψ_3^1	$X_1(s,t) = \frac{\partial}{\partial s}[sX_n(s,t)]$	$\widetilde{X}_1(k,t) = \frac{k^2}{\pi}\widetilde{X}_3(k,t)$
T_1^2	$X_2(s,t) = -\frac{1}{\pi}\int_s^\infty \frac{du}{\sqrt{u^2-s^2}} \frac{\partial X_1(u,t)}{\partial u}$	$\widetilde{X}_2(k,t) = \widetilde{X}_1(k,t)$
T_1^3	$X_3(s,t) = -\frac{1}{2\pi s}\frac{\partial X_1(s,t)}{\partial s}$	$\widetilde{X}_3(k,t) = \widetilde{X}_1(k,t)$

Proposition 3.3

The T_3^1 and Ψ_3^1 of the partial spatial derivative of $X_3(\boldsymbol{s},t)$ are given by, respectively,

$$T_3^1\left[\frac{\partial}{\partial s_i}X_3\right](\sigma,\boldsymbol{\theta},t)=\theta_i\frac{\partial}{\partial\sigma}\widehat{X}_{1,\boldsymbol{\theta}}(\sigma,t), \tag{3.27}$$

and

$$\Psi_3^1\left[\frac{\partial}{\partial s_i}X_3\right](\sigma,\boldsymbol{\theta},t)=\Omega T_3^1\left[\frac{\partial}{\partial s_i}X_3(\boldsymbol{s},t)\right](\sigma,\boldsymbol{\theta},t)=\theta_i\frac{\partial}{\partial\sigma}X_{1,\boldsymbol{\theta}}(\sigma,t). \tag{3.28}$$

As it might be anticipated, the expression for the gradient of a generalized S/TRF is somehow more involved.

Example 3.4

The T_3^1 of the gradient of $X_3(\boldsymbol{s},t)\delta(\sigma-\boldsymbol{s}\cdot\boldsymbol{\theta})$ is

$$T_3^1\left[\frac{\partial}{\partial s_i}(X_3(\boldsymbol{s},t)\delta(\sigma-\boldsymbol{s}\cdot\boldsymbol{\theta}))\right](p,\boldsymbol{\theta}',t)=\theta_i'\frac{\partial}{\partial p}\widehat{X}_1(\sigma,\boldsymbol{\theta},p,\boldsymbol{\theta}',t), \tag{3.29}$$

where $\widehat{X}_1(\sigma,\boldsymbol{\theta},p,\boldsymbol{\theta}',t)=T_3^1[X_3(\boldsymbol{s},t)\delta(\sigma-\boldsymbol{s}\cdot\boldsymbol{\theta})](p,\boldsymbol{\theta}',t)$ is a function of a double projection on two separate direction vectors. Eq. (3.29) is very useful in the study of SPDEs.

All ST applications involve at some point an inversion operation that reconstructs the three-dimensional spatial structure of the S/TRF from its projections on all possible directions in S^3. The inverse T_1^3 of the function $\widehat{X}_{1,\boldsymbol{\theta}}(\sigma,t)$ consists—according to Eq. (3.8)—of two steps: application of the Ω operator leads to $X_{1,\boldsymbol{\theta}}(\sigma,t)$, and the Ψ_3^1 operator reconstructs $X_3(\boldsymbol{s},t)$. It follows from Eq. (3.2) that the Ω operator in three dimensions is given by

$$X_{1,\boldsymbol{\theta}}(\sigma,t)=\Omega\left[\widehat{X}_{1,\boldsymbol{\theta}}(\sigma,t)\right]=-\frac{1}{2\pi}\frac{\partial^2}{\partial\sigma^2}\widehat{X}_{1,\boldsymbol{\theta}}(\sigma,t), \tag{3.30}$$

which is also a useful expression in SPDE studies.

Example 3.5

The unit vectors of the transform lines in three dimensions can be represented in terms of the polar angle χ and the azimuthal angle ϕ as follows

$$\boldsymbol{\theta}=(\theta_1,\theta_2,\theta_3)=(\sin\chi\,\cos\phi,\sin\chi\,\sin\phi,\cos\chi). \tag{3.31}$$

A single covering is obtained by considering unit vectors $\boldsymbol{\theta}$ in one hemisphere while leaving the projection length σ unconstrained. In terms of the polar and azimuthal angles,

$$X_3(\boldsymbol{s},t)=\Psi_1^3[X_{1,\boldsymbol{\theta}}(\sigma,t)](\boldsymbol{s})=\frac{1}{4\pi}\int_0^\pi d\phi\int_0^\pi d\chi\sin\chi\left[X_{1,\boldsymbol{\theta}}(\sigma,t)+X_{1,\boldsymbol{\theta}}(-\sigma,t)\right]_{\sigma=\boldsymbol{s}\cdot\boldsymbol{\theta}}, \tag{3.32}$$

which is the image of the operator Ψ_3^1 in three dimensions.

The following proposition concerns the ST of the product of two S/TRFs in the physical and the spectral domains (Christakos and Hristopulos, 1997).

Proposition 3.4

Consider two S/TRFs, $X_3(\boldsymbol{s},t)$ and $Y_3(\boldsymbol{s},t)$. The T_3^1 transformation of their product is given by

$$T_3^1[X_3Y_3](\sigma,\boldsymbol{\theta},t) = \int dz\, X_3(z,t)\, \widehat{Y}_{1,\boldsymbol{\theta},z}(\sigma,t). \tag{3.33}$$

The inverse operator T_1^3 of the product $\widehat{X}_{1,\boldsymbol{\theta}}(\sigma,t)\widehat{Y}_{1,\boldsymbol{\theta}}(\sigma,t)$ in the spectral domain is as follows

$$FT\left\{T_1^3\left[\widehat{X}_{1,\boldsymbol{\theta}}\widehat{Y}_{1,\boldsymbol{\theta}}\right]\right\}(\boldsymbol{k},t) = \frac{1}{4\pi}\int dp\widetilde{X}_3\left(\frac{\boldsymbol{k}}{|\boldsymbol{k}|}p,t\right)\widetilde{Y}_3\left(\boldsymbol{k}-\frac{\boldsymbol{k}}{|\boldsymbol{k}|}p,t\right), \tag{3.34}$$

which relates the spectral T_1^3 of the product of two spatial T_3^1 with the one-dimensional integration of the product of the corresponding spectral T_1^3.

In certain real-world applications, approximations based on series expansion of STs may turn out to be useful.

Proposition 3.5

The T_3^1 operator of the product of $X_3(\boldsymbol{s},t)$ and $Y_3(\boldsymbol{s},t)$ in $R^{3,1}$ that admits individually the ST representation is given in terms of the series expansion

$$\int d\sigma\widehat{X}_{1,\boldsymbol{\theta}}(\sigma,t)Y_{1,\boldsymbol{\theta}}(\sigma,t) = \sum_{k=0}^{\infty}\widehat{X}_{1,\boldsymbol{\theta}}^{(k)}(\alpha,t)\int d\sigma\frac{(\sigma-\alpha)^k}{k!}Y_{1,\boldsymbol{\theta}}(\sigma,t), \tag{3.35}$$

where $\widehat{X}_{1,\boldsymbol{\theta}}^{(k)}(\alpha,t) = \frac{\partial^k}{\partial\sigma^k}\widehat{X}_{1,\boldsymbol{\theta}}(\sigma,t)|_{\sigma=\alpha}$. Eq. (3.35) is exact, assuming that the Taylor expansion of $\widehat{X}_{1,\boldsymbol{\theta}}(\sigma,t)$ converges to $\widehat{X}_{1,\boldsymbol{\theta}}(\sigma,t)$.

An interesting point to be made regarding the Taylor expansions of STs is that a meaningful expansion of $\widehat{X}_{1,\boldsymbol{\theta}}(\sigma,t)$ must contain more higher-order terms than a local expansion of $X_3(\boldsymbol{s},t)$, because T_3^1 assigns to each value of σ the integral of $X_3(\boldsymbol{s},t)$ over a planar domain of points $\boldsymbol{s}\in R^3$.

I conclude this section on ST-based random field modeling by reminding the readers that several SPDEs representing random fields $X_n(\boldsymbol{s},t)$ can be conveniently reduced to SODEs representing $X_1(s,t)$ by means of suitably chosen STs. The SPDE solution can then be reconstructed from the SODE solutions along a finite set of transform lines. The application of the ST approach in the solution of SPDE representing physical laws is discussed in Section 3.5 and in Chapter XIV.

3.3 SPACE TRANSFORMATION FOR SPATIOTEMPORAL VARIABILITY FUNCTIONS

Naturally, all relevant ST formulas introduced earlier apply in the case of the space–time covariance, $c_n(\boldsymbol{h},\tau)$, $(\boldsymbol{h},\tau)\in R^{n,1}$, see Table 3.3. As usual, the $\widetilde{c}_n(\boldsymbol{k},\tau)$ and $\widetilde{\widehat{c}}_{1,\boldsymbol{\theta}}(k,\tau)$ are called the wavevector–temporal and wavenumber–temporal spectra corresponding to the covariances $c_n(\boldsymbol{h},\tau)$ and $\widehat{c}_{1,\boldsymbol{\theta}}(h,\tau)$, respectively.

Remark 3.2

As the readers observe, the basic STs are T_n^1 and Ψ_1^n, whereas the others can be defined in terms of these two. In particular, in the physical domain, $\Psi_n^1 = \Omega T_n^1$ and $T_1^n = \Psi_1^n\Omega$; and in the frequency domain, $T_1^n = \left(T_n^1\right)^{-1}$ and $T_n^1 = \left(\Psi_1^n\right)^{-1}$.

Table 3.3 Covariance Space Transformations (STs) in $R^{n,1}$

ST	Physical Domain ($h = h\theta$)	Frequency Domain ($k = k\theta$)
T_n^1	$\int dh\, c_n(\boldsymbol{h},\tau)\delta(h - \boldsymbol{h}\cdot\boldsymbol{\theta}) = \widehat{c}_{1,\theta}(h,\tau)$	$\widetilde{c}_n(\boldsymbol{k},\tau) = \widetilde{\widehat{c}}_{1,\theta}(k,\tau)$
Ψ_n^1	$\Omega T_n^1[c_n](h,\boldsymbol{\theta},\tau) = c_{1,\theta}(h,\tau)$	$\frac{S_n k^{n-1}}{(2\pi)^{n-1}}\widetilde{c}_n(\boldsymbol{k},\tau) = \widetilde{c}_{1,\theta}(k,\tau)$
Ψ_1^n	$S_n^{-1}\int_{S_n} d\boldsymbol{\theta}\, c_{1,\theta}(\boldsymbol{h}\cdot\boldsymbol{\theta},\tau) = c_n(\boldsymbol{h},\tau)$	$\frac{(2\pi)^{n-1}}{S_n k^{n-1}}\widetilde{c}_{1,\theta}(k,\tau) = \widetilde{c}_n(\boldsymbol{k},\tau)$
T_1^n	$\Psi_1^n\Omega[\widehat{c}_{1,\theta}](\boldsymbol{h},\tau) = c_n(\boldsymbol{h},\tau)$	$\widetilde{\widehat{c}}_{1,\theta}(k,\tau) = \widetilde{c}_n(\boldsymbol{k},\tau)$

Example 3.6

The formulas in Table 3.2 apply in the case of the isotropic covariance and spectral functions with $c_n(h,\tau) = c_n(|\boldsymbol{h}|,\tau)$, $\widetilde{c}_n(k,\tau) = \widetilde{c}_n(|\boldsymbol{k}|,\tau)$ $(n = 2, 3)$, by replacing X and $\widetilde{X}$ with c and $\widetilde{c}$, respectively.

Example 3.7

In certain physical applications, the projection of a random field on a lower dimensionality domain is observed, and one needs to calculate the original field in the higher dimensionality domain (e.g., atmospheric refractive index fluctuations in $R^{3,1}$ projected onto a wavefront in $R^{2,1}$). Consider such a random field $X_3(\boldsymbol{s},t)$, $\boldsymbol{s} = (s_1,s_2,s_3)$, and its transformation from $R^{3,1}$ to $R^{2,1}$ as follows,

$$X_2(s_1,s_2,t) = \int ds_3 u(s_3) X_3(\boldsymbol{s},t),$$

where $u(s_3)$ is a box function of width δs_3 and height δs_3^{-1} (physically, this means that the projection X_2 represents the X_3 average through a plate). In these applications, it is important that the observed covariance or spectrum in $R^{2,1}$ can be properly connected with the unobserved ones in $R^{3,1}$. The corresponding covariance functions are related by

$$c_{X_2}(h_1,h_2,t) = \iint ds_3 ds_3' u(s_3) u(s_3') c_{X_3}\left(\left(h_1^2 + h_2^2 + (s_3 - s_3')^2\right)^{\frac{1}{2}}, t\right).$$

For illustration, assume that $r = \left(h_1^2 + h_2^2\right)^{\frac{1}{2}} << \delta s_3$, and the simultaneous two-point covariance function has the form of a power law $c_{X_k}(r,t) \propto r^{-b_k}$, with $k = 2$, 3 and $b_3 = b_2 + 1 > 1$. Then, the covariance equation becomes

$$c_{X_2}(r,t) \approx \frac{1}{\delta s_3}\int ds_3 c_{X_3}\left(\left(r^2 + s_3^2\right)^{\frac{1}{2}}, t\right) = \frac{r}{\delta s_3} c_{X_3}(r,t).$$

For the associated spectral functions, where the projection through the plate of thickness δs_3 in the physical domain implies that in the spectral domain it holds that $\delta s_3 \delta k_3 \leq 1$ (see, also, uncertainty principle in Section 6 of Chapter VIII), it is valid that

$$\widetilde{c}_{X_2}(k,t) \approx \frac{1}{\delta s_3}\widetilde{c}_{X_3}(k,t),$$

where $\delta s_3 \delta k_3 \leq 1$, and the coefficient $\frac{1}{\delta s_3}$ is consistent with the physical condition of this application that $\widetilde{c}_{X_n}$ should be characterized by its spatial dimensions $(length)^n$, $n = 2$, 3.

The next proposition is a straightforward application of the analysis above (Christakos and Panagopoulos, 1992).

Proposition 3.6

The covariance $c_n(\boldsymbol{h},\tau)$, $(\boldsymbol{h},\tau) \in R^{n,1}$, is uniquely determined by means of its T_1^n or Ψ_n^1.

Example 3.8

In $R^{3,1}$ the $c_3(\boldsymbol{h},\tau)$ is uniquely defined if the values of $\widehat{c}_{1,\boldsymbol{\theta}}(h,\tau)$ are known along all lines passing through a given point, or if the values of $\widehat{c}_{2,\boldsymbol{\theta}}(h,\tau)$ are known on all planes passing through a given line. Similarly, $c_3(\boldsymbol{h},\tau)$ is uniquely determined if the $c_{1,\boldsymbol{\theta}}(h,\tau)$ or $c_{2,\boldsymbol{\theta}}(h,\tau)$ are known for all $\boldsymbol{\theta}$.[14]

Assuming that $c_n(\boldsymbol{h},\tau)$ is everywhere continuous, and since it forms an FT pair with $\widetilde{c}_n(\boldsymbol{w},\tau)$ in $R^{n,1}$, there exists an RV $\boldsymbol{u}$ in R^n that has $c_n(\boldsymbol{h},\tau)/c_n(\boldsymbol{0},\tau)$ as a CF; that is,

$$\frac{c_n(\boldsymbol{h},\tau)}{c_n(\boldsymbol{0},\tau)} = \overline{e^{i\boldsymbol{h}\cdot\boldsymbol{u}}}, \tag{3.36}$$

where the PDF of $\boldsymbol{u}$ is $f_{n,\boldsymbol{u}}(\boldsymbol{k},\tau) = \frac{\widetilde{c}_n(\boldsymbol{k},\tau)}{c_n(\boldsymbol{0},\tau)}$. In the isotropic case, the PDF will be a function of $k = |\boldsymbol{k}|$.

An interesting situation is described in the following proposition (Christakos and Panagopoulos, 1992).

Proposition 3.7

If $f_{1,\boldsymbol{u}}(k,\tau)$ is the one-dimensional PDF obtained by applying Ψ_n^1 on the PDF $f_{n,\boldsymbol{u}}(\boldsymbol{k},\tau)$, it is valid that

$$f_{1,\boldsymbol{u}}(k,\tau) = \Psi_n^1\left[f_{n,\boldsymbol{u}}(\boldsymbol{k},\tau)\right] = \frac{1}{2}f(k,\tau), \tag{3.37}$$

where $\Psi_n^1[\cdot] = \frac{S_n k^{n-1}}{(2\pi)^{n-1}}[\cdot]$ is the spectral domain expression of Ψ_n^1, and $f(k,\tau)$ is the one-dimensional PDF of k.

3.4 SPACE TRANSFORMATION IN THE SIMULATION OF SPATIOTEMPORAL RANDOM FIELDS

Let us take stock: On the basis of our ST results so far, it became evident that it is generally possible to apply the ST approach to find solutions of n-dimensional-time problems of S/TRF theory by transferring analysis to one-dimensional-time time problems of S/TRF theory by transferring analysis to one-dimensional-time problems. Typical among these problems is the simulation of random fields in the $R^{n,1}$ domain. Here, the fundamental practical issue is summarized as follows: "When can an S/TRF in $R^{n,1}$ be represented as a linear combination of statistically uncorrelated random fields in $R^{1,1}$?" Regarding this practical issue, two results have been derived (Christakos and Panagopoulos, 1992).

---

[14]Note that uniqueness requires an infinite number of hyperplanes; that is, $c_3(\boldsymbol{h},\tau)$ is not uniquely determined by any finite set of $c_{1,\boldsymbol{\theta}}(h,\tau)$ or $c_{2,\boldsymbol{\theta}}(h,\tau)$.

Proposition 3.8

(S/TRF representation; finite case): The random field $\widehat{X}_n(\boldsymbol{s},t)$, $(\boldsymbol{s},t) \in R^{n,1}$, admits a linear representation

$$\widehat{X}_n(\boldsymbol{s},t) = \sum_{i=1}^{N} u_i X_{1,i}(\boldsymbol{s}\cdot\boldsymbol{\theta}_i, t) \tag{3.38}$$

of pairwise statistical uncorrelated unidimensional-time fields $X_{1,i}$ along lines $\boldsymbol{\theta}_i$, $i = 1, 2,\ldots, N$, passing through a given point and with $u_i \in R^1$, if and only if the corresponding covariance admits the linear representation

$$\widehat{c}_n(\boldsymbol{h} = \boldsymbol{s} - \boldsymbol{s}', \tau = t - t') = \sum_{i=1}^{N} u_i u_i' c_{1,i}(\boldsymbol{h}\cdot\boldsymbol{\theta}_i, \tau), \tag{3.39}$$

where $c_{1,i}(\boldsymbol{h}\cdot\boldsymbol{\theta}_i, \tau)$ is the unidimensional covariance of the $X_{1,i}(\boldsymbol{s}\cdot\boldsymbol{\theta}_i, \boldsymbol{t})$, $i = 1, 2,\ldots, N$.[15]

In the general case, STs appear in the scene.

Proposition 3.9

(S/TRF representation; general case): The $X_n(\boldsymbol{s},t)$, $(\boldsymbol{s},t) \in R^{n,1}$, admits the representation

$$X_n(\boldsymbol{s},t) = \Psi_1^n\left[X_{1,\boldsymbol{\theta}}(s,t)\right] \tag{3.40}$$

with weight function $u(\boldsymbol{s},\boldsymbol{\theta})$, $s = \boldsymbol{s}\cdot\boldsymbol{\theta}$, and $X_{1,\boldsymbol{\theta}}$ are pairwise uncorrelated fields along lines $\boldsymbol{\theta}$ passing through a given point, if and only if the corresponding covariances are related by

$$c_n(\boldsymbol{h},\tau) = \Psi_1^n\left[c_{1,\boldsymbol{\theta}}(h,\tau)\right] \tag{3.41}$$

with weight function $u(\boldsymbol{s},\boldsymbol{\theta})u'(\boldsymbol{s}',\boldsymbol{\theta})$, where $\boldsymbol{h} = \boldsymbol{s}-\boldsymbol{s}'$, $h = \boldsymbol{h}\cdot\boldsymbol{\theta}$.

Some useful results—immediate consequence of the S/TRF representation results above and the central limit theorem—are introduced by the following corollaries.

Corollary 3.1

Assumptions of Proposition 3.8. Moreover, in Eq. (3.38) let $u_{ki} = \frac{1}{\sqrt{N}}$, that is,

$$\widehat{X}_n(\boldsymbol{s},t) = \frac{1}{\sqrt{N}} \sum_{i=1}^{N} X_{1,i}(\boldsymbol{s}\cdot\boldsymbol{\theta}_i, t). \tag{3.42}$$

The corresponding covariance will be given by, see Eq. (3.39),

$$\widehat{c}_n(\boldsymbol{h},\tau) = \frac{1}{N} \sum_{i=1}^{N} c_{1,i}(\boldsymbol{h}\cdot\boldsymbol{\theta}_i, \tau). \tag{3.43}$$

Corollary 3.2

In view of Eq. (3.42), let

$$X_n(\boldsymbol{s},t) = \lim_{N\to\infty} \widehat{X}_n(\boldsymbol{s},t). \tag{3.44}$$

[15]For simplicity, it is usually assumed that $X_{1,i}(\boldsymbol{s}_k\cdot\boldsymbol{\theta}_i, \boldsymbol{t})$ have zero mean and unit variance.

If the limit is finite, then $X_n(\boldsymbol{s},t)$ is an S/TRF with *Gaussian* univariate PDF and covariance

$$c_n(\boldsymbol{h},\tau) = \lim_{N\to\infty} \widehat{c}_n(\boldsymbol{h},\tau) = \Psi_1^n\left[c_{1,\theta}(h,\tau)\right], \tag{3.45}$$

where $h = \boldsymbol{h}\cdot\boldsymbol{\theta}$ and the weight function is S_n^{-1}.

On the strength of Corollaries 3.1 and 3.2, given an S/TRF $X_n(\boldsymbol{s},t)$ with covariance $c_n(\boldsymbol{h},\tau)$, the simulation problem in $R^{n,1}$ may be seen as the factorization of $X_n(\boldsymbol{s},t)$ as a weighted sum of N statistically uncorrelated random fields $X_{1,i}(\boldsymbol{s}\cdot\boldsymbol{\theta}_i,t)$, $i=1, 2,\ldots, N$, which is equivalent to a description of the factorization of $c_n(\boldsymbol{h},\tau)$ in the form of a weighted sum of the one-dimensional covariances $c_{1,\theta}(h,\tau)$. The latter factorization tends to the ST $\Psi_1^n\left[c_{1,\theta}(h,\tau)\right]$ as $N\to\infty$. The ST-based random field simulation technique has been termed the "turning bands method" is geostatistics. More details on the subject, including a variety of other random field simulation techniques, can be found in Christakos (1992).

Remark 3.4

In some situations of practical importance, deviations from the assumptions of Corollaries 3.1 and 3.2 may apply, such as, (*i*) the finite number of lines N considered and (*ii*) the nonexistence of the second-order moments of $X_{1,i}(\boldsymbol{s}\cdot\boldsymbol{\theta}_i,t)$, $i=1,\ldots,N$. Regarding (*i*), obviously all the integral relations defined in the preceding sections hold exactly as long as an infinite set of projection angles θ_i is considered. Yet, in practice only a finite number of θ_i is assumed, so that certain of the results obtained are only approximations of the integral relations above (which is not a novel fact, since all types of integral transforms—like the Fourier, the Abel, and the Radon ones—experience similar problems). Nevertheless, it turns out that in most applications of practical importance these approximations are very satisfactory. In the case that the univariate densities of $X_{1,i}(\boldsymbol{s}\cdot\boldsymbol{\theta}_i,t)$ are non-Gaussian and $N<\infty$, the random field of Eq. (3.42) may be only approximately Gaussian. Yet, in most situations in the simulation practice of natural attributes, such an approximation is satisfactory. Regarding (*ii*), an interesting situation arises when the $X_{1,i}(\boldsymbol{s}\cdot\boldsymbol{\theta}_i,t)$, $i=1, 2,\ldots, N$, all have *Cauchy* densities, $\left[\pi\left(1+X_{1,i}^2(\boldsymbol{s}\cdot\boldsymbol{\theta}_i,t)\right)\right]^{-1}$. Second moments do not exist and the $X_n^*(\boldsymbol{s},t)$ of Eq. (3.42) is distributed with density $\left[N^{\frac{1}{2}}\pi\left(1+\frac{X^2}{N}\right)\right]^{-1}$, which under no circumstances tends to the Gaussian density. This aspect may be important in practice, when experimental results show that several natural attributes do not satisfy the Gaussian assumption.[16]

In the case of space−time heterogeneous S/TRF, the variance might not exist, and then the variogram $\gamma_n(\boldsymbol{h},\tau)$ or the generalized covariance $\kappa_n(\boldsymbol{h},\tau)$ (Chapter XIII) are used instead of the ordinary covariance $c_n(\boldsymbol{h},\tau)$. As a matter of fact, the ST results derived above still apply by simply replacing $\{c_n(\boldsymbol{h},\tau),\ c_{1,\theta}(h,\tau)$ and $\widehat{c}_{1,\theta(h,\tau)}\}$ with

$$\left\{\gamma_n(\boldsymbol{h},\tau),\ \gamma_{1,\theta}(h,\tau) = \Psi_n^1[\gamma_n(\boldsymbol{h},\tau)],\ \widehat{\gamma}_{1,\theta}(h,\tau) = T_n^1[\gamma_n(\boldsymbol{h},\tau)]\right\},$$

or with

$$\left\{\kappa_n(\boldsymbol{h},\tau), \kappa_{1,\theta}(h,\tau) = \Psi_n^1[\kappa_n(\boldsymbol{h},\tau)],\ \widehat{\kappa}_{1,\theta}(h,\tau) = T_n^1[\kappa_n(\boldsymbol{h},\tau)]\right\},$$

[16] In general, a certain amount of experimentation with various numerical approaches and different algorithms is essential for the efficient application of the ST approach in simulation practice.

respectively. Several practical applications of the ST are discussed in the relevant literature (Christakos, 1992; Christakos and Panagopoulos, 1992; Christakos and Hristopulos, 1997, 1998; Hristopulos et al., 1999; Kolovos et al., 2004).

3.5 SPACE TRANSFORMATION IN THE SOLUTION OF STOCHASTIC PARTIAL DIFFERENTIAL EQUATION

As far as SPDE with space–time-dependent, in general, coefficients are concerned, the ST approach consists in reducing the original multidimensional problem to a unidimensional problem, usually an SODE. The solutions derived for the simpler unidimensional problem can be then used to obtain solutions of the multidimensional problem by means of inverse ST. Here I will focus on the T_n^1 transform. However, similar results can be obtained by means of the Ψ_n^1 transform.

Example 3.9

The subsurface flow law in $R^{2,1}$ governing the random hydraulic head field $X_2(\boldsymbol{s})$ is

$$\sum_{j=1}^{2}\left[\frac{\partial}{\partial s_j}+2s_j\right]X_2(\boldsymbol{s})=0, \tag{3.46}$$

subject to the boundary condition (BC) $X_2(\boldsymbol{s}=\boldsymbol{0})=1$. Using the ST approach, the exact solution is found to be

$$X_2(\boldsymbol{s})=\mathrm{e}^{-s^2+\alpha(s_1-s_2)}, \tag{3.47}$$

where α is an arbitrary real constant. The ST-based solution of Eq. (3.47) coincides with that obtained by the standard SPDE-solving method of separation of variables (Bialynicki-Birula et al., 1992). If $\alpha=0$, one recovers the isotropic solution

$$X_2(\boldsymbol{s})=\mathrm{e}^{-s^2}, \tag{3.48}$$

the T_2^1 of which is

$$\widehat{X}_{1,\boldsymbol{\theta}}(\sigma)=\sqrt{\pi}\mathrm{e}^{-\sigma^2}. \tag{3.49}$$

Above, the terms isotropic and anisotropic solution are used in reference to the spatial dependence of the solution. The isotropic solution depends entirely on the magnitude of the position vector, and it has no directional dependence.

The application of the ST approach in the solution of SPDE representing natural laws is discussed in more detail in Chapter XIV.

4. THE TRAVELING TRANSFORMATION

The idea behind the TT is that, from a modeling viewpoint, it may be attractive to project the composite space–time representation of a phenomenon on a pseudospatial domain of reduced dimensionality, and then take advantage of this projection in the study of the phenomenon.

4.1 BASIC NOTIONS

I start with the definition of a random field model that fits the above TT requirement of reduced dimensionality in terms of a change of space–time coordinates.

Definition 4.1

For any $\boldsymbol{p} = (\boldsymbol{s}, t)$, the *TT* is generally defined as a change of coordinates of the form

$$(\boldsymbol{s}, t) \in R^{n,1} \mapsto \widehat{\boldsymbol{s}} = \varpi(\boldsymbol{s}, t) \in R^n, \tag{4.1a}$$

where ϖ is a suitable function of the original space–time coordinates such that the S/TRF $X(\boldsymbol{p})$ is represented as

$$X(\boldsymbol{p}) = \widehat{X}(\widehat{\boldsymbol{s}}), \tag{4.1b}$$

with the resulting random field $\widehat{X}(\widehat{\boldsymbol{s}})$ called a *traveling random field* (*TRF*).

On practical grounds, of key importance in Definition 4.1 is the term "suitable function," that is, the function ϖ must be selected so that $\varpi(\boldsymbol{s},t) \in R^n$ and the calculations involving $\widehat{X}(\widehat{\boldsymbol{s}})$ are considerably simpler than those involving the original $X(\boldsymbol{p})$. Accordingly, the goal of TT is to project the study of a physical phenomenon in a convenient domain with certain attractive features (reduced dimensionality, improved representation of space–time variation, higher estimation accuracy, and lower computational cost).

I continue with a particularly simple yet fruitful selection of the ϖ function. I assume that for any $\boldsymbol{p} = (\boldsymbol{s}, t) \in R^{n,1}$ the TT

$$(\boldsymbol{s}, t) \in R^{n,1} \mapsto \widehat{\boldsymbol{s}} = \varpi(\boldsymbol{s}, t) = \boldsymbol{s} - \boldsymbol{\upsilon} t \in R^n \tag{4.2a}$$

can be defined with the TT velocity vector $\boldsymbol{\upsilon} = (\upsilon_1 \ldots \upsilon_n) \in R^n$ being generally a function of space–time, in which case the TRF can be written as

$$X(\boldsymbol{p}) = X(\boldsymbol{s} - \boldsymbol{\upsilon} t,\ 0) = \widehat{X}(\widehat{\boldsymbol{s}}), \tag{4.2b}$$

which also has some physical background.

Remark 4.1

Indeed, an interesting feature of TT of Eqs. (4.2a–b) should be pointed out, which is also the motivation behind this specific selection of the ϖ function. A similar to the random field equality of Eq. (4.2b) has been used in physical sciences to introduce the so-called *frozen* random field model (discussed in Section 2 of Chapter X). However, it should be stressed that in the TT context above a different interpretation of Eq. (4.2b) is used: while the frozen field notion assumes that the phenomenon satisfies Eq. (4.2b) and a physically meaningful justification of this assumption is sought, depending on the phenomenon under study, in Definition 4.1, Eq. (4.2a) is seen as a TT that is established so that Eq. (4.2b) "projects" $X(\boldsymbol{p})$ values assigned to points in the space–time domain onto $\widehat{X}(\widehat{\boldsymbol{s}})$ values in a traveling pseudospatial domain, where $\boldsymbol{s}$, t, and $\boldsymbol{\upsilon}$ are intrinsically linked and the $\boldsymbol{\upsilon}$ is specified in a self-consistent manner.

In operational terms, an advantage of the TT of Eqs. (4.2a–b) is that it requires the specification of a single parameter $\boldsymbol{\upsilon}$. Empirically, Eqs. (4.2a–b) introduce a certain link between the original, fully spatiotemporal, random field $X(\boldsymbol{p})$ and the traveling, pseudospatial, random field $\widehat{X}(\widehat{\boldsymbol{s}})$. This link can be viewed in different ways offering useful insights. Specifically, the space–time variation of $X(\boldsymbol{p})$ can be seen as determined by the pseudospatial attribute distribution $\widehat{X}(\widehat{s})$ that "travels" along the direction of the TT velocity $\boldsymbol{\upsilon}$ with speed $\upsilon = |\boldsymbol{\upsilon}|$. Although velocity $\boldsymbol{\upsilon}$ is, in general, a function of space–time, in several cases of practical interest, its speed υ may be considered constant (see $PM_{2.5}$ distribution in Case Study II.2.1).

Example 4.1

Consider a region in which the mean speed of disease spread is constant. From an epidemiologic viewpoint, this implies that the heterogeneities of the disease attribute distribution $X(\boldsymbol{p})$ move along the direction of $\boldsymbol{\upsilon}$ with speed $|\boldsymbol{\upsilon}|$ without changes.

Case Study II.2.1 (cont.)

Regarding Case Study II.2.1, the heterogeneities of $PM_{2.5}$ distribution in the Shandong region (China) during January 1–31, 2014 move along the direction of pollutant spread velocity that measures pollutant divergence from a region to another (Christakos et al., 2017c). The transformation from the original to the traveling $PM_{2.5}$ distribution involves an interconnection between pollutant, space, time, and velocity that can be specified accordingly.

A direct consequence of the projection hypothesis, Eqs. (4.1a–b), in the quantitative characterization of attribute mean trend and correlation structure (e.g., the amount by which a natural attribute values change together in a composite space–time domain) is as follows.

Proposition 4.1

Under the TRF hypothesis of Definition 4.1, the corresponding mean, covariance, and variogram functions of $X(\boldsymbol{p})$ and $\widehat{X}(\widehat{\boldsymbol{s}})$ are related by

$$\begin{aligned} \overline{X(\boldsymbol{p})} &= \overline{\widehat{X}(\widehat{\boldsymbol{s}})}, \\ c_X(\boldsymbol{h},\tau) &= c_{\widehat{X}}(\widehat{\boldsymbol{h}}), \\ \gamma_X(\boldsymbol{h},\tau) &= \gamma_{\widehat{X}}(\widehat{\boldsymbol{h}}), \end{aligned} \tag{4.3a–c}$$

where $(\boldsymbol{h},\tau) \in R^n \times T$ and $\widehat{\boldsymbol{s}} = \varpi(\boldsymbol{s},t)$, $\widehat{\boldsymbol{h}} = \varpi(\boldsymbol{h},t) \in R^n$.

The interrelations between the means, covariances, and variograms introduced by Eqs. (4.3a–c) may be associated to different types of attribute distributions. The observed interrelations need to be quantified for various purposes, including testing for the presence of significant attribute correlations, assessing space–time attribute dependence ranges, selecting theoretical models of dependency or attribute laws of change to summarize the observed spatiotemporal attribute structure, and inferring about the underlying attribute spread mechanism (spread velocity components, dispersal distances, and directional differences). On a related note, the covariance pair $(c_X, c_{\widehat{X}})$ of Eq. (4.3b) and the variogram pair $(\gamma_X, \gamma_{\widehat{X}})$ of Eq. (4.3c) establish *multidomain* representations of spatiotemporal variation, in the sense that the domain of c_X (γ_X) is $R^{n,1}$ and that of $c_{\widehat{X}}$ ($\gamma_{\widehat{X}}$) is R^n. Herein, I will focus on the specific TT of Eqs. (4.2a–b), in which case Eqs. (4.3a–c) become (see, also, Proposition 2.1 in Chapter X)

$$\begin{aligned}
&\overline{X(\boldsymbol{p})} = \overline{X(\boldsymbol{s},t)} = \overline{\widehat{X}(\widehat{\boldsymbol{s}})},\\
&c_X(\boldsymbol{h},\tau) = c_X(\boldsymbol{h}-\upsilon\tau,0) = c_{\widehat{X}}(\widehat{\boldsymbol{h}}),\\
&\gamma_X(\boldsymbol{h},\tau) = \gamma_X(\boldsymbol{h}-\upsilon\tau,0) = \gamma_{\widehat{X}}(\widehat{\boldsymbol{h}}),
\end{aligned} \quad (4.4\text{a–c})$$

where now $(\boldsymbol{h},\tau) \in R^n \times T$ and $\widehat{\boldsymbol{s}} = (\boldsymbol{s} - \upsilon t)$, $\widehat{\boldsymbol{h}} = (\boldsymbol{h} - \upsilon\tau) \in R^n$.

Case Study II.2.1 (cont.)

In the Shandong region of Fig. 3.1 of Chapter II, certain quantitative relationships were established between the original and the traveling $PM_{2.5}$ distribution by means of their respective variogram functions (offering a measure of pollutant relatedness across space–time) and the partial differential equation (PDE) laws these functions must satisfy (providing information about the direction in which pollutant's magnitude changes the most). These relationships identify pollutant distribution characteristics across space–time that can contribute to our understanding of $PM_{2.5}$ spread.

A direct result of Eqs. (4.4b–c) is that in the purely spatial case, $\tau = 0$, and the purely temporal case, $\boldsymbol{h} = \boldsymbol{0}$, it is valid, respectively (Exercise V.9),

$$\begin{aligned}
&c_X(\boldsymbol{h}) = c_{\widehat{X}}(\boldsymbol{h}),\\
&c_X(\tau) = c_{\widehat{X}}(\upsilon\tau),
\end{aligned} \quad (4.5\text{a–b})$$

where $c_X(\boldsymbol{h}) = c_X(\boldsymbol{h}, 0)$ and $c_X(\tau) = c_X(0,\tau)$.

Example 4.2

Consider the S/TRF $X(\boldsymbol{p})$, $\boldsymbol{p} \in R^{n,1}$, with covariance

$$c_X(\boldsymbol{h},\tau) = e^{-(\boldsymbol{h}+\tau^2)}, \quad (4.6\text{a})$$

where $(\boldsymbol{h},\tau) \in R^{n,1}$. Then, υ, $\widehat{\boldsymbol{h}} = (\boldsymbol{h} - \upsilon\tau) \in R^n$, and Eq. (4.4b) lead to,

$$e^{-(\boldsymbol{h}+\tau^2)} = e^{-[(\boldsymbol{h}-\upsilon\tau)+0^2]} = e^{-(\widehat{\boldsymbol{h}})}. \quad (4.6\text{b})$$

Eq. (4.6b) displays the different functional forms of the covariance functions of the fields X and $\widehat{X}$,

$$c_X(\cdot,\cdot) = e^{-(\cdot)-(\cdot)^2} \text{ and } c_{\widehat{X}}(\cdot) = e^{-(\cdot)},$$

respectively. As noted earlier, covariance interrelations introduced by Eq. (4.6b) may be associated with different kinds of space–time attribute distributions.

In the special case of a spatially isotropic covariance, the multidomain representation of the covariance pair $\left(c_X, c_{\widehat{X}}\right)$ of Eq. (4.4b) is conveniently modified as follows.

Proposition 4.2

In the spatially isotropic case, the covariances $c_X(|\boldsymbol{h}|,\tau)$ of the original attribute $X(\boldsymbol{p})$ and $c_{\widehat{X}}(\widehat{h})$ of the traveling attribute $\widehat{X}(\widehat{s})$ are related by the multidomain representation,

$$c_X(|\boldsymbol{h}|,\tau) = c_X(|\boldsymbol{h}| - \upsilon\tau, 0) = c_{\widehat{X}}\left(\widehat{h}\right), \tag{4.7}$$

where $|\boldsymbol{h}| = \left(\sum_{i=1}^{n} h_i^2\right)^{\frac{1}{2}}$ and $\upsilon = |\boldsymbol{\upsilon}| = \left(\sum_{i=1}^{n} \upsilon_i^2\right)^{\frac{1}{2}}$, with $(|\boldsymbol{h}|,\tau) \in R^1 \times T$, and $\widehat{h} = (|\boldsymbol{h}| - \upsilon\tau) \in R^1$.

Example 4.3

Consider the random field $X(\boldsymbol{p})$, $\boldsymbol{p} \in R^{n,1}$, with the spatially isotropic covariance function

$$c_X(\boldsymbol{h},\tau) = e^{-\left(|\boldsymbol{h}|^2+\tau^2\right)^{\frac{1}{2}}}, \tag{4.8}$$

where $(|\boldsymbol{h}|,\tau) \in R^{1,1}$. Then, υ, $\widehat{h} = (|\boldsymbol{h}| - \upsilon\tau) \in R^1$, and Eq. (4.4b) leads to,

$$e^{-\left(|\boldsymbol{h}|^2+\tau^2\right)^{\frac{1}{2}}} = e^{-\left[(|\boldsymbol{h}|-\upsilon\tau)^2+0^2\right]^{\frac{1}{2}}} = e^{-\widehat{\boldsymbol{h}}}, \tag{4.9}$$

which clearly distinguishes between the multidomain covariance representations.

Example 4.4

For numerical illustration, a database is considered that is the simulated unidimensional space–time ($R^{1,1}$) mortality distribution shown in Fig. 4.1A. A total of 100 (10 × 10) grid points with coordinates s, t were considered, and the mortality values were generated at these points using the sequential simulation technique ("simuseq.m" function of the *BMELib 2.0* software; Yu et al., 2007). The histogram of the simulated mortality values is plotted in Fig. 4.1B. Besides the fact that this is a computationally convenient study serving the purpose of illustrating numerically certain features of the TT model, it is also a reasonable representation of a real-world situation in which a study's focus is the temporal variation of mortality along the dominant spatial direction of epidemic spread. The correlation space–time structure of mortality distribution is represented by the covariance model

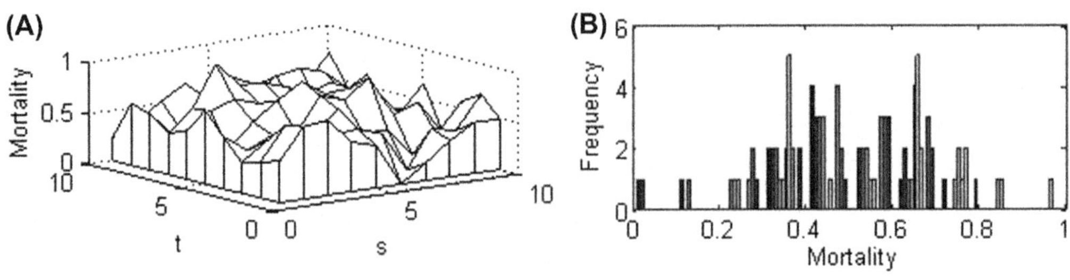

FIGURE 4.1

(A) Simulated mortality distribution in $R^1 \times T$; and (B) the corresponding histogram.

$c_X(h,\tau)$ of Eq. (4.8) with $n = 1$, i.e., $c_X(h, \tau) = e^{-(h^2+\tau^2)^{\frac{1}{2}}}$. This covariance, which is plotted in Fig. 4.2A measures the strength of the dependency between mortality values across space–time (the stronger the spatiotemporal dependence, the smaller the variance between nearby pairs of mortality observations). The $c_X(h,\tau)$ accounts for different components of mortality variance, and, due to its shape, it represents adequately the space–time correlations of the spatially homogeneous/temporally stationary mortality pattern of Fig. 4.1A. The covariance magnitude drops rather fast, implying that strong mortality correlations appear at short space and time lags, whereas they are negligible at longer lags. Moreover, the sharp slope of $c_X(h,\tau)$ near the origin implies a mortality pattern with considerable variability, which is, indeed, the case of the distribution of Fig. 4.1A.

Given the close formal link between the TT of Eqs. (4.2a–b) and the frozen random field model, several of the results obtained in the context of the latter can be fruitfully used in applications of the former. Moreover, some special cases of the TT may apply in space–time modeling, which can simplify matters considerably. According to one such special case, the $\boldsymbol{v}$ of the TT may be specified so that an STHS random field $X(\boldsymbol{p})$, $\boldsymbol{p} \in R^{n,1}$, admits a covariance of the form

$$c_X(\mathbf{0}, \tau) = c_X(\boldsymbol{v}\tau, 0), \tag{4.10}$$

where the TT velocity vector $\boldsymbol{v}$ is specified so that Eq. (4.10) is valid. Clearly, Eq. (4.10) can be derived from the TT of Eq. (4.4b), see, also, Eq. (2.3) of Chapter X.

Example 4.5

A class of STIS covariance functions members of which satisfy Eq. (4.10) is the following

$$c_X(\boldsymbol{h}, \tau) = c_0 e^{-\beta(|\boldsymbol{h}|^\alpha + \tau^\alpha)} \tag{4.11}$$

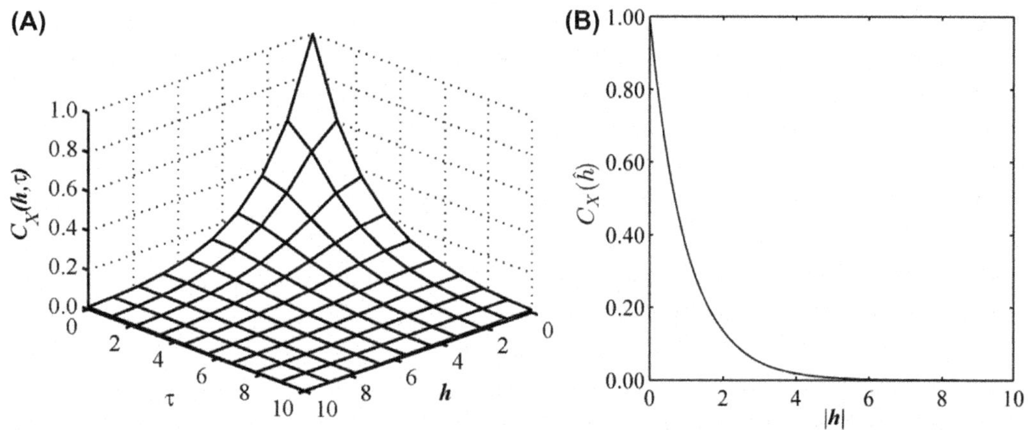

FIGURE 4.2

(A) Space–time covariance of the mortality distribution ($R^1 \times T$); and (B) projected covariance of the traveling mortality distribution $c_{\widehat{X}}(\widehat{h})$ in R^1.

where $\alpha \in (0,2]$, and $\beta > 0$. The corresponding TT velocity vector is simply given by

$$v_1 = 1, \ \ v_2 = \ldots = v_n = 0, \tag{4.12}$$

i.e., the attribute moves along the direction of v_1 in the region of interest. Obviously, the covariance of the mortality distribution (Fig. 4.2A) does not belong to the class of Eq. (4.10). In fact, as we shall see below, the covariance function of the mortality distribution could satisfy Eq. (4.10) only if it could be assumed that the velocity vector is constant, which, though, is not the case of the mortality distribution of Fig. 4.1A.

Furthermore, when a vector $\boldsymbol{v}$ is defined so that Eqs. (4.2a–b) holds, the TT leads to PDE representations of the random field and its covariance function, just as in the case of a space–time frozen random field (a thorough discussion of these representations can be found in Section 2.3 of Chapter X).

Example 4.6

The covariance function of Eq. (4.4b) admits the representations

$$\begin{aligned} &\left[\frac{\partial}{\partial \tau} + \sum_{i=1}^{n} v_i \frac{\partial}{\partial h_i}\right] c_X(\boldsymbol{h}, \tau) = \left[\frac{\partial}{\partial \tau} + \boldsymbol{v}^T \nabla\right] c_X(\boldsymbol{h}, \tau) = 0, \\ &\left[\frac{\partial^2}{\partial h_j \partial \tau} + \sum_{i=1}^{n} v_i \frac{\partial^2}{\partial h_i \partial h_j}\right] c_X(\boldsymbol{h}, \tau) = \nabla \frac{\partial c_X}{\partial \tau} + \boldsymbol{H}_c \boldsymbol{v} = 0, \end{aligned} \tag{4.13a–b}$$

where $j = 1,\ldots, n$, and $\boldsymbol{H}_c$ is the Hessian matrix (Eq. 2.14d of Chapter X). From Eqs. (4.13a–b) it is valid

$$\frac{\partial c_X}{\partial \tau} - \left(\boldsymbol{H}_c^{-1} \nabla \frac{\partial c_X}{\partial \tau}\right)^T \nabla c_X = 0. \tag{4.14}$$

For illustration, in $R^{1,1}$ Eqs. (4.13a–b) and (4.14) give

$$v = \begin{cases} -\dfrac{\partial c_X}{\partial \tau} \left(\dfrac{\partial c_X}{\partial h}\right)^{-1} \\ -\dfrac{\partial^2 c_X}{\partial h \partial \tau} \left(\dfrac{\partial^2 c_X}{\partial h^2}\right)^{-1} \end{cases}, \tag{4.15a–b}$$

which is an equation that should be kept in mind in the context of the traveling vector determination, which is the subject of the next section.

4.2 DETERMINATION OF THE TRAVELING VECTOR

The traveling vector $\boldsymbol{v}$ is, generally, a function of space–time, in which case it is essential to establish ways of determining $\boldsymbol{v}$. In fact, there exists an intrinsic dependence between $\boldsymbol{v}$, $\boldsymbol{h}$, and τ that plays an important role in the self-consistent selection of these parameters. I start by noticing that TT self-consistency implies that the change of space–time coordinates

$$(\boldsymbol{h}, \tau) \in R^{n,1} \mapsto (\boldsymbol{h} - \boldsymbol{v}\tau) \in R^n \tag{4.16}$$

is one-to-one. In certain cases an exact approach for determining $\boldsymbol{v}$ is possible (i.e., one can specify an explicit relationship between $\boldsymbol{v}$, $\boldsymbol{h}$, and τ, the form of which depends on the shape of the S/TVF considered), whereas in some other cases an approximate approach may be needed.

Approach 1

As regards the exact approach, given the S/TVF (covariance or the variogram) form, Eqs. (4.4b–c) can specify a relationship between the values of $\boldsymbol{v}$, $\boldsymbol{h}$, and τ, say,

$$V(\boldsymbol{h}, \tau, \boldsymbol{v}) = 0, \tag{4.17}$$

where the form of V depends on that of the S/TVF. By definition, $\widehat{\boldsymbol{h}} \in R^n$ enters into the picture by means of the relationship

$$\widehat{\boldsymbol{h}} = \boldsymbol{h} - \boldsymbol{v}\tau. \tag{4.18}$$

Given $(\boldsymbol{h}, \tau) \in R^{n,1}$, Eq. (4.17) determines the value of $\boldsymbol{v}$ in terms of the spatial and temporal correlation ranges. Then, Eq. (4.18) provides the corresponding $\widehat{\boldsymbol{h}}$. In this way, each set of $(\boldsymbol{h}, \tau)$ values is related to a unique set of $\left(\boldsymbol{v}, \widehat{\boldsymbol{h}}\right)$ values.

Example 4.7

In $R^{n,1}$ consider the random field $X(\boldsymbol{p})$, with space–time covariance

$$c_X(\boldsymbol{h}, \tau) = c_0 e^{-\left(\frac{|\boldsymbol{h}|^2}{\beta_s^2} + \frac{\tau^2}{\beta_t^2}\right)^{\frac{1}{2}}}. \tag{4.19}$$

In view of Eq. (4.19), Eq. (4.4b) gives

$$c_0 e^{-\left(\frac{|\boldsymbol{h}|^2}{\beta_s^2} + \frac{\tau^2}{\beta_t^2}\right)^{\frac{1}{2}}} = c_0 e^{-\left(\frac{(|\boldsymbol{h}| - v\tau)^2}{\beta_s^2}\right)^{\frac{1}{2}}}, \tag{4.20}$$

which leads to a V relationship of the form

$$\beta_t^2 \tau\, v^2 - 2\beta_t^2 |\boldsymbol{h}| v - \beta_s^2 \tau = 0. \tag{4.21}$$

The solution of Eq. (4.21) determines the TT velocity values along different directions in space as

$$v_1 = \frac{|\boldsymbol{h}|}{\tau} \pm \left(\frac{|\boldsymbol{h}|^2}{\tau^2} + \frac{\beta_s^2}{\beta_t^2}\right)^{\frac{1}{2}}, \quad v_2 = \ldots = v_n = 0. \tag{4.22}$$

The covariance function of Eq. (4.19) represents an attribute distribution that travels along direction s_1 with varying speed, in which case Eq. (4.18) becomes

$$\widehat{h} = |\boldsymbol{h}| - \upsilon\tau = \mp\left(|\boldsymbol{h}|^2 + \frac{\beta_s^2}{\beta_t^2}\tau^2\right)^{\frac{1}{2}}, \tag{4.23}$$

which links the lags in the two domains.

Example 4.8

To obtain some more specific results, consider the covariance functions Eqs. (4.11) and (4.19) that satisfy Eq. (4.10) with TT velocity vectors given by, respectively,

$$\begin{aligned} \upsilon_1 &= 1, \quad \upsilon_2 = \ldots = \upsilon_n = 0, \\ \upsilon_1 &= \frac{\beta_s}{\beta_t}, \quad \upsilon_2 = \ldots = \upsilon_n = 0. \end{aligned} \tag{4.24a–b}$$

For these specific velocity choices, the above two covariance functions, indeed, satisfy Eq. (4.10). Yet, an interesting observation is that for the same covariance function of Eq. (4.19), the $\boldsymbol{\upsilon}$ value for the hypothesis of Eq. (4.10) is space independent and time independent, see Eq. (4.24b), whereas for the more general TT hypothesis it is space dependent and time dependent, see Eq. (4.22).

Example 4.9

In the case of the simulated mortality data (Fig. 4.1A), the space- and time-dependent velocity of mortality spread is given by Eq. (4.22) with $\beta_s = \beta_t = 1$, $h = |\boldsymbol{h}| \in R^1$, and

$$\upsilon_1 = \upsilon^{(i)} = \frac{h}{\tau} \pm \left(\frac{h^2}{\tau^2} + 1\right)^{\frac{1}{2}}, \tag{4.25}$$

where the superscript $i = 1$ corresponds to the "+" sign, and $i = 2$ to the "−" sign in the equation. Accordingly, the covariance of the traveling mortality distribution is given by

$$c_X(h,\tau) = c_{\widehat{X}}\left(\widehat{h}\right) = e^{-\left(h-\upsilon^{(i)}\tau\right)},$$

where $\widehat{h} = \left(h - \upsilon^{(i)}\tau\right)$, directly implying that $\widehat{h}^2 = h^2 + \left(\upsilon^{(i)}\tau\right)^2 - 2\upsilon^{(i)}h\tau = h^2 + \tau^2$. The plots in Fig. 4.3 show how the $\upsilon^{(i)}$ values of Eq. (4.22) vary with space and time for various space h and time τ lag values. The plot of Fig. 4.3A corresponds to the option $\upsilon^{(1)} \geq 0$, and that of Fig. 4.3B corresponds to the option $\upsilon^{(2)} \leq 0$. A visual inspection of Fig. 4.3A and B concludes that $\upsilon^{(1)}$ increases with space lag h and decreases with time lag τ, whereas, in absolute values, $\upsilon^{(2)}$ decreases with h and increases with τ (although at different rates). The choice between the two options of mortality velocity should depend on the particular features of the disease pattern in the region of interest. To gain additional insight, the temporal and spatial rates of change of the mortality velocity, viz.,

$$\begin{aligned} \frac{\partial\upsilon^{(i)}}{\partial\tau} &= -\frac{1}{\tau^2}\left[h \pm \frac{h^2}{\tau}\left(\frac{h^2}{\tau^2}+1\right)^{-\frac{1}{2}}\right], \\ \frac{\partial\upsilon^{(i)}}{\partial h} &= \frac{1}{\tau}\left[1 \pm \frac{h}{\tau}\left(\frac{h^2}{\tau^2}+1\right)^{-\frac{1}{2}}\right] \end{aligned} \tag{4.26a–b}$$

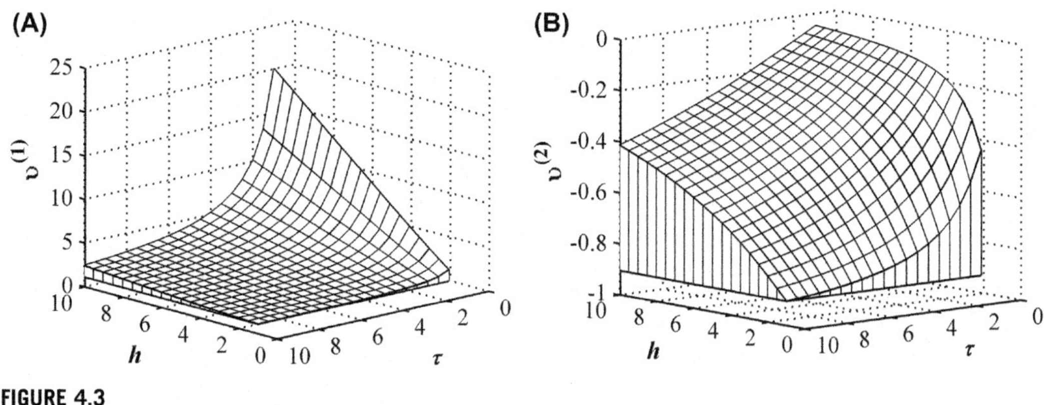

FIGURE 4.3

Mortality velocity plots as a function of space lag, h, and time lag, τ: (A) $\upsilon^{(1)}$ and (B) $\upsilon^{(2)}$.

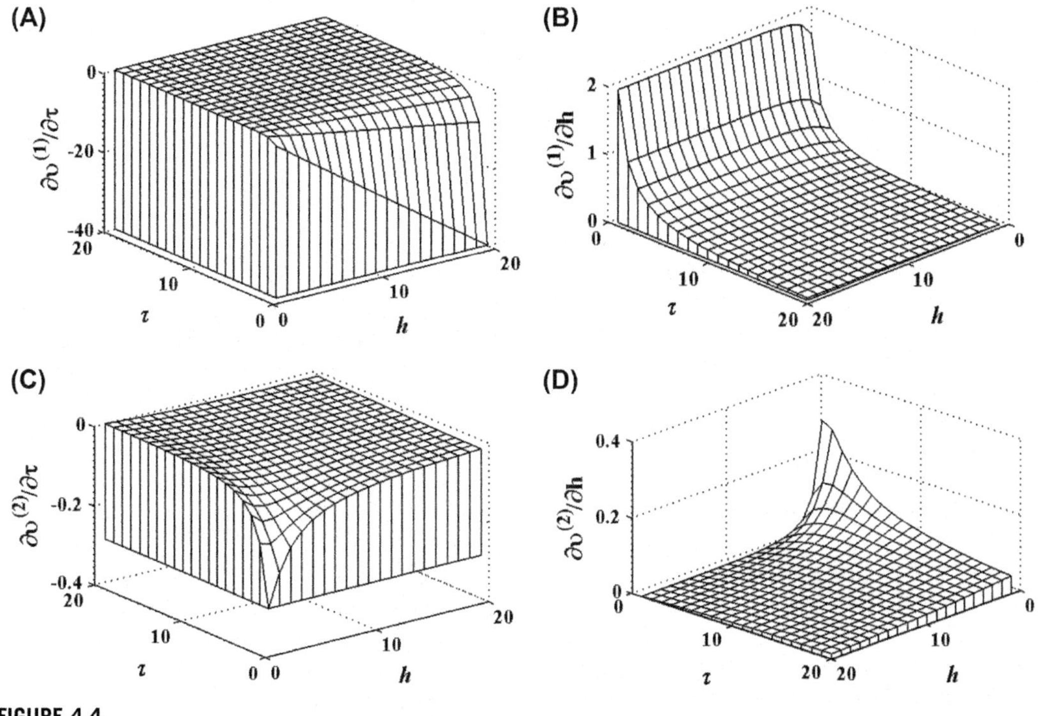

FIGURE 4.4

Plots of temporal and spatial mortality velocity rates as a function of space lag, h, and time lag, τ: (A) $\frac{\partial\upsilon^{(1)}}{\partial\tau}$, (B) $\frac{\partial\upsilon^{(1)}}{\partial h}$, (C) $\frac{\partial\upsilon^{(2)}}{\partial\tau}$, and (D) $\frac{\partial\upsilon^{(2)}}{\partial h}$.

($i = 1, 2$) are plotted in Fig. 4.4A–D. A positive rate of TT velocity change may imply that the disease spreads out of a given region with time, whereas a negative rate may denote a disease entering the region with time. The temporal rate of $\upsilon^{(1)}$ change is initially negative but it drops to zero very fast with increasing τ lag, whereas it initially increases quickly to reach a certain value but then also tends to zero quickly with increasing $|\boldsymbol{h}|$ lag. The spatial rate of $\upsilon^{(1)}$ change is positive, decreasing very slowly (or even staying constant) with $|\boldsymbol{h}|$ lag but much faster with τ lag. As regards $\upsilon^{(2)}$, its temporal rate of change is negative, dropping to zero fast with increasing τ and h lags. The spatial rate of $\upsilon^{(2)}$ change is positive and it also decreases with increasing τ and $|\boldsymbol{h}|$ lag (although the decrease is somehow faster with increasing $|\boldsymbol{h}|$). The projected covariance of the traveling mortality distribution is plotted in Fig. 4.2B (notice that only the $\upsilon^{(2)}$ value is used in this plot, since $\upsilon^{(1)}$ gives negative projected lags).

Approach 2

In the case of a space/time-independent traveling vector $\boldsymbol{\upsilon}$, another approach is based on the PDE representations of the TT. Indeed, the traveling vector $\boldsymbol{\upsilon}$ can be calculated directly from Eq. (4.13b) as

$$\boldsymbol{\upsilon} = -\boldsymbol{H}_c^{-1}\frac{\partial}{\partial\tau}\nabla c_X, \tag{4.27}$$

anywhere on the $c_X(\boldsymbol{h},\tau)$ space–time surface, except at the points in which the determinant of the Hessian (det $\boldsymbol{H}_c$) vanishes.

Example 4.10

In $R^{1,1}$ consider the covariance function

$$c_X(h,\tau) = c_0 e^{-\left(\frac{h}{\beta_s} + \frac{\tau}{\beta_t}\right)}. \tag{4.28}$$

It is easily found that

$$\upsilon = -\frac{\partial c_X}{\partial\tau}\left(\frac{\partial c_X}{\partial h}\right)^{-1} = -\frac{\partial^2 c_X}{\partial h\partial\tau}\left(\frac{\partial^2 c_X}{\partial h^2}\right)^{-1} = -\beta_s\beta_t^{-1}, \tag{4.29}$$

i.e., the condition of Eq. (4.15a–b) is satisfied, and the traveling parameter υ can be calculated as in Eq. (4.27), and the covariance of $\widehat{X}(s - \upsilon t)$ will be

$$c_{\widehat{X}}(h - \upsilon\tau) = c_0 e^{-\frac{h-\upsilon\tau}{\beta_s}} = c_0 e^{-\frac{h+\beta_s\beta_t^{-1}\tau}{\beta_s}}. \tag{4.30}$$

On the other hand, for the covariance of Eq. (4.19) with $n = 1$ it is valid that

$$\begin{aligned} &\frac{\partial c_X}{\partial\tau}\left(\frac{\partial c_X}{\partial h}\right)^{-1} = \beta_s^2\beta_t^{-2}\tau\, h^{-1}, \\ &\frac{\partial^2 c_X}{\partial h\partial\tau}\left(\frac{\partial^2 c_X}{\partial h^2}\right)^{-1} = \frac{1 + (\beta_s^{-2}h^2 + \beta_t^{-2}\tau^2)^{-\frac{1}{2}}}{1 + (\beta_s^{-2}h^2 + \beta_t^{-2}\tau^2)^{-\frac{1}{2}} - \beta_s^2 h^{-2}(\beta_s^{-2}h^2 + \beta_t^{-2}\tau^2)^{\frac{1}{2}}}\beta_s^2\beta_t^{-2}\tau\, h^{-1}, \end{aligned} \tag{4.31a–b}$$

i.e., the condition of Eq. (4.15a–b) is not satisfied for a space- and time-dependent traveling vector, as was expected.

Example 4.11

In the $R^{2,1}$ domain, Eq. (4.13b) leads to the set of equations,

$$\begin{aligned}&\frac{\partial^2 c_X}{\partial h_1 \partial \tau} + v_1 \frac{\partial^2 c_X}{\partial h_1^2} + v_2 \frac{\partial^2 c_X}{\partial h_1 \partial h_2} = 0,\\&\frac{\partial^2 c_X}{\partial h_2 \partial \tau} + v_1 \frac{\partial^2 c_X}{\partial h_2 \partial h_1} + v_2 \frac{\partial^2 c_X}{\partial h_2^2} = 0.\end{aligned} \tag{4.32a–b}$$

In matrix formulation, Eqs. (4.32a–b) become

$$\begin{bmatrix} \dfrac{\partial^2 c_X}{\partial h_1 \partial \tau} \\[2ex] \dfrac{\partial^2 c_X}{\partial h_2 \partial \tau} \end{bmatrix} + \boldsymbol{H}_c \begin{bmatrix} v_1 \\ v_2 \end{bmatrix} = \mathbf{0},$$

where

$$\boldsymbol{H}_c = \begin{bmatrix} \dfrac{\partial^2 c_X}{\partial h_1^2} & \dfrac{\partial^2 c_X}{\partial h_1 \partial h_2} \\[2ex] \dfrac{\partial^2 c_X}{\partial h_1 \partial h_2} & \dfrac{\partial^2 c_X}{\partial h_2^2} \end{bmatrix}.$$

The vector $\boldsymbol{v} = (v_1\ v_2)^T$ can be directly calculated from Eqs. (4.32a–b), i.e.,

$$\begin{aligned}\begin{bmatrix} v_1 \\ v_2 \end{bmatrix} &= -\boldsymbol{H}_c^{-1} \begin{bmatrix} \dfrac{\partial^2 c_X}{\partial h_1 \partial \tau} \\[2ex] \dfrac{\partial^2 c_X}{\partial h_2 \partial \tau} \end{bmatrix} = -\frac{1}{\dfrac{\partial^2 c_X}{\partial h_1^2}\dfrac{\partial^2 c_X}{\partial h_2^2} - \left(\dfrac{\partial^2 c_X}{\partial h_1 \partial h_2}\right)^2} \begin{bmatrix} \dfrac{\partial^2 c_X}{\partial h_2^2} & -\dfrac{\partial^2 c_X}{\partial h_1 \partial h_2} \\[2ex] -\dfrac{\partial^2 c_X}{\partial h_1 \partial h_2} & \dfrac{\partial^2 c_X}{\partial h_1^2} \end{bmatrix} \begin{bmatrix} \dfrac{\partial^2 c_X}{\partial h_1 \partial \tau} \\[2ex] \dfrac{\partial^2 c_X}{\partial h_2 \partial \tau} \end{bmatrix},\\ &= \frac{1}{\left(\dfrac{\partial^2 c_X}{\partial h_1 \partial h_2}\right)^2 - \dfrac{\partial^2 c_X}{\partial h_1^2}\dfrac{\partial^2 c_X}{\partial h_2^2}} \begin{bmatrix} \left(\dfrac{\partial^2 c_X}{\partial h_2^2}\dfrac{\partial^2 c_X}{\partial h_1 \partial \tau} - \dfrac{\partial^2 c_X}{\partial h_2 \partial \tau}\dfrac{\partial^2 c_X}{\partial h_1 \partial h_2}\right) \\[2ex] \left(\dfrac{\partial^2 c_X}{\partial h_1^2}\dfrac{\partial^2 c_X}{\partial h_2 \partial \tau} - \dfrac{\partial^2 c_X}{\partial h_1 \partial h_2}\dfrac{\partial^2 c_X}{\partial h_1 \partial \tau}\right) \end{bmatrix},\end{aligned} \tag{4.33a–b}$$

which satisfy Eq. (4.13a–b) in the $R^{2,1}$-domain.[17]

[17]Geometrically, det $\boldsymbol{H}_c$ is the Gaussian curvature of the c_X surface at a fixed time. Points having large curvature are those exhibiting large directional contrasts, and, hence, they are adequate for calculating vector $\boldsymbol{v}$.

Case Study II.2.1 (cont.)

Case Study II.2.1 considered a $PM_{2.5}$ data set in the Shandong province (China) with empirical and theoretical $PM_{2.5}$ variograms plotted in Fig. 3.2 of Chapter II. Let $\boldsymbol{v} = (v_1,\ v_2)$ be the TT velocity vector of pollutant spread, where $X(\boldsymbol{p})$ now represents the $PM_{2.5}$ distribution with

$$\boldsymbol{p} = (\boldsymbol{s}, t) = (s_1, s_2, t) \in R^{2,1} \mapsto \widehat{\boldsymbol{s}} = \boldsymbol{s} - \boldsymbol{v}t \in R^2. \tag{4.34}$$

The TT of Eq. (4.2b) introduces a link between the original $X(\boldsymbol{p})$ in $R^{2,1}$ and the traveling $\widehat{X}(\widehat{\boldsymbol{s}})$ in R^2. The time variation of the $PM_{2.5}$ distribution is specified by the pseudospatial distribution $\widehat{X}(\widehat{\boldsymbol{s}})$ that "travels" along the $\boldsymbol{v}$ direction with speed $v = |\boldsymbol{v}|$. Direct consequence of the TT-based characterization of $PM_{2.5}$ dependence (the amount by which pollutant values change together in space–time) is that the $PM_{2.5}$ variogram $\gamma_X(|\boldsymbol{h}|,\tau)$ of Eq. (3.10) of Chapter II determining the space–time dependence of the original $X(\boldsymbol{p})$ distribution and the variogram $\gamma_{\widehat{X}}(\widehat{h})$ of the TT-based distribution $\widehat{X}(\widehat{\boldsymbol{s}})$ are related by the two-domain variogram representation

$$\gamma_X(|\boldsymbol{h}|, \tau) = \gamma_X(|\boldsymbol{h}| - v\tau, 0) = \gamma_{\widehat{X}}(\widehat{h}), \tag{4.35}$$

where $(|\boldsymbol{h}|,\tau) \in R^{1,1}$, $\widehat{h} = (|\boldsymbol{h}| - v\tau) \in R^1$, $|\boldsymbol{h}| = \left(h_1^2 + h_2^2\right)^{\frac{1}{2}} = h_E$, $v = |\boldsymbol{v}| = \left(v_1^2 + v_2^2\right)^{\frac{1}{2}}$ (a function of space–time). Interrelations between γ_X and $\gamma_{\widehat{X}}$ are linked to different kinds of $PM_{2.5}$ patterns. The observed interrelations may be quantified for various purposes, including testing for the presence of significant $PM_{2.5}$ correlations, assessing space–time $PM_{2.5}$ dependence ranges, selecting theoretical models (variogram class) or pollutant laws of change to summarize the observed spatiotemporal pollutant structure, and inferring about the $PM_{2.5}$ spread mechanism (spread velocity components, dispersal distances, directional differences). By combining the theoretical variogram of Eq. (3.10) of Chapter II with Eq. (4.35), the TT velocity is calculated as

$$v = \frac{1}{\tau}\left[|\boldsymbol{h}| - \left(\frac{|\boldsymbol{h}|^{2r}}{(b\tau^{\alpha}+1)^{\beta r}} + \frac{\beta d \ln(b\tau^{\alpha}+1)}{2\kappa}\right)^{\frac{1}{2r}}\right] \tag{4.36}$$

(km/day), which expresses the interdependence between $\boldsymbol{v}$, $\boldsymbol{h}$, and τ. The v distribution of Eq. (4.36) is plotted in Fig. 4.5. Self-consistency of the $PM_{2.5}$ representation generally implies that the TT $(|\boldsymbol{h}|, \tau) \mapsto \widehat{h}$ is one-to-one, whereas $\widehat{h}$ enters the picture by means of the relationship

$$\widehat{h} = |\boldsymbol{h}| - v\tau = \left(\frac{|\boldsymbol{h}|^{2r}}{(b\tau^{\alpha}+1)^{\beta r}} + \frac{\beta d \ln(b\tau^{\alpha}+1)}{2\kappa}\right)^{\frac{1}{2r}}. \tag{4.37}$$

In this way, each pair of $(|\boldsymbol{h}|,\tau)$ values is related to a unique pair of $(v, \widehat{h})$ values. By applying Eq. (4.37) on the original space–time empirical variogram of Fig. 3.2 of Chapter II, I obtained the traveling empirical variogram in Fig. 4.6. To this empirical variogram, I fit the theoretical traveling model

$$\gamma_{\widehat{X}}(\widehat{h}) = c_0 + c\begin{cases} 1 - e^{-a^{-1}\widehat{h}} & (\widehat{h} \le 3a), \\ 1 & (\widehat{h} > 3a), \end{cases} \tag{4.38}$$

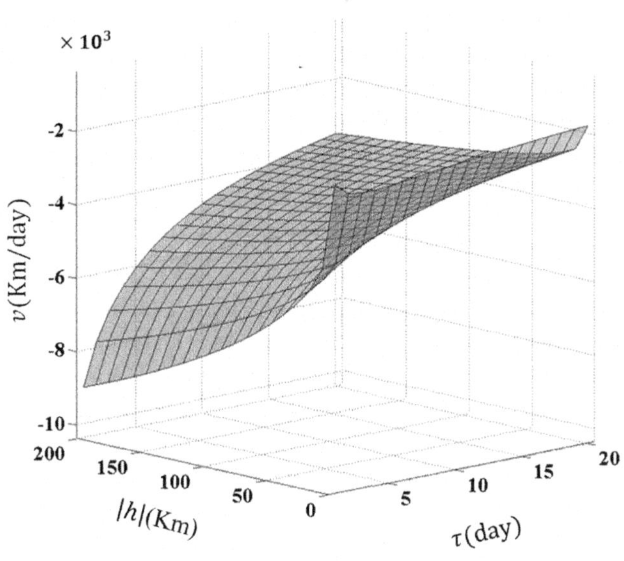

FIGURE 4.5

Plot of the $PM_{2.5}$ spread velocity (in km/day) as function of space lag, $|\boldsymbol{h}|$, and time lag, τ.

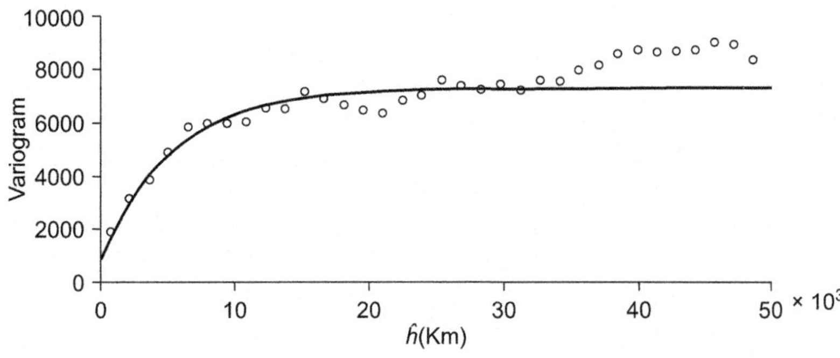

FIGURE 4.6

Empirical variogram (*dotted line*, in $(\mu g/m^3)^2$) and theoretical variogram (*continuous line*) of the traveling $PM_{2.5}$ distribution.

with $\widehat{h} \in R^1$, $c_0 = 857.77$ $(\mu g/m^3)^2$, $c = 6428.4$ $(\mu g/m^3)^2$, and $a = 5346.3$ km. The $\widehat{h}$ lag beyond which $PM_{2.5}$ observations are independent (i.e., variogram range beyond which the correlation between pollutant values is practically negligible) is about 16,039 km. The shape of $\gamma_{\widehat{X}}(\widehat{h})$ represents the manner in which the $PM_{2.5}$ values are interrelated or manifest dependence,[18] whereas the $\gamma_{\widehat{X}}(\widehat{h})$

[18]Dependence between $PM_{2.5}$ values decreases with $\widehat{h}$.

magnitude measures the strength of this dependence.[19] Since the variogram measures a particular kind of pollutant dependence, the degree of linear relationship between $PM_{2.5}$ values across space–time, I can assess the strength of this relationship by comparing the $PM_{2.5}$ variogram magnitude with the largest possible value (=7286.17(μg/m^3)^2, Fig. 4.6). A noteworthy feature of the TT-derived pollutant model is that I can select more adequate $PM_{2.5}$ variogram models $\gamma_{\widehat{X}}\left(\widehat{h}\right)$ in the reduced dimensionality TT domain than in the original space–time domain, in which it is often much harder to select $\gamma_X(|\boldsymbol{h}|,\tau)$ models that reflect adequately the composite space–time $PM_{2.5}$ variation structure. The $\left(\gamma_X, \gamma_{\widehat{X}}\right)$ interrelation introduced by Eq. (4.35) leads to a law of pollutant change that summarizes the observed spatiotemporal pollutant structure and infers about the underlying $PM_{2.5}$ spread mechanism (e.g., spread velocity components, dispersal distances, and directional differences):

$$\frac{\partial \gamma_X}{\partial \tau} + \theta \sum_{i=1}^{2} \frac{\partial \gamma_X}{\partial h_i} = 0, \tag{4.39}$$

where

$$\theta = \frac{v + \tau \dfrac{\partial v}{\partial \tau}}{\sum_{k=1}^{2} \left[h_k |\boldsymbol{h}|^{-1} - \tau \dfrac{\partial v}{\partial h_k} \right]} = \frac{\alpha \beta b |\boldsymbol{h}|^{2-2r} \tau^{\alpha-1}}{2(b\tau^{\alpha}+1) \sum_{k=1}^{2} h_k} \left[|\boldsymbol{h}|^{2r} - \frac{d(b\tau^{\alpha}+1)^{\beta r}}{2r\kappa} \right] \tag{4.40}$$

is the pollutant spread velocity function for the $PM_{2.5}$ variogram of Eq. (3.10) of Chapter II. Eq. (4.39) is a useful $PM_{2.5}$ variogram law that expresses quantitatively the relationship between the rates of variogram changes under the specified conditions ($PM_{2.5}$ data set and variogram shape), which, in turn, reflect the corresponding changes of dependence between $PM_{2.5}$ values at different spatial locations and time periods. Specifically (*a*) in Eq. (4.39) the temporal rate of $PM_{2.5}$ variogram change is explained as the spatial rates of $PM_{2.5}$ variogram change multiplied by the distance-weighted TT velocity function θ; and (*b*) as is expected, in light of the $PM_{2.5}$ variogram of Eq. (3.10) of Chapter II, the solution of Eqs. (4.39) and (4.40) with respect to v also leads to the velocity expression of Eq. (4.36) (thus, further confirming the validity of the variogram representation of Eq. (4.39) in the Shandong case).[20]

[19]Variogram magnitude indicates how the $PM_{2.5}$ concentration at one point responds in relation to $PM_{2.5}$ concentration changes at another point.

[20]The variogram law of Eq. (4.39) is valid for any pollutant variogram that satisfies the multidomain representation of Eq. (4.35), and is not restricted to the $PM_{2.5}$ variogram model of Eq. (3.10) of Chapter II. Otherwise said, Eq. (4.39) is a general variogram law whose applicability extends beyond the $PM_{2.5}$ spread velocity $v = v(|\boldsymbol{h}|,\tau)$ of Eq. (4.36). In the same context, the variogram PDE can be further extended to even more general variogram classes.

4.3 TRAVELING TRANSFORMATION IN SPATIOTEMPORAL RANDOM FIELD ESTIMATION: THE SPACE–TIME PROJECTION TECHNIQUE

The TT approach also introduces a space–time estimation technique with some interesting characteristics. I start by considering that the TT of Eq. (4.2a) "projects" $X(\boldsymbol{p})$ values assigned to points in the space–time attribute domain onto $\widehat{X}(\widehat{\boldsymbol{s}})$ values in a traveling domain. Otherwise said, in the space–time mapping context the TT goal is to develop a space–time mapping technique that "projects" $X(\boldsymbol{p})$ values assigned to points in the space–time attribute domain $\boldsymbol{p} = (\boldsymbol{s},t) \in R^{n,1}$ onto $\widehat{X}(\widehat{\boldsymbol{s}})$ values in a traveling pseudospatial domain $\widehat{\boldsymbol{s}} = \boldsymbol{s} - \boldsymbol{v}t \in R^n$. Specifically, let $X(\boldsymbol{p})$ be characterized by the covariance $c_X(\boldsymbol{h},\tau)$ (or the variogram $\gamma_X(\boldsymbol{h},\tau)$). In light of the random field TT of Eq. (4.2b), the associated pseudospatial attribute distribution $\widehat{X}(\widehat{\boldsymbol{s}})$ is determined by

$$\widehat{X}(\boldsymbol{p}) = \widehat{\widehat{X}}(\widehat{\boldsymbol{s}}). \tag{4.41}$$

Assume that the attribute $X(\boldsymbol{p})$ is observed at m space–time points ($R^{n,1}$) and that attribute estimates are sought at k additional (mapping) points. The mapping technique suggested by TT, herein termed as the *space–time projection* (STP) technique, transfers attribute estimation from the $R^{n,1}$ domain onto the R^n domain of reduced dimensionality. The main methodological steps of the STP technique of space–time mapping are described in Table 4.1, whereas the STP technique is illustrated next by means of a real-world case study.

Case Study II.2.1 (cont.)

The TT-based STP mapping technique has some interesting characteristics. Specifically, in the Shandong region during the time period January 1–31, 2014, $PM_{2.5}$ values were observed at $m = 2163$ space–time points (Table 3.1 of Chapter II), and $PM_{2.5}$ estimates were subsequently sought at $k = 69 \times 47 \times 31 = 100{,}533$ mapping (grid) points with cell size 10 km × 10 km × 1 day ($R^{2,1}$). STP transfers $PM_{2.5}$ estimation from the $R^{2,1}$ domain onto the R^2 domain. To obtain the space–time $PM_{2.5}$

Table 4.1 The Space–Time Projection Mapping Technique

Step No.	Description
1	On the basis of the analysis of Section 4.2, I determine the traveling transformation velocity vector $\boldsymbol{v}$ of the attribute distribution.
2	Given the $(\boldsymbol{s}, t)$ coordinates of the original $m + k$ space–time points, I calculate the corresponding $\widehat{\boldsymbol{s}} = \boldsymbol{s} - \boldsymbol{v}t$ coordinates (R^n) of the projected $m + k$ space points.
3	I use the $\boldsymbol{v}$ values above to find the values of the traveling disease attribute covariance $c_{\widehat{X}}(\widehat{\boldsymbol{h}})$ (or variogram $\gamma_{\widehat{X}}(\widehat{\boldsymbol{h}})$) with $\widehat{\boldsymbol{h}} = \boldsymbol{h} - \boldsymbol{v}\tau$ (R^n).
4	I derive estimates $\widehat{\widehat{X}}(\widehat{\boldsymbol{s}})$ in R^n of the traveling attribute values $\widehat{X}(\widehat{\boldsymbol{s}})$ at the k projected mapping points using the attribute values at the m projected data points (and other relevant information, if available).
5	In view of Eq. (4.41), I determine the estimates $\widehat{X}(\boldsymbol{p})$ of the attribute $X(\boldsymbol{p})$ in the original $R^{n,1}$ domain from the estimates $\widehat{\widehat{X}}(\widehat{\boldsymbol{s}})$ in the projected R^n domain.

estimation map in the Shandong province by means of the STP technique of Table 4.1, I proceeded as follows (Christakos et al., 2017c):

- In Steps 1–2, I used the υ expression of Eq. (4.36) to compute the R^2-projected $\widehat{s}_i$ $(i = 1, 2)$ coordinates for each one of the $m + k = 102{,}696$ points with original coordinates $(\boldsymbol{s}, t) = (s_1, s_2, t) \in R^{2,1}$, as follows:

$$\widehat{s}_i = s_i - \upsilon_i t = \left(\frac{s_i^{2r}}{(bt^\alpha + 1)^{\beta r}} + \frac{\beta d \ln(bt^\alpha + 1)}{2\kappa} \right)^{\frac{1}{2r}} \tag{4.42a–b}$$

  $(i = 1, 2)$.

- In Step 3, the transformation calculations led to the $\gamma_{\widehat{X}}(\widehat{h})$ plots shown in Fig. 4.6.
- In Step 4, I generated the TT-projected $\widehat{\widehat{X}}(\widehat{\boldsymbol{s}})$ map in R^2, see Fig. 4.7A.[21]
- In Step 5, the $\widehat{X}_{stp}(\boldsymbol{p})$ map was generated in the $R^{2,1}$ domain of the $PM_{2.5}$ distribution in Shandong province during the time period January 1–31, 2014, which is shown in Fig. 4.7B.

Fig. 4.8A and B offer another visual interpretation of the idea behind the dimensionality reduction achieved by the STP technique. Specifically, Fig. 4.8A shows a plot of the original coordinates s_1 and s_2 in $R^{2,1}$, and Fig. 4.8B shows a plot of the transformed coordinates $\widehat{s}_1 = s_1 - \upsilon_1 t$ and $\widehat{s}_2 = s_2 - \upsilon_2 t$ in R^2. Clearly, in the original $R^{2,1}$ domain the coordinates s_1 and s_2 are entirely uncorrelated, whereas in the transformed R^2 domain these coordinates $\widehat{s}_1$ and $\widehat{s}_2$ are strongly correlated. Hence, working with coordinates $\widehat{s}_1$ and $\widehat{s}_2$ rather than with s_1 and s_2 (and t) would represent the situation more concisely and reduce the number of coordinates.[22] For numerical comparison purposes, the $PM_{2.5}$ estimation in the Shandong province was also carried out using the standard *space–time ordinary kriging (STOK)* technique based on the same pollutant data set and variogram (Figs. 3.1 and 3.2 of Chapter II).[23] The space–time $PM_{2.5}$ estimation map, $\widehat{X}_{stok}(\boldsymbol{p})$, generated by the STOK technique is plotted in Fig. 4.7C.[24] The visual comparison of the $\widehat{X}_{stp}(\boldsymbol{p})$ and $\widehat{X}_{stok}(\boldsymbol{p})$ maps (Fig. 4.7B and C) led to the conclusion that the STP map was more informative than the STOK map, as it provided a more detailed and realistic representation of the $PM_{2.5}$ distribution in the Shandong province during the period January 1–31, 2014. Moreover, the numerical accuracy of the above pollutant mapping techniques was tested by means of two commonly used accuracy indicators: mean absolute error (MAE), and root mean square error (RMSE). For accuracy comparison purposes, a total of $v = 433$ validation points were assumed.[25] The obtained pollutant estimation accuracy results for the two techniques are listed in

---

[21]At most 16 neighboring points located by a searching radius of 20×10^3 km were used to get an estimate at each one of the mapping (grid) points $\widehat{\boldsymbol{s}} = (\widehat{s}_1, \widehat{s}_2)$ in R^2.

[22]In practical terms, the efficient application of the STP technique would assume a strong correlation between $\widehat{s}_1$ and $\widehat{s}_2$ (Fig. 4.8B). More precisely, to optimize the implementation of the STP technique, we would like the calculated TT velocity to be such that the TT coordinates covary as close as possible.

[23]Technical details about the STOK technique are found in Christakos (1992) and Christakos and Hristopulos (1998).

[24]A maximum of 16 neighboring points located by a 1000 km/15 days searching radius were used to derive $PM_{2.5}$ estimates at each mapping (grid) point in $R^{2,1}$ (i.e., the same as those used by the STP technique).

[25]That is, space–time data points at which the actual $PM_{2.5}$ concentrations were known and could be compared to the $PM_{2.5}$ concentrations estimated by the STP and STOK techniques.

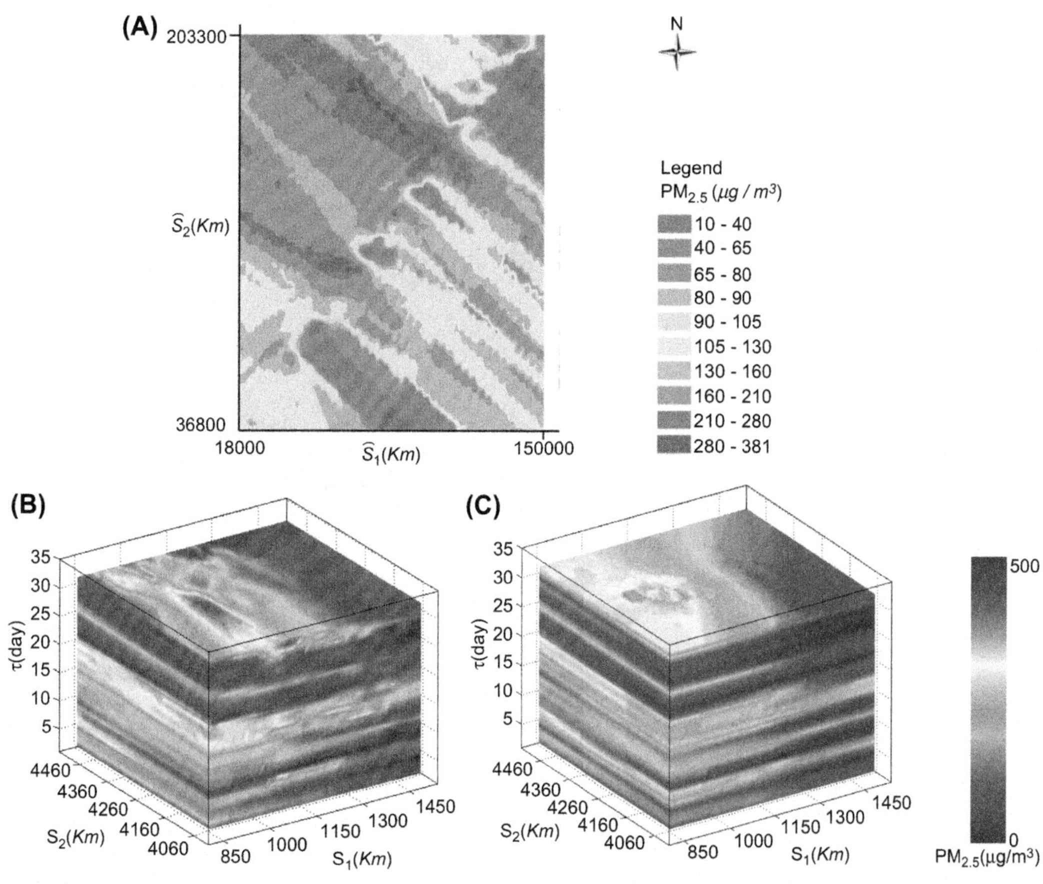

FIGURE 4.7

(A) Traveling map. And, space–time estimation maps of the $PM_{2.5}$ distribution (μg/m^3) in Shandong province during January 1–31, 2014, using (B) the space–time projection technique and (C) the space–time ordinary kriging technique.

Table 4.2. Clearly, the STP technique was more accurate than the STOK technique (by 26.21% and 28.47%, respectively, in terms of the MAE and the RMSE criteria). As regards computational time, the STP technique again performed better than the STOK technique in the Shandong study (about 19% less computational time was required in the implementation of the STP technique compared to that of the STOK technique, Table 4.2). As an additional accuracy test, Fig. 4.9 shows plots of the daily-averaged $PM_{2.5}$ concentrations (actual, STP estimated, and STOK estimated) at the $v = 433$ validation points for the period January 1–31, 2014. Both techniques performed well, but the STP technique again performed better than the STOK technique generating estimates of the daily-averaged $PM_{2.5}$ concentrations that provided an almost perfect fit to the actual daily-averaged $PM_{2.5}$ values. The calculated MAE values were found to be very small: 1.68 μg/m^3 (STP) *versus* 6.03 μg/m^3 (STOK). An illustrative outline of the STP technique is given in Fig. 4.10: Starting with the multidomain variogram

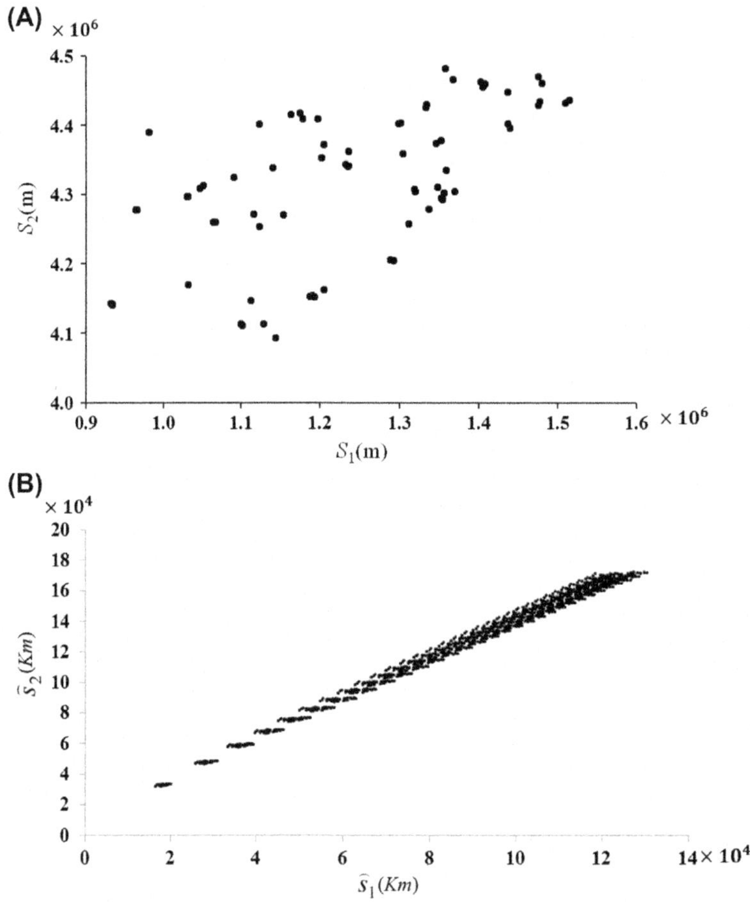

FIGURE 4.8

Plots of the arrangements of (A) the original coordinates s_1 and s_2, and (B) the transformed coordinates $\widehat{s}_1$ and $\widehat{s}_2$.

Eq. (4.35), the pollutant spread velocity υ was computed and plotted as a function of $|\boldsymbol{h}|$ and τ.[26] Given υ, the traveling variogram $\gamma_{\widehat{X}}(\widehat{h})$ was calculated from the original space–time $PM_{2.5}$ variogram $\gamma_X(|\boldsymbol{h}|,\tau)$, using the TT-derived coordinates:

$$\boldsymbol{p} = (s_1, s_2, t) \in R^{2,1} \mapsto \widehat{s} = (s_1 - \upsilon\, t, s_2 - \upsilon\, t) \in R^2.$$

Next, using the calculated $\gamma_{\widehat{X}}(\widehat{h})$ the traveling map $\widehat{\widehat{X}}(\widehat{s})$ was generated in the R^2 domain. Lastly, the space–time map of the $PM_{2.5}$ distribution in the Shandong province (January 1–31, 2014) was

[26]Note that the same result can be obtained using the useful PDE pollutant representation of Eq. (4.39).

Table 4.2 Accuracy Indicators and Computational Time of Space−Time Projection (STP) Versus Space−Time Ordinary Kriging (STOK) Techniques

Technique	Mean Absolute Error (μg/m³)	Root Mean Square Error (μg/m³)	Computational Time (s)
$\widehat{X}_{stok}$	24.95	38.07	965
$\widehat{X}_{stp}$	18.41	27.23	759

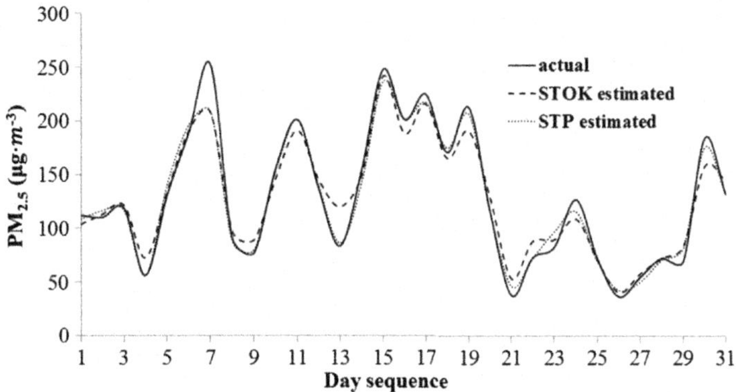

FIGURE 4.9

Plots of daily-averaged $PM_{2.5}$ values at 433 validation points for the period January 1−31, 2014: actual values (*continuous line*), space−time projection (STP) estimates (*dotted line*), and space−time ordinary kriging (STOK) estimates (*dashed line*).

generated in $R^{2,1}$ by means of the equality $\widehat{X}_{stp}(\boldsymbol{p}) = \widehat{\widehat{X}}(\widehat{\boldsymbol{s}})$, where the following inverse TT-derived coordinates were used:[27]

$$\widehat{\boldsymbol{s}} = (s_1 - \upsilon\, t, s_2 - \upsilon\, t) \in R^2 \mapsto \boldsymbol{p} = (s_1, s_2, t) \in R^{2,1}.$$

The fact that the STP technique transforms the original $PM_{2.5}$ distribution into a TT distribution enables a computationally more efficient pollutant mapping that takes advantage of the lower dimensionality of the transformed domain while avoiding some of the practical complexities of the original higher-dimensionality domain. Specifically, in the STOK technique the neighborhood search was based on the original set of coordinates (s_1, s_2, t), whereas in the STP technique it was based on the projected set of coordinates ($\widehat{s}_1$, $\widehat{s}_2$). In the domain of the STOK technique the spatiotemporal distances were calculated using the formula

$$|\Delta\boldsymbol{p}| = \left(h_1^2 + h_2^2 + \varepsilon\tau^2\right)^{\frac{1}{2}},$$

[27]The $\widehat{\widehat{X}}(\widehat{\boldsymbol{s}})$ map constitutes an intermediate (transformation) stage, which rather serves as a modeling "vehicle" allowing us to perform TT analysis in the convenient R^2 domain of reduced computational complexity, and then move the obtained results onto the actual $R^2 \times T$ domain of the Shandong pollution study.

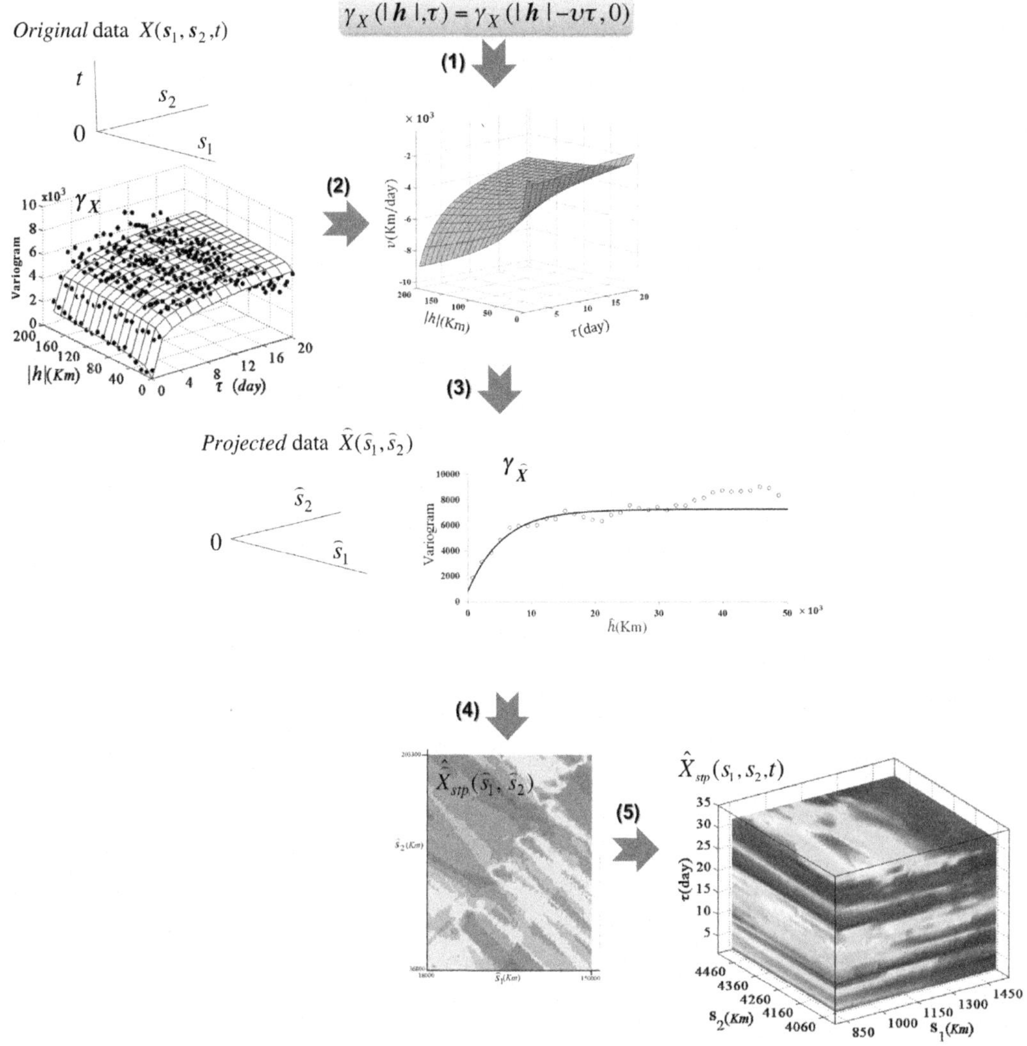

FIGURE 4.10

An outline of the implementation of the space–time projection technique in Shandong province (China).

where the parameter ε needed to be properly selected. Often, this choice is not an easy matter. In light of the $PM_{2.5}$ variogram plots of Fig. 3.2 of Chapter II, I have chosen the approximation

$$\varepsilon = \left(\frac{\text{spatial correlation range}}{\text{temporal correlation range}}\right)^2 = \left(\frac{200,000 \text{ m}}{20 \text{ days}}\right)^2 = 10^8 \text{ m/day},$$

so that all terms of $|\Delta \boldsymbol{p}|$ have the same units.[28] On the other hand, in the STP technique the projected spatiotemporal distances of the traveling pollutant model were directly calculated using the formula $\left(\widehat{h}_1^2 + \widehat{h}_2^2\right)^{\frac{1}{2}}$ that does not involve any approximations in terms of the ε parameter. In sum, the TT-based random field relations link two different perspectives of pollutant mapping, one in the $R^{2,1}$ domain with one in the R^2 domain. While these two perspectives are different, they communicate with its other through the TT. This is an element that makes the STP technique methodologically attractive: it can transfer the mathematical modeling of pollution into a more convenient domain and then properly link the modeling results in this domain to the pollutant mapping in the original domain.

5. CLOSING COMMENTS

Concluding this chapter, Table 4.3 provides a summary comparison of the three transformation categories (FT, ST, and TT) based on three criteria: dimensionality, mapping domain, and corresponding coordinates. The FT transfers the study of the attribute into a frequency domain, where derivatives of all orders are eliminated turning a calculus problem into an algebra problem; the ST reduces the study into the lowest dimensionality space; and the TT is essentially a redefinition of the attribute arguments. This makes TT the simplest mathematical operator among the three transformations. Naturally, every transformation has its own *pros* and *cons*, depending on the situation considered, whereas certain trade-offs naturally apply in each case. Adequately assessing these *pros* and *cons* are a matter of experience with the application of each transformation in real-world situations.

Table 4.3 A Comparison of Transformations

Name	From	To	Operator
Fourier transformation	$\boldsymbol{p} = (\boldsymbol{s},t) \in R^{n,1}$	$\boldsymbol{w} = (\boldsymbol{k},\omega) \in R^{n,1}$	$\int d\boldsymbol{w}\ e^{i\boldsymbol{w}\cdot\boldsymbol{p}}$
Space transformation	$\boldsymbol{p} = (\boldsymbol{s},t) \in R^{n,1}$	$(\boldsymbol{s}\cdot\boldsymbol{\theta}, t) \in R^{1,1}$	$\int d\boldsymbol{s}\ \delta(\sigma - \boldsymbol{s}\cdot\boldsymbol{\theta})$
Traveling transformation	$\boldsymbol{p} = (\boldsymbol{s},t) \in R^{n,1}$	$\widehat{\boldsymbol{s}} \in R^n$	$\boldsymbol{s} - \boldsymbol{v}t$

[28]This procedure of defining space–time distances is used in the well-known SEKS-GUI library, as well as in other space–time mapping libraries, when searching for neighboring points in the space–time domain (Yu et al., 2007).

CHAPTER

GEOMETRICAL PROPERTIES OF SPATIOTEMPORAL RANDOM FIELDS

CHAPTER OUTLINE

1. INTRODUCTION

In applied stochastics, the study of many natural systems needs the development of a *stochastic calculus* for random fields, which can address the notions of random field *geometry* or *regularity* (continuity, differentiability, integrability). Of course, since a spatiotemporal random field (S/TRF) consists of a large number of sample paths (realizations), applying ordinary calculus to each and every one of them is rather impractical and overly restrictive. Therefore, the stochastic versions of the above notions refer to probabilities or average behavior of the sample paths. Subsequently, the study of the behavior of an S/TRF with respect to the mathematical notions of random field regularity can lead to valuable conclusions regarding the geometry of the S/TRF and, consequently, the homogeneity and smoothness of the natural attribute distribution modeled by the S/TRF.

To start with, I need to reemphasize that the manner in which the argument $\boldsymbol{p}$ of an S/TRF enters stochastic calculus depends on the specific objectives of the study and the interpretability of its results. So, in mathematical manipulations directly involving space−time coordinates, the argument may be seen as an $n+1$-dimensional vector $\boldsymbol{p} = (s_1,\ldots, s_n, s_0)$ in the R^{n+1} domain (in which case, the calculus in R^{n+1} follows parallel paths with that in R^n); or, alternatively, it can be viewed as a vector−scalar pair $\boldsymbol{p} = (\boldsymbol{s}, t)$, where $\boldsymbol{s} = (s_1,\ldots, s_n)$ and t form a Cartesian product $R^{n,1}$. Moreover, the space−time metric

Spatiotemporal Random Fields. http://dx.doi.org/10.1016/B978-0-12-803012-7.00006-4

$m(\boldsymbol{p},\boldsymbol{p}')$ enters in the space–time analysis, either as a single scalar $m(\boldsymbol{p},\boldsymbol{p}') = |\Delta\boldsymbol{p}| = \Delta p$ in the R^{n+1} domain, or as a pair of scalars $m(\boldsymbol{p},\boldsymbol{p}') = (|\Delta\boldsymbol{s}|, \tau)$ in the $R^{n,1}$ domain. The vector $\Delta\boldsymbol{p}$ and the number $|\Delta\boldsymbol{p}| = \Delta p$ are related via a function g that accounts for the physical differences between space and time as well as for their structural interrelations. In particular, as was suggested in Chapter I, the Riemannian space–time metric (Eq. 2.14, Chapter I) and the Pythagorean space–time metric (Eq. 2.15a, Chapter I) are often assumed in space–time analysis in the R^{n+1} domain, whereas the pairs "Euclidean spatial distance-absolute time difference," Eq. (2.18a) of Chapter I, and "Manhattan spatial distance-absolute time difference," Eq. (2.18b) of Chapter I, are often assumed in the $R^{n,1}$ domain.

According to the discussion in Chapter IV, we can view an S/TRF $X(\boldsymbol{p}, u)$ as producing for each random event $u \in \Omega$ and at each point $\boldsymbol{p} = (\boldsymbol{s}, t)$ of the $n+1$-dimensional space–time a sample path or realization $\chi(\boldsymbol{p})$. Then, just as for a deterministic function $\chi(\boldsymbol{p})$, where questions about the geometrical properties of $\chi(\boldsymbol{p})$ are transferred to questions about the convergence of the series

$$\{\chi(\boldsymbol{p}_m)\}, \text{ with } \boldsymbol{p}_m \underset{m\to\infty}{\rightarrow} \boldsymbol{p}, \tag{1.1}$$

space–time geometrical questions concerning the random field $X(\boldsymbol{p})$ reduce to the concept of stochastic convergence of the corresponding random variable (RV) sequence

$$\{x_m(u) = X(\boldsymbol{p}_m)\}, \text{ with } m = 1, 2\ldots \tag{1.2}$$

This key concept is discussed in Section 2.

I conclude this introduction with a postulate that is, basically, an extension of Postulate 2.3 of Chapter I. When applicable, this postulate can clarify certain theoretical and practical aspects of space–time analysis.

Postulate 1.1

Many random field geometry results that have been derived for spatial random fields can be formally extended to spatiotemporal random fields as long as (1) the time argument does not change during the derivations so that they carry through as before, or (2) a valid space–time metric of a specified form can be assumed in the analysis leading from R^n to $R^{n,1}$.

The two conditions described in the postulate allow the formal derivation of many useful results in the $R^{n,1}$ domain directly from the corresponding results in the R^n domain. As usual, in the following an effort is made to maintain a balance between mathematical rigor and physical interpretability. This is necessary, especially in cases in which the purely mathematical results seem rather "unphysical" and the notation used rather "anti-intuitive". Then, it may be worth relaxing mathematical rigor in favor of a presentation that can be translated into meaningful statements about the real-world.

2. STOCHASTIC CONVERGENCE

As with ordinary calculus, the study of S/TRF geometry in the context of stochastic calculus should involve the notion of limit of a random field. I start with a brief discussion of the main types of convergence in a stochastic framework. This brings us back to Section 3 of Chapter I on RV convergence. Generally, a random field $X(\boldsymbol{p})$ is said to converge to $X(\boldsymbol{p}_0)$ in a *certain sense* when

$\boldsymbol{p} \to \boldsymbol{p}_0$ if the corresponding RV sequence converges according to the discussion in Section 3 of Chapter I. Specifically:

Definition 2.1

Let $X(\boldsymbol{p})$ be an S/TRF in $L_2(\Omega, \mathcal{F}, P)$, or simply L_2, and $\mathcal{S} \subset R^{n+1}$. The $X(\boldsymbol{p})$ is said to converge to $X(\boldsymbol{p}_0)$ in $\mathcal{S}$ as $\boldsymbol{p} \to \boldsymbol{p}_0$ when either of the following two equivalent conditions is valid:

(i) If given an $\varepsilon > 0$, there exists a $\delta > 0$ so that

$$\Phi(X(\boldsymbol{p}) - X(\boldsymbol{p}_0)) < \varepsilon \text{ when } \boldsymbol{p} \in \mathcal{S} \text{ and } |\boldsymbol{p} - \boldsymbol{p}_0| < \delta,$$

where the function Φ takes a suitable form depending on the type of convergence considered.

(ii) If the RV sequence

$$\{x_m = X(\boldsymbol{p})\} \text{ at } \boldsymbol{p} = \boldsymbol{p}_1, \ldots, \boldsymbol{p}_m \text{ tends to } x_0 = X(\boldsymbol{p}_0) \text{ as } m \to \infty,$$

where for every $u \in \Omega$ the sequence $\{x_m(u)\}$, or simply $\{x_m\}$, $m = 1, 2, \ldots$ is a set of real numbers.

I will apply this general definition to develop various types of S/TRF convergence. Yet, it should be noticed that, (a) of the above two equivalent conditions, the second one is most often used in the literature, and (b) different authors have different views as regards the definition of certain types of stochastic convergence and their relation.

Definition 2.2

The $X(\boldsymbol{p})$ is said to converge to $X(\boldsymbol{p}_0)$ in $\mathcal{S}$:

(a) *Surely* or in the *ordinary calculus* (*o.c.*) sense,

$$X(\boldsymbol{p}, u) \xrightarrow{o.c.} X(\boldsymbol{p}_0, u) \text{ as } \boldsymbol{p} \to \boldsymbol{p}_0, \tag{2.1a}$$

if for all $u \in \Omega$

$$\lim_{\boldsymbol{p} \to \boldsymbol{p}_0} X(\boldsymbol{p}) = X(\boldsymbol{p}_0). \tag{2.1b}$$

The o.c. convergence is sometimes linked to convergence with *probability one* (*w.p.1.*), i.e.,

$$X(\boldsymbol{p}) \xrightarrow{P.1} X(\boldsymbol{p}_0) \text{ as } \boldsymbol{p} \to \boldsymbol{p}_0 \tag{2.2a}$$

if

$$P[\lim_{\boldsymbol{p} \to \boldsymbol{p}_0} X(\boldsymbol{p}) = X(\boldsymbol{p}_0) \text{ for all } \boldsymbol{p} \in \mathcal{S}] = 1. \tag{2.2b}$$

(b) *Almost surely* (*a.s.*),

$$X(\boldsymbol{p}, u) \xrightarrow{a.s.} X(\boldsymbol{p}_0, u) \text{ as } \boldsymbol{p} \to \boldsymbol{p}_0 \tag{2.3a}$$

if for all $\boldsymbol{p} \in \mathcal{S}$

$$P[\lim_{\boldsymbol{p} \to \boldsymbol{p}_0} X(\boldsymbol{p}) = X(\boldsymbol{p}_0)] = 1; \tag{2.3b}$$

or, equivalently,

$$\text{for any } \mathcal{V} \subseteq \Omega \text{ satisfying } P[\mathcal{V}] = 0, \text{ and } \lim_{\boldsymbol{p} \to \boldsymbol{p}_0} |X(\boldsymbol{p}) - X(\boldsymbol{p}_0)| = 0 \text{ for all } u \notin \mathcal{V}. \tag{2.3c–d}$$

(c) *In probability* (*P*),

$$X(\boldsymbol{p}) \xrightarrow{P} X(\boldsymbol{p}_0) \text{ as } \boldsymbol{p} \to \boldsymbol{p}_0, \text{ or } \underset{\boldsymbol{p} \to \boldsymbol{p}_0}{\text{l.i.p.}}\, X(\boldsymbol{p}) = X(\boldsymbol{p}_0), \tag{2.4a}$$

if for all $\varepsilon > 0$,

$$\lim_{\boldsymbol{p} \to \boldsymbol{p}_0} P[|X(\boldsymbol{p}) - X(\boldsymbol{p}_0)| > \varepsilon] = 0; \tag{2.4b}$$

or, equivalently, given an $\varepsilon > 0$, there exists a $\delta > 0$ so that

$$P[X(\boldsymbol{p}) - X(\boldsymbol{p}_0) \geq \varepsilon] < \varepsilon \text{ when } \boldsymbol{p} \in \mathcal{S} \text{ and } |\boldsymbol{p} - \boldsymbol{p}_0| < \delta. \tag{2.4c}$$

(d) *Weakly* or *in distribution* (*F*),

$$X(\boldsymbol{p}) \xrightarrow{F} X(\boldsymbol{p}_0) \text{ as } \boldsymbol{p} \to \boldsymbol{p}_0 \tag{2.5a}$$

if

$$\lim_{\boldsymbol{p} \to \boldsymbol{p}_0} F_X(\chi, \boldsymbol{p}) = F_X(\chi, \boldsymbol{p}_0) \tag{2.5b}$$

on the continuity set χ of F_X; or, equivalently, given an $\varepsilon > 0$, there exists a $\delta > 0$ so that

$$F_X(\chi, \boldsymbol{p}) - F_X(\chi, \boldsymbol{p}_0) < \varepsilon \text{ when } \boldsymbol{p} \in \mathcal{S} \text{ and } |\boldsymbol{p} - \boldsymbol{p}_0| < \delta, \tag{2.5c}$$

on the continuity set χ of F_X.

(e) *Mean squarely* (*m.s.*),

$$X(\boldsymbol{p}) \xrightarrow{m.s.} X(\boldsymbol{p}_0), \text{ or } \underset{\boldsymbol{p} \to \boldsymbol{p}_0}{\text{l.i.m.}}\, X(\boldsymbol{p}) = X(\boldsymbol{p}_0), \tag{2.6a}$$

if

$$\lim_{\boldsymbol{p} \to \boldsymbol{p}_0} \overline{|X(\boldsymbol{p}) - X(\boldsymbol{p}_0)|^2} = 0; \tag{2.6b}$$

or, equivalently, given an $\varepsilon > 0$, there exists a $\delta > 0$ so that

$$\overline{[X(\boldsymbol{p}) - X(\boldsymbol{p}_0)]^2} < \varepsilon \text{ when } \boldsymbol{p} \in \mathcal{S} \text{ and } |\boldsymbol{p} - \boldsymbol{p}_0| < \delta. \tag{2.6c}$$

In principle, the stochastic formulas of Eqs. (2.1a), (2.2a), (2.3a), (2.4a), (2.5a), and (2.6a), should be interpreted as meaning the corresponding deterministic Eqs. (2.1b), (2.2b), (2.3b), (2.4b), (2.5b), and (2.6b). This correspondence is useful, because it links a less familiar notion (stochastic convergence) with a commonly used notion (deterministic or ordinary convergence).

Remark 2.1

In view of Definition 2.2, the main types of convergence above are related to each other. Particularly,

$$X(\boldsymbol{p}) \xrightarrow{o.c.} X(\boldsymbol{p}_0) \Rightarrow \left\{ \begin{array}{l} X(\boldsymbol{p}) \xrightarrow{m.s.} X(\boldsymbol{p}_0) \\ X(\boldsymbol{p}) \xrightarrow{a.s.} X(\boldsymbol{p}_0) \end{array} \right\} \Rightarrow X(\boldsymbol{p}) \xrightarrow{P} X(\boldsymbol{p}_0) \Rightarrow X(\boldsymbol{p}) \xrightarrow{F} X(\boldsymbol{p}_0),$$

which is a relationship to keep in mind for future use.

The deterministic (o.c.) convergence, Eqs. (2.1a) and (2.1b), is of little or no use in practice. Of rather limited use are the a.s., *P*, and *F* convergence types. Admittedly, the most useful type of convergence is the m.s. convergence of Eqs. (2.6a)–(2.6c). In this case, the following two criteria are of considerable importance.

Theorem 2.1

(*Cauchy's convergence criterion*): The m.s. limit

$$\underset{\boldsymbol{p},\boldsymbol{p}'\to\boldsymbol{p}_0}{\text{l.i.m.}} [X(\boldsymbol{p}) - X(\boldsymbol{p}')] = \underset{\Delta\boldsymbol{p},\Delta\boldsymbol{p}'\to 0}{\text{l.i.m.}} [X(\boldsymbol{p}_0 + \Delta\boldsymbol{p}) - X(\boldsymbol{p}_0 + \Delta\boldsymbol{p}')] = 0, \tag{2.7a}$$

exists if and only if

$$\lim_{\boldsymbol{p},\boldsymbol{p}'\to\boldsymbol{p}_0} \overline{\left|X(\boldsymbol{p}) - X(\boldsymbol{p}')\right|^2} = \lim_{\Delta\boldsymbol{p},\Delta\boldsymbol{p}'\to\boldsymbol{0}} \overline{\left|X(\boldsymbol{p}_0 + \Delta\boldsymbol{p}) - X(\boldsymbol{p}_0 + \Delta\boldsymbol{p}')\right|^2} = 0, \tag{2.7b}$$

where the equations must hold regardless of how $\boldsymbol{p}$, $\boldsymbol{p}'$ approach $\boldsymbol{p}_0$, or $\Delta\boldsymbol{p}$, $\Delta\boldsymbol{p}'$ approach $\boldsymbol{0}$.

In some cases, a useful expression of the criterion of Theorem 2.1 is as follows.

Theorem 2.2

(*Loeve's convergence criterion*): The limit in Eq. (2.7b) exists if and only if

$$\lim_{\boldsymbol{p},\boldsymbol{p}'\to\boldsymbol{p}_0} \overline{X(\boldsymbol{p})X(\boldsymbol{p}')} = c, \tag{2.8}$$

where c is a constant independent of how $\boldsymbol{p}$, $\boldsymbol{p}'$ approach $\boldsymbol{p}_0$.

For $X(\boldsymbol{p}) \in L_2$, the L_2-convergence is m.s. convergence. Let $X(\boldsymbol{p}) \xrightarrow{m.s.} X(\boldsymbol{p}_0)$ and $Y(\boldsymbol{p}) \xrightarrow{m.s.} Y(\boldsymbol{p}_0)$. Working along similar lines as in Eqs. (3.32a–b) of Chapter I, it can be shown that the proposition below is valid.

Proposition 2.1

If for $\boldsymbol{p} \to \boldsymbol{p}_0$, $\boldsymbol{p}' \to \boldsymbol{p}_0'$, the m.s. limits hold respectively

$$\begin{aligned} X(\boldsymbol{p}) &\xrightarrow{m.s.} X(\boldsymbol{p}_0), \\ Y(\boldsymbol{p}') &\xrightarrow{m.s.} Y(\boldsymbol{p}_0'), \end{aligned} \tag{2.9a–b}$$

then

$$\begin{aligned} &\overline{X(\boldsymbol{p})} \to \overline{X(\boldsymbol{p}_0)}, \\ &\overline{Y(\boldsymbol{p}')} \to \overline{Y(\boldsymbol{p}_0')}, \\ &\overline{X(\boldsymbol{p})Y(\boldsymbol{p}')} \to \overline{X(\boldsymbol{p}_0)Y(\boldsymbol{p}_0')}. \end{aligned} \tag{2.10a–c}$$

Proposition 2.1 concerns the continuity of random field expectations and is the basis of many useful stochastics results.

Remark 2.2

Two interesting consequences of this proposition are as follows: the operators "l.i.m." and "$\overline{[\cdot]}$" commute, i.e.,

$$\overline{\underset{p\to p_0}{\text{l.i.m.}}\, X(p)} = \lim_{p\to p_0} \overline{X(p)},$$

and

$$X(p) \xrightarrow{m.s.} const \text{ if and only if } \overline{X(p)X(p')} \to const \text{ as } p, p' \to p_0.$$

An attractive feature of m.s. convergence, which is one of the reasons for its principal importance in applications, is that it can be linked to covariance convergence, and often, while the former cannot be translated into physical statements, the latter can. This leads to the m.s. convergence criterion described by the following proposition.

Proposition 2.2

Let $X(\boldsymbol{p})$ be an S/TRF and let $\boldsymbol{p}_0$ be a fixed point in R^{n+1}. Then, $X(\boldsymbol{p}) \xrightarrow{m.s.} X(\boldsymbol{p}_0)$ if and only if

$$\overline{X(\boldsymbol{p})X(\boldsymbol{p}')} \to \overline{X(\boldsymbol{p}_0)^2} \text{ as } \boldsymbol{p}, \boldsymbol{p}' \to \boldsymbol{p}_0. \tag{2.11}$$

What is interesting about Proposition 2.2 is that while the convergence of $X(\boldsymbol{p})$ is in the m.s. (stochastic) sense, that of $\overline{X(\boldsymbol{p})X(\boldsymbol{p}')}$ is in the ordinary (deterministic) sense. Also, this proposition states that L_2-random fields coincide with those fields that have second-order moments.

Next, moving to another type of convergence, and recalling that a random field $X(\boldsymbol{p}, u)$ can be viewed as producing at each $u \in \Omega$ a sample path or realization $\chi(\boldsymbol{p})$, the following proposition holds concerning the a.s. convergence of the random field.

Proposition 2.3

(*a.s. convergence criterion*): If for some $k > 0$ it is valid that

$$\sum_{i=1}^{\infty} \overline{\left|X(\boldsymbol{p}_i) - X(\boldsymbol{p}_0)\right|^k} < \infty, \tag{2.12a}$$

then

$$X(\boldsymbol{p}) \xrightarrow{a.s.} X(\boldsymbol{p}_0). \tag{2.12b}$$

To conclude, it must be borne in mind that as regards random field convergence, those in the m.s. and a.s. sense are the two most important types. In much of what follows, it would be more precise to write $X(\boldsymbol{p}, u)$, but for simplicity in presentation the u will be usually suppressed.

3. STOCHASTIC CONTINUITY

In this section, I turn to stochastic continuity. In physical terms, the continuity of a random field $X(\boldsymbol{p})$ implies that the difference

$$\delta X(\boldsymbol{p}) = X(\boldsymbol{p} + \Delta \boldsymbol{p}) - X(\boldsymbol{p}) \tag{3.1}$$

is small, for small $\Delta \boldsymbol{p}$. But the $X(\boldsymbol{p})$ is a random field, which means that the difference $\delta X(\boldsymbol{p})$ is itself a random field that follows a probability law, thus complicating matters significantly. Only in special cases, like a space−time homostationarity (STHS) Gaussian $X(\boldsymbol{p}) \sim N(0, \sigma_X^2)$, we also know the probability law of $\delta X(\boldsymbol{p})$, i.e.,

$$\delta X(\boldsymbol{p}) \sim N\left(0, \sigma_{\delta X}^2(\Delta \boldsymbol{p}) = \sigma_X^2 - c_X(\Delta \boldsymbol{p})\right), \tag{3.2}$$

where $c_X(\boldsymbol{p}, \boldsymbol{p}') = c_X(\Delta \boldsymbol{p})$. In this case, if the covariance is continuous at the origin, it is continuous everywhere, and the same is valid for the random field. Many of the results in this section are straightforward extensions of the analysis in the R^n domain to the R^{n+1} domain ($n \geq 1$).

3.1 BASIC TYPES OF STOCHASTIC CONTINUITY

Since mathematically $X(\boldsymbol{p})$ is viewed as producing for each $u \in \Omega$ a (deterministic) sample path $\chi(\boldsymbol{p})$, we can use the conventional notions and techniques of classical calculus to study its continuity features. For instance, we can say that

$X(\boldsymbol{p})$ is continuous at $\boldsymbol{p}^*$ if for any $u \in \Omega$ the difference $|\chi(\boldsymbol{p}) - \chi(\boldsymbol{p}^*)| \xrightarrow{\boldsymbol{p} \to \boldsymbol{p}^*} 0$.

In some simple cases of random fields, continuity is rather easily established.

Example 3.1

For illustration, let us assume a random field representation of the periodic variation

$$X(s,t) = a(t)\cos(s + \theta),$$

where $a(t)$ is a continuous deterministic function of t, and the RV $\theta = 0$ or π with equal probability. The sample paths can take two forms,

$$\chi(s,t) = \pm a(t)\cos(s),$$

which are continuous at each point.

Matters are not so obvious in the vast majority of cases, where the continuity of a random field is directly linked to the convergence issue considered earlier. Accordingly, it seems natural that to the various kinds of random field convergence we can associate the corresponding kinds of continuity.

Definition 3.1

The $X(\boldsymbol{p})$ has *continuous sample paths with probability* 1 in $\mathcal{S} \subset R^{n+1}$ if $X(\boldsymbol{p}) \xrightarrow{P.1} X(\boldsymbol{p}_0)$ as $\boldsymbol{p} \to \boldsymbol{p}_0$, or more conveniently,

$$X(\boldsymbol{p} + \delta \boldsymbol{p}) \xrightarrow{P.1} X(\boldsymbol{p}) \text{ as } \delta \boldsymbol{p} \to \boldsymbol{0} \tag{3.3}$$

for all $u \in \Omega$, i.e., with probability 1 there are no discontinuities within $\mathcal{S}$. In this case, we also talk about *sample path continuity* (or *continuity everywhere*).

The sample path continuity is the strongest type of stochastic continuity that is concerned with all sample paths of the random field. A weaker sense of continuity is obtained by eliminating a set of sample paths linked to the space $\mathcal{V} \subseteq \Omega$ such that $P[\mathcal{V}] = 0$. In this case, the following definition is introduced.

Definition 3.2

The $X(\boldsymbol{p})$ is *a.s. sample path continuous* at $\boldsymbol{p} \in \mathcal{S}$ if

$$X(\boldsymbol{p}+\delta\boldsymbol{p}) \xrightarrow{a.s.} X(\boldsymbol{p}) \text{ as } \delta\boldsymbol{p} \to \mathbf{0}. \tag{3.4}$$

When Eq. (3.4) holds for all $\boldsymbol{p}$ in the domain $\mathcal{S}$ of $X(\boldsymbol{p})$, the latter is a.s. continuous everywhere.[1]

On occasion, if physically justified, random field continuity considerations may focus on the spatial or the temporal coordinate suppressing the time or space argument, respectively.

Example 3.2

For a fixed time, the random field $X(s, t)$, $s \in [a, b]$, is a.s. continuous in space if there exist some κ, λ_t, $\mu > 0$ so that

$$\overline{|X(s+h,t) - X(s,t)|^{\kappa}} < \lambda_t |h|^{1+\mu}.$$

for sufficiently small h. A similar result is valid regarding a.s. continuity in time.

Otherwise said, in the case of a.s. continuity Eq. (3.4) holds everywhere except on the space of events $\mathcal{V}$ with $P[\mathcal{V}] = 0$.[2] In the context of Definition 3.1, the a.s. continuity of Definition 3.2 is essentially piecewise continuity, that is, it may allow discontinuities within $\mathcal{S}$ (although, according to Definition 3.1 the probability of a discontinuity at a specific $\boldsymbol{p} \in \mathcal{S}$ is 0, nevertheless, the random field realization may have discontinuities). Another, still weaker, kind of continuity is described next, although it is implied by the m.s. continuity to be discussed later.

Definition 3.3

The $X(\boldsymbol{p})$ is *continuous in probability* (P) at $\boldsymbol{p}$ if

$$X(\boldsymbol{p}+\delta\boldsymbol{p}) \xrightarrow{P} X(\boldsymbol{p}) \text{ as } \delta\boldsymbol{p} \to \mathbf{0}. \tag{3.5}$$

Because of the restrictions of the previous types of continuity that are often difficult to satisfy in applications, yet another type of "mean" continuity is introduced that avoids focusing on each and every sample path of a random field.

Definition 3.4

The $X(\boldsymbol{p})$ is said to be *continuous in the m.s. sense* (or L_2 continuous) at $\boldsymbol{p}$ if

$$X(\boldsymbol{p}+\delta\boldsymbol{p}) \xrightarrow{m.s.} X(\boldsymbol{p}) \text{ as } \delta\boldsymbol{p} \to \mathbf{0}, \text{ or } \underset{\delta\boldsymbol{p}\to\mathbf{0}}{\text{l.i.m.}}\, X(\boldsymbol{p}+\delta\boldsymbol{p}) = X(\boldsymbol{p}). \tag{3.6}$$

Eq. (3.6) is essentially equivalent to Eq. (2.6a). When Eq. (3.6) holds for all $\boldsymbol{p}$ in the domain of the random field $X(\boldsymbol{p})$, the latter is m.s. continuous everywhere. Being the most significant by far in applications, m.s. continuity considers the set of random field sample paths us a whole.

---

[1]As noted earlier, the a.s. continuity is sometimes linked to the p.1 continuity. Yet, certain authors consider them as two different kinds of continuity, emphasizing the relative importance of the latter in applied stochastics.

[2]Equivalently, there exists a space $\mathcal{V}^C$ with $P[\mathcal{V}^C] = 1$ so that Eq. (3.4) holds for all $u \in \mathcal{V}^C$.

Stochastic continuity conditions can be expressed in terms of covariance functions (in fact, the most fruitful analysis is possible in special cases, like Gaussian and/or STHS random fields). Indeed, Definition 3.4 implies that $X(\boldsymbol{p})$ is m.s. continuous at $\boldsymbol{p}$ if

$$\lim_{\boldsymbol{p}\to\boldsymbol{p}_0} \overline{|X(\boldsymbol{p}) - X(\boldsymbol{p}_0)|^2} = 0 \tag{3.7a}$$

(this is usually called L_2-continuity). On the strength of Proposition 2.2, a useful implication of Eq. (3.7a) is that

$$\lim_{\boldsymbol{p}\to\boldsymbol{p}_0} \overline{|X(\boldsymbol{p}) - X(\boldsymbol{p}_0)|^2} = \lim_{\boldsymbol{p}\to\boldsymbol{p}_0} [c_X(\boldsymbol{p},\boldsymbol{p}) - c_X(\boldsymbol{p}_0,\boldsymbol{p}) - c_X(\boldsymbol{p},\boldsymbol{p}_0) + c_X(\boldsymbol{p}_0,\boldsymbol{p}_0)] = 0, \tag{3.7b}$$

i.e., if the covariance $c_X(\boldsymbol{p},\boldsymbol{p}')$ is continuous at $(\boldsymbol{p}_0,\boldsymbol{p}_0)$, the right side of the last equation will indeed become 0 as $\boldsymbol{p}\to\boldsymbol{p}_0$, and Eq. (3.7b) is satisfied. Eq. (3.7b) is most useful in practice, since it shows that the definition of m.s. continuity, Eq. (3.7a), is equivalent to an expression involving covariance functions, Eq. (3.7b).[3] Furthermore, in this case

$$\overline{[X(\boldsymbol{p}) - X(\boldsymbol{p}_0)]^2} \geq [\overline{X(\boldsymbol{p}) - X(\boldsymbol{p}_0)}]^2,$$

so that if Eq. (3.7a) holds as $\boldsymbol{p}\to\boldsymbol{p}_0$, the last equation above implies that $\overline{X(\boldsymbol{p}) - X(\boldsymbol{p}_0)}\to 0$ as $\boldsymbol{p}\to\boldsymbol{p}_0$. In other words, the mean $\overline{X(\boldsymbol{p})}$ must be also m.s. continuous, i.e.,

$$\lim_{\boldsymbol{p}\to\boldsymbol{p}_0} \overline{X(\boldsymbol{p}) - X(\boldsymbol{p}_0)} = 0. \tag{3.8}$$

Accordingly, the fundamental proposition below is valid.

Proposition 3.1

The $X(\boldsymbol{p})$ is continuous in the m.s. sense at $\boldsymbol{p}$, i.e., $X(\boldsymbol{p}) \in L_2$, if and only if (1) its mean value is m.s. continuous at $\boldsymbol{p}$ and (2) its covariance $c_X(\boldsymbol{p},\boldsymbol{p}')$ is continuous at $(\boldsymbol{p},\boldsymbol{p}'=\boldsymbol{p})$. The $X(\boldsymbol{p})$ is everywhere continuous if $c_X(\boldsymbol{p},\boldsymbol{p}')$ is continuous at every diagonal point $(\boldsymbol{p},\boldsymbol{p}'=\boldsymbol{p})$.

Proposition 3.1 suggests that the m.s. continuity of a second-order random field is determined by the ordinary continuity of the corresponding covariances. A nice result that links sample path continuity with covariance features can be proven in the case of an STHS random field (see Chapter VII). The two cases of the following example correspond to an m.s. continuous random field that is also a.s. continuous and to an m.s. continuous random field that is not.

Example 3.3

Consider a zero mean *Wiener* field $W(s, t)$ with the simultaneous two-point covariance of the space–time separable form

$$c_W(s,s',t) = a(t)\min(s,s'). \tag{3.9}$$

where $a(t) > 0$ is a continuous function of t. According to Postulate 1.1, the analysis can be simplified as follows. Let $s = s^* + h_1,\ \ s' = s^* + h_2$, so that

$$|c_W(s^*+h_1, s^*+h_2, t) - c_W(s^*, s^*, t)| = a(t)|\min\ (s^*+h_1, s^*+h_2) - s^*| \leq a(t)\max(h_1,h_2).$$

[3] As we will see below, continuity in $\boldsymbol{p}$ and $\boldsymbol{p}_0$ provides sufficient m.s. continuity conditions.

As $h_1, h_2 \to 0$, $|c_W(s^* + h_1, s^* + h_2, t) - c_W(s^*, s^*, t)| \to 0$, meaning that $W(s, t)$ is m.s. continuous in s. In addition, almost all its sample paths are continuous (i.e., $W(s, t)$ is a.s. continuous). The *Poisson* field $\Pi(s, t)$ also has a covariance of the form of Eq. (3.9), and, hence, it is m.s. continuous too. However, almost all of its realizations have discontinuities.

Remark 3.1

Recalling Remark 2.1, and given that one stochastic convergence type does not, in general, imply the other, likewise one stochastic continuity type does not necessarily imply the other. Yet, the following statements are valid: (1) The a.s. continuity implies P continuity, and so does m.s. continuity. (2) If $X(\boldsymbol{p})$ is a bounded random field, then a.s. continuity implies m.s. continuity. (3) The a.s. continuity does not generally imply m.s. continuity (see an illustration in Example 3.4). (4) Although m.s. continuity is the one usually applied in second-order random field studies, it does not generally imply sample path continuity (a classical special case is discussed in Example 3.5).

Example 3.4

Consider the random field

$$X(s,t) = \left(s - \frac{1}{2}\right)^{-1} 1_{\left(\frac{1}{2}, s\right)}(t) \text{ if } s \in \left(\frac{1}{2}, 1\right], = 0 \text{ if } s \in \left[0, \frac{1}{2}\right), \tag{3.10}$$

where $s \in [0, 1]$, $t \sim U(0, 1)$, and $1_{\left(\frac{1}{2}, s\right)}(t)$ is an indicator function. It can be shown that $X(s,t) \xrightarrow{a.s.} 0$ as $s \to \frac{1}{2}$, but the random field is not m.s. continuous in s (Exercise VI.1).

Concerning the practical significance of stochastic continuity, most of the random field models considered in applied stochastics are m.s. continuous. Yet, discontinuities are also encountered in the space–time variation of certain natural attributes. Permeabilities in porous media, e.g., jump discontinuously at lithotype boundaries. In such cases, it may be more appropriate to use a continuous random field across the boundary, and simulations should account for the discontinuity (Christakos and Hristopulos, 1998).

3.2 EQUIVALENCE, MODIFICATION, AND SEPARABILITY[4]

As was discussed in Section 3.1, a central issue with the sample path continuity of an S/TRF $X(\boldsymbol{p})$ is that the probability of occurrence P of the sample path set of interest (i.e., the set of sample paths $\chi(\boldsymbol{p})$ that are continuous functions) may not be defined, because of the spatiotemporal complexity of $\chi(\boldsymbol{p})$ for the different $u \in \Omega$, and the structure of $(\Omega, \mathcal{F})$. And, even if this set of sample paths is defined, it may not be uniquely determined by the finite-dimensional cumulative distribution function (CDF) of $X(\boldsymbol{p})$. In this context, the notions of equivalence and modification are studied because sample path features, such as continuity, are of considerable importance in applied stochastics.

Definition 3.5

Two random fields $X_1(\boldsymbol{p})$ and $X_2(\boldsymbol{p})$ defined on two, generally different, probability spaces $(\Omega, \mathcal{F}, P)$ and $(\Omega', \mathcal{F}', P')$ are said to be *equivalent* or *versions* of one another, i.e.,

[4]To be distinguished from covariance separability.

$$X_1(\boldsymbol{p})\overset{e}{=}X_2(\boldsymbol{p}), \tag{3.11a}$$

if and only if for all $m = 1, 2,..., \boldsymbol{p} \in R^{n+1}$ and $A_i \in \mathcal{F}$, it is valid that

$$P[u: X_1(\boldsymbol{p}_i, u) \in A_i, i = 1, ..., m] = P[u': X_2(\boldsymbol{p}_i, u') \in A_i, i = 1, ..., m]. \tag{3.11b}$$

I.e., the $X_1(\boldsymbol{p})$ and $X_2(\boldsymbol{p})$ have the same finite dimensional probability distributions.

Definition 3.6

The $X_1(\boldsymbol{p})$ and $X_2(\boldsymbol{p})$ defined on the same probability space $(\Omega, \mathcal{F}, P)$ are called a *modification* of one another if and only if for all $\boldsymbol{p} \in R^{n+1}$ and $u \in \Omega$, it is valid that

$$P[u: X_1(\boldsymbol{p}, u) = X_2(\boldsymbol{p}, u)] = 1. \tag{3.12}$$

Clearly, Eq. (3.12) implies Eq. (3.11b). The construction of a modification may or may not be possible, depending on the shape of the finite-dimensional CDF under consideration (e.g., it is possible in the case of a Brownian field, but it is not possible in the case of a Poisson field). Generally, the significance of the notions of equivalence and modification lies in the fact that two equivalent random fields do not always have the same sample path features (continuity, differentiability, etc.). This means that it may be appropriate that all results concerned with sample path features should clarify that, "there exists a version or a modification of the random field under consideration that has the stated sample path features." For illustration let us discuss two examples. The first example (Cramer and Leadbetter, 1967) shows that two equivalent random fields, although they have identical probability distributions, they may have different continuity features.

Example 3.5

Given that $s \in [0, 1]$ and $\Omega = [0, 1]$, two random fields are defined by

$$X_1(s, u) = 0 \quad (\text{for all } s, \text{ and all } u \in \Omega), \tag{3.13}$$

and

$$X_2(s, u) = 1 \ (s = u), \ = 0 \ (s \neq u). \tag{3.14}$$

The fields have identical CDF,

$$F_{s_1,...,s_m}(\chi_1, ..., \chi_m) = 1 \quad \text{if } \chi_i \geq 0 \ (i = 1, ..., m), \quad = 0 \text{ otherwise.}$$

Since for all $s \in [0, 1]$,

$$P[u: X_1(s, u) = X_2(s, u)] = P[[0, 1] - \{s\}] = 1,$$

the two random fields are a modification of one another. This also implies that the two random fields are a.s. continuous and m.s. continuous. But the sample path behavior of the two fields are completely different, one is continuous, $P[u : X_1(s, u)$ continuous in $[0, 1]] = 1$, whereas the other one is not continuous, $P[u : X_2(s, u)$ continuous in $[0, 1]] = 0$. Otherwise said, while $X_1(s)$ has continuous realizations with probability 1, $X_2(s)$ has not.

The above simple example demonstrates that the multivariate CDF does not necessarily determine the continuity features of a random field. The second example shows that there are cases in which it cannot be concluded that the random field of interest has the specified sample path features.

Example 3.6

Given that $s \in [0, 1]$, $\Omega = R^{[0,1]}$, and $\mathcal{F}$ is a σ-field generated by cylinder sets $\{u \in \Omega : (u(s_1), \ldots, u(s_n)) \in \mathcal{B}_n\}$, a random field is defined as $X(s, u) = u(s)$. Now, the $\{u \in \Omega$: sample path $\chi(s, u)$ is continuous$\} \notin \mathcal{F}$, and, hence, one cannot say whether the random field has continuous sample paths with positive probability or not.

Definition 3.7

The $X(\boldsymbol{p})$ is called a *sample path separable* random field if there exists a countable set $S \subset R^{n+1}$ and an event $E \subset \Omega$ for which $P[E] = 0$, such that for any closed interval $I_C \subset R^1$ and open interval $I_O \subset R^1$ the sets

$$\begin{aligned} &\{u: X(\boldsymbol{p}, u) \in I_C, \boldsymbol{p} \in I_O\}, \\ &\{u: X(\boldsymbol{p}, u) \in I_C, \boldsymbol{p} \in I_O \cap S\} \end{aligned} \quad (3.15\text{a–b})$$

differ by a subset of E.

Example 3.7

A random field is sample path separable if there exists a countable set $S \subset R^1$ such that, with probability 1, the random field behaves more irregularly on R^1 than on S. In this case, the conditions of Definition 3.7 reduce to

$$\begin{aligned} \sup\nolimits_{s \in S \cap \mathcal{S}} X(s, u) &= \sup\nolimits_{s \in \mathcal{S}} X(s, u), \\ \inf\nolimits_{s \in S \cap \mathcal{S}} X(s, u) &= \inf\nolimits_{s \in \mathcal{S}} X(s, u), \end{aligned} \quad (3.16\text{a–b})$$

where $\mathcal{S} \subset R^1$. In the light of Eqs. (3.16a–b), the random field X_2 in Example 3.5 is not separable, since its sample paths are not determined by their values on a countable subset $S \subset R^1$.

In sum, a.s. continuity does not guarantee sample path continuity, for which additional conditions are required. Although comprehensive conditions may not be available in the general case, R^{n+1} and non-Gaussian $X(\boldsymbol{p})$, one well-known condition is sample path separability, which ensures that finite dimensional CDF determine sample path features by requiring that sample paths are determined by their values on an everywhere dense, countable set of points in R^{n+1}, say $S \subset R^{n+1}$ (i.e., if $X(\boldsymbol{p})$ is separable, then the set of CDFs $F_{\boldsymbol{p}_1,\ldots,\boldsymbol{p}_m}$ for all m, determine the random field's sample path features). To put it in more words, once we can credibly assume that we are dealing with the separable version of the random field $X(\boldsymbol{p})$, a.s. continuity in the sense discussed earlier can be extended to sample path continuity, since the probability 1 statement at individual points is simultaneously valid on a dense countable set of points S, and the sample path properties of $X(\boldsymbol{p})$ are specified by its properties on S. Given then that to every random field there corresponds a separable field whose properties are determined by its finite-dimensional probability distributions, it is reasonable that the study of separability focuses on sample path continuity.

In the light of Postulate 1.1, several spatial random field regularity results can be formally extended to spatiotemporal random fields, as long as either of the conditions (1) or (2) of the postulate can be assumed to hold. For sure, the point of the postulate's rationale is an interesting one that should be considered in terms of what is displaced and what gets interpreted. With this important note in mind, Proposition 3.2 provides some sufficient conditions for sample path continuity.[5]

---

[5]Relevant proofs can be found, e.g., in Cramer and Leadbetter (1967), Belyaev (1972), Adler (1981), Adler and Taylor (2007); and references therein.

Proposition 3.2

Let $X(s, t)$ be a zero mean random field in the $R^{n,1}$ domain. Assuming that the time argument does not change during derivations (Postulate 1.1):

(a) If for $\alpha_t > 0$, $\beta > \lambda > 0$ it holds that

$$\overline{|X(\boldsymbol{s}+\boldsymbol{h},t) - X(\boldsymbol{s},t)|^{\lambda}} \leq \frac{\alpha_t \boldsymbol{h}^{2n}}{\left|\log|\boldsymbol{h}|\right|^{1+\beta}}, \tag{3.17}$$

then $X(s, t)$ has continuous sample paths in space with probability 1 over any compact set $\mathcal{S} \subset R^n$.

(b) Given the simultaneous two-point covariance $c_X(\boldsymbol{s},\boldsymbol{s}',t)$, if for all $\boldsymbol{h} \in \mathcal{S}$ it is true that

$$c_X(\boldsymbol{s}+\boldsymbol{h},\boldsymbol{s}+\boldsymbol{h},t) - c_X(\boldsymbol{s}+\boldsymbol{h},\boldsymbol{s},t) - c_X(\boldsymbol{s},\boldsymbol{s}+\boldsymbol{h},t) + c_X(\boldsymbol{s},\boldsymbol{s},t) \leq \frac{\alpha_t \boldsymbol{h}^{2n}}{\left|\log|\boldsymbol{h}|\right|^{1+\beta}}, \tag{3.18}$$

where $\beta > 2$, the $X(s, t)$ has continuous sample paths in space with probability 1 over any $\mathcal{S}$ (this result is obtained from Eq. 3.17 for $\lambda = 2$).

(c) If $X(s, t)$ is a centered Gaussian field with a continuous covariance function, and, if for some $\alpha_t < \infty$ and $\beta > 0$ it is valid that

$$\overline{|X(\boldsymbol{s}+\boldsymbol{h},t) - X(\boldsymbol{s},t)|^{2}} \leq \frac{\alpha_t}{\left|\log|\boldsymbol{h}|\right|^{1+\beta}}, \tag{3.19}$$

then the random field $X(s, t)$ has continuous sample paths in space with probability 1 over any $\mathcal{S}$.

(d) If the simultaneous two-point covariance $c_X(\boldsymbol{s} + \boldsymbol{h}, \boldsymbol{s} - \boldsymbol{h}, t)$ is an n-times continuously differentiable function with respect to $\boldsymbol{h}$, and for some $\alpha > 0$ it holds that

$$\sup\left|c_X(\boldsymbol{s}+\boldsymbol{h},\boldsymbol{s}-\boldsymbol{h},t) - \eta_{n,t}(\boldsymbol{h},\boldsymbol{s})\right| = O_t\left(|\boldsymbol{h}|^{n+\alpha}\right), \tag{3.20}$$

as $\boldsymbol{h} \to \boldsymbol{0}$, where $\eta_{n,t}$ is a Taylor polynomial of degree n in $\boldsymbol{h}$ for each $\boldsymbol{s}$ and with time-dependent coefficients, in general, and the supremum is taken over $\boldsymbol{s}$ lying in each compact subset of $R^{n,1}$, then there exists a version of the random field with continuous realizations in space.

An example may help us clarify some conditions of the theoretical statements of Proposition 3.2. Yet, the interest here lies not so much on theory but rather on the practical significance of these statements that apply in a range of cases in the physical world. Some of these cases are mentioned next.

Example 3.8

In the case of attributes characterized by a space–time separable covariance $c_X(\boldsymbol{s},\boldsymbol{s}',t) = c_{X(1)}(\boldsymbol{s},\boldsymbol{s}')c_{X(2)}(t,t), c_{X(2)}(t,t) > 0$, we find that Eq. (3.18) holds with $\alpha_t = \alpha c_{X(2)}(t, t)$, $\alpha > 0$ (Exercise VI.14).

The next example examines the possibility that the findings of Proposition 3.2 are stated in the R^{n+1} domain.

Example 3.9

If we define $\Delta \boldsymbol{p}_0 = (h_1,\ldots, h_n, h_0 = 0) = (\boldsymbol{h}, 0)$, Eq. (3.19) could hold rather trivially by simply replacing $\boldsymbol{h}$ with $\Delta \boldsymbol{p}_0$. A more interesting generalization is possible if a specific space−time metric is used (e.g., the Pythagorean metric $r_p = |\Delta \boldsymbol{p}|$, see Table 1.1 of Chapter III). Such a metric generalization would raise interesting theoretical questions worthy of investigation, but I will not get into this issue here.

By a way of a summary, the following conclusions can be drawn based on our discussion of stochastic continuity so far:

(i) The m.s. continuity depends on stochastic expectation, i.e., on a certain average over all sample paths (realizations) of the random field. The m.s. random field properties are usually considered, in part because they can be determined from the corresponding covariance functions.[6]

(ii) The a.s. (or the sample path) continuity depends on the behavior of each individual sample path (i.e., it involves conditions on an infinite number of points), which makes it more difficult to study.

(iii) Yet, useful a.s. continuity conditions are available, especially for Gaussian random fields. Some of these conditions involve statistical moments of random field differences, which assure the existence of a random field version with continuous sample paths.

(iv) Since the finite dimensional CDF of a random field do not determine its sample path properties, demonstrating sample path continuity (as well as differentiability, see Section 4 next) relies on the notion of separability (i.e., separable random fields whose finite-dimensional CDF can specify sample path properties).

We will revisit these important random field geometry issues in Chapter VII in the context of key hypotheses [like random field STHS and space−time isostationarity (STIS)], where additional insight can be gained regarding the space−time variation features of the phenomenon of interest. At this point, the question naturally emerges whether a continuous random field is also differentiable. As it turns out, it is physically possible that continuity implies differentiability (this is, e.g., often the case with sea surface), yet mathematically it is a more complicated issue to determine the conditions so that random field differentiability holds. This leads us to the topic of the next section.

4. STOCHASTIC DIFFERENTIATION

Understanding S/TRF differentiability plays a central role in the adequate mathematical representation of many physical phenomena. For example, the differentiability of a random field representing ocean surface is directly related to surface roughness and its consequences (on the reliability of marine structures, the economic feasibility of offshore wind power utilization etc.). Mathematically, the differentiation of a random field $X(\boldsymbol{p})$ generates new random fields that are the derivatives of $X(\boldsymbol{p})$ along various directions in space and in time. Beyond its purely mathematical interest, stochastic differentiation has significant implications in scientific practice in which random field differentiability often has a definite physical meaning and considerable real-world consequences.

[6]Jumping ahead, in a following section it will be shown that a random field is space−time (ν, μ) m.s. differentiable at $\boldsymbol{p}$ if and only if the $(2\nu, 2\mu)$ derivatives of its covariance exists and are finite at $(\boldsymbol{p}, \boldsymbol{p})$.

4.1 BASIC NOTATION AND DEFINITIONS

Although the spatial and temporal derivatives of $X(\boldsymbol{p})$ can be formally considered in the same setting, that is, with respect to each component of the space–time point vector $\boldsymbol{p} = (s_1, \ldots, s_n, s_0) \in R^{n+1}$, for applied stochastics purposes,[7] it may be advantageous to work in the $R^{n,1}$ domain. In this case, we should distinguish between differentiation of $X(\boldsymbol{p})$ with respect to time $s_0 = t \in R^1$ and differentiation

(a) either along arbitrary spatial directions defined by $\boldsymbol{\varepsilon} = (\boldsymbol{\varepsilon}_{d_1}, \ldots, \boldsymbol{\varepsilon}_{d_\rho})$, where $\boldsymbol{\varepsilon}_{d_i}$ is the unit vector along direction d_i $(i = 1, \ldots, \rho)$ at location $\boldsymbol{s} = (s_1, \ldots, s_n) \in R^n$;[8]

(b) or along the spatial coordinate directions $(s_1, \ldots, s_n)$ defined by $\boldsymbol{\varepsilon} = (\boldsymbol{\varepsilon}_1, \ldots, \boldsymbol{\varepsilon}_n)$, where $\boldsymbol{\varepsilon}_i$ $(i = 1, \ldots, n)$ are unit vectors along the (orthogonal) directions of the space coordinates s_i of $\boldsymbol{s}$.[9]

The added value of the above setup is that the R^{n+1} analysis and the $R^{n,1}$ analysis can be linked operationally by means of the vector–scalar notation introduced in Chapter I. In particular, in the R^{n+1} domain let $\boldsymbol{p} = (s_1, \ldots, s_n, s_0) = (\boldsymbol{s}, s_0)$ be the space–time point vector with the multi–indexes $\boldsymbol{d}^* = (d_1, \ldots, d_\rho, d_0)$ and $\boldsymbol{\nu}^* = (\nu_1, \ldots, \nu_\rho, \nu_0)$ denoting direction and order of differentiation, respectively $\left(\sum_{i=1}^{\rho,0} \nu_i = |\boldsymbol{\nu}^*| = \nu^*\right)$. In the $R^{n,1}$ domain, on the other hand, we consider separately in R^n the spatial location vector $\boldsymbol{s} = (s_1, \ldots, s_n)$ with the multi–indexes of nonnegative integers $\boldsymbol{d} = (d_1, \ldots, d_\rho)$ and $\boldsymbol{\nu} = (\nu_1, \ldots, \nu_\rho)$ $\left(\sum_{i=1}^{\rho} \nu_i = |\boldsymbol{\nu}| = \nu\right)$, and separately in R^1 the time scalar t with the nonnegative integers d_0 and $\nu_0 = \mu$. Then, we can link the analyses in the R^{n+1} and $R^{n,1}$ domains by letting $\boldsymbol{d}^* = (\boldsymbol{d}, d_0)$, and $\boldsymbol{\nu}^* = (\boldsymbol{\nu}, \nu_0)$ such that $\nu^* = \nu + \mu$.

Example 4.1

A space–time monomial in the R^{n+1} domain, $\boldsymbol{p}^{\boldsymbol{\nu}^*} = \Pi_{i=1}^{\rho,0} s_i^{\nu_i^*}$, could be expressed in the $R^{n,1}$ domain as $\boldsymbol{p}^{(\boldsymbol{\nu},\mu)} = \boldsymbol{s}^{\boldsymbol{\nu}} s_0^{\mu}$, where $\boldsymbol{s}^{\boldsymbol{\nu}} = \Pi_{i=1}^{\rho} s_i^{\nu_i}$ is a spatial monomial, and $s_0^{\mu} = t^{\mu}$ is a temporal monomial (Section 2 of Chapter I; and Appendix).

I continue with some basic space–time differentiation notation for an S/TRF $X(\boldsymbol{p})$. For any subsets of multiindexes $\boldsymbol{d}$ and $\boldsymbol{\nu}$ as above, we can define the spatial (arbitrary directions)/temporal partial derivatives of $X(\boldsymbol{p})$ as

$$X_{\boldsymbol{d}^*}^{(\boldsymbol{\nu}^*)}(\boldsymbol{p}) = \frac{\partial^{\nu^*} X(\boldsymbol{p})}{\partial \boldsymbol{p}_{\boldsymbol{d}^*}^{\boldsymbol{\nu}^*}} \quad (R^{n+1}),$$

$$X_{\boldsymbol{d},0}^{(\boldsymbol{\nu},\mu)}(\boldsymbol{s}, s_0) = \frac{\partial^{\nu+\mu} X(\boldsymbol{s}, s_0)}{\partial \boldsymbol{s}_{\boldsymbol{d}}^{\boldsymbol{\nu}} \partial s_0^{\mu}} \quad (R^{n,1}), \tag{4.1a–b}$$

[7]Concerning, e.g., the view that a natural attribute may depend on the space argument in a different way than it depends on the time argument.

[8]With a few notable exceptions, in the majority of cases considered in applied stochastics it is assumed that $\rho = n$.

[9]Differentiation along the coordinate directions is a special case of differentiation along arbitrary directions, under the condition $\boldsymbol{\varepsilon}_{d_i} = \boldsymbol{\varepsilon}_i, i = 1, \ldots n (= \rho)$.

where $\partial \boldsymbol{p}_{\boldsymbol{d}^*}^{\boldsymbol{\nu}^*} = \Pi_{i \in I_{\nu^*}} \partial s_{d_i}^{\nu_i}$, $\partial \boldsymbol{s}_{\boldsymbol{d}}^{\boldsymbol{\nu}} = \Pi_{i \in I_\nu} \partial s_{d_i}^{\nu_i}$, and the "$i \in I_\nu$" and "$i \in I_{\nu^*}$" denote that for a given ρ the i may assume values in any subset $I_\nu \subseteq \{1, \ldots, \rho\}$ and $I_{\nu^*} \subseteq \{1, \ldots, \rho, 0\}$, respectively.[10]

Example 4.2

If i spans all directions, i.e., $I_{\nu^*} = \{1, \ldots, \rho, 0\}$ and $I_\nu = \{1, \ldots, \rho\}$, the $\partial \boldsymbol{p}_{\boldsymbol{d}^*}^{\boldsymbol{\nu}^*}$ and $\partial \boldsymbol{s}_{\boldsymbol{d}}^{\boldsymbol{\nu}}$ in the denominator of Eqs. (4.1a–b) become, respectively, $\Pi_{i=1}^{\rho,0} \partial s_{d_i}^{\nu_i}$ and $\Pi_{i=1}^{\rho} \partial s_{d_i}^{\nu_i}$. If, in addition, $\nu_i = \mu = 1$, the $\partial \boldsymbol{p}_{\boldsymbol{d}^*}^{\boldsymbol{\nu}^*}$ and $\partial \boldsymbol{s}_{\boldsymbol{d}}^{\boldsymbol{\nu}}$ reduce to $\Pi_{i=1}^{\rho,0} \partial s_{d_i}$ and $\Pi_{i=1}^{\rho} \partial s_{d_i}$, respectively.

Similarly, for any subset of the multiindexes $\boldsymbol{i} = (1, \ldots, i)$ referring to the s_i directions ($i \leq n$) and the associated $\boldsymbol{\nu} = (\nu_1, \ldots, \nu_i)$ we can define the spatial (coordinate directions)/temporal partial derivatives of $X(\boldsymbol{p})$ by

$$X_{\boldsymbol{i},0}^{(\boldsymbol{\nu}^*)}(\boldsymbol{p}) = \frac{\partial^{\nu^*} X(\boldsymbol{p})}{\partial \boldsymbol{p}^{\boldsymbol{\nu}^*}} \quad (R^{n+1}),$$

$$X_{\boldsymbol{i},0}^{(\boldsymbol{\nu},\mu)}(\boldsymbol{s}, s_0) = \frac{\partial^{\nu+\mu} X(\boldsymbol{s}, s_0)}{\partial \boldsymbol{s}^{\boldsymbol{\nu}} \partial s_0^{\mu}} \quad (R^{n,1}), \tag{4.2a–b}$$

where $\partial \boldsymbol{p}^{\boldsymbol{\nu}^*} = \Pi_{i \in I_{\nu^*}} \partial s_i^{\nu_i}$, $I_{\nu^*} \subseteq \{1, \ldots, n, 0\}$, $\partial \boldsymbol{s}^{\boldsymbol{\nu}} = \Pi_{i \in I_\nu} \partial s_i^{\nu_i}$, $I_\nu \subseteq \{1, \ldots, n\}$. In the following, random field differentiations will be usually performed in terms of the definition in Eq. (4.2b).

Example 4.3

Commonly encountered special cases of Eq. (4.2b) are

$$X_{i_1,i_2,0}^{(\nu,\mu)}(\boldsymbol{s}, s_0) = X_{i_1,i_2,0}^{(\nu_1,\nu-\nu_1,0)}(\boldsymbol{s}, s_0) = \frac{\partial^{\nu+\mu} X(\boldsymbol{s}, s_0)}{\partial s_{i_1}^{\nu_1} \partial s_{i_2}^{\nu-\nu_1} \partial s_0^{\mu}},$$

$$X_{i,0}^{(\nu,\mu)}(\boldsymbol{s}, s_0) = \frac{\partial^{\nu+\mu} X(\boldsymbol{s}, s_0)}{\partial s_i^{\nu} \partial s_0^{\mu}},$$

$$X_{i_1,i_2,0}^{(\nu,0)}(\boldsymbol{s}, s_0) = X_{i_1,i_2}^{(\nu_1,\nu-\nu_1)}(\boldsymbol{s}, s_0) = \frac{\partial^{\nu} X(\boldsymbol{s}, s_0)}{\partial s_{i_1}^{\nu_1} \partial s_{i_2}^{\nu-\nu_1}}, \tag{4.3a–d}$$

$$X_{i,0}^{(\nu,0)}(\boldsymbol{s}, s_0) = X_i^{(\nu)}(\boldsymbol{s}, s_0) = \frac{\partial^{\nu} X(\boldsymbol{s}, s_0)}{\partial s_i^{\nu}}$$

$$(i_1, i_2 = 1, \ldots, n:\ \nu, \mu = 0, 1, \ldots).$$

As before, if i spans all directions in Eqs. (4.2b), i.e., $I_\nu = \{1, \ldots, n\}$, then $\partial \boldsymbol{s}^{\boldsymbol{\nu}}$ reduces to $\Pi_{i=1}^{n} \partial s_i^{\nu_i}$. In this case, the derivative may be simply denoted as, respectively,

$$X^{(\boldsymbol{\nu},\mu)}(\boldsymbol{s}, s_0) \text{ or } \partial_{\boldsymbol{s},s_0}^{\nu+\mu} X(\boldsymbol{s}, s_0).$$

---

[10]Surely, for a given ν there are many combinations of ν_i such that $\sum_{i \in I_\nu} \nu_i = \nu$ or $\sum_{i \in I_{\nu^*}} \nu_i = \nu^*$.

The latter is, as a matter of fact, the kind of random field differentiation that will be mostly used herein. If, in addition, $n = \nu$ and $\nu_i = 1$, the notation in Eqs. (4.2b) simplifies further to

$$X^{(n,1)}(\boldsymbol{s}, s_0) = \frac{\partial^{n+1} X(\boldsymbol{s}, s_0)}{\partial \boldsymbol{s}^n \partial s_0} \quad (R^{n,1}), \tag{4.4}$$

where $\partial \boldsymbol{s}^n = \Pi_{i=1}^n \partial s_i$.[11]

In Eq. (4.4), the differentiation of $X(\boldsymbol{s}, s_0)$ may generate n random fields, which are its first derivatives along the directions of the n coordinates in space combined with one derivative in time, denoted as

$$X_{1,0}^{(1,1)}(\boldsymbol{s}, s_0), \ldots, X_{n,0}^{(1,1)}(\boldsymbol{s}, s_0), \tag{4.5}$$

i.e., the subscripts 1,…, n correspond to the spatial coordinates $s_1, \ldots, s_n$, and the subscript 0 to the time coordinate $s_0 = t$. The set of these derivatives is sometimes called the *gradient* random field set. S/TRFs are also generated by considering higher-order derivatives along the coordinate directions and/or in time. When referring to random field derivatives, one should be careful to clarify if partial derivatives along all directions are considered or only along a few of them. As regards their interpretation, the derivatives represent the rates of change of $X(\boldsymbol{p})$ as the space−time arguments change. Of course, what the physical meaning of the derivative is depends on what physical quantity $X(\boldsymbol{p})$ represents. For instance, if $X(\boldsymbol{p})$ represents momentum, the $X_{i,0}^{(1,0)}(\boldsymbol{p})$ is the spatial gradient along the direction s_i.

Example 4.4

To shed light on the implementation of the general random field differentiation expressions presented above, some specific cases of $X(\boldsymbol{p})$ derivatives are listed in Table 4.1.

Table 4.1 Examples of Random Field Derivatives in $R^{n,1}$

ν	μ	I_ν	Derivatives of $X(\boldsymbol{p})$
2	2	{1, 3}	$X_{\boldsymbol{d},0}^{(2,2)}(\boldsymbol{p}) = \frac{\partial^{2+2} X(\boldsymbol{p})}{\Pi_{i\in\{1,3\}} \partial s_{d_i}^{\nu_i} \partial s_0^2} = \frac{\partial^4 X(\boldsymbol{p})}{\partial s_{d_1} \partial s_{d_3} \partial s_0^2}$
2	2	{1, 3}	$X_{i,0}^{(2,2)}(\boldsymbol{p}) = X^{(2,2)}(\boldsymbol{p}) = \frac{\partial^{2+2} X(\boldsymbol{p})}{\Pi_{i\in\{1,3\}} \partial s_i^{\nu_i} s_0^2} = \frac{\partial^4 X(\boldsymbol{p})}{\partial s_1 \partial s_3 \partial s_0^2}$
2	1	{1}	$X_{i,0}^{(2,1)}(\boldsymbol{p}) = X^{(2,1)}(\boldsymbol{p}) = \frac{\partial^{2+1} X(\boldsymbol{p})}{\Pi_{i\in\{1\}} \partial s_i^{\nu_i} \partial s_0} = \frac{\partial^3 X(\boldsymbol{p})}{\partial s_1^2 \partial s_0}$
1	0	{2}	$X_{i,0}^{(1,0)}(\boldsymbol{p}) = X^{(1,0)}(\boldsymbol{p}) = \frac{\partial X(\boldsymbol{p})}{\Pi_{i\in\{2\}} \partial s_i^{\nu_i}} = \frac{\partial X(\boldsymbol{p})}{\partial s_2}$

[11]This is also written simply as, $\partial \boldsymbol{s} = \Pi_{i=1}^{n} \partial s_i$.

Herein, I will focus on the spatial (coordinate directions)/temporal partial derivatives[12] of $X(\boldsymbol{p})$, and I will define some of the derivatives in the m.s. sense using limits of generalized differences, starting with the definition of the derivative of $X^{(\boldsymbol{\nu},\mu)}(\boldsymbol{p})$ in the m.s. sense.

Definition 4.1

The νth-order in the sequence of space directions and μth-order in time *partial derivative* in the m.s. sense is defined by

$$X^{(\boldsymbol{\nu},\mu)}(\boldsymbol{p}) = \frac{\partial^{\nu+\mu}X(\boldsymbol{p})}{\partial \boldsymbol{s}^{\boldsymbol{\nu}}\partial s_0^{\mu}} = \underset{\boldsymbol{h}\to\boldsymbol{0},h_0\to 0}{\text{l.i.m.}} \frac{\Delta_{\boldsymbol{h},h_0}^{\nu+\mu}X(\boldsymbol{p})}{\boldsymbol{h}^{\boldsymbol{\nu}}h_0^{\mu}}, \tag{4.6a}$$

where the limit $\boldsymbol{h} \to \boldsymbol{0}$ operates sequentially ($h_1 \to 0$, then $h_2 \to 0$, etc.), and

$$\Delta_{\boldsymbol{h},h_0}^{\nu+\mu}X(\boldsymbol{p}) = \sum_{\eta_0\in\{0,1\}^{\mu}} (-1)^{\mu-\eta_0}\left[\sum_{\boldsymbol{\eta}\in\{0,1\}^{\nu}} (-1)^{\nu-\sum_{k=1}^{\nu}\eta_k} X\left(\boldsymbol{s}+\sum_{k=1}^{\nu}\eta_k h_k \boldsymbol{\varepsilon}_{i_k}, s_0+\eta_0 h_0\right)\right], \tag{4.6b}$$

$\boldsymbol{\eta} = (\eta_1,\ldots, \eta_k)$, is a *generalized space–time difference* operator.

Otherwise said, an S/TRF $X(\boldsymbol{p})$ has a νth/μth-order partial derivative in the m.s. sense if there exists a random field $X^{(\boldsymbol{\nu},\mu)}(\boldsymbol{p})$ defined as in Eq. (4.6a) (it converges in L_2). To put it in more symbolic terms,

$$\frac{\Delta_{\boldsymbol{h},h_0}^{\nu+\mu}X(\boldsymbol{p})}{\boldsymbol{h}^{\boldsymbol{\nu}}h_0^{\mu}} \xrightarrow{L_2} \left(\text{or}, \xrightarrow{m.s.}\right) X^{(\boldsymbol{\nu},\mu)}(\boldsymbol{p}) \text{ as } h_1\to 0,\ldots,h_\nu\to 0, h_0\to 0. \tag{4.6c}$$

As the readers observe, Definition 4.1, although formally rigorous, it is not always useful as a practical random field differentiability criterion, since it includes the unknown random field $X^{(\boldsymbol{\nu},\mu)}(\boldsymbol{p})$. An alternative, and admittedly more useful, criterion of differentiability is as follows.

Definition 4.2

The $X(\boldsymbol{p})$ is νth/μth-differentiable in the m.s. sense in R^{n+1} if the 2νth/2μth-*generalized derivative*

$$\lim_{\boldsymbol{h},\boldsymbol{h}'\to\boldsymbol{0},h_0,h_0'\to 0} \overline{\frac{\Delta_{\boldsymbol{h},h_0}^{\nu+\mu}X(\boldsymbol{p})\Delta_{\boldsymbol{h}',h_0'}^{\nu+\mu}X(\boldsymbol{p}')}{\boldsymbol{h}^{\boldsymbol{\nu}}\boldsymbol{h}'^{\boldsymbol{\nu}}h_0^{\mu}h_0'^{\mu}}} \tag{4.7}$$

exists for all $\boldsymbol{p}, \boldsymbol{p}'\in R^{n+1}$, all directions, and all $\boldsymbol{h}$, $\boldsymbol{h}'$, h_0, h_0'.

The above result is a direct consequence of the m.s. convergence properties discussed earlier. As we shall see in the following section, Eq. (4.7) provides the 2νth/2μth-derivative of the corresponding covariance function.

Remark 4.1

In light of the above definitions, if $X(\boldsymbol{p})$ is a *Gaussian* random field, its m.s. derivatives, if they exist, are Gaussian as well. Clearly, if the S/TRF has a constant mean, then $\overline{X^{(\boldsymbol{\nu},\mu)}(\boldsymbol{p})} = 0$.

It should be pointed out that the m.s. differentiability of $X(\boldsymbol{p})$ concerns the average behavior of its sample paths and does not prevent the existence of certain sample paths that are not differentiable in

[12]For the corresponding formulas in the case of spatial (arbitrary directions)/temporal partial derivatives, see the relevant exercises of this chapter.

the ordinary sense. Next, derivatives along the directions of the coordinates of a point will be considered, in which case the following results are basically special cases of Definition 4.1.

Example 4.5

Let the random field $X(\boldsymbol{p})$ represent ocean wave slope. The $X(\boldsymbol{p})$ is m.s. differentiable of the 1st-order with respect to both the spatial coordinate s_i ($i = 1,\ldots, n$) and the time instant $t = s_0$ of the point $\boldsymbol{p}$ (i.e., 1/1-order m.s. differentiable) if the $X_{i,0}^{(1,1)}(\boldsymbol{p}) = \frac{\partial^2 X(\boldsymbol{p})}{\partial s_i \partial s_0}$ exists so that

$$\frac{X(\boldsymbol{s}+h_i\boldsymbol{\varepsilon}_i, s_0+h_0) - X(\boldsymbol{s}, s_0+h_0) + X(\boldsymbol{s}, s_0) - X(\boldsymbol{s}+h_i\boldsymbol{\varepsilon}_i, s_0)}{h_i h_0} \xrightarrow{m.s.} X_{i,0}^{(1,1)}(\boldsymbol{p}) \text{ as } h_i, h_0 \to 0, \quad (4.8a)$$

or

$$\begin{aligned} X_{i,0}^{(1,1)}(\boldsymbol{p}) &= \frac{\partial^2 X(\boldsymbol{p})}{\partial s_i \partial s_0} = \underset{h_i,h_0\to 0}{\text{l.i.m.}} \frac{\Delta^2_{h_i,h_0} X(\boldsymbol{p})}{h_i h_0} \\ &= \underset{h_i,h_0\to 0}{\text{l.i.m.}} \frac{X(\boldsymbol{s}+h_i\boldsymbol{\varepsilon}_i, s_0+h_0) - X(\boldsymbol{s}, s_0+h_0) + X(\boldsymbol{s}, s_0) - X(\boldsymbol{s}+h_i\boldsymbol{\varepsilon}_i, s_0)}{h_i h_0}, \end{aligned} \quad (4.8b)$$

$i = 1,\ldots, n$. Each random field $X_{i,0}^{(1,1)}(\boldsymbol{p})$ is also referred to as the ith-partial/temporal derivative of $X(\boldsymbol{p})$ at $\boldsymbol{p}$. Eq. (4.8b) is obtained from Eqs. (4.6a) and (4.6b) for $\boldsymbol{\varepsilon} = \boldsymbol{\varepsilon}_{i_1} = \boldsymbol{\varepsilon}_i$.[13] The limit of Eq. (4.8b) existing in the m.s. sense by definition indicates that,

$$\lim_{h_i,h_0\to 0} \overline{\left[\frac{\Delta^2_{h_i,h_0} X(\boldsymbol{p})}{h_i h_0} - \frac{\partial^2 X(\boldsymbol{p})}{\partial s_i \partial s_0}\right]^2} = 0. \quad (4.9)$$

When Eq. (4.8a) holds for all $\boldsymbol{p}$ in the domain of $X(\boldsymbol{p})$, the random field is m.s. differentiable everywhere. Since Eq. (4.9) includes the unknown $X_{i,0}^{(1,1)}(\boldsymbol{p})$, another way to look at Eqs. (4.8a) and (4.8b) is in terms of the Cauchy's convergence criterion discussed earlier. In these terms, the random field $\frac{\Delta^2_{h_i,h_0} X(\boldsymbol{p})}{h_i h_0}$ has an m.s. limit if

$$\underset{h_i,h_0,h_i',h_0'\to 0}{\text{l.i.m.}} \left[\frac{\Delta^2_{h_i,h_0} X(\boldsymbol{p})}{h_i h_0} - \frac{\Delta^2_{h_i',h_0'} X(\boldsymbol{p})}{h_i' h_0'}\right] = 0, \quad (4.10)$$

---

[13]The implementation of the operator of Eq. (4.6b) gives for $\nu=1$,

$$\Delta^1_{h_i} X(\boldsymbol{p}) = \sum_{\eta_i=0,1} (-1)^{1-\eta_i} X(\boldsymbol{s}+\eta_i h_i \boldsymbol{\varepsilon}_i, s_0+\eta_0 h_0) = X(\boldsymbol{s}+h_i\boldsymbol{\varepsilon}_i, s_0+\eta_0 h_0) - X(\boldsymbol{s}, s_0+\eta_0 h_0);$$

and for $\mu = 1$,

$$\begin{aligned} \Delta^2_{h_i,h_0} X(\boldsymbol{p}) &= \Delta^1_{h_0}\left[\Delta^1_{h_i} X(\boldsymbol{p})\right] = \sum_{\eta_0=0,1} (-1)^{1-\eta_0} \Delta^1_{h_i} X(\boldsymbol{p}) \\ &= \sum_{\eta_0=0,1} (-1)^{1-\eta_0} [X(\boldsymbol{s}+h_i\boldsymbol{\varepsilon}_i, s_0+\eta_0 h_0) - X(\boldsymbol{s}, s_0+\eta_0 h_0)] \\ &= [X(\boldsymbol{s}+h_i\boldsymbol{\varepsilon}_i, s_0+h_0) - X(\boldsymbol{s}, s_0+h_0)] - [X(\boldsymbol{s}+h_i\boldsymbol{\varepsilon}_i; s_0) - X(\boldsymbol{s}, s_0)], \end{aligned}$$

as in Eq. (4.8b).

which means that in place of Eqs. (4.8a) and (4.8b) a sufficient condition of differentiability involving expectation may be used: The random field $X(\boldsymbol{p})$ is (1, 1) m.s. differentiable with respect to the coordinates s_i and s_0 (time) of the point $\boldsymbol{p}$ if

$$\lim_{h_i,h_0,h_i',h_0'\to 0}\overline{\left[\frac{\Delta^2_{h_i,h_0}X(\boldsymbol{p})}{h_ih_0}-\frac{\Delta^2_{h_i',h_0'}X(\boldsymbol{p})}{h_i'h_0'}\right]^2}=0, \tag{4.11}$$

which holds independent of the way h_i, h_0, h_i', h_0' approach 0. As we shall see in the following section, Eq. (4.11) plays a key role in the derivation of a necessary and sufficient condition of m.s. differentiability in terms of the covariance function.

Remark 4.2

Consider a *linear* and *homogeneous* space–time differential operator

$$\mathcal{L}_{\boldsymbol{p}}X(\boldsymbol{s},s_0)=\sum_{|\boldsymbol{\rho}|=\nu}\alpha_{\boldsymbol{\rho},\mu}\frac{\partial^{\rho+\mu}}{\partial \boldsymbol{s}^{\boldsymbol{\rho}}\partial s_0^{\mu}}X(\boldsymbol{s},s_0), \tag{4.12}$$

where $s_0=t$, $\alpha_{\boldsymbol{\rho},\mu}$ are known coefficients, $\boldsymbol{\rho}$ is the set of integers $(\rho_1,\ldots,\rho_n)$ such that $|\boldsymbol{\rho}|=\sum_{i=1}^n\rho_i=\nu$, ρ_i denotes the order of the partial derivative with respect to s_i, and $\partial\boldsymbol{s}^{\boldsymbol{\rho}}=\Pi_{i=1}^n\partial s_i^{\rho_i}$.

For any $\mathcal{L}_{\boldsymbol{p}}$, the stochastic expectation and differentiation operators commute, i.e.,

$$\begin{aligned}
&\overline{\mathcal{L}_{\boldsymbol{p}}X(\boldsymbol{p})}=\mathcal{L}_{\boldsymbol{p}}\overline{X(\boldsymbol{p})},\\
&\operatorname{cov}(\mathcal{L}_{\boldsymbol{p}}X(\boldsymbol{p}),X(\boldsymbol{p}'))=\mathcal{L}_{\boldsymbol{p}}c_X(\boldsymbol{p},\boldsymbol{p}'),\\
&\operatorname{cov}(\mathcal{L}_{\boldsymbol{p}}X(\boldsymbol{p}),\mathcal{L}_{\boldsymbol{p}'}X(\boldsymbol{p}'))=\mathcal{L}_{\boldsymbol{p}}\mathcal{L}_{\boldsymbol{p}'}c_X(\boldsymbol{p},\boldsymbol{p}'),\\
&\operatorname{var}(\mathcal{L}_{\boldsymbol{p}}X(\boldsymbol{p}))=\mathcal{L}_{\boldsymbol{p}}\mathcal{L}_{\boldsymbol{p}'}c_X(\boldsymbol{p},\boldsymbol{p}')\big|_{\boldsymbol{p}=\boldsymbol{p}'}
\end{aligned}$$

for all $\boldsymbol{p}$, $\boldsymbol{p}'$. These formulas are rather routinely applied in random field modeling.

Beyond the partial space–time derivatives and the differential operators, other kinds of derivatives of random fields also emerge in applications, such as the directional derivative. Specifically, the *directional derivative* of $X(\boldsymbol{p})$ at a given point $\boldsymbol{p}$ along a given direction determined by the unit vector $\boldsymbol{\varepsilon}$, $D_{\boldsymbol{\varepsilon}}X(\boldsymbol{p})$, is the inner product of the gradient $\nabla X(\boldsymbol{p})$ and $\boldsymbol{\varepsilon}$ at a given point $\boldsymbol{p}$, i.e.,

$$D_{\boldsymbol{\varepsilon}}X(\boldsymbol{p})=\nabla X(\boldsymbol{p})\cdot\boldsymbol{\varepsilon}, \tag{4.13}$$

and is interpreted as the rate of change of $X(\boldsymbol{p})$ moving through $\boldsymbol{p}$ with a velocity specified by $\boldsymbol{\varepsilon}$. Hence, $D_{\boldsymbol{\varepsilon}}X(\boldsymbol{p})$ generalizes the notion of the partial derivatives defined above (in which the rate of change was taken along specific coordinates, say s_i, all other coordinates s_j $(j\neq i)$ being held constant).

Example 4.6

Let $X(\boldsymbol{p})=(e^{s_0s_1}+s_2)v$, $\boldsymbol{p}=(s_1,s_2,s_0)$, where the RV $v\sim N(0,1)$. The directional derivative along the unit vector $\boldsymbol{\varepsilon}=(\varepsilon_1,\varepsilon_2)$ is defined in terms of the partial space–time random field derivatives, i.e.,

$$D_{\varepsilon_1,\varepsilon_2}X(\boldsymbol{p})=\varepsilon_1X^{(1,0,0)}_{1,2,0}(\boldsymbol{p})+\varepsilon_2X^{(0,1,0)}_{1,2,0}(\boldsymbol{p})=[\varepsilon_1s_0e^{s_0s_1}+\varepsilon_2]v,$$

which is the rate of change of $X(\boldsymbol{p})$ in the direction $\boldsymbol{\varepsilon}$ if s_1 and s_2 change simultaneously. At $\boldsymbol{p} = \boldsymbol{0}$, $D_{\varepsilon_1,\varepsilon_2}X(\boldsymbol{0}) = \varepsilon_2 \upsilon$. Furthermore,

$$\frac{\Delta^{1,1}_{\varepsilon_1,\varepsilon_2}X(\boldsymbol{p})}{h} = \frac{\upsilon}{h}\left[e^{s_0(s_1+h\varepsilon_1)} + h\varepsilon_2 - e^{s_0 s_1}\right].$$

At $\boldsymbol{p} = \boldsymbol{0}$, $\dfrac{\Delta^{1,1}_{\varepsilon_1,\varepsilon_2}X(\boldsymbol{0})}{h} = \varepsilon_2 \upsilon$. In this case,

$$\underset{h\to 0}{\text{l.i.m.}}\ \frac{\Delta^{1,1}_{\varepsilon_1,\varepsilon_2}X(\boldsymbol{0})}{h} = \varepsilon_2\upsilon = D_{\varepsilon_1,\varepsilon_2}X(\boldsymbol{0}),$$

i.e., the directional derivative exists in the m.s. sense. It is also valid that,

$$\lim_{s_1\to 0, s_2\to 0, s_0\to 0} \overline{[X(s_1,s_2,s_0) - X(0,0,0)]^2} = \lim_{s_1\to 0, s_2\to 0, s_0\to 0} [(e^{s_0 s_1} + s_2) - 1]^2 = 0,$$

i.e., in this case, the random field is also m.s. continuous at $\boldsymbol{p} = \boldsymbol{0}$. The maximum of $D_{\varepsilon_1,\varepsilon_2}X(\boldsymbol{p})$ is obtained in the direction $\boldsymbol{\varepsilon} = \frac{\nabla X(\boldsymbol{p})}{|\nabla X(\boldsymbol{p})|}$ (Exercise VI.15).

Mathematically, the direction of the directional derivative can be arbitrary. In applications, however, there may be some physical reason to select a specific direction that has a scientific interpretation (e.g., the rate of temperature change in the direction of air flow may need to be calculated). The directional derivative (a vector) is the projection of the gradient (a scalar) in a given direction, i.e., the latter contains more information than the former. The existence of the directional derivative of a random field does not guarantee that it is m.s. continuous. Indeed, random fields can be found whose directional derivatives exists in the m.s. sense, but, nevertheless, the random fields are not m.s. continuous (Exercise VI.1).

4.2 COVARIANCES OF RANDOM FIELD DERIVATIVES

Since the m.s. differentiability of a random field is directly related to the existence of covariance derivatives, it is appropriate to specify these derivatives explicitly. Fortunately, space−time covariances of the random field derivatives can be defined in a straightforward manner. So, the covariances of the random field derivatives of Eqs. (4.1b) and (4.2b) in $R^{n,1}$ are given by, respectively,

$$\text{cov}\left(X^{(\boldsymbol{\nu},\mu)}_{\boldsymbol{d},0}(\boldsymbol{p}), X^{(\boldsymbol{\nu}',\mu')}_{\boldsymbol{d},0}(\boldsymbol{p}')\right) = \frac{\partial^{\nu+\nu'+\mu+\mu'} c_X(\boldsymbol{p},\boldsymbol{p}')}{\partial s^{\boldsymbol{\nu}}_{\boldsymbol{d}} \partial s'^{\boldsymbol{\nu}'}_{\boldsymbol{d}} \partial s^{\mu}_0 \partial s'^{\mu'}_0}, \tag{4.14a}$$

and

$$\text{cov}\left(X^{(\boldsymbol{\nu},\mu)}_{i,0}(\boldsymbol{p}), X^{(\boldsymbol{\nu}',\mu')}_{i,0}(\boldsymbol{p}')\right) = \frac{\partial^{\nu+\nu'+\mu+\mu'} c_X(\boldsymbol{p},\boldsymbol{p}')}{\partial \boldsymbol{s}^{\boldsymbol{\nu}} \partial \boldsymbol{s}'^{\boldsymbol{\nu}'} \partial s^{\mu}_0 \partial s'^{\mu'}_0}. \tag{4.14b}$$

If all directions are considered, the products $\Pi_{i\in I_\nu}$ and $\Pi_{i\in I_{\nu'}}$, should be replaced by $\Pi^{\rho}_{i=1}$ (Eq. 4.14a) and $\Pi^{n}_{i=1}$ (Eq. 4.14b). In applications, one may be interested about the derivatives of the covariances of measurements of various phenomena, such as the convective boundary layers obtained during tropical ocean-global atmosphere experiments. As with random field derivatives, when we refer to covariance derivatives we should be careful to clarify whether partial derivatives along all directions are involved or only along a few of them. In this setting, the decision over which covariance models are the most useful for interpretive purposes is linked to the mathematical behavior of the space−time derivatives of these models.

Table 4.2 Examples of Covariances of Random Field Derivatives in $R^{n,1}$

ν	μ	ν'	μ'	I_ν	$I_{\nu'}$	Covariance of Derivatives of $X(p)$, $h = s' - s$, $\tau = t' - t$
1	1	1	1	{1}	{2}	$\text{cov}\left(X^{(1,1)}(\boldsymbol{p}), X^{(1,1)}(\boldsymbol{p}')\right) = \frac{\partial^{1+1+1+1} c_X(\boldsymbol{p},\boldsymbol{p}')}{\Pi_{i\in\{1\}}\partial s_i^{\nu_i}\Pi_{i\in\{2\}}\partial s_i^{\prime\nu'_i}\partial s_0 \partial s'_0} = \frac{\partial^4 c_X(\boldsymbol{p},\boldsymbol{p}')}{\partial s_1 \partial s'_2 \partial s_0 \partial s'_0}$
2	1	1	0	{2, 3}	{1}	$\text{cov}\left(X^{(2,1)}(\boldsymbol{p}), X^{(1,0)}(\boldsymbol{p}')\right) = \frac{\partial^{2+1+1} c_X(\boldsymbol{p},\boldsymbol{p}')}{\Pi_{i\in\{2,3\}}\partial s_i^{\nu_i}\Pi_{i\in\{1\}}\partial s_i^{\prime\nu'_i}\partial s_0} = \frac{\partial^4 c_X(\boldsymbol{p},\boldsymbol{p}')}{\partial s_2 \partial s_3 \partial s'_1 \partial s_0}$
2	0	0	1	{2}	∅	$\text{cov}\left(X^{(2,0)}(\boldsymbol{p}), X^{(0,1)}(\boldsymbol{p}')\right) = \frac{\partial^{2+1} c_X(\boldsymbol{p},\boldsymbol{p}')}{\Pi_{i\in\{2\}}\partial s_i^{\nu_i}\Pi_{i\in\varnothing}\partial s_i^{\prime\nu'_i}\partial s'_0} = \frac{\partial c_X^3(\boldsymbol{p},\boldsymbol{p}')}{\partial s_2^2 \partial s'_0}$

Example 4.7

The examples listed in Table 4.2 below illustrate the implementation of Eqs. (4.14a) and (4.14b) in a few selected cases in $R^{n,1}$. Other cases are considered in the relevant exercises of this chapter.

Although m.s. differentiability allows the existence of sample paths that are not differentiable in the ordinary sense, if, nevertheless, all the random field sample paths happen to be differentiable at a point, the m.s. derivative exists since Eqs. (4.10) or (4.11) are satisfied.

Example 4.8

Consider the first-order spatial/first-order temporal derivative $X_{i,0}^{(1,1)}(\boldsymbol{p})$ at $\boldsymbol{p}$ and along the direction $\boldsymbol{\varepsilon}_i$ of the coordinate s_i ($i = 1,\ldots, n$) and at time $t = s_0$ defined by Eq. (4.8b). A sufficient condition for the existence of $X_{i,0}^{(1,1)}(\boldsymbol{p})$ is the existence of the (2/2) generalized derivative (see Eq. 4.7 with $\nu = \mu = 1$),

$$\frac{\partial^4 c_X(\boldsymbol{p},\boldsymbol{p}')}{\partial s_i \partial s'_i \partial s_0 \partial s'_0} = \lim_{h_i,h_0,h'_i,h'_0 \to 0} \frac{\overline{\Delta^2_{h_i,h_0} X(\boldsymbol{p}) \Delta^2_{h'_i,h'_0} X(\boldsymbol{p}')}}{h_i h'_i h_0 h'_0}, \tag{4.15}$$

which is a condition of covariance differentiability. A point worth our attention is that from Eq. (4.6c) we also find

$$\frac{\Delta^2_{h_i,h_0} X(\boldsymbol{p})}{h_i h_0} \frac{\Delta^2_{h'_i,h'_0} X(\boldsymbol{p}')}{h'_i h'_0} \xrightarrow{L_2} X_{i,0}^{(1,1)}(\boldsymbol{p}) X_{i,0}^{(1,1)}(\boldsymbol{p}') \text{ as } h_0 \to 0, h_i \to 0,$$

which implies

$$\overline{\left| \frac{\Delta^2_{h_i,h_0} X(\boldsymbol{p})}{h_i h_0} \frac{\Delta^2_{h'_i,h'_0} X(\boldsymbol{p}')}{h'_i h'_0} - X_{i,0}^{(1,1)}(\boldsymbol{p}) X_{i,0}^{(1,1)}(\boldsymbol{p}') \right|} \xrightarrow{L_1} 0.$$

The last limit, in turn, implies Eq. (4.15). If $X_{i,0}^{(1,1)}(\boldsymbol{p})$ exists, then

$$\overline{X(\boldsymbol{p}) X_{i,0}^{(1,1)}(\boldsymbol{p}')} = \frac{\partial^2 c_X(\boldsymbol{p},\boldsymbol{p}')}{\partial s'_i \partial s'_0},$$

$$\overline{X_{i,0}^{(1,1)}(\boldsymbol{p}) X_{j,0}^{(1,1)}(\boldsymbol{p}')} = \frac{\partial^4 c_X(\boldsymbol{p},\boldsymbol{p}')}{\partial s_i \partial s'_j \partial s_0 \partial s'_0}. \tag{4.16a–b}$$

Also, the 2/0 m.s. derivative (i.e., second-order partial derivative in the m.s. sense in the directions i, j of the coordinates s_i, s_j, and no temporal derivative), is given by

$$X_{i,j,0}^{(2,0)}(\boldsymbol{p}) = \frac{\partial^2 X(\boldsymbol{p})}{\partial s_i \partial s_j} = \underset{h_i, h_j \to 0}{\text{l.i.m.}} \frac{X(\boldsymbol{s} + h_i \boldsymbol{\varepsilon}_i + h_j \boldsymbol{\varepsilon}_j, s_0) - X(\boldsymbol{s} + h_i \boldsymbol{\varepsilon}_i, s_0) + X(\boldsymbol{s}, s_0) - X(\boldsymbol{s} + h_j \boldsymbol{\varepsilon}_j, s_0)}{h_i h_j}, \tag{4.17}$$

where $i, j = 1, \ldots, n$.[14]

Remark 4.3

The following point should be stressed: For methodological and operational reasons, we should distinguish between

(a) the 2νth/2μth-orders partial derivative of the space−time covariance, $\text{cov}\left(X^{(\nu,\mu)}(\boldsymbol{p}), X^{(\nu,\mu)}(\boldsymbol{p}')\right)$, which is given by

$$\begin{aligned} \frac{\partial^{2(\nu+\mu)} c_X(\boldsymbol{p}, \boldsymbol{p}')}{\partial \boldsymbol{s}^\nu \partial \boldsymbol{s}'^\nu \partial s_0^\mu \partial s_0'^\mu} &= \frac{\partial^{\nu+\mu}}{\partial \boldsymbol{s}^\nu \partial s_0^\mu} \left(\frac{\partial^{\nu+\mu} c_X(\boldsymbol{p}, \boldsymbol{p}')}{\partial \boldsymbol{s}'^\nu \partial s_0'^\mu} \right) \\ &= \lim_{\boldsymbol{h} \to \boldsymbol{0}, h_0 \to 0} \left\{ \frac{1}{\boldsymbol{h}^\nu h_0^\mu} \Delta_{\boldsymbol{h}, h_0}^{\nu+\mu} \left[\lim_{\boldsymbol{h}' \to \boldsymbol{0}, h'_0 \to 0} \frac{\Delta_{\boldsymbol{h}', h'_0}^{\nu+\mu} c_X(\boldsymbol{p}, \boldsymbol{p}')}{\boldsymbol{h}'^\nu h_0'^\mu} \right] \right\}, \end{aligned} \tag{4.18a}$$

and

(b) the 2νth/2μth-orders generalized derivative of the covariance given by

$$\frac{\partial^{2(\nu+\mu)} c_X(\boldsymbol{p}, \boldsymbol{p}')}{\partial \boldsymbol{s}^\nu \partial \boldsymbol{s}'^\nu \partial s_0^\mu \partial s_0'^\mu} = \lim_{\boldsymbol{h}, \boldsymbol{h}' \to \boldsymbol{0}, h_0, h'_0 \to 0} \frac{\overline{\Delta_{\boldsymbol{h}, h_0}^{\nu+\mu} X(\boldsymbol{p}) \Delta_{\boldsymbol{h}', h'_0}^{\nu+\mu} X(\boldsymbol{p}')}}{\boldsymbol{h}^\nu \boldsymbol{h}'^\nu h_0^\mu h_0'^\mu} \tag{4.18b}$$

(see also, Eq. 4.7). Particularly, we start by calculating the first limit (as $\boldsymbol{h}' \to \boldsymbol{0}$, $h'_0 \to 0$) of the part of Eq. (4.18a) within [•], in which case the part within {•} reduces to a function of the form $\frac{1}{\boldsymbol{h}^\nu h_0^\mu} f(\boldsymbol{h}, h_0)$, and then the second limit (as $\boldsymbol{h} \to \boldsymbol{0}$, $h_0 \to 0$) is taken of this function. To better illustrate this difference, we consider an example.

Example 4.9

A random field is defined by (the time argument is suppressed for simplicity of presentation, as it does not affect the derivations)

$$X(s) = \begin{cases} 0 \text{ at } s = 0 \\ v_j \text{ at } s \in \left(\frac{1}{2^j}, \frac{1}{2^{j-1}} \right], j = 1, 2, \ldots \\ X(-s) \text{ at } s \in [-1, 0], \end{cases} \tag{4.19}$$

[14] Eq. (4.17) is also obtained from Eqs. (4.6a) and (4.6b) for $\boldsymbol{\varepsilon} = (\boldsymbol{\varepsilon}_i, \boldsymbol{\varepsilon}_j)$.

where v_j are independent, identically distributed RVs of zero mean and unit variance, and $s \in [-1, 1]$. The covariance is

$$c_X(s,s') = \begin{cases} 1 \text{ for } s,s' \in \left(\frac{1}{2^j}, \frac{1}{2^{j-1}}\right], & j = 1,2,\ldots \\ 0 \text{ otherwise} \end{cases} \tag{4.20a}$$

and at $s = s'$,

$$c_X(s,s) = \begin{cases} 1 \text{ for } s \in (0,1] \\ 0 \text{ otherwise.} \end{cases} \tag{4.20b}$$

The covariance partial derivative of Eq. (4.18a) in one dimension ($\nu = 1$, $\mu = 0$) reduces to

$$\begin{aligned} \left.\frac{\partial^2 c_X(s,s')}{\partial s \partial s'}\right|_{s=s'=0} &= \left[\frac{\partial}{\partial s}\left(\frac{\partial c_X(s,s')}{\partial s'}\right)_{s'=0}\right]_{s=0} = \frac{\partial}{\partial s}\left[\lim_{h'\to 0}\frac{1}{h'}\Delta^1_{h'}c_X(s,s')_{s'=0}\right]_{s=0} \\ &= \frac{\partial}{\partial s}\left[\lim_{h'\to 0}\frac{1}{h'}0\right]_{s=0} = 0. \end{aligned} \tag{4.21}$$

On the other hand, taking the corresponding limit as in Eq. (4.18b) we find the generalized derivative

$$\left.\frac{\partial^2 c_X(s,s')}{\partial s \partial s'}\right|_{s=s'=0} = \lim_{h\to 0}\frac{1}{h^2}\left[\Delta^1_h\Delta^1_h c_X(s,s')\right]_{s=s'=0} = \lim_{h\to 0}\frac{1}{h^2}\left[c_X(h,h)\right] = \lim_{h\to 0}\frac{1}{h^2} = \infty, \tag{4.22}$$

which contradicts the result of Eq. (4.21).

Example 4.10

For further illustration, let us consider the second/zeroth-orders generalized derivative obtained from Eq. (4.18b), with $\nu = 1$, $\mu = 0$, i.e.,

$$\operatorname{cov}\left(\frac{\partial X(\boldsymbol{p})}{\partial s_i}, \frac{\partial X(\boldsymbol{p}')}{\partial s'_i}\right) = \lim_{h_i, h'_i \to 0} \frac{\Delta^1_{h_i}\Delta^1_{h'_i} c_X(\boldsymbol{p},\boldsymbol{p}')}{h_i h'_i}, \tag{4.23}$$

where

$$\frac{\Delta^1_h\Delta^1_{h'} c_X(\boldsymbol{p},\boldsymbol{p}')}{h_i h'_i} = \frac{c_X(\boldsymbol{s}+h_i\boldsymbol{\varepsilon}_i, s_0; \boldsymbol{s}'+h'_i\boldsymbol{\varepsilon}_i, s'_0) - c_X(\boldsymbol{s}, s_0; \boldsymbol{s}'+h'_i\boldsymbol{\varepsilon}_i, s'_0) - c_X(\boldsymbol{s}+h_i\boldsymbol{\varepsilon}_i, s_0; \boldsymbol{s}', s'_0) + c_X(\boldsymbol{p},\boldsymbol{p}')}{h_i h'_i}.$$

And the corresponding second/zeroth-order partial derivative of the covariance,

$$\operatorname{cov}\left(\frac{\partial X(\boldsymbol{p})}{\partial s_i}, \frac{\partial X(\boldsymbol{p}')}{\partial s'_i}\right) = \frac{\partial^2 c_X(\boldsymbol{p},\boldsymbol{p}')}{\partial s_i \partial s'_i} = \lim_{h_i\to 0}\frac{1}{h_i}\Delta^1_{h_i}\left[\lim_{h'_i\to 0}\frac{1}{h'_i}\Delta^1_{h'_i}c_X(\boldsymbol{p},\boldsymbol{p}')\right]. \tag{4.24}$$

Although existence of the second/zeroth-order generalized derivative implies existence of the second/zeroth-order partial derivative of the covariance, existence of the former is not equivalent to that of the latter, since existence of the second/zeroth-order partial derivative does not require joint continuity of $c_X(\boldsymbol{p},\boldsymbol{p}')$ in $(\boldsymbol{p},\boldsymbol{p}')$. This comment is better appreciated in the context of the Proposition 4.1 of the following section, which is one of the important consequences of the condition of Eq. (4.11).

Remark 4.4

I may digress for a moment and recall that if a random field is Gaussian, the CDFs of its derivatives are Gaussian, and so are the joint CDFs of the random field with its derivatives.

4.3 MEAN SQUARELY DIFFERENTIABILITY CONDITIONS

As noted earlier, random field differentiability conditions are important in practice since random fields are widely used to represent the space-time variations of many physical phenomena. Modeling water surface roughness, e.g., involves derivatives of random wave fields. Hence, I continue with an important proposition regarding m.s. differentiability that is, essentially, an extension in the space–time domain of a classical earlier result due to Loeve (1953).

Proposition 4.1

A random field $X(\boldsymbol{p})$ is νth/μth-order m.s. differentiable, if and only if, the mean value $\overline{X(\boldsymbol{p})}$ is differentiable,[15] and the 2νth/2μth-order generalized derivative of Eq. (4.18b) exists and is finite at all points $\boldsymbol{p}=\boldsymbol{p}'$.

The above is a necessary and sufficient m.s. differentiability condition. To shed light on the theoretical result of the above proposition and illustrate some of its particularities an example is considered next.

Example 4.11

Let $X(\boldsymbol{p})$ be a 1/1-order m.s. differentiable S/TRF. Eq. (4.11) holds and can be written in an expanded form as

$$0=\lim_{h_i,h_0,h_i',h_0'\to 0}\left\{\overline{\left[\frac{\Delta^2_{h_i,h_0}X(\boldsymbol{p})}{h_ih_0}\right]^2}+\overline{\left[\frac{\Delta^2_{h'_i,h'_0}X(\boldsymbol{p})}{h_i'h_0'}\right]^2}-2\frac{\overline{\Delta^2_{h_i,h_0}X(\boldsymbol{p})\Delta^2_{h'_i,h'_0}X(\boldsymbol{p})}}{h_ih_i'h_0h_i'}\right\}=2(C_1-C_2),\tag{4.25}$$

where $C_1=\lim_{h_i,h_0\to 0}\frac{\Delta^2_{h_i,h_0}\Delta^2_{h_i,h_0}c_X(\boldsymbol{p},\boldsymbol{p})}{h_i^2h_0^2}$, and $C_2=\lim_{h_i,h_0,h'_i,h'_0\to 0}\frac{\Delta^2_{h_i,h_0}\Delta^2_{h'_i,h'_0}c_X(\boldsymbol{p},\boldsymbol{p})}{h_ih_i'h_0h_0'}$. But C_2 is, in fact, Eq. (4.18b) for $\nu=\mu=1$ at $\boldsymbol{p}=\boldsymbol{p}'$, which means that Eq. (4.18b) exists and is finite at $\boldsymbol{p}=\boldsymbol{p}'$. Conversely, if Eq. (4.18b) exists and is finite at $\boldsymbol{p}=\boldsymbol{p}'$, then $C_1=C_2$, which means that Eq. (4.11) holds, i.e., the $X(\boldsymbol{p})$ is 1/1-m.s. differentiable.

We should bear in mind that, generally, the existence of the limit in Eq. (4.18b) implies the existence of the corresponding covariance of Eq. (4.18a) at $\boldsymbol{p}=\boldsymbol{p}'$, but the converse is not necessarily valid. Underlying Eq. (4.18b) is the assumption of independent sample paths, whereas

[15]Since a zero mean random field is usually assumed in theoretical analysis, this part of the condition may be relaxed.

the way h_i, h_0, h'_i, h'_0 approach 0 has no effect. This is not the same with the 2/2-order covariance derivative

$$\frac{\partial^4 c_X(\boldsymbol{p},\boldsymbol{p}')}{\partial s_i \partial s_0 \partial s'_i \partial s'_0} = \lim_{h_i,h_0\to 0} \frac{1}{h_i h_0} \Delta^2_{h_i,h_0} \left[\lim_{h'_i,h'_0\to 0} \frac{1}{h'_i h'_0} \Delta^2_{h'_i,h'_0} c_X(\boldsymbol{p},\boldsymbol{p}') \right], \tag{4.26}$$

where the limit of the operator $\frac{1}{h'_i,h'_0}\Delta^2_{h'_i,h'_0} c_X(\boldsymbol{p},\boldsymbol{p}')$ as $h'_i, h'_0 \to 0$ is first applied, followed by the limit of $\frac{1}{h_i,h_0}\Delta^2_{h_i,h_0}$ as h_i, $h_0 \to 0$. Existence of Eq. (4.18b) at a point $\boldsymbol{p}$ implies that of Eq. (4.18a) at $\boldsymbol{p}$. However, existence of Eq. (4.18a) does not imply that of Eq. (4.18b), and, consequently, of the m.s. differentiability of $X(\boldsymbol{p})$. In view of these observations, the following sufficient condition of m.s. differentiability holds.

Proposition 4.2

The random field $X(\boldsymbol{p})$ is νth/μth-order m.s. differentiable, if the $2(\nu+\mu)$th-order partial derivative of Eq. (4.18a) exists, it is finite and continuous at $\boldsymbol{p} = \boldsymbol{p}'$.[16]

To help us further illustrate the application of Propositions 4.1 and 4.2, we examine a few more simple examples.

Example 4.12

Consider wind speed variation with height s and at time t taken from shipboard observations and represented by the random field

$$X(s,t) = v(t)s, \tag{4.27}$$

where $s \in R^1$, $v(t)$ is an RV with zero mean and variance $\sigma_v^2(t)$. The simultaneous two-point covariance is $c_X(s,s',t) = \sigma_v^2(t)ss'$. Accordingly, the partial covariance derivative is

$$\frac{\partial^2 c_X(s,s',t)}{\partial s \partial s'} = \sigma_v^2(t)\frac{\partial^2 (ss')}{\partial s \partial s'} = \sigma_v^2(t),$$

i.e., it exists, is finite and continuous, and, hence, $X(s, t)$ is m.s. differentiable. The corresponding generalized derivative is expressed by

$$\frac{\partial^2 c_X(s,s',t)}{\partial s \partial s'} = \lim_{h,h'\to 0} \frac{1}{hh'} \Delta^1_h \Delta^1_{h'} c_X(s,s',t) = \sigma_v^2(t),$$

which confirms random field differentiability.

Example 4.13

Unlike Example 4.12, in Example 4.9, the covariance derivatives obtained by Eqs. (4.21) and (4.22) differed, which disconfirms random field differentiability. In other words, the generalized derivative, Eq. (4.22), does not exist at $s = s' = 0$, and, therefore, the $X(s)$ does not have an m.s. derivative at $s = 0$ (according to Proposition 4.1), even that the partial covariance derivative, Eq. (4.21), exists, but is not continuous at $s = s'$.

[16]The last condition is emphasized because the fact that the second covariance derivative exists does not necessarily imply that it is also continuous.

Water surface roughness generally depends on wind speed and the spectrum of waves. For wind speeds below a certain limit, the water surface is considered aerodynamically smooth, and may be represented by continuous and differentiable random fields. In mathematical terms, the m.s. (1/1) differentiability of $X(\boldsymbol{p})$ at $\boldsymbol{p}$ implies m.s. continuity of $X(\boldsymbol{p})$ at $\boldsymbol{p}$, since

$$
\begin{aligned}
&\lim_{h_i,h_0\to 0}\overline{|X(\boldsymbol{s}+h_i\boldsymbol{\varepsilon}_i,s_0+h_0)-X(\boldsymbol{s},s_0)|^2}=\lim_{h_i,h_0\to 0}h_i^2h_0^2\lim_{h_i,h_0\to 0}\frac{\overline{|X(\boldsymbol{s}+h_i\boldsymbol{\varepsilon}_i,s_0+h_0)-X(\boldsymbol{s},s_0)|^2}}{h_i^2h_0^2}\\
&=0\times\lim_{h_i,h_0\to 0}\frac{c_X(\boldsymbol{s}+h_i\boldsymbol{\varepsilon}_i,s_0+h_0;\boldsymbol{s}+h_i\boldsymbol{\varepsilon}_i,s_0+h_0)-2c_X(\boldsymbol{s}+h_i\boldsymbol{\varepsilon}_i,s_0+h_0;\boldsymbol{s},s_0)+c_X(\boldsymbol{s},s_0;\boldsymbol{s},s_0)}{h_i^2h_0^2}\\
&=0\times\frac{\partial^4 c_X(\boldsymbol{p},\boldsymbol{p})}{\partial s_i^2\partial s_0^2}=0,
\end{aligned}
\tag{4.28}
$$

where the 2/2-order derivative of the covariance is finite by hypothesis. However, the converse is not generally valid, and an example is discussed next.

Example 4.14

Consider the Wiener random field $W(s, t)$ studied in Example 3.3. Since the Wiener covariance of Eq. (3.9) does not possess a second partial derivative at $s = s'$,[17] $W(s, t)$ is not m.s. differentiable in space. However, as was proven in Example 3.3, the Wiener random field is m.s. continuous. We notice that the second derivative of $W(s, t)$ exists in the generalized sense, since the corresponding simultaneous two-point covariance is given by

$$
\frac{\partial^2 c_W(s,s',t)}{\partial s\partial s'}=a(t)\delta(s,s')=\begin{cases}\infty & \text{for } s=s'\\ 0 & \text{for } s\neq s'.\end{cases}
\tag{4.29a–b}
$$

In view of Eq. (4.29a), the m.s. Wiener derivative $\frac{\partial W(s,t)}{\partial s}$ is infinite too. Also, the $W(s, t)$ and $W(s', t)$ are uncorrelated for all $s \neq s'$, regardless of how close to each other s and s' are. The latter feature implies that the Wiener random field varies very fast with s. Lastly, the readers may notice that the random field

$$
B(s,t)=\frac{\partial W(s,t)}{\partial s}
\tag{4.30}
$$

is the so-called *white-noise* random field (which follows a Gaussian probability distribution).

In some applications, it is especially interesting to study the stochastic dependence between a random field and its derivatives, as well as between the random field derivatives themselves (of various orders). This happens, e.g., with phenomena represented by the heat equation in which case the correlation between temperature and its derivatives are calculated.

---

[17]Indeed, the first derivative of the Wiener covariance is $\frac{\partial c_W(s,s',t)}{\partial s'}=a(t)\theta(s-s')$, where θ is a step function ($=1$ for $s'<s=0$ for $s'>s$. The θ-derivative does not exist at its discontinuity point, and, hence, the second derivative of the Wiener covariance does not exist too.

Example 4.15

If the variance of $X(\boldsymbol{p})$ is constant, then

$$c_{X,X^{(1,0)}_{d_j,0}}(\boldsymbol{p},\boldsymbol{p}) = \frac{\partial c_X(\boldsymbol{p},\boldsymbol{p}')}{\partial s'_{d_j}}\bigg|_{\boldsymbol{p}=\boldsymbol{p}'} = 0, \tag{4.31}$$

i.e., the random fields $X(\boldsymbol{p})$ and $X^{(1,0)}_{d_j,0}(\boldsymbol{p})$ are uncorrelated (for all j). If the random field is Gaussian, then $X(\boldsymbol{p})$ and $X^{(1,0)}_{d_j,0}(\boldsymbol{p})$ are stochastically independent. This is not necessarily valid for $X(\boldsymbol{p})$ and $X^{(1,0)}_{d_j,0}(\boldsymbol{p}')$ for $\boldsymbol{p} \neq \boldsymbol{p}'$. If the variance of $X^{(1,0)}_{d_i,0}(\boldsymbol{p})$ is constant, then

$$c_{X^{(1,0)}_{d_i,0},X^{(2,0)}_{d_i,d_j,0}}(\boldsymbol{p},\boldsymbol{p}) = 0. \tag{4.32}$$

It is not necessarily true, however, that Eq. (4.32) holds if the first subscript d_i differs from the second subscripts d_i and d_j. The above result can be generalized to higher orders. If the variance of $X^{(\nu,0)}_{\boldsymbol{d},0}(\boldsymbol{p})$ is constant,[18] then

$$c_{X^{(\nu,0)}_{\boldsymbol{d},0},X^{(\nu+1,0)}_{\boldsymbol{d}',0}}(\boldsymbol{p},\boldsymbol{p}) = 0, \tag{4.33}$$

where $d_i = d'_i$ for some $i = 1,\ldots,\nu$. A similar result is also valid for any $\mu \neq 0$.

Another useful property of m.s. differentiation is that it is a linear operator, i.e., if the random fields $X_l(\boldsymbol{p})$, $l = 1,\ldots, k$, are m.s. differentiable, their linear combination

$$Y(\boldsymbol{p}) = \sum_{l=1}^{k} a_l X_l(\boldsymbol{p}) \tag{4.34a}$$

(a_l are real coefficients) is also an m.s. differentiable random field. In addition, it is valid that

$$\frac{\partial^{\nu+\mu}}{\partial \boldsymbol{s}^{\nu}\partial s_0^{\mu}} \sum_{l=1}^{k} a_l X_l(\boldsymbol{p}) = \sum_{l=1}^{k} a_l \frac{\partial^{\nu+\mu} X_l(\boldsymbol{p})}{\partial \boldsymbol{s}^{\nu}\partial s_0^{\mu}}. \tag{4.34b}$$

Extensions of the above results are rather straightforward.

Since the random field m.s. continuity and differentiability conditions involve covariance functions, some interesting results are obtained in the case of random fields with *separable* covariance functions (Section 2.7 of Chapter II), i.e.,

$$c_X(\boldsymbol{s},t,\boldsymbol{s}',t') = c_{X(1)}(\boldsymbol{s},\boldsymbol{s}')c_{X(2)}(t,t'). \tag{4.35}$$

Indeed, in the case of separability the basic m.s. continuity condition becomes

$$\begin{aligned}\overline{|X(\boldsymbol{p}) - X(\boldsymbol{p}_0)|^2} &= \lim_{\boldsymbol{s}\to\boldsymbol{s}_0, t\to t_0}\left[c_{X(1)}(\boldsymbol{s},\boldsymbol{s})c_{X(2)}(t,t) - 2c_{X(1)}(\boldsymbol{s},\boldsymbol{s}_0)c_{X(2)}(t,t_0)\right] \\ &\quad + c_{X(1)}(\boldsymbol{s}_0,\boldsymbol{s}_0)c_{X(2)}(t_0,t_0) = 0,\end{aligned} \tag{4.36}$$

[18] Note that this condition by itself does not necessarily imply that the random field is STHS.

i.e., if the purely spatial covariance component $c_{X(1)}(\boldsymbol{s}, \boldsymbol{s})$ is continuous at $(\boldsymbol{s}_0, \boldsymbol{s}_0)$ and the purely temporal component $c_{X(2)}(t, t)$ is continuous at (t_0, t_0), the right side of the last equation will indeed become 0 as $\boldsymbol{s} \rightarrow \boldsymbol{s}_0$, $t \rightarrow t_0$. Eq. (4.36) permits the consideration of the spatial and temporal limits separately, and then their joint effect can be studied. Accordingly, the following corollary is valid.

Corollary 4.1

The random field $X(\boldsymbol{p})$ with a separable covariance function is continuous in the m.s. sense at $\boldsymbol{p}$ if its mean value is continuous at $\boldsymbol{p}$, and its spatial and temporal covariance components are continuous at $(\boldsymbol{s}, \boldsymbol{s}' = \boldsymbol{s})$ and $(t, t' = t)$, respectively.

Example 4.16

The zero random field $X(\boldsymbol{p})$ with a separable covariance function as in Eq. (4.35) is continuous in the m.s. sense at $\boldsymbol{p}$ if $c_{X(1)}(\boldsymbol{s}, \boldsymbol{s}')$ and $c_{X(2)}(t, t')$, are continuous at $\boldsymbol{s}' = \boldsymbol{s}$ and $t' = t$, respectively.

Thus, in the case of space–time covariance separability any continuity conditions imposed on the space–time covariance are transferred separately on its spatial and temporal components. Similar conclusions hold for m.s. differentiability.

Example 4.17

A zero mean random field $X(\boldsymbol{p})$ with a separable covariance function is (1, 1)-order differentiable in the m.s. sense at $\boldsymbol{p}$, if $\frac{\partial^2 c_{X(1)}(\boldsymbol{s},\boldsymbol{s}')}{\partial s_i^2}$ and $\frac{\partial^2 c_{X(2)}(t,t')}{\partial t^2}$ exist and are continuous at $\boldsymbol{s}' = \boldsymbol{s}$ and $t' = t$, respectively.

Before leaving this section, I would like to bring to the readers attention yet another practical issue. We studied above the theoretical conditions that a covariance function must satisfy for the corresponding space–time attribute distribution to be m.s. continuous or m.s. differentiable. A practical issue is that in many cases the actual attribute distribution may be so irregular that cannot be considered itself m.s. continuous or differentiable. Nevertheless, as it turns out the following postulate often makes sense in practice.

Postulate 4.1

When observations of the real-world attribute are obtained using a scientific equipment, an unavoidable averaging of the attribute values usually takes place that results to a recorded attribute distribution that is regularly varying (i.e., for practical modeling purposes the resulting attribute distribution can be considered as m.s. continuous and differentiable).

In many applications, such postulates can provide sufficient justification for using meaningfully the theoretical results in practice.

4.4 ALMOST SURELY DIFFERENTIABILITY CONDITIONS

We recall that the notion of m.s. differentiability discussed above, neither implies nor requires that the random field sample paths are individually differentiable. Nevertheless, sample path regularity or smoothness properties may have a significant physical interpretation. Accordingly, just as in the case of stochastic continuity, another important type of random field stochastic differentiability is the differentiability of the realizations of the random field, and the related almost surely differentiability condition.

Definition 4.3

A random field $X(\boldsymbol{p})$ has a νth/μth-order partial derivative in the *a.s.* sense if there exists a random field $X^{(\nu,\mu)}(\boldsymbol{p})$ so that

$$\frac{\Delta_{\boldsymbol{h},h_0}^{\nu+\mu} X(\boldsymbol{p})}{\boldsymbol{h}^{\nu} h_0^{\mu}} \xrightarrow{a.s.} X^{(\nu,\mu)}(\boldsymbol{p}) \tag{4.37}$$

as $h_0 \to 0, h_1 \to 0, \ldots, h_\nu \to 0$.

The example that follows is based on the same premises introduced in Proposition 3.2 (i.e., the results hold as long as either of the conditions (1) or (2) of Postulate 1.1 hold.).

Example 4.18

The $X(\boldsymbol{p})$ is 1/1-order a.s. differentiable at $\boldsymbol{p}$, if there exists a random field $X_{i,0}^{(1,1)}(\boldsymbol{p})$ such that

$$\frac{\Delta_{h_i,h_0}^{2} X(\boldsymbol{p})}{h_i h_0} = \frac{X(\boldsymbol{s}+h_i\boldsymbol{\varepsilon}_i, s_0+h_0) - X(\boldsymbol{s}, s_0+h_0) + X(\boldsymbol{s}, s_0) - X(\boldsymbol{s}+h_i\boldsymbol{\varepsilon}_i, s_0)}{h_i h_0} \xrightarrow{a.s.} X_{i,0}^{(1,1)}(\boldsymbol{p}) \text{ as } h_0, h_i \to 0. \tag{4.38}$$

Some a.s. differentiability conditions involve the specification of metrics (e.g., $|\Delta\boldsymbol{p}| = r_E$ or r_p; Chapter III). The m.s. spatial derivative $X_i^{(1,0)}(\boldsymbol{p}) = \frac{\partial}{\partial s_i} X(\boldsymbol{p})$ is a.s. continuous (i.e., it has continuous realizations) if for all $\boldsymbol{h}$ the spatial derivatives of simultaneous two-point covariances satisfy

$$c_{X_i^{(1,0)}}(\boldsymbol{s}+\boldsymbol{h}, \boldsymbol{s}+\boldsymbol{h}, t) - c_{X_i^{(1,0)}}(\boldsymbol{s}+\boldsymbol{h}, \boldsymbol{s}, t) - c_{X_i^{(1,0)}}(\boldsymbol{s}, \boldsymbol{s}+\boldsymbol{h}, t) + c_{X_i^{(1,0)}}(\boldsymbol{s}, \boldsymbol{s}, t) \leq \frac{\alpha_t \boldsymbol{h}^{2n}}{\left|\log|\boldsymbol{h}|\right|^{1+\beta}} \tag{4.39a}$$

where $\alpha_t > 0$; or for all $\Delta\boldsymbol{p}_0 = (\boldsymbol{h}, 0)$ it holds that

$$\begin{aligned} & c_{X_i^{(1,0)}}(\boldsymbol{p}+\Delta\boldsymbol{p}_0, \boldsymbol{p}+\Delta\boldsymbol{p}_0) - c_{X_i^{(1,0)}}(\boldsymbol{p}+\Delta\boldsymbol{p}_0, \boldsymbol{p}) - c_{X_i^{(1,0)}}(\boldsymbol{p}, \boldsymbol{p}+\Delta\boldsymbol{p}_0) + c_{X_i^{(1,0)}}(\boldsymbol{p}, \boldsymbol{p}) \\ & \leq \frac{\alpha \Delta\boldsymbol{p}_0^{2n}}{\left|\log|\Delta\boldsymbol{p}_0|\right|^{1+\beta}}, \end{aligned} \tag{4.39b}$$

where $\alpha > 0$, $\beta > 2$, $c_{X_i^{(1,0)}}(\boldsymbol{p}, \boldsymbol{p}') = \frac{\partial^2}{\partial s_i \partial s'_i} c_X(\boldsymbol{p}, \boldsymbol{p}')$. An alternative approach is to use a suitable space–time metric (e.g., $r_p = |\Delta\boldsymbol{p}|$ in Table 1.1 of Chapter III).

Sufficient conditions with regard to the sample function continuity of higher-order derivatives can be obtained in a similar way. Just as it happens with sample function continuity, m.s. differentiability does not imply a.s. differentiability. Yet, in the case of Gaussian random fields, m.s. differentiability practically implies that the respective sample path derivatives of the sample states exist in the a.s. sense. Concluding this section, it should be stressed that random field differentiability conditions are especially important in the modeling of phenomena characterized by rough and erratic surfaces observed in geodesy, geomorphology, atmospheric physics, biology etc. These surfaces may correspond to random fields that are differentiable only in a certain sense, or they are nondifferentiable in any stochastic sense (and this can happen even if the random fields are continuous). More results concerning the a.s. differentiability of STHS and STIS are discussed in Chapter VII.

5. THE CENTRAL LIMIT THEOREM

The *central limit theorem* (*CLT*) is, perhaps, one of the most renowned theorem in statistics and probability theory. Generally speaking, it states that under certain conditions a probability distribution (typically the distribution of the sum of a large number of independent RVs) will tend to approach the Gaussian (or normal) probability distribution.

More specifically, let us consider the CLT theorem associated with the convergence of probability distributions.

Theorem 5.1

Assume that $\{x_k\}$, $k = 1,2,\ldots$ is a sequence of independent RVs with mean values $\overline{x}_k$, variances $\sigma_k^2 \neq 0$, and $\overline{|x_k - \overline{x}_k|^3} < \infty$. Let

$$\left\{ y_m = \frac{\sum_{k=1}^{m} (x_k - \overline{x}_k)}{\left(\sum_{k=1}^{m} \sigma_k^2 \right)^{\frac{1}{2}}} \right\}, \tag{5.1}$$

$m = 1,2,\ldots$ be an RV sequence with probability distributions $\{F_m(\psi)\}$. If

$$\lim_{m \to \infty} \frac{\left(\sum_{k=1}^{m} \overline{(x_k - \overline{x}_k)^3} \right)^{\frac{1}{3}}}{\left(\sum_{k=1}^{m} \sigma_k^2 \right)^{\frac{1}{2}}} = 0, \tag{5.2}$$

then the $\{F_m(\psi)\}$ tends to the standard Gaussian distribution $N(0,\ 1)$ as $m \to \infty$.

The above is one type of CLT. For a detailed treatment of the subject see Cramer (1946), Loeve (1953), and Rosenblatt (1956).

Example 5.1

Let $x_1, \ldots, x_k$ be identically distributed independent RVs with mean $\overline{x}$ and variance σ^2. If $S_k = \sum_{i=1}^{k} x_i$, then

$$\lim_{k \to \infty} P\left[\frac{S_k - k\overline{x}}{\sigma\sqrt{k}} \leq \chi \right] = \frac{1}{\sqrt{2\pi}} \int_{-\infty}^{\chi} du e^{-\frac{u^2}{2}} \tag{5.3}$$

for all $\chi \in R^1$. Hence,

$$\frac{S_k - k\overline{x}}{\sigma\sqrt{k}} \underset{k \to \infty}{\overset{F}{\longrightarrow}} N(0, 1) \tag{5.4}$$

i.e., the $\frac{S_k - k\overline{x}}{\sigma\sqrt{k}}$ tends to a Gaussian RV with zero mean and unit variance as $k \to \infty$.

The above version of the CLT involves convergence of probability distribution function. Due to its importance, there exist several other versions of the CLT in the literature (in terms of convergence of probability density functions, involving vector RVs etc., Fisz, 1964; Renyi, 2007; Dudley, 2014; McKean, 2014).

6. STOCHASTIC INTEGRATION

The notion of stochastic integration can emerge in a variety of contexts, including the averaging of natural attributes within a specified space−time domain, the solution of physical differential equations, and the transformation between different domains (e.g., from real to spectral domains). We will be concerned with random field integrability in the m.s. sense.

I will first look at the problem purely formally. Let us consider the fundamental extension of a classical Riemann integral in the stochastic context.

Definition 6.1

For the random field $X(\boldsymbol{p})$ and a deterministic bounded and piecewise continuous function $\alpha(\boldsymbol{u}, \boldsymbol{p})$, the *m.s. Riemann integral* is given by

$$Z(\boldsymbol{p}) = \int_V d\boldsymbol{u}\alpha(\boldsymbol{u}, \boldsymbol{p})X(\boldsymbol{u}) \tag{6.1}$$

$\left(V \subset R^{n+1}\right)$, which exists in the m.s. sense if the limit of the sum

$$Z_m(\boldsymbol{p}) = \sum_{i=1}^{m} \alpha(\boldsymbol{p}_i, \boldsymbol{p})X(\boldsymbol{p}_i)\Delta\boldsymbol{p}_i \tag{6.2}$$

exists for $m \rightarrow \infty$ (also in the m.s. sense). The above limit is called the *m.s. Riemann integral* of $X(\boldsymbol{p})$. Here, $\Delta\boldsymbol{p}_i$ is an infinitesimal volume, where the measure of the largest $\Delta\boldsymbol{p}_i$ tends to zero, and the sum of all $\Delta\boldsymbol{p}_i$ equals V (clearly, both $Z(\boldsymbol{p})$ and $Z_m(\boldsymbol{p})$ are random fields). Then, it is valid that

$$Z(\boldsymbol{p}) = \underset{m \rightarrow \infty}{\text{l.i.m.}}\, Z_m(\boldsymbol{p}), \tag{6.3}$$

and the random field $X(\boldsymbol{p})$ is said to be *integrable* in the m.s. sense.

Furthermore, given the mean and covariance function of a random field in some applications, we want to find the mean and covariance function of an integral of the random field. In the case of an integral of the above form, the means and covariances are related by

$$\begin{aligned} \overline{Z(\boldsymbol{p})} &= \int_V d\boldsymbol{u}\alpha(\boldsymbol{u}, \boldsymbol{p})\overline{X(\boldsymbol{u})}, \\ c_Z(\boldsymbol{p}, \boldsymbol{p}') &= \int_V\int_V d\boldsymbol{u}d\boldsymbol{u}'\alpha(\boldsymbol{u}, \boldsymbol{p})\alpha^*(\boldsymbol{u}', \boldsymbol{p}')c_X(\boldsymbol{u}, \boldsymbol{u}'). \end{aligned} \tag{6.4a−b}$$

where α^* denotes the complex conjugate of α, and $\boldsymbol{u} = (\boldsymbol{v}, \tau)$. An interesting implication of Eq. (6.4b) is that if an uncorrelated random field has finite variance, the covariance of the integral of this random field is identically zero. Then, the following corollary presents a special case.

Corollary 6.1

For the integral of an uncorrelated random field to have a nonzero covariance, it is necessary that it is a white-noise random field.

Example 6.1

First, let $X(s, t)$ be a zero mean white-noise random field in $R^{1,1}$ with $c_X(s,t,s',t') = W(s,t,s',t')$ $\delta(s-s',t-t')$. Then, Eq. (6.4b) reduces to

$$c_Z(s,t,s',t') = \iint dv d\tau \alpha(s,t,v,\tau)\alpha^*(s',t',v,\tau)W(v,\tau). \tag{6.5}$$

That is, the covariance of the integral of a white-noise random field is expressed by a double integral and not a quadruple integral. Next, let $X(s, t)$ be a zero mean Brownian random field in $R^{1,1}$ with $c_X(s,t,s,t') = \overline{X(s,t)X(s,t')} = a(s)\min\ (t,t')$ so that

$$Z(s) = \int_0^T dt X(s,t) \tag{6.6}$$

is a Gaussian random field. The integral in Eq. (6.6) is a Riemann integral. Indeed, the Riemann sum approximation of Eq. (6.2) is

$$Z_m(s) = \Delta t \sum_{i=1}^{m} X(s,t_i), \tag{6.7}$$

where $\Delta t = \frac{T}{m}$ and $t_i = i\Delta t$. It is valid that

$$Z(s) = \underset{m\to\infty}{\text{l.i.m.}}\ Z_m(s), \tag{6.8}$$

since $X(s, t)$ is a continuous function of its arguments. The random field $Z(s)$ is Gaussian as the limit of a summation of Gaussian fields, and its variance is

$$\overline{Z_m^2(s)} = \Delta t^2 \sum_{i,j=1}^{m} \overline{X(s,t_i)X(s,t_j)}, \tag{6.9}$$

which, as $\Delta t \to 0$, converges to

$$\overline{Z^2(s)} = \int_0^T\int_0^T dt dt' \overline{X(s,t)X(s,t')}. \tag{6.10}$$

Furthermore,

$$\overline{Z^2(s)} = a(s)\int_0^T\int_0^T dt dt' \min\ (t,t') = \frac{a(s)T^3}{3} \tag{6.11}$$

in the case of the one-point/two-times Brownian covariance. Also, for the integral of an uncorrelated random field to have a nonzero covariance, it is necessary that it is a white-noise random field.

Remark 6.1

A worth-noticing special case of the integral of Eq. (6.1) is when we let

$$\alpha(\boldsymbol{w};\boldsymbol{p}) = \frac{1}{(2\pi)^{\frac{n+1}{2}}} e^{i\boldsymbol{w}\cdot\boldsymbol{p}}, \tag{6.12}$$

in which case Eq. (6.1) defines the FT already discussed in Chapter V.

Working along the lines of Definition 6.1 the proposition below can be proven (Gihman and Skorokhod, 1974a).

Proposition 6.1

The random field $X(\boldsymbol{p})$ is integrable in the m.s. Riemann sense if and only if

$$\overline{Z^2(\boldsymbol{p})} = \int_V \int_V d\boldsymbol{u} d\boldsymbol{u}' \alpha(\boldsymbol{p},\boldsymbol{u})\alpha^*(\boldsymbol{p},\boldsymbol{u}') c_X(\boldsymbol{u},\boldsymbol{u}') < \infty, \tag{6.13}$$

where α^* denotes the complex conjugate of α.

Under certain circumstances, we may need to define another kind of integral that is a generalization of the Riemann integral above.

Definition 6.2

Consider the integral

$$Z(\boldsymbol{p}) = \int_V d\mathcal{M}_X(\boldsymbol{w})\alpha(\boldsymbol{w},\boldsymbol{p}), \tag{6.14}$$

where $\mathcal{M}_X$ is a random additive set function associated with the random field $X(\boldsymbol{p})$ and defined on some class A of sets $S_i \in R^{n+1}$. Let $V = \cup_{i=1}^{n_m} S_i^{(m)}$ $(m = 1, 2, \ldots)$ be a sequence of partitions $\{P_m\}$ of V so that

$$\Delta_m = \max_i \sup_{\boldsymbol{p},\boldsymbol{p}' \in S_i^{(m)}} |\boldsymbol{p} - \boldsymbol{p}'| \underset{m\to\infty}{\longrightarrow} 0.$$

In this case, the counterpart of the sum in Eq. (6.2) is

$$Z_m(\boldsymbol{p}) = \sum_{i=1}^{m} \alpha(\boldsymbol{p},\boldsymbol{p}_i^{(m)}) \mathcal{M}_X\left(S_i^{(m)}\right), \tag{6.15}$$

where $\boldsymbol{p}_i^{(m)}$ is any point of the region $S_i^{(m)}$. Then, the m.s. Riemann–Stieltjes integral (Eq. 6.14) is defined as in Eq. (6.3).

Proposition 6.1 may also hold in the m.s. Riemann–Stieltjes sense where, though, the integral in Eq. (6.13) must be replaced by one in the ordinary Riemann–Stieltjes sense (Loeve, 1953; Pugachev and Sinitsyn, 1987). Of significant importance is the construction of the so-called integral canonical representation of a random field $X(\boldsymbol{p})$, namely

$$X(\boldsymbol{p}) = \overline{X(\boldsymbol{p})} + \int_V d\mathcal{M}_X(\boldsymbol{w})\alpha(\boldsymbol{w},\boldsymbol{p}), \tag{6.16}$$

where $\mathcal{M}_X$ has now zero mean. To Eq. (6.16) we can associate an integral canonical representation of its covariance. In many applications, it is useful to express stochastic integrals as integrals containing white-noise random field, viz.

$$\int_V d\mathcal{M}_X(\boldsymbol{p}')\alpha(\boldsymbol{p}', \boldsymbol{p}) = \int_V d\boldsymbol{p}'\eta(\boldsymbol{p}')\alpha(\boldsymbol{p}', \boldsymbol{p}), \tag{6.17}$$

in which η is a zero-mean white-noise random field. On the basis of Eq. (6.17) the following integral canonical representation of the field $X(\boldsymbol{p})$ may be constructed.

$$X(\boldsymbol{p}) = \overline{X(\boldsymbol{p})} + \int_V d\boldsymbol{p}'\eta(\boldsymbol{p}')\alpha(\boldsymbol{p}', \boldsymbol{p}). \tag{6.18}$$

Representation (Eq. 6.18), when possible, has important consequences in applications related to the stochastic differential equation modeling of physical systems. To conclude, stochastic integration can make especially useful and well-defined contributions. The m.s. integrals have the formal properties of ordinary integrals. Some other results of stochastic integration will be discussed later in the context of specific classes of S/TRFs.

AUXILIARY HYPOTHESES OF SPATIOTEMPORAL VARIATION

CHAPTER VII

CHAPTER OUTLINE

1. INTRODUCTION

As stated in earlier chapters, the argument of a spatiotemporal random field (S/TRF) $X(\boldsymbol{p})$ may be seen as a unique ($n+1$-dimensional) vector $\boldsymbol{p} = (s_1,\ldots, s_n, s_0)$ in the R^{n+1} domain (in this convention, $s_1,\ldots, s_n$ denote space coordinates and s_0 denotes time); or, alternatively, as a pair $\boldsymbol{p} = (\boldsymbol{s},t)$ of an n-dimensional

Spatiotemporal Random Fields. http://dx.doi.org/10.1016/B978-0-12-803012-7.00007-6

spatial vector $\boldsymbol{s} \in R^n$ and a time scalar $t \in T(\subseteq R^1)$ in the Cartesian product $R^n \times T$ or, simply, $R^{n,1}$. Similarly, the composite space–time lag $\Delta\boldsymbol{p} = \boldsymbol{p}' - \boldsymbol{p}$ may be viewed either as a unique vector

$$\Delta\boldsymbol{p} = (\Delta s_1, \ldots, \Delta s_n, \Delta s_0) \tag{1.1a}$$

in R^{n+1}, where $\Delta s_i = s_i' - s_i (i = 1, \ldots, n, 0)$ denote combined space–time lag components; or, alternatively, as a vector–scalar pair

$$\Delta\boldsymbol{p} = (\boldsymbol{h}, \tau) \tag{1.1b}$$

in $R^{n,1}$, where $\boldsymbol{h} = (h_1, \ldots, h_n) \in R^n$ and $\tau = t' - t = \Delta t \in T$ denote separate spatial vector and time scalar lags.[1] In both domains the physically distinct features of space and time are taken into account as well as their interconnection. Also, a direct correspondence can be established between Eqs. (1.1a) and (1.1b) by observing that $\Delta s_i = h_i (i = 1, \ldots, n)$ and $\Delta s_0 = h_0 = \tau\,(i = 0)$ [i.e., by setting $h_0 = \tau$ in Eq. (1.1a) we directly obtain Eq. (1.1b), and *vice versa*, by setting $\tau = h_0$ in Eq. (1.1b) we obtain Eq. (1.1a)]. This kind of formal "correspondence," however, does not apply in the case of the space–time metrics to be discussed next.

As was suggested in Chapters I and III, the distinction between analysis in the R^{n+1} domain and analysis in the $R^{n,1}$ domain is most consequential in studies involving the space–time metric. Specifically, in R^{n+1} the metric admits a unified space–time representation

$$m(\boldsymbol{p}, \boldsymbol{p}') = |\Delta\boldsymbol{p}| = g(\Delta\boldsymbol{p}), \tag{1.2a}$$

in terms of a single function g that accounts for the physics of composite space–time, whereas in $R^{n,1}$ the metric is a pair of distinct space and time representations

$$m(\boldsymbol{p}, \boldsymbol{p}') = (|\Delta\boldsymbol{s}|, |\Delta t|) = (g_1(\boldsymbol{h}), g_2(\tau)) \tag{1.2b}$$

in terms of two generally different functions g_1 and g_2 that account individually for the physics of space and time. Many applications simply assume that $g_2(\tau) = \tau$, although this does not have to be always the case, at least in theory. Obviously, the metric representations of Eqs. (1.2a) and (1.2b) are not equivalent.

The deeper issue of whether random fields should be presented in the R^{n+1} or in the $R^{n,1}$ domain calls for a decision to be made on theoretical modeling as well as on case-specific grounds. In many cases the formal treatment of random fields may be analytically tractable in the R^{n+1} domain, whereas in some other cases the specific distinction of space and time introduced by the $R^{n,1}$ analysis may be appropriate due to physical considerations.[2]

The study of an S/TRF in terms of spatiotemporal variability functions (S/TVFs) represented by statistical moments up to second-order (mean, covariance, structure, and variogram functions; Chapter IV) eventually requires the introduction of certain *auxiliary hypotheses* concerning the manner these moments vary as functions of $\Delta\boldsymbol{p}$ or $|\Delta\boldsymbol{p}|$, so that the purely mathematical S/TRF model is compatible with the variability features of the phenomenon it describes and, at the same time, this model is applicable and interpretable under practical circumstances.

In the context of group theory (Helgason, 1984), these auxiliary hypotheses refer to classes of random fields that are invariant with respect to some transformation group: the hypotheses involve

---

[1]For convenience, in the following we will usually assume that $t' > t$, i.e., $|\tau| = \tau$.

[2]For example, it may make more sense to consider the domain $R^{n,1}$ when we want to isolate the dynamical character of the phenomenon, or to focus on its spatial physical features, or when we deal with separable random fields.

some kind of a *homogeneous space*, i.e., a space S on which a group $\mathcal{G}$ acts transitively, as was discussed in Section 2 of Chapter I. More specifically, in group-theoretic terms, we can define an S/TRF $X(\boldsymbol{p})$ over the group space $\mathcal{G}$ as a function on $\mathcal{G}$ with values in the Hilbert space $\mathcal{H}_2 = L_2(\Omega, \mathcal{F}, P)$. Then, the $\mathcal{G}$-homogeneity property of the $X(\boldsymbol{p})$ implies that its S/TVFs (mean, covariance, variogram, etc.) remain invariant on the application of the transformation $\mathcal{g} \in \mathcal{G}$ on $\boldsymbol{p}$, $\mathcal{g}\boldsymbol{p} = \boldsymbol{p}'$, so that

$$\overline{X(\boldsymbol{p})} = \overline{X(\mathcal{g}\boldsymbol{p})} = \overline{X}, \tag{1.3}$$

and

$$c_X(\boldsymbol{p}, \boldsymbol{p}') = c_X(\mathcal{g}\boldsymbol{p}, \mathcal{g}\boldsymbol{p}').^3 \tag{1.4}$$

If $\mathcal{g} = U_\Delta$, i.e., translation by $\Delta\boldsymbol{p}$, we talk about a space−time homostationary random field. If $\mathcal{g} = \Lambda_\perp$, i.e., orthogonal transformation (rotation[4] and reflection, Section 2 of Chapter I), we talk about a space−time isostationary random field.

In this chapter we consider $\mathcal{G}$-homogeneity hypotheses for *scalar* random fields. The relevant hypotheses for *vector* random fields will be discussed in Chapter IX.[5] I shall distinguish between alternative auxiliary hypotheses and their respective links. These hypotheses are theoretical constructs, which, if they are going to be of any use in real-world applications, should satisfy two conditions: (a) they should be either substantiable or falsifiable on the basis of physical knowledge and empirical evidence, and (b) they should be interpretable and computationally feasible. Although these hypotheses are presented below in terms of covariance functions, similar hypotheses may hold in terms of the variogram or structure functions, as well. Lastly, I remind the readers that some introductory comments about these hypotheses were made in Section 7.4 of Chapter II.

1.1 HYPOTHESIS 1: HOMOSTATIONARITY

Under Hypothesis 1, which is a constitutive hypothesis necessary for inference, a natural attribute is modeled (a) as a *wide sense* (*w.s.*) *space−time homogeneous* (STH) random field or (b) as a *w.s. space−time homostationary* (STHS) random field.[6]

In terminology a, the STH hypothesis means that a group of translation operators U_Δ acting on R^{n+1} exists such that the invariance under U_Δ holds for the mean and covariance of the random field $X(\boldsymbol{p})$, i.e.,

$$\overline{X(\boldsymbol{p})} = \overline{X(U_{\Delta\boldsymbol{p}}\boldsymbol{p})} = \overline{X}, \tag{1.5a}$$

[3]As was stressed in Section 2 of Chapter I, this kind of random field homogeneity should be distinguished from domain homogeneity according to which a natural attribute will occur the same way under identical conditions independent of place, or an experimental setup will generate the same result whether performed at one time instant or another.
[4]The rotation group is also denoted as $\mathcal{G} \equiv SO(n)$. In dimensions $n = 2$ and $n = 3$, the group's elements are the usual rotations around a point ($n = 2$) and around a line ($n = 3$).
[5]Both, U_Δ and $\Lambda_\perp$ are rigid motions (i.e., mappings of domains that preserve the Euclidean or the Pythagorean distance, depending on the case).
[6]STH should be seen in the light of $\mathcal{G}$-homogeneity, Eqs. (1.3) and (1.4), above with $\mathcal{g} = U_\Delta$, whereas STHS means space homogeneous/time stationarity. In a certain respect, homogeneity here is analogous to stationarity except that position is the argument, and not time.

and

$$c_X(\boldsymbol{p},\boldsymbol{p}') = c_X(U_{\Delta p}\boldsymbol{p}, U_{\Delta p}\boldsymbol{p}'). \tag{1.5b}$$

By letting $\Delta\boldsymbol{p} = -\boldsymbol{p}$ in the above equation, we find that

$$c_X(\boldsymbol{p},\boldsymbol{p}') = c_X(0,\boldsymbol{p}' - \boldsymbol{p}) = c_X(\Delta\boldsymbol{p}). \tag{1.6}$$

In more words, in the case of a w.s. STH random field, the mean field value is a constant and its covariance depends only on the single vector lag $\Delta\boldsymbol{p}$ between two space–time points $\boldsymbol{p}$ and $\boldsymbol{p}'$. Clearly, under this hypothesis the function $c_X(\boldsymbol{p},\boldsymbol{p}')$ defined on $R^{n+1} \times R^{n+1}$ can be represented by a function $c_X(\Delta\boldsymbol{p})$ defined on R^{n+1}. Notice that we can also write,

$$c_X(\boldsymbol{p},\boldsymbol{p} + \Delta\boldsymbol{p}) = c_X(\boldsymbol{p} + \Delta\boldsymbol{p},\boldsymbol{p}) = c_X(\boldsymbol{p},\boldsymbol{p} - \Delta\boldsymbol{p}) \tag{1.7}$$

for all $\Delta\boldsymbol{p} \in R^{n+1}$.

In terminology b, which is rather common in physical applications, the hypothesis implies that in the case of a w.s. STHS random field in $R^{n,1}$ the mean field value is a constant, as in Eq. (1.5a), but its covariance now depends separately on the space vector lag $\boldsymbol{h}$ and the time separation τ between two points $\boldsymbol{p}$ and $\boldsymbol{p}'$, i.e.,

$$c_X(\boldsymbol{p},\boldsymbol{p}') = c_X(\boldsymbol{h}, \tau). \tag{1.8}$$

This is a direct consequence of the fact that, as we saw in Eqs. (1.1a) and (1.1b), the $\Delta\boldsymbol{p}$ can be written as a unique vector in R^{n+1} as well as a vector–scalar pair in the Cartesian product $R^{n,1}$, thus leading to Eq. (1.8). By virtue of the convention

$$\begin{aligned} \Delta\boldsymbol{p} &= (h_1, \ldots, h_n, h_0), \\ \boldsymbol{h} &= (h_1, \ldots, h_n), \\ h_0 &= \tau, \end{aligned} \tag{1.9a–c}$$

Eqs. (1.6) and (1.8) are formally (although not necessarily physically) equivalent representations of the space–time covariance.

Remark 1.1

In many applications, the physical interpretation of the STHS hypothesis is that the large-scale characteristics (macrostructure) of the underlying natural attribute do not change over space–time. Such phenomena abound in Nature and are encountered in oceanography, climatology, biology, ecology, and earth and atmospheric sciences.

Furthermore, an S/TRF $X(\boldsymbol{p})$ is called *strict sense* (*s.s.*) STH in the R^{n+1} domain if the probability distributions of the random variable (RV) sequence $x(\boldsymbol{p}_1),\ldots, x(\boldsymbol{p}_m)$ for any integer m and all points $\boldsymbol{p}_1,\ldots, \boldsymbol{p}_m$ remain the same when all points $\boldsymbol{p}_1,\ldots, \boldsymbol{p}_m$ are translated by an arbitrary vector $\Delta\boldsymbol{p} \in R^{n+1}$, i.e.,

$$F_X[\chi_1(\boldsymbol{p}_1), \ldots, \chi_m(\boldsymbol{p}_m)] = F_X[\chi_1(U_{\Delta p}\boldsymbol{p}_1), \ldots, \chi_m(U_{\Delta p}\boldsymbol{p}_m)] \tag{1.10a}$$

for all $m = 1, 2,\ldots$, and all spatiotemporal lags $\Delta\boldsymbol{p}$. Equivalently, $X(\boldsymbol{p})$ is called s.s. STHS in the $R^{n,1}$ domain if these probabilities remain the same when the space–time points are translated by arbitrary spatial lag-temporal separation pairs $(\boldsymbol{h},\tau) \in R^{n,1}$, i.e.,

$$F_X[\chi_1(\boldsymbol{s}_1, t_1), \ldots, \chi_N(\boldsymbol{s}_m, t_m)] = F_X[\chi_1(U_{\boldsymbol{h}}\boldsymbol{s}_1, t_1 + \tau), \ldots, \chi_m(U_{\boldsymbol{h}}\boldsymbol{s}_m, t_m + \tau)] \tag{1.10b}$$

for all $m = 1, 2, \ldots,$ and all space–time lags $(\boldsymbol{h},\tau)$. For reasons that will become obvious, herein I will usually refer to STHS random fields.

Clearly, if the mean and covariance of an s.s. STHS random field exist, they will also satisfy Eqs. (1.5a) and (1.8), i.e., an s.s. S/TRF is w.s. STHS too. A w.s. STHS field, however, is not necessarily an s.s. STHS field. An important special case of a random field where w.s. and s.s STHS imply each other is the Gaussian S/TRF. Nevertheless, s.s. STHS is rarely applicable in practical situations. In the following, w.s. STHS random fields will be simply called STHS random fields.

Given that unlike time, which is a scalar, location is a vector, it is also possible to have only *partial* homogeneity. As a matter of fact, in many real-world cases it is physically impossible to have phenomena (e.g., turbulent flows) that are homogeneous in all coordinate directions and stationary as well, but the notion is methodologically useful, nonetheless. In particular, it is possible to have homogeneous turbulence in a flow field that is nonhomogeneous on large scales. Furthermore, it is possible to deal with random fields that are spatially homogeneous/temporally nonstationary or spatially nonhomogeneous/temporally stationary. The properties of such S/TRFs can be obtained from expressions that are similar to the ones considered above. The following example considers some illustrative specific cases.

Example 1.1

A random field in $R^{3,1}$ may be homogeneous, say, along the s_1 and s_2 directions but not along the s_3 direction, in which case the covariance is written as

$$c_X(\boldsymbol{p},\boldsymbol{p}') = c_X(h_1, h_2, s_3, s_3', \tau).$$

In the case of a homogeneous/nonstationary random field, the mean and covariance are given by

$$\begin{aligned} \overline{X}(\boldsymbol{p}) &= \overline{X}(t), \\ c_X(\boldsymbol{p},\boldsymbol{p}') &= c_X(\boldsymbol{h}, t, t'), \end{aligned}$$

i.e., the mean is a function of time, and the covariance is a function of the space lag and both time instants considered. In the case of a nonhomogeneous/stationary field, we can write

$$\begin{aligned} \overline{X}(\boldsymbol{p}) &= \overline{X}(\boldsymbol{s}) \\ c_X(\boldsymbol{p},\boldsymbol{p}') &= c_X(\boldsymbol{s}, \boldsymbol{s}', \tau), \end{aligned}$$

i.e., the mean is a function of space, and the covariance is a function of the time lag and both locations considered.

The following hypothesis is a restriction of STHS in the sense that all directions in space are essentially taken as equivalent.

1.2 HYPOTHESIS 2: ISOSTATIONARITY

A natural attribute may be modeled, either (a) as a w.s. space–time *isotropic* (STI) random field in the R^{n+1} domain or (b) as a w.s. space–time *isostationary* (STIS) random field in the $R^{n,1}$ domain. These two terminologies are not always equivalent (as was the case with the corresponding terminologies in Hypothesis 1). Instead, the use of the (a) *versus* the (b) terminology depends on the structure of the covariance arguments (e.g., how the space and time arguments are physically linked) and the group transformation applied.[7]

According to terminology (a), the w.s. STI hypothesis implies that the random field is w.s. STH, and, in addition, a group of orthogonal transformations $\Lambda_{\perp}$ acting on R^{n+1} exists such that the invariance under $\Lambda_{\perp}$ holds for the random field covariance, i.e.,

$$c_X(\boldsymbol{p},\boldsymbol{p}') = c_X(\Lambda_{\perp}(\boldsymbol{p}),\Lambda_{\perp}(\boldsymbol{p}')). \tag{1.11a}$$

Since isostationarity presupposes homostationarity, it is valid by definition that $c_X(\boldsymbol{p},\boldsymbol{p}') = c_X(\Delta\boldsymbol{p})$ and $c_X(\Lambda_{\perp}(\boldsymbol{p}),\Lambda_{\perp}(\boldsymbol{p}')) = c_X(\Lambda_{\perp}(\Delta\boldsymbol{p}))$, which imply that $c_X(\Delta\boldsymbol{p}) = c_X(\Lambda_{\perp}(\Delta\boldsymbol{p}))$. Specifically, in the case of rotational invariance the covariance depends only on $|\Delta\boldsymbol{p}|$, i.e.,

$$c_X(\boldsymbol{p},\boldsymbol{p}') = c_X(|\Delta\boldsymbol{p}|), \tag{1.11b}$$

where $|\Delta\boldsymbol{p}| = \Delta p$ is a mixed space–time metric as discussed in Chapter III. In more words, the covariance function depends on a single scalar: the absolute value of the vector distance $\Delta\boldsymbol{p}$ between any two points $\boldsymbol{p}$ and $\boldsymbol{p}'$ in space–time. Of particular interest in applications is the case of a real-valued $c_X(|\Delta\boldsymbol{p}|)$ defined on $[0,\infty)$.

Terminology (b) refers to the $R^{n,1}$ domain, where it introduces the w.s. STIS hypothesis according to which

$$c_X(\boldsymbol{p},\boldsymbol{p}') = c_X(r,\tau), \tag{1.11c}$$

where the covariance depends on two separate arguments: the absolute value $r = |\boldsymbol{h}|$ of the space distance $\boldsymbol{h}$ and the time separation τ between $\boldsymbol{p}$ and $\boldsymbol{p}'$. Put differently, w.s. STIS should be clearly distinguished from the w.s. STI hypothesis in the sense that Eq. (1.11b) is replaced by Eq. (1.11c).

In view of the above considerations, at its core, the challenge is how to adequately determine the corresponding metric: Should it be represented as the single argument $|\Delta\boldsymbol{p}| = \Delta p$ in the case of STI (Eq. 1.11b) or as the pair of arguments $(r = |\boldsymbol{h}|, \tau)$ in the case of STIS (Eq. 1.11c)? I have already expressed some thoughts regarding this important metric determination issue in Chapters I and III. In any case, it should be kept in mind that unlike the covariance representations of Eqs. (1.6) and (1.8), the covariance representations of Eqs. (1.11b) and (1.11c) are not equivalent, in general.

Example 1.2

To further illustrate the distinction between the representations of Eqs. (1.11b) and (1.11c), I would like to notice that in the case of the Pythagorean space–time metric of Eq. (2.19b) of Chapter I, Eq. (1.11b) implies a covariance that is a function of the single argument $|\Delta\boldsymbol{p}| = r_p$, so that

$$c_X(|\Delta\boldsymbol{p}|) = c_X(r_p). \tag{1.12a}$$

[7]To avoid possible confusion regarding the meaning of isotropy, what I consider here is stochastic isotropy, i.e., a property that characterizes the space–time statistical moments under rotations of the coordinate system. As earlier, STI should be seen in the abstract sense of Eqs. (1.3) and (1.4) with $g = \Lambda_{\perp}$, whereas STIS means space isotropic/time stationary.

On the other hand, in the case of the separate space−time metric of Eq. (2.19a) of Chapter I, Eq. (1.11c) refers to a covariance that is a function of the two arguments, the Euclidean metric $r = r_E$ and the time lag τ, i.e.,

$$c_X(r, \tau) = c_X(r_E, \tau). \tag{1.12b}$$

Eqs. (1.12a) and (1.12b) are termed, respectively, the *nonseparate* (composite) space−time metric covariance form and the *separate* space−time metric covariance form (more details in following sections).

In the remaining of this chapter, as well as in the following chapters, the discussion will focus mainly on STHS and STIS random fields. This is, at least in part, due to the fact that these kinds of random fields offer adequate quantitative representations of many real-world situations, where they acquire physically meaningful and informative interpretations. The STIS hypothesis, in particular, can make especially useful and well-defined contributions in the context of vector S/TRFs (Chapter IX).

Remark 1.2

STIS random fields are used to model natural attributes that exhibit some kind of isotropy, space−time separate or otherwise, in their distribution. In earth sciences, e.g., physical symmetry conditions may make it appropriate for the covariance function to depend (regarding its spatiotemporal behavior) only on some kind of "distance" of the coordinates. STIS covariance functions are commonly used in turbulence (McComb, 1990), in statistical topography (Isichenko, 1992), in flow and transport (Dagan, 1989), in atmospheric pollution (Christakos and Hristopulos, 1996), and in public health studies (Christakos and Hristopulos, 1998). The rationale for the use of these covariances is that in the absence of macroscopic directional trends, averaging over the stochastic ensemble eliminates directional preferences. Although isotropy may be not always an appropriate modeling assumption (many natural attributes are anisotropic at large scales, e.g., in the case of flow in stratified media the hydraulic conductivity covariance depends significantly on flow direction), nonetheless, it is possible to transform a stochastically anisotropic random field into an isotropic one by a simple rescaling of the axes (Christakos and Hristopulos, 1998).

Similarly to the s.s. STHS discussed above, STIS random fields in the s.s. (i.e., in terms of probability functions) can be also considered. As with s.s. STHS, the s.s. STIS hypothesis is rather rarely realized in practice. Therefore, I will not get any further into this issue here.

Our third hypothesis below serves as an alternative in the case that Hypotheses 1 and 2 do not apply in the real-world phenomenon of interest, that is, when the S/TRF under study is space−time heterogeneous.

1.3 HYPOTHESIS 3: HETEROGENEITY

In physical applications, the requirements that a random field is STHS or STIS sometimes turn out to be too restrictive. Then, a natural attribute is modeled as an S/TRF with *incremental homostationarity* or, more precisely, with *homostationary increments of order* ν *in space* and μ *in time* (S/TRF-ν/μ).[8] In general terms, this hypothesis means that although the random field $X(\boldsymbol{p})$ itself is space−time

[8]A detailed presentation of this class of S/TRFs, which has its origin in the theory of random distributions (Ito, 1954; Gel'fand, 1955), is given in Chapter XIII.

heterogeneous (space nonhomogeneous/time nonstationary), there exists a transformation $\mathcal{T}$ such that the

$$Y(\boldsymbol{p}) = \mathcal{T}[X(\boldsymbol{p})] \tag{1.13}$$

is an STHS random field. As regards the specific forms that $\mathcal{T}$ can assume, a well-known case is discussed in the following example.

Example 1.3

A useful transformation, for the purpose of the above hypothesis, is of the form

$$\mathcal{T}[X(\boldsymbol{p})] = X(\boldsymbol{p}) - \eta_{\nu,\mu}(\boldsymbol{p}), \tag{1.14}$$

where $\eta_{\nu/\mu}$ is a polynomial of degree ν in $\boldsymbol{s}$ and μ in t, with random coefficients (already introduced in Section 2 of Chapter I). The physical interpretation of $\eta_{\nu/\mu}$ is that it expresses quantitatively the space–time trends of the phenomenon represented by the random field $X(\boldsymbol{p})$. It can be also shown that if Eq. (1.14) holds, the space–time heterogeneous covariance function of $X(\boldsymbol{p})$ can be written as

$$c_X(\boldsymbol{p},\boldsymbol{p}') = \kappa_X(\Delta\boldsymbol{p}) + \eta_{\nu/\mu}(\boldsymbol{p},\boldsymbol{p}'), \tag{1.15}$$

i.e., $c_X(\boldsymbol{p},\boldsymbol{p}')$ consists of an STHS part $\kappa_X(\Delta\boldsymbol{p})$, called a *generalized spatiotemporal covariance* of order ν/μ (the term "generalized" is used here to distinguish κ_X from the ordinary covariance c_X), and a space–time polynomial part $\eta_{\nu/\mu}$ in $\boldsymbol{p}$ and $\boldsymbol{p}'$.

Remark 1.3

It can be also proven that, in theory, an STHS random field is also an S/TRF-ν/μ for any values of ν and μ, but the converse is not generally true (Christakos, 1991b). In certain cases of space–time heterogeneity, one may make use of alternative tools of stochastic inference such as the structure or the variogram function (already introduced in Chapter II), and their higher-order extensions (Chapter XIII).

1.4 HYPOTHESIS 4: ERGODICITY

Under this hypothesis, the natural attribute is represented by a so-called *w.s. ergodic* S/TRF, which implies that its ensemble moments[9] (mean, covariance, variogram, structure function) that in theory are expressed in terms of the random field probability density function (PDF) coincide with the sample moments calculated in practice on the basis of the single available realization. Ergodicity, which is a term borrowed from statistical mechanics (Khinchin, 1949), generally refers to STHS random fields (space–time heterogeneous random fields are nonergodic). A stricter ergodicity hypothesis, that of an *s.s. ergodic* S/TRF, assumes that all the ensemble and sample multivariate probability distributions are equal to each other, but is rarely considered in real-world applications.

[9]To be sure, all statistical moments defined so far as stochastic averages are also termed ensemble moments, ensemble averages, or, more commonly in applications, spatiotemporal variability functions (S/TVFs).

When studying w.s. ergodicity in terms of the spatiotemporal moments (or the associated S/TVFs), the ensemble moments considered earlier in a stochastic convergence sense are properly linked to limits of the corresponding space and/or time sample averages, as follows:

(a) If an STHS random field $X(\boldsymbol{p}) = X(\boldsymbol{s},t)$ is *t-ergodic*, i.e., ergodic as a function of time, then the ensemble mean can be expressed in terms of the sample average (denoted by $\overline{\overline{}}$), i.e.,

$$\overline{X(\boldsymbol{s},t)} = \lim_{\Delta T\to\infty}\overline{\overline{X(\boldsymbol{s},t)}} = \lim_{\Delta T\to\infty}\frac{1}{|\Delta T|}\int_{\Delta T} dt\, X(\boldsymbol{s},t), \tag{1.16a}$$

where ΔT is a suitable time interval that varies between T_1 and T_2, or between 0 and T (in which case, $|\Delta T| = T$). In practice, we can restrict ΔT to a finite value that is, though, considerably larger than the temporal correlation (or dependence) range ε_t ($|\Delta T| >> \varepsilon_t$).

(b) If $X(\boldsymbol{p})$ is *s-ergodic*, i.e., ergodic as a function of space, then the sample mean is given by

$$\overline{X(\boldsymbol{s},t)} = \lim_{|S|\to\infty}\overline{\overline{X(\boldsymbol{s},t)}} = \lim_{|S|\to\infty}\frac{1}{|S|}\int_{S} d\boldsymbol{s}X(\boldsymbol{s},t), \tag{1.16b}$$

where S is the spatial region of averaging. Sometimes, we let S be the n-dimensional cube $S = \{\boldsymbol{s} : |s_i| \le L,\ \ i = 1,2,\ldots,n\}$ with $|S| = (2L)^n$, and $|L| \to \infty$ in the limit. In practice, we can limit S to a finite value such that $|S|^{\frac{1}{n}} >> \varepsilon_s$, where ε_s is the spatial correlation range (several formulas for the calculation of the sample moments can be found in Christakos, 1992).[10]

(c) Lastly, the $X(\boldsymbol{p})$ is termed space−time (*st*) *ergodic* if Eqs. (1.16a) and (1.16b) are satisfied simultaneously.

In a similar manner, the ensemble covariance and variogram functions can be expressed in terms of the sample covariance and variogram functions by means of equations analogous to Eqs. (1.16a) and (1.16b), see, also, Section 7 of Chapter VIII.

Remark 1.4

In a relevant note, since the ergodicity hypothesis emerges from the fact that in most practical problems we have only one sequence of measurements across space−time on the basis of which the sample moments need to be calculated, it may be considered a working or methodological hypothesis (i.e., a hypothesis that, while not necessarily always verified in practice, can be tested on the basis of the successes to which it leads in practice).

So with all this being said, I would also like to notice that in many practical applications the validity of Hypotheses 1−4 is restricted to limited domains over space−time. Then, Hypotheses 1−4 may be called, respectively, *quasihomostationarity*, *quasiisostationarity*, *quasiheterogeneity*, and *quasiergodicity* (or *microergodicity*).

---

[10] The notation $X_T(\boldsymbol{s})$ and $X_S(t)$ is sometimes used in the literature to denote the averages of Eqs. (1.16a) and (1.16b), respectively.

1.5 HYPOTHESIS 5: SEPARABILITY

This hypothesis presupposes that a physical attribute is represented by an S/TRF $X(\boldsymbol{p})$ characterized by a *space–time separable covariance* function, such that (see Eq. 2.32 of Chapter II),

$$c_X(\boldsymbol{s}, t, \boldsymbol{s}', t') = c_{X(1)}(\boldsymbol{s}, \boldsymbol{s}')c_{X(2)}(t, t'), \tag{1.17}$$

where $c_{X(1)}(\boldsymbol{s}, \boldsymbol{s}')$ is a purely spatial and $c_{X(2)}(t, t')$ is a purely temporal covariance component. Obviously, this hypothesis could be combined with one of the previous hypotheses. If we are dealing with an STHS or an STIS random field, it is valid that

$$\begin{aligned} c_X(\boldsymbol{h}, \tau) &= c_{X(1)}(\boldsymbol{h})c_{X(2)}(\tau), \\ c_X(r, \tau) &= c_{X(1)}(r)c_{X(2)}(\tau), \end{aligned} \tag{1.18a–b}$$

respectively ($r = |\boldsymbol{h}|$). In the case of generalized space–time covariances, the c_X in Eqs. (1.18a–b) should be replaced by κ_X.

Example 1.4

Several variations of the separability condition of Eq. (1.17) are possible in practice, such as

$$\begin{aligned} c_X(\boldsymbol{h}, \tau) &= c_{X(2)}(\tau)\prod_{i=1}^{n} c_{X(1)}(h_i), \\ c_X(\boldsymbol{h}, \tau) &= c_{X(2)}(\tau)c_{X(1)}(h_j, h_k)\prod_{i \neq j,k} c_{X(1)}(h_i), \\ c_X(\boldsymbol{h}, \tau) &= c_{X(2)}(\tau)c_{X(1)}(h_j, h_k, h_l)\prod_{i \neq j,k,l} c_{X(1)}(h_i), \end{aligned} \tag{1.19a–c}$$

etc., depending on the physical conditions of the study in $R^{n,1}$. For illustration, in $R^{3,1}$ the separation

$$c_X(\boldsymbol{h}, \tau) = c_{X(2)}(\tau)c_{X(1)}(h_1, h_2)c_{X(1)}(h_3)$$

is physically meaningful if $c_{X(1)}(h_1, h_2)$ represents horizontal attribute variation and $c_{X(1)}(h_3)$ represents vertical attribute variation. Another kind of separation that is rather common in ocean science studies is specified by

$$c_X(r, \theta, \tau) = a(\theta)c_{X(1)}(r)c_{X(2)}(\tau),$$

where θ is a directional angle and $a(\theta)$ is the so-called spreading function that is used, e.g., to parameterize the directional spreading of wind seas or to represent the direction-dependent spatial correlation of ocean wave heights (Forristall and Ewans, 1998).

Remark 1.5

Generally, on physical grounds Hypothesis 5 makes sense when the study's primary concern is twofold: (a) how the attribute's correlation at a specified time varies across space, and, separately, (b) how the correlation at a specified location varies with time. Then, separable covariance models of the form of Eqs. (1.18a–b) are physically justified, where $c_{X(1)}(\boldsymbol{h})$ accounts for concern (a) and $c_{X(2)}(\tau)$ for concern (b).

1.6 HYPOTHESIS 6: SYMMETRY

The standard property of *symmetric* (*even*) STHS covariance functions is usually stated in the literature as

$$c_X(\Delta \boldsymbol{p}) = c_X(-\Delta \boldsymbol{p}). \tag{1.20}$$

Beyond the standard symmetry of Eq. (1.20), there are other symmetry hypotheses, most of which are not commonly used in practice because they are difficult to interpret on physical grounds. These hypotheses include spatially and temporally symmetric covariances of the form, respectively,

$$c_X(\boldsymbol{s}, t; \boldsymbol{s}', t') = c_X(\boldsymbol{s}', t; \boldsymbol{s}, t'), \tag{1.21a–b}$$

and

$$c_X(\boldsymbol{s}, t; \boldsymbol{s}', t') = c_X(\boldsymbol{s}, t'; \boldsymbol{s}', t)$$

for all $\boldsymbol{s}, \boldsymbol{s}' \in R^n$ and $t, t' \in T$. In the case of STHS random fields, Eqs. (1.21a–b) may be replaced by the *full symmetry* conditions

$$\begin{aligned} c_X(\boldsymbol{h}, \tau) &= c_X(-\boldsymbol{h}, -\tau), \\ &= c_X(-\boldsymbol{h}, \tau), \\ &= c_X(\boldsymbol{h}, -\tau), \end{aligned} \tag{1.22a–c}$$

with $(\boldsymbol{h},\tau) \in R^{n,1}$. The corresponding spectral density functions (SDFs), if they exist, they are also fully symmetric.

Although interesting in theory (e.g., certain space–time covariance construction techniques assume full symmetry; Gneiting, 2002, and Chapter XVI), full covariance symmetry needs to be checked carefully in real-world applications (e.g., in Example 2.9 we will see that the full symmetry covariance conditions may be physically justified in terms of the Lagrangian space–time coordinates). This fact is often related to the property of separability implying full symmetry (i.e., a separable STHS covariance is fully symmetric too), although the converse is not necessarily valid. The implication is that if a covariance is not fully symmetric, then it is not separable (this observation may serve as a separability test, when one is needed).

Lastly, another kind of symmetry assumes that the space–time covariance function is even with respect to each individual component of the space lag vector and the time lag, i.e., the equalities

$$c_X(h_1, \ldots, h_i, \ldots, h_n, \tau) = c_X(h_1, \ldots, -h_i, \ldots, h_n, \tau) = c_X(h_1, \ldots, h_i, \ldots, h_n, -\tau). \tag{1.23}$$

are satisfied for all $i = 1, \ldots, n$. This kind of condition is sometimes called the space–time *lag-component* symmetry (some investigators seem to have reservations with the apparent time symmetry of Eqs. (1.22a), (1.22c), and (1.23), on physical grounds).

1.7 HYPOTHESIS 7: LOCATIONAL DIVERGENCE

Depending on the physical conditions of the phenomenon encountered in a study, some kind of case-specific hypothesis could be useful. One such hypothesis is the *locational divergence* (i.e., location-specific attribute variation), which is stated as

$$c_X(\boldsymbol{s}, \boldsymbol{s}', t, t') = c_X(\boldsymbol{s} + a_1\boldsymbol{s}', \boldsymbol{s} + a_2\boldsymbol{s}', t, t'), \tag{1.24}$$

where a_1 and a_2 are real numbers. In Eq. (1.24), the values of the coefficients a_1 and a_2 can be selected so that the resulting covariance reflects different degrees of variation at the locations $\boldsymbol{s}+a_1\boldsymbol{s}'$ and $\boldsymbol{s}+a_2\boldsymbol{s}'$ (e.g., a high covariance variation may be associated with one point, and a low or moderate variation may be associated with another point). A similar assumption can be made with respect to the time argument.

Example 1.5

Suppose that $a_1=-a_2=-\frac{1}{2}$ in Eq. (1.24), in which case this equation gives

$$c_X(\boldsymbol{s},\boldsymbol{s}',t,t')=c_X\left(\boldsymbol{s}-\frac{1}{2}\boldsymbol{s}',\boldsymbol{s}+\frac{1}{2}\boldsymbol{s}',t,t'\right). \tag{1.25}$$

In Eq. (1.25) the focus of the space−time variability analysis is the correlation of the attribute values between the pair of points $\boldsymbol{s}-\frac{1}{2}\boldsymbol{s}'$ and $\boldsymbol{s}+\frac{1}{2}\boldsymbol{s}'$, where $\boldsymbol{s}$ is the point in the middle of the above pair of points, and $\boldsymbol{s}'$ is the separation distance between this pair of points.

The combination of STHS with the case-specific hypothesis above yields some interesting results. In particular, Eq. (1.24) gives

$$c_X(\boldsymbol{s},\boldsymbol{s}',t,t')=c_X((a_1-a_2)\boldsymbol{s}',\tau), \tag{1.26}$$

which describes a modification of the STHS hypothesis such that

$$\left.\begin{array}{l}\boldsymbol{h}=\boldsymbol{s}'-\boldsymbol{s}\\ \tau=t'-t\end{array}\right\}\mapsto\left\{\begin{array}{l}\boldsymbol{h}=(a_1-a_2)\boldsymbol{s}'\\ \tau=t'-t.\end{array}\right.$$

Eq. (1.26) may be interpreted as representing a phenomenon with correlations between locations $\boldsymbol{s},\boldsymbol{s}'$ and times t, t' depending on a different location $(a_1-a_2)\boldsymbol{s}'$ and the time separation τ. Also, Eq. (1.25) may be further simplified to give

$$c_X(\boldsymbol{s},\boldsymbol{s}',t,t')=c_X(\boldsymbol{s}',\tau), \tag{1.27}$$

i.e., in this case the correlation structure between $(\boldsymbol{s},t)$ and $(\boldsymbol{s}',t')$ is only a function of the second location $\boldsymbol{s}'$ and the time separation τ.

The majority of the results to be discussed in the following sections are valid under one or more of the Hypotheses 1−7. Concerning hypothesis choice, when a hypothesis is simply chosen to be "representative" of the real-world phenomenon, its value as a model of the phenomenon depends completely on whether it is truly representative.

2. SPACE−TIME HOMOSTATIONARITY

The STHS is, perhaps, the most widely used hypothesis in real-world applications, with the corresponding set of S/TRFs denoted by $\mathcal{K}_0\subset\mathcal{K}$, where $\mathcal{K}$ is the class of all S/TRFs. Not surprisingly, then, that the space−time correlation theory has found its widest applicability in the context of STHS random fields. Below I will concentrate on the key tools of this theory, namely, the S/TVFs (covariance, variogram, structure functions), and their frequency domain counterparts.

2.1 OMNIDIRECTIONAL SPATIOTEMPORAL VARIABILITY FUNCTIONS

The STHS hypothesis is often seen as a modeling choice based on sound theory, on empirical evidence, or both (as with all empirical evidence, we may need to consider multiple interpretations and link them to the appropriate hypothesis). When justified, the STHS hypothesis can simplify matters considerably.

Example 2.1

On physical grounds, STHS has significant consequences for the equations governing averaged attributes, e.g., turbulent flow motion, since the spatial derivative of the averaged flow motion should be zero. As it turns out, homogeneity along just a single direction simplifies the physical study significantly. For instance, in the case of the Reynolds stress transport equation, if the field is homogeneous, the turbulence transport is zero.

It is important to emphasize certain issues regarding the internal consistency of STHS covariance functions starting with some useful comments about the role of the vector lag $\boldsymbol{\Delta p} \in R^{n+1}$. As was stated in Section 2 of Chapter IV, the covariance expressions associated with the pairs of spatiotemporal points

$$(\boldsymbol{p} = \boldsymbol{p}' + \boldsymbol{\Delta p}, \boldsymbol{p}') \ \ vs. \ \ (\boldsymbol{p}, \boldsymbol{p}' = \boldsymbol{p} + \boldsymbol{\Delta p}) \ \ vs. \ \ (\boldsymbol{p}, \boldsymbol{p}' = \boldsymbol{p} - \boldsymbol{\Delta p})$$

are different from one another. In the STHS case, in particular, the direction of the vector lag $\boldsymbol{\Delta p}$ (i.e., whether it is defined as $\boldsymbol{\Delta p} = \boldsymbol{p}' - \boldsymbol{p}$ or as $\boldsymbol{\Delta p} = \boldsymbol{p} - \boldsymbol{p}'$) still plays a key role when our analytical calculations proceed from two independent arguments $(\boldsymbol{p}, \boldsymbol{p}')$ to a single one $(\boldsymbol{\Delta p})$. The frequency domain analysis of a random field discussed in Section 2 of Chapter II is valid in the STHS case, although in this case a distinct spectral representation applies (see, Section 3).

Furthermore, the following definitions of *omnidirectional*, i.e., involving all directions, covariance functions are considered in the study of STHS random fields along the lag $\boldsymbol{\Delta p}$:

$$c_X(\boldsymbol{\Delta p}) := \begin{cases} \overline{X'(\boldsymbol{p}' + \boldsymbol{\Delta p})X'(\boldsymbol{p}')}, \\ \overline{X'(\boldsymbol{p})X'(\boldsymbol{p} + \boldsymbol{\Delta p})}, \\ \overline{X'(\boldsymbol{p})X'(\boldsymbol{p} - \boldsymbol{\Delta p})}, \end{cases} \tag{2.1a–c}$$

where, as before, $X'(\boldsymbol{p}) = X(\boldsymbol{p}) - \overline{X(\boldsymbol{p})}$ is the random fluctuation field.

Example 2.2

As regards the interpretation of the STHS covariance $c_X(\boldsymbol{\Delta p})$, it can represent various physical situations, such as waves occurring in a practically infinite ocean over which a statistically uniform wind blows continuously or the space–time distribution of antibiotics in surface soil, among many other phenomena.

The choice between the above covariance definitions to represent the spatiotemporal variability of a phenomenon should be internally consistent, i.e., it should satisfy any physical equation or empirical model governing the phenomenon. In the case of the opposite lag, $-\boldsymbol{\Delta p}$, the following definitions are also used,

$$c_X(-\boldsymbol{\Delta p}) := \begin{cases} \overline{X'(\boldsymbol{p}' - \boldsymbol{\Delta p})X'(\boldsymbol{p}')}, \\ \overline{X'(\boldsymbol{p})X'(\boldsymbol{p} - \boldsymbol{\Delta p})}, \\ \overline{X'(\boldsymbol{p})X'(\boldsymbol{p} + \boldsymbol{\Delta p})}, \end{cases} \tag{2.2a–c}$$

where $c_X(\Delta\boldsymbol{p}) = c_X(-\Delta\boldsymbol{p})$. These different definitions should be kept in mind, especially, when the derivatives of covariance functions are considered, in which case the results may differ, depending on the choice of the covariance definition above. Also, when dealing with *complex* random fields, some authors avoid using the definition of Eq. (2.1b), because it seems to be inconsistent with the definition of the covariance function of a complex random field

$$c_X(\Delta\boldsymbol{p}) := \begin{cases} \overline{X'(\boldsymbol{p}+\Delta\boldsymbol{p})X'^*(\boldsymbol{p})}, \\ \overline{X'(\boldsymbol{p})X'^*(\boldsymbol{p}-\Delta\boldsymbol{p})}, \\ \left[\overline{X'(\boldsymbol{p})X'^*(\boldsymbol{p}+\Delta\boldsymbol{p})}\right]^*, \\ \overline{X'(\boldsymbol{p})X'^*(\boldsymbol{p}+\Delta\boldsymbol{p})}, \end{cases} \neq \overline{X'(\boldsymbol{p})X'^*(\boldsymbol{p}+\Delta\boldsymbol{p})} \tag{2.3a–d}$$

(see, also, Section 6 of Chapter IV). I remind the readers that the units of the covariance functions are those of the random field squared. And, in the special case that the random field is dimensionless, so is the covariance function.

This is a good point to introduce the spatiotemporal *correlation function* of an STHS random field $X(\boldsymbol{p})$, which is commonly defined as the normalized covariance function

$$\rho_X(\Delta\boldsymbol{p}) = \frac{c_X(\Delta\boldsymbol{p})}{c_X(\boldsymbol{0})}.$$

The correlation between the values of $X(\boldsymbol{p})$ and $X(\boldsymbol{p}')$ typically tends to zero, i.e., they become uncorrelated as they become widely separated in space and/or time. Yet, since the location $\boldsymbol{s}$ is a vector, correlation may reduce (or tend to zero) at different rates along different directions in space. Hence, both the directions of the location vectors ($\boldsymbol{s}$, $\boldsymbol{s}'$) and the direction of the separation vector ($\boldsymbol{h}$) between $\boldsymbol{s}$ and $\boldsymbol{s}'$ should be carefully considered in space−time analysis. Keeping in mind that for STHS random fields the variance is a constant, $c_X(\boldsymbol{0}) = \sigma_X^2 = c_0$, the $\rho_X(\Delta\boldsymbol{p})$ is sometimes used in place of the covariance function $c_X(\Delta\boldsymbol{p})$, because of

(a) its fixed range of variation, $\rho_X(\Delta\boldsymbol{p}) \in [-1,1]$,
(b) its attractive scale-independency property, and
(c) the fact that it is dimensionless.

For future reference, in Table 2.1 we summarize some of the most important classes of STHS correlation functions in R^{n+1}. Clearly, the following hierarchies are valid in terms of m.s. continuity and m.s. differentiability,

$$\mathcal{H}_{n+1,C} \subset \mathcal{H}_{n+1,C-\{0\}} \subset \mathcal{H}_{n+1}, \text{ and } \mathcal{H}_{n+1,D} \subset \mathcal{H}_{n+1,D-\{0\}} \subset \mathcal{H}_{n+1}.$$

Certain properties of the correlation function in the R^{n+1} domain are presented in Table 2.2. The counterparts of these properties in the $R^{n,1}$ domain are obtained in a straightforward manner (for a detailed review, see Christakos, 1992). According to Property no. 1, the correlation function is an even (symmetric) and bounded function. Property no. 2 expresses some limit features of the correlation

Table 2.1 Classes of Space–Time Homostationary (STHS) Correlation Functions

Symbol	Description
$\mathcal{H}_{n+1}$	Class of STHS, in general, correlation functions.
$\mathcal{H}_{n+1,C-\{0\}}$	Class of STHS correlation functions that are everywhere mean square (m.s.) continuous except, perhaps, at the origin (nugget effect).
$\mathcal{H}_{n+1,C}$	Class of STHS correlation functions that are m.s. continuous everywhere.
$\mathcal{H}_{n+1,D}$	Class of STHS correlation functions that are everywhere m.s. differentiable.
$\mathcal{H}_{n+1,D-\{0\}}$	Class of STHS correlation functions that are everywhere m.s. differentiable except, perhaps, at the origin.

Table 2.2 Properties of Space–Time Homostationary Correlation Functions

No.	Formulation
1	$\rho_X(\Delta\boldsymbol{p}) = \rho_X(-\Delta\boldsymbol{p})$ Even (symmetric) functions $\lvert\rho_X(\Delta\boldsymbol{p})\rvert \leq 1$ Bounded function
2	$\rho_X(\Delta\boldsymbol{p}) \in \mathcal{H}_{n+1,C}$ $\therefore \lim_{\lvert\Delta\boldsymbol{p}\rvert\to\infty} \rho_X(\Delta\boldsymbol{p}) = 0$
3	$\rho_X(\Delta\boldsymbol{p}) \in \mathcal{H}_{n+1,D}$ $\therefore \frac{\partial\rho_X}{\partial h_i}(\mathbf{0}) = 0 \quad (i = 0, 1, \ldots, n)$
4	$\rho_X(\Delta\boldsymbol{p}) \in \mathcal{H}_{n+1,C-\{0\}}$ $\therefore \rho_X(\Delta\boldsymbol{p}) = \alpha\delta_{\Delta\boldsymbol{p}} + \beta\rho_b(\Delta\boldsymbol{p})$ $X(\boldsymbol{p}) = X_\alpha(\boldsymbol{p}) + X_b(\boldsymbol{p})$
5	$\rho_X(\Delta\boldsymbol{p}) \in \mathcal{H}_{n+1,C}$ $\therefore \rho_X(\Delta\boldsymbol{p}) = \overline{e^{i\eta\cdot\Delta\boldsymbol{p}}}$
6	$\rho_X(\Delta\boldsymbol{p})$ permissible in $D \subseteq R^{n+1}$ $\therefore \rho_X(\Delta\boldsymbol{p})$ permissible in $D' \subset D$
7	$\lvert\rho_X(\Delta\boldsymbol{p})\rvert \leq 1$ Cauchy – Schwartz inequality

functions, where $|\Delta\boldsymbol{p}| \to \infty$ is here equivalent to $|\boldsymbol{h}|, |\tau| \to \infty$. The physical meaning of this equation is that the random field values at two points that are widely separated in either space or time are assumed uncorrelated. Property no. 3 states that the correlation derivative vanishes at the space–time origin. In Property no. 4 the correlation function is decomposed into two functions: Kronecker delta,[11] $\delta_{\Delta\boldsymbol{p}}$,

[11] Associated with the nugget effect phenomenon encountered in applied sciences are the mathematical notions of the *delta* (or *Dirac*) function and the *Kronecker delta* (Remark 1.1 of Chapter I). The delta function models nugget effects in the continuous-parameter case; the Kronecker delta is used to model nugget effects of discrete-parameter random fields.

and a function $\rho_b(\Delta\boldsymbol{p}) \in \mathcal{H}_{n+1,C}$ (α and β are nonnegative coefficients). This decomposition represents the so-called *nugget effect* phenomenon and implies that the corresponding STHS field $X(\boldsymbol{p})$ can be also decomposed into a "chaotic" component $X_\alpha(\boldsymbol{p})$ and an m.s. continuous component $X_b(\boldsymbol{p})$. Property no. 5 notices that, if a correlation function belongs to the class $\mathcal{H}_{n+1,C}$ of STHS functions, then it has the specified exponential form, where $\boldsymbol{\eta}$ is an $n+1$-dimensional random vector. That is, the $\rho_X(\Delta\boldsymbol{p})$ can be considered as the characteristic function of $\boldsymbol{\eta}$. According to Property no. 6, if a function is a permissible correlation in a space−time domain, it is so in any other space−time domain of lower spatial dimensionality. Some authors call this the *hereditary* property (Gaspari and Cohn, 1999). Lastly, Property no. 7 presents the celebrated Cauchy−Schwartz inequality for STHS random fields. These properties can be also expressed in terms of covariance functions in a rather straightforward manner. Minor differences may exist, such as, e.g., the Cauchy−Schwartz inequality should be written as $|c_X(\Delta\boldsymbol{p})| \leq \sigma_X^2$, where the upper limit of the correlation value is the variance, and the correlation function decreases monotonically with the space (time) lag when the time (space) is kept fixed.

Working in the $R^{n,1}$ domain, and assuming a zero mean, an alternative expression of the covariance function of Eq. (2.1b) is given by

$$c_X(\boldsymbol{h},\tau) = \overline{X(\boldsymbol{s},t)X(\boldsymbol{s}+\boldsymbol{h},t+\tau)}. \tag{2.4}$$

This is an expression that is appropriate for the definition of other useful covariance forms, which have the distinctive quality, that are usually easier to interpret physically. The matter is illustrated in the following examples.

Example 2.3

Based on the space−time covariance function of Eq. (2.4), several physically important quantities can be computed. Setting the time argument equal to zero, Eq. (2.4) immediately yields the simultaneous two-point (two-point/one-time) covariance

$$c_X(\boldsymbol{h},0) = \overline{X(\boldsymbol{s},0)X(\boldsymbol{s}+\boldsymbol{h},0)},$$

where time is considered unchanged. In such cases, we often simply write $c_X(\boldsymbol{h})$ to put emphasis on two-point spatial correlation and suppress time. In atmospheric studies, the Fourier transformation (FT) of $c_X(\boldsymbol{h})$ is interpreted as the magnetic energy spectrum. The effective distance and period can be directly calculated from the covariance (Section 6 of Chapter VIII), which is interpreted as the energy-containing scale. The temporal covariance resulting from Eq. (2.4) by setting the space argument equal to zero is the single-point/two-time covariance function

$$c_X(\boldsymbol{0},\tau) = \overline{X(\boldsymbol{0},t)X(\boldsymbol{0},t+\tau)},$$

or simply, $c_X(\tau)$. Physically, this covariance form is closely linked to intermittency, accelerations, and sweeping effects in turbulence, and it can also impact energetic particle scattering (Matthaeus et al., 2016).

Example 2.4

Let $X(\boldsymbol{s},t)$ represent sea depth in $R^{2,1}$. Based on physical considerations, the covariance function can be written as

$$c_X\left(h_{\|},h_{\perp},\tau\right) = c_0\frac{\sin(a_{\perp}|h_{\perp}|)}{a_{\perp}|h_{\perp}|}e^{-a_{\|}|h_{\|}|-a_{\tau}|\tau|},$$

where $h_{||}$ is the longshore component of the spatial lag (i.e., parallel to the shoreline) and $h_{\perp}$ its cross-shore component (i.e., orthogonal to the shoreline), $c_0 > 0$ is the variance, and $a_{||}$, $a_{\perp}$, and $a_{\tau} > 0$ are empirical coefficients (Bourgine et al., 2001).

2.2 DIRECTION-SPECIFIC SPATIOTEMPORAL COVARIANCE FUNCTION

In addition to the covariance functions above considered along all coordinate directions, the *direction-specific* covariance functions are limited along specified directions in space, i.e., they focus on correlations along these directions. Therefore, a direction-specific covariance is typically obtained by measuring the correlation between attribute values at a pair of points along a line in the s_j direction, and writing the covariance function as

$$c_X\left(\boldsymbol{s},\boldsymbol{h}\cdot\boldsymbol{\varepsilon}_j,t,t+\tau\right) = c_X\left(\boldsymbol{h}\cdot\boldsymbol{\varepsilon}_j,\tau\right) = \overline{X(\boldsymbol{s},t)X\left(\boldsymbol{s}+h\boldsymbol{\varepsilon}_j,t+\tau\right)} = c_{X,j}\left(h_j,\tau\right), \tag{2.5}$$

where $\boldsymbol{\varepsilon}_j$ specifies the unit vector along direction s_j. Eq. (2.5) is the STHS case of the direction-specific covariance introduced in Eq. (2.22) of Chapter IV.

Example 2.5

The goal of the covariance function of Eq. (2.5) is to focus on correlations between attribute values along a direction of physical significance. In $R^{3,1}$, this direction-specific covariance function is given by

$$c_X(\boldsymbol{h}\cdot\boldsymbol{\varepsilon}_1,\tau) = \overline{X(s_1,s_2,s_3,t)X(s_1+h_1,s_2,s_3,t+\tau)} = c_{X,1}(h_1,\tau),$$

which is a covariance expression used in fluid mechanics studies (Monin and Yaglom, 1971) and elsewhere.

2.3 ANISOTROPIC FEATURES

An STHS random field $X(\boldsymbol{p})$ is, in general, *anisotropic* (*nonisotropic*), which is a property of random fields in $R^{n,1}$ with $n > 1$. Large-scale anisotropic structures in porous media, e.g., are modeled using scalar random fields with anisotropic covariance functions. In formal terms, anisotropy implies that the covariance $c_X(\Delta\boldsymbol{p})$ depends on the vector space–time lag $\Delta\boldsymbol{p}$ (i.e., the vector magnitude $|\Delta\boldsymbol{p}|$ and the vector direction $\boldsymbol{\varepsilon}_{\Delta p}$ of $\Delta\boldsymbol{p}$), and can be expressed as

$$c_X(\Delta\boldsymbol{p}) = c_X(|\Delta\boldsymbol{p}|,\boldsymbol{\varepsilon}_{\Delta p}) = c_X\left(\left(\Delta\boldsymbol{p}^T\boldsymbol{A}\Delta\boldsymbol{p}\right)^{\frac{1}{2}}\right), \tag{2.6}$$

where $\boldsymbol{A}$ is a nonnegative matrix. Under certain circumstances, this situation is called *geometric* anisotropy. Note that if $\boldsymbol{A}$ is the identity matrix, Eq. (2.6) reduces to Eq. (1.11b) and the random field is STI (in the sense described earlier). Two special cases are of practical interest:

(a) space–time $\mathcal{A}_{\alpha,\alpha_0}$ geometric anisotropy, where the covariance is invariant under the transformation $\mathcal{A}_{\alpha,\alpha_0}$ of Eq. (2.29) of Chapter I, i.e.,

$$c_X(\Delta\boldsymbol{p}) = c_X\left(\left(\sum_{i=1}^{n,0}\alpha_i^2h_i^2\right)^{\frac{1}{2}}\right), \tag{2.6a}$$

where $\alpha_i > 0$, $i = 1,\ldots, n$ and 0.

(b) spatial $\mathcal{A}_{\alpha,0}$ geometric anisotropy, where the covariance is invariant under the transformation $\mathcal{A}_{\alpha,0}$ of Eq. (2.27) of Chapter I, i.e.,

$$c_X(\boldsymbol{h},\tau) = c_X\left(\left(\sum_{i=1}^{n}\alpha_i^2 h_i^2\right)^{\frac{1}{2}}, \tau\right), \tag{2.6b}$$

where $\alpha_i > 0$, $i = 1,\ldots, n$.

Rather typical covariance models of STHS but anisotropic random fields are those having the following two general forms in R^{n+1}: The form

$$c_X(\Delta\boldsymbol{p}) = c_X\left(\frac{|h_1|}{a_1},\ldots,\frac{|h_n|}{a_n},\frac{|h_0|}{a_0}\right), \tag{2.7a}$$

where the coefficients $a_i > 0$ ($i = 1,\ldots, n$ and 0) characterize the correlation scale in the n spatial directions and in time. And, the form

$$c_X(\Delta\boldsymbol{p}) = c_X\left(\frac{\sum_{i=1}^{n,0} c_i h_i}{a}\right), \tag{2.7b}$$

where the coefficients $c_i > 0$ ($i = 1,\ldots, n$ and 0), and the coefficient $a > 0$ is considered separately because it may be interpreted in terms of the effective correlation ranges of the random field $X(\boldsymbol{p})$ (Section 6 of Chapter VIII).

Example 2.6

An interesting special case of Eq. (2.6a) is the space−time anisotropic Gaussian (or double exponential) covariance function in the R^{n+1} domain,[12]

$$c_X(\Delta\boldsymbol{p}) = c_X(\boldsymbol{0})e^{-\frac{1}{2}\sum_{i=1}^{n,0}\frac{h_i^2}{a_i^2}}. \tag{2.8a}$$

A straightforward generalization of Eq. (2.8a) leads to the covariance function

$$c_X(\Delta\boldsymbol{p}) = c_X(\boldsymbol{0})e^{-\frac{1}{2}\sum_{i,j=1}^{n,0} b_{ij}h_i h_j}, \tag{2.8b}$$

where b_{ij} ($i, j = 1,\ldots, n$ and 0) are the elements of a matrix $\boldsymbol{B}$ with det $\boldsymbol{B} \neq 0$. As the readers observe, the space−time metric of the covariance of Eq. (2.8a) is of the Pythagorean form of Eq. (2.15a) of Chapter I, whereas that of the covariance of Eq. (2.8b) is of the Riemannian form of Eq. (2.14) of Chapter I. Both space−time metrics are commonly assumed in analytical derivations in the R^{n+1} domain.

[12]The equivalent expression of this model in the $R^{n,1}$ domain is simply written as $c_X(\Delta\boldsymbol{p}) = c_X(\boldsymbol{0})e^{-\frac{1}{2}\left(\sum_{i=1}^{n}\frac{h_i^2}{a_i^2}+\frac{\tau^2}{a_\tau^2}\right)}$.

2.4 SPATIOTEMPORAL VARIOGRAM AND STRUCTURE FUNCTIONS: OMNIDIRECTIONAL AND DIRECTION SPECIFIC

As we saw in Section 2 of Chapter II, in addition to the covariance function, the space–time variation features of $X(\boldsymbol{p})$ can be represented by the omnidirectional *space–time variogram* function $\gamma_X(\boldsymbol{p},\boldsymbol{p}')$ of Eq. (2.17a) of Chapter II. If the variogram function $\gamma_X(\boldsymbol{p},\boldsymbol{p}')$ and the associated mean function of the field increment $\Delta X(\boldsymbol{p}',\boldsymbol{p}) = X(\boldsymbol{p}') - X(\boldsymbol{p})$ depend only on the vector lag $\Delta\boldsymbol{p} = \boldsymbol{p}' - \boldsymbol{p}$, then we can write

$$\gamma_X(\boldsymbol{p}',\boldsymbol{p}) = \begin{cases} \gamma_X(\Delta\boldsymbol{p}) = \dfrac{1}{2}\overline{[X(\boldsymbol{p}+\Delta\boldsymbol{p}) - X(\boldsymbol{p})]^2} & \text{in } R^{n+1}, \\ \gamma_X(\boldsymbol{h},\tau) = \dfrac{1}{2}\overline{[X(\boldsymbol{s}+\boldsymbol{h},t+\tau) - X(\boldsymbol{s},t)]^2} & \text{in } R^{n,1}, \end{cases} \tag{2.9a–b}$$

and

$$\overline{\Delta X(\boldsymbol{p}',\boldsymbol{p})} = \begin{cases} \overline{\Delta X(\Delta\boldsymbol{p})} & \text{in } R^{n+1}, \\ \overline{\Delta X(\boldsymbol{h},\tau)} & \text{in } R^{n,1}. \end{cases} \tag{2.10}$$

Under these conditions, the $X(\boldsymbol{p})$ is sometimes called a random field with STHS *increments*. As a matter of fact, it is usually a more reasonable assumption in the real-world that a natural attribute has STHS increments than that it is STHS itself. Moreover, the notion of a random field with STHS increments is sometimes linked to *local* STHS, which refers to homostationarity in a spatial area or over a time period that are considerably smaller than the corresponding spatial and temporal integral scales (see Section 6 of Chapter VIII for detailed definitions of these scales).

Remark 2.1

Noticeable features of the class of random field models with STHS increments include the following:

(a) It imposes restrictions on the moments of $\Delta X(\Delta\boldsymbol{p})$, and not on the moments of $X(\boldsymbol{p})$ itself. This makes this class of fields larger than that of STHS fields. Hence, in the early stages of a real-world investigation, when we are not certain whether or not the random field is an STHS one, it may be preferable to use the variogram function.

(b) It requires that $\overline{X(\boldsymbol{p})}$ varies linearly with $\boldsymbol{p}$.

(c) Its variogram does not require knowledge of the mean, as the covariance does.

(d) Its variogram exists for attributes characterized by practically unlimited variability, whereas the covariance of an STHS field only exists for attributes with limited variability.

The following proposition introduces some basic results concerning the variogram function, which are widely used in practice.

Proposition 2.1
The space–time variogram function, (i) is related to the space–time covariance through the following expression

$$\gamma_X(\Delta\boldsymbol{p}) = c_X(\boldsymbol{0}) - c_X(\Delta\boldsymbol{p}), \tag{2.11a}$$

and (ii) satisfies the inequality

$$\gamma_X(\boldsymbol{p}+\boldsymbol{p}')^{\frac{1}{2}} \leq \gamma_X(\boldsymbol{p})^{\frac{1}{2}} + \gamma_X(\boldsymbol{p}')^{\frac{1}{2}} \tag{2.11b}$$

for $\boldsymbol{p},\boldsymbol{p}' \in R^{n+1}$.

In more words, while the covariance $c_X(\Delta\boldsymbol{p})$ expresses correlation, the variogram $\gamma_X(\Delta\boldsymbol{p})$ in Eq. (2.11a) expresses *decorrelation*, i.e., the decrease of the covariance from the variance. Eqs. (2.11a–b) are direct consequences of Eqs. (2.9a–b). The class of random fields with STHS increments is generalized further in Chapter XIII. Whatever the differences in definition and classification are among the covariance and the variogram functions, it is the accurate assessment of the physical attribute's spatiotemporal variability that underpins their goals and means.

As noted (Section 2 in both Chapters II and IV), the space–time structure function is simply twice the variogram function, which is why these two terms are used interchangeably throughout the book. Historically, the structure function is linked to developments in statistical turbulence theory during the early 1940s and the variogram in developments in geostatistics about 20 years later. Indeed, the structure function

$$\xi_X(\boldsymbol{h},t) = \overline{[X(\boldsymbol{s},t) - X(\boldsymbol{s}+\boldsymbol{h},t)]^2}, \tag{2.12}$$

($=2\gamma_X(\boldsymbol{h},t)$) has been a key element of Kolmogorov's theory of turbulence (Kolmogorov, 1941). To gain some insight concerning the matter, let $X(\boldsymbol{s},t)$ denote an attribute varying in space and time, and assume that the distance r is short so that the low-order Taylor series expansion is approximately valid,

$$X(\boldsymbol{s},t) - X(\boldsymbol{s}+\boldsymbol{h},t) \approx \frac{\partial X}{\partial s_j} r,$$

$r = |\boldsymbol{h}|$. Since it corresponds to an attribute change over a short distance, this expression may be viewed as a fluctuation representing small-scale behavior. On the other hand, the structure function, $\xi_X(r,t)$, may be seen as representing fluctuation energy. Given that wavenumbers (k) in the spectral domain are related to distances (r) in the physical domain, if an expression for ξ_X as a function of r can be derived, the energy distribution can be found as a function of k.

The structure function of Eq. (2.12) is specifically a second-order structure function, and for this reason, it is sometimes indicated as $\xi_X^{[2]}(r,t)$. Moreover, in certain applications (statistical fluid mechanics, random sea waves, etc.), interpretive scientific insights can be extended in terms of *higher-order* structure functions of the form

$$\xi_X^{[\rho]}(\boldsymbol{p},\boldsymbol{p}') = \overline{[X(\boldsymbol{p}') - X(\boldsymbol{p})]^{\rho}}, \tag{2.13}$$

where ρ is a positive integer (see, also, Eq. 2.20 of Chapter IV). This is called the structure function of the ρth *order*. In fact, structure functions can be defined for any integer order ρ corresponding to the power to which the field differences above are raised (such higher-order structure functions constitute a central part of the Kolmogorov theory assessing how these functions should scale with distance between data points, say, in a power law form).

Even-order space–time structure functions, when physically justified, are combined with the assumption of a Gaussian PDF, since, obviously, it simplifies considerably their calculation. The statistical independence assumption, on the other hand, is used for the valuable insight it offers regarding higher-order statistics. Within the framework of such assumptions, certain of the higher-order structure functions can be actually computed experimentally and physically interpreted, as is discussed in the following example.

Example 2.7

In fluid dynamics applications, data from wind tunnel experiments are routinely used to study the behavior of fourth-order velocity structure functions for nearly isotropic turbulence. By assuming that flow velocities at several points are jointly Gaussian, whereas locally averaged velocities are statistically independent of velocity differences, the fourth-order velocity statistics can be related to second-order structure functions and velocity covariances. The predictions of the jointly Gaussian and statistical independence assumptions can be then compared with data. These comparisons can quantify how accurately the fourth-order statistics follow the scaling dependence on second-order velocity structure functions and on velocity covariances as predicted by the jointly Gaussian and statistical independence assumptions.

As was the case with the covariance function, in addition to the variogram functions considered along all coordinate directions, variogram functions are often limited along specified directions. Hence, *direction-specific* structure functions are obtained in many applications by measuring the differences between attribute values at pairs of points along a line in the s_j direction, i.e.,

$$\xi_{X,j}(h_j,\tau) = \overline{[X(\boldsymbol{s}+h\boldsymbol{\varepsilon}_j, t+\tau) - X(\boldsymbol{s},t)]^2}, \tag{2.14}$$

where $\boldsymbol{\varepsilon}_j$ is the unit vector along s_j. Also, the direction-specific structure function of the ρth *order* can be defined,

$$\xi^{[\rho]}_{X,j}(h_j,\tau) = \overline{[X(\boldsymbol{s}+h\boldsymbol{\varepsilon}_j, t+\tau) - X(\boldsymbol{s},t)]^\rho}. \tag{2.15}$$

Given that the two functions are particularly germane, the variogram function can be defined from the structure function in a straightforward manner.

Example 2.8

The space–time variogram function of hydraulic conductivity and along a line in the s_3 direction within in a three-dimensional porous medium is given in terms of the corresponding conductivity structure function as

$$\gamma_X(0,0,h_3,\tau) = \frac{1}{2}\xi_X(0,0,h_3,\tau) = \frac{1}{2}\overline{[X(s_1,s_2,s_3,t) - X(s_1,s_2,s_3+h_3,t+\tau)]^2} = \gamma_{X,3}(h_3,\tau), \tag{2.16}$$

in the $R^{3,1}$ domain. The satisfactory characterization of the porous medium usually requires the calculation of several direction-specific variograms as functions of time.

Up to this point, the S/TVFs were expressed in Eulerian space–time coordinates. We will conclude this section by comparing these expressions with S/TVFs derived in Lagrangian space–time coordinates (Remark 2.1 of Chapter I). For illustration, the covariance function of Eq. (2.4) can be written in Lagrangian space–time coordinates as

$$c_X(\boldsymbol{h}, \tau) = \overline{X(\boldsymbol{s}, t)X(\boldsymbol{s} + \boldsymbol{h}(t+\tau))}, \tag{2.17}$$

where the displacement vector $\boldsymbol{h}(t+\tau)$ describes the location at times $t+\tau$. The formulation of Eq. (2.17) would increase richness of interpretation by determining correlations of attributes such as the properties of fluid particles passing through locations $\boldsymbol{s}$ (in homogeneous planes of the flow) at times t, traveling along certain trajectories and arriving at locations $\boldsymbol{s} + \boldsymbol{h}(t+\tau)$ at times $t+\tau$. In this setting, the displacement vector $\boldsymbol{h}(t+\tau)$ may represent an RV describing the locations at times $t+\tau$ of the particles in the ensemble averaging with respect to the initial locations $\boldsymbol{s}$ at times t.

Example 2.9

For certain phenomena, such as storm systems represented by STHS random fields in $R^{2,1}$, the full symmetry covariance conditions of Eqs. (1.22a–c) may be physically justified in terms of the Lagrangian space–time coordinates. Specifically, in Lagrangian coordinates, Eq. (1.22a) holds as a direct consequence of the assumption that the random field is STHS. Eq. (1.22b) represents an S/TRF with no mean motion. Lastly, Eq. (1.22c) holds as a result of the previous two Eqs. (1.22a–b). We will revisit Lagrangian covariances in Chapter X.

3. SPECTRAL REPRESENTATIONS OF SPACE–TIME HOMOSTATIONARITY

To begin with, to study STHS random fields in the $R^{n,1}$ domain we usually need to define an important type of complex-valued random field (the material that follows is closely linked to Section 2 of Chapter V).

Definition 3.1

The complex-valued random field with *orthogonal (uncorrelated) increments*, $\mathcal{M}_X(\Delta\boldsymbol{w})$, is defined so that for any pair of disjoint sets $\Delta\boldsymbol{w}, \Delta\boldsymbol{w}'$ it is valid that

$$\begin{aligned} &\overline{\mathcal{M}_X(\Delta\boldsymbol{w})\mathcal{M}_X^*(\Delta\boldsymbol{w}')} = 0, \\ &\mathcal{M}_X(d\boldsymbol{w}) = \mathcal{M}_X([\boldsymbol{w}, \boldsymbol{w} + d\boldsymbol{w}]) = d\mathcal{M}_X(\boldsymbol{w}). \end{aligned} \tag{3.1a–b}$$

These considerations imply that an STHS random field admits the Fourier–Stieltjes representations

$$X(\boldsymbol{p}) = \int d\mathcal{M}_X(\boldsymbol{w}) e^{i\boldsymbol{w}\cdot\boldsymbol{p}}, \tag{3.2}$$

where a key property of $d\mathcal{M}_X(\boldsymbol{w})$ is that it determines a measure $dQ_X(\boldsymbol{w}) = \overline{|d\mathcal{M}_X(\boldsymbol{w})|^2}$, which, combined with Eqs. (3.1a–b), yields

$$dQ_X(\boldsymbol{w})d\boldsymbol{w}'\delta(\boldsymbol{w} - \boldsymbol{w}') = \overline{d\mathcal{M}_X(\boldsymbol{w})d\mathcal{M}_X^*(\boldsymbol{w}')}.^{13} \tag{3.3}$$

Notice that the random field $\mathcal{M}_X(\boldsymbol{w})$ is not necessarily differentiable. If it is differentiable, the Fourier–Stieltjes integral of Eq. (3.2) reduces to the Riemann integral of Eq. (2.5) of Chapter V. But the functions $\mathcal{M}_X(\boldsymbol{w})$ associated with STHS random fields generally do not satisfy this property.

[13] Other properties include $d\mathcal{M}_X(-\boldsymbol{w}) = d\mathcal{M}_X^*(\boldsymbol{w})$ and $\overline{d\mathcal{M}_X(\boldsymbol{w})} = 0$.

Example 3.1
In ocean sciences, the profile of a wave traveling at an angle θ to the direction s_1 in $R^{2,1}$ can be represented as

$$X(\boldsymbol{s},t) = \int d\mathcal{M}_X(\boldsymbol{k})\, e^{i(\boldsymbol{k}\cdot\boldsymbol{s}-\omega t)} = \int_{-\infty}^{\infty}\int_{-\pi}^{\pi} d\mathcal{M}_X(\omega,\theta)e^{i(\boldsymbol{k}\cdot\boldsymbol{s}-\omega t)}, \tag{3.4a}$$

where $\boldsymbol{s} = (s_1,s_2)$, $\boldsymbol{k} = (k_1,k_2) = (k\cos\theta,k\sin\theta)$.[14] The wavenumber $k = |\boldsymbol{k}|$ is related to the wave frequency ω by the dispersion relation

$$\omega^2 = gk\tanh(kd), \tag{3.4b}$$

where g is the acceleration due to gravity, and d denotes water depth. As it happens in such cases, the dispersion relation (3.4b) imposes a restriction on k and ω, which affects the calculation of the integrals in Eq. (3.4a)—more about the dispersion relation can be found in Section 3.4.

Example 3.2
In communication sciences, on the other hand, the weakly harmonizable STHS field has the following representation in $R^{n,1}$,

$$X(\boldsymbol{s},t) = \sum_{l=0}^{\infty}\sum_{l'=0}^{h(l,n)} S_l^{l'}\left(\frac{\boldsymbol{s}}{|\boldsymbol{s}|}\right)\int_{-\infty}^{\infty}\int_{0}^{\infty} d\mathcal{M}_l^{l'}(k,\omega)\frac{J_{\frac{n}{2}}(k|\boldsymbol{s}|)}{(k|\boldsymbol{s}|)^{\frac{n}{2}}-1}e^{-i\omega t},$$

where $S_l^{l'}$ denotes spherical harmonics on the unit n-sphere, $J_{\frac{n}{2}}$ denotes the Bessel function of the first kind and $n/2$th order (Gradshteyn and Ryzhik, 1965), and $h(l,n) = \frac{(2l+n-2)(l+n-3)!}{(n-2)!\,l!}$.

3.1 SPECTRAL FUNCTIONS OF SPACE–TIME HOMOSTATIONARY RANDOM FIELDS

A similar classification with that of complete versus partial S/TVFs suggested in Remarks 2.3 and 2.4 of Chapter II applies in the case of spectral functions. So, we have complete and partial spectral functions, as well. In this section, in particular, we will discuss the complete (or full) SDF, which has useful and well-defined contributions in physical applications (partial spectral functions is the concern of Section 3.3 that follows).

From the space–time covariance definition we obtain the following expression of the STHS covariance function in R^{n+1},

[14] In many ocean wave studies, only the real part is considered, i.e.,

$$X(\boldsymbol{s},t) = \mathrm{Re}\int_{-\infty}^{\infty}\int_{-\pi}^{\pi} d\mathcal{M}_X(\omega,\theta)e^{i(\boldsymbol{k}\cdot\boldsymbol{s}-\omega t)} = \int_{-\infty}^{\infty}\int_{-\pi}^{\pi} d\mathcal{M}_X(\omega,\theta)\cos(\boldsymbol{k}\cdot\boldsymbol{s}-\omega t).$$

$$\begin{aligned} c_X(\boldsymbol{\Delta p}) &= \overline{[X(\boldsymbol{p}) - \overline{X(\boldsymbol{p})}]\left[X(\boldsymbol{p}+\boldsymbol{\Delta p}) - \overline{X(\boldsymbol{p}+\boldsymbol{\Delta p})}\right]} \\ &= \iint d\chi_{\boldsymbol{p}} d\chi_{\boldsymbol{p}+\boldsymbol{\Delta p}} \left[\chi_{\boldsymbol{p}} - \overline{X(\boldsymbol{p})}\right]\left[\chi_{\boldsymbol{p}+\boldsymbol{\Delta p}} - \overline{X(\boldsymbol{p}+\boldsymbol{\Delta p})}\right] f_X(\chi_{\boldsymbol{p}}, \chi_{\boldsymbol{p}+\boldsymbol{\Delta p}}). \end{aligned} \quad \text{(3.5a–b)}$$

Moreover, by setting $\boldsymbol{p} = \boldsymbol{p}' + \boldsymbol{\Delta p}$, Eqs. (2.12b–c) of Chapter V yield

$$\begin{aligned} c_X(\boldsymbol{\Delta p}) &= \iint \overline{d\mathcal{M}_X(\boldsymbol{w}) d\mathcal{M}_X^*(\boldsymbol{w}')} e^{i(\boldsymbol{w}\cdot(\boldsymbol{p}'+\boldsymbol{\Delta p}) - \boldsymbol{w}'\cdot\boldsymbol{p}')} \\ &= \iint dQ_X(\boldsymbol{w}) d\boldsymbol{w}' \delta(\boldsymbol{w} - \boldsymbol{w}') e^{i(\boldsymbol{\Delta w}\cdot\boldsymbol{p}' + \boldsymbol{w}\cdot\boldsymbol{\Delta p})}, \end{aligned} \quad \text{(3.5c–d)}$$

where the representation of Eq. (3.3) has been used in Eq. (3.5d). Finally, assuming that the $Q_X(\boldsymbol{w})$ is differentiable, Eq. (3.3) is reduced to

$$\widetilde{c}_X(\boldsymbol{w}')\delta(\boldsymbol{w} - \boldsymbol{w}') d\boldsymbol{w}\, d\boldsymbol{w}' = \overline{d\mathcal{M}_X(\boldsymbol{w}) d\mathcal{M}_X^*(\boldsymbol{w}')}, \quad \text{(3.6a)}$$

where $\delta(\boldsymbol{w} - \boldsymbol{w}') = \prod_{i=1}^{n,0} \delta(w_i - w_i')$, and

$$\widetilde{c}_X(\boldsymbol{w}) = \frac{\partial^{n+1}}{\partial \boldsymbol{w}^{n+1}} Q_X(\boldsymbol{w}) \quad \text{(3.6b)}$$

is the corresponding SDF (originally introduced in Definition 2.5 of Chapter II).

Remark 3.1

A relevant observation is that in the physical domain, different space–times ($\boldsymbol{p} \neq \boldsymbol{p}'$) are generally correlated, Eq. (3.5a), but in the spectral domain, different wave frequencies ($\boldsymbol{w} \neq \boldsymbol{w}'$) are uncorrelated, Eq. (3.6a), which is a result of the STHS (translational invariance) hypothesis, and the field is non-STHS, e.g., the $\overline{d\mathcal{M}_X^2(\boldsymbol{w})}$ is not constant but a function of the point $\boldsymbol{w}$, although $\overline{X(\boldsymbol{p})^2}$ is a constant.

In light of the above considerations, Eqs. (3.5c–d) can be reduced to the following expressions (Exercise VII.2)

$$\begin{aligned} c_X(\boldsymbol{\Delta p}) &= \int dQ_X(\boldsymbol{w}) e^{i\boldsymbol{w}\cdot\boldsymbol{\Delta p}} \\ &= \int d\boldsymbol{w}\, \widetilde{c}_X(\boldsymbol{w}) e^{i\boldsymbol{w}\cdot\boldsymbol{\Delta p}}, \end{aligned} \quad \text{(3.6c–d)}$$

where the Q_X is a continuous, nonnegative, bounded, and nondecreasing function.[15] By inverting Eq. (3.6d) one gets the well known *Khinchin–Bochner* representation

$$\widetilde{c}_X(\boldsymbol{w}) = \frac{1}{(2\pi)^{n+1}} \int d\boldsymbol{\Delta p}\; c_X(\boldsymbol{\Delta p}) e^{-i\boldsymbol{w}\cdot\boldsymbol{\Delta p}} \geq 0 \quad \text{(3.7)}$$

(Khinchin, 1949; Bochner, 1959). These representations lead to the following result, known as *Khinchin*'s theorem (Khinchin, 1934).

[15] In fact, the spectral covariance representation of Eq. (3.6d) follows from the condition $\int d\boldsymbol{\Delta p} |c_X(\boldsymbol{\Delta p})| < \infty$.

Theorem 3.1

An STHS random field $X(\boldsymbol{p})$ admits a spectral representation of the form of Eq. (3.2), where $\mathcal{M}_X(\boldsymbol{w})$ is a complex-valued random field with orthogonal increments. To $X(\boldsymbol{p})$ one can attach an SDF $\widetilde{c}_X(\boldsymbol{w})$ of the covariance $c_X(\Delta\boldsymbol{p})$, Eqs. (3.6d) and (3.7), provided that Eq. (3.6a) is satisfied.

Theorem 3.1 establishes a powerful result as regards the spectral representation of an S/TRF and its space–time covariance function. If we are dealing with real fields where both $c_X(\boldsymbol{p})$ and $\widetilde{c}_X(\boldsymbol{w})$ are even functions, they are related by the *Fourier cosine transform* pair

$$\begin{aligned} c_X(\Delta\boldsymbol{p}) &= \int d\boldsymbol{w}\ \widetilde{c}_X(\boldsymbol{w})\cos(\boldsymbol{w}\cdot\Delta\boldsymbol{p}), \\ \widetilde{c}_X(\boldsymbol{w}) &= \frac{1}{(2\pi)^{n+1}}\int d\Delta\boldsymbol{p}\ c_X(\Delta\boldsymbol{p})\cos(\boldsymbol{w}\cdot\Delta\boldsymbol{p}), \end{aligned} \tag{3.8a–b}$$

in the R^{n+1} domain. The Fourier cosine transform pair may be favored over the Fourier exponential transform (or simply FT) pair in certain applications, such as dealing with real random fields, under symmetric conditions, variogram spectral representations, or in studies with considerable computational components.

Remark 3.2

Interestingly, in view of Eqs. (3.7) or (3.8b), the physical units of $\widetilde{c}_X$ are those of c_X multiplied by the units of $\Delta\boldsymbol{p}$. If the c_X is dimensionless, the units of $\widetilde{c}_X$ will directly depend on the form of the composite space–time lag $\Delta\boldsymbol{p}$.

When it is convenient to work in the $R^{n,1}$ domain, the previous spectral representation can be rewritten to account for w.s. STHS. Specifically, the random field $X(\boldsymbol{s},t)$ admits the Fourier–Stieltjes representation

$$X(\boldsymbol{s},t) = \iint d\mathcal{M}_X(\boldsymbol{k},\omega)e^{i(\boldsymbol{k}\cdot\boldsymbol{s}-\omega\,t)}, \tag{3.9}$$

where $\boldsymbol{k} = (k_1,\ldots, k_n)$ and ω denote the spatial frequency (wavevector) and temporal frequency, respectively. The spectral amplitude increment $d\mathcal{M}_X(\boldsymbol{k},\omega)$ is determined by the requirements reflecting the space–time statistics of the random field. So, from the STHS requirement of a constant mean, Eq. (3.9) gives

$$\overline{d\mathcal{M}_X(\boldsymbol{k},\omega)} = c;$$

(c denotes a constant), and from the STHS covariance requirement we get

$$\overline{d\mathcal{M}_X(\boldsymbol{k},\omega)d\mathcal{M}_X^*(\boldsymbol{k}',\omega')} = \widetilde{c}_X(\boldsymbol{k},\omega)\delta(\boldsymbol{k}-\boldsymbol{k}')\delta(\omega-\omega'),$$

where $\widetilde{c}_X(\boldsymbol{k},\omega)$ is the SDF in $R^{n,1}$. Also, the STHS requirement $c_X(\boldsymbol{s},t,\boldsymbol{s}',t') = c_X(\boldsymbol{h},\tau)$ implies

$$\iint\iint d\mathcal{M}_X(\boldsymbol{k},\omega)d\mathcal{M}_X^*(\boldsymbol{k}',\omega')e^{i(\boldsymbol{k}\cdot\boldsymbol{s}-\omega\,t-\boldsymbol{k}'\cdot\boldsymbol{s}'+\omega'\,t')} = \iint d\boldsymbol{k}d\omega\ e^{i(\boldsymbol{k}\cdot\boldsymbol{h}-\omega\tau)}\widetilde{c}_X(\boldsymbol{k},\omega).$$

In other words, the following result holds, which is, perhaps, the most frequently used spectral representation of the covariance function.

Proposition 3.1

The SDF $\widetilde{c}_X(\boldsymbol{k}, \omega)$ satisfies the spectral representation of the covariance function $c_X(\boldsymbol{h},\tau)$, as follows

$$c_X(\boldsymbol{h}, \tau) = \iint dkd\omega \, e^{i(\boldsymbol{k}\cdot\boldsymbol{h}-\omega\tau)} \widetilde{c}_X(\boldsymbol{k}, \omega),$$

$$\widetilde{c}_X(\boldsymbol{k}, \omega) = \frac{1}{(2\pi)^{n+1}} \iint dhd\tau \, e^{-i(\boldsymbol{k}\cdot\boldsymbol{h}-\omega\tau)} c_X(\boldsymbol{h}, \tau). \tag{3.10a–b}$$

Eqs. (3.10a–b) are the space–time extensions of Khinchin's (1949) "stationary covariance–SDF" pair of functions stating that for STHS random fields the Fourier coefficients at different wavevectors and frequencies are uncorrelated. The wavevector–frequency spectrum $\widetilde{c}_X(\boldsymbol{k}, \omega)$ and the space–time covariance $c_X(\boldsymbol{h},\tau)$ are FT pairs and possess the same symmetries.

Remark 3.3

An alternative derivation of the above spectral domain results has some instructional features. As the readers recall, in Remark 2.1 of Chapter V, I defined the general spectral representation (i.e., no STHS or STIS assumptions were made),

$$\widetilde{c}_X(\boldsymbol{k}, \omega, \boldsymbol{s}, t) dkd\omega = \iint e^{i(\boldsymbol{k}'\cdot\boldsymbol{s}-\omega' t)} \overline{d\mathcal{M}_X(\boldsymbol{k}, \omega) d\mathcal{M}_X^*(\boldsymbol{k}-\boldsymbol{k}', \omega-\omega')}. \tag{3.11a}$$

The $\widetilde{c}_X(\boldsymbol{k}, \omega, \boldsymbol{s}, t)$ is a general *unrestricted* spectrum that is, though, difficult to physically measure in practice. If the S/TRF can be assumed to be STHS, the above expressions should be independent of $\boldsymbol{s}$ and t. Specifically, by omitting $\boldsymbol{s}$ and t from the left side of the above equation implies that the right side of Eq. (3.11a) must be also independent of $\boldsymbol{s}$ and t. For this to happen, the $\overline{d\mathcal{M}_X d\mathcal{M}_X^*}$ must take the form of delta (generalized) functions, i.e.,

$$\overline{d\mathcal{M}_X(\boldsymbol{k}, \omega) d\mathcal{M}_X^*(\boldsymbol{k}-\boldsymbol{k}', \omega-\omega')} = \delta(\boldsymbol{k}', \omega') \widetilde{c}_X(\boldsymbol{k}, \omega) dk' d\omega' = \delta(\boldsymbol{k}') \delta(\omega') \widetilde{c}_X(\boldsymbol{k}, \omega) dk' d\omega', \tag{3.11b}$$

which means that the only contributions to the integral of Eq. (3.11a) occur when $\boldsymbol{k}' = \boldsymbol{0}$ and $\omega' = 0$ (i.e., the only components making contributions to the covariance are those at the same wavevector and frequency). By virtue of Eq. (3.11b), the right side of Eq. (3.11a) properly reduces to

$$\iint dk' d\omega' \, e^{i(\boldsymbol{k}'\cdot\boldsymbol{s}-\omega' t)} \delta(\boldsymbol{k}') \delta(\omega') \widetilde{c}_X(\boldsymbol{k}, \omega) = \widetilde{c}_X(\boldsymbol{k}, \omega),$$

as expected. In Remark 2.1 of Chapter V, we also saw that[16]

[16]Here $\widetilde{c}_X(\boldsymbol{k}, \omega, \boldsymbol{s}, t) = \iint dk' d\omega' \, e^{i(\boldsymbol{k}'\cdot\boldsymbol{s}-\omega' t)} \widetilde{c}_X(\boldsymbol{k}, \boldsymbol{k}', \omega, \omega')$.

$$c_X(\boldsymbol{s}, \boldsymbol{h}, t, \tau) = \iint \iint dkdk'd\omega\, d\omega'\, e^{i(\boldsymbol{k}'\cdot \boldsymbol{s}-\omega' t)+i(\boldsymbol{k}\cdot \boldsymbol{h}-\omega\tau)} \widetilde{c}_X(\boldsymbol{k}, \boldsymbol{k}', \omega, \omega'). \tag{3.11c}$$

If the random field is STHS, the above expressions should be independent of $\boldsymbol{s}$ and t. Specifically, by omitting $\boldsymbol{s}$ and t from the left side of the last equation implies that the right side of the equation must be also independent of $\boldsymbol{s}$ and t. For this to happen, the $\widetilde{c}_X(\boldsymbol{k}, \boldsymbol{k}', \omega, \omega')$ in the right side of Eq. (3.11c) must take the form of delta functions, i.e.,

$$\widetilde{c}_X(\boldsymbol{k}, \boldsymbol{k}', \omega, \omega') = \delta(\boldsymbol{k}', \omega')\widetilde{c}_X(\boldsymbol{k}, \omega) = \delta(\boldsymbol{k}')\delta(\omega')\widetilde{c}_X(\boldsymbol{k}, \omega),$$

which means that contributions to the covariance emerge only when $\boldsymbol{k}' = \boldsymbol{0}$ and $\omega' = 0$, in which case they take the form of wavevectors and frequencies. By virtue of the above assumptions, Eq. (3.11c) properly reduces to

$$\begin{aligned} c_X(\boldsymbol{h}, \tau) &= \iint \iint dkdk'd\omega\, d\omega'\, e^{i(\boldsymbol{k}'\cdot \boldsymbol{s}-\omega' t)+i(\boldsymbol{k}\cdot \boldsymbol{h}-\omega\tau)}\, \delta(\boldsymbol{k}')\delta(\omega')\widetilde{c}_X(\boldsymbol{k}, \omega) \\ &= \iint dkd\omega e^{i(\boldsymbol{k}\cdot \boldsymbol{h}-\omega\tau)} \widetilde{c}_X(\boldsymbol{k}, \omega), \end{aligned}$$

which brings us back to Eq. (3.10a).

The starting point of Remark 3.3 was the unrestricted spectrum that was introduced in Chapter V. The unrestricted spectrum is a nonnegative function, i.e., $\widetilde{c}_X(\boldsymbol{k}, \omega, \boldsymbol{s}, t) \geq 0$, because it is the transform of the mean of the product of complex conjugate numbers. This spectrum has six arguments: two for wavevector, one for frequency, two for spatial position, and one for time. Generally, there will be a different spectrum for each position $\boldsymbol{s}$ and each time instant t.

Example 3.3

As suggested earlier, in oceanography the $\widetilde{c}_X(\boldsymbol{k}, \omega, \boldsymbol{s}, t)$ can model surface displacement of ocean waves. Specifically, the $\widetilde{c}_X(\boldsymbol{k}, \omega, \boldsymbol{s}, t)$ is interpreted as the potential energy spectrum associated with sea surface displacement in $R^{2,1}$. The mean potential energy of the ocean wave surface is directly expressed in terms of this spectrum, i.e.,

$$\frac{1}{2}\rho g\overline{X^2(\boldsymbol{s}, t)} = \iint dkd\omega\, \widetilde{c}_X(\boldsymbol{k}, \omega, \boldsymbol{s}, t),$$

where ρ is the fluid density, and $\overline{X^2(\boldsymbol{s}, t)}$ is the mean square of the sea surface displacement.

As regards the complete SDF $\widetilde{c}_X(\boldsymbol{k}, \omega)$, it must be borne in mind that it is the most restricted but easier to interpret spectrum compared to some of the partial spectra used in applications (Section 3.3). Below I discuss another application of SDF in ocean sciences.

Example 3.4
The $\widetilde{c}_X(\boldsymbol{k},\omega)$ can be used to model the interior of a 3- or 4-day North Atlantic storm that covers an area of about 500 × 500 nautical miles. But, it will not be an adequate model in the case of transient conditions, of which the most important may be the generation or growth phase of the sea (Kinsman, 1984).

3.2 PROPERTIES OF THE SPECTRAL DENSITY FUNCTION

The SDF, $\widetilde{c}_X(\boldsymbol{k},\omega)$, has certain interesting properties in the $R^{n,1}$ domain, from both a mathematical and a physical viewpoint:

(a) Since the covariance $c_X(\boldsymbol{h},\tau)$ is a nonnegative-definite (NND) function in $R^{n,1}$, and in light of Bochner's theorem, we have the convenient inequality that must be satisfied by the corresponding SDF[17]

$$\widetilde{c}_X(\boldsymbol{k},\omega) \geq 0 \tag{3.12}$$

for all $\boldsymbol{k}$,ω. Hence, the covariance functions of m.s. continuous random fields are NND functions. The converse can also be shown, namely that every NND function can be a covariance function of an m.s. continuous random field.

(b) The SDF is a function that gives the power (amplitude) of the S/TRF as a function of the wavevector $\boldsymbol{k}$ and the frequency ω.

(c) The SDF provides information about the distribution of the S/TRF over the wavevector $\boldsymbol{k}$ and frequency ω. Indeed, from Eq. (3.10a) for $(\boldsymbol{h},\tau)=(\boldsymbol{0},0)$, we get

$$c_X(\boldsymbol{0},0) = \iint d\boldsymbol{k}d\omega\, \widetilde{c}_X(\boldsymbol{k},\omega),$$

i.e., the field variance is obtained by integrating the SDF.

(d) From Eq. (3.10b) we see that for any random field X, the SDF has units of L^nTX^2. So, if the field is dimensionless, so is the covariance, in which case the SDF has areal × time or volume × time units, depending on whether $n=2$ or 3, respectively.[18]

(e) The full SDF admits another noticeable interpretation. The corresponding spectral random field $\widetilde{X}(\boldsymbol{k},\omega)$ satisfies the expressions

$$\overline{\left[\widetilde{X}(\boldsymbol{k}_{i'},\omega_{i'}) - \widetilde{X}(\boldsymbol{k}_i,\omega_i)\right]^2} = \frac{1}{(2\pi)^{n+1}} \int_{\boldsymbol{k}_i}^{\boldsymbol{k}_{i'}} d\boldsymbol{k} \int_{\omega_i}^{\omega_{i'}} d\omega\, \widetilde{c}_X(\boldsymbol{k},\omega),$$

$$\overline{\left[\widetilde{X}(\boldsymbol{k}_{i'},\omega_{i'}) - \widetilde{X}(\boldsymbol{k}_i,\omega_i)\right]\left[\widetilde{X}\left(\boldsymbol{k}_{j'},\omega_{j'}\right) - \widetilde{X}(\boldsymbol{k}_j,\omega_j)\right]} = 0 \tag{3.13a–b}$$

for $((\boldsymbol{k}_i,\boldsymbol{k}_{i'})\times(\omega_i,\omega_{i'}))\cap((\boldsymbol{k}_j,\boldsymbol{k}_{j'})\times(\omega_j,\omega_{j'}))=\emptyset$.[19] The added value of Eq. (3.13a) is that it forms the basis of many developments regarding the spectral or frequency representation of S/TRFs

[17] Permissibility criteria for covariance functions are described in Chapter XV.

[18] This observation may be considered in the light of Remark 3.2 regarding a similar finding in the $R^{n,1}$ domain.

[19] In more formal terms, Eq. (3.13a) can be written as $\overline{\left|d\widetilde{X}(\boldsymbol{k},\omega)\right|^2} = \frac{1}{(2\pi)^{n+1}}\widetilde{c}_X(\boldsymbol{k},\omega)d\boldsymbol{k}d\omega$.

that can be found in applications of atmospheric sciences, meteorology, communications, etc. Eq. (3.13b) implies that in the case of disjoint wavevectors–frequencies the $\widetilde{X}(\boldsymbol{k}, \omega)$ increments are uncorrelated, and, hence, the random fields can be composed of a superposition of $(\boldsymbol{k},\omega)$-plane waves.

For illustration, below I discuss an example of spectral random field representations generated by the stochastic partial differential equation (SPDE) of a physical law. Chapter XIV will develop these representations more fully.

Example 3.5

Consider the natural attributes $X(\boldsymbol{s},t)$ and $Y(\boldsymbol{s},t)$ with joint space–time distributions in $R^{2,1}$ governed by the SPDE

$$\left[a\frac{\partial^2}{\partial t^2} + b\sum_{i=1}^{2}\frac{\partial^4}{\partial s_i^4} + 2b\frac{\partial^4}{\partial s_1^2 \partial s_2^2}\right]X(\boldsymbol{s},t) = Y(\boldsymbol{s},t), \tag{3.14}$$

where $Y(\boldsymbol{s},t)$ is the observable attribute, and a and b are known coefficients. In light of the analysis above, and assuming that a solution of Eq. (3.14) exists in the m.s. sense, the latter can be written in terms of the spectral representation

$$\widetilde{X}(\boldsymbol{k},\omega) = \frac{1}{b\boldsymbol{k}^4 - a\omega^2}\widetilde{Y}(\boldsymbol{k},\omega). \tag{3.15}$$

Consequently, the space–time dependence structure of $X(\boldsymbol{s},t)$ can be expressed in terms of the known SDF of $Y(\boldsymbol{s},t)$ as

$$c_X(\boldsymbol{h},\tau) = \iint d\boldsymbol{k}d\omega\frac{e^{i(\boldsymbol{k}\cdot\boldsymbol{h}-\omega\tau)}}{b\boldsymbol{k}^4 - a\omega^2}\widetilde{c}_Y(\boldsymbol{k},\omega). \tag{3.16}$$

This example illustrates the use of the basic FT property of transforming differentiation operations into multiplications (Table 2.1 of Chapter V), thus greatly simplifying the solution of SPDE like Eq. (3.14).

As suggested earlier, for real-valued STHS covariances that are even functions in $R^{n,1}$, the following symmetry is valid,

$$c_X(\boldsymbol{h},\tau) = c_X(-\boldsymbol{h},-\tau), \tag{3.17a}$$

and the exponential factor in the spectral representation can be conveniently replaced by $\cos(\boldsymbol{k}\cdot\boldsymbol{h} - \omega\tau)$. When physically justified, the class of STHS random fields largely simplifies computations. It also allows the derivation of simple yet useful relationships between the various S/TVFs. Such is, e.g., the covariance–variogram relationship for STHS random fields

$$c_X(\boldsymbol{h},\tau) = c_X(0,0) - \gamma_X(\boldsymbol{h},\tau), \tag{3.17b}$$

which allows us to readily move from covariance to variogram representations and *vice versa*. Combined with Eq. (3.10a), the above relationship leads to the following useful result.

Proposition 3.2
The variogram function can be represented in $R^{n,1}$ as

$$\gamma_X(\boldsymbol{h},\tau) = \iint dkd\omega[1-\cos(\boldsymbol{k}\cdot\boldsymbol{h}-\omega\tau)]\tilde{c}_X(\boldsymbol{k},\omega). \tag{3.18}$$

The spectral variogram representation of Eq. (3.18) is used in many applications, e.g., in situations where linear local trends are present.

Example 3.6
Consider the variogram function in $R^{3,1}$,

$$\gamma_X(r,\tau) = c_0\left(1-e^{-\left(\frac{1}{a^2}r^2+\frac{1}{b^2}\tau^2\right)^{\frac{1}{2}}}\right).$$

where $r = |\boldsymbol{h}|$ and $c_0 > 0$. This variogram function characterizes the space–time variation patterns of an STIS random field, and it admits the spectral representation of Eq. (3.18), with

$$\tilde{c}_X(k,\omega) = \frac{abc_0}{2\pi(1+a^2k^2+b^2\omega^2)^{\frac{3}{2}}}.$$

The class of STIS random fields is discussed in detail in Chapter VIII.

3.3 PARTIAL SPECTRAL REPRESENTATIONS

In several real-world applications, physical considerations suggest that there is no need to transform all the arguments of an S/TRF, in which case the following expressions of spatiotemporal variation prove to be valuable.

We recall that the complete or full SDF, $\tilde{c}_X(\boldsymbol{k},\omega)$, see Eqs. (3.10a–b), forms a $(\boldsymbol{k},\omega)$-FT pair with the space–time covariance function $c_X(\boldsymbol{h},\tau)$ and can be characterized as a *wavevector–frequency* spectrum. In addition, alternative representations of the covariance function[20] are often physically justified or even required in terms of *partial* spectral (Fourier) operations with respect to either space or time. Partial spectra explicitly include space lag or time separation dependence, and may be presented as partially spectral and partially covariance functions. Such partial spectra are defined below.

Definition 3.2
The *space–frequency* spectrum $\tilde{c}_X(\boldsymbol{h},\omega)$ is a partial spectral function defined in two ways:

(a) as a (ω)-FT pair with $c_X(\boldsymbol{h},\tau)$, i.e.,

$$\begin{aligned} c_X(\boldsymbol{h},\tau) &= \int d\omega e^{-i\omega\tau}\tilde{c}_X(\boldsymbol{h},\omega), \\ \tilde{c}_X(\boldsymbol{h},\omega) &= \frac{1}{2\pi}\int d\tau e^{i\omega\tau}c_X(\boldsymbol{h},\tau); \end{aligned} \tag{3.19a–b}$$

[20]Or of other S/TVFs such as the variogram and the structure function, for that matter.

and
(b) as a ($\boldsymbol{k}$)-FT pair with $\widetilde{c}_X(\boldsymbol{k},\omega)$, i.e.,

$$\widetilde{c}_X(\boldsymbol{h},\omega) = \int d\boldsymbol{k} e^{i\boldsymbol{k}\cdot\boldsymbol{h}} \widetilde{c}_X(\boldsymbol{k},\omega),$$

$$\widetilde{c}_X(\boldsymbol{k},\omega) = \frac{1}{(2\pi)^n} \int d\boldsymbol{h} e^{-i\boldsymbol{k}\cdot\boldsymbol{h}} \widetilde{c}_X(\boldsymbol{h},\omega). \tag{3.20a–b}$$

If the S/TRF $X(\boldsymbol{s},t)$ is real, the covariance $c_X(\boldsymbol{h},\tau)$ is real too, and so is $\widetilde{c}_X(\boldsymbol{k},\omega)$, but $\widetilde{c}_X(\boldsymbol{h},\omega)$ may be complex, in general. Another partial spectrum is defined next, in which inclusion of an explicit time separation dependence allows representation of the dynamical component of the phenomenon represented by the spectrum.

Definition 3.3

The *wavevector–time* spectrum[21] $\widetilde{c}_X(\boldsymbol{k},\tau)$ can be defined in two ways:

(a) as a ($\boldsymbol{k}$)-FT pair with $c_X(\boldsymbol{h},\tau)$, i.e.,

$$c_X(\boldsymbol{h},\tau) = \int d\boldsymbol{k} e^{i\boldsymbol{k}\cdot\boldsymbol{h}} \widetilde{c}_X(\boldsymbol{k},\tau),$$

$$\widetilde{c}_X(\boldsymbol{k},\tau) = \frac{1}{(2\pi)^n} \int d\boldsymbol{h} e^{-i\boldsymbol{k}\cdot\boldsymbol{h}} c_X(\boldsymbol{h},\tau), \tag{3.21a–b}$$

where the $\widetilde{c}_X(\boldsymbol{k},\tau)$ is also called the *angular spectrum*; and
(b) as a (ω)-FT pair with $\widetilde{c}_X(\boldsymbol{k},\omega)$, i.e.,

$$\widetilde{c}_X(\boldsymbol{k},\tau) = \int d\omega e^{-i\omega\tau} \widetilde{c}_X(\boldsymbol{k},\omega),$$

$$\widetilde{c}_X(\boldsymbol{k},\omega) = \frac{1}{2\pi} \int d\tau e^{i\omega\tau} \widetilde{c}_X(\boldsymbol{k},\tau). \tag{3.22a–b}$$

In many applications, to isolate further the dynamical or the spatial characters of the spectrum, we can introduce the following concepts derived from partial spectra.

Definition 3.4

The *normalized angular spectrum* (also called the *dynamical correlation function* in certain physical sciences) is defined as

$$\widetilde{\rho}_X(\boldsymbol{k},\tau) = \frac{\widetilde{c}_X(\boldsymbol{k},\tau)}{\widetilde{c}_X(\boldsymbol{k},0)},$$

[21]For the readers information, in some applied sciences this is also called the *wavevector–time covariance*.

where $\widetilde{c}_X(\boldsymbol{k}, \tau = 0) = \int d\omega\, \widetilde{c}_X(\boldsymbol{k}, \omega) = \widetilde{c}_X(\boldsymbol{k})$. Similarly, the *normalized frequency spectrum* is defined by

$$\widetilde{\rho}_X(\boldsymbol{h}, \omega) = \frac{\widetilde{c}_X(\boldsymbol{h}, \omega)}{\widetilde{c}_X(\boldsymbol{0}, \omega)},$$

where $\widetilde{c}_X(\boldsymbol{h} = \boldsymbol{0}, \omega) = \int d\boldsymbol{k}\, \widetilde{c}_X(\boldsymbol{k}, \omega) = \widetilde{c}_X(\omega)$.

Some interesting relationships can be obtained with the help of the normalized angular and frequency spectra.

Example 3.7

After some straightforward manipulations we find that the following equations are valid,

$$c_X(\tau) = \int d\boldsymbol{k}\, \widetilde{c}_X(\boldsymbol{k})\widetilde{\rho}_X(\boldsymbol{k}, \tau),$$

$$c_X(\boldsymbol{h}) = \int d\omega\, \widetilde{c}_X(\omega)\widetilde{\rho}_X(\boldsymbol{h}, \omega),$$

where $\widetilde{c}_X(\boldsymbol{k})$ and $\widetilde{c}_X(\omega)$ are called the wavevector and frequency spectrum, respectively (Exercise VII.22). The interpretation of these equations is that the purely temporal covariance $c_X(\tau)$ is a combination of normalized angular spectra weighted by the wavevector spectrum $\widetilde{c}_X(\boldsymbol{k})$, whereas the purely spatial covariance $c_X(\boldsymbol{h})$ is a combination of normalized frequency spectra weighted by the frequency spectrum $\widetilde{c}_X(\omega)$.

Notice that in the Definition 3.4 the $\widetilde{\rho}_X(\boldsymbol{k}, \tau)$ is considered a correlation function with respect to the time separation τ. If $\widetilde{\rho}_X(\boldsymbol{k}, \tau) = e^{-a|\boldsymbol{k}|\tau}$, the $a|\boldsymbol{k}|$ is a parameterization that allows several kinds of timescales to be considered in temporal decorrelation modeling, where correlation decays exponentially with a timescale $(a|\boldsymbol{k}|)^{-1}$. Another case is $\widetilde{\rho}_X(\boldsymbol{k}, \tau) = e^{-a^2k^2\tau^2}$, where the parameter a allows us to adjust for the strength of dynamical effects, i.e., to choose a physically motivated decorrelation timescale.

Among their many other applications (in oceanology, geophysics, radio, optics, acoustics, materials, and radiation sciences), the normalized angular and frequency spectra have been used to generate useful classes of space−time covariance models (see, Section 3 of Chapter XVI).

The above spectra and their relationships are outlined in Fig. 3.1. Generally, by pursuing a study of the random fields in the frequency domain, a considerable amount of insight is often gained into the operations of these fields.

In addition to the full and partial spectra shown in Fig. 3.1, certain *marginal* spectra of a random field can be derived as follows:

$$\widetilde{c}_X(\boldsymbol{k}) = \frac{1}{(2\pi)^n} \int d\boldsymbol{h}\, e^{-i\boldsymbol{k}\cdot\boldsymbol{h}} c_X(\boldsymbol{h}),$$

$$c_X(\boldsymbol{h}) = \int d\boldsymbol{h}\, e^{i\boldsymbol{k}\cdot\boldsymbol{h}} \widetilde{c}_X(\boldsymbol{k}), \tag{3.23a–b}$$

and

$$\widetilde{c}_X(\boldsymbol{k}) = \int d\omega \widetilde{c}_X(\boldsymbol{k}, \omega). \tag{3.23c}$$

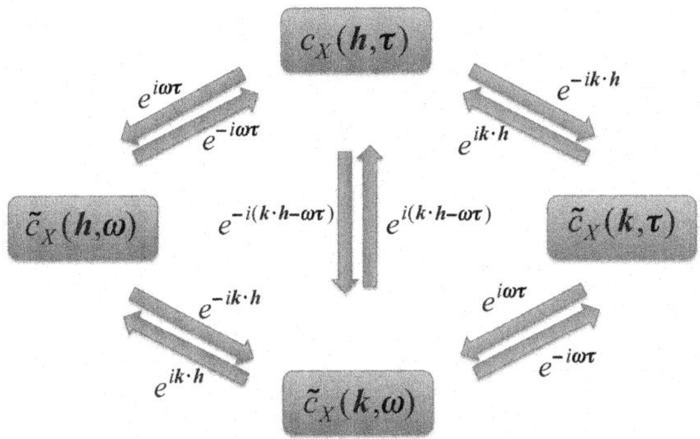

FIGURE 3.1

Spectra (complete and partial) and their relationships for STHS random fields.

Similarly,

$$\tilde{c}_X(\omega) = \frac{1}{2\pi}\int d\tau\, e^{i\omega\tau} c_X(\tau),$$

$$c_X(\tau) = \int d\tau\, e^{-i\omega\tau}\tilde{c}_X(\omega), \tag{3.24a–b}$$

and

$$\tilde{c}_X(\omega) = \int d\boldsymbol{k}\, \tilde{c}_X(\boldsymbol{k}, \omega) \tag{3.24c}$$

is the relationship between the corresponding spectra.

Example 3.8

In ocean science, see Example 3.1, the following spectral representations have been also used ($R^{2,1}$),

$$\begin{aligned} c_X(\boldsymbol{h}, \tau) &= \int d\boldsymbol{k}\, \tilde{c}_X(\boldsymbol{k})\, e^{i(\boldsymbol{k}\cdot\boldsymbol{h}+\omega\tau)} \\ &= \int_{-\infty}^{\infty}\int_{-\pi}^{\pi} d\omega d\theta\, \tilde{\sigma}_X(\omega, \theta)\, e^{i(\boldsymbol{k}\cdot\boldsymbol{h}+\omega\tau)}, \end{aligned}$$

where $\tilde{c}_X(\boldsymbol{k})$ is called the wavevector spectrum, $\tilde{\sigma}_X(\omega, \theta)$ is interpreted as a two-dimensional SDF of surface waves, $\boldsymbol{k} = (k_1,k_2) = (k\cos\theta,\ k\sin\theta)$, and the wavenumber $k = |\boldsymbol{k}|$ is related to the wave frequency ω by the dispersion relation of Eq. (3.4b).

Furthermore, in the case of random fields with STHS increments in $R^{n,1}$ (Section 2.4), the spectrum of a variogram viewed as a simultaneous two-point function can be defined in terms of partial spectra.

Example 3.9

For a random field $X(\boldsymbol{p})$ with STHS increments in $R^{3,1}$, the partial spectral representation of the simultaneous two-point variogram is such that

$$\gamma_X(\boldsymbol{h},t) = \int d\boldsymbol{k}[1-\cos(\boldsymbol{k}\cdot\boldsymbol{h})]\widetilde{c}_X(\boldsymbol{k},t),$$

$$\widetilde{c}_X(\boldsymbol{k},t) = \frac{1}{(2\pi)^3 k^2}\int d\boldsymbol{h}\sin(\boldsymbol{k}\cdot\boldsymbol{h})\boldsymbol{k}\cdot\nabla\gamma_X(\boldsymbol{h},t), \tag{3.25a–b}$$

where Eq. (3.25b) is derived by first differentiating Eq. (3.25a) with respect to $\boldsymbol{h}$, and then considering the inverse FT of the differentiation outcome.

I end this section by reemphasizing that the varying conditions of real-world investigations and physical intuition often require that random field modeling distinguishes between simultaneous and nonsimultaneous covariances, one- and two-point variograms, as well as between full and partial spectral functions, and use them as appropriate.

3.4 MORE ON DISPERSION RELATIONS

This section will bring us back to the importance of dispersion relations in random field modeling. In particular, the following point should be stressed: In the case of physical laws representing wave phenomena, a link is imposed between $\boldsymbol{k}$ and ω by the so-called *dispersion relation* of the phenomenon, $D_R(\boldsymbol{k},\omega)$, i.e., $\boldsymbol{k}$ and ω are not independent but linked by the dispersion relation. This requirement clearly affects the mathematical analysis of the corresponding spectral representations, since the $\widetilde{c}_X(\boldsymbol{k},\omega)$ in the spectra of Eq. (3.10a) must be multiplied by a delta function $\delta(D_R(\boldsymbol{k},\omega))$ that assures that the dispersion relation $D_R(\boldsymbol{k},\omega)$ is satisfied, thus implying that the multiplicity of the Fourier integral is reduced. Below we examine two classical cases of dispersion relations in ocean sciences.

Example 3.10

As we briefly saw in Example 2.3 of Chapter I, in the case of the wave law

$$\left[\upsilon^{-2}\frac{\partial^2}{\partial t^2} - \nabla^2\right]X(\boldsymbol{s},t) = 0 \tag{3.26a}$$

of underground acoustics (υ is the local sound velocity), the dispersion relation linking $\boldsymbol{k}$ and ω is

$$D_R(\boldsymbol{k},\omega): \boldsymbol{k}^2 - \upsilon^{-2}\omega^2 = 0, \tag{3.26b}$$

where $\boldsymbol{k}^2 = \sum_{i=1}^{3} k_i^2$. Then, $\widetilde{c}_X(\boldsymbol{k},\omega)$ is multiplied by $\delta(\omega \mp \upsilon|\boldsymbol{k}|)$ and Eq. (3.10a) yields

$$c_X(\boldsymbol{h},\tau) = \int d\boldsymbol{k}\, e^{i(\boldsymbol{k}\cdot\boldsymbol{h}\mp\upsilon|\boldsymbol{k}|\tau)}\widetilde{c}_X(\boldsymbol{k},\omega),$$

i.e., the integral with respect to $\boldsymbol{k}$ and ω reduces to an integral with respect to $\boldsymbol{k}$ only (for a given ω, $\widetilde{c}_X$ is nonzero only on a circle of radius $|\boldsymbol{k}| = \frac{\omega}{\upsilon}$).[22] This, obviously, affects the shape of the covariance c_X.

[22] In $R^{3,1}$ the $\widetilde{c}_X$ is nonzero on a sphere, and in $R^{1,1}$ it is nonzero at two points opposite in sign.

Example 3.11

The study of surface waters often assumes uniform motion in the horizontal direction s_2, in which case the velocity potential is a function of s_1 (horizontal), s_3 (depth), and t (time), i.e., $X(s_1,s_3,t)$. This is an assumption that simplifies the calculations greatly, and, yet, the model still yields valuable results that improve our understanding of the ocean problem. The dynamic boundary condition on $X(s_1,s_3,t)$ implies a dispersion relation as

$$D_R(k,\omega): k\left(k^2\sigma + \rho g\right)\tanh(kh) - \rho\omega^2 = 0, \tag{3.27}$$

where σ is the surface tension force, ρ is fluid density, and g denotes gravity, as usual. The dispersion Eq. (3.27), which expresses a crucial relationship between the temporal behavior of the model (represented by the angular frequency ω) and the spatial behavior of the model (represented by the physical wavenumber k) imposes a definite restriction of the way k and ω covary, which should be taken into consideration when considering S/TRF representations in the spectral domain.

3.5 SEPARABILITY ASPECTS

As was stated in Section 1, the STHS covariance $c_X(\boldsymbol{h},\tau)$ in $R^{n,1}$ will be termed a space–time separable covariance if[23]

$$c_X(\boldsymbol{h},\tau) = c_{X(1)}(\boldsymbol{h})c_{X(2)}(\tau). \tag{3.28a}$$

Clearly, this implies that

$$\tilde{c}_X(\boldsymbol{k},\omega) = \tilde{c}_{X(1)}(\boldsymbol{k})\tilde{c}_{X(2)}(\omega) \geq 0. \tag{3.28b}$$

If $c_{X(1)}$ is functionally equal to $c_{X(2)}$, Eq. (3.28a) simplifies to $c_X(\boldsymbol{h},\tau) = c_{X(1)}(\boldsymbol{h})c_{X(1)}(\tau)$. I have already stressed the point that, when physically justified, separability is a very convenient property in applications.

One more thing to remember is that, if one of the arguments (space or time) is assumed fixed, certain conditions apply for the function to be a valid covariance of the other argument according to Eq. (3.28b). For instance, assuming a fixed $\tau = \tau^*$, for the function $c_X(\boldsymbol{h},\tau^*)$ to be a valid spatial covariance, it is necessary that $c_{X(2)}(\tau^*) \geq 0$. Analogous is the case for a fixed $\boldsymbol{h} = \boldsymbol{h}^*$. Moreover, as Stein (2005) has noticed, separable space–time covariance functions are not smoother away from the origin than at the origin, and they tend to have "ridges" along their axes.

Example 3.12

In hydrogeology, observations and their interpretation have suggested that the space–time variation of point precipitation intensity $X(\boldsymbol{s},t)$ at location $\boldsymbol{s}$ during time t can be represented by an S/TRF with a separable covariance of the form

$$c_X(\boldsymbol{h},\tau) = \sigma_X^2\rho_{X(1)}(\boldsymbol{h})\rho_{X(2)}(\tau), \tag{3.29}$$

where σ_X^2 is the point variance of $X(\boldsymbol{s},t)$, $\rho_{X(1)}(\boldsymbol{h})$ is the spatial correlation, and $\rho_{X(2)}(\tau)$ is the temporal correlation of precipitation intensity. By taking advantage of covariance separability, comprehensive charts for rainfall network design can be constructed (Bras and Rodriguez-Iturbe, 1985).

[23] An equation analogous to Eq. (3.28a) is not necessarily valid in terms of the variogram function.

Remark 3.4

I noticed in Section 2 of Chapter II that the fact that a temporal random process is a unidimensional random field readily implies that most of the results of general random field theory can be expressed in a straightforward manner in terms of temporal random processes. Valid covariance functions for spatial fields, e.g., remain valid for temporal processes by replacing the spatial lag $|\boldsymbol{h}|$ with the temporal lag τ (>0). This is justified if one observes that the values of a random field $X(s_1,s_2,s_3)$ on the line $s_2 = s_3 = 0$ represent a homogeneous field on this line, i.e., a stationary random process in the coordinate s_1. Of course, there are random field properties (e.g., spatial anisotropy) that do not have a direct analog in the context of random processes.

As was commented on several occasions, a large variety of known purely spatial and purely temporal covariance models can be combined in a straightforward manner, to produce valid space–time models. In light of Eq. (3.28a), we consider below a few examples of space–time covariances that are combinations of valid models belonging to families of spatially homogeneous covariances, $c_X(\boldsymbol{h})$, and temporally stationary covariances, $c_X(\tau)$. Spatial and temporal covariances of higher continuity orders can be constructed based on homogeneous and stationary residuals, respectively (Christakos and Hristopulos, 1998). This procedure can be used in the R^{n+1} and in the $R^{n,1}$ domain.

Example 3.13

Let us consider the space–time separable covariance function in $R^{2,1}$,

$$c_X(\boldsymbol{h},\tau) = c_0 e^{-\frac{\boldsymbol{h}^2}{a^2}-\tau} J_0(b|\boldsymbol{h}|), \tag{3.30a}$$

where $c_0 > 0$, $\boldsymbol{h} \in R^2$, with metric pair ($|\boldsymbol{h}| = r_E$, τ). The corresponding SDF is also separable,

$$\widetilde{c}_X(\boldsymbol{k},\omega) = \widetilde{c}_{X(1)}(\boldsymbol{k})\widetilde{c}_{X(2)}(\omega) = \frac{c_0 a^2}{4\pi} I_0\left(\frac{a^2 b|\boldsymbol{k}|}{2}\right) e^{-\frac{a^2(b^2+\boldsymbol{k}^2)}{4}} \frac{1}{\pi(1+\omega^2)}, \tag{3.30b}$$

where I_0 denotes a modified Bessel function of the first kind. Clearly, the SDF satisfies the condition of Eq. (3.12). Also, the existence of the integral representation of the temporal part of the covariance requires that the limits hold,

$$\lim_{|\omega|\to 0}\left[\omega \widetilde{c}_{X(2)}(\omega)\right] = \lim_{|\omega|\to\infty}\left[\omega \widetilde{c}_{X(2)}(\omega)\right] = 0, \tag{3.31}$$

which are clearly both valid for $\widetilde{c}_{X(2)}(\omega) = \pi^{-1}\left(1+\omega^2\right)^{-1}$. As another illustration, the following STHS covariance in R^{n+1} can be derived as the product of unidimensional von Kármán or Matern covariances,[24]

$$c_X(\Delta\boldsymbol{p}) = \frac{2^{1-\nu}}{\Gamma(\nu)} \prod_{i=1}^{n,0} (a_i|h_i|)^{\nu} K_\nu(a_i|h_i|), \tag{3.32}$$

where $\Delta\boldsymbol{p} = (h_1,\ldots, h_n, h_0)$, K_ν denotes the modified Bessel function of the second kind and νth order, and $a_i > 0$.

[24]This covariance was proposed by von Kármán (1948) to characterize the random velocity of turbulent media and in the spatial statistics literature is also known as the Matern covariance.

I continue with a comment that also provides a link with the following section. Certain random field properties (e.g., convergence and differentiation conditions) may depend on the manner in which the corresponding metric is defined. In the case of an STHS random field with $\Delta \boldsymbol{p} \in R^{n+1}$ and a Pythagorean space–time metric $|\Delta \boldsymbol{p}| = r_p$, a set of convergence conditions may be considered that differ from the corresponding conditions in the case of an STHS random field with $(\boldsymbol{h},\tau) \in R^{n,1}$ and a Euclidean spatial metric $|\boldsymbol{h}| = r_E$.

Example 3.14

In the latter case described above, i.e., $(\boldsymbol{h},\tau) \in R^{n,1}$ and $|\boldsymbol{h}| = r_E$, the counterparts of the convergence conditions involving the integral representations of Eqs. (3.6c–d) lead to the conclusion that if the covariance $c_X(\boldsymbol{h},\tau)$ tends to zero fast enough with $|\boldsymbol{h}|, \tau \to \infty$ (rapidly falling-off covariance), then

$$\iint d\boldsymbol{h} d\tau |c_X(\boldsymbol{h}, \tau)| < \infty, \tag{3.33}$$

and the $c_X(\boldsymbol{h},\tau)$ can be represented by Eq. (3.10a), in which case the $\widetilde{c}_X(\boldsymbol{k}, \omega)$ is given by Eq. (3.10b). In other words, if $c_X(\boldsymbol{h},\tau)$ is an absolutely integrable function in $R^{n,1}$, then the SDF $\widetilde{c}_X(\boldsymbol{k}, \omega)$ exists. In this case, obviously,

$$\int d\boldsymbol{k} d\omega \widetilde{c}_X(\boldsymbol{k}, \omega) = c_X(\boldsymbol{0}, 0) < \infty, \tag{3.34}$$

that is, the SDF is integrable and Eq. (3.12) is valid (in the particular case of rapidly falling-off covariance functions, $c_X(\boldsymbol{h},\tau)$ is a covariance function if and only if Eq. (3.12) is valid).

Similarly to Eqs. (3.8a–b) in R^{n+1}, when we are dealing with real-valued fields where both $c_X(\boldsymbol{h},\tau)$ and $\widetilde{c}_X(\boldsymbol{k}, \omega)$ are even functions, they are related by the Fourier cosine transform in $R^{n,1}$,

$$\begin{aligned} c_X(\boldsymbol{h}, \tau) &= \iint d\boldsymbol{k} d\omega \cos(\boldsymbol{k}\cdot\boldsymbol{h} - \omega\tau)\widetilde{c}_X(\boldsymbol{k}, \omega), \\ \widetilde{c}_X(\boldsymbol{k}, \omega) &= \frac{1}{(2\pi)^{n+1}} \iint d\boldsymbol{h} d\tau \cos(\boldsymbol{k}\cdot\boldsymbol{h} - \omega\tau) c_X(\boldsymbol{h}, \tau) \end{aligned} \tag{3.35a–b}$$

(it is always assumed that these integrals exist). As noted earlier, separability implies full symmetry, that is, a separable STHS covariance is fully symmetric too (the converse is not necessarily valid). Based on Eq. (3.35a), the following result holds in the case of full symmetry.

Proposition 3.3

A space–time covariance $c_X(\boldsymbol{h},\tau)$ is fully symmetric if and only if

$$c_X(\boldsymbol{h}, \tau) = \iint d\boldsymbol{k} d\omega \cos(\boldsymbol{k}\cdot\boldsymbol{h})\cos(\omega\tau)\widetilde{c}_X(\boldsymbol{k}, \omega) \tag{3.36}$$

in $R^{n,1}$.

The covariance representation of Eq. (3.36) is also useful in the modeling of separable functions. As a final note to this section, an STHS hypothesis may also apply to multipoint S/TVFs. For example, the trivariance $\overline{X(\boldsymbol{s},t)X(\boldsymbol{s}',t')X(\boldsymbol{s}'',t'')}$ depends only on $(\boldsymbol{s}-\boldsymbol{s}', t-t')$ and $(\boldsymbol{s}-\boldsymbol{s}'', t-t'')$, or $(\boldsymbol{s}'-\boldsymbol{s}'', t'-t'')$ in $R^{n,1}$.

4. THE GEOMETRY OF SPACE–TIME HOMOSTATIONARITY

In Chapter VI we saw that the study of many natural systems needs the development of a stochastic calculus and that the key feature of stochastic calculus that renders it particularly useful is that it addresses the notions of S/TRF continuity, differentiability, and integrability (e.g., sea surface roughness depends on the steepness of the waves that is closely linked to the notion of differentiability of the random fields modeling the waves). Naturally, these notions refer to probabilities or average behavior of the sample paths (random field realizations), which can lead to informative conclusions regarding the S/TRF geometry, and, consequently, the regularity, homogeneity, and smoothness of the natural attribute modeled by the random field. In Chapter VI it became also clear that S/TRF continuity, differentiability, and integrability should involve the notion of the "limit of a random field." In this chapter, these geometrical properties are considered in the particular case of STHS random fields. Accordingly, I start with the presentation of some basic differentiation formulas that are valid in the STHS case.

4.1 DIFFERENTIATION FORMULAS: PHYSICAL AND SPECTRAL DOMAINS

Mathematically, the differentiation of an S/TRF $X(\boldsymbol{p})$ generates new S/TRFs that are the derivatives of $X(\boldsymbol{p})$ along various directions in space and in time. As was suggested in Chapter VI, the stochastic differentiation of an S/TRF with respect to the arguments $\boldsymbol{p},\boldsymbol{p}'$ is uniquely linked to the deterministic differentiation of the corresponding S/TVF with respect to the argument $\Delta\boldsymbol{p}$. The transformation

$$\boldsymbol{p},\boldsymbol{p}' \mapsto \Delta\boldsymbol{p}$$

entails some mathematical manipulations of the S/TVFs basically in terms of the chain rule of standard mathematical analysis.

Proposition 4.1

For a linear and homogeneous differential operator $\mathcal{L}_{\boldsymbol{p}}$, in general (specific cases will be considered later) it is valid that

$$\begin{aligned} \mathcal{L}_{\boldsymbol{p}} c_X(\boldsymbol{p},\boldsymbol{p}') &= \pm \mathcal{L}_{\Delta\boldsymbol{p}} c_X(\Delta\boldsymbol{p}), \\ \mathcal{L}_{\boldsymbol{p}'}\mathcal{L}_{\boldsymbol{p}}[c_X(\boldsymbol{p},\boldsymbol{p}')] &= \pm \mathcal{L}_{\boldsymbol{p}'}\mathcal{L}_{\Delta\boldsymbol{p}}[c_X(\Delta\boldsymbol{p})], \\ \mathcal{L}_{\boldsymbol{p}}\mathcal{L}_{\boldsymbol{p}'}[c_X(\boldsymbol{p},\boldsymbol{p}')] &= \mp \mathcal{L}_{\boldsymbol{p}}\mathcal{L}_{\Delta\boldsymbol{p}}[c_X(\Delta\boldsymbol{p})], \end{aligned} \qquad (4.1\text{a–c})$$

where the indicated sign in Eqs. (4.1a–c) refers to the corresponding sign of the argument $\Delta\boldsymbol{p} = \pm(\boldsymbol{p}-\boldsymbol{p}')$.

Essentially, given that both the random field $X(\boldsymbol{p})$ and its space–time derivatives are functions in the R^{n+1} domain, higher-order $X(\boldsymbol{p})$ derivatives are formally defined in terms of a recursive process involving the differentiation of lower-order $X(\boldsymbol{p})$ derivatives. Beyond its mathematical interest, stochastic differentiation has significant implications in scientific practice in which S/TRF differentiability often has a physical meaning and real-world implications.

Methodologically, the implementation of Eqs. (4.1a–c) often boils down to specifying the $\mathcal{L}_{\Delta\boldsymbol{p}}$ operator that corresponds to a given $\mathcal{L}_{\boldsymbol{p}}$ operator. That is, to the argument transformation $\boldsymbol{p},\boldsymbol{p}' \mapsto \Delta\boldsymbol{p}$, the corresponding operator transformation must be established,

$$\mathcal{L}_{\boldsymbol{p}} \mapsto \mathcal{L}_{\Delta\boldsymbol{p}}$$

(taking into consideration, of course, the sign of the space–time lag $\boldsymbol{\Delta p}$). I start by presenting some differentiation formulas for STHS random fields in the R^{n+1} domain that are direct consequences of the general stochastic differentiation results of Section 4 of Chapter VI. These expressions link the covariance functions of various combinations of S/TRF derivatives.

Example 4.1

Let us consider some specific cases of the operator $\mathcal{L}_p$ (in all cases below, $\boldsymbol{\Delta p} = \boldsymbol{p} - \boldsymbol{p}'$). I start with the simple operator transformation

$$\mathcal{L}_p = \frac{\partial}{\partial s_i} \mapsto \mathcal{L}_{\Delta p} = \frac{\partial}{\partial h_i}, \tag{4.2a}$$

which is here applied on the covariance function $c_X(\boldsymbol{p} - \boldsymbol{p}') = c_X(\boldsymbol{\Delta p})$. This is a case of Eq. (4.1a) of Proposition 4.1, using the chain rule

$$\frac{\partial c_X(\boldsymbol{p} - \boldsymbol{p}')}{\partial s_i} = \frac{\partial(\boldsymbol{p} - \boldsymbol{p}')}{\partial s_i} \cdot \nabla c_X(\boldsymbol{\Delta p}),$$

thus leading to the covariance relation

$$\frac{\partial c_X(\boldsymbol{p} - \boldsymbol{p}')}{\partial s_i} = \frac{\partial c_X(\boldsymbol{\Delta p})}{\partial h_i} \tag{4.2b}$$

(Exercise VII.8). For further illustration, next I consider the operator transformation

$$\mathcal{L}_p = \frac{\partial^{\nu+\mu}}{\partial \boldsymbol{s}^\nu \partial s_0^\mu} \mapsto \mathcal{L}_{\Delta p} = \frac{\partial^{\nu+\mu}}{\partial \boldsymbol{h}^\nu \partial h_0^\mu}, \tag{4.3}$$

where, according to the multi-index notation introduced in Section 4 of Chapter VI, $\partial \boldsymbol{s}^\nu = \prod_{i=1}^n \partial s_i^{\nu_i}$ and $\partial \boldsymbol{h}^\nu = \prod_{i=1}^n \partial h_i^{\nu_i}$, $h_0 = \tau$ (see also Appendix). In light of Eq. (4.3), the covariance relation is given by

$$\frac{\partial^{\nu+\mu} c_X(\boldsymbol{p} - \boldsymbol{p}')}{\partial \boldsymbol{s}^\nu \partial s_0^\mu} = \frac{\partial^{\nu+\mu} c_X(\boldsymbol{\Delta p})}{\partial \boldsymbol{h}^\nu \partial h_0^\mu}, \tag{4.4}$$

which, of course, is a generalization of Eq. (4.2b). Working along the same lines it can be shown that, in general, the following covariance relationships hold,

$$\begin{aligned}
&\mathcal{L}_p c_X(\boldsymbol{p} - \boldsymbol{p}') = \mathcal{L}_{\Delta p} c_X(\boldsymbol{\Delta p}),\\
&\mathcal{L}_{p'} c_X(\boldsymbol{p} - \boldsymbol{p}') = (-1)^{\nu+\mu} \mathcal{L}_{\Delta p} c_X(\boldsymbol{\Delta p}),\\
&\mathcal{L}_{p'} \mathcal{L}_{\Delta p} c_X(\boldsymbol{\Delta p}) = (-1)^{\nu+\mu} \mathcal{L}_{\Delta p} \mathcal{L}_{\Delta p} c_X(\boldsymbol{\Delta p})
\end{aligned} \tag{4.5a–c}$$

(Exercise VII.7).[25] Lastly, for an even $\mathcal{L}_{\Delta p}$ operator,

$$\mathcal{L}_{\Delta p} c_X(\Delta\boldsymbol{p}) = \mathcal{L}_{\Delta p} c_X(-\Delta\boldsymbol{p}),$$
$$\mathcal{L}_p \mathcal{L}_{\Delta p} c_X(\Delta\boldsymbol{p}) = \mathcal{L}_p \mathcal{L}_{\Delta p}[c_X(-\Delta\boldsymbol{p})] = \mathcal{L}_{\Delta p} \mathcal{L}_{\Delta p} c_X(\Delta\boldsymbol{p}), \tag{4.6a–b}$$

which is also equal to $\mathcal{L}_p \mathcal{L}_{p'}[c_X(\boldsymbol{p},\boldsymbol{p}')]$.

I reemphasize that when we deal with S/TRF and S/TVF derivatives, we should be careful to clarify whether partial derivatives along all directions or only along a few of the directions are considered.

Corollary 4.1

Assume that the operators $\mathcal{L}_p$ express S/TRF derivatives in the R^{n+1} domain as follows,[26]

$$\mathcal{L}_p X(\boldsymbol{p}) = X_{i,0}^{(\nu,\mu)}(\boldsymbol{p}) = \frac{\partial^{\nu+\mu} X(\boldsymbol{p})}{\partial \boldsymbol{s}^{\nu} \partial s_0^{\mu}},$$
$$\mathcal{L}_{p,d} X(\boldsymbol{p}) = X_{\boldsymbol{d},0}^{(\nu,\mu)}(\boldsymbol{p}) = \frac{\partial^{\nu+\mu} X(\boldsymbol{p})}{\partial \boldsymbol{s}_{\boldsymbol{d}}^{\nu} \partial s_0^{\mu}}, \tag{4.7a–b}$$

where $\partial \boldsymbol{s}^{\nu} = \prod_{i \in I_\nu} \partial s_i^{\nu_i}$, $I_\nu \subseteq \{1,\ldots, n\}$, and $\partial \boldsymbol{s}_{\boldsymbol{d}}^{\nu} = \prod_{i \in I_\nu} \partial s_{d_i}^{\nu_i}$, $I_\nu \subseteq \{1,\ldots, \rho\}$. The covariances of Eqs. (4.7a–b) with $\Delta\boldsymbol{p} = \boldsymbol{p} - \boldsymbol{p}'$ are, respectively,[27]

$$\operatorname{cov}(\mathcal{L}_p X(\boldsymbol{p}), \mathcal{L}'_{p'} X(\boldsymbol{p}')) = (-1)^{\nu'+\mu'} \mathcal{L}_{\Delta p} \mathcal{L}'_{\Delta p} c_X(\Delta\boldsymbol{p}),$$
$$\operatorname{cov}(\mathcal{L}_{p,d} X(\boldsymbol{p}), \mathcal{L}'_{p',d} X(\boldsymbol{p}')) = (-1)^{\nu'+\mu'} \mathcal{L}_{\Delta p,d} \mathcal{L}'_{\Delta p,d} c_X(\Delta\boldsymbol{p}), \tag{4.8a–b}$$

where,

$$\mathcal{L}'_{\Delta p} = \frac{\partial^{\nu'+\mu'}}{\partial \boldsymbol{h}^{\nu'} \partial h_0^{\mu'}},$$
$$\mathcal{L}_{\Delta p,d} = \frac{\partial^{\nu+\mu}}{\partial \boldsymbol{h}_{\boldsymbol{d}}^{\nu} \partial h_0^{\mu}}, \tag{4.9a–c}$$
$$\mathcal{L}'_{\Delta p,d} = (-1)^{\nu'+\mu'} \frac{\partial^{\nu'+\mu'}}{\partial \boldsymbol{h}_{\boldsymbol{d}}^{\nu'} \partial h_0^{\mu'}},$$

with $\partial \boldsymbol{h}^{\nu'} = \prod_{i \in I_{\nu'}} \partial h_i^{\nu_i'}$, $\partial \boldsymbol{h}_{\boldsymbol{d}}^{\nu} = \prod_{i \in I_\nu} \partial h_{d_i}^{\nu_i}$, and $\partial \boldsymbol{h}_{\boldsymbol{d}}^{\nu'} = \prod_{i \in I_{\nu'}} \partial h_{d_i}^{\nu_i'}$.

---

[25] A noticeable difference between Eqs. (4.1a–c) and Eqs. (4.5a–c) is that the left side of the former is a function of $\boldsymbol{p},\boldsymbol{p}'$, whereas that of the latter a function of $\boldsymbol{p} - \boldsymbol{p}'$.

[26] Herein, for simplicity, the $X_{i,0}^{(\nu,\mu)}(\boldsymbol{p})$ will be usually written as $X^{(\nu,\mu)}(\boldsymbol{p})$.

[27] The derivation is left as an exercise (see, Exercise VII.11).

Eqs. (4.8a–b) are general expressions of covariance differentiation that are valid for any operator of the type $\mathcal{L}_p$. Their specific implementation in practice is facilitated by some analytical formulations as described below.

Example 4.2

A straightforward implementation of Eq. (4.8a) gives the covariance derivative expression

$$\operatorname{cov}\left(X^{(\nu,\mu)}(\boldsymbol{p}),X^{(\nu',\mu')}(\boldsymbol{p}')\right)=(-1)^{\nu+\mu}\mathcal{L}_{\Delta p}\mathcal{L}'_{\Delta p}c_X(\Delta\boldsymbol{p})=(-1)^{\nu+\mu}\frac{\partial^{\nu+\nu'+\mu+\mu'}c_X(\Delta\boldsymbol{p})}{\partial\boldsymbol{h}^{\nu+\nu'}\partial h_0^{\mu+\mu'}},\qquad(4.10a)$$

where now $\Delta\boldsymbol{p}=\boldsymbol{p}'-\boldsymbol{p}$, $\partial\boldsymbol{h}^{\nu+\nu'}=\partial\boldsymbol{h}^{\nu}\partial\boldsymbol{h}^{\nu'}$. An obvious consequence of the STHS assumption is that it reduces the number of covariance derivatives. When the order of the space–time points changes, i.e., $\Delta\boldsymbol{p}=\boldsymbol{p}-\boldsymbol{p}'$, Eq. (4.8a) becomes

$$\operatorname{cov}\left(X^{(\nu,\mu)}(\boldsymbol{p}),X^{(\nu',\mu')}(\boldsymbol{p}')\right)=(-1)^{\nu'+\mu'}\frac{\partial^{\nu+\nu'+\mu+\mu'}c_X(\Delta\boldsymbol{p})}{\partial\boldsymbol{h}^{\nu+\nu'}\partial h_0^{\mu+\mu'}},\qquad(4.10b)$$

i.e., $(-1)^{\nu+\mu}$ is replaced by $(-1)^{\nu'+\mu'}$.

Eqs. (4.10a) and (4.10b) are useful operational instruments in the development of space–time covariance derivatives for STHS random fields.

Example 4.3

The cases of space–time covariance derivative listed in Table 4.1 further illustrate the step-by-step implementation of Eqs. (4.10a) and (4.10b) in $R^{n,1}$ (for comparison with the general case, see, also Table 4.2 of Chapter VI).

It may be instructive to look in a bit more detail into the mathematical procedure that leads from the differentiation formulas of general covariance functions $c_X(\boldsymbol{p},\boldsymbol{p}')$ to those of STHS covariance functions $c_X(\Delta\boldsymbol{p})$, where the sign of $\Delta\boldsymbol{p}$ affects the calculations, and then compare the results with Eqs. (4.10a) and (4.10b) for confirmation.

Example 4.4

Let us start our study of the differentiation process with an example in $R^{1,1}$, i.e., for the random field $X^{(1,0)}(s,t)=\frac{\partial X(s,t)}{\partial s}$, consider the product

$$X^{(1,0)}(s,t)X^{(1,0)}(s',t)=\frac{\partial^2}{\partial s\partial s'}X(s,t)X(s',t).$$

By taking the stochastic expectation of the above equation, and interchanging the operations of differentiation and expectation (which are commutable), we find

$$\overline{X^{(1,0)}(s',t)X^{(1,0)}(s,t)}=\frac{\partial^2}{\partial s\partial s'}\overline{X(s,t)X(s',t)}=\frac{\partial^2}{\partial s\partial s'}c_X(s'-s,t),$$

where due to STHS the argument in the covariance is $h=s'-s$. Next, by letting $u=s'+s$, transforming the derivatives involving s and s' to h and u, and by using the chain rule we get

$$\frac{\partial^2}{\partial s\partial s'}=\frac{\partial^2}{\partial u^2}-\frac{\partial^2}{\partial h^2}.$$

Table 4.1 Covariance Cases of Random Field Derivatives in $R^{n,1}$

ν	μ	ν'	μ'	I_ν	$I_{\nu'}$	Covariance of Derivatives of $X(p)$
1	1	1	1	{1}	{2}	$\boldsymbol{h}=\boldsymbol{s}'-\boldsymbol{s}, \tau=t'-t$ $\operatorname{cov}\left(X^{(1,1)}(\boldsymbol{p}), X^{(1,1)}(\boldsymbol{p}')\right)=\frac{\partial^{1+1+1+1}c_X(\boldsymbol{p},\boldsymbol{p}')}{\prod_{i\in\{1\}}\partial s_i^{\nu_i}\prod_{i\in\{2\}}\partial s_i'^{\nu_i'}\partial s_0\partial s_0'}=\frac{\partial^4 c_X(\boldsymbol{p},\boldsymbol{p}')}{\partial s_1\partial s_2'\partial s_0\partial s_0'}$ $=(-1)^{1+1}\frac{\partial^4 c_X(\boldsymbol{h})}{\partial h_1\partial h_2\partial h_0^2}=\frac{\partial^4 c_X(\boldsymbol{h})}{\partial h_1\partial h_2\partial h_0^2}$
2	1	1	0	{2,3}	{1}	$\boldsymbol{h}=\boldsymbol{s}'-\boldsymbol{s}, \tau=t'-t$ $\operatorname{cov}\left(X^{(2,1)}(\boldsymbol{p}), X^{(1,0)}(\boldsymbol{p}')\right)=\frac{\partial^{2+1+1}c_X(\boldsymbol{p},\boldsymbol{p}')}{\prod_{i\in\{2,3\}}\partial s_i^{\nu_i}\prod_{i\in\{1\}}\partial s_i'^{\nu_i'}\partial s_0}=\frac{\partial^4 c_X(\boldsymbol{p},\boldsymbol{p}')}{\partial s_2\partial s_3\partial s_1'\partial s_0}$ $=(-1)^{2+1}\frac{\partial^4 c_X(\Delta\boldsymbol{p})}{\partial h_1\partial h_2 h_3\partial h_0}=-\frac{\partial^4 c_X(\Delta\boldsymbol{p})}{\partial h_1\partial h_2 h_3\partial h_0}$
2	0	0	1	{2}	∅	$\boldsymbol{h}=\boldsymbol{s}'-\boldsymbol{s}, \tau=t'-t$ $\operatorname{cov}\left(X^{(2,0)}(\boldsymbol{s}), X^{(0,1)}(\boldsymbol{s}')\right)=\frac{\partial^{2+1}c_X(\boldsymbol{p},\boldsymbol{p}')}{\prod_{i\in\{2\}}\partial s_i^{\nu_i}\prod_{i\in\varnothing}\partial s_i'^{\nu_i'}\partial s_0'}=\frac{\partial^3 c_X(\boldsymbol{p},\boldsymbol{p}')}{\partial s_2^2\partial s_0'}$ $=(-1)^{2+0}\frac{\partial^3 c_X(\Delta\boldsymbol{p})}{\partial h_2^2\partial h_0}=\frac{\partial^3 c_X(\Delta\boldsymbol{p})}{\partial h_2^2\partial h_0}$
0	0	1	1	∅	{2}	$\boldsymbol{h}=\boldsymbol{s}-\boldsymbol{s}', \tau=t-t'$ $\operatorname{cov}\left(X(\boldsymbol{p}), X^{(1,1)}(\boldsymbol{p}')\right)=\frac{\partial^{1+1}c_X(\boldsymbol{p},\boldsymbol{p}')}{\prod_{j\in\{2\}}\partial s_j'^{\nu_j'}\partial s_0'}=\frac{\partial^2 c_X(\boldsymbol{p}-\boldsymbol{p}')}{\partial s_2'\partial s_0'}$ $=(-1)^{1+1}\frac{\partial^2 c_X(\Delta\boldsymbol{p})}{\partial h_2\partial h_0}=\frac{\partial^2 c_X(\Delta\boldsymbol{p})}{\partial h_2\partial h_0}$
1	1	1	1	{1}	{2}	$\boldsymbol{h}=\boldsymbol{s}-\boldsymbol{s}', \tau=t-t'$ $\operatorname{cov}\left(X^{(1,1)}(\boldsymbol{p}), X^{(1,1)}(\boldsymbol{p}')\right)=\frac{\partial^{1+1+1+1}c_X(\boldsymbol{p},\boldsymbol{p}')}{\prod_{i\in\{1\}}\partial s_i^{\nu_i}\prod_{j\in\{2\}}\partial s_j'^{\nu_j'}\partial s_0\partial s_0'}=\frac{\partial^4 c_X(\boldsymbol{p}-\boldsymbol{p}')}{\partial s_1\partial s_2'\partial s_0\partial s_0'}$ $=(-1)^{1+1}\frac{\partial^4 c_X(\Delta\boldsymbol{p})}{\partial h_1\partial h_2\partial h_0^2}=\frac{\partial^4 c_X(\Delta\boldsymbol{p})}{\partial h_1\partial h_2\partial h_0^2}$
1	1	0	2	{1}	∅	$\boldsymbol{h}=\boldsymbol{s}-\boldsymbol{s}', \tau=t-t'$ $\operatorname{cov}\left(X^{(1,1)}(\boldsymbol{p}), X^{(0,2)}(\boldsymbol{p}')\right)=\frac{\partial^{1+1+0+2}c_X(\boldsymbol{p},\boldsymbol{p}')}{\prod_{i\in\{1\}}\partial s_i^{\nu_i}\partial s_0\partial s_0'^2}=\frac{\partial^4 c_X(\boldsymbol{p}-\boldsymbol{p}')}{\partial s_1\partial s_0\partial s_0'^2}$ $=(-1)^{0+2}\frac{\partial^4 c_X(\Delta\boldsymbol{p})}{\partial h_1\partial h_0^3}=\frac{\partial^4 c_X(\Delta\boldsymbol{p})}{\partial h_1\partial h_0^3}$

Then, the covariance derivation above becomes,

$$\overline{X^{(1,0)}(s',t)X^{(1,0)}(s,t)} = \left[\frac{\partial^2}{\partial u^2} - \frac{\partial^2}{\partial h^2}\right] c_X(h,t),$$

and since $c_X(h,t)$ is a function of h but not of u, we obtain

$$\mathrm{cov}\left(X^{(1,0)}(s,t), X^{(1,0)}(s',t)\right) = \overline{X^{(1,0)}(s,t)X^{(1,0)}(s',t)} = -\frac{\partial^2}{\partial h^2} c_X(h,t), \tag{4.11a}$$

which is the desired result. This is, also, obtained from the general expression of Eqs. (4.10a) and (4.10b) in $R^{1,1}$, for $\nu = \nu' = 1$, $\mu = \mu' = 0$. In sum, the two basic mathematical tricks used in the derivation of the derivative were: first, to replace the covariance argument from (s', s) to $h = s' - s$, and second, to define a new argument $u = s' + s$, and use the chain rule. A similar differentiation procedure applies in terms of other S/TVFs, such as the variogram or the structure function. As another illustration, for the product in $R^{n,1}$

$$X(\boldsymbol{s},t)X^{(1,0)}(\boldsymbol{s},t) = X(\boldsymbol{s},t)\frac{\partial X(\boldsymbol{s},t)}{\partial s_j},$$

we find that

$$\overline{X(\boldsymbol{s}',t)X^{(1,0)}(\boldsymbol{s},t)} = \frac{\partial}{\partial s_j}\overline{X(\boldsymbol{s},t)X(\boldsymbol{s}',t)} = \frac{\partial}{\partial s_j} c_X(\boldsymbol{h},t),$$

where due to STHS the argument in the covariance is $\boldsymbol{h} = \boldsymbol{s}' - \boldsymbol{s}$. By letting $u_j = s'_j + s_j$ and $h_j = s'_j - s_j$, or $s'_j = \frac{u_j + h_j}{2}$ and $s_j = \frac{u_j - h_j}{2}$, and then using the chain rule we get

$$\frac{\partial}{\partial s_j} = \frac{\partial u_k}{\partial s_j}\frac{\partial}{\partial u_k} - \frac{\partial h_k}{\partial s_j}\frac{\partial}{\partial h_k} = \delta_{kj}\frac{\partial}{\partial u_k} - \delta_{kj}\frac{\partial}{\partial h_k} = \frac{\partial}{\partial u_j} - \frac{\partial}{\partial h_j}.$$

Finally, the covariance derivation procedure becomes,

$$\mathrm{cov}\left(X(\boldsymbol{p}'), X^{(1,0)}(\boldsymbol{p})\right) = \overline{X(\boldsymbol{s}',t)X^{(1,0)}(\boldsymbol{s},t)} = \left[\frac{\partial}{\partial u_j} - \frac{\partial}{\partial h_j}\right] c_X(\boldsymbol{h},t) = -\frac{\partial}{\partial h_j} c_X(\boldsymbol{h},t), \tag{4.11b}$$

since $c_X(\boldsymbol{h},t)$ is a function of $\boldsymbol{h}$ but not of u_j. As a confirmation, the same result is obtained from the general expression of Eq. (4.10a) in $R^{n,1}$, for $\nu = 1$ and $\nu' = \mu = \mu' = 0$.

Example 4.5

In the R^{n+1} domain assume that

$$\mathrm{cov}\left(X^{(n,1)}(\boldsymbol{p}), X^{(n,1)}(\boldsymbol{p}')\right) = \frac{\partial^{2n+2} c_X(\boldsymbol{p},\boldsymbol{p}')}{\prod_{i=1}^{n,0}\partial s_i \prod_{j=1}^{n,0}\partial s_j}, \tag{4.12}$$

where $\Delta\boldsymbol{p} = \boldsymbol{p} - \boldsymbol{p}'$, and $X^{(n,1)}(\boldsymbol{p}) = \frac{\partial^{n+1}X(\boldsymbol{p})}{\prod_{i=1}^{n,0}\partial s_i}$, $X^{(n,1)}(\boldsymbol{p}') = \frac{\partial^{n+1}X(\boldsymbol{p}')}{\prod_{i=1}^{n,0}\partial s_i'}$. In view of Eq. (4.10a), we can write,

$$\operatorname{cov}\left(\frac{\partial^{n+1}X(\boldsymbol{p})}{\prod_{i=1}^{n,0}\partial s_i}, \frac{\partial^{n+1}X(\boldsymbol{p}')}{\prod_{i=1}^{n,0}\partial s_i'}\right) = (-1)^{n+1}c_X^{(2n+2)}(\Delta\boldsymbol{p}) \tag{4.13}$$

(Exercise VII.12).

Example 4.6

As noted earlier, the covariance derivatives obtain a physical meaning that depends on the phenomenon under study. Let $X(\boldsymbol{p})$ be an STHS random field in $R^{n,1}$, and $\Delta\boldsymbol{p} = \boldsymbol{p} - \boldsymbol{p}'$. The following partial derivatives are defined

$$c_{X^{(1)}X}(\Delta\boldsymbol{p}) = \frac{\partial}{\partial s_i}c_X(s_1 - s_1', \ldots, s_n - s_n', s_0 - s_0') = \frac{\partial}{\partial h_i}c_X(\Delta\boldsymbol{p}),$$

$$c_{XX^{(1)}}(\Delta\boldsymbol{p}) = \frac{\partial}{\partial s_i'}c_X(s_1 - s_1', \ldots, s_n - s_n', s_0 - s_0') = -\frac{\partial}{\partial h_i}c_X(\Delta\boldsymbol{p}),$$

$$c_{X^{(1)}X^{(1)}}(\Delta\boldsymbol{p}) = \frac{\partial}{\partial s_i}\left[\frac{\partial}{\partial s_i'}c_X(s_1 - s_1', \ldots, s_n - s_n', s_0 - s_0')\right] = \frac{\partial}{\partial s_i}\left[-\frac{\partial}{\partial h_i}c_X(\Delta\boldsymbol{p})\right] = -\frac{\partial^2}{\partial h_i^2}c_X(\Delta\boldsymbol{p}),$$

(4.14a–c)

which are used in the study of ionic fluids (Dennis, 2003).

The existence of the S/TRF derivatives (m.s. sense) imposes certain requirements on the SDF of $X(\boldsymbol{p})$, and *vice versa*.

Proposition 4.2

In $R^{n,1}$, the existence condition for the derivative

$$X^{(\boldsymbol{\nu},\mu)}(\boldsymbol{p}) = \frac{\partial^{\nu+\mu}X(\boldsymbol{p})}{\partial \boldsymbol{s}^{\boldsymbol{\nu}}\partial s_0^{\mu}}$$

($\partial \boldsymbol{s}^{\boldsymbol{\nu}} = \prod_{i=1}^{n}\partial s_i^{\nu_i}$, $\sum_{i=1}^{n}\nu_i = |\boldsymbol{\nu}| = \nu$) is that the corresponding SDF ($\widetilde{c}_X$) must be such that

$$\overline{\left[X^{(\boldsymbol{\nu},\mu)}(\boldsymbol{p})\right]^2} = (-1)^{\nu+\mu}c_X^{(2\boldsymbol{\nu},2\mu)}(\boldsymbol{0},0) = \iint dkd\omega\, \boldsymbol{k}^{2\boldsymbol{\nu}}\omega^{2\mu}\widetilde{c}_X(\boldsymbol{k},\omega) < \infty. \tag{4.15}$$

The last inequality requires that $\widetilde{c}_X(\boldsymbol{k},\omega)$ decreases as $|\boldsymbol{k}|, \omega \to \infty$ faster than $|\boldsymbol{k}|^{-n-2\nu}\omega^{-1-2\mu}$.

Alternatively, in the R^{n+1} domain, if the $\boldsymbol{w}^{\boldsymbol{\nu}^*}\widetilde{c}_X(\boldsymbol{w})$ is integrable, $\boldsymbol{\nu}^* = (\boldsymbol{\nu},\mu)$, the $c_X^{(2\boldsymbol{\nu},2\mu)}(\Delta\boldsymbol{p})$ exists and is given by

$$c_X^{(2\boldsymbol{\nu},2\mu)}(\Delta\boldsymbol{p}) = (-1)^{\nu+\mu}\int d\boldsymbol{w}\boldsymbol{w}^{2(\boldsymbol{\nu},\mu)}e^{i\boldsymbol{w}\cdot\Delta\boldsymbol{p}}\widetilde{c}_X(\boldsymbol{w}), \tag{4.16a}$$

where the multi-index notation $\boldsymbol{\Delta p} = (h_1,\ldots, h_n, h_0)$, $\boldsymbol{w} = (w_1,\ldots, w_n, w_0)$, $\boldsymbol{w}^{\nu^*} = \boldsymbol{w}^{(\nu,\mu)} = \prod_{i=1}^{n} w_i^{\nu_i} w_0^{\mu}$, and $\sum_{i=1}^{n,0} \nu_i = \nu^* = \nu + \mu$ has been used. For $\boldsymbol{\Delta p} = \boldsymbol{0}$, Eq. (4.16a) reduces to

$$c_X^{(2\nu,2\mu)}(\boldsymbol{0}) = (-1)^{\nu+\mu} \int d\boldsymbol{w}\, \boldsymbol{w}^{2(\nu,\mu)} \widetilde{c}_X(\boldsymbol{w}). \tag{4.16b}$$

The expressions in $R^{n,1}$ and R^{n+1} above are equivalent. So, from Eq. (4.16a), we directly obtain Eq. (4.15) if we let $\boldsymbol{w}^{2(\nu,\mu)} = |\boldsymbol{k}|^{2\nu} \omega^{2\mu}$ and $d\boldsymbol{w} = d\boldsymbol{k}d\omega$.

Example 4.7

For a zero-mean STHS random field $X(s,t)$ in $R^{1,1}$, its space–time derivative

$$X^{(1,1)}(s,t) = \frac{\partial^2}{\partial s \partial t} X(s,t) \tag{4.17a}$$

is an STHS field too. By differentiating the integrable representation of Eq. (3.9) with $n = 1$, we find that

$$X^{(1,1)}(s,t) = \iint d\mathcal{M}_{X^{(1,1)}}(k,\omega) e^{i(ks-\omega t)}, \tag{4.17b}$$

where $d\mathcal{M}_{X^{(1,1)}}(k,\omega) = -k\omega\, d\mathcal{M}_X(k,\omega)$. Assuming that these spectral measures are differentiable, the corresponding spectral random field representations are related by

$$\widetilde{X}^{(1,1)}(k,\omega) = -k\omega\, \widetilde{X}(k,\omega). \tag{4.18}$$

The covariance, on the other hand, is given by

$$c_{X^{(1,1)}}(h,\tau) = \iint dQ_{X^{(1,1)}}(k,\omega)\, e^{i(kh-\omega\tau)}, \tag{4.19}$$

where $dQ_{X^{(1,1)}}(k,\omega) = k^2\omega^2 dQ_X(k,\omega)$. The integral of Eq. (4.19) exists if the condition

$$\iint dQ_X(k,\omega) k^2\omega^2 < \infty \tag{4.20}$$

holds. Assuming measure differentiability, Eq. (4.20) becomes

$$\iint dk d\omega\, k^2\omega^2\, \widetilde{c}_X(k,\omega) < \infty, \tag{4.21}$$

where $\widetilde{c}_{X^{(1,1)}}(k,\omega) = k^2\omega^2\, \widetilde{c}_X(k,\omega)$, which is a special case of Eq. (4.15) with $\nu = \mu = 1$. Eq. (4.21) is equivalent to the existence of the variance

$$\sigma^2_{X^{(1,1)}} = \frac{\partial^4}{\partial h^2 \partial \tau^2} c_X(h,\tau)\Big|_{h=\tau=0} < \infty, \tag{4.22}$$

which, in turn, is equivalent to the existence of $X^{(1,1)}(s,t)$ in the m.s. sense. Notice that Eq. (4.15) requires that the $\widetilde{c}_X(k,\omega)$ decreases as $k,\omega \to \infty$ faster than $k^{-1-2\nu}\omega^{-1-2\mu}$. For instance, this is valid if $X(s,t)$ has a Gaussian space–time covariance. On the other hand, if $X(s,t)$ has an exponential space–time covariance in $R^{1,1}$,

$$c_X(h,\tau) = e^{-|h|-\tau}, \tag{4.23}$$

the corresponding SDF

$$\widetilde{c}_X(k,\omega) = \frac{1}{\pi^2(1+k^2)(1+\omega^2)} \tag{4.24}$$

does not satisfy the above condition for $\nu = \mu = 1$, since the SDF falls off with k and ω as $k^{-2}\omega^{-2}$, which is slower than the required $k^{-3}\omega^{-3}$. Hence, the random field is not m.s. differentiable. For the space–time exponential covariance, this result is valid in higher dimensions. Indeed, assume that a random field $X(s,t)$ has the exponential covariance of Eq. (4.23) extended in the $R^{3,1}$ domain, i.e.,

$$c_X(r_E,\tau) = e^{-r_E-\tau}, \tag{4.25}$$

where r_E is the Euclidean spatial metric. In $R^{3,1}$ and for $\nu = \mu = 1$, the condition of Eq. (4.15) requires that the SDF decreases as $|\boldsymbol{k}|, \omega \to \infty$ faster than $|\boldsymbol{k}|^{-5}\omega^{-3}$. However, the SDF of Eq. (4.25) is

$$\widetilde{c}_X(\boldsymbol{k},\omega) = \frac{1}{\pi^3\left(1+|\boldsymbol{k}|^2\right)^2(1+\omega^2)}, \tag{4.26}$$

where $|\boldsymbol{k}| = \left(\sum_{i=1}^{3} k_i^2\right)^{\frac{1}{2}}$. This SDF falls off with $|\boldsymbol{k}|$, ω as $|\boldsymbol{k}|^{-4}\omega^{-2}$, i.e., it does not satisfy the above condition. Hence, the random field with the exponential space–time covariance is not differentiable in $R^{3,1}$.

Remark 4.1

An interesting observation can be made here. Consider the derived random field $Y(s,t) = X^{(1,1)}(s,t)$ in $R^{1,1}$ as in Example 4.7, where it is assumed that $X(s,t)$ has the exponential covariance of Eq. (4.23). In view, of Eq. (4.24), the SDF of $Y(s,t)$ is

$$\widetilde{c}_Y(k,\omega) = k^2\omega^2\widetilde{c}_X(k,\omega) = \frac{k^2\omega^2}{\pi^2(1+k^2)(1+\omega^2)} \geq 0, \tag{4.27}$$

which, though, does not satisfy Eq. (4.21). This is a warning that the condition $\widetilde{c}_Y(k,\omega) \geq 0$ by itself is not generally sufficient for the permissibility of $c_Y(h,\tau)$.

Example 4.8

For a space–time random field $X(\boldsymbol{p})$ in $R^{3,1}$ we find

$$\frac{\partial^{\nu+\mu} c_X(\boldsymbol{h},\tau)}{\partial h_1^{\nu_1}\partial h_2^{\nu_2}\partial h_3^{\nu_3}\partial\tau^{\mu}} = (-1)^{\mu} i^{\nu+\mu} \iint dkd\omega \prod_{l=1}^{3} k_l^{\nu_l}\omega^{\mu}\, e^{i(\boldsymbol{k}\cdot\boldsymbol{h}-\omega\tau)}\widetilde{c}_X(\boldsymbol{k},\omega), \tag{4.28}$$

where $\nu = \sum_{l=1}^{3} \nu_l$. In light of Eq. (4.28), if

$$\iint dkd\omega \prod_{l=1}^{3} k_l^{2\nu_l}\omega^{2\mu}\widetilde{c}_X(\boldsymbol{k},\omega) < \infty,$$

the corresponding random field derivative

$$\frac{\partial^{(\nu,\mu)} X(\boldsymbol{s},t)}{\partial s_1^{\nu_1}\partial s_2^{\nu_2}\partial s_3^{\nu_3}\partial t^{\mu}} = (-1)^{\mu} i^{\nu+\mu} \iint d\mathcal{M}_X(\boldsymbol{k},\omega) \prod_{l=1}^{3} k_l^{\nu_l}\omega^{\mu}\, e^{i(\boldsymbol{k}\cdot\boldsymbol{s}-\omega t)}$$

will exist in the m.s. sense (and all lower-order derivatives will exist too). Moreover, it holds that

$$\overline{\frac{\partial^{(\nu,\mu)}X(\boldsymbol{s},t)}{\partial s_1^{\nu_1}\partial s_2^{\nu_2}\partial s_3^{\nu_3}\partial t^{\mu}}\frac{\partial^{(\nu',\mu')}X(\boldsymbol{s}',t')}{\partial s'^{\nu'_1}_1\partial s'^{\nu'_2}_2\partial s'^{\nu'_3}_3\partial t'^{\mu'}}} = (-1)^{\nu+\mu}\frac{\partial^{\nu+\nu'+\mu+\mu'}c_X(\boldsymbol{h},\tau)}{\partial h_1^{\nu_1+\nu'_1}\partial h_2^{\nu_2+\nu'_2}\partial h_3^{\nu_3+\nu'_3}\partial\tau^{\mu+\mu'}}, \tag{4.29}$$

where $\boldsymbol{h} = \boldsymbol{s}' - \boldsymbol{s}$ was assumed.

The link between random field differentiability and SDF discussed above raises interesting questions worthy of investigation. Exercise VII.5 presents a useful list of space–time random field derivatives and the corresponding covariance and variance derivatives. Also, Exercise VII.14 is particularly instructive, since it refers to alternative approaches of calculating the covariances of random field derivatives in the m.s. sense. Lastly, in Exercise VIII.16 it is stated that in $R^{1,1}$ the SDF, $\widetilde{c}_X(k,\omega) = ck^{-\ln k}\omega^{-\ln\omega}$, satisfies the existence condition

$$\iint dkd\omega\, k^{2\nu}\omega^{2\mu}\widetilde{c}_X(k,\omega) < \infty$$

for any ν, μ; i.e., the SDF decays faster than $k^{-2\nu}\omega^{-2\mu}$.

4.2 STOCHASTIC CONTINUITY AND DIFFERENTIABILITY

To start with, I need to reemphasize that stochastic continuity and differentiability are two important notions of random field modeling. So, although these notions have already been discussed in various parts of this book (particularly, Chapter VI), it is worth summarizing the key points for the purpose of comparison:

- S/TRF continuity may be considered individually along different directions in space, in time, or in a combined manner involving several directions.
- S/TRF continuity and differentiability are closely linked.
- S/TRF continuity and differentiability in the m.s. sense are usually considered, because they are specified from the covariance functions.[28] Generally, if the covariance is continuously differentiable and, hence, continuous at the origin, the corresponding S/TRF is m.s. continuous (i.e., m.s. field differentiability implies m.s continuity).
- The m.s. properties are useful for specifying a.s. properties (e.g., many studies focus on the specification of a.s. differentiability for m.s. differentiable S/TRFs).
- The a.s. continuity and differentiability depends on the behavior of each individual sample path, which makes it more difficult to study (yet, useful a.s. continuity and differentiability conditions are available for Gaussian S/TRFs).
- Since the finite-dimensional probability distributions of an S/TRF do not determine its sample path properties, demonstrating sample path continuity and differentiability rely on the notion of S/TRF separability.[29]

[28] Or the SDF, in the case of STHS fields, as we will see below.

[29] To be distinguished from covariance separability discussed earlier.

Example 4.9

If the separability assumption holds, the random field $X(\boldsymbol{p})$ realizations are absolutely continuous and the random field derivatives exist almost everywhere, the m.s. derivative random field is equal to the sample path derivative field with probability 1. If, in addition, the random field is Gaussian, the derivative random field is Gaussian too.

As noted earlier, some special random field continuity conditions apply in the STHS case. These conditions are particularly easy to describe in the case of Gaussian random fields.

Example 4.10

As we saw in Section 3 of Chapter VI, for an STHS Gaussian random field, $X(\boldsymbol{p})\sim N(0,\sigma_X^2)$, the probability law of the corresponding random field increment, $\delta X(\boldsymbol{p})=X(\boldsymbol{p}+\Delta\boldsymbol{p})-X(\boldsymbol{p})$, is given by

$$\delta X(\boldsymbol{p})\sim N(0,\sigma_{\delta X}^2(\Delta\boldsymbol{p})),$$

where $c_X(\boldsymbol{p},\boldsymbol{p}')=c_X(\Delta\boldsymbol{p})$, $\sigma_{\delta X}^2(\Delta\boldsymbol{p})=\sigma_X^2-c_X(\Delta\boldsymbol{p})$. In this case, if the covariance is continuous at the origin, it is continuous everywhere, and the same is valid for the random field.

I reiterate that just as in general random field modeling (Chapter VI), the regularity of STHS fields may be considered in the R^{n+1} domain or in the $R^{n,1}$ domain. The focus of a regularity study in the R^{n+1} domain involves a formal mathematical analysis of regularity (viewed, more or less, as an extension of the corresponding R^n analysis), whereas in the $R^{n,1}$ domain the analysis is carried out in the Cartesian domain $R^n\times T$ in which regularity is studied analytically assuming some explicit split of time from space. The regularity of an S/TRF in R^{n+1} is related to the behavior of the covariance at the origin defined by the composite space–time lag $\Delta\boldsymbol{p}=\boldsymbol{0}$. Some regularity requirements may be less strict due to properties such as the $\Delta\boldsymbol{p}$ dependence of the covariance, and the limits depending only on whether $\boldsymbol{p}$ approaches zero with a positive or a negative sign. With this in mind, the following proposition summarizes some standard results concerning S/TRF geometry.

Proposition 4.3

Assume that $X(\boldsymbol{p})$ is an STHS random field in L_2.

(a) If its covariance function $c_X(\Delta\boldsymbol{p})$ is continuous at the space–time origin of the R^{n+1} domain, it is continuous everywhere in R^{n+1}, in which case the $X(\boldsymbol{p})$ is m.s. continuous. If $c_X(\Delta\boldsymbol{p})$ is not continuous at the origin, $X(\boldsymbol{p})$ is not m.s. continuous at any point $\boldsymbol{p}$.

(b) If the $(2\nu, 2\mu)$th-order partial covariance derivative, $c_X^{(2\nu,2\mu)}(\Delta\boldsymbol{p})$, exists and is finite at the space–time origin, then and only then the partial random field derivative $X^{(\nu,\mu)}(\boldsymbol{p})$ exists in the m.s sense. In this case, the $2(\nu+\mu)$th-order generalized derivative

$$\lim_{\Delta\boldsymbol{p},\Delta\boldsymbol{p}'\to\boldsymbol{0}}\frac{\Delta_{\Delta\boldsymbol{p}}^{\nu+\mu}\Delta_{\Delta\boldsymbol{p}'}^{\nu+\mu}c_X}{\Delta\boldsymbol{p}^{(\nu,\mu)}\Delta\boldsymbol{p}'^{(\nu,\mu)}} \tag{4.30}$$

exists for all $\Delta\boldsymbol{p}$, $\Delta\boldsymbol{p}'$

(c) As a particular case of (b) above, the derivative $X_{i,j}^{(\nu,\mu)}(\boldsymbol{p})$ exists in the m.s. sense if and only if the covariance derivative

$$c_{X_i}^{(2\nu,2\mu)}(\Delta\boldsymbol{p}) = \frac{\partial^{2(\nu+\mu)}}{\partial h_i^{2\nu}\partial h_0^{2\mu}} c_X(\Delta\boldsymbol{p}) \tag{4.31}$$

exists and is finite at $\Delta\boldsymbol{p} = \boldsymbol{0}$.

Generally, the regularity of an S/TRF in $R^{n,1}$ is related to the behavior of the covariance at the origin defined by the pair of space and time lags $\boldsymbol{h} = \boldsymbol{0}$, $\tau = 0$. Yet, under the STHS assumption some regularity requirements may be less strict due to the covariance's dependence only on the pair of space and time lags $\boldsymbol{h}$ and τ. A version of Proposition 4.3 can be stated in the domain $R^{n,1}$ by replacing the single vector $\Delta\boldsymbol{p}$ with the vector–scalar pair $\boldsymbol{h}$,τ, in which case the relevant continuity and differentiability conditions apply separately on $\boldsymbol{h}$ and τ. We may then distinguish between certain characteristic shapes of the covariance c_X near the space–time origin:

(i) If c_X is discontinuous at the space (time) origin,[30] the S/TRF is not m.s. continuous in space (time) and, thus, it exhibits a very irregular pattern; the random field becomes more irregular if the covariance is discontinuous at the space (time) origin, and then drops immediately to zero for $\boldsymbol{h} > \boldsymbol{0}$ ($\tau > 0$).[31]

(ii) If c_X behaves linearly near the space (time) origin, that is, the c_X is continuous and once differentiable, the S/TRF is continuous in the m.s. differentiable in space (time).

(iii) If the shape of c_X near the space (time) origin is parabolic, the c_X is continuous and twice differentiable in space (time), and the corresponding S/TRF is also m.s. continuous and once differentiable.

The links between the covariance function characteristics and the natural attribute's space–time variation structure (smoothness, homogeneity, anisotropy etc.) are reasonably well known, and some of them are listed above. This knowledge can be used to select the appropriate covariance model for specific applications. Thus, whether the model reliably represents the space–time variability features of the actual attribute is determined, to a considerable extent, by its geometrical features (shape, continuity, differentiability etc.). Even if it seems to fit the data closely, it is inappropriate to select a covariance model whose features correspond to a noncontinuous random field across space and/or time, and then use it to represent an attribute that has been observed to vary smoothly in space–time (the relative value of data-fitness as regards attribute representational adequacy is further discussed in Chapter XVI in a covariance model construction setting).

I will now revisit the case where physical considerations suggest to spatially and/or temporally conditionalize the analysis (Section 2.1 of Chapter I), i.e., to focus individually on the spatial and the temporal behavior of the random field. Accordingly, we may assume a fixed time (i.e., formally suppress time) and study random field continuity with respect to space, and/or to assume a fixed space (formally suppress space) and study continuity with respect to time. Then, the conclusions drawn can be interpreted separately for the spatial and for the temporal continuity.

[30]A discontinuity in the space (time) origin is sometimes called the "nugget effect" (this term is used mainly in geostatistics).
[31]This situation is sometimes called the "pure nugget effect" in geostatistics.

Example 4.11

Suppose that physical interpretation focuses on the spatial pattern of a natural attribute $X(s,t)$ in $R^{1,1}$, which requires to suppress time and consider the spatial derivative of a zero-mean STHS random field representing the rate of change of the attribute in space,

$$X^{(1,0)}(s,t) = \frac{\partial}{\partial s} X(s,t).$$

Consider the simultaneous two-point covariance function given by

$$\frac{\partial}{\partial h} c_X(h,t) = \frac{\partial}{\partial h} \overline{X(s,t)X(s+h,t)} = \overline{X(s,t)\frac{\partial}{\partial h} X(s+h,t)} = \overline{X(s,t)\frac{\partial}{\partial s} X(s+h,t)}.$$

Under these conditions, it is found that

$$\frac{\partial}{\partial h} c_X(h,t)\big|_{h=0} = \overline{X(s,t)\frac{\partial}{\partial s} X(s,t)} = \frac{1}{2}\frac{\partial}{\partial s}\overline{X^2(s,t)} = \frac{1}{2}\frac{\partial}{\partial s}\sigma_X^2 = 0,$$

since for an STHS field, σ_X^2 is constant. Hence, if $X^{(1,0)}(s,t)$ exists, i.e., the $X(s,t)$ is m.s. differentiable, then $\frac{\partial}{\partial h} c_X(h,t)\big|_{h=0} = 0$.[32]

Postulate 1.1 of Chapter VI, concerning the validity in the $R^{n,1}$ domain of certain random field results directly obtained from known results in R^n, also applies to covariance functions. The corollary below is a rather straightforward extension of a similar result for spatial random fields, taking into account the general expressions

$$\frac{\partial^{\nu+\mu} c_X(\boldsymbol{h},\tau)}{\partial h_i^{\nu}\partial \tau^{\mu}} = \overline{X(\boldsymbol{s},t)\frac{\partial^{\nu+\mu} X(\boldsymbol{s}+h_i\boldsymbol{\varepsilon}_i, t+\tau)}{\partial h_i^{\nu}\partial \tau^{\mu}}} = \overline{X(\boldsymbol{s},t)\frac{\partial^{\nu+\mu} X(\boldsymbol{s}+h_i\boldsymbol{\varepsilon}_i, t+\tau)}{\partial s_i^{\nu}\partial t^{\mu}}} = c_{XX^{(\nu,\mu)}}(\boldsymbol{h},\tau),$$

$$\frac{\partial^{\nu+\mu} c_X(\boldsymbol{h},\tau)}{\partial h_i^{\nu}\partial \tau^{\mu}}\bigg|_{\boldsymbol{h}=\boldsymbol{0},\tau=0} = c_{XX^{(\nu,\mu)}}(\boldsymbol{0},0),$$

(4.32a–b)

and the fact that the odd-ordered derivatives of covariances (which are symmetric functions) are zero.

Corollary 4.2

In view of Postulate 1.1 of Chapter VI, if the STHS random field $X(\boldsymbol{p})$ in $R^{n,1}$ is m.s. differentiable of the νth order in space, the simultaneous $(2\nu-1)$th-order spatial derivative of the corresponding covariance function is such that

$$\frac{\partial^{2\nu-1} c_X(\boldsymbol{h},t)}{\partial h_i^{2\nu-1}}\bigg|_{\boldsymbol{h}=0} = c_{XX^{(2\nu-1,0)}}(\boldsymbol{0},t) = 0, \tag{4.33}$$

where $c_{XX^{(2\nu-1,0)}}(\boldsymbol{h},t) = \overline{X(\boldsymbol{s},t)\frac{\partial^{2\nu-1} X(\boldsymbol{s}+h_i\boldsymbol{\varepsilon}_i,t)}{\partial h_i^{2\nu-1}}}$.

[32] Working long the same lines it is found that a similar result is also valid in terms of the nonsimultaneous two-point covariance.

Similar expressions are valid in terms of the separation time argument (τ) and, under certain conditions, for the combined space–time (h_i, τ) arguments. Section 2 of Chapter VIII will address the topic more fully. At the moment, a simple example may clarify some conditions as regards the applicability of Corollary 4.2.

Example 4.12

In the case of the double exponential (Gaussian) covariance model

$$c_X(h,\tau) = c_0 e^{-h^2-\tau^2}$$

in $R^{1,1}$, its third-order spatial derivative

$$c_{XX^{(3,0)}}(h,\tau) = \frac{\partial^3}{\partial h^3} c_X(h,\tau) = 4c_0 h(3-2h^2)e^{-h^2-\tau^2},$$

is equal to zero at the space origin (and, consequently, at the space–time origin), $c_{X^{(3,0)}}(0,\tau) = c_{X^{(3,0)}}(0,0) = 0$. This is expected according to Corollary 4.2, given the odd-order space derivative of the covariance. But the second-order space derivative

$$c_{XX^{(2,0)}}(h,\tau) = \frac{\partial^2}{\partial h^2} c_X(h,\tau) = 2c_0(2h^2-1)e^{-h^2-\tau^2}$$

is nonzero at the space and the time origins, $\frac{\partial^2}{\partial h^2} c_X(0,\tau), \frac{\partial^2}{\partial h^2} c_X(h,0) \neq 0$. Again, this result is expected according to Corollary 4.2, given the even-order space derivative of the covariance. Analogous conclusions can be drawn in terms of time derivatives or the combined space–time derivatives of the covariance. For illustration, the 2/1 space–time derivative

$$c_{XX^{(2,1)}}(h,\tau) = \frac{\partial^3}{\partial h^2 \partial \tau} c_X(h,\tau) = 4c_0\tau(1-2h^2)e^{-h^2-\tau^2}$$

is zero at the time and the space–time origin, $\frac{\partial^3}{\partial h^2 \partial \tau} c_X(h,0) = \frac{\partial^3}{\partial h^2 \partial \tau} c_X(0,0) = 0$, but it is nonzero at the space origin, $\frac{\partial^3}{\partial h^2 \partial \tau} c_X(0,\tau) \neq 0$. That is, the even-order space derivative is linked to the nonzero covariance value at the space origin, and the odd-order time covariance derivative is linked to the zero covariance value at the time origin. Also, the 3/1 space–time derivative

$$c_{XX^{(3,1)}}(h,\tau) = \frac{\partial^4}{\partial h^3 \partial \tau} c_X(h,\tau) = 8c_0 h\tau(2h^2-3)e^{-h^2-\tau^2}$$

is zero at the space–time origin, $\frac{\partial^4}{\partial h^3 \partial \tau} c_X(0,\tau) = \frac{\partial^4}{\partial h^3 \partial \tau} c_X(h,0) = \frac{\partial^4}{\partial h^3 \partial \tau} c_X(0,0) = 0$. That is, the odd-order space/odd-order time derivatives [which is the even (fourth)-order combined space–time derivative] of the covariance is zero at the space, time, or space–time origin. On the other hand, for the exponential covariance model

$$c_X(h,\tau) = c_0 e^{-h-\tau},$$

Eq. (4.33) is not valid, and the corresponding S/TRF is not m.s. differentiable.

Since m.s. random field geometrical properties are linked to the differentiability of the covariance function and the corresponding spectral density, the following proposition (Stein, 2005) is of theoretical interest, yet not easily implemented in practice.

Proposition 4.4

Assume that $\frac{\partial^l}{\partial w_j^l}\widetilde{c}_X(\boldsymbol{w})$ ($\boldsymbol{w} = (w_1,\ldots, w_n, w_0)$, $j = 1,\ldots, n, 0$) exists and is integrable for $l \leq k$, and that $|\boldsymbol{w}|^{\nu+\mu}\frac{\partial^k}{\partial w_j^k}\widetilde{c}_X(\boldsymbol{w})$ is integrable, $\nu = |\boldsymbol{\nu}| = \sum_{j=1}^n \nu_j$. Then, for all space–time lags $\Delta\boldsymbol{p} \neq \boldsymbol{0}$ the

$$c_X^{(\nu,\mu)}(\Delta\boldsymbol{p}) = \frac{\partial^{\nu+\mu} c_X(\Delta\boldsymbol{p})}{\partial\boldsymbol{h}^{\nu}\partial h_0^{\mu}}, \tag{4.34a}$$

where $\partial\boldsymbol{h}^{\nu} = \prod_{i=1}^n \partial h_i^{\nu_i}$ and $\boldsymbol{\nu} = (\nu_1,\ldots, \nu_n)$, exists for all $\boldsymbol{\nu}$ such that $\sum_{j=1}^{n,0} \nu_j \leq \nu + \nu_0 = \nu + \mu$ ($\nu_0 = \mu$ according to earlier notation). In addition, if $h_j \neq 0$, the $c_X^{(\nu,\mu)}(\Delta\boldsymbol{p})$ is given by the series representation

$$c_X^{(\nu,\mu)}(\Delta\boldsymbol{p}) = \sum_{\rho=0}^{\nu_j}\binom{\nu_j}{\rho} i^k(-1)^{\rho}(k)_{\rho} h_j^{-k-\rho}\int_{R^{n+1}} d\boldsymbol{w} e^{i\Delta\boldsymbol{p}\cdot\boldsymbol{w}}\frac{(i\boldsymbol{w})^{(\nu,\mu)}}{(iw_j)^{\rho}}\frac{\partial^k}{\partial w_j^k}\widetilde{c}_X(\boldsymbol{w}), \tag{4.34b}$$

where $(k)_{\rho} = k(k+1)\ldots(k+\rho-1)$, $(k)_0 = 1$.

Example 4.13

If in $R^{1,1}$ the SDF is of the form

$$\widetilde{c}_X(w_1, w_0) = \left[\left(1+w_1^2\right)\left(1+w_1^2+w_0^2\right)\right]^{-1},$$

the covariance derivative $c_X^{(\nu,0)}(h_1, h_0)$ exists, whereas the $c_X^{(0,\mu)}(h_1, h_0)$ does not. If in $R^{n,1}$ the SDF is of the form

$$\widetilde{c}_X(\boldsymbol{k}, \omega) = \left[\eta_1\left(\boldsymbol{k}^2\right) + \eta_2\left(\omega^2\right)\right]^{-\nu},$$

with $\nu > 0$ and properly selected polynomials η_1 and η_2 (for the detailed conditions, see Proposition 4.6), all partial derivatives of all orders of the corresponding covariance $c_X(\boldsymbol{h},\tau)$ exist for $(\boldsymbol{h},\tau) \neq (\boldsymbol{0},0)$. As an illustration, the model no. 30 of Table 5.1 of Chapter VIII corresponds to an S/TRF that is once m.s. differentiable along any spatial direction (time origin), and not differentiable in time (space origin), although the model is infinitely differentiable away from the origin.

Once more, the hypothesis of space–time covariance separability (Section 3.5) has some interesting consequences as regards the study of random field regularity properties. Taking advantage of the separability hypothesis, we can conclude that when the decomposition

$$c_X(\boldsymbol{h}, \tau) = c_{X(1)}(\boldsymbol{h}) c_{X(2)}(\tau)$$

is valid, random field continuity results may be derived separately for the space and time domains, and subsequently combined, as appropriate. In many cases, it makes sense to study the covariances $c_{X(1)}(\boldsymbol{h})$ and $c_{X(2)}(\tau)$ separately. So, we can consider time regularity by fixing the spatial lag argument $\boldsymbol{h}$ in the purely spatial covariance component $c_{X(1)}(\boldsymbol{h})$ and focusing on $c_{X(2)}(\tau)$. Similarly, we can study space regularity by fixing the time separation argument τ in the purely temporal component $c_{X(2)}(\tau)$ and focusing on $c_{X(1)}(\boldsymbol{h})$. Moreover, for a fixed τ, for $c_X(\boldsymbol{h},\tau)$ to be a permissible covariance function of $\boldsymbol{h}$, sometimes denoted as $c_{X,\tau}(\boldsymbol{h})$, it must be that $c_{X(2)}(\tau) \geq 0$. Assuming that the last condition holds, we can focus on $c_{X(1)}(\boldsymbol{h})$. Obviously, a similar approach applies if $\boldsymbol{h}$ is assumed fixed, in which case for $c_X(\boldsymbol{h},\tau)$ to be a permissible covariance function of τ, $c_{X,\boldsymbol{h}}(\tau)$, it must hold that $c_{X(1)}(\boldsymbol{h}) \geq 0$.

Example 4.14

Consider a zero-mean random field $X(s,t)$ that is STHS and m.s. differentiable in $R^{1,1}$ with a separable covariance, $c_X(h,\tau) = c_{X(1)}(h)c_{X(2)}(\tau)$. According to the discussion above, we can study $c_{X(1)}(h)$ and $c_{X(2)}(\tau)$ separately. If the analysis involves only spatial operations, we can focus on $c_{X(1)}(h)$ (the $c_{X(2)}(\tau)$ will remain unchanged). For illustration, consider spatial differentiation, i.e., $c_X^{(1,0)}(h,\tau) = c_{X(2)}(\tau)c_{X(1)}^{(1)}(h)$, or

$$\frac{\partial c_X(h,\tau)}{\partial h} = c_{X(2)}(\tau)\frac{dc_{X(1)}(h)}{dh},$$

i.e., the partial derivative of $c_X(h,\tau)$ is equal to the ordinary derivative of $c_{X(1)}(h)$ times $c_{X(2)}(\tau)$, and similarly for higher-order derivatives. By definition, $c_{X(1)}(h)$ is an even function, which means that $\frac{dc_{X(1)}(h)}{dh}$ is an odd function. Also, by virtue of Eq. (4.33), it is true that

$$\frac{dc_{X(1)}(h)}{dh}\Big|_{h=0} = c_{X(1)}^{(1)}(0) = 0, \tag{4.35a}$$

which, in view of the above comments, also implies that

$$\frac{dc_X(h,\tau)}{dh}\Big|_{h=0} = 0. \tag{4.35b}$$

The second-order generalized derivative is[33]

$$\begin{aligned}\lim_{h_1,h_1'\to 0}\frac{\Delta_{h_1}^1\Delta_{h_1'}^1 c_{X(1)}(h)}{h_1h_1'} &= \lim_{h_1,h_1'\to 0}\frac{1}{h_1'}\left[\frac{c_{X(1)}(h_1-h_1')-c_{X(1)}(-h_1')}{h_1}-\frac{c_{X(1)}(h_1)-c_{X(1)}(0)}{h_1}\right]\\ &= \lim_{h_1'\to 0}\frac{1}{h_1'}\left[\frac{\partial}{\partial h_1}c_{X(1)}(-h_1')-\frac{\partial}{\partial h_1}c_{X(1)}(0)\right] = -\frac{\partial^2}{\partial h^2}c_{X(1)}(0),\end{aligned} \tag{4.35c}$$

which is finite. Hence, the second-order generalized derivative exists at all $h \in R^1$. If $X^{(1)}(s)$ exists,

$$\frac{\partial c_X(h,\tau)}{\partial h} = c_{X(1)^{(1)}X(1)}(h)c_{X(2)}(\tau),\ \text{with}\ c_{X(1)^{(1)}X(1)}(h) = \frac{\partial}{\partial h}\overline{X(s'+h)X(s')} = \frac{dc_{X(1)}(h)}{dh},$$

[33]Since the covariance $c_{X(1)}(h)$ is an even function, the expression is independent of the order in which the intervals are taken, e.g., this limit is equivalent to $\lim_{h_1,h_1'\to 0}\frac{1}{h_1}\left[\frac{c_{X(1)}(h_1-h_1')-c_{X(1)}(h_1)}{h_1'}-\frac{c_{X(1)}(-h_1')-c_{X(1)}(0)}{h_1'}\right]$, etc.

and, since the $\sigma^2_{X(1)}$ is constant,

$$\frac{dc_{X(1)}(h)}{dh}\Big|_{h=0} = \frac{1}{2}\frac{d}{ds'}\overline{X(s')^2} = \frac{1}{2}\frac{\partial \sigma^2_{X(1)}}{\partial s'} = 0,$$

i.e., the same result as Eq. (4.35a). The readers may recall that the existence of $c^{(2)}_{X(1)}(h)$ is a necessary and sufficient condition for the existence of $X^{(1)}(s)$. Now, for $c^{(2)}_{X(1)}(h)$ to be continuous for all h, it is only needed that it is so at $h = 0$ (and, of course, Eq. 4.35a must hold). To see if this condition applies, I start from

$$\sigma^2_{X^{(1)}(1)} = -c^{(2)}_{X(1)}(0) = -\frac{d^2}{dh^2}c_{X(1)}(h)\big|_{h=0},$$

and notice that since $\sigma^2_{X^{(1)}(1)} > 0$, it must hold that $c^{(2)}_{X(1)}(0) < 0$. For the last inequality to be satisfied, the covariance must have a maximum at $h = 0$, i.e., $c^{(1)}_{X(1)}(0) = 0$. By way of a summary, at $h = 0$, the covariance expressions are

$$\begin{aligned}
c_{X^{(1)}(1)X(1)}(0) &= -c_{X(1)X^{(1)}(1)}(0) = \frac{dc_{X(1)}(h)}{dh}\Big|_{h=0} = 0,\\
c_{X^{(1)}(1)X^{(1)}(1)}(0) &= -\frac{d^2c_{X(1)}(h)}{dh^2}\Big|_{h=0} = \sigma^2_{X^{(1)}(1)} > 0,\\
c_{X^{(2)}(1)X^{(2)}(1)}(0) &= \frac{d^4c_{X(1)}(h)}{dh^4}\Big|_{h=0} = \sigma^2_{X^{(2)}(1)} > 0,
\end{aligned} \tag{4.36a–c}$$

which are frequently used in applications involving space–time separable covariances. Similar results can be derived for the purely temporal covariance $c_{X(2)}(\tau)$, and then properly combined with those for $c_{X(1)}(h)$ above.

The notions of equivalence and modification were introduced in Section 3 of Chapter VI, because sample path features, such as continuity, are of considerable importance in applications. The significance of the above notions lies in the fact that two equivalent random fields do not always have the same sample path features. In the same chapter it was suggested that it may be appropriate that all results concerned with sample path features clarify that, "there exists a version or a modification of the random field under consideration that has the stated sample path features." It was also shown in Chapter VI that the multivariate cumulative distribution function (CDF) does not necessarily determine the continuity features of a random field, and that there are cases in which it cannot be concluded that the random field of interest has the specified sample path features. The a.s. continuity does not guarantee sample path continuity, for which additional conditions are required. Although comprehensive conditions may not be available in the general case (i.e., the case of non-Gaussian random fields in R^{n+1}), one well-known condition is sample path separability, which ensures that finite-dimensional CDF determines sample path features by requiring that sample paths are determined by their values on an everywhere dense, countable set of points.

Also, as was noted (Postulate 1.1 and Remark 3.2 of Chpater VI), although certain random field geometry results available in the literature were derived for spatial random fields (e.g., Kent, 1989; Adler and Taylor, 2007; and references therein), they can be extended to space–time random fields under either of the following two conditions: (i) the derivation of these results that involved only spatial operations (e.g., spatial derivatives) carry through as before if the time argument does not change (i.e., random fields that are also functions of time may be also considered, if as long as time is held constant the derivations are unchanged), or (ii) a space—time metric of a specific form is assumed (e.g., the random field geometry results involve a metric of the general space—time Pythagorean form). With this in mind, the next proposition provides some sufficient conditions concerning the sample path continuity of STHS random fields.[34]

Proposition 4.5

Let $X(\boldsymbol{s},t)$ be a zero mean random field in $R^{n,1}$. Assuming that time does not change during the derivations (i.e., condition (i) applies):

(a) If the simultaneous two-point covariance $c_X(\boldsymbol{h},t)$ is an n-times continuously differentiable function in space, and for some $\alpha > 0$ it holds that

$$\left|c_X(\boldsymbol{h},t) - \eta_{n,t}(\boldsymbol{h})\right| = O_t\left(\left|\boldsymbol{h}\right|^{n+\alpha}\right), \tag{4.37}$$

as $\boldsymbol{h} \to \boldsymbol{0}$, where $\eta_{n,t}$ is a Taylor polynomial of degree n with time-dependent coefficients, in general (as denoted by the subscript t), then there exists a version of the random field with continuous realizations. In addition, since $c_X(\boldsymbol{h},t)$ is continuous at $\boldsymbol{h} = \boldsymbol{0}$, the random field $X(\boldsymbol{s},t)$ is m.s. continuous in space.

(b) If $X(\boldsymbol{s},t)$ is a centered Gaussian random field with a continuous covariance, the validity of the inequality

$$\overline{\left|X(\boldsymbol{s}+\boldsymbol{h},t) - X(\boldsymbol{s},t)\right|^2} \leq \frac{\alpha_t}{\left|\log\left|\boldsymbol{h}\right|\right|^{1+\beta}}, \tag{4.38}$$

for some $\alpha_t < \infty$ and some $\beta > 0$ implies that the random field $X(\boldsymbol{s},t)$ has continuous sample paths with probability 1. Given that the denominator in the right side of Eq. (4.38) tends to ∞ as $\boldsymbol{h} \to \boldsymbol{0}$, and that m.s. continuity requires that the left side of Eq. (4.38) tends to 0 as $\boldsymbol{h} \to \boldsymbol{0}$, for Eq. (4.38) to hold it is required that the latter happens at least as fast as the former. For a homogeneous Gaussian random field, Eq. (4.38) can be written as

$$c_X(\boldsymbol{0},t) - c_X(\boldsymbol{h},t) \leq \frac{\alpha_t}{\left|\log\left|\boldsymbol{h}\right|\right|^{1+\beta}}, \tag{4.39}$$

for some $\alpha_t < \infty$ and some $\beta > 0$ (Eq. (4.39) is also obtained from Eq. (3.18) of Chapter VI assuming STHS).

An example may illustrate some conditions as regards the applicability of the theoretical results presented in Proposition 4.5.

[34]Proposition 4.4 is essentially a consequence of Proposition 3.2 of Chapter VI.

Example 4.15

In the case of a space–time separable covariance $c_X(\boldsymbol{h},\tau) = c_{X(1)}(\boldsymbol{h})c_{X(2)}(\tau)$, Eq. (4.37) holds with $\eta_{n,t}(\boldsymbol{h}) = \eta_n(\boldsymbol{h})c_{X(2)}(\tau)$ and $O_t\left(|\boldsymbol{h}|^{n+\alpha}\right) = O\left(|\boldsymbol{h}|^{n+\alpha}\right)c_{X(2)}(\tau)$. In a similar way, it is found that in the case of Eq. (4.39) we have $\alpha_t = \alpha c_{X(2)}(\tau)$. (Exercise VII.23).

Remark 4.1

Taking into consideration Postulate 1.1 of Chapter VI, other mathematical expressions of the conditions presented in Proposition 4.5 are generally possible. Some of them are briefly examined here.

(a) Certain important random field continuity results obtained by Kent (1989) in the R^n domain may be extended in the $R^{n,1}$ domain in terms of partial covariance derivatives. Specifically, if for some $\alpha > 0$ it is valid that

$$\left|\frac{\partial^n}{\partial \boldsymbol{h}^n}\left(c_X(\boldsymbol{h},t) - \eta_{n+1,t}(\boldsymbol{h})\right)\right| = O_t\left(|\boldsymbol{h}|^{-3-\alpha}\right)$$

as $\boldsymbol{h} \to \boldsymbol{0}$, where $\partial \boldsymbol{h}^n = \prod_{i=1}^n \partial h_i$, there exists a version of the random field with continuous realizations.

(b) In the case of a Gaussian random field in the $R^{n,1}$ domain, as a special case of the Kolmogorov's continuity theorem, if there exists an $\alpha > 0$ and a constant $C_t > 0$ such that

$$\overline{|X(\boldsymbol{s}+\boldsymbol{h},t) - X(\boldsymbol{s},t)|^2} \leq C_t|\boldsymbol{h}|^{2\alpha} \tag{4.40}$$

is valid, then for every $\alpha' < \alpha$, the $X(\boldsymbol{s},t)$is a.s. α'-Hölder continuous.[35]

(c) Proposition 4.5 may be expressed in the R^{n+1} domain if a suitable metric is assumed (e.g., $|\Delta \boldsymbol{p}| = r_p$). In this case, if the covariance is a continuously differentiable function, and for some $\alpha > 0$ it holds that

$$|c_X(\Delta \boldsymbol{p}) - \eta_{n+1}(\Delta \boldsymbol{p})| = O\left(|\Delta \boldsymbol{p}|^{n+1+\alpha}\right),$$

as $\Delta \boldsymbol{p} \to \boldsymbol{0}$, then there exists a version of the random field with continuous realizations. Similar modification in the R^{n+1} domain is valid for Eqs. (4.38) and (4.39).

In certain applications, it may be appropriate to isolate the study of random field time regularity by setting $\boldsymbol{s} = \boldsymbol{s}'$ (or $\boldsymbol{h} = \boldsymbol{0}$) in the covariance function, and that of space regularity by letting $t = t'$ (or $\tau = 0$), highlighting similarities and differences as they emerge. This perspective acknowledges the essential role of physical evidence and interpretation within the modeling process. An illustration is given in the example below.

Example 4.16

Many marine observing systems allow environmental sensing and nearly real-time data streaming at selected monitoring locations. The data time resolution and space coverage offer improved

[35] If $\alpha > 0$, the condition implies that the function is continuous; if $\alpha = 0$ the function need not be continuous, but it is bounded.

understanding and managing of marine systems in coastal oceans. Suppose that the space–time covariance of a water quality attribute in $R^{1,1}$ is given by

$$c_X(h,t,t') = \frac{1}{4}\int_{|t-t'|}^{t+t'} du\, u^{-1/2} e^{-\frac{h^2}{4u}},$$

(Exercise VII.27). The regularity of the corresponding Gaussian random field $X(s,t)$ is determined by the behavior of the covariance at $(s,t)=(s',t')$. To study time regularity, let $h=0$, in which case the covariance expression gives

$$c_X(0,t,t') = \frac{1}{4}\int_{|t-t'|}^{t+t'} du\, u^{-1/2} = \frac{1}{2}\left(|t+t'|^{1/2} - |\tau|^{1/2}\right),$$

where $|\tau| = |t-t'|$. For $t \to t'$, $\overline{|X(0,t') - X(0,t)|^2} = |\tau|^{1/2}$. Hence, in light of Eq. (4.40) the random field is a.s. α'-Hölder continuous in time with $\alpha' < \frac{1}{4}$. Similarly, for space regularity, let $t = t'$ (or $\tau = 0$), in which case the covariance expression gives

$$c_X(h,t,t) = \frac{1}{4}\int_0^{2t} du\, u^{-1/2} e^{-\frac{h^2}{4u}} = t^{1/2} + \frac{\pi^{1/2}}{4}|h| + O\left(\frac{h^2}{8} t^{-1/2}\right),$$

i.e., $\overline{|X(h',t) - X(h,t)|^2} = \frac{\pi^{1/2}}{2}|h|$ for $h \to h'$, and, hence, the random field is a.s. α'-Hölder continuous in space with $\alpha' < \frac{1}{2}$.

The sufficient condition of Proposition 4.5 (case a) for a.s. sample path continuity is independent of the shape of the probability law. Even though the condition of Eq. (4.37) is very sharp, significantly more covariance regularity is required compared to Eq. (4.38) to ensure the same degree of sample path smoothness. By virtue of Proposition 4.5 (case b), sample path smoothness is primarily determined by covariance smoothness. The Gaussian assumption is critical here. If $X(\boldsymbol{p})$ is, e.g., an STHS Poisson field, it has discontinuous sample paths although the condition is satisfied.

Another practical consequence of Eq. (4.39) is as follows. If for the simultaneous two-point covariance function of an STHS Gaussian random field $X(\boldsymbol{p})$ it is valid that

$$\frac{\alpha_{1,t}}{(-\log|\boldsymbol{h}|)^{1+\beta_1}} \le c_X(\boldsymbol{0},t) - c_X(\boldsymbol{h},t) \le \frac{\alpha_{2,t}}{(-\log|\boldsymbol{h}|)^{1+\beta_2}} \tag{4.41}$$

for $|\boldsymbol{h}|$ small enough, then the random field will be sample path continuous in space if $\beta_2 > 0$ and discontinuous if $\beta_1 < 0$.

Example 4.17

Eq. (4.39) is valid for many covariance models $c_X(\boldsymbol{h})c_X(t)$, including the class of power exponential functions

$$c_X(\boldsymbol{h}) = e^{-\left(\frac{\boldsymbol{h}}{a}\right)^{\lambda}},$$

where $\lambda \in (0,2]$. Also, consider the random field $X(s,t)$, $s \in [a,b]$. The field is a.s. space continuous if $\overline{|X(s+h,t)-X(s,t)|^{\kappa}} \leq \alpha_t |h|^{1+\varepsilon}$, for some α_t, κ, $\varepsilon > 0$ and sufficiently small h (Exercise VII.16).

If $g(\bullet)$: $L_2 \rightarrow R^1$ is a continuous function and $X(\boldsymbol{p})$ is an a.s. continuous random field, the random field

$$Y(\boldsymbol{p}) = g(X(\boldsymbol{p})) \tag{4.42}$$

is also a.s. continuous. Matters are not that straightforward as regards m.s. continuity. In fact, by extending in the R^{n+1} domain a result by Banerjee and Gelfand (2003), we find that the $g(\bullet)$: $R^1 \rightarrow R^1$ must be a continuous Lipschitz of order 1 function,[36] and $X(\boldsymbol{p})$ an m.s. continuous random field, for the resulting field $Y(\boldsymbol{p})$ of Eq. (4.42) to be m.s. continuous too (Exercise VII.17).

Example 4.18

Interesting cases of $g(\bullet)$ functions are the finite summations of m.s. (resp. a.s.) continuous random fields $X_i(\boldsymbol{p})$ multiplied by continuous deterministic weighting functions, in which case the resulting random fields $Y(\boldsymbol{p})$ are also m.s (resp. a.s.) continuous (Exercise VII.18). The result remains basically valid if the summation is replaced by integration.

The random field regularity (continuity and differentiability) criteria discussed above in a theoretical setting and under rather strict conditions should be tested systematically for sensitivity and robustness across a range of application contexts. For sure, to select a criterion for a specific use, we must have a fairly complete understanding of the underlying assumptions and applicability range, and also an appreciation of the interpretability of the conclusions to be drawn by means of the criterion.

4.3 SPATIOTEMPORAL RANDOM FIELD INTEGRABILITY

I will now briefly revisit the topic of random field integrability, initially introduced in Section 6 of Chapter VI. As regards the stochastic integration of STHS random fields, in particular, the condition expressed by Eq. (6.4b) of Chapter VI becomes

$$\overline{Z^2(\boldsymbol{p})} = \int_V \int_V d\boldsymbol{u} d\boldsymbol{u}' \alpha(\boldsymbol{u},\boldsymbol{p})\, \alpha^*(\boldsymbol{u}',\boldsymbol{p}) c_X(\boldsymbol{u}'-\boldsymbol{u}) < \infty, \tag{4.43}$$

where $\boldsymbol{p} = (s_1,\ldots, s_n, s_0)$ and $V \subseteq R^{n+1}$. The function can take many forms, depending on the goals of the analysis or the in situ conditions of the attribute.

Example 4.19

Assuming $\alpha(\boldsymbol{u},\boldsymbol{p}) = 1$, the stochastic integral of Eq. (6.1) of Chapter VI and the condition of Eq. (4.43) can be written as

$$Z = \int d\boldsymbol{p}\, I(\boldsymbol{p}) X(\boldsymbol{p}), \tag{4.44}$$

[36]That is, $|g(X(\boldsymbol{p}+\Delta\boldsymbol{p})) - g(X(\boldsymbol{p}))| \leq c|X(\boldsymbol{p}+\Delta\boldsymbol{p}) - X(\boldsymbol{p})|$, where c is a constant.

and

$$\overline{Z^2} = \int_V d\Delta\boldsymbol{p}\,\Lambda(\Delta\boldsymbol{p})c_X(\Delta\boldsymbol{p}) < \infty, \tag{4.45}$$

respectively, where

$$I(\boldsymbol{p}) = \begin{cases} 1 & \text{if } \boldsymbol{p} \in V \\ 0 & \text{otherwise} \end{cases}$$

is the indicator function, and $\Lambda(\Delta\boldsymbol{p}) = \int d\boldsymbol{p}\, I(\boldsymbol{p})I(\boldsymbol{p}+\Delta\boldsymbol{p})$. If $c_X(\Delta\boldsymbol{p})$ is continuous in R^{n+1}, the stochastic integral of Eq. (4.45) exists in the m.s. sense.

The following result is valid concerning the integrability of the SDF $\widetilde{c}_X(\boldsymbol{k},\omega)$ in $R^{n,1}$ (Stein, 2005).

Proposition 4.6

Consider a bounded SDF of the form

$$\widetilde{c}_X(\boldsymbol{k},\omega) = \left[\eta_1\left(\boldsymbol{k}^2\right) + \eta_2\left(\omega^2\right)\right]^{-\nu}, \tag{4.46}$$

where $\nu > 0$, and η_1 and η_2 are nonnegative polynomials on $[0, \infty)$ of positive degree α_1 and α_2, respectively. Then $\widetilde{c}_X(\boldsymbol{k},\omega)$ is integrable if and only if $\frac{n}{\alpha_1\nu} + \frac{1}{\alpha_2\nu} < 2$ and, if so, its FT, i.e., the covariance, $c_X(\boldsymbol{h},\tau)$, possesses all partial derivatives of all orders for $(\boldsymbol{h},\tau) \neq (\boldsymbol{0},0)$.

In addition, Proposition 4.6 describes a general SDF for which the conditions of Proposition 4.3 concerning the existence of $c_X^{(\nu,\mu)}(\Delta\boldsymbol{p})$ hold.

5. SPECTRAL MOMENTS AND LINEAR RANDOM FIELD TRANSFORMATIONS

I return now to the spectral moments introduced in earlier chapters. Interpretive physical insights can be extended and useful expressions for m.s. differentiable STHS random fields can be obtained in terms of these moments.

Definition 5.1

The *spectral moments* of an STHS random field are defined as follows

$$\beta_X^{(\nu,\mu,\nu',\mu')} = \int d\boldsymbol{k} \int d\omega \prod_{\lambda=1}^{m} k_\lambda^{\nu_\lambda}\omega^{\mu+\mu'}\widetilde{c}_X(\boldsymbol{k},\omega), \tag{5.1}$$

where $\sum_{\lambda=1}^{m}\nu_\lambda = \nu + \nu'$.

Under the homostationarity and differentiability conditions associated with Eq. (4.10a), the validity of the following proposition is rather straightforward.

Proposition 5.1

The space–time variance of the random field $X(\boldsymbol{p})$ derivatives are expressed in terms of the spectral moments by

$$\left.\frac{\partial^{\nu+\nu'+\mu+\mu'}c_X(\boldsymbol{h},\tau)}{\partial\boldsymbol{h}^{\nu+\nu'}\partial\tau^{\mu+\mu'}}\right|_{\boldsymbol{h}=\boldsymbol{0},\tau=0} = (-1)^{\mu+\mu'}\imath^{\nu+\nu'+\mu+\mu'}\beta_X^{(\nu,\nu',\mu,\mu')} \tag{5.2}$$

In physical applications characterized by simultaneous spatial variation, Eq. (5.2) reduces to an expression in terms of the wavenumber–time spectrum $\widetilde{c}_X(\boldsymbol{k},t)$.

Corollary 5.1

The simultaneous spatial variance of the random field $X(\boldsymbol{p})$ derivatives is expressed in terms of the spectral moments by

$$\begin{aligned}\beta_X^{(\nu,0,\nu',0)} &= \int d\boldsymbol{k}\, \widetilde{c}_X(\boldsymbol{k},\tau)\boldsymbol{k}^{\nu+\nu'} \\ &= i^{-(\nu+\nu')}\frac{\partial^{\nu+\nu'} c_X(\boldsymbol{h},t)}{\partial \boldsymbol{h}^{\nu+\nu'}}\bigg|_{\boldsymbol{h}=0}\end{aligned} \qquad (5.3\text{a–b})$$

Given the symmetry of the spectrum, if $\nu+\nu'$ is an odd number, it is valid that

$$\beta_X^{(\nu,0,\nu',0)} = 0. \qquad (5.3\text{c})$$

Eq. (5.3c) is an interesting result that has been proven to be useful in the study of random field continuity and differentiability. In the following example, some useful spectral moments and their interpretations in ocean studies are discussed.

Example 5.1

In the case of spectral representations frequently used in ocean sciences, see Example 3.1, the spectral moments can be written in $R^{2,1}$ as

$$\beta_X^{(\nu,\mu,\nu',\mu')} = \int\int\int dk_1 dk_2 d\omega k_1^{\nu} k_2^{\nu'} \omega^{\mu+\mu'} \widetilde{c}_X(k_1,k_2,\omega), \qquad (5.4)$$

where $\boldsymbol{h} = (h_1,h_2)$, and

$$\text{cov}\left(X^{(\nu,\mu)}(\boldsymbol{p}), X^{(\nu',\mu')}(\boldsymbol{p}')\right) = (-1)^{\nu+\mu}\frac{\partial^{\nu+\nu'+\mu+\mu'} c_X(\boldsymbol{h},\tau)}{\partial h_1^{\nu}\partial h_2^{\nu'}\partial \tau^{\mu+\mu'}}.$$

As should be expected, for $\nu=\nu'$, $\mu=\mu'$, $\boldsymbol{p}=\boldsymbol{p}'$, the above equation gives Eq. (4.15). Another interesting expression is obtained if we assume that $\nu+\nu'=2\lambda$ and $\mu=\mu'=0$, in which case

$$\beta_X^{(\nu,0,\nu',0)} = \beta_X^{(\nu,\nu',0)} = (-1)^{\lambda}\frac{\partial^{2\lambda} c_X(\boldsymbol{h},\tau)}{\partial h_1^{\nu}\partial h_2^{\nu'}}\bigg|_{\boldsymbol{h}=\boldsymbol{0}}. \qquad (5.5)$$

We can also write that

$$c_X(\Delta\boldsymbol{p}) = c_X(\boldsymbol{h},\tau) = \text{Re}\int\int dkd\omega\, \widetilde{c}_X(\boldsymbol{k},\tau)e^{i(\boldsymbol{k}\cdot\boldsymbol{h}-\omega\tau)} = \int\int dkd\omega\, \widetilde{c}_X(\boldsymbol{k},\tau)\cos(\boldsymbol{k}\cdot\boldsymbol{h}-\omega\tau), \qquad (5.6)$$

where $\boldsymbol{h} = (h_1,h_2)$, $\tau = h_0$, and $\boldsymbol{k} = (k_1,k_2)$. In the following, when two space dimensions but no time are considered in the spectral moments, for simplicity the notation $\beta_X^{(\nu,\nu',0)}$ will be used, where ν and ν' refer to h_1 and h_2, respectively, and 0 denotes that no time derivative is considered. A testimony to the interpretive value of these spectral moments is the fact that some important oceanographic conditions can be expressed in terms of the spectral moments, in particular:

(a) For the wave energy to travel in one direction, the SDF of Eq. (5.6) should be reduced to a single unidimensional spectral function. For this to happen it must be valid that

$$\beta_X^{(2,0,0)}\beta_X^{(0,2,0)} = \left(\beta_X^{(1,1,0)}\right)^2. \tag{5.7}$$

(b) For the wave energy to consist of wave components of the same length but possibly different directions we must have

$$\left(\beta_X^{(4,0,0)} + 2\beta_X^{(2,2,0)} + \beta_X^{(0,4,0)}\right)\beta_X^{(0,0,0)} = \left(\beta_X^{(2,0,0)} + \beta_X^{(0,2,0)}\right)^2. \tag{5.8}$$

(c) For the wave energy to be situated at two diametrically opposite points of the spectrum, Eqs. (5.7) and (5.8) must be satisfied simultaneously. And, for the wave energy to be concentrated on lines parallel to the k_1 and the k_2 axis, the conditions are

$$\begin{aligned} \beta_X^{(0,2,0)}\beta_X^{(0,0,0)} &= \left(\beta_X^{(0,1,0)}\right)^2, \\ \beta_X^{(2,0,0)}\beta_X^{(0,0,0)} &= \left(\beta_X^{(1,0,0)}\right)^2, \end{aligned} \tag{5.9a–b}$$

respectively.

Example 5.2

Consider in R^{n+1} the case, $i, j, k, l = 0, 1, \ldots, \nu$. Then,

$$\begin{aligned} \beta_X^{(\nu_i,\nu_j,\nu_k,\nu_l,0)} &= \int dQ_X(\mathbf{w}) w_i^{\nu_i} w_j^{\nu_j} w_k^{\nu_k} w_l^{\nu_l} \\ &= i^{-(\nu_i+\nu_j+\nu_k+\nu_l)} \frac{\partial^{\nu_i+\nu_j+\nu_k+\nu_l} c_X(\mathbf{p})}{\partial s_i^{\nu_i}\partial s_j^{\nu_j}\partial s_k^{\nu_k}\partial s_l^{\nu_l}}\bigg|_{\mathbf{p}=0} \\ &= (-1)^{-(\nu_i+\nu_j)} i^{-(\nu_i+\nu_j+\nu_k+\nu_l)} \overline{\frac{\partial^{\nu_i+\nu_j} X(\mathbf{p})}{\partial s_i^{\nu_i}\partial s_j^{\nu_j}} \frac{\partial^{\nu_k+\nu_l} X(\mathbf{p})}{\partial s_k^{\nu_k}\partial s_l^{\nu_l}}}, \end{aligned} \tag{5.10a–c}$$

which involves only even spectral moments—all odd spectral moments and odd covariance derivatives are zero.

Another interesting feature of the spectral moments is that they are directly linked to the random field variance and the statistical moments of its derivatives.

Example 5.3

In $R^{2,1}$ with $\mathbf{p} = (s_1,s_2,t)$, the spectral moments of the random field $X(\mathbf{p})$ representing ocean wave heights reduce to the unidimensional integral of the simultaneous two-point spectrum, i.e.,

$$\beta_X^{(\nu_1,\nu_2,0)} = \int dk_1 \int dk_2\, k_1^{\nu_1} k_2^{\nu_2} \tilde{c}_X(k_1, k_2, t). \tag{5.11}$$

The variance of the zero mean random field $X(s_1,s_2,t)$ is

$$\overline{X^2(s_1,s_2,t)} = \iint dk_1 dk_2\, \tilde{c}_X(k_1,k_2,t) = c_X(0,0,t). \tag{5.12}$$

Let $X^{(1,0,0)}(s_1,s_2,t) = \frac{\partial X(s_1,s_2,t)}{\partial s_1}$.[37] The corresponding wavevector spectra are linked by $\tilde{c}_{X^{(1,0,0)}}(k_1,k_2,t) = k_1^2 \tilde{c}_X(k_1,k_2,t)$, and the covariance of $X^{(1,0,0)}(s_1,s_2,t)$ is

$$\overline{X^{(1,0,0)}(s_1,s_2,t)X^{(1,0,0)}(s_1',s_2',t)} = c_{X^{(1,0,0)}}(h_1,h_2,t) = -c_X^{(2,0,0)}(h_1,h_2,t) \tag{5.13}$$

(it should be reminded that the signs of the covariance differentiation equations depend on the sign of the spatial and/or temporal lag assumed). For the wavevector–temporal spectra we observe that $k_1^2\, \tilde{c}_X(k_1,k_2,t) = -(ik_1)^2 \tilde{c}_X(k_1,k_2,t)$ is minus the spectrum of $c_X^{(2,0,0)}(h_1,h_2,t) = \frac{\partial^2 c_X(h_1,h_2,t)}{\partial h_1^2}$, i.e.,

$$\iint dk_1 dk_2 k_1^2\, \tilde{c}_X(k_1,k_2,t) = \beta_X^{(2,0,0)} = -\frac{\partial^2 c_X(h_1,h_2,t)}{\partial h_1^2}\Big|_{h_1=h_2=0}. \tag{5.14}$$

We can also express other covariance functions in terms of the spectral moments. Indeed, the covariance function

$$\overline{X(s_1,s_2,t)\frac{\partial X(s_1',s_2',t)}{\partial s_1'}} = \frac{\partial c_X(h_1,h_2,t)}{\partial h_1} = c_X^{(1,0,0)}(h_1,h_2,t)$$

is proportional to $\iint dk_1 dk_2\, k_1\, \tilde{c}_X(k_1,k_2,t) = \beta_X^{(1,0,0)}$. Given that $\tilde{c}_X(k_1,k_2,t)$ is an even function and the power of k_1 is odd, we have

$$\beta_X^{(1,0,0)} = 0. \tag{5.15}$$

On the other hand, for the covariance

$$\overline{X(s_1,s_2,t)\frac{\partial^2 X(s_1',s_2',t)}{\partial s_1'^2}} = \frac{\partial^2 c_X(h_1,h_2,t)}{\partial h_1^2} = c_X^{(2,0,0)}(h_1,h_2,t)$$

it is valid that

$$-c_X^{(2,0,0)}(h_1,h_2,t)\big|_{h_1=h_2=0} = \iint dk_1 dk_2 k_1^2\, \tilde{c}_X(k_1,k_2,t) = \beta_X^{(2,0,0)}. \tag{5.16}$$

i.e., the same result as in Eq. (5.14).

I will conclude the discussion in this chapter with some elementary results on linear random field transformations. As earlier, let $Y(\boldsymbol{p}) = \mathfrak{I}[X(\boldsymbol{p})]$, and assume that both $\mathcal{M}_X(\boldsymbol{w})$ and $\mathcal{M}_Y(\boldsymbol{w})$ are random

[37]In the spectral domain, $\tilde{X}^{(1,0,0)}(k_1,k_2,t) = ik_1\, \tilde{X}(k_1,k_2,t)$.

fields with uncorrelated increments, and such that $\overline{|\mathcal{M}_X(\mathbf{w})|^2} = Q_X(\mathbf{w})$ and $\overline{|\mathcal{M}_Y(\mathbf{w})|^2} = Q_Y(\mathbf{w})$. In the STHS case, the covariance function of $Y(\mathbf{p})$ can be represented by

$$\begin{aligned} c_Y(\Delta\mathbf{p}) &= \int \overline{d\mathcal{M}_Y(\mathbf{w})d\mathcal{M}_Y^*(\mathbf{w})}\, e^{i\mathbf{w}\cdot\Delta\mathbf{p}} \\ &= \int \overline{d\mathcal{M}_X(\mathbf{w})d\mathcal{M}_X^*(\mathbf{w})}\left|\tilde{\eta}(\mathbf{w})\right|^2 e^{i\mathbf{w}\cdot\Delta\mathbf{p}}, \end{aligned} \tag{5.17a–b}$$

where $\overline{d\mathcal{M}_Y(\mathbf{w})d\mathcal{M}_Y^*(\mathbf{w})} = \left|\tilde{\eta}(\mathbf{w})\right|^2 \overline{d\mathcal{M}_X(\mathbf{w})d\mathcal{M}_X^*(\mathbf{w})}$, or

$$dQ_Y(\mathbf{w}) = \left|\tilde{\eta}(\mathbf{w})\right|^2 dQ_X(\mathbf{w}). \tag{5.18}$$

As usual, if the functions $Q_X(\mathbf{w})$ and $Q_Y(\mathbf{w})$ are differentiable, Eq. (5.18) can be expressed by means of the SDF, viz.,

$$\tilde{c}_Y(\mathbf{w}) = \left|\tilde{\eta}(\mathbf{w})\right|^2 \tilde{c}_X(\mathbf{w}), \tag{5.19}$$

i.e., the SDF of the random field $Y(\mathbf{p})$ is obtained from that of $X(\mathbf{p})$ by simply multiplying it by $\left|\tilde{\eta}(\mathbf{w})\right|^2$.

The spectral moments have many applications. *Inter alia*, the spectral moments are used extensively in the study of random vibrations due to seismic activity, etc. Also, they have been used in the signal processing of the output from photoacoustic instruments monitoring crude oil in water. Spectral moments offer an efficient way to analyze the signal, they work directly on the spectrum, and they can be limited within the required frequency ranges. The zeroth-order moment can replace the peak-to-peak value. The second-order moment also shows ability to contribute new information when modeling photoacoustic responses. Yet, little is known of the influence of the higher-order moments, but there might be important information that can be used to get a better representation of the oil concentration.

ISOSTATIONARY SCALAR SPATIOTEMPORAL RANDOM FIELDS

CHAPTER VIII

CHAPTER OUTLINE

1. INTRODUCTION

This chapter starts with some fundamental considerations and mathematical formulations of the notion of space–time isostationarity. It is often argued that isotropy implies that correlation is invariant to coordinate system rotations or reflections about itself. Yet, some important distinctions are worth pointing out regarding the concept of isotropy in a space–time setting (some initial thoughts about the matter were presented in Chapter VII).

1.1 BASIC CONSIDERATIONS

Common sense seems to imply that spatial isotropy cannot be readily extended to space–time, because we are apparently dealing with two physically different quantities. Yet, even if space and time are not directly comparable, they are, nevertheless, linked via physical laws and empirical evidence. As a consequence, it was pointed out in Section 1 of Chapter VII that, as regards an S/TRF $X(\boldsymbol{p})$, two kinds of "isotropic" behavior may be considered in the space–time domain, depending on the manner in which space is linked to time through the associated metric:

(a) The $X(\boldsymbol{p})$, $\boldsymbol{p} \in R^{n+1}$, is called wide sense (w.s.) space–time isotropic (STI) if it has a constant mean and its covariance function depends only on the single argument $|\Delta\boldsymbol{p}|$, viz.,

Spatiotemporal Random Fields. http://dx.doi.org/10.1016/B978-0-12-803012-7.00008-8

$$c_X(\boldsymbol{p},\boldsymbol{p}') = c_X(|\Delta\boldsymbol{p}|), \tag{1.1a}$$

where $\Delta\boldsymbol{p}$ denotes the vector lag between any two space–time points $\boldsymbol{p}$ and $\boldsymbol{p}'$, and $|\Delta\boldsymbol{p}|$ is a composite space–time metric (see, e.g., Table 1.1 of Chapter III).

(b) The $X(\boldsymbol{p})$, $\boldsymbol{p} \in R^{n,1}$, is called w.s. space isotropic/time stationary or, simply, w.s. space–time isostationary (STIS) if it has a constant mean and its covariance function depends on the argument pair $(|\boldsymbol{h}|,\tau)$, viz.,

$$c_X(\boldsymbol{p},\boldsymbol{p}') = c_X(|\boldsymbol{h}|,\tau), \tag{1.1b}$$

where $|\boldsymbol{h}|$ is a spatial distance and τ is a time separation.

Case (b) assumes an explicit way of splitting time from space, whereas case (a) allows several implicit ways of linking time with space depending on the functional form of $|\Delta\boldsymbol{p}|$. Eqs. (1.1a) and (1.1b) are termed the *metric-nonseparate* (*composite*) and the *metric-separate*[1] covariance forms, respectively, and they introduce some further restrictions on the space–time homostationary (STHS) hypothesis discussed in Chapter VII. In both cases (a) and (b), the metrics $|\Delta\boldsymbol{p}|$ and $|\boldsymbol{h}|$ may assume various forms (see Chapter III). We also noticed in Section 1 of Chapter VII that the STI covariance representation of Eq. (1.1a) and the STIS representation of Eq. (1.1b) are not equivalent, in general.

To facilitate our discussion, I first assume that the covariance function under study is an STI one in R^{n+1}, i.e.,

$$c_X(|\Delta\boldsymbol{p}|) = c_X(r_p), \tag{1.2a}$$

where r_p is a space–time Pythagorean metric, see Eq. (2.15a) of Chapter I with $\varepsilon_{ii} = a_i^{-2}$ $(i = 1, \ldots, n$ and $0)$, i.e., $r_p = \left(\sum_{i=1}^{n,0} \frac{h_i^2}{a_i^2}\right)^{\frac{1}{2}}$. Then, I assume that the STIS covariance in $R^{n,1}$ is given by

$$c_X(r,\tau) = c_X\left(r_E, \frac{\tau}{a_0}\right), \tag{1.2b}$$

where r_E is a spatial Euclidean metric, Eq. (2.18a) of Chapter I with $\varepsilon_{ii} = a_i^{-2}$ $(i = 1, \ldots, n)$, i.e., $r_E = \left(\sum_{i=1}^{n} \frac{h_i^2}{a_i^2}\right)^{\frac{1}{2}}$. The two covariances obviously have different mathematical formulations: in Eq. (1.2a) the space–time metric is defined as a single positive real number, r_p, whereas in Eq. (1.2b) the metric is defined as a pair of positive real numbers, r_E and $a_0^{-1}\tau$. I continue with some examples of covariance functions in R^{n+1} or $R^{n,1}$, with different space–time metrics. Certain space–time covariance functions admit more than one kind of metrics, as was explained in Chapter III.

Example 1.1

I start with an STI covariance in R^{n+1} that has a wide range of applications in earth and atmospheric sciences, namely, the *von Kármán* model

$$c_X(r_p) = 2^{1-\nu}\Gamma^{-1}(\nu)(br_p)^{\nu}K_{\nu}(br_p), \tag{1.3a}$$

[1] To be distinguished from the notion of a *separable* covariance function (Section 1.5 of Chapter VII).

where $b, \nu > 0$. Its spectral density function (SDF) is of the form $\widetilde{c}_X(w_p) = c\left(b^2 + w_p^2\right)^{-\nu-\frac{n+1}{2}}$, where $w_p^2 = \mathbf{w}^2$, and c is a coefficient of proportionality. Eq. (1.3a) can be also expressed in terms of other space–time metrics (some of which were reviewed in Chapter III). So, the covariance of Eq. (1.3a) remains valid if the Pythagorean metric r_p is replaced by the traveling metric $r_T = \left|\sum_{i=1}^{n} (|h_i| + \varepsilon_{0i}\tau)^2\right|^{\frac{1}{2}}$, the plane wave metric $r_W = \varepsilon_{00}^{\frac{1}{2}}\tau + \varepsilon_{00}^{-\frac{1}{2}}\sum_{i=1}^{n}\varepsilon_{0i}|h_i|$, or their combination (e.g., the metric $(r_T^2 + r_W^2)^{\frac{1}{2}}$). In applications, these covariance models have distinct physical interpretations. In stationary turbulence studies, the following formulation of Eq. (1.3a) in terms of the r_T metric has been suggested,

$$c_X(r_T) = 8.6 \times 10^{-2}\left(\frac{L_0}{r_0}\right)^{\frac{5}{3}}\left(\frac{2\pi}{L_0}r_T\right)^{\frac{5}{3}} K_{\frac{5}{3}}\left(\frac{2\pi r}{L_0}r_T\right), \tag{1.3b}$$

where L_0 and r_0 are atmospheric parameters. Lastly, an STIS formulation of the von Kármán covariance function in the $R^{n,1}$ domain is

$$c_X(r,\tau) = \frac{b^\nu}{2^{\nu-1}\Gamma(\nu)}\left(r^2 + \frac{a_1}{a_2}\tau^2\right)^{\frac{\nu}{2}} K_\nu\left(b\left(r^2 + \frac{a_1}{a_2}\tau^2\right)^{\frac{1}{2}}\right), \tag{1.3c}$$

where $a_1, a_2, b, \nu > 0$, and the corresponding SDF is $\widetilde{c}_X(k,\omega) = c\left[a_1\left(b^2 + k^2\right) + a_2\omega^2\right]^{-\nu-\frac{n+1}{2}}$, where c is a constant of proportionality. The spatiotemporal correlations quantified by models such as the above are instrumental in the understanding of the dynamic coupling between spatial and temporal scales of motion in turbulent flows.

What characterizes the above hypotheses is that their validity implies that certain S/TRF notions and properties (statistical moments, spectral functions, continuity, differentiability, integrability) depend closely on the space–time metric considered (separate or nonseparate). So, for example, the SDF derivation in the isotropic case depends on the manner in which the metrics $|\Delta \boldsymbol{p}|$ (in R^{n+1}) or $|\boldsymbol{h}|$ (of $R^{n,1}$) are defined. Put differently, the metric form determines what will be interpreted and what will be neglected.

Next, I focus on STIS fields in the $R^{n,1}$ domain with the (r_E, τ) metric, in which case many of the results of Section 3 of Chapter VII can be transformed into their isotropic counterparts by applying the n-dimensional spherical coordinate transformations (these transformations were defined in Eq. (2.34) of Chapter I and also in the Appendix),

$$\begin{aligned} k_1 &= k\cos\theta_1, \\ k_i &= k\cos\theta_i\prod_{j=1}^{i-1}\sin\theta_j \quad (i = 2, 3, \ldots, n-1), \\ k_n &= k\prod_{j=1}^{n-1}\sin\theta_j, \\ \boldsymbol{k}\cdot\boldsymbol{h} &= k\,r\cos\theta_1, \end{aligned} \tag{1.4a–d}$$

where $k \geq 0$, $\theta_i \in [0,\pi]$ $(i = 1, \ldots, n-2)$, $\theta_{n-1} \in [0,2\pi]$. Specifically, the fundamental STHS spectral representations of Eqs. (3.10a–b) of Chapter VII in the STIS case become (Exercise VIII.1),

$$\begin{aligned} c_X(r,\tau) &= (2\pi)^{\frac{n}{2}} r^{1-\frac{n}{2}} \int_0^\infty \int_0^\infty dk\, d\omega\, k^{\frac{n}{2}} J_{\frac{n}{2}-1}(kr) e^{-i\omega\tau} \tilde{c}_X(k,\omega), \\ \tilde{c}_X(k,\omega) &= \frac{1}{(2\pi)^{\frac{n}{2}}} k^{1-\frac{n}{2}} \int_0^\infty \int_0^\infty dr\, d\tau\, r^{\frac{n}{2}} J_{\frac{n}{2}-1}(kr) e^{i\omega\tau} c_X(r,\tau), \end{aligned} \tag{1.5a–b}$$

where $J_{\frac{n}{2}-1}$ is the usual Bessel function of the first kind and $(\frac{n}{2}-1)$th order. The wavenumbers in the FT domain are related to distances in physical space, $k \sim r^{-1}$, $\omega \sim \tau^{-1}$. Furthermore, the covariance is an even and symmetric function, in which case Eqs. (1.5a–b) may be also written as

$$\begin{aligned} c_X(r,\tau) &= (2\pi)^{\frac{n}{2}} r^{1-\frac{n}{2}} \int_0^\infty \int_0^\infty dk\, d\omega\, k^{\frac{n}{2}} J_{\frac{n}{2}-1}(kr) \cos(\omega\tau) \tilde{c}_X(k,\omega), \\ \tilde{c}_X(k,\omega) &= (2\pi)^{-\frac{n}{2}} k^{1-\frac{n}{2}} \int_0^\infty \int_0^\infty dr\, d\tau\, r^{\frac{n}{2}} J_{\frac{n}{2}-1}(kr) \cos(\omega\tau) c_X(r,\tau), \end{aligned} \tag{1.6a–b}$$

in $R^{n,1}$.[2] For $c_X(r,\tau)$ to be a covariance function of an STIS field, it is necessary and sufficient that this function admits a representation of the form of Eqs. (1.5a) or (1.6a), where $\tilde{c}_X(k,\omega)$ is a nonnegative bounded function (some conditions apply here, as discussed in Chapter VII).

Example 1.2

In $R^{2,1}$, Eqs. (1.5a and b) yield, respectively,

$$\begin{aligned} c_X(r,\tau) &= 2\pi \int_0^\infty \int_0^\infty dk\, d\omega k\, J_0(kr) e^{-i\omega\tau} \tilde{c}_X(k,\omega), \\ \tilde{c}_X(k,\omega) &= \frac{1}{2\pi} \int_0^\infty \int_0^\infty dr\, d\tau r\, J_0(kr) e^{i\omega\tau} c_X(r,\tau). \end{aligned} \tag{1.7a–b}$$

Similarly, in $R^{3,1}$,

$$\begin{aligned} c_X(r,\tau) &= 4\pi \int_0^\infty \int_0^\infty dk\, d\omega\, k \frac{\sin(kr)}{r} e^{-i\omega\tau} \tilde{c}_X(k,\omega), \\ \tilde{c}_X(k,\omega) &= \frac{1}{2\pi^2} \int_0^\infty \int_0^\infty dr\, d\tau\, r \frac{\sin(kr)}{k} e^{i\omega\tau} c_X(r,\tau) \end{aligned} \tag{1.8a–b}$$

(Exercise VIII.2). Since most real-world phenomena are studied in two- or three-dimensional physical spaces, the above expressions are standard tools in these spaces.

---

[2]For even functions, $c_X(r,\tau) = c_X(-r,-\tau)$, such as for real or STIS random fields, the SDF is even too, $\tilde{c}_X(k,\omega) = \tilde{c}_X(-k,-\omega)$. In this case, the $e^{i\omega\tau}$ and $e^{-i\omega\tau}$ in Eqs. (1.5a–b) can be replaced by $\cos(\omega\tau)$.

Example 1.3

The distribution of contaminant concentration, $X(\boldsymbol{p}), \boldsymbol{p} = (\boldsymbol{s},t) \in R^{3,1}$, in the subsurface, is represented by a zero mean STIS random field that is mean square (m.s.) continuous but not differentiable, As such, the concentration covariance has an exponential shape

$$c_X(\boldsymbol{h},\tau) = c_0 e^{-|\boldsymbol{h}|-\tau}, \tag{1.9a}$$

where the space distance $|\boldsymbol{h}|$ is the standard Euclidean. The corresponding SDF is

$$\widetilde{c}_X(\boldsymbol{k},\omega) = \frac{c_0}{\pi^3(1+k^2)^2(1+\omega^2)}, \tag{1.9b}$$

where $k = \left(\sum_{i=1}^{3} k_i^2\right)^{\frac{1}{2}}$. Eq. (1.9b) is a wavevector isostationary function, in the sense that $\widetilde{c}_X(\boldsymbol{k},\omega) = \widetilde{c}_X(k,\omega)$. Interestingly, a different result is obtained if another space distance $|\boldsymbol{h}|$ is used in Eq. (1.9a), say the Manhattan distance, in which case the derived covariance

$$\widetilde{c}_X(\boldsymbol{k},\omega) = \frac{c_0}{\pi^4\left(1+k_1^2\right)\left(1+k_2^2\right)\left(1+k_3^2\right)(1+\omega^2)}, \tag{1.9c}$$

is an anisotropic function.

Analogous to the above results and permissibility conditions are valid in terms of the space–time variogram and structure functions of the random field $X(\boldsymbol{p})$. The readers are reminded that the variogram $\gamma_X(r,\tau)$ and the structure function $\xi_X(r,\tau)$ are essentially the same mathematical tool of space–time variability assessment, since the latter is simply twice the former. The STIS variogram may be defined as

$$\gamma_X(r,\tau) = \frac{1}{2}\overline{\left[X(\boldsymbol{s}+\boldsymbol{\varepsilon}r,t+\tau) - X(\boldsymbol{s},t)\right]^2} = \frac{1}{2}\xi_X(r,\tau), \tag{1.10a}$$

where $\boldsymbol{\varepsilon}$ is a unit vector. Certain advantages of the variogram (structure) function over the covariance function have been discussed in Remark 2.1 of Chapter VII. In $R^{3,1}$, which is a common space–time domain encountered in applications, the partial spectral representations of the simultaneous two-point variogram are

$$\begin{aligned} \gamma_X(r,\tau) &= \int_0^\infty dk\left[1-\frac{\sin(kr)}{kr}\right]k^2\widetilde{c}_X(k,\tau), \\ \widetilde{c}_X(k,\tau) &= \frac{1}{(2\pi)^2k^3}\int_0^\infty dr\left[\sin(kr) - kr\cos(kr)\right]\frac{\partial}{\partial r}\gamma_X(r,\tau), \end{aligned} \tag{1.10b–c}$$

which are derived from the general Eqs. (3.25a–b) of Chapter VII, see Example 3.8 of Chapter VII. Given the shape of the covariance or the variogram function, the correlation ranges and scales of the attribute they represent can be readily specified. In the following examples, the standard Euclidean space distance is assumed, unless otherwise noted.

Example 1.4

The space–time variogram function presented in Example 3.6 of Chapter VII, i.e.,

$$\gamma_X(r,\tau) = c_0\left(1 - e^{-\left(\frac{1}{a^2}r^2 + \frac{1}{b^2}\tau^2\right)^{\frac{1}{2}}}\right),$$

was found to have a nonnegative SDF. The spatial range of this variogram function is defined as the r value for which $\gamma_X(r,0) = 0.95c_0$, i.e., $r = \varepsilon_r = 3^{1/2}a$. Similarly, its time range is defined as the τ value for which $\gamma_X(0,\tau) = 0.95c$, i.e., $r = \varepsilon_\tau = 3^{1/2}b$ (a more detailed analysis of space and time ranges and scales is presented in Section 6 in this chapter).

Example 1.5

A valid variogram model representing STIS variations in $R^{n,1}$ is given by (Gneiting, 2002),

$$\gamma_X(r,\tau) = \sigma^2\left[1 - \frac{1}{(b\tau^\alpha + 1)^\beta}e^{-\frac{\kappa r^\lambda}{(b\tau^\alpha+1)^\beta}}\right], \tag{1.11}$$

where σ, α, β, b, κ, λ are suitable coefficients.

At this point, it may be instructive to consider random field isotropy and the derivation of the corresponding variogram function using physical arguments.

Example 1.6

In fluid dynamics, the theory of statistical turbulence is formulated on the basis of two hypotheses (Kraichnan, 1974): (*a*) the distributions of the velocity differences $X(\boldsymbol{s}+\boldsymbol{h},t) - X(\boldsymbol{s},t)$ are isotropic functions solely of the lag vectors $\boldsymbol{h}$, the kinematic viscosity ν (in m^2/s units) and the mean rate of energy dissipation per unit mass ε (in m^2/s^3 units), provided that all the $\boldsymbol{h}$ are small compared with the turbulence macroscales; and (*b*) when the vectors $\boldsymbol{h}$ are large compared with dissipation-range scales, the velocity distributions are independent of ν. These two hypotheses lead, by dimensional analysis, to explicit functional forms for the variogram function of $X(\boldsymbol{s},t)$ that depend only on $r = |\boldsymbol{h}|$, i.e., the simultaneous two-point variogram is given by (for simplicity, the time argument in suppressed)

$$\gamma_X(\boldsymbol{h}) = \beta_1\gamma\left(\frac{|\boldsymbol{h}|}{\beta_2}\right), \tag{1.12}$$

where, based on the two hypotheses above, it is found that $\beta_1 = (\varepsilon\nu)^{1/2}$ (in $(m/s)^2$ units) and $\beta_2 = \left(\frac{\nu^3}{\varepsilon}\right)^{1/4}$ (in m units). Moreover, for the variogram function to be independent of ν in the inertial range (hypothesis (*b*)), it must have the functional form

$$\gamma\left(\frac{|\boldsymbol{h}|}{\beta_2}\right) = \beta_3\left(\frac{|\boldsymbol{h}|}{\beta_2}\right)^{2/3}, \tag{1.13}$$

where the constant β_3 should be dimensionless, because only in this case the dependence on ν drops out in Eq. (1.12). Accordingly, Eqs. (1.12)–(1.13) yield

$$\gamma_X(\boldsymbol{h}, t) = \alpha|\boldsymbol{h}|^{2/3}, \tag{1.14}$$

where $\alpha = \beta_1\beta_3(\beta_2)^{2/3} = (\varepsilon\nu)^{1/2}\beta_3\left(\left(\frac{\nu^3}{\varepsilon}\right)^{-1/4}\right)^{2/3} = \beta_3\varepsilon^{2/3}$ represents turbulence strength.

To conclude this section, it should be noticed that several covariance properties of the spatial domain have direct analogues in the spatiotemporal domain.

Example 1.7

For illustration, consider $n+1$ space–time points $(\boldsymbol{s}_i, t_i) \in R^{n,1}$ such that $(\boldsymbol{s}_i - \boldsymbol{s}_j, t_i - t_j) = (\boldsymbol{h}, \tau)$ for all $i \neq j$. It is valid that

$$\begin{aligned}\overline{\left[\sum_{i=1}^{n+1} X(\boldsymbol{s}_i, t_i)\right]^2} &= \sum_{i,j=1}^{n+1} \overline{X(\boldsymbol{s}_i, t_i)X(\boldsymbol{s}_j, t_j)} = (n+1)c_X(\boldsymbol{0}, 0) + \sum_{i\neq j=1}^{n} c_X(\boldsymbol{s}_i - \boldsymbol{s}_j, t_i - t_j) \\ &= (n+1)c_X(\boldsymbol{0}, 0) + \sum_{i\neq j=1}^{n+1} c_X(\boldsymbol{h}, \tau) = (n+1)c_X(\boldsymbol{0}, 0) + n(n+1)c_X(\boldsymbol{h}, \tau) \geq 0,\end{aligned} \tag{1.15}$$

which, implies that $c_X(\boldsymbol{h}, \tau) \geq -\frac{1}{n}c_X(\boldsymbol{0}, 0)$.

1.2 POWER-LAW CORRELATIONS

Short-range (spatial or temporal) covariance models have correlations that decay to zero fast enough for the integral of the STIS covariance function to exist. Examples include the spherical, the exponential, the cubic, and the Gaussian covariance models. The correlations of short-range models become negligible when the lag increases beyond a certain value. For many of these models the decays of the correlation functions are determined by two scales, the space and the time correlation lengths.

Dimensionless lags can be defined by dividing the actual lags with the corresponding correlation lengths, in which case the SDF of a large class of two-scale covariance models admit the representation

$$\widetilde{c}_X(k, \omega) = a\,\varepsilon_r^n\varepsilon_\tau\widetilde{f}_r(\varepsilon_r k)\widetilde{f}_\tau(\varepsilon_\tau\omega), \tag{1.16}$$

where a is a constant, ε_r and ε_τ denote spatial and temporal correlation ranges, respectively ($\varepsilon_r k$ and $\varepsilon_\tau \omega$ are dimensionless arguments), and $\widetilde{f}_r$ and $\widetilde{f}_\tau$ are suitable functions (see Example 1.8).

Example 1.8

The exponential space–time covariance model is a short-range model in $R^{n,1}$ given by[3]

$$c_X(r,\tau) = c_0\, e^{-\frac{r}{\varepsilon_r} - \frac{\tau}{\varepsilon_\tau}}, \tag{1.17}$$

with SDF of the form of Eq. (1.16), where $a = \dfrac{c_0}{\pi^{\frac{n+2}{2}} \Gamma\left(2 - \frac{n}{2}\right)}$, $\widetilde{f}_r(\varepsilon_r k) = \left(1 + \varepsilon_r^2 k^2\right)^{-\frac{n+1}{2}}$, and $\widetilde{f}_\tau(\varepsilon_\tau \omega) = \left(1 + \varepsilon_\tau^2 \omega^2\right)^{-1}$ (the exponential model is also called *Markovian*, because it can be shown that a Gaussian stationary process is Markovian only if the covariance function is exponential).

Certain random fields, including fractional noises and fractional Brownian motions, are characterized by power-law correlations, i.e.,

$$c_X(\lambda r, \lambda \tau) = \lambda_r^{2H_r} \lambda_\tau^{2H_\tau} c_{X(1)}(r) c_{X(2)}(\tau), \tag{1.18}$$

where H_r and H_τ are the scaling (fractal) exponents in space and time, respectively, and λ_r and λ_τ are multiplication constants (Mandelbrot, 1982; Mandelbrot and Van Ness, 1968). Many data sets that were previously thought to represent incoherent and structureless noise have been well characterized by means of fractal correlation models (Feder, 1988; Bak and Chen, 1989). The apparent irregularity of such fields was shown to derive from the long-range nature of the power-law correlations among individual events.

Example 1.9

Power-law correlations have been observed in environmental and economic processes. A recent analysis of the epidemic size and duration of measles in small, isolated communities showed that they also exist in biological systems and have important implications for human health (Rhodes and Anderson, 1996).

Power-law correlations can be classified as short range, if they are integrable and do not change the asymptotic scaling properties of the system, and *long range* when they lead to new types of asymptotic behavior at large distances or times.

Example 1.10

In the case of diffusion in random media, long-range correlations in the velocity field lead to anomalous, non-Fickian diffusion (Bouchaud and Georges, 1990). Anomalous diffusion has been observed in porous rocks, and it has been studied by means of a two-dimensional layered media model (Matheron and de Marsily, 1980).

---

[3]This is a generalization in the $R^{n,1}$ domain of an earlier result obtained in Example 1.3. Indeed, for $n = 1$ and $\varepsilon_r = \varepsilon_r = 1$, Eq. (1.17) reduces to Eq. (1.9b).

Example 1.11

The STIS fractal fields are self-similar; therefore, they are characterized by a symmetric SDF,

$$\widetilde{c}_X(k,\omega)=\frac{A}{k^{\xi_r}\omega^{\xi_\tau}}, \tag{1.19}$$

for $k\in[k_m,\,k_0]$, $\omega\in[\omega_m,\,\omega_0]$, where k and ω are the spatial and temporal frequency, respectively, ξ_r and ξ_τ are the corresponding spectral exponents, and A is a constant. This leads to a finite range of distances and periods within which the field is characterized by fractal behavior. If $\widetilde{c}_X(k,\omega)$ is integrable, it represents the SDF of an STIS covariance. The real-domain covariance function can be obtained from the inverse FT of Eq. (1.19), and it behaves as a power law within the fractal range. The integrability of $\widetilde{c}_X(k,\omega)$ depends on the values of the exponents ξ_r and ξ_τ (Christakos and Hristopulos, 1998). If $\widetilde{c}_X(k,\omega)$ is not integrable, it represents the variogram spectrum of a random field with STHS increments. Furthermore, The STIS fractal field covariances exhibit power-law behavior in $R^{n,1}$ of the decompositional form

$$c_X(r,\tau)=A\zeta_1\,\zeta_n\,r^{2H_r}\tau^{2H_\tau}, \tag{1.20a}$$

where $k_0^{-1}<<r_0\le r\le r_m<<k_m^{-1}$, $\omega_0^{-1}<<\tau_0\le\tau\le\tau_m<<\omega_m^{-1}$, the SDF exponents in Eq. (1.19) are within the ranges $\xi_r\in\left(\frac{n-1}{2},n\right)$, $\xi_\tau\in(0,1)$, and the ζ_n and ζ_1 are constant numbers. The parameters H_r and H_τ represent the *Hurst exponents*, which are related to the SDF exponents via

$$H_r=\frac{\xi_r-n}{2}\ \text{ and }\ H_\tau=\frac{\xi_\tau-1}{2}. \tag{1.20b}$$

The SDF of Eq. (1.20a) exists only if the Hurst exponents lie in the ranges[4]

$$H_r\in\left(-\frac{n+1}{4},0\right)\ \text{ and }\ H_\tau\in\left(-\frac{1}{2},0\right). \tag{1.20c}$$

Note that this equation holds within the fractal windows $r\in[r_0,\,r_m]$, $\tau\in[\tau_0,\,\tau_m]$ above. If it is extrapolated outside the scaling range, it leads to a divergence near zero, due to the negative value of the Hurst exponent. This behavior is not meaningful, unless the field is in a critical state that has correlations extending throughout its domain. Covariance models with nonsingular behavior at the origin and asymptotic power-law behavior can be constructed; such a model is given by

$$c_X(r,\tau)=c_0\left(1+\frac{r^2}{\varepsilon_r^2}\right)^{H_r}\left(1+\frac{\tau^2}{\varepsilon_\tau^2}\right)^{H_\tau}, \tag{1.21}$$

which is an asymptotically scaling model, i.e., its validity requires that $r>>\varepsilon_r$, $\tau>>\varepsilon_\tau$.

Anisostationary separable covariance functions are generally used to model directional heterogeneity. The SDF of such models may be expressed as

$$\widetilde{c}_X(\boldsymbol{k},\omega)=c_0\left(\varepsilon_\tau\prod_{i=1}^{n}\varepsilon_i\right)\widetilde{f}_r(k_1\varepsilon_1,\ldots,k_n\varepsilon_n)\widetilde{f}_\omega(\omega\varepsilon_\tau), \tag{1.22}$$

[4]Power-law fields with Hurst exponents that satisfy this range are called fractional noises.

where ε_i, $i = 1, \ldots, n$ are the spatial correlation ranges along the principal directions and ε_τ is the temporal correlation range. By means of the rescaling transformations $\frac{\varepsilon_i}{\varepsilon_1} = \lambda_i$, and $k_1' = k_1$, $k_i' = \lambda_i\, k_i$ for $i = 2, \ldots, n$, Eq. (1.22) becomes

$$\tilde{c}_X(k', \omega) = c_0 \left(\varepsilon_1^n \varepsilon_\tau \prod_{i=2}^{n} \lambda_i \right) \tilde{f}_r'(k' \varepsilon_1) \tilde{f}_\omega(\omega \varepsilon_\tau), \tag{1.23}$$

where $k' = \left(\sum_{i=1}^{n} k_i^2 \right)^{\frac{1}{2}}$. Obviously, such STIS covariance functions are more convenient for analytical calculations.

Example 1.12

For illustration, let us apply the rescaling transformation to the anisostationary exponential covariance function in $R^{3,1}$ with SDF

$$\tilde{c}_X(\boldsymbol{k}, \omega) = \frac{2 c_0 \varepsilon_\tau \prod_{i=1}^{3} \varepsilon_i}{\pi^2 \left(1 + \sum_{i=1}^{3} k_i^2 \varepsilon_i^2 \right)^2 \left(1 + \omega^2 \varepsilon_\tau^2 \right)^2}. \tag{1.24}$$

As before, we set $k_1' = k_1$, $\frac{\varepsilon_2}{\varepsilon_1} = \lambda_2$, $\frac{\varepsilon_3}{\varepsilon_1} = \lambda_3$, $k_2' = \lambda_2\, k_2$, and $k_3' = \lambda_2\, k_3$, in which case,

$$\tilde{c}_X(k', \omega) = \frac{2 c_0 \lambda_2 \lambda_3 \varepsilon_1^3 \varepsilon_\tau}{\pi^2 \left(1 + k'^2 \varepsilon_1^2 \right)^2 \left(1 + \omega^2\, \varepsilon_\tau^2 \right)^2}, \tag{1.25}$$

where $k'^2 = \sum_{i=1}^{3} k_i^2$. An interesting application of the rescaling technique in the calculation of the effective hydraulic conductivity in porous media can be found in Hristopulos and Christakos (1997).

In a formal sense, anisostationary covariance models are direct extensions of the STIS models. A rather straightforward extension of the model of Eq. (1.21) leads to the following anisostationary covariance model

$$c_X(\boldsymbol{h}, \tau) = c_0 \left(1 + \sum_{i=1}^{n} \frac{h_i^2}{\varepsilon_i^2} \right)^{H_r} \left(1 + \frac{\tau^2}{\varepsilon_\tau^2} \right)^{H_\tau}, \tag{1.26}$$

where $H_r \in \left(-\frac{n+1}{4}, 0 \right)$ and $H_\tau \in \left(-\frac{1}{2}, 0 \right)$. As expected, the two exponents characterize asymptotic scaling in Eq. (1.26). It is, also, possible for the power-law exponents to vary in different directions, such as in the covariance function

$$c_X(\boldsymbol{h}, \tau) = c_0 \left(1 + \tau^2 \right)^{-\frac{1}{2}\beta_\tau} \sum_{i=1}^{n} \left(\frac{1}{n} + h_i^2 \right)^{-\frac{1}{2}\beta_i}, \tag{1.27}$$

where β_i are (positive) directional exponents and β_τ is a temporal exponent. The model (Eq. 1.26) is asymptotically scaling if distances are measured in the rescaled coordinate system. This is not true of the model of Eq. (1.27), because the power-law exponents vary with the direction, i.e., it is a self-affine model. Lastly, interesting is the case of fractal models with STHS increments.

Example 1.13

For fractal models with STHS increments, the covariance offers a useful two-point correlation function for a negative Hurst exponent, i.e., as in the case of Eq. (1.20c). If the Hurst exponent is positive (i.e., $\xi_r > n$, $\xi_\tau > 1$ in Eq. 1.20b), the spectrum is not integrable due to the infrared divergence. When the SDF exponents satisfy $\xi_r \in (n,n+2)$, $\xi_\tau \in (1,3)$, the $\widetilde{c}_X(\boldsymbol{k},\tau)$ is a permissible spectrum for the variogram of a random field with STHS increments. The variogram function in $R^{n,1}$ is then obtained from this spectrum as

$$\gamma_X(\boldsymbol{h},\tau) = \iint d\boldsymbol{k}d\omega \left(1 - e^{i\,\boldsymbol{k}\cdot\boldsymbol{h}}\right)\left(1 - e^{i\,\omega\tau}\right)\widetilde{c}_X(\boldsymbol{k},\tau). \tag{1.28}$$

The integral in Eq. (1.28) converges at both limits if the exponents of the spectrum satisfy the inequalities $\xi_r \in (n,n+2)$, $\xi_\tau \in (1,3)$. Hence, the corresponding ranges of the Hurst exponents are H_r, $H_\tau \in (0,1)$. Hurst exponents in the above range are characteristic of fractional Brownian motions. The space–time variogram obtained from the spectrum is a power law

$$\gamma_X(r,\tau) = A\eta_1\,\eta_n\,r^{2H_r}\tau^{2H_\tau}, \tag{1.29}$$

where $r_0 \le r = |\boldsymbol{h}| \le r_m$, and the η_1 and η_n are constant numbers.

In all the above models, the joint consideration of space and time in random field modeling introduces considerable methodological complications, in which case we need to make adequate assessments of space and time interconnections and develop rigorous quantitative interpretations of these interconnections.

1.3 PHYSICAL CONSIDERATIONS OF VARIOGRAM FUNCTIONS

In applications, we often use *ad hoc* formulations of variogram functions, depending on the modeling needs of the particular physical study. These formulations can be of various kinds, one of which is the (time or space) difference operator formulation discussed below.

In many real-world phenomena, some of the spectral functions can be measured relatively accurately or reasonably hypothesized. This is the case with the frequency spectrum $\widetilde{c}_X(\omega)$, which is often directly measurable and interpretable. To take advantage of this fact, we define the time difference operator $\Delta_1(\boldsymbol{s},\tau) = X(\boldsymbol{s},t+\tau) - X(\boldsymbol{s},t)$, based on which several variograms can be derived, such as the single point/two-time variograms

$$\begin{aligned}\gamma_X(\tau) &= \frac{1}{2}\overline{\Delta_1(\boldsymbol{s},\tau)^2},\\ \gamma_{\Delta_1 X}(\tau) &= \frac{1}{2}\overline{[\Delta_1(\boldsymbol{s},t+\tau) - \Delta_1(\boldsymbol{s},t)]^2}.\end{aligned} \tag{1.30a–b}$$

These variograms can be represented in terms of the frequency spectrum $\widetilde{c}_X(\omega)$, as

$$\begin{aligned}\gamma_X(\tau) &= \int d\omega[1 - \cos(\omega\tau)]\widetilde{c}_X(\omega),\\ \gamma_{\Delta_1 X}(\tau) &= \int d\omega[3 - 4\cos(\omega\tau) + \cos(2\omega\tau)]\widetilde{c}_X(\omega).\end{aligned} \tag{1.31a–b}$$

Given $\widetilde{c}_X(\omega)$, the variograms can be calculated as above.

Example 1.14

Assume that the available frequency spectrum is of the form

$$\widetilde{c}_X(\omega) = c\omega^{-k},$$

where c is a constant of proportionality and $k > 0$. By virtue of Eqs. (1.31a–b), it is easily found that

$$\frac{\gamma_{\Delta_1 X}(\tau)}{\gamma_X(\tau)} = 4\left(1 - 2^{k-3}\right),$$

i.e., the variogram ratio depends on the form of the frequency spectrum only. Also, in this case, given $\gamma_X(\tau)$, the $\gamma_{\Delta_1 X}(\tau)$ is obtained by simply multiplying $\gamma_X(\tau)$ by $4(1-2^{k-3})$.

2. RELATIONSHIPS BETWEEN COVARIANCE DERIVATIVES AND SPACE–TIME ISOSTATIONARITY

The STIS being a restriction of STHS, many of the STHS random field geometry results discussed in Chapter VII can be properly modified for STIS fields. Moreover, under certain conditions, isotropic random field geometry results that have been derived in a purely spatial domain may be extended in the STIS domain.

In the case of STIS random fields, partial covariance derivatives with respect to the coordinate directions can be related to covariance derivatives with respect to the corresponding metric. Indeed, in $R^{2,1}$ and with an Euclidean $|\boldsymbol{h}|$, the following relationships are valid for the derivatives of the random field $X(\boldsymbol{s},t)$ and the corresponding covariances,

$$\overline{\frac{\partial X(s_1+h_1, s_2+h_2, t+\tau)}{\partial s_i} X(s_1, s_2, t)} = \frac{\partial c_X(r,\tau)}{\partial h_i} = -\frac{h_i}{r}\frac{\partial}{\partial r} c_X(r,\tau),$$

$$\overline{\frac{\partial X(s_1+h_1, s_2+h_2, t+\tau)}{\partial s_i}\frac{\partial X(s_1, s_2, t)}{\partial s_j}} = -\frac{\partial^2 c_X(r,\tau)}{\partial h_i \partial h_j} = \frac{1}{r}\left[\frac{h_i h_j}{r}\left(-\frac{\partial^2}{\partial r^2} + r^{-1}\frac{\partial}{\partial r}\right) - \delta_{ij}\frac{\partial}{\partial r}\right] c_X(r,\tau),$$

$$\overline{\frac{\partial^2 X(s_1+h_1, s_2+h_2, t+\tau)}{\partial s_i \partial s_j}\frac{\partial^2 X(s_1, s_2, t)}{\partial s_k \partial s_l}} = \frac{\partial^4 c_X(r,\tau)}{\partial h_i \partial h_j h_k \partial h_l}$$

$$= \frac{1}{r^2}\left[\left(-\frac{\alpha_1}{r} + \frac{3\alpha_2}{r^3} - \frac{15\alpha_3}{r^5}\right)\frac{\partial}{\partial r} - \left(-\alpha_1 + \frac{3\alpha_2}{r^2} - \frac{15\alpha_3}{r^4}\right)\frac{\partial^2}{\partial r^2} + \left(\frac{\alpha_2}{r} - \frac{6\alpha_3}{r^3}\right)\frac{\partial^3}{\partial r^3} + \frac{\alpha_3}{r^2}\frac{\partial^4}{\partial r^4}\right] c_X(r,\tau),$$

(2.1a–c)

where $\alpha_1 = \delta_{ij}\delta_{kl} + \delta_{ik}\delta_{jl} + \delta_{il}\delta_{jk}$, $\alpha_2 = \delta_{ij}h_k h_l + \delta_{ik}h_j h_l + \delta_{il}h_j h_k + \delta_{jl}h_i h_k + \delta_{jk}h_i h_l + \delta_{kl}h_i h_j$, and $\alpha_3 = h_i h_j h_k h_l$, $i, j, k, l = 1, 2$ (similar expressions can be derived for the different kinds of space–time metrics discussed in Chapter III). To gain some insight about how Eqs. (2.1a–c) may be used in applications, I present below a few examples.

Table 2.1 Examples of Covariance Derivatives in $R^{2,1}$ Obtained From Eq. (2.1b)

i	j	h_1	h_2	τ	Eq. (2.1b) at $(s,t) = (s_1,s_2,t) = (0,0,0)$
2	2	r	0	0	$\overline{\frac{\partial X(r,0,0)}{\partial s_2}\frac{\partial X(0,0,0)}{\partial s_2}} = -\frac{1}{r}\frac{\partial}{\partial r}c_X(r,0)$
2	2	0	r	0	$\overline{\frac{\partial X(0,r,0)}{\partial s_2}\frac{\partial X(0,0,0)}{\partial s_2}} = -\frac{\partial^2}{\partial r^2}c_X(r,0)$
2	1	r	0	0	$\overline{\frac{\partial X(r,0,0)}{\partial s_2}\frac{\partial X(0,0,0)}{\partial s_1}} = 0$
2	2	0	0	τ	$\overline{\frac{\partial X(0,0,\tau)}{\partial s_2}\frac{\partial X(0,0,0)}{\partial s_2}} = -\frac{1}{r}\frac{\partial}{\partial r}c_X(0,\tau)$
1	1	0	0	τ	$\overline{\frac{\partial X(0,0,\tau)}{\partial s_1}\frac{\partial X(0,0,0)}{\partial s_1}} = -\frac{1}{r}\frac{\partial}{\partial r}c_X(0,\tau)$
2	1	0	0	τ	$\overline{\frac{\partial X(0,0,\tau)}{\partial s_2}\frac{\partial X(0,0,0)}{\partial s_1}} = 0$

Example 2.1

A list of cases concerning the application of Eq. (2.1b) is given in Table 2.1. The first covariance of Table 2.1 is termed the longitudinal covariance function in physical sciences and is denoted by $c_{\ell\ell} = -\frac{1}{r}\frac{\partial}{\partial r}c_X$, whereas the second one is termed the transverse (or lateral) covariance function and is denoted by $c_{\eta\eta} = -\frac{\partial^2}{\partial r^2}c_X$. In the case of STIS random fields, the covariance function between the field derivatives can be expressed in terms of the longitudinal and transverse covariance functions above, which are not independent, since they are both derived from the random field $X(s,t)$. Furthermore, at $(s,t) = (s_1,s_2,t) = (0,0,0)$ if we let $i = l = 1$, $j = k = 2$, $h_1 = r$, and $h_2 = \tau = 0$ in Eq. (2.1c), we find

$$\overline{\frac{\partial^2 X(r,0,0)}{\partial s_1 \partial s_2}\frac{\partial^2 X(0,0,0)}{\partial s_1 \partial s_2}} = \frac{1}{r}\left[\frac{2}{r^2}\frac{\partial}{\partial r} - \frac{2}{r}\frac{\partial^2}{\partial r^2} + \frac{\partial^3}{\partial r^3}\right]c_X(r,0).$$

Also, using Eq. (2.1c) we find that the covariance of $Y(s,t) = -\frac{\partial^2 X(s,t)}{\partial s_2^2} - \frac{\partial^2 X(s,t)}{\partial s_1^2}$ is given by

$$\overline{Y(r,0,0)Y(0,0,0)} = \left[\frac{1}{r^3}\frac{\partial}{\partial r} - \frac{1}{r^2}\frac{\partial^2}{\partial r^2} + \frac{2}{r}\frac{\partial^3}{\partial r^3} + \frac{\partial^4}{\partial r^4}\right]c_X(r,0)$$

for $i = k = 2$, $j = l = 1$, $h_1 = r$, and $h_2 = \tau = 0$.

Random field differentiability in the m.s. sense is linked to covariance differentiability at the origin in the straightforward manner described in the following proposition.

Proposition 2.1

In view of Postulate 1.1 of Chapter VI, let $X(\boldsymbol{p})$ be an STIS random field in $R^{n,1}$ that is m.s. differentiable. If any of $\nu_i + \nu_i'$ $(i = 1,\ldots, n)$ is odd with $\sum_{i=1}^{n}(\nu_i + \nu_i') = \nu + \nu'$, the derivative of the simultaneous two-point covariance is

$$\frac{\partial^{\nu+\nu'}}{\partial \boldsymbol{h}^{\nu+\nu'}}c_X(r,t) = \frac{\partial^{\nu+\nu'}}{\prod_{i=1}^{n}\partial h_i^{\nu_i+\nu_i'}}c_X(r,t) = 0 \tag{2.2}$$

at the space origin, where $r = |\boldsymbol{h}|$ is the Euclidean distance.

A similar result involving the time argument is valid (see examples below). A direct consequence of Proposition 2.1 is the following corollary, which is the STIS version of Corollary 4.2 of Chapter VII discussed in a previous chapter.

Corollary 2.1

If the STIS random field $X(\boldsymbol{p})$ is m.s. differentiable, then its covariance satisfies the relationship

$$\frac{\partial^{2\nu-1}}{\partial h_i^{2\nu-1}} c_X(r,t) = 0 \tag{2.3}$$

at the origin.

It is instructive to notice that Proposition 2.1 and Corollary 2.1 are based on Postulate 1.1 of Chapter VI and on two premises: (*a*) because of STIS, we can write in $R^{n,1}$,

$c_X(r,t) = c_X\left(\left(r^2\right)^{\frac{1}{2}}, t\right) = \widehat{c}_X\left(r^2, t\right)$, where $r^2 = \sum_{i=1}^n h_i^2$ and (*b*) with the change of variables $r^2 = u$

so that $\widehat{c}_X^{(2\nu-1-j,0)}\left(r^2, t\right) = \frac{\partial^{2\nu-1-j}}{\partial u^{2\nu-1-j}} \widehat{c}_X(u,t)$, it is valid that

$$\begin{aligned}
\frac{\partial^{2\nu-1}}{\partial h_i^{2\nu-1}} \widehat{c}_X\left(r^2, t\right) &= \sum_{j=0}^{\lfloor (2\nu-1)/2 \rfloor} \frac{(2\nu-1)!}{(2\nu-1-2j)!j!} (2h_i)^{2\nu-1-2j} \widehat{c}_X^{(2\nu-1-j,0)}(u,t) \\
&= 2h_i \sum_{j=0}^{\lfloor (2\nu-1)/2 \rfloor} \frac{(2\nu-1)!}{(2\nu-1-2j)!j!} (2h_i)^{2\nu-2-2j} \widehat{c}_X^{(2\nu-1-j,0)}(u,t).
\end{aligned} \tag{2.4a–b}$$

The right side of Eq. (2.4b) is equal to 0 at the origin ($h_i \to 0$, regardless of the t value). An interesting point about the above formulation is that the behavior at the space origin does not change if time derivatives are involved. We will illustrate the implementation of these useful theoretical expressions with a few examples.

Example 2.2

We can calculate the $\frac{\partial^3}{\partial h_i^3} \widehat{c}_X\left(r^2, t\right)$ in two ways. First, in terms of the usual partial differentiation, we get

$$\frac{\partial^3}{\partial h_i^3} \widehat{c}_X\left(r^2, t\right) = 4h_i \left[3 \frac{\partial^2}{\partial u^2} \widehat{c}_X(u,t) + 2h_i^2 \frac{\partial^3}{\partial u^3} \widehat{c}_X(u,t)\right] = 4h_i \left[3 \widehat{c}_X^{(2,0)} \widehat{c}_X(u,t) + 2h_i^2 \widehat{c}_X^{(3,0)} \widehat{c}_X(u,t)\right],$$

where $u = r^2$. Second, the same result is obtained by using the expressions of Eqs. (2.4a–b) with $\nu = 2$, i.e.,

$$\begin{aligned}
\frac{\partial^3}{\partial h_i^3} \widehat{c}_X\left(r^2, t\right) &= 2h_i \sum_{j=0}^{\lfloor (3/2) \rfloor} \frac{3!}{(3-2j)!j!} (2h_i)^{3-2j-1} \frac{\partial^{3-j}}{\partial u^{3-j}} \widehat{c}_X(u,t) \\
&= 2h_i \sum_{j=0}^{1} \frac{3!}{(3-2j)!j!} (2h_i)^{2-2j} \widehat{c}_X^{(3-j,0)}\left(r^2, t\right) = 4h_i \left[2h_i^2 \widehat{c}_X^{(3,0)}\left(r^2, t\right) + 3 \widehat{c}_X^{(2,0)}\left(r^2, t\right)\right],
\end{aligned}$$

as should be expected.

Example 2.3

In $R^{2,1}$, let $\nu_1 = 2$, $\nu_2 = \nu_1' = \mu' = 0$, and $\nu_2' = \mu = 1$, so that $\nu = \nu_1 + \nu_2 = 2$, and $\nu' = \nu_1' + \nu_2' = 1$. Due to the STIS assumption, we can write $c_X(r, t) = \widehat{c}_X(r^2, t)$, in which case,

$$\frac{\partial^4}{\prod_{i=1}^{2} \partial h_i^{\nu_i+\nu_i'} \partial t} \widehat{c}_X\left(r^2, t\right) = \frac{\partial^4}{\partial h_1^2 \partial h_2 \partial t} \widehat{c}_X\left(h_1^2 + h_2^2, t\right) = 4h_2\left[\frac{\partial^3}{\partial u^2 \partial t} \widehat{c}_X(u, t) + 4h_1^2 \frac{\partial^4}{\partial u^3 \partial t} \widehat{c}_X(u, t)\right].$$

Clearly, the right side of the above expression is 0 at the space origin (regardless of the t value), which is expected on the basis of Proposition 2.1, since $\nu_2' + \nu_2' = 1$ is odd (this result is valid here although a time derivative is involved). On the other hand, consider the case, $\nu_1 = 2$ and $\nu_2 = \nu_1' = \nu_2' = 0$, so that $\nu = \nu_1 + \nu_2 = 2$ and $\nu' = \nu_1' + \nu_2' = 0$. Then,

$$\frac{\partial^2}{\partial h_1^2} \widehat{c}_X\left(h_1^2 + h_2^2, t\right) = 2 \frac{\partial}{\partial u} \widehat{c}_X(u, t) + 4h_1^2 \frac{\partial^2}{\partial u^2} \widehat{c}_X(u, t),$$

which is not necessarily 0 at the origin. This is also expected on the basis of Proposition 2.1, since $\nu_1' + \nu_1' = 2$ and $\nu_2' + \nu_2' = 0$ are not odd.

I continue with a direct consequence of Corollary 2.1 described in the following corollary.

Corollary 2.2

For an m.s. differentiable random field $X(\boldsymbol{p})$ the simultaneous two-point STIS covariance satisfies the equality

$$\frac{\partial^{2\nu-1}}{\partial r^{2\nu-1}} c_X(0, t) = 0 \tag{2.5}$$

at the origin.

Example 2.4

In $R^{2,1}$, and by letting $c_X(r, t) = \widehat{c}_X\left(r^2, t\right)$, it is immediately found that

$$\frac{\partial^3}{\partial r^3} \widehat{c}_X\left(r^2, t\right) = 4r\left[\frac{\partial^2}{\partial u^2} + 2\frac{\partial}{\partial u} + 2r^2 \frac{\partial^2}{\partial u^2}\right] \widehat{c}_X(u, t),$$

which is equal to 0 at the space origin. On the other hand,

$$\frac{\partial^2}{\partial r^2} \widehat{c}_X\left(r^2, t\right) = 2\left[\frac{\partial}{\partial u} + 2r^2 \frac{\partial}{\partial u}\right] \widehat{c}_X(u, t),$$

$u = r^2$, which is not necessarily zero at the space origin (again, regardless of the τ value).

On the basis of the fundamental Eq. (1.5a), some useful expressions of the STIS covariance derivatives are obtained.

Proposition 2.2

In $R^{n,1}$ the following expressions of the time and space derivatives of an STIS covariance function hold,

$$\frac{\partial^{\mu}}{\partial\tau^{\mu}}c_X(r,\tau) = (-i)^{\mu}(2\pi)^{\frac{n}{2}}r^{1-\frac{n}{2}}\int_0^{\infty}\int_0^{\infty} dk\, d\omega\, k^{\frac{n}{2}}\omega^{\mu}J_{\frac{n}{2}-1}(kr)e^{-i\omega\tau}\tilde{c}_X(k,\omega),$$
$$\left(\frac{1}{r}\frac{\partial}{\partial r}\right)^{\nu}c_X(r,\tau) = (-1)^{\nu}(2\pi)^{\frac{n}{2}}r^{1-\frac{n}{2}-\nu}\int_0^{\infty}\int_0^{\infty} dk\, d\omega\, k^{\frac{n}{2}+\nu}J_{\frac{n}{2}-1+\nu}(kr)e^{-i\omega\tau}\tilde{c}_X(k,\omega). \tag{2.6a–b}$$

Some comments are in order here concerning Eqs. (2.6a–b). The derivation of Eq. (2.6a) is rather trivial, whereas the derivation of Eq. (2.6b) used the recursive expression

$$\left(\frac{1}{\zeta}\frac{\partial}{\partial\zeta}\right)^{\nu}f(\zeta,\tau) = \left(\frac{1}{\zeta}\frac{\partial}{\partial\zeta}\right)\left[\left(\frac{1}{\zeta}\frac{\partial}{\partial\zeta}\right)^{\nu-1}f(\zeta,\tau)\right], \tag{2.7}$$

where

$$\left(\frac{1}{\zeta}\frac{\partial}{\partial\zeta}\right)^{1}f(\zeta,\tau) = \left(\frac{1}{\zeta}\frac{\partial}{\partial\zeta}\right)f(\zeta,\tau),$$
$$\left(\frac{1}{\zeta}\frac{\partial}{\partial\zeta}\right)^{\nu}\left[\frac{J_{\mu}(\zeta)}{\zeta^{\mu}}\right] = (-1)^{\nu}\frac{J_{\mu+\nu}(\zeta)}{\zeta^{\mu+\nu}} \tag{2.8a–b}$$

(see also Appendix). The implementation of Eq. (2.6b) is illustrated in the following example.

Example 2.5

For $n = \nu = 1$, Eq. (2.6b) gives[5]

$$\frac{\partial}{\partial r}c_X(r,\tau) = -2\int_0^{\infty}\int_0^{\infty} dk\, d\omega\, k\sin(kr)e^{-i\omega\tau}\tilde{c}_X(k,\omega). \tag{2.9a}$$

Similarly, for $n = 2$, $\nu = 1$, Eq. (2.6b) gives

$$\frac{\partial}{\partial r}c_X(r,\tau) = -2\pi\int_0^{\infty}\int_0^{\infty} dk\, d\omega\, k^2J_1(kr)e^{-i\omega\tau}\tilde{c}_X(k,\omega), \tag{2.9b}$$

where J_1 denotes a Bessel function of the first kind and first order. Furthermore, using Eq. (2.8b) and the Bessel inequality $|J_{\nu}(kr)| < a(kr)^{-\frac{1}{2}}$ (a is a constant), it can be shown that the simultaneous two-point covariance function of an STIS in $R^{3,1}$ is space differentiable (Yadrenko, 1983).

[5]Notice that $J_{\frac{1}{2}}(r) = \sqrt{\frac{2}{\pi r}}\sin r$.

3. HIGHER-ORDER SPATIOTEMPORAL VARIOGRAM AND STRUCTURE FUNCTIONS

Higher-order variogram and structure functions that are particularly useful in ocean and atmospheric sciences are defined in a rather direct way. So, the νth-*order* structure function is given by

$$\xi_X^{[\nu]}(r,\tau) = \overline{[X(\boldsymbol{s}+\boldsymbol{\varepsilon}r, t+\tau) - X(\boldsymbol{s},t)]^\nu}, \tag{3.1}$$

where ν is a positive integer. Higher-order structure functions such as the above can provide additional information (compared to the standard second-order structure functions) about the space–time variation nature of the phenomenon. In this respect, the order ν may obtain a deeper physical meaning, as illustrated in this example.

Example 3.1

The following νth-order simultaneous two-point structure function is known to be very useful in fluid mechanics studies,

$$\xi_X^{[\nu]}(r) = \alpha_\nu (\upsilon r)^{\frac{\nu}{3}} \left(\frac{r}{\ell}\right)^{\lambda_\nu - \frac{\nu}{3}}, \tag{3.2}$$

where $r = |\boldsymbol{h}|$, and λ_ν, ℓ, α_ν, and υ are physical parameters (the time argument is routinely suppressed in these studies because of simultaneity). The order ν of the structure function plays a key role. A well-known special case of the above expression is the Kolmogorov νth-order structure function with $\lambda_\nu = \frac{\nu}{3}$. Other λ_ν values that have been also used in applications are $\lambda_\nu = \frac{\nu}{9} + 2\left[1 - \left(\frac{2}{3}\right)^{\frac{\nu}{3}}\right]$ and $\lambda_\nu = 1 - \ln_2\left(0.7^{\frac{\nu}{3}} + 0.3^{\frac{\nu}{3}}\right)$, which are associated with different fluid flow phenomena.

It should be noticed that a higher-order space–time structure function can be expressed in terms of the standard binomial theorem as follows

$$\xi_X^{[2\nu]}(r,\tau) = \sum_{i=0}^{2\nu} (-1)^i \binom{2\nu}{i} \overline{X(\boldsymbol{s}+\boldsymbol{\varepsilon}r, t+\tau)^{2\nu-i} X(\boldsymbol{s},t)^i}, \tag{3.3}$$

where $\binom{2\nu}{i} = C^i_{2\nu} = \frac{(2\nu)!}{i!(2\nu-i)!}$. Some special cases of Eq. (3.3) are of interest. For large lags r and τ, the $X(\boldsymbol{s},t)$ and $X(\boldsymbol{s}+\boldsymbol{\varepsilon}r,\ t+\tau)$ become statistically independent, i.e.,

$$\overline{X(\boldsymbol{s}+\boldsymbol{\varepsilon}r, t+\tau)^{2\nu-i} X(\boldsymbol{s},t)^i} = \overline{X(\boldsymbol{s},t)^{2\nu-i}}\ \overline{X(\boldsymbol{s},t)^i}$$

(the odd moments are zero), in which case Eq. (3.3) reduces to

$$\xi_X^{[2\nu]} \approx \sum_{i=0}^{\nu} \binom{2\nu}{2i} \overline{X(\boldsymbol{s},t)^{2\nu-2i}}\ \overline{X(\boldsymbol{s},t)^{2i}}.$$

Let us examine some specific yet insightful cases of the above results.

Example 3.2

For $\nu = 1$, Eq. (3.3) becomes the standard second-order structure function $\xi_X^{[2]}$, which for large r and τ, gives $\xi_X^{[2]} = \xi_X \approx 2\overline{X^2}$, and for $\nu = 2$, it gives $\xi_X^{[4]} \approx 2\overline{X^4} + 6\left(\overline{X^2}\right)^2$. Both of them are useful structure function approximations in practice.

4. SEPARABLE CLASSES OF SPACE–TIME ISOSTATIONARY COVARIANCE MODELS

I now turn to some of the most encountered space–time separable STIS covariance models, i.e., models that split the covariance into separate spatial and temporal components. One of the most common separable models is the (multiplicative) separable one of the form

$$c_X(r, \tau) = c_{X(1)}(r) c_{X(2)}(\tau). \tag{4.1}$$

We recall that if $X(\boldsymbol{p})$ is an STIS random field, its partial derivatives are STHS but not necessarily STIS random fields. Not surprisingly, in the case of separability, many geometrical properties of the STIS random fields (continuity, differentiability, integrability) are immediate consequences of the corresponding properties of STHS fields, taking into account, of course, the well-known space–time metric effects discussed in earlier parts of the book.

As we saw in Chapters VI and VII, under certain circumstances, spatial and temporal geometry can be considered separately, followed by the analysis of their joint effects. So, on occasion we may consider time regularity by fixing the space argument in the purely spatial covariance component $c_{X(1)}$ and focusing on $c_{X(2)}$. Similarly, we can study space regularity by fixing the time argument in the purely temporal component $c_{X(2)}$ and focusing on $c_{X(1)}$. We should always keep in mind that it is not possible, in principle, to indicate a complete analogue of isotropy in the time domain, with the exception, perhaps, of a limited analogue in the case of real-valued temporal fields with an even covariance, so that $c_{X(2)}(\tau) = c_{X(2)}(|\tau|)$.

Example 4.1

For the multiplicative separable random field $X(s,t)$ in $R^{1,1}$, the simultaneous m.s derivative $\frac{\partial^\nu}{\partial s^\nu} X(s, t)$, $\nu = 1, 2, \ldots$, exists if and only if the $(-1)^\nu c_X^{(2\nu,0)}(0, \tau)$ exists and is finite. Moreover, the random field derivative exists in the m.s. sense if and only if $c_X^{(2\nu-1,0)}(0, \tau) = 0$. Similar comments hold for the derivative $\frac{\partial^\mu}{\partial t^\mu} X(s, t)$, $\mu = 1, 2, \ldots$ (fixed s).

The list of Table 4.1 presents covariance models considered separately in space and time. These models have been used in many studies in oceanography, atmospheric physics, forestry, mining, and earth sciences, in general. In this list, the $c_X(\tau)$ models are formally the unidimensional versions of the $c_X(r)$ models. It is rather common knowledge that various valid separable space–time covariance functions can be constructed by using proper combinations of the $c_{X(1)}$ and $c_{X(2)}$ models of Table 4.1 (see, also, Chapter XVI). A brief summary of the main characteristics of these models and their physical interpretation is outlined next:

- The exponential model no. 1 is associated with (m.s.) continuous but not differentiable random fields. As regards its interpretation, this model represents attributes with a correlation (dependence) range $\varepsilon_\zeta \approx 3a_\zeta$ ($\zeta = r$ or τ) and is associated with not particularly smooth attribute variations. Since small- and large-scale heterogeneities are considerably suppressed, this model cannot explain satisfactorily, e.g., seismic wave scattering and travel-time variation data (Frankel and Clayton, 1986).
- The Gaussian (or squared exponential) covariance model no. 2 is continuous at $\zeta = 0$, where all its derivatives are finite. Hence, the random field is m.s. continuous and differentiable of any

Table 4.1 Separable Space–Time Covariance Models in $R^{n,1}$

No.	Name	Form of $c_{X(i)}(\zeta)$ ($\zeta = r$ (i = 1), $= \tau$ (i = 2)), $c_0 \geq 0$
1	Exponential (n = 1, 2, 3)	$c_0 e^{-\frac{\zeta}{a_\zeta}}$ $(a_\zeta > 0)$
2	Gaussian (n = 1, 2, 3)	$c_0 e^{-\frac{\zeta^2}{a_\zeta^2}}$ $(a_\zeta > 0)$
3	Expo-cosine (n = 1, 2, 3)	$c_0 e^{-\frac{\zeta}{a_\zeta}} \cos(b_\zeta \zeta)$ $(a_\zeta, b_\zeta > 0)$
4	Spherical (n = 1, 2, 3)	$\begin{cases} c_0\left[1 - \frac{3\zeta}{2a_\zeta} + \frac{\zeta^3}{2a_\zeta^3}\right], & \zeta \in [0, a_\zeta] \\ 0, & \zeta > a_\zeta \end{cases}$ $(a_\zeta > 0)$
5	von Kármán/Matérn (n = 1, 2, 3)	$\frac{c_0}{2^{\nu-1}\Gamma(\nu)}\left(\frac{\zeta}{a_\zeta}\right)^\nu K_\nu\left(\frac{\zeta}{a_\zeta}\right)$ $(a_\zeta, \nu > 0)$
6	Gamma-Bessel (n = 1, 2, 3)	$c_0 \Gamma(\nu+1)\left(\frac{2}{a_\zeta \zeta}\right)^\nu J_\nu(a_\zeta \zeta)$ $(a_\zeta > 0)$
7	Rational quadratic (n = 1, 2, 3)	$c_0\left(1 + \frac{\zeta^2}{a_\zeta^2}\right)^{-b_\zeta}$ $(a_\zeta, b_\zeta > 0)$
8	Gaussian-Bessel (n = 1, 2)	$c_0 e^{-\frac{\zeta^2}{a_\zeta^2}} J_0(b_\zeta \zeta)$ $(a_\zeta, b_\zeta > 0)$
9	Linear (n = 1)	$\begin{cases} c_0[1 - a_\zeta \zeta], & \zeta \leq a_\zeta^{-1} \\ 0, & \zeta \geq a_\zeta^{-1} \end{cases}$ $(a_\zeta > 0)$
10	Cauchy (n = 1, 2, 3)	$c_0(1 + (c_\zeta \zeta)^{\alpha_\zeta})^{-\nu}(\nu,\ c_\zeta > 0, \alpha_\zeta \in (0, 2])$
11	Powered exponential (n = 1, 2, 3)	$c_0\ e^{-\left(\frac{\zeta}{\alpha_\zeta}\right)^{b_\zeta}}$ $(a_\zeta > 0,\ b_\zeta \in (0, 2])$
12	Wave (n = 1, 2, 3)	$c_0 \frac{\sin a_\zeta \zeta}{\zeta}$ $(a_\zeta > 0)$
13	De Wijsian (n = 1, 2, 3)	$c_0 - \ln \zeta^{\alpha_\zeta}$ $c_0 - \ln(1 + \zeta^{\alpha_\zeta})$ $(\alpha_\zeta \in (0, 2))$
14	Linear exponential (n = 1, 2)	$c_0 e^{-\frac{\zeta}{\alpha_\zeta}}\left(1 + \frac{\zeta}{\alpha_\zeta}\right)$ $(\alpha_\zeta > 0)$

order. Its range is $\varepsilon_\zeta \approx \sqrt{3}a_\zeta$, and the model is associated with very smooth variations. Due to its considerable smoothness, model no. 2 cannot represent the inhomogeneities of many earth processes (e.g., Wu and Aki, 1985).

- The expo-cosine model no. 3 has the particularity that in R^3 it is valid only if $a_\zeta b_\zeta < 0.577$. This model too is associated with (m.s.) continuous but not differentiable random fields.

- The range of the spherical model no. 4 is practically $\varepsilon_\zeta \approx 0.8a_\zeta$. This model, which has been rather popular in geostatistics, is permissible in the case of a Normal (Gaussian) random field but not for a lognormal one.
- In the von Kármán or Matérn covariance model no. 5 (von Kármán, 1948), the Γ and K_ν denote the Gamma and the modified Bessel function, respectively. The correlation range is a function of ν, which controls the smoothness of the corresponding attribute (the ν controls how many times differentiable the model is at the origin; which, in turn, controls how many times m.s. differentiable a Gaussian random field with this covariance model is) . For $\mu = 1$, this model is sometimes called the Whittle model. For $\nu = \frac{1}{2}$, the model gives the exponential model no. 1, and when $\nu \rightarrow \infty$, it becomes the Gaussian model no. 2. If $\nu = \lambda + \frac{1}{2}$ (λ is a nonnegative integer), model no. 5 can be expressed as the product of a polynomial of degree λ in $\frac{\zeta}{a_\zeta}$ and $e^{-\frac{\zeta}{a_\zeta}}$.[6] If the range parameter a_ζ is large, model no. 5 approximates the power covariance model ($\nu > 0$) and the de Wijs model (when $\nu \rightarrow 0$). When ν is small ($\nu \rightarrow 0$), the variation of the corresponding natural attribute is rough. An alternative parameterization of model no. 5 is obtained by replacing a_ζ with $a_\zeta\left(2\mu^{\frac{1}{2}}\right)^{-1}$. The Kármán model is commonly used in geophysical studies of self-affine random media (Klimes, 2002a).
- In the Gamma–Bessel model no. 6, J_ν denotes the Bessel function of the first kind and νth order, with $\nu \geq \frac{n-2}{2}$. For $\nu = 0.5$ and $\nu = -0.5$ this model reduces, respectively, to the wave model (see model no. 12) $c_0\frac{\sin(\zeta)}{\zeta}$ (valid for $n = 1, 2, 3$), and the hole effect model $c_0\frac{\cos(\zeta)}{\zeta}$ (valid for $n = 1$).
- In the rational quadratic covariance model no. 7, the range coefficient α_ζ controls how fast the model approaches zero with ζ, and the roughness parameter b_ζ controls the smoothness of the random field realizations. This model corresponds to an m.s. differentiable random field for all values of a_ζ, $b_\zeta > 0$. The model has been used in applications, such as the prediction of wax precipitation in crude oil systems.
- Model no. 8 is basically a combination of Gaussian–Bessel functions (J_0 is the Bessel function of the first kind and zeroth order). Such combinations have been tested, inter alia, in medical applications, including corneal endothelium studies.
- The linear model no. 9 is a special case of the power variogram (with exponents between 0 and 2), which is used in petroleum engineering. In this more general model, the closer the exponent is to 0, the more random the variation. The model corresponds to random fields with homogeneous increments and is also linked to fractals.
- The Cauchy model no. 10 has been suggested as a suitable model to describe gravity or magnetic data (yet, generally, this is not widely used). The model is a completely monotone function if $\alpha_\zeta \in (0,1)$, $\nu > 0$.
- The powered exponential covariance model no. 11 is continuous at the origin for $b_\zeta \in (0,1]$. The corresponding random field is m.s. continuous but not m.s. differentiable. For $b_\zeta \in (0,2]$, the covariance is continuous at the origin, but only in the case $b_\zeta = 2$ it is continuous everywhere, and,

[6]A few examples: for $\mu = \frac{1}{2}$, model no. 5 gives $c_0 e^{-\frac{\zeta}{a_\zeta}}$; for $\mu = \frac{3}{2}$, it gives $c_0\left(\frac{\zeta}{a_\zeta} + 1\right)e^{-\frac{\zeta}{a_\zeta}}$; and for $\mu = \frac{5}{2}$, it gives $c_0\left[\left(\frac{\zeta}{a_\zeta}\right)^2 + 3\frac{\zeta}{a_\zeta} + 3\right]e^{-\frac{\zeta}{a_\zeta}}$.

Table 4.2 Properties of Correlation Functions

No.	Formulation
1	$\rho_X(r) \in I_n : \rho_X(r) \geq -\frac{1}{n}$ for all r $\rho_X(r) \in I_{n,0} : \rho_X(r) \geq \inf_{u \geq 0}\left[\left(\frac{n}{2}-1\right)!\left(\frac{2}{u}\right)^{\frac{n}{2}-1} J_{\frac{n}{2}-1}(u)\right]$
2	$\left.\begin{matrix}\rho_X(r) \in I_n \quad (n>1) \\ \rho_X(r^*) = 1 \quad \text{for some } r^* > 0\end{matrix}\right\} : \rho_X(r) = 1$ for all r
3	$\rho_X(r) \in I_{n,c} \quad (n>1)$: $\rho_X(r) = a + (1-a)\rho_X^*(r)$, $\left(a \in [0,1], \rho_X^*(r) \text{ continuous}, \lim_{r \to \infty} \rho_X^*(r) = 0\right)$
4	$I_n - I_{n,0} = \varnothing \quad (n > 1)$

also, it is m.s. differentiable. Similar geometrical properties exhibit the exponential covariance (model no. 1) and the expo—cosine covariance (model no. 3).

- The wave model no. 12 characterizes attributes exhibiting strong periodicities, and it has been associated with patterns where points at longer distances may be more correlated than points at shorter distances.
- The De Wijsian model no. 13 has been used to describe spatial patterns of ore grades in mining applications, and it is linked to fractal processes.
- The linear exponential model no. 14 is an interesting special case of model no. 5. Model no. 14 is widely used in geopotential and wind-field correlation studies (Thiébaux, 1985).

In the case of separability, certain properties of the spatial component $c_X(r)$ of Eq. (4.1) should be kept in mind. Let $\rho_X(r) = \frac{c_X(r)}{c_0}$ be the associated isotropic correlation function. Three important classes of correlation functions are considered: (*a*) the class I_n of isotropic, in general, correlation functions in R^n, (*b*) the class $I_{n,0}$ of isotropic correlation functions in R^n that are everywhere continuous except, perhaps, at the origin, and (*c*) the class $I_{n,c}$ of isotropic correlation functions in R^n which are continuous everywhere. Clearly, $I_{n,c} \subset I_{n,0} \subset I_n$. By means of the above classification, a number of interesting properties of $\rho_X(r)$ can be derived, as shown in Table 4.2 (for a more detailed review with relevant references, see Christakos, 1992).

Property no. 1 is a direct consequence of the representation of Eq. (1.5a) restricted in the purely spatial domain. Concerning property no. 3, the $\rho_X(r)$ can be decomposed into three parts (specifically, an additional delta function component is included in the right side of the formula shown in Table 4.2).[7] Property no. 4 is the celebrated Schoenberg conjecture. Some useful practical results can be obtained from these properties.

Example 4.2

For $n = 1, 2, 3$, and 4, property no. 1 gives $\rho_X(r) \geq \inf_{u \geq 0} \cos(u) = -1$, $\rho_X(r) \geq \inf_{u \geq 0} J_0(u) \approx -0.403$, $\rho_X(r) \geq \inf_{u \geq 0}\frac{\sin(u)}{u} \approx -0.218$, and $\rho_X(r) \geq \text{Inf}_u\left[\frac{2}{u}J_1(u)\right] \approx -0.133$, respectively.

I conclude this section by stressing the following points: Separability is a modeling assumption that is often used since it produces results that are valid (provided that their purely spatial and purely temporal

[7]Otherwise said, all isotropic correlations are a mixture of a nugget effect and a continuous isotropic correlation.

components hold independently) and it is also easy to analyze and interpret. On the other hand, the main drawback of separability is that it may fail to incorporate key space–time interconnections.

5. A SURVEY OF SPACE–TIME COVARIANCE MODELS

Several space–time nonseparable covariance models are listed in Table 5.1,[8] including models for STHS and STIS random fields. The step-by-step techniques used for constructing these space–time covariance models are discussed in Chapter XVI. An important reminder here is that a function that is a valid covariance model in R^{n+1} ($R^{n,1}$), it is also a valid model in $R^{n'+1}$ ($R^{n',1}$) for $n' < n$ (the converse is not necessarily true). As noted earlier, this is sometimes called the hereditary property. Most of the models in Table 5.1 have closed-form analytical expressions, whereas some others are expressed in terms of an integral that needs to be calculated numerically. Notice that in Table 5.1, $r = |\boldsymbol{h}|$, $E_n = \frac{2\Gamma(n)}{\sqrt{\pi}\Gamma\left(\frac{n-1}{2}\right)}$, and $B_n = (4\alpha\pi\,\tau)^{-\frac{1}{2}}E_n$. More specifically:

- In the space–time nonseparable covariance model no. 1, the permissibility condition is simply $b > 0$. Model no. 2 ($\alpha > 0$) is inspired from the diffusion equation, and it tends to a delta function as $\tau \to 0$. The symbol "$\cong$" denotes that the model presents this functional form asymptotically (as $r \to \infty$ and $\tau \to \infty$) but not close to the origin. To obtain permissible covariance models, the singularity at zero lag must be tamed, e.g., by means of a short-range cutoff. The cutoff can be implemented either in real space or in frequency space (in the frequency space it is a high-frequency cutoff, Hristopulos, 2003). In either case, the modified function needs to be checked for permissibility. The shape of the covariance changes with the n and α values. Clearly, the same is true for the correlation ranges and the behavior near the space–time origin. Other formulations and extensions of model no. 2 are possible (to deal with the singularity at zero or to account for physical features of the underlying process). Hristopulos (2002) and Gneiting (2002) proposed formulations that involve the addition of constants after the time lag.
- Model no. 3 is a modification of model no. 2 with $0 \leq \beta \leq 1$ and $0 < \gamma \leq 1$. Applications of this covariance class can be found in fluid mechanics studies (Monin and Yaglom, 1971). In model no. 4, $\tau > 0$, and the KummerM function is a solution to the Kummer's differential equation. And in model no. 5, Erfc(x) is a complementary error function, and $r, \tau > 0$.
- In model no. 6, the a and c are constant coefficients. The model tends to $2Erfc\left(a\left(\frac{\tau}{c}\right)^{\frac{1}{2}}\right)$ as $r \to 0$, and to σ^2 as $r, \tau \to 0$. The values of the coefficients affect the correlation ranges and shape of the model. The covariance model, e.g., decreases faster for increasing values of the $\frac{a}{c}$ ratio. Another observation is that the model declines faster in the spatial direction than with time.
- In model no. 7, $A = e^{0.25D}\left[4\left(\pi D^3\right)^{\frac{1}{2}}\right]^{-1}$ and $\alpha, z, D > 0$. The shape and the correlation range of the model depend on the parameters α and z. The magnitude of the model's slope increases with the ratio α/z.

[8]Surely, this list is by no means exhaustive.

Table 5.1 Space–Time Covariance Models in $R^{n,1}$

No.	c_X	n
1	$e^{-\frac{r^2}{2(1+b\,\tau^2)}}\begin{cases}(1+b\,\tau^2)^{-1.5}\left[1-0.5r^2(1+b\,\tau^2)^{-1}\right] \\ (1+b\,\tau^2)^{-2.5}\left\{1-r^2(1+b\,\tau^2)^{-1}+r^4\left[8(1+b\,\tau^2)^2\right]^{-1}\right\}\end{cases}$	1,2
2	$\cong(4\alpha\pi\,\tau)^{-0.5n}e^{-\frac{r^2}{4\alpha\tau}}$	1,2,3
3	$(\beta\,\tau^{2\gamma}+1)^{-0.5n}e^{-r^2(\beta\tau^{2\gamma}+1)^{-1}}$	1,2,3
4	$\frac{1}{8}\left(\frac{\pi}{\alpha\tau}\right)^{0.5}E_2\,KummerM\left[0.5,2,-r^2(4\alpha\tau)^{-1}\right]$	1,2
5	$0.5r^{-1}E_3\left[\left(1-2\frac{\alpha\tau}{r^2}\right)Erfc\left(\frac{r}{2}\left(\frac{1}{\alpha\tau}\right)^{0.5}\right)+\frac{2}{r}\left(\frac{\alpha\tau}{\pi}\right)^{0.5}e^{-r^2(4\alpha\tau)^{-1}}\right]$	1,2,3
6	$0.5\left[e^{-a\,r}Erfc\left(a(c^{-1}\,\tau)^{0.5}-0.5\,r(c\,\tau^{-1})^{0.5}\right)+e^{a\,r}Erfc\left(a(c^{-1}\,\tau)^{0.5}+\frac{1}{2}\,r(c\,\tau^{-1})^{0.5}\right)\right]$	1,2,3
7	$\cong A\left(r^{z-2\alpha}\tau^{-1}\right)^{0.5}\;e^{-r^{-z}\,\tau}$	1
8	$\beta_n\,r^{1-0.5n}\int_0^\infty dk\,k^{0.5n}\,e^{-c^{-1}(k^2+\alpha^2)^p\tau}(k^2+\alpha^2)^{-p}J_{\frac{n}{2}-1}(kr)$	1,2,3
9	$e^{-\frac{\lvert \boldsymbol{r}\pm\boldsymbol{v}\,\tau\rvert}{\alpha}}$	1,2,3
10	$e^{-\frac{(\boldsymbol{r}\pm\boldsymbol{v}\,\tau)^2}{\alpha^2}}$	1,2,3
11	$\left[1+\frac{(\boldsymbol{r}\pm\boldsymbol{v}\,\tau)^2}{\beta^2}\right]^{0.5\nu}e^{-\frac{\lvert\boldsymbol{r}\pm\boldsymbol{v}\,\tau\rvert}{\alpha}}$	1,2,3
12	$\frac{\sin(\beta\tau)}{\pi^4\tau}\prod_{i=1}^{3}\frac{\sin(a_ih_i)}{h_i}$	1,2,3
13	$\sigma^2\,\widehat{f}_z(\tau/r^\beta;u_c)\widehat{f}_\alpha(r;w_c)$	1,2,3
14	$\sum_{j,k=0}^{\infty}\left\{\begin{matrix}c_{jk}\chi_{1j}(\boldsymbol{s})\,\chi_{1k}(\boldsymbol{s}') \\ \frac{A_jA_kc_{\chi(j,k)}(\boldsymbol{s},\boldsymbol{s}')}{A_jA_k\,\chi_{1j}(\boldsymbol{s})\,\chi_{1k}(\boldsymbol{s}')}\end{matrix}\right\}\chi_{2j}(t)\,\chi_{2k}(t')-\overline{X(\boldsymbol{s},t)}\,\overline{X(\boldsymbol{s}',t')}$	1

Continued

Table 5.1 Space–Time Covariance Models in $R^{n,1}$—cont'd

No.	c_X	n
15	$\cong r^{-1} g\left(\frac{r^z}{\tau}\right)$	1,2,3
16	$r^{\alpha z-1}\tau^{-\alpha} \quad (r^z\tau^{-1} << 1)$ $\tau^b r^{-bz-1} \quad (r^z\tau^{-1} >> 1)$	1,2
17	$\left(\frac{B}{(\tau-\zeta)^\lambda}\right)\phi\left(\frac{r^2}{\chi(\tau-\zeta)}\right)$	1,2,3
18	$\cong (b\,\tau)^{-m} e^{-\frac{r^2}{\alpha\,\tau}}$	1,2,3
19	$\left(1+b\,\tau^2\right)^{-1.5}\left[1-0.5r^2\left(1+b\,\tau^2\right)^{-1}\right]e^{-0.5r^2(1+b\,\tau^2)^{-1}}$	1,2
20	$\left(1+b\,\tau^2\right)^{-2.5}\left\{1-r^2\left(1+b\,\tau^2\right)^{-1}+r^4\left[8\left(1+b\,\tau^2\right)^2\right]^{-1}\right\}e^{-0.5r^2(1+b\,\tau^2)^{-1}}$	1,2
21	$\left(1+\frac{\tau^2}{\nu^2}\right)^{-0.5\nu} e^{-\frac{\boldsymbol{h}}{a}}$	1,2,3
22	$e^{-\left(\frac{r^2}{a^2}+\frac{\tau^2}{b^2}\right)^{0.5}}$	1
23	$\frac{\pi\eta_0}{2\eta_1^{0.5}}\left[e^{-\eta_1^{-0.5}\xi^{-1}r}\,Erfc\left(\left(\widetilde{D}\lvert\tau\rvert\right)^{0.5}-0.5\xi^{-1}r\left(\widetilde{D}\eta_1\lvert\tau\rvert\right)^{-0.5}\right)+e^{\eta_1^{-0.5}\xi^{-1}r}\,Erfc\left(\left(\widetilde{D}\lvert\tau\rvert\right)^{0.5}+0.5\xi^{-1}r\left(\widetilde{D}\eta_1\lvert\tau\rvert\right)^{-0.5}\right)\right]$	1
24	$\frac{\pi^2\eta_0\xi}{\eta_1 r}\left[e^{-\eta_1^{-0.5}\xi^{-1}r}\,Erfc\left(\left(\widetilde{D}\lvert\tau\rvert\right)^{0.5}-0.5\xi^{-1}r\left(\widetilde{D}\eta_1\lvert\tau\rvert\right)^{-0.5}\right)-e^{\eta_1^{-0.5}\xi^{-1}r}\,Erfc\left(\left(\widetilde{D}\lvert\tau\rvert\right)^{0.5}+0.5\xi^{-1}r\left(\widetilde{D}\eta_1\lvert\tau\rvert\right)^{-0.5}\right)\right]$	1,2,3
25	$a_1\delta(\boldsymbol{h},\tau)+a_2\delta(\boldsymbol{h})+a_3\delta(\tau)$	1,2,3
26	$c_0\left(1+\frac{r^{\lambda_1}}{a_1^2}+\frac{\tau^{\lambda_2}}{a_2^2}\right)^{-\lambda_3}$	1,2,3
27	$\frac{\left[1+\left(a_1\tau^{2\lambda}\right)^{\lambda\mu}\right]^\mu}{\left(1+a_1\tau^{2\lambda}\right)^\mu\left[1+\left(a_1\tau^{2\lambda}\right)^{\lambda\mu}+a_2r^{2\lambda}\right]^\mu}$	1,2
28	$e^{-a\lvert\tau\rvert}Erfc\left(\gamma_X(\boldsymbol{h})^{0.5}-\frac{a\lvert\tau\rvert}{2\gamma_X(\boldsymbol{h})^{0.5}}\right)+e^{a\lvert\tau\rvert}Erfc\left(\gamma_X(\boldsymbol{h})^{0.5}+\frac{a\lvert\tau\rvert}{2\gamma_X(\boldsymbol{h})^{0.5}}\right)$	1,2,3

29	$c_X(h,0) + \sum_{i=1}^{k} w_i(h)\lvert\tau\rvert^{a_i} + u_h(\tau)$	1
30	$\frac{\pi^2}{16a^6} e^{br}\mathrm{Erfc}\left(ab\lvert\tau\rvert^{0.5} + \frac{r}{2a\lvert\tau\rvert^{0.5}}\right)\left(\frac{1}{b^3} - \frac{r}{b^2} + \frac{4a^4\tau^2}{r}\right)$ $+\frac{\pi^2}{16a^6} e^{-br}\mathrm{Erfc}\left(ab\lvert\tau\rvert^{0.5} - \frac{r}{2a\lvert\tau\rvert^{0.5}}\right)\left(\frac{1}{b^3} + \frac{r}{b^2} - \frac{4a^4\tau^2}{r}\right) + \frac{\pi^{1.5}\lvert\tau\rvert^{0.5}}{4a^5b^2} e^{-a^2b^2\lvert\tau\rvert - \frac{r^2}{4a^2\lvert\tau\rvert}}$	1,2,3
31	$a[(2\nu + n + 1)M_\nu(r_p) - 2b\beta_1(\boldsymbol{h}\cdot\boldsymbol{u})\beta_2\tau M_{\nu-1}(r_p)]$	1,2,3
32	$\frac{\alpha^n \pi^{\frac{n}{2}}}{2^{n+\gamma(\tau)-1} 2\Gamma\left(\nu + \gamma(\tau) + \frac{1}{2}n\right)} M_{\nu+\gamma(\tau)}(r_T)$	1,2,3
33	$\frac{1}{\gamma_{X(2)}(\tau) - \gamma_{X(1)}(\boldsymbol{h})}\left[e^{-\gamma_{X(1)}(\boldsymbol{h})} - e^{-\gamma_{X(2)}(\tau)}\right]$	1,2,3
34	$\frac{1}{\gamma_{X(1)}(\boldsymbol{h}) + \gamma_{X(2)}(\tau)}\left[1 - e^{-\gamma_{X(1)}(\boldsymbol{h}) - \gamma_{X(2)}(\tau)}\right]$ $(=1 \text{ if } (\boldsymbol{h},\tau) = (\boldsymbol{0},0))$	1,2,3
35	$\frac{\alpha_1\alpha_2 e^{-\gamma_{X(1)}(\boldsymbol{h}) - \gamma_{X(2)}(\tau)}}{1 - \alpha_0 e^{-\gamma_{X(1)}(\boldsymbol{h}) - \gamma_{X(2)}(\tau)}} \frac{e^{-\gamma_{X(1)}(\boldsymbol{h})} + e^{-\gamma_{X(2)}(\tau)} - (1+\alpha_0)e^{-\gamma_{X(1)}(\boldsymbol{h}) - \gamma_{X(2)}(\tau)}}{\left[1 - (1-\alpha_1)e^{-\gamma_{X(1)}(\boldsymbol{h})}\right]\left[1 - (1-\alpha_2)e^{-\gamma_{X(2)}(\tau)}\right]}$	1,2,3
36	$\frac{a}{2\sigma^2} r\, I_0\left(\frac{c}{2\sigma^2} r\tau\right) e^{-\frac{r^2 + c^2\tau^2}{4\sigma^2}}$	1,2
37	$c_0\left[1 + \frac{\gamma_{X(1)}(\boldsymbol{h}) + \gamma_{X(2)}(\tau)}{\alpha_0}\right]^{-\beta_0}\left[1 + \frac{\gamma_{X(1)}(\boldsymbol{h})}{\alpha_1}\right]^{-0.5\beta_1}\left[1 + \frac{\gamma_{X(2)}(\tau)}{\alpha_2}\right]^{-\beta_2} \frac{K_{\beta_1}\left(2\left(\left(\alpha_1 + \gamma_{X(1)}(\boldsymbol{h})\right)\varepsilon\right)^{0.5}\right)}{K_{\beta_1}\left(2(\alpha_1\varepsilon)^{0.5}\right)}$	1,2,3
38	$v^2\left(\frac{\tau_0}{\tau}\right)^{0.5}\left(\delta(h - v\tau) + \delta(h + v\tau) - \frac{\theta(v\tau - \lvert h\rvert)}{2v\tau}\right)$	1
39	$0.5\left[e^{-\beta r} Erfc\left(\frac{2\beta\tau - \alpha r}{2\sqrt{\alpha\tau}}\right) + e^{\beta r} Erfc\left(\frac{2\beta\tau + \alpha r}{2\sqrt{\alpha\tau}}\right)\right]$	1

- In model no. 8, $\beta_n = \frac{(2\pi)^{\frac{n}{2}+1}}{2c}$, and $J_{\frac{n}{2}-1}$ denotes the Bessel function of the first kind and order $\frac{n}{2}-1$. The integral in this model needs to be calculated numerically. Kolovos et al. (2004) considered the case $p = n = 2$ for varying values of the parameters α and c. The plotted covariance was shown to decline faster along the time direction than along the space direction.
- In model nos. 9–11, $\alpha, \beta, \nu > 0$. These models are useful in several physical applications, e.g., in cases in which the "frozen" random field hypothesis applies.
- In model no. 12, as $\alpha_i, \beta \rightarrow \infty$, the covariance is proportional to $\delta(\boldsymbol{h},\tau)$ and the random field tends to a white noise. Fractal models exhibit a power-law asymptotic decay of the correlations with a noninteger exponent. The models can be fractal either in space (in which case the asymptotic behavior refers to $r \rightarrow \infty$) or in time (in which case the asymptotic behavior refers to $\tau \rightarrow \infty$).
- The model no. 13 has fractal properties, where $\widehat{f}_\nu(r; u_c) = f_\nu(r; u_c) f_\nu^{-1}(0; u_c)$ and $f_\nu(r; u_c) = \Gamma(-\nu)^{-1} \int_0^{u_c} du\, e^{-ur}\, u^{-(\nu+1)}$. As is shown in Christakos et al. (2000), the function $\widehat{f}_z\left(\frac{\tau}{r^\beta}; u_c\right)$ has an unusual dependence on the space and time lags through $\frac{\tau}{r^\beta}$. For large τ and r, $\frac{\tau}{r^\beta} \rightarrow 0$ and $\widehat{f}_z\left(\frac{\tau}{r^\beta}; u_c\right) \rightarrow 1$. With regard to $\widehat{f}_z\left(\frac{\tau}{r^\beta}; u_c\right)$, two pairs of spatiotemporal points are "equidistant" if $\tau_1 r_1^{-\beta} = \tau_2 r_2^{-\beta}$ (in contrast with, e.g., a Gaussian spatiotemporal covariance function where equidistant lags satisfy the equation $r^2 a_r^{-2} + \tau^2 a_\tau^{-2} = c$).[9] Permissibility conditions for this model imply that $-1 < z < 0$ and $-0.5(n+1) < \alpha - \beta z < 0$ in $R^{n,1}$. These conditions can be relaxed by cutting off the short- and long-range behavior of the model using the methods in Hristopulos (2003). Numerical illustrations of this model are plotted in Kolovos et al. (2004) for several values of the parameters z, β and α.
- Model no. 14 represents space–time heterogeneous attributes, due to a number of reasons, including the boundary and initial condition (BIC) effects. The χ_{1j} and χ_{2j} are partial differential equation modes with amplitudes A_j (deterministic or random) determined from the BIC of the physical law; in the first equation, c_{ij} are mode coefficient correlations; in the second equation, $c_{\chi(j,k)}$ are the mode correlations $\overline{\chi_{1j}(\boldsymbol{s})\chi_{1k}(\boldsymbol{s}')}$ and A_j are deterministic mode amplitudes; and in the third equation, A_j are random variables determined from the BIC.[10]
- In model nos. 15 and 16, z, α, b and $g(x)$ are suitable coefficient and function (for details, see Christakos and Hristopulos, 1998). In model no. 17, B, ζ, and λ are physical coefficients, and ϕ is a physically determined function (Monin and Yaglom, 1971).
- In model 18, the coefficients m, α, and b obtain physical meaning in the context of the turbulence study considered. This model may be seen as an extension of the covariance model 2.
- In model nos. 19 and 20, $b \geq 0$. For $b = 0$ among the noticeable features of the covariance is the presence of "hole effects," mainly, along the space direction. And, in model no. 21, $a, \nu > 0$, and in model no. 22, $a, b > 0$.
- In model nos. 23 and 24, $\widetilde{D} = \frac{D}{\xi^n \eta_0}$, where D is a diffusion coefficient, $\eta_0, \eta_1 > 0$, ξ are coefficients, and $\mu \geq 0$ is a normalization constant.

---

[9]In contrast with, e.g., a Gaussian spatiotemporal covariance function where equidistant lags satisfy the equation $r^2 a_r^{-2} + \tau^2 a_\tau^{-2} = c$.

[10]That is, randomness in the models can be introduced by the BIC leading to random coefficients A_j, the differential operator $\mathcal{L}_s$ leading to random eigenfunctions χ_{1j}, and by both of the above.

- In the nugget model no. 25 the coefficients $a_i \geq 0$ $(i = 1, 2, 3)$. In model no. 26, $\lambda_1, \lambda_2 \in (0, 2)$, a_1, a_2, c_0, $\lambda_3 > 0$. In model no. 27, λ, $\mu \in (0, 1)$ and $a_1, a_2 > 0$. And, in model no. 28, $a > 0$ and $\gamma_X(\boldsymbol{h})$ is a continuous variogram model in R^n.
- In model no. 29,[11] $w_i(h)$ are even functions with $w_i(0) \neq 0$, and $u_h(\tau) = O(\tau^2)$ as $\tau \to 0$,[12] with a bounded second derivative. And, model no. 30 $(a, b > 0)$ is infinitely differentiable away from the origin. The corresponding random field is once m.s. differentiable along any spatial direction (time origin), and not differentiable in time (space origin).
- In model no. 31, a, v, β_1, $\beta_2 > 0$, $b \in [0,1]$, $M_\nu(u) = u^\nu K_\nu(u)$, the vector $\boldsymbol{u} \in R^n$ has unit length, and $r_p = \left(\beta_1^2 r^2 + \beta_2^2 \tau^2\right)^{\frac{1}{2}}$ is a Pythagorean metric in $R^{n,1}$, see Eq. (2.15a) of Chapter I with $\varepsilon_{ii} = \beta_1^2$ $(i = 1, \ldots, n)$ and $\varepsilon_{00} = \beta_2^2$. This model has the same smoothness features in space and in time.
- Model no. 32 can have any differentiability degree across space and time, $r_T = \alpha|\boldsymbol{h} - \theta\tau\boldsymbol{u}|$, where $\alpha > 0$, $\theta \geq 0$, $\boldsymbol{u} \in R^n$ is a unit vector, and $\gamma(\tau)$ is a valid temporal variogram model.
- In models nos. 33 and 34, $\gamma_{X(1)}(\boldsymbol{h})$ and $\gamma_{X(2)}(\tau)$ are a purely spatial and a purely temporal variogram, respectively (in addition, in model no. 33, $\gamma_{X(1)} \neq \gamma_{X(2)}$). And, in model no. 35, $\alpha_i \in [0,1]$, $i = 1, 2, 3$, $\sum_{i=1}^{3} \alpha_i = 1$. Despite its theoretical interest, this model has not yet found considerable real-world applications.
- In model no. 36, a, c, $\sigma > 0$, and I_0 is the modified Bessel function. A dominant feature of this model is the tilted ridge of high correlation values that extend from the origin and diagonally in the space−time domain. In model no. 37, c_0, α_0, α_2, $\beta_2 > 0$ if $\beta_0 \geq 0$; $\alpha_1 > 0$ and $\boldsymbol{\varepsilon} \geq 0$ if $\beta_1 > 0$; α_1, $\boldsymbol{\varepsilon} > 0$ if $\beta_1 = 0$; and $\alpha_1 \geq 0$ and $\boldsymbol{\varepsilon} > 0$ if $\beta_1 < 0$. And, in model no. 38, the υ, $\tau_0 > 0$ are empirically calculated physical constants, and θ here denotes the step function. Lastly, model no. 39 is similar in structure to the models nos. 6, 23, and 24.

As we will see in Chapter XVI, starting from any covariance model of lower dimensionality in Table 5.1 and using space transformations, the corresponding models in higher dimensionality can be derived. As the list of Table 5.1 shows, it is not particularly difficult to construct a plethora of space-time covariance models that are formally valid; what is often challenging is to make sure that these models are physically meaningful representations of the space-time variability features of the phenomenon.

6. SCALES OF SPATIOTEMPORAL DEPENDENCE AND THE UNCERTAINTY PRINCIPLE

In many applications, the physical conditions require that some stages of the investigation focus on specified spatial and temporal properties of the phenomenon (e.g., varying dependency levels along space and time, anisotropic properties of space, and principal differences between spatial and temporal regularity characteristics of the phenomenon). Focusing on the space dependence of a random field brings out the features due to its dependence on several spatial coordinates, whereas when we are concerned with the time dependence of a random field, we seek to bring out the features due to its dynamic variation. In this setting, the notions of *spatiotemporal dependency scales* provide measures of the extent of long- and short-range correlations in space and interrelationships in time, and measures of the degrees of attribute distance and period of influence in space and time, respectively. Each of

[11]Model nos. 29−35 have been studied in Ma (2002b, 2003b), and Stein (2005).

[12]That is, $u_h(\tau) = O(\tau^2)$ as $\tau \to 0$ if and only if there exist positive numbers ε and λ such that $u_h(\tau) \leq \lambda|\tau|$ for $|\tau| < \varepsilon$.

these scales, while focusing either on a spatial dimension or on time, it does not lose track of its temporal or spatial links, respectively. Furthermore, in the spatial domain, and since location $\boldsymbol{s}$ is a vector, the dependence or correlation scales should depend on the directions of the location vector and the direction of the separation vector ($\boldsymbol{h}$) between any pair of locations $\boldsymbol{s}$ and $\boldsymbol{s}'$. In the general case, the following scales are directly obtained from the covariance function.

Definition 6.1

The space–time-dependent *effective distance* $r_{e,j}(\boldsymbol{s},t)$ of a random field $X(\boldsymbol{s},t)$ along the s_j direction, and the corresponding *effective period* $\tau_{e,j}(\boldsymbol{s},r)$ are defined as, respectively,

$$\begin{aligned} r_{e,j}(\boldsymbol{s},t) &= \int_0^\infty dr\, \frac{c_{X,j}(r,\boldsymbol{s},t)}{c_{X,j}(0,\boldsymbol{s},t)}, \\ \tau_{e,j}(\boldsymbol{s},r) &= \int_0^\infty dt\, \frac{c_{X,j}(r,\boldsymbol{s},t)}{c_{X,j}(r,\boldsymbol{s},0)}, \end{aligned} \tag{6.1a–b}$$

where $c_{X,j}(r,\boldsymbol{s},t) = c_X(\boldsymbol{h}\cdot\boldsymbol{\varepsilon}_j,\boldsymbol{s},t)$ is the direction-specific covariance of Eq. (2.22) of Chapter IV specified by measuring correlations between field values at pair of points along the s_j direction with $r = h_j$, and the integration in Eq. (6.1a) is along the direction specified by the unit vector $\boldsymbol{\varepsilon}_j$.

In Definition 6.1 it is assumed, of course, that the integrals exist. To gain additional insight, we will consider Eqs. (6.1a–b) under the usual restrictions on space–time variability (e.g., STHS or STIS), in which case some practically interesting scales of spatial and temporal variation emerge.

6.1 SCALES FOR SPATIOTEMPORAL RANDOM FIELDS WITH RESTRICTED SPACE–TIME VARIABILITY

Let us start with the case of STHS random fields such that the expression of the direction-specific covariance above reduces to

$$c_{X,j}(r,\tau) = c_X\left(\boldsymbol{h}\cdot\boldsymbol{\varepsilon}_j,\tau\right) = c_X\left(0,\ldots,h_j,\ldots,0,\tau\right), \tag{6.2}$$

where the covariance is independent of the specific location $\boldsymbol{s}$ and depends only on the lag magnitude $r = h_j$. I continue with the definition of the effective distance and the effective period for STHS random fields, which are special cases of Eqs. (6.1a–b).

Definition 6.2

The time-dependent *effective distance* $r_{e,j}(\tau)$ of an STHS $X(\boldsymbol{s},t)$ along the s_j direction, and the space-dependent *effective period* $\tau_{e,j}(r)$ are defined as, respectively,

$$\begin{aligned} r_{e,j}(\tau) &= \int_0^\infty dr\, \frac{c_{X,j}(r,\tau)}{c_{X,j}(0,\tau)}, \\ \tau_{e,j}(r) &= \int_0^\infty d\tau\, \frac{c_{X,j}(r,\tau)}{c_{X,j}(r,0)}, \end{aligned} \tag{6.3a–b}$$

where, as before, the integration in Eq. (6.3a) is along the direction $\boldsymbol{\varepsilon}_j$, and it is assumed that the integrals exist. The $r_{e,j}(\tau)$ and $\tau_{e,j}(r)$ are also termed the *effective ranges* or *integral scales*.

Let me make a few comments concerning the interpretation of these ranges or scales. For each index j, the $r_{e,j}(\tau)$ and $\tau_{e,j}(r)$ are, respectively, the effective spatial and temporal ranges over which significant correlations prevail. These effective ranges offer useful measures of how long a spatial (temporal) lag needs to be in order for the largest part of the S/TRF to become uncorrelated in space (time). Typically, this requires a few integral scales. Each effective range will be different for each j, unless stricter assumptions, beyond STHS, are valid for the random field, such as STIS. In other words, it is important to specify which effective range (integral scale) we refer to, which components of the vector locations are considered, and along which direction the integration in Eq. (6.3a) is carried out.

Example 6.1

In turbulence studies an effective distance (also called a correlation length) may express the energy-containing scale (Matthaeus et al., 2016). In the $R^{2,1}$ domain, in particular, and focusing at the space–time origin ($r = 0$ and $\tau = 0$), Eqs. (6.3a–b) become

$$\begin{aligned} r_{e,1} &= \frac{1}{c_0}\int_0^{\infty} dh_1\, c_X(h_1,0,0), \\ r_{e,2} &= \frac{1}{c_0}\int_0^{\infty} dh_2\, c_X(0,h_2,0), \\ \tau_e &= \frac{1}{c_0}\int_0^{\infty} d\tau\, c_X(0,0,\tau), \end{aligned} \tag{6.4a–c}$$

where $c_0 = c_X(0,0,0)$. In terms of the energy spectrum, the $r_{e,j}$ ($i = 1, 2$) may be linked with the wavenumber of maximum energy (a similar observation is valid for τ_e and its relation with frequency). An analogous timescale can also be constructed. Obviously, for $r_{e,j}$ and τ_e to exist, the integrals above must exist. For this to happen, some necessary conditions are

$$h_1 c_X(h_1,0,0) \xrightarrow{|h_1|\to\infty} 0, \quad h_2 c_X(0,h_2,0) \xrightarrow{|h_2|\to\infty} 0, \quad \tau c_X(0,0,\tau) \xrightarrow{|\tau|\to\infty} 0.$$

Physically, it is possible that

$$c_X(h_1,0,0) \xrightarrow{|h_1|\to\infty} |h_1|^{-\lambda_1}, \quad c_X(0,h_2,0) \xrightarrow{|h_2|\to\infty} |h_2|^{-\lambda_2}, \quad c_X(0,0,\tau) \xrightarrow{|\tau|\to\infty} |\tau|^{-\lambda_0},$$

in which cases the $r_{e,1}$, $r_{e,2}$, and τ_e exist if $\lambda_1 > 1$, $\lambda_2 > 1$ and $\lambda_0 > 1$, respectively. The $r_{e,j}$ ($i = 1, 2$) and τ_e are measures, respectively, of how far in space and how deep in time the field values remain correlated. In practice, this is a certain fraction of the distance h_1, h_2 or the time τ to the first zero of c_X $(h_1,0,0)$, c_X $(0,h_2,0)$, or c_X $(0,0,\tau)$, respectively. In some cases, the $r_{e,1}$, $r_{e,2}$, and τ_e may be defined so that

$$c_X\left(r_{e,1},0,0\right) = c_X\left(0,r_{e,2},0\right) = c_X(0,0,\tau_e) = \frac{c_0}{e}. \tag{6.5}$$

For illustration, in the case of an anisotropic exponential covariance function,

$$c_X(h_1,h_2,\tau) = c_0\, e^{-\left(\frac{h_1}{a_1}+\frac{h_2}{a_2}+\frac{\tau}{a_0}\right)},$$

Eq. (6.5) gives $r_{e,1} = a_1$, $r_{e,2} = a_2$, and $\tau_e = a_0$.

Example 6.2

Consider the anisotropic Gaussian covariance model of Eq. (2.8a) of Chapter VII for $n = 3$, i.e.,

$$c_X(\Delta \boldsymbol{p}) = c_X(\boldsymbol{0}) e^{-\frac{1}{2}\sum_{i=1}^{3,0}\frac{h_i^2}{a_i^2}}.$$

The coefficients a_1, a_2, and a_3 characterize the scales of spatial dependence along the directions h_1, h_2, and h_3, respectively, and the coefficient a_0 characterizes the scale of temporal dependence. With covariances of the form of Eq. (2.7b) of Chapter VII for $n = 3$, i.e.,

$$c_X(\Delta \boldsymbol{p}) = c_X\left(\frac{\sum_{i=1}^{3,0} c_i h_i}{a}\right),$$

the coefficient a may under certain conditions characterize the correlation distance in the direction perpendicular to the Euclidean plane defined by $\sum_{i=1}^{3} c_i h_i = 0$. On that Euclidean plane, as well as on all planes parallel to it, spatial correlation occurs up to infinity.

In the case of an STIS random field such that $c_{X,j}(r,\tau) = c_X(r,\tau)$, Eqs. (6.3a–b) reduce to the expressions

$$\begin{aligned} r_e(\tau) = r_{e,j}(\tau) &= \int_0^\infty dr \frac{c_X(r,\tau)}{c_X(0,\tau)}, \\ \tau_e(r) = \tau_{e,j}(r) &= \int_0^\infty d\tau \frac{c_X(r,\tau)}{c_X(r,0)}, \end{aligned} \tag{6.6a–b}$$

for all j.

Example 6.3

In an analogous manner to the analysis of Example 6.1, and focusing on the origin $(r_e(0), \tau_e(0))$ in the $R^{2,1}$ domain, Eqs. (6.6a–b) become $r_e = c_0^{-1}\int_0^\infty dr\, c_X(r,0)$ and $\tau_e = c_0^{-1}\int_0^\infty d\tau\, c_X(0,\tau)$, where $c_X(0,0) = c_0$, i.e., the effective distance and effective period become constants.

Some additional insight regarding the effective ranges is gained if we assume that we are dealing with the special case of space–time separable covariances, i.e., $c_X(r,\tau) = c_{X(1)}(r)c_{X(2)}(\tau)$. Then, Eqs. (6.3a–b) simply yield

$$\begin{aligned} r_e = r_{e,j} &= \frac{1}{c_{X(1)}(0)}\int_0^\infty dr\, c_{X(1)}(r), \\ \tau_e = \tau_{e,j} &= \frac{1}{c_{X(2)}(0)}\int_0^\infty d\tau\, c_{X(2)}(\tau), \end{aligned} \tag{6.7a–b}$$

which are time independent and space independent, respectively, for all j. Recalling that the $r_{e,j}$ ($i = 1$, 2) and the τ_e are measures of, respectively, how far in space and how long in time the random field values remain correlated, in physical applications these "how far" and "how long" are usually determined according to some practical rule. One such rule is as follows: the spatial range r_e may be defined so that the $c_{X(1)}(r_e)$ value is approximately 50% of the value of the corresponding variance

$c_{X(1)}(0)$; in a similar manner, the temporal range τ_e is defined so that the $c_{X(2)}(\tau_e)$ value is approximately 50% of the value of the corresponding variance $c_{X(2)}(0)$. As noticed earlier, another rule replaces 50% with e^{-1}.

Example 6.4

Table 6.1 lists the r_e and τ_e values for different combinations of three commonly used models of $c_{X(1)}(r)$ and $c_{X(2)}(\tau)$, where a_ζ ($\zeta = r, \tau$) are known model parameters (for the mathematical expressions and interpretations of these covariance models, see Table 4.1 in this chapter). Note that there exist covariance models that do not have effective ranges because the integrals in Eqs. (6.3a–b) diverge. A typical example is the covariance

$$c_X(r,\tau) = c_0\left(1+\frac{r^2}{a_r^2}\right)^{-b} e^{-\frac{\tau}{a_\tau}}, \tag{6.8}$$

in $R^{3,1}$, where $b < 0.5$.

The effective ranges $r_e(\tau)$ and $\tau_e(r)$ should be distinguished from the dependence (corellation) ranges defined next, which constitute limiting kinds of scales of spatial and temporal correlation.

Definition 6.3

The *spatial dependence range* $\varepsilon_{r,j}$ and the *temporal dependence range* $\varepsilon_{\tau,j}$ are defined as the ranges beyond which the covariance can be considered approximately equal to zero.

While $r_{e,j}(\tau)$ and $\tau_{e,j}(r)$ are the ranges over which significant correlations prevail across space and time, respectively, the $\varepsilon_{r,j}$ and $\varepsilon_{\tau,j}$ are the ranges over which nonzero correlations prevail across space and time, respectively. Practically, in the case of space–time separability, the dependence ranges are taken to be the distance and the period at which the value of the covariance is approximately equal to 5% of the corresponding variances, $c_{X(1)}(0)$ and $c_{X(2)}(0)$, respectively.

Example 6.5

Table 6.2 presents the ε_r (isotropic case) and ε_τ values for different combinations of $c_{X(1)}(r)$ and $c_{X(2)}(\tau)$ models, where, as before, a_ζ ($\zeta = r, \tau$) are known model parameters. On the basis of these values, an obvious conclusion is that, among the covariance models considered in Table 6.2, the exponential covariance has the longest correlations and the spherical the shortest ones.

Table 6.1 Spatial r_e and Temporal τ_e Effective Ranges for Different Covariance Models

$\downarrow c_{X(2)}$ $\quad c_{X(1)} \rightarrow$	Gaussian	Exponential	Spherical
Gaussian	$\frac{\sqrt{\pi}}{2}a_r, \frac{\sqrt{\pi}}{2}a_\tau$	$a_r, \frac{\sqrt{\pi}}{2}a_\tau$	$\frac{3}{8}a_r, \frac{\sqrt{\pi}}{2}a_\tau$
Exponential	$\frac{\sqrt{\pi}}{2}a_r, a_\tau$	a_r, a_τ	$\frac{3}{8}a_r, a_\tau$
Spherical	$\frac{\sqrt{\pi}}{2}a_r, \frac{3}{8}a_\tau$	$a_r, \frac{3}{8}a_\tau$	$\frac{3}{8}a_r, \frac{3}{8}a_\tau$

Table 6.2 Spatial ε_r and Temporal ε_τ Dependence Ranges for Different Covariance Models

$\downarrow c_{X(2)}$ $\quad c_{X(1)} \rightarrow$	Gaussian	Exponential	Spherical
Gaussian	$\sqrt{3}a_r, \sqrt{3}a_\tau$	$3a_r, \sqrt{3}a_\tau$	$a_r, \sqrt{3}a_\tau$
Exponential	$\sqrt{3}a_r, 3a_\tau$	$3a_r, 3a_\tau$	$a_r, 3a_\tau$
Spherical	$\sqrt{3}a_r, a_\tau$	$3a_r, a_\tau$	a_r, a_τ

With space–time anisostationary (i.e., homogeneous but anisotropic) covariances, the ε_r range varies with direction in space, which makes the determination of the dependency range a more complicated matter. Also, the fact that for many covariance functions the important region of variation is characterized by finite ranges (say, a_r or a_τ) allows the direct calculation of the order of magnitude of its derivatives. So, the magnitude of the derivative $\frac{d}{dr}c_{X(1)}(r)$ is simply of the order $a_r^{-1}c_{X(1)}(a_r)$. Similarly, the magnitude of the derivative $\frac{d}{d\tau}c_{X(2)}(\tau)$ is of the order $a_\tau^{-1}c_{X(2)}(a_\tau)$.

Example 6.6

For illustration, let us consider the Gaussian space–time separable covariance model

$$c_X(r,\tau) = c_{X(1)}(r)c_{X(2)}(\tau) = e^{-\frac{r^2}{a_r^2}-\frac{\tau^2}{a_\tau^2}},$$

where $c_{X(1)}(r) = e^{-\frac{r^2}{a_r^2}}$ and $c_{X(2)}(\tau) = e^{-\frac{\tau^2}{a_\tau^2}}$. Then, $\frac{d}{dr}c_{X(1)}(r) = -\frac{2r}{a_r^2}c_{X(1)}(r)$, so that $\frac{d}{dr}c_{X(1)}(a_r) \sim a_r^{-1}c_{X(1)}(a_r)$, as expected. Similarly, $\frac{d}{d\tau}c_{X(2)}(a_\tau) \sim a_\tau^{-1}c_{X(2)}(a_\tau)$.

6.2 RELATIONSHIPS BETWEEN PHYSICAL AND SPECTRAL DOMAINS: THE UNCERTAINTY PRINCIPLE

We saw in previous sections that the space–time covariance function and the spectral function form an FT pair. As a consequence, an STIS covariance $c_X(r,\tau)$ is uniquely determined by means of the spectral density $\tilde{c}_X(k,\omega)$, and vice versa. Some interesting results are obtained in the case of separability when the space–time covariance is the product of a spatial and a temporal function. A noticeable feature is the inverse relationship between the widths of the two functions. More precisely, a wide $c_{X(1)}(r)$ (which implies a spatially long-correlated random field) corresponds to a narrow $\tilde{c}_{X(1)}(k)$. Conversely, a narrow $c_{X(1)}(r)$ (short-correlated field) corresponds to a wide $\tilde{c}_{X(1)}(k)$. Similar observations are valid for the pair $c_{X(2)}(\tau)$ and $\tilde{c}_{X(2)}(\omega)$.

This relationship between $c_{X(1)}(r)$ and $\tilde{c}_{X(1)}(k)$, and between $c_{X(2)}(\tau)$ and $\tilde{c}_{X(2)}(\omega)$, can be expressed statistically by means of some principle. To show how this works, for modeling purposes let the distance r and the time interval τ be viewed as RVs with probability density functions (PDFs), respectively,

$$f_r = \frac{c_{X(1)}(r)}{r_e\, c_{X(1)}(0)},$$
$$f_\tau = \frac{c_{X(2)}(\tau)}{\tau_e\, c_{X(2)}(0)}. \tag{6.9a–b}$$

Similarly, the k and ω can be considered as RVs with PDFs, respectively,

$$\begin{aligned} f_k &= \frac{\tilde{c}_{X(1)}(k)}{c_{X(1)}(0)}, \\ f_\omega &= \frac{\tilde{c}_{X(2)}(\omega)}{c_{X(2)}(0)}. \end{aligned} \tag{6.10a–b}$$

Then, using some well-known FT properties, it can be shown that there exist certain relationships that link statistically the r and k, and the τ and ω, thus leading to the introduction of the so-called uncertainty principles.

Definition 6.4

The *uncertainty principles* between the RVs r and k, and between τ and ω are

$$\begin{aligned} \sigma_r \sigma_k &\geq 1, \\ \sigma_\tau \sigma_\omega &\geq 1, \end{aligned} \tag{6.11a–b}$$

respectively.

An interpretation of σ_r and σ_k is that they are the standard deviations of the distance (r) and the wavenumber (k) marginal PDFs, respectively, of the joint space–wavenumber probability density. Similarly, the σ_τ and σ_ω are the standard deviations of the time (τ) and frequency (ω) marginal probability densities, respectively, of the joint time–frequency PDF.[13] Moreover, in light of Eqs. (6.9a–b)–(6.10a–b), the σ_r and σ_k can be seen as the uncertainties (widths) of $c_{X(1)}(r)$ and $\tilde{c}_{X(1)}(k)$, respectively; and, similarly, the σ_τ and σ_ω can be seen as the uncertainties (widths) of $c_{X(2)}(\tau)$ and $\tilde{c}_{X(2)}(\omega)$, respectively. The uncertainty principle of Eq. (6.11a) implies that we cannot reduce the spatial spread and the wavenumber spread simultaneously. Similar is the relationship between time spread and frequency spread based on the principle of Eq. (6.11b). To orient the readers about such "physical–spectral" domain associations, once more we use a simple example that examines a special case of the uncertainty principle.

Example 6.7

In R^1, the Gaussian spatial covariance of Eq. (2.8a) of Chapter VII with $n = 1$ and its SDF are given by

$$\begin{aligned} c_{X(1)}(r) &= c_{X(1)}(0) e^{-\frac{r^2}{2a^2}}, \\ \tilde{c}_{X(1)}(k) &= \frac{a c_{X(1)}(0)}{(2\pi)^{\frac{1}{2}}} e^{-\frac{a^2 k^2}{2}}, \end{aligned} \tag{6.12a–b}$$

respectively. The spatial uncertainty σ_r is proportional to a, whereas the frequency uncertainty σ_k is inversely proportional to a so that $\sigma_r \sigma_k = 1$, in this case. Hence, the equality in Eq. (6.11a) occurs when the $c_{X(1)}(r)$ is Gaussian, and similarly for Eq. (6.11b) and $c_{X(2)}(\tau)$.

[13]Mathematically, these standard deviations are tightly linked via the FT stretching or scaling property (Table 2.1 of Chapter V).

When the random field of interest is anisotropic in R^n $(n \geq 2)$, the correlation radii change with different directions in space. This basically means that there exist n spatial uncertainty principles,

$$\sigma_{r_i}\sigma_{k_i} \geq 1 \tag{6.13}$$

for all $i = 1, \ldots, n$; r_i are distances along directions $s_1, \ldots, s_n$ in space, and k_i are the associated wavenumbers in the spectral domain. As regards time, the uncertainty principle of Eq. (6.11b) remains, of course, valid.

Example 6.8

In R^3, the Gaussian covariance model of Eq. (2.8a) of Chapter VII with $n = 3$, and its SDFs are given by

$$\begin{aligned} c_{X(1)}(h_1, h_2, h_3) &= c_0 e^{-\frac{1}{2}\sum_{i=1}^{3} a_i^{-2} h_i^2}, \\ \widetilde{c}_{X(1)}(k_1, k_2, k_3) &= \frac{c_0 \prod_{i=1}^{3} a_i}{(2\pi)^{\frac{3}{2}}} e^{-\frac{1}{2}\sum_{i=1}^{3} a_i^2 k_i^2}, \end{aligned} \tag{6.14a–b}$$

respectively. Obviously, the widths σ_{k_i} are inversely proportional to a_i $(i = 1,2,3)$, and, hence, inversely proportional to σ_{r_i} so that Eq. (6.13) holds.

7. ON THE ERGODICITY HYPOTHESES OF SPATIOTEMPORAL RANDOM FIELDS

The ensemble moment-based determination of S/TVFs involving multivariate probability distributions makes perfect sense on theoretical grounds. Yet, in the real-world, only one sequence of measurements (realizations) is usually available, in which case we need to employ the notion of an *ergodic* S/TRF, i.e., a random field whose sample moments (that is, spatiotemporal averages calculated over the single realization) converge to the corresponding ensemble moments. In certain cases this convergence can be rigorously proved with the help of the branch of mathematics that is known as ergodic theory. Nevertheless, in the majority of applications, this option is not a real possibility, and then the *ergodic hypothesis* is adopted, instead, that assumes that the equivalence between ensemble and sample moments is valid (see Hypothesis 4 of Chapter VII). This is, in fact, the reason that some authors view ergodicity as a working or a modeling hypothesis.

Otherwise said, the ergodicity hypothesis was created with regard to the experimental or empirical determination of S/TVFs in practice (i.e., this is a hypothesis that is needed when only one sequence of measurements is realistically available). Under the ergodicity hypotheses, certain expressions for the *empirical* S/TVFs have been proposed. In particular, commonly used expressions for the empirical space–time mean, covariance (centered and noncentered) and variogram functions are as follows

$$\overline{\overline{X(s,t)}} = \frac{1}{|S|\Delta T}\iint_{S\times\Delta T} ds dt X(s,t) = \overline{\overline{X}}, \tag{7.1}$$

$$\overline{\overline{c_X(\boldsymbol{h},\tau)}} = \frac{1}{|S|\Delta T}\iint_{S\times\Delta T} ds dt \left[X(\boldsymbol{s},t) - \overline{\overline{X(\boldsymbol{s},t)}}\right]\left[X(\boldsymbol{s}+\boldsymbol{h},t+\tau) - \overline{\overline{X(\boldsymbol{s},t)}}\right],$$
$$\overline{\overline{X(\boldsymbol{s},t)X(\boldsymbol{s}+\boldsymbol{h},t+\tau)}} = \frac{1}{|S|\Delta T}\iint_{S\times\Delta T} ds dt X(\boldsymbol{s},t)X(\boldsymbol{s}+\boldsymbol{h},t+\tau) \tag{7.2a–b}$$

$$\overline{\overline{\gamma_X(\boldsymbol{h},\tau)}} = \frac{1}{2|S|\Delta T}\iint_{S\times\Delta T} ds dt [X(\boldsymbol{s}+\boldsymbol{h},t+\tau) - X(\boldsymbol{s},t)]^2, \tag{7.3}$$

where the integration runs over the spatial region S and the time period ΔT.

The following example presents an empirical correlation function with an important interpretation in weather forecasting.

Example 7.1

In weather studies, the distribution of lightning events is a natural indicator of thunderstorms. Based on the observed lightning data at a set of positions $\boldsymbol{s}_i$ and time instants t_i ($i = 1, \ldots, m$) associated with the detected lightning return strokes for the series of m lightning events, the distribution of lightning events in $R^{2,1}$ can be described by

$$X(\boldsymbol{s},t) = \sum_{i=1}^{m} \delta(\boldsymbol{s}-\boldsymbol{s}_i)\delta(t-t_i),$$

where the $\boldsymbol{s}$ and t may be viewed as random vector and random variable, respectively, and $\overline{X(\boldsymbol{s},t)} = 0$ represents the background state of zero lightning distribution. In light of this interpretation, and by virtue of Eq. (7.2b), the corresponding empirical space–time correlation function of the lightning events is

$$\overline{\overline{\rho_X(\boldsymbol{h},\tau)}} = c\overline{\overline{X(\boldsymbol{s},t)X(\boldsymbol{s}+\boldsymbol{h},t+\tau)}} = c\sum_{i,j=1}^{m} \delta(\boldsymbol{h}-(\boldsymbol{s}_i-\boldsymbol{s}_j))\delta(\tau-(t_i-t_j))$$

(c is a normalization constant[14]), which may represent the distribution of spatial distances and time separations between the observed lightning events.

Among the various forms of stochastic convergence discussed in earlier chapters, we will consider mainly convergence in the *m.s.* sense, although some results will be presented with regard to almost sure (a.s.) convergence, as well. Moreover, as we will see below, different types of ergodicity can be established for S/TRFs. Specifically, we distinguish between *t*-ergodic, *s*-ergodic, and *st*-ergodic random fields, corresponding to random fields that are ergodic as a function of time, space, and space–time, respectively. The notion of *t*-ergodicity applies when we are not concerned about the dependence of the random field on spatial coordinates to bring out the random field features due primarily to its time dependence, whereas *s*-ergodicity applies when we suppress time dependence to bring out the random field features mainly due to dependency on several spatial coordinates. As a consequence, below we study the conditions of *t*-ergodicity, followed by those of *s*-ergodicity. The conditions of *st*-ergodicity are a direct combination of the above, i.e., in this case *t*- and *s*-ergodicity are

[14]Typically, the inverse of the variance, $\overline{\overline{X(\boldsymbol{s},t)^2}} = \sum_{i,j=1}^{m} \delta(\boldsymbol{s}_i-\boldsymbol{s}_j)\delta(t_i-t_j)$.

satisfied simultaneously. When the *t*-ergodicity or the *s*-ergodicity is considered in the following paragraphs, the corresponding covariance or variogram functions are usually regarded as simultaneous two-point or nonsimultaneous one-point functions, respectively (see, also, Remark 2.4 of Chapter II).

Naturally, the ergodic hypothesis places certain restrictions on the S/TVFs, which means that for ergodicity to hold it is necessary that one of the auxiliary hypotheses of Section 1 of Chapter VII holds too. Specifically, ergodicity in the mean requires that the random field has a constant mean (i.e., it is STHS in the mean), whereas ergodicity in the covariance requires that the field has a covariance function that depends only on the space and time lags (i.e., it is STHS in the covariance). As we saw in Section 1 of Chapter VII, *t*-ergodicity in the mean requires that the random field $X(\boldsymbol{s},t)$ has a constant mean denoted as $\overline{X(\boldsymbol{s},t)}$ so that Eq. (1.16a) of Chapter VII holds, *viz.*,

$$\overline{X(\boldsymbol{s},t)} = \lim_{T\to\infty} \overline{\overline{X(\boldsymbol{s},t)}} = \lim_{T\to\infty} \frac{1}{T}\int_0^T dtX(\boldsymbol{s},t),$$

where the time interval is chosen so that $\Delta T = T$, in this case (this choice is convenient in computations). Similarly, *s*-ergodicity in the mean requires that $X(\boldsymbol{s},t)$ has a mean $\overline{X(\boldsymbol{s},t)}$ so that Eq. (1.16b) of Chapter VII holds, i.e.,

$$\overline{X(\boldsymbol{s},t)} = \lim_{|S|\to\infty} \overline{\overline{X(\boldsymbol{s},t)}} = \lim_{|S|\to\infty} \frac{1}{|S|}\int_S d\boldsymbol{s}X(\boldsymbol{s},t),$$

where S is the spatial region of averaging (for convenience, we often let $|S| = (2L)^n$ with the integration region denoted by $S = U_L$, see below).

In the above equations, the sample average $\overline{\overline{X(\boldsymbol{s},t)}}$ may be seen as an RV depending on T and S (or U_L), but not on ($\boldsymbol{s}$, t), and is such that the mean of $\overline{\overline{X(\boldsymbol{s},t)}}$ equals to that of $\overline{X(\boldsymbol{s},t)}$, i.e., $\overline{\overline{\overline{X(\boldsymbol{s},t)}}} = \overline{X(\boldsymbol{s},t)}$,[15] in which the limit operator in front of $\overline{\overline{X(\boldsymbol{s},t)}}$ is replaced by the mean value operator over $\overline{\overline{X(\boldsymbol{s},t)}}$.

Example 7.2

For illustration, let $\overline{X(\boldsymbol{s},t)} = \overline{X}$, i.e., a constant. By virtue of $\lim_{T\to\infty}[\cdot] \mapsto \overline{[\cdot]}$, we also get $\overline{\overline{\overline{X(\boldsymbol{s},t)}}} = (T)^{-1}\int_0^T dt\,\overline{X} = \overline{X}$, in the case of *t*-ergodicity. A similar result holds in the case of *s*-ergodicity.

Real-world observations confirm the existence of spatial and temporal averages of natural attributes as defined in Eqs. (1.16a and b) of Chapter VII. At a given time instant, a distinct spatial pattern is repeated regularly (i.e., the attribute variation exhibits the same overall structure throughout the domain of interest), whereas at a given spatial location a distinct attribute pattern is repeated in time. Moreover, there exist considerable differences between observed attribute variations with distinct patterns, which imply that a quantitative description of these patterns requires the introduction of the notion of attribute scale or range (see, also, the preceding Section 6). In practice, these scales may be

---

[15]It is a standard convention that, depending on whether *t*- or *s*-ergodicity is considered, one of the arguments is considered as a dummy variable under integration.

represented by the temporal and the spatial correlation (or dependence) ranges ε_τ and ε_r, respectively. Then, to obtain ensemble averages in terms of sample averages, as ergodicity investigations require, the time interval T is restricted to a value that is considerable larger than the temporal correlation range, i.e., $T >> \varepsilon_\tau$, whereas the area of the spatial region S is limited to a finite value that is considerable larger than the spatial correlation range, say, $|S|^{\frac{1}{n}} >> \varepsilon_r$.

Example 7.3

In the case of STHS phenomena in geophysics, oceanography, and meteorology, under the ergodicity hypothesis the sample spatial mean and the sample temporal mean are assumed to yield the same result (Hinze, 1975). If the sample temporal mean velocity of turbulent flow contains very slow variations that we do not regard as belonging to the particular flow, we need to consider T as a finite time interval that is though sufficiently large compared with the timescale represented by the temporal range ε_τ of the phenomenon. The same T, on the other hand, should be small compared with the period ϑ_τ of any slow variations in the flow field that do not belong to the turbulence phenomenon, i.e., in this case $\varepsilon_\tau << T << \vartheta_\tau$. Surely, there is a certain level of arbitrariness in the selection of the variations that we regard as not belonging to the particular flow, yet, as it turns out, physical understanding allows a proper selection most of the time in practice.

We return now to the case of m.s. convergence, where Eqs. (1.16a and b) of Chapter VII lead to the following conditions of t-ergodicity and s-ergodicity in the mean in terms of the one-point/two-time and the two-point/one-time (simultaneous two-point) covariance functions, respectively,

$$\lim_{T\to\infty} \overline{\left[\overline{X(\boldsymbol{s},t)} - X(\boldsymbol{s},t')\right]^2} = \lim_{T\to\infty} \frac{1}{T^2}\int_0^T\int_0^T dtdt' c_X(\boldsymbol{s},t;\boldsymbol{s},t') = 0,$$
$$\lim_{L\to\infty} \overline{\left[\overline{X(\boldsymbol{s},t)} - X(\boldsymbol{s}',t)\right]^2} = \lim_{L\to\infty} \frac{1}{(2L)^{2n}}\int_{U_L}\int_{U_L} d\boldsymbol{s}d\boldsymbol{s}' c_X(\boldsymbol{s},t;\boldsymbol{s}',t) = 0. \tag{7.4a–b}$$

If the random field $X(\boldsymbol{s},t)$ is homogeneously distributed, the condition of Eq. (7.4b) becomes accordingly

$$\lim_{L\to\infty} \frac{1}{(2L)^{2n}}\int_{U_L} d\boldsymbol{h} c_X(\boldsymbol{h},t) = 0. \tag{7.5a}$$

A result similar to that of Eq. (7.5a) is valid if the random field is stationary, i.e.,

$$\lim_{T\to\infty} \frac{1}{T^2}\int_0^T d\tau c_X(\boldsymbol{s},\tau) = 0. \tag{7.5b}$$

Apart from the analysis in the physical domain above, ergodicity investigations can be carried out in the spectral domain too, where Eq. (7.5b) is equivalent to the requirement that the SDF is continuous at the origin. Further, a sufficient condition for Eq. (7.5a) to hold is that the limit of the simultaneous two-point covariance vanishes, i.e.,

$$\lim_{u\to\infty} c_X(u\boldsymbol{\varepsilon},t) = 0 \tag{7.6}$$

for all $\boldsymbol{\varepsilon} \neq \mathbf{0}$. Lastly, in the case of an STIS random field, the sufficient condition of Eq. (7.6) simply reduces to

$$\lim_{r \to \infty} c_X(r,t) = 0, \tag{7.7}$$

where $r = |\boldsymbol{h}|$. Let us look at an application of the above analysis in the case of a particular covariance model.

Example 7.4

The STIS exponential covariance model

$$c_X(r,\tau) = c_0 e^{-ar-\beta\tau}, \tag{7.8}$$

where a and β are suitable coefficients as described earlier, satisfies the condition of Eq. (7.7) and it corresponds to a random field that is s-ergodic in the mean. Obviously, if we are dealing with a space–time separable covariance function, $c_X(r,t) = c_{X(1)}(r)c_{X(2)}(\tau)$, the condition of Eq. (7.7) reduces to $\lim_{r\to\infty} c_{X(1)}(r) = 0$. Other interesting results are obtained in terms of the variogram function. For an STIS random field, a sufficient but not necessary condition for s-ergodicity is

$$\lim_{r \to \infty} \gamma_X(r,\tau) = c_0. \tag{7.9}$$

The variogram corresponding to the covariance model of Eq. (7.8) obviously satisfies Eq. (7.9) and, hence, it corresponds to a random field that is s-ergodic in the mean.

As we saw earlier, in many cases in practice we encounter random fields with limited (spatial or temporal) correlation ranges, which means that their covariance functions are negligible for spatial lags $r >> \varepsilon_r$ or time lags $\tau >> \varepsilon_\tau$ for certain ε_r and ε_τ values. Under these conditions, some interesting results can be derived from the ergodicity expressions presented above.

Example 7.5

If we assume that $X(\boldsymbol{s},t)$ is a homogeneous random field in $R^{n,1}$, with a zero mean (for simplicity) and a simultaneous covariance function, and we subsequently define the random process

$$Y(t) = \frac{1}{(2L)^n} \int_{U_L} d\boldsymbol{s} X(\boldsymbol{s},t), \tag{7.10}$$

the spatial average considered in Eq. (7.4b) can be written as

$$\overline{Y^2(t)} = \frac{1}{(2L)^{2n}} \int_{U_L}\int_{U_L} d\boldsymbol{s} d\boldsymbol{s}' c_X(\boldsymbol{s},t;\boldsymbol{s}',t) = \frac{1}{(2L)^{2n}} \int_{U_L} d\boldsymbol{h}\, c_X(\boldsymbol{h},t)$$

Then, an approximation is obtained if it is assumed that $\varepsilon_r << 2L$, in which case,

$$\overline{Y^2(t)} \approx \frac{\varepsilon_r^n}{(2L)^{2n}} \overline{X^2(\boldsymbol{s},t)},$$

that is, the standard deviation of *Y*(*t*) is approximately equal to that of *X*(*s*,*t*) divided by $a_{L,\varepsilon_r} = \frac{(2L)^n}{\varepsilon_r^{\frac{n}{2}}}$. The a_{L,ε_r} can be seen as the number of correlation (ε_r) subdomains within the domain U_L. In this case, the condition of Eq. (7.4b) is clearly satisfied, since

$$\lim_{L\to\infty} \overline{\left[\overline{\overline{X(s,t)}} - \overline{X(s,t)}\right]^2} = \lim_{L\to\infty} \frac{\varepsilon_r^n}{(2L)^{2n}} \overline{X^2(s,t)} = 0.$$

A similar analysis is valid in terms of the temporal average of Eq. (7.4a).

We now turn our attention to the case of *s*-ergodicity in the covariance or in the variogram function. The conditions for this kind of *s*-ergodicity may involve higher-order moments. The *s*-ergodicity in the covariance also requires that the random field is STHS, whereas for *s*-ergodicity in the variogram it is necessary that the random field has spatially homogeneous increments. More specifically, if the sample covariance is defined as

$$\overline{\overline{c_X(\boldsymbol{h},t)}} = \frac{1}{(2L)^n}\int_{U_L} d\boldsymbol{s}\, [X(\boldsymbol{s},t) - \overline{\overline{X(\boldsymbol{s},t)}}][X(\boldsymbol{s}+\boldsymbol{h},t) - \overline{\overline{X(\boldsymbol{s},t)}}], \tag{7.11}$$

a random field is *s*-ergodic in the covariance if Eq. (7.12) of Table 7.1 holds. In the light of m.s. convergence, Eq. (7.12) leads to the condition of Eq. (7.13a) of Table 7.1, which may also be expressed in the more tractable form as in Eq. (7.13b) of Table 7.1, where

$$S_L[\cdot] = \frac{1}{(2L)^n}\int_{U_L} d\boldsymbol{h}[\cdot],$$

and $c_Z(\boldsymbol{h},t)$ is the covariance function of Eq. (7.13c) of Table 7.1 with

$$\Delta_{s'}X(\boldsymbol{h},t) = \left[X(\boldsymbol{h}+\boldsymbol{s}',t) - \overline{\overline{X}}\right], \text{ and } \Delta_0 X(\boldsymbol{h},t) = \left[X(\boldsymbol{h},t) - \overline{\overline{X}}\right].$$

In the special case of a simultaneous homogeneous Gaussian random field, the condition of Eq. (7.13b) reduces to that of Eq. (7.14) of Table 7.1. In the spectral domain, the condition of Eq. (7.14) is equivalent to the requirement that the SDF is continuous everywhere. The covariance condition described in Eq. (7.14) can be expressed in terms of the variogram, as well, see Eq. (7.15) of Table 7.1.

Example 7.6

Let us consider the random field in $R^{1,1}$,

$$X(s,t) = a\cos(ks + \omega t + u),$$

where a is a constant and u is a uniformly distributed RV, i.e., its PDF is given by $f_u(v) = \frac{1}{2\pi}$ for $0 \le v \le 2\pi$, and $= 0$, otherwise. The ensemble mean of the random field is

$$\overline{X(s,t)} = a\frac{1}{2\pi}\int_0^{2\pi} du\, \cos(ks + \omega t + u) = 0,$$

Table 7.1 A List of Ergodicity Conditions

Domain	In the Covariance	In the Variogram
$\boldsymbol{h}$	$\lim_{L\to\infty} \overline{\overline{c_X(\boldsymbol{h},t)}} = c_X(\boldsymbol{h},t)$ (7.12) $\lim_{L\to\infty} \overline{\left[\overline{\overline{c_X(\boldsymbol{h},t)}} - c_X(\boldsymbol{h},t)\right]^2} = 0$ (7.13a) $\lim_{L\to\infty} S_L[c_Z(\boldsymbol{h},t)] = 0$ (7.13b) $Z_{s'}(\boldsymbol{h},t) = \Delta_{s'}X(\boldsymbol{h},t)\Delta_0 X(\boldsymbol{h},t)$ $c_Z(\boldsymbol{h},t) = \overline{[Z_{s'}(\boldsymbol{s}+\boldsymbol{h},t) - c_X(\boldsymbol{s}',t)][Z_{s'}(\boldsymbol{s},t) - c_X(\boldsymbol{s}',t)]}$ (7.13c–d) $X \sim G$: $\lim_{L\to\infty} S_L\left[c_X(\boldsymbol{h},t)^2\right] = 0$ (7.14) $\lim_{L\to\infty} S_L\left[\gamma_X(\boldsymbol{h},t)\left(2 - \frac{\gamma_X(\boldsymbol{h},t)}{c_0}\right)\right] = c_0$ (7.15)	$\lim_{L\to\infty} \overline{\overline{\gamma_X(\boldsymbol{h},t)}} = \gamma_X(\boldsymbol{h},t)$ (7.18) $\lim_{L\to\infty} S_L'[\gamma_X(\boldsymbol{h}'+\boldsymbol{h},t) + \gamma_X(\boldsymbol{h}'-\boldsymbol{h},t)$ $-2\gamma_X(\boldsymbol{h}',t)]^2 = 0$ (7.19)
τ	$\lim_{T\to\infty} \overline{\overline{c_X(\boldsymbol{s},\tau)}} = c_X(\boldsymbol{s},\tau)$ (7.21) $\lim_{T\to\infty} \overline{\left[\overline{\overline{c_X(\boldsymbol{s},\tau)}} - c_X(\boldsymbol{s},\tau)\right]^2} = 0$ (7.22a) $\lim_{T\to\infty} S_T[c_Z(\boldsymbol{s},\tau)] = 0$ (7.22b) $Z_{t'}(\boldsymbol{s},\tau) = \Delta_{t'}X(\boldsymbol{s},t+\tau)\Delta_0 X(\boldsymbol{s},\tau)$ $c_Z(\boldsymbol{s},\tau) = \overline{[Z_{t'}(\boldsymbol{s},t+\tau) - c_X(\boldsymbol{s},t')][Z_{t'}(\boldsymbol{s},t) - c_X(\boldsymbol{s},t')]}$ (7.22c–d) $X \sim G$: $\lim_{T\to\infty} S_T\left[c_X(\boldsymbol{s},t)^2\right] = 0$ (7.23) $\lim_{T\to\infty} S_T\left[\gamma_X(\boldsymbol{s},t)\left(2 - \frac{\gamma_X(\boldsymbol{s},t)}{c_0}\right)\right] = c_0$ (7.24)	$\lim_{T\to\infty} \overline{\overline{\gamma_X(\boldsymbol{s},\tau)}} = \gamma_X(\boldsymbol{s},\tau)$ (7.26) $\lim_{T\to\infty} S_T'[\gamma_X(\boldsymbol{s},\tau'+\tau) + \gamma_X(\boldsymbol{s},\tau'-\tau)$ $-2\gamma_X(\boldsymbol{s},\tau')]^2 = 0$ (7.27)

and the sample average with respect to time is

$$\lim_{T\to\infty}\overline{\overline{X(s,t)}} = a\lim_{T\to\infty}\frac{1}{T}\int_0^T dt\cos(ks+\omega t+u) = a\lim_{T\to\infty}\frac{1}{T\omega}\sin(ks+\omega t+u) = 0,$$

Hence, the random field is t-ergodic in the mean. Moreover, the ensemble covariance is

$$\begin{aligned}\overline{X(s,t)X(s+h,t+\tau)} &= a^2\frac{1}{2\pi}\int_0^{2\pi} du\cos(ks+\omega t+u)\cos(k(s+h)+\omega(t+\tau)+u)\\ &= \frac{a^2}{2}\cos(kh+\omega\tau)\end{aligned}$$

whereas the sample covariance with respect to time is

$$\begin{aligned}\lim_{T\to\infty}\overline{\overline{X(s,t)X(s+h,t+\tau)}} &= a^2\lim_{T\to\infty}\frac{1}{T}\int_0^T dt\cos(ks+\omega t+u)\cos(k(s+h)+\omega(t+\tau)+u)\\ &= \frac{a^2}{2}\cos(kh+\omega\tau)\end{aligned}$$

Hence, the random field is t-ergodic in the covariance. Similar calculations with respect to the space argument show that the random field is s-ergodic in the mean and the covariance, as well.

Next, consider the case where the random field $X(\boldsymbol{p})$ has simultaneous homogeneous increments $Y_{\boldsymbol{h}}(\boldsymbol{s},t) = X(\boldsymbol{s}+\boldsymbol{h},t) - X(\boldsymbol{s},t)$ that are also Gaussian with covariance

$$c_{Y_h}(\boldsymbol{h}',t) = \overline{Y_{\boldsymbol{h}}(\boldsymbol{s},t)Y_{\boldsymbol{h}}(\boldsymbol{s}',t)} = \gamma_X(\boldsymbol{h}'+\boldsymbol{h},t) + \gamma_X(\boldsymbol{h}'-\boldsymbol{h},t) - 2\gamma_X(\boldsymbol{h}',t), \tag{7.16}$$

where $\boldsymbol{h}' = \boldsymbol{s}-\boldsymbol{s}'$. The sample variogram function is given by

$$\overline{\overline{\gamma_X(\boldsymbol{h},t)}} = \frac{1}{2(2L)^n}\int_{U_L} d\boldsymbol{s}[X(\boldsymbol{s},t) - X(\boldsymbol{s}+\boldsymbol{h},t)]^2, \tag{7.17}$$

and a random field is s-ergodic in the variogram if Eq. (7.18) of Table 7.1 holds. As a matter of fact, assuring ergodicity in the variogram can be an important issue in real-world applications, since it often turns out that it is preferable to work with variograms.[16] By applying condition (7.14) of Table 7.1, where c_X is replaced by c_{Y_h} of Eq. (7.16), we find that the condition for s-ergodicity in the variogram can be written as in Eq. (7.19) of Table 7.1, where

$$S_L'[\cdot] = \frac{1}{2(2L)^n}\int_{U_L} d\boldsymbol{h}'[\cdot].$$

As regards t-ergodicity in the covariance, if we define the sample covariance function as the time integral

[16]The readers may recall that the number of data points necessary to obtain covariance and variogram functions with similar accuracy is much smaller for the latter than for the former function, etc.

$$\overline{\overline{c_X(s,\tau)}} = \frac{1}{T}\int_0^T dt \left[X(s,t) - \overline{\overline{X(s,t)}}\right]\left[X(s,t+\tau) - \overline{\overline{X(s,t+\tau)}}\right], \tag{7.20}$$

then an STHS field is t-ergodic in the covariance if Eq. (7.21) of Table 7.1 holds. By virtue of m.s. convergence, Eq. (7.21) leads to the condition of Eq. (7.22a) of Table 7.1, which may also be expressed in the more tractable form as in Eq. (7.22b), where $c_Z(\boldsymbol{h},\tau)$ is the covariance of Eq. (7.13c) of Table 7.1 with $\Delta_{t'}X(s,\tau) = X(s,t'+\tau) - \overline{\overline{X}}$, and $\Delta_0 X(s,\tau) = X(s,\tau) - \overline{\overline{X}}$. In the special case of a Gaussian STHS field, condition (7.22b) reduces to Eq. (7.23) of Table 7.1, where

$$S_T[\cdot] = \frac{1}{T}\int_0^T d\tau.$$

In the spectral domain, condition (7.23) of Table 7.1 is equivalent to the requirement that the SDF $\widetilde{c}_X(\boldsymbol{k},\omega)$ is continuous everywhere. Again, Eq. (7.23) can be expressed in terms of the variogram function, viz., Eq. (7.24) of Table 7.1.

In an analogous manner, as regards the t-ergodicity in the variogram we proceed as follows. If the sample variogram function is given by

$$\overline{\overline{\gamma_X(s,\tau)}} = \frac{1}{2T}\int_0^T dt\, [X(\boldsymbol{h},t) - X(\boldsymbol{h},t+\tau)]^2, \tag{7.25}$$

a random field is t-ergodic if Eq. (7.26) of Table 7.1 holds. The condition for t-ergodicity in the variogram can be then written as in Eq. (7.27) of Table 7.1, where now

$$S_T'[\cdot] = \frac{1}{2T}\int_0^T d\tau'.$$

Example 7.7

The variogram $\gamma_X(\boldsymbol{h},\tau) = ah\tau$ satisfies the conditions of Eqs. (7.19) and (7.27) and, hence, the underlying random field is s- and t-ergodic in the variogram function.

The treatment of ergodicity in the space–time context presents no technical problems, so that the appropriate extensions hardly need a detailed treatment. Indeed, by combining the above results on s- and t-ergodicity, the $X(\boldsymbol{p})$ is space–time ergodic (*st-ergodic*) in the corresponding space–time statistical moment, if the associated s-ergodic and t-ergodic conditions are satisfied simultaneously. For instance, an S/TRF is st-ergodic in the mean if Eqs. (1.16a and b) of Chapter VII are satisfied simultaneously.

An interesting aspect of ergodicity emerges in applications in which we are interested merely in the behavior of the covariance or the variogram function near the origin. It then suffices to verify some sort of ergodicity only near the origin. Practically, this means that the random field is only s-ergodic in volumes significantly small as compared with the scales of spatial correlation. This sort of s-ergodicity has been assigned the names of *quasiergodicity* and *microergodicity*. For quasiergodicity to make sense, the S must satisfy certain conditions. In the STIS case in $R^{3,1}$, e.g., we usually require that $\varepsilon_r << S^{\frac{1}{3}} << L_s$, where L_s denotes a characteristic spatial scale.

Example 7.8

Consider a random field $X(\boldsymbol{p})$ with STIS Gaussian increments and a variogram $\gamma_X(r,\tau)$. Assume that the variogram $\gamma_X(r,\tau)$ expansion near the origin can be characterized by the term with the lowest degree, say $r^{l^*}\tau^{l'^*}$, where $l^*, l'^* \leq 2$. In order that the actual variogram be determined accurately in terms of the sample variogram $\overline{\overline{\gamma_X(r,\tau)}}$, which is here an RV, we can consider the m.s. limit,

$$\underset{r,\tau\to 0}{\text{l.i.m.}} \frac{\overline{\overline{\gamma_X(r,\tau)}}}{\gamma_X(r,\tau)} = 1. \tag{7.28}$$

Under these conditions, the m.s. convergence of Eq. (7.28) entails that

$$\lim_{r,\tau\to 0} \overline{\left[\frac{\overline{\overline{\gamma_X(r,\tau)}}}{\gamma_X(r,\tau)} - 1\right]^2} = \lim_{r,\tau\to 0} \frac{Var\left[\overline{\overline{\gamma_X(r,\tau)}}\right]}{\gamma_X^2(r,\tau)} = 0. \tag{7.29}$$

Suppose further that for small lags we can write that

$$\frac{Var\left[\overline{\overline{\gamma_X(r,\tau)}}\right]}{\gamma_X^2(r,\tau)} = r^{4-2l^*}\tau^{4-2l'^*} + \beta\left(r^n\tau^{n'}\right), \tag{7.30}$$

in which case the analysis holds only for $l^*, l'^* < 2$. The latter means that in order that the micro-ergodicity in the variogram to be a sound assumption it is required that certain conditions be imposed on the m.s. differentiability of the random field.

CHAPTER

VECTOR AND MULTIVARIATE RANDOM FIELDS

IX

CHAPTER OUTLINE

1. INTRODUCTION

When several correlated spatiotemporal random fields (S/TRFs) are physically combined in applications, many important results of earlier chapters need to be carefully extended in a "multi-scalar" setting. As usual, analysis is restricted in terms either of a single space–time vector lag $\Delta\boldsymbol{p}$ in R^{n+1}, or of a pair of "space vector–time scalar" lags $(\boldsymbol{h}, \tau)$ in $R^{n,1}$. Put differently, space–time homostationary (STHS) random fields will be mainly assumed in this chapter. Other assumptions, like space–time isostationarity (STIS) and separability, will be also considered. As regards the physical justification of these assumptions, some of them have sharper testable implications than the others, which could be tested through a range of quantitative evidence and physical constraints. To orient our presentation, I will start with an elementary definition.

Spatiotemporal Random Fields. http://dx.doi.org/10.1016/B978-0-12-803012-7.00009-X

Definition 1.1

A *vector* S/TRF (VS/TRF)

$$\boldsymbol{X}(\boldsymbol{p}) = [X_1(\boldsymbol{p})\ldots X_k(\boldsymbol{p})]^T, \tag{1.1}$$

consists of k interconnected components, $X_l(\boldsymbol{p})$, $l = 1,\ldots, k$, each of which is a scalar random field.

The readers are reminded that the general notion of a VS/TRF was briefly introduced in Section 5 of Chapter II, where some of its most important properties were outlined. Trivially, for $k = 1$ the VS/TRF of Eq. (1.1) reduces to a scalar random field. It is useful to notice that in many physical applications the notion of a VS/TRF is commonly associated with the case $k = n$ in Eq. (1.1). Yet, Definition 1.1 is operationally convenient, because Eq. (1.1) formally includes both the notion of a VS/TRF ($k = n$) and that of a multivariate S/TRF ($k \neq n$, in general) to be considered next. To phrase it in more words, although on formal grounds the $\boldsymbol{X}(\boldsymbol{p})$ in Eq. (1.1) can be considered as a collection of scalar random fields $X_l(\boldsymbol{p})$ ($l = 1,\ldots, k$), on physical grounds it rather makes sense to distinguish between three possible scenarios in which $\boldsymbol{X}(\boldsymbol{p})$ represents:

(a) a set of different scalar physical attributes $X_l(\boldsymbol{p})$ (e.g., for $l = 1, 2, 3$ the components $X_l(\boldsymbol{p})$ may denote pressure, temperature, and density),

(b) a single (composite) physical attribute with components $X_l(\boldsymbol{p})$ (e.g., $\boldsymbol{X}(\boldsymbol{p})$ may denote velocity in $R^{3,1}$ with directional components $X_l(\boldsymbol{p})$, $l = 1, 2, 3$), or

(c) a combination of scalar and composite physical attributes (e.g., $\boldsymbol{X}(\boldsymbol{p})$ may include a temperature and three velocity components).

In light of the above considerations, in scenario (a) we substantively refer to a *multivariate* S/TRF (i.e., one consisting of a set of scalar fields that are physically distinct), in scenario (b) about a VS/TRF (i.e., one consisting of a set of scalar fields that belong to the same physical vector attribute), and a combination of the above in scenario (c). In all these scenarios the relationships between the $\boldsymbol{X}(\boldsymbol{p})$ components are usually specified by means of physical considerations (scientific theories, physical laws, empirical equations). To shed light on the definition of $\boldsymbol{X}(\boldsymbol{p})$ above (which formally represents all three cases (a)–(b) above) a visual representation of the VS/TRF and its components is given in the following example.

Example 1.1

Fig. 1.1 displays a two-component (bivariate) VS/TRF

$$\boldsymbol{X}(\boldsymbol{p}) = [X_1(\boldsymbol{p})\; X_2(\boldsymbol{p})]^T$$

in the $R^{2,1}$ domain, particularly at the space–time points $\boldsymbol{p} = (\boldsymbol{s}, t)$ and $\boldsymbol{p}' = (\boldsymbol{s}', t')$, with $\boldsymbol{s} = (s_1, s_2)$ and $\boldsymbol{s}' = (s'_1, s'_2)$. The vector spatial lag $\boldsymbol{h} = \boldsymbol{s}' - \boldsymbol{s}$ is also shown in the same figure.

Herein, for convenience I will use the standard V/STRF notation $\boldsymbol{X}(\boldsymbol{p})$ to denote both the case $k = n$ (physically motivated vector random field) and $k \neq n$ (multivariate random field), and which one of the two I refer to will be obvious from the context. On formal grounds, and since each scalar random field component $X_l(\boldsymbol{p})$ of $\boldsymbol{X}(\boldsymbol{p})$ represents an attribute varying in space–time, each $X_l(\boldsymbol{p})$ has its own spectral representation in R^{n+1} or $R^{n,1}$, of the form, respectively,

$$\begin{aligned} X_l(\boldsymbol{p}) &= \int d\mathcal{M}_{X_l}(\boldsymbol{w})e^{i\boldsymbol{w}\cdot\boldsymbol{p}}, \\ X_l(\boldsymbol{s}, t) &= \iint d\mathcal{M}_{X_l}(\boldsymbol{k}, \omega)e^{i(\boldsymbol{k}\cdot\boldsymbol{s}-\omega t)}, \end{aligned} \tag{1.2}$$

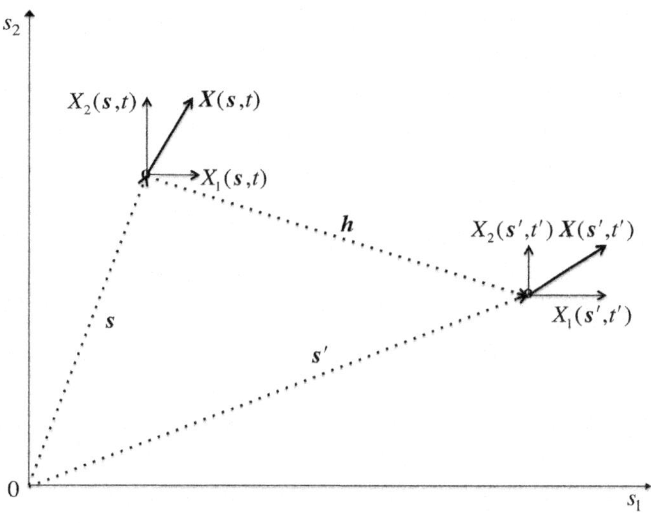

FIGURE 1.1

A two-component (bivariate) vector random field in $R^{2,1}$ (in this and the following figures, the time axis is suppressed for simplicity of presentation).

where $\mathcal{M}_{X_l}$ satisfies Eq. (3.3) of Chapter VII. The readers may recall that the notations $\boldsymbol{p}, \boldsymbol{w} \in R^{n+1}$ and $(\boldsymbol{s}, t)$, $(\boldsymbol{k}, \omega) \in R^{n,1}$ are formally equivalent. The $c_{X_l X_{l'}}(\boldsymbol{p},\boldsymbol{p}')$ will denote the cross-covariance of the components $X_l(\boldsymbol{p})$ and $X_{l'}(\boldsymbol{p}')$ of the VS/TRF $\boldsymbol{X}(\boldsymbol{p})$ (say, fluid velocity vector), whereas the $c_{X_l Y}(\boldsymbol{p},\boldsymbol{p}')$ will denote the cross-covariance of the component $X_l(\boldsymbol{p})$ of $\boldsymbol{X}(\boldsymbol{p})$ and the scalar random field $Y(\boldsymbol{p})$ (say, atmospheric temperature).

Remark 1.1

Vectorial and multivariate random field modeling is often limited by the fact that far fewer classes of suitable covariance functions are available compared to scalar random field modeling. In the vectorial or multivariate notation introduced above, the covariances represent relationships between the attributes $X_l(\boldsymbol{p})$, $l = 1,\ldots, k$, which are the components of the vector $\boldsymbol{X}(\boldsymbol{p})$. The situation is more involved in this case, for the additional reason that each $X_l(\boldsymbol{p})$ has a distinct physical meaning that is reflected in the corresponding cross-covariance function. Then, the known relationships between $X_l(\boldsymbol{p})$ may represent complete or (usually) partial knowledge about the attributes in the form of physical laws, scientific theories, empirical associations, etc. Accordingly, we can confront the problem of limited covariance classes by generating covariance models directly from the corresponding physical laws governing the phenomenon of interest, or, generally, having a form consistent with these physical laws or any other form of empirical knowledge about the phenomenon (Christakos, 1992; Christakos and Hristopulos, 1998; and Chapter XIV).

2. HOMOSTATIONARY AND HOMOSTATIONARILY CONNECTED CROSS–SPATIOTEMPORAL VARIABILITY FUNCTIONS AND CROSS–SPECTRAL DENSITY FUNCTIONS

Several statistical moments can be defined among the components of a VS/TRF $\boldsymbol{X}(\boldsymbol{p})$. In practice, we limit ourselves to up to second-order statistical moments, also termed spatiotemporal variability functions (S/TVFs) in applied stochastics. In our discussion so far, the S/TVFs include the mean function (which are often assumed that they have been subtracted from the random field realizations), and the covariance function (alternatively, the variogram or the structure functions) of each component random field. Here we will add the *cross*-S/TVFs, i.e., the cross-covariance (cross-variograms or cross-structure) functions between the component random fields.

2.1 BASIC NOTIONS AND INTERPRETATIONS

Specifically, we continue our discussion with the definition of the space−time cross-covariance function for jointly considered random fields. This may be the case, e.g., of the covariance between the wind speed in the horizontal and the vertical direction, or the covariance between the temperature and the vertical wind speed fluctuation.

Definition 2.1

Let $X_l(\boldsymbol{p})$ and $X_{l'}(\boldsymbol{p}'), \boldsymbol{p}, \boldsymbol{p}' \in R^{n+1}$, be any pair of STHS random fields (meaning that the random fields are homostationary and homostationarily connected). The space−time *cross-covariance* is defined as

$$\begin{aligned} c_{X_l X_{l'}}(\Delta\boldsymbol{p}) &= \overline{[X_l(\boldsymbol{p}) - \overline{X_l(\boldsymbol{p})}][X_{l'}(\boldsymbol{p}+\Delta\boldsymbol{p}) - \overline{X_{l'}(\boldsymbol{p}+\Delta\boldsymbol{p})}]}, \\ &= \iint d\chi_l d\chi_{l'} [\chi_l - \overline{X_l(\boldsymbol{p})}][\chi_{l'} - \overline{X_{l'}(\boldsymbol{p}+\Delta\boldsymbol{p})}] f_{\boldsymbol{p},\boldsymbol{p}+\Delta\boldsymbol{p}}(\chi_l, \chi_{l'}), \end{aligned} \tag{2.1a–b}$$

where $l, l' = 1, \ldots, k$, and $f_{\boldsymbol{p},\boldsymbol{p}+\Delta\boldsymbol{p}}(\chi_l, \chi_{l'})$ is the joint probability density function (PDF) of $X_l(\boldsymbol{p})$ and $X_{l'}(\boldsymbol{p}')$.

Since the real-world applicability of the VS/TRF theory is a principal concern of the book, it is appropriate to bring to the readers' attention that some variations of the space−time covariance expression above have been considered in applications, depending on the physical situation under consideration. One such expression in R^{n+1} is

$$c_{X_l X_{l'}}(\Delta\boldsymbol{p}) = \frac{1}{2}\left[\overline{X_l(\boldsymbol{p})X_{l'}(\boldsymbol{p})} + \overline{X_l(\boldsymbol{p}+\Delta\boldsymbol{p})X_{l'}(\boldsymbol{p}+\Delta\boldsymbol{p})}\right], \tag{2.1c}$$

which is used in oceanic studies (Hill and Wilczak, 2001). Also, it is sometimes more convenient to work in the $R^{n,1}$ domain, in which case Eq. (2.1a) can be written as

$$c_{X_l X_{l'}}(\boldsymbol{h}, \tau) = \overline{\left[X_l(\boldsymbol{s}, t) - \overline{X_l(\boldsymbol{s}, t)}\right]\left[X_{l'}(\boldsymbol{s}+\boldsymbol{h}, t+\tau) - \overline{X_{l'}(\boldsymbol{s}+\boldsymbol{h}, t+\tau)}\right]}, \tag{2.2}$$

where $l, l' = 1, \ldots, k$.

Example 2.1

Fig. 2.1 presents the specific pair of components in the case of the two-component VS/TRF $\boldsymbol{X}(\boldsymbol{p}) = [X_1(\boldsymbol{p})\, X_2(\boldsymbol{p})]^T$ discussed in Example 1.1. These component pairs are considered in the corresponding four STHS cross-covariance functions $c_{X_1X_1}(\boldsymbol{h}, \tau)$, $c_{X_2X_2}(\boldsymbol{h}, \tau)$, $c_{X_1X_2}(\boldsymbol{h}, \tau)$, and $c_{X_2X_1}(\boldsymbol{h}, \tau)$.

As was the case with the scalar covariances, the cross-covariance functions between any pair of STHS random fields

$$X_l(\boldsymbol{p}) \text{ and } X_{l'}(\boldsymbol{p}'), \quad l, l' = 1, \ldots, k,$$

typically tend to zero as they become widely separated in space or time. We should keep in mind, however, that, since the location $\boldsymbol{s}$ is a vector, the correlation may tend to zero at different rates along

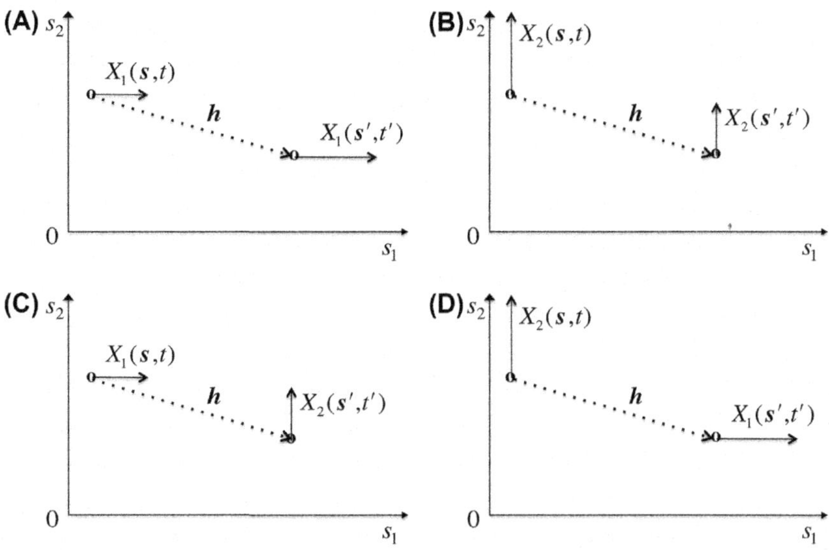

FIGURE 2.1

The components of the vector random field in $R^{2,1}$ considered in the covariance (A) $c_{X_1X_1}(\boldsymbol{h},\tau)$, (B) $c_{X_2X_2}(\boldsymbol{h},\tau)$, (C) $c_{X_1X_2}(\boldsymbol{h},\tau)$, and (D) $c_{X_2X_1}(\boldsymbol{h},\tau)$.

different directions in space. Hence, the direction notion (and this includes the directions of the location vectors $\boldsymbol{s}$, $\boldsymbol{s}'$, and the direction of the separation vector $\boldsymbol{h}$ between $\boldsymbol{s}$ and $\boldsymbol{s}'$) should be an important constituent of any assessment of space−time variability based on cross-covariance functions. Next, without any complication, the coefficient of spatiotempotal cross-correlation can be written as

$$\rho_{X_lX_{l'}}(\Delta\boldsymbol{p}) = \frac{c_{X_lX_{l'}}(\Delta\boldsymbol{p})}{\sigma_{X_l}(\boldsymbol{p})\sigma_{X_{l'}}(\boldsymbol{p}+\Delta\boldsymbol{p})}, \tag{2.3}$$

where $\sigma_{X_l}(\boldsymbol{p})$ and $\sigma_{X_{l'}}(\boldsymbol{p}+\Delta\boldsymbol{p})$ are the standard deviations of the random fields $X_l(\boldsymbol{p})$ and $X_{l'}(\boldsymbol{p}')$, respectively.

Remark 2.1

I remind the readers of an important property of correlation and cross-correlation functions, known as scale independency (Remark 4.1 of Chapter IV). Linked to this property is the STHS-*perseverance* property of correlation functions as follows. If $X_l(\boldsymbol{p})$ is an STHS random field with mean $\overline{X_l(\boldsymbol{p})}$ and constant variance $\sigma^2_{X_l}$, let us define a new random field

$$X_{l'}(\boldsymbol{p}) = a(\boldsymbol{p})X_l(\boldsymbol{p}),$$

where $a(\boldsymbol{p})$ is a known deterministic function. The corresponding statistical moments will be

$$\overline{X_{l'}(\boldsymbol{p})} = a(\boldsymbol{p})\overline{X_l(\boldsymbol{p})},\ \ \sigma^2_{X_{l'}}(\boldsymbol{p}) = a^2(\boldsymbol{p})\sigma^2_{X_l}\ \text{ and}$$

$$c_{X_{l'}}(\boldsymbol{p},\boldsymbol{p}') = a(\boldsymbol{p})a(\boldsymbol{p}')c_{X_l}(\boldsymbol{p}-\boldsymbol{p}'),$$

i.e., the new random field $X_{l'}(\boldsymbol{p})$ is space–time heterogeneous, in general, and its covariance depends on both points $\boldsymbol{p}$ and $\boldsymbol{p}'$. However, in terms of the correlation functions we find that

$$\rho_{X_{l'}}(\boldsymbol{p},\boldsymbol{p}') = \rho_{X_l}(\boldsymbol{p}-\boldsymbol{p}'),$$

i.e., the correlation function of $X_{l'}(\boldsymbol{p})$ depends only on the space–time lag $\boldsymbol{p}-\boldsymbol{p}'$. Naturally, in such cases it may be preferable to work in terms of correlation functions.

The example that follows demonstrates the fundamental role of cross-S/TVFs (like the cross-covariance and the cross-correlation functions above) and their spectral equivalents (like the spectral functions, full or partial) in the development of physical theories.

Example 2.2

Let $X_l(\boldsymbol{s}, t)$ denote flow field velocity fluctuations in the $R^{3,1}$ domain, and assume that the simultaneous two-point functions

$$c_{X_lX_{l'}}(\boldsymbol{h},t) = \overline{X_l(\boldsymbol{s},t)X_{l'}(\boldsymbol{s}+\boldsymbol{h},t)}$$

$(l, l' = 1, 2, 3)$ represent the corresponding velocity fluctuation cross-covariance functions (linked to Reynolds shear stress). These cross-covariance functions are fundamental quantities of high interpretive value, from which several physically significant variables can be determined. In particular, the turbulent kinetic energy is defined as

$$K_X = \frac{1}{2}\mu\sum_{l=1}^{3}\sigma^2_{X_l},$$

where μ is the field density and $\sigma^2_{X_l} = \overline{X_l(\boldsymbol{s},t)^2}$ (the σ_{X_l} is a measure of the amplitude of the velocity fluctuation components). The energy spectrum $E_{X_lX_{l'}}$ is given in terms of the flow velocity partial spectrum of the cross-covariance function, i.e.,

$$E_{X_lX_{l'}}(\boldsymbol{k},t) = \widetilde{c}_{X_lX_{l'}}(\boldsymbol{k},t) = \frac{1}{(2\pi)^3}\int d\boldsymbol{h}e^{-i\boldsymbol{k}\cdot\boldsymbol{h}}c_{X_lX_{l'}}(\boldsymbol{h},t).$$

Formally, the cross-covariance is the inverse of the energy spectrum

$$c_{X_lX_{l'}}(\boldsymbol{h},t) = \int d\boldsymbol{k}e^{i\boldsymbol{k}\cdot\boldsymbol{h}}E_{X_lX_{l'}}(\boldsymbol{k},t),$$

where $d\boldsymbol{k} = dk_1dk_2dk_3$ is the differential volume in wavevector space. For $l = l'$ and $\boldsymbol{h} = \boldsymbol{0}$, we obtain the turbulent kinetic energy spectrum defined simply as one-half the cross-covariance function,

$$K_X(t) = \frac{1}{2}c_{X_lX_l}(\boldsymbol{0},t) = \frac{1}{2}\int d\boldsymbol{k}E_{X_lX_l}(\boldsymbol{k},t).$$

In spherical coordinates, we obtain

$$K_X(t) = \frac{1}{2}\int_0^\infty dk\int_{|\boldsymbol{k}|=k} d\varpi E_{X_lX_l}(\boldsymbol{k},t),$$

where $d\boldsymbol{k} = d\varpi dk$, and $d\varpi$ is an element of solid angle in $\boldsymbol{k}$-space. Then, the energy spectrum is expressed as

$$E_X(k,t) = \frac{1}{2}\int_{|\boldsymbol{k}|=k} d\varpi E_{X_l X_l}(\boldsymbol{k},t),$$

where $k = |\boldsymbol{k}|$, in which case the turbulent kinetic energy spectrum is given by

$$K_X(t) = \int_0^{\infty} dk E_X(k,t),$$

where $E_X(k, t)$ represents the energy in a spherical shell in $\boldsymbol{k}$-space located at $k = |\boldsymbol{k}|$. Additional quantities that are physically significant can be obtained in the purely temporal domain. Let

$$\rho_{E,s}(\tau) = \frac{\overline{X(t)X(t+\tau)}}{\overline{X(t)^2}}$$

be the temporal correlation function of a single random field at a fixed location $\boldsymbol{s}$ in space and correlated as a function of time $\left(\text{without loss of generality, let } \overline{X(t)^2} = \sigma_X^2\right)$. The integral time scale is given by

$$\lambda_E = \int_{R^1} d\tau\, \rho_{E,s}(\tau).$$

In the frequency domain, the temporal correlation function is given in terms of the correlation function as

$$\rho_{E,s}(\tau) = \frac{1}{2\pi}\int_{R^1} d\omega' e^{-i\omega'\tau}\widetilde{\rho}_{E,s}(\omega'),$$

which at $\tau = 0$ defines the frequency spectrum

$$E_{11}(\omega) = 2\sigma_X^2 \widetilde{\rho}_{E,s}(2\pi\omega),$$

and leads to the following expression for the variance,

$$\sigma_X^2 = \frac{1}{2}\int_{R^1} d\omega E_{11}(\omega)$$

(if $\rho_{E,s}(\tau)$ is symmetric, the coefficient $\frac{1}{2}$ is dropped and the integration is from zero to infinity). A practical benefit of the analysis is that $E_{11}(\omega)$ can be easily specified from experimental time series data. In isotropic turbulence studies involving Taylor's hypothesis, the estimated $E_{11}(\omega)$ can be used to calculate several important physical quantities, like $\rho_{E,s}(\tau)$, $K_X(t)$, and $E_{X_l X_l}(\boldsymbol{k},t)$. As our discussion above showed, the space–time analysis may be performed in different domains, ($\boldsymbol{h}$, t), ($\boldsymbol{k}$, t), (k, t), ($\boldsymbol{k}$, ω), where important physical quantities are determined on the basis of the covariance and correlation functions.

The next step is to introduce the *matrix* of the spatiotemporal cross-covariance functions between the component random fields as

$$\boldsymbol{c}_X = \left[c_{X_l X_{l'}}(\Delta\boldsymbol{p})\right]$$

for all $l, l' = 1, 2, \ldots, k$ and $\boldsymbol{\Delta p} = (\boldsymbol{h}, \tau)$. The $\boldsymbol{c_X}$ represents all possible combinations of spatial, temporal, and spatiotemporal correlations. Trivially, for $\boldsymbol{\Delta p} = \boldsymbol{0}$ the diagonal elements of $\boldsymbol{c_X}$ are random field variances and the off-diagonal elements are space–time covariances. For $\boldsymbol{h} \neq \boldsymbol{0}$ and $\Delta t = 0$, the $\boldsymbol{c_X}$ elements are purely spatial cross-covariances; and for $\boldsymbol{h} = \boldsymbol{0}$ and $\tau \neq 0$, the $\boldsymbol{c_X}$ elements are purely temporal cross-covariances. The properties below are straightforward consequences of the preceding analysis.

Property 2.1

The matrix $\boldsymbol{c_X}$ is nonnegative-definite (NND), namely,

$$\boldsymbol{\lambda}^T \boldsymbol{c_X} \boldsymbol{\lambda} \geq 0 \tag{2.4}$$

for all $m \in N$ and all deterministic vectors $\boldsymbol{\lambda} = \{\boldsymbol{\lambda}_i\}$ $(i = 1,\ldots, m)$, $\boldsymbol{\lambda}_i = [\lambda_{i1} \ldots \lambda_{ik}]^T$. Just as for scalar random fields, this is an immediate consequence of the fact that if

$$Y(\boldsymbol{p}) = \sum_{i=1}^{m} \boldsymbol{\lambda}_i^T \boldsymbol{X}(\boldsymbol{p} + \boldsymbol{p}_i), \tag{2.5a}$$

it must hold that

$$\text{var}\ [Y(\boldsymbol{p})] = \sum_{i=1}^{m} \sum_{j=1}^{m} \boldsymbol{\lambda}_i^T \boldsymbol{c_X}\left(\boldsymbol{\Delta p}_{ij}\right) \boldsymbol{\lambda}_j \geq 0 \tag{2.5b}$$

for all $m \in N$, $\boldsymbol{\lambda}_i \in R^k$ $(i = 1,\ldots, m)$, and all $\boldsymbol{\Delta p}_{ij}$.

Property 2.2

The $\boldsymbol{c_X}$ is a symmetric matrix in the sense that

$$c_{X_l X_{l'}}\left(\boldsymbol{\Delta p}_{ij}\right) = c_{X_{l'} X_l}\left(\boldsymbol{\Delta p}_{ji}\right), \tag{2.6a}$$

where $\boldsymbol{\Delta p}_{ij} = \boldsymbol{p}_i - \boldsymbol{p}_j$, $\boldsymbol{\Delta p}_{ij} = -\boldsymbol{\Delta p}_{ji}$. However, in general,

$$c_{X_l X_{l'}}\left(\boldsymbol{\Delta p}_{ij}\right) \neq c_{X_{l'} X_l}\left(\boldsymbol{\Delta p}_{ij}\right). \tag{2.6b}$$

Property 2.3

A straightforward application of the Cauchy–Schwartz inequality yields

$$\left|c_{X_l X_{l'}}\left(\boldsymbol{\Delta p}_{ij}\right)\right| \leq \sigma_{X_l}(\boldsymbol{p}_i) \sigma_{X_{l'}}(\boldsymbol{p}_j) \tag{2.6c}$$

for all $l, l' = 1, \ldots, k$ and $i, j = 1,\ldots, m$.

Given the matrix $\boldsymbol{c_X}$ of STHS cross-covariances, the definition of the corresponding matrix of cross–spectral density functions (cross-SDFs) is rather straightforward, namely, it is the symmetric matrix

$$\widetilde{\boldsymbol{c}}_{\boldsymbol{X}} = \left[\widetilde{c}_{X_l X_{l'}}(\boldsymbol{\Delta w})\right], \tag{2.7a}$$

for all $\boldsymbol{\Delta w}$ and $l, l' = 1, \ldots, k$, and the component cross-SDFs $\widetilde{c}_{X_l X_{l'}}(\boldsymbol{\Delta w})$ are expressed in terms of the corresponding cross-covariance functions as

$$\widetilde{c}_{X_l X_{l'}}(\boldsymbol{\Delta w}) = \frac{1}{(2\pi)^{n+1}} \int d\boldsymbol{\Delta p}\, e^{-i\boldsymbol{\Delta w} \cdot \boldsymbol{\Delta p}} c_{X_l X_{l'}}(\boldsymbol{\Delta p}), \tag{2.7b}$$

assuming that these integrals exist. More specifically, the diagonal elements of the matrix $\widetilde{c}_X$ represent the SDFs of the component S/TRFs, and the off-diagonal elements represent their cross-SDFs. The matrix $\widetilde{c}_X$ is Hermitian for all Δw. Partial cross-SDFs (i.e., space–frequency and wavenumber–time cross-SDFs) can be also defined using the transformations shown in Fig. 3.1 of Chapter VII. The spectral functions above manifest statistical relationships between the components of $X(p)$ at the same or different locations and times. To these relationships we may assign different physical interpretations, depending on the real-world phenomenon.

Example 2.3

In the special case, e.g., of plane wave random fields propagating in a certain direction with a certain speed (Section 3 of Chapter X) the elements of the spectral matrix of such random fields essentially consist of a common amplitude and a set of phase shifts between points.

On the basis of the spectral representations above, we can derive the relevant criteria of permissibility for space–time covariance matrices (see Chapter XV).

2.2 GEOMETRY OF VECTOR SPATIOTEMPORAL RANDOM FIELDS

The geometrical properties of VS/TRFs, e.g., differentiation and integration, can be considered in a manner analogous to that for scalar S/TRFs. So, the derivatives of the component random fields are given by

$$X_l^{(\nu,\mu)}(p) = \mathcal{L}_p X_l(p),$$

where $l = 1,\ldots, k$, $s_0 = t$ and $\mathcal{L}_p = \frac{\partial^{\nu+\mu}}{\partial s^\nu \partial s_0^\mu}$. These derivatives exists in $R^{n,1}$ if

$$\int dk d\omega \prod_{i=1}^{n} k_i^{2\nu_i} \omega^{2\mu} \widetilde{c}_{X_l X_{l'}}(\boldsymbol{k}, \omega) < \infty, \tag{2.8a}$$

where $\nu = \sum_{i=1}^{n} \nu_i$. Then,

$$\begin{aligned} \overline{\frac{\partial^{(\nu,\mu)} X_l(s,t)}{\partial s^\nu \partial t^\mu} \frac{\partial^{(\nu',\mu')} X_{l'}(s',t')}{\partial s'^{\nu'} \partial t'^{\mu'}}} &= (-1)^{\nu+\mu} i^{\nu+\nu'-\mu-\mu'} \int dk\, d\omega \prod_{i=1}^{n} k_i^{\nu_i+\nu_i'} \omega^{\mu+\mu'} e^{i(\boldsymbol{k}\cdot\boldsymbol{h}-\omega\tau)} \widetilde{c}_{X_l X_{l'}}(\boldsymbol{k}, \omega) \\ &= (-1)^{\nu+\mu} \frac{\partial^{\nu+\nu'+\mu+\mu'} c_{X_l X_{l'}}(\boldsymbol{h}, \tau)}{\partial \boldsymbol{h}^{\nu+\nu'} \partial \tau^{\mu+\mu'}} \end{aligned} \tag{2.8b}$$

with $\boldsymbol{h} = s' - s$, $\tau = t' - t$, $\partial s^\nu = \prod_{i=1}^{n} \partial s_i^{\nu_i}$, and $\partial \boldsymbol{h}^{\nu+\nu'} = \prod_{i=1}^{n} \partial \boldsymbol{h}_i^{\nu_i+\nu_i'}$ (the readers are reminded that the imaginary coefficient in the right hand side of Eq. (2.8b) is due to the space–time FT definition introduced in Chapter V).

Example 2.4

Consider a VS/TRF, $X(s, t) = [X_1(s, t)\ X_2(s, t)\ X_3(s, t)]^T$, in $R^{3,1}$. The divergence law in the case of this vector field gives

$$\operatorname{div} X(s,t) = \nabla \cdot X(s,t) = \sum_{l=1}^{3} \frac{\partial X_l(s,t)}{\partial s_l} = i \sum_{l=1}^{3} \int d\mathcal{M}_{X_l}(\boldsymbol{k}, \omega) k_l\, e^{i(\boldsymbol{k}\cdot s - \omega t)}, \tag{2.9a}$$

and

$$\overline{[\nabla\cdot X(s,t)\nabla\cdot X(s+h,t)]} = \iint dk d\omega\, e^{i(k\cdot h-\omega t)} \sum_{l,l'=1}^{3} k_l k_{l'} \widetilde{c}_{X_l X_{l'}}(k,\omega)$$
$$= -\sum_{l,l'=1}^{3} \frac{\partial^2 c_{X_l X_{l'}}(h,t)}{\partial h_l \partial h_{l'}}, \qquad (2.9b\text{–}c)$$

i.e., the covariance of $\nabla\cdot X(s,t)$ is given by Eq. (2.9b), and its spectrum by $\sum_{l,l'=1}^{3} k_l k_{l'} \widetilde{c}_{X_l X_{l'}}(k,\omega)$.

Random field integrability issues emerge when the core knowledge is available in the form of integral equations. Covariance functions and matrices estimated from the data using statistical techniques should satisfy any core knowledge concerning attribute variation. Physical laws (usually in the form of differential or integral equations) are the prime source of constraints imposed on the estimated space–time covariance functions. Otherwise said, the estimated covariance functions are not solely data dependent but they must also conform to all physical constraints relevant to the phenomenon under study.

Example 2.5

Assume that the space–time variation of the observed random vector $X(s, t)$ is governed in $R^{n,1}$ by the integrodifferential equation

$$\frac{\partial}{\partial t} X(s,t) + I_s[b(s), X(s,t)] = a(s,t), \qquad (2.10a)$$

where

$$I_s[Y(s), X(s,t)] = \int_{\Theta} ds' b(s-s') X(s',t).$$

The $b(s)$ expresses the dynamics of the phenomenon in the spatial domain Θ, and $a(s, t)$ is a vector of space–time white-noise random fields, independent of $X(s, t)$, and such that $\overline{a(s,t)a(s,t')^T} = A(s-s')\delta(t-t')$, $t \geq t'$. The cross-covariance equation associated with Eq. (2.10a) is of the form

$$c_X(s-s', t-t') - I_s[G(s,t), c_X(s',t')] = 0, \qquad (2.10b)$$

where $c_X(s, t)$ is the covariance matrix, $G(s, t)$ is the Green's function matrix of Eq. (2.10a) with $a(s, t) = 0$, $G(s, t) = 0$ $(t < 0)$, $G(s) = I\delta(s)$ $(t = 0)$, and I is the identity matrix.[1] The covariance model generated from the data should satisfy Eq. (2.10b) for all space–time lags.

3. SOME SPECIAL CASES OF COVARIANCE FUNCTIONS

Two covariance functions that play a special role in physical applications are the *longitudinal* and the *lateral* (or *transverse*) covariance functions. For the longitudinal covariance function, the distance $r = |h|$ is taken parallel to the direction of the lag h between each pair of locations s and s', i.e.,

$$c_{\ell\ell}(r,\tau) = \overline{\left[\left(X(s,t) - \overline{X(s,t)}\right)\left(X(s+h,t+\tau) - \overline{X(s+h,t+\tau)}\right)\right]}\cdot \varepsilon_l, \qquad (3.1a)$$

[1]That is, $I_s[G(s,t), c_X(s',t')] = \int_{\Theta} ds'' G(s-s'', t-t'') c_X(s''-s', t''-t')$, and Eq. (2.10b) is a recursive expression with $t \geq t'' \geq t'$.

where $\boldsymbol{\varepsilon}_l$ is a unit vector in the longitudinal (parallel) direction with respect to the direction of the vector random field $\boldsymbol{X}(\boldsymbol{s}, t)$. In an analogous manner, for the transverse covariance function the distance is taken perpendicular to the direction of the lag $\boldsymbol{h}$, i.e.,

$$c_{\eta\eta}(r,\tau) = \overline{\left[\left(\boldsymbol{X}(\boldsymbol{s},t) - \overline{\boldsymbol{X}(\boldsymbol{s},t)}\right)\left(\boldsymbol{X}(\boldsymbol{s}+\boldsymbol{h},t+\tau) - \overline{\boldsymbol{X}(\boldsymbol{s}+\boldsymbol{h},t+\tau)}\right)\right]} \cdot \boldsymbol{\varepsilon}_\eta, \tag{3.1b}$$

where $\boldsymbol{\varepsilon}_\eta$ is a unit vector in the transverse (perpendicular) direction.

As we shall see later, the cross-covariance function $c_{X_l X_{l'}}(\boldsymbol{h},\tau)$ can be expressed in terms of the above two scalar functions, the longitudinal covariance function $c_{\ell\ell}(r, \tau)$ and the transverse covariance function $c_{\eta\eta}(r, \tau)$. Examples of these two types of covariance functions are discussed next to shed some light on the general expressions above.

Example 3.1

Let $X_l(\boldsymbol{s}, t)$ denote zero mean flow field velocity fluctuations in $R^{3,1}$, and assume that $c_{X_l X_{l'}}(\boldsymbol{h},t) = \overline{X_l(\boldsymbol{s},t)X_{l'}(\boldsymbol{s}+\boldsymbol{h},t)}$, $l,l' = 1,2,3$, are the corresponding velocity fluctuation cross-covariance functions. In this case, the longitudinal and transverse covariance functions are derived from the simultaneous two-point cross-covariance $c_{X_l X_{l'}}(\boldsymbol{h},t)$ as follows (see also, Fig. 3.1),

$$\begin{aligned} c_{\ell\ell}(r,t) &= \overline{X_1(s_1,s_2,s_3,t)X_1(s_1+r,s_2,s_3,t)} = c_{X_1X_1}(r,0,0,t) = c_{X_1X_1}(r\boldsymbol{\varepsilon}_1,t), \\ c_{\eta\eta}(r,t) &= c_{X_2X_2}(r\boldsymbol{\varepsilon}_1,t) = c_{X_3X_3}(r\boldsymbol{\varepsilon}_1,t), \end{aligned} \tag{3.2a–b}$$

where $r = |\boldsymbol{h}|$, $\boldsymbol{\varepsilon}_1$ is the unit vector in the coordinate direction. In more words, the longitudinal covariance function $c_{\ell\ell}(r, \tau)$ is associated with the correlation between the same components (X_1) of the vector random field $\boldsymbol{X}(\boldsymbol{s}, t)$ at two different points on a line parallel to these components. The transverse covariance function $c_{\eta\eta}(r, \tau)$, on the other hand, is linked to the correlation between the

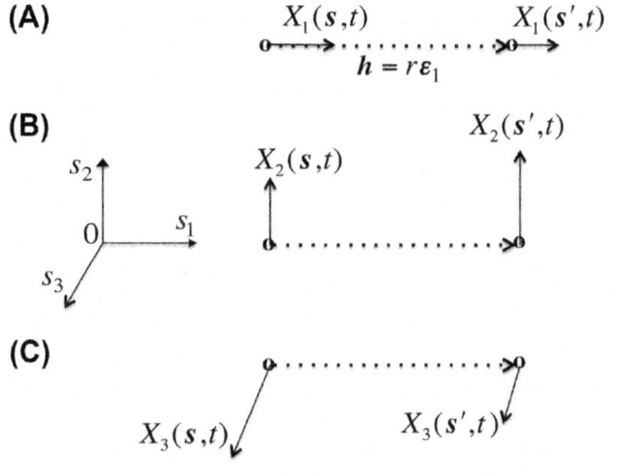

FIGURE 3.1

The components of the vector random field in $R^{3,1}$ considered in the covariance (A) $c_{\ell\ell}(r,t) = c_{X_1X_1}(r,t)$, and (B), (C) $c_{\eta\eta}(r,t) = c_{X_2X_2}(r,t) = c_{X_3X_3}(r,t)$.

same components (X_2) of $\boldsymbol{X}(\boldsymbol{s}, t)$ at two different points on a line perpendicular to these components. In many cases, the corresponding longitudinal and transverse correlation functions are used as follows,

$$\rho_{\ell\ell}(r,t) = \frac{c_{X_1X_1}(r\boldsymbol{\varepsilon}_1,t)}{\sigma_1^2},$$
$$\rho_{\eta\eta}(r,t) = \frac{c_{X_2X_2}(r\boldsymbol{\varepsilon}_1,t)}{\sigma_2^2} = \frac{c_{X_3X_3}(r\boldsymbol{\varepsilon}_1,t)}{\sigma_3^2}, \tag{3.3a–b}$$

where, for simplicity, we usually let $\sigma_1^2 = \sigma_2^2 = \sigma_3^2 = 1$. Also, we notice that $\rho_{\ell\ell}(r, t)$ and $\rho_{\eta\eta}(r, t)$ are nondimensional.

The longitudinal and the transverse covariance functions are closely linked to parameters with considerable physical meaning. Among them are the scales at which turbulent dissipation occurs, as discussed in the following example.

Example 3.2

The linear scale λ_1 offers a measure of the scale at which turbulent dissipation occurs, and is defined from the Taylor series expansion of $c_{\ell\ell}(r, t)$ at $r = 0$, i.e.,[2]

$$c_{\ell\ell}(r,t) = c_{\ell\ell}(0,t) + r\frac{\partial c_{\ell\ell}}{\partial r}(0,t) + \frac{r^2}{2!}\frac{\partial^2 c_{\ell\ell}}{\partial r^2}(0,t) + \ldots \tag{3.4}$$

Consider the value $r = r^*$ so that $c_{\ell\ell}(r^*, t) = 0$, and since $r^*\frac{\partial c_{\ell\ell}}{\partial r}(0,t) = 0$, it is found that $0 \approx c_{\ell\ell}(0,t) + \frac{r^{*2}}{2}\frac{\partial^2 c_{\ell\ell}}{\partial r^2}(0,t)$. Then, by letting $\lambda_1^2 = \frac{1}{2}r^{*2}$, we define the linear scale in terms of the longitudinal covariance function, i.e.,

$$\lambda_1^2 = -c_{\ell\ell}(0,t)\left[\frac{\partial^2 c_{\ell\ell}}{\partial r^2}(0,t)\right]^{-1}. \tag{3.5}$$

From Eq. (3.5), the longitudinal covariance function satisfies the approximation

$$c_{\ell\ell}(r,t) \approx c_{\ell\ell}(0,t) - \frac{r^2}{2\lambda_1^2}. \tag{3.6}$$

Working along similar lines, we find that the linear scale λ_2 is expressed in terms of the transverse covariance function

$$\lambda_2^2 = -c_{\eta\eta}(0,t)\left[\frac{\partial^2 c_{\eta\eta}}{\partial r^2}(0,t)\right]^{-1}. \tag{3.7}$$

And the transverse covariance function is such that

$$c_{\eta\eta}(r,t) \approx c_{\eta\eta}(0,t) - \frac{r^2}{2\lambda_2^2} \tag{3.8}$$

(compare it with Eq. 3.6).

[2]Here $\frac{\partial c_{\ell\ell}}{\partial r}(0,t)$ is a simpler notation for $\frac{\partial}{\partial r}c_{\ell\ell}(r,t)\big|_{r=0}$ etc.

We will revisit the longitudinal and the transverse covariance functions in the following section, in the context of the solenoidal and potential VS/TRF.

As we saw in previous sections, another S/TVF that is popular in earth and atmospheric science applications where heterogeneous, in general, random fields are involved, is the *space–time cross-structure* function defined in two ways, namely

$$\begin{aligned} \xi_{X_l X_{l'}}(\Delta \boldsymbol{p}) &= \overline{[X_l(\boldsymbol{p}) - X_l(\boldsymbol{p}+\Delta\boldsymbol{p})][X_{l'}(\boldsymbol{p}) - X_{l'}(\boldsymbol{p}+\Delta\boldsymbol{p})]}, \\ \xi_{X_l X_{l'}}(\Delta \boldsymbol{p}) &= \overline{[X_l(\boldsymbol{p}) - X_{l'}(\boldsymbol{p}+\Delta\boldsymbol{p})]^2}, \end{aligned} \tag{3.9a–b}$$

for $l, l' = 1, \ldots, k$. Eq. (3.9a), in particular, defines the cross-structure function as the covariance of the random field differences between two points $\boldsymbol{p}$ and $\boldsymbol{p}+\Delta\boldsymbol{p}$.

Similarly to the longitudinal and the transversal covariance functions presented earlier, the longitudinal and transversal structure functions can be also defined by restricting the general expressions of Eqs. (3.9a–b) along the parallel ($\boldsymbol{\varepsilon}_l$) and the perpendicular ($\boldsymbol{\varepsilon}_\eta$) directions, i.e.,

$$\begin{aligned} \xi_{\ell\ell}(r,\tau) &= \overline{[(\boldsymbol{X}(\boldsymbol{s},t) - \boldsymbol{X}(\boldsymbol{s}+\boldsymbol{h},t+\tau))\cdot\boldsymbol{\varepsilon}_l]^2}, \\ \xi_{\eta\eta}(r,\tau) &= \overline{\left[(\boldsymbol{X}(\boldsymbol{s},t) - \boldsymbol{X}(\boldsymbol{s}+\boldsymbol{h},t+\tau))\cdot\boldsymbol{\varepsilon}_\eta\right]^2}, \end{aligned} \tag{3.10a–b}$$

respectively, [compare with Eqs. (3.1a) and (3.1b)]. The readers may find it interesting that the longitudinal structure function is more often measured in laboratory experiments.

Example 3.1 (cont.)

As in Example 3.1, let $X_l(\boldsymbol{s}, \boldsymbol{t})$ denote zero mean flow field velocity fluctuations in $R^{3,1}$, and

$$\xi_{X_l X_{l'}}(\boldsymbol{h},t) = \overline{[X_l(\boldsymbol{s}+\boldsymbol{h},t) - X_l(\boldsymbol{s},t)][X_{l'}(\boldsymbol{s}+\boldsymbol{h},t) - X_{l'}(\boldsymbol{s},t)]}, \tag{3.11}$$

$l, l' = 1, 2, 3$, are the corresponding velocity fluctuation cross-structure functions. The longitudinal and transverse structure functions are derived from the simultaneous two-point cross-structure function $c_{X_l X_{l'}}(\boldsymbol{h},t)$ as follows (see also, Fig. 3.1),

$$\begin{aligned} &\xi_{\ell\ell}(r,t) = \overline{[X_1(s_1+r,s_2,s_3,t) - X_1(s_1,s_2,s_3,t)]^2} = \xi_{X_1X_1}(r,0,0,t) = \xi_{X_1X_1}(r\boldsymbol{\varepsilon}_1,t), \\ &\xi_{\eta\eta}(r,t) = \xi_{X_2X_2}(r\boldsymbol{\varepsilon}_1,t) = \xi_{X_3X_3}(r\boldsymbol{\varepsilon}_1,t), \end{aligned} \tag{3.12a–b}$$

where $r = |\boldsymbol{h}|$, $\boldsymbol{\varepsilon}_1$ is the unit vector in the coordinate direction. Just as with the cross-covariance function, the cross-structure function $\xi_{X_l X_{l'}}(\boldsymbol{h},t)$ can be expressed in terms of two scalar structure functions, $\xi_{\ell\ell}(r, t)$ and $\xi_{\eta\eta}(r, t)$.

In Section 4 that follows, the longitudinal and transversal covariance functions will be related to a specific class of VS/TRFs of considerable importance in applied stochastics studies. The following example is an introduction to this important class.

Example 3.2

In $R^{3,1}$, consider the continuity equation for incompressible fluids

$$\nabla\cdot\boldsymbol{X}(\boldsymbol{p}) = 0, \tag{3.13}$$

where $\boldsymbol{X}(\boldsymbol{p}) = [X_1(\boldsymbol{p})\, X_2(\boldsymbol{p})\, X_3(\boldsymbol{p})]^T$ is the velocity vector. By multiplying the continuity equation, in turn, by $\frac{\partial X_{l'}(\boldsymbol{s},t)}{\partial s_{l'}}$ ($l' = 1, 2, 3$), and taking expectations, we find

$$\sum_{l=1}^{3} \overline{\frac{\partial X_{l'}(\boldsymbol{s},t)}{\partial s_{l'}} \frac{\partial X_l(\boldsymbol{s},t)}{\partial s_l}} = 0,$$

$l' = 1, 2, 3$. This equation set can be solved to give the new set of equations

$$\begin{aligned}
\overline{\frac{\partial X_1(\boldsymbol{s},t)}{\partial s_2} \frac{\partial X_2(\boldsymbol{s},t)}{\partial s_1}} &= -\frac{1}{2}\left\{ \overline{\left[\frac{\partial X_1(\boldsymbol{s},t)}{\partial s_1}\right]^2} + \overline{\left[\frac{\partial X_2(\boldsymbol{s},t)}{\partial s_2}\right]^2} - \overline{\left[\frac{\partial X_3(\boldsymbol{s},t)}{\partial s_3}\right]^2} \right\}, \\
\overline{\frac{\partial X_1(\boldsymbol{s},t)}{\partial s_3} \frac{\partial X_3(\boldsymbol{s},t)}{\partial s_1}} &= -\frac{1}{2}\left\{ \overline{\left[\frac{\partial X_1(\boldsymbol{s},t)}{\partial s_1}\right]^2} - \overline{\left[\frac{\partial X_2(\boldsymbol{s},t)}{\partial s_2}\right]^2} + \overline{\left[\frac{\partial X_3(\boldsymbol{s},t)}{\partial s_3}\right]^2} \right\}, \\
\overline{\frac{\partial X_2(\boldsymbol{s},t)}{\partial s_3} \frac{\partial X_3(\boldsymbol{s},t)}{\partial s_2}} &= -\frac{1}{2}\left\{ -\overline{\left[\frac{\partial X_1(\boldsymbol{s},t)}{\partial s_1}\right]^2} + \overline{\left[\frac{\partial X_2(\boldsymbol{s},t)}{\partial s_2}\right]^2} + \overline{\left[\frac{\partial X_3(\boldsymbol{s},t)}{\partial s_3}\right]^2} \right\},
\end{aligned} \qquad (3.14a–c)$$

The added value of Eqs. (3.14a–c) is that the cross-covariance derivatives are obtained in terms of the variance derivatives. Random fields that satisfy Eq. (3.13) constitute a special class of fields, the solenoidal random fields, which will be discussed in the following section.

In applied sciences, we often encounter systems of physical equations that are coupled in several attributes (e.g., displacements and velocities). Under certain conditions, these equations can be uncoupled by expressing the components of the vector attribute in terms of derivatives of scalar and vector fields in the form of the celebrated Helmholtz decomposition (Stewart, 2011).

Helmholtz Decomposition Theorem[3]

Any well-behaved vector field can be decomposed into the sum of a longitudinal (diverging, noncurling, irrotational) vector field and a transverse (solenoidal, curling, rotational, nondiverging) vector field.

As regards random fields, the Helmholtz decomposition implies that a VS/TRF $\boldsymbol{X}(\boldsymbol{p})$ in $R^{n,1}$ ($n = 2$, 3), which is sufficiently smooth (e.g., twice continuously differentiable and vanishes at infinity), is determinate when we know the values of two random fields: the scalar $W(\boldsymbol{p})$ and the vector $\boldsymbol{U}(\boldsymbol{p})$ fields such that

$$\begin{aligned}
\text{div } \boldsymbol{X}(\boldsymbol{p}) &= \nabla \cdot \boldsymbol{X}(\boldsymbol{p}) = W(\boldsymbol{p}), \\
\text{curl } \boldsymbol{X}(\boldsymbol{p}) &= \nabla \times \boldsymbol{X}(\boldsymbol{p}) = \boldsymbol{U}(\boldsymbol{p}).
\end{aligned} \qquad (3.15a–b)$$

---

[3]Also known as the fundamental theorem of vector calculus.

If $\boldsymbol{X}(\boldsymbol{p})$ is limited to a simply connected region with boundary S_B, the $\boldsymbol{X}(\boldsymbol{p})$ is determinate by $W(\boldsymbol{p})$ and $\boldsymbol{U}(\boldsymbol{p})$, and the $\boldsymbol{X}(\boldsymbol{p})$ values normal to the boundary $(\boldsymbol{X}(\boldsymbol{p})\cdot\boldsymbol{n}, \boldsymbol{p}\in S_B)$.[4]

Example 3.3

Consider the VS/TRF in $R^{2,1}$, $\boldsymbol{X}(\boldsymbol{p}) = [X_1(\boldsymbol{p})\, X_2(\boldsymbol{p})]^T$. According to Helmholtz decomposition, the $\boldsymbol{X}(\boldsymbol{p})$ can be separated into two scalar components, a divergent (irrotational) component and a non-divergent (rotational) component, i.e.,

$$\begin{aligned} \nabla\cdot\boldsymbol{X}(\boldsymbol{p}) &= \frac{\partial X_1(\boldsymbol{p})}{\partial s_1} + \frac{\partial X_2(\boldsymbol{p})}{\partial s_2} = \Theta(\boldsymbol{p}), \\ \nabla\times\boldsymbol{X}(\boldsymbol{p}) &= \frac{\partial X_2(\boldsymbol{p})}{\partial s_1} - \frac{\partial X_1(\boldsymbol{p})}{\partial s_2} = \Psi(\boldsymbol{p}). \end{aligned} \quad (3.16\text{a--b})$$

Assuming that the random field possesses suitable convergence properties, the $\boldsymbol{X}(\boldsymbol{p})$ is determinate if the $\Theta(\boldsymbol{p})$ and $\Psi(\boldsymbol{p})$ are known. In certain cases, it is physically meaningful to express the $\boldsymbol{X}(\boldsymbol{p})$ components in terms of the continuous and differentiable scalar random fields $V_1(\boldsymbol{p})$ and $V_2(\boldsymbol{p})$,

$$\begin{aligned} X_1(\boldsymbol{p}) &= \frac{\partial}{\partial s_1}V_1(\boldsymbol{p}) - \frac{\partial}{\partial s_2}V_2(\boldsymbol{p}), \\ X_2(\boldsymbol{p}) &= \frac{\partial}{\partial s_2}V_1(\boldsymbol{p}) + \frac{\partial}{\partial s_1}V_2(\boldsymbol{p}). \end{aligned} \quad (3.17\text{a--b})$$

If the $\boldsymbol{X}(\boldsymbol{p})$ represents, say, fluid flow, the random fields $V_1(\boldsymbol{p})$ and $V_2(\boldsymbol{p})$ may denote velocity potential and stream function, respectively. Then, it can be shown that

$$\begin{aligned} \Psi(\boldsymbol{p}) &= \nabla^2 V_2(\boldsymbol{p}), \\ \Theta(\boldsymbol{p}) &= \nabla^2 V_1(\boldsymbol{p}). \end{aligned} \quad (3.18\text{a--b})$$

In addition, given that the $V_1(\boldsymbol{p})$ and $V_2(\boldsymbol{p})$ are observed to be uncorrelated in a real-world homogeneous flow situation, the cross-covariances of the VS/TRF $\boldsymbol{X}(\boldsymbol{p})$ can be conveniently expressed in terms of those of the scalar random fields $V_1(\boldsymbol{p})$ and $V_2(\boldsymbol{p})$, i.e., the simultaneous two-point covariance

$$c_{X_1X_2}(\boldsymbol{s},\boldsymbol{s}',t) = \frac{\partial^2}{\partial h_2^2}c_{V_2V_1}(\boldsymbol{h},t) + \frac{\partial^2}{\partial h_2\partial h_1}c_{V_2}(\boldsymbol{h},t) - \frac{\partial^2}{\partial h_1^2}c_{V_1V_2}(\boldsymbol{h},t) - \frac{\partial^2}{\partial h_1\partial h_2}c_{V_1}(\boldsymbol{h},t), \quad (3.19)$$

---

[4]We recall that For $n = 2$, the curl operator in Eq. (3.15b) is sometimes also denoted as rot $\boldsymbol{X}$, and is given by curl $\boldsymbol{X} = $ rot $\boldsymbol{X} = \frac{\partial X_2}{\partial s_1} - \frac{\partial X_1}{\partial s_2}$; whereas for $n = 3$, it is given by

$$\text{curl } \boldsymbol{X} = \left(\frac{\partial X_3}{\partial s_2} - \frac{\partial X_2}{\partial s_3}\right)\boldsymbol{\varepsilon}_1 + \left(\frac{\partial X_1}{\partial s_3} - \frac{\partial X_3}{\partial s_1}\right)\boldsymbol{\varepsilon}_2 + \left(\frac{\partial X_2}{\partial s_1} - \frac{\partial X_1}{\partial s_2}\right)\boldsymbol{\varepsilon}_3 = \begin{bmatrix} 0 & -\frac{\partial}{\partial s_3} & \frac{\partial}{\partial s_2} \\ \frac{\partial}{\partial s_3} & 0 & -\frac{\partial}{\partial s_1} \\ -\frac{\partial}{\partial s_2} & \frac{\partial}{\partial s_1} & 0 \end{bmatrix} \boldsymbol{X},$$

where $\boldsymbol{\varepsilon}_l$ denote the unit vectors along directions s_l ($l = 1, 2, 3$). Notice that div $\boldsymbol{X} = $ rot $\boldsymbol{X}^\perp$, where $\boldsymbol{X}^\perp = (-X_2, X_1)$.

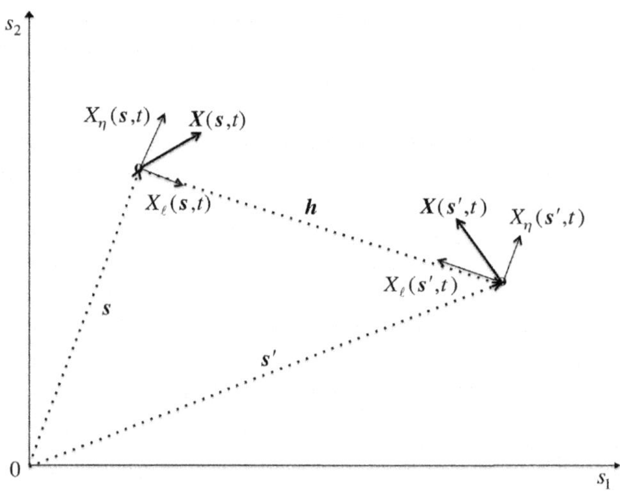

FIGURE 3.2

The longitudinal and transverse components of the vector spatiotemporal random fields $X(s, t)$ in $R^{2,1}$.

where $h = s - s'$. Hence, we do not need to assume homogeneity for the vector random field $X(p)$ but only for the scalar velocity potential and stream function. Furthermore, another physical assumption is that the covariance of velocity potential and stream function is isotropic, and, since $V_1(p)$ and $V_2(p)$ are observed to be uncorrelated, it holds that $c_{V_1V_2}(h,t) = 0$. In this case, for the longitudinal and transverse components of $X(p) = X(s, t)$ (Fig. 3.2) we can define the corresponding covariances as,

$$\begin{aligned}
c_{\ell\ell}(r,t) &= -r^{-1}\frac{\partial}{\partial r}c_{V_2}(r,t) - \frac{\partial^2}{\partial r^2}c_{V_1}(r,t),\\
c_{\eta\eta}(r,t) &= -\frac{\partial^2}{\partial r^2}c_{V_2}(r,t) - r^{-1}\frac{\partial}{\partial r}c_{V_1}(r,t),\\
c_{\ell\eta}(r,t) &= c_{\eta\ell}(r,t) = 0,
\end{aligned} \tag{3.20a–c}$$

where $r = |h|$. Hence, given the covariances of the longitudinal and transverse velocity components in practice, we can obtain the velocity covariance functions.

4. SOLENOIDAL AND POTENTIAL VECTOR SPATIOTEMPORAL RANDOM FIELDS

According to another form of the Helmholtz decomposition, a sufficiently smooth vector field $X(p)$ (i.e., possessing certain convergence properties at infinity) can be decomposed into two special kinds of vector fields: the *solenoidal* (divergence-free or incompressible) field, $X^s(p)$, and the *potential* (curl-free or irrotational) field, $X^p(p)$. These fields, which play a major role in atmospheric studies (e.g., Monin and Yaglom, 1971), are formally defined next.

Definition 4.1

A *solenoidal* VS/TRF, $X^s(p) = [X_1(p)...X_n(p)]^T$,[5] is defined as a vector random field that is divergence free, i.e., it satisfies the law

$$\text{div } X^s(p) = \nabla \cdot X^s(p) = \sum_{k=1}^{n} \frac{\partial}{\partial s_k} X_k(p) = 0. \tag{4.1a}$$

The corresponding *potential* VS/TRF $X^p(p)$ satisfies the gradient law

$$X^p(p) = \nabla Y(p) = \left[\frac{\partial}{\partial s_1} Y(p) \dots \frac{\partial}{\partial s_n} Y(p) \right]^T, \tag{4.1b}$$

where $Y(p)$ is an STIS scalar random field.

Obviously, the velocity vector field of the continuity Eq. (3.13) is a solenoidal VS/TRF. In light of Eqs. (3.15a–b), a vector random field $Z(p)$ can be defined so that

$$\nabla \cdot X^s(p) = \nabla \cdot (\nabla \times Z(p)) = 0.$$

Also, a necessary and sufficient condition for the existence of a vector potential field $X^p(p)$ is that $X^p(p)$ is curl free, i.e.,

$$\text{curl } X^p(p) = \nabla \times X^p(p) = \nabla \times (\nabla Y(p)) = 0, \tag{4.2}$$

which is indeed the case, and, also, $\nabla \cdot X^p(p) = \nabla^2 Y(p)$.

In view of the above definitions and properties, the Helmholtz decomposition can be now expressed as

$$X(p) = X^s(p) + X^p(p) = \nabla \times Z(p) + \nabla Y(p), \tag{4.3}$$

where $p = (s, t)$, and the $Y(p)$ and $Z(p)$ are given in $R^{3,1}$ by

$$Y(s,t) = -\frac{1}{4\pi} \int ds' \frac{W(s',t)}{|s - s'|},$$

$$Z(s,t) = \frac{1}{4\pi} \int ds' \frac{U(s',t)}{|s - s'|}$$

(the vanishing condition at infinity mentioned earlier is imposed to ensure that the integrals above converge). Note that originally the above results were derived for spatial fields, but they can be extended to space–time fields, assuming that their derivation involves only spatial derivatives and these carry through as before if the time argument does not change (Stewart, 2003). Helmholtz decomposition theorem has been also studied in different domains, such as smooth spaces, convex domains and antisymmetric second-rank tensor fields, and under specific or general conditions and domains (Fujiwara and Morimoto, 1977; Kobe, 1984; Hauser, 1971; Sprössig, 2010).

The cross-covariances of the vector random field $X^p(p)$ components are related to the covariance of the scalar random field $Y(p)$ by

$$c^p_{X_l X_{l'}}(s, s', t) = \frac{\partial^2}{\partial s_l \partial s'_{l'}} c_Y(s, s', t). \tag{4.4}$$

[5]Usually in practice, $n = 2, 3$.

If the scalar random field $Y(\boldsymbol{p})$ is STHS, so is the vector random field $\boldsymbol{X}^p(\boldsymbol{p})$. The readers should recall that space–time correlations are formed when we let $\boldsymbol{s}$ and $\boldsymbol{s}'$ denote different locations and let the time be different for the two attributes being correlated. By definition of the solenoidal VS/TRF $\boldsymbol{X}^s(\boldsymbol{p})$, its cross-covariances and cross-spectra are such that

$$\sum_{l=1}^{n} \frac{\partial}{\partial h_l} c^s_{X_l X_{l'}}(\boldsymbol{h}, t) = \sum_{l'=1}^{n} \frac{\partial}{\partial h_{l'}} c^s_{X_l X_{l'}}(\boldsymbol{h}, t) = 0,$$
$$\sum_{l=1}^{n} k_l \widetilde{c}^s_{X_l X_{l'}}(\boldsymbol{k}, t) = \sum_{l=1}^{n} k_{l'} \widetilde{c}^s_{X_l X_{l'}}(\boldsymbol{k}, t) = 0, \tag{4.5a–b}$$

which are equivalent expressions in the physical and the spectral domains. To shed some light in the formal analysis above, it may be useful to review some real-world examples.

Example 4.1

In atmospheric turbulence studies, a measure of the propagation direction of an optical wave on arrival at a receiver is provided by the angle of arrival (AA). This is the angle between the actual plane of the phase front at the receiver and a plane of reference (Rasouli et al., 2014). The AA covariances are evaluated for pairs of points with a given horizontal separation on a single trace (normal to the grating rulings) or vertically between separate traces. A coordinate system is chosen in $R^{2,1}$ such that the s_1 component of the AA fluctuations can be measured, and longitudinal and lateral directions are referenced to the direction of the measured AA component. Then, given the covariance function of the AA fluctuation $X(s_1, s_2, t)$ determined by

$$c_X(h_1, h_2, t) = \overline{X(s_1, s_2, t) X(s_1 + h_1, s_2 + h_2, t)}, \tag{4.6a}$$

and its relation to the corresponding structure function by the physical relationship

$$c_X(h_1, h_2, t) = -\frac{\lambda^2}{8\pi^2} \frac{\partial^2}{\partial h_2^2} \xi_X(h_1, h_2, t), \tag{4.7b}$$

two distinct structure functions of $X(s_1, s_2, t)$ can be defined, accordingly: the simultaneous longitudinal and the transverse (lateral) covariance functions, which are calculated by

$$c_{\ell\ell}(h_1, 0) = A(\beta - 3) h_1^{\beta-4},$$
$$c_{\eta\eta}(0, h_2) = A h_2^{\beta-4}, \tag{4.8a–b}$$

where $A = 0.0127 \gamma_\beta \lambda^2 r_0^{2-\beta} (\beta - 2)$; the parameters λ and φ denote the wavelength and the wave-front phase, respectively, γ_β is a structural coefficient, β is the exponent of the power spectrum for the index of refraction fluctuations, r_0 is a length defined on the wave plane, and d is the grating's period. Notice that for $h_1 = h_2$, the ratio of the covariances in Eq. (4.8a–b) is equal to $\beta - 3$. The above covariance functions can be measured experimentally between pairs of spatial points, thus allowing a quantitative measure of anisotropy in the atmospheric surface layer.

Example 4.2

Interestingly, other random vectors with sound physical meaning can be derived from the above special classes of VS/TRFs. In $R^{3,1}$, let us define the vector field $\boldsymbol{Z}(\boldsymbol{p}) = \nabla \times \boldsymbol{X}^s(\boldsymbol{p})$, where $\boldsymbol{X}^s(\boldsymbol{p})$ is the velocity vector field of fluid flow, and $\boldsymbol{Z}(\boldsymbol{p})$ denotes the vorticity vector field. The cross-covariances and the cross-spectra between components of the two vectors are then given by, respectively,

$$\begin{aligned} c_{Z_l X_l}(\boldsymbol{h}, \tau) &= -\nabla^2 c^s_{X_l X_l}(\boldsymbol{h}, \tau), \\ \widetilde{c}_{Z_l X_l}(\boldsymbol{h}, \tau) &= k^2 \widetilde{c}^s_{X_l X_l}(\boldsymbol{k}, \omega). \end{aligned} \tag{4.9a–b}$$

The physical interpretation of Eq. (4.9a) is that the vorticity-velocity cross-covariance is equal to the negative Laplacian of the velocity covariance, whereas the interpretation of Eq. (4.9b) is that the vorticity-velocity cross-spectrum is related to the velocity spectrum by a squared wavenumber.

5. PARTIAL CROSS-COVARIANCE AND CROSS-SPECTRAL FUNCTIONS

As in the scalar case, partial cross-covariance and cross-spectral functions can be defined without difficulty in the case of VS/TRFs. The wavevector–time cross-spectrum, e.g., is defined by

$$\begin{aligned} \widetilde{c}_{X_l X_{l'}}(\boldsymbol{k}, t) &= \frac{1}{(2\pi)^n} \int d\boldsymbol{h}\, e^{-i\boldsymbol{k}\cdot\boldsymbol{h}} c_{X_l X_{l'}}(\boldsymbol{h}, t), \\ c_{X_l X_{l'}}(\boldsymbol{h}, t) &= \int d\boldsymbol{k}\, e^{i\boldsymbol{k}\cdot\boldsymbol{h}} \widetilde{c}_{X_l X_{l'}}(\boldsymbol{k}, t). \end{aligned} \tag{5.1a–b}$$

The space-frequency spectrum $\widetilde{c}_{X_l X_{l'}}(\boldsymbol{h}, \omega)$ can be similarly defined. As usual, the space and time arguments have distinct effects in the VS/TRF description. By setting $\boldsymbol{h} = \boldsymbol{0}$, e.g., Eq. (5.1b) becomes

$$c_{X_l X_{l'}}(\boldsymbol{0}, t) = \int d\boldsymbol{k}\, \widetilde{c}_{X_l X_{l'}}(\boldsymbol{k}, t), \tag{5.2}$$

which is a measure of the contribution to the cross-covariance of the components $X_l(\boldsymbol{p})$ with wavevector $\boldsymbol{k}$. Important physical quantities can be determined in terms of the partial spectral functions, as the following example demonstrates.

Example 5.1

In a fluid flow situation in $R^{3,1}$, let $X_l(\boldsymbol{p})$, $l = 1, 2, 3$, represent the corresponding velocity components of the VS/TRF. The energy spectrum is given by

$$E(\kappa, t) = \frac{1}{2} \int d\boldsymbol{k}\, \delta(|\boldsymbol{k}| - \kappa) \widetilde{c}_{X_l X_{l'}}(\boldsymbol{k}, t), \tag{5.3}$$

which may be seen as a wavevector–time spectrum linked to the directional information contained in $\widetilde{c}_{X_l X_{l'}}(\boldsymbol{k}, t)$. Furthermore, Eq. (5.3) yields

$$\int d\kappa E(\kappa, t) = \frac{1}{2} c_{X_l X_{l'}}(0, t), \tag{5.4}$$

which is the contribution to the covariance of the components $X_l(\boldsymbol{p})$, $l = 1, \ldots, k$, with wavenumber $|\boldsymbol{k}| \in [\kappa, \kappa + d\kappa]$.

The partial cross-spectra introduce a partialization analysis in the frequency domain that can isolate and reveal key components of the random field. This can be achieved, e.g., by explicitly including space-lag or time-separation dependences that allow representations of the spatial and dynamical component of the phenomenon represented by the spectrum. Furthermore, these spectra can detect and interpret peaks and troughs associated with excitatory and inhibitory connections and discriminate direct and indirect attribute interconnections across space and time.

6. HIGHER-ORDER CROSS—SPATIOTEMPORAL VARIABILITY FUNCTIONS

Higher-order cross-S/TVFs can be specified in a vectorial or a multivariate context. Depending on the physical considerations, various forms of higher-order S/TVFs (cross-covariance, cross-variogram, and cross-structure functions) can be defined. As with all modeling tools, there are both dominant and alternative theoretical interpretations of these S/TVFs.

Among the most interesting higher-order S/TVFs are the fourth-order cross-covariance and cross-structure functions

$$
\begin{aligned}
c_{X_iX_jX_kX_l}(\boldsymbol{p},\boldsymbol{p}') &= \overline{X_i(\boldsymbol{p})X_j(\boldsymbol{p})X_k(\boldsymbol{p}')X_l(\boldsymbol{p}')},\\
\xi_{X_iX_jX_kX_l}(\boldsymbol{p},\boldsymbol{p}') &= \overline{[X_i(\boldsymbol{p})X_j(\boldsymbol{p}) - X_i(\boldsymbol{p}')X_j(\boldsymbol{p}')][X_k(\boldsymbol{p})X_l(\boldsymbol{p}) - X_k(\boldsymbol{p}')X_l(\boldsymbol{p}')]},\\
\xi'_{X_iX_jX_kX_l}(\boldsymbol{p},\boldsymbol{p}') &= \overline{[X_i(\boldsymbol{p}) - X_i(\boldsymbol{p}')][X_j(\boldsymbol{p}) - X_j(\boldsymbol{p}')][X_k(\boldsymbol{p}) - X_k(\boldsymbol{p}')][X_l(\boldsymbol{p}) - X_l(\boldsymbol{p}')]}.
\end{aligned}
\tag{6.1a–c}
$$

for any combination of i, j, k, l. Surely, before fourth-order cross-covariance and cross-structure functions such as above are used, the underlying assumptions (normality, independence, homostationarity, isostationarity, separability, etc.) must be validated.

Example 6.1

Experimental fluid dynamics observations are sometimes derived based on the assumption that the streamwise velocity components for several spacings within the energy-containing range have a joint normal (Gaussian) PDF. This implies that the PDF of any component of velocity difference is normal too. An alternative assumption is that for locally homogeneous turbulence the volume-averaged velocities are statistically independent of local velocity differences, where the averaging volume has a size comparable with this spacing and contains the two points. Both the above assumptions enable considerable simplifications in the in situ calculation of the corresponding velocity S/TVFs. The jointly normal and the statistically independency assumptions are very useful because they allow scaling predictions for the fourth-order velocity statistics in terms of the velocity covariance and structure functions.

Methodologically speaking, cross-S/TVF modeling may involve two kinds of generalization:

(a) empirical generalization inferred from inductive models and substantiated or falsified on the basis of empirical data, and

(b) analytical generalization not seeking to match theory and reality as per empirical generalization, instead, the main idea is abstracting away from empirics (depart from data) to reach a conceptual level that may not be directly validated through empirical testing, as if holding a mirror between models and data.

The cross-S/TVF models of analytical generalization are neither true nor false, but useful or not in making sense of complicated higher-order cross-correlations. As a result, certain applications in earth and atmospheric sciences may require the consideration of some analytical generalizations of the cross-S/TVFs, as follows,

$$
\begin{aligned}
c_{X_1\ldots X_{2\rho}}(\boldsymbol{p},\boldsymbol{p}') &= \overline{\prod_{l=1}^{\rho} X_l(\boldsymbol{p}) \prod_{l=\rho+1}^{2\rho} X_l(\boldsymbol{p}')},\\
\xi_{X_1\ldots X_{2\rho}}(\boldsymbol{p},\boldsymbol{p}') &= \overline{\prod_{l=1,3}^{2\rho-1} [X_l(\boldsymbol{p})X_{l+1}(\boldsymbol{p}) - X_l(\boldsymbol{p}')X_{l+1}(\boldsymbol{p}')]},\\
\xi'_{X_1\ldots X_{2\rho}}(\boldsymbol{p},\boldsymbol{p}') &= \overline{\prod_{l=1}^{2\rho} [X_l(\boldsymbol{p}) - X_l(\boldsymbol{p}')]}.
\end{aligned}
\tag{6.2a–c}
$$

The product operators in the definition of $c_{X_1\ldots X_{2\rho}}$ and $\xi'_{X_1\ldots X_{2\rho}}$ have 2ρ terms (each of which involves a single random field), whereas the product operator in the definition of $\xi_{X_1\ldots X_{2\rho}}$ has ρ terms (each of which involves two S/TRFs). As it turns out, the cross-structure function of Eq. (6.2c) is often easier to formulate in practice than the cross-covariance function of Eq. (6.2a).

As a matter of fact, going even further with analytical generalization, the higher-order cross-S/TVFs of Eqs. (6.2a–c) are not the only ones possible. Depending on the situation, other forms may also emerge that are useful in the description of physical observations.

Example 6.2

Other cross-S/TVFs can be derived, like

$$
\begin{aligned}
c'_{X_1X_2X_3X_4}(\boldsymbol{p},\boldsymbol{p}') &= \overline{[X_1(\boldsymbol{p}) - X_1(\boldsymbol{p}')][X_2(\boldsymbol{p})X_3(\boldsymbol{p})X_4(\boldsymbol{p}) - X_2(\boldsymbol{p}')X_3(\boldsymbol{p}')X_4(\boldsymbol{p}')]},\\
c''_{X_1X_2}(\boldsymbol{p},\boldsymbol{p}') &= \frac{1}{2}[c_{X_1X_2}(\boldsymbol{p},\boldsymbol{p}) + c_{X_1X_2}(\boldsymbol{p}',\boldsymbol{p}')].
\end{aligned}
\tag{6.3a–b}
$$

Eq. (6.3b) is simply an average of two cross-covariance functions. Table 6.1 presents the second-order ($\rho = 1$) and fourth-order ($\rho = 2$) space–time cross-covariance and cross-structure functions. Moreover, the above cross-S/TVFs (cross-covariances and cross-structure functions) are related by the useful identities

$$
\begin{aligned}
\xi_{X_1X_2X_3X_4}(\boldsymbol{p},\boldsymbol{p}') &= \overline{X_1(\boldsymbol{p})X_2(\boldsymbol{p})X_3(\boldsymbol{p})X_4(\boldsymbol{p}) + X_1(\boldsymbol{p}')X_2(\boldsymbol{p}')X_3(\boldsymbol{p}')X_4(\boldsymbol{p}')}\\
&\quad - c_{X_1X_2X_3X_4}(\boldsymbol{p},\boldsymbol{p}') - c_{X_3X_4X_1X_2}(\boldsymbol{p},\boldsymbol{p}'),\\
\xi'_{X_1X_2X_3X_4}(\boldsymbol{p},\boldsymbol{p}') &= \bar{c}_{X_1X_2X_3X_4}(\boldsymbol{p},\boldsymbol{p}') - \xi_{X_1X_2X_3X_4}(\boldsymbol{p},\boldsymbol{p}') - \xi_{X_1X_3X_2X_4}(\boldsymbol{p},\boldsymbol{p}') - \xi_{X_1X_4X_2X_3}(\boldsymbol{p},\boldsymbol{p}').
\end{aligned}
\tag{6.4a–b}
$$

where $\bar{c}_{X_1X_2X_3X_4}(\boldsymbol{p},\boldsymbol{p}') = c'_{X_1X_2X_3X_4}(\boldsymbol{p},\boldsymbol{p}') + c'_{X_2X_1X_3X_4}(\boldsymbol{p},\boldsymbol{p}') + c'_{X_3X_1X_2X_4}(\boldsymbol{p},\boldsymbol{p}') + c'_{X_4X_1X_2X_3}(\boldsymbol{p},\boldsymbol{p}')$.

By way of a summary, higher-order cross-covariance and cross-structure functions are of considerable interest in the study of pressure and velocity statistics, acceleration correlations, turbulence phenomena, and acoustic generation (Batchelor, 1951; Lin, 1953; Tennekes, 1975; Hill, 1996; Zhou and Rubinstein, 1996). The higher-order cross-S/TVFs become analytically more tractable if certain simplifying assumptions can be justified. If the Gaussian assumption, e.g., can be used, certain higher-order cross-S/TVF can be expressed in terms of second-order space–time statistics. Some interesting expressions of the above higher-order cross-S/TVFs are considered in the exercise section of this chapter.

Table 6.1 Examples of Space–Time Cross–Spatiotemporal Variability Functions

ρ	$c_{X_1X_2\ldots X_{2\rho}}(\boldsymbol{p},\boldsymbol{p}')$	$\xi_{X_1X_2\ldots X_{2\rho}}(\boldsymbol{p},\boldsymbol{p}')$	$\xi'_{X_1X_2\ldots X_{2\rho}}(\boldsymbol{p},\boldsymbol{p}')$
1	$c_{X_1X_2}(\boldsymbol{p},\boldsymbol{p}') = \overline{X_1(\boldsymbol{p})X_2(\boldsymbol{p}')}$	$\xi_{X_1X_2}(\boldsymbol{p},\boldsymbol{p}') = \overline{X_1(\boldsymbol{p})X_2(\boldsymbol{p})} - \overline{X_1(\boldsymbol{p}')X_2(\boldsymbol{p}')}$	$\xi'_{X_1X_2}(\boldsymbol{p},\boldsymbol{p}') = \overline{[X_1(\boldsymbol{p}) - X_1(\boldsymbol{p}')][X_2(\boldsymbol{p}) - X_2(\boldsymbol{p}')]}$
2	$c_{X_1X_2X_3X_4}(\boldsymbol{p},\boldsymbol{p}') = \overline{X_1(\boldsymbol{p})X_2(\boldsymbol{p})X_3(\boldsymbol{p}')X_4(\boldsymbol{p}')}$	$\xi_{X_1X_2X_3X_4}(\boldsymbol{p},\boldsymbol{p}') = \overline{[X_1(\boldsymbol{p})X_2(\boldsymbol{p}) - X_1(\boldsymbol{p}')X_2(\boldsymbol{p}')][X_3(\boldsymbol{p})X_4(\boldsymbol{p}) - X_3(\boldsymbol{p}')X_4(\boldsymbol{p}')]}$	$\xi'_{X_1X_2X_3X_4}(\boldsymbol{p},\boldsymbol{p}') = \overline{[X_1(\boldsymbol{p}) - X_1(\boldsymbol{p}')][X_2(\boldsymbol{p}) - X_2(\boldsymbol{p}')][X_3(\boldsymbol{p}) - X_3(\boldsymbol{p}')][X_4(\boldsymbol{p}) - X_4(\boldsymbol{p}')]}$

7. ISOSTATIONARY VECTOR SPATIOTEMPORAL RANDOM FIELDS

I now turn to what is a more challenging issue, that is, the consideration of space–time isostationary VS/TRFs. Interestingly, the generalization of the scalar random field notion of STIS into that of a VS/TRF may be considered in two ways with physically distinct motivations:

(a) the *direct* (lag-based) generalization, and
(b) the *composite* (lag-field–based) generalization.

To phrase it in more words: in case (a), STIS refers to a physically motivated multivariate random field (i.e., a random field characterized as multivariate based on physical considerations, see Section 1), and is defined in terms of its space–time arguments (so that the cross-covariances depend only on the lag magnitude). In case (b), STIS refers to a physically motivated vectorial random field and is defined in terms of group transformations of both the space–time arguments and the VS/TRF components (so that the cross-covariances are functions of the lag direction, as well). As it turns out, under certain conditions the STIS of case (a) may be seen as a special case of the STIS of case (b).

7.1 DIRECT (LAG-BASED) SPACE–TIME ISOSTATIONARITY

This generalization of the scalar random field notion of isotropy refers to the invariance of the mean and cross-covariance functions under arbitrary transformations of points in $R^{n,1}$ and a fixed coordinate system with respect to which the components of the VS/TRF are specified. This kind of direct generalization of scalar STIS fields, which is used in most spatial statistics and mainstream geostatistics applications, usually applies to VS/TRF representations with components scalar physical attributes. In this case, $X_1(\boldsymbol{p})$, $X_2(\boldsymbol{p})$, and $X_3(\boldsymbol{p})$ may denote, e.g., temperature, pressure, and moisture, that is, the vector components refer to three different physical quantities in $R^{3,1}$.

The following developments will generally refer to the VS/TRF

$$\boldsymbol{X}(\boldsymbol{p}) = [X_1(\boldsymbol{p}), \ldots, X_k(\boldsymbol{p})]^T,$$

with mean vector field $\overline{\boldsymbol{X}(\boldsymbol{p})}$, and space–time cross-covariance matrix

$$\boldsymbol{c}_X = [c_{X_l X_{l'}}(\boldsymbol{p}, \boldsymbol{p}')],$$

$l, l' = 1, \ldots, m$. Also, some results are closely linked to the analysis in Section 1 of Chapter VII.

Definition 7.1

In the direct notion of STIS, the VS/TRF is simply a vector of isostationary random fields, isostationarily correlated to each other. The means of the component random fields are constant,

$$\overline{\boldsymbol{X}(\boldsymbol{p})} = \overline{\boldsymbol{X}}, \tag{7.1}$$

and the corresponding matrix of cross-covariance functions between the component random fields is such that

$$\boldsymbol{c}_X = [c_{X_l X_{l'}}(\boldsymbol{s}, t, \boldsymbol{s}', t')] = [c_{X_l X_{l'}}(g\boldsymbol{s}, U_\tau t, g\boldsymbol{s}', U_\tau t')], \tag{7.2a}$$

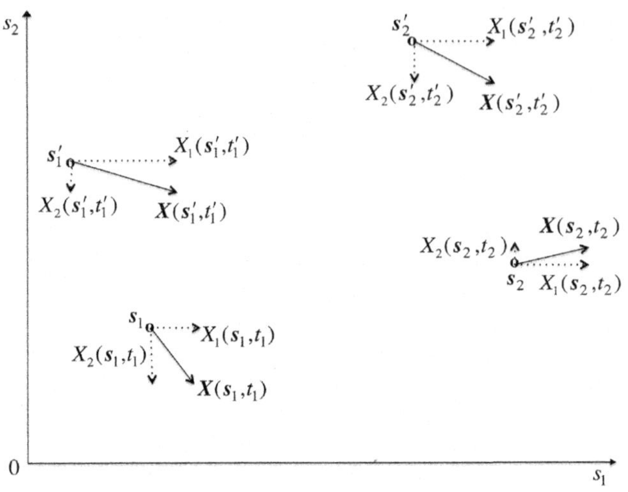

FIGURE 7.1

The direct multivariate notion of a space–time isostationary vector random field in $R^{2,1}$ (as usual, the time axis is suppressed for visualization simplicity).

for all $g \in SO(n)$, all U_τ, and all $l, l' = 1, \ldots, k$. In simpler terms, this means that

$$\boldsymbol{c_X} = [c_{X_l X_{l'}}(\boldsymbol{s}, t, \boldsymbol{s}', t')] = [c_{X_l X_{l'}}(r, \tau)], \tag{7.2b}$$

where $r = |\boldsymbol{s} - \boldsymbol{s}'|$, $\tau = |t - t'|$ for all $l, l' = 1, \ldots, k$, i.e., all scalar cross-covariances functions are isotropic (depend only the distance between pairs of points in space and separations in time).

So with all this being said, it may be worth noticing at this point that some authors prefer to express the STIS condition as,

$$\boldsymbol{c_X} = [c_{X_l X_{l'}}(\boldsymbol{p}, \boldsymbol{p}')] = [c_{X_l X_{l'}}(\boldsymbol{R_\theta s}, U_\tau t, \boldsymbol{R_\theta s}', U_\tau t')]$$

for all rotation matrices $\boldsymbol{R_\theta}$, where $r = |\boldsymbol{s} - \boldsymbol{s}'|$ for all $l, l' = 1, \ldots, k$. The following example illustrates STIS matters in the direct multivariate sense.[6]

Example 7.1

I will now focus on the $R^{2,1}$ domain, within which I will consider four different space–time points $\boldsymbol{p}_1 = (\boldsymbol{s}_1, t_1)$, $\boldsymbol{p}_2 = (\boldsymbol{s}_2, t_2)$, $\boldsymbol{p}'_1 = (\boldsymbol{s}'_1, t'_1)$, and $\boldsymbol{p}'_2 = (\boldsymbol{s}'_2, t'_2)$. In the case of a bivariate random field $\boldsymbol{X}(\boldsymbol{s}) = [X_1(\boldsymbol{p})\, X_2(\boldsymbol{p})]^T$, the direct STIS condition is illustrated in Fig. 7.1 (for visual convenience, only the spatial coordinates are shown). The corresponding cross-covariance functions can be written as

$$c_{X_l X_{l'}}(\boldsymbol{p}_1, \boldsymbol{p}_2) = c_{X_l X_{l'}}(\boldsymbol{p}'_1, \boldsymbol{p}'_2) = c_{X_l X_{l'}}(r, \tau),$$

[6]It was noted earlier that as regards STIS, translation may apply in both R^n and T domains (i.e., $R^{n,1}$), whereas rotation and reflection is usually considered in the R^n domain.

where $r = |s_1 - s_2| = |s_1' - s_2'|$, $\tau = |t_1 - t_2| = |t_1' - t_2'|$ for $l, l' = 1, 2$. To phrase it in more words, the $c_{X_lX_{l'}}$ are invariant to transformations $\mathcal{g}$ replacing $\boldsymbol{s}_1$ and $\boldsymbol{s}_2$ by, respectively, $\boldsymbol{s}_1' = \mathcal{g}\boldsymbol{s}_1$ and $\boldsymbol{s}_2' = \mathcal{g}\boldsymbol{s}_2$, ($\mathcal{g}$ denotes translation, rotation, or reflection of the set of points in R^2).

Under the direct STIS condition, the results concerning scalar covariance functions can be directly extended to the case of the component STIS cross-covariance functions in Eq. (7.2b). So, the cross-covariance function and the cross-spectrum in $R^{n,1}$ are given by, respectively,

$$\begin{aligned} c_{X_lX_{l'}}(r,\tau) &= (2\pi)^{\frac{n}{2}}r^{1-\frac{n}{2}}\int_0^\infty\int_0^\infty dk\, d\omega k^{\frac{n}{2}}J_{\frac{n}{2}-1}(kr)e^{-i\omega\tau}\widetilde{c}_{X_lX_{l'}}(k,\omega), \\ \widetilde{c}_{X_lX_{l'}}(k,\omega) &= (2\pi)^{-\frac{n}{2}}k^{1-\frac{n}{2}}\int_0^\infty\int_0^\infty dr d\tau r^{\frac{n}{2}}J_{\frac{n}{2}-1}(kr)e^{i\omega\tau}c_{X_lX_{l'}}(r,\tau), \end{aligned} \tag{7.3a–b}$$

$l, l' = 1, \ldots, k$. Also, many of the results of vector STHS random fields remain valid in the case of STIS vector random fields above. The cross-spectra in Eqs. (7.3a–b) satisfy the standard requirements

$$\begin{aligned} &\widetilde{c}_{X_lX_{l'}}(k,\omega) = \widetilde{c}_{X_{l'}X_l}(k,\omega), \\ &\widetilde{c}_{X_lX_{l'}}(k,\omega) \geq 0, \\ &\sum_{l=1}^{k}\sum_{l'=1}^{k}\lambda_l\lambda_{l'}\widetilde{c}_{X_lX_{l'}}(\boldsymbol{k},\omega) \geq 0 \end{aligned} \tag{7.4a–c}$$

for all real coefficients λ_l and $\lambda_{l'}$ $(l, l' = 1, \ldots, k)$. And, any functions $\widetilde{c}_{X_lX_{l'}}(k,\omega)$ that can be written as in Eqs. (7.3a–b) and satisfy Eqs. (7.4a–c) are components of the spectral matrix, $\widetilde{\boldsymbol{c}}_{\boldsymbol{X}}$, of an STIS vector random field. In some physical situations the term "local isotropy" is used to indicate isotropy only on small scales, identified by their (high) wavenumbers.

Nevertheless, the above definition of direct isotropy for VS/TRFs can turn out to be very restrictive, on both theoretical and practical grounds. This definition applies, indeed, to scalar physical attributes (e.g., the VS/TRF elements are temperature, pressure, and moisture). However, in many other real-world situations involving vector physical attributes (e.g., the VS/TRF elements are fluid velocity components), it is possible that the cross-covariances of the VS/TRF in different coordinate systems (defined from the translation/rotation of the coordinate axes) are not isotropic, although the VS/TRF itself is defined as isotropic. The following example explains why this may be the case.

Example 7.2

Consider a scalar S/TRF, $X(\boldsymbol{s}, t)$, in $R^{3,1}$. STIS typically implies that its covariance does not depend on the orientation of the lag vector $\boldsymbol{h}$. This is not, however, the case with a VS/TRF $\boldsymbol{X}(\boldsymbol{s}, t)$ where the cross-covariances $c_{X_lX_{l'}}(\boldsymbol{h},\tau)$ of the matrix $\boldsymbol{c}_{\boldsymbol{X}}$ depend on the orientation of the coordinates. For illustration, assume that the point $\boldsymbol{s}$ is the origin of a coordinate system in $R^{n,1}$, and let point $\boldsymbol{s}'$ be located on the s_1-axis of $\boldsymbol{s}$, so that $\boldsymbol{s}' = \boldsymbol{s} + \boldsymbol{h}$. I will focus on the component field $X_1(\boldsymbol{s}, t)$ that is the projection of the vector field $\boldsymbol{X}(\boldsymbol{s}, t)$ on the s_1-axis, in which case, according to our discussion in Section 3, $c_{X_1X_1}(\boldsymbol{h},\tau)$ is the longitudinal covariance. Next, I rotate $\boldsymbol{h}$ by 90 degrees about $\boldsymbol{s}$ so that the points $\boldsymbol{s}$ and $\boldsymbol{s}'$ become parallel to the s_2-axis, in which case the above covariance function becomes the lateral or transverse (relative to $\boldsymbol{h}$) covariance function $c_{X_2X_2}(\boldsymbol{h},\tau)$ of $\boldsymbol{X}(\boldsymbol{s}, t)$, i.e., the covariance of the component field $X_2(\boldsymbol{s}, t)$ that is the projection of $\boldsymbol{X}(\boldsymbol{s}, t)$ on the s_2-axis (perpendicular to s_1-axis). By the standard definition of isotropy, $c_{X_1X_1}(\boldsymbol{h})$ should be the same as $c_{X_2X_2}(\boldsymbol{h},\tau)$ that resulted by the 90 degrees rotation described above. This cannot be valid, which means that the cross-covariances $c_{X_l X_{l'}}(\boldsymbol{h}, \tau)$ depend on the orientation of $\boldsymbol{h}$.

As a matter of fact, the situation described in Example 7.2, provides the motivation for the following notion of vector random field STIS.

7.2 COMPOSITE LAG-FIELD—BASED SPACE—TIME ISOSTATIONARITY

The generalization of the STIS hypothesis motivated by Example 7.2 is both physically desirable and possible. This is a vectorial STIS generalization that applies in the case where the study of a phenomenon focuses on vector physical attributes (or combinations of vector and scalar physical attributes, see scenarios (b) and (c) at the beginning of Section 1). This is the case, e.g., of the vector physical attribute $\boldsymbol{X}(\boldsymbol{p}) = [X_1(\boldsymbol{p}), X_2(\boldsymbol{p}), X_3(\boldsymbol{p})]^T$, where $\boldsymbol{X}(\boldsymbol{p}), \boldsymbol{p} = (\boldsymbol{s}, t)$, is a velocity vector in $R^{3,1}$ with $X_1(\boldsymbol{p})$, $X_2(\boldsymbol{p})$, and $X_3(\boldsymbol{p})$ denoting the velocity components along the corresponding directions (i.e., the vector components refer to the same physical quantity, the velocity).

Consider the VS/TRF $\boldsymbol{X}(\boldsymbol{p})$ with $k = n$. In view of the above considerations, the definition of STIS refers to group invariance of the means and cross-covariances of $\boldsymbol{X}(\boldsymbol{p})$, (1) under arbitrary transformations $\mathcal{g}$ ($=U_\Delta$, $\Lambda_\perp$, i.e., translations, rotations, and reflections) of points in R^{n+1}, which (2) are carried out simultaneously with transformations of the coordinate system with respect to which the components of $\boldsymbol{X}(\boldsymbol{p})$ are specified (projected). Condition (2) assumes the existence of a matrix $\mathcal{U}_g$ describing the linear transformation of the components of the vector random field $\boldsymbol{X}(\boldsymbol{p})$ corresponding to the transformations $\mathcal{g}$ of domain R^n under the condition (1) above. It is valid that, $\mathcal{U}_{g_1}\mathcal{U}_{g_2} = \mathcal{U}_{g_1 g_2}$ and $\mathcal{U}_{g^{-1}} = [\mathcal{U}_g]^{-1}$, i.e., the set of matrices $\mathcal{U}_g$ constitutes a k-dimensional representation of the group $\mathcal{G}$ of all orthogonal transformations. A formal definition of the STIS vector random field is given next.

Definition 7.2

The VS/TRF $\boldsymbol{X}(\boldsymbol{p})$ is *wide sense composite* (*lag-field*—based) *STIS* if and only if the mean vector is such that[7]

$$\mathcal{U}_g \overline{\boldsymbol{X}(\mathcal{g}\boldsymbol{p})} = \overline{\boldsymbol{X}}, \tag{7.5a}$$

and the covariance matrix satisfies

$$\mathcal{U}_g \boldsymbol{c}_X(\mathcal{g}\Delta\boldsymbol{p})\,\mathcal{U}_g^* = \boldsymbol{c}_X(\Delta\boldsymbol{p}), \tag{7.5b}$$

for all $\mathcal{g} \in \mathcal{G}$, where $\Delta\boldsymbol{p} = (\boldsymbol{h}, \tau)$, $\boldsymbol{c}_X$ is the covariance matrix of $\boldsymbol{X}(\boldsymbol{p})$, and $\mathcal{U}_g^*$ is the Hermitian conjugate of the matrix $\mathcal{U}_g$ denoting transformation of the random vector components under the transformation $\mathcal{g}$ of points on R^{n+1}.[8]

Definition 7.2 implies that the mean vector and the covariance matrix do not change by the combined transformations $\mathcal{g}$ and $\mathcal{U}_g$. The deeper issue here is that while the cross-covariances in Definition 7.1 are functions of the scalar distance $r = |\boldsymbol{h}|$ only, in Definition 7.2 the isotropy is defined in a manner such that the cross-covariances depend on the orientation of the vector distance $\boldsymbol{h}$, see also,

[7]In fact, the condition that the mean vector must be invariant under arbitrary rotations may imply that the mean value is zero.
[8]A similar definition is valid in terms of probability distributions (strict sense, s.s., isotropy).

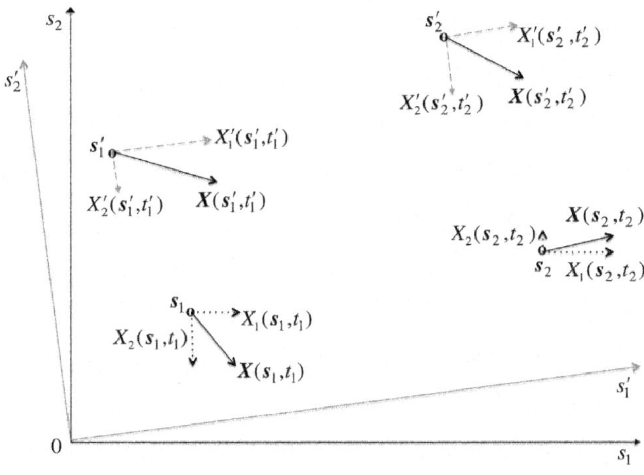

FIGURE 7.2

The composite (lag-field) notion of the space–time isostationary vector random field in $R^{2,1}$.

Examples 7.1(cont.), 7.3, and 7.4. In the special case that $\mathcal{U}_g$ is the unit $k \times k$-matrix for all $g \in \mathcal{G}$, Eqs. (7.5a) and (7.5b) reduce to Eqs. (7.2a) and (7.2b), respectively, in which case $\boldsymbol{X}(\boldsymbol{p})$ is simply a vector of STIS random fields that are isostationarilly correlated to each other.[9]

Example 7.1 (cont.)

For the vector random field $\boldsymbol{X}(\boldsymbol{p}) = [X_1(\boldsymbol{p}), X_2(\boldsymbol{p})]^T$ studied in Example 7.1, the composite STIS vector random field situation is visually illustrated in Fig. 7.2, where

$$c_{X_l X_{l'}}(\boldsymbol{s}_1, \boldsymbol{s}_2, \tau) = c_{X'_l X'_{l'}}(\boldsymbol{s}'_1, \boldsymbol{s}'_2, \tau),$$

$l, l' = 1, 2$ and $\boldsymbol{s}_1 - \boldsymbol{s}_2 = \boldsymbol{s}'_1 - \boldsymbol{s}'_2 = \boldsymbol{h}$. That is, the cross-covariance of $X_1(\boldsymbol{s}_1, t_1)$ and $X_2(\boldsymbol{s}_2, t_2)$ is the same as that of $X'_1(\boldsymbol{s}'_1, t'_1)$ and $X'_2(\boldsymbol{s}'_2, t'_2)$ ($X'_l(\boldsymbol{s}'_1, t'_1)$ and $X'_l(\boldsymbol{s}'_2, t'_2)$, $l = 1, 2$, are the projections of $\boldsymbol{X}(\boldsymbol{s}'_1, t'_1)$ and $\boldsymbol{X}(\boldsymbol{s}'_2, t'_2)$, respectively, on to the (s'_1, s'_2) coordinate system, and the angles between $\boldsymbol{h}$ and the vector random field must be maintained as the coordinate system is rotated and reflected). In more formal terms, the $c_{X_l X_{l'}}(\boldsymbol{s}_1, \boldsymbol{s}_2, \tau)$ are invariant to transformations g that replace $\boldsymbol{s}_1$ and $\boldsymbol{s}_2$ with, respectively, $\boldsymbol{s}'_1 = g\boldsymbol{s}_1$ and $\boldsymbol{s}'_2 = g\boldsymbol{s}_2$, and simultaneously carry out the linear transformation u of the VS/TRF components $X'_l(\boldsymbol{s}'_1, t'_1) = uX_l(g\boldsymbol{s}_1, t_1)$ and $X'_l(\boldsymbol{s}'_2, t'_2) = u\, X_l(g\boldsymbol{s}_2, t_2)$ corresponding to the transformation g of the coordinates. The readers may find it interesting to compare the visual illustrations of Figs. 7.1 and 7.2.

An important result of the composite generalization of the notion of STIS is presented by the following theorem (Robertson, 1940).

[9]A simpler version of Eq. (7.5b) is obtained when only rotations are considered, i.e., $R_\theta^T \boldsymbol{c}_X(R_\theta \boldsymbol{p}, R_\theta \boldsymbol{p}')R_\theta = \boldsymbol{c}_X(\Delta\boldsymbol{p})$, for all rotation matrices R_θ.

Theorem 7.1

Let $\boldsymbol{X}(\boldsymbol{p})$ satisfy the composite STIS conditions of Definition 7.2. Then, the cross-covariances $c_{X_l X_{l'}}(\boldsymbol{h}, \tau)$, $l, l' = 1, \ldots, k$, of the matrix $\boldsymbol{c}_X$ are generally expressed in $R^{n,1}$ by

$$c_{X_l X_{l'}}(\boldsymbol{h}, \tau) = A(r, \tau) h_l h_{l'} + B(r, \tau)\delta_{ll'}, \tag{7.6}$$

where $A(r, \tau)$ and $B(r, \tau)$ are real-valued functions of the lag magnitude $r = |\boldsymbol{h}|$, and $\delta_{ll'}$ is the usual Kronecker delta.

According to Theorem 7.1, in the case of an STIS vector random field, the most general form of the cross-covariance is given by Eq. (7.6), in which case the STIS condition reduces to two independent components, $A(r, \tau)$ and $B(r, \tau)$. It is convenient to express these two components in terms of the covariance functions that are commonly measured in physical applications, viz., the longitudinal and the transverse (or lateral) covariance functions (Section 3). For this purpose, let the s_1-axis be along the lag vector $\boldsymbol{h} = (1, 0, 0)$ in R^3, which means that,

$$A(r, \tau) + B(r, \tau) = c_{X_1 X_1}(r, \tau) = c_{\ell\ell}(r, \tau),$$

i.e., a longitudinal covariance function (the two vector field components are aligned with the lag vector), and

$$B(r, \tau) = c_{X_2 X_2}(r, \tau) = c_{X_3 X_3}(r, \tau) = c_{\eta\eta}(r, \tau),$$

i.e., a lateral or transverse covariance function (the pairs of the vector field components are selected so that they are in the same plane but perpendicular to the lag vector). Then, the following proposition is a direct consequence of Theorem 7.1.

Proposition 7.1

Let $\boldsymbol{X}(\boldsymbol{p})$ satisfy the composite STIS conditions of Definition 7.2. Then, the cross-covariances in $R^{n,1}$ are given by

$$c_{X_l X_{l'}}(\boldsymbol{h}, \tau) = \frac{h_l h_{l'}}{r^2}\left[c_{\ell\ell}(r, \tau) - c_{\eta\eta}(r, \tau)\right] + c_{\eta\eta}(r, \tau)\delta_{ll'}, \tag{7.7}$$

$l, l' = 1, \ldots, k$, where $c_{\ell\ell}(r, \tau)$ and $c_{\eta\eta}(r, \tau)$ are the longitudinal and lateral (or transverse) scalar (isotropic) covariances, respectively, with $c_{\ell\ell}(0, \tau) = c_{\eta\eta}(0, \tau) = \frac{1}{3}\overline{X(\boldsymbol{p})^2}$.

Eq. (7.7) is a remarkable results, since it implies that the cross-covariances $c_{X_l X_{l'}}(\boldsymbol{h}, \tau)$ are uniquely specified by two scalar functions, $c_{\ell\ell}(r, \tau)$ and $c_{\eta\eta}(r, \tau)$. Obviously, due to the term $h_l h_{l'}$, the covariance $c_{X_l X_{l'}}(\boldsymbol{h}, \tau)$ is not STIS, although the VS/TRF $\boldsymbol{X}(\boldsymbol{p})$ it represents is defined as STIS (in the composite sense). In particular, $c_{\ell\ell}(r, \tau)$ is the covariance of $X_l(\boldsymbol{s}, t)$, which is the projection of $\boldsymbol{X}(\boldsymbol{s}, t)$ on a line along the direction of $\boldsymbol{h}$; and $c_{\eta\eta}(r, \tau)$ is the covariance of $X_\eta(\boldsymbol{s}, t)$, which is the projection of $\boldsymbol{X}(\boldsymbol{s}, t)$ on a line perpendicular to $\boldsymbol{h}$, so that $\frac{h_l}{r}$ are the components of the unit vector $\frac{\boldsymbol{h}}{r}$. In other words, if the coordinate system in $R^{3,1}$ is chosen so that $\boldsymbol{h}$ is in the s_1 direction (say, $\boldsymbol{h} = r\boldsymbol{\varepsilon}_1$), Eq. (7.7) gives,

$$\begin{aligned}
&c_{X_1 X_1}(r, t) = c_{X_1 X_1}(r\boldsymbol{\varepsilon}_1, t) = c_{\ell\ell}(r, t),\\
&c_{X_l X_l}(r, t) = c_{X_l X_l}(r\boldsymbol{\varepsilon}_1, t) = c_{\eta\eta}(r, t) \quad (l = 2, 3),\\
&c_{X_l X_{l'}}(r, t) = 0 \quad (l \neq l')
\end{aligned}$$

[recall, Eqs. (3.2a–b)]. These relationships, in combination with Fig. 7.2, convincingly demonstrate the importance of $c_{\ell\ell}(r, \tau)$ and $c_{\eta\eta}(r, \tau)$.

Example 7.3

In Eq. (7.7), let us consider one of the diagonal matrix $\boldsymbol{c_X}$ components in $R^{3,1}$, which is the covariance

$$c_{X_lX_l}(\boldsymbol{h},\tau) = \frac{h_l^2}{r^2}\left[c_{\ell\ell}(r,\tau) - c_{\eta\eta}(r,\tau)\right] + c_{\eta\eta}(r,\tau), \tag{7.8a}$$

i.e., the covariances $c_{X_lX_l}(\boldsymbol{h},\tau)$, $l = 1, 2, 3$, of the STIS vector random field are not STIS themselves, since they depend on the direction of the lag coordinate h_l. Matters can be developed further in certain physical situations. One of them is the situation in which the field is divergent free, $\nabla\cdot\boldsymbol{X}(\boldsymbol{p}) = 0$ (say, the phenomenon under study involves a continuity equation (Eq. 3.13) for homogeneous fields of incompressible fluids). It is then valid that

$$\sum_{l=1}^{3}\frac{\partial}{\partial h_l}c_{X_lX_{l'}}(\boldsymbol{h},\tau) = \sum_{l'=1}^{3}\frac{\partial}{\partial h_{l'}}c_{X_lX_{l'}}(\boldsymbol{h},\tau) = 0, \tag{7.8b}$$

in which case Eq. (7.7) leads to[10]

$$c_{\eta\eta}(r,\tau) = c_{\ell\ell}(r,\tau) + \frac{r}{2}\frac{\partial}{\partial r}c_{\ell\ell}(r,\tau), \tag{7.8c}$$

i.e., the longitudinal covariance can be obtained from the transverse covariance, and vice versa.

Similarly to Eq. (7.7), in the frequency domain we have the following general expression of the wavevector–time cross-spectrum

$$\widetilde{c}_{X_lX_{l'}}(\boldsymbol{k},\tau) = \frac{k_lk_{l'}}{k^2}\left[\widetilde{c}_{\ell\ell}(k,\tau) - \widetilde{c}_{\eta\eta}(k,\tau)\right] + \widetilde{c}_{\eta\eta}(k,\tau)\delta_{ll'}, \tag{7.9}$$

where

$$\widetilde{c}_{X_lX_{l'}}(\boldsymbol{k},\tau) = \frac{1}{(2\pi)^n}\int d\boldsymbol{h}\, e^{-i\boldsymbol{k}\cdot\boldsymbol{h}}c_{X_lX_{l'}}(\boldsymbol{h},\tau),$$

$k = |\boldsymbol{k}|$ is the wavenumber, and $\widetilde{c}_{\ell\ell}(k,\tau)$ and $\widetilde{c}_{\eta\eta}(k,\tau)$ are the corresponding longitudinal and lateral spectra. If the covariances $c_{\ell\ell}(r,\tau)$ and $c_{\eta\eta}(r,\tau)$ decrease fast enough as $r\to\infty$, the $\widetilde{c}_{\ell\ell}(k,\tau)$ and $\widetilde{c}_{\eta\eta}(k,\tau)$ can be determined accordingly.

Example 7.4

Consider the case in which $\boldsymbol{X}(\boldsymbol{p}) = [X_1(\boldsymbol{p})\, X_2(\boldsymbol{p})\, X_3(\boldsymbol{p})]^T$ denotes an STIS atmospheric velocity vector field in $R^{3,1}$, where $X_1(\boldsymbol{p})$, $X_2(\boldsymbol{p})$, and $X_3(\boldsymbol{p})$ are the velocity components along the corresponding directions. The spectra $\widetilde{c}_{X_lX_{l'}}(\boldsymbol{k},\tau)$, $l, l' = 1, 2, 3$, in Eq. (7.9) are then the three-dimensional spectra of the velocity vector field $\boldsymbol{X}(\boldsymbol{p})$. In physical applications, it is easier to measure the unidimensional longitudinal and lateral spectra of $\boldsymbol{X}(\boldsymbol{p})$ by

$$\begin{aligned}\widetilde{c}_{\ell}(k,\tau) &= (2\pi)^{-1}\int dr e^{-ikr}c_{\ell\ell}(r,\tau),\\ \widetilde{c}_{\eta}(k,\tau) &= (2\pi)^{-1}\int dr e^{-ikr}c_{\eta\eta}(r,\tau).\end{aligned} \tag{7.10a–b}$$

[10] Recall that $\frac{\partial}{\partial h_l}c_{\ell\ell}(r,\tau) = \frac{\partial}{\partial r}c_{\ell\ell}(r,\tau)\frac{\partial r}{\partial h_l} = \frac{\partial}{\partial r}c_{\ell\ell}(r,\tau)\frac{h_l}{r}$.

We recall that $X_1(s_1, 0, 0)$ and $X_2(s_1, 0, 0)$ will be STHS fields along the line $s_2 = s_3 = 0$. We also notice that

$$\begin{aligned}\widetilde{c}_\ell(k_1,\tau) &= \iint dk_2 dk_3 \widetilde{c}_{X_1X_1}(k_1,k_2,k_3,\tau),\\ \widetilde{c}_\eta(k_1,\tau) &= \iint dk_2 dk_3 \widetilde{c}_{X_2X_2}(k_1,k_2,k_3,\tau)\end{aligned} \tag{7.11a–b}$$

(in practice, $\widetilde{c}_\ell$ and $\widetilde{c}_\eta$ are often generated from tower and aircraft velocity measurements). By substituting the spectra $\widetilde{c}_{X_lX_{l'}}$ expressions of Eq. (7.9) into Eqs. (7.11a–b), it is found that

$$\begin{aligned}\widetilde{c}_\ell(k_1,\tau) &= 2\pi \int dk\, k^{-1}\left[k_1^2 \widetilde{c}_{\ell\ell}(k,\tau) + (k^2 - k_1^2)\widetilde{c}_{\eta\eta}(k,\tau)\right],\\ \widetilde{c}_\eta(k_1,\tau) &= \pi \int dk\, k^{-1}\left[\left(k^2 - k_1^2\right)\widetilde{c}_{\ell\ell}(k,\tau) + \left(k^2 + k_1^2\right)\widetilde{c}_{\eta\eta}(k,\tau)\right].\end{aligned} \tag{7.11c–d}$$

These equations imply that $\widetilde{c}_\ell, \widetilde{c}_\eta \geq 0$, if $\widetilde{c}_{\ell\ell}, \widetilde{c}_{\eta\eta} \geq 0$. It is noteworthy that, by inverting Eqs. (7.10a–b), the corresponding longitudinal and lateral covariances can be expressed as

$$\begin{aligned}c_{\ell\ell}(r,\tau) &= 2\int_0^\infty dk \cos(kr)\widetilde{c}_\ell(k,\tau),\\ c_{\eta\eta}(r,\tau) &= 2\int_0^\infty dk \cos(kr)\widetilde{c}_\eta(k,\tau).\end{aligned} \tag{7.12a–b}$$

Lastly, by combining Eqs. (7.7) and (7.12a–b), one can generate space–time cross-covariance models in $R^{3,1}$, starting from valid models in $R^{1,1}$ (see also, Chapter XVI).

Example 7.5

Interestingly, when both a scalar and a vector random field are involved (scenario (c) discussed in the beginning of Section 1), i.e.,

$$\boldsymbol{X}(\boldsymbol{s},t) = [Y(\boldsymbol{s},t)\ X_1(\boldsymbol{s},t)\ X_2(\boldsymbol{s},t)\ X_3(\boldsymbol{s},t)]^T,$$

Eq. (7.7) in $R^{3,1}$ reduces to

$$\begin{aligned}c_{YX_l}(\boldsymbol{h},\tau) &= \frac{h_l}{r} c_{Y\ell}(r,\tau),\\ c_{X_lY}(\boldsymbol{h},\tau) &= \frac{h_l}{r} c_{\ell Y}(r,\tau) = -\frac{h_l}{r} c_{Y\ell}(r,\tau),\end{aligned}$$

i.e., the cross-covariance of any vector random field component and the scalar random field is equal to minus the cross-covariance of the scalar random field and any vector random field component.

I will conclude this section by noticing that much of the analysis above is valid in terms of the structure function. So, Eq. (7.7) can be written as

$$\xi_{X_lX_{l'}}(\boldsymbol{h},\tau) = \frac{h_l h_{l'}}{r^2}\left[\xi_{\ell\ell}(r,\tau) - \xi_{\eta\eta}(r,\tau)\right] + \xi_{\eta\eta}(r,\tau)\delta_{ll'}, \tag{7.13}$$

where $\xi_{\ell\ell}(r, \tau)$ and $\xi_{\eta\eta}(r, \tau)$ are the longitudinal and transverse (lateral) structure functions. By differentiating the above relationship, we find that

$$\frac{\partial}{\partial h_l}\xi_{X_lX_{l'}}(\boldsymbol{h},\tau) = \frac{h_{l'}}{r^2}\left\{2[\xi_{\ell\ell}(r,\tau) - \xi_{\eta\eta}(r,\tau)] + r\frac{\partial}{\partial r}\xi_{\ell\ell}(r,\tau)\right\},$$

which, in the case of a divergent-free field $\boldsymbol{X}(\boldsymbol{p})$, yields

$$\xi_{\eta\eta}(r,t) = \xi_{\ell\ell}(r,t) + \frac{r}{2}\frac{\partial}{\partial r}\xi_{\ell\ell}(r,t).$$

If the coordinate system is such that in $\xi_{X_lX_{l'}}(\boldsymbol{h},\tau)$ the space argument is $\boldsymbol{h} = r\boldsymbol{\varepsilon}_1$, then it is found that

$$\begin{aligned}
&\xi_{X_1X_1}(r,\tau) = \xi_{\ell\ell}(r,\tau),\\
&\xi_{X_lX_l}(r,\tau) = \xi_{\eta\eta}(r,\tau) \quad (l = 2,3),\\
&\xi_{X_lX_{l'}}(r,\tau) = 0 \quad (l \neq l').
\end{aligned}$$

And, if $\boldsymbol{X}(\boldsymbol{p})$ is divergent free, the structure function is related to the simultaneous two-point covariance function by

$$\xi_{X_lX_{l'}}(\boldsymbol{h},t) = 2c_{X_lX_{l'}}(0,t) - c_{X_lX_{l'}}(\boldsymbol{h},t) - c_{X_{l'}X_l}(\boldsymbol{h},t),$$

where $c_{X_{l'}X_l}(\boldsymbol{h},t) = c_{X_lX_{l'}}(-\boldsymbol{h},t)$. In addition,

$$\xi_{\ell\ell}(r,t) = 2c_{\ell\ell}(0,t) - 2c_{\ell\ell}(r,t),$$

i.e., the longitudinal structure function is directly related to the longitudinal covariance function. (Exercise IX.6).

Example 7.6

Assume that $\xi_{\ell\ell}(r, t) = (\varepsilon r)^a c(t)$ in $R^{3,1}$ with $a > 0$. From Eq. (7.8c) expressed in terms of the structure function, we find

$$\xi_{\eta\eta}(r,t) = \xi_{\ell\ell}(r,t) + \frac{r}{2}\frac{\partial}{\partial r}\xi_{\ell\ell}(r,t) = \frac{a+2}{2}(\varepsilon r)^a c(t) = \frac{a+2}{2}\xi_{\ell\ell}(r,t).$$

Also, from Eq. (7.7) expressed in terms of the structure function, we get

$$\begin{aligned}
\xi_{X_lX_{l'}}(\boldsymbol{h},t) &= \frac{h_lh_{l'}}{r^2}\left[\xi_{\ell\ell}(r,t) - \xi_{\eta\eta}(r,t)\right] + \xi_{\eta\eta}(r,t)\delta_{ll'}\\
&= \frac{1}{2}\left[(a+2)\delta_{ll'} - \frac{a}{r^2}h_lh_{l'}\right]\xi_{\ell\ell}(r,t) = \frac{1}{2}\left[(a+2)\delta_{ll'} - \frac{a}{r^2}h_lh_{l'}\right](\varepsilon r)^a c(t).
\end{aligned}$$

That is, in this case the $\xi_{X_lX_{l'}}(\boldsymbol{h},t)$, which is STHS but generally not STIS, is determined by the single scalar function $\xi_{\ell\ell}(r, t)$, which is STIS. Moreover, for $\boldsymbol{h} = r\boldsymbol{\varepsilon}_1$, we get

$$\begin{aligned}
&\xi_{X_1X_1}(r,t) = \xi_{\ell\ell}(r,t) = (\varepsilon r)^a c(t),\\
&\xi_{X_2X_2}(r,t) = \frac{a+2}{2}\xi_{\ell\ell}(r,t) = \frac{a+2}{2}(\varepsilon r)^a c(t) = \xi_{X_3X_3}(r,t),\\
&\xi_{X_lX_{l'}}(r,t) = 0 \quad (l \neq l'),
\end{aligned}$$

which concludes this example.

7.3 LINKS WITH SOLENOIDAL AND POTENTIAL SPATIOTEMPORAL RANDOM FIELDS

In this section, I will discuss some interesting relationships between the longitudinal $c_{\ell\ell}(r, \tau)$ and the lateral $c_{\eta\eta}(r, \tau)$ covariances for some special kinds of VS/TRFs. In this setting, the following point should be stressed: The longitudinal and lateral covariances of an isostationary VS/TRF $\boldsymbol{X}(\boldsymbol{p})$ in Eq. (7.7) can be decomposed as

$$\begin{aligned} c_{\ell\ell}(r,\tau) &= c^{s}_{\ell\ell}(r,\tau) + c^{p}_{\ell\ell}(r,\tau), \\ c_{\eta\eta}(r,\tau) &= c^{s}_{\eta\eta}(r,\tau) + c^{p}_{\eta\eta}(r,\tau), \end{aligned} \tag{7.14a–b}$$

where $c^{s}_{\eta\eta}$ and $c^{s}_{\ell\ell}$ are the lateral and longitudinal covariances of the solenoidal VS/TRF $\boldsymbol{X}^{s}(\boldsymbol{p})$, and $c^{p}_{\eta\eta}$ and $c^{p}_{\ell\ell}$ are the lateral and longitudinal covariances of the potential VS/TRF $\boldsymbol{X}^{p}(\boldsymbol{p})$.

To continue with our analysis, we notice that the $c^{s}_{\ell\ell}$ and $c^{s}_{\eta\eta}$ covariances of an STIS $\boldsymbol{X}^{s}(\boldsymbol{p})$ jointly satisfy the following equation in $R^{n,1}$,

$$\frac{\partial}{\partial r} c^{s}_{\ell\ell}(r,t) + \frac{n-1}{r}\left[c^{s}_{\ell\ell}(r,t) - c^{s}_{\eta\eta}(r,t)\right] = 0, \tag{7.15}$$

where $r = |\boldsymbol{h}|$. Along the lines of the discussion in Example 7.4, the $c^{s}_{\ell\ell}$ and $c^{s}_{\eta\eta}$ are even functions, have their maximum values at $r = 0$, $c^{s}_{\ell\ell}(0,t) = c^{s}_{\eta\eta}(0,t)$, and they can be approximated by

$$\begin{aligned} c^{s}_{\ell\ell}(r,t) &\approx c^{s}_{\ell\ell}(0,t) - \frac{r^2}{2\lambda_1^2}, \\ c^{s}_{\eta\eta}(r,t) &\approx c^{s}_{\eta\eta}(0,t) - \frac{r^2}{2\lambda_2^2}, \end{aligned} \tag{7.16a–b}$$

where

$$\begin{aligned} \lambda_1^2 &= -c^{s}_{\ell\ell}(0,t)\left[\frac{\partial^2 c^{s}_{\ell\ell}}{\partial h^2}(0,t)\right]^{-1}, \\ \lambda_2^2 &= -c^{s}_{\eta\eta}(0,t)\left[\frac{d^2 c^{s}_{\eta\eta}}{dh^2}(0,t)\right]^{-1} \end{aligned} \tag{7.17a–b}$$

(see also, Exercise IX.7).

I will now return to the fundamental for the purposes of our analysis Eq. (7.7), by virtue of which the covariances of the solenoidal VS/TRF $\boldsymbol{X}^{s}(\boldsymbol{p})$ can be written as

$$c^{s}_{X_l X_{l'}}(\boldsymbol{h},t) = \frac{1}{r^2}\left[c^{s}_{\ell\ell}(r,t) - c^{s}_{\eta\eta}(r,t)\right] h_l h_{l'} + c^{s}_{\eta\eta}(r,t)\delta_{ll'}, \tag{7.18}$$

where $r = |\boldsymbol{h}|$, and, as before, the $c^{s}_{\ell\ell}$ and $c^{s}_{\eta\eta}$ correspond to the VS/TRF components in the direction of $\boldsymbol{h}$ and normal to $\boldsymbol{h}$, respectively. These covariances are even functions and have their maximum values at $r = 0$. To phrase it in more words, the covariances of $\boldsymbol{X}^{s}(\boldsymbol{p})$ are uniquely expressed in terms of only two STIS covariances, $c^{s}_{\ell\ell}$ and $c^{s}_{\eta\eta}$. Similarly, the covariances of $\boldsymbol{X}^{p}(\boldsymbol{p})$ can be expressed as

$$c^{p}_{X_l X_{l'}}(\boldsymbol{h},t) = \frac{1}{r^2}\left[c^{p}_{\ell\ell}(r,t) - c^{p}_{\eta\eta}(r,t)\right] h_l h_{l'} + c^{p}_{\eta\eta}(r,t)\delta_{ll'}. \tag{7.19}$$

In applied sciences, the longitudinal and lateral covariances can be experimentally calculated, providing a measure of the physical behavior of the attribute of interest.

Example 7.7

By virtue of Eq. (7.15), the following expressions are obtained for the longitudinal and lateral covariances of the solenoidal random field,

$$\begin{aligned} \frac{\partial}{\partial r}\left[r c^{s}_{\ell\ell}(r,t)\right] &= c^{s}_{\eta\eta}(r,t) \qquad (n=2), \\ \frac{\partial}{\partial r}\left[r^{2} c^{s}_{\ell\ell}(r,t)\right] &= 2r c^{s}_{\eta\eta}(r,t) \quad (n=3). \end{aligned} \tag{7.20a–b}$$

Eq. (7.20b) is known as the von Karman equation. Note that at $r=0$,

$$\begin{aligned} \lambda_1^2 &= 3\lambda_2^2 \quad (n=2), \\ \lambda_1^2 &= 2\lambda_2^2 \quad (n=3) \end{aligned} \tag{7.21a–b}$$

(Exercise IX.8). On the other hand, the $c^{p}_{\ell\ell}$ and $c^{p}_{\eta\eta}$ covariances of a potential vector field $\boldsymbol{X}^{p}(\boldsymbol{s})$ satisfy the Obukhov–Yaglom equation,

$$c^{p}_{\ell\ell}(r,t) = \frac{\partial}{\partial r}\left[r c^{p}_{\eta\eta}(r,t)\right], \tag{7.22}$$

where now

$$\begin{aligned} c^{p}_{\ell\ell}(r,t) &= -\frac{\partial^2 c_Y(r,t)}{\partial r^2}, \\ c^{p}_{\eta\eta}(r,t) &= -\frac{1}{r}\frac{\partial c_Y(r,t)}{\partial r}. \end{aligned} \tag{7.23a–b}$$

Eq. (7.22) is a necessary and sufficient condition for realizing $\boldsymbol{X}^{p}(\boldsymbol{p})$. In this case, and in view of Eqs. (7.23a–b), it is valid that

$$c_{X_l X_{l'}}(\boldsymbol{h},t) = -\frac{\partial^2}{\partial h_l \partial h_{l'}} c_Y(\boldsymbol{h},t) = -\frac{1}{r}\frac{\partial}{\partial r}\left[\frac{1}{r}\frac{\partial c_Y(r,t)}{\partial r}\right] h_l h_{l'} - \frac{1}{r}\frac{\partial c_Y(r,t)}{\partial r}\delta_{ll'}, \tag{7.24}$$

$r=|\boldsymbol{h}|$. Also, by virtue of Eq. (7.20a–b), Eq. (7.18) gives

$$c_{X_l X_{l'}}(\boldsymbol{h},t) = \frac{1}{r}\frac{\partial c^{p}_{\eta\eta}(r,t)}{\partial r} h_l h_{l'} + c^{p}_{\eta\eta}(r,t)\delta_{ll'}, \tag{7.25}$$

i.e., the cross-covariances of the initial VS/TRF can be expressed in terms of $c^{p}_{\eta\eta}$, in this case.

In sum, in the case of an STIS solenoidal VS/TRF, Eq. (7.7) gives Eq. (7.15) that relates $c^{s}_{\ell\ell}$ and $c^{s}_{\eta\eta}$ (in this case, $\tilde{c}^{s}_{\ell\ell}=0$ in the frequency domain, see below); and, in the case of a potential VS/TRF, Eq. (7.7) gives Eq. (7.22) that relates $c^{p}_{\ell\ell}$ and $c^{p}_{\eta\eta}$ (in this case, $\tilde{c}^{p}_{\eta\eta}=0$). Accordingly, the covariance $c^{s}_{X_l X_{l'}}(\boldsymbol{h},\tau)$ of a solenoidal vector random field is fully determined by a single function $c^{s}_{\ell\ell}$ or $c^{s}_{\eta\eta}$; and, the covariance $c^{p}_{X_l X_{l'}}(\boldsymbol{h},\tau)$ of a potential vector random field is fully determined by a single function $c^{p}_{\ell\ell}$ or $c^{p}_{\eta\eta}$.

In the frequency domain, we have the following general expressions of the spectral functions for a solenoidal and a potential VS/TRFs, respectively,

$$\widetilde{c}^{s}_{X_lX_{l'}}(\boldsymbol{k},\tau) = \left(\delta_{ll'} - \frac{k_l k_{l'}}{k^2}\right)\widetilde{c}^{s}_{\eta\eta}(k,\tau),$$
$$\widetilde{c}^{p}_{X_lX_{l'}}(\boldsymbol{k},\tau) = \frac{k_l k_{l'}}{k^2}\widetilde{c}^{p}_{\ell\ell}(k,\tau), \tag{7.26a–b}$$

where $\widetilde{c}^{s}_{\ell\ell} = \widetilde{c}^{p}_{\eta\eta} = 0$. Furthermore,

$$\int_0^\infty dr\, r^{n-2} c^{s}_{\eta\eta}(r,\tau) = 0,$$
$$\int_0^\infty dr\, c^{p}_{\ell\ell}(r,\tau) = 0, \tag{7.27a–b}$$

i.e., $c^{s}_{\eta\eta}$ is not everywhere positive, and $c^{p}_{\ell\ell}$ would change sign.

Example 7.8

In light of Eqs. (7.20a–b), $c^{s}_{X_lX_{l'}}(\boldsymbol{h},\tau)$ can be specified in terms of the corresponding longitudinal covariance $c^{s}_{\ell\ell}(r,\tau)$ as

$$c^{s}_{X_lX_{l'}}(\boldsymbol{h},\tau) = -\frac{1}{r}\frac{\partial c^{s}_{\ell\ell}(r,\tau)}{\partial r}h_l h_{l'} + \left[c^{s}_{\ell\ell}(r,\tau) + r\frac{\partial c^{s}_{\ell\ell}(r,\tau)}{\partial r}\right]\delta_{ll'} \quad (n=2),$$
$$c^{s}_{X_lX_{l'}}(\boldsymbol{h},\tau) = -\frac{1}{2r}\frac{\partial c^{s}_{\ell\ell}(r,\tau)}{\partial r}h_l h_{l'} + \left[c^{s}_{\ell\ell}(r,\tau) + \frac{r}{2}\frac{\partial c^{s}_{\ell\ell}(r,\tau)}{\partial r}\right]\delta_{ll'} \quad (n=3). \tag{7.28a–b}$$

So, Eqs. (7.28a–b) can generate covariance models for the multidimensional $c^{s}_{X_lX_{l'}}(\boldsymbol{h},\tau)$ starting from unidimensional models of $c^{s}_{\ell\ell}(r,\tau)$. Further, in light of Eq. (7.22), $c^{p}_{X_lX_{l'}}(\boldsymbol{h},\tau)$ can be written in terms of $c^{p}_{\eta\eta}(r,\tau)$ as

$$c^{p}_{X_lX_{l'}}(\boldsymbol{h},\tau) = \frac{1}{r}\frac{\partial c^{p}_{\eta\eta}(r,\tau)}{\partial r}h_l h_{l'} + c^{p}_{\eta\eta}(r,\tau)\delta_{ll'}. \tag{7.29}$$

Lastly, it can be shown that $c^{p}_{\ell\ell}(r,\tau) \le c^{p}_{\eta\eta}(r,\tau)$ (the equality holds when $r=0$), and $c^{p}_{\ell\ell}(0,\tau) = c^{p}_{\eta\eta}(0,\tau) = -\frac{\partial^2}{\partial r^2}c_Y(0,\tau)$.

Interestingly, the cross-covariances of STHS vector random fields can be expressed in terms of cross-covariances satisfying the composite STIS conditions of Definition 7.2. Here is an example.

Example 7.9

Let the vector random field $\boldsymbol{X}(\boldsymbol{p})$ be an STHS Gaussian field that is spherically but not reflectionally symmetric in space. Then, its cross-covariances can be written as

$$c_{X_lX_{l'}}(\boldsymbol{h},\tau) = A(r,\tau)h_l h_{l'} + B(r,\tau)\delta_{ll'} + C(r,\tau)\varepsilon_{ll'l''}h_{l''}, \tag{7.30}$$

where $\varepsilon_{ll'l''}$ is the Levi-Civita symbol defined in Table 1.1 of Chapter I. Hence, these covariances consist of the isostationary part (according to Definition 7.2) introduced in Eq. (7.6) of Theorem 7.1, plus an additional function $C(r,\tau)$ of r and τ. The corresponding spectra can be decomposed as

$$\widetilde{c}_{X_lX_{l'}}(\boldsymbol{k},\tau) = \widetilde{c}^{s}_{X_lX_{l'}}(\boldsymbol{k},\tau) + \widetilde{c}^{p}_{X_lX_{l'}}(\boldsymbol{k},\tau), \tag{7.31}$$

where $\tilde{c}^{s}_{X_lX_{l'}}(\boldsymbol{k},\tau) = \left(\delta_{ll'} - \frac{k_lk_{l'}}{k^2}\right)\tilde{c}^{s}_{\eta\eta}(k,\tau)$, and $\tilde{c}^{p}_{X_lX_{l'}}(\boldsymbol{k},\tau) = \frac{k_lk_{l'}}{k^2}\tilde{c}^{p}_{\ell\ell}(k,\tau)$. Moreover, we can write

$$c_{X_lX_{l'}}(\boldsymbol{h}) = \int_0^\infty d\tau\, c_{X_lX_{l'}}(\boldsymbol{h},\tau) = A(r)h_lh_{l'} + B(r)\delta_{ll'} + C(r)\varepsilon_{ll'l''}h_{l''},$$
$$\overline{X(\boldsymbol{p})^2} = \int d\boldsymbol{k}\left[(n-1)\tilde{c}^{s}_{\eta\eta}(k) + \tilde{c}^{p}_{\ell\ell}(k)\right] \tag{7.32a–b}$$

From the last equations we can obtain some interesting relationships. For illustration, Eq. (7.32a) gives at $\boldsymbol{h} = \boldsymbol{0}$, $\frac{\partial}{\partial h_{l''}}c_{X_lX_{l'}}(0,\tau) = C(0,\tau)\varepsilon_{ll'l''}$.

8. EFFECTIVE DISTANCES AND PERIODS

I will revisit here the notions of effective distances and effective period introduced in Section 6 of Chapter VIII for scalar S/TRFs. As we pass from scalar to vector S/TRFs, we find that the required generalizations are mathematically rather obvious. Indeed, for each of the cross-covariances $c_{X_lX_{l'}}$ of the matrix $\boldsymbol{c_X}$ the corresponding effective distances and periods can be defined, once a direction for the lag vector has been selected.

Definition 8.1

For VS/TRFs, the space–time-dependent *effective distance* $r_{e,j}(\boldsymbol{s}, t)$ along the s_j direction as well as the *effective period* $\tau_{e,j}(\boldsymbol{s}, r)$ can be defined as, respectively,

$$r_{e,j}(\boldsymbol{s},t) = \frac{1}{c_{X,j}(0,\boldsymbol{s},t)}\int_0^\infty dr\, c_{X_lX_{l'},j}(r,\boldsymbol{s},t),$$
$$\tau_{e,j}(\boldsymbol{s},r) = \frac{1}{c_{X,j}(r,\boldsymbol{s},0)}\int_0^\infty dt\, c_{X_lX_{l'},j}(r,\boldsymbol{s},t), \tag{8.1a–b}$$

where

$$c_{X_lX_{l'},j}(r,\boldsymbol{s},t) = c_{X_lX_{l'}}(h\boldsymbol{\varepsilon}_j,\boldsymbol{s},t) = c_{X_lX_{l'}}(s_1,\ldots,s_j+h_j,\ldots,s_n,t), \tag{8.2}$$

$\Delta\boldsymbol{p} = (r,\boldsymbol{s},t)$ with $r = h_j$, and the integration in Eq. (8.1a) is along the direction $\boldsymbol{\varepsilon}_j$.

The above expressions of the effective distances and the effective period that hold for VS/TRFs are similar in form to respective properties of scalar S/TRFs (discussed in Section 6 of Chapter VIII). To gain additional insight concerning the matter, let us look at some simple examples.

Example 8.1

Let us revisit Example 6.1 of Chapter VIII, assuming a two-component vector random field with $l, l' = 1,2$ in $R^{2,1}$. For the cross-correlation $\rho_{X_1X_2}$, the effective scales along the main direction corresponding to Eqs. (6.4a–c) of Chapter VIII are

$$r_{e,1}^{[1,2]} = \int_0^\infty dh_1\, \rho_{X_1X_2}(h_1,0,0),$$
$$r_{e,2}^{[1,2]} = \int_0^\infty dh_2\, \rho_{X_1X_2}(0,h_2,0), \tag{8.3a–c}$$
$$\tau_e^{[1,2]} = \int_0^\infty d\tau\, \rho_{X_1X_2}(0,0,\tau).$$

The effective scales above, generally differ from each other unless some additional restrictive assumption (e.g., isotropy) is imposed.

Given that the $r_{e,i}^{[1,2]}$ ($i = 1, 2$) and $\tau_e^{[1,2]}$ are measures of, respectively, how far in space and how long in time the field values remain correlated, in practice the $r_{e,1}^{[1,2]}$, $r_{e,2}^{[1,2]}$, or $\tau_e^{[1,2]}$ may correspond to certain fractions of the distances h_1, h_2 or the time τ to the first zero of $\rho_{X_1X_2}(h_1, 0, 0)$, $\rho_{X_1X_2}(0, h_2, 0)$, or $\rho_{X_1X_2}(0, 0, \tau)$, respectively. A practical rule is that the effective distances and the effective time are obtained so that the following equation holds

$$\rho_{X_1X_2}\left(r_{e,1}^{[1,2]}, 0, 0\right) = \rho_{X_1X_2}\left(0, r_{e,2}^{[1,2]}, 0\right) = \rho_{X_1X_2}\left(0, 0, \tau_e^{[1,2]}\right) \approx \frac{1}{2}.$$

Generally, the $r_{e,i}^{[1,2]}$ ($i = 1, 2$) and $\tau_e^{[1,2]}$ are different from each other. Other practical rules also exist, like replacing the coefficient $\frac{1}{2}$ in the above equation with e^{-1}.

I will conclude this chapter by looking at the formulation of the effective distances and effective time in the case that longitudinal and transverse covariance functions are used.

Example 8.2

In the particular case of longitudinal and transverse covariances, the corresponding effective distances and effective period are given by

$$\begin{aligned} r_{e,\ell}(t) &= \int_0^\infty dr\, \rho_{\ell\ell}(r,t), \\ r_{e,\eta}(t) &= \int_0^\infty dr\, \rho_{\eta\eta}(r,t), \\ \tau_{e,\ell}(r) &= \int_0^\infty dt\, \rho_{\ell\ell}(r,t), \\ \tau_{e,\eta}(r) &= \int_0^\infty dt\, \rho_{\eta\eta}(r,t). \end{aligned} \tag{8.4a–d}$$

Furthermore, in view of Eq. (7.8c) it is easily shown that in the case of divergence-free random fields (e.g., homogeneous incompressible fluids) the following simple relationships hold (Exercise IX.9),

$$\begin{aligned} r_{e,\eta}(t) &= \frac{1}{2} r_{e,\ell}(t), \\ \tau_{e,\eta}(t) &= \frac{1}{2} \tau_{e,\ell}(t) \end{aligned} \tag{8.5a–b}$$

i.e., the lateral effective distance (resp., lateral effective period) is one-half the longitudinal effective distance (resp., longitudinal effective period). It is also found from Eq. (7.8c) that

$$\int_0^\infty dr\, r\rho_{\eta\eta}(r,\tau) = 0, \tag{8.6}$$

if, for large r, the $\rho_{\ell\ell}(r, \tau)$ decays faster than r^{-2} (Exercise IX.11).

CHAPTER

SPECIAL CLASSES OF SPATIOTEMPORAL RANDOM FIELDS

CHAPTER OUTLINE

1. INTRODUCTION

This chapter is devoted to the study of certain special classes of spatiotemporal random fields (S/TRFs) representing attributes with specified physical nature. Indeed, these classes are introduced as a result of either physical observations justifying their use or theoretical analysis that seeks to simplify mathematical manipulations, or both. As a result, these random field classes have a wide applicability in sciences. It would be instructive to start our discussion with the description of some real-world phenomena and the special S/TRF model characteristics required to describe them adequately.

Example 1.1

In meteorology, many attributes occur over a very large global region so that they can be assumed to be fairly homogeneous in space, and they also decay over time. Moreover, the attributes may be initially in some steady state, and then a disturbance (e.g., a rain storm or hurricane) happens that perturbs the

Spatiotemporal Random Fields. http://dx.doi.org/10.1016/B978-0-12-803012-7.00010-6

environment. Hence, a special class of S/TRF models is needed to represent the above features that are common among many meteorological attributes.

Example 1.2

In oceanography, special classes of S/TRF models is needed to accommodate the main features of the dynamic distribution of ocean waves in space−time. These features include the fact that the waves are mostly wind generated, whereas other wave-generating mechanisms include earthquakes, which are the major cause of tsunamis (while rare, waves can be catastrophic if the earthquake occurs near or on the coast), and planetary forces (which drive tides and cause long period waves of the order of 12−24 h). These S/TRF classes should also account for the fact that a smooth water surface is associated with wind speeds below 2.5 m/s, a fully rough water surface is observed in the case of speeds exceeding 10 m/s, and for intermediate speeds, the flow is smooth over some parts of the water surface but rough around and in the lee of the breaking whitecaps.

Generally speaking, and based on the material presented in the book so far, the S/TRF models can be classified into distinguishable classes that differ by:

- **(i)** the formal and physical characteristics of the space−time domain within which they vary,
- **(ii)** the different techniques of measuring or observing natural attributes,
- **(iii)** the tools—such as spatiotemporal variability functions (S/TVFs) (covariance, variogram, structure functions) and domain transformation techniques (usually from the real space to a frequency, a reduced dimensionality, or a traveling space)—used to analyze the available data and other information sources,
- **(iv)** the S/TVF parameters that need to be estimated (space and time covariance ranges, anisotropy ratios, variogram slopes at the space−time origin and along different directions or in time, spectra peaks, and behavior at specified frequencies etc.),
- **(v)** the equations relating the S/TVF parameters to physical characteristics of the real phenomenon, and
- **(vi)** finally, the assumptions which are used for deriving these equations.

Each S/TRF tool provides an important piece of information about the phenomenon of interest, whereas in many cases it is the combination of these tools that offers the optimal attribute modeling approach in conditions of uncertainty. In reality, multiple data sources provide a considerable amount of raw information, and the use of each tool has its advantages and shortcomings (e.g., under certain conditions the structure function may have some advantages with respect to covariance function and spectra, or, working in the spectral domain may generate valuable information that is not available in the physical domain). The general objective of S/TRF modeling is to provide a representation of reality that extracts as much useful information as possible concerning the phenomenon of interest. Each tool extracts a part of useful information from multiple data sources, and their combined implementation may increase data informational efficiency by supplementing each other.

In the end, we should keep in mind that whatever S/TRF model, technique, or tool we decide to use, our ultimate goal always is to keep in touch with reality. In this respect, special S/TRF classes based on physically meaningful modeling assumptions that considerably simplify the mathematical analysis or conveniently relocate the mathematical difficulties are often appropriate if not necessary.

2. FROZEN SPATIOTEMPORAL RANDOM FIELDS AND TAYLOR'S HYPOTHESIS

Frozen random fields constitute a very useful class of physically motivated models with a wide range of applicability in oceanography, marine and coastal engineering, atmospheric sciences, weather forecasting and climatology, and health geography. I start with some basic notions concerning this special class of S/TRFs.

2.1 BASIC NOTIONS

To start with, a primary feature that should be encapsulated in the S/TRF modeling of a certain type of sea surface is that at each time instant the previous spatial distribution of the sea surface moves to a new location with specified velocity. I now present a formal definition of this special class of space–time homostationary (STHS) random fields that can model this sea surface feature.

Definition 2.1

A *frozen* random field, $X(\boldsymbol{s}, t)$, is an S/TRF whose temporal variation is determined by the random field

$$\begin{aligned} &\widehat{X}(\widehat{\boldsymbol{s}}), \\ &\widehat{\boldsymbol{s}} = \boldsymbol{s} - \boldsymbol{v}t, \end{aligned} \tag{2.1a–b}$$

which travels with velocity $\boldsymbol{v} = (v_1, \ldots, v_n)$, i.e.,

$$X(\boldsymbol{s}, t) = \widehat{X}(\widehat{\boldsymbol{s}}). \tag{2.1c}$$

The $\widehat{X}(\widehat{\boldsymbol{s}})$ may be seen as a *traveling* (pseudospatial) random field so that the $X(\boldsymbol{s}, t)$ value at $\boldsymbol{s}$, t is that of $\widehat{X}(\widehat{\boldsymbol{s}})$ at $\widehat{\boldsymbol{s}} = \boldsymbol{s} - \boldsymbol{v}t$.[1]

Sometimes $\boldsymbol{v}$ is interpreted as a convection velocity, in which case the S/TRF that satisfies Eq. (2.1c) is termed a frozen random field because such a field is convected with $\boldsymbol{v}$ as if it were frozen. The following examples describe phenomena, the basic features of which are encapsulated in the frozen random field model as described in Definition 2.1.

Example 2.1

In atmospheric turbulence, the streamwise velocity component at location $\boldsymbol{s} = (s_1, s_2, s_3)$ and time t is $X(\boldsymbol{s}, t)$, then the streamwise velocity $X(\boldsymbol{s} + \boldsymbol{h}, t + \tau)$ at the downstream location $\boldsymbol{s} + \boldsymbol{h} = (s_1 + r, s_2, s_3)$, and the later time $t + \tau$ can be expressed as

$$X(\boldsymbol{s} + \boldsymbol{h}, t + \tau) = X(\boldsymbol{s} + \boldsymbol{h} - \boldsymbol{v}\tau, t),$$

where $\boldsymbol{v} = (v, 0, 0)$ is the convection velocity with v being a constant. Also, a significant part of remote sensing literature is based on the assumption that the index refraction field is frozen and moving with a constant wind velocity $\boldsymbol{v}$.

Example 2.2

In ocean dynamics, during the initial stage of sea wave generation, the surface pressure field $X(\boldsymbol{s}, t)$ is convected as a rigid pattern across the water, and it can be represented as in Eq. (2.1c) with $\boldsymbol{s} = (s_1, s_2)$, and $\boldsymbol{v} = (v_1, v_2)$ denoting the convection velocity, i.e.,

$$X(s_1, s_2, t) = \widehat{X}(s_1 - v_1 t, s_2 - v_2 t).$$

---

[1]We remind the readers that the frozen field concept is the motivation for the development of a special case of the traveling transformation technique in Chapter V, although in a different conceptual context.

For further illustration, the most elementary way to represent a free wave surface of the above form is to let $\widehat{X}(\cdot) = a\cos(\cdot)$, where a is called the amplitude representing half the vertical distance between a peak and a trough. More complicated phenomena are studied in terms of superpositions (linear combinations) of cosine waves. As we will see in Chapter XIV, many natural attributes are governed by stochastic partial differential equation (SPDE) that admits solutions of the form of Eqs. (2.1a–b) and (2.1c) in the space–time domain $R^{n,1}$.

The added value of the random frozen field representation of Eqs. (2.1a–b) lies in reducing the number of arguments from $\boldsymbol{s}, t$ to $\widehat{\boldsymbol{s}}$. The determination of the velocity vector $\boldsymbol{\upsilon}$ depends on the specifics of the attribute considered. Some techniques for specifying $\boldsymbol{\upsilon}$ have been already discussed in Chapter V in the context of traveling transformations. If $c_{\widehat{X}}(\boldsymbol{h})$ is the covariance function of $\widehat{X}(\widehat{\boldsymbol{s}})$ in Eqs. (2.1a–b) and (2.1c), the following proposition can be proven (Exercise X.1).

Proposition 2.1

The mean, covariance, and variogram functions of the original random field, $X(\boldsymbol{s}, t)$, and the traveling random field, $\widehat{X}(\widehat{\boldsymbol{s}})$, are related by

$$\begin{aligned} \overline{X(\boldsymbol{s},t)} &= \overline{\widehat{X}(\widehat{\boldsymbol{s}})}, \\ c_X(\boldsymbol{h},\tau) &= c_{\widehat{X}}(\widehat{\boldsymbol{h}},0) = c_{\widehat{X}}(\widehat{\boldsymbol{h}}), \\ \gamma_X(\boldsymbol{h},\tau) &= \gamma_{\widehat{X}}(\widehat{\boldsymbol{h}},0) = \gamma_{\widehat{X}}(\widehat{\boldsymbol{h}}), \end{aligned} \tag{2.2a–c}$$

where $(\boldsymbol{s}, t), (\boldsymbol{h}, \tau) \in R^{n,1}$ and $\widehat{\boldsymbol{s}} = (\boldsymbol{s} - \boldsymbol{\upsilon}t)$, $\widehat{\boldsymbol{h}} = (\boldsymbol{h} - \boldsymbol{\upsilon}\tau) \in R^n$.[2]

Proposition 2.1 involves two domains: The original domain $R^{n,1}$, and the traveling domain R^n. Furthermore, the assumed linear transformation in Eqs. (2.2a–c) implies that the isomean, isocovariance, and isovariogram contours are straight lines, $|\boldsymbol{s}| - \upsilon t = |\widehat{\boldsymbol{s}}|$ and $|\boldsymbol{h}| - \upsilon\tau = |\widehat{\boldsymbol{h}}|$, respectively, where $|\widehat{\boldsymbol{s}}|$ and $|\widehat{\boldsymbol{h}}|$ are viewed as contour levels. The covariance model $c_{\widehat{X}}(\widehat{\boldsymbol{h}},0)$ of Eq. (2.2b) expresses space–time correlation in terms of spatial correlation ($\tau = 0$) and the linear transformation $(\boldsymbol{h} - \boldsymbol{\upsilon}\tau)$. Interestingly, this transformation seems to imply that on the lines $|\boldsymbol{h}| - \upsilon\tau = c$ (c is a constant) the space–time covariance never declines for any $\boldsymbol{h}$ and τ, which seems to contradict a basic characteristic of covariance functions (this observation justifies the development of an elliptic space–time isostationarity (STIS) representation of the covariance in Section 2.4.3 later).

Example 2.3

An illustration of a frozen field mean distribution $\overline{X(s,t)}$ in the $R^{1,1}$ domain is shown in Fig. 2.1. In this case, the mean Eq. (2.2a) is $\overline{X(s,t)} = \overline{\widehat{X}(\widehat{s})}$, with $\widehat{s} = s - \upsilon t$ and $\frac{ds}{dt} = \upsilon$. At the time origin, $\overline{X(s,0)} = \overline{\widehat{X}\left(\widehat{s} = s\right)}$. By virtue of the frozen field definition, the material derivative is

$$\frac{d\overline{X(s,t)}}{dt} = \frac{\partial\overline{X(s,t)}}{\partial t}\frac{dt}{dt} + \frac{\partial\overline{X(s,t)}}{\partial s}\frac{ds}{dt} = -\upsilon\frac{\partial\overline{\widehat{X}(\widehat{s})}}{\partial\widehat{s}} + \frac{\partial\overline{\widehat{X}(\widehat{s})}}{\partial\widehat{s}}\upsilon = 0,$$

[2] Some authors use the notation $\widehat{\boldsymbol{h}} = \boldsymbol{h} + \boldsymbol{\upsilon}\tau$ instead of $\widehat{\boldsymbol{h}} = \boldsymbol{h} - \boldsymbol{\upsilon}\tau$.

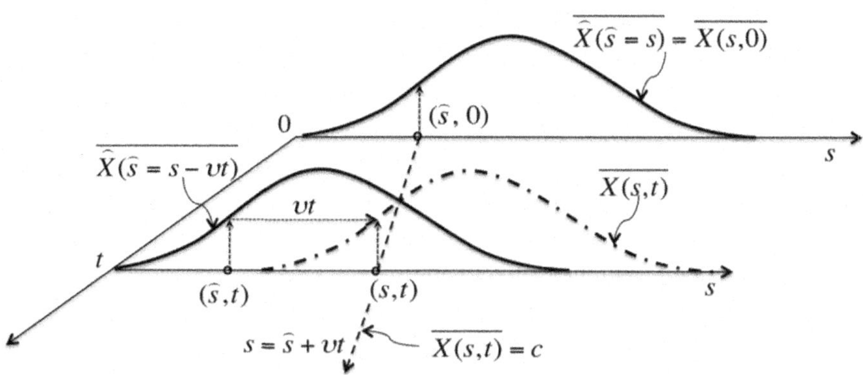

FIGURE 2.1

An illustration of the frozen field notion in $R^{1,1}$.

which means that $\overline{X(s,t)} = c$ (constant) along the line $s = \widehat{s} + vt$. Interestingly, the frozen field mean $\overline{X(s,t)}$ satisfies the partial differential equation (PDE)

$$\frac{\partial \overline{X(s,t)}}{\partial t} + v\frac{\partial \overline{X(s,t)}}{\partial s} = 0,$$

with initial condition $\overline{X(s,0)} = \overline{\widehat{X}(s)}$. If, say, $\overline{X(s,0)} = \overline{\widehat{X}(s)} = \cos(s)$ and $v = -c$, a solution to the PDE is of the form $\overline{X(s,t)} = \cos(s + ct)$. In Fig 2.1, the spatial plot of $\overline{X(s,t)}$ at time t is simply that of $\overline{\widehat{X}(s - vt)}$ shifted to the right by vt ($v > 0$), i.e., as expected, the $\overline{X(s,t)}$ values are the same as the $\overline{\widehat{X}(\widehat{s})}$ values at $\widehat{s} = s - vt$. A similar analysis is valid in terms of covariance and variogram functions in Eqs. (2.2b) and (2.2c).

A special case of the covariance and variogram functions of Eqs. (2.2b) and (2.2c) is encountered in some atmospheric optics applications, as described in the following corollary.

Corollary 2.1

If the lag $\boldsymbol{h}$ and velocity $\boldsymbol{v}$ vectors are parallel to each other, Eqs. (2.2b) and (2.2c) give

$$c_{\widehat{X}}(\boldsymbol{h} - \boldsymbol{v}\tau) = c_{\widehat{X}}\left(\boldsymbol{v}\left(\frac{h}{v} - \tau\right)\right),$$

$$\gamma_{\widehat{X}}(\boldsymbol{h} - \boldsymbol{v}\tau) = \gamma_{\widehat{X}}\left(\boldsymbol{v}\left(\frac{h}{v} - \tau\right)\right),$$

where $\boldsymbol{h} = \boldsymbol{v}\frac{h}{v}$, $h = |\boldsymbol{h}|$, and $v = |\boldsymbol{v}|$.

Another well-known special case of the covariance function of Eq. (2.2b) is described by the celebrated *Taylor's hypothesis* (introduced by Taylor, 1938), which is such that

$$c_X(\mathbf{0}, \tau) = c_X(\boldsymbol{v}\tau, 0). \quad (2.3)$$

The covariance function of Eq. (2.3) is assumed in earth and atmospheric studies when it is consistent with physical knowledge about the attribute's space–time distribution. The frozen field hypothesis can be used, e.g., to deduce spatial information about fluid turbulent fluctuations based on time series of measurements at a single location, or at a set of locations where measurements have not been made simultaneously (Tatarskii, 1971). Otherwise said, based on the velocity υ, Eq. (2.3) can be used to obtain the spatial covariance function from the temporal covariance function (and vice versa). Nevertheless, this straightforward link between the spatial and the temporal covariance functions does not necessarily extend to the corresponding data (i.e., usually it is not an obvious matter how we can use Eq. (2.3) to transform temporal data directly into spatial data; see also Section 6 later).

There exist alternative versions of Taylor's hypothesis, e.g., one of them links spatial and temporal derivatives (Heskestad, 1965), see also Section 2.4 on possible extensions of the hypothesis. While Taylor's hypothesis can be derived as a special case of the frozen random field, the reverse is not necessarily true. Typically, Eq. (2.3) involves Eulerian coordinates. Its counterpart in Lagrangian coordinates may be simply written as $c_X(\upsilon\tau, \tau) = c_X(\upsilon\tau, 0)$ (Section 6).

Example 2.4

Assume that an attribute is characterized by the following space–time nonseparable covariance model in $R^{n,1}$,

$$c_X(\boldsymbol{h}, \tau) = c_0\left(1 + \frac{|\boldsymbol{h}|^{\kappa}}{a_r^{\kappa}} + \frac{|\tau|^{\lambda}}{a_\tau^{\lambda}}\right)^{-\nu}, \tag{2.4}$$

where c_0, a_r, a_τ, $\boldsymbol{\nu} > 0$, κ, $\lambda \in (0,2]$. This a flexible model that satisfies Eq. (2.3) for $\kappa = \lambda$ and

$$\boldsymbol{\upsilon}^T = \left[\frac{a_r}{a_\tau} 0 \ldots 0\right] \in R^n. \tag{2.5}$$

The space–time separable covariance model

$$c_X(\boldsymbol{h}, \tau) = c_0 e^{-\left(\frac{|\boldsymbol{h}|^{\mu}}{a_r^{\mu}} + \frac{|\tau|^{\mu}}{a_\tau^{\mu}}\right)}, \tag{2.6}$$

where $\mu = 1, 2$ and c_0, a_r, $b_\tau > 0$, satisfies Eq. (2.3) for $\boldsymbol{\upsilon}^T = [1\ \ 0 \ldots 0] \in R^n$. For the model

$$c_X(\boldsymbol{h}, \tau) = c_0 e^{-\left(\frac{r^2}{a_r^2} + \frac{\tau^2}{a_\tau^2}\right)^{\frac{1}{2}}},$$

where $r = |\boldsymbol{h}|$, it is found that υ is of the form of Eq. (2.5). Other cases were discussed in Chapter V in the context of the traveling transformation.

Remark 2.1

The space–time variability representations introduced by Eqs. (2.2b) and (2.2c) surely allow especially useful and well-defined simplifications in space–time analysis and modeling, such as the derivation of spatial variability functions, simply by temporally averaging observed attribute values at pairs of spatial locations.

The covariance of a random field $X(\boldsymbol{s}, t)$ can be decomposed into a frozen and a temporal part. Assume that the S/TRF is generated by a recursive equation of the form

$$X(\boldsymbol{s}, t) = \lambda X(\boldsymbol{s} - \boldsymbol{\upsilon}\tau, t - \tau) + (1 - \lambda^2)^{\frac{1}{2}} Y_t(\boldsymbol{s}),$$

where $\lambda = \lambda(t)$ is a given correlation coefficient, $Y_t(\boldsymbol{s})$ are independent random fields with covariance $c_{Y_t}(\boldsymbol{h}) = e^{-\frac{|\boldsymbol{h}|^2}{2a^2}}$ (for all $t = l\tau$, $l = 0,1,\ldots$, and a is a constant), and $X(\mathbf{0},0) = Y_0(\mathbf{0})$. Then the corresponding space–time covariance is of the form

$$c_X(\boldsymbol{h}, \tau) = \lambda(\tau) e^{-\frac{|\boldsymbol{h}-\boldsymbol{v}\tau|^2}{2a^2}}$$

(Baxevani et al., 2006).

2.2 SPECTRAL DOMAIN ANALYSIS

Beyond the physical domain of an attribute distribution, interpretable information can be obtained by studying the mathematical behavior of the attribute distribution in the spectral domain. To orient the reader, I start the discussion with an example.

Example 2.5

One can study certain effects of the frozen random field model introduced in the representation of the sea surface pressure field in Example 2.2 by taking the Fourier transformation of $X(\boldsymbol{s}, t) = \widehat{X}(\boldsymbol{s} - \boldsymbol{v}t)$, where $\boldsymbol{s} = (s_1, s_2)$ and $\boldsymbol{v} = (v_1, v_2)$, and setting $\widehat{\boldsymbol{s}} = \boldsymbol{s} - \boldsymbol{v}t$, i.e.,

$$\widetilde{X}(\boldsymbol{k}, t) = \frac{1}{2\pi}\int ds\, e^{-i\boldsymbol{k}\cdot\boldsymbol{s}}\widehat{X}(\boldsymbol{s} - \boldsymbol{v}t) = e^{-i\boldsymbol{k}\cdot\boldsymbol{v}t}\left[\frac{1}{2\pi}\int d\widehat{\boldsymbol{s}}\, e^{-i\boldsymbol{k}\cdot\widehat{\boldsymbol{s}}}\widehat{X}(\widehat{\boldsymbol{s}})\right] = e^{-i\boldsymbol{k}\cdot\boldsymbol{v}t}\widetilde{X}(\boldsymbol{k}, 0).$$

The physical interpretation of this equation is that in conditions of rigid convection, the spectrum of the pressure field can be decomposed into a time-dependent exponential function and a time-invariant spectrum of the pressure field.

Furthermore, the following corollary concerns the behavior of the corresponding covariance functions in the spectral domain, and it is a straightforward implication of Proposition 2.1 (or, an extension of the analysis in Example 2.5, for that matter).

Corollary 2.2

The full spectrum $\widetilde{c}_X(\boldsymbol{k}, \omega)$ and the wavevector–time spectrum $\widetilde{c}_X(\boldsymbol{k}, \tau)$ associated with the space–time covariance function of Eq. (2.2b) are given in terms of the spectra of the traveling covariance function, viz.,

$$\begin{aligned} \widetilde{c}_X(\boldsymbol{k}, \omega) &= \widetilde{c}_{\widehat{X}}(\boldsymbol{k})\delta(\omega - \boldsymbol{k}\cdot\boldsymbol{v}), \\ \widetilde{c}_X(\boldsymbol{k}, \tau) &= \widetilde{c}_{\widehat{X}}(\boldsymbol{k})e^{-i\tau(\boldsymbol{k}\cdot\boldsymbol{v})} \end{aligned} \quad (2.7a\text{–}b)$$

These are frequency domain representations that can considerably simplify attribute space–time modeling.

An interesting fact about Eq. (2.7a) is that it connects the space–time spectrum of the original random field $X(\boldsymbol{p})$ with the spectrum of the traveling random field $\widehat{X}(\widehat{\boldsymbol{s}})$. Then, it holds that

$$\widetilde{c}_X(\boldsymbol{k}) = \widetilde{c}_{\widehat{X}}(\boldsymbol{k}), \quad (2.8a)$$

i.e., the spectrum of the space–time random field $X(\boldsymbol{p})$ at $\boldsymbol{h} = \mathbf{0}$ is the same as that of the traveling random field $\widehat{X}(\widehat{\boldsymbol{s}})$. Similarly, it is found that

$$\widetilde{c}_X(\omega) = \frac{2\pi}{v}\int_{\omega/v}^{\infty} dkk\widetilde{c}_{\widehat{X}}(k), \quad (2.8b)$$

where $v = |\boldsymbol{v}|$. An example of Eq. (2.8b) in the isotropic case is as follows.

Example 2.6

If $\widehat{X}(\widehat{s})$ is an isotropic frozen random field such that $\widetilde{c}_X(k) = \widetilde{c}_{\widehat{X}}(k)$, $k = |\boldsymbol{k}|$, it is found that

$$\widetilde{c}_X(k) = \begin{cases} \upsilon\, \widetilde{c}_X(\omega)|_{\omega=\upsilon k} & (R^{1,1}), \\ -\dfrac{\upsilon^2}{2\pi k}\dfrac{d}{d\omega}\widetilde{c}_X(\omega)|_{\omega=\upsilon k} & (R^{3,1}). \end{cases} \tag{2.9a–b}$$

The $\widetilde{c}_X(\omega)$ may be interpreted as the frequency spectrum on a time series $X(\boldsymbol{s},t)$ obtained by a fixed observer. Eqs. (2.9a–b) relate the wavenumber (spatial) spectrum to the frequency (temporal) spectrum of an isotropic frozen random field. For illustration, let the traveling covariance form in R^3 be given by

$$c_{\widehat{X}}(r) = c_0 e^{-\frac{r^2}{a^2}}, \tag{2.9c}$$

$r = |\boldsymbol{h}|$, in which case,

$$\begin{aligned} \widetilde{c}_X(\omega) &= \frac{c_0 a}{2\sqrt{\pi}\upsilon} e^{-\frac{a^2\omega^2}{4\upsilon^2}}, \\ \widetilde{c}_X(k) &= \frac{c_0 a^3}{(2\sqrt{\pi})^3} e^{-\frac{a^2k^2}{4}}\, \widetilde{c}_X(k) = \widetilde{c}_{\widehat{X}}(k), \end{aligned} \tag{2.9d–e}$$

are the corresponding frequency and wavenumber spectra, respectively (Exercise X.3).

A noteworthy use of the frozen random field model is to study weather attributes such as rainfall intensity at the ground surface. In these situations the distribution of the S/TRF $X(s, t)$ is determined by the variability of the random field $\widehat{X}(\widehat{s})$, $\widehat{s} = s - \upsilon t$, which travels with a constant velocity vector υ. As a consequence, a frozen S/TRF is defined as $X(s,t) = \widehat{X}(\widehat{s})$. The following case looks at some more properties of a frozen random field that is also STIS.

Example 2.7

In the STIS case, the corresponding frozen random field covariances and spectra are related by

$$\begin{aligned} c_X(r,\tau) &= c_{\widehat{X}}(r - \upsilon\tau), \\ \widetilde{c}_X(k,\omega) &= \widetilde{c}_{\widehat{X}}(k)\delta(\omega - \upsilon k), \end{aligned} \tag{2.10a–b}$$

where $\widetilde{c}_X(k) = \widetilde{c}_{\widehat{X}}(k)$. Assuming that the random field $\widehat{X}(\widehat{s})$ has the isotropic covariance of Eq. (2.9c) in R^3, we obtain the wavenumber spectrum of Eq. (2.9e), and the frequency spectrum of Eq. (2.9d), in which case the wavenumber–frequency spectrum is given by

$$\widetilde{c}_X(k,\omega) = \frac{c_0 a^3}{(2\sqrt{\pi})^3} e^{-\frac{a^2k^2}{4}} \delta(\omega - \upsilon k),$$

which is a generalized function.

In some optical scattering applications, partial variograms of frozen random fields have been proven to be useful, such as the variogram functions

$$\gamma_X(v\tau) = \frac{1}{2}\overline{[X(\boldsymbol{h}-v\tau,t)-X(\boldsymbol{h},t)]^2},$$

$$\gamma'_X(v\tau) = \frac{1}{2}\overline{[X(\boldsymbol{s},t)-X(\boldsymbol{s}+\boldsymbol{h},t)][X(\boldsymbol{s}-v\tau,t)-X(\boldsymbol{s}+\boldsymbol{h}-v\tau,t)]}.$$

Interestingly, the latter variogram can be conveniently expressed in terms of the former, i.e.,[3]

$$\gamma'_X(v\tau) = \frac{1}{2}[\gamma_X(\boldsymbol{h}-v\tau)+\gamma_X(\boldsymbol{h}+v\tau)] - \gamma_X(v\tau),$$

which implies that as soon as γ_X has been experimentally calculated, the γ'_X is directly obtained using the above relationship, without the need to calculate it experimentally. This is useful in practice, because it is computationally more demanding to obtain γ'_X than γ_X.

2.3 DIFFERENTIAL EQUATION REPRESENTATIONS

As it turns out, the frozen random field hypothesis leads to a differential equation representation of the S/TRF $X(\boldsymbol{p})$ with important consequences in the study of the space–time variability patterns of a phenomenon (Christakos et al., 2017a; also, Exercise X.4).

Proposition 2.2

The frozen S/TRF model $X(\boldsymbol{p})$ obeys the SPDE in $R^{n,1}$,

$$\frac{\partial X}{\partial t} + \boldsymbol{\theta}^T \nabla X = 0, \tag{2.11}$$

where

$$\boldsymbol{\theta} = [\theta_1 \ldots \theta_n]^T,$$

$$\nabla X = \left[\frac{\partial X}{\partial h_1} \cdots \frac{\partial X}{\partial h_n}\right]^T, \tag{2.12a–b}$$

and the vector $\boldsymbol{\theta}$ is the velocity vector function of the attribute with directional components

$$\theta_i = \frac{v_i + \tau \dfrac{\partial v_i}{\partial t}}{1 - \tau \dfrac{\partial v_i}{\partial s_i}} \tag{2.12c}$$

($i = 1, \ldots, n$) in the space–time domain.

The following proposition, which is a direct consequence of Eq. (2.11), provides the PDE representations of the space–time attribute covariance (Christakos et al., 2017a; Exercise X.5).

[3]Here, the identity $2(\chi_1-\chi_2)(\chi_3-\chi_4) = (\chi_1-\chi_4)^2 + (\chi_2-\chi_3)^2 - (\chi_1-\chi_3)^2 - (\chi_2-\chi_4)^2$ has been used.

Proposition 2.3

The space–time mean and covariance functions of the frozen random field $X(\boldsymbol{p})$ satisfy the set of PDEs,

$$\frac{\partial c_X}{\partial \tau} + \boldsymbol{\theta}^T \nabla c_X = 0,$$
$$\nabla \frac{\partial c_X}{\partial \tau} + \boldsymbol{\Theta}^T \nabla c_X + \boldsymbol{H}_c \boldsymbol{\theta} = 0, \tag{2.13a–b}$$

where

$$\nabla c_X = \left[\frac{\partial c_X}{\partial h_1} \cdots \frac{\partial c_X}{\partial h_n}\right]^T,$$

$$\nabla \frac{\partial c_X}{\partial \tau} = \left[\frac{\partial^2 c_X}{\partial h_1 \partial \tau} \cdots \frac{\partial^2 c_X}{\partial h_n \partial \tau}\right]^T,$$

$$\boldsymbol{\Theta} = \begin{bmatrix} \frac{\partial \theta_1}{\partial h_1} & \cdots & \frac{\partial \theta_1}{\partial h_n} \\ \vdots & & \\ \frac{\partial \theta_n}{\partial h_1} & \cdots & \frac{\partial \theta_n}{\partial h_n} \end{bmatrix}, \tag{2.14a–d}$$

$$\boldsymbol{H}_c = \begin{bmatrix} \frac{\partial^2 c_X}{\partial h_1^2} & \cdots & \frac{\partial^2 c_X}{\partial h_1 \partial h_n} \\ \vdots & & \\ \frac{\partial^2 c_X}{\partial h_n \partial h_1} & \cdots & \frac{\partial^2 c_X}{\partial h_n^2} \end{bmatrix}.$$

The $\boldsymbol{H}_c$ is the Hessian matrix, and the vector $\boldsymbol{\theta}$ has the directional components of Eq. (2.12c) in the $(\boldsymbol{h}, \boldsymbol{\tau})$ domain.

Interpretation of the space–time covariance PDEs above can offer valuable insight regarding the actual attribute distribution in a real-world setting. Generally, this set of PDEs expresses quantitatively the relationship between the rates of covariance changes, which reflect the corresponding changes of space–time dependence between attribute values at different points. Specifically, in Eq. (2.13a) the temporal rate of covariance change is explained as the spatial rate of covariance change multiplied by the corresponding velocity function $\boldsymbol{\theta}$. Eq. (2.13b) involves $\boldsymbol{\theta}$ derivatives (see, matrix $\boldsymbol{\Theta}$) requiring some sort of consistency between neighboring velocity functions. Each θ_i is a function of the attribute velocity along direction s_i, its spatial and temporal rates of change, and the spatial and temporal lags considered. Another implication of the PDE representations of space–time attribute dependence is that a frozen covariance function must, generally, satisfy Eqs. (2.13a–b), which is why, as was shown in Chapter III, this kind of PDEs can play a key role in the determination of a physically meaningful space–time metric.

Remark 2.2

In more detailed analytical terms, the PDE representations of Eqs. (2.13a–b) take the following forms, respectively,

$$\left[\frac{\partial}{\partial \tau}+\sum_{i=1}^{n}\theta_i\frac{\partial}{\partial h_i}\right]c_X(\boldsymbol{h},\tau)=0,$$
$$\left[\frac{\partial^2}{\partial h_j\partial \tau}+\sum_{i=1}^{n}\frac{\partial \theta_i}{\partial h_j}\frac{\partial}{\partial h_i}+\sum_{i=1}^{n}\theta_i\frac{\partial^2}{\partial h_i\partial h_j}\right]c_X(\boldsymbol{h},\tau)=0, \tag{2.15a–b}$$

where $j = 1,\ldots, n$ and θ_i ($i = 1,\ldots, n$) are given by Eq. (2.12c). As the following illustration shows, the usefulness of these PDE formulations becomes more evident when specific covariance models are involved in the space–time analysis.

Example 2.8

In the $R^{1,1}$ domain, the covariance PDEs of Eqs. (2.15a–b) yield

$$\left[\frac{\partial}{\partial \tau}+\theta\frac{\partial}{\partial h}\right]c_X(h,\tau)=0,$$
$$\left[\frac{\partial^2}{\partial h\partial \tau}+\frac{1}{1-\tau\frac{\partial \upsilon}{\partial h}}\left(\frac{\partial \upsilon}{\partial h}+\tau\frac{\partial^2 \upsilon}{\partial h\partial \tau}+\tau\theta\frac{\partial^2 \upsilon}{\partial h^2}\right)\frac{\partial}{\partial h}+\theta\frac{\partial^2}{\partial h^2}\right]c_X(h,\tau)=0, \tag{2.16a–b}$$

where now we have a single velocity function $\theta=\left(\upsilon+\tau\frac{\partial \upsilon}{\partial \tau}\right)\left(1-\tau\frac{\partial \upsilon}{\partial h}\right)^{-1}$. For illustration, in the case of the space–time covariance model

$$c_X(h,\tau)=e^{-(h^2+\tau^2)^{\frac{1}{2}}}, \tag{2.17}$$

it is found that $\theta=-\frac{\tau}{h}$, $\upsilon=\frac{h}{\tau}\pm\left(\frac{h^2}{\tau^2}+1\right)^{\frac{1}{2}}$ (Exercise X.6).[4] We recall that the covariance model of Eq. (2.17) was used in the mortality simulation of Example 4.4 of Chapter V. In this case, the PDE covariance representations of Eqs. (2.16a–b) reduce to

$$\left[\frac{\partial}{\partial \tau}-\frac{\tau}{h}\frac{\partial}{\partial h}\right]c_X(h,\tau)=0,$$
$$\left[\frac{\partial^2}{\partial h\partial \tau}+\frac{\tau}{h^2}\frac{\partial}{\partial h}-\frac{\tau}{h}\frac{\partial^2}{\partial h^2}\right]c_X(h,\tau)=0. \tag{2.18a–b}$$

In the case of the mortality simulation plotted in Fig. 4.1A of Chapter V, the interpretation of Eq. (2.18a) implies that the rate of temporal correlation changes of mortality values can be explained by the rate of spatial correlation changes of mortality values multiplied by the corresponding ratio of the temporal lag over the spatial lag.

[4]The derivation of the υ expression was also discussed in Section 4.2 of Chapter V in the traveling transformation context.

Naturally, matters simplify considerably if the traveling vector $\boldsymbol{v}$ can be assumed to be space- and time-independent, i.e., the attribute travels with a constant velocity in the region of interest, and Eq. (2.12c) yields $\theta_i = \upsilon_i$ (i.e., the velocity function reduces to the velocity itself). In this case, Eq. (2.11) yields

$$\left[\frac{\partial}{\partial t} + \sum_{i=1}^{n} \upsilon_i \frac{\partial}{\partial s_i}\right] X(\boldsymbol{p}) = 0, \tag{2.19}$$

and the covariance representations become considerably simpler. This is further demonstrated in the following corollary.

Corollary 2.2

In the case of space- and time-independent traveling vector, $\boldsymbol{\theta} = \boldsymbol{v}$ and $\boldsymbol{\Theta} = \mathbf{0}$ in Eqs. (2.13a–b), and the PDE representations of the space–time covariance function of $X(\boldsymbol{p})$ are reduced to

$$\begin{aligned} &\frac{\partial c_X}{\partial \tau} + \boldsymbol{v}^T \nabla c_X = 0, \\ &\nabla \frac{\partial c_X}{\partial \tau} + \boldsymbol{H}_c \boldsymbol{v} = 0, \end{aligned} \tag{2.20a–b}$$

where $\boldsymbol{v} = [\upsilon_1 \ldots \upsilon_n]^T$.

For subsequent mathematical manipulations the set of Eqs. (2.20a–b) can be written in more detailed analytical terms,

$$\begin{aligned} &\left[\frac{\partial}{\partial \tau} + \sum_{i=1}^{n} \upsilon_i \frac{\partial}{\partial h_i}\right] c_X(\boldsymbol{h}, \tau) = 0, \\ &\left[\frac{\partial^2}{\partial h_j \partial \tau} + \sum_{i=1}^{n} \upsilon_i \frac{\partial^2}{\partial h_i \partial h_j}\right] c_X(\boldsymbol{h}, \tau) = 0 \\ &(j = 1, \ldots, n). \end{aligned} \tag{2.21a–b}$$

Corollary 2.3

In the case of a space- and time-independent traveling vector $\boldsymbol{v}$, the following condition holds concerning the covariance function,

$$\frac{\partial c_X}{\partial \tau} - \left(\boldsymbol{H}_c^{-1} \nabla \frac{\partial c_X}{\partial \tau}\right)^T \nabla c_X = 0 \tag{2.22}$$

(see, also, Exercise X.7).

Example 2.9

In the $R^{1,1}$ domain, the condition of Eq. (2.22) gives,

$$\frac{\partial c_X}{\partial \tau} - \left(\frac{\partial^2 c_X}{\partial h^2}\right)^{-1} \frac{\partial^2 c_X}{\partial h \partial \tau} \frac{\partial c_X}{\partial h} = 0,$$

or, in light of Eqs. (2.20a–b),

$$\frac{\frac{\partial c_X}{\partial \tau}}{\frac{\partial c_X}{\partial h}} = \frac{\frac{\partial^2 c_X}{\partial h \partial \tau}}{\frac{\partial^2 c_X}{\partial h^2}} = -\upsilon, \tag{2.23}$$

which is an equation that the covariance c_X of the frozen random field must satisfy, in the case of space- and time-independent velocity.

Lastly, some interesting covariance function representations in the case of the frozen random field are obtained in terms of Taylor series expansions, as follows

$$\begin{aligned} c_X(\boldsymbol{h} - \boldsymbol{\upsilon}\tau, 0) = c_X(\boldsymbol{h}, \tau) &- \tau \left[\sum_{i=1}^{n} \upsilon_i \frac{\partial c_X(\boldsymbol{h}, \tau)}{\partial h_i} + \frac{\partial c_X(\boldsymbol{h}, \tau)}{\partial \tau} \right] \\ &+ \frac{1}{2}\tau^2 \left[\sum_{i,j=1}^{n} \upsilon_i \upsilon_j \frac{\partial^2 c_X(\boldsymbol{h}, \tau)}{\partial h_i \partial h_j} + 2 \sum_{i=1}^{n} \upsilon_i \frac{\partial^2 c_X(\boldsymbol{h}, \tau)}{\partial h_i \partial \tau} + \frac{\partial^2 c_X(\boldsymbol{h}, \tau)}{\partial \tau^2} \right] - \ldots \end{aligned} \tag{2.24}$$

(Exercise X.8), which leads to

$$\left(-\tau + \frac{\tau^2}{2}\right) \left[\sum_{i=1}^{n} \upsilon_i \frac{\partial}{\partial h_i} + \frac{\partial}{\partial \tau} \right] c_X(\boldsymbol{h}, \tau) + \frac{\tau^2}{2} \sum_{i=1}^{n} \upsilon_i \left[\sum_{j=1}^{n} \upsilon_j \frac{\partial^2}{\partial h_i \partial h_j} + \frac{\partial^2}{\partial h_i \partial \tau} \right] c_X(\boldsymbol{h}, \tau) - \ldots = 0. \tag{2.25}$$

It is worth noticing that by setting equal to 0 the first term within the brackets of the series expansion of Eq. (2.25) we get Eq. (2.21a), obtained earlier. And, similarly, by setting equal to 0 the second term within the brackets of the series expansion we get Eq. (2.21b), also obtained earlier.

2.4 EXTENSIONS OF THE FROZEN RANDOM FIELD MODEL

Several extensions of the frozen random field model have been considered, depending on the modeling needs of the phenomenon of interest. Some of these extensions are briefly presented next, noticing that in most cases their further development is the subject of ongoing research.

2.4.1 Spectral Domain Decomposition

This extension of the frozen random field model is obtained if in place of the physical $R^{n,1}$ domain of Eq. (2.1c), a spectral domain decomposition of $X(\boldsymbol{s}, t)$ is introduced that assumes that at each wavevector $\boldsymbol{k}$ the field is moving with velocity $\boldsymbol{\upsilon}(\boldsymbol{k})$, as follows

$$X(\boldsymbol{s}, t) = \int d\boldsymbol{k}\, e^{i(\boldsymbol{k} \cdot \boldsymbol{s} - t\upsilon(\boldsymbol{k}))} \widetilde{X}(\boldsymbol{k}), \tag{2.26}$$

where $\widetilde{X}(\boldsymbol{k})$ is a random field in the frequency domain. What is special about Eq. (2.26) is that it introduces a decomposition of the random field in the frequency domain that depends on the form of $\boldsymbol{\upsilon}(\boldsymbol{k})$.

2.4.2 Nonstationary Representation

Another extension is useful in applications such as studies in which the advection velocity includes a certain acceleration part that cannot be physically ignored. Then, the velocity $\boldsymbol{\upsilon}$ needs to be considered as a function of the acceleration $\boldsymbol{a}$ and an average time $\overline{\tau}$, i.e., $\boldsymbol{\upsilon} = \boldsymbol{\upsilon}(\boldsymbol{a}, \overline{\tau})$, where $\boldsymbol{a}$ is the acceleration and $\bar{\tau}$ is the time in the middle of the interval τ. This type of nonstationarity can be taken into account by means of an extended form of the covariance Eq. (2.2b) where now the traveling coordinates are given by

$$\widehat{\boldsymbol{h}} = \boldsymbol{h} - \boldsymbol{v}(\boldsymbol{a}, \bar{\tau})\tau \tag{2.27}$$

Often, a linear form of the velocity is assumed, $\boldsymbol{v}(\boldsymbol{a}, \bar{\tau}) = \boldsymbol{v}_0 + \boldsymbol{a}\bar{\tau}$, where $\boldsymbol{v}_0$ is the velocity at time $t = 0$. In this respect, Eq. (2.27) introduces a kind of an accelerated form of the frozen random field hypothesis.

2.4.3 Elliptic Representation

As was emphasized earlier, the fact that the covariance values of Eq. (2.2b) may remain constant for sufficiently large lags $|\boldsymbol{h}|$ or τ in the linear contours $|\boldsymbol{h}| - \upsilon\tau = |\widehat{\boldsymbol{h}}|$ apparently violates a basic covariance property (*viz.*, the covariance should decrease with increasing $|\boldsymbol{h}|$ or τ). Therefore, the frozen random field model might not apply at larger lags. This observation suggests the development of an elliptic STIS representation of the covariance Eq. (2.2b) with

$$|\widehat{\boldsymbol{h}}| = \left[(|\boldsymbol{h}| - \upsilon\tau)^2 + (\upsilon'\tau)^2\right]^{\frac{1}{2}}, \tag{2.28}$$

where

$$\upsilon = -\frac{\partial^2 c_X(0,0)}{\partial r \partial \tau}\left(\frac{\partial^2 c_X(0,0)}{\partial r^2}\right)^{-1},$$

$$\upsilon' = \left[\frac{\partial^2 c_X(0,0)}{\partial \tau^2}\left(\frac{\partial^2 c_X(0,0)}{\partial r^2}\right)^{-1} - \upsilon^2\right]^{\frac{1}{2}}.$$

where $r = |\boldsymbol{h}|$. The isocovariance contours defined by Eq. (2.28) have an elliptic shape. The readers may find it interesting to compare the above equations with Eq. (2.23).

2.4.4 Random Velocity

In some other applications, the real situation is better represented by assuming that the velocity $\boldsymbol{v}(\boldsymbol{s}, t)$ is an STHS random field statistically independent to $X(\boldsymbol{s}, t)$, so that $X(\boldsymbol{s}, t) = \widehat{X}(\boldsymbol{s} - \boldsymbol{v}(\boldsymbol{s}, t)t)$. In this case, Eq. (2.7a) becomes

$$\tilde{c}_X(\boldsymbol{k}, \omega) = \frac{1}{2\pi}\tilde{c}_{\widehat{X}}(\boldsymbol{k})\int d\tau\, e^{-i\omega\tau}\phi_\upsilon(\boldsymbol{k}\tau) \tag{2.29a}$$

where $\phi_\upsilon(-\boldsymbol{k}\tau) = \overline{e^{-i\tau(\boldsymbol{k}\cdot\boldsymbol{v})}}$ is the characteristic function (CF) of the velocity. The velocity is a function of the space−time point coordinates $(\boldsymbol{s}, t)$, but, given that it is an STHS random field, its CF is independent of $(\boldsymbol{s}, t)$.

Example 2.10

Lets us assume that the velocity in the $R^{3,1}$ domain follows a Gaussian law with mean $\bar{\boldsymbol{v}}$ and variance $\overline{(\upsilon_i - \bar{\upsilon}_i)^2} = \frac{1}{3}\sigma_\upsilon^2$ $(i = 1, 2, 3)$. Then

$$\phi_\upsilon(-\boldsymbol{k}\tau) = e^{-i\tau(\boldsymbol{k}\cdot\bar{\boldsymbol{v}}) - \frac{1}{6}(\tau\boldsymbol{k})^2\sigma_\upsilon^2},$$

and

$$\widetilde{c}_X(\boldsymbol{k},\omega) = a^{-\frac{1}{2}}\widetilde{c}_{\widehat{X}}(\boldsymbol{k})e^{-\frac{\pi}{a}(\omega+\boldsymbol{k}\cdot\overline{\boldsymbol{v}})^2},$$

where $a = \frac{2}{3}\pi k^2\sigma_v^2$. This result generalizes Eq. (2.7a). If $\sigma_v^2 = 0$, it gives Eq. (2.7a).

To improve the flexibility of the model of Eq. (2.2b), the case of a random velocity has been also studied by Cox and Isham (1988), who, after assuming that $\boldsymbol{v}$ is random variable, extended Eq. (2.2b) as

$$c_X(\boldsymbol{h},\tau) = \overline{c_{\widehat{X}}(\boldsymbol{h}-\boldsymbol{v}\tau)}. \tag{2.29b}$$

The determination of the probability law of the random velocity vector $\boldsymbol{v}$ depends on the physics of the phenomenon under study. The frozen covariance model of Eq. (2.2b) is an STHS covariance that may be seen as a special case of the covariance of Eq. (2.29b) under the simplification that $\boldsymbol{v}$ is the mean value of the directional attribute (e.g., dominant wind velocity).

2.4.5 Heterogeneous Representation

It is an interesting development that frozen random fields with varying physical nature can be associated with structure functions of any order in the $R^{n,1}$ domain (Section 2.2 of Chapter IV). Indeed, structure functions of the ρth order can be specified having the general heterogeneous (space nonhomogeneous/time stationary) form

$$\xi_X^{[\rho]}(\boldsymbol{s},\boldsymbol{h},\tau) = \overline{[X(\boldsymbol{s}+\boldsymbol{h},t+\tau) - X(\boldsymbol{s},t)]^\rho}, \tag{2.30}$$

i.e., by definition, the structure function is the ρth-order moment of the random field difference (or, increment) in space–time. Additional information is often gained by deriving structure function equations in the physical or the spectral domain to analyze, e.g., multiple received signals for spaced antenna radars (Praskovsky and Praskovskaya, 2003).

Specifically, to derive structure function equations of any order ρ for a frozen random field $X(\boldsymbol{s}, t)$ in $R^{n,1}$, first the velocity vector is decomposed into a time-independent mean $\overline{\boldsymbol{v}}(\boldsymbol{s})$ and a zero-mean fluctuation $\boldsymbol{v}'(\boldsymbol{s},t)$,

$$\boldsymbol{v} = \boldsymbol{v}(\boldsymbol{s},t) = \overline{\boldsymbol{v}}(\boldsymbol{s}) + \boldsymbol{v}'(\boldsymbol{s},t), \tag{2.31}$$

which is assumed to be locally stationary at each location $\boldsymbol{s}$. Then, in the expansion

$$X(\boldsymbol{s}+\boldsymbol{h},t+\tau) = X(\boldsymbol{s},t) + \nabla X(\boldsymbol{s},t)\cdot\boldsymbol{h} + \frac{dX(\boldsymbol{s},t)}{dt}\tau + O(\boldsymbol{h}^2,\tau^2), \tag{2.32}$$

it is assumed that the $X(\boldsymbol{s}, t)$ is mean square (m.s.) continuous and differentiable. The time derivative is given by

$$\frac{dX(\boldsymbol{s},t)}{dt} = X(\boldsymbol{s},t) - \nabla X(\boldsymbol{s},t)\cdot\boldsymbol{v}. \tag{2.33}$$

where the local Taylor hypothesis, $X(\boldsymbol{s} + \boldsymbol{v}\tau,\ t + \tau) = X(\boldsymbol{s},\ t)$, for small τ has been used. In light of Eqs. (2.31)–(2.33), Eq. (2.30) gives

$$\xi_X^{[\rho]}(\boldsymbol{s},\boldsymbol{h},\tau) = (-1)^\rho\varphi_\rho(\boldsymbol{s},\boldsymbol{h}) + (-1)^{\rho+1}\rho\chi_\rho(\boldsymbol{s},\boldsymbol{h})\tau + (-1)^\rho\frac{\rho(\rho-1)}{2}\psi_\rho(\boldsymbol{s},\boldsymbol{h})\tau^2 + O(\tau^3), \tag{2.34}$$

where

$$\varphi_\rho(\boldsymbol{s},\boldsymbol{h}) = \overline{(\nabla X(\boldsymbol{s},t)\cdot\boldsymbol{h})^\rho},$$

$$\chi_\rho(\boldsymbol{s},\boldsymbol{h}) = \overline{(\nabla X(\boldsymbol{s},t)\cdot\boldsymbol{h})^{\rho-1}\left(\nabla X(\boldsymbol{s},t)\cdot\overline{\boldsymbol{v}(\boldsymbol{s})}\right)},$$

$$\psi_\rho(\boldsymbol{s},\boldsymbol{h}) = \overline{(\nabla X(\boldsymbol{s},t)\cdot\boldsymbol{h})^{\rho-2}\left[\left(\nabla X(\boldsymbol{s},t)\cdot\overline{\boldsymbol{v}(\boldsymbol{s})}\right)^2 + (\nabla X(\boldsymbol{s},t)\cdot\boldsymbol{v}'(\boldsymbol{s},t))^2\right]} + \overline{(\nabla X(\boldsymbol{s},t)\cdot\boldsymbol{h})^{\rho-2}\left(\frac{\partial}{\partial t}X(\boldsymbol{s},t)\right)^2}. \tag{2.35a–c}$$

The coefficients φ_ρ depend only on the field gradient moments and not on the velocity moments, the coefficients χ_ρ depend on the field gradient moments and the mean velocity but not on the velocity fluctuations, and the coefficients ψ_ρ depend on the field gradient moments and the velocity mean and fluctuations. Also, since, in this case, the random field is time stationary the mean of $X(\boldsymbol{s}, t)$, and its gradient are only functions of space, i.e., $\overline{X(\boldsymbol{s},t)} = \overline{X(\boldsymbol{s})}$ and $\overline{\nabla X(\boldsymbol{s},t)} = \overline{\nabla X(\boldsymbol{s})}$, respectively.

Other kinds of structure functions can be defined, such as the one-point/time stationary structure function of any order ρ,

$$\xi_X^{[\rho]}(\boldsymbol{s},\tau) = \overline{[X(\boldsymbol{s},t+\tau) - X(\boldsymbol{s},t)]^\rho}, \tag{2.36}$$

which is the ρth-order moment of the temporal random field difference (or increment). Following a similar approach as above it is found that

$$\xi_X^{[\rho]}(\boldsymbol{s},\tau) = \varpi_\rho(\boldsymbol{s})\tau^\rho + O(\tau^{\rho+1}), \tag{2.37}$$

where

$$\varpi_\rho(\boldsymbol{s}) = \overline{\left[\nabla X(\boldsymbol{s},t)\cdot\overline{\boldsymbol{v}(\boldsymbol{s})} + \nabla X(\boldsymbol{s},t)\cdot\boldsymbol{v}'(\boldsymbol{s},t) - \frac{\partial}{\partial t}X(\boldsymbol{s},t)\right]^\rho}. \tag{2.38}$$

The example below is an illustration of the particular forms the parameters φ_ρ, χ_ρ, and ψ_ρ can take, and their physical interpretation.

Example 2.11

In the $R^{3,1}$ domain we let $\rho = 2$ so that we focus on the structure functions $\xi_X^{[2]}(\boldsymbol{s},\boldsymbol{h},\tau) = \xi_X(\boldsymbol{s},\boldsymbol{h},\tau)$ and $\xi_X^{[2]}(\boldsymbol{s},\tau) = \xi_X(\boldsymbol{s},\tau)$. Then,

$$\phi_2(\boldsymbol{s},\boldsymbol{h}) = \sum_{i=1}^{3} h_i^2\overline{\left(X_i^{(1,0)}\right)^2} + \sum_{i\neq i'=1}^{3} h_i h_{i'}\overline{X_i^{(1,0)}X_{i'}^{(1,0)}},$$

$$\chi_2(\boldsymbol{s},\boldsymbol{h}) = \sum_{i=1}^{3} h_i\overline{v_i(\boldsymbol{s})}\,\overline{\left(X_i^{(1,0)}\right)^2} + \sum_{i\neq i'=1}^{3} h_i\overline{v_{i'}(\boldsymbol{s})}\,\overline{X_i^{(1,0)}X_{i'}^{(1,0)}},$$

$$\psi_2(\boldsymbol{s}) = \sum_{i=1}^{3}\overline{\left(X_i^{(1,0)}\right)^2}\left[\left(\overline{v_i(\boldsymbol{s})}\right)^2 + \overline{v_i'^2(\boldsymbol{s},t)}\right] + \sum_{i\neq i'=1}^{3}\overline{X_i^{(1,0)}X_{i'}^{(1,0)}}\left[\overline{v_i(\boldsymbol{s})}\,\overline{v_{i'}(\boldsymbol{s})} + \overline{v_i'(\boldsymbol{s},t)v_{i'}'(\boldsymbol{s},t)}\right]$$

$$+\overline{\left(X^{(0,1)}\right)^2} = \varpi_2(\boldsymbol{s}), \tag{2.39a–c}$$

where $X_i^{(1,0)} = \frac{\partial}{\partial s_i}X$, $X_{i'}^{(1,0)} = \frac{\partial}{\partial s_{i'}}X$, and $X^{(0,1)} = \frac{\partial}{\partial t}X$ (Exercise X.10). That is, the structure functions are expressed in terms of $X_i^{(1,0)}$ and $X^{(0,1)}$ that describe the spatial and temporal rates of changes of the $X(s, t)$ values at location s and at instant t. In many cases, the $X_i^{(1,0)}$, $X^{(0,1)}$, and $\boldsymbol{\upsilon}$ describe physically different features of $X(s, t)$ so that these rates and the velocity can be assumed to be statistically independent random fields.

2.5 INTEGRALS OF FROZEN SPATIOTEMPORAL RANDOM FIELDS

In some applications, the physical interpretation of an attribute modeled as a frozen S/TRF may require the introduction of a partial integration of the covariance function leading to new covariance functions with different properties. Among the most common integrations is over all directions of the attribute's $R^{2,1}$ domain, i.e.,

$$c_X(\boldsymbol{h}, \tau) = r\int_0^{2\pi} d\varphi\, c_X(r, \varphi, \tau), \tag{2.40}$$

where (r, φ) are polar coordinates.

Example 2.12

In the $R^{3,1}$ domain consider the space–time covariance

$$c_X(\boldsymbol{h}, \tau) = \frac{c_0}{4\pi\sigma_X^2} e^{-\frac{(\boldsymbol{h}-\boldsymbol{\upsilon}\tau)^2}{4\sigma^2}}, \tag{2.41}$$

where c_0 is a constant. By applying Eq. (2.40) it is found that

$$c_X(r, \tau) = \frac{c_0}{2\sigma_X^2} r\, I_0\left(\frac{c_0 r\tau}{2\sigma_X^2}\right) e^{-\frac{r^2+\upsilon^2\tau^2}{4\sigma^2}}, \tag{2.42}$$

$r = |\boldsymbol{h}|$, which is a covariance with different features than that of Eq. (2.41). So, the covariance model of Eq. (2.42) is characterized by tilted ridges of high correlation values.

2.6 VECTOR FROZEN SPATIOTEMPORAL RANDOM FIELDS

The extension of the analysis above to frozen vector S/TRFs presents no technical problems. Let $X_l(\boldsymbol{p})$ and $X_{l'}(\boldsymbol{p}')$, $\boldsymbol{p}, \boldsymbol{p}' \in R^{n+1}$, $l, l' = 1, \ldots, k$, be pairs of random fields that are frozen and frozenly connected. In this case, the corresponding cross-covariance and cross-variogram functions are expressed as, respectively,

$$\begin{aligned} c_{X_l X_{l'}}(\boldsymbol{h}, \tau) &= c_{\widehat{X}_l \widehat{X}_{l'}}(\boldsymbol{h} - \boldsymbol{\upsilon}\tau), \\ \gamma_{X_l X_{l'}}(\boldsymbol{h}, \tau) &= \gamma_{\widehat{X}_l \widehat{X}_{l'}}(\boldsymbol{h} - \boldsymbol{\upsilon}\tau). \end{aligned} \tag{2.43a–b}$$

Obviously, for $\tau = 0$ Eqs. (2.43a–b) reduce to $c_{X_l X_{l'}}(\boldsymbol{h}, 0) = c_{\widehat{X}_l \widehat{X}_{l'}}(\boldsymbol{h})$ and $\gamma_{X_l X_{l'}}(\boldsymbol{h}, 0) = \gamma_{\widehat{X}_l \widehat{X}_{l'}}(\boldsymbol{h})$.

Furthermore, if the velocity $\boldsymbol{v}$ is random, Eqs. (2.43a–b) become the integral covariance representations

$$c_{X_l X_{l'}}(\boldsymbol{h},\tau) = \int d\boldsymbol{v} f_{\boldsymbol{v}}\, c_{\widehat{X}_l \widehat{X}_{l'}}(\boldsymbol{h} - \boldsymbol{v}\tau),$$
$$\gamma_{X_l X_{l'}}(\boldsymbol{h},\tau) = \int d\boldsymbol{v} f_{\boldsymbol{v}}\, \gamma_{\widehat{X}_l \widehat{X}_{l'}}(\boldsymbol{h} - \boldsymbol{v}\tau), \tag{2.44a–b}$$

where $f_{\boldsymbol{v}}$ is the probability density of $\boldsymbol{v}$ ($f_{\boldsymbol{v}}$ may depend on the direction of $\boldsymbol{v}$, in general). The $c_{X_l X_{l'}}$, $c_{\widehat{X}_l \widehat{X}_{l'}}$, $\gamma_{X_l X_{l'}}$, and $\gamma_{\widehat{X}_l \widehat{X}_{l'}}$ are all symmetric in $l, l' = 1, \ldots, k$, even if $f_{\boldsymbol{v}}$ is an anisotropic function. These equations admit a certain interpretation, depending on the physical context. An illustration is given below.

Example 2.13

In fluid dynamics studies, Eq. (2.44b) distinguishes the purely kinematic (convective) effect of large-scale eddies from the intrinsic small-scale properties of turbulence. Also, for STIS random fields in R^3, with $\gamma_{X_l}(r,\tau) = \sum_{l'=1}^{k} \gamma_{X_{l'}}(\boldsymbol{h},\tau)$, $r = |\boldsymbol{h}|$, Eq. (2.44b) leads to

$$\gamma_{X_l}(r,\tau) = \int d\boldsymbol{v} f_{\boldsymbol{v}} \gamma_{\widehat{X}_l}(|\boldsymbol{h} - \boldsymbol{v}\tau|), \tag{2.45}$$

where $f_{\boldsymbol{v}}(\boldsymbol{v}) = 3\left(\frac{6}{\pi}\right)^{\frac{1}{2}} a^{-3} e^{-1.5\frac{v^2}{a^2}}$, a is a physical constant (Tatarskii, 1971).

3. PLANE-WAVE SPATIOTEMPORAL RANDOM FIELDS

I now turn to another special class of random fields with a considerable number of applications in electromagnetic radiation, optics and microlasers, elastic solids, and acoustic wave in gas or fluid.

Definition 3.1

The *plane-wave* random field propagating in direction $\boldsymbol{\eta}$ with speed υ is defined by

$$X(\boldsymbol{s},t) = \widehat{X}(\widehat{t}), \tag{3.1}$$

where $\widehat{t} = t - \frac{\boldsymbol{\eta}\cdot\boldsymbol{s}}{\upsilon}$ introduces a time transformation.

For sure, there is an analogy between the frozen and the plane-wave hypotheses: The former hypothesis involves a space transformation that is time dependent, whereas the latter hypothesis involves a time transformation that is space dependent. This analogy is reflected in the following proposition, which is essentially the counterpart of Proposition 2.1 (its validity is a straightforward consequence of the Definition 3.1).

Proposition 3.1

Assuming temporal stationarity, the mean and covariance functions of the original random field, $X(\boldsymbol{p})$, and the corresponding temporal random process, $\widehat{X}(\widehat{t})$, are related by

$$\overline{X(\boldsymbol{p})} = \overline{\widehat{X}(\widehat{t})},$$
$$c_X(\boldsymbol{h},\tau) = c_{\widehat{X}}(\widehat{\tau}), \tag{3.2a–b}$$

where $\widehat{t} = \left(t - \frac{\eta \cdot s}{\upsilon}\right)$, $\widehat{\tau} = \left(\tau - \frac{\eta \cdot h}{\upsilon}\right) \in T(\subseteq R^1)$.[5]

As before, the space–time variability representations of Eqs. (3.2a–b) allow considerable simplifications in space–time analysis. Beyond the physical domain, useful expressions can be derived in the spectral domain, as well (Exercise X.11).

Corollary 3.1

The SDF $\widetilde{c}_X(\boldsymbol{k}, \omega)$ and the space–frequency function $\widetilde{c}_X(\boldsymbol{h}, \omega)$ associated with the covariance function of Eq. (3.2b) are given by, respectively,

$$\begin{aligned} \widetilde{c}_X(\boldsymbol{k}, \omega) &= \widetilde{c}_{\widehat{X}}(\omega)\delta\left(\boldsymbol{k} - \frac{\omega}{\upsilon}\boldsymbol{\eta}\right), \\ \widetilde{c}_X(\boldsymbol{h}, \omega) &= \widetilde{c}_{\widehat{X}}(\omega)e^{i\omega\frac{\boldsymbol{\eta}\cdot\boldsymbol{h}}{\upsilon}} \end{aligned} \tag{3.3a–b}$$

(compare with Eqs. 2.7a–b).

Eq. (3.3a) links the spectrum of the original field $X(\boldsymbol{p})$ with that of the temporal field $\widehat{X}\left(\widehat{t}\right)$. In physical terms, the spectrum of a plane curve at two different locations is linked to that at one location by means of a linear phase shift reflecting the propagation between these two locations. Furthermore, $\widetilde{c}_X(\omega) = \widetilde{c}_{\widehat{X}}(\omega)$, i.e., the spectrum of the S/TRF $X(\boldsymbol{p})$ at $\boldsymbol{h} = \boldsymbol{0}$ and that of the temporal field $\widehat{X}\left(\widehat{t}\right)$ coincide. From a phenomenological perspective, there exists a considerable body of theoretical, numerical, and experimental evidence of phenomena with the distinctive features of the plane-wave S/TRF model.

Combinations of the above special kind of random fields is possible leading to random fields with space–time covariances of the forms

$$c_X(\boldsymbol{h}, \tau) = \begin{cases} c_X(r_{TW}), \\ c_X\left(\boldsymbol{h} - \upsilon\tau, \tau - \dfrac{\boldsymbol{\eta}\cdot\boldsymbol{h}}{\upsilon}\right) \end{cases} \tag{3.4a–b}$$

where $r_{TW} = \left(\widehat{\boldsymbol{h}}^2 + \upsilon^2\widehat{\tau}^2\right)^{\frac{1}{2}}$, $\widehat{\boldsymbol{h}} = (\boldsymbol{h} - \upsilon\tau)$, and $\widehat{\tau} = \left(\tau - \frac{\boldsymbol{\eta}\cdot\boldsymbol{h}}{\upsilon}\right)$.

Example 3.1

A space–time covariance function of the form described above is given by

$$c_X(r_{TW}) = a r_{TW}^{\nu - \frac{n+1}{2}} K_{\nu - \frac{n+1}{2}}(r_{TW}), \tag{3.5}$$

where $a > 0, \nu > \frac{n+1}{2}$.

Lastly, as is shown in Chapter XVI, a combination of the above equations leads to a simple approach of generating new covariance models.

---

[5] The derivation of Eq. (3.2a) is a straightforward expectation of Eq. (3.1). For Eq. (3.2b), one observes that $c_X(\boldsymbol{h}, \tau) = \overline{X\left(t - \frac{\eta\cdot s}{\upsilon}\right) X\left(t - \tau - \frac{\eta\cdot(s-h)}{\upsilon}\right)} = c_X\left(\tau - \frac{\eta\cdot h}{\upsilon}\right) = c_{\widehat{X}}(\widehat{\tau})$.

4. LOGNORMAL SPATIOTEMPORAL RANDOM FIELDS

This special and very useful in practice class of S/TRFs was introduced in Section 7 of Chapter II. As we saw in this section, a lognormal S/TRF $Y(\boldsymbol{p})$ is defined in terms of a Gaussian random field $X(\boldsymbol{p})$ as

$$\begin{aligned} &Y(\boldsymbol{p}) \sim L^m\left(\overline{Y}, \sigma_Y^2\right), \\ &\text{s.t. } \ln Y(\boldsymbol{p}) = X(\boldsymbol{p}) \sim G^m\left(\overline{X}, \sigma_X^2\right). \end{aligned} \tag{4.1a–b}$$

where G^m and L^m denote the m-variate Gaussian and the corresponding lognormal probability law, respectively. Otherwise said, a random field $Y(\boldsymbol{p})$ is a lognormal S/TRF if the logarithm of $Y(\boldsymbol{p})$ is a Gaussian random field $X(\boldsymbol{p})$, that is, lognormal random fields are of the form $e^{X(\boldsymbol{p})}$.

The relationships between the statistics of $Y(\boldsymbol{p})$ and $X(\boldsymbol{p})$ were described in Table 7.1 of Chapter II. These relationships show that a lognormal random field can be characterized by defining the mean field and the covariance function of the underlying Gaussian random field. In applications where the hypotheses of STHS and STIS apply, the covariance function arguments $\boldsymbol{p}, \boldsymbol{p}'$ should be replaced by $\Delta\boldsymbol{p}$ and $|\Delta\boldsymbol{p}|$, respectively.

Example 4.1

In view of Eqs. (4.1a–b), we can derive the following variogram function expressions

$$\begin{aligned} \gamma_Y(\Delta\boldsymbol{p}) &= \frac{1}{2}\overline{[Y(\boldsymbol{p}+\Delta\boldsymbol{p}) - Y(\boldsymbol{p})]^2}, \\ \gamma_X(\Delta\boldsymbol{p}) &= \frac{1}{2}\overline{[\ln Y(\boldsymbol{p}+\Delta\boldsymbol{p}) - \ln Y(\boldsymbol{p})]^2}. \end{aligned} \tag{4.2a–b}$$

Moreover, by virtue of the basic relationships of Table 7.1 of Chapter II, we find the covariance expression for the lognormal random field,

$$c_Y(\Delta\boldsymbol{p}) = \overline{Y}^2\left[e^{\sigma_X^2 - \gamma_X(\Delta\boldsymbol{p})} - 1\right]. \tag{4.3}$$

These equations lead to a remarkable result. Assuming that $\sigma_X^2 >> \gamma_X(\Delta\boldsymbol{p})$, Eq. (4.3) reduces to the simpler expression (Exercise X.12),

$$c_Y(\Delta\boldsymbol{p}) = \sigma_Y^2 e^{-\gamma_X(\Delta\boldsymbol{p})}. \tag{4.4}$$

Therefore, if $\gamma_X(\Delta\boldsymbol{p})$ is a valid variogram model, the covariance model $c_Y(\Delta\boldsymbol{p})$ specified by Eq. (4.4) is valid too. This interesting expression can be used to generate new covariance models from known variogram ones (Chapter XVI).

5. SPHERICAL SPATIOTEMPORAL RANDOM FIELDS

As noted earlier (Chapters I and III), in many applications the natural attribute varies on a surface (e.g., the earth) that is viewed as a spherical surface in R^n, i.e., $S^{n-1} = \{\boldsymbol{s} \in \boldsymbol{R}^n : |\boldsymbol{s}| = \rho\}$,[6] where ρ is the

[6]Some authors use the notation $S_n = \{\boldsymbol{s} \in R^n : |\boldsymbol{s}| = \rho\}$, instead. In this book we will use both notations, depending on the context.

spherical radius, and $|s| = \left(\sum_{i=1}^{n} s_i^2\right)^{\frac{1}{2}}$. In such cases, the space−time metric is of the form (r_s, τ), where[7]

$$r_S = |\boldsymbol{h}| = \rho\theta = 2\rho \sin^{-1}\left(\frac{r_E}{2\rho}\right) \tag{5.1}$$

is the arc length between any two points $\boldsymbol{s}$ and $\boldsymbol{s}'$ on the surface of the sphere defined earlier by Eq. (2.18c) of Chapter I,

$$\theta = \arccos(\boldsymbol{s}\cdot\boldsymbol{s}') \in [0, \pi] \tag{5.2}$$

is the angle of the arc, $r_E = |\boldsymbol{s} - \boldsymbol{s}'|$ is the Euclidean distance in R^n, and τ is the time lag in $T \subseteq R^1$. Also, it is easily seen that $\boldsymbol{s}\cdot\boldsymbol{s}' = \rho^2 - 0.5r_E^2$ (Exercise X.13), in which case

$$\theta = \arccos\left(\rho^2 - 0.5r_E^2\right),$$

i.e., the arc angle θ, the radius ρ, and the Euclidean distance r_E are also related.

Example 5.1

Many global phenomena in oceanography, radiophysics, geophysics, and climatology (e.g., mixed layer depths over the global ocean, paleoceanographic temperatures, satellite and meteorological observations, atmospheric pollution data, and large-scale fire spread) need to be represented in terms of random field models on $S^{n-1} \times T$.

An m.s. continuous random field $X(\boldsymbol{s}, t)$ is wide sense (w.s.) STIS on $S^{n-1} \times T$, or simply $S^{n-1,1}$, if its mean is constant (say, zero for simplicity), and its covariance between locations $\boldsymbol{s}$ and $\boldsymbol{s}'$ and between time instants t and t' is a function of r_s and $\tau = t' - t$, i.e.,

$$\overline{X(\boldsymbol{s},t)X(\boldsymbol{s}',t')} = c_X(r_s,\tau), \tag{5.3}$$

where r_s is given by Eq. (5.1). A link between the spherical random field model of Eq. (5.3) and the underlying physics can be established via some geodesy considerations.

Example 5.2

In the common case of the earth considered as a spherical surface

$$S^2 = \left\{\boldsymbol{s} \in R^3 : |\boldsymbol{s}| = 1\right\}, \tag{5.4}$$

one recognizes the metric of Eq. (5.1) as the geodetic distance between points on the surface of the sphere. In this case, a space−time point on $S^{2,1}$ is denoted as

$$\boldsymbol{p} = (\boldsymbol{s}, t) = (\vartheta, \varphi, t), \tag{5.5}$$

where $\vartheta \in [0, \pi]$ is the colatitude and $\varphi \in [0, 2\pi]$ is the longitude. A random field on $S^{2,1}$ is written as

$$X(\boldsymbol{s}, t) = X(\vartheta, \varphi, t), \tag{5.6}$$

[7]In many cases, for simplicity, it is assumed that $\rho = 1$, in which case $r_s = \theta \in [0, \pi]$.

which is an STIS random field with covariance

$$c_X(\boldsymbol{s}, t, \boldsymbol{s}', t') = c_X(\vartheta, \varphi, t, \vartheta', \varphi', t') \tag{5.7}$$

that depends on the arc distance θ between locations $\boldsymbol{s} = (\vartheta, \varphi)$ and $\boldsymbol{s}' = (\vartheta', \varphi')$, i.e., $\cos\theta = \boldsymbol{s} \cdot \boldsymbol{s}'$. The covariance function can be written as

$$c_X(\boldsymbol{s}, t, \boldsymbol{s}', t') = c_X(\mathcal{R}_\theta \boldsymbol{s}, N, \tau) = c_X(r_s, \tau),$$

where $r_s = \theta$, and $\mathcal{R}_\theta$ denotes the rotation that takes $\boldsymbol{s}'$ onto the North Pole N and $\boldsymbol{s}$ onto the plane $s_1 s_3$.

The following Proposition was proven by Mokljacuk and Jadrenko (1979) together with several other useful results on the subject.

Proposition 5.1

The $X(s, t)$ can be expressed in $S^{n-1,1}$ by means of the series expansion

$$X(\boldsymbol{s}, t) = \sum_{l=0}^{\infty} \sum_{k=1}^{h(l,n)} Y_l^k(t) S_l^k(\boldsymbol{s}), \tag{5.8}$$

where $S_l^k(\boldsymbol{s})$ are orthonormal spherical harmonics of degree l on the $S^{n-1,1}$ domain, and $h(l,n) = (2l + n - 2)\frac{(l+n-3)!}{(n-2)!l!}$ is the number of linearly independent spherical harmonics of degree l, and $Y_l^k(t)$ is a random process sequence such that

$$\overline{Y_l^k(t) Y_{l'}^{k'}(t')} = \delta_{ll'} \delta_{kk'} b_l(t' - t),$$

where $\delta_{ll'}$ and $\delta_{kk'}$ denote Kronecker delta, and $b_l(t' - t) = b_l(\tau)$ is a stationary covariance sequence such that $\sum_{l=0}^{\infty} h(l,n) b_l(0) < \infty$. Then, the covariance function of $X(s, t)$ is

$$c_X(r_s, \tau) = \sum_{l=0}^{\infty} \frac{h(l,n)}{S_n G_{l,\frac{n-2}{2}}(1)} G_{l,\frac{n-2}{2}}(\cos r_s) b_l(\tau), \tag{5.9}$$

where $G_{l,\frac{n-2}{2}}$ are Gegenbauer polynomials (Appendix), and $S_n = \frac{2\pi^{\frac{n}{2}}}{\Gamma(\frac{n}{2})}$, as usual.

It would be instructive to study some examples that could clarify the application of Proposition 5.1.

Example 5.3

If $n = 3$, Eq. (5.9) gives the space–time covariance

$$c_X(r_s, \tau) = \sum_{l=0}^{\infty} (2l + 1) L_l(\cos r_s) b_l(\tau), \tag{5.10}$$

where L_l are Legendre polynomials of degree l. If $n > 2$, the

$$c_X(r_s, \tau) = \left[1 - 2b(\tau)\cos r_s + b^2(\tau)\right]^{-\frac{n-2}{2}}, \tag{5.11}$$

$b(\tau)$ is a positive-definite kernel, is a covariance of an STIS random field.

Example 5.4

Oceanographic studies (tide analysis, seasonal variations, ocean geophysical oscillations) use random field models in $S^2 \times T$ (with $\rho = 1$) and $t \in Z$. Assume that $X(s, t)$ is an STHS and m.s. continuous random field. Then, it follows that

$$X(s,t) = \sum_{l=0}^{\infty} \sum_{k=-l}^{l} Y_l^k(t) S_l^k(s) \tag{5.12}$$

is a series that is m.s. convergent, where

$$Y_l^k(t) = \int_{S^2} dsX(s,t) S_l^k(s) ds$$

is a stochastic integral in the m.s. sense, and

$$b_l(\tau) = \int_{-\pi}^{\pi} dQ_l(\omega) e^{i\omega\tau},$$

where Q_l is the frequency spectral distribution of $X(s, t)$. Then, the covariance is given by

$$c_X(r_s, \tau) = \frac{1}{4\pi} \sum_{l=0}^{\infty} (2l+1) L_l(\cos r_s) \int_{-\pi}^{\pi} dQ_l(\omega)\, e^{i\omega\tau}, \tag{5.13}$$

which is a model based on spherical random field considerations that is insightful and able to capture the essential physics of the situation.

Proposition 5.1 can be extended in the case of a vector S/TRF, as is discussed in the following example.

Example 5.5

The covariance matrix of an m-variate Gaussian vector random field $\boldsymbol{X}(s, t)$ is given by

$$\boldsymbol{c}_{\boldsymbol{X}}(r_s, \tau) = \sum_{l=0}^{\infty} \frac{h(l,n)}{S_n G_{l, \frac{n-2}{2}}(1)} G_{l, \frac{n-2}{2}}(\cos r_s) \boldsymbol{b}_l(\tau), \tag{5.14}$$

where $\boldsymbol{b}_l(\tau)$ is a stationary covariance $(m \times m)$ matrix sequence such that $\sum_{l=0}^{\infty} h(l,n)\boldsymbol{b}_l(0)$ converges.

A large family of space–time covariance models in $S^{n-1,1}$ can be constructed using the properties of the Gegenbauer polynomials (for a more detailed discussion, see Chapter XVI). Furthermore, under certain conditions, several of the results regarding a random field $X(s, t)$ in $R^{n,1}$ remain valid for a random field in $S^{n-1,1}$. One example is the nonnegative-definiteness (NND) property (Section 2 of Chapter IV), i.e., a covariance function must be NND on $S^{n-1,1}$. Put differently, the NND property remains valid by replacing r_E with r_S. Another example is the extension in the space–time domain of many of the useful results obtained by Yadrenko (1983). This extension is rather straightforward and implies that given a covariance function in the $R^{n,1}$ domain, $c_X(r_E, \tau)$, there can be defined a valid covariance function on the spherical domain $S^{n-1,1}$, $c_X(r_S, \tau)$. In the case of separability, $c_X(r_S, \tau) = c_X(r_S)c_X(\tau)$, one needs to focus on $c_X(r_S)$ in S^{n-1} and use some earlier results. In particular, Huang et al. (2011) and Gneiting (2013) have suggested a list of functions in the R^n domain ($n = 1,2,3$) that can also be easily extended in the S^{n-1} domain, subject to certain conditions on their coefficient ranges, as is shown in Table 5.1. On the other hand, common covariance models in R^n that are not valid in S^{n-1} include the Gaussian, the hole effect, and the Cauchy model for $\alpha = \kappa = 2$ (for mathematical expressions of these models, see Table 4.1 of Chapter VIII).

Table 5.1 Covariance Models in S^{n-1}

Covariance Model	$c_X(r_s)$ $(\rho = 1, r_s = \theta)$	Range of Validity
Power exponential or stable	$c_0 e^{-\frac{r_s^\alpha}{b^\alpha}}$	$c_0,\ b > 0$ $\alpha \in (0,\ 1]$
Spherical	$c_0\left(1 - \frac{3r_s}{2a} + \frac{r_s^3}{2a}\right)$ $(r_s \in [0, a]$ 0 $(r_s > a)$	$c_0,\ a > 0$
von Kármán/Matern	$\frac{2^{\nu-1} c_0 r_s^\nu}{\Gamma(\nu) a^\nu} K_\nu\left(\frac{r_s}{a}\right)$	$c_0,\ a > 0$ $\nu \in \left(0, \frac{1}{2}\right]$
Power	$c_0 - \left(\frac{r_s}{a}\right)^\alpha$	$\alpha \in (0,\ 1]$ $c_0 \geq \int_0^\pi dr_s \left(\frac{r_s}{a}\right)^\alpha \sin r_s$
Generalized Cauchy	$c_0\left(1 + \left(\frac{r_s}{b}\right)^\alpha\right)^{-\frac{\kappa}{\alpha}}$	$c_0,\ b,\ \kappa > 0$ $\alpha \in (0,\ 1]$
Power sine	$c_0\left[1 - \left(\sin\frac{r_s}{2}\right)^\alpha\right]$	$\alpha \in (0,\ 2)$

By combining the covariance models of Table 5.1, valid covariance models can be easily constructed in $S^{n-1,1}$, as illustrated in the following example.

Example 5.6

The functions

$$c_X(r_s, \tau) = c_0 e^{-\frac{r_s^\lambda + \tau^\lambda}{a^\lambda}},$$

$$c_X(r_s, \tau) = \frac{2^{\nu-1} c_0 r_s^\nu}{\Gamma(\nu) a^{\nu-b} \left(a^\lambda + \tau^\lambda\right)^{\frac{b}{\lambda}}} K_\nu\left(\frac{r_s}{a}\right),$$

are valid covariance on $S^{n-1,1}$ for $c_0,\ a,\ b > 0$, $\nu \in \left(0, \frac{1}{2}\right]$, and $\lambda \in (0,\ 1]$.

Remark 5.1

Given that in the analysis above a sphere of radius one ($\rho = 1$, so that $r_S = \theta$) was assumed, in real-world applications we must account for the actual earth radius ($\rho = 6371$ km) within the covariance function.

Lastly, as a useful approximation, in some cases a covariance function on $S^{n-1,1}$ is considered a function of the chordal distance

$$r_c = r_E = 2\rho \sin\left(\frac{\theta}{2}\right), \tag{5.15}$$

and the time lag τ. Under the condition that r_s is small compared to $\rho\pi$, one can assume that $r_c \approx r_s$, in which case using r_c instead of r_s will produce a covariance $c_X(r_c, \tau)$ that is not significantly different from $c_X(r_s, \tau)$. If, however, the above condition is not valid, and since the difference between r_c and r_s increases as the distance between two points on S^{n-1} increases, a covariance function defined in terms of r_c can yield physically nonsensical results.

6. LAGRANGIAN SPATIOTEMPORAL RANDOM FIELDS

Up to this point, the analysis involved Eulerian coordinates. In certain applications (e.g., describing fluid motion), however, it may be advantageous to work in Lagrangian space–time coordinates specified as follows,

$$\begin{aligned} r_L &= r_E - \upsilon\tau_E, \\ \tau_L &= \tau_E = \tau \end{aligned} \tag{6.1}$$

where υ is a constant speed. Let c_X and $c_{\breve{X}}$ denote the covariance function in Eulerian and Lagrangian coordinates, respectively. Then, the following equality holds between the two,

$$c_X(r_E, \tau) = c_{\breve{X}}(r_E - \upsilon\tau, \tau). \tag{6.2}$$

In certain applications, the $c_X(r_E, \tau)$ is a symmetric function about the line $r_E - \upsilon\tau = 0$, so that $c_X(\upsilon\tau - r_E, \tau) = c_X(\upsilon\tau + r_E, \tau)$.

In this setting, Taylor's hypothesis may allow temporal data to be transformed into spatial data by means of the velocity. Furthermore, Taylor's hypothesis leads to

$$c_{\breve{X}}(\boldsymbol{\upsilon}\tau, \tau) = c_{\breve{X}}(\boldsymbol{\upsilon}\tau, 0). \tag{6.3}$$

Assuming that the velocity $\boldsymbol{\upsilon}$ is uncorrelated to the Lagrangian space–time correlations, Eq. (6.3) can be used to determine the validity of Taylor's hypothesis in terms of $\boldsymbol{\upsilon}$.

Example 6.1

Several interesting conclusions can be drawn by assuming a specific space–time covariance model. So, let us consider the following space–time covariance model in Lagrangian space–time coordinates

$$c_{\breve{X}}(r_L, \tau) = c_0 e^{-\left(\frac{r_L^2}{a_r^2} + \frac{\tau^2}{a_\tau^2}\right)^{\frac{1}{2}}}. \tag{6.4}$$

If one defines the speed as the ratio $\upsilon' = \frac{a_r}{a_\tau}$, Eq. (6.4) gives

$$c_{\breve{X}}(r_L, \tau) = c_0 e^{-\left(\frac{r_L^2 + (\upsilon'\tau)^2}{a_r^2}\right)^{\frac{1}{2}}}. \tag{6.5}$$

which satisfies the relationship

$$c_{\breve{X}}(0, \tau) = c_{\breve{X}}(\upsilon'\tau, 0), \tag{6.6}$$

i.e., in this case, Taylor's hypothesis in Lagrangian coordinates has the same form as the Taylor's hypothesis in Eulerian coordinates, Eq. (2.3). Let us define the time lag τ_L^* so that both Eqs. (6.3) and (6.6) apply. This leads to

$$r_E = \left(\upsilon^2 + \upsilon'^2\right)^{\frac{1}{2}}\tau, \tag{6.7}$$

in which case Taylor's hypothesis in Eulerian coordinates gives

$$c_X(0,\tau) = c_X\left(\left(\upsilon^2 + \upsilon'^2\right)^{\frac{1}{2}}\tau, 0\right). \tag{6.8}$$

This is a remarkable relationship that applies in the case that the corresponding covariance function in Lagrangian coordinates is of the general form

$$c_{\breve{X}}(r_L, \tau) = c_{\breve{X}}\left(\tau_L^*\right), \tag{6.9}$$

where $r_E = \upsilon'\tau_L^*$ and

$$\tau_L^{*2} = \left(\frac{r_L}{\upsilon'}\right)^2 + \tau^2, \tag{6.10}$$

with the corresponding $r_E = \frac{a_r}{a_\tau}\left(\left(\frac{\upsilon}{\upsilon'}\right)^2 + 1\right)^{\frac{1}{2}}\tau$. Based on the above general results, Potvin (1993) has noticed that: (a) assuming that for a frozen field we have $a_\tau \rightarrow \infty$, then $\upsilon' \rightarrow 0$, and Eq. (6.7) reduces to $r_E = \upsilon\tau$; (b) if $\upsilon \rightarrow 0$ but $\upsilon' \neq 0$ (e.g., the space–time correlations are dominated by the internal activity of the storm system), Eq. (6.7) reduces to $r_E = \upsilon'\tau$; and (c) if υ' is not constant, i.e., $\upsilon'\left(\tau_L^*\right)$, then Eqs. (6.8) and (6.10) hold in terms of $\upsilon'\left(\tau_L^*\right)$, and Taylor's hypothesis becomes a nonlinear relationship.

CONSTRUCTION OF SPATIOTEMPORAL PROBABILITY LAWS

XI

CHAPTER OUTLINE

1. INTRODUCTION

As was mentioned in several places throughout the book, stochastic modeling is concerned with natural attributes that vary across space–time following a certain law of change. While the second-order space–time statistics offer a partial characterization of an S/TRF in terms of correlation theory (Chapter IV), the random field is, nevertheless, fully characterized stochastically in terms of probability functions, such as the *multivariate* probability density function (M-PDF) or the multivariate cumulative distribution function (M-CDF). Accordingly, this chapter focuses on techniques for constructing M-PDF models that can be used in a variety of scientific applications, together with a comparative discussion of these models, highlighting similarities and differences as they emerge.

Basically, the M-PDF model construction techniques can be classified into two major groups: the group of *formal* M-PDF model construction techniques and the group of *substantive* model construction techniques. Each one of these techniques has its merits and limitations. But before I continue with the discussion of formal and substantive techniques, it may be instructive to digress for a moment and stress the difference between the formal and the substantive notions.

Spatiotemporal Random Fields. http://dx.doi.org/10.1016/B978-0-12-803012-7.00011-8

Example 1.1

To provide an initial illustration of this difference, consider the simple yet frequently occurring situation in real-world applications that involves the physical condition expressed by the inequality $X_2(\boldsymbol{p}) > X_1(\boldsymbol{p})$ between two related attributes (e.g., rainfall depth and the corresponding critical depth used in "depth–duration–frequency" curves). A standard modeling approach is to write

$$X_2(\boldsymbol{p}) = X_1(\boldsymbol{p}) + \varepsilon,$$

where ε is a positive variable, and then proceed with modeling the joint probability distribution of $X_1(\boldsymbol{p})$ and ε. In such simple examples, the physical bound is rather well defined a priori. Yet, the analysis also points out the basic fact that possibly complex physical relationships could occur in the real-world that need to be incorporated in M-PDF model construction, because they represent substantive knowledge.

Returning to formal M-PDF model construction, the first thing to say is that it includes models that are speculative, analytically tractable, or ready-made. The following are well-known formal techniques:

(a) the *direct* techniques,
(b) the *factora* techniques, and
(c) the *copula* techniques.

More specifically, direct M-PDF techniques focus on suitable transformations or combinations of existing PDF models. A common feature of such M-PDF models is that the corresponding *univariate* PDFs (U-PDFs) are of the same kind (e.g., if the M-PDF is Gaussian, so is the U-PDF). The inverse, however, is not generally true, i.e., a U-PDF (say, Gaussian) may be associated with an M-PDF of a different kind (non-Gaussian). These facts, obviously, can cause serious problems in many applications in which we deal with non-Gaussian fields associated with different kinds of U-PDF models.

In many cases in practice, a recurring problem is how to extend a specified U-PDF model, which is usually available in practice, to an M-PDF that fits the natural attribute of interest. This kind of problem is a prime reason for the systematic development of the copula- and factora-based representations of the M-PDF. In particular, the main idea behind factoras and copulas is to express an unknown function, the M-PDF, in terms of a better-known function (the factora or the copula), and the corresponding U-PDF. The factora concept can be traced back to the Gaussian *tetrachoric series* expansion of Pearson (1901), whereas the copula technology has its origins in the multivariate probability analysis of Sklar (1959). Although factora is apparently an older concept than copula, both concepts share some common features.

Next, I continue with substantive M-PDF model construction, which includes techniques that take into account the contentual and contextual domain of the situation at hand. To this group of techniques belong:

(a) the *Bayesian maximum entropy* (*BME*) techniques, and
(b) the *stochastic differential equation* (*SDE*) techniques.

A BME technique, which is fundamentally different than the factora and the copula techniques, is essentially based on a synthesis of core and in situ knowledge bases, which, subsequently, serves as the input to a set of equations involving various space–time points. The solution of these equations provides the M-PDF model (Christakos, 2000, 2010). The basic idea of the SDE techniques, on the other hand, is that from the SDE governing the S/TRF of interest, the corresponding differential equations of the M-PDF can be derived in several manners, including the classical, the characteristic function (CF), and the functional ones.

In sum, M-PDF model construction techniques are used in various branches of applied sciences, where their choice depends on the specifics of the in situ situation. In the following, the $f_X(\boldsymbol{p}) = f_X(\boldsymbol{p}_1,\ldots,\boldsymbol{p}_k)$ denotes an M-PDF, whereas the $f_X(\boldsymbol{p}_i)$, $i = 1, 2,\ldots$ denotes a set of U-PDFs.

2. DIRECT PROBABILITY DENSITY MODEL CONSTRUCTION TECHNIQUES

As already noted, direct M-PDF construction techniques basically rely on valid combinations or transformations of known models, which are considered as the component PDF models. For instance, based on our understanding of the phenomenon, we may be able to justifiably postulate that the corresponding U-PDF has a known form, and then we use direct techniques to construct a valid M-PDF model in terms of such U-PDFs. Naturally, direct techniques select models with analytically tractable properties. Popular component U-PDF models include the Gaussian (normal), the Student, the exponential, the lognormal, the elliptical, the Cauchy, the beta, the gamma, the logistic, the Liouville, and the Pareto models (for the detailed mathematical expressions of these models, see, e.g., Kotz et al., 2000). A list of well-known direct techniques is presented in Table 2.1. Next, I will discuss its one of these models to some detail.

Table 2.1 Direct Probability Density Function (PDF) Model Construction Techniques

PDF Construction Technique	PDF Form	
Full independency	$\prod_{i=1}^{k} f_X(\boldsymbol{p}_i)$	(2.1)
Partial independency	$\prod_{\text{some } i} f_X(\boldsymbol{p}_i)$	(2.2)
Spherical symmetry	$f_X\left(\left(\sum_{i=1}^{k} \chi_{\boldsymbol{p}_i}^2\right)^{\frac{1}{2}}\right)$	(2.3)
Transformation	$f_X(g_1^{-1},\ldots,g_k^{-1})\lvert J\rvert$	(2.4)

2.1 THE INDEPENDENCY TECHNIQUES

One of the simplest direct techniques assumes *full independency*, in which case the derived M-PDF model $f_X(\boldsymbol{p})$ is the product of the component U-PDFs, see Table 2.1. This model essentially describes phenomena that do not transmit knowledge across space−time, i.e., our knowledge of the attribute's state at point $\boldsymbol{p}_i$ does not affect our knowledge of the state at point $\boldsymbol{p}_j$. Although mathematically convenient, the model of Eq. (2.1) is of rather limited use in real-world situations. Perhaps, a more useful model in practice is that of *partial independency* of Eq. (2.2).

Example 2.1

The M-PDF model

$$f_X(\boldsymbol{p}_1,\boldsymbol{p}_2,\boldsymbol{p}_3) = \prod_{i=1}^{3} f_X(\boldsymbol{p}_i) \tag{2.5}$$

determines a fully independent trivariate PDF (T-PDF) model, whereas the model defined by the pair of conditions

$$\begin{aligned} f_X(\boldsymbol{p}_1,\boldsymbol{p}_2) &= \prod_{i=1}^{2} f_X(\boldsymbol{p}_i), \\ f_X(\boldsymbol{p}_1,\boldsymbol{p}_2,\boldsymbol{p}_3) &\neq \prod_{i=1}^{3} f_X(\boldsymbol{p}_i), \end{aligned} \tag{2.6a–b}$$

is a partially independent PDF model.

2.2 THE SPHERICAL SYMMETRY TECHNIQUE

M-PDF models can be also derived in cases when a specific relationship is known to exist between the random field realizations. In this respect, an interesting model is the *spherically symmetric* M-PDF model of Eq. (2.3), which is an even function that is symmetric with respect to $\chi_{\boldsymbol{p}_i}$, $i = 1, \ldots, k$. All U-PDFs are here the same, $f_X(\boldsymbol{p}_i) = f_0(\chi)$. Then, the class of M-PDFs in Eq. (2.3) is obtained by assuming any U-PDF model f_0 in the integral equation

$$f_0(\chi) = 2\pi^{\frac{k-1}{2}}\Gamma^{-1}\left(\frac{k-1}{2}\right)\int_0^{\infty} d\xi\,\xi^{k-2} f_X\left(\left(\chi^2+\xi^2\right)^{\frac{1}{2}}\right), \tag{2.7}$$

and subsequently inverting the integral equation for f_X, which is the M-PDF model constructed by this technique.

Example 2.2

In the case of stochastic independence, Eq. (2.7) simply yields

$$f_X(\boldsymbol{p}) = f_X(\xi) = \prod_{i=1}^{k} f_X(\boldsymbol{p}_i), \tag{2.8}$$

where $f_X(\boldsymbol{p}_i) = c_0 e^{c_1 \chi_i^2}$ is a Gaussian model with suitable coefficients c_0 and c_1 (i.e., the combination of spherical and independent assumptions produces a Gaussian M-PDF model).

2.3 THE TRANSFORMATION TECHNIQUE

In the *transformation* technique (also listed in Table 2.1), one starts with an S/TRF $X(\boldsymbol{p})$ that has a known M-PDF $f_X(\boldsymbol{p}_1,\ldots,\boldsymbol{p}_k)$, and then uses suitable one-to-one functions g_i, $i = 1,\ldots, k$, to derive a new S/TRF

$$Y(\boldsymbol{p}_i) = g_i(X(\boldsymbol{p}_1), \ldots, X(\boldsymbol{p}_k)) \tag{2.9}$$

(particularly, the g_i functions in Eq. (2.9) may include atmost k elements, depending on the case considered). The M-PDF model is subsequently constructed by means of Eq. (2.4), where

$$|J| = \left| \frac{\partial\left(g_1^{-1}, \ldots, g_k^{-1}\right)}{\partial(Y_1, \ldots, Y_k)} \right| \tag{2.10}$$

is the corresponding Jacobian with

$$X(\boldsymbol{p}_i) = g_i^{-1}(Y(\boldsymbol{p}_1), \ldots, Y(\boldsymbol{p}_k)) \tag{2.11}$$

for $i = 1,\ldots, k$. The transformation technique is considered as the most general and potentially useful direct technique.

Example 2.3

Let $Y(\boldsymbol{p}_i) = g_i(X(\boldsymbol{p}_i)) = F_{Y_i}^{-1}(X(\boldsymbol{p}_i))$, $i = 1, 2$, where the F_{Y_i} have the shapes of known CDFs, i.e., $F_{Y_i} = F_Y(\boldsymbol{p}_i)$, $i = 1, 2$. Also, let us assume that the Gaussian bivariate PDF (B-PDF) $f_X(\boldsymbol{p}_1, \boldsymbol{p}_2)$ is spherically symmetric, i.e.,

$$f_X(\boldsymbol{p}_1, \boldsymbol{p}_2) = f_X\left(\left(\chi_1^2 + \chi_2^2\right)^{\frac{1}{2}}\right) = f_X\left(\left(F_{Y_1}^2 + F_{Y_2}^2\right)^{\frac{1}{2}}\right).$$

By virtue of Example 2.2, this implies that

$$f_X(\boldsymbol{p}_1, \boldsymbol{p}_2) = f_X(\boldsymbol{p}_1) f_X(\boldsymbol{p}_2) = a e^{b\chi_1^2} a e^{b\chi_2^2}.$$

Then, the new B-PDF is given by

$$f_Y(\boldsymbol{p}_1, \boldsymbol{p}_2) = a^2 e^{b\left(\chi_1^2 + \chi_2^2\right)} f_Y(\boldsymbol{p}_1) f_Y(\boldsymbol{p}_2), \tag{2.12}$$

since $X(\boldsymbol{p}_i) = g_i^{-1}(Y(\boldsymbol{p}_i)) = F_Y(\boldsymbol{p}_i)$, $i = 1, 2$, and the $f_Y(\boldsymbol{p}_1) = \dfrac{\partial F_Y(\boldsymbol{p}_1)}{\partial \psi_1}$ and $f_Y(\boldsymbol{p}_2) = \dfrac{\partial F_Y(\boldsymbol{p}_2)}{\partial \psi_2}$ are known U-PDFs.

In some other situations, the M-PDF $f_X(\boldsymbol{p})$ model can be expressed in terms of its U-PDF $f_X(\boldsymbol{p}_i)$ models ($i = 1,\ldots, k$) and a set of functions of $\chi_{\boldsymbol{p}_i}$. This kind of PDF model building could be of considerable interest, because one often has good knowledge of $f_X(\boldsymbol{p}_i)$ and seeks to construct a $f_X(\boldsymbol{p}) = f_X(\boldsymbol{p}_1,\ldots, \boldsymbol{p}_k)$ that is both physically meaningful and its parameters can be estimated accurately in practice. Two noteworthy cases of this PDF model building approach are considered in the following sections: copulas and factoras.

3. FACTORA-BASED PROBABILITY DENSITY MODEL CONSTRUCTION TECHNIQUES

A key question that leads to the development of the copula- and factora-based PDF representations is related to the extension of U-PDFs that are commonly available in practice to valid M-PDFs. The basic argument is as follows: Under certain general conditions in theory, an M-PDF, $f_X(\boldsymbol{p}) = f_X(\boldsymbol{p}_1,\ldots,\boldsymbol{p}_k)$, can be expressed in terms of its U-PDF $f_X(\boldsymbol{p}_i)$ and a multivariate function of $\chi(\boldsymbol{p}_i)$ ($i = 1,\ldots,k$). A basic feature of the derived M-PDF (e.g., Gaussian, lognormal, or gamma) often is that the corresponding U-PDFs are all of the same kind (i.e., Gaussian, lognormal, or gamma, respectively). The inverse, however, is often not true, that is, a U-PDF (say, Gaussian) may be associated with an M-PDF of a different kind (non-Gaussian, in general). These aspects can cause some problems in many in situ applications in which one deals with a non-Gaussian attribute $X(\boldsymbol{p})$ that has different kinds of U-PDF $f_X(\boldsymbol{p}_i)$ (e.g., the $f_X(\boldsymbol{p}_1,\boldsymbol{p}_2)$ is non-Gaussian, whereas the $f_X(\boldsymbol{p}_1)$ is Gaussian and the $f_X(\boldsymbol{p}_2)$ is gamma). In such cases, the critical question is how to properly extend the U-PDF that may be available in the particular application to an M-PDF that fits the probabilistic features of the attribute of interest. This kind of a problem constitutes a prime reason for the development of the factora representations, which derive M-PDF models starting from U-PDF ones.

Pearson's (1901) original insight on tetrachoric models can be extended in a non-Gaussian random field context, leading to the class of factorable S/TRFs, as follows (Christakos, 1986b, 1989). Let $\theta_{X;\boldsymbol{p}} = \theta(\chi_{\boldsymbol{p}})$, $\chi_{\boldsymbol{p}} = \left(\chi_{\boldsymbol{p}_1},\ldots,\chi_{\boldsymbol{p}_k}\right)$, be a multivariate function of $L_2\left(R^k, \prod_{i=1}^{k} f_X(\boldsymbol{p}_i)\right)$ with

$$r_k = \int d\chi_{\boldsymbol{p}} \prod_{i=1}^{k} f_X(\boldsymbol{p}_i)\theta^2\left(\chi_{\boldsymbol{p}}\right) < \infty,$$

in which case one can write

$$f_X(\boldsymbol{p}) = f_X(\boldsymbol{p}_1,\ldots,\boldsymbol{p}_k) = \left[\prod_{i=1}^{k} f_X(\boldsymbol{p}_i)\right]\theta\left(\chi_{\boldsymbol{p}}\right). \tag{3.1}$$

And, let $\varpi_{j_i}\left(\chi_{\boldsymbol{p}_i}\right)$ be complete sets of polynomials of degrees $j_i = 0, 1,\ldots$ ($i = 1,\ldots,k$) in $L_2(R^1, f_X(\boldsymbol{p}_i))$ that are orthonormal with respect to $f_X(\boldsymbol{p}_i)$.

Definition 3.1

The *factora* function is defined as

$$\theta\left(\chi_{\boldsymbol{p}}\right) = \theta_{X;\{\boldsymbol{p}_i\}} = \left[\prod_{i=1}^{k}\sum_{j_i=0}^{\infty}\right]\left(\theta_{j_1\ldots j_k}\prod_{i=1}^{k}\varpi_{j_i}\left(\chi_{\boldsymbol{p}_i}\right)\right), \tag{3.2}$$

where the coefficients $\theta_{j_1\ldots j_k}$ satisfy the corresponding completeness relationship

$$\left[\prod_{i=1}^{k}\sum_{j_i=0}^{\infty}\right]\theta^2_{j_1\ldots j_k} = r_k,$$

which assures that the series expansions converges, and the M-PDF of Eq. (3.1) is written as

$$f_X(\boldsymbol{p}) = \left[\prod_{i=1}^{k} f_X(\boldsymbol{p}_i)\right]\theta_{X;\{\boldsymbol{p}_i\}}. \tag{3.3}$$

A random field $X(\boldsymbol{p})$ that satisfies Eq. (3.3) is called a *factorable* of order k random field (Christakos, 1992). Essentially, Eq. (3.3) decomposes the modeling of the M-PDF, $f_X(\boldsymbol{p})$, into a product of U-PDFs $f_X(\boldsymbol{p}_i)$ and factoras $\left(\theta_{X;\boldsymbol{p}} = \theta_{X;\{\boldsymbol{p}_i\}}\right)$ that express interactions between univariate functions of $\chi_{\boldsymbol{p}_i}$.[1] Further, the factoras may offer a measure of the deviation of the M-PDF model from the product of U-PDF models. If $\theta_{X;\{\boldsymbol{p}_i\}} \equiv 1$, then $f_X(\boldsymbol{p}) = \left[\prod_{i=1}^{k} f_X(\boldsymbol{p}_i)\right]$. Departure from 1 can be seen as deviation from independence. In fact, Eq. (3.3) can be written as

$$\theta_{X;\{\boldsymbol{p}_i\}} \equiv f_X(\boldsymbol{p})\left[\prod_{i=1}^{k} f_X(\boldsymbol{p}_i)\right]^{-1},$$

and, hence,

$$\overline{[\log \theta_{X;\{\boldsymbol{p}_i\}}]} = \overline{\log\left\{f_X(\boldsymbol{p})\left[\prod_{i=1}^{k} f_X(\boldsymbol{p}_i)\right]^{-1}\right\}},$$

where the expectation is with respect to $f_X(\boldsymbol{p})$. This can be interpreted as "mutual information" between the $\chi_{\boldsymbol{p}_i}$'s (a k-dimensional extension of the usual concept of mutual information between two random variables), a measure that quantifies departure from independence (Christakos et al., 2011).

Remark 3.1

Similar quantities can be obtained in terms of *Tsallis* mutual information–dependent version (Tsallis, 1998), by replacing "log" with the corresponding approximation used by Tsallis entropy.

Certain variations of the basic Eqs. (3.1)–(3.3) have been considered in the relevant literature, as discussed next.

Example 3.1

Kotz and Seeger (1991) considered Eq. (3.1) for $k = 2$ and replaced the symbol $\theta_{X;\boldsymbol{p}_1,\boldsymbol{p}_2}$ with the symbol $\psi_{X;\boldsymbol{p}_1,\boldsymbol{p}_2}$, which was termed the density weighting function (DWF). The $\theta_{X;\boldsymbol{p}_1,\boldsymbol{p}_2}$ (or $\psi_{X;\boldsymbol{p}_1,\boldsymbol{p}_2}$) can be determined in several ways, one of which is in terms of Eq. (3.2). In the bivariate case, Eq. (3.3) can be reduced to the B-PDF expansion

$$f_X(\boldsymbol{p}_1,\boldsymbol{p}_2) = f_X(\boldsymbol{p}_1)f_X(\boldsymbol{p}_2)\sum_{j=0}^{\infty}\theta_j\varpi_j(\chi_{\boldsymbol{p}_1})\varpi_j(\chi_{\boldsymbol{p}_2}), \tag{3.4}$$

[1]This is also an advantage of the way factoras are defined over that of copulas (Section 4).

for all $\boldsymbol{p}_1, \boldsymbol{p}_2$. In this case,

$$\theta_{X;\boldsymbol{p}_1,\boldsymbol{p}_2} = \sum_{j=0}^{\infty} \theta_j \varpi_j(\chi_{\boldsymbol{p}_1}) \varpi_j(\chi_{\boldsymbol{p}_2});$$

e.g., $\theta_0 = 1$, $\theta_1 = \rho_{X;\boldsymbol{p}_1,\boldsymbol{p}_2}$ (correlation coefficient), and $\theta_j \delta_{jj'} = \overline{\varpi_j(\chi_{\boldsymbol{p}_1}) \varpi_{j'}(\chi_{\boldsymbol{p}_2})}$, with $\varpi_0(\chi_{\boldsymbol{p}_i}) = 1$ and $\varpi_1(\chi_{\boldsymbol{p}_i}) = \left(\chi_{\boldsymbol{p}_i} - \overline{X_{\boldsymbol{p}_i}}\right) \sigma_{\boldsymbol{p}_i}^{-1}$ for all space−time points. In Eq. (3.4) the marginals have the same form, and the same set of polynomials (and corresponding spaces) for $i = 1$ and $i = 2$ are considered. This equation is simpler than Eq. (3.2) particularized for $k = 2$, since cross-terms (involving different polynomials) are discarded, that is, $\theta_{j_1 j_2}$ is assumed to be 0 for $j_1 \neq j_2$, and renamed as θ_j for $j_1 = j_2$.

Remark 3.2

It is interesting that in Eq. (3.4) knowledge of lower-order statistics is linked to the first terms of the series, whereas that of higher-order statistics is linked to later terms of the series. Using first-order (Hermite, Laguerre, etc.) polynomial expansions, Eq. (3.4) yields

$$f_X(\boldsymbol{p}_1, \boldsymbol{p}_2) = f_X(\boldsymbol{p}_1) f_X(\boldsymbol{p}_2) \left[1 + \theta_1 \ \chi_{\boldsymbol{p}_1} \chi_{\boldsymbol{p}_2}\right],$$

which is also what one obtains by performing a Taylor expansion of a B-PDF around a fixed value of $\theta_1 = \theta_1^*$. An example is given in Sungur (1990). The approximation error depends on θ_1. Since usually $\theta_1^* = 0$, the approximation is satisfactory around the independence case and worsens as dependence increases.

Another technique proposed in the literature is that to build $\psi_{X;\boldsymbol{p}_1,\boldsymbol{p}_2}$ (essentially, $\theta_{X;\boldsymbol{p}_1,\boldsymbol{p}_2}$), one can use the following DWF representation (Long and Krzysztofowicz, 1995)

$$\theta_X(\boldsymbol{p}_1, \boldsymbol{p}_2) = \psi_X(\boldsymbol{p}_1, \boldsymbol{p}_2) = 1 + \lambda \omega \left[F_X(\chi_{\boldsymbol{p}_1}), F_X(\chi_{\boldsymbol{p}_2})\right], \tag{3.5}$$

where ω is called the covariance characteristic and λ the covariance scaler. Using Eq. (3.4), it can be expressed in terms of the expansion

$$\lambda \omega = \sum_{j=1}^{\infty} \theta_j \varpi_j(\chi_{\boldsymbol{p}_1}) \varpi_j(\chi_{\boldsymbol{p}_2}).$$

Other authors (de la Peña et al., 2006) have extended the work of Kotz and Seeger (1991) and Long and Krzysztofowicz (1995) in a multivariate setting so that Eq. (3.5) can be generalized as

$$\theta_{X;\{\boldsymbol{p}_i\}} = \left[1 + U_{X;\{\boldsymbol{p}_i\}}\right], \tag{3.6}$$

where

$$U_{X;\{\boldsymbol{p}_i\}} = \sum_{c=2}^{k} \sum_{1 \leq i_1 < \ldots < i_c \leq i_k} s_{i_1 \ldots i_c}(\chi_{\boldsymbol{p}_1} \cdots \chi_{\boldsymbol{p}_c}),$$

and the $s_{i_1 \ldots i_c}$ satisfy three general conditions (integrability, degeneracy, and positive definiteness).

A key step in Eq. (3.4) is to calculate ϖ_j that are orthogonal with respect to a U-PDF. There exist several methods for this purpose, where the ϖ_j include Hermite, Laguerre, Generalized Laguerre,

Legendre, Gegenbauer, Jacobi, and Stieltjes—Wigert polynomials (e.g., the Gaussian, gamma, or Poisson U-PDF is associated with Hermite, Laguerre, or Charlier polynomials). In Eq. (3.4), one needs to define factoras $\theta_{X;\{p_i\}}$ with the prescribed mathematical properties and associated complete sets of orthogonal polynomials (the difficulty increases with $k > 2$). For this purpose, a widely applicable method is based on the formula

$$\varpi_j(\chi_{p_i}) = f_X^{-1}(p_i)\frac{d^j}{d\chi_{p_i}^j}\left[\upsilon(\chi_{p_i})^j f_X(p_i)\right],$$

where $\upsilon(\chi_{p_i})$ is a function that satisfies specific conditions (Christakos, 1986b, 1992). This formula has been used to find $\varpi_j(\chi)$ for a wide range of continuous functions $f_X(\chi)$, including the Gaussian, exponential, and Pearson (Type I).

Example 3.2

If we consider the U-PDF models specified by

$$f_X(p_i) = \frac{1}{\sqrt{2\pi}}e^{\frac{1}{2}\chi_{p_i}^2}$$

$\left(-\infty \leq \chi_{p_i} \leq \infty\right)$, the $\varpi_j = H_{a(j)}$ are Hermite polynomials, and the bivariate factora is

$$\theta_X(p_1,p_2) = \sum_{j=0}^{\infty}\rho_X^{a(j)}H_{a(j)}(\chi_{p_1})H_{a(j)}(\chi_{p_2}),$$

where $\theta_j = \rho_X^{a(j)}$ and ρ_X is a correlation coefficient. For $a(j) = j$, the $f_X(p_1,p_2)$ is a Gaussian B-PDF; but for $a(j) = 2j$, it is non-Gaussian (Christakos, 1992). This is not surprising, since to a given U-PDF one may associate more than one B-PDFs. Many other examples can be found in the cited literature.

The next proposition has been shown to be particularly useful in many cases of PDF model construction.

Proposition 3.1

If $X(\boldsymbol{p})$ is a factorable S/TRF, and $\phi(\cdot)$ is a strictly monotonic function, the random field $Z(\boldsymbol{p}) = \phi(X(\boldsymbol{p}))$ is also factorable. The corresponding B-PDF is given by

$$f_Z(p_1,p_2) = f_Z(p_1)f_Z(p_2)\sum_{j=0}^{\infty}\theta_j\varpi_j\left[\phi^{-1}(\zeta_{p_1})\right]\varpi_j\left[\phi^{-1}(\zeta_{p_2})\right]. \tag{3.7}$$

Proposition 3.1 implies that starting from known classes of factorable random fields $X(\boldsymbol{p})$, new classes $Z(\boldsymbol{p})$ can be constructed using different kinds of functions ϕ. The structure of the factora is preserved under strictly monotonic transformations (as in the copula framework to be discussed later). In Eq. (3.7), the values of the parameters θ_j change, since they have to be computed on the transformed S/TRF. Another interesting property of the factora model is discussed in the following proposition.

Proposition 3.2

The factora model satisfies the relationship

$$\int d\chi_{p_1}\left(\chi_{p_1}-\overline{X(p_1)}\right)f_X(p_1,p_2)=c_X(p_1,p_2)\sigma_{X;p_2}^{-2}\left(\chi_{p_2}-\overline{X(p_2)}\right)f_X(p_2), \tag{3.8}$$

for all p_1, p_2.[2]

Example 3.3

In the special case that $\overline{X(p_i)}=\overline{X}=\mu$ and $f_X(p_i)=f_X$ (for all p_i), Eq. (3.8) reduces to a rather more tractable form,

$$\int d\chi_{p_1}(\chi_{p_1}-\mu)f_X(p_1,p_2)=\rho_X(\Delta p)(\chi_{p_2}-\mu)f_X. \tag{3.9}$$

where $\Delta p=(h,\tau)$. A direct consequence of Eq. (3.9) is,

$$\overline{X(p_1)X^m(p_2)}=\rho_X(\Delta p)\overline{X^{m+1}(p_2)}-\mu(\rho_X(\Delta p)-1)\overline{X^m(p_2)}; \tag{3.10}$$

i.e., a higher-order, two-point dependence is conveniently expressed in terms of one-point functions.

Yet another interesting property of the factoras is that they generate estimators of nonlinear state–nonlinear measurement systems that can be superior to those of the Kalman filter (Christakos, 1989, 1992). For example, the Kalman filter estimates include only linear correlations, whereas the factora estimates include linear and nonlinear correlations; also, the Kalman filter is limited to the estimation of lower-order moments (mean and variance), whereas the factora estimator can provide lower- and higher-order moments.

Remark 3.3

A special case of factora analysis is the so-called disjunctive kriging technique (Matheron, 1976). Note that this technique involves orthogonal polynomials of the random field rather than the original field itself (Christakos, 1992).

The implementation of the factora-based PDF model construction technique relies on our knowing the U-PDFs at different locations and times across the entire study area. Since this kind of information may not be available in many applications, the U-PDFs are assumed to be Gaussian, which may be unrealistic in some cases. Also, factoras involve infinite series that have to be truncated. Lastly, as regards the numerical implementation of the factora technique, sometimes the choice is made to use indicator kriging, which could cause some problems (see Example 7.1 later).

4. COPULA-BASED PROBABILITY DENSITY MODEL CONSTRUCTION TECHNIQUES

The starting point of copulas is that a continuous M-CDF, $F_{X;\{p_i\}}=F_X(p_1,\ldots,p_k)$, can be generally written in terms of a special kind of functions, which are defined as follows (Sklar, 1959; Genest and Rivest, 1993; Nelsen, 1999).

[2]As a matter of fact, Eq. (3.8) is valid for random field classes other than the factorable ones (Christakos, 2010).

Definition 4.1

The multivariate *copula* function is given by

$$\begin{aligned} F_{X;\{p_i\}} &= C_X(F_X(p_1),\ldots,F_X(p_k)) = P_X\left[F_X(p_1)\le \upsilon_{p_1},\ldots,F_X(p_k)\le \upsilon_{p_k}\right] \\ &= C_{X;\{p_i\}}\left(\upsilon_{p_1},\ldots,\upsilon_{p_k}\right), \end{aligned} \tag{4.1}$$

where $F_X(\boldsymbol{p}_i)$ are U-CDFs, and $\upsilon_{\boldsymbol{p}_i}$ are realizations of uniformly distributed random fields

$$U_{p_i} = F_X^{-1}(p_i) \sim U(0,1),$$
$$i = 1,\ldots,k.$$

In a few words, a copula is an M-CDF with uniform marginals. As a matter of fact, any M-CDF with support on $[0,1]^k$ and uniform marginals is considered a copula (Mikosch, 2006a,b). The corresponding multivariate *copula density* $\varsigma_{X;\{p_i\}}$ is defined by

$$\varsigma_{X;\{p_i\}} d\upsilon = dC_{X;\{p_i\}}$$

(assuming copula continuity and differentiability).

Proposition 4.1

Since continuous functions are assumed, the M-PDF $f_X(\boldsymbol{p})$ can be reformulated in terms of its U-PDFs and the multivariate copula density as

$$f_X(\boldsymbol{p}) = f_{X;\{p_i\}} = \left[\prod_{i=1}^{k} f_X(\boldsymbol{p}_i)\right] \varsigma_{X;\{p_i\}}. \tag{4.2}$$

Eq. (4.2) basically decomposes the M-PDF into the product of the U-PDFs and the multivariate copula density that expresses a certain form of interaction between the U-PDFs. It has been argued that this decomposition may offer some modeling flexibility. Indeed, several assumptions about the shape of the copula density can be made, including the elliptic, the Archimedian, the Marshall–Olkin, the Pareto, the Gaussian, and the *t*-copulas. In other words, assumptions concerning the shape of the M-PDF have been replaced by assumptions concerning the shapes of the U-PDFs and the copula density. Proponents of the copula technique seem to claim that a copula density can be assumed in a more rigorous and realistic manner than an M-PDF, although critics have argued that there is no logical reason for choosing one copula over the other. Instead, one makes such a modeling decision based purely on mathematical convenience. Also, it must be kept in mind that a suitable copula should be chosen as well as the corresponding U-PDF. In other words, assuming a priori a Gaussian copula is like assuming Gaussian marginals without any theoretical reason or empirical evidence.

As is the case with all mathematical tools, the copulas have their pros and cons. On modeling grounds, copula is a tool to develop non-Gaussian M-PDFs that are suitable for some applications, but not for some others (Joe, 2006). Eq. (4.2) makes it possible to separate modeling multivariate-dependent models into two parts: fitting unidimensional marginal distributions and fitting joint dependence across marginal CDFs. However, Eq. (4.2) may increase calibration errors due to the extra step in estimation. Under certain conditions, copulas yield useful parametric descriptions of non-Gaussian M-PDF (Scholzel and Friederichs, 2008). The copula approach has the flexibility of modeling a multivariate distribution given different marginal behaviors. This feature may help avoid misspecification of the marginal distributions and focus directly on the dependence structure.

Proposition 4.2

Copulas are scale invariant in the sense that the copula of $Z(\boldsymbol{p}) = \phi(X(\boldsymbol{p}))$ is equal to the copula of $X(\boldsymbol{p})$ if $\phi(\cdot)$ is a strictly monotonic function.

The copula property described in Proposition 4.2 is particularly relevant to financial research. Since copulas are simply joint probability distributions with uniform marginals, the representations of Eqs. (4.1)−(4.2) can be also applied to them. On the other hand, we should keep in mind that the copula approach mainly applies to continuous-valued attributes so that the marginals are uniform according to the so-called probability integral transform theorem. No general approach exists to construct the most appropriate copula for a random field, whereas the choice of a copula family for an in situ problem is often based not on substantive reasoning, but on mathematical convenience (Mikosch, 2006a,b). If construction methods are available for component-wise maxima, not unique approaches can be established for a set of attributes that are not all extremes. This is also the case of univariate analysis, where distribution functions are usually chosen on the basis of theoretical observations and goodness-of-fit criteria. Direct interpretation of the copula alone does not offer insight about the complete stochastic nature of the random field, and there is no dependence separately from the marginals. In addition, copulas do not solve satisfactorily the dimensionality problem (Scholzel and Friederichs, 2008).

Interpretive issues concerning the in situ applications of the copulas may emerge too. There are many real-world attributes that are not continuous valued but rather discrete valued or mixed valued (e.g., daily rainfall), which means that the integral transform theorem (on which the copula technology of continuous variables relies) cannot be implemented, since the $F_X(\boldsymbol{p}_i)$ are no longer uniformly distributed on the interval (0,1), thus giving rise to so-called unidentifiability issues (Genest and Nešlehová, 2007). In this respect, although copulas can be used in simulation and robustness studies, they have to be implemented with caution because some properties do not hold in the discrete case. Attracted by the possibility to select arbitrary marginals, one sometimes forgets that a suitable copula should be chosen as well as marginals. Lastly, certain critics argue that, as defined, copulas are "static"-dependent measures; that is, they are only snapshots of the components' marginal distributions at a specific time. It may be then futile to try to obtain the marginal distributions of dependent time series models, so copulas may fail in time-dependent structures.

Copulas can be directly related to factoras. Let us consider a random field $X(\boldsymbol{p})$ having an absolutely continuous B-CDF $F_X(\boldsymbol{p}_1, \boldsymbol{p}_2)$ and the corresponding B-PDF is given by $f_X(\boldsymbol{p}_1,\boldsymbol{p}_2) = f_X(\boldsymbol{p}_1)f_X(\boldsymbol{p}_2)\theta(\chi_{\boldsymbol{p}_1}, \chi_{\boldsymbol{p}_2})$. The

$$F_Z(\zeta_{\boldsymbol{p}_1}, \zeta_{\boldsymbol{p}_2}) = F_X\left(F_X^{-1}(F_Z(\zeta_{\boldsymbol{p}_1})), F_X^{-1}(F_Z(\zeta_{\boldsymbol{p}_2}))\right), \tag{4.3}$$

is a B-CDF of $(\zeta_{\boldsymbol{p}_1}, \zeta_{\boldsymbol{p}_2})$ with marginals (U-CDF) $F_Z(\boldsymbol{p}_1)$ and $F_Z(\boldsymbol{p}_2)$. Under these conditions, the following proposition holds (Christakos et al., 2011).

Proposition 4.3
Let the CDFs $F_X(\boldsymbol{p}_1)$, $F_X(\boldsymbol{p}_2)$, $F_Z(\boldsymbol{p}_1)$, and $F_Z(\boldsymbol{p}_2)$ be absolutely continuous functions with U-PDFs $f_X(\boldsymbol{p}_1)$, $f_X(\boldsymbol{p}_2)$, $f_Z(\boldsymbol{p}_1)$, and $f_Z(\boldsymbol{p}_2)$. Then, the M-PDF of the random field $Z(\boldsymbol{p})$ is given by

$$f_Z(\zeta_{p_1},\zeta_{p_2}) = f_Z(\zeta_{p_1})f_Z(\zeta_{p_2})\theta\left(F_X^{-1}(F_Z(\zeta_{p_1})),F_X^{-1}(F_Z(\zeta_{p_2}))\right). \tag{4.4}$$

Proposition 4.3 implies that if we assume that $F_X(\boldsymbol{p}_1,\boldsymbol{p}_2)$ is a B-CDF with marginals $F_X(\boldsymbol{p}_1)$ and $F_X(\boldsymbol{p}_2)$ and factora $\theta(\chi_{p_1},\chi_{p_2})$, then the $\theta(F_X^{-1}(\upsilon_{p_1}),F_X^{-1}(\upsilon_{p_2}))$ is also a B-PDF in $[0,1]^2$ with uniform U-PDFs, i.e., a bivariate copula function $\varsigma(F_X^{-1}(\upsilon_{p_1}),F_X^{-1}(\upsilon_{p_2}))$.

Example 4.1
Consider a B-PDF $F_Z(\boldsymbol{p}_1,\boldsymbol{p}_2)$ with arbitrary marginals $F_Z(\boldsymbol{p}_1)$ and $F_Z(\boldsymbol{p}_2)$. One can apply a suitable strictly monotonic function ϕ such that $\zeta_{p_i} = \phi(\chi_{p_i})$. To ensure that the transformed attributes follow the desired marginals, the simplest method may be to set up $\phi(\cdot) = F_Z^{-1}(F_X(\cdot))$ (under the hypothesis of continuous functions). Then, Eq. (4.4) leads to

$$f_Z(\boldsymbol{p}_1,\boldsymbol{p}_2) = f_Z(\boldsymbol{p}_1)f_Z(\boldsymbol{p}_2)\sum_{j=0}^{\infty}\theta_j\varpi_j\left[F_X^{-1}(F_Z(\zeta_{p_1}))\right]\varpi_j\left[F_X^{-1}(F_Z(\zeta_{p_2}))\right], \tag{4.5}$$

where $\zeta_{p_l} = \phi(\chi_{p_l}) = F_Z^{-1}(F_X(\chi_{p_l}))$, $l = 1,2$, imply $\chi_{p_l} = \phi^{-1}(\zeta_{p_l})$. And, the combination of the two yields $\phi^{-1}(\zeta_{p_l}) = F_X^{-1}(F_Z(\zeta_{p_l}))$.

To phrase it in more words, ϕ, F_X, and F_Z are linked: If we use an arbitrary function ϕ without knowing the corresponding F_Z, it is not possible to build a B-PDF with known marginals, whereas if we seek a B-PDF with specified marginals F_Z, this becomes possible by letting $\phi(\cdot) = F_Z^{-1}(F_X(\cdot))$.

Example 4.2
On the basis of the above analysis it follows that the

$$\sum_{j=0}^{\infty}\theta_j\varpi_j\left[F_X^{-1}(\upsilon_{p_1})\right]\varpi_j\left[F_X^{-1}(\upsilon_{p_2})\right]$$

is a B-PDF with uniform U-PDF in $[0,1]^2$, namely, a copula density.

Lastly, by means of factoras we can build B-PDF with equal nonuniform marginals and apply monotone transformations to change the marginals. However, in doing so we pass implicitly through copula representation.

5. STOCHASTIC DIFFERENTIAL EQUATION–BASED PROBABILITY DENSITY MODEL CONSTRUCTION TECHNIQUES

Experience shows that the consequences of using the wrong PDF in real-world applications can have severe consequences, which is why substantive modeling incorporates in the PDF model construction as much physical knowledge as possible. As a result, the substantive approach of M-PDF model construction adopts a science-based viewpoint, in which a prime source of core physical knowledge is provided by natural laws and scientific theories. The incorporation of such core knowledge in the derivation of the M-PDF models constitutes a definite advantage of substantive model construction (e.g., prior probability problems of the so-called objective and subjective Bayesian analyses are avoided). In this section, we study substantive PDF model construction based on SDE techniques, ordinary (SODE) and partial (SPDE). In Section 6 that follows substantive PDF model construction based on BME techniques will be discussed.

To start with, I assume that a vector S/TRF $\boldsymbol{X}(\boldsymbol{p}) = [X_1(\boldsymbol{p})\ldots X_k(\boldsymbol{p})]^T$ represents a set of natural attributes whose variations in space and time are governed by an SDE generally expressed as,

$$\mathcal{L}_X[\boldsymbol{X}, \boldsymbol{w}; \boldsymbol{p}] = 0, \tag{5.1a}$$

with boundary and initial condition (BIC), $\boldsymbol{X}(\boldsymbol{p}_0) = [X_1(\boldsymbol{p}_0)\ldots X_k(\boldsymbol{p}_0)]^T$, where $\boldsymbol{w}(\boldsymbol{p}) = [w_1(\boldsymbol{p})\ldots w_m(\boldsymbol{p})]^T$ are parameters with known probability distributions or summary statistics. Then, from Eq. (5.1a) the partial differential equation (PDE) of the corresponding PDF

$$\mathcal{L}_f[f_X, f_w; \boldsymbol{p}] = 0 \tag{5.1b}$$

can be, in principle, derived with known BIC, $f_{X_0}(\chi_0) = f_{X_0}$, and with $f_w(\omega) = f_w$. Eq. (5.1b) describes the manner in which the PDF f_X changes in space and time, and, simultaneously, it introduces a method of PDF model construction.

Remark 5.1

In some cases, it may be easier, analytically, instead of the PDF Eq. (5.1b) to derive from Eq. (5.1a) the differential equation

$$\mathcal{L}_\Phi[\Phi_X, \Phi_w; \boldsymbol{p}] = 0, \tag{5.2}$$

satisfied by the corresponding characteristic functionals (CFLs) Φ_X and Φ_w.[3] After the Φ_X has been obtained in terms of Eq. (5.2), the associated PDF model f_X is readily available, since (f_X, Φ_X) constitute an FT pair.

What remains to be discussed is how the PDF Eq. (5.1b) or the CFL Eq. (5.2) is derived from the random field SDE of Eq. (5.1a). In this respect, three distinct approaches will be presented next, namely, the transformation of variables approach, the CF approach, and the functional approach.

5.1 THE TRANSFORMATION OF VARIABLES APPROACH

This is a rather straightforward approach, the basic feature of which is that it assumes that an explicit solution of the SDE of Eq. (5.1a) exists in the m.s. sense, say,

$$\boldsymbol{X}(\boldsymbol{p}) = \mathcal{S}[\boldsymbol{w}; \boldsymbol{p}], \tag{5.3}$$

where $X_l(\boldsymbol{p}) = \mathcal{S}_l(w_1, \ldots, w_m)$, $w_q(\boldsymbol{p}) = \mathcal{S}_q^{-1}(X_1, \ldots, X_k)$, $l = 1, \ldots, k$, $q = 1, \ldots, m$. Then, a suitable transformation of variables is used to obtain the PDF model of $\boldsymbol{X}(\boldsymbol{p})$, as follows,

$$f_X(\chi_1, \ldots, \chi_k) = f_w(w_q)|\boldsymbol{J}| \tag{5.4}$$

for $q = 1, \ldots, m$,

[3]For more information about the CFL notion and its basic features, see Chapter XII.

$$|\boldsymbol{J}| = \left|\frac{\partial(\mathcal{S}^{-1})}{\partial(\boldsymbol{\chi})}\right| = \left|\frac{\partial(\mathcal{S}_1^{-1},\ldots,\mathcal{S}_k^{-1})}{\partial(\chi_1,\ldots,\chi_k)}\right| = \begin{vmatrix} \frac{\partial \mathcal{S}_1^{-1}}{\partial \chi_1} \cdots \frac{\partial \mathcal{S}_1^{-1}}{\partial \chi_k} \\ \vdots \\ \frac{\partial \mathcal{S}_k^{-1}}{\partial \chi_1} \cdots \frac{\partial \mathcal{S}_k^{-1}}{\partial \chi_k} \end{vmatrix} \tag{5.5}$$

is the Jacobian of the transformation (for simplicity, above we let $\chi_{p_i} = \chi_i$, $i = 1,\ldots, k$).

To obtain some more specific results, I suppose that the vector S/TRF $\boldsymbol{X}(\boldsymbol{p})$ obeys the SPDE

$$\frac{\partial \boldsymbol{X}(\boldsymbol{p})}{\partial s_l} = \mathcal{L}_X[\boldsymbol{X}(\boldsymbol{p}),\boldsymbol{p}], \tag{5.6}$$

$l = 1,\ldots, k$, with known BIC, $\boldsymbol{X}(\boldsymbol{p}_0) = \boldsymbol{X}_0 \sim f_X(\boldsymbol{\chi},\boldsymbol{p}_0)$. Assume that the solution of the SPDE of Eq. (5.6) is

$$\boldsymbol{X}(\boldsymbol{p}) = \mathcal{S}[\boldsymbol{X}_0,\boldsymbol{p}_0,\boldsymbol{p}], \tag{5.7}$$

and its inverse is given by

$$\boldsymbol{X}_0 = \mathcal{S}^{-1}[\boldsymbol{X}(\boldsymbol{p}),\boldsymbol{p}]. \tag{5.8}$$

Then, in light of Eq. (5.4) the M-PDF model of $\boldsymbol{X}(\boldsymbol{p})$ is constructed by means of

$$f_X(\boldsymbol{\chi},\boldsymbol{p}) = f_0(\boldsymbol{\chi}_0,\boldsymbol{p})|\boldsymbol{J}|, \tag{5.9}$$

where $\boldsymbol{\chi}_0 = \mathcal{S}^{-1}[\boldsymbol{\chi}]$.

I continue our discussion with some simple yet instructive examples, which are special cases of Eqs. (5.3)–(5.4) with $k = m = 1$.

Example 5.1

In real-world applications, a random process $X(t)$ is obtained if the variability of an S/TRF $X(s,t)$ is due mainly to temporal behavior, or if the spatial fluctuations are eliminated by averaging (in such cases, the space argument is suppressed, for simplicity). Assume further that the $X(t)$ obeys the following SODE

$$\frac{dX(t)}{dt} = aX(t), \tag{5.10}$$

with a known deterministic coefficient a and a random initial condition $X(t_0) = X_0$ obeying the PDF law

$$f_X(\chi, t_0) = f_0 = e^{\sum_{i=0}^{2} \nu_i \chi^i}, \tag{5.11}$$

where $\nu_0 = -\frac{1}{2}\left(\overline{X_0}^2 \sigma_0^{-2} + \ln\left(2\,\pi\sigma_0^2\right)\right)$, $\nu_1 = \overline{X_0}\sigma_0^{-2}$, and $\nu_2 = -\frac{1}{2}\sigma_0^{-2}$, with $\overline{X_0} = \overline{X(t_0)}$ and $\sigma_0^2 = \sigma_X^2(t_0)$ denoting, respectively, the random process mean and variance at the time origin $t = t_0$. Following the transformation of variables approach, the well-known classical solution of the SODE of Eq. (5.10) is

$$X(t) = X_0 e^{at} = \mathcal{S}(X_0), \tag{5.12}$$

in which case its inverse is given by

$$X_0 = \mathcal{S}^{-1}(X) = X(t)e^{-at}. \tag{5.13}$$

Accordingly,

$$\frac{\partial \mathcal{S}^{-1}}{\partial X} = e^{-at}, \tag{5.14}$$

and the following PDF model is constructed

$$f_X(\chi, t) = f_0(\chi_0 = \chi e^{-at})e^{-at} = \frac{1}{\sqrt{2\pi}\sigma_0} e^{-\sigma_0^{-2}\left(\frac{1}{2}\overline{X_0}^2 - \overline{X_0}e^{-at}\chi + \frac{1}{2}e^{-2at}\chi^2\right) - at}, \tag{5.15}$$

which is a result to be used in subsequent examples, for comparison purposes.

The next example deals with a situation where the objective of the physical study is the time variation of the attribute along a specified direction s_i in space (that is, with the exception of s_i, all other coordinates are assumed constant).

Example 5.2

Suppose that the attribute of interest is represented by a unidirectional space–time random field $X(s_i,t)$. Assume further that this direction-dependent random field obeys the law

$$\frac{\partial X(s_i, t)}{\partial t} = w(s_i, t), \tag{5.16}$$

where $w(s_i,t) = a(s_i)t$, and $a(s_i) \sim f_w$. Following the transformation of variables approach above, the corresponding solution

$$X(s_i, t) = \frac{t}{2} w(s_i, t) = \mathcal{S}(w), \tag{5.17}$$

is first obtained, with its inverse given by

$$w(s_i, t) = \mathcal{S}^{-1}(X) = \frac{2}{t} X(s_i, t). \tag{5.18}$$

Accordingly,

$$\frac{\partial \mathcal{S}^{-1}}{\partial X} = \frac{2}{t}, \tag{5.19}$$

and the PDF model

$$f_X(\chi, t) = f_w\left(\omega = \frac{\chi}{2t}\right)\frac{2}{t} \tag{5.20}$$

is constructed, as desired.

Example 5.3

In the case of Eqs. (5.3)–(5.4) let $k = m = 2$, i.e., two random fields, $X_1(\boldsymbol{p})$ and $X_2(\boldsymbol{p})$, are now considered with their B-PDF given by

$$f_X(\chi_1, \chi_2) = f_w\left(w_1 = \mathcal{S}_1^{-1}(\chi_1, \chi_2), w_2 = \mathcal{S}_2^{-1}(\chi_1, \chi_2)\right) \begin{vmatrix} \frac{\partial \mathcal{S}_1^{-1}}{\partial \chi_1} & \frac{\partial \mathcal{S}_1^{-1}}{\partial \chi_2} \\ \frac{\partial \mathcal{S}_2^{-1}}{\partial \chi_1} & \frac{\partial \mathcal{S}_2^{-1}}{\partial \chi_2} \end{vmatrix} \tag{5.21}$$

$$= f_w\left(\mathcal{S}_1^{-1}, \mathcal{S}_2^{-1}\right)\left[\frac{\partial \mathcal{S}_1^{-1}}{\partial \chi_1}\frac{\partial \mathcal{S}_2^{-1}}{\partial \chi_2} - \frac{\partial \mathcal{S}_1^{-1}}{\partial \chi_2}\frac{\partial \mathcal{S}_2^{-1}}{\partial \chi_1}\right].$$

Next, assume that the given random fields $w_1(\boldsymbol{p})$ and $w_2(\boldsymbol{p})$ are spherically symmetric, i.e.,

$$f_w(\omega_1, \omega_2) = f_w\left(\left(\omega_1^2 + \omega_2^2\right)^{\frac{1}{2}}\right), \tag{5.22}$$

and the B-PDF is Gaussian, in which case the random fields $w_1(\boldsymbol{p})$ and $w_2(\boldsymbol{p})$ are independent, i.e.,

$$f_w(\omega_1, \omega_2) = f_{w_1}(\omega_1) f_{w_2}(\omega_2) = a^2 e^{b\left(\omega_1^2 + \omega_2^2\right)}.$$

Also, let us assume that the above random fields are expressed as $w_i = F_{X_i}(X_i), i = 1, 2$ (F_{X_i} are CDFs), which implies that $\mathcal{S}_i^{-1} = F_{X_i}$, and

$$X_i(\boldsymbol{p}) = \mathcal{S}_i(w_i) = F_{X_i}^{-1}(w_i). \tag{5.23}$$

Therefore, the B-PDF of the random fields $X_1(\boldsymbol{s})$ and $X_2(\boldsymbol{s})$ is given by

$$f_X(\chi_1, \chi_2, \boldsymbol{p}) = a^2 e^{b\left(F_{X_1}^2(\chi_1) + F_{X_2}^2(\chi_2)\right)} f_{X_1}(\chi_1, \boldsymbol{p}) f_{X_2}(\chi_2, \boldsymbol{p}) \tag{5.24}$$

(Exercise XI.1).

5.2 THE CHARACTERISTIC FUNCTION APPROACH

One more attractive feature of the notion of characteristic function is that it provides a useful approach of PDF model building. It may be instructive to introduce the CF approach with the help of a simple example.

Example 5.4

Let us look at an attribute whose temporal rate of change is equal to a known function of the attribute. Under this physical condition, the attribute may be represented by a random field $X(\boldsymbol{s},t)$ that obeys the SPDE

$$\frac{\partial}{\partial t} X(\boldsymbol{s}, t) = g(X(\boldsymbol{s}, t)), \tag{5.25}$$

with the random initial condition $X(\boldsymbol{s}, t_0) = X_0 \sim f_{X_0}(\chi_0)$, where g is a known function. Typically, the corresponding CF is given by (see, Definition 2.3 of Chapter II)

$$\phi_X(u) = \overline{e^{iuX}},$$

and by differentiation,

$$\frac{\partial}{\partial t}\phi_X(u) = iu\overline{g\,e^{iuX}} = iu\int d\chi\, e^{iu\chi}(gf_X).$$

Since (f_X,ϕ_X) form an FT pair, it is valid that

$$gf_X = \int du\, e^{-iu\chi}\frac{1}{iu}\frac{\partial}{\partial t}\phi_X,$$

and after taking the derivative with respect to χ, we find (Exercise XI.2)

$$\frac{\partial}{\partial t}f_X + \frac{\partial}{\partial \chi}(f_X g) = 0. \tag{5.26}$$

That is, in this case the PDF model $f_X(\chi)$ is constructed as the solution of the deterministic Eq. (5.26) with initial condition f_{X_0}.

The analysis of Example 5.4 can be extended to the case of the vector S/TRF, $\boldsymbol{X}(\boldsymbol{s},t) = [X_1(\boldsymbol{s},t)\ldots X_k(\boldsymbol{s},t)]^T$, that satisfies the law

$$\frac{\partial}{\partial t}\boldsymbol{X}(\boldsymbol{s},t) = \boldsymbol{g}(\boldsymbol{X}(\boldsymbol{s},t)), \tag{5.27}$$

with random vector initial conditions $\boldsymbol{X}(\boldsymbol{s},t_0) = \boldsymbol{X}_0$, where $\boldsymbol{g}(\boldsymbol{\chi}) = [g_1(\chi)\ldots g_k(\chi)]^T$. Accordingly, Eq. (5.26) is extended to the more general PDF equation

$$\frac{\partial}{\partial t}f_X + \sum_{l=1}^{k}\frac{\partial}{\partial \chi_l}(f_X g_l) = 0 \tag{5.28}$$

($l = 1,\ldots, k$), which is also known as the Liouville equation. I will not get into more details here, though, a systematic exposition of the Liouville equation can be found in the relevant literature.

5.3 THE FUNCTIONAL APPROACH

The functional treatment of random fields has been found to be particularly useful in the study of differential equations modeling a variety of physical systems (e.g., Hopf, 1952; Lewis and Kraichnan, 1962; Beran, 1968).[4] In principle, we start from the SPDE $\mathcal{L}_X$ governing the random fields of interest, and using functional analysis we can derive the corresponding PDE, $\mathcal{L}_f$, or $\mathcal{L}_\Phi$ obeyed by the PDF or the CFL of the random field, respectively.[5] Then, the PDE (CFL), which completely defines the random field, can be constructed by solving this differential equation. We continue with an example that refers to some CFL results described in Chapter XII.

[4]The material presented in this section is closely linked to the theory of random functionals to be discussed in the following Chapter XII, in which case the readers may find it appropriate to properly consult Chapter XII, while reading the present section, or to read the present section after reading Chapter XII.

[5]For a presentation of the CFL theory, see Chapter XII.

Example 5.5

Let us assume that the vector random field $\boldsymbol{X}(t)$ with components $X_l(t)$, $l = 1,\ldots, k$, satisfies the set of differential Eq. (2.8) of Chapter XII. The PDF of $\boldsymbol{X}(t)$ is given by the space–time delta function (Eq. (2.10b) of Chapter XII),

$$f_X(\boldsymbol{p}) = \overline{\delta(\boldsymbol{X}(t) - \boldsymbol{s})}, \tag{5.29}$$

where, as usual, $\boldsymbol{p} = (\boldsymbol{s},t)$. By integrating Eq. (2.8) of Chapter XII we get

$$X_i(t) = X_i(0) + \int_0^t dt'' \int d\boldsymbol{s}\delta(\boldsymbol{X}(t'') - \boldsymbol{s})[W_i(\boldsymbol{s},t'') + Z_i(\boldsymbol{s},t'')],^6$$

which, by letting $\boldsymbol{p}' = (\boldsymbol{s}', t')$, implies that (Eq. (2.11) of Chapter XII),

$$\frac{\delta X_i(t)}{\delta Z_k(\boldsymbol{p}')} = 0 \tag{5.30}$$

if $t' < 0$ or $t' > t$. From Eq. (5.29), we find that (Exercise XI.3)

$$\frac{\partial}{\partial t} f_X(\boldsymbol{p}) = -\sum_{i=1}^{k} \frac{\partial}{\partial s_i}[W_i(\boldsymbol{X}(t),t)f_X(\boldsymbol{s},t)] + \sum_{i=1}^{k} \frac{\partial}{\partial s_i}\left[\overline{\delta(\boldsymbol{X}(t) - \boldsymbol{s})Z_i(\boldsymbol{X}(t),t)}\right]. \tag{5.31}$$

Finally, in light of Eq. (2.16) of Chapter XII, Eqs. (2.9a–b) of the same chapter yield the following equation governing the PDF of $\boldsymbol{X}(t)$,

$$\frac{\partial}{\partial t} f_X(\boldsymbol{p}) + \sum_{i=1}^{k} \frac{\partial}{\partial s_i} \Omega_i(\boldsymbol{p}) f_X(\boldsymbol{p}) - \sum_{i=1}^{k}\sum_{p=1}^{k} \frac{\partial^2 \zeta_{ip}(\boldsymbol{s},\boldsymbol{s},t)}{\partial s_i \partial s_p} f_X(\boldsymbol{p}) = 0, \tag{5.32}$$

where

$$\Omega_i(\boldsymbol{p}) = W_i(\boldsymbol{p}) + \sum_{p=1}^{k} \left[\frac{\partial \zeta_{ip}(\boldsymbol{s},\boldsymbol{s}',t)}{\partial s_p}\right]_{s=s'}. \tag{5.33}$$

The solution of Eq. (5.33) allows the construction of the PDF models $f_X(\boldsymbol{p})$. This equation is also known as the probabilistic Fokker–Planck equation associated with the set of dynamic stochastic Eq. (2.8) of Chapter XII.

Example 5.6

Consider the set of stochastic differential Eqs. (1.37a–b) of Chapter XII. In light of Eq. (1.38) of Chapter XII, and using the functional properties discussed in the same chapter, we obtain the functional equation

$$\frac{\delta}{\delta q_Y(\boldsymbol{p})}\left[\frac{\delta}{\delta q_l(\boldsymbol{p})} \Phi_{X,Y}(q_1,q_2,q_3,q_Y)\right] = -\int d\chi(\boldsymbol{p}) d\psi(\boldsymbol{p}) e^{i\sum_{l=1}^{3}\int dp q_l \chi_l + i\int dp q_Y \psi} \chi_l \psi f_{X,Y}(\chi,\psi). \tag{5.34}$$

[6] Here the identity $\delta(\boldsymbol{X}(t) - \boldsymbol{s})W(\boldsymbol{X}(t),t) = \delta(\boldsymbol{X}(t) - \boldsymbol{s})W(\boldsymbol{s},t)$, for any function W, has been used.

Differentiating with respect to s_l, and then taking into consideration Eq. (1.37a) of Chapter XII we get the differential equation governing the CFL,

$$\sum_{l=1}^{3}\frac{\partial}{\partial s_l}\left[\frac{\delta}{\delta q_Y(\boldsymbol{p})}\frac{\delta}{\delta q_l(\boldsymbol{p})}\Phi_{X,Y}(q_1,q_2,q_3,q_Y)\right]=0. \tag{5.35}$$

Working along the same lines, starting from Eq. (1.37b) of Chapter XII, we find that

$$\sum_{j=1}^{3}\sum_{k=1}^{3}\varepsilon_{ljk}\frac{\partial}{\partial s_j}\left[\frac{\delta}{\delta q_k(\boldsymbol{p})}\Phi_{X,Y}(q_1,q_2,q_3,q_Y)\right]=0 \tag{5.36}$$

($l = 1, 2, 3$). Lastly, subject to the appropriate boundary conditions, linked to the particular physical application, Eqs. (5.35) and (5.36) should be solved with respect to the CFL $\Phi_{X,Y}(q_1, q_2, q_3, q_Y)$.

6. BAYESIAN MAXIMUM ENTROPY—BASED MULTIVARIATE PROBABILITY DENSITY MODEL CONSTRUCTION TECHNIQUES

The BME technique (Christakos, 1990, 2000) provides a broad knowledge synthesis framework for constructing M-PDF models in a manner that incorporates the general or core knowledge base, *G*-KB (consisting of natural laws, theoretical models, scientific theories, empirical relationships), and the specificatory knowledge base, *S*-KB (site-specific knowledge such as hard data, uncertain information, secondary sources) of the in situ situation. In real-world applications, the *S*-KB is usually available at a set of points $\boldsymbol{p}_D = (\boldsymbol{p}_H,\boldsymbol{p}_S)$, where $\boldsymbol{p}_H = \{\boldsymbol{p}_i, i = 1,\ldots, m_H\}$, and $\boldsymbol{p}_S = \{\boldsymbol{p}_i, i = m_H + 1,\ldots, m\}$ are the set of points where hard (exact) measurements and soft (uncertain) data are available, respectively. The BME technique quantifies uncertainty in multiple ways, including in terms of interval values and probability functions, as part of the *S*-KB. This technique allows uncertainty quantification in a natural and powerful manner, and requires fragments of information that give valuable insights about the PDF form.

The BME-based construction of M-PDF models is compactly expressed in terms of the equations introduced in the following proposition (Christakos, 1990, 2000).

Proposition 6.1

Given the *G*- and *S*-KB as described above, the M-PDF of the S/TRF at a set of space–time points $\boldsymbol{p}_k = \left(\boldsymbol{p}_{k_1},\ldots,\boldsymbol{p}_{k_p}\right)$ is given by

$$f_X(\boldsymbol{p}_k) = A^{-1}\int d\chi_S \xi_S e^{\boldsymbol{\mu}_G\cdot \boldsymbol{g}_G}, \tag{6.1}$$

where $\boldsymbol{g}_G$ is a vector with elements representing the *G*-KB, ξ_S represents the *S*-KB available, $\boldsymbol{\mu}_G$ is a space–time vector with elements that assign proper weights to the elements of $\boldsymbol{g}_G$ and are the solutions of the system of equations

$$\int d\chi_G\left(\boldsymbol{g}_G - \overline{\boldsymbol{g}}_G\right)e^{\boldsymbol{\mu}_G\cdot \boldsymbol{g}_G} = 0, \tag{6.2}$$

and A is a normalization parameter.

Eqs. (6.1) and (6.2) introduce a process that integrates S with G in a physically and logically consistent manner, and it accounts for local and nonlocal attribute dependencies across space–time. The above are noticeable features of the BME approach of M-PDF model building. The following corollaries (Christakos, 2000) present some analytical cases of M-PDF model construction in the light of Proposition 6.1. Specifically, in these corollaries, uncertainty is quantified in terms of interval values $I_S = [l,u]$ (the l and u denote lower and upper bounds, respectively), and probability functions f_S (Corollary 6.1 and Corollary 6.2, respectively).

Corollary 6.1

Assume that the available G-KB includes the S/TRF mean vector

$$\overline{X_G} = \left[\overline{X(\boldsymbol{p}_1)}\ldots\overline{X(\boldsymbol{p}_m)}\ \overline{X(\boldsymbol{p}_{k_1})}\ldots\overline{X\left(\boldsymbol{p}_{k_\rho}\right)}\right]^T, \tag{6.3}$$

and the (centered) covariance matrix

$$\boldsymbol{c}_{X_G} = \overline{\left(\boldsymbol{X}_G - \overline{\boldsymbol{X}_G}\right)\left(\boldsymbol{X}_G - \overline{\boldsymbol{X}_G}\right)^T}. \tag{6.4}$$

The S-KB includes hard data at points $\boldsymbol{p}_H$ and soft data of the interval type I_S at points $\boldsymbol{p}_S$. Then, Eq. (6.1) gives the M-PDF model

$$f_X(\boldsymbol{p}_k) = A^{-1}\phi\left(\boldsymbol{\chi}_k; \boldsymbol{B}_{k|H}\boldsymbol{\chi}_H, \boldsymbol{c}_{k|H}\right)\int_{\boldsymbol{l}-\boldsymbol{B}_{S|kH}\boldsymbol{\chi}_{kH}}^{\boldsymbol{u}-\boldsymbol{B}_{S|kH}\boldsymbol{\chi}_{kH}} d\boldsymbol{\chi}_S\phi\left(\boldsymbol{\chi}_S; \boldsymbol{0}, \boldsymbol{c}_{S|kH}\right), \tag{6.5}$$

where $\boldsymbol{\chi}_{kH} = (\boldsymbol{\chi}_k, \boldsymbol{\chi}_H)$, $\boldsymbol{B}_{k|H} = \boldsymbol{c}_{k,H}\,\boldsymbol{c}_{H,H}^{-1}$, $\boldsymbol{c}_{k|H} = \boldsymbol{c}_{k,k} - \boldsymbol{B}_{k|H}\,\boldsymbol{c}_{H,k}$, $\boldsymbol{B}_{S|kH} = \boldsymbol{c}_{S,kH}\,\boldsymbol{c}_{kH,kH}^{-1}$, $\boldsymbol{c}_{S|kH} = \boldsymbol{c}_{S,S} - \boldsymbol{B}_{S|kH}\,\boldsymbol{c}_{kH,S}$, $\boldsymbol{l} = (l_{m_H+1}, \ldots, l_m)$, and $\boldsymbol{u} = (u_{m_H+1}, \ldots, u_m)$; the $\varphi(\boldsymbol{\chi}; \overline{\boldsymbol{x}}, \boldsymbol{c})$ denotes a Gaussian distribution with mean vector $\overline{\boldsymbol{x}}$ and covariance matrix $\boldsymbol{c}$; and $A = \int_{\boldsymbol{l}-\boldsymbol{B}_{S|H}\boldsymbol{\chi}_H}^{\boldsymbol{u}-\boldsymbol{B}_{S|H}\boldsymbol{\chi}_H} d\boldsymbol{\chi}_S\varphi\left(\boldsymbol{\chi}_S; \boldsymbol{0}, \boldsymbol{c}_{S|H}\right)$.[7]

Corollary 6.2

Assume that the G-KB is as in Corollary 6.1, but the S-KB now includes hard data at points $\boldsymbol{p}_H$ and soft data of the probabilistic type $f_S(\boldsymbol{\chi}_S)$ at points $\boldsymbol{p}_S$. In this case, Eq. (6.1) gives the M-PDF model

$$f_X(\boldsymbol{p}_k) = A^{-1}\phi\left(\boldsymbol{\chi}_k; \boldsymbol{B}_{k|H}\boldsymbol{\chi}_H, \boldsymbol{c}_{k|H}\right)\int d\boldsymbol{\chi}_S f_H(\boldsymbol{\chi}_S)\phi\left(\boldsymbol{\chi}_S; \boldsymbol{B}_{s|kH}\boldsymbol{\chi}_{kH}, \boldsymbol{c}_{S|kH}\right), \tag{6.6}$$

where $A = \int d\boldsymbol{\chi}_S f_S(\boldsymbol{\chi}_S)\phi(\boldsymbol{\chi}_S; \boldsymbol{B}_{s|H}\boldsymbol{\chi}_H, \boldsymbol{c}_{s|H})$.

Let us continue with an example that shows that the BME approach is, in a sense, a generalization of the SDE approach of M-PDF model construction, because in addition to the physical law it also accounts for other kinds of site-specific information, such as uncertain measurements. This is important, since most available measurements are indeed uncertain rather than exact, and almost every measuring instrument is characterized by a finite level of precision. For demonstration purposes, the BME results are compared below with those of the SDE analysis of Example 5.1.

[7]More technical details concerning these mathematical expressions can be found in Christakos (2000).

Example 6.1

Consider again the random process $X(t)$ satisfying the physical law expressed by the stochastic differential Eq. (5.10) in Example 5.1. Site-specific information, *S*-KB, about $X(t)$ is available in the form of the probability distribution $f_S(\chi,t)$, where the possible attribute values at each time instant t belong to the interval $I \in [\chi_l,\chi_u]$, χ_l and χ_u denote lower and upper values. This is a realistic situation because a measurement process is usually performed with a limited accuracy and the measurement of a physical attribute is uncertain. This uncertainty is here expressed by the probability distribution above. The initial conditions are as in Example 5.1. In the BME setting, the physical Eqs. (5.10)–(5.11) constitute the *G*-KB of the situation, in which case the corresponding BME equations associated with Eqs. (5.10)–(5.11) are

$$\int d\chi\, e^{\sum_{i=0}^{2}\mu_{i,t}\chi^i} = 1,$$

$$\int d\chi\ \chi \frac{\partial}{\partial t} e^{\sum_{i=0}^{2}\mu_{i,t}\chi^i} = a \int d\chi\ \chi e^{\sum_{i=0}^{2}\mu_{i,t}\chi^i}, \tag{6.7a-c}$$

$$\int d\chi\ \chi^2 \frac{\partial}{\partial t} e^{\sum_{i=0}^{2}\mu_{i,t}\chi^i} = 2a \int d\chi\ \chi^2 e^{\sum_{i=0}^{2}\mu_{i,t}\chi^i},$$

where in light of the condition of Eq. (5.11), we find that $\mu_{i,0} = \nu_i\ (i = 0, 1, 2)$. Eq. (6.7a) is obviously a normalization constraint. The solution of Eqs. (6.7a–c) with respect to the three unknown coefficients $\mu_{i,t}\ (i = 0, 1, 2)$ gives

$$\mu_{0,t} = \nu_0 - at,\ \ \mu_{1,t} = \nu_1 e^{-at},\ \ \mu_{2,t} = \nu_2 e^{-2at}, \tag{6.8a–c}$$

and the PDF model derived based on the *G*-KB is given by

$$f_G(\chi,t) = e^{\sum_{i=0}^{2}\mu_{i,t}\chi^i} = \frac{1}{\sqrt{2\pi}\sigma_0} e^{-\sigma_0^{-2}\left(\frac{1}{2}\overline{X}_0^2 - \overline{X}_0 e^{-at}\chi + \frac{1}{2}e^{-2at}\chi^2\right) - at}. \tag{6.9}$$

Remarkably, Eq. (6.9) is the same as the classical solution $f_X(\chi,t)$ of Eq. (5.15) of the stochastic differential Eq. (5.10) subject to the condition of Eq. (5.11). The present BME implementation included only two equations, Eqs. (6.7b and c), in addition to the normalization Eq. (6.7a), because the condition of Eq. (5.11) involved a Gaussian PDF f_0 with up to second-order statistical moments. Also, we notice that the solution so far accounts for the *G*-KB in the form of the differential Eq. (5.10), but it does not yet account for the *S*-KB in the form of f_S. As a matter of fact, the classical transformation of variables approach discussed in Section 5.1 cannot incorporate the *S*-KB into the solution. BME, on the other hand, can account for many kinds of site-specific information. In the present case, BME updates the solution of Eq. (6.9) to account for the site-specific probability model f_S expressing measurement uncertainty by means of the equation

$$f_X(\chi,t) = A^{-1} f_S(\chi,t) f_G(\chi,t), \tag{6.10}$$

where $A = \int_I dv f_S(v,t) f_G(v,t)$ is a normalization constant, and both the *G*-KB and the *S*-KB are incorporated in the final PDF model of Eq. (6.10). Furthermore, if instead of the Bayesian conditional we use the stochastic logic (material) conditional proposed in Christakos (2002, 2010), the PDF model becomes

$$f_X(\chi, t) = \frac{1}{2A - 1}[2f_S(\chi, t) - 1]f_G(\chi, t). \tag{6.11}$$

Obviously, the PDF model of Eq. (6.11) is different than that of Eq. (6.10). Which one of the two models to use depends on factors associated with the phenomenon under study.

7. METHODOLOGICAL AND TECHNICAL COMMENTS

Comparison is a basic principle of inductive inference to the point that thinking without comparison is unthinkable. When different methods are compared under the same conditions, emerging similarities and differences are highlighted, and their relative performance is assessed by means of indicators such as predictive accuracy or uncertainty. Following this principle, practical insight about the M-PDF model construction techniques can be gained with the help of comparative simulation studies (one such study was presented in Christakos et al., 2011).

Underlying both the copula- and the factora-based techniques is the basic idea of replacing an unknown entity (original M-PDF) with another unknown entity (factora or copula) that is supposedly easier to infer from the available data and manipulate analytically. Whether this is actually a valid claim of practical significance depends on a number of technical and substantive issues, a few of which are discussed in the example that follows.

Example 7.1

As was mentioned earlier, the implementation of the factora- and the copula-based PDF model construction techniques generally requires knowledge of the U-PDFs at different times and locations across the entire study area. Yet, this information is not usually available in practice, in which case Gaussian U-PDFs are assumed. Obviously, this may be an unrealistic assumption. Also, when the choice is made to use indicator kriging in the numerical implementation of the factora technique, it can generate negative PDF values as a result of indicator kriging's tendency to produce pseudo-PDFs rather than proper PDFs.

These caveats notwithstanding, the above techniques are very useful tools for generating valid PDF models. In technical terms, a prime advantage of the copula technique is its analytical tractability, although this is mainly valid in low dimensions (2–4). While factoras involve infinite series that have to be truncated, many copulas are available in a closed form. This comes at the cost of some restrictive assumptions, such as low dimensionality, uniform marginals, and the applicability of the integral transform theorem. Attempts to involve transforms of uniform marginals are rather *ad hoc* and can add considerable complexity to the process. On the other hand, potential advantages of factoras include the elimination of restrictive requirements (uniform marginals, integral transform theorem, etc.), and the rich classes of PDFs derived by taking advantage of the ϕ-property and the generalization formulas. The functional form of $\theta_{X;\{pi\}}$ (in the case of factoras) is explicitly given in terms of known polynomials, whereas the explicit form of $\varsigma_{X;\{p_i\}}$ (in the case of copulas) is generally unknown and needs to be derived every time.

The basic idea of the SDE-based construction of an M-PDF is that the latter can be derived from the differential equation governing the attribute of interest using the classical, the characteristic function, or the functional techniques. In principle, these techniques account for core physical knowledge but not for site-specific information. The BME-based PDF model construction technique is conceptually a very different approach.[8] According to the BME technique, (*a*) the problem of constructing an appropriate PDF model should be tackled by exploiting all sources of knowledge (core and site-specific) and synthesizing them into a coherent framework, and (*b*) the basic properties of the phenomenon should be encapsulated in a PDF model construction technique. In other words, BME's viewpoint is that there is not a single or universal PDF model construction technique, but rather that PDF model construction is inescapably grounded in the physical conditions and relevant knowledge sources of the situation of interest.

Numerical simulations studied in Christakos et al. (2011) show that the PDF models generated by the factoras, the copulas, and the BME techniques may differ from each other. There are certain reasons for this, as follows. The factoras and copulas use different PDF expansions. They also use different spatiotemporal estimation schemes: while the numerical implementation of the factora technique is often combined with indicator kriging, BME does not need to make such an assumption. Also, the three methods incorporate the available knowledge bases in different ways. Lastly, the above techniques can be used to generate realistic non-Gaussian simulations of environmental attributes (e.g., rainfall assessment, climate, and watersheds) and to estimate extreme values in Nature (e.g., droughts and hazards) without making any restrictive assumptions such as normality, linearity, etc. On the other hand, physical studies using the mainstream statistics assumptions (normality, linearity, space–time separation, etc.) can yield unrealistic results.

In the end, a key factor in selecting among different PDF model construction techniques is intimately linked to the interpretive values of the models these techniques produce, an issue which ultimately comes down to their dynamics of physical meaning making.

[8]In fact, the only similarity between the BME technique, on the one side, and the factora and copula techniques, on the other, is methodological, i.e., at certain stages of their implementation, questions will be asked and data will be collected.

SPATIOTEMPORAL RANDOM FUNCTIONALS

CHAPTER XII

CHAPTER OUTLINE

1. CONTINUOUS LINEAR RANDOM FUNCTIONALS IN THE SPACE–TIME DOMAIN

This chapter describes spatiotemporal random fields (S/TRFs) in the R^{n+1} (or $R^{n,1}$) domain by means of the so-called *functional*, i.e., a "function of a function." The functional description naturally involves more complex mathematics, but it has its rewards on both theoretical and applied grounds. These include the ability to carry out important operations that cannot be performed with standard tools (e.g., it allows integral transformations that do not exist in the ordinary sense but they are very important in applications), as well as to offer a more complete and consistent presentation of the S/TRF theory (e.g., as it was shown in Chapter XI, it allows the study of differential equations representing the evolution of probability functions, and as it will be discussed in Chapter XIII, it enables the rigorous modeling of attributes with space–time heterogeneous variation). This chapter presents the basics of the theory of spatiotemporal random functionals, with a special emphasis on its notions and results that are useful for the purposes of the book.

1.1 BASIC NOTIONS

To start with, a functional is generally defined in a domain of functions if a rule exists so that each function belonging to the domain is referred to a number termed the value of the functional of this function. For illustration, I select a few simple cases of functionals.

Example 1.1

Classical examples of functionals are the definite Riemann integral of a function over a finite support, and the delta function that is a generalized function defined in terms of a functional

Spatiotemporal Random Fields. http://dx.doi.org/10.1016/B978-0-12-803012-7.00012-X

instead of its point values. For an S/TRF $X(\boldsymbol{p})$ we can define a cumulative distribution functional (CDFL) that is a natural generalization of the cumulative distribution function (CDF), as

$$F_X(\chi(\boldsymbol{p})) = P[X(\boldsymbol{p}) < \chi(\boldsymbol{p})].$$

The value of $F_X(\chi(\boldsymbol{p}))$ is a functional, since it depends on the form of $\chi(\boldsymbol{p})$. The functional takes as input a function $\chi(\boldsymbol{p})$ on a domain, i.e., not the value of the function at a specific point $\boldsymbol{p}$, but all the values of $\chi(\boldsymbol{p})$ at all the $\boldsymbol{p}$'s in the domain. Its output is a number. A similar functional formalism is possible for higher-order CDFLs. Another example is the exponential random functional

$$\Lambda_X(t) = \int_0^t du e^{-X(u)},$$

where $X(t)$ is a Levy process ($t \geq 0$). This functional represents important random dynamical systems of mathematical physics and, also, it expresses key quantities in stochastic risk theory. Lastly, let us look at the functional

$$S_X(\mathcal{L}) = \int d\boldsymbol{p} \mathcal{L}\left(X, \frac{\partial}{\partial t}X, \frac{\partial}{\partial s}X\right),$$

where $X(\boldsymbol{p}) = X(s, t)$ denotes the displacement of a point s on a string at time t, and $\mathcal{L}$ denotes the Lagrangian density, which is a function of $X(s, t)$ and its temporal and spatial derivatives. The functional S_X represents the action and is used in the study of elastic media (Exercise XII.14).

Apparently, a functional can take many different forms, but we are arguably interested about those particular forms that have considerable theoretical and practical significance in the context of S/TRF modeling, which is the subject of this book. With this goal in mind, I continue with the definition of a particular case of a functional that is, indeed, of great value in S/TRF characterization, and, hence, it will be at the center of this chapter's developments.

Definition 1.1

Let $q(\boldsymbol{p})$ be a nonrandom function such that[1]

$$X(q) = \int_V d\boldsymbol{p}\, q(\boldsymbol{p})X(\boldsymbol{p}) = \langle q(\boldsymbol{p}), X(\boldsymbol{p})\rangle, \tag{1.1}$$

where $V \in R^{n+1}$, exists for almost all realizations of the random field $X(\boldsymbol{p})$, $\boldsymbol{p} \in R^{n+1}$. The $X(q)$ is called a continuous linear *random functional.*

Before proceeding further, one brief note should be made concerning terminology. The $q(\boldsymbol{p})$ is often called the *test* or *support* function. Generally, in applications, one needs to use random functionals on spaces of test or support functions $q(\boldsymbol{p})$ that satisfy two basic conditions:

(i) they are well-behaved functions, and
(ii) they are interpretable and lead to representations with physical meaning.

[1]Herein, for simplicity the V will be dropped but implicitly implied.

Concerning condition (i), $q(\boldsymbol{p})$ should belong to spaces of functions that, in general, are continuous, integrable, infinitely differentiable, and all their derivatives vanish outside a certain interval in R^{n+1} (Gel'fand, 1955). Such are the spaces already defined in Section 2.4 of Chapter I:

- space C^{∞} of continuous and infinitely differentiable functions with compact support, and
- space C_0^{∞} of continuous and infinitely differentiable functions which, together with their derivatives of all orders, approach zero rapidly at infinity (e.g., faster than $|\boldsymbol{s}|^{-1}$ and t^{-1} as $|\boldsymbol{s}| \rightarrow \infty$ and $t \rightarrow \infty$).

These spaces of functions are also known as Schwartz spaces (Schwartz, 1950–51). Choosing test functions that are space–time separable, $q(\boldsymbol{s}, t) = q_1(\boldsymbol{s})q_2(t)$, is often a convenient choice. Examples of test functions belonging to the above spaces of functions are described next.

Example 1.2

A test function belonging to the C^{∞} space is given by

$$q(\boldsymbol{s}, t) = \begin{cases} e^{-\left(\frac{s^2}{R^2 - s^2} + \frac{t^2}{T^2 - t^2}\right)}, & \text{if } |\boldsymbol{s}| < R, \ |\tau| < T \\ 0, & \text{if } |\boldsymbol{s}| \geq R, \ |\tau| \geq T \end{cases}$$

where R and T are specified space and time intervals (all derivatives of $q(\boldsymbol{s}, t)$ are continuous at the support boundary). A test or support function in C_0^{∞} is given by

$$q(\boldsymbol{s}, t) = e^{-\left(\frac{s^2}{R^2} + \frac{t^2}{T^2}\right)},$$

which, although it has an infinite support, it can be used to model a finite support filter, because it decays fast outside the range determined by R and T. Generally, the form of the test or support function $q(\boldsymbol{s}, t)$ depends on the geometry of the physical domain considered. In some subsurface porous media geometries, $q(\boldsymbol{s}, t)$ may be assumed to be nonzero on some finite domain (e.g., deep electromagnetic experiments) or zero everywhere except on a line between a source and a receiver (e.g., in seismic traveling experiments).

Concerning condition (ii), some elementary choices of test functions are discussed in the following example, which also offers an assessment of their potential uses.

Example 1.3

Assume that the random field $X(\boldsymbol{p}) = X(\boldsymbol{s}, t)$ represents the concentration of an aerosol substance in the atmosphere. By choosing the test function

$$q(\boldsymbol{p}) = \delta(\boldsymbol{s}, \boldsymbol{s}^*)\delta(t, t^*),$$

Eq. (1.1) provides the value of the substance at the space location–time instant $\boldsymbol{p}^* = (\boldsymbol{s}^*, t^*)$. Delta test functions are also used when the measurements are localized in space and/or time. If, on the other hand, we chose the test function

$$q(\boldsymbol{s}, t) = \begin{cases} 1, & \text{if } \boldsymbol{s} \in U, \ t \in [t_1, t_2] \\ 0, & \text{otherwise,} \end{cases}$$

then Eq. (1.1) provides the total amount of substance in the volume U during the time period $[t_1, t_2]$. Also, test functions with finite support provide appropriate models for the observation effect (Christakos, 1992). Generally, when choices must be made from a suite of candidate test functions, it is desirable to make the selection on objective grounds and based on physical considerations regarding the phenomenon of interest.

It is noteworthy that the functional characterization of a random field is closely related to the theory of *generalized random fields* to be discussed in Chapter XIII, which is why the present chapter may be also viewed as an introduction to the following chapter. Particularly, if $q(\boldsymbol{p})$ belongs to the Schwartz space $C_0^\infty\left(R^{n+1}\right)$, the functional of Eq. (1.1) is a generalized S/TRF as will be presented in Chapter XIII.

Remark 1.1

The notion of a random functional also emerges in the context of the so-called *functional data analysis* (FDA, Ramsay and Silverman, 2002) that considers a set of data functions rather than a matrix of observations. In FDA, the q-function is chosen so that it properly represents the main features of the functional data of interest. Examples of q-functions include truncated power series, B-splines, exponential, Fourier, and wavelet functions. For instance, in many practical applications a q-function of the Fourier form is often an appropriate choice for periodic data functions. I will not get into a more detailed discussion of FDA here.

1.2 GENERALIZED FOURIER TRANSFORM

One of the most important features of the functional characterization of an S/TRF is that it leads to the definition of an extended formulation of the Fourier transformation (FT) operation (discussed in Chapter V), the so-called *generalized* FT (Gel'fand and Shilov, 1964), which resolves the key issue of existence of spectral representations encountered in many physical applications, noticeably in ocean and atmospheric sciences. Indeed, the powerful notion of a random functional makes it possible to perform S/TRF operations of principal importance in physical studies that are impossible to perform in terms of ordinary functions.

More specifically, as the discussion in Chapter V brought to our attention, in many physical studies (e.g., unlimited oceans) the spectral representations of many random fields do not exist in the ordinary FT sense. Fortunately, as it turns out, these representations exist in the generalized FT sense defined in terms of the random functional above (which is why the transform is sometimes called the *q-sense* FT). Operationally, the approach for obtaining the generalized FT of an S/TRF is rather simple and involves three main steps:

(a) Select a suitable test function. Generally, this can be any function that makes the corresponding Fourier integral operation to exist. One possible class of test functions is the ε-parametered class of functions $q_\varepsilon(s, t)$ such that $q_\varepsilon(s, t)$ goes to zero with exponential rapidity as its arguments (s, t) go to infinity, and $\lim_{\varepsilon \to \infty} q_\varepsilon(s, t) = 1$.[2]

[2]A good choice is a function of the form $q_\varepsilon(s,t) = e^{-\frac{s^2+t^2}{\varepsilon^2}}$, but other possibilities obviously exist.

(b) Multiply the random field $X(s, t)$ by $q_\varepsilon(s, t)$ so that the corresponding FT

$$\widetilde{X}_\varepsilon(\boldsymbol{k},\omega) = \frac{1}{(2\pi)^{\frac{n+1}{2}}} \iint ds dt\, e^{-i(\boldsymbol{k}\cdot\boldsymbol{s} - \omega t)} X(\boldsymbol{s},t) q_\varepsilon(\boldsymbol{s},t) \tag{1.2a}$$

can be determined in the standard sense.

(c) Lastly, make the passage to the limit, to obtain the desired spectrum,

$$\widetilde{X}(\boldsymbol{k},\omega) = \lim_{\varepsilon\to\infty} \widetilde{X}_\varepsilon(\boldsymbol{k},\omega). \tag{1.2b}$$

A similar approach leads to the integral expression

$$X(\boldsymbol{s},t) = \lim_{\varepsilon\to\infty} X_\varepsilon(\boldsymbol{s},t) = \lim_{\varepsilon\to\infty} \iint dk d\omega\, e^{i(\boldsymbol{k}\cdot\boldsymbol{s} - \omega t)} \widetilde{X}(\boldsymbol{k},\omega) q_\varepsilon(\boldsymbol{k},\omega). \tag{1.2c}$$

Thus, the above integrals exist only in the limit q-sense of the generalized FT.

The generalized FT approach was presented above in terms of an S/TRF, but it is also valid in terms of its spatiotemporal variability functions (S/TVFs) (e.g., the functional formalism above allows the determination of nonsimultaneous as well as simultaneous covariance functions). In fact, as we will see in Chapter XIII, many space–time covariance and spectral functions exist only in the q-sense. Until then, let us examine a typical situation in the real-world that is among those that have motivated the development of generalized FT.

Example 1.4

Consider the random sea surface displacement $X(s, t)$ in $R^{2,1}$. As is well known (e.g., Dean and Dalrymple, 1991), under the common assumption of an unlimited ocean the $X(s, t)$ cannot be integrated in the ordinary sense because the integral is unbounded. Accordingly, the FT of $X(s, t)$, which requires that $\int ds |X(\boldsymbol{s},t)| < \infty$, does not exist in the standard sense. However, the FT exists in the generalized (q) sense described above. Next, I will show how the generalized Fourier integral can be applied on spectral functions, leading to the derivation of the corresponding space–time covariance functions. For this purpose, let the spectral density function (SDF) of $X(s, t)$ be $\widetilde{c}_X(\boldsymbol{k},\omega) = c_0$. As was argued earlier, its Fourier integral yielding the water surface displacement covariance exists only in the q-sense. Specifically, by letting the test function q be the ε-parametered function

$$q_\varepsilon(\boldsymbol{k},\omega) = e^{-\frac{k^2+\omega^2}{2\varepsilon^2}},$$

such that $\lim_{\varepsilon\to\infty} q_\varepsilon(\boldsymbol{k}, \omega) = 1$, the SDF can be defined as

$$\widetilde{c}_X(\boldsymbol{k},\omega) = \lim_{\varepsilon\to\infty} \left[q_\varepsilon(\boldsymbol{k},\omega)\widetilde{c}_X(\boldsymbol{k},\omega)\right].$$

The corresponding space–time covariance function is determined by the Fourier integral in the q-sense,

$$c_X(\boldsymbol{h},\tau) = \iint dk d\omega e^{i(\boldsymbol{k}\cdot\boldsymbol{h} - \omega\tau)} \widetilde{c}_X(\boldsymbol{k},\omega) = \lim_{\varepsilon\to\infty} \iint dk d\omega e^{i(\boldsymbol{k}\cdot\boldsymbol{h} - \omega\tau)} e^{-\frac{k^2+\omega^2}{2\varepsilon^2}} c_0$$

$$= c_0 \lim_{\varepsilon\to\infty} (2\pi)^2 \varepsilon^4\, e^{-\frac{h^2+\tau^2}{2}\varepsilon^2}.$$

The last expression above is a definition of a basic generalized function, the Dirac delta (or, simply, delta) function[3]

$$\delta(\boldsymbol{h},\tau) = \lim_{\varepsilon\to\infty} (2\pi)^2 \varepsilon^4 \, e^{-\frac{h^2+\tau^2}{2}\varepsilon^2},$$

in which case the

$$c_X(\boldsymbol{h},\tau) = c_0\delta(\boldsymbol{h},\tau).$$

is the corresponding space–time covariance function (also known as space–time pure nugget effect).

In view of the above considerations, when appropriate, in the following sections the FT will be interpreted in the generalized q-sense. In this setting, a useful generalization can be made concerning the analysis of Examples 1.3 and 1.4 that makes analysis more rigorous and efficient. Let FT and FT_q denote the FT in the ordinary and in the q-sense, respectively. Then, for a function f in the $R^{n,1}$ domain, it is valid that

$$FT_q[f] = \lim_{\varepsilon\to\infty} FT[q_\varepsilon f],$$

where q_ε is an ε-parametered test function. In this sense, several well-known results can be derived, such as, $FT_q[1] = \delta(\boldsymbol{h},\tau)$, or

$$\delta(\boldsymbol{h},\tau) = \iint d\boldsymbol{k} d\omega e^{i(\boldsymbol{k}\cdot\boldsymbol{h} - \omega\tau)}.$$

In the following discussion, the subscript q will be dropped when it is obvious from the context that the integral is considered in the generalized sense.

Remark 1.2

The application of generalized FTs often involves a Gamma function for negative values of its argument. In this case, the formula

$$\Gamma\left(-\frac{\rho}{2}\right) = \begin{cases} \dfrac{(-1)^\ell}{\ell!}, & \rho = 2\ell \\[2ex] \dfrac{(-1)^{\ell+1}2^{\ell+1}\pi^{\frac{1}{2}}}{(2\ell+1)!!}, & \rho = 2\ell+1 \end{cases},$$

is very useful in calculations (see also Appendix).

Since polynomial functions are not bounded, their FTs do not exist in the ordinary sense, but generalized FTs can be defined and evaluated.

Example 1.5

The generalized FT of $\eta_{\rho/\zeta}(\boldsymbol{s},t) = s^\rho t^\zeta$ in the $R^{n,1}$ domain, where $s = |\boldsymbol{s}|$, is

$$\widetilde{\eta}_{\rho/\zeta}(\boldsymbol{k},\omega) = \widetilde{\eta}_{\rho(1)}(\boldsymbol{k})\widetilde{\eta}_{\zeta(2)}(\omega),$$

---

[3]More formally, the delta function is defined so that $\iint d\boldsymbol{h}d\tau\delta(\boldsymbol{h},\tau) = lim_{\varepsilon\to\infty} \iint d\boldsymbol{h}d\tau(2\pi)^2\varepsilon^4 e^{-\frac{h^2+\tau^2}{2}\varepsilon^2}$. Yet, in physical sciences the shorter definition is usually favored.

where

$$\widetilde{\eta}_{\rho(1)}(\boldsymbol{k}) = 2^{\rho+n}\pi^{\frac{n}{2}}\frac{\Gamma\left(\frac{\rho+n}{2}\right)}{\Gamma\left(-\frac{\rho}{2}\right)k^{\rho+n}},$$

$$\widetilde{\eta}_{\zeta(2)}(\omega) = 2^{\zeta+1}\pi^{\frac{1}{2}}\frac{\Gamma\left(\frac{\zeta+1}{2}\right)}{\Gamma\left(-\frac{\zeta}{2}\right)\omega^{\zeta+1}},$$

with $k = |\boldsymbol{k}|$.

The generalized FTs of space–time polynomials have various useful applications, e.g., they are used in the permissibility conditions for generalized random functionals (Chapters XIII and XV).

1.3 SPACE–TIME CHARACTERISTIC FUNCTIONALS

In light of the fundamental Definition 1.1, a special kind of a functional can be associated with a random field, which has many attractive properties, and, hence, it is of great practical significance.

Definition 1.2

The *characteristic functional* (CFL) of the S/TRF $X(\boldsymbol{p})$ is defined as

$$\Phi_X(q) = \overline{e^{iX(q)}} = \int d\chi_{\boldsymbol{p}}\, e^{i\int d\boldsymbol{p}q(\boldsymbol{p})\chi_{\boldsymbol{p}}} f_X(\chi_{\boldsymbol{p}}), \tag{1.3}$$

where $X(q)$ is given by Eq. (1.1).

Clearly, the $\Phi_X(q)$ must be known for any $q(\boldsymbol{p})$. The principal value of the CFL of Eq. (1.3) is that it contains all the required information about the random field representation of the natural attribute.

Remark 1.3

The CFL notion can be viewed as an extension of the characteristic function (CF) notion, where now the concern is with random fields in space–time rather than with random variables (RVs). Specifically, in the case of $X(\boldsymbol{p})$, each realization (sample function) corresponds to a vector $\boldsymbol{\chi}$ in an infinite-dimensional Hilbert space, so the $\boldsymbol{u}$ in Eqs. (2.14a–b) of Chapter II is replaced by an infinite-dimensional vector $\boldsymbol{q}$ in the same Hilbert space as $\boldsymbol{\chi}$. That is, $\boldsymbol{q}$ describes a function $q(\boldsymbol{p})$ so that the CF becomes a CFL $\Phi_X(q)$. Since the random field $X(\boldsymbol{p})$ replaces the discrete set of variables of the CF discussed in Chapter II, the CF is generalized from a sum over products of discrete χ_i and u_i values in the exponential of Eqs. (2.14a–b) of Chapter II to a space–time function of $\chi(\boldsymbol{p})$ and $q(\boldsymbol{p})$ in the exponential of Eq. (1.3). Thus, the CF of the RVs x_i becomes the CFL of the random field $X(\boldsymbol{p})$.

Assuming that the $\Phi_X(q)$ is available, several important functions can be derived from it, such as the CF, the corresponding probability density function (PDF), and the statistical moments of $X(\boldsymbol{p})$. Specifically, this can be done by letting

$$q_m(\boldsymbol{p}) = \sum_{k=1}^{m} q_k\delta(\boldsymbol{p}-\boldsymbol{p}_k),$$

in which case

$$\int dp\, q_m(\boldsymbol{p})X(\boldsymbol{p}) = \sum_{k=1}^{m} q_k X(\boldsymbol{p}_k),$$

and the CFL will be reduced to the CF of Eqs. (2.14a–b) of Chapter II. Then, any space–time statistical moment can be calculated from the CF (see, Eqs. 2.7a–b of Chapter IV), whereas the PDF is obtained as the FT of the CF. The CFL $X(q)$ completely defines the S/TRF $X(\boldsymbol{p})$, by uniquely defining the PDF $f_X(\chi_1,\ldots,\chi_m)$ for any m and selection of space–time points $\boldsymbol{p}_1,\ldots,\boldsymbol{p}_m$. Other interesting properties of the CFL are as follows:

(a) Let $\Phi_{X_1}[q]$ and $\Phi_{X_2}[q]$ denote the CFLs of the random fields $X_1(\boldsymbol{p})$ and $X_2(\boldsymbol{p})$. If $a_1, a_2 \geq 0$ are such that $a_1 + a_2 = 1$, then the

$$a_1 \Phi_{X_1}[q] + a_2 \Phi_{X_2}[q], \tag{1.4}$$

and the

$$\Phi_{X_1}[q]\Phi_{X_2}[q] \tag{1.5}$$

are CFLs, as well.

(b) Let $\Phi_u[q]$ be a family of CFL that depend on the RV u. Then, the

$$\Phi_X[q] = \int_U \Phi_u[q] dF_u(v), \tag{1.6}$$

where $F_u(v)$ is the probability distribution function of u in a space U, is a CFL too.

(c) In the case of a space–time homostationary (STHS) random field we have that

$$\Phi_X(q(\boldsymbol{p})) = \Phi_X(q(U_{\Delta p}\boldsymbol{p})), \tag{1.7}$$

i.e., the CFL is invariant under translation. And, in the case of a space–time isostationary (STIS) random field it holds that

$$\Phi_X(q(\boldsymbol{p})) = \Phi_X\left(q\left(R_\theta^{-1}\boldsymbol{p}\right)\right), \tag{1.8}$$

i.e., the CFL is invariant under rotation. Notice that if the random field is STHS (STIS) in a subset of coordinates only, then Eq. (1.7) (resp., Eq. 1.8) applies strictly to this subset.

Example 1.6

Consider the atmospheric permittivity attribute in $R^{3,1}$ represented by the random field $X(s, t)$ with a CFL of the form

$$\Phi_X(q_1, q_2) = \overline{e^{i\int ds[q_1(\boldsymbol{s})X(\boldsymbol{s},t) + q_2(\boldsymbol{s})X^*(\boldsymbol{s},t)]}},$$

and let $q_1(\boldsymbol{s}) = -a\sum_{j=1}^{k} \delta(\boldsymbol{s}_\eta - \boldsymbol{s}_{\eta,j})$ and $q_2(\boldsymbol{s}) = a\sum_{j=k+1}^{k+m} \delta(\boldsymbol{s}_\eta - \boldsymbol{s}_{\eta,j})$, where $a > 0$, and the $\boldsymbol{s} \in R^3$ is decomposed into its longitudinal coordinate $s_\ell \in R^1$ and its transverse coordinate vector $\boldsymbol{s}_\eta \in R^2$, i.e., $\boldsymbol{s} = (s_\ell, \boldsymbol{s}_\eta)$. Then, the CFL of the atmospheric permittivity field is

$$\Phi_X(q_1, q_2) = \overline{e^{ia\int_0^{s_\ell} ds'\left[-\sum_{j=1}^{k} X\left(s', \boldsymbol{s}_{\eta,j}, t\right) + \sum_{j=k+1}^{k+m} X^*\left(s', \boldsymbol{s}_{\eta,j}, t\right)\right]}},$$

in terms of longitudinal and transverse coordinates.

1.4 FUNCTIONAL DERIVATIVES

Generally, the formal definition of Eq. (1.1) applies to nonrandom (deterministic) or random functionals $X(q)$ associated with a field $X(\boldsymbol{p})$. And, just as any field $X(\boldsymbol{p})$ has a derivative with respect to its argument $\boldsymbol{p}$, a functional $X(q)$ can have a functional derivative with respect to its input function $q(\boldsymbol{p})$. Indeed, while in the case of a regular derivative, the idea is to change the argument $\boldsymbol{p}$ by $d\boldsymbol{p}$ and see how the function $X(\boldsymbol{p})$ changes in response, a functional derivative changes the entire input function $q(\boldsymbol{p})$ by a small amount $\delta q(\boldsymbol{p})$ and observes how the functional changes in response.

Specifically, consider the values of the continuous linear functional $X(q)$ in connection with the functions $q(\boldsymbol{p})$ and $q(\boldsymbol{p})+\delta q(\boldsymbol{p})$, where $\delta q(\boldsymbol{p})$ is only nonzero in a certain region $D\boldsymbol{p}$ at the point $\boldsymbol{p}_0$. This leads to the following functional derivative definition.

Definition 1.3

The *functional derivative* of the continuous linear functional $X(q)$ evaluated at $\boldsymbol{p}_0$ is the limit

$$\frac{\delta}{\delta q(\boldsymbol{p}_0)}X(q)=\lim_{\substack{|D\boldsymbol{p}|\to 0\\ \max|\delta q|\to 0}}\frac{X(q+\delta q)-X(q)}{\int_{D\boldsymbol{p}}d\boldsymbol{p}\,\delta q(\boldsymbol{p})},\tag{1.9}$$

assuming that this limit exists and depends neither on the form of $\delta q(\boldsymbol{p})$ nor on the way in which $|D\boldsymbol{p}|$ and $\max|\delta q(\boldsymbol{p})|$ tend to 0.

Example 1.7

The functional derivative of $X(q)$, Eq. (1.1), at a point $\boldsymbol{p}_0$ is obtained by means of the limit[4]

$$\frac{\delta}{\delta q(\boldsymbol{p}_0)}\int d\boldsymbol{p}\,q(\boldsymbol{p})X(\boldsymbol{p})=\lim_{\substack{|D\boldsymbol{p}|\to 0\\ \max|\delta q|\to 0}}\frac{\int_{D\boldsymbol{p}}d\boldsymbol{p}X(\boldsymbol{p})[q(\boldsymbol{p})+\delta q(\boldsymbol{p})-q(\boldsymbol{p})]}{\int_{D\boldsymbol{p}}d\boldsymbol{p}\,\delta q(\boldsymbol{p})}.\tag{1.10}$$

The derivatives of more complicated functionals are obtained in the same way. As should be expected, explicit analytical expressions of the CFL and its derivatives exist for Gaussian and Poisson random fields.

A point to be stressed is that the functional derivative of $X(q)$ above differs from its partial derivative that plays a key role in many applications where $X(q)$ is considered a random functional and $X(s,t)$ a random field (see Eqs. (1.12) and (1.13) of Chapter XIII). When we apply Definition 1.3 in particular cases, we need to specify δq. There is an infinite number of ways we could change q in the functional. Obviously, to get a consistent definition of a functional derivative, we clearly need a more definite specification of δq. A well-known specification, which is also used in the context of generalized S/TRF theory (Chapter XIII), is to choose

$$\delta q(\boldsymbol{p})=\varepsilon\delta(\boldsymbol{p}-\boldsymbol{p}_0)=\varepsilon\prod_{i=1}^{n,0}\delta\left(s_i-s_{0,i}\right),\tag{1.11}$$

[4]Here, the mean value theorem, $\int_{D\boldsymbol{p}}d\boldsymbol{p}X(\boldsymbol{p})\delta q(\boldsymbol{p})=X(\boldsymbol{p}_1)\int_{D\boldsymbol{p}}d\boldsymbol{p}\,\delta q(\boldsymbol{p})$, has been used, where $\boldsymbol{p}_1$ lies in the neighborhood of $\boldsymbol{p}_0$, and then the limit is taken as $|D\boldsymbol{p}|\to 0$.

$\boldsymbol{p} \in R^{n+1}$, $s_0 = t$, and $s_{0,0} = t_0$ (by convention), and the ε is real valued. In this case,

$$\begin{aligned}\frac{\delta}{\delta q(\boldsymbol{p}_0)} X(q) &= \lim_{\varepsilon \to 0} \frac{X(q(\boldsymbol{p}) + \varepsilon\delta(\boldsymbol{p} - \boldsymbol{p}_0)) - X(q(\boldsymbol{p}))}{\varepsilon} \\ &= \frac{\partial}{\partial \varepsilon} X(q(\boldsymbol{p}) + \varepsilon\delta(\boldsymbol{p} - \boldsymbol{p}_0))\bigg|_{\varepsilon=0} \\ &= \frac{\partial X(q)}{\partial q(\boldsymbol{p})} \delta(\boldsymbol{p}_0 - \boldsymbol{p}),\end{aligned} \tag{1.12a–c}$$

where $\int d\boldsymbol{p}\delta q(\boldsymbol{p}) = \varepsilon$, and the rule is introduced that the limit $\varepsilon \to 0$ has to be taken first, before any other possible limiting operations. As Eqs. (1.12a–c) demonstrate, the functional derivative has been reduced to a conventional derivative. It is also worth-noticing that the functional derivative notation of Eq. (1.12b) is most often used in physical sciences. The above process is, perhaps, better appreciated with the help of a few examples.

Example 1.8

Let us assume the following functional form for the field $X(s, t)$ in $R^{1,1}$,

$$X(q) = \int_a^b dt X(s, t) q(t),$$

where a and b are real limits. Then, letting $\delta q(t) = \varepsilon\delta(t - t_0)$ based on Eq. (1.11), the functional derivative is given by

$$\begin{aligned}\frac{\delta}{\delta q(t_0)} X(q) &= \lim_{\varepsilon \to 0} \frac{1}{\varepsilon}\left[\int_a^b dt X(s, t)(q(t) + \varepsilon\delta(t - t_0)) - \int_a^b dt X(s, t) q(t)\right] \\ &= \int_a^b dt X(s, t)\delta(t - t_0) = \begin{cases} X(s, t_0) & \text{if } t_0 \in [a, b], \\ 0 & \text{otherwise,}\end{cases}\end{aligned}$$

i.e., the functional derivation is reduced to the original field. Also, the derivative does not depend on the form of $q(t)$. As another illustration, let

$$X(q) = \int_a^b du \left(\frac{\partial q(u)}{\partial u}\right)^2,$$

where, as before, the a and b are real limits. Then, the functional derivative is given by

$$\begin{aligned}\frac{\delta}{\delta q(t)} X(q) &= \lim_{\varepsilon \to 0} \frac{1}{\varepsilon} \int_a^b du \left[\left(\frac{\partial(q(u) + \varepsilon\delta(u - t))}{\partial u}\right)^2 - \left(\frac{\partial q(u)}{\partial u}\right)^2\right] \\ &= 2\int_a^b du \frac{\partial q(u)}{\partial u} \frac{\partial \delta(u - t)}{\partial u} = 2\left[\frac{\partial q(u)}{\partial u}\delta(u - t)\bigg|_a^b - \int_a^b du \frac{\partial^2 q(u)}{\partial u^2}\delta(u - t)\right] \\ &= \begin{cases} -2\dfrac{\partial^2 q(t)}{\partial t^2} & \text{if } t_0 \in [a, b], \\ 0 & \text{otherwise,}\end{cases}\end{aligned}$$

i.e., the functional derivation is, again, reduced to a regular derivation.

Analogous to the basic rules of conventional differentiation are the basic rules of functional differentiation, some of which are listed in Table 1.1 (obviously, I refer to linear operations). Essentially, most cases of functional differentiation encountered in this book will be handled using the equations of Table 1.1. These equations show that in most cases a functional can be differentiated under the sign of standard differentiation or integration. There are, however, several cases in which the functional

Table 1.1 Rules of Functional Differentiation

Assumptions	Functional Derivative
a_i $(i = 1,\ldots, k)$ independent of q	$\frac{\delta}{\delta q(\boldsymbol{p})}\sum_{i=1}^{k} a_i X_i(q) = \sum_{i=1}^{k} a_i \frac{\delta}{\delta q(\boldsymbol{p})} X_i(q)$ (1.13)
Direct implementation of product rule	$\frac{\delta}{\delta q(\boldsymbol{p})}[X_1(q)X_2(q)] = X_1(q)\frac{\delta}{\delta q(\boldsymbol{p})}X_2(q) + X_2(q)\frac{\delta}{\delta q(\boldsymbol{p})}X_1(q)$ (1.14)
$\frac{d}{d\zeta}g(\zeta)$ is conventional derivative at $\zeta = X(q)$	$\frac{\delta}{\delta q(\boldsymbol{p})}g(X(q)) = \frac{\delta}{\delta q(\boldsymbol{p})}X(q)\frac{d}{d\zeta}g(\zeta)\Big\vert_{\zeta=X(q)}$ (1.15)
$X(\boldsymbol{p}) = \delta(\boldsymbol{p} - \boldsymbol{p}_1)$ in Eq. (1.10) $s_{1,0} = t_1$, $s_{0,0} = t_0$	$\frac{\delta q(\boldsymbol{p}_1)}{\delta q(\boldsymbol{p}_0)} = \delta(\boldsymbol{p}_1 - \boldsymbol{p}_0)$ (1.16)
Any pair of functions q_j, q_k	$\frac{\delta q_j(\boldsymbol{p}')}{\delta q_k(\boldsymbol{p})} = \delta_{jk}\delta(\boldsymbol{p} - \boldsymbol{p}')$ (1.17)
$X(q) = q^n(\boldsymbol{p})$ at $\boldsymbol{p}_0$	$\frac{\delta}{\delta q(\boldsymbol{p}_0)}X(q) = nq^{n-1}(\boldsymbol{p})\delta(\boldsymbol{p} - \boldsymbol{p}_0)$ (1.18)
$X(q) = \int d\boldsymbol{p}q^n(\boldsymbol{p})$	$\frac{\delta}{\delta q(\boldsymbol{p}_0)}X(q) = nq^{n-1}(\boldsymbol{p})\big\vert_{\boldsymbol{p}=\boldsymbol{p}_0}$ (1.19)
$X(q) = \int d\boldsymbol{p}\left(\frac{\partial q(\boldsymbol{p})}{\partial s_i}\right)^n$ $(i = 1,\ldots, n, 0)$	$\frac{\delta}{\delta q(\boldsymbol{p}_0)}X(q) = -n\frac{\partial}{\partial s_i}\left(\frac{\partial q(\boldsymbol{p})}{\partial s_i}\right)^{n-1}\Big\vert_{\boldsymbol{p}=\boldsymbol{p}_0}$ (1.20)
$X(q) = \int d\boldsymbol{p}g\left(\frac{\partial q(\boldsymbol{p})}{\partial s_i}\right)$	$\frac{\delta}{\delta q(\boldsymbol{p}_0)}X(q) = -\frac{\partial}{\partial s_i}\frac{\partial g}{\partial\left(\frac{\partial q}{\partial s_i}\right)}\Bigg\vert_{\boldsymbol{p}=\boldsymbol{p}_0}$ (1.21)

differentiation differs from ordinary differentiation. To illustrate these points, some examples are discussed next.

Example 1.9

Suppose that in an application the functional is defined as the gradient of a test function specified by the physics of the phenomenon, i.e.,

$$X(q) = \nabla q(\boldsymbol{p}). \tag{1.22}$$

Then, the functional derivative of Eq. (1.22) at $\boldsymbol{p}_0$ in R^{n+1} is given by

$$\frac{\delta}{\delta q(\boldsymbol{p}_0)} X(q) = \lim_{\varepsilon \to 0} \frac{\nabla (q(\boldsymbol{p}) + \varepsilon \delta(\boldsymbol{p} - \boldsymbol{p}_0) - q(\boldsymbol{p}))}{\varepsilon} = \nabla \delta(\boldsymbol{p} - \boldsymbol{p}_0). \tag{1.23}$$

This differs from the partial derivative, which is such that $\frac{\delta}{\delta q} X = 0$.

Example 1.10

Assume that we chose the test function $q(s, t)$ to denote the displacement of a string at location s and time t, so that the action of the string is given by the associated functional

$$X(q) = \mu \iint ds' dt' \left\{ \frac{1}{2} \left(\frac{\partial}{\partial t'} q(s', t') \right)^2 - \frac{v^2}{2} \left(\frac{\partial}{\partial s'} q(s', t') \right)^2 \right\} \tag{1.24}$$

(μ is the string mass per unit length). Then, using the stationarity condition of the phenomenon, we find that

$$0 = \frac{\delta}{\delta q(s, t)} X(q(s', t')) = \mu \iint ds' dt' \left\{ -\frac{\partial q(s', t')}{\partial t'} + v^2 \frac{\partial^2 q(s', t')}{\partial s'^2} \right\} \delta(s - s') \delta(t - t'), \tag{1.25}$$

where (see Table 1.1)

$$\frac{\delta q(s', t')}{\delta q(s, t)} = \delta(s - s') \delta(t - t'), \tag{1.26}$$

and the property of delta functions

$$g(s) \frac{\partial \delta(s - s')}{\partial s} = -\frac{\partial g(s)}{\partial s} \delta(s - s') \tag{1.27}$$

(g is a known function) have been used. Lastly, Eq. (1.25) leads to the wave equation

$$\frac{\partial^2 q(s, t)}{\partial t^2} = v^2 \frac{\partial^2 q(s, t)}{\partial s^2}, \tag{1.28}$$

which is the well-known equation of motion, in this case.

Some other interesting functional expressions, for future use, are discussed next. An exponential expression is as follows,

$$X(q_1 + q_2) = \left[e^{\int d\boldsymbol{p} q_2(\boldsymbol{p}) \frac{\delta}{\delta q_1(\boldsymbol{p})}} \right] X(q_1), \tag{1.29}$$

where the operator on the right-hand side of Eq. (1.29) is linked to the Taylor series expansion by definition, i.e.,

$$\left[e^{\int d\boldsymbol{p} q_2(\boldsymbol{p})\frac{\delta}{\delta q_1(\boldsymbol{p})}}\right] X(q_1) = \left[1 + \int d\boldsymbol{p} q_2(\boldsymbol{p})\frac{\delta}{\delta q_1(\boldsymbol{p})} + \frac{1}{2!}\left(\int d\boldsymbol{p} q_2(\boldsymbol{p})\frac{\delta}{\delta q_1(\boldsymbol{p})}\right)^2 \ldots\right] X(q_1), \quad (1.30)$$

where the differential operator $\frac{\delta}{\delta q_1(\boldsymbol{p})}$ is treated as a variable.

The application of the functional derivative on the CFL notion produces some very important results. The partial derivatives that give the RV moments in the case of the CF (Eqs. 2.7a–b of Chapter IV) now become the CFL derivatives that give the random field moments. Starting from the CFL definition in Eq. (1.3), it can be shown that the functional derivative with respect to q evaluated at the space–time point $\boldsymbol{p}_1$ is given by

$$\frac{1}{i}\frac{\delta}{\delta q(\boldsymbol{p}_1)}\Phi_X(q) = \overline{X(\boldsymbol{p}_1)e^{i\langle q(\boldsymbol{p}_1), X(\boldsymbol{p}_1)\rangle}}. \quad (1.31a)$$

Repeating the above operator and evaluating Eq. (1.31a) in the case that $q(\boldsymbol{p}_1) = 0$ it is found that

$$\left.\frac{1}{i^k}\frac{\delta^k \Phi_X(q)}{\delta q^k(\boldsymbol{p}_1)}\right|_{q=0} = \frac{1}{i^k}\frac{\delta^k \Phi_X}{\delta q^k(\boldsymbol{p}_1)}(0) = \overline{X^k(\boldsymbol{p}_1)}. \quad (1.31b)$$

Notice that the space–time point $\boldsymbol{p}_1$ where the q function is evaluated (i.e., with which the functional derivative is performed) is contained in the domain over which the integral involved in Eqs. (1.31a–b) is defined.

Working along similar lines, the corresponding expressions for the kth-order spatiotemporal moment is

$$\left.\frac{1}{i^k}\frac{\delta^k}{\delta q(\boldsymbol{p}_1)\ldots\delta q(\boldsymbol{p}_k)}\Phi_X(q)\right|_{q=0} = \overline{X(\boldsymbol{p}_1)\ldots X(\boldsymbol{p}_k)}, \quad (1.32)$$

i.e., the right side of Eq. (1.32) is the k-point space–time covariance of the random field $X(\boldsymbol{p})$; and

$$\frac{1}{i^k}\frac{\delta^k}{\delta q(\boldsymbol{p}_1)\ldots\delta q(\boldsymbol{p}_k)}\ln\Phi_X(q)|_{q=0} = \vartheta_{X,k}(\boldsymbol{p}_1, \ldots, \boldsymbol{p}_k), \quad (1.33)$$

where $\vartheta_{X,k}(\boldsymbol{p}_1,\ldots,\boldsymbol{p}_k)$ is called the kth-*order cumulant function.* Using Eq. (1.33), we can derive the following useful expressions involving spatiotemporal moments

$$\begin{aligned}
\vartheta_{X,1}(\boldsymbol{p}_1) &= \overline{X(\boldsymbol{p}_1)},\\
\vartheta_{X,2}(\boldsymbol{p}_1, \boldsymbol{p}_2) &= \overline{X(\boldsymbol{p}_1)X(\boldsymbol{p}_2)} - \overline{X(\boldsymbol{p}_1)}\,\overline{X(\boldsymbol{p}_2)} = c_X(\boldsymbol{p}_1, \boldsymbol{p}_2),\\
\vartheta_{X,3}(\boldsymbol{p}_1, \boldsymbol{p}_2, \boldsymbol{p}_3) &= \overline{X(\boldsymbol{p}_1)X(\boldsymbol{p}_2)X(\boldsymbol{p}_3)} - \overline{X(\boldsymbol{p}_1)}\,\overline{X(\boldsymbol{p}_2)X(\boldsymbol{p}_3)}\\
&\quad - \overline{X(\boldsymbol{p}_2)}\,\overline{X(\boldsymbol{p}_1)X(\boldsymbol{p}_3)} - \overline{X(\boldsymbol{p}_3)}\,\overline{X(\boldsymbol{p}_1)X(\boldsymbol{p}_2)} + 2\overline{X(\boldsymbol{p}_1)}\,\overline{X(\boldsymbol{p}_2)}\,\overline{X(\boldsymbol{p}_3)}.
\end{aligned} \quad (1.34a\text{–}c)$$

In light of the above developments, we can draw the conclusion that the q functions may serve as auxiliary, intermediate developments that vanish from the final outcome of the analysis. For example, after functional differentiation the q functions are set equal to zero to derive random field moments in which the q no longer appear. In other cases, the form of the q function is dictated by the particular situation to which the functional is adopted (Example 1.10), or it is chosen so that it serves a specified purpose, e.g., to eliminate complex trends in the space–time distribution of the S/TRF (Chapter XIII).

In some applications, products of S/TRFs with random functionals emerge. For illustration, consider the random field $X(\boldsymbol{p})$ and the functional $\Theta(X)$. Then, it is valid that

$$\overline{X(\boldsymbol{p})\Theta(X)} = \sum_{j=1}^{\infty} \frac{1}{(j-1)!} \int d\boldsymbol{p}_1 \ldots d\boldsymbol{p}_{j-1} \vartheta_{X,j}\left(\boldsymbol{p}, \boldsymbol{p}_1, \ldots, \boldsymbol{p}_{j-1}\right) \overline{\frac{\delta^{(j-1)}}{\delta X(\boldsymbol{p}_1) \ldots \delta X\left(\boldsymbol{p}_{j-1}\right)} \Theta(X)}. \tag{1.35}$$

As regards applications, it is particularly useful that the Definition 1.2 can be easily extended to include several random fields [say, a vector S/TRF (VS/TRF), Chapter IX] by means of the following definition.

Definition 1.4

The CFL of the VS/TRF $\boldsymbol{X}(\boldsymbol{p}) = [X_1(\boldsymbol{p}) \ldots X_k(\boldsymbol{p})]^T$ is defined as

$$\Phi_X(q_1, \ldots, q_k) = \overline{e^{i \sum_{l=1}^{k} \int d\boldsymbol{p} q_l X_l}}, \tag{1.36}$$

where $X_l(\boldsymbol{p})$ $(l = 1,\ldots,k)$ are the scalar random field components of the VS/TRF $\boldsymbol{X}(\boldsymbol{p})$.

Many of the previous results derived for scalar S/TRFs can be extended accordingly. For instance, Eqs. (1.13)–(1.16) remain valid in the case of a vector random field $\boldsymbol{X}(\boldsymbol{p})$ by properly replacing $\frac{\delta}{\delta q(\boldsymbol{p})}$ with $\frac{\delta}{\delta q_l(\boldsymbol{p})}$ $(l = 1,\ldots,k)$.

Example 1.11

Consider the set of stochastic partial differential equations (SPDEs) commonly encountered in applications (see, Chapter IX),

$$\begin{aligned} \nabla \cdot [Y\, \boldsymbol{X}] &= \sum_{i=1}^{3} \frac{\partial}{\partial s_i} [Y(\boldsymbol{p}) X_i(\boldsymbol{p})] = 0, \\ \nabla \times \boldsymbol{X} &= \sum_{j=1}^{3} \sum_{k=1}^{3} \varepsilon_{ijk} \frac{\partial}{\partial s_j} [X_k(\boldsymbol{p})] = 0, \end{aligned} \tag{1.37a–b}$$

where $i = 1, 2, 3$ and ε_{ijk} is the Levi-Civita symbol defined in Table 1 of Chapter I. Physically, these equations may represent, e.g., Maxwell's equations in the absence of sources and under static conditions (i.e., constant with time), where the vector $\boldsymbol{X} = [X_1(\boldsymbol{p})\, X_2(\boldsymbol{p})\, X_3(\boldsymbol{p})]^T$ denotes the electric field and $Y(\boldsymbol{p})$ is the permittivity. The corresponding CFL, Eq. (1.36), should be written as

$$\Phi_{X,Y}(q_1, q_2, q_3, q_Y) = \int d\chi d\psi e^{i \sum_{l=1}^{3} \int d\boldsymbol{p} q_l \chi_l + i \int d\boldsymbol{p} q_Y \psi} f_{X,Y}(\chi(\boldsymbol{p}), \psi(\boldsymbol{p})), \tag{1.38}$$

where $d\chi = d\chi_1 d\chi_2 d\chi_3$. As the readers may recall, a specific reference to Eqs. (1.37a–b) was made in Chapter XI.

Complex-valued random fields (Section 6 of Chapter II) use the representation $X(\boldsymbol{p}) = X_R(\boldsymbol{p}) + iX_I(\boldsymbol{p})$, and, similarly, it can be written that $q(\boldsymbol{p}) = q_R(\boldsymbol{p}) + iq_I(\boldsymbol{p})$. In the CFL setting, $X_R(\boldsymbol{p})$ and

$X_I(\boldsymbol{p})$ are associated with the corresponding $q_R(\boldsymbol{p})$ and $q_I(\boldsymbol{p})$. Alternatively, two independent test functions $q_1(\boldsymbol{p})$ and $q_2(\boldsymbol{p})$ associated with $X(\boldsymbol{p})$ and $X^*(\boldsymbol{p})$ can be used, respectively, so that

$$\Phi_X(q_1, q_2) = \overline{e^{i[X(q_1) + X^*(q_2)]}} = \overline{e^{i\int d\boldsymbol{p}[q_1(\boldsymbol{p})X(\boldsymbol{p}) + q_2(\boldsymbol{p})X^*(\boldsymbol{p})]}}. \tag{1.39}$$

Example 1.12

In light of Eq. (1.39), the first-order statistical moments (mean values) for complex random fields are given by

$$\begin{aligned} \frac{1}{i}\frac{\delta}{\delta q_1(\boldsymbol{p})}\Phi_X(0,0) &= \overline{X(\boldsymbol{p})}, \\ \frac{1}{i}\frac{\delta}{\delta q_2(\boldsymbol{p})}\Phi_X(0,0) &= \overline{X^*(\boldsymbol{p})}. \end{aligned} \tag{1.40}$$

Similarly, the second-order statistical moments (S/TVFs) are

$$\begin{aligned} \frac{1}{i^2}\frac{\delta^2}{\delta q_1(\boldsymbol{p}_1)\delta q_2(\boldsymbol{p}_2)}\Phi_X(0,0) &= \overline{X(\boldsymbol{p}_1)X^*(\boldsymbol{p}_2)}, \\ \frac{1}{i^2}\frac{\delta^2}{\delta q_1^2(\boldsymbol{p}_1)}\Phi_X(0,0) &= \overline{X(\boldsymbol{p}_1)^2}, \\ \frac{1}{i^2}\frac{\delta^2}{\delta q_2^2(\boldsymbol{p}_2)}\Phi_X(0,0) &= \overline{X^*(\boldsymbol{p}_2)^2} \end{aligned} \tag{1.41}$$

i.e., all moments above are specified in terms of the CFL, as expected.

As a final thought for this section, the functional formalism described above has been used extensively in the study of many important physical phenomena. One of them is turbulent flow satisfying the Navier–Stokes equations, where the CFL of the complete space–time probability distribution of fluid velocity amplitudes was first derived, and, then, the covariance functions of the velocity field (which may involve simultaneous or nonsimultaneous time arguments) were obtained from the CFL by functional differentiation (Lewis and Kraichnan, 1962).

2. GAUSSIAN FUNCTIONALS

As noted earlier, the notion of a CFL can be properly extended to several S/TRFs, and such an extension is particularly interesting in the case of Gaussian vector random fields. But before we introduce the CFLs of a Gaussian VS/TRF $\boldsymbol{X}(\boldsymbol{p})$, let us first look at some appealing features of the CFL of a Gaussian scalar S/TRF.

Definition 2.1

For a Gaussian S/TRF $X(\boldsymbol{p})$, the corresponding *Gaussian* CFL (GCFL) is defined by

$$G_X(q) = e^{i\overline{X(q)} - \frac{1}{2}\sigma_X^2}, \tag{2.1}$$

with mean value

$$\overline{X(q)} = \int dp\, q(s)\overline{X(p)},$$

and variance

$$\sigma_X^2 = \overline{[X(q) - \overline{X(q)}]^2} = \int\int dp_1 dp_2 q(p_1)q(p_2)c_X(p_1,p_2).$$

The examples that follow examine some differentiation cases of the GCFL.[5]

Example 2.1

Using Eqs. (1.13)–(1.16), the functional derivative of the GCFL $G_X(q)$ is found to be

$$\frac{\delta}{\delta q(p)}G_X(q) = G_X(q)\left[i\overline{X(p)} - \int dp_1 q(p_1)c_X(p,p_1)\right], \tag{2.2}$$

where the symmetry $c_X(p, p_1) = c_X(p_1, p)$ has been used.[6]

Example 2.2

Eq. (2.1) gives

$$\ln G_X(q) = i\int dpq(p)\overline{X(p)} - \frac{1}{2}\int\int dp_1 dp_2 q(p_1)q(p_2)c_X(p_1,p_2),$$

and also,

$$\begin{aligned}\ln G_X(q) &= \sum_{k=1}^{\infty}\frac{i^k}{k!}\int dp_1 ... dp_k q(p_1)...q(p_k)\vartheta_{X,k}(p_1,...,p_k)\\ &= i\int dp_1 q(p_1)\overline{X(p_1)} - \frac{1}{2}\int dp_1 dp_2 q(p_1)q(p_2)c_X(p_1,p_2)\\ &\quad - \frac{i}{6}\int dp_1 dp_2 dp_3 q(p_1)q(p_2)q(p_3)\vartheta_{X,3}(p_1,p_2,p_3) + ...\end{aligned}$$

By comparing the last two equations, the well-known result is obtained that for a Gaussian random field all the cumulant functions of order higher than 2 are 0, viz.,

$$\vartheta_{X,k\geq 3}(p_1,...,p_k) = 0. \tag{2.3a}$$

[5]The readers may find it interesting to compare Eq. (2.1) with Eq. (1.3).

[6]Actually, the nonsymmetric part of $c_X(p_1, p_2)$, if it exists, has zero contribution to the integral since it is multiplied by a symmetric function $q(p_1)q(p_2) = q(p_2)q(p_1)$.

As a consequence of this basic result, the corresponding multipoint spatiotemporal moments are as follows:

$$\begin{aligned}&\overline{X'(p_1)...X'(p_{2i+1})} = 0,\\&\overline{X'(p_1)...X'(p_{2i})} = \sum_{DP}\overline{X'(p_{\alpha_1})X'(p_{\alpha_2})}...\overline{X'(p_{\alpha_{2k-1}})X'(p_{\alpha_{2k}})},\end{aligned} \tag{2.3b–c}$$

where $i = 1,2,\ldots,X'(p_i) = X(p_i) - \overline{X(p_i)}$, and Σ_{DP} denotes summation over all different permutations of pairs from $1,2,\ldots,2k$. In fact, the number of terms in the summation is $(2k-1)!! = 1 \times 3 \times \ldots \times (2k-1)$. For illustration,

$$\begin{aligned}\overline{X'(p_1)X'(p_2)X'(p_3)X'(p_4)} &= \overline{X'(p_1)X'(p_2)}\,\overline{X'(p_3)X'(p_4)} + \overline{X'(p_1)X'(p_3)}\,\overline{X'(p_2)X'(p_4)}\\&\quad + \overline{X'(p_1)X'(p_4)}\,\overline{X'(p_2)X'(p_3)},\end{aligned} \tag{2.4}$$

$$\begin{aligned}&\overline{X'(p_1)X'(p_2)X'(p_3)X'(p_4)X'(p_5)X'(p_6)}\\&= \overline{X'(p_1)X'(p_2)}\left[\overline{X'(p_3)X'(p_4)}\;\overline{X'(p_5)X'(p_6)} + \overline{X'(p_3)X'(p_5)}\;\overline{X'(p_4)X'(p_6)}\right.\\&\left. + \overline{X'(p_3)X'(p_6)}\;\overline{X'(p_4)X'(p_5)}\right] + \overline{X'(p_1)X'(p_3)}\left[\overline{X'(p_2)X'(p_4)}\;\overline{X'(p_5)X'(p_6)}\right.\\&\left. + \overline{X'(p_2)X'(p_5)}\;\overline{X'(p_4)X'(p_6)} + \overline{X'(p_2)X'(p_6)}\;\overline{X'(p_4)X'(p_5)}\right] + \overline{X'(p_1)X'(p_4)}\\&\times\left[\overline{X'(p_2)X'(p_3)}\;\overline{X'(p_5)X'(p_6)} + \overline{X'(p_2)X'(p_5)}\;\overline{X'(p_3)X'(p_6)} + \overline{X'(p_2)X'(p_6)}\;\overline{X'(p_3)X'(p_5)}\right]\\&+ \overline{X'(p_1)X'(p_5)}\left[\overline{X'(p_2)X'(p_3)}\;\overline{X'(p_4)X'(p_6)} + \overline{X'(p_2)X'(p_4)}\;\overline{X'(p_3)X'(p_6)}\right.\\&\left. + \overline{X'(p_2)X'(p_6)}\;\overline{X'(p_3)X'(p_4)}\right] + \overline{X'(p_1)X'(p_6)}\left[\overline{X'(p_2)X'(p_3)}\;\overline{X'(p_4)X'(p_5)}\right.\\&\left. + \overline{X'(p_2)X'(p_4)}\;\overline{X'(p_3)X'(p_5)} + \overline{X'(p_2)X'(p_5)}\;\overline{X'(p_3)X'(p_4)}\right].\end{aligned} \tag{2.5}$$

Eqs. (2.4) and (2.5) are, respectively, the fourth- and sixth-order covariance functions of the random field.

Example 2.3

In the case of a Gaussian random field, Eq. (1.35) can be simplified as

$$\overline{X(p)\Theta(X)} = \overline{X(p)}\;\overline{\Theta(X)} + \int dp_1 c_X(p,p_1)\overline{\frac{\delta}{\delta X(p_1)}\Theta(X)}, \tag{2.6}$$

where $c_X(p,p_1) = \overline{\left[X(p) - \overline{X(p)}\right]\left[X(p_1) - \overline{X(p_1)}\right]}$, and the convenient properties described by Eqs. (2.3a–c) have been taken into account.

The formal treatment of Gaussian VS/TRFs presents no significant technical problems. Indeed, many of the previous results can be extended without much difficulty in the case of a Gaussian vector random field $\boldsymbol{X}(\boldsymbol{p}) = [X_1(\boldsymbol{p})\ldots X_k(\boldsymbol{p})]^T$. In this respect, Eq. (2.6) can be properly extended to

$$\overline{X_l(\boldsymbol{p})\Theta(\boldsymbol{X})} = \overline{X_l(\boldsymbol{p})}\,\overline{\Theta(\boldsymbol{X})} + \sum_{l'=1}^{k} \int d\boldsymbol{p}_1 c_{X_lX_{l'}}(\boldsymbol{p},\boldsymbol{p}_1) \overline{\frac{\delta}{\delta X_{l'}(\boldsymbol{p}_1)}\Theta(\boldsymbol{X})}, \tag{2.7}$$

where, $\Theta(\boldsymbol{X}) = \Theta(X_1(\boldsymbol{p}),\ldots, X_k(\boldsymbol{p}))$, and $c_{X_lX_{l'}}(\boldsymbol{p},\boldsymbol{p}_1) = \overline{\left[X_l(\boldsymbol{p}) - \overline{X_l(\boldsymbol{p}_1)}\right]\left[X_{l'}(\boldsymbol{p}) - \overline{X_{l'}(\boldsymbol{p}_1)}\right]}$.

Example 2.4

Consider the vector random process $\boldsymbol{X}(t) = [X_1(t)\ldots X_k(t)]^T$ varying in time so that its components $X_l(t)$, $l = 1,\ldots,k$, satisfy the set of differential equations

$$\frac{d}{dt}X_l(t) - W_l(\boldsymbol{X}(t),t) - Z_l(\boldsymbol{X}(t),t) = 0, \tag{2.8}$$

with initial condition $X_l(0)$. The W_l are deterministic functions, and the $Z_l(\boldsymbol{p})$, $\boldsymbol{p} = (\boldsymbol{s}, t)$, are Gaussian S/TRFs such that

$$\begin{aligned} &\overline{Z_l(\boldsymbol{p})} = 0, \\ &\overline{Z_l(\boldsymbol{p})Z_{l'}(\boldsymbol{p}')} = 2\zeta_{ll'}(\boldsymbol{s},\boldsymbol{s}',t)\delta(t-t') \end{aligned} \tag{2.9a–b}$$

$(l, l' = 1, \ldots, k)$, where $\zeta_{ll'}(\boldsymbol{s},\boldsymbol{s}',t)$ are known deterministic functions. Also, let

$$\begin{aligned} \Theta(\boldsymbol{X}) &= \delta(\boldsymbol{X}(t) - \boldsymbol{s}), \\ f_X(\boldsymbol{s},t) &= \overline{\delta(\boldsymbol{X}(t) - \boldsymbol{s})}, \end{aligned} \tag{2.10a–b}$$

where each $X_l(t)$ is a function of $Z_l(\boldsymbol{p})$ via Eq. (2.8). Notice that by integrating Eq. (2.8) we get[7]

$$\begin{aligned} X_l(t) &= X_l(0) + \int_0^t dt''[W_l(\boldsymbol{X}(t''),t'') + Z_l(\boldsymbol{X}(t''),t'')] \\ &= X_l(0) + \int_0^t dt'' \int d\boldsymbol{s}\delta(\boldsymbol{X}(t'') - \boldsymbol{s})[W_l(\boldsymbol{s},t'') + Z_l(\boldsymbol{s},t'')], \end{aligned}$$

with

$$\frac{\delta X_l(t)}{\delta Z_{l'}(\boldsymbol{s}',t')} = 0, \tag{2.11}$$

[7]Here the identity $\delta(\boldsymbol{X}(t) - \boldsymbol{s})W(\boldsymbol{X}(t), t) = \delta(\boldsymbol{X}(t) - \boldsymbol{s})W(\boldsymbol{s}, t)$, for any function W, has been used.

if $t' < 0$ or $t' > t$. In light of Eq. (2.7),

$$\begin{aligned}\overline{Z_l(\boldsymbol{X}(t),t)\delta(\boldsymbol{X}(t)-\boldsymbol{s})} &= \sum_{l'=1}^{k}\int ds'\int dt' 2\zeta_{ll'}(\boldsymbol{s},\boldsymbol{s}',t)\delta(t-t')\overline{\frac{\delta}{\delta Z_{l'}(\boldsymbol{s}',t')}\delta(\boldsymbol{X}(t)-\boldsymbol{s})}\\ &= \sum_{l'=1}^{k}\int ds'\zeta_{ll'}(\boldsymbol{s},\boldsymbol{s}',t)\overline{\frac{\delta}{\delta Z_{l'}(\boldsymbol{s}',t)}\delta(\boldsymbol{X}(t)-\boldsymbol{s})},\end{aligned} \tag{2.12}$$

where $c_{Z_lZ_{l'}}(\boldsymbol{p},\boldsymbol{p}') = 2\zeta_{ll'}(\boldsymbol{s},\boldsymbol{s}',t)\delta(t-t')$.[8] Moreover, by virtue of the functional properties discussed earlier, we find that

$$\frac{\delta X_l(t)}{\delta Z_j(\boldsymbol{s}',t')} = \delta_{lj}\delta(\boldsymbol{X}(t')-\boldsymbol{s}') + \int_{t'}^{t} dt''\int ds\delta(\boldsymbol{X}(t'')-\boldsymbol{s})\left\{\sum_{l'=1}^{k}\frac{\delta X_{l'}(t'')}{\delta Z_j(\boldsymbol{s}',t')}\frac{\partial}{\partial s_{l'}}[W_l(\boldsymbol{s},t'')+Z_l(\boldsymbol{s},t'')]\right\}, \tag{2.13}$$

In light of Eq. (2.11), $\frac{\delta X_{l'}(t'')}{\delta Z_j(\boldsymbol{s}',t')} = 0$ at $t'' < t'$, which has allowed to replace the zero lower limit of the integral of t'' by t'.[9] At the limit $t' \to t$, the integral becomes 0, and Eq. (2.13) reduces to

$$\frac{\delta X_l(t)}{\delta Z_j(\boldsymbol{s}',t)} = \delta_{lj}\delta(\boldsymbol{X}(t)-\boldsymbol{s}'). \tag{2.14}$$

Then, using Eq. (2.14),

$$\begin{aligned}\frac{\delta}{\delta Z_{l'}(\boldsymbol{s}',t)}\delta(\boldsymbol{X}(t)-\boldsymbol{s}) &= -\sum_{l=1}^{k}\frac{\partial\delta(\boldsymbol{X}(t)-\boldsymbol{s})}{\partial s_l}\frac{\delta X_l(t)}{\delta Z_{l'}(\boldsymbol{s}',t)} = -\sum_{l=1}^{k}\frac{\partial\delta(\boldsymbol{X}(t)-\boldsymbol{s})}{\partial s_l}\delta_{ll'}\delta(\boldsymbol{X}(t)-\boldsymbol{s}')\\ &= -\frac{\partial}{\partial s_{l'}}[\delta(\boldsymbol{s}-\boldsymbol{s}')\delta(\boldsymbol{X}(t)-\boldsymbol{s})].\end{aligned} \tag{2.15}$$

Subsequently,

$$\overline{Z_l(\boldsymbol{X}(t),t)\delta(\boldsymbol{X}(t)-\boldsymbol{s})} = \sum_{l'=1}^{k}\left\{-\frac{\partial}{\partial s_{l'}}[\zeta_{ll'}(\boldsymbol{s},\boldsymbol{s},t)f_X(\boldsymbol{s},t)] + f_X(\boldsymbol{s},t)\left[\frac{\partial\zeta_{ll'}(\boldsymbol{s},\boldsymbol{s}',t)}{\partial s_{l'}}\right]_{\boldsymbol{s}=\boldsymbol{s}'}\right\}, \tag{2.16}$$

The results above play a key role in derivations concerning probability equations of multivariate PDF model construction (Chapter XI).

Example 2.5

Consider the heat conduction equation in an isotropic medium,

[8]Here, $\int_0^t dt'\delta(t-t') = \frac{1}{2}$, and $t' \in (0,t)$, since δ is taken to be the limit of an even covariance.

[9]Eq. (1.17) has been used, i.e., $\frac{\delta Z_l(\boldsymbol{s},t'')}{\delta Z_j(\boldsymbol{s}',t')} = \delta_{lj}\delta(\boldsymbol{s}-\boldsymbol{s}')\delta(t''-t')$.

$$\frac{\partial}{\partial t}X(\boldsymbol{p}) = D\nabla^2 X(\boldsymbol{p}), \tag{2.17}$$

where $X(\boldsymbol{p}) = X(\boldsymbol{s}, t)$ denotes a zero mean temperature field and D is heat diffusivity, with boundary and initial condition

$$\begin{aligned} X(\boldsymbol{p}) &= X(\boldsymbol{s}) \ \ \boldsymbol{s} \in A, \\ X(\boldsymbol{p})|_{t=0} &= X_0(\boldsymbol{s}), \end{aligned} \tag{2.18}$$

where $X_0(\boldsymbol{s})$ is a Gaussian random field with known statistics. The corresponding CFL is

$$\Phi_X = \overline{e^{i\int ds q(\boldsymbol{s})X(\boldsymbol{s},t)}}. \tag{2.19}$$

The time derivative of this functional satisfies the equation

$$\frac{\partial}{\partial t}\Phi_X = D\int dsq\nabla_s^2\frac{\delta}{\delta q}\Phi_X, \tag{2.20}$$

where $\nabla_s^2 = \sum_{j=1}^{3}\frac{\partial^2}{\partial s_j^2}$. Eq. (2.20) has a Gaussian solution

$$\Phi_X = e^{-\frac{1}{2}\left[\iint dsds' q(\boldsymbol{s})q(\boldsymbol{s}')c_X(\boldsymbol{s},\boldsymbol{s}',t)\right]}. \tag{2.21}$$

By substituting Eq. (2.20) into Eq. (2.19), we find that

$$\iint dsds' q(\boldsymbol{s})q(\boldsymbol{s}')\left[\frac{\partial}{\partial t}c_X(\boldsymbol{s},\boldsymbol{s}',t) - D\nabla_s^2 c_X(\boldsymbol{s},\boldsymbol{s}',t) - D\nabla_{s'}^2 c_X(\boldsymbol{s},\boldsymbol{s}'',t)\right] = 0. \tag{2.22}$$

Eq. (2.22) implies that

$$\frac{\partial}{\partial t}c_X(\boldsymbol{s},\boldsymbol{s}',t) = D\left[\nabla_s^2 c_X(\boldsymbol{s},\boldsymbol{s}',t) + \nabla_{s'}^2 c_X(\boldsymbol{s},\boldsymbol{s}',t)\right]. \tag{2.23}$$

Assuming homogeneity, i.e., $c_X(\boldsymbol{s},\boldsymbol{s}',t) = c_X(\boldsymbol{h},t)$, $\boldsymbol{h} = \boldsymbol{s}' - \boldsymbol{s}$, Eq. (2.23) reduces to

$$\frac{\partial}{\partial t}c_X(\boldsymbol{h},t) = 2D\nabla_h^2 c_X(\boldsymbol{h},t), \tag{2.24}$$

with solution the space−time covariance function

$$c_X(\boldsymbol{h},t) = \frac{H(t)}{D^{\frac{1}{2}}(2\pi t)^{\frac{3}{2}}}\int d\boldsymbol{h}' e^{-\frac{\sum_{i=1}^{3}(h_i-h_i')^2}{8Dt}} c_{X_0}(\boldsymbol{h}'), \tag{2.25}$$

where $H(t)$ is a Heaviside unit step function. The same results as above (derived in terms of the CFL) can also be derived by the classical stochastic partial differential equation (SPDE) approach, see Example 3.9 of Chapter XIV.

I will conclude this chapter with some final thoughts. Depending on the test function selected, the notion of a random functional can become physically meaningful and very flexible, admitting several interpretations. In a very real sense, the test function maintains a close link between the mathematical description and the physical phenomenon described. This means that the decision over which test functions are the most useful for interpretive purposes are by necessity context bound, and they can be influenced by the physical knowledge and empirical evidence available regarding the phenomenon under study, and in some cases of the ways experiments are set up and performed, as well (e.g., the choice of a test function should be such that the functional it specifies can describe adequately what is known, observed, and experienced about the phenomenon). As a result, the range of applications of the random functional theory introduced in this chapter is very wide and continues to increase.

GENERALIZED SPATIOTEMPORAL RANDOM FIELDS

CHAPTER OUTLINE

1. BASIC NOTIONS

There are several important phenomena and natural attributes in the real-world that exhibit heterogeneous space–time variations and complex patterns (local trends, global irregularities, fluctuations of varying magnitude, unbounded domains). In other cases, the attribute data that provide information about physical process or mechanism are discontinuous with no well-defined point values and, instead, they consist of coarse-grained measurements (i.e., obtained by averaging over specific space–time windows due to finite instrument bandwidth). These attributes cannot be described using tools of ordinary mathematical analysis [e.g., the ordinary Fourier transformation (FT) may not exist], and they cannot be studied in terms of the restrictive theories of space–time homostationarity (STHS) or space–time isostationarity (STIS) random fields discussed in Chapters VII and VIII.

Spatiotemporal Random Fields. http://dx.doi.org/10.1016/B978-0-12-803012-7.00013-1

As a matter of fact, when we are confronted with natural attributes characterized by heterogeneous (space nonhomogeneous/time nonstationary) variations, complex patterns, or coarse-grained data, it will be useful or even necessary to rely on *generalized spatiotemporal random fields* (GS/TRFs). These are random fields defined in terms of the random functional theory introduced in Chapter XII for suitable choices of the test or support functions. Some real-world situations indicating the need to use GS/TRFs and their tools (generalized FTs, delta functions, etc.) are briefly described in the following example.

Example 1.1

The space–time distribution of many air pollutants (fine particles, ozone, sulfur, and nitrogen dioxide, etc.) are characterized by complicated patterns and varying local trends that cannot be represented adequately by any of the standard space–time variability hypotheses (such as STHS and STIS). Also, the Fourier–Stieltjes representation of Eq. (3.2) of Chapter VII is useful in certain turbulence studies, but it is completely inadequate to model water wave phenomena generated by winds blowing off a shore. Furthermore, as was noted in the previous chapter, in ocean studies an unlimited ocean is often assumed, in which case, if $X(\boldsymbol{s},t)$ denotes surface displacement of ocean water, the FT of $X(\boldsymbol{s},t)$ does not exist in the ordinary sense but only in the generalized sense.

Methodologically then, the notion of GS/TRFs is a consequence of the recognition that as contextualized as the study of practice may be, modeling should constantly look beyond specific cases, toward generalizations that include a wider range of phenomena. The GS/TRF theory is, indeed, considerably richer than the STHS random field theory of Chapter VII, in the sense that the former theory can be linked to a larger number of phenomena than the latter theory. As we will see below, a very useful class of GS/TRFs is characterized by its spatial and temporal heterogeneity orders that satisfy lawful conditions of change in the space–time domain (moreover, other interesting random field models, such as fractals and wavelets, can be derived as special cases of this class for a suitable choice of the heterogeneity orders and test or support functions).

1.1 THE NOTION OF GENERALIZED SPATIOTEMPORAL RANDOM FIELD

The notion of a GS/TRF is closely linked to the notion of a random functional introduced in Chapter XII in the following respect. Let Q be a specified linear space of elements q, termed the test or support functions in the theory of functionals. As was suggested in Section 1.2 of Chapter XII, among the best-known Q spaces are the Schwartz spaces of functions, C^{∞} and C_0^{∞} (Schwartz, 1950–51), and the elements $q \in Q$ are in $R^{n,1}$, that is, $q = q(\boldsymbol{p}) = q(\boldsymbol{s},t)$. Also, let $\mathcal{H}_{2,q}$ be the Hilbert space of all random variables (RVs) $x(q)$ on Q endowed with the scalar product

$$(x(q_1), x(q_2)) = \overline{x(q_1)x(q_2)} = \iint dF(\chi_1, \chi_2)\chi_1\chi_2, \tag{1.1}$$

where $F(\chi_1,\chi_2)$ denotes the joint cumulative distribution function (CDF) of the RVs $x(q_1)$, $x(q_2)$ with

$$||x(q)||^2 = \overline{|x(q)|^2} < \infty, \tag{1.2}$$

and satisfying the linearity condition

$$x\left(\sum_{i=1}^{m}\lambda_i q_i\right)=\sum_{i=1}^{m}\lambda_i x(q_i) \tag{1.3}$$

for all test or support functions $q_i \in Q$, and all (real or complex) numbers λ_i ($i = 1,\ldots, m$). Then, the following definition is introduced.

Definition 1.1

A GS/TRF on Q, denoted as the functional $X(q)$, is the random mapping

$$X(q): Q \rightarrow \mathcal{H}_{2,q}. \tag{1.4a}$$

If $X(\boldsymbol{p})$, $\boldsymbol{p} = (\boldsymbol{s},t)$, is an ordinary spatiotemporal random field (S/TRF),[1] we can associate with it a GS/TRF by means of the continuous linear functional

$$X(q)=\left\langle q(\boldsymbol{p}),X(\boldsymbol{p})\right\rangle=\int_V d\boldsymbol{p}\,q(\boldsymbol{p})X(\boldsymbol{p}), \tag{1.4b}$$

such that[2]

$$\overline{\left|X(q_n)-X(q)\right|^2}\rightarrow 0 \quad \text{as} \quad q_n \underset{n\rightarrow\infty}{\rightarrow} q, \tag{1.4c}$$

where q, $q_n \in Q$ and $V \subseteq R^{n+1}$. The set of all continuous GS/TRFs on Q will be denoted by $\mathcal{G}$.

From a physical modeling perspective, via the test or support functions, Eqs. (1.4a–c) can provide mathematical representations of the coarse graining involved in measurements of natural attributes represented by random fields (i.e., this is the case where the actual point values of the S/TRF are inaccessible to observers, because the measurement process involves a certain averaging of the random field over the characteristic scales of the measuring equipment). I will have to say more about the physical interpretation of GS/TRFs in following sections. At this point, an example may clarify interpretational matters by linking the GS/TRF model to specific applications.

Example 1.2

G/STRFs can be seen as coarse-grained functions that represent averaging effects in human exposure measurements due to finite instrument bandwidth or numerical averaging (Christakos and Hristopulos, 1998). Also, although intermittent processes (say, rainfall) are discontinuous with no well-defined values at all points, coarse-grained values can be obtained by averaging over specific space–time windows, and then modeled by GS/TRFs. Many other attributes with complex space–time variations can be represented by GS/TRF, including river discharge records (Robert and Roy, 1990), financial processes (Mandelbrot, 1982), and fluid turbulence dynamics (Hinze, 1975).

The probability laws and the second-order space–time moments of the GS/TRF $X(q)$ can be defined in a straightforward manner by replacing the point values χ of the ordinary S/TRF by

[1] In the sense discussed in Chapter II.

[2] The $q_n \underset{n\rightarrow\infty}{\rightarrow} q$ means that all test functions q_n, q vanish outside a compact support and all the partial derivatives of q_n converge to the corresponding partial derivatives of q on this support.

the functionals $X(q)$. In particular, by virtue of Eqs. (1.1)–(1.4a–c), the second-order moments of the GS/TRF are defined as the spatiotemporal *mean functional*

$$\overline{X(q)} = \int dF_{X(q)}(\chi)\chi, \tag{1.5}$$

where $F_{X(q)}(\chi)$ denotes the CDF of $X(q)$, the spatiotemporal *variance functional*

$$\sigma_X^2(q) = \overline{\left[X(q) - \overline{X(q)}\right]^2}, \tag{1.6}$$

and the (centered) spatiotemporal *covariance functional*

$$c_X(q_1, q_2) = \overline{\left[X(q_1) - \overline{X(q_1)}\right]\left[X(q_2) - \overline{X(q_2)}\right]} \tag{1.7}$$

of the GS/TRF $X(q)$.

The $\overline{X(q)}$ is a generalized function (or distribution) on Q (in the sense of Schwartz, 1950–51). Like $X(q)$, the mean functional is also linear, since for any N it holds

$$\overline{X\left(\sum_{i=1}^{N}\lambda_i q_i\right)} = \overline{\sum_{i=1}^{N}\lambda_i X(q_i)} = \sum_{i=1}^{N}\lambda_i \overline{X(q_i)}. \tag{1.8}$$

Due to this linearity, the $c_X(q_1,q_2)$ is a bilinear functional on Q, and

$$c_X(q, q) = \overline{\left|X(q) - \overline{X(q)}\right|^2} \geq 0. \tag{1.9}$$

These functionals belong to the dual space of Q, the Q', i.e., one often writes that $c_X(q, q) \in Q'$. Also, both the mean and the covariance functionals will be assumed to be real-valued and continuous relative to the topology of Q, in the sense that

$$\overline{X(q_n)} \to \overline{X(q)} \quad \text{when } q_n \underset{n \to \infty}{\to} q, \tag{1.10a}$$

and

$$c_X\left(q_n, q_n'\right) \to c_X(q, q') \quad \text{when } q_n \underset{n \to \infty}{\to} q,\ q_n' \underset{n \to \infty}{\to} q' \tag{1.10b}$$

for q, q', q_n, $q_n' \in Q$.

Other useful second-order characteristics are the spatiotemporal *structure functional* (ξ_X) and the spatiotemporal *variogram functional* (γ_X) defined by

$$\xi_X(q_1, q_2) = 2\gamma_X(q_1, q_2) = \overline{[X(q_1) - X(q_2)]^2}. \tag{1.11}$$

Apart from the physical (real) domain considered in Eqs. (1.6)–(1.11), mathematically equivalent second-order space–time functionals may be also constructed in the spectral (frequency) domain by taking the FT of the covariance, structure, and variogram functionals (Section 3.3, and Exercise XIII.1).

Remark 1.1

Before proceeding further, I would like to make a few comments regarding notation. First, $\boldsymbol{p} = (\boldsymbol{s}, s_0) = (\boldsymbol{s}, t)$, where $s_0 = t$ by convention (in the following, s_0 and t will be used interchangeably, as convenient). Also, for convenience the following notations in the R^{n+1} and $R^{n,1}$ domains will be used interchangeably:

$$\Delta\boldsymbol{p} \leftrightarrow (\boldsymbol{h}, \tau), \quad \Delta\boldsymbol{w} \leftrightarrow (\boldsymbol{k}, \omega), \quad \Delta\boldsymbol{p}^{(\nu,\mu)} \leftrightarrow \boldsymbol{h}^{\nu}\tau^{\mu}, \quad \Delta\boldsymbol{w}^{(\nu,\mu)} \leftrightarrow \boldsymbol{k}^{\nu}\omega^{\mu}$$

(the operational usefulness of this notational equivalence has been discussed in Section 2.1 of Chapter I; see also Appendix).

Since the partial derivatives of the test functions also belong to the same spaces of functions (C^{∞} and C_0^{∞}), the partial derivatives of the GS/TRF $X(q)$ of any combined order ρ in space and order ζ in time exist and are defined by means of the random functional

$$X^{(\rho,\zeta)}(q) = (-1)^{\rho+\zeta} X\left(q^{(\rho,\zeta)}\right), \tag{1.12}$$

where ζ is a nonnegative integer, $\boldsymbol{\rho} = (\rho_1,\ldots, \rho_n)$ is a multiindex of nonnegative integers, $\rho = |\boldsymbol{\rho}|$, and

$$X\left(q^{(\rho,\zeta)}\right) = \left\langle q^{(\rho,\zeta)}(\boldsymbol{p}), X(\boldsymbol{p})\right\rangle = \int d\boldsymbol{p}\, q^{(\rho,\zeta)}(\boldsymbol{p}) X(\boldsymbol{p}). \tag{1.13}$$

That is, the key point here is that the GS/TRF derivatives are evaluated in terms of the derivatives of the test or support function. An important conclusion of the above analysis is that, although the derivative of an ordinary S/TRF may not exist in the standard sense, they can still be defined in terms of GS/TRFs. This result, in addition to being of pure theoretical interest, has significant consequences in many physical applications (see discussion in following Section 1.2).

Remark 1.2

Eq. (1.13) is easily obtained if we assume that $Q = C^{\infty}$ and use integration by parts in the expression

$$X^{(\rho,\zeta)}(q) = \left\langle q(\boldsymbol{p}), X^{(\rho,\zeta)}(\boldsymbol{p})\right\rangle = \int d\boldsymbol{p}\, q(\boldsymbol{p}) X^{(\rho,\zeta)}(\boldsymbol{p}),$$

where (as defined in Chapter VI)

$$X^{(\rho,\zeta)}(\boldsymbol{p}) = \frac{\partial^{\rho+\zeta}}{\partial \boldsymbol{s}^{\boldsymbol{\rho}} \partial t^{\zeta}} X(\boldsymbol{s}, t),$$

with $\partial \boldsymbol{s}^{\boldsymbol{\rho}} = \prod_{i=1}^{n} \partial s_i^{\rho_i}$, $\rho = |\boldsymbol{\rho}| = \sum_{i=1}^{n} \rho_i$.

The following example from Christakos and Hristopulos (1998) illustrates the application of Eq. (1.12) in terms of a random field with a generalized time derivative.

Example 1.3

A Brownian motion $X(t)$ (Section 6.1 of Chapter VI), also known as Wiener process, provides a physical model for small-scale dynamics with apparently noisy behavior at larger scales. Brownian motion increments are stationary and uncorrelated. Its realizations (sample paths) in the normalized time interval [0,1] are obtained by means of the infinite series

$$X(t) = 2^{\frac{1}{2}}\pi^{-1} \sum_{i=1}^{\infty} \frac{Z_i}{i - \frac{1}{2}} \left[\sin\left(i - \frac{1}{2}\right)\pi t\right],$$

where Z_i are $N(0,1)$ RVs. In many applications it is useful to calculate the velocity of Brownian motion, i.e., the time derivative $\frac{d}{dt}X(t)$. However, the realizations are almost everywhere continuous but nondifferentiable functions of time, and, hence, the derivative $\frac{d}{dt}X(t)$ does not exist. In fact, if we take the derivatives of the $X(t)$ series terms above we get a series that diverges, because the fast fluctuations (high orders) are undamped. The issue is resolved by evaluating the derivative of the generalized random field

$$X(q) = \int_{-\infty}^{\infty} dt' X(t - t') q(t'),$$

instead of the point field $X(t)$. The test function $q(t)$ filters out the fast fluctuations that cause series divergence and, hence, the derivative $Y(q) = X^{(1)}(q)$ exists and is stationary. Particularly, if the Gaussian test function $q(t) = (2\pi)^{-\frac{1}{2}}\sigma^{-1}e^{-\frac{t^2}{2\sigma^2}}$ is used, the generalized derivative is given by

$$X^{(1)}(q) = 2^{\frac{1}{2}} \sum_{i=1}^{\infty} Z_i \cos\left[\left(i - \frac{1}{2}\right)\pi t\right] e^{-\frac{1}{2}\left(i-\frac{1}{2}\right)^2 \pi^2 \sigma^2}.$$

The high-order terms in $X^{(1)}(q)$ are damped by the exponential that ensures convergence of the series, and hence the derivative is well defined.

Since a GS/TRF $X(q)$ cannot be assigned values at isolated spatial points−time instances $(\boldsymbol{s},t)$, unless the test function q is a delta function, the following random field has been introduced that is useful in the mathematical analysis and application of random functionals.

Definition 1.2

A *convoluted S/TRF* (CS/TRF) is defined as the random field

$$Y_q(\boldsymbol{p}) = \left\langle q(\boldsymbol{p}'), U_{\boldsymbol{p}} X(\boldsymbol{p}') \right\rangle = \int_V d\boldsymbol{p}' q(\boldsymbol{p}') U_{\boldsymbol{p}} X(\boldsymbol{p}') = q(\boldsymbol{p}) * X(\boldsymbol{p}), \tag{1.14}$$

where $U_{\boldsymbol{p}} X(\boldsymbol{p}') = X(U_{\boldsymbol{p}}(\boldsymbol{p}')) = X(\boldsymbol{p}' + \boldsymbol{p})$ is a translation transformation, and $*$ denotes convolution.

While Eq. (1.4b) can be interpreted as an average of $X(\boldsymbol{p})$ over the support V, Eq. (1.14) rather represents a nonlocal average of $X(\boldsymbol{p})$ over a window determined by $q(\boldsymbol{p})$. The partial differentiation of orders $\boldsymbol{\rho}$ in space and ζ in time of the CS/TRF is

$$Y_q^{(\boldsymbol{\rho},\zeta)}(\boldsymbol{p}) = (-1)^{\rho+\zeta} U_{\boldsymbol{p}} X\left(q^{(\boldsymbol{\rho},\zeta)}\right), \tag{1.15}$$

where

$$U_{\boldsymbol{p}} X\left(q^{(\boldsymbol{\rho},\zeta)}\right) = \left\langle q^{(\boldsymbol{\rho},\zeta)}(\boldsymbol{p}'), U_{\boldsymbol{p}} X(\boldsymbol{p}')\right\rangle = \int d\boldsymbol{p}' q^{(\boldsymbol{\rho},\zeta)}(\boldsymbol{p}') U_{\boldsymbol{p}} X(\boldsymbol{p}').$$

We will revisit the differentiation issue in following parts of the book, notably in the context of spectral analysis.

1.2 GENERALIZED SPATIOTEMPORAL RANDOM FIELD PROPERTIES AND PHYSICAL SIGNIFICANCE

An ordinary S/TRF $X(\boldsymbol{s},t)$ admits a linear extension that is the GS/TRF $X(q)$ defined by Eq. (1.4a). Depending on the choice of the test function q, the continuous linear functional of Eq. (1.4b) may admit a variety of physical interpretations. The following example connects the GS/TRF model with applications considered in previous chapters.

Example 1.4

As we saw in Example 1.3 of Chapter XII, depending on the choice of the test or support function, Eq. (1.4b) may determine point or volume attribute values. So, for a delta test function, Eq. (1.4b) represents the concentration value of an aerosol substance at a single location in the atmosphere and at a specific time instant; whereas for a step test function, Eq. (1.4b) represents the total amount of hazardous substance within a certain volume during the time period considered.

At this point, it is appropriate to comment further on the physical significance of the representations of Eqs. (1.4b) and (1.14) and describe applications where it may be more realistic to develop mathematical models in terms of $X(q)$ and $Y_q(\boldsymbol{p})$ rather than in terms of $X(\boldsymbol{p})$:

Physical significance A: In applications, statements such as "the value of the natural attribute X at spatial location $\boldsymbol{s}$ and time instant t," are purely mathematical, and what one actually observes or measures is "the value of the natural attribute X averaged over some neighborhood of $\boldsymbol{p} = (\boldsymbol{s},t)$," as in Eqs. (1.4b) or (1.14).

Physical significance B: The proper choice of q in Eqs. (1.4b) or (1.14) may assure that the random fields $X(q)$ and $Y_q(\boldsymbol{p})$ possess certain desirable properties from a modeling and a physical interpretation perspective.

Physical significance C: Eq. (1.4b) may specify a filter whose input is the ordinary S/TRF $X(\boldsymbol{p})$ and output the CS/TRF $Y_q(\boldsymbol{p})$.

Physical significance D: Analysis in terms of $X(q)$ and $Y_q(\boldsymbol{p})$ can solve problems not tractable otherwise, such as the study of space nonhomogeneous/time nonstationary random fields, or the development of differential equation representations that capture essential features of the phenomena they describe.

Example 1.5

A case of physical significance A is when the actual point values of the attribute represented by $X(\boldsymbol{p})$ are inaccessible to the observer, because the measurement process involves an averaging of the attribute values over the characteristic scales of the measuring apparatus (e.g., Yaglom, 1986). Then, Eqs. (1.4b) and (1.14) provide mathematical representations of the coarse graining involved in such measurements by linking the test or support function q to the measuring device or the instrument's window. The coarse-graining process can significantly modify the properties of random fields, a situation sometimes termed the observation effect. If the apparatus function is known, the point values of the S/TRF are determined by the deconvolution of Eqs. (1.4b) and (1.14). Such techniques have been used successfully for enhanced information recovery by spectrum deconvolution in atomic spectroscopy (e.g., Fisher et al., 1997). A case of physical significance B, is when the attribute represented by $X(\boldsymbol{p})$ has rather rough geometrical characteristics [characterized in terms of mean square (m.s.) continuity and differentiability], in which case $X(q)$ and $Y_q(\boldsymbol{p})$ often have smoother geometrical features than $X(\boldsymbol{p})$, and, thus, they provide better representations of spatiotemporal variability. A case of physical significance *C* is when, depending on the choice of the test function, we can develop a filter that removes noise and other useless quantities, and emphasizes only the properties of interest (e.g., a filter that yields maps containing high-frequency information that enhances details of the attribute pattern, Christakos, 1992). Lastly, the kind of space−time heterogeneity described in physical significance *D* has been observed in applications of climatology, oceanography, atmospheric sciences, meteorology, and weather forecasting. This is the case, e.g., when rainfall nonstationarity in area *B* is due to rainfall in area *A* and the dominant wind direction being from area *A* to area *B*. The source of rainfall spatial nonhomogeneity is the different variation patterns of cloud density, temperature, and humidity. Then Eqs. (1.4b) and (1.14) offer the theoretical means to study such phenomena.

The random fields $X(q)$ and $Y_q(\boldsymbol{p})$ are linked together and also with the original random field $X(\boldsymbol{p})$ through the interesting properties they share, which are as follows.

Property 1.1

The space $\mathcal{K}$ of ordinary S/TRFs is a subset of the space $\mathcal{G}$ of GS/TRF, viz., $\mathcal{K} \subset \mathcal{G}$.

Property 1.2

The CS/TRF of Eq. (1.14) is such that

$$Y_q(\mathbf{0}) = X(q) \tag{1.16}$$

for all $q \in Q$, and

$$Y_q(\boldsymbol{p}) = U_{\boldsymbol{p}} X(q) = X\left(U_{-\boldsymbol{p}} q\right) \tag{1.17}$$

for all $q \in Q$ and all $\boldsymbol{p} \in R^{n+1}$.

Property 1.3

The means and covariances of $X(q)$ and $Y_q(\boldsymbol{p})$ are written as the inner products

$$\overline{X(q)} = \left\langle \overline{X(\boldsymbol{p})}, q(\boldsymbol{p}) \right\rangle, \tag{1.18}$$

$$\overline{Y_q(\boldsymbol{p})} = \langle \overline{U_{\boldsymbol{p}} X(\boldsymbol{p}')}, q(\boldsymbol{p}') \rangle = q(\boldsymbol{p}) * \overline{X(\boldsymbol{p})}, \tag{1.19}$$

and

$$c_X(q_1, q_2) = \overline{\left[X(q_1) - \overline{X(q_1)}\right]\left[X(q_2) - \overline{X(q_2)}\right]} = \langle\langle c_X(\boldsymbol{p};\boldsymbol{p}'), q_1(\boldsymbol{p})\rangle, q_2(\boldsymbol{p}')\rangle, \tag{1.20}$$

$$\begin{aligned} c_Y(\boldsymbol{p};\boldsymbol{p}') &= \overline{\left[Y_{q_1}(\boldsymbol{p}) - \overline{Y_{q_1}(\boldsymbol{p})}\right]\left[Y_{q_2}(\boldsymbol{p}') - \overline{Y_{q_2}(\boldsymbol{p}')}\right]} \\ &= \langle\langle c_X(U_{\boldsymbol{p}} X(\boldsymbol{p}''), U_{\boldsymbol{p}'} X(\boldsymbol{p}''')), q_1(\boldsymbol{p}'')\rangle, q_2(\boldsymbol{p}''')\rangle. \end{aligned} \tag{1.21}$$

Thus, the mean and covariance functionals of the GS/TRF and CS/TRF are linearly related to the mean and covariance functions of the corresponding ordinary S/TRF. From Eqs. (1.18) and (1.20), we confirm that

$$\begin{aligned} \overline{X(q)} &= \overline{Y_q(\mathbf{0})}, \\ \sigma_X^2(q) &= c_X(q, q) = c_Y(\mathbf{0}, \mathbf{0}) \end{aligned} \tag{1.22a–b}$$

for all $q \in Q$, i.e., the mean and variance of $X(q)$ are equal to that of $Y_q(\boldsymbol{p})$ at $\boldsymbol{p} = \mathbf{0}$.

Property 1.4

The covariance functional of $X(q)$ is an nonnegative-definite (NND) bilinear functional so that

$$c_X(q_1, q_2) = \overline{\left|X(q) - \overline{X(q)}\right|^2} \geq 0 \tag{1.23}$$

for all $q \in Q$. Conversely, every continuous NND bilinear functional $c_X(q_1,q_2)$ in Q is a covariance functional of some GS/TRF $X(q)$.[3]

Property 1.5

As noted earlier, $X(q)$ and $Y_q(\boldsymbol{p})$ are always differentiable, even when $X(\boldsymbol{p})$ is not. The partial differentiation of orders $\boldsymbol{\rho} = (\rho_1,\ldots, \rho_n)$ in space and ζ in time was defined by Eq. (1.12), which can be also written as

$$\frac{\partial^{\rho+\zeta}}{\partial s^{\rho} \partial t^{\zeta}} X(q) = (-1)^{\rho+\zeta} X\left(\frac{\partial^{\rho+\zeta} q(\boldsymbol{p})}{\partial s^{\rho} \partial t^{\zeta}}\right). \tag{1.24}$$

Similarly for the space–time derivatives of the CS/TRF, Eq. (1.15) is valid, which can be also written as

$$\frac{\partial^{\rho+\zeta}}{\partial s^{\rho} \partial t^{\zeta}} Y_q(\boldsymbol{p}) = (-1)^{\rho+\zeta} U_{\boldsymbol{p}} X\left(\frac{\partial^{\rho+\zeta} q(\boldsymbol{p})}{\partial s^{\rho} \partial t^{\zeta}}\right). \tag{1.25}$$

Thus, although there may exist no $X^{(\rho,\zeta)}(\boldsymbol{p})$ as such, the $X^{(\rho,\,\zeta)}(q)$ and $Y_q^{(\rho,\zeta)}(\boldsymbol{p})$ can be always obtained in the sense defined above. An important consequence is that, although the derivative of an ordinary S/TRF may not exist in the standard sense, it can still be defined in terms of GS/TRFs. This feature of the GS/TRF and CS/TRF has several other interesting consequences. It leads to a more realistic evaluation of the microscale properties that are closely related to the space–time pattern of the random field. And the behavior of the spatiotemporal variability functions (S/TVFs) near the space–time origin governs some key geometrical characteristics of random fields, such as m.s. continuity and differentiability (Chapters VI and VII).

Property 1.6

By applying the Riesz–Radon theorem in terms of generalized functions, we find that the mean $\overline{X(q)}$ can be written as

$$\overline{X(q)} = \left\langle \sum_{\rho \leq \nu} \sum_{\zeta \leq \mu} q^{(\rho,\zeta)}(\boldsymbol{p}), f_{\rho,\zeta}(\boldsymbol{p}) \right\rangle, \tag{1.26}$$

where the ν, μ are nonnegative integers, $q(\boldsymbol{p}) \in C^{\infty}$, and the $f_{\rho,\zeta}(\boldsymbol{p})$ are continuous functions, only a finite number of which are different from zero on any given finite support V of C^{∞}. Integration by parts yields

[3]The covariance function $c_Y(\boldsymbol{p},\boldsymbol{p}')$ is also an NND function in the ordinary sense defined earlier (Section 3 of Chapter IV).

$$\overline{X(q)} = \left\langle \sum_{\rho \le \nu} \sum_{\zeta \le \mu} (-1)^{\rho+\zeta} f_{\rho,\zeta}^{(\rho,\zeta)}(\boldsymbol{p}), q(\boldsymbol{p}) \right\rangle. \tag{1.27}$$

A similar expression may be derived for the mean $\overline{Y_q(\boldsymbol{p})}$, namely,

$$\overline{Y_q(\boldsymbol{p})} = \left\langle \sum_{\rho \le \nu} \sum_{\zeta \le \mu} (-1)^{\rho+\zeta} U_{\boldsymbol{p}} f_{\rho,\zeta}^{(\rho,\zeta)}(\boldsymbol{p}'), q(\boldsymbol{p}') \right\rangle. \tag{1.28}$$

For convenience in the subsequent analysis, let

$$g_{\rho,\zeta}(\boldsymbol{p}) = f_{\rho,\zeta}^{(\rho,\zeta)}(\boldsymbol{p}). \tag{1.29}$$

I will conclude this section by noticing that closely related to Property 1.6 is the following subsection.

1.3 HOMOSTATIONARY GENERALIZED SPATIOTEMPORAL RANDOM FIELDS

Ordinary random field homostationarity in the group transformation sense (Section 1 of Chapter VII) can be considered for GS/TRFs, as well. The random functional $X(q)$, $q(\boldsymbol{p}) \in C^\infty$, $\boldsymbol{p} = (\boldsymbol{s}, t) \in R^{n,1}$ will be called wide sense (w.s.) STHS if its mean value $\overline{X(q)}$ and covariance functional $c_X(q_1, q_2)$ are invariant with respect to any translation transformation of the parameters, that is,

$$\overline{X(q)} = \overline{X(U_{\Delta \boldsymbol{p}} q)}, \tag{1.30}$$

$$c_X(q_1, q_2) = c_X(U_{\Delta \boldsymbol{p}} q_1, U_{\Delta \boldsymbol{p}} q_2) \tag{1.31}$$

for any $\Delta \boldsymbol{p} = (\boldsymbol{h}, \tau) \in R^{n,1}$. Clearly, when $X(q)$ is STHS, the $c_X(q_1, q_2)$ is a translation invariant NND bilinear functional, in which case the following proposition can be proven (Christakos, 1991b,c).

Proposition 1.1

If $X(q)$ is STHS on Q, there exists one and only one covariance functional $c_X(q_1, q_2)$ such that

$$\langle X(q_1), X(q_2) \rangle = c_X(q_1, q_2) \tag{1.32}$$

for $q_1, q_2 \in Q$.

I shall denote by $\mathcal{G}_0$ the set of all STHS generalized random fields. Note that $\mathcal{K}_0 \subset \mathcal{G}_0 \subset \mathcal{G}$. Similarly, the CS/TRF $Y_q(\boldsymbol{p})$ is called STHS if

$$\overline{Y_q(\boldsymbol{p})} = ct, \tag{1.33}$$

and

$$c_Y(\boldsymbol{p}, \boldsymbol{p}') = c_Y(\Delta \boldsymbol{p}) \tag{1.34}$$

for any $\Delta \boldsymbol{p} = \boldsymbol{p} - \boldsymbol{p}' = (\boldsymbol{h}, \tau) \in R^{n,1}$. The $c_Y(\boldsymbol{p}, \boldsymbol{p}')$ is an NND function.

Remark 1.3

In view of Eq. (1.28) and the condition of Eq. (1.30), it follows that in the STHS case the functions $f_{\rho,\zeta}(\boldsymbol{p})$ are constants. Therefore,

$$g_{\rho,\zeta}(\boldsymbol{p}) = \begin{cases} f_{\rho,\zeta}^{(\rho,\zeta)}(\boldsymbol{p}) = 0 & \text{for} \quad \rho, \zeta \geq 1, \\ f_{0,0}^{(0,0)}(\boldsymbol{p}) = a & \text{for} \quad \rho = \zeta = 0, \end{cases} \tag{1.35a–b}$$

where a is a constant, in which case the $\overline{X(q)}$ will have the form

$$\overline{X(q)} = a \int d\boldsymbol{p} \; q(\boldsymbol{p}) = a\langle q(\boldsymbol{p}), 1 \rangle, \tag{1.36}$$

i.e., the random functional mean is proportional to the test function integral.

The generalized covariance $c_X(q_1,q_2) \in Q'$ can be expressed in terms of the corresponding ordinary $c_X(\Delta\boldsymbol{p})$ as follows,

$$c_X(q_1, q_2) = \left\langle c_X(\Delta\boldsymbol{p}), q_1 * \breve{q}_2(\Delta\boldsymbol{p}) \right\rangle = c_X\left(q_1 * \breve{q}_2\right) \tag{1.37}$$

for all $q_1,q_2 \in Q$, where the symbol $\smile$ here denotes inversion, $\breve{q}_2(\Delta\boldsymbol{p}) = \breve{q}_2(-\Delta\boldsymbol{p})$. An example is considered below to simplify the presentation of the theoretical discussion above.

Example 1.6

Let us define in $R^{1,1}$ a zero mean Wiener random field $W(s,t)$, $s \in [s_1,s_2]$, $t \in [0,\infty)$, as a Gaussian S/TRF with covariance function

$$c_W(s, t; s', t') = \min(s - s_1, s' - s_2)\min(t, t'). \tag{1.38}$$

As we saw earlier, the $X(s,t) = \frac{\partial^2}{\partial s \partial t} W(s,t)$ will be a zero mean white-noise random field with covariance function

$$c_X(h, \tau) = \delta(h, \tau), \tag{1.39}$$

where $\Delta\boldsymbol{p} = (h,\tau) \in R^{1,1}$ and $\iint dh d\tau \delta(h,\tau) = \iint dh d\tau \delta(h)\delta(\tau)$. The corresponding GS/TRF has the covariance functional

$$c_X(q_1, q_2) = \left\langle c_x(\Delta p), q_1 * \breve{q}_2(\Delta p) \right\rangle = \left\langle \delta, q_1 * \breve{q}_2 \right\rangle = \left(q_1 * \breve{q}_2\right)(0) = \delta\left(q_1 * \breve{q}_2\right). \tag{1.40}$$

The above results can be generalized to more than one dimension. Specifically, one may define in $R^{n,1}$ the so-called Brownian sheet $B(\boldsymbol{p})$, which is a zero mean Gaussian S/TRF with covariance function

$$c_B(\boldsymbol{p},\boldsymbol{p}') = \prod_{i=1}^{n} min(s_i, s_i')min(t, t'). \tag{1.41}$$

The Brownian sheet has important applications in the context of stochastic partial differential equation (SPDE) (Chapter XIV).

Since $c_X(q_1,q_2)$ is a translation invariant bilinear functional, it will have the form $c_X(q_1,q_2) = \langle c_X^{\text{o}},q_1 * q_2\rangle$, where c_X^{o} is an NND generalized function that is the FT of some positive-tempered measure $\phi(\boldsymbol{w})$, $\boldsymbol{w} = (\boldsymbol{k},\omega)$, that is,

$$c_X^{\text{o}}(q) = \int d\phi(\boldsymbol{w})\widetilde{q}(\boldsymbol{w}),$$

where $\widetilde{q}(\boldsymbol{w})$ is the FT of $q(\boldsymbol{p})$, and $\int d\phi(\boldsymbol{w})\left(1+\boldsymbol{w}^2\right)^{-\lambda}<\infty$ for some $\lambda>0$. Moreover, in view of the FT properties of generalized functions, it is valid that

$$c_X(q_1,q_2)=\left\langle c_X^{\mathrm{o}},q_1*\breve{q}_2\right\rangle=\left\langle\phi,\widetilde{q}_1\widetilde{q}_2\right\rangle,$$

which leads to the following result (Christakos, 1991b).

Proposition 1.2

Let $X(q)$ be a GS/TRF. The covariance functional can be written as

$$c_X(q_1,q_2)=\int d\phi(\boldsymbol{w})\widetilde{q}_1(\boldsymbol{w})\widetilde{q}_2(\boldsymbol{w}), \tag{1.42}$$

where $\widetilde{q}_1(\boldsymbol{w})$ and $\widetilde{q}_2(\boldsymbol{w})$ are the FT of the test functions $q_1(\boldsymbol{p})$ and $q_2(\boldsymbol{p})$, respectively, and $\phi(\boldsymbol{w})$ is some positive-tempered measure. In this case, the $\phi(\boldsymbol{w})$ may be called the *spectral measure* of the GS/TRF.

Example 1.7

Consider once more the Example 1.6. Since

$$c_X(q_1,q_2)=\left\langle c_X^{\mathrm{o}},q_1*\breve{q}_2\right\rangle=\left\langle\delta,q_1*\breve{q}_2\right\rangle=\left\langle\phi,\widetilde{q}_1\widetilde{q}_2\right\rangle,$$

and the FT of $c_X^{\mathrm{o}}=\delta$ is $d\boldsymbol{w}$ (Lebesgue measure), we conclude that the spectral measure of $X(q)$ in this case is $d\phi(\boldsymbol{w})=d\boldsymbol{w}$.

The generalized STHS analysis has an interesting implication, as presented in the property below.

Property 1.7

The $Y_q(\boldsymbol{s},t)$ can be a zero mean STHS random field even when the associated ordinary S/TRF $X(\boldsymbol{s},t)$ is space nonhomogeneous/time nonstationary. This convenient result holds under certain conditions concerning the choice of the test functions $q(\boldsymbol{p})=q(\boldsymbol{s},t)$ and the form of the function $g_{\rho,\zeta}(\boldsymbol{p})=g_{\rho,\zeta}(\boldsymbol{s},t)$. Specifically, the q-space must be defined as

$$Q_{\nu/\mu}=\left\{q\in Q\colon\left\langle q(\boldsymbol{p}),g_{\rho,\zeta}(\boldsymbol{p})\right\rangle=0\quad\text{for all }\rho\le\nu,\zeta\le\mu\right\}, \tag{1.43}$$

and the g-space as

$$C_{\nu/\mu}=\left\{g_{\rho,\zeta}(\boldsymbol{p})\in C\colon<q(\boldsymbol{p}),g_{\rho,\zeta}(\boldsymbol{p})>=0\Rightarrow<q(\boldsymbol{p}),U_{\Delta\boldsymbol{p}}g_{\rho,\zeta}(\boldsymbol{p})>=0\quad\text{for all }\rho\le\nu,\zeta\le\mu\right\}, \tag{1.44}$$

where C is the space of continuous functions in R^{n+1} with compact support, in which case, the new space $Q_{\nu/\mu}$ is termed an *admissible space* of orders ν/μ (AS-ν/μ).

Eq. (1.43) assures a zero mean value for $Y_q(\boldsymbol{p})$ at $\boldsymbol{p}=(\boldsymbol{s},t)=(\boldsymbol{0},0)$, whereas the closeness of $C_{\nu/\mu}$ to translation is necessary in order that inferences about $X(q)$ make sense (i.e., in order that the correlation properties of $X(q)$ remain unaffected by a shift $U_{\Delta\boldsymbol{p}}$ of the space–time origin). Space–time functions $g_{\rho,\zeta}(\boldsymbol{p})$ that satisfy these conditions are of the form

$$g_{\rho,\zeta}(\boldsymbol{p})=g_{\rho,\zeta}(\boldsymbol{s},t)=\boldsymbol{s}^{\rho}t^{\zeta}e^{\alpha\cdot\boldsymbol{s}+\beta t}, \tag{1.45}$$

where $\boldsymbol{\alpha}$ and β are (real or complex) vector and number, respectively. To confirm this suggestion, suppose that the $g_{\boldsymbol{\rho},\zeta}(\boldsymbol{p})$, $\boldsymbol{p} = (s,t)$, is of the form (Eq. 1.45) and let

$$\iint dsdt\, q(\boldsymbol{s},t)\boldsymbol{s}^{\boldsymbol{\rho}}t^{\zeta}e^{\boldsymbol{\alpha}\cdot\boldsymbol{s}+\beta t} = 0.$$

Then, by applying a shift $U_{\Delta\boldsymbol{p}} = U_{\boldsymbol{h},\tau}$, and using the series expansions of Eqs. (2.8c–d) of Chapter I (see also Appendix), it can be shown that (Exercise XIII.3)

$$\iint dsdt\, q(\boldsymbol{s},t)U_{\boldsymbol{h},\tau}(\boldsymbol{s}^{\boldsymbol{\rho}}t^{\zeta})e^{U_{\boldsymbol{h},\tau}(\boldsymbol{\alpha}\cdot\boldsymbol{s}+\beta t)} = \iint dsdt\, q(\boldsymbol{s},t)(\boldsymbol{s}+\boldsymbol{h})^{\boldsymbol{\rho}}(t+\tau)^{\zeta}e^{\boldsymbol{\alpha}\cdot(\boldsymbol{s}+\boldsymbol{h})+\beta(t+\tau)} = 0.$$

In other words, the condition of Eq. (1.44) is fulfilled.

Example 1.8

Let us choose the function $\overset{\circ}{q}(\boldsymbol{s},t) \in C_0^{\infty}$ so that its FT $\widetilde{q}^{\circ}(\boldsymbol{k},\omega)$ satisfies the equation

$$[1-\widetilde{q}^{\circ}(\boldsymbol{k},\omega)]^{(\boldsymbol{\rho},\zeta)}\Big|_{(\boldsymbol{k},\omega)=(\boldsymbol{0},0)} = 0$$

for all ρ up to $2\nu' > \nu$ and all ζ up to $2\mu' > \mu$.

It can be shown that (Exercise XIII.4),

$$\iint dsdt g_{\boldsymbol{\rho},\zeta}(\boldsymbol{s},t)[\delta(\boldsymbol{s},t) - q^{\circ}(\boldsymbol{s},t)] = (-1)^{\rho+\zeta}[1-\widetilde{q}^{\circ}(\boldsymbol{k},\omega)]^{(\boldsymbol{\rho},\zeta)}\Big|_{(\boldsymbol{k},\omega)=(\boldsymbol{0},0)} = 0$$

for all ρ up to $2\nu' > \nu$ and all ζ up to $2\mu' > \mu$, by definition of $\widetilde{q}^{\circ}(\boldsymbol{k},\omega)$. Therefore,

$$q(\boldsymbol{s},t) = \delta(\boldsymbol{s},t) - q^{\circ}(\boldsymbol{s},t) \in Q_{\nu/\mu},$$

i.e., it is an appropriate test function.

From a practical point of view, both the modeling of spatiotemporal variations and the space–time attribute estimation are easier and more efficiently carried out when the $g_{\boldsymbol{\rho},\zeta}$ are pure space–time polynomials (Section 2.1 of Chapter I),

$$g_{\boldsymbol{\rho},\zeta}(\boldsymbol{s},t) = \boldsymbol{p}^{(\boldsymbol{\rho},\zeta)} = \boldsymbol{s}^{\boldsymbol{\rho}}t^{\zeta} = s_1^{\rho_1}s_2^{\rho_2}\ldots s_n^{\rho_n}t^{\zeta}. \tag{1.46a}$$

This is due mainly to the convenient invariance and linearity properties that these polynomials have, i.e.,

$$\langle q(\boldsymbol{p}), U_{\Delta\boldsymbol{p}}g_{\boldsymbol{\rho},\zeta}(\boldsymbol{p})\rangle = 0. \tag{1.46b}$$

for all $\rho \leq \nu, \zeta \leq \mu$. The associated q-space is an AS-ν/μ space, $Q_{\nu/\mu}$, as defined earlier. In sum, the "derived" fields $X(q)$ and $Y_q(\boldsymbol{p})$ have a very convenient mathematical structure. Interpretationally, this means that even if $X(\boldsymbol{p})$ represents a natural attribute that has, in general, very irregular space–time heterogeneous features, we can derive random fields $X(q)$ and $Y_q(\boldsymbol{p})$ that have regular STHS features. Hence, data analysis and modeling become much easier.

I will close this section with a final remark. As we saw in Section 1.3 of Chapter XII, a complete stochastic characterization of the GS/TRF $X(q)$ is provided by its characteristic functional (CFL) defined as

$$\Phi(q) = \overline{e^{iX(q)}}. \tag{1.47}$$

The $\Phi(q)$ must be known for any $q(\boldsymbol{p})$, such that the integral of Eq. (1.4b) exists for all possible realizations of $X(\boldsymbol{p})$. When the $\Phi(q)$ is available, one may derive the CF, probability density function (PDF), and the space–time moments of $X(\boldsymbol{p})$, see Eqs. (1.31a–b) of Chapter XII. Most of the CFL properties discussed in Chapter XII remain valid here too.

2. SPATIOTEMPORAL RANDOM FIELDS OF ORDERS ν/μ

Finding ways to probe the credibility of modeling assumptions is a critical part of successful scientific inference and interpretation. In this section, I will start with an intuitive assessment of the STHS assumption leading to the introduction of a class of heterogeneous S/TRFs that is obtained from the GS/TRF theory, and then I will present some technical details of the mathematical theory.

2.1 DEPARTURE FROM SPACE–TIME HOMOSTATIONARITY

As was discussed in Section 1.2, particularly, regarding the "Physical significance D" of GS/TRF, these random fields can be used in the composite space–time modeling of natural attributes with complicated distributional patterns (e.g., varying space nonhomogeneous/time nonstationary trends). In this respect, random field determination from a single realization is not well defined mathematically, since there is no a priori criterion for distinguishing between space–time attribute trends and fluctuations. In light of this statistical indeterminacy, trend-free (i.e., space homogeneous/time stationary, STHS, or space isotropic/time stationary, STIS) random fields have been assumed widely in scientific modeling, because they are very efficient models for explicit and numerical calculations (such models were introduced in Chapters VII and VIII). However, trend-free models may not be an adequate option for phenomena with large and complicated space–time variabilities. In view of these concerns, a class of GS/TRF is discussed in this section that is considerably more general than the restricted class of trend-free random fields. This class is capable of handling complicated space–time variations of any magnitude in a mathematically rigorous and physically meaningful manner.

The mathematical theory of a space–time heterogeneous random field presented here is based on the central idea that the variability of a random field can be characterized by means of its *degree of departure* from the STHS state. Put differently, this idea considers that the STHS hypothesis is very useful for explicit modeling calculations, and it is worth exploiting it further, even in the study of random fields that lack such features. This is accomplished by means of a mathematical operation on the heterogeneous random field that eliminates any nonhomogeneous and nonstationary parts, in which case the departure of a random field from STHS determines the order of this operation, in a sense that will be rigorously defined later.

I now come to the important class of generalized random fields with STHS increments of orders $\boldsymbol{\nu}$ in space and μ in time. That is, the variability of a random field in space–time is characterized by two integers: the vector $\boldsymbol{\nu} = (\nu_1,\ldots, \nu_n)$ for space and the scalar μ for time. In this setting, the orders $\boldsymbol{\nu}$ and μ are called:

(a) *heterogeneity orders*, referring to the levels of departure of the *original* S/TRF variation from the state of STHS or

(b) *continuity orders*, referring to the smoothness of the *derived* S/TRF variation.

In other words, the higher the values of $\boldsymbol{\nu}$ and μ: (i) the greater the departure of the original field from space homogeneity and time stationarity, respectively (i.e., the more complicated the spatial or temporal variation features, respectively) and (ii) the greater the spatial or temporal continuity of the derived field (i.e., the smoother the spatial or temporal variation features, respectively) compared to the original field.

In certain applications, the orders ν and μ may be functions of space and time, i.e., different space—time neighborhoods may have different degrees of space—time heterogeneity.

Remark 2.1

Two particular cases of the ν, μ values are worth mentioning: (a) By standard convention, the values $\nu = \mu = -1$ correspond to STHS random fields (no trends). (b) The case $\nu = \mu = 0$ denotes a random field with STHS increments represented by $Y_q(\boldsymbol{p})$. Random fields of this type may involve linear trends in space and time that are due to the space—time mean function $\overline{X(\boldsymbol{p})}$. If the mean $\overline{Y_q(\boldsymbol{p})}$ is zero, then the $\overline{X(\boldsymbol{p})}$ is a constant.

GS/TRF share the stochastic symmetries of ordinary S/TRF, i.e., homogeneous and stationary GS/TRF are defined by means of the second-order moment functionals (w.s.) or the PDFs [strict sense (s.s.)]. Random fields with higher heterogeneity orders $\boldsymbol{\nu}$ and μ demonstrate these symmetries by means of appropriately defined generalized increments (produced by difference operators that remove trends from the initial S/TRF thus producing STHS residuals, discussed later).

In this important class of random fields, the space Q in the linear mapping of Eq. (1.4) is replaced by the AS-ν/μ space, $Q_{\nu/\mu}$, defined in Eqs. (1.43)—(1.44). Otherwise said, the idea that S/TRF heterogeneity can be characterized by means of its degree of departure from STHS is linked to the key Property 1.7 of Section 1. This critical observation leads to the following definition.

Definition 2.1

A GS/TRF $X(q)$ with *space homogeneous of order* $\boldsymbol{\nu}$ and *time stationary of order* μ *increments* in $R^{n,1}$ (GS/TRF-ν/μ) is a linear mapping as in Eqs. (1.4a—b), where $Q \equiv Q_{\nu/\mu}$ and the corresponding CS/TRF $Y_q(\boldsymbol{p})$ of Eq. (1.14) is a zero mean STHS random field for all $q \in Q_{\nu/\mu}$ and all $\Delta\boldsymbol{p} = (\boldsymbol{h},\tau) \in R^{n,1}$.

The zero mean condition is imposed for convenience and does not restrict generality. Beyond the generalized random field notion of $X(q)$, the ordinary random field notion of $X(\boldsymbol{p})$ associated with the space $\mathcal{G}_{\nu/\mu}$ of all continuous GS/TRF-ν/μ is often more useful in applications. Hence, the following definition.

Definition 2.2

The $X(\boldsymbol{p})$ is called an *ordinary S/TRF of order* ν/μ (S/TRF-ν/μ) if for all $q \in Q_{\nu/\mu}$, the corresponding CS/TRF $Y_q(\boldsymbol{p})$ is zero mean STHS.

The perspective introduced by the notion of S/TRF-ν/μ raises interesting questions worthy of investigation. In this respect, I will examine below the link of the $Q_{\nu/\mu}$ space with the notion of space–time detrending. The link of the $Q_{\nu/\mu}$ space with lawfulness (e.g., derived from physical laws) will be examined in Section 2.4.

2.2 SPACE–TIME DETRENDING

The transformation of the space–time heterogeneous random field $X(\boldsymbol{p})$ into an STHS residual random field $Y(\boldsymbol{p})$[4] can be made in terms of the space–time *detrending* operator $Q_{\nu/\mu}$ that removes spatial and temporal trends from the original random field. If the multivariate PDF of $X(\boldsymbol{p})$ is known a priori, it is possible, at least in principle, to construct the transformation $Q_{\nu/\mu}$ that yields an s.s. STHS random field $Y(\boldsymbol{p})$. In reality, of course, the multivariate probability distribution is rarely if ever available, and in most applications a second-order representation of $X(\boldsymbol{p})$ is employed. Then, the $Q_{\nu/\mu}$ operator generates a w.s. STHS $Y(\boldsymbol{p})$.

The detrending operator is not unique. If one detrending operator exists, then differential operators of higher-order will also generate STHS residuals on acting on $X(\boldsymbol{p})$. Linear operators are more convenient to work with, although other $Q_{\nu/\mu}$-operators are also possible. In view of the above considerations, the following definition of an ordinary S/TRF-ν/μ is introduced below that is more convenient to implement in applications.

Definition 2.3

Given the random field $X(\boldsymbol{p})$, if an operator $Q_{\nu/\mu}$ exists so that all S/TRFs

$$Y(\boldsymbol{p}) = Q_{\nu/\mu}[X(\boldsymbol{p})] \tag{2.1}$$

are STHS, the $X(\boldsymbol{p})$ is an ordinary S/TRF-ν/μ. Without loss of generality, the residual random field $Y(\boldsymbol{p})$ is usually assumed to have zero mean.

As we will see later, the operator in Eq. (2.1) imposes some restrictions on the corresponding space–time statistics (or S/TVFs). It is worth noticing that, if the $Q_{\nu/\mu}$ operator expresses the dynamical laws that govern the natural attribute, the $X(\boldsymbol{p})$ is fully determined. For continuous random fields of the space $\mathcal{G}_{\nu/\mu}$, the detrending operator can be represented by a linear space–time differential operator associated with the corresponding space $Q_{\nu/\mu}$ that eliminates space–time polynomial trends in the mean function.

As a result, several space–time differential operators $Q_{\nu/\mu}$ exist that cover many physical applications. In particular, consider the linear homogeneous differential operator of order ν/μ defined as

$$Q_{\nu/\mu}X(\boldsymbol{p}) = \nabla_{\alpha}^{(\nu+1,\mu+1)}X(\boldsymbol{p}) = \sum_{|\rho|=\nu+1} \alpha_{\rho,\mu+1}X^{(\rho,\mu+1)}(\boldsymbol{p}), \tag{2.2}$$

where $\alpha_{\rho,\mu+1}$ are constant coefficients,

$$X^{(\rho,\mu+1)}(\boldsymbol{p}) = \frac{\partial^{|\rho|+\mu+1}}{\partial s^{\rho}\,\partial t^{\mu+1}}X(\boldsymbol{p}), \tag{2.3}$$

[4]This is the random field $Y_q(\boldsymbol{p})$, where q is herein dropped for simplicity.

$\boldsymbol{\rho}$ is the set of integers $(\rho_1,\ldots,\rho_n)$ such that $|\boldsymbol{\rho}| = \sum_{i=1}^{n} \rho_i = \nu + 1$, ρ_i denotes the order of the partial derivative with respect to s_i, and $|\boldsymbol{\rho}| = \rho$ is the order of the spatial differential operator $\frac{\partial^{|\rho|}}{\partial s^{\rho}} = \frac{\partial^{|\rho|}}{\prod_{i=1}^{n} \partial s_i^{\rho_i}}$ (Section 4 of Chapter VI). Notice that for an ordinary S/TRF-ν/μ, the differential operator $Q_{\nu/\mu}$ in Eq. (2.2) involves partial derivatives of order $\nu + 1$ in space and ordinary derivatives of order $\mu + 1$ in time, which technically eliminate space–time polynomial trends of degree ν/μ when operating on the original random field $X(\boldsymbol{p})$.[5] If the partial point derivatives of the $X(\boldsymbol{p})$ do not exist, the definition of Eq. (2.2) can be extended using generalized random fields.

For generalized random field purposes, among the space–time detrending operators $Q_{\nu/\mu}$ that are useful special cases of Eq. (2.2) are the following:

(a) The space–time *Ito-Gel'fand* differential operator

$$Q_{\nu/\mu} = D^{(\nu+1,\mu+1)} = \frac{\partial^{\nu+\mu+2}}{\partial s^{\nu} \partial t^{\mu+1}}, \tag{2.4}$$

where $\partial s^{\nu} = \prod_{i=1}^{n} \partial s_i^{\nu_i}$, and for any combination of $\nu_1,\ldots,\nu_n$ such that $\sum_{i=1}^{n} \nu_i = \nu + 1$.

(b) The space–time *delian* differential operator

$$Q_{\nu/\mu} = \nabla^{(\nu+1,\mu+1)} = \sum_{i=1}^{n} \frac{\partial^{\nu+\mu+2}}{\partial s_i^{\nu+1} \partial t^{\mu+1}}. \tag{2.5}$$

(c) The space–time *additive* differential operator

$$Q_{\nu/\mu} = \frac{\partial^{\mu+1}}{\partial t^{\mu+1}} + \nabla^{(\nu+1)} = \frac{\partial^{\mu+1}}{\partial t^{\mu+1}} + \sum_{i=1}^{n} \frac{\partial^{\nu+1}}{\partial s_i^{\nu+1}}. \tag{2.6}$$

All the above operators eliminate space–time polynomial trends of degree ν/μ. Obviously, these operators are not uniquely defined: If $Q_{\nu/\mu}$ is a differential operator of spatial order ν and temporal order μ, all operators $Q_{\nu'/\mu'}$ such that $\nu' > \nu$ and $\mu' > \mu$ are also admissible operators. On physical grounds, the choice of $Q_{\nu/\mu}$ could be linked to the space–time changes of the random field $X(\boldsymbol{p})$ whose statistical variation properties (covariance shape, spatial and/or temporal correlation ranges, spectral density peaks, etc.) are of particular interest to the investigation, and are, thus, distinguished from other changes that are uninteresting, albeit unavoidable, like mean value trends, drifts and slow field fluctuations. Lastly, the operators are interpretable, because they may be associated with SPDE representing physical laws (also, Chapter XIV).

In the context of the Ito-Gel'fand differential operator of Eq. (2.4), a special case of a GS/TRF is defined as follows.

Definition 2.4

The $X(q)$ is called a GS/TRF of order ν/μ (*GS/TRF-ν/μ*), if the random field $Y(q)$ determined by the Ito-Gel'fand differential operator, i.e.,

[5] I.e., by convention, in Eq. (2.2) the space–time derivatives are one order higher than the indicated subscript of $Q_{\nu/\mu}$.

$$
\begin{aligned}
Y(q) &= D^{(\nu+1,\mu+1)} X(q) = X^{(\nu+1,\mu+1)}(q) \\
&= (-1)^{\nu+\mu} X\left(q^{(\nu+1,\mu+1)}\right)
\end{aligned}
\tag{2.7a–b}
$$

are zero mean STHS (generalized) random fields for any given μ and any combination of $\nu_1,\ldots, \nu_n$ for which $\sum_{i=1}^{n} \nu_i = \nu + 1$.

In relation to the above Ito-Gel'fand notion of a GS/TRF-ν/μ, and in view of Definition 2.4, the ordinary S/TRF associated with the space $\mathcal{G}_{\nu/\mu}$ can be defined as follows.

Definition 2.5

The $X(\boldsymbol{p})$ is called an *ordinary* S/TRF of order ν/μ (S/TRF-ν/μ) if the random fields

$$
Y(\boldsymbol{p}) = X^{(\nu+1,\mu+1)}(\boldsymbol{p})
\tag{2.8}
$$

exist and are zero mean STHS for any given μ and any combination of $\nu_1,\ldots, \nu_n$ for which $\sum_{i=1}^{n} \nu_i = \nu + 1$.

An interesting feature of Eq. (2.8) is that given the derived field $Y(\boldsymbol{p})$, the original field $X(\boldsymbol{p})$ can be, in principle, determined in terms of the solution of Eq. (2.8). In connection with this observation, a useful proposition can be proven (Christakos, 1991b,c).

Proposition 2.1

The following random fields are ordinary S/TRF-ν/μ:

(a) The solutions $X(\boldsymbol{p})$ of the SPDE of Eq. (2.8), in which case the

$$
Y(q) = \langle q(\boldsymbol{p}), Y(\boldsymbol{p})\rangle = \left\langle q(\boldsymbol{p}), X^{(\nu+1,\mu+1)}(\boldsymbol{p})\right\rangle = Y_q^{(\nu+1,\mu+1)}(\boldsymbol{0})
\tag{2.9}
$$

is an STHS generalized random field.[6]

(b) The random field

$$
X(\boldsymbol{p}) = \sum_{\rho=0}^{\nu} \sum_{\zeta=0}^{\mu} \beta_{\rho,\zeta} g_{\rho,\zeta}(\boldsymbol{p}),
\tag{2.10}
$$

where $\beta_{\rho,\zeta}$ ($\rho \le \nu$ and $\zeta \le \mu$) are RVs in the associated Hilbert space.

Below, I will examine some interesting random field representations that are obtained by inverting the detrending $Q_{\nu/\mu}$-operators.

Example 2.1

Consider the partial residuals $Y_i(\boldsymbol{s},t)$, $i = 1,\ldots, n$, of the random field $X(\boldsymbol{s},t)$ generated by

$$
Y_i(\boldsymbol{s}, t) = Q_{\nu/\mu}[X(\boldsymbol{s}, t)] = \frac{\partial^{\nu+\mu+2}}{\partial s_i^{\nu+1} \partial t^{\mu+1}} X(\boldsymbol{s}, t).
\tag{2.11}
$$

[6] Notice that $Y_q^{(\nu+1,\mu+1)}(\boldsymbol{0}) = (-1)^{\nu+\mu+2} \langle q^{(\nu+1,\mu+1)}(\boldsymbol{p}), X(\boldsymbol{p})\rangle$.

By exploiting the linearity of the $Q_{\nu/\mu}$-operator, the following relation is obtained,

$$\frac{\partial^{\mu+1}}{\partial t^{\mu+1}} \sum_{i=1}^{n} \frac{\partial^{\nu+1}}{\partial s_i^{\nu+1}} X(s,t) = \sum_{i=1}^{n} Y_i(s,t) = Y(s,t). \tag{2.12}$$

Given the residual $Y(s,t)$, the random field $X(s,t)$ is defined by solving the above SPDE, i.e.,

$$X(s,t) = \beta \overline{Y} \left(\sum_{i=1}^{n} \theta_i^2 s_i^{\nu+1} \right) t^{\mu+1} + \eta_{\nu/\mu}(s,t) + \int ds' G_0^{(\nu+1)}(s,s') z(s',t), \tag{2.13}$$

where the θ_i are the direction cosines of an arbitrary unit vector $\boldsymbol{\theta}$ $\left(\sum_{i=1}^{n} \theta_i^2 = 1 \right)$, $\beta = \frac{1}{(\mu+1)!(\nu+1)!}$, $G_0^{(\nu+1)}(s,s')$ is the spatial Green's function of the $Q_{\nu/\mu}$-operator that satisfies

$$\sum_{i=1}^{n} \frac{\partial^{\nu+1}}{\partial s_i^{\nu+1}} G_0^{(\nu+1)}(s,s') = \delta(s-s'),$$

$\overline{Y}$ is the constant mean of the residual $Y(s,t)$ with fluctuations $Y'(s,t)$, and the

$$z(s,t) = \frac{1}{\mu!} \int_{-\infty}^{t} dt' (t-t')^{\mu} Y'(s,t')$$

represents the contribution due to the zero mean residual. Some comments are in order here: The first term in the right side of Eq. (2.13) is a monomial of degree $\nu + 1/\mu + 1$ generated by the mean of the residual $Y(s,t)$. This term includes an arbitrary spatial dependence through the direction cosines θ_i that is filtered out, and thus lost in the residual. The second term, the polynomial $\eta_{\nu/\mu}(s,t)$, offers a partial representation of space–time trends. The third term indicates that additional polynomial terms or combinations of polynomials with other functions that result from the fluctuation integral.[7]

Example 2.2

The inversion of the space–time additive $Q_{\nu/\mu}$-operator of Eq. (2.6), i.e., the solution of the equation

$$\left[\frac{\partial^{\mu+1}}{\partial t^{\mu+1}} + \sum_{i=1}^{n} \frac{\partial^{\nu+1}}{\partial s_i^{\nu+1}} \right] X(s,t) = Y(s,t), \tag{2.14}$$

gives

$$\begin{aligned} X(s,t) = \overline{Y} & \left[\frac{1}{(\nu+1)!} \sum_{i=1}^{n} \theta_i^2 s_i^{\nu+1} + \frac{1}{(\mu+1)!} \theta_{n+1}^2 t^{\mu+1} \right] \\ & + \iint ds' dt' G_0^{(\nu+1/\mu+1)}(s,s';t,t') Y'(s',t') + \eta_{\nu/\mu}(s,t), \end{aligned} \tag{2.15}$$

[7]Different S/TRF models $X(s,t)$ can be derived using different $Q_{\nu/\mu}$-operators.

where the associated Green's function satisfies the equation

$$\left[\frac{\partial^{\mu+1}}{\partial t^{\mu+1}}+\sum_{i=1}^{n}\frac{\partial^{\nu+1}}{\partial s_i^{\nu+1}}\right]G_0^{(\nu+1/\mu+1)}(s,s';t,t')=\delta(s-s')\delta(t-t'),$$

and $\sum_{i=1}^{n}\theta_i^2=1$.

2.3 ORDINARY SPATIOTEMPORAL RANDOM FIELD-ν/μ REPRESENTATIONS OF THE GENERALIZED SPATIOTEMPORAL RANDOM FIELD-ν/μ

In view of Eqs. (1.4a–b) and (1.14), one comes to the logical conclusion that to each generalized random field $X(q)$ correspond various ordinary random fields $X^a(\boldsymbol{p})$, $a=1,2,\ldots$, all having the same CS/TRF $Y_q(\boldsymbol{p})$. That is, one can write schematically that

$$\begin{array}{ccc} X(q) & \leftrightarrow & X^a(\boldsymbol{p}) \\ \downarrow\uparrow & & \downarrow\uparrow \\ X(U_{-\boldsymbol{p}}q) & = & Y_q(\boldsymbol{p}) \end{array} \tag{2.16}$$

This observation, in turn, leads to the following definition.

Definition 2.6

The set

$$\mathcal{R}_q=\{X^a(\boldsymbol{p})\}, \tag{2.17}$$

$a=1,2,\ldots$, of all ordinary S/TRF-ν/μ that have the same CS/TRF-ν/μ $Y_q(\boldsymbol{p})$ in $Q_{\nu/\mu}$ will be termed the *generalized representation set* of order ν/μ (GRS-ν/μ). Each member of the GRS-ν/μ will be considered a representation of the random functional $X(q)$.

I can now present the proposition below that generates ordinary S/TRF-ν/μ representations in terms of space–time functions $g_{\rho,\zeta}(\boldsymbol{p})$ (Christakos, 1991b).

Proposition 2.2

Let $\overset{\circ}{X}(\boldsymbol{p})$ be a representation belonging to set $\mathcal{R}_q$ of $X(q)$. The ordinary S/TRF $X^a(\boldsymbol{p})$, $a=1,2,\ldots$, is another representation if and only if it can be expressed as

$$X^a(\boldsymbol{p})=\overset{\circ}{X}(\boldsymbol{p})+\sum_{\rho\le\nu}\sum_{\zeta\le\mu}c_{\rho,\zeta}g_{\rho,\zeta}(\boldsymbol{p}), \tag{2.18}$$

where the $c_{\rho,\zeta}$ ($\rho\le\nu$ and $\zeta\le\mu$) are RVs such that

$$c_{\rho,\zeta}=\langle\eta_{\rho,\zeta}(\boldsymbol{p}),X^a(\boldsymbol{p})\rangle, \tag{2.19}$$

and the $\eta_{\rho,\zeta}$ satisfy the equation

$$\left\langle\eta_{\rho,\zeta}(\boldsymbol{p}),g_{\rho',\zeta'}(\boldsymbol{p})\right\rangle=\begin{cases}1 & \text{if}\quad \rho=\rho' \text{ and } \zeta=\zeta' \\ 0 & \text{otherwise.}\end{cases} \tag{2.20}$$

Example 2.3

By virtue of Eq. (2.18), the $Q_{\nu/\mu}$ operator can eliminate polynomial expressions, $\eta_{\nu/\mu}$, such that

$$\begin{aligned} Q_{\nu/\mu}\left[\eta_{\nu/\mu}(\boldsymbol{p})\right] &= 0, \\ Q_{\nu/\mu}\left[X^{\circ}(\boldsymbol{p})\right] &= Y(\boldsymbol{p}), \end{aligned} \tag{2.21a–b}$$

where $Y(\boldsymbol{p})$ is an STHS field, and

$$\eta_{\nu/\mu}(\boldsymbol{s},t) = \sum_{\zeta=0}^{\mu}\sum_{|\rho|=0}^{\nu} c_{\rho,\zeta}\boldsymbol{s}^{\rho}t^{\zeta} = \sum_{\zeta=0}^{\mu}\sum_{|\rho|=0}^{\nu} c_{\rho,\zeta}s_1^{\rho_1}\ldots s_n^{\rho_n}t^{\zeta}, \tag{2.21c}$$

where ν, μ, ζ, and ρ_i are integers, and the coefficients $c_{\rho,\zeta}$ are, in general, RVs with mean value $\overline{c_{\rho,\zeta}}$ and correlation $\overline{c_{\rho,\zeta}c_{\rho',\zeta'}}$. In addition, the polynomial coefficients are uncorrelated with the random field $X^{\circ}(\boldsymbol{p})$ for all $\boldsymbol{\rho}$ and ζ. As was suggested in Section 2.1 of Chapter I, an alternative notation that is more efficient than Eq. (2.21c) is given by Eq. (2.12b) of Chapter I.

As noted earlier, an ordinary S/TRF-ν/μ is not always differentiable. It can, however, be expressed in terms of a differentiable S/TRF-ν/μ, as shown in the next proposition (Christakos, 1991b).

Proposition 2.3

If $X(\boldsymbol{p})$ is a continuous S/TRF-ν/μ, the decomposition

$$X(\boldsymbol{p}) = X^{id}(\boldsymbol{p}) + Y(\boldsymbol{p}) \tag{2.22}$$

is valid, where $X^{id}(\boldsymbol{p})$ is an infinitely differentiable S/TRF-ν/μ, and $Y(\boldsymbol{p})$ is an STHS random field.

2.4 DETERMINATION OF THE OPERATOR $Q_{\nu/\mu}$ AND ITS PHYSICAL SIGNIFICANCE

The prime significance of the operator $Q_{\nu/\mu}$ is the elimination of space–time trends from the random field representing the natural attribute of interest. As was noticed in Christakos (1992) and Christakos and Hristopulos (1998), the $Q_{\nu/\mu}$ can be specified in practice, either

(a) *lawfully*, i.e., on the basis of physical laws (when available) or
(b) *empirically*, i.e., by fitting random field models locally to the data.

Underlying the techniques of case (a) is a scientific method rather than a purely data-driven perspective. Indeed, much more than data analysis is needed to make random field modeling compelling. While data-driven techniques are satisfied with the mere description of data across space–time, the operator $Q_{\nu/\mu}$ has an explanatory character as a result of its potential connection with the physical laws describing the mechanisms underlying the data. Knowledge produced from these physical laws is used in the definition of the S/TRF model and the derivation of the corresponding S/TVF models. This leads to an exact specification of attribute variation models (about which limited or no information exists) in terms of other attribute models about which sufficient information is available.

Example 2.4

It is common practice in stochastic hydrology that the hydraulic head covariance is determined from the conductivity covariance using the continuity equation and Darcy's law (Dagan, 1989). Further, as regards groundwater flow in a heterogeneous aquifer, the principle of mass conservation is expressed as

$$\nabla \cdot [K(\boldsymbol{p})\nabla X(\boldsymbol{p})] = S_s \frac{\partial}{\partial t} X(\boldsymbol{p}) + U(\boldsymbol{p}), \tag{2.23}$$

where $K(\boldsymbol{p})$ is the hydraulic conductivity field, $X(\boldsymbol{p})$ is the hydraulic head, S_s is the specific storage coefficient, and $U(\boldsymbol{p})$ is a source. In steady state (t-independent) flow with no source term, the following decomposition of the steady state hydraulic head is considered $X(\boldsymbol{s}) = \overline{X(\boldsymbol{s})} + X'(\boldsymbol{s})$, where the $\overline{X(\boldsymbol{s})}$ and $X'(\boldsymbol{s})$ denote the head mean and fluctuations, respectively.[8] Furthermore, that $\overline{X(\boldsymbol{s})}$ is not renormalized by hydraulic conductivity fluctuations implies that it satisfies the Laplace equation $\nabla^2 \overline{X(\boldsymbol{s})} = 0$, and the head field is space nonhomogeneous. Also, $\nabla^2 X(\boldsymbol{s}) = Y(\boldsymbol{s})$, where $Y(\boldsymbol{s}) = \nabla^2 X'(\boldsymbol{s})$ is a homogeneous residual. As a result of this analysis, it holds that $Q_{\nu/\mu} = \nabla^2$, i.e., for the steady state flow with no source term, the $Q_{\nu/\mu}$ operator is the Laplacian one.

In case (b), the operator $Q_{\nu/\mu}$ is chosen so that it satisfies application-related requirements (e.g., $Q_{\nu/\mu}$ is a detrending operator that annihilates empirical trend functions with space−time coordinates). Hence, the S/TRFs defined by Eq. (2.1) are capable of handling complicated space−time patterns based on the intuitive idea that the variability of an attribute can be characterized by means of its degrees of departure from STHS. Unlike case (a), in case (b) the operator $Q_{\nu/\mu}$ is not directly related to an explicit physical mechanism (e.g., diffusion or convection). In fact, it may account for the effects of more than one mechanisms acting simultaneously. In such cases, when prior information regarding the laws that govern the process is insufficient, the only feasible approach is to initially let the data speak for themselves, so to speak. At a later stage, it may be possible by examining the spatiotemporal patterns revealed by the data to formulate a mechanistic hypothesis. This approach of starting with a mathematical structure that is later given physical meaning is common in science (Stratton, 1941; Longair, 1984).

Yet, scientists who use the generalized random field theory in practice know that in many cases determining uniquely the detrending operator over the entire physical domain of interest may not be always possible. In certain applications the attribute trends vary considerably over large regions or time intervals due to the interaction of various physical mechanisms, complicated boundary conditions, and time-varying inputs. Then, the form of the detrending operator $Q_{\nu/\mu}$ changes in space−time, and it can be considered invariant only within local neighborhoods. As a consequence of this natural complexity, it is impossible to determine these neighborhoods and the form of the detrending operator without taking the data into account. Hence, in practice the neighborhoods and the local form of the operator $Q_{\nu/\mu}$ are provided by the model that leads to the optimal fit with the control data (the approach is discussed in more detail in Christakos and Hristopulos, 1998). In statistics this is usually call a "nonparametric approach," stressing the fact that no assumptions regarding the form of the $Q_{\nu/\mu}$-operator are made.

[8]The reader may notice that the twofold decomposition introduced in Eq. (2.18) puts on a more formal basis the decomposition of the hydraulic head into mean and fluctuation components as suggested here.

Example 2.5

The above approach is powerful in exploring relationships between environmental exposure and health effects, and it may also provide useful diagnostic tools based on these relationships (e.g., Christakos and Kolovos, 1999; Yu and Christakos, 2006).

In the same context, it is possible that the $Q_{\nu/\mu}$-operator enhances our knowledge about the original attribute represented by $X(\boldsymbol{p})$. Assume that the $X(\boldsymbol{p})$ leads to $Y_q(\boldsymbol{p})$ via $Q_{\nu/\mu}$. Then, the inverse operation $Q_{\nu/\mu}^{-1}$ may yield a new S/TRF representation of $X(\boldsymbol{p})$ that may contain more information than the original one (Christakos, 2010).

Example 2.6

For illustration, let us examine the very simple case where the phenomenon in $R^{1,1}$ is originally represented by $X(s,t) = as^3t^2$ (a is an RV), and let us consider the operator $Q_{2/1} = \frac{d^3}{ds^2dt}[\cdot]$. In this case $Q_{2/1}X(s,t) = 12ast = Y(s,t)$. Now, if the inverse operator, $Q_{2/1}^{-1}$, is applied on $Y(s,t)$ we find $X(s,t) = a\left(\frac{1}{6}s^3 + bs + c\right)\left(\frac{1}{2}t^2 + d\right)$, where b, c, and d are coefficients to be calculated from the available auxiliary conditions. So, it is possible that the new and more general $X(s,t)$ model offers a more complete representation of the phenomenon than the original representation does. This perspective would raise interesting questions worthy of further investigation.

Lastly, making the connection with physical significance C of Section 1.2, the $Q_{\nu/\mu}$ operation can be viewed as a spatiotemporal filter at the observation scale determined by the local neighborhood. In particular, the operation $Y = Q_{\nu/\mu}[X]$ is a high-pass filter that annihilates trends and enhances detail of the space–time pattern. Conversely, the $Q'_{\nu/\mu}[X] = X - Q_{\nu/\mu}[X]$ is a low-pass filter (containing long-term trends and seasonal effects).

3. THE CORRELATION STRUCTURE OF SPATIOTEMPORAL RANDOM FIELD-ν/μ

This section is devoted to the study of the essential spatiotemporal variation characteristics (trend and correlation structure) of S/TRF-ν/μ, both generalized and ordinary.

3.1 SPACE–TIME FUNCTIONAL STATISTICS

To start with, and in view of the preceding results, the space–time functional statistics of the generalized random field

$$X^{(\nu,\mu)}(q) = (-1)^{\nu+\mu}X\left(q^{(\nu,\mu)}\right)$$

can be defined without any technical difficulty. Specifically, $X^{(\nu,\mu)}(q)$ has a constant mean functional

$$\overline{X^{(\nu,\mu)}(q)} = (-1)^{\nu+\mu}\overline{X\left(q^{(\nu,\mu)}\right)} = a\tilde{q}(\mathbf{0}), \tag{3.1}$$

where a is a constant; and the noncentered and centered covariance functionals of $X^{(\nu,\mu)}(q)$, respectively, are as follows

$$\overline{X\left(q_1^{(\nu+1,\mu+1)}\right)X\left(q_2^{(\nu+1,\mu+1)}\right)} = \overline{X^{(\nu+1,\mu+1)}(q_1)X^{(\nu+1,\mu+1)}(q_2)} = \overline{Y(q_1)Y(q_2)},$$
$$c_X\left(q_1^{(\nu+1,\mu+1)}, q_2^{(\nu+1,\mu+1)}\right) = c_Y(q_1,q_2). \tag{3.2a–b}$$

As noted earlier, the $c_Y(q_1,q_2)$ in Eq. (3.2b) is a translation-invariant bilinear functional and, therefore, so is $c_X\left(q_1^{(\nu+1,\mu+1)}, q_2^{(\nu+1,\mu+1)}\right)$.

Taking into account the properties of bilinear functionals (for the relevant theory, see Gel'fand and Vilenkin, 1964), Eqs. (3.1) and (3.2a–b) lead to an interesting result presented next.

Proposition 3.1

Let $X(q)$ be a GS/TRF-ν/μ in $R^{n,1}$. Its mean value and covariance functional have the following forms

$$\overline{X(q)} = \sum_{0\le|\boldsymbol{\rho}|\le\nu}\sum_{0\le\zeta\le\mu} a_{\boldsymbol{\rho},\zeta}\langle s^{\boldsymbol{\rho}}t^{\zeta}, q(\mathbf{s},t)\rangle = \sum_{\rho_1}\cdots\sum_{\rho_n}\sum_{\zeta} a_{\rho_1,\ldots,\rho_n,\zeta}\langle s_1^{\rho_1}\ldots s_n^{\rho_n}t^{\zeta}, q(\mathbf{s},t)\rangle, \tag{3.3}$$

where $\alpha_{\boldsymbol{\rho},\zeta}$ are suitable coefficients, $0 \le \rho = |\boldsymbol{\rho}| = \sum_{i=1}^n \rho_i \le \nu$, and

$$c_X(q_1,q_2) = \iint_{R^{n,1}_{-\{0,0\}}} d\phi_X(\mathbf{k},\omega)\widetilde{q}_1(\mathbf{k},\omega)\widetilde{q}_2(\mathbf{k},\omega) + G\left[\widetilde{q}_1^{(\nu+1,\mu+1)}(\mathbf{0},0), \widetilde{q}_2^{(\nu+1,\mu+1)}(\mathbf{0},0)\right], \tag{3.4}$$

where $R^{n,1}_{-\{0,0\}} = (R^n - \{\mathbf{0}\}) \times (T - \{0\})$, ϕ_X is a positive-tempered measure, and G is a function in $\widetilde{q}_1^{(\nu+1,\mu+1)}(\mathbf{0},0)$ and $\widetilde{q}_2^{(\nu+1,\mu+1)}(\mathbf{0},0)$.

Example 3.1

Consider in $R^{2,1}$ the case $\nu = \mu = 1$. According to Proposition 3.1, the mean value of $X(q)$ will be

$$\overline{X(q)} = \sum_{0\le\rho\le1}\sum_{0\le\zeta\le1} a_{\rho_1\rho_2\zeta}\langle s_1^{\rho_1}s_2^{\rho_2}t^{\zeta}, q(\mathbf{s},t)\rangle$$
$$= a_{000} + a_{100}\langle s_1,q\rangle + a_{010}\langle s_2,q\rangle + a_{001}\langle t,q\rangle + a_{101}\langle s_1t,q\rangle + a_{011}\langle s_2t,q\rangle.$$

The mean value of $Y(q)$ is

$$\overline{Y(q)} = \overline{X^{(2,2)}(q)} = \frac{\partial^4}{\partial s_1^{\rho_1}\partial s_2^{\rho_2}\partial t^2}\overline{X(q)},$$

where $\rho_1 + \rho_2 = 2$, and it is valid that

$$\frac{\partial^4}{\partial s_1^2\partial t^2}\overline{X(q)} = \frac{\partial^4}{\partial s_2^2\partial t^2}\overline{X(q)} = \frac{\partial^4}{\partial s_1\partial s_2\partial t^2}\overline{X(q)} = 0.$$

In other words, $Y(q)$ has zero mean as expected. Obviously, the mean

$$\overline{X^{(1,1)}(q)} = \frac{\partial^2}{\partial s_1^{\rho_1}\partial s_2^{\rho_2}\partial t}\overline{X(q)}.$$

is constant for all possible combinations of ρ_1 and ρ_2 such that $\rho_1 + \rho_2 = 1$. That is, the $X^{(1,1)}(q)$ is a constant mean, space–time heterogeneous GS/TRF.

3.2 GENERALIZED SPATIOTEMPORAL COVARIANCE FUNCTIONS

As noted earlier, it is unavoidable that the operator $Q_{v/\mu}$ in Eq. (2.1) imposes some restrictions on the space–time statistics of $X(\boldsymbol{p})$. I proceed with the analysis of the spatiotemporal correlation structure of an ordinary S/TRF-v/μ, by introducing the following definition.

Definition 3.1

Given a continuous ordinary S/TRF v/μ $X(\boldsymbol{s},t)$, a continuous and symmetric function $\kappa_X(\Delta\boldsymbol{p})$ is termed a *generalized spatiotemporal covariance* function of order v in space and μ in time (GS/TC-v/μ) if and only if

$$(X(q_1),X(q_2)) = \langle \kappa_X(\Delta\boldsymbol{p}), q_1(\boldsymbol{p})q_2(\boldsymbol{p}')\rangle \geq 0 \tag{3.5}$$

for all $q_1,q_2 \in Q_{v/\mu}$, where $\Delta\boldsymbol{p} = \boldsymbol{p} - \boldsymbol{p}'$.

In other words, in order that a given function be a valid model of a GS/TC-v/μ it is necessary and sufficient that the condition of Eq. (3.5) is satisfied (more on the subject in Chapter XV). The κ_X can be viewed as a generalized function defined on the spaces C^∞ or C_0^∞ by means of a linear functional, as follows

$$\left\langle q(\boldsymbol{p}), \kappa_X(\boldsymbol{p})\right\rangle = \int d\boldsymbol{p}\, q(\boldsymbol{p})\kappa_X(\boldsymbol{p}), \tag{3.6}$$

where $\kappa_X(\Delta\boldsymbol{p}) = \kappa_X(\boldsymbol{h},\tau)$ and $q(\boldsymbol{p})$ is a test function in one of the above spaces (for simplicity, it is usually assumed that $\kappa_X(\boldsymbol{h},\tau)$ is a real function).

We saw earlier that with a particular GS/TRF-v/μ $X(q)$ we can associate a GRS v/μ, $\mathcal{R}_q$, whose elements are the corresponding ordinary S/TRF-v/μ. Similarly, with a particular $X(q)$ we can associate a set of GS/TC-v/μ, satisfying Definition 3.1. This set will be called the generalized spatiotemporal covariance representation set of order v/μ, and will be denoted by $\mathcal{W}_{v/\mu}$.[9] We will see latter that some interesting properties of $\mathcal{W}_{v/\mu}$ may be obtained by assuming that the GS/TC-v/μ is STIS, that is,

$$\kappa_X(\Delta\boldsymbol{p}) = \kappa_X(r,\tau),$$

where $r = |\boldsymbol{h}|$.

Remark 3.1

Let us now explore Eq. (3.2b) a little further in the light of Definition 3.1. It is valid that

$$\begin{aligned} c_X\left(q_1^{(v+1,\mu+1)}, q_2^{(v+1,\mu+1)}\right) &= \left\langle c_X(\boldsymbol{p},\boldsymbol{p}'), q_1^{(v+1,\mu+1)}(\boldsymbol{p})q_2^{(v+1,\mu+1)}(\boldsymbol{p}')\right\rangle \\ &= \left\langle c_X^{(2v+2,2\mu+2)}(\boldsymbol{p},\boldsymbol{p}'), q_1(\boldsymbol{p})q_2(\boldsymbol{p}')\right\rangle \\ &= \langle c_Y(\Delta\boldsymbol{p}), q_1(\boldsymbol{p})q_2(\boldsymbol{p}')\rangle, \end{aligned} \tag{3.7a–c}$$

[9]As was noticed in Christakos (1992), the concept of the GS/TC-v/μ is the space–time extension of the purely spatial generalized covariance in the sense of Matheron (1973).

where

$$c_X^{(2\nu+2,2\mu+2)}(\boldsymbol{p},\boldsymbol{p}') = c_Y(\Delta\boldsymbol{p}). \tag{3.7d}$$

The above PDE can be solved with respect to $c_X(\boldsymbol{p},\boldsymbol{p}')$.

To orient the readers about the preceding developments, it would be instructive to consider the $R^{1,1}$ case first.

Example 3.2

As was shown in Christakos (1991c), the solutions of the SDPE

$$X^{(\nu+1,\mu+1)}(s,t) = Y(s,t)$$

in $R^{1,1}$, where the $Y(s,t)$ is a zero mean STHS, are S/TRF-ν/μ, and

$$Y(q) = < q(s,t), Y(s,t) > = < q(s,t), X^{(\nu+1,\mu+1)}(s,t) > = Y_q^{(\nu+1,\mu+1)}(0,0).$$

Moreover, the corresponding covariance functions are related by

$$c_X^{(2\nu+2,2\mu+2)}(s,t;s',t') = c_Y(r,\tau),$$

where $r = |s' - s|$ and $\tau = t' - t$. The solution of the last PDE is

$$c_X(s,t;s',t') = \kappa_X(r,\tau) + \eta_{\nu/\mu}(s,t;s',t'), \tag{3.8}$$

where

$$\kappa_X(r,\tau) = (-1)^{\nu+\mu}\int_0^r\int_0^\tau dudu' \frac{(r-u)^{2\nu+1}(\tau-u')^{2\mu+1}}{(2\nu+1)!(2\mu+1)!}c_Y(u,u') \tag{3.9}$$

is the corresponding GS/TC-ν/μ, and $\eta_{\nu/\mu}(s,t;s',t')$ is a polynomial of degree ν in s, s' and μ in t, t'. In fact, Eq. (3.9) can be solved with respect to $c_Y(r,\tau)$, viz.,

$$c_Y(r,\tau) = (-1)^{\nu+\mu}\kappa_X^{(2\nu+2,2\mu+2)}(r,\tau), \tag{3.10}$$

where the κ_X derivatives may be obtained in the generalized function sense. As we will see later, the added value of Eq. (3.10) is that its validity can be properly extended in the $R^{n,1}$ domain to provide a useful permissibility condition for GS/TC-ν/μ, see Corollary 3.2.

Generalizations of the above results are desirable and possible. In fact, the following proposition has been shown to hold (Christakos, 1991b,c).

Proposition 3.2

Let $X(\boldsymbol{p})$ be an S/TRF-ν/μ. Its covariance function can be expressed as the decomposition relationship

$$c_X(\boldsymbol{p},\boldsymbol{p}') = \kappa_X(\Delta\boldsymbol{p}) + \eta_{\nu,\mu}(\boldsymbol{p},\boldsymbol{p}'), \tag{3.11}$$

where $\kappa_X(\Delta\boldsymbol{p})$, $\Delta\boldsymbol{p} = (\boldsymbol{h},\tau) = (\boldsymbol{s}'-\boldsymbol{s}, t'-t)$, is the associated GS/TC-ν/μ, the orders ν and μ depend on the spatial and temporal lags $\boldsymbol{h}$ and τ, and the $Q_{\nu/\mu}$-operator filters out the polynomials with variable coefficients of degree ν in $\boldsymbol{s}$, $\boldsymbol{s}'$, and degree μ in t, t', $\eta_{\nu,\mu}(\boldsymbol{p},\boldsymbol{p}')$.

Proposition 3.2 together with the definition of the GS/TRF-ν/μ lead to the following result that links the spatiotemporal covariance functional with GS/TC-ν/μ.

Corollary 3.1
If $X(q)$ is a GS/TRF-ν/μ, it is valid that

$$c_X(q_1, q_2) = \langle c_X(\boldsymbol{p},\boldsymbol{p}'), q_1(\boldsymbol{p})q_2(\boldsymbol{p}')\rangle = \langle \kappa_X(\Delta\boldsymbol{p}), q_1(\boldsymbol{p})q_2(\boldsymbol{p}')\rangle. \tag{3.12}$$

In view of Corollary 3.1, the condition of Eq. (3.5) satisfied by all $\kappa_X(\Delta\boldsymbol{p})$ also emerges from the fact that $c_X(q_1,q_2)$ is an NND functional in $Q_{\nu/\mu}$ that satisfies Eq. (3.12). In connection to the above, a continuous and symmetric function $\kappa_X(\Delta\boldsymbol{p})$ is a permissible GS/TC-ν/μ if and only if the condition of Eq. (3.5) holds.

The $\kappa_X(\Delta\boldsymbol{p})$ is termed a conditionally NND function of order ν/μ (see also, Chapter XV).

3.3 GENERALIZED SPECTRAL REPRESENTATIONS AND PERMISSIBILITY OF GENERALIZED COVARIANCES

S/TRF representations in the spectral (frequency) domain are obtained by means of FT using spectral densities of residual S/TVFs. Spectral random field representations are particularly useful, because they take advantage of the STHS properties of the residual field. If $X(\boldsymbol{s},t)$ is a differentiable S/TRF-ν/μ, the spectral representation of the covariance of the residual $Y(\boldsymbol{p})$ is written as

$$c_Y(\boldsymbol{h}, \tau) = \int d\phi_Y(\boldsymbol{k}, \omega)e^{i(\boldsymbol{k}\cdot\boldsymbol{h} - \omega\tau)}, \tag{3.13}$$

where $\phi_Y(\boldsymbol{k},\omega)$ is a positive summable measure without atom at the origin (Loeve, 1953, i.e., integrable and without a point discontinuity at the origin involving a delta function; some authors refer to an absolutely continuous measure).

Even if the ordinary FT of the S/TRF $X(\boldsymbol{s},t)$ does not exist, spectral densities for its increments of order ν/μ can be defined. Thus, for STHS random fields the spectral density is used for the covariance function, whereas for fields with only STHS increments, it is used for the variogram or structure functions.

Example 3.3
In the case of a random field $X(\boldsymbol{s},t)$ with orders $\nu/\mu = 0/0$ (i.e., STHS increments), the following representation is possible

$$X(\boldsymbol{s}, t) = X(\boldsymbol{0}, 0) + c_1 t + \boldsymbol{d}_1 \cdot \boldsymbol{s} + \int d\mathcal{M}_X(\boldsymbol{w}), \tag{3.14}$$

where c_1 and $\boldsymbol{d}_1$ are scalar and vector coefficients, respectively, and $\mathcal{M}_X(\boldsymbol{w})$ is a random density that satisfies the conditions

$$\overline{\mathcal{M}_X(\boldsymbol{w})} = 0$$

$$\overline{\mathcal{M}_X^*(\boldsymbol{w}_1)\mathcal{M}_X(\boldsymbol{w}_2)} = (2\pi)^{n+1}\delta(\boldsymbol{w}_1 + \boldsymbol{w}_2)|\mathcal{M}_X(\boldsymbol{w}_1)|^2.$$

For a real-valued S/TRF $X(\boldsymbol{s},t)$ the density is symmetric $\mathcal{M}_X^*(\boldsymbol{w}) = \mathcal{M}_X(-\boldsymbol{w})$. If the $Q_{\nu/\mu}$-operator is of the weighted additive form of Eq. (2.6) with $\nu/\mu = 0/0$, i.e.,

$$Q_{0/0} = \left[a_0\frac{\partial}{\partial t} + \sum_{i=1}^{n} a_i\frac{\partial}{\partial s_i}\right],$$

the 0/0 increment has the form

$$Y(\boldsymbol{s},t) = \overline{Y} + \int d\mathcal{M}_X(\boldsymbol{w})e^{i\boldsymbol{w}\cdot\boldsymbol{p}}(i\boldsymbol{a}\cdot\boldsymbol{w}), \tag{3.15}$$

where $\boldsymbol{a} = (a_0,\ldots, a_n)$ and $\overline{Y} = c_1 + \boldsymbol{a}\cdot\boldsymbol{d}_1$.
Also, expressions of the S/TRFs and their S/TVFs in the frequency domain are directly derived by means of the summation process.

Example 3.4

Let us assume that $X(\boldsymbol{s},t)$ is a differentiable S/TRF-ν/μ, and then let $A = \left\{\boldsymbol{a} = (\nu_1,\ldots,\nu_n) : \sum_{i=1}^{n}\nu_i = \nu+1\right\}$. By definition, the

$$Y_{\boldsymbol{a}}(\boldsymbol{p}) = X_{\boldsymbol{a}}^{(\nu+1,\mu+1)}(\boldsymbol{p}) = \frac{\partial^{\nu+\mu+2}X(\boldsymbol{p})}{\partial\boldsymbol{s}^{\boldsymbol{a}}\partial t^{\mu+1}}$$

is a zero mean STHS random field for all $\boldsymbol{a} \in A$. The spectral representation of the covariance of each $Y_{\boldsymbol{a}}(\boldsymbol{p})$ is written as (see also, Eq. 3.13)

$$c_{Y_a}(\boldsymbol{h},\tau) = \int d\phi_{Y_a}(\boldsymbol{k},\omega)e^{i(\boldsymbol{k}\cdot\boldsymbol{h}-\omega\tau)},$$

where ϕ_{Y_a}, $\boldsymbol{a} \in A$, are positive summable measures without atom at the origin. By summation of the individual covariances we define the covariance function

$$\begin{aligned} c_Y(\boldsymbol{h},\tau) &= \sum_{\boldsymbol{a}\in A} c_{Y_a}(\boldsymbol{h},\tau) = \sum_{\boldsymbol{a}\in A} \overline{X_{\boldsymbol{a}}^{(\nu+1,\mu+1)}(\boldsymbol{p})X_{\boldsymbol{a}}^{(\nu+1,\mu+1)}(\boldsymbol{p}')} \\ &= \sum_{\boldsymbol{a}\in A} c_{X_a}^{(2\nu+2,2\mu+2)}(\boldsymbol{p},\boldsymbol{p}') = \int d\phi_Y(\boldsymbol{k},\omega)e^{i(\boldsymbol{k}\cdot\boldsymbol{h}-\omega\tau)}, \end{aligned}$$

where $\phi_Y = \sum_{\boldsymbol{a}\in A}\phi_{Y_a}$.

The generalized FT $\widetilde{\kappa}_X(\boldsymbol{k},\omega)$ of $\kappa_X(\boldsymbol{h},\tau)$ is defined from the functional relationship between the real and the frequency domain

$$< q(\boldsymbol{p}), \kappa_X(\boldsymbol{p}) > = < \widetilde{q}(\boldsymbol{w}), \widetilde{\kappa}_X(\boldsymbol{w}) > = \int d\boldsymbol{w}\widetilde{q}(\boldsymbol{w})\widetilde{\kappa}_X(\boldsymbol{w}),$$

where, as usual, $\widetilde{q}(\boldsymbol{w}) = \widetilde{q}(\boldsymbol{k},\omega)$ is the FT of the test function. Explicit expressions for generalized FTs are given in Gel'fand and Shilov (1964).

In applications, an important issue is obtaining valid GS/TC-ν/μ, $\kappa_X(\Delta\boldsymbol{p})$. In principle, the solution of Eq. (2.1) determines $\kappa_X(\Delta\boldsymbol{p})$ in terms of the residual random field, and this GS/TC-ν/μ is permissible by construction. However, an explicit solution of the random field Eq. (2.1) is not always possible (this is an issue even with linear homogeneous differential operators of the form of Eq. (2.2) often used in this analysis). Instead, it is often easier to solve the expression that links κ_X with the covariance c_Y (Christakos and Hristopulos, 1998): multiplying Eq. (2.1) by $Q^*_{\nu/\mu}[X(\boldsymbol{s}',t')] = Y(\boldsymbol{s}',t')$,[10] and then taking the expected value of the product, the following covariance expression is obtained

$$U_Q \kappa_X(\Delta \boldsymbol{p}) = c_Y(\Delta \boldsymbol{p}), \tag{3.16}$$

where $U_Q = Q_{\nu/\mu} Q^*_{\nu/\mu}$ is a linear space–time operator with $Q^*_{\nu/\mu}$ being the complex conjugate of the $Q_{\nu/\mu}$ operator of Eq. (2.1).[11] The GS/TC-ν/μ obtained from the solution of Eq. (3.16) involves a number of constants (e.g., polynomial term coefficients) that are not determined by the solution. Permissibility criteria (PC, Chapter XV) are then needed to ensure that $\kappa_X(\Delta \boldsymbol{p})$ is a valid GS/TC-ν/μ. The need to derive PC that are independent of the above solutions, requires the introduction of the following definition.

Definition 3.2

A generalized function $\kappa_X(\Delta \boldsymbol{p})$ is termed conditionally NND if $U_Q \kappa_X(\Delta \boldsymbol{p})$ is an NND function.

An NND function is also conditionally NND (Gel'fand and Vilenkin, 1964), but the converse is not always true. Moreover, the following theorem provides a general PC for GS/TC-ν/μ (Christakos, 1991b, 1992).

Theorem 3.1

A function $\kappa_X(\Delta \boldsymbol{p})$ is a permissible GS/TC-ν/μ if and only if it is conditionally NND.

The next corollary presents some interesting results that are direct consequences of the preceding analysis.

Proposition 3.3

If $\kappa_X(\Delta \boldsymbol{p}) \in \mathcal{W}_{\nu/\mu}$, then

$$\kappa_X(\Delta \boldsymbol{p}) + \eta_{2\nu,2\mu}(\Delta \boldsymbol{p}) \in \mathcal{W}_{\nu/\mu} \tag{3.17}$$

too (i.e., if the former is a permissible GS/TC-ν/μ, so is the latter). And, a GS/TC-ν/μ is also a GS/TC-ν'/μ' for all $\nu' \geq \nu$ and $\mu' \geq \mu$.

Another direct consequence of the preceding analysis is described in the following corollary that has many uses in the random field modeling of heterogeneous (non-STHS) natural attributes.

Corollary 3.2

A permissible GS/TC-ν/μ satisfies the relationship

$$\nabla^{2\nu+2} \frac{\partial^{2\mu+2}}{\partial \tau^{2\mu+2}} \kappa_X(\Delta \boldsymbol{p}) = (-1)^{\nu+\mu} c_Y(\Delta \boldsymbol{p}). \tag{3.18}$$

where $c_Y(\Delta \boldsymbol{p})$ is the covariance of an STHS random field $Y(\boldsymbol{p})$.

---

[10] Here $Q^*_{\nu/\mu} = \sum_{|\rho|=\nu+1} \alpha^*_{\rho,\mu+1} \frac{\partial^{|\rho|+\mu+1}}{\partial s^{|\rho|} \partial t^{\mu+1}}$ is the complex conjugate operator of the spatiotemporal differential operator $Q_{\nu/\mu}$ defined in Eq. (2.2).

[11] In deriving Eq. (3.16), the property of the $Q_{\nu/\mu}$-operator to filter out polynomials $\eta_{\nu/\mu}(\boldsymbol{s},t)$ and $\eta_{\nu/\mu}(\boldsymbol{s}',t')$ has been taken into account.

A formulation of Eq. (3.18) in $R^{1,1}$ was derived analytically earlier, see Eq. (3.10). In relation to Eq. (3.18), the measure $\phi_Y(\boldsymbol{w})$ is the FT of $(-1)^{\nu+\mu}\frac{\partial^{2\mu+2}}{\partial\tau^{2\mu+2}}\nabla^{2\nu+2}\kappa_X(\Delta\boldsymbol{p})$. In the case that $\phi_Y(\boldsymbol{w})$ is differentiable, we can define the *generalized spectral density function of order* ν/μ, $\widetilde{\kappa}_X(\boldsymbol{w})$, as the space–time generalized FT of $\kappa_X(\Delta\boldsymbol{p})$.

Some useful PCs for the GS/TC-ν/μ in terms of spectral domain representations are discussed in considerable detail in Chapter XV. The FTs of polynomials are used in the PCs of κ_X (as was suggested in Section 1.2 of Chapter XII, the FTs of polynomial functions do not exist in the ordinary sense, because polynomial functions are not bounded, but generalized FT can be defined and evaluated). If the ordinary FT of the function $\kappa_X(\boldsymbol{h},\tau)$ exists, it is identical to the generalized FT.

Further, the properties of the ordinary FT of partial derivatives remain valid for the generalized FTs of $\kappa_X(\boldsymbol{h},\tau)$. For instance,

$$\left\langle \frac{\partial\kappa_X(\boldsymbol{h},\tau)}{\partial h_j}, q(\boldsymbol{h},\tau)\right\rangle = i\left\langle k_j\widetilde{\kappa}_X(\boldsymbol{k},\omega), \widetilde{q}(\boldsymbol{k},\omega)\right\rangle$$

$(j = 1,\ldots, n)$, and

$$\left\langle \frac{\partial\kappa_X(\boldsymbol{h},\tau)}{\partial \tau}, q(\boldsymbol{h},\tau)\right\rangle = -i\langle \omega\widetilde{\kappa}_X(\boldsymbol{k},\omega), \widetilde{q}(\boldsymbol{k},\omega)\rangle.$$

Similarly, for the operator U_Q we have

$$\langle U_Q\kappa_X(\boldsymbol{h},\tau), q(\boldsymbol{h},\tau)\rangle = \left\langle \widetilde{U}_Q\widetilde{\kappa}_X(\boldsymbol{k},\omega), \widetilde{q}(\boldsymbol{k},\omega)\right\rangle,$$

where $\widetilde{U}_Q$ is the U_Q equivalent in Fourier space. In general, the operator $\widetilde{U}_Q$ is obtained from U_Q by means of the transformations

$$\frac{\partial}{\partial h_j} \mapsto ik_j, \frac{\partial}{\partial \tau} \mapsto -i\omega.$$

For example, if $U_Q = \nabla^{2\nu+2}\frac{\partial^2}{\partial\tau^{2\mu+2}}$, then $\widetilde{U}_Q = k^{2\nu+2}\omega^{2\mu+2}$, where $k = |\boldsymbol{k}|$.

3.4 GENERALIZED COVARIANCE FUNCTION MODELS

Some comments are in order here concerning the types of functions that can be used as generalized covariance models. Similarly to ordinary covariances, GS/TC-ν/μ models can be derived that are either space–time separable or nonseparable, and I will present examples from both groups here.

First, it is noteworthy that if the covariance of $Y(\boldsymbol{p})$ is spatially isotropic and space–time separable, i.e., $c_Y(r,\tau) = c_{Y(1)}(r)c_{Y(2)}(\tau)$, $r = |\boldsymbol{h}|$, then the $\kappa_X(r,\tau)$ is separable too, i.e.,

$$\kappa_X(r,\tau) = \kappa_{X(1)}(r)\kappa_{X(2)}(\tau). \tag{3.19}$$

Below, I will examine a series of generalized covariance functions of this type. Specifically, I will start with GS/TC-ν/μ that have a space–time polynomial form. There is a good reason for this: the polynomial terms arise naturally in GS/TC-ν/μ models, because they represent space–time trends or fluctuation correlations. The distinction between trends and fluctuations is important: Brownian motion is a zero mean random process, but its variance increases linearly with the support size.

Example 3.5

Consider the SPDE of Eq. (2.8), where $Y(s,t)$ is a zero mean white-noise random field in $R^{1,1}$, so that

$$Y(s,t) = X^{(\nu+1,\mu+1)}(s,t) = \frac{\partial^{\nu+\mu+2}}{\partial s^{\nu+1}\partial t^{\mu+1}}X(s,t), \tag{3.20}$$

with covariance $c_Y(r,\tau) = \delta(r)\delta(\tau)$. Then, Eq. (3.9) gives the GS/TC-ν/μ model of the space–time polynomial form,

$$\kappa_X(r,\tau) = (-1)^{\nu+\mu}\frac{r^{2\nu+1}\tau^{2\mu+1}}{(2\nu+1)!(2\mu+1)!}, \tag{3.21}$$

which has been proven very useful in applications.

The next proposition (Christakos and Hristopulos, 1998) introduces a useful nonseparable generalization of Eq. (3.21), which can be used in many applications.

Proposition 3.4

A valid GS/TC-ν/μ function in $R^{n,1}$ is given by

$$\kappa_X(r,\tau) = \sum_{\rho=0}^{2\nu+1}\sum_{\zeta=0}^{2\mu+1}(-1)^{s(\rho)+s(\zeta)}a_{\rho\zeta}r^{\rho}\tau^{\zeta}, \tag{3.22}$$

where the coefficients $a_{\rho\zeta}$ should satisfy certain PCs so that the $\kappa_X(r,\tau)$ is a conditionally NND function in the sense of Eq. (3.5) (explicit PCs are derived in Section 4.3 of Chapter XV, and are not presented here, to avoid repetition); the $s(\rho)$ and $s(\zeta)$ are sign functions such that $s(u) = 0.5(u - \delta_{u,2p+1})$, $u = \rho, \zeta$; $p = \nu, \mu$.

Based on the above proposition, two more useful corollaries can be derived in connection with the property of Eq. (3.19).

Corollary 3.3

In many cases the polynomial GS/TC-ν/μ of Eq. (3.22) can be conveniently expressed as a product of separable space and time polynomials, i.e.,

$$\kappa_X(r,\tau) = \kappa_{X(1)}(r)\kappa_{X(2)}(\tau) = \left[\sum_{\rho=0}^{2\nu+1}(-1)^{s(\rho)}a_{\rho}r^{\rho}\right]\left[\sum_{\zeta=0}^{2\mu+1}(-1)^{s(\zeta)}b_{\zeta}\tau^{\zeta}\right], \tag{3.23}$$

where the polynomial coefficients are related via $a_{\rho\zeta} = a_{\rho}b_{\zeta}$. If both the space and time components are valid covariances, their product is also a valid covariance (specific permissibility conditions for Eq. (3.23) are also derived in Section 4.3 of Chapter XV).

Corollary 3.4

The nonseparable function of Eq. (3.22) can be further generalized to include additional, non-polynomial, terms, as follows

$$\kappa_X(r,\tau) = \alpha_0\delta(r)\delta(\tau) + \delta(r)\sum_{\zeta=0}^{\mu}(-1)^{\zeta+1}a_{\zeta}\tau^{2\zeta+1} + \delta(\tau)\sum_{\rho=0}^{\nu}(-1)^{\rho+1}b_{\rho}r^{2\rho+1}$$

$$+\sum_{\rho=0}^{\nu}\sum_{\zeta=0}^{\mu}(-1)^{\rho+\zeta}d_{\rho/\zeta}r^{2\rho+1}\tau^{2\zeta+1} + \delta_{n,2}r^{2\nu}\log r\sum_{\zeta=0}^{\mu}(-1)^{\zeta}c_{\zeta}\tau^{2\zeta+1}, \tag{3.24}$$

where $\delta_{n,2}$ is Kronecker's delta (Table 1.1 of Chapter I), and α_0, a_ζ, b_ρ, c_ζ, and $d_{\rho/\zeta}$ are suitable coefficients.

The first three terms in Eq. (3.24) represent discontinuities at the space–time origin; the fourth term is purely polynomial; the fifth term, which is logarithmic in the space lag, is obtained only in $R^{2,1}$. In Kolovos et al. (2004) the readers may find some interesting plots of the model of Eq. (3.24). Theoretically, this model is useful for natural attributes that have white-noise residuals, $Y(\boldsymbol{p})$, but due to its simplicity it has been widely used in applications.[12]

Based now on the observation that an ordinary S/TRF-ν/μ that satisfies Eq. (3.20) can be assigned a GS/TC-ν/μ of the polynomial form of Eq. (3.21), the following proposition holds (Christakos, 1991b).

Proposition 3.5

Assume that an ordinary S/TRF-ν/μ in $R^{1,1}$ can be expressed by

$$X(s,t) = \sum_{\rho=0}^{\nu}\sum_{\zeta=0}^{\mu}\frac{a_{\rho\zeta}}{\rho!\zeta!}\int_0^s\int_0^t dudu'(s-u)^\rho(t-u')^\zeta Y(u,u'), \tag{3.25}$$

where $a_{\rho\zeta}$, $\rho = 0,1,\ldots,\nu$ and $\zeta = 0,1,\ldots,\mu$ are suitable coefficients, and $Y(s,t)$ is a zero mean white-noise random field in $R^{1,1}$. Then, its GS/TC-ν/μ is of the form of Eq. (3.22).

Example 3.6

Working along lines similar to those of Example 3.5, it is found that starting with a nonseparable ordinary covariance function

$$c_Y(r,\tau) = ae^{-br-c\tau}, \tag{3.26}$$

the corresponding space–time separable GS/TC-ν/μ in $R^{1,1}$ is (Exercise XIII.5)

$$\kappa_X(r,\tau) = a\frac{(-1)^{\nu+\mu}}{(2\nu+1)!(2\mu+1)!}\frac{\gamma(2\nu+2,-br)\gamma(2\mu+2,-c\tau)}{b^{2\nu+2}c^{2\mu+2}}e^{-(br+c\tau)}, \tag{3.27}$$

where $\gamma(\cdot,\cdot)$ denotes here the incomplete gamma function. After some manipulations, Eq. (3.27) may also be written as

$$\kappa_X(r,\tau) = a\frac{(-1)^{\nu+\mu}}{b^{2\nu+2}c^{2\mu+2}}e^{-(br+c\tau)}\left[1-e^{br}\sum_{i=0}^{2\nu+1}\frac{(-br)^i}{i!}\right]\left[1-e^{c\tau}\sum_{i=0}^{2\mu+1}\frac{(-c\tau)^i}{i!}\right]. \tag{3.28}$$

Consider, for instance, the case $\nu = \mu = 0$. Then, Eq. (3.28) gives

$$\kappa_X(r,\tau) = \frac{a}{b^2c^2}\left[1+(br-1)e^{br}\right]\left[1+(c\tau-1)e^{c\tau}\right]e^{-(br+c\tau)}, \tag{3.29}$$

which is of a separable form.

---

[12]Some authors have argued that the assumption of white-noise residual is restrictive, since more flexible models can be obtained using residuals with finite range correlations.

The next proposition considers another class of separable GS/TC-ν/μ functions, which are solutions of Eq. (3.18) (Christakos and Hristopulos, 1998; Kolovos et al., 2004).

Proposition 3.6

A valid GS/TC-ν/μ function in $R^{n,1}$ is given by

$$\kappa_X(r,\tau) = [G_1(r) + \eta_{2\nu+1}(r)]\,[G_2(\tau) + \eta_{2\mu+1}(\tau)], \tag{3.30}$$

where $G_1(r)$ and $G_2(\tau)$ are Green's functions defined as

$$G_1(r) = \begin{cases} \dfrac{r^{2\nu}\log r}{2^{2\nu+1}\pi(\nu!)^2}, & n=2 \\ \dfrac{(-1)^{\nu+1}\Gamma\left(\dfrac{1}{2}-\nu\right)r^{2\nu-1}}{2^{2\nu+1}\pi^{\frac{3}{2}}\nu!}, & n=3 \end{cases} \tag{3.31a–b}$$

and

$$G_2(\tau) = \frac{(-1)^{\mu}\tau^{2\mu+1}}{(2\mu+1)!}\theta(\tau), \tag{3.31c}$$

where $\theta(\tau)$ is a step function (=1 if $\tau > 0$, =0 otherwise). The spatial and temporal polynomial are, respectively,

$$\begin{aligned} \eta_{2\nu+1}(r) &= \sum_{\rho=0}^{\nu}(-1)^{\rho}a_{\rho}r^{2\rho+1}, \\ \eta_{2\nu+1}(\tau) &= \sum_{\zeta=0}^{\nu}(-1)^{\zeta}b_{\zeta}\tau^{2\zeta+1}. \end{aligned} \tag{3.32a–b}$$

Certain PCs apply on the coefficients a_ρ and b_ζ (again, see Section 4.3 of Chapter XV).

The following corollary is a direct consequence of Proposition 3.6.

Corollary 3.5

A useful class of GS/TC-ν/μ functions is described by,

$$\kappa_X(\boldsymbol{h},\tau) = \kappa_{X(p)}(\boldsymbol{h},\tau) + \eta_{2\nu+1}(\boldsymbol{h})\eta_{2\mu+1}(\tau), \tag{3.33}$$

where

$$\begin{aligned} \kappa_{X(p)}(\boldsymbol{h},\tau) &= \int_{-\infty}^{\tau} d\tau' \int d\boldsymbol{h}' K(\boldsymbol{h}-\boldsymbol{h}',\tau-\tau')c_Y(\boldsymbol{h}',\tau'), \\ K(\boldsymbol{h}-\boldsymbol{h}',\tau-\tau') &= G_1(\boldsymbol{h}-\boldsymbol{h}')G_2(\tau-\tau'), \end{aligned} \tag{3.34a–b}$$

and

$$G_2(\tau-\tau') = \frac{(-1)^{\mu}}{(2\mu+1)!}(\tau-\tau')^{2\mu+1}\theta(\tau-\tau'),$$

$$G_1(\boldsymbol{h}-\boldsymbol{h}') = \begin{cases} \dfrac{1}{2^{2\nu+1}\pi(\nu!)^2}|\boldsymbol{h}-\boldsymbol{h}'|^{2\nu}\log|\boldsymbol{h}-\boldsymbol{h}'|, & n=2 \\[2ex] \dfrac{(-1)^{\nu+1}\Gamma\left(\dfrac{1}{2}-\nu\right)}{2^{2\nu+1}\pi^{\frac{3}{2}}\nu!}|\boldsymbol{h}-\boldsymbol{h}'|^{2\nu-1}, & n=3, \end{cases} \tag{3.35a–c}$$

and the polynomials $\eta_{2\nu+1}(r)$ and $\eta_{2\nu+1}(\tau)$ are given by Eqs. (3.32a–b).

The following corollaries suggest some general GS/TC-ν/μ models with interesting features that can be further exploited.

Corollary 3.6

A valid GS/TC-ν/μ function in $R^{3,1}$ is given by

$$\kappa_X(r,\tau) = \left[\kappa_{X(1)}(r) + \eta_{2\nu+1}(r)\right]\left[\kappa_{X(2)}(\tau) + \eta_{2\mu+1}(\tau)\right],$$

$$\kappa_{X(1)}(r) = \frac{(-1)^{2\nu+1}}{2^{\nu+1}\nu!\prod_{i=1}^{\nu+1}(2i-1)}\int_0^{\infty} dr'\frac{r'}{r}\left[(r+r')^{2\nu+1} - |r-r'|^{2\nu+1}\right]c_{Y(1)}(r'), \tag{3.36a–c}$$

$$\kappa_{X(2)}(\tau) = \int_{-\infty}^{\tau} d\tau' G_1(\tau-\tau')c_{Y(2)}(\tau'),$$

where $G_1(\tau-\tau')$ is given by Eq. (3.35a), and $c_Y(r,\tau) = c_{Y(1)}(r)c_{Y(2)}(\tau)$, $r = |\boldsymbol{h}|$, is known STIS (separable) covariance function.

Corollary 3.7

A valid GS/TC-ν/μ function in $R^{3,1}$ is given by

$$\kappa_X(r,\tau) = (-1)^{\mu+\nu+1}c_0\, a^{2\nu+2}\, b^{2\mu+2}\Lambda_\nu\left(\frac{r}{a}\right)\Psi_\mu\left(\frac{\tau}{b}\right) \tag{3.37}$$

where

$$\Lambda_\nu(u) = \frac{(-1)^{\nu}}{2^{\nu+1}\nu!\prod_{i=1}^{\nu+1}(2i-1)}u^{-1}\int_0^{\infty} du' u' e^{-u'}\left[(u+u')^{2\nu+1} - |u-u'|^{2\nu+1}\right],$$

$$\Psi_\mu(w) = e^{-|w|} + \theta(w)\sum_{i=0}^{2\mu+1}(i!)^{-1}\left[(w)^i - (-w)^i\right], \tag{3.38a–b}$$

$$u = \frac{r}{a},\ \ w = \frac{\tau}{b},\ \ |w| = \frac{|\tau|}{b},\ \ a>0,\ b>0,\ \text{ and }\ r = |\boldsymbol{h}|.$$

Table 3.1 The Functions $\Lambda_\nu(u)$, $u = r/a$, and $\Psi_\mu(w)$, $w = \tau/b$, in $R^{3,1}$

ν	μ	$\Lambda_\nu(u)$	$\Psi_\mu(w)$
0	0	$2u^{-1}(1-e^{-u})-e^{-u}$	$e^{-\|w\|}+2w\theta(w)$
1	1	$-4u^{-1}(1-e^{-u})+e^{-u}-u$	$e^{-\|w\|}+\theta(w)\left(2w+\frac{w^3}{3}\right)$
2	2	$\frac{6}{u}(1-e^{-u})-e^{-u}+2u\left(1+\frac{u^3}{24}\right)$	$e^{-\|w\|}+\theta(w)\left(2w+\frac{w^3}{3}+\frac{w^5}{60}\right)$

The integrals in Eqs. (3.38a–b) can be evaluated explicitly for different ν and μ values as shown in Table 3.1. The Λ_ν values at zero lag depend on the continuity order ν. For $\nu = 0$, the Λ_0 lacks a polynomial term and it decreases monotonically, whereas for $\nu > 0$ the Λ_ν increases monotonically due to the polynomial terms. For negative lags, the Ψ_μ increases exponentially and do not depend on the continuity order μ, whereas for positive lags, the Ψ_μ increases faster with increasing μ.

Corollary 3.8

A valid GS/TC-ν/μ function in $R^{3,1}$ is given by

$$\kappa_X(r,\tau) = (-1)^{\mu+\nu+1} c_0\, a^{2\nu+2}\, b^{2\mu+2} \Lambda_\nu\left(\frac{r}{a}\right)\Psi_\mu\left(\frac{\tau}{b}\right) \tag{3.39}$$

where (the readers may notice the difference with Eqs. (3.38a–b))

$$\begin{aligned}\Lambda_\nu(u) &= \frac{(-1)^\nu}{2^{\nu+1}\nu!\prod_{i=1}^{\nu+1}(2i-1)} u^{-1}\int_0^\infty du' u' e^{-u'^2}\left[(u+u')^{2\nu+1} - \left|u-u'\right|^{2\nu+1}\right],\\ \Psi_\mu(w) &= \frac{1}{(2\mu+1)!}\int_{-\infty}^{w} dw' e^{-w'^2}(w-w')^{2\mu+1},\end{aligned} \tag{3.40a–b}$$

$$u = \frac{r}{a},\quad w = \frac{\tau}{b},\quad \left|w\right| = \frac{|\tau|}{b},\quad a>0,\ b>0,\quad \text{and}\quad r = \left|\boldsymbol{h}\right|.$$

The integrals in Eqs. (3.40a–b) can be evaluated explicitly for different ν and μ values as shown in Table 3.2.

Table 3.2 The Functions $\Lambda_\nu(u)$, $u = r/a$, and $\Psi_\mu(w)$, $w = \tau/b$, in $R^{3,1}$

ν	μ	$\Lambda_\nu(u)$	$\Psi_\mu(w)$
0	0	$\pi^{\frac{1}{2}}\frac{erf(u)}{4u}$	$\frac{1}{2}\left\{\pi^{\frac{1}{2}}u[1+erf(u)]+e^{-u^2}\right\}$
1	1	$-\left(\pi^{\frac{1}{2}}erf(u)\frac{1+2u^2}{16u}+\frac{e^{-u^2}}{8}\right)$	$\frac{1}{4}\left\{\pi^{\frac{1}{2}}(3u+2u^3)[1+erf(u)] + 2(1+u^2)e^{-u^2}\right\}$
2	2	$\pi^{\frac{1}{2}}\frac{3+12u^2+4u^4}{384u}erf(u)+\frac{10+4u^2}{384}e^{-u^2}$	$\frac{1}{8}\left\{\pi^{\frac{1}{2}}(15w+20w^3+4w^5)[1+erf(w)] + (8+18w^2+4w^4)e^{-w^2}\right\}$

Remark 3.2

I reemphasize that a clear distinction should be made regarding the role of Eqs. (2.1) and (3.16) in GS/TC-ν/μ determination. An explicit expression for the covariance involves solving Eq. (2.1), i.e., $Q_{\nu/\mu}[X(s,t)] = Y(s,t)$. However, solutions of this equation are not easily obtained for $n \geq 2$ and $\nu, \mu \geq 0$. In contrast, explicit solutions for the generalized covariance Eq. (3.16), i.e., $U_Q\kappa_X(\boldsymbol{h}, \tau) = c_Y(\boldsymbol{h}, \tau)$, can be obtained, as we saw earlier. The price to be paid for bypassing the $Q_{\nu/\mu}$-equation is the indeterminacy of the polynomial coefficients, that is, the polynomial terms of a GS/TC-ν/μ often are not determined from the solution of the U_Q-equation. These polynomial coefficients need to be constrained by the corresponding PCs of the generalized covariance.

The usefulness of the space–time generalized covariance functions above is demonstrated, inter alia, by the fact that many useful models, like fractals and wavelets, are special cases of the GS/TRF theory (Christakos, 2000). Furthermore, note that after the generalized covariance models κ_X have been constructed as above (other model construction techniques are discussed in Chapter XVI), ordinary space–time heterogeneous covariance models can be derived by means of Eq. (3.11). In light of Eq. (3.11), even if one starts with a separable generalized covariance function, like Eq. (3.19), the resulting ordinary space–time covariance (Eq. 3.11) is nonseparable, nevertheless. Hence, Eq. (3.11) offers the means for generating a large class of potentially useful nonseparable spatiotemporal covariance models, which can be used to represent space–time heterogeneous natural attributes.

4. DISCRETE LINEAR REPRESENTATIONS OF SPATIOTEMPORAL RANDOM FIELDS

A key element in passing from abstract theory to the practical analysis of spatiotemporal data is the development of suitable discrete linear representations of the S/TRF model. This is necessary because real data are usually discretely distributed in space–time.

4.1 SPACE–TIME RANDOM INCREMENTS

Let $X(\boldsymbol{p}_i)$, $i = 1,\ldots, m$, be a discrete-parameter ordinary S/TRF. Assume that $q \in Q = \mathcal{Q}$, where $\mathcal{Q}$ is the space of real measures with finite support and such that

$$q(\boldsymbol{p}) = \sum_{i=1}^{m} q(\boldsymbol{p}_i)\delta(\boldsymbol{p}_i - \boldsymbol{p}) = \sum_{i=1}^{m} q_i\delta_i(\boldsymbol{p}), \tag{4.1}$$

where $\boldsymbol{p}_i = (s_{i1},\ldots, s_{in}, s_{i0} = t_i)$, and m denotes the number of space–time points considered. In the $R^{n,1}$ domain, Eq. (4.1) can be written in terms of the space–time point conditional expressions (time conditioned to space and space conditioned to time, see Eqs. (2.4b–c) of Chapter I) as

$$q(\boldsymbol{s}, t) = \sum_{i=1}^{m}\sum_{j=1_i}^{m_i} q(\boldsymbol{s}_i, t_i)\delta(\boldsymbol{s}_i - \boldsymbol{s}, t_j - t) = \sum_{i=1}^{m}\sum_{j=1_i}^{m_i} q_{ij}\delta_{ij}(\boldsymbol{p}), \tag{4.2}$$

where $q \in \mathcal{Q}$, and m_i denotes the number of time instances t_j ($j = 1_i,\ldots, m_i$) used given that we are at the spatial position $\boldsymbol{s}_i$ ($i = 1,\ldots, m$). The discrete CS/TRF corresponding to Eq. (4.1) is

$$Y_q(\boldsymbol{p}) = \left\langle \sum_{i=1}^{m} q_i \delta_i(\boldsymbol{p}'), U_{\boldsymbol{p}} X(\boldsymbol{p}') \right\rangle = \sum_{i=1}^{m} q_i U_{\boldsymbol{p}} X(\boldsymbol{p}_i) \tag{4.3}$$

with composite ordinates in $R^{n,1}$, whereas the discrete CS/TRF corresponding to Eq. (4.2) may assume either of the following forms

$$\begin{aligned} Y_q(\boldsymbol{s}, t) &= \sum_{i=1}^{m} \sum_{j=1_i}^{m_i} q_{ij} U_{\boldsymbol{s}_i, t_j} X(\boldsymbol{s}, t), \\ Y_q(\boldsymbol{s}, t) &= \sum_{i=0}^{m(\nu)} \sum_{j=0}^{m(\mu)} q_{ij} U_{\delta \boldsymbol{s}_i, \delta t_j} X(\boldsymbol{s}, t) \end{aligned} \tag{4.4a–b}$$

with coordinates in $R^{n,1}$. Both Eqs. (4.4a) and (4.4b) are used in computational random field modeling, as appropriate. Eq. (4.4b) involves linear combinations of attribute values at points indicated by $i = 0,\ldots, m(\nu)$, $j = 0,\ldots, m(\mu)$, within a space–time neighborhood around ($\boldsymbol{s}$,t) with $\delta \boldsymbol{s}_0 = \delta t_0 = 0$. Eqs. (4.3–4.4a–b) are discrete representations of the continuous operator of Eq. (1.14), in which case the weights q_{ij} represent the discretized detrending operator $Q_{\nu/\mu}$. In the following, the residual field is in certain places denoted by Y_q to emphasize the dependence on the detrending operator.

As we pass from continuous to discrete representations, I remind the readers that in the continuous case considered in the previous sections the operator $Q_{\nu/\mu}$ properly eliminated the polynomial trends of the continuous S/TRF $X(\boldsymbol{s},t)$. Linear combinations of the S/TRF values $X(\boldsymbol{s}_i,t_j)$ weighted by q_{ij} have a similar same effect on the discretized system. The following example illustrates this effect.

Example 4.1

Let us calculate the weights q_{ij} for an S/TRF with continuity orders $\nu/\mu = 1/1$ in $R^{1,1}$. The detrending operator has the continuum representation

$$Q_{1/1} = \frac{\partial^4}{\partial s^2\, \partial t^2}. \tag{4.5}$$

On the other hand, the discretized difference operator to $O(\delta s^2 \delta t^2)$ accuracy is given by

$$\frac{\partial^4 X_{0,0}}{\partial s^2 \partial t^2} \cong \frac{X_{1,1} + X_{-1,1} + X_{1,-1} + X_{-1,-1} - 2X_{1,0} - 2X_{-1,0} - 2X_{0,1} - 2X_{0,-1} + 4X_{0,0}}{\delta s^2 \delta t^2}, \tag{4.6}$$

where $X_{k,l} \equiv X(s + k\,\delta s, t + l\,\delta t)$ and δs, δt denote the space and time spacings. In the approximation of Eq. (4.6), the weights are given by

$$q_{1,1} = q_{-1,1} = q_{1,-1} = q_{-1,-1} = 1, \quad q_{1,0} = q_{-1,0} = q_{0,-1} = q_{0,1} = -2, \quad q_{0,0} = 4. \tag{4.7}$$

By definition, these weights eliminate linear polynomials of the space–time trends, i.e., they satisfy the sum rules

$$\sum_{i=-1}^{1} q_{ij} i^{\rho} = \sum_{j=-1}^{1} q_{ij} j^{\zeta} = \sum_{i=-1}^{1} \sum_{j=-1}^{1} q_{ij} i^{\rho} j^{\zeta} = 0, \tag{4.8}$$

$\rho,\zeta = 0,1$. In order to see how these sum rules are obtained, consider the monomials $\eta_{\rho/\zeta}(s,t) = s^\rho\, t^\zeta$ with $0 \leq \rho,\, \zeta \leq 1$. The effect of the detrending operator on these monomials is such that

$$Q_{1/1}\left[\eta_{\rho/\zeta}(s,t)\right] \cong \sum_{i=-1}^{1}\sum_{j=-1}^{1} q_{ij}(s+i\delta s)^\rho (t+j\delta t)^\zeta + O\left(\delta s^2 \delta t^2\right), \tag{4.9}$$

$\rho,\zeta = 0,1$. If one solves the equation $Q_{1/1}\left[\eta_{\rho/\zeta}(s,t)\right] = 0$ up to $O\left(\delta s^2 \delta t^2\right)$ for all the combinations $0 \leq \rho,\, \zeta \leq 1$, one finds that

$$\begin{aligned}
Q_{1/1}\left[\eta_{0/0}(s,t)\right] &= \sum_{i=-1}^{1}\sum_{j=-1}^{1} q_{ij} = 0 \\
Q_{1/1}\left[\eta_{1/0}(s,t)\right] &= \sum_{i=-1}^{1}\sum_{j=-1}^{1} q_{ij}(s+i\delta s) = 0 \\
Q_{1/1}\left[\eta_{0/1}(s,t)\right] &\cong \sum_{i=-1}^{1}\sum_{j=-1}^{1} q_{ij}(t+j\delta t) = 0 \\
Q_{1/1}\left[\eta_{1/1}(s,t)\right] &= \sum_{i=-1}^{1}\sum_{j=-1}^{1} q_{ij}(s+i\delta s)(t+j\delta t) = 0.
\end{aligned} \tag{4.10a–d}$$

Eqs. (4.10a–d) lead to the sum rules in Eq. (4.8). Similar sum rules are satisfied for higher-order derivatives and their discretized representation.

The following example illustrates the distinct yet equivalent ways Eqs. (4.3)–(4.4a–b) are formulated.

Example 4.2

Consider four points with composite space–time coordinates $\boldsymbol{p}_1 = \delta_{\boldsymbol{h},\tau} = (\boldsymbol{h},\tau)$, $\boldsymbol{p}_2 = \delta_{\boldsymbol{h},0} = (\boldsymbol{h},0)$, $\boldsymbol{p}_3 = \delta_{\boldsymbol{0},\tau} = (\boldsymbol{0},\tau)$, and $\boldsymbol{p}_4 = \boldsymbol{0} = (0,0)$. From the definition of $Y_q(\boldsymbol{p})$, Eq. (4.3), with $m = 4$, and $q_1 = q_4 = 1$, $q_2 = q_3 = -1$, one finds that

$$Y_q(\boldsymbol{p}) = \sum_{i=1}^{4} q_i X(\boldsymbol{p}+\boldsymbol{p}_i) = X\left(\boldsymbol{p}+\delta_{\boldsymbol{h},\tau}\right) - X\left(\boldsymbol{p}+\delta_{\boldsymbol{h},0}\right) - X\left(\boldsymbol{p}+\delta_{\boldsymbol{0},\tau}\right) + X(\boldsymbol{p}). \tag{4.11}$$

Similarly, consider the same four points as above but this time with separately defined spatial coordinates $\boldsymbol{s}_1 = \boldsymbol{s}_2 = \boldsymbol{h}$, $\boldsymbol{s}_3 = \boldsymbol{s}_4 = \boldsymbol{s}_1 - \boldsymbol{h} = 0$, and space-conditioned temporal instants $t_{1_1} = t_{1_3} = \tau$, $t_{1_2} = t_{1_4} = 0$. From the definition of $Y_q(\boldsymbol{s},t)$, Eq. (4.4a), with $m_i = 1_i$ denoting that one time instance t_j $(j = 1_i)$ is used given that we are at the spatial position $\boldsymbol{s}_i$ $(i = 1,\ldots,4)$, and $q_{11_1} = q_{41_4} = 1$, $q_{21_2} = q_{31_3} = -1$, it is found that

$$Y_q(\boldsymbol{s},t) = \sum_{i=1}^{4} q_{i1_i} X(\boldsymbol{s}_i+\boldsymbol{s}, t_{1_i}+t) = X(\boldsymbol{h}+\boldsymbol{s},\tau+t) - X(\boldsymbol{h}+\boldsymbol{s},t) - X(\boldsymbol{s},\tau+t) + X(\boldsymbol{s},t), \tag{4.12}$$

i.e., the same as the expression of $Y_q(\boldsymbol{p})$ in Eq. (4.11).[13]

[13]This is because Eq. (4.12) can be also written as $Y_q(\boldsymbol{s},t) = X(\boldsymbol{p}+\Delta\boldsymbol{p}) - X(\boldsymbol{p}+(\boldsymbol{h},0)) - X(\boldsymbol{p}+(0,\tau)) + X(\boldsymbol{p}) = X(\boldsymbol{p}+\Delta\boldsymbol{p}) - X\left(\boldsymbol{p}+\delta_{\boldsymbol{h},0}\right) - X\left(\boldsymbol{p}+\delta_{\boldsymbol{0},\tau}\right) + X(\boldsymbol{p})$.

The analysis of the discrete S/TRF representations so far, acknowledges the essential role of the discrete S/TRF $Y_q(\boldsymbol{p})$ of Eq. (4.3) and $Y_q(\boldsymbol{s},t)$ of Eq. (4.4a–b), and justifies the introduction of the following definition.

Definition 4.1

The $Y_q(\boldsymbol{p})$ or $Y_q(\boldsymbol{s},t)$ of Eqs. (4.4a–b) will be called *spatiotemporal increments of order v in space and μ in time* (*S/TI-v/μ*) on $Q_{v/\mu}$ if it holds that

$$\sum_{i=1}^{m} q_i \boldsymbol{p}_i^{(\rho,\zeta)} = 0, \tag{4.13a}$$

or either of the following conditions is valid

$$\begin{aligned}&\sum_{i=1}^{m}\sum_{j=1_i}^{m_i} q_{ij}\boldsymbol{s}_i^{\rho} t_j^{\zeta} = 0,\\ &\sum_{i=0}^{m(\nu)}\sum_{j=0}^{m(\mu)} q_{ij} U_{\delta s_i,\delta t_j}\eta_\alpha(\boldsymbol{s},t) = 0\end{aligned} \tag{4.13b–c}$$

for all $\boldsymbol{\rho} \leq v$ and $\zeta \leq \mu$, or all monomials η_α with $\alpha = 1,\ldots, N_n(\nu/\mu)$, where $N_n(\nu/\mu)$ is the number of η_α as defined in Eq. (2.12c) of Chapter I. In this case the coefficients $\{q_i\},\{q_{ij}\} \in Q_{v/\mu} \subset \mathcal{Z}$, $i = 1,\ldots, m$ and $j = 1_i,\ldots, m_i$, will be termed *admissible sets of coefficients of order v/μ* (AC-v/μ).

Example 4.3

Let $X(\boldsymbol{s},t)$ be an S/TRF-1/1 in $R^{1,1}$, and $\frac{\partial^4}{\partial s^2 \partial t^2} X(s,t) = Y(s,t)$, where $Y(\boldsymbol{s},t)$ is an STHS residual field. As we saw in Example 4.1 the discrete representation of $Q_{1/1}[X]$ is given by Eq. (4.6). Eqs. (4.10a–d) then establish that the weights q_{ij} satisfy the conditions in Eq. (4.13b). Therefore, the residual $Y(s,t)$ is a spatiotemporal increment of order 1/1, according to Definition 4.1.

In light of Definition 4.1 the next one follows rather naturally.

Definition 4.2

The discrete representation of the ordinary S/TRF $X(\boldsymbol{s},t)$ will be called an ordinary S/TRF-ν/μ on $Q_{\nu/\mu}$ if the corresponding S/TI-ν/μ $Y_q(\boldsymbol{s},t)$ is a zero mean STHS random field.

Example 4.4

In Example 4.3, since the residual random field $Y(\boldsymbol{s},t)$ is STHS, the $X(\boldsymbol{s},t)$ is an S/TRF-1/1, according to Definition 4.2. As another illustration, consider the case (Christakos, 1992) in which $\boldsymbol{p} = (s_1, s_2, t) \in R^{2,1}$, and $h,\tau \geq 0$. Also, let

$$\begin{aligned}Y_q(s_1,s_2,t) &= \sum_{i=1}^{5}\sum_{j=1}^{3} q_{ij}X(s_{i1},s_{i2},t_j)\\ &= X(s_1+h,s_2,t+\tau) - 2X(s_1+h,s_2,t) + X(s_1+h,s_2,t-\tau) + X(s_1+h,s_2,t+\tau)\\ &\quad - 2X(s_1,s_2+h,t) + X(s_1,s_2+h,t-\tau) + X(s_1-h,s_2,t+\tau) - 2X(s_1-h,s_2,t)\\ &\quad + X(s_1-h,s_2,t-\tau) + X(s_1,s_2-h,t+\tau) - 2X(s_1,s_2-h,t) + X(s_1,s_2-h,t-\tau)\\ &\quad - 4[X(s_1,s_2,t+\tau) - 2X(s_1,s_2,t) + X(s_1,s_2,t-\tau)].\end{aligned} \tag{4.14}$$

It is easily shown that

$$\sum_{i=1}^{5}\sum_{j=1}^{3} q_{i_1 i_2} s_{i_1}^{\rho_1} s_{i_2}^{\rho_2} s_j^{\zeta} = 0$$

for all $\boldsymbol{\rho} = (\rho_1, \rho_2)$ such that $|\boldsymbol{\rho}| = \rho_1 + \rho_2 \leq 1$ and $\zeta \leq 1$. Therefore, the $Y_q(\boldsymbol{p})$ above is an S/TI-l/l. If, in addition, it is a STHS random field, the corresponding $X(\boldsymbol{p})$ is an S/TRF-1/1 with mean value $\overline{X(s_1, s_2, t)} = \sum_{|\boldsymbol{\rho}|, \zeta \leq 1} a_{\rho_1, \rho_2, \zeta} s_1^{\rho_1} s_2^{\rho_2} t^{\zeta}$.

Proposition 4.1

An ordinary S/TRF-ν/μ can be represented by the *spatiotemporal autoregressive model of order* $\nu + 1$ *in space and* $\mu + 1$ *in time*, S/TAR $(\nu + 1, \mu + 1)$,

$$\Delta_{\boldsymbol{s},t}^{(\nu+1,\mu+1)} X(\boldsymbol{p}) = Y_q(\boldsymbol{p}), \tag{4.15}$$

where

$$\Delta_{\boldsymbol{s},t}^{(\nu+1,\mu+1)} X(\boldsymbol{p}) = \sum_{\rho=0}^{\nu+1}\sum_{\zeta=0}^{\mu+1} (-1)^{\rho+\zeta} C_{\nu+1}^{\rho} C_{\mu+1}^{\zeta} X(\boldsymbol{s} + (\nu + 1 - \rho)\boldsymbol{h}, t + (\mu + 1 - \zeta)\tau) \tag{4.16}$$

is the finite difference of order $\nu + 1$ in space and $\mu + 1$ in time, $Y_q(\boldsymbol{p})$ is an STHS random field, and $C_i^j = \binom{i}{j}$ (Appendix).

Among the noticeable features of Eq. (4.15) is its direct link with the generalized space–time difference operator of Eq. (4.6b) of Chapter XI used in random field differentiation. Another interesting feature of Eq. (4.15) is that it may be considered as the discrete counterpart of the continuous representation of Eq. (2.8). A comparison of the representations of Eq. (2.8) and Eq. (4.15) concludes that all discrete-parameter S/TRF-ν/μ admit a representation of the form of Eq. (4.15), whereas for a continuous-parameter S/TRF-ν/μ to be represented by Eq. (2.8), it is necessary that it is $\nu + 1$ times differentiable in space and $\mu + 1$ times in time. The discrete representations are also used for space–time estimation purposes.

Proposition 4.2

Let $X(\boldsymbol{s},t)$ be an ordinary S/TRF on $Q_{\nu/\mu}$, and let

$$\widehat{X}(\boldsymbol{s}_0, t_0) = \sum_{i=1}^{m}\sum_{j=1_i}^{m_i} \lambda_{ij} X(\boldsymbol{s}_i, t_j) \tag{4.17}$$

be the linear estimator of $X(\boldsymbol{s}_0,t_0)$ at location/instant $(\boldsymbol{s}_0,t_0)$ such that

$$\overline{\widehat{X}(\boldsymbol{s}_0, t_0) - X(\boldsymbol{s}_0, t_0)} = 0, \tag{4.18}$$

and

$$\overline{X(\boldsymbol{s}_0, t_0)} = \sum_{|\boldsymbol{\rho}| \leq \nu, \zeta \leq \mu} \eta_{\boldsymbol{\rho}\zeta} \boldsymbol{s}_0^{\boldsymbol{\rho}} t_0^{\zeta}, \tag{4.19}$$

where $\eta_{\rho\zeta}$ are suitable coefficients. Then, the difference

$$Y_q(s_0, t_0) = \widehat{X}(s_0, t_0) - X(s_0, t_0) \tag{4.20}$$

is an S/TI-ν/μ on $Q_{\nu/\mu}$.

Notice that if the $Y_q(\boldsymbol{p})$ of Eq. (4.20) is STHS, the $X(\boldsymbol{p})$ is by definition an S/TRF-ν/μ. Conversely, if $X(\boldsymbol{p})$ is an S/TRF-ν/μ, the S/TI-ν/μ $Y_q(\boldsymbol{p})$ of Eq. (4.20) is STHS. In the discrete framework, the condition of Eq. (3.5) implies that a function κ_X, is a GS/TC-ν/μ if and only if for all AC-ν/μ it holds that

$$\overline{X(q)^2} = \overline{Y_q(\mathbf{0})^2} = \overline{\left[\sum_{i=1}^{m}\sum_{j=1_i}^{m_i} q_{ij}X(s_i, t_j)\right]^2} = \sum_{i=1}^{m}\sum_{j=1_i}^{m_i}\sum_{i'=1}^{m}\sum_{j'=1_{i'}}^{m_{i'}} q_{ij}q_{i'j'}\kappa_X(\boldsymbol{h}_{ii'}, \tau_{jj'}) \geq 0 \tag{4.21}$$

where $\boldsymbol{h}_{ii'} = \boldsymbol{s}_i - \boldsymbol{s}_{i'}$, and $\tau_{jj'} = t_j - t_{j'}$. In practical applications where a finite number of discretely distributed data are available, it is convenient to use GS/TC-ν/μ models of the space–time polynomial form of Eq. (3.22). The parameters of these models (i.e., the orders ν and μ as well as the coefficients $a_{\rho\zeta}$, $\rho = 0, 1,\ldots, \nu$ and $\zeta = 0, 1,\ldots, \mu$) can be estimated on the basis of the available data by means of parameter estimation techniques such as least squares or maximum likelihood (e.g., Rao, 1973). Note that the estimated values of the coefficients $a_{\rho\zeta}$ should satisfy the corresponding PCs (Section 3).

4.2 SPACE–TIME VARIOGRAM ANALYSIS

A special case of the S/TRF-ν/μ theory that is worth further study is the limiting case $\nu/\mu = 0/0$. This case, which is often encountered in practice, may be associated with the space–time detrending operators of Eqs. (4.13a–c) above with $\nu = \mu = 0$. In the following developments, we will freely pass from the continuous to the discrete domain, and vice versa, as appropriate.

Let us start with the Ito-Gel'fand differential operator of Eq. (2.4) with $\nu = \mu = 0$, i.e., the (1,1) space–time derivatives of the S/TRF $X(\boldsymbol{p})$ at point $\boldsymbol{p} \in R^{n,1}$ with respect to both the spatial coordinate s_i and the time instant $t = s_0$ of $\boldsymbol{p}$,[14]

$$X_{i,0}^{(1,1)}(\boldsymbol{p}) = \frac{\partial^2 X(\boldsymbol{p})}{\partial s_i \partial s_0} = \underset{h_i,\tau \to 0}{\text{l.i.m.}} \frac{\Delta_{i,0}^2 X(\boldsymbol{p})}{h_i \tau}, \tag{4.22}$$

where $i = 1,\ldots, n$, and

$$\Delta_{i,0}^2 X(\boldsymbol{p}) = X(\boldsymbol{p} + \delta_{h_i,\tau}) - X(\boldsymbol{p} + \delta_{0,\tau}) + X(\boldsymbol{p}) - X(\boldsymbol{p} + \delta_{h_i,0}) \tag{4.23}$$

with $\boldsymbol{p} = (\boldsymbol{s},t)$, $\delta_{h_i,\tau} = (h_i\boldsymbol{\varepsilon}_i,\tau)$, $\delta_{0,\tau} = (0,\tau)$, $\delta_{h_i,0} = (h_i\boldsymbol{\varepsilon}_i,0)$. If $X(\boldsymbol{p})$ is an S/TRF-$\nu/\mu = 0/0$, by definition in the *continuous* domain

$$X_{i,0}^{(1,1)}(\boldsymbol{p}) = Y_i(\boldsymbol{p}), \tag{4.24}$$

where $Y_i(\boldsymbol{p})$ $(i = 1,\ldots, n)$ is a zero mean STHS random field. The ordinary covariance functions of $X(\boldsymbol{p})$ and $Y_i(\boldsymbol{p})$ are related by

[14]According to the notation adapted in Chapter VI, the subscripts i and 0 refer to s_i and $s_0 = t$, respectively.

$$\frac{\partial^4 c_X(\boldsymbol{p},\boldsymbol{p}')}{\partial s_i \partial s'_i \partial t \partial t'} = c_{Y_i}(\boldsymbol{p}-\boldsymbol{p}') \tag{4.25}$$

in the continuous domain along the direction s_i, with $i = 1,\ldots, n$.

In view of Eqs. (4.22)–(4.25), it seems quite appropriate to study in more detail the random field 0/0-increments in the *discrete* domain, in particular,

$$Y_q(\boldsymbol{p}) = \left\langle \sum_{k=1}^{4} q_k \delta_k(\boldsymbol{p}'), U_{\Delta\boldsymbol{p}} X(\boldsymbol{p}') \right\rangle = \sum_{k=1}^{4} q_k U_{\Delta\boldsymbol{p}} X(\boldsymbol{p}_k) = \Delta^2 X(\boldsymbol{p}), \tag{4.26}$$

where $q_1 = q_3 = 1$, $q_2 = q_4 = -1$, and $\sum_{k=1}^{4} q_k = 0$. The $Y_q(\boldsymbol{p})$ plays a central role in variogram analysis. The $Y_q(\boldsymbol{p})$ of Eq. (4.26) and the $Y_i(\boldsymbol{p})$ of Eq. (4.24) explicitly refer to the discrete and the continuous domain, respectively. In certain cases, e.g., when the direction of the increment needs to be explicitly denoted, the notation $Y_{q,i}(\boldsymbol{p})$ will be also used in the discrete domain. Accordingly, the mean of the 0/0 random increment along the direction s_i will be

$$\overline{Y_{q,i}(\boldsymbol{p})} = \overline{\Delta_{i,0}^2 X(\boldsymbol{p})} = \overline{X(\boldsymbol{p}+\delta_{h_i,\tau})} - \overline{X(\boldsymbol{p}+\delta_{0,\tau})} + \overline{X(\boldsymbol{p})} - \overline{X(\boldsymbol{p}+\delta_{h_i,0})}, \tag{4.27}$$

and the variogram along the direction s_i, will be

$$\begin{aligned} \gamma_X(\boldsymbol{p},\boldsymbol{p}\prime;\Delta\boldsymbol{p},\Delta\boldsymbol{p}\prime) &= \frac{1}{2}\overline{Y_{q,i}(\boldsymbol{p})Y_{q,i}(\boldsymbol{p}\prime)} \\ &= \frac{1}{2}\overline{\left[X(\boldsymbol{p}+\delta_{h_i,\tau}) - X(\boldsymbol{p}+\delta_{0,\tau}) + X(\boldsymbol{p}) - X(\boldsymbol{p}+\delta_{h_i,0})\right]} \\ &\quad \overline{\left[X(\boldsymbol{p}'+\delta_{h\prime_i,\tau\prime}) - X(\boldsymbol{p}'+\delta_{0,\tau\prime}) + X(\boldsymbol{p}') - X(\boldsymbol{p}'+\delta_{h\prime_i,0})\right]} \end{aligned} \tag{4.28}$$

To distinguish it from the classical variogram function, the variogram of Eq. (4.28) is sometimes called the *extended space–time variogram*. To better appreciate the difference between the two, Eq. (4.28) is expressed in terms of the corresponding covariance function $c_{Y_{q,i}}(\boldsymbol{p},\boldsymbol{p}')$ as

$$\gamma_X(\boldsymbol{p},\boldsymbol{p}';\Delta\boldsymbol{p},\Delta\boldsymbol{p}') = \frac{1}{2}c_{Y_{q,i}}(\boldsymbol{p},\boldsymbol{p}') = \frac{1}{2}\Delta_{\Delta\boldsymbol{p}}^2\Delta_{\Delta\boldsymbol{p}'}^2 c_X(\boldsymbol{p},\boldsymbol{p}') \tag{4.29a}$$

(Exercise XIII.8). This variogram function is related to the corresponding structure function by

$$\gamma_X(\boldsymbol{p},\boldsymbol{p}';\Delta\boldsymbol{p},\Delta\boldsymbol{p}') = \frac{1}{2}\xi_X(\boldsymbol{p},\boldsymbol{p}';\Delta\boldsymbol{p},\Delta\boldsymbol{p}'). \tag{4.29b}$$

Furthermore, the covariance of $X(\boldsymbol{p})$ is related to that of $Y_q(\boldsymbol{p})$ in terms of

$$\frac{\partial^4 c_X(\boldsymbol{p},\boldsymbol{p}')}{\partial s_i \partial s'_i \partial t \partial t'} = \lim_{h_i,h'_i,\tau,\tau' \to 0} \frac{c_{Y_{q,i}}(\boldsymbol{p},\boldsymbol{p}')}{h_i h'_i \tau \tau'}, \tag{4.30}$$

where $c_{Y_{q,i}}(\boldsymbol{p},\boldsymbol{p}')$ is given by the right hand side of Eq. (4.29a). Since $X(\boldsymbol{p})$ is m.s. continuous, the c_{Y_i}, γ_X, ξ_X, and c_X of Eqs. (4.29a–b)–(4.30) are continuous functions of their arguments.

Example 4.5

Let us now assume that $\boldsymbol{p} = \boldsymbol{p}'$ and $\Delta\boldsymbol{p} = \Delta\boldsymbol{p}'$. The corresponding space–time variogram function is given by (assuming a constant $X(\boldsymbol{p})$ mean)

$$\gamma_X(\boldsymbol{p}, \Delta\boldsymbol{p}) = \frac{1}{2}\overline{Y_{q,i}(\boldsymbol{p})^2} = \frac{1}{2}\overline{\left[X\left(\boldsymbol{p} + \delta_{h_i,\tau}\right) - X\left(\boldsymbol{p} + \delta_{0,\tau}\right) + X(\boldsymbol{p}) - X\left(\boldsymbol{p} + \delta_{h_i,0}\right)\right]^2}, \tag{4.31}$$

which is a space–time extension of the classical, purely spatial form of the geostatistical variogram (Olea, 1999).

By further exploring the duality principles between $X(\boldsymbol{p})$ and $Y_q(\boldsymbol{p})$, we arrive at the definition below, which is essentially a special case of the Definition 4.1.

Definition 4.3

A random field $X(\boldsymbol{p})$ will be called an S/TRF with STHS 0/0 *increments* $Y_i(\boldsymbol{p})$ along the direction s_i $(i = 1,\ldots, n)$, if the mean value and the variogram function depend only on the vector space–time lag $\Delta\boldsymbol{p}$, that is,

$$\overline{Y_{q,i}(\boldsymbol{p})} = \overline{\Delta_{i,0}^2 X(\Delta\boldsymbol{p})}, \tag{4.32}$$

and

$$\gamma_X(\Delta\boldsymbol{p}) = \frac{1}{2}\overline{Y_{q,i}(\boldsymbol{p})^2} = \frac{1}{2}c_{Y_{q,i}}(\mathbf{0}). \tag{4.33}$$

Example 4.6

In the case of the $Y_{q,i}(\boldsymbol{p})$ of Eq. (4.27), Eq. (4.33) yields

$$\gamma_X(\Delta\boldsymbol{p}) = 2\left[c_X(\mathbf{0}) - c_X\left(\delta_{h_i,0}\right) - c_X\left(\delta_{0,\tau}\right)\right] + c_X\left(\delta_{h_i,\tau}\right) + c_X\left(\delta_{-h_i,\tau}\right). \tag{4.34}$$

If $X(\boldsymbol{p})$ is an STHS random field, it is valid that

$$\overline{Y_{q,i}(\boldsymbol{p})} = 0, \tag{4.35}$$

and

$$\begin{aligned}\gamma_X(\Delta\boldsymbol{p}) &= \frac{1}{2}\overline{\left[X\left(\boldsymbol{p} + \delta_{h_i,\tau}\right) - X\left(\boldsymbol{p} + \delta_{0,\tau}\right)\right]^2} + \frac{1}{2}\overline{\left[X(\boldsymbol{p}) - X\left(\boldsymbol{p} + \delta_{h_i,0}\right)\right]^2} \\ &\quad + \overline{\left[X\left(\boldsymbol{p} + \delta_{h_i,\tau}\right) - X\left(\boldsymbol{p} + \delta_{0,\tau}\right)\right]\left[X(\boldsymbol{p}) - X\left(\boldsymbol{p} + \delta_{h_i,0}\right)\right]} \\ &= 2\left[\gamma_X\left(\delta_{h_i,0}\right) + \gamma_X\left(\delta_{0,\tau}\right) - \gamma_X\left(\delta_{h_i,\tau}\right)\right],\end{aligned} \tag{4.36}$$

i.e., the extended space–time variogram $\gamma_X(\Delta\boldsymbol{p})$ is here expressed in terms of the standard variograms $\gamma_X\left(\delta_{h_i,\tau}\right)$, $\gamma_X\left(\delta_{h_i,0}\right)$, and $\gamma_X\left(\delta_{0,\tau}\right)$.

Proposition 4.3

Let $X(\boldsymbol{p})$ be an S/TRF in $R^{n,1}$ with STHS increments 0/0. It can be shown that

$$\gamma_X(2^m\Delta\boldsymbol{p}) \leq 4^{2m}\gamma_X(\Delta\boldsymbol{p}), \tag{4.37}$$

where m is an integer. Furthermore,

$$\lim_{h,\tau \to \infty} \frac{\gamma_X(\boldsymbol{h},\tau)}{\boldsymbol{h}^2\tau^2} = 0, \tag{4.38}$$

where a is a constant (Exercise XIII.18).[15]

As regards the space–time geometry of the S/TRF $X(\boldsymbol{p})$ with STHS 0/0 increments $Y_q(\boldsymbol{p})$ and its relation with the associated variogram $\gamma_X(\Delta\boldsymbol{p})$, a series of interesting results are available including the following.

Proposition 4.4

A continuous in the m.s. sense S/TRF $X(\boldsymbol{p})$ with STHS 0/0-increments $Y_q(\boldsymbol{p})$ implies a continuous variogram function $\gamma_X(\Delta\boldsymbol{p})$. Conversely, a continuous variogram function implies an m.s. continuous S/TRF. Also, an S/TRF is 1/1-differentiable in the m.s. sense if and only if the corresponding variogram function is (2/2)-differentiable.

Under certain conditions, the result of proposition 4.4 can be extended to the case of $X(\boldsymbol{p})$ with STHS ν/μ-increments. Furthermore, let $X(\boldsymbol{p})$ be a heterogeneous, in general, S/TRF but with STHS 0/0-increments. We assume that the $X(\boldsymbol{p})$ is m.s. differentiable, i.e., the $X_{i,0}^{(1,1)}(\boldsymbol{p})$ of Eq. (4.22) exist for all $i = 1, 2, \ldots, n$. By definition, these derivatives will be STHS and such that

$$\lim_{h_i,\tau \to 0} \overline{\left[\frac{Y_q(\boldsymbol{p})}{h_i\tau} - X_{i,0}^{(1,1)}(\boldsymbol{p})\right]^2} = 0,$$

which, after some manipulations, implies that,

$$\left.\frac{\partial^2\gamma_X(\boldsymbol{h},\tau)}{\partial h_i \partial\tau}\right|_{(h_i,\tau)=(0,0)} = 0 \tag{4.39}$$

for all $i = 1, 2, \ldots, n$. The analysis above leads to the following useful corollary.

Corollary 4.1

An S/TRF $X(\boldsymbol{p})$ with STHS 0/0-increments is m.s. differentiable of order (1,1), i.e., the $X_{i,0}^{(1,1)}(\boldsymbol{p})$ exist in the m.s. sense for all $i = 1, 2, \ldots, n$, if and only if Eq. (4.39) holds for all $i = 1, 2, \ldots, n$. Under certain conditions, this result also can be extended to the case of $X(\boldsymbol{p})$ with STHS increments of orders ν/μ.

The variogram function derivatives are related to the corresponding covariance function derivatives, when they exist. We recall that the variogram and structure functions exist for a more general class of random fields, i.e., random fields with STHS increments, whereas the covariance function requires that the random field is itself STHS, which is a more restrictive requirement.

Also, some interesting results can be obtained in terms of the Taylor expansion of a variogram around the origin, viz.

$$\gamma_X(\Delta\boldsymbol{p}) \approx \gamma_X(\boldsymbol{0}) + \sum_{\alpha\geq 1} \frac{1}{\alpha!} \sum_{i_1,\ldots,i_\alpha=0}^{n} a_{i_1,\ldots i_\alpha} \prod_{j=i_1}^{i_\alpha} h_j, \tag{4.40}$$

[15] Compare this result with that of Eq. (2.7) of Chapter XV with $\nu = \mu = 0$.

where $\Delta\boldsymbol{p} = (h_1,\ldots, h_n, h_0 = \tau)$, and $a_{i_1,\ldots,i_\alpha} = \frac{\partial^\alpha}{\partial \boldsymbol{h}^\alpha}\gamma_X(\Delta\boldsymbol{p})\big|_{\Delta\boldsymbol{p}=0}$, with $\partial \boldsymbol{h}^\alpha = \prod_{j=i_1}^{i_\alpha} h_j$. The expansion of (Eq. 4.40) may provide useful information regarding the geometry of the random field. In particular, since by definition the $X(\boldsymbol{p})$ is m.s. continuous if and only if the $\gamma_X(\Delta\boldsymbol{p})$ is continuous at origin, Eq. (4.40) shows that this can happen if and only if $\gamma_X(\mathbf{0}) = 0$; and, according to Eq. (4.39), the $X(\boldsymbol{p})$ is differentiable in the m.s. sense if and only if $a_{i_1} = 0$ in the expansion of Eq. (4.40). In the special case of an STIS variogram $\gamma_X(r,\tau)$, the expansion of Eq. (4.40) leads to some interesting results concerning the existence of the space−time derivatives of orders (ν,μ) of the random field $X(\boldsymbol{p})$ (see, Exercise XIII.21).

Example 4.7

For illustration, the expansion of Eq. (4.40) in $R^{1,1}$ with $(h_1,h_0) = (h,\tau)$ gives under certain conditions (Exercise XIII.22)

$$\begin{aligned}\gamma_X(h,\tau) \approx \gamma_X(0,0) + \frac{\partial\gamma_X(0,0)}{\partial\tau}\tau + \frac{\partial\gamma_X(0,0)}{\partial h}h + \frac{\partial^2\gamma_X(0,0)}{\partial h\partial\tau}h\tau \\ + \frac{1}{2}\left(\frac{\partial^2\gamma_X(0,0)}{\partial\tau^2}\tau^2 + \frac{\partial^2\gamma_X(0,0)}{\partial h^2}h^2\right).\end{aligned} \tag{4.41}$$

In the STHS case, $\frac{\partial\gamma_X(0,0)}{\partial\tau} = \frac{\partial\gamma_X(0,0)}{\partial h} = 0$. In practice, the derivatives can be sometimes calculated from physical considerations. For instance, in homogeneous turbulent shear flow in $R^{3,1}$ under STIS conditions with $r = |\boldsymbol{h}|$, these derivatives are estimated by

$$\begin{aligned}\frac{\partial^2\gamma_X(0,0)}{\partial r^2} &= 0.75\int_0^\infty dk\, k^2 E(k),\\ \frac{\partial^2\gamma_X(0,0)}{\partial r\partial\tau} &= -u_1\frac{\partial^2\gamma_X(0,0)}{\partial r^2},\\ \frac{\partial^2\gamma_X(0,0)}{\partial\tau^2} &= \left(u_1^2 + u_0^2\right)\frac{\partial^2\gamma_X(0,0)}{\partial r^2} + 0.75S^2\int_0^\infty dkE(k),\end{aligned}$$

where u_1 is a mean velocity component, S is a shear rate, u_0 is the sweeping velocity, and $E(k)$ is the energy spectrum calculated from experimental data.

With all this being discussed in this chapter, there is something really interesting and I think underappreciated about the GS/TRF theory: This may be a very good modeling choice to improve the representation of space−time heterogeneous variations encountered in the real-world.

CHAPTER

PHYSICAL CONSIDERATIONS

CHAPTER OUTLINE

1. SPATIOTEMPORAL VARIATION AND LAWS OF CHANGE

The study of spatiotemporal variation, i.e., the manner in which the values of a natural attribute are linked across space and time, generally involves *physical considerations*, including scientific laws and theories, testable hypotheses, and empirical associations. This is a vital body of information about the attributes that is often more significant than the uncertain data sets usually available in space−time. In this respect, prominent is the role of the physical laws that govern the change of the attributes across space and time. To phrase it in more words, it is desirable and possible that an attribute's spatiotemporal variation is quantitatively expressed in terms of quantitative relations that codify lawful regularities of Nature that science has been able to discover. Among other things, this means that we need to investigate approaches that make it possible to derive spatiotemporal variability functions (S/TVFs) in the physical (real) domain (covariance, variogram, and structure functions) or in the spectral domain (spectral functions, complete or partial) using physical considerations about the attribute.

Before a more technical discussion of random field modeling based on physical law is presented, it may be fruitful to digress for a moment and briefly review the matter from a broader perspective. Generally, the foundational physical laws scientists believe in at the present time include Newton's laws, conservation of momentum and energy laws, thermodynamics laws, and Maxwell's laws. Surely, a long list of other laws (physical, biological, social, etc.) has been developed over the years, many of which are not as important as the above ones, yet they offer very valuable information about several aspects of Nature. For illustration, a partial list of such laws is given in Table 1.1 (Christakos, 2010). The list covers a wide range of disciplines, including ocean, earth and atmospheric sciences, life sciences, and economics (a detailed discussion of these laws can be found in the relevant scientific

Spatiotemporal Random Fields. http://dx.doi.org/10.1016/B978-0-12-803012-7.00014-3

Table 1.1 Partial List of Natural Laws

Physics	Abney, Archimedes, Bernoulli–Euler, Biot, Boltzmann, Bose–Einstein, Clausius, Coulomb, Curie, Euler, Faraday, Fick, Fresnel–Arago, Heisenberg, Hooke, Joule, Kirchhoff, Lambert, Maxwell, Newton, Ohm, Planck, Rayleigh, Schrodinger, Snell, Steinmetz, Wien
Chemistry	Avogadro, Beer–Lambert, Bouguer–Lambert, Boyle, Coppet, Dalton, Einstein–Stark, Fajans–Soddy, Gay–Lussac, Humboldt, Maxwell–Boltzmann, Nernst, Ostwald, Proust, Raoult, Retger, Sommerfeld, Wenzel, Wullner
Earth and atmospheric sciences	Archie, Bernoulli, Braggs, Buys–Ballot, Darcy, Dittus–Boelter, Drude, Egnell, Glen, Hack, Hale, Hazen, Hilt, Hopkins, Jordan, King, Kramer, MacArthur–Wilson, Richards, Steno, Stokes, Wake, Walther, Werner, Young–Laplace
Life sciences	Behring, Bowditch, Courvoisier, Dastre–Morat, Dollo, Du Bois, Elliott, Edinger, Emmert, Farr, Gloger, Gogli, Gompertz, Haeckel, Hardy–Weinberg, Liebig, Mendel, Reed–Frost, Wallace, van Valen, von Baer, Yoda, Zeune
Psychology	Bell–Magendie, Charpentier, Ebbinghaus, Fechner, Fitt, Fullerton–Cattell, Hick–Hyman, Horner, Jackson, Jost, Korte, Merkel, Piper, Ricco, Talbot–Plateau, Vierdot, Weber
Economics	Engel, Goodhart, Gresham, Hotelling, Okun, Pareto, Say, Verdoom, Wagner, Wald

literature). When studying this list, it should be kept in mind that a general difference between a *fundamental* physical law and an *empirical* physical law is that the former is a law that is derived logically from deeper principles and often has explanatory power, whereas the latter is a quantitative relationship that somehow fits the data, but it is not necessarily understood why it works.

In random field modeling, it is generally recognized that investigators often know considerably more about the phenomenon of interest than is conveyed to them by the available site-specific data (experimental, numerical, statistical, etc.). This is called *core knowledge*, and includes physical laws and theories of the same or relevant scientific discipline (core knowledge is basically derived by means of analytical or logical thinking rather than in terms of statistical data analysis). *Empirical testing* relies fundamentally on the use of lawful regularities expressed by physical laws. Data are inescapably grounded in the underlying natural processes and mechanisms. Specifically, scientific measurements and observations implicitly depend on the laws governing the underlying physical mechanisms and phenomena, which make them reproducible under the same conditions. If we cannot rely on these laws, we cannot reasonably use the measurements and observations as scientific evidence. Physical laws are also very useful in the case of *inaccessible attributes*. This happens when an attribute, say X, is difficult, expensive or even impossible to measure or calculate experimentally. This attribute, however, is often related to a measurable attribute, say Y, via a physical, health, or social law. Using this law, the statistics of X can be adequately calculated from those of Y. Natural laws, phenomenological models, and empirical associations across various disciplines are often available in the form of differential equations (ordinary and partial), and algebraic equations (polynomial, exponential, etc.).

To clarify the meaning and the substantive content of the equations expressing mathematically the physical laws, it may be appropriate to distinguish between physical laws that are *material* laws (e.g., Ohm's, Hooke's, or Timoshenko's laws), those that express *conservation* (e.g., conservation of charge, mass, or energy), and those that are *equilibrium* laws (e.g., of fluid or chemical equilibrium). In random field modeling, basically two groups of equations are obtained from these and other physical laws, a direct (stochastic) one and a derived (deterministic) one, as follows:

(i) from the physical law governing the attribute, a *stochastic* equation is obtained that should be obeyed by the spatiotemporal random field (S/TRF) representing it, and
(ii) from this stochastic equation, the corresponding *deterministic* equation that the S/TVF must satisfy can be derived.

Deriving the S/TVF, in particular, from the physical law and not solely on the basis of data (as is often done in practice) improves the physical interpretation of the S/TVF, and, also, avoids the self-reference problem of statistical techniques (e.g., in Kriging or statistical regression the data are used initially to calculate the covariance function and then to derive spatiotemporal estimates as weighted averages of the same data, which is an approach that creates some profound logical problems).[1]

Let us look at the matter in a little more technical detail. The behavior of a phenomenon, i.e., its law of change, is quantitative described by a mathematical relationship between the pertinent attributes, i.e., by an equation. It is my intention in this chapter to investigate, in a relevant random field context, the issues surrounding physical laws that can be mathematically represented in terms of a stochastic equation of the general form [case (i), above]

$$\mathcal{L}_{\boldsymbol{p}}[X(\boldsymbol{p}), Y_l(\boldsymbol{p}), a_{l'}(\boldsymbol{p})] = 0, \tag{1.1}$$

$l = 1, \ldots, k$, $l' = 1, \ldots, m$, where $\mathcal{L}_{\boldsymbol{p}}$ is a mathematical operator in the space–time domain (usually differential, although integral and algebraic operators are also considered, see below), $X(\boldsymbol{p})$ is the random field representing the unknown attribute, $Y_l(\boldsymbol{p})$, $l = 1, \ldots, k$, are known random fields representing observable or measurable attributes, and $a_{l'}(\boldsymbol{p})$, $l' = 1, \ldots, m$, denote model parameters. Each one of the attributes and parameters has specific physical dimensions and they can only be combined by means of dimensionally consistent operations. In physical terms, Eq. (1.1) then describes the constraints between attributes linked by the laws of Nature in conditions of uncertainty.

In the vast majority of scientific applications encountered in the real-world the $\mathcal{L}_{\boldsymbol{p}}$ denotes a differential operator so that Eq. (1.1) represents a *stochastic partial differential equation* (SPDE). The SPDE specifies attribute's local behavior during small space and time changes and in conditions of uncertainty. In some other cases $\mathcal{L}_{\boldsymbol{p}}$ denotes an integral operator so that Eq. (1.1) represents a *stochastic integral equation* (SIE), and in a smaller number of applications $\mathcal{L}_{\boldsymbol{p}}$ is an algebraic operator, in which case Eq. (1.1) represents a *stochastic algebraic equation* (SAE). In the majority of interesting cases, a space–time argument $\boldsymbol{p} = (\boldsymbol{s}, t) \in R^{n,1}$ is considered. When the argument reduces to $s \in R^1$ or $t \in T$, we are dealing with a stochastic *ordinary differential equation* (SODE), which, though, is of rather little interest in the context of spatiotemporal random field modeling. Uncertainty generally enters Eq. (1.1) in three ways: random boundary and/or initial conditions (BIC), random $Y_l(\boldsymbol{p})$, and random coefficients $a_{l'}(\boldsymbol{p})$. These three possibilities are not mutually exclusive. In some cases, we deal with a mixture of these possibilities.

Subsequently, and very importantly, from the SPDE obeyed by the random fields representing the natural attributes, the corresponding *deterministic* partial differential equations (PDE) obeyed by the attributes' S/TVFs can be derived subject to the associated BIC (group of equations (ii), above).

[1]See Christakos (2000, 2010).

In many cases, an analytical solution of the PDE is not available, in which case an approximate solution is sought, and when this is not possible, we search for a numerical solution by using computational techniques based on the discretization of the PDE. *Computational* random field modeling is concerned with physical quantities, which distinguishes it from computational mathematics and statistics that are concerned with numbers. Surely, physical quantities are represented by numbers in computational random field modeling, but they also possess physical meaning and content (which are usually deliberately ignored by computational mathematics and statistics), and, in addition, they are associated with case-specific space–time arguments (coordinates, domains, and metrics).

By way of a summary, Eq. (1.1) is a mathematical representation of the space–time distribution of a natural phenomenon (usually in terms of SPDE and less frequently in terms of SAE) starting from prescribed BIC. Whatever specific form Eq. (1.1) takes, we want to make sure that the mathematical representation we select does not oversimplifies the physical structure of the phenomenon under study to allow tractable computations and make a statistical technique applicable (e.g., this is what often happens with hierarchical Bayesian techniques; Wikle et al., 1998, 2001). Surely, even if we restrict the presentation to natural laws that can be mathematically represented by Eq. (1.1), it is impossible to present here every natural law that falls into this category. Rather the choice of laws to be presented here is, of necessity, representative yet limited in number. Surely, many of the issues (theoretical and interpretational) surrounding the use of SPDE representations of physical laws in random field modeling are many and challenging. Some of them are touched on in this book, whereas several others continue to be a matter of research and even controversy.

Remark 1.1

At this point, it is important to point out that many of the results of Sections 2 and 3 that follow on the representation of random fields in terms of SAE and SPDE can be fruitfully used in the construction of space–time covariance models (to be discussed in Chapter XVI, particularly, Section 5). This approach has the great merit of maintaining close contact between the mathematical description and the physical phenomenon described. In particular, the covariance models thus constructed are permissible by construction, they are interpretable, and they have considerable dynamics of physical meaning-making.

2. EMPIRICAL ALGEBRAIC EQUATIONS

In many practical applications, the S/TVFs can be directly derived from empirical SAE governing the joint variation of the attributes of interest. In these applications, the operator in Eq. (1.1) has a linear algebraic form, which leads to the following SAE model solution of Eq. (1.1) with $l = l' = 1, \ldots, m$, $k = m$,

$$X(\boldsymbol{p}) = \mathcal{L}_{\boldsymbol{p}}^{-1}[Y_l(\boldsymbol{p}), a_l] = \sum_{l=1}^{m} a_l Y_l(\boldsymbol{p}). \tag{2.1}$$

Eq. (2.1) links the attribute of interest $X(\boldsymbol{s})$, about which, say, we do not have sufficient information, or it is physically difficult or expensive to measure, with other attributes $Y_l(\boldsymbol{p})$, ($l = 1, \ldots, m$) about which a significant amount of easily collected information is available, and a set of coefficients a_l. Usually, uncertainty enters the law representation of Eq. (2.1) through the modeling of $Y_l(\boldsymbol{p})$ or a_l.

Simple empirical models of the form of Eq. (2.1) are used in many real-world situations, such as, subsurface pollution assessment in which the cost of analysis for inorganic contaminants $Y_l(\boldsymbol{p})$ is usually much lower than that for an organic contaminant $X(\boldsymbol{s})$. Also, index tests are correlated with engineering soil properties via empirical relationships of the form of Eq. (2.1). Naturally, there is a wide variability of SAE models, ranging from the very simple to the very complex. In fact, even the simplest SAE models can have a considerable contribution in practical attribute modeling.

Example 2.1

In Seismology, the space–time-dependent epidemic-type aftershock sequence model is given by

$$X(\boldsymbol{s},t) = X_b(\boldsymbol{s}) + \sum_{t_l<t} a_l(t)Y(\boldsymbol{s}_l), \tag{2.2}$$

where $X(\boldsymbol{s}, t)$ is the seismicity rate, $X_b(\boldsymbol{s})$ is the background rate due to the tectonic loading, $Y(\boldsymbol{s}_l)$ is a function related to the spatial decay of aftershocks with respect to the mainshock epicenter,

$$a_l(t) = \frac{Ke^{\alpha\Delta m_l}}{(t-t_l+c)^p},$$

Δm_l is the difference between the magnitude m_l of the lth earthquake and the smallest magnitude (threshold magnitude) m_0 of earthquakes to be treated in the data set, t_l is the occurrence time of the lth earthquake, and the α, K, c, and p are aftershock parameters. In Zoology, an empirical law that relates air temperature and the rate of crickets chirp (also known as Dolbear's law) has the algebraic form

$$X(\boldsymbol{s},t) = 50 + 0.25(Y(\boldsymbol{s},t) - 4), \tag{2.3}$$

where the coefficients have been calculated empirically, $X(\boldsymbol{s}, t)$ denotes temperature ($^\circ F$), and $Y(\boldsymbol{s}, t)$ is the number of chirps per minute. Lastly, in soil mechanics a well-known empirical relationship between standard penetration resistance, $Y(\boldsymbol{s}, t)$, and vertical stress, $X(\boldsymbol{s}, t)$, for a cohesionless soil is as follows,

$$X(\boldsymbol{s},t) = a + bY(\boldsymbol{s},t), \tag{2.4}$$

where the coefficients a and b are uniformly distributed random variables (RVs) that are assumed to be independent of each other and of $X(\boldsymbol{s}, t)$ and $Y(\boldsymbol{s}, t)$. Commonly used empirical statistics of these coefficients are $\overline{a} = 1.35$, $\sigma_a = 1.05$, $\overline{b} = 7.9$, $\sigma_b = 1.76$. The above empirical law is useful in soil mechanics applications, since in practice vertical stress $X(\boldsymbol{s}, t)$ is much more difficult to measure than standard penetration resistance $Y(\boldsymbol{s}, t)$. In all the above cases, the empirical laws can be used to calculate the corresponding S/TVFs, see Section 5.2 of Chapter XVI.

Undoubtedly, the S/TVFs estimated from the data using statistical techniques should also encapsulate the existing core physical knowledge about space–time attribute variation. For instance, natural laws may impose certain physical *constraints*[2] on the estimated covariance functions. Otherwise said, the estimated covariance functions are not solely data-dependent but they must also conform to physical constraints relevant to the phenomenon under study. One example is discussed here, and some more can be found in the last section of this chapter.

[2]Physical constraints are discussed in Section 5 at the end of this chapter.

Example 2.2

Let the attributes of a natural system be represented by the vector S/TRF $\boldsymbol{X}(\boldsymbol{p}) = [X_1(\boldsymbol{p})\ldots X_k(\boldsymbol{p})]^T$, and assume that the following empirical relationships exist among the attributes $X_l(\boldsymbol{p})$, $l = 1,\ldots, k$,

$$\boldsymbol{X}(\boldsymbol{p})^T \boldsymbol{a}_\rho(\boldsymbol{p}) = b_\rho(\boldsymbol{p}), \tag{2.5}$$

where $\boldsymbol{a}_\rho^T(\boldsymbol{p}) = [a_1(\boldsymbol{p})\ldots a_k(\boldsymbol{p})]_\rho$ and $b_\rho(\boldsymbol{p})$ are known physical parameters ($\rho = 1,\ldots,m$). It seems reasonable that $\boldsymbol{X}(\boldsymbol{p})$ modeling and data processing should account for the empirical constraints of Eq. (2.5) in a rigorous manner. Interestingly, Eq. (2.5) leads to another set of constraints that must be satisfied by the corresponding cross-covariance matrix $\boldsymbol{c_X} = [c_{X_l X_{l'}}]$ of the attributes (and by any other S/TVF, in general). This is an important issue as regards the adequate selection of a covariance matrix $\boldsymbol{c_X}$ model, which is the topic to be discussed in Section 12 of Chapter XVI.

Many situations similar to those discussed in Examples 2.1 and 2.2, and of varying levels of sophistication, can be found throughout this book, and even more of them in the relevant scientific literature.

3. PHYSICAL DIFFERENTIAL EQUATIONS

A large class of SPDE across space–time has the general form of Eq. (1.1), where the notion of randomness enters the SPDE models via the random BIC, a random forcing function, the random coefficients of the operator $\mathcal{L}_{\boldsymbol{p}}$, or combinations of the above. The analysis below is mainly concerned with the general linear and homogeneous space–time differential operator of the form introduced in Remark 4.2 of Chapter VI, i.e.,

$$\mathcal{L}_{\boldsymbol{p}} X(\boldsymbol{p}) = \sum_{|\boldsymbol{\rho}|=\nu} \alpha_{\boldsymbol{\rho},\mu} \frac{\partial^{\boldsymbol{\rho}+\mu}}{\partial \boldsymbol{s}^{\boldsymbol{\rho}} \partial t^{\mu}} X(\boldsymbol{p}),$$

where $\boldsymbol{p} = (\boldsymbol{s}, t)$, $\alpha_{\boldsymbol{\rho},\mu}$ are coefficients, $\boldsymbol{\rho}$ is the set of integers $(\rho_1,\ldots, \rho_n)$ such that $|\boldsymbol{\rho}| = \sum_{i=1}^n \rho_i = \nu$, ρ_i denotes the order of the partial derivative with respect to s_i, and $\partial \boldsymbol{s}^{\boldsymbol{\rho}} = \prod_{i=1}^n \partial s_i^{\rho_i}$ (multi-index notation). While many of the basic theoretical problems in the study of SPDEs are essentially the same as those for classical (deterministic) PDE (existence and uniqueness of solutions, stability, dependence of solutions on coefficients, BIC), there are considerable differences, as well. These differences naturally arise from the study of the random fields described by the SPDE. For instance, the interpretation of an SPDE will depend on whether the random field is viewed as a collection of RVs (mean square sense) or as a family of realizations (sample path sense). As a consequence, different types of solutions to an SPDE may be obtained:

(a) In terms of random field representations of the natural attributes (e.g., in the mean square or the sample path sense).

(b) By determining the probability density functions, or the characteristic functions, of the random fields involved (in more complicated situations, the characteristic functionals, may be needed).

(c) By means of the deterministic PDEs that govern the corresponding S/TVFs, or the deterministic algebraic equations relating the corresponding spectral functions.

In this section, we will focus on approach (a) above, i.e., only on certain cases of physical laws represented as SPDE, which are closely related to the random field models considered in the book (and which can be used to derive the corresponding S/TVFs, if needed). Approaches (b) and (c) will be discussed in Chapter XVI. The examples that follow describe several kinds of SPDE models of attributes that need to be accounted for when studying the random fields representing these attributes.

Example 3.1

Several natural laws in conditions of in situ uncertainty are expressed by SPDEs of the evolutionary form,

$$\frac{\partial}{\partial t}X(\boldsymbol{s},t) = aX(\boldsymbol{s},t) + bY(\boldsymbol{s},t), \tag{3.1a}$$

where a and b are known coefficients, and $Y(\boldsymbol{s}, t)$ is a zero mean white-noise random field. A general solution of Eq. (3.1a) is given by the random field expression

$$X(\boldsymbol{s},t) = e^{at}X(\boldsymbol{s},0) + b\int_0^t du e^{a(t-u)}Y(\boldsymbol{s},u), \tag{3.1b}$$

subject to the necessary BIC (Exercise XIV.3). On average, $\overline{X(\boldsymbol{s},t)} = e^{at}\overline{X(\boldsymbol{s},0)}$, i.e., the space–time variation depends on the space-dependent initial condition and on time.

Example 3.2

A relatively large class of scientific models is represented by the νth-order SPDE along the direction of the unit vector $\boldsymbol{\varepsilon}_i$ (fixed time), i.e., $\mathcal{L}_{\boldsymbol{p}} = \sum_{l=0}^{\nu} a_l \frac{\partial^l}{\partial s_i^l}$, where a_l ($l = 1,\ldots,\nu$) are known coefficients, in which case the SPDE and the associated boundary conditions are given by, respectively,

$$\begin{aligned} &\sum_{l=0}^{\nu} a_l \frac{\partial^l}{\partial s_i^l} X(\boldsymbol{s},t) = Y(\boldsymbol{s},t), \\ &\frac{\partial^l}{\partial s_i^l} X(\boldsymbol{s},t)\bigg|_{\boldsymbol{s}=\boldsymbol{0}} = \frac{\partial^l}{\partial s_i^l} X(\boldsymbol{0},t) = 0 \end{aligned} \tag{3.2a–b}$$

($l = 0, 1,\ldots, \nu - 1$), where the attribute of interest is represented by the random field $X(\boldsymbol{p}) \in L_2$, i.e., it is mean square (m.s.) differentiable with respect to the coordinate s_i of the point $\boldsymbol{p}$, $Y(\boldsymbol{p})$ represents a measurable attribute, and a_i are known coefficients (see also, Section 5 of Chapter XVI).

The following examples examine random field representations of physical laws encountered in ocean, earth, and atmospheric sciences (these examples will be revisited in Chapter XVI, in the context of space–time covariance development).

Example 3.3

The following SPDE has been used in applied sciences to represent several physical phenomena in the $R^{n,1}$ domain,

$$\nabla\cdot[a(\boldsymbol{s},t)\nabla X(\boldsymbol{s},t)] = -Y(\boldsymbol{s},t), \tag{3.3}$$

where, to account for real-world uncertainty (due to parameter measurement errors etc.), the $Y(\boldsymbol{s}, t)$ and $a(\boldsymbol{s}, t) > 0$ are modeled as random fields. For each $Y(\boldsymbol{s}, t)$ and $a(\boldsymbol{s}, t)$, Eq. (3.3) is solved for $X(\boldsymbol{s}, t)$ subject the BIC that are necessary for the uniqueness of the solution. The particular physical meaning of $X(\boldsymbol{s}, t)$, $Y(\boldsymbol{s}, t)$, and $a(\boldsymbol{s}, t)$, as well as the interpretation of Eq. (3.3) depend on the application (e.g., in electrodynamics, the SPDE may be seen as the Gauss law, where $X(\boldsymbol{s}, t) = E(\boldsymbol{s}, t)$, $Y(\boldsymbol{s}, t) = \rho(\boldsymbol{s}, t)$, and $a(\boldsymbol{s}, t) = \varepsilon_0$ denote, respectively, the electric field, charge density rate, and medium permittivity). For illustration, in the $R^{2,1}$ domain, and assuming that $a(t) = e^{W(t)}$ with $W(t) \sim G(0, 1)$ for all t, and $Y(\boldsymbol{s}, t) = \sin(\pi s_1)\sin(\pi s_2)$, the random field solution of Eq. (3.3) is given by

$$X(\boldsymbol{s}, t) = \frac{1}{2\pi^2} e^{-W(t)} \sin(\pi s_1)\sin(\pi s_2), \tag{3.4}$$

where $\boldsymbol{s} = (s_1, s_2)$, see Exercise XIV.9. From Eq. (3.4), we conclude that the average time-variation of $X(\boldsymbol{s}, t)$ depends on the statistics of $W(t)$, and it varies periodically in space.

Example 3.4

Several kinds of random fields modeling meteorological attributes obey SPDEs of the heat conduction (diffusion) and fractional types. Specifically Whittle (1954, 1963) and Jones and Zhang (1997) studied random fields $X(\boldsymbol{p})$ satisfying a fractional SPDE of the form

$$\left[\left(\nabla^2 - a^2\right)^p - b\frac{\partial}{\partial t}\right] X(\boldsymbol{p}) = Y(\boldsymbol{p}), \tag{3.5}$$

where a and b are known coefficients, $p = \frac{2\mu+n}{4}$ ($\mu > 0$), and $Y(\boldsymbol{p})$ is a zero mean white-noise random field with variance σ_Y^2. The covariance model corresponding to Eq. (3.5) is presented in case no. 4 of Table 5.2 of Chapter XVI.

General hydrodynamic laws constitute another important category of physical laws represented by random fields and SPDE. An example is discussed next.

Example 3.5

Transient single-phase fluid flow in $R^{n,1}$ is governed by the continuity equation and Darcy's law in their stochastic form (Zhang, 2002)

$$\begin{aligned} S_s(\boldsymbol{s})\frac{\partial}{\partial t}X(\boldsymbol{p}) + \nabla\cdot\boldsymbol{Q}(\boldsymbol{p}) &= Y(\boldsymbol{p}), \\ K_S(\boldsymbol{s})\frac{\partial}{\partial s_j}X(\boldsymbol{p}) &= -Q_i(\boldsymbol{p}), \end{aligned} \tag{3.6a–b}$$

subject to the BIC

$$\begin{aligned} X(\boldsymbol{s}, 0) &= X_0(\boldsymbol{s}), & \boldsymbol{s} \in \Omega, \\ X(\boldsymbol{s}, t) &= X_B(\boldsymbol{s}, t), & \boldsymbol{s} \in \Gamma_D, \\ \boldsymbol{Q}(\boldsymbol{s}, t)\cdot\boldsymbol{\eta}(\boldsymbol{s}) &= Q(\boldsymbol{s}, t), & \boldsymbol{s} \in \Gamma_N, \end{aligned} \tag{3.7a–c}$$

where $\boldsymbol{p} = (\boldsymbol{s}, t) \in R^{n,1}$, $X(\boldsymbol{p})$ is the random hydraulic head field, $\boldsymbol{Q}$ is the specific discharge vector, $S_s(\boldsymbol{s})$ is the specific storage, $Y(\boldsymbol{p})$ denotes the source/sink function, $K_S(\boldsymbol{s})$ is the random hydraulic conductivity (locally isotropic), $X_0(\boldsymbol{s})$ is the initial head, $X_B(\boldsymbol{s}, t)$ is the prescribed head on Dirichlet boundary segments Γ_D, the scalar Q denotes the prescribed flux across Neumann boundary segments Γ_N, and $\boldsymbol{\eta}(\boldsymbol{s})$ is an outward unit vector normal to the boundary. Often the log-transformed hydraulic conductivity, $f(\boldsymbol{s}) = \ln K_S(\boldsymbol{s})$, is used, in which case, the following decomposition is valid $f(\boldsymbol{s}) = \overline{f(\boldsymbol{s})} + f'(\boldsymbol{s})$, where $\overline{f(\boldsymbol{s})}$ and $f'(\boldsymbol{s})$ are the corresponding mean and random fluctuation.

Moreover, since the source of the $X(\boldsymbol{p})$ uncertainty is the $f'(\boldsymbol{s})$ randomness, the hydraulic head can be expanded as

$$X(\boldsymbol{p}) = \sum_{\lambda=0}^{\infty} X^{(\lambda)}(\boldsymbol{p}), \tag{3.8}$$

with $X^{(\lambda)}(\boldsymbol{p})$ denoting a term of λth-order in σ_f [standard deviation of $f(\boldsymbol{s})$], i.e., $X^{(\lambda)}(\boldsymbol{p}) = O\left(\sigma_f^{(\lambda)}\right)$. Accordingly, by letting $K_G(\boldsymbol{s}) = e^{\overline{f(\boldsymbol{s})}}$, and substituting Eq. (3.8) into Eqs. (3.6a–b) we find the corresponding equation satisfied by each $X^{(\lambda)}(\boldsymbol{p})$. For illustration, for the $X^{(0)}(\boldsymbol{p})$ it is valid that

$$\left[\frac{\partial^2}{\partial s_i^2} + \frac{\partial}{\partial s_i}\overline{f(\boldsymbol{s})}\frac{\partial}{\partial s_i} - \frac{S_s(\boldsymbol{s})}{K_G(\boldsymbol{s})}\frac{\partial}{\partial t}\right] X^{(0)}(\boldsymbol{p}) = -\frac{Y(\boldsymbol{s})}{K_G(\boldsymbol{s})}, \tag{3.9}$$

with BIC

$$\begin{aligned} X^{(0)}(\boldsymbol{s},0) &= X_0(\boldsymbol{s}), & \boldsymbol{s} \in \Omega, \\ X^{(0)}(\boldsymbol{s},t) &= X_B(\boldsymbol{s},t), & \boldsymbol{s} \in \Gamma_D, \\ \eta_i(\boldsymbol{s})\frac{\partial}{\partial s_i}X^{(0)}(\boldsymbol{s},t) &= -Q(\boldsymbol{s},t), & \boldsymbol{s} \in \Gamma_N, \end{aligned} \tag{3.10a–c}$$

where $\overline{X^{(0)}(\boldsymbol{p})} = X^{(0)}(\boldsymbol{p})$. For higher $\lambda \geq 1$–orders we find

$$\begin{aligned} \left[\frac{\partial^2}{\partial s_i^2} + \frac{\partial}{\partial s_i}\overline{f(\boldsymbol{s})}\frac{\partial}{\partial s_i}\right] X^{(\lambda)}(\boldsymbol{p}) = &-\frac{\partial}{\partial s_i}f'(\boldsymbol{s})\frac{\partial}{\partial s_i}X^{(\lambda-1)}(\boldsymbol{p}) - \frac{(-1)^\lambda}{\lambda!}\frac{Y(\boldsymbol{s})}{K_G(\boldsymbol{s})}[f'(\boldsymbol{s})]^\lambda \\ &+ \sum_{m=0}^{\lambda}\frac{S_s(\boldsymbol{s})}{K_G(\boldsymbol{s})}\frac{(-1)^{\lambda-m}}{(\lambda-m)!}[f'(\boldsymbol{s})]^{\lambda-m}\frac{\partial}{\partial t}X^{(\lambda)}(\boldsymbol{p}), \end{aligned} \tag{3.11}$$

with BIC

$$\begin{aligned} X^{(\lambda)}(\boldsymbol{s},0) &= 0, & \boldsymbol{s} \in \Omega, \\ X^{(\lambda)}(\boldsymbol{s},t) &= 0, & \boldsymbol{s} \in \Gamma_D, \\ \eta_i(\boldsymbol{s})\frac{\partial}{\partial s_i}X^{(\lambda)}(\boldsymbol{s},t) &= 0, & \boldsymbol{s} \in \Gamma_N, \end{aligned} \tag{3.12a–c}$$

where $\overline{X^{(1)}(\boldsymbol{p})} = 0$, and is the head fluctuation to first-order in σ_f.

The following example discusses a subsurface flow law and the solution of the associated SPDE in terms of the space transformation (ST) method presented in Chapter V.

Example 3.6

Consider the SPDE representation of the subsurface flow law governing the hydraulic head field $X(\boldsymbol{p})$, $\boldsymbol{p} = (\boldsymbol{s}, t)$ in $R^{3,1}$, i.e.,

$$\sum_{j=1}^{3}\left[\frac{\partial}{\partial s_j} + w_j(\boldsymbol{p})\right] X(\boldsymbol{p}) = 0, \tag{3.13}$$

where the scalar field $K(\boldsymbol{p})$ denotes hydraulic conductivity, and $w_j(\boldsymbol{p})$ are the random components of the log-conductivity gradient. The log conductivity, $\ln K(\boldsymbol{p})$, is assumed to be a random field with known correlation structure,[3] and a distinction is made between isotropic and anisotropic dependence. As regards BIC, for illustration we can consider the case of an infinite flow domain, where Eq. (3.13) is conditioned by a set of point-like conditions to be specified in the flow domain. Using the ST technique (Section 3 of Chapter V), the general solution of Eq. (3.13) is found to be the random field (Christakos and Hristopulos, 1997)

$$X(\boldsymbol{p}) = -\frac{1}{2(2\pi)^2}\int d\boldsymbol{\theta}\widehat{X}_{1,\boldsymbol{\theta}}(0,t)\; m(\sigma,\boldsymbol{\theta}), \tag{3.14}$$

at $\sigma = \boldsymbol{s}\cdot\theta$, where $\widehat{X}_{1,\boldsymbol{\theta}}$ is the directional distribution, and $m(\sigma,\boldsymbol{\theta})$ is the directional head generator. These should be determined by means of the *self-consistency* rule, i.e., so that the three-dimensional head solution of Eq. (3.14) satisfies the original SPDE of Eq. (3.13) and the point-like conditions (measurements).

Remark 3.1

The ST analysis above can be easily extended to the case of an anisotropic hydraulic conductivity tensor, in which case Darcy's law relates the constant specific discharge vector $\boldsymbol{Q}$ with the gradient of the hydraulic head and the hydraulic conductivity of the porous medium by means of

$$K(\boldsymbol{s},t)\nabla X(\boldsymbol{s},t) = -\boldsymbol{Q} \tag{3.15}$$

in $R^{3,1}$ (Christakos and Hristopulos, 1997). This form of Darcy's law implies that the equipotentials are the planes $\boldsymbol{Q}\cdot\boldsymbol{s} = const.$, which, in turn, imposes certain restrictions on the functional form of the solution $X(\boldsymbol{s}, t)$.

Example 3.7

Consider the following SPDE in $R^{1,1}$ governing the distribution of the attributes $X(s, t)$ and $Y(s, t)$,

$$\left[\frac{\partial}{\partial t} - \frac{\partial^2}{\partial s^2}\right]X(s,t) = Y(s,t), \tag{3.16}$$

where $Y(s, t)$ is a known Gaussian random field with $\overline{Y(s,t)Y(s',t')} = \delta(t'-t)\delta(s'-s)$, and $X(0, 0) = X_0$. The solution of Eq. (3.16) is of the form

$$X(s,t) = \int_{-\infty}^{\infty} ds' \int_0^t dt' \frac{1}{4\pi|t-t'|^{\frac{1}{2}}} e^{-\frac{|s-s'|^2}{4\pi(t-t')}} Y(s',t'), \tag{3.17}$$

which is also a Gaussian random field (Exercise XIV.4).

[3]Although here hydraulic conductivity is viewed as a scalar, the analysis can be extended to the case of an anisotropic (tensor) hydraulic conductivity.

Example 3.8

The advection-reaction equation used in various applications (e.g., water contamination, Kolovos et al., 2002) is generally an SPDE in $R^{2,1}$ of the form

$$\left[\frac{\partial}{\partial t} + \boldsymbol{v}\cdot\nabla - \nabla\cdot\boldsymbol{a}\nabla + b\right]X(\boldsymbol{p}) = Y(\boldsymbol{p}), \tag{3.18a}$$

where $\boldsymbol{p} = (s_1, s_2, t)$, $\boldsymbol{v}$ is a velocity vector, $\boldsymbol{a}$ is a diffusion matrix, $b > 0$ is a dumping coefficient, and $Y(\boldsymbol{p})$ is a forcing or source–sink term, often represented by a Gaussian random field. The physical interpretation of $X(\boldsymbol{p})$ depends on the application (e.g., contaminant concentration in the case of river contamination). The advection term $\boldsymbol{v}\cdot\nabla X$ accounts for transport effects, the term $\nabla\cdot\boldsymbol{a}\nabla X$ represents diffusion, and the term bX represents damping effects. The covariance model corresponding to Eq. (3.18a) is presented in case no. 5 of Table 5.2 of Chapter XVI.

Example 3.9

For the heat conduction law described in Example 2.5 of Chapter XII, the corresponding PDE governing the covariance function $c_X(\boldsymbol{s}, \boldsymbol{s}', t)$ of the temperature field $X(\boldsymbol{p})$ is derived by first multiplying equation Eq. (2.17) of Chapter XII at $\boldsymbol{p}'$ by $X(\boldsymbol{p})$ and taking the expectation, i.e.,

$$\frac{\partial}{\partial t}\overline{X(\boldsymbol{p}')X(\boldsymbol{p})} = D\Delta_{s'}\overline{X(\boldsymbol{p}')X(\boldsymbol{p})}, \tag{3.19}$$

then multiplying Eq. (2.17) of Chapter XII at $\boldsymbol{p}$ by $X(\boldsymbol{p}')$ and taking the expectation, i.e.,

$$\frac{\partial}{\partial t}\overline{X(\boldsymbol{p})X(\boldsymbol{p}')} = D\Delta_{s}\overline{X(\boldsymbol{p})X(\boldsymbol{p}')}, \tag{3.20}$$

and, lastly, adding them to find that

$$\frac{\partial}{\partial t}c_X(\boldsymbol{s}, \boldsymbol{s}', t) = D[\Delta_s c_X(\boldsymbol{s}, \boldsymbol{s}', t) + \Delta_{s'} c_X(\boldsymbol{s}, \boldsymbol{s}', t)], \tag{3.21}$$

which is the same covariance expression as the one derived in Chapter XII by means of the random functional method, see Eq. (2.23) of Chapter XII.

Example 3.10

The SPDE that describes heat transfer phenomena in the Nea Kessani study was studied in Yu et al. (2007). An isotropic three-dimensional region of uniform thermal conductivity with no heat generation was assumed (i.e., the amount of energy that comes in and out of an elementary unit volume is constant/steady state, since no information about the heat source or thermal conductivities of the geological formations is available), in which case the heat transfer law at a fixed time t can be expressed by the SPDE

$$\nabla^2 X(\boldsymbol{p}) = 0, \tag{3.22}$$

where $\boldsymbol{p} = (\boldsymbol{s}, t) \in R^{3,1}$, and $X(\boldsymbol{p})$ is the temperature random field. The corresponding boundary conditions are

$$\begin{aligned} &X(\boldsymbol{p}) = A(\boldsymbol{p}) + \varepsilon_A(\boldsymbol{p}), && X(\boldsymbol{p}) \in B_D, \\ &\frac{\partial}{\partial s_i}X(\boldsymbol{p}) = B(\boldsymbol{p}) + \varepsilon_B(\boldsymbol{p}), \quad i = 1, 2, 3, && X(\boldsymbol{p}) \in B_N, \end{aligned} \tag{3.23a–b}$$

where the B_D and B_N denote the stochastic Dirichlet and Neumann boundary conditions, respectively; the A and B are deterministic trends; and the random fluctuations ε_A and ε_B express uncertainty in our knowledge of the boundary conditions. We do not merely seek the direct mathematical solution of Eq. (3.22), isolated from other influences. Instead, we seek composite solutions of the geothermal situation that are physically consistent with Eq. (3.22) in a stochastic sense that accounts for the uncertainty expressed by Eqs. (3.23a–b) and, in addition, they incorporate empirically important site-specific information sources. Let the temperature random field $X(\boldsymbol{p})$ at a certain time t be expressed as the sum of a mean $\overline{X}(\boldsymbol{p})$ and a random fluctuation field $X'(\boldsymbol{p})$, i.e., $X(\boldsymbol{p}) = \overline{X}(\boldsymbol{p}) + X'(\boldsymbol{p})$. The $\overline{X}(\boldsymbol{p})$ and $X'(\boldsymbol{p})$ satisfy the PDE and SPDE, respectively,

$$\nabla^2 \overline{X}(\boldsymbol{p}) = 0, \tag{3.24}$$

with boundary conditions

$$\begin{aligned} &\overline{X}(\boldsymbol{p}) = A(\boldsymbol{p}), && \overline{X}(\boldsymbol{p}) \in B_D, \\ &\frac{\partial}{\partial s_i}\overline{X}(\boldsymbol{p}) = B(\boldsymbol{p}), \quad i = 1,2,3, && \overline{X}(\boldsymbol{p}) \in B_N; \end{aligned} \tag{3.25}$$

and

$$\nabla^2 X'(\boldsymbol{p}) = 0, \tag{3.26}$$

with boundary conditions

$$\begin{aligned} &X'(\boldsymbol{p}) = \varepsilon_A(\boldsymbol{p}), && X'(\boldsymbol{p}) \in B_D, \\ &\frac{\partial}{\partial s_i}X'(\boldsymbol{p}) = \varepsilon_B(\boldsymbol{p}), \quad i = 1,2,3, && X'(\boldsymbol{p}) \in B_N. \end{aligned} \tag{3.27}$$

Because of the assumption of homogeneous-isotropic thermal conductivity, the uncertainty associated with the temperature field in the above equations is due to the boundary condition uncertainty.

To conclude this section, I remind the readers of two important general laws we discussed in Chapter VIII. In one case, the solenoidal vector random field, $\boldsymbol{X}(\boldsymbol{p}) = [X_1(\boldsymbol{p})\ldots X_k(\boldsymbol{p})]^T$, satisfies the divergence physical law

$$\nabla \cdot \boldsymbol{X}(\boldsymbol{p}) = 0. \tag{3.28}$$

Solenoidal fields play a major role in studies of atmospheric turbulence. Another class of random fields used in meteorological studies is the potential vector random field, $\boldsymbol{X}(\boldsymbol{p})$ (also, Chapter VIII) satisfying the gradient physical law

$$\boldsymbol{X}(\boldsymbol{p}) = \nabla Y(\boldsymbol{p}), \tag{3.29}$$

where $Y(\boldsymbol{p})$ is a scalar random field. As we will see in Chapter XVI, several of the techniques presented above can be used to construct new space–time covariance models.

4. LINKS BETWEEN STOCHASTIC PARTIAL DIFFERENTIAL EQUATION AND GENERALIZED RANDOM FIELDS

In this section, our attention will be focused on SPDE aspects that display certain connections with the *generalized* S/TRF (GS/TRF) models considered in Chapter XIII. The relevance of such "GS/TRF-SPDE" links owe to the fact that a variety of natural attributes are governed by such SPDEs (e.g., flow through porous media, hydroclimatic systems, and transport and diffusion in the atmosphere). Furthermore, these links can be valuable tools in the improvement of existing physical models. For example, in flood prediction, an important problem is quantitative precipitation forecasting. By studying the spatiotemporal residual series of model errors, it is possible to develop corrections to the model to account for persistent errors. In groundwater contaminant transport modeling, there often arise structural errors in model predictions due to complexities in the subsurface system that cannot reasonably be modeled deterministically. The ability to model stochastically the resulting space−time processes offers the potential of developing corrections to the model predictions to better reflect the true system.

4.1 LINKS IN TERMS OF THE RANDOM FUNCTIONAL

The GS/TRF formalism discussed in the previous chapters provides simple solutions to SPDE problems that are difficult to solve or cannot be solved in terms of ordinary S/TRF. Consider an SPDE of the general form of Eq. (1.1), with known BIC. Eq. (1.1), which generates an S/TRF $X(s, t)$, will be called the "basic SPDE." To the SPDE we associate the functional

$$\langle \mathcal{L}_{p}[X(\boldsymbol{p})], X^{*}(\boldsymbol{p})\rangle = \langle Y(\boldsymbol{p}), X^{*}(\boldsymbol{p})\rangle, \tag{4.1}$$

where $X^{*}(\boldsymbol{p})$ is an S/TRF whose meaning will become clear shortly. Next, assume that by using the random functional properties introduced in Chapter XIII, Eq. (4.1) is transformed into the form

$$\left\langle \mathcal{L}_{\boldsymbol{p}}^{*}[X^{*}(\boldsymbol{p})], X(\boldsymbol{p})\right\rangle = F[X(\boldsymbol{p}), X^{*}(\boldsymbol{p})] + \langle Y(\boldsymbol{p}), X^{*}(\boldsymbol{p})\rangle, \tag{4.2}$$

where F is a suitable function of the S/TRF. If we define the "adjoint SPDE"

$$\mathcal{L}_{\boldsymbol{p}}^{*}[X^{*}(\boldsymbol{p})] = q(\boldsymbol{p}), \tag{4.3}$$

Eq. (4.2) becomes

$$X(q) = F[X(\boldsymbol{p}), X^{*}(\boldsymbol{p})] + \langle Y(\boldsymbol{p}), X^{*}(\boldsymbol{p})\rangle, \tag{4.4}$$

where $X(q) = \langle X(\boldsymbol{p}), q(\boldsymbol{p})\rangle$ is a GS/TRF in the sense described in Chapter XIII. Therefore, the "basic SPDE" has been transformed into the functional Eq. (4.4), where $X^{*}(\boldsymbol{p})$ is the solution of the "adjoint SPDE" (Eq. 4.3).

To illustrate the practical implications of this transformation, we must emphasize that the physical interpretation of the generalized random functional $X(q)$ depends on the choice of the test (or support) function $q(\boldsymbol{p}) = q(s, t)$. If, for example, we choose $q(\boldsymbol{s}, t) = \delta(\boldsymbol{s} - \boldsymbol{s}')\delta(t - t')$, then

$$X(q) = X(\boldsymbol{s}', t'). \tag{4.5}$$

Eq. (4.5) gives the value of $X(\boldsymbol{s}', t')$ at the point $(\boldsymbol{s}', t')$. Consider the space−time domain $S = V \times T$, and let

$$q(\boldsymbol{s}, t) = \begin{cases} |S|^{-1} & \text{if } (\boldsymbol{s}, t) \in S \\ 0 & \text{otherwise} \end{cases}. \tag{4.6}$$

Then, we can define the functional

$$X(q) = |S|^{-1}\int_{S} d\boldsymbol{s}dt\, X(\boldsymbol{s}, t), \tag{4.7}$$

that is, the mean value of $X(\boldsymbol{p})$ within the space–time domain $V \times T$. Moreover, a class of important practical problems in sciences may be solved faster and more efficiently by means of the GS/TRF formulation. For illustration, below we study an air pollution situation (Christakos, 1992, 2005).

Example 4.1

Let $X(\boldsymbol{s}, t)$ in the $R^{3,1}$ domain denote the concentration of aerosol substance in the atmosphere in the region of interest A within a time period T. Suppose that A is approximately cylindrical with total surface $S = S_B + S_T + S_L$, where S_B, S_T, and S_L denote the base, top, and lateral surfaces of A. Substance transport and diffusion within A is governed by

$$\left[\frac{\partial}{\partial t} + \nabla \cdot \boldsymbol{v} - \frac{\partial}{\partial s_3}\zeta\frac{\partial}{\partial s_3} - \xi\nabla^2 + \upsilon\right]X(\boldsymbol{p}) = w\delta(\boldsymbol{s}, \boldsymbol{s}_0), \tag{4.8}$$

with BIC

$$\begin{aligned} X(\boldsymbol{p}) &= 0 && (\boldsymbol{s} \in S_L) \\ \frac{\partial}{\partial s_3}X(\boldsymbol{p}) &= \alpha X(\boldsymbol{p}) && (\boldsymbol{s} \in S_B) \\ \frac{\partial}{\partial s_3}X(\boldsymbol{p}) &= 0 && (\boldsymbol{s} \in S_T) \\ X(\boldsymbol{s}, T) &= X(\boldsymbol{s}, 0). \end{aligned} \tag{4.9a–d}$$

In the above equations, $\boldsymbol{v}$ is the velocity vector of air particles with components v_i $(i = 1, 2, 3)$ along the horizontal directions s_1, s_2, and the vertical direction s_3; ξ, and ζ are the horizontal and vertical diffusion coefficients, respectively; υ is a quantity that has an inverse time dimension, $\alpha \geq 0$ is a parameter determining the interaction of the impurities with the underlying surface; w is the intensity of the aerosol discharge; and $\boldsymbol{s}_0$ is the location of the aerosol source (e.g., industrial plant). Next, we consider the random functional

$$X(q) = \int_A \int_0^T dsdt\, q(\boldsymbol{p})X(\boldsymbol{p}), \tag{4.10}$$

where

$$q(\boldsymbol{p}) = \begin{cases} T^{-1} + \beta\delta(s_3) & \text{if } \boldsymbol{s} \in A \\ 0 & \text{if } \boldsymbol{s} \notin A \end{cases}, \tag{4.11}$$

β is a coefficient that accounts for the aerosol fraction that gets into the soil. In view of the functional formulation, the "adjoint SPDE" with respect to Eqs. (4.8) and (4.9a–d) is

$$\left[\upsilon - \frac{\partial}{\partial t} + \nabla \cdot \boldsymbol{v} - \frac{\partial}{\partial s_3}\zeta\frac{\partial}{\partial s_3} - \xi\nabla^2\right]X^*(\boldsymbol{p}) = q(\boldsymbol{s}, t), \tag{4.12}$$

and

$$
\begin{aligned}
&X^*(\boldsymbol{p}) = 0 && (\boldsymbol{s} \in S_L) \\
&\frac{\partial}{\partial s_3} X^*(\boldsymbol{p}) = \alpha X^*(\boldsymbol{p}) && (\boldsymbol{s} \in S_B) \\
&\frac{\partial}{\partial s_3} X^*(\boldsymbol{p}) = 0 && (\boldsymbol{s} \in S_T) \\
&X^*(\boldsymbol{s}, T) = X^*(\boldsymbol{s}, 0).
\end{aligned}
\qquad (4.13\text{a–d})
$$

After solving Eqs. (4.12) and (4.13a–d), we substitute $X^*(\boldsymbol{p})$ into Eq. (4.10) to find

$$
X(q) = w \int_0^T dt\, X^*(\boldsymbol{s}, T) = X_q(\boldsymbol{s}). \qquad (4.14)
$$

It is worth noticing that certain important practical problems can be studied in terms of the functional solution of Eq. (4.14). For illustration, assume that the problem is to find a region $U \subset A$ where a new industrial plant can be located, so that for all $\boldsymbol{s}_0 \in A$ the resulting pollution over a nearby populated area $D \subset A$ during the time period T does not exceed a permissible level c, imposed by global and local sanitary requirements, viz., $X(\boldsymbol{s}) < c$ for all $\boldsymbol{s} \in D$. Assuming that all necessary information about the wind fields in the region is available, from the above condition we can find the locations $\boldsymbol{s}_0$: $X_q(\boldsymbol{s}_0) < c$, which determines the region U.

4.2 LINKS IN TERMS OF THE DETRENDING OPERATOR

We saw in earlier chapters that, by definition, a continuous parameter S/TRF-ν/μ obeys certain SPDEs, and the corresponding covariance functions (ordinary and generalized) satisfy the corresponding deterministic PDEs. In the following, we study ordinary S/TRF representations that are obtained by inverting the detrending operator $Q_{\nu/\mu}$ of the general form (Section 2.2 of Chapter XIII)

$$
Q_{\nu/\mu}[X(\boldsymbol{p})] = Y(\boldsymbol{p}), \qquad (4.15)
$$

where $X(\boldsymbol{p})$ is an S/TRF-ν/μ, and $Y(\boldsymbol{s}, t)$ is an STHS random field. In principle, $X(\boldsymbol{p})$ can be generated from $Y(\boldsymbol{p})$ by inverting Eq. (4.15), which is a solution that can contribute to clarifying the physical content of the SPDE of the attribute represented by the random field.

Based on the definition of an S/TRF-ν/μ, the $Q_{\nu/\mu}$-operators, which involve partial derivatives of the space–time coordinates, generate STHS random fields $Y(\boldsymbol{p})$ when they operate on $X(\boldsymbol{p})$. Specifically, if $X(\boldsymbol{p})$ is an S/TRF-ν/μ, by definition, all

$$
Y_i(\boldsymbol{p}) = \frac{\partial^{\mu+\nu+2}}{\partial s_i^{\nu+1} \partial t^{\mu+1}} X(\boldsymbol{p}) \qquad (4.16)
$$

($i = 1, \ldots, n$) are STHS random fields. Let us now consider the space–time delian differential operator of Eq. (2.5) of Chapter XIII, i.e.,

$$
Q_{\nu/\mu}[\cdot] = \nabla^{(\nu+1,\mu+1)}[\cdot] = \sum_{i=1}^{n} \frac{\partial^{\mu+\nu+2}}{\partial s_i^{\nu+1} \partial t^{\mu+1}}[\cdot]. \qquad (4.17)
$$

The random field

$$Y(\boldsymbol{p}) = \sum_{i=1}^{n} Y_i(\boldsymbol{p}) = \nabla^{(\nu+1,\mu+1)} X(\boldsymbol{p}), \tag{4.18}$$

is STHS too, and Eq. (4.18) is interpreted in the m.s. sense.[4] This observation leads to the following result (Christakos and Hristopulos, 1998).

Proposition 4.1

Let $Y(\boldsymbol{p})$ be an STHS random field. There is one and only one ordinary S/TRF-ν/μ, $X(\boldsymbol{p})$, with representations satisfying the differential Eq. (4.18). The solution of Eq. (4.18) is

$$\begin{aligned} X(\boldsymbol{p}) &= Q_{\nu/\mu}^{-1}[Y(\boldsymbol{p})] = \left(\nabla^{(\nu+1,\mu+1)}\right)^{-1}[Y(\boldsymbol{p})], \\ &= \frac{1}{\mu!}\int d\boldsymbol{s}' \int_{-\infty}^{t} dt' G_0^{(\nu+1)}(\boldsymbol{s},\boldsymbol{s}')(t-t')^{\mu}\left[Y(\boldsymbol{s},t') - \overline{Y}\right] + \overline{Y}\,\vartheta_{\nu/\mu}(\boldsymbol{p}) + \eta_{\nu/\mu}(\boldsymbol{p}), \end{aligned} \tag{4.19a–b}$$

where $G_0^{(\nu+1)}(\boldsymbol{s},\boldsymbol{s}')$ is the spatial Green's function that satisfies

$$\sum_{i=1}^{n} \frac{\partial^{\nu+1}}{\partial s_i^{\nu+1}} G_0^{(\nu+1)}(\boldsymbol{s},\boldsymbol{s}') = \delta(\boldsymbol{s}-\boldsymbol{s}'), \tag{4.20a}$$

$$\vartheta_{\nu/\mu}(\boldsymbol{s},t) = \beta\, t^{\mu+1} \sum_{i=1}^{n} \theta_i^2\, s_i^{\nu+1}, \tag{4.20b}$$

θ_i are the direction cosines of an arbitrary unit vector $\boldsymbol{\theta}\left(\sum_{i=1}^{n}\theta_i^2 = 1\right)$, $\beta = [(\mu+1)!(\nu+1)!]^{-1}$, and $\eta_{\nu/\mu}(\boldsymbol{p})$ is a polynomial of degree ν in space and μ in time.

One can interpret the Green's function as a device that is used to handle a probabilistic range of possible solutions when a discontinuity is present. An immediate consequence of Proposition 4.1 is the next corollary.

Corollary 4.1

If $X(\boldsymbol{p})$ is an ordinary S/TRF-ν/μ, there exists an S/TRF-$(\nu+2k)/(\mu+2\lambda)$, $Z(\boldsymbol{s}, t)$, such that

$$X(\boldsymbol{p}) = \nabla^{(2k,2l)} Z(\boldsymbol{p}), \tag{4.21}$$

where $\nabla^{(2k,2l)} = \sum_{i=1}^{n} \frac{\partial^{2(k+l)}}{\partial s_i^{2k}\partial t^{2l}}$.

The space–time covariance functions in $R^{n,1}$ associated with the differential operators of Eqs. (4.16) and (4.18) are, respectively,

$$c_{Y_i}(\Delta\boldsymbol{p}) = \frac{\partial^{2\mu+2\nu+4}}{\partial s_i^{\nu+1}\partial s_i'^{\nu+1}\partial t^{\mu+1}\partial t'^{\mu+1}} c_X(\boldsymbol{p},\boldsymbol{p}'), \tag{4.22}$$

[4]Under certain conditions, say, $Y(\boldsymbol{p})$ is a white-noise random field, Eq. (4.18) may be also interpreted as an equation for the sample path of the random field.

where $\Delta \boldsymbol{p} = (\boldsymbol{h}, \tau)$, and

$$\begin{aligned} c_Y(\Delta \boldsymbol{p}) &= \frac{\partial^{2\mu+2}}{\partial t^{\mu+1} \partial t'^{\mu+1}} \sum_{i=1}^{n} \sum_{j=1}^{n} \frac{\partial^{2\nu+2}}{\partial s_i^{\nu+1} \partial s_j'^{\nu+1}} c_X(\boldsymbol{p}, \boldsymbol{p}'), \\ &= \nabla^{(\nu+1,\mu+1)} \nabla'^{(\nu+1,\mu+1)} c_X(\boldsymbol{p}, \boldsymbol{p}'), \\ &= (-1)^{\nu+\mu} \nabla^{(2\nu+2,2\mu+2)} \kappa_X(\Delta \boldsymbol{p}), \end{aligned} \quad (4.23\text{a–c})$$

where the operators $\nabla^{(\nu+1,\mu+1)}$ and $\nabla^{(2\nu+2,2\mu+2)}$ were defined earlier.

Example 4.2

Suppose that the S/TRF-1/1 $X(\boldsymbol{p})$ and the STHS random field $Y(\boldsymbol{s}, t)$ are related by the SPDE

$$\nabla^{(2,2)} X(\boldsymbol{p}) = Y(\boldsymbol{p}), \quad (4.24)$$

where $\nabla^{(2,2)} = \sum_{i=1}^{n} \frac{\partial^4}{\partial s_i^2 \partial t^2}$. We may now study several properties of the random field $X(\boldsymbol{p})$ by means of those of $Y(\boldsymbol{p})$. For instance, the corresponding covariance functions are related by

$$\begin{aligned} c_Y(\Delta \boldsymbol{p}) &= \nabla^{(2,2)} \nabla'^{(2,2)} c_X(\boldsymbol{p}, \boldsymbol{p}'), \\ &= \nabla^{(4,4)} \kappa_X(\Delta \boldsymbol{p}), \end{aligned} \quad (4.25\text{a–b})$$

where $\nabla'^{(2,2)} = \sum_{j=1}^{n} \frac{\partial^4}{\partial s_j'^2 \partial t'^2}$, and $\nabla^{(4,4)} = \sum_{i=1}^{n} \frac{\partial^8}{\partial h_i^4 \partial \tau^4}$.

To further clarify the link between SPDEs and GS/TRFs, we assume next that the operator $Q_{\nu/\mu}$ is of the space–time additive differential form of Eq. (2.6) of Chapter XIII. In this case, the following proposition is valid (Christakos and Hristopulos, 1998).

Proposition 4.2

Let $Y(\boldsymbol{p})$ be an STHS random field. There is one and only one S/TRF-ν/μ, $X(\boldsymbol{p})$, with representations satisfying the SPDE[5]

$$Q_{\nu/\mu} X(\boldsymbol{p}) = \left[\frac{\partial^{\mu+1}}{\partial t^{\mu+1}} + \nabla^{(\nu+1)} \right] X(\boldsymbol{p}) = Y(\boldsymbol{p}). \quad (4.26)$$

The solution of Eq. (4.26) is

$$\begin{aligned} X(\boldsymbol{p}) = Q_{\nu/\mu}^{-1}[Y(\boldsymbol{p})] &= \left[\frac{\partial^{\mu+1}}{\partial t^{\mu+1}} + \nabla^{(\nu+1)} \right]^{-1} Y(\boldsymbol{p}) \\ &= \int ds' dt' \psi(s', t') G_0^{(\nu+1/\mu+1)}(\boldsymbol{p}, \boldsymbol{p}') + \overline{Y} \vartheta_{\nu/\mu}(\boldsymbol{p}), \end{aligned} \quad (4.27)$$

[5]This is Eq. (2.1) of Chapter XIII with the operator $Q_{\nu/\mu}$ of Eq. (2.6) of Chapter XIII.

where $G_0^{(\nu+1,\mu+1)}(s,s',t,t')$ is the spatial Green's function that satisfies

$$\left[\frac{\partial^{\mu+1}}{\partial t^{\mu+1}}+\nabla^{(\nu+1)}\right]\left[G_0^{(\nu+1,\mu+1)}(\boldsymbol{p},\boldsymbol{p}')\right]=\delta(s-s')\delta(t-t'), \tag{4.28a}$$

where $\delta(s-s')$ and $\delta(t-t')$ are delta functions in space and time, respectively; and

$$\vartheta_{\nu/\mu}(s,t)=\frac{1}{(\nu+1)!}\sum_{i=1}^{n}\theta_i^2 s_i^{\nu+1}+\frac{1}{(\mu+1)!}t^{\mu+1}, \tag{4.28b}$$

with $\sum_{i=1}^{n}\theta_i^2=1$.

In sum, the study of the links between SPDEs and GS/TRFs has the considerable merit of maintaining close contact between the mathematical GS/TRFs modeling and the physical phenomenon described by the SPDEs. Surely, there are interesting issues surrounding the use and nature of these links some of which are addressed in the relevant literature.

5. PHYSICAL CONSTRAINTS IN THE FORM OF INTEGRAL RELATIONSHIPS, DOMAIN RESTRICTIONS, AND DISPERSION EQUATIONS

In the previous sections, we used physical laws describing the space–time distribution of the attributes of interest to derive equations of the corresponding random field representations and the S/TVFs (mainly covariance functions). This formulation has the great merit of maintaining close contact between the mathematical description and the physical phenomenon. In the context of this formulation, the following points should be stressed:

(i) The attributes are represented quantitatively by numbers, but they also have a physical meaning.
(ii) Physical reasons require the association of attributes with well-defined space–time arguments (points, surfaces or volumes, time instants or time intervals, coordinates and metrics).
(iii) On account of their physical meanings, attributes have a significant information content that must be taken into consideration.

Conditions (i)–(iii) are very influential in scientific modeling, in general, and in random field modeling, in particular. Among other things, they may impose certain *physical constraints* on the S/TVFs of an attribute, some of which have already been mentioned above, and some others are discussed below. Furthermore, in the context of computational random field modeling these conditions forbid certain choices in the process of differential equation discretization and, at the same time, they may suggest other choices (e.g., the use of a particular discretization mesh or grid is dictated by physical reasons). Simplifications are possible in the computational process, but should be made with caution. It may be true that to understand we must simplify, but it is also true that every simplification has the potential to lead us away from reality.

Using rigorous formulations of core physical knowledge (like the lawful differential equation representations discussed in previous sections) may not be always possible, for various reasons (e.g., the case-specific BIC associated with the differential equation representations may be not available). Instead, in certain cases other forms of valuable information may be available about the phenomenon of interest that act as physical constraints on random field modeling and may lead to S/TVF models that depend on both the physics and the site-specific data of the real-world situation. Before we examine some examples, let us notice that constraint modeling applies in both cases of scalar and vector (multivariate) random fields.

In any case, an adequate understanding of the physical conditions of the situation of interest is central to any S/TVF modeling or theorization approach that represents accurately those features of the phenomena that it is intended to describe, model, or theorize. I continue with a discussion of physical constraints in the form of *integral* relationships. What is interesting about this kind of constraint is that they may take a particularly tractable form in the frequency domain.

Example 5.1

As in Example 2.2 in this chapter, we consider the physical attributes represented by the vector random field

$$\boldsymbol{X}(\boldsymbol{p}) = [X_1(\boldsymbol{p})...X_k(\boldsymbol{p})]^T,$$

and we assume that the following integral relationships exist among the attributes $X_l(\boldsymbol{p})$, $l = 1,..., k$:

$$\int d\boldsymbol{p}\boldsymbol{X}(\boldsymbol{p})^T \boldsymbol{a}_\rho(\boldsymbol{p}-\boldsymbol{p}') = b_\rho(\boldsymbol{p}'), \tag{5.1}$$

where $\boldsymbol{a}_\rho^T(\boldsymbol{p}) = [a_1(\boldsymbol{p})...a_k(\boldsymbol{p})]_\rho$ and $b_\rho(\boldsymbol{p})$ are known physical parameters ($\rho = 1,...,m$). These equations show that the calculated cross-S/TVFs are not solely data-dependent but they must also conform to the constraints of Eq. (5.1). Eq. (5.1) impose certain physical constraints on the space−time distribution of the attribute vector $\boldsymbol{X}(\boldsymbol{p})$, which must be taken into consideration in modeling and data processing (see also, Chapter XVI).

Several other constraints, of varying level of sophistication, can be found in the relevant literature. In many applications the covariance equations associated with the physical law of the phenomenon under consideration may impose definite restrictions on the space−time covariance *domain* as described by its dependence and influence features.

Example 5.2

Assume that the space−time covariance function $c_X(h, \tau)$ of the random field $X(s, t)$ satisfies the wave equation in $R^{1,1}$,

$$\frac{\partial^2 c_X(h,\tau)}{\partial \tau^2} - \upsilon^2 \frac{\partial^2 c_X(h,\tau)}{\partial h^2} = 0, \tag{5.2}$$

where υ is the velocity with initial conditions

$$\begin{aligned} c_X(h,0) &= c_{X,0}(h), \\ \frac{\partial c_X}{\partial \tau}(h,0) &= c_{X,\tau}(h). \end{aligned} \tag{5.3a–b}$$

The solution to this equation is well known and is given by

$$c_X(h,\tau) = \frac{1}{2}\left[c_{X,0}(h-\upsilon\tau) + c_{X,0}(h+\upsilon\tau)\right] + \frac{1}{2\upsilon}\int_{h-\upsilon\tau}^{h+\upsilon\tau} dh' c_{X,\tau}(h'). \tag{5.4}$$

The covariance solution at any space–time lags (h_k, τ_k) depends on the values of the covariance boundary condition $c_{X,0}(h)$ at $h_k - v\tau_k$ and $h_k + v\tau_k$, and also on the values of the covariance initial condition $c_{X,\tau}(h)$ in the interval $\Delta_k = [h_k - v\tau_k, h_k + v\tau_k]$. This interval forms the base of an isosceles triangle with vertices (h_k, τ_k), $(h_k - v\tau_k, 0)$, and $(h_k + v\tau_k, 0)$. The base Δ_k is called the *domain of dependence* of the covariance at (h_k, τ_k), and the velocity v over Δ_k contributes to the solution $c_X(h, \tau)$ of the wave equation above. If the initial conditions of Eqs. (5.3a–b) vanish within the domain of dependence, then $c_X(h_k, \tau_k) = 0$. Reversing the concept of domain of dependence, the *region of influence* of an interval $[h_1, h_2]$ consists of those lags (h_k, τ_k) in the $h \times \tau$ plane whose domains of dependence overlap with $[h_1, h_2]$. It follows that if (h_k, τ_k) is outside the region of influence of $[h_1, h_2]$, then no initial covariance conditions within $[h_1, h_2]$ can determine $c_X(h_k, \tau_k)$. To specify whether any lags (h_k, τ_k) lie within the region of influence of $[h_1, h_2]$, we use its domain of dependence Δ_k to conclude that (h_k, τ_k) is within the region of influence if and only if

$$\begin{aligned} h_k - v\tau_k &\leq h_2, \\ h_k + v\tau_k &\geq h_1 \end{aligned} \tag{5.5a–b}$$

(i.e., the initial covariance conditions affect $c_X(h, \tau)$ at those space–time lags (h_k, τ_k) that satisfy these two inequalities). So, the region of influence forms a truncated characteristic cone bounded by $[h_1, h_2]$ on the h-axis, and the lines $h_k + v\tau_k = h_1$ and $h_k - v\tau_k = h_2$ for $\tau_k > 0$. Otherwise put, if the initial covariance conditions are supported in an interval $\{h: |h - h_k| \leq r\}$, the covariance is supported in the region $\{(h, \tau): \tau \geq 0, h \in [h_k - r - v\tau, h_k + r + v\tau]\}$.

Lastly, constraints on the space–time covariance function are also imposed by physical *dispersion equations*. Since the subject of physical dispersion has been addressed in previous chapters, and to simplify the theoretical presentation, we refer to an example already discussed in Chapter VII.

Example 5.3

In Example 3.10 of Chapter VII, we saw that the acoustic pressure of an ocean wave propagating in space–time is governed by a wave law characterized by the dispersion equation,

$$\boldsymbol{k}^2 - v^{-2}\omega^2 = 0.$$

($\boldsymbol{k}$ and ω denote wavevector and time-frequency, respectively, and v is the local sound velocity). This dispersion equation introduces a physical constraint on the shape of the S/TVF characterizing the space–time dependency pattern of the attribute. Specifically, the spectral covariance representation associated with the underwater acoustics law reduces to an integral with respect to $\boldsymbol{k}$ only (for a given ω, the corresponding spectral density function is nonzero only on a circle of radius $\frac{\omega}{v}$). This, obviously, affects the shape of the space–time covariance function.

To conclude, in court, as well as in science, ignorance of the law is not acceptable. We will revisit the subject of physical considerations and their critical role in the random field modeling of natural phenomena in the context of the systematic covariance construction techniques to be discussed in Chapter XVI.

PERMISSIBILITY IN SPACE–TIME

CHAPTER OUTLINE

1. CONCERNING PERMISSIBILITY

The spatiotemporal variability functions (S/TVFs) (ordinary and generalized covariance functions, variogram, and structure functions) represent and measure primary aspects of the space–time variation of a natural attribute. In many applications, the S/TVFs are often obtained empirically from the data. This is usually an ill-posed problem, since the complete empirical determination of an S/TVF requires infinitely many points, while the available data provide only a finite set. In practice, the issue is simply resolved by fitting various empirical S/TVF models to the data and choosing the one that provides the best fit. For the purpose of determining various models for the fit, sufficient knowledge of the mathematical properties of correlations based on a physical analysis of the phenomenon of interest can be very helpful.

In the above setting, one of the most recurring and consequential features of rigorous S/TVF modeling concerns the *permissibility conditions* (PCs) that must constrain the parameters of the empirical S/TVF models so that these models are valid space–time moments. A distinction should be made between sufficient and necessary PCs: Sufficient PCs indicate that given the set of conditions a conclusion is obtained (regarding the validity or not of the S/TVF), whereas necessary conditions indicate that given a conclusion that conclusion satisfies the stated conditions.

It is worthwhile to devote an entire chapter on the important role the PCs play in the random field modeling of natural attributes. The PCs constitute necessary and/or sufficient mathematical conditions

Spatiotemporal Random Fields. http://dx.doi.org/10.1016/B978-0-12-803012-7.00015-5

that must be satisfied by the S/TVFs. In the case of space–time homostationary (STHS) fields, PCs are obtained by means of the *nonnegative-definiteness* (NND) conditions based on Bochner's analysis, also known as the Khinchin–Bochner analysis (Khinchin, 1949; Bochner, 1959). In the case of space–time heterogeneous (non-STHS) random fields, the NND conditions are based on the Bochner–Schwartz theorem (Gel'fand and Vilenkin, 1964).[1]

A few worth-noticing statements concerning NND, and the associated PCs, that the readers should keep in mind are as follows:

(a) The NND conditions frequently are neither adequately understood nor rigorously applied.
(b) Some of these conditions are not as widely applicable as they are commonly thought to be.
(c) Their implications in real-world modeling are not sufficiently appreciated, although they can have serious consequences.

Another point to be stressed is that the PCs would not be necessary if the S/TVF could be determined from first principles. When possible in practice, this approach focuses on obtaining explicit expressions for the unknown random field $X(\boldsymbol{p})$ in terms of a known random field $Y(\boldsymbol{p})$ via physical modeling. Solutions of such models provide scientifically lawful relations between different attributes of a phenomenon, which determine the S/TVF of $X(\boldsymbol{p})$ in terms of the known S/TVF of $Y(\boldsymbol{p})$. The $Y(\boldsymbol{p})$ may represent a natural gradient, e.g., temperature, and $X(\boldsymbol{p})$ the resulting flux. In several situations, however, the random field $Y(\boldsymbol{p})$ has multiple physical components, or the field $X(\boldsymbol{p})$ satisfies a complicated dynamic equation, which can make explicit representation of $X(\boldsymbol{p})$ in terms of a well-characterized $Y(\boldsymbol{p})$ a rather difficult task. In any case, physical modeling can be extremely valuable in providing information about the form of the S/TVF that can also be used in the context of empirical S/TVF calculations. The following postulate, then, offers a useful physical PC.

Postulate 1.1
Fitting a physical law–based function to the data is a much better approach than using an arbitrary or *ad hoc* function that merely provides a statistically good fit to the data. Hence, when the available knowledge makes it possible, a combination of both approaches, law- and data-based, is strongly recommended.

In this chapter we discuss the important role the PCs play in space–time variability studies. We consider both scalar and vector spatiotemporal random fields (S/TRFs), in the ordinary and the generalized sense (although we focus on covariance functions, the analysis can be extended to other S/TVFs too, such as variogram or structure functions). The development of comprehensive PCs is surely a helpful step forward as regards rigorous spatiotemporal variability analysis. Yet, it is just as critical to point out that these formal PCs should be distinguished from the physical PCs.

2. BOCHNERIAN ANALYSIS

I will start with a review of the most fundamental perspective regarding NND and the associated PC. The mathematical underpinning of this perspective is Bochner's celebrated theorem.

[1]Both theorems have been discussed in previous chapters.

2.1 MAIN RESULTS

As it has already been noticed in earlier chapters, in the case of a scalar S/TRF $X(\boldsymbol{p})$ a standard property of its covariance function $c_X(\boldsymbol{p}_i, \boldsymbol{p}_j)$ is that it must be of the NND type, that is,

$$\sum_{i=1}^{m}\sum_{j=1}^{m} q_i q_j\, c_X(\boldsymbol{p}_i,\boldsymbol{p}_j) \geq 0 \tag{2.1a}$$

for all nonnegative integers m, all points $\boldsymbol{p}_i, \boldsymbol{p}_j \in R^{n+1}$ and all real (or complex) numbers q_i and q_j $(i,j = 1,\ldots,m)$. Or, alternatively, we can use conditional coordinates (Eqs. 2.4b–c of Chapter I), so that the NND restriction is written as

$$\sum_{i=1}^{m}\sum_{j=1_i}^{k_i}\sum_{i'=1}^{m}\sum_{j'=1_{i'}}^{k_{i'}} q_{ij} q_{i'j'}\, c_X(\boldsymbol{s}_i, t_j; \boldsymbol{s}_{i'}, t_{j'}) \geq 0 \tag{2.1b}$$

for all m, k_i, $k_{i'}$ $(= 1, 2, \ldots)$, all locations–instants $(\boldsymbol{s}_i, t_j) \in R^{n,1}$, and all real (or complex) numbers q_{ij}, $q_{i'j'}$ (k_i denotes the number of time instants t_j, $j = 1_i, 2_i, \ldots k_i$, used, given that we are at the spatial position $\boldsymbol{s}_i$). The above inequalities are a direct consequence of the classical inequality,

$$\overline{X'(\boldsymbol{p}_i)^2} = \sum_{i=1}^{m}\sum_{j=1}^{m} q_i q_j\, \overline{X'(\boldsymbol{p}_i) X'(\boldsymbol{p}_j)} \geq 0, \tag{2.2}$$

where, as usual, $X' = X - \overline{X}$ denotes random field fluctuation. Otherwise said, the space of covariance functions is that of NND functions.

What is interesting about the inequalities presented in Eqs. (2.1a) and (2.1b) is that they are completely general, i.e., they are satisfied by space–time covariance functions regardless of space–time variability hypotheses (i.e., they hold for STHS and non-STHS random fields). Yet, confirming the validity of these inequalities in practical applications is surely a complicated affair. Fortunately, a famous theorem by *Bochner* (1933) provides useful conditions in the frequency domain for functions to be NND. Bochner's theorem was initially formulated for functions $f(\boldsymbol{h})$ defined in a subspace of R^n with a Euclidean distance $r_E = |\boldsymbol{h}|$ (see, Eq. 2.18a of Chapter I), but it can be extended to space–time functions because its validity does not depend on the definition of a space–time distance.

Specifically, Bochner's theorem has been primarily used for STHS [and space–time isostationary (STIS)] random fields. In this case, the following theorem holds in terms of the ordinary covariance function[2] (Gnedenko, 1962).

Theorem 2.1

A continuous function $c_X(\Delta\boldsymbol{p})$ in R^{n+1} is an NND function if and only if it can be expressed as in[3]

$$c_X(\Delta\boldsymbol{p}) = \int dQ_X(\Delta\boldsymbol{w}) e^{i\Delta\boldsymbol{w}\cdot\Delta\boldsymbol{p}}, \tag{2.3}$$

[2]Analogous results are valid in terms of variogram or structure functions.
[3]See, also, Eq. (3.6c) of Chapter VII.

where $Q_X(\Delta \boldsymbol{w})$ is a nonnegative bounded nondecreasing function (sometimes termed the spectral distribution function).

It is usually assumed that the $Q_X(\Delta \boldsymbol{w})$ is absolutely continuous, and the $c_X(\Delta \boldsymbol{p})$ is absolute integrable on R^{n+1} so that the conditions are satisfied for the more comprehensive covariance representations

$$\tilde{c}_X(\Delta \boldsymbol{w}) = \frac{1}{(2\pi)^{n+1}} \int_{R^{n+1}} d\Delta \boldsymbol{p}\, c_X(\Delta \boldsymbol{p}) e^{-i\Delta \boldsymbol{w} \cdot \Delta \boldsymbol{p}} \geq 0,$$
$$c_X(\Delta \boldsymbol{p}) = \int_{R^{n+1}} d\Delta \boldsymbol{w}\, \tilde{c}_X(\Delta \boldsymbol{w}) e^{i\Delta \boldsymbol{w} \cdot \Delta \boldsymbol{p}} \tag{2.4a–b}$$

to be valid for all $\Delta \boldsymbol{p}, \Delta \boldsymbol{w} \in R^{n+1}$.[4] As were the cases with many representations in the R^{n+1} domain, those of Eqs. (2.3) and (2.4a–b) remain valid in the $R^{n,1}$ domain by simply replacing $\Delta \boldsymbol{p}$ and $\Delta \boldsymbol{w}$ with $(\boldsymbol{h}, \tau)$ and $(\boldsymbol{k}, \omega)$, respectively (see, e.g., Theorem 4.1 later).

Example 2.1

Consider the random field $X(s, t)$ introduced in Example 2.1 of Chapter V, with the covariance representation

$$c_X(h, \tau) = \int dQ_X(\alpha) e^{i(h + c\tau)\alpha}.$$

This function satisfies the conditions of Theorem 2.1, and, hence, it is a permissible covariance function.

For space–time heterogeneous (non-STHS), random fields the Bochner–Schwartz theorem can be used in terms of the space–time generalized covariance functions $\kappa_X(\boldsymbol{h}, \tau)$. Then, the following result is valid (Christakos and Hristopulos, 1998).

Theorem 2.2

A function $\kappa_X(\boldsymbol{h}, \tau)$ is conditionally NND if and only if its spectral density function (SDF) $\tilde{\kappa}_X(\boldsymbol{k}, \omega)$ is a real-valued and nonnegative function such that the tempered integral

$$\int\int \frac{d\boldsymbol{k}\, d\omega}{(1 + \boldsymbol{k}^2)^{\rho_1}(1 + \omega^2)^{\rho_2}} \tilde{U}_Q\, \tilde{\kappa}_X(\boldsymbol{k}, \omega) \tag{2.5}$$

converges for some numbers $\rho_1, \rho_2 \geq 0$, where $\tilde{U}_Q$ is the U_Q equivalent in Fourier space.[5]

One must be aware that the PCs described by Theorem 2.2 are satisfied if the following conditions hold:

(i) $\tilde{\kappa}_X(\boldsymbol{k}, \omega) \geq 0$;
(ii) $\tilde{U}_Q\, \tilde{\kappa}_X(\boldsymbol{k}, \omega)$ does not have any nonintegrable singularities around $k = |\boldsymbol{k}| = 0, \ \omega = 0$; and
(iii) as $k, \omega \to \infty$, $\tilde{U}_Q\, \tilde{\kappa}_X(\boldsymbol{k}, \omega)$ increases slower than $k^{2\rho_1}\, \omega^{2\rho_2}$ for some numbers $\rho_1, \rho_2 \geq 0$.

The expression $\tilde{U}_Q = k^{2\nu+2}\omega^{2\mu+2}$ is commonly used for the detrending operator in the spectral domain, in which case the PCs are described by the following proposition (Christakos, 1991c).

---

[4]See, also, Eq. (3.7) of Chapter VII.
[5]The operator U_Q was defined in Section 3.3 of Chapter XIII.

Proposition 2.1
The conditions for $\kappa_X(\boldsymbol{h}, \tau)$ to be a generalized spatiotemporal covariance of order ν/μ (GS/TC-ν/μ) are[6]

$$k^{2\nu+2+(n-1)}\omega^{2\mu+2}\,\widetilde{\kappa}_X(\boldsymbol{k}, \omega) \geq 0, \tag{2.6}$$

and[7]

$$\lim_{|\boldsymbol{h}|\to\infty} \frac{\kappa_X(\boldsymbol{h}, \tau)}{\boldsymbol{h}^{2\nu+2}} = \lim_{\tau\to\infty} \frac{\kappa_X(\boldsymbol{h}, \tau)}{\tau^{2\mu+2}} = 0. \tag{2.7}$$

The PCs introduced by Proposition 2.1 have the significant merit that they are rather easy to implement in practice. It is, also, worth noticing that in the case of space–time covariance separability in the $R^{n,1}$ domain, i.e.,

$$c_X(\boldsymbol{h}, \tau) = c_{X(1)}(\boldsymbol{h})\, c_{X(2)}(\tau),$$

the composite space–time integral of Bochner's theorem, Eq. (2.4b), conveniently reduces to the product of two separate integrals, one in space and one in time. Specifically, the following corollary is a rather straightforward consequence of Theorem 2.1.

Corollary 2.1
A separable continuous function $c_X(\boldsymbol{h}, \tau)$ is an NND function in $R^{n,1}$ if and only if it can be expressed as

$$c_X(\boldsymbol{h}, \tau) = \int d\boldsymbol{k} e^{i\boldsymbol{k}\cdot\boldsymbol{h}}\,\widetilde{c}_{X(1)}(\boldsymbol{k}) \int d\omega\, e^{-i\,\omega\,\tau}\,\widetilde{c}_{X(2)}(\omega), \tag{2.8}$$

where the SDFs $\widetilde{c}_{X(1)}(\boldsymbol{k})$ and $\widetilde{c}_{X(2)}(\omega)$ are real-valued, integrable, and nonnegative functions of the wavevector $\boldsymbol{k}$ and the frequency ω, respectively.

Similar expressions are obtained in the case of space–time separable generalized covariances $\kappa_X(\boldsymbol{h}, \tau)$, as well. Yet, not all are ideal in paradise, as the following discussion on the limitations of formal PC analysis points out.

2.2 LIMITATIONS OF BOCHNERIAN ANALYSIS

A cautionary note is due here. There are some more restrictions underlying the validity of Theorem 2.1: A function $c_X(\boldsymbol{p}_i, \boldsymbol{p}_j)$ is a permissible covariance of a multi-Gaussian (say, m-variate Gaussian, G^m), in particular, random field $X(\boldsymbol{p})$ if and only if it is an NND function. To demonstrate the necessary part, one must simply show that all combinations $\sum_{i=1}^{k} q_i X(\boldsymbol{p}_i)$ must have a nonnegative variance. To prove the sufficient part, one must essentially construct a G^m-S/TRF that admits this covariance. The *Gaussian restriction* appears naturally, since the proof is based on certain properties of the G^m probability distribution (Gnedenko, 1962).

[6]The $\boldsymbol{k}^{n-1}$ is included because the integration is in R^n.

[7]The polynomial terms arise naturally in $\kappa_X(\boldsymbol{h}, \tau)$ models: They represent trends or fluctuation correlations. The distinction between trends and fluctuations is important: Brownian motion is a zero-mean random process, but its variance increases linearly with the support size.

Insofar as NND validity is concerned, some issues have emerged regarding the range of applicability of the NND condition and when this condition is adequate. As it turns out, two major conclusions can be drawn regarding the matter:

(a) Bochnerian criteria guarantee that a G^m-distributed S/TRF exists with the NND function as a space–time covariance, but not necessarily that the function is a permissible covariance for a non-G^m field.

(b) Permissibility of a space–time covariance function depends on the space–time metric, i.e., a function that is permissible covariance for one metric may be not so for another.

Let us examine these conclusions in more detail, starting with (a). According to Bochner's analysis, a function $c_X(\Delta\boldsymbol{p})$ in R^{n+1} (or, $c_X(\boldsymbol{h}, \tau)$ in $R^{n,1}$) is a permissible covariance function of a G^m-STHS random field $X(\boldsymbol{p})$ if and only if it is an NND function. Yet, this analysis does not always guarantee that there is a non-G^m random field that admits as a covariance an NND function (in fact, this is not valid, in general). One way around this problem is to link the covariance of the non-G^m random field of interest to the covariance function of a G^m-STHS random field, and check if the latter is permissible. This approach is described in the following proposition.

Proposition 2.2

Let $c_Y(\Delta\boldsymbol{p})$ be the covariance function of a non-G^m random field $Y(\boldsymbol{p})$ that is related to the covariance function $c_X(\Delta\boldsymbol{p})$ of the STHS G^m random field $X(\boldsymbol{p})$ by means of

$$c_Y(\Delta\boldsymbol{p}) = \mathcal{F}[c_X(\Delta\boldsymbol{p})], \tag{2.9}$$

where $\mathcal{F}$ is a known function. Then, for the permissibility of c_Y it is not sufficient that it is an NND function itself, but a stronger condition must be satisfied, namely, that the c_X related to c_Y via Eq. (2.9) must be an NND function.

The function $\mathcal{F}$ may assume various forms, depending on the situation. A rather typical case is examined in the following example.

Example 2.2

If $c_Y(\Delta\boldsymbol{p})$ is the covariance of a *lognormal* S/TRF $Y(\boldsymbol{p})$ with mean $\overline{Y}$, the function $\mathcal{F}$ in Eq. (2.9) is necessarily of the form

$$c_Y(\Delta\boldsymbol{p}) = \mathcal{F}[c_X(\Delta\boldsymbol{p})] = \overline{Y}^2\left(e^{c_X(\Delta\boldsymbol{p})} - 1\right), \tag{2.10}$$

where c_X is the covariance of a G^m random field $X(\boldsymbol{p})$. Hence, it is not sufficient that c_Y is an NND function, instead, it is the function c_X within Eq. (2.10) that must be NND. These two fields are related in terms of the classical results listed in Table 2.1 (see, also, Table 7.1 of Chapter II). An implication of the last equation of Table 2.1 is that one must test whether or not the $\ln\left[\eta^2\frac{c_Y}{\sigma_Y^2} + 1\right]$ is an NND function, instead of attempting to do the same merely for c_Y.

Table 2.1 Classical Results of Lognormality

$X(\boldsymbol{p}) \sim G^m(\overline{X}, \sigma_X^2)$	$Y(\boldsymbol{p}) = e^{X(\boldsymbol{p})}$	$Y(\boldsymbol{p}) \sim L^m(\overline{Y}, \sigma_Y^2)$
$\overline{Y} = e^{\overline{X}+\frac{1}{2}\sigma_X^2}$		$\overline{X} = \ln \frac{\overline{Y}^2}{\sqrt{\overline{Y}^2+\sigma_Y^2}}$
$\overline{Y^k} = e^{k\overline{X}+\frac{1}{2}k^2\sigma_X^2},\ k \geq 1$		
$\sigma_Y^2 = e^{2\overline{X}+\sigma_X^2}\left[e^{\sigma_X^2} - 1\right]$		$\sigma_X^2 = \ln\left[\frac{\sigma_Y^2}{\overline{Y}^2} + 1\right]$
$c_Y(\Delta\boldsymbol{p}) = \frac{\sigma_Y^2}{\eta^2}\left[e^{c_X(\Delta\boldsymbol{p})} - 1\right],\ \eta = \frac{\sigma_Y}{\overline{Y}}$		$c_X(\Delta\boldsymbol{p}) = \ln\left[\eta^2 \frac{c_Y(\Delta\boldsymbol{p})}{\sigma_Y^2} + 1\right]$

Example 2.3

To show that the spherical model is not a permissible covariance for a lognormal (L^m) random field, consider the

$$X(\boldsymbol{p}) \sim G^m\left(\overline{X} = 1, \sigma_X^2 = 0.1^2\right).$$

In view of Table 2.1, a lognormal field $Y(\boldsymbol{p}) = e^{X(\boldsymbol{p})}$ can be defined such that

$$Y(\boldsymbol{p}) \sim L^m\left(\overline{Y} = e^{1.005}, \sigma_Y^2 = e^{2.01}\left(e^{0.01} - 1\right)\right).$$

For simplicity of the numerical computations, let us suppress the time argument and consider the spherical covariance model in R^2 (Table 4.1 of Chapter VIII), $c_Y(|\boldsymbol{h}_{ij}|) = sph(c_0, \boldsymbol{\varepsilon}_r) = sph(2, 3)$. Also, select a set of six points $\boldsymbol{s}_i$ $(i = 1,\ldots, 6)$ so that $\boldsymbol{s}_1 = (1, 1)$, $\boldsymbol{s}_2 = (2, 4)$, $\boldsymbol{s}_3 = (3, 1)$, $\boldsymbol{s}_4 = (3, 4)$, $\boldsymbol{s}_5 = (4, 2)$, and $\boldsymbol{s}_6 = (5, 6)$. Using the expressions in Table 2.1 we obtain the corresponding relationship between covariance functions,

$$c_X\left(|\boldsymbol{h}_{ij}|\right) = \ln\left[e^{-4.02}\left(e^{0.01} - 1\right)^{-2} c_Y\left(|\boldsymbol{h}_{ij}|\right) + 1\right].$$

Then, the $c_X(|\boldsymbol{h}_{ij}|)$ covariance values shown in Table 2.2 are obtained for each pair of points (Exercise XV. 1), on the basis of which it can be shown that there exist combinations $\sum_{i=1}^{k} \lambda_i X(\boldsymbol{p}_i)$ that violate the NND condition. Indeed, the combination of weights $\lambda_1 = \lambda_2 = \lambda_3 = -10$, $\lambda_4 = 0.477$, $\lambda_5 = 5.858$, $\lambda_6 = 24.665$ together with the corresponding covariance values from Table 2.2 gives

Table 2.2 Values of $c_X(|\boldsymbol{h}_{ij}|)$

Location	s_1	s_2	s_3	s_4	s_5	s_6
s_1	0.2374					
s_2	0.2374	0.2374				
s_3	0.0389	0.2374	0.2374			
s_4	0.2374	0.1301	0.2374	0.2374		
s_5	0.2374	0.0013	0.0885	0.0236	0.2374	
s_6	0.2374	0.2374	0.2374	0.0013	0.2374	0.2374

$Var\left[\sum_{i=1}^{6}\lambda_i X(s_i)\right] = -0.0827 < 0$.[8] Hence, the spherical covariance model is permissible for a normal (Gaussian) random field but not for a lognormal one. It can also be shown that other functions are not permissible covariance models for a lognormal random field (e.g., the cosine model, Exercise XV. 2).

An important point made earlier should be stressed: not all formally permissible covariance functions are good candidates to represent the space–time variation of a natural attribute. Covariance models describe how attribute correlations change in space and time and, therefore, they are inherently connected to the physical laws governing the attribute.

3. METRIC DEPENDENCE

The main issue of this section concerns metric-dependent PCs, i.e., with the fact that a covariance that is valid for one space–time metric may be not necessarily so for another. This issue emerges with increasing frequency in applications. I remind the readers that Table 5.1 of Chapter VIII presented a number of space–time covariance models, the permissibility of which depended on the metric considered. Among the space–time metrics associated with these models were the Pythagorean metric (r_P), the Minkowski metric (r_{Mi}), the traveling metric (r_T), and the mixed traveling–plane wave metric (r_{TW}). Some more results about metric dependence are discussed next.

Example 3.1

In his study of covariance permissibility, Ma (2007) has shown that the space–time covariance model

$$c_X(|\boldsymbol{h}|,\tau) = (1 + a\,|\boldsymbol{h}| + b\,\tau + c\,|\boldsymbol{h}|\,\tau)e^{-|\boldsymbol{h}|-\tau}, \tag{3.1}$$

is permissible in $R^{n,1}$ for both the Euclidean r_E and the Manhattan (absolute) r_M metrics, the difference being the conditions on the model coefficients, i.e., $|b-c| \le 1-a$, $|b+cn| \le 1+an$ for r_E, and $|b-cn| \le 1-an$, $|b+cn| \le 1+an$ for r_M. Also, focusing on the purely spatial case Banerjee (2005) seems to agree with earlier results that many well-known covariance functions (exponential, spherical, and Matern functions) that are valid in terms of the Euclidean metric are not necessarily so if geodetic metrics are used, instead.[9]

Moreover, the following proposition has been proven in the spatial case by Christakos and Papanicolaou (2000), but it can be extended to simultaneous covariance functions in the space–time ($|\boldsymbol{h}|$,τ) domain.

Proposition 3.1

(1) Assume that the covariance function $c_X(|\boldsymbol{h}|)$ has an even extension on R^1 that is twice differentiable in an open neighborhood of 0, while $|\boldsymbol{h}|$ is a norm on R^n. Then, $|\boldsymbol{h}|$ must be the spatial Euclidean metric

[8] On the other hand, $Var\left[\sum_{i=1}^{6}\lambda_i Y(s_i)\right] = 0.0461 > 0$, which could lead to misinterpretation of the PCs.

[9] Although Benarjee admits that "ambiguity prevails among practicing statisticians about distance metrics", nevertheless, he seems to be completely unaware of the existing substantial literature on the subject.

$|\boldsymbol{h}| = r_E$. (2) When the conditions of case (1) are not satisfied, a function $c_X(|\boldsymbol{h}|)$ may be a permissible covariance for a non-Euclidean metric of the form

$$|\boldsymbol{h}| = r_\rho = \sqrt[\rho]{\sum\nolimits_{i=1}^{n} |h_i|^\rho}, \tag{3.2}$$

where $0 < \rho < 2$.

Some covariance models are considered next to illustrate the implementation of the theoretical result above.

Example 3.2

The spatial component of the separable space–time Gaussian covariance model of Eq. (2.8a) of Chapter VII satisfies the conditions of Proposition 3.1, and, hence it is permissible only for r_E. On the other hand, the exponential covariance model of Eq. (4.23) of Chapter VII does not satisfy the conditions of Proposition 3.1 (i.e., the even extension of c_X is not twice differentiable in an open neighborhood of zero), and, r_E is not necessarily the only choice. In fact, the exponential c_X in $(|\boldsymbol{h}|, \tau)$ is permissible for both the Euclidean $|\boldsymbol{h}| = r_E$ ($\rho = 2$ in Eq. 3.2) and the Manhattan $|\boldsymbol{h}| = r_M$ ($\rho = 1$ in Eq. 3.2) metrics.

In sum, the PCs play an important role in determining covariance dependence on the space–time metric considered. In the following section we present a list of comprehensive PCs that cover a considerable number of classes of space–time covariance models.

4. FORMAL AND PHYSICAL PERMISSIBILITY CONDITIONS FOR COVARIANCE FUNCTIONS

With the considerations of the previous sections in mind, I proceed with the presentation of a series of PCs for covariance functions that are widely used in applications. When they refer to the corresponding spectral function, rather than the covariance function itself, these PCs usually turn out to be necessary and sufficient. Another set of PCs although they are only sufficient, they have the advantage that they refer directly to the covariance function.

As was mentioned in the introduction to the present chapter, in principle, a covariance model is permissible if it is determined from first principles. Hence, in addition to the formal PCs presented in the previous sections, a physical PC is valid that is a direct result of the fundamental physical characteristics of the phenomenon represented by the random field.

Proposition 4.1

If a continuous and symmetric function is the solution of an equation derived from the physical law, it is a permissible by construction covariance model, so long as all the covariance functions that pertain to the equation inputs (e.g., sources, boundary and initial condition) are valid.

Proposition 4.1 introduces a substantive rather than formal PC, which is a sufficient permissibility condition that is also necessary, in the sense that a physically valid covariance of a natural attribute must be consistent with the physical law governing the attribute. In Chapter XIV we presented several kinds of physical laws and the associated statistical equations describing the change of the attribute covariance in space–time. In Chapter XVI, we will discuss several techniques of constructing permissible covariance models from physical laws and their covariance equations. Another PC result of general applicability is described next.

Proposition 4.2

If a function is a permissible covariance function in $R^{n,1}$, it is also permissible in $R^{n',1}$ for any $n' < n$.

In fact, since permissibility in $n = 3$ implies permissibility in $n = 2$, in some cases it may be more convenient mathematically to investigate the permissibility of a covariance model in $R^{3,1}$, even if the domain of interest is $R^{2,1}$.

4.1 PERMISSIBILITY CONDITIONS FOR SPACE–TIME HOMOSTATIONARY COVARIANCE FUNCTIONS

In most of the PCs to be discussed below, a Euclidean metric is assumed. As was argued in previous sections, the standard procedure is that, since the application of Eq. (2.1a) or Eq. (2.1b) is practically very difficult or even impossible, one often implements Bochner's theorem (Theorem 2.1). In this setting, Theorem 2.1 leads to the following formulation.

Theorem 4.1

A continuous space–time function $c_X(\boldsymbol{h}, \tau)$ in $R^{n,1}$ is NND if and only if it can be expressed as

$$c_X(\boldsymbol{h}, \tau) = \iint d\boldsymbol{k} d\omega \, e^{i(\boldsymbol{k} \cdot \boldsymbol{h} - \omega\tau)} \, \widetilde{c}_X(\boldsymbol{k}, \omega), \tag{4.1}$$

where the SDF $\widetilde{c}_X(\boldsymbol{k}, \omega)$ is a real-valued, integrable, and nonnegative function of the spatial frequency (wavevector) $\boldsymbol{k}$ and the temporal frequency ω.

As a matter of fact, the SDF must be integrable but not necessarily bounded, i.e., it may have singular points. Theorem 4.1 can also be expressed in terms of the *spectral distribution function* $\widetilde{F}_X(\boldsymbol{\kappa}, \omega)$ instead of the SDF (Adler, 1981). Accordingly, the conditions of Bochner's theorem for the spectral distribution function are that it should be real valued and bounded, and that the measure

$$\mu(A) = \int_A d\widetilde{F}_X(\boldsymbol{k}, \omega) \tag{4.2}$$

is a nondecreasing function. This formulation is useful if the SDF does not exist, e.g., when the spectral distribution function has a jump; however, even in these cases the SDF can be defined by means of generalized functions. Lastly, in the special case of the rather restrictive, full symmetry hypothesis of Eqs. (1.22a–c) of Chapter VII, the $\widetilde{c}_X(\boldsymbol{k}, \omega)$ is also symmetric.

Example 4.1

SDFs that behave, e.g., as $\widetilde{c}_X(\boldsymbol{k}, \omega) \propto |\boldsymbol{k}|^{-\alpha}$ for $\boldsymbol{k}$ near zero are singular for $\alpha > 0$, but they are integrable if $n > \alpha$, where n denotes the space dimension.

It is worth noticing that many of the results to be presented below are consequences of Theorem 4.1, although the following formulation is easier to implement in applications.

Proposition 4.3

A continuous and symmetric function $c_X(\boldsymbol{h}, \tau)$ in $R^{n,1}$ will be a permissible covariance of an STHS random field if and only if it is the $n+1$-fold Fourier transformation (FT) of a nonnegative bounded function[10]

$$\tilde{c}_X(\boldsymbol{k}, \omega) \geq 0 \tag{4.3}$$

for all $(\boldsymbol{k}, \omega) \in R^{n,1}$.

I now continue with some examples of space–time covariance functions that satisfy the conditions of Proposition 4.3.

Example 4.2

Let us define the generalized function

$$c_X(\boldsymbol{h}, \tau) = a\delta(\boldsymbol{h}, \tau), \tag{4.4}$$

where $a > 0$, and $\delta(\boldsymbol{h}, \tau) = \prod_{i=1}^{n} \delta(h_i)\delta(\omega)$ is the $n+1$-dimensional delta function. The function of Eq. (4.4) is such that $\iint d\boldsymbol{h} d\tau\, c_X(\boldsymbol{h}, \tau) = a < \infty$, and it can be represented by the integral of Eq. (4.1), where the generalized FT of Eq. (4.4) gives the spectrum

$$\tilde{c}_X(\boldsymbol{k}, \omega) = \frac{a}{(2\pi)^{n+1}} \geq 0. \tag{4.5}$$

This implies that Eq. (4.4) is a permissible covariance function (the corresponding random field is a white noise).

Example 4.3

Consider in $R^{1,1}$ the space–time double exponential (Gaussian) covariance function

$$c_X(h, \tau) = e^{-h^2 - \tau^2}. \tag{4.6}$$

It has been shown earlier in this book that the corresponding spectrum is such that

$$\tilde{c}_X(k, \omega) = \frac{1}{4\pi} e^{-\frac{1}{4}(k^2 + \omega^2)} \geq 0 \tag{4.7}$$

for all $(k, \omega) \in R^{1,1}$. Therefore, the space–time function of Eq. (4.6) is a permissible covariance.

Some useful covariance relationships that can serve as necessary PC conditions are presented in the following proposition (Ma, 2003a).

Proposition 4.4

Let $c_X(\boldsymbol{h}, \tau)$ be a space–time covariance function in $R^{n,1}$. Then, it is necessary that:

(a) the inequality holds,

$$c_X(\boldsymbol{h} + \boldsymbol{h}', \tau + \tau') + c_X(\boldsymbol{h} - \boldsymbol{h}', \tau - \tau') \geq 2\left[c_X(\boldsymbol{h}, \tau) + c_X(\boldsymbol{h}', \tau') - c_X(\boldsymbol{0}, 0)\right]; \tag{4.8}$$

---

[10]The condition of a bounded $\tilde{c}_X(\boldsymbol{k}, \omega)$ is critical here.

(b) for a fixed time lag $\tau = \tau^* \in T$, the functions

$$c_X(\boldsymbol{h}, 0) \pm \frac{1}{2}[c_X(\boldsymbol{h}, \tau^*) + c_X(\boldsymbol{h}, -\tau^*)] \tag{4.9}$$

are valid spatial covariances with $\boldsymbol{h} \in R^n$; and

(c) for a fixed space lag $\boldsymbol{h} = \boldsymbol{h}^* \in R^n$, the functions

$$c_X(\boldsymbol{0}, t) \pm \frac{1}{2}[c_X(\boldsymbol{h}^*, \tau) + c_X(\boldsymbol{h}^*, -\tau)] \tag{4.10}$$

are valid temporal covariances with $\tau \in T$.

4.2 PERMISSIBILITY CONDITIONS FOR SPACE–TIME ISOSTATIONARY COVARIANCE FUNCTIONS

This section considers the covariance permissibility of STIS random fields (Chapter VIII). In this respect, the following PC is valid (for reasons explained earlier, in the STIS case it is generally more convenient to present our results in the $R^{n,1}$ domain).

Theorem 4.2

For $c_X(r, \tau)$, $r = r_E$, to be a covariance function of a STIS random field in $R^{n,1}$, it is necessary and sufficient that this function admits a representation of the form (see, also, Eqs. 1.6a–b of Chapter VIII)

$$\begin{aligned} c_X(r,\tau) &= (2\pi)^{\frac{n}{2}} r^{1-\frac{n}{2}} \int_0^\infty \int_0^\infty dk\, d\omega\, k^{\frac{n}{2}} J_{\frac{n}{2}-1}(kr)\, e^{-i\omega\tau}\, \widetilde{c}_X(k,\omega), \\ \widetilde{c}_X(k,\omega) &= (2\pi)^{-\frac{n}{2}} k^{1-\frac{n}{2}} \int_0^\infty \int_0^\infty dr\, d\tau\, r^{\frac{n}{2}} J_{\frac{n}{2}-1}(kr)\, e^{i\omega\tau} c_X(r,\tau), \end{aligned} \tag{4.11a–b}$$

where

$$\widetilde{c}_X(k,\omega) \geq 0 \tag{4.12}$$

with $k,\ \omega \geq 0$.

For illustration, Table 4.1 displays a list of space–time covariance functions and the associated SDFs (some of these results have already been studied in previous chapters of the book, although in a different context). In all cases with dimensionality $n > 1$, it is assumed that $r = r_E$ and $k = \left(\sum_{i=1}^n k_i^2\right)^{\frac{1}{2}}$.

Remark 4.1

In the case of a vector S/TRF (VS/TRF), $\boldsymbol{X}(\boldsymbol{p}) = [X_1(\boldsymbol{p})\ldots X_k(\boldsymbol{p})]^T$, Eqs. (4.11a–b) remain formally valid if we replace the scalars c_X and $\widetilde{c}_X$ with the $k \times k$ covariance matrices $\boldsymbol{c}_X$ and $\widetilde{\boldsymbol{c}}_X$, respectively.

Some specific formulations of Theorem 4.2 have been proven to be very useful in applications. One such formulation is described next, which involves space transformation (ST) operators in the frequency domain (Chapter V).

Table 4.1 Examples of Space–Time Covariances and Associated Spectral Density Functions

$c_X(r, \tau)$			$\widetilde{c}_X(k, \omega)$	
$e^{-r-\tau}$	$(R^{3,1})$	(4.13a)	$\frac{1}{\pi^3(1+k^2)^2(1+\omega^2)}$	(4.13b)
$e^{-a^{-2}(r-\upsilon\tau)^2}$	$(R^{3,1})$	(4.14a)	$\frac{a^3}{8\sqrt{\pi^3}}e^{-\frac{1}{4}(ak)^2}\delta(\omega-\upsilon k)$	(4.14b)
$e^{-\frac{r^2}{a^2}-\tau}J_0(br)$	$(R^{2,1})$	(4.15a)	$\frac{a^2}{4\pi^2(1+\omega^2)}I_0\left(\frac{a^2 b}{2}k\right)e^{-\frac{a^2(b^2+k^2)}{4}}$	(4.15b)
$\frac{1}{\cosh(a^{-1}r)e^{\tau}}$	$(R^{1,1})$	(4.16a)	$\frac{a}{2\pi\cosh\left(\frac{1}{2}\pi ak\right)(1+\omega^2)}$	(4.16b)
$e^{-\sqrt{\frac{1}{a^2}r^2+\frac{1}{b^2}\tau^2}}$	$(R^{1,1})$	(4.17a)	$\frac{ab}{2\pi\sqrt{(1+a^2k^2+b^2\omega^2)^3}}$	(4.17b)

Proposition 4.5

Given a valid covariance model $c_{X,1}$ in $R^{1,1}$, its permissibility in $R^{n,1}$ can be tested by deriving the corresponding SDF as follows:

$$\widetilde{c}_{X,n}(k,\omega) = \begin{cases} \Psi_1^n\left[\widetilde{c}_{X,1}\right](k,\omega), \\ T_1^n\left[\widetilde{c}_{X,1}\right](k,\omega), \end{cases} \tag{4.18a–b}$$

and then testing if the condition of Eq. (4.12) is satisfied for the function $\widetilde{c}_{X,n}(k,\omega)$.[11]

Remark 4.2

To reemphasize the effect of metric, I remind the readers that different results may be obtained for the covariance models of Table 4.1 if another metric is used. For instance, as we saw in Example 1.3 of Chapter VIII, if the Manhattan metric r_M ($n = 3$) is used in Eq. (4.13a) of Table 4.1, the corresponding SDF is given by Eq. (1.9c) of Chapter VIII, which, unlike Eq. (4.13b), is an anisotropic function.

[11]The same ST approach applies in the case of variogram and generalized covariance models (Christakos, 1984b).

To be sure, if the space–time separability hypothesis is valid for the covariance function of interest, some comprehensive sufficient conditions can be derived (Christakos, 1984b). The readers may recall that separability is often not valid in applications, yet, when it can be seen as a reasonably valid hypothesis it gives crucial insights into the space–time variability structure of the phenomenon.

Proposition 4.6

A continuous separable function $c_X(r,\tau) = c_{X(1)}(r)c_{X(2)}(\tau)$ in $R^{n,1}$ $(n = 1, 2, 3)$, with $c_{X(1)}(0), c_{X(2)}(0) > 0$, is a permissible covariance if the conditions listed in Table 4.2 hold, depending on the dimensionality of the function.

The conditions of Proposition 4.6, despite the fact that they are only sufficient and deal with rather specific classes of covariance models, they are, nevertheless, very convenient, since they refer directly to the covariance function and not to the SDF.

Example 4.4

In $R^{3,1}$, the exponential model

$$c_X(r,\tau) = e^{-\frac{r}{a}-\frac{\tau}{b}},$$

where $a, b > 0$, gives at the origin,

$$\frac{\partial}{\partial r} c_X(r,0)\Big|_{r=0} = -a^{-1} < 0, \quad \frac{\partial}{\partial \tau} c_X(0,\tau)\Big|_{\tau=0} = -b^{-1} < 0;$$

Table 4.2 Sufficient Conditions of Space–Time Covariance Permissibility

$\frac{\partial}{\partial r} c_X(r,0)\Big\|_{r=0} < 0$ $\frac{\partial}{\partial \tau} c_X(0,\tau)\Big\|_{\tau=0} < 0$ $\frac{\partial^2}{\partial \tau^2} c_X(0,\tau) \geq 0$	(4.19a–c)
$\lim_{r\to\infty} r^{\frac{1}{2}(n-1)} c_X(r,0) = 0$ $\lim_{\tau\to\infty} c_X(0,\tau) = 0$	(4.19d–e)
$c''_X(r,0) = \frac{\partial^2}{\partial r^2} c_X(r,0) \geq 0 \quad (R^{1,1})$ $\int_r^\infty dc''_X(r,0)\, u(u^2 - r^2)^{-\frac{1}{2}} \geq 0 \quad (R^{2,1})$ $\left(1 - r\frac{\partial}{\partial r}\right) c''_X(r,0) \geq 0 \quad (R^{3,1})$	(4.19f–h)

at infinity, $\lim_{r\to\infty} r\, c_X(r,0) = 0$, $\lim_{\tau\to\infty} c_X(0,\tau) = 0$; and also,

$$\frac{\partial^2}{\partial\tau^2} c_X(0,\tau) = b^{-2} e^{-\frac{\tau}{b}} \geq 0, \quad \left(\frac{\partial^2}{\partial r^2} - r\frac{\partial^3}{\partial r^3}\right) c_X(r,0) = (r+a)a^{-3} e^{-\frac{r}{a}} \geq 0.$$

Therefore, the exponential function above satisfies the conditions of Eqs. (4.19a–c), (4.19d–e), and (4.19h), which implies that it is a permissible covariance model in $R^{3,1}$ (of course, the same applies in $R^{2,1}$ and in $R^{1,1}$).

Lastly, in the case of random fields varying on the surface of a sphere (see, Section 5 of Chapter X), the corresponding covariance function is written as $c_X(r_s, \tau)$, where r_s is the arc length between any two points on a sphere. Then, the covariance PC described by the following proposition is valid.

Proposition 4.7

If a real-valued continuous function $c_X(r_s, \tau)$ in $S^{n-1} \times T$ satisfies Eq. (5.9) of Chapter X, then it is a permissible covariance model.

As is rather customary in any presentation that seeks a proper blend of the requirements for mathematical rigor and physical meaningfulness, I will conclude the discussion in this section with the customary warning: The formal PCs presented above should not be viewed as the singular acid test that can be applied to any real-world application. This is because, as it was argued in Postulate 2.1 of Chapter IV, in real-world applications a covariance function is not enough to be formally permissible (i.e., to satisfy a relevant PC), but it also needs to be physically permissible (i.e., to be consistent with physical considerations and interpretable).

4.3 PERMISSIBILITY CONDITIONS FOR GENERALIZED SPATIOTEMPORAL COVARIANCE FUNCTIONS

I turn in this section to the discussion of PCs that apply to space–time heterogeneously varying attributes represented by generalized spatiotemporal random fields of orders ν/μ (GS/TRF-ν/μ). As we saw in Chapter XIII, a general model for GS/TC-ν/μ is of the form of Eq. (XIII.3.22), i.e.,

$$\kappa_X(r,\tau) = \sum_{\rho=0}^{2\nu+1} \sum_{\zeta=0}^{2\mu+1} (-1)^{s(\rho)+s(\zeta)} a_{\rho\zeta} r^{\rho} \tau^{\zeta}. \qquad \text{(XIII.3.22)}$$

where $a_{\rho\zeta} > 0$, and the $s(\rho)$ and $s(\zeta)$ are sign functions such that $s(u) = \frac{1}{2}\left(u - \delta_{u,2l+1}\right)$, $u = \rho, \zeta$; $l = \nu, \mu$. The sign functions compensate the negative signs introduced by the generalized FT when using the Bochner–Schwartz theorem. Hence, the generalized FTs of all the monomials involved in Eq. (XIII.3.22) are positive, and the right-hand side of Eq. (XIII.3.22) satisfies by construction the permissibility condition (2.6). This leads to the following comprehensive PC for generalized space–time covariance functions.

Proposition 4.8

The necessary and sufficient conditions for the $\kappa_X(\boldsymbol{h}, \tau)$ of Eq. (XIII.3.22) to be a permissible GS/TC-ν/μ are that the coefficients $a_{\rho\zeta}$ satisfy the conditions of Eq. (2.7).

The PCs derived for the GS/TC-ν/μ models in the remainder of this section are based on Proposition 4.8. In the case of the polynomial GS/TC-ν/μ expressed as the product of separable space and time polynomials in Eq. (XIII.3.23), i.e.,

$$\kappa_X(r,\tau) = \kappa_{X(1)}(r)\kappa_{X(2)}(\tau) = \left[\sum_{\rho=0}^{2\nu+1} (-1)^{s(\rho)} a_\rho\, r^\rho\right]\left[\sum_{\zeta=0}^{2\mu+1} (-1)^{s(\zeta)} b_\zeta\, \tau^\zeta\right], \tag{XIII.3.23}$$

the PCs are expressed as

$$\begin{aligned}\widetilde{\kappa}_{X(1)}(k) &= \sum_{\rho=0}^{2\nu+1} c_\rho a_\rho\, k^{2\nu+1-\rho} \geq 0,\\ \widetilde{\kappa}_{X(2)}(\omega) &= \sum_{\zeta=0}^{2\mu+1} d_\zeta\, b_\zeta \omega^{2\mu+1-\zeta} \geq 0\end{aligned} \tag{4.20a–b}$$

for all $k \geq 0$ and $\omega \geq 0$, where the coefficients c_ρ and d_ζ of the generalized FT are given by

$$\begin{aligned}c_\rho &= (-1)^{s(\rho)} 2^{\rho+n}\, \pi^{\frac{n}{2}} \Gamma\left(\frac{\rho+n}{2}\right)\Gamma^{-1}\left(-\frac{\rho}{2}\right),\\ d_\zeta &= (-1)^{s(\zeta)} 2^{\zeta+1}\, \pi^{\frac{1}{2}} \Gamma\left(\frac{\zeta+1}{2}\right)\Gamma^{-1}\left(-\frac{\zeta}{2}\right).\end{aligned} \tag{4.21a–b}$$

The PCs are satisfied if the coefficients of the leading powers of the polynomials $\widetilde{\kappa}_{X(1)}(k)$ and $\widetilde{\kappa}_{X(2)}(\omega)$ in Eqs. (4.20a–b) are positive, and their roots are either complex conjugate or real and negative.

Another class of GS/TC-ν/μ models which, as we saw in Chapter XIII, are solutions of Eq. (3.18) of Chapter XIII, is of the space–time separable form (Christakos and Hristopulos, 1998; Kolovos et al., 2004)

$$\kappa_X(r,\tau) = \left[G_1(r) + \eta_{2\nu+1}(r)\right]\left[G_2(\tau) + \eta_{2\mu+1}(\tau)\right], \tag{XIII.3.30}$$

where $G_1(r)$ and $G_2(\tau)$ are Green's functions, and the polynomials, $\eta_{2\nu+1}(r)$, and $\eta_{2\nu+1}(\tau)$ are defined in Eqs. (3.32a–b) of Chapter XIII. Depending on the heterogeneity (or continuity) orders ν, $\mu = 0, 1, 2$, certain PCs apply on the coefficients a_ρ and b_ζ (some examples in the case of polynomial generalized covariances that involve only odd powers are shown in Table 4.3), where, depending on the dimensionality, the coefficients c_i and d_i ($i = 0, 1, 2, 3$) may assume different values (Table 4.4). Some plots of the generalized covariance models of Eq. (XIII.3.30) for various ν, μ values can be found in Kolovos et al. (2004).

Table 4.3 Permissibility Conditions for Coefficients a_ρ and b_ζ

ν, μ	a_ρ	b_ζ
0	$a_0 \geq 0$	$b_0 \geq 0$
1	$a_0,\ a_1 \geq 0$	$b_0,\ b_1 \geq 0$
2	$a_0,\ a_2 \geq 0,\ a_1 \geq \frac{2}{c_1}\sqrt{a_0 a_2 c_0 a_2}$	$b_0,\ b_2 \geq 0,\ b_1 \geq \frac{2}{d_1}\sqrt{b_0 b_2 d_0 d_2}$

Table 4.4 Values of Coefficients c_i and d_i (i = 0, 1, 2, 3)

	$R^{2,1}$	$R^{3,1}$
c_0, d_0	2π, 2	8π, 2
c_1, d_1	18π, 12	96π, 12
c_2, d_2	450π, 240	2880π, 240
c_3, d_3	22050π, 10080	161280π, 10080

The following section extends the PC analysis to vector random fields (VS/TRFs).

4.4 PERMISSIBILITY CONDITIONS FOR SPATIOTEMPORAL COVARIANCE MATRICES

The PCs derived in Sections 4.1–4.3 for scalar space–time covariance functions (STHS, STIS, and generalized covariance functions) can be extended to PCs that apply in the case of space–time covariance matrices introduced in Section 2 of Chapter IX in the context of VS/TRFs.

To start with, in the case of a homostationary and homostationarily connected VS/TRF, $\boldsymbol{X}(\boldsymbol{p}) = [X_1(\boldsymbol{p})\ldots X_k(\boldsymbol{p})]^T$, the corresponding matrix of spatiotemporal cross-covariance functions between the component random fields of the $\boldsymbol{X}(\boldsymbol{p})$ is given by

$$\boldsymbol{c}_X = \left[c_{X_l X_{l'}}(\Delta\boldsymbol{p})\right] \tag{4.22}$$

for all $l,\ l' = 1, 2, \ldots, k$. The matrix $\boldsymbol{c}_X$ should be NND, namely,

$$\boldsymbol{\lambda}^T \boldsymbol{c}_X \boldsymbol{\lambda} = \sum_{i=1}^{m}\sum_{j=1}^{m} \boldsymbol{\lambda}_i^T \boldsymbol{c}_X(\Delta\boldsymbol{p}_{ij})\boldsymbol{\lambda}_j \geq 0 \tag{4.23}$$

for all $m \in N$, deterministic vectors $\boldsymbol{\lambda}_i \in R^k$ ($i = 1,\ldots, m$), and all $\Delta\boldsymbol{p}_{ij}$.

Given the matrix of the STHS cross-covariances of Eq. (4.22), the definition of the corresponding matrix of cross-SDFs (cross-spectra) is rather straightforward, namely, it is the symmetric matrix

$$\widetilde{\boldsymbol{c}}_X = \left[\widetilde{c}_{X_l X_{l'}}(\Delta\boldsymbol{w})\right], \tag{4.24}$$

for all $\Delta\boldsymbol{w}$ and $l, l' = 1, \ldots, k$. The component cross-SDFs $\widetilde{c}_{X_l X_{l'}}(\Delta\boldsymbol{w})$ are expressed in terms of the corresponding cross-covariance functions as

$$\widetilde{c}_{X_l X_{l'}}(\Delta\boldsymbol{w}) = \frac{1}{(2\pi)^{n+1}} \int d\Delta\boldsymbol{p}\; e^{-i\Delta\boldsymbol{w}\cdot\Delta\boldsymbol{p}} c_{X_l X_{l'}}(\Delta\boldsymbol{p}), \tag{4.25}$$

assuming that these integrals exist. Obviously, the diagonal elements of the matrix $\widetilde{\boldsymbol{c}}_X$ represent the SDFs of the component S/TRFs, and the off-diagonal elements represent their cross-SDFs. The space–time covariance $c_Y(\Delta\boldsymbol{p}_{ij})$ of the random field $Y(\boldsymbol{p}) = \sum_{i=1}^{m} \boldsymbol{\lambda}_i^T \boldsymbol{X}(\boldsymbol{p}+\boldsymbol{p}_i)$ must be NND, which means that its SDF needs to satisfy $\widetilde{c}_Y(\Delta\boldsymbol{w}_{ij}) \geq 0$ for all $\Delta\boldsymbol{w}_{ij} \in R^{n+1}$. This analysis, in turn, implies that

$$\sum_{i=1}^{m}\sum_{j=1}^{m} \boldsymbol{\lambda}_i^T \widetilde{\boldsymbol{c}}_X(\Delta\boldsymbol{p}_{ij})\boldsymbol{\lambda}_j \geq 0 \tag{4.26}$$

for all $\boldsymbol{\lambda}_i$, $\boldsymbol{\lambda}_j$. Hence, the symmetric matrix of Eq. (4.24) is NND. Conversely, for any vector of deterministic coefficients $\boldsymbol{\lambda}_i$ it is valid that

$$\int dw \boldsymbol{\alpha}^T \tilde{\boldsymbol{c}}_X \boldsymbol{\alpha} = \sum_{i=1}^{m} \sum_{j=1}^{m} \boldsymbol{\lambda}_i^T \boldsymbol{c}_X(\Delta \boldsymbol{p}_{ij}) \boldsymbol{\lambda}_j \geq 0, \tag{4.27}$$

i.e., Eq. (4.26), where $\boldsymbol{\alpha} = \sum_{i=1}^{m} e^{-i\, \boldsymbol{w}_i \cdot \boldsymbol{p}_i} \boldsymbol{\lambda}_i$ (Exercise XV. 4). On the basis of these considerations, we obtain the relevant PC for space–time covariance matrices, as follows.

Theorem 4.3

For the matrix of Eq. (4.22) to be the matrix of permissible cross-covariance functions for a VS/TRF, it is necessary and sufficient that the matrix of Eq. (4.24) of the corresponding cross-SDF is NND for all $\Delta \boldsymbol{w}$. By virtue of the matrix theory, the latter means that the principal minor determinants of the matrix of Eq. (4.24) must be nonnegative.

The PC introduced by Theorem 4.3 for VS/TRFs could be better appreciated with the help of a simple example.

Example 4.5

Let us consider the case $k = 2$, i.e., $\boldsymbol{X}(\boldsymbol{p}) = [X_1(\boldsymbol{p})\ X_2(\boldsymbol{p})]^T$, in which case Eq. (4.24) reduces to

$$\tilde{\boldsymbol{c}}_X = \begin{bmatrix} \tilde{c}_{X_1}(\Delta \boldsymbol{w}) & \tilde{c}_{X_1X_2}(\Delta \boldsymbol{w}) \\ \tilde{c}_{X_2X_1}(\Delta \boldsymbol{w}) & \tilde{c}_{X_2}(\Delta \boldsymbol{w}) \end{bmatrix}, \tag{4.28}$$

which must be NND. In terms of the principal minor determinants of the above matrix the latter requirement implies that

$$\begin{aligned} &\tilde{c}_{X_1}(\Delta \boldsymbol{w}), \tilde{c}_{X_2}(\Delta \boldsymbol{w}) \geq 0, \\ &\left|\tilde{c}_{X_1X_2}(\Delta \boldsymbol{w})\right|^2 = \left|\tilde{c}_{X_2X_1}(\Delta \boldsymbol{w})\right|^2 \leq \tilde{c}_{X_1}(\Delta \boldsymbol{w}) \tilde{c}_{X_2}(\Delta \boldsymbol{w}). \end{aligned} \tag{4.29a–b}$$

It is noteworthy that by using the theory of matrices (e.g., Horn and Johnson, 1985), one can derive several useful results regarding the permissibility of the covariance matrix $\boldsymbol{c}_X$.

As is rather evident, most of the properties of a VS/TRF can be derived from those of its component scalar S/TRFs. Indeed, the vector random field $\boldsymbol{X}(\boldsymbol{p})$ is said to be space–time heterogeneous (spatially nonhomogeneous and temporally nonstationary), if its vector mean value $\overline{\boldsymbol{X}}(\boldsymbol{p})$ is a function of the space–time point $\boldsymbol{p}$, and its centered matrix covariance

$$\boldsymbol{c}_X(\boldsymbol{p}, \boldsymbol{p}') = \overline{[\boldsymbol{X}(\boldsymbol{p}) - \overline{\boldsymbol{X}}(\boldsymbol{p})]\,[\boldsymbol{X}(\boldsymbol{p}') - \overline{\boldsymbol{X}}(\boldsymbol{p}')]^T} \tag{4.30}$$

is a function of $\boldsymbol{p}$ and $\boldsymbol{p}'$. STHS (in the w.s.) occur when the mean $\overline{\boldsymbol{X}}(\boldsymbol{p})$ is constant and the matrix covariance $\boldsymbol{c}_X(\boldsymbol{p}, \boldsymbol{p}')$ depends only on the space–time lags $\boldsymbol{h}$ and τ. A more general class of VS/TRFs can be defined as follows.

Definition 4.1

A VS/TRF-$\boldsymbol{\nu}/\boldsymbol{\mu}$ $\boldsymbol{X}(\boldsymbol{p})$ consists of k scalar S/TRF-ν_l/μ_l, $X_l(\boldsymbol{p})$, $l = 1, \ldots, k$.

Higher heterogeneity orders ν_l and μ_l impose fewer restrictions on the data, allowing the application of the random field concepts to a wider range of natural attributes than classical statistical models permit. Given the VS/TRF-$\boldsymbol{\nu}/\boldsymbol{\mu}$ $\boldsymbol{X}(\boldsymbol{p})$ of dimension k, we can define the matrix

$$\kappa_X(\boldsymbol{h},\tau) = \left[\kappa_{X,ll'}(\boldsymbol{h},\tau)\right], \quad (l,l' = 1,\ldots,k), \tag{4.31}$$

i.e., with elements the GS/TC-ν_l/μ_l $\kappa_{X,ll}(\boldsymbol{h}, \tau)$ of the $X_l(\boldsymbol{s}, t)$ $(l = 1,\ldots, k)$, and the cross – GS/TC – $\nu_l, \nu_{l'}/\mu_l, \mu_{l'}$ $\kappa_{X,ll'}(\boldsymbol{h},\tau)$ of $X_l(\boldsymbol{s}, t)$ and $X_{l'}(\boldsymbol{s}',t')$ with $l \neq l'$ $(l,l' = 1,\ldots,k)$. The latter are defined as follows: Consider the detrending operator for each $X_l(\boldsymbol{s}, t)$ such that

$$Q_{\nu/\mu,l}\, X_l(\boldsymbol{s},t) = Y_l(\boldsymbol{s},t),$$

where $Y_l(\boldsymbol{s}, t)$ denotes STHS residuals. The $\kappa_{ll'}(\boldsymbol{h},\tau)$ are the solutions of the set of partial differential equations (PDEs)

$$U_{Q(ll')}\kappa_{X,ll'}(\boldsymbol{h},\tau) = c_{Y,ll'}(\boldsymbol{h},\tau), \quad (l,l' = 1,\ldots,k), \tag{4.32}$$

where $U_{Q(ll')} = Q_l^* Q_{l'}$, and $c_{Y,ll'}(\boldsymbol{h},\tau) = \overline{Y_l(\boldsymbol{s},t)Y_{l'}(\boldsymbol{s}',t')} - \overline{Y_l(\boldsymbol{s},t)}\;\overline{Y_{l'}(\boldsymbol{s}',t')}$. The permissibility criterion can be formulated as an extension of the Bochner–Schwartz theorem in terms of the generalized FT covariance matrix $\widetilde{\kappa}_X(\boldsymbol{k},\omega)$ that has elements $\widetilde{\kappa}_{X,ll'}(\boldsymbol{k},\omega)$ $(l,l' = 1,\ldots,k)$.

Definition 4.2

The Hermitian matrix $\widetilde{\kappa}_X(\boldsymbol{k},\omega)$ is NND if every characteristic root of the polynomial

$$\eta(\lambda) = Det\left[\lambda \boldsymbol{I} - \widetilde{\kappa}_X(\boldsymbol{k},\omega)\right] \tag{4.33}$$

is nonnegative for all $(\boldsymbol{k}, \omega)$.

A consequence of this definition is that both the trace and the determinant of $\widetilde{\kappa}_X(\boldsymbol{k},\omega)$ are positive for all $(\boldsymbol{k}, \omega)$. In relation to this, the following proposition holds (Christakos and Hristopulos, 1998).

Proposition 4.9

The matrix $\widetilde{\kappa}_X(\boldsymbol{k},\omega)$ is NND if and only if all the principal submatrices $\widetilde{\kappa}_X[\alpha|\alpha](\boldsymbol{k},\omega)$ are NND.

In matrix theory (Marcus and Minc, 1992), principal submatrices are obtained from the initial matrix by excluding rows and columns designated by the set α, which contains $0 \leq l \leq k-1$ elements. If, e.g., $\alpha = \{1, 2\}$ the principal submatrix is obtained by excluding the first two rows and columns. This brings us to the final result of this section.

Proposition 4.10

The $\kappa_X(\boldsymbol{h},\tau)$ is a permissible generalized covariance matrix if the corresponding spectral matrix $\widetilde{\kappa}_X(\boldsymbol{k},\omega)$ is Hermitian NND, and if there are numbers p_1, $p_2 \geq 0$ such that the following tempered integrals converge

$$\iint dkd\omega \frac{\widetilde{U}_{Q(ll')}\widetilde{\kappa}_{ll'}(\boldsymbol{k},\omega)}{(1+\boldsymbol{k}^2)^{p_1}(1+\omega^2)^{p_2}} < \infty \tag{4.34}$$

for all $l,l' = 1,\ldots,k$.

In the following section, I will present some more consequences of the PC analysis discussed above.

5. MORE CONSEQUENCES OF PERMISSIBILITY

Using a nonpermissible covariance function or the wrong covariance metric can have considerable consequences in applications. As we saw earlier, not all formally permissible covariances are good candidates to describe the spatiotemporal variability of a natural attribute. It should also be taken into consideration that covariance models describe how the correlations behave in space and time and, therefore, they are inherently connected to the scientific laws governing the attribute.

Example 5.1

Physically, there is a significant difference between short-range (e.g., exponential or Gaussian covariance models) and power law correlations. The latter indicate the existence of scaling in the system, which may arise due to a number of different physical causes. It could possibly denote that the system is near so-called critical points, where correlation functions become invariant under scale transformations (Christakos and Hristopulos, 1998).

Example 5.2

Percolation type models, which have been successfully applied to various natural phenomena—including flow and transport in porous media and the spread of epidemics—are among the systems that exhibit critical behavior. For illustration, assume that in lattice percolation, two phases (say, red and blue) occupy the lattice sites with probabilities p_r (red) and $1 - p_r = p_b$ (blue), respectively. At the critical threshold p_c a connected cluster of red sites spans the whole lattice. If the red phase represents the void space, the lattice becomes permeable at p_c. Near the percolation threshold the correlation functions behave like power laws with fractal exponents (Stauffer and Aharony, 1992), and the geometric structures on the lattice are fractal objects.

Important consequences of the failure to satisfy the conditions of covariance permissibility are encountered in several other applications, such as extreme value analysis (EVA, Coles, 2001). EVA is concerned with the quantification of extreme situations for the purpose of environmental disaster planning (severe weather, earthquakes, floods, fires, tornados, climate change, etc.), health risk assessment (epidemics, human exposure, heat waves, air pollution, etc.), or financial asset management (portfolio risk analysis, insurance, stock performance, etc.). Hence, in most cases the main EVA objective is to extrapolate beyond the range of the available observations and calculate the probability of occurrence of events that have never (or rarely) been observed.

Example 5.3

Let us revisit Example 2.3 and consider the same normal and lognormal random fields $X(\boldsymbol{s}, t)$ and $Y(\boldsymbol{s}, t)$, respectively, together with their covariance functions (the readers are reminded of the covariance permissibility problems detected in that example). Then, in Table 5.1 we chose three points and the corresponding distances between them. In EVA we are generally interested about probabilities of the form

$$P[Y(\boldsymbol{p}_i) > y] = P[X(\boldsymbol{p}_i) > \ln y] = (2\pi)^{-\frac{3}{2}}|\boldsymbol{c}|^{-\frac{1}{2}}\left[\prod_{i=1}^{3}\int_{\ln y}^{\infty} d\chi_i\right] e^{-\frac{1}{2}(\boldsymbol{\chi}-\bar{\boldsymbol{x}})^T \boldsymbol{c}^{-\frac{1}{2}}(\boldsymbol{\chi}-\bar{\boldsymbol{x}})}. \tag{5.1}$$

Table 5.1 Distances $|h_{ij}|$, $i, j = 1, 2, 3$

Location	s_1	s_2	s_3
s_1	0		
s_2	$\sqrt{10}$	0	
s_3	2	$\sqrt{10}$	0

Using the covariance function of Example 2.3 in the present case, we find that

$$|c|^{-\frac{1}{2}} = -0.0094, \quad (i = 1, 2, 3), \tag{5.2}$$

which is obviously an unacceptable result and can lead to the inaccurate estimation of the extreme part of a sample.

To conclude this section, the phenomena studied by EVA provide yet another demonstration of how inappropriate can be to rely on data alone, as well as to use a nonpermissible covariance function or the wrong covariance metric.

CHAPTER

CONSTRUCTION OF SPATIOTEMPORAL COVARIANCE MODELS

CHAPTER OUTLINE

1. INTRODUCTION

Several techniques of constructing valid models for spatiotemporal variability functions (S/TVFs) have been suggested in the literature. As its title indicates, the focus of this chapter is primarily on space–time covariance models. Yet, techniques for constructing other kinds of S/TVFs and their

Spatiotemporal Random Fields. http://dx.doi.org/10.1016/B978-0-12-803012-7.00016-7

spectral (frequency) domain counterparts will be also discussed briefly. In particular, this chapter describes covariance model construction techniques many of which have been proven to be efficient and fruitful tools in the development of new and useful models that are permissible, physically consistent, interpretable, and flexible in their use. In this fourfold context, several of the theoretical results obtained in previous chapters of the book can be used to build valid covariance models. After all, as inductively derived as they may be in many cases, the S/TVFs models remain, to a considerable extent, theoretical constructs. It is also noticed that working in the spectral domain may have certain benefits, like the simplification of analytical calculations. This means that, in some cases, it may be easier to construct first a valid spectral function and then transform it to the corresponding covariance model.

Before proceeding any further with our discussion of the covariance model construction techniques, I would like to remind the readers that it was a deliberate choice that the theoretical developments in the previous chapters focused on random fields with physical nature motivated by real-world phenomena. This is because, as many studies have noticed, it is not unusual that covariance models inadequate to describe the specific real-world phenomenon of interest are frequently used in the literature, simply relying on the false premise that for a formally valid covariance model to be an adequate representation of the phenomenon it is enough that it fits reasonably well the available data set. Also, as was emphasized in Postulate 2.1 of Chapter IV, in real-world applications a covariance model is not enough to be formally permissible (i.e., to satisfy the nonnegative definiteness or any relevant permissibility condition (PC) as discussed in Chapter XV). Beyond data fitting and formal permissibility, an adequate covariance model also needs to be physically permissible (i.e., consistent with physical considerations concerning the phenomenon). This situation is more serious, as well as more common, in the case of cross-covariance functions of vectorial or multivariate random fields representing different attributes connected via empirical models, auxiliary relationships etc., where the generation of suitable cross-covariance models is often limited by the fact that far fewer classes of suitable models and established guidelines exist compared to the case of scalar random field modeling (see, Remark 1.1 of Chapter IX).

Example 1.1

In the case of a meteorological attribute that decays with time and occurs over a very large global region within which it varies homogeneously, the physical characteristics that an adequate covariance model should capture include: (1) how fast the attribute correlation declines with increasing distance between points, (2) any inherent symmetry in the data, and (3) the space−time regularity (continuity and differentiability) properties of the attribute. A physically valid covariance model should be also interpretable. For instance, a reasonable interpretation of the fact that the functional form of the covariance model of seismic attributes depends on shorter and longer separation distances could be that different factors control data decorrelation at shorter and longer separation lags.

Perhaps not surprising for those readers with considerable real-world experience, there may exist reasons that in certain applications the same phenomenon may be characterized by different S/TVFs. This is the case described in my second example.

Example 1.2

In seismology there is a plethora of S/TVFs describing seismic data variability. There are several real and technical reasons for this situation, such as: (1) the observed variability of seismic data recorded during different events at different sites, (2) the technical differences between the numerical data processing techniques used, and (3) the distinct functional forms assumed in regression data-fitting in cases of data with large scatter.

Accordingly, it is highly desirable or even necessary that the generated covariance models also satisfy the core and site-specific knowledge about the natural attribute variation that these models purport to represent, and that they are feasible computationally and flexible in their use.

Let us take stock: Arguably, it is not difficult to construct numerous space−time covariance models that are formally valid (the discussion in this chapter presents several techniques of covariance model construction). What is usually difficult is to make sure that these models are physically meaningful representations of the actual space−time variability features of the phenomenon. Accordingly, an adequate understanding of the physical laws and conditions pertinent to the situation of interest is central to any covariance modeling or theorization approach that seeks to represent accurately those features of the phenomena that it is intended to describe, model, or theorize. This perspective of covariance model construction enables us to account satisfactorily for any disagreements between the actual phenomenon and the generated covariance models of the phenomenon. In the remaining of this chapter I discuss a variety of space−time covariance model construction techniques, highlighting similarities and differences as they emerge. Clearly the collections of techniques presented below exhaust neither all theoretically possible covariance models nor the large number of potential physical interpretations.

2. PROBABILITY DENSITY FUNCTION—BASED AND RELATED TECHNIQUES

The basic idea behind the probability density function (PDF) technique is that there is a fundamental link between a wide class of space−time covariance models and the family of known PDFs, ordinary or generalized, which can be taken advantage of to construct new space−time covariance models starting from known PDFs. This is discussed in Section 2.1 below. Furthermore, it is noteworthy that several other covariance model construction techniques have been developed over the years based on the basic idea of the PDF technique. Some of these techniques are discussed in Sections 2.2 and 2.3.

2.1 LINKING DIRECTLY COVARIANCE MODELS AND PROBABILITY DENSITY FUNCTIONS

The PDF technique of space−time covariance model construction (Christakos, 1984b) is outlined in Table 2.1. By virtue of Eqs. (2.1a−b) of this table, a rich class of covariance models can be generated in the $R^{n,1}$ domain directly from the family of known PDFs using the spectral operators H_n and $H_{n,1}$ of Eqs. (2.2a−b). Given that these PDFs can be also linked to spectral densities (complete and partial), the class of space−time covariance models thus generated can be further enriched (this possibility is also discussed later in this section). To orient the readers, several examples will be examined below, which may clarify certain key aspects of the technique described in Table 2.1.

Table 2.1 The Probability Density Function Technique of Space–Time Covariance Model Construction

Step	Description
1	Select a valid probability density function f for the wavenumber and the frequency.
2	Insert the probability density function f in $$c_X(r,\tau)=\begin{cases}\int_0^\infty\int_0^\infty dk d\omega f_{k,\omega}(k,\omega)H_{n,1},\\ \int_0^\infty dk f_k(k,\tau)H_n,\end{cases} \quad (2.1a–b)$$ where $r=\lvert\boldsymbol{h}\rvert$, and $$H_n=c_0(2\pi)^{\frac{n}{2}}r^{1-\frac{n}{2}}k^{\frac{n}{2}}J_{\frac{n}{2}-1}(kr), \quad H_{n,1}=e^{-i\omega\tau}H_n, \quad (2.2a–b)$$ (c_0 is a constant) to obtain the corresponding space–time covariance model in $R^{n,1}$.

Example 2.1

For illustration, if one selects an ordinary PDF with an exponential shape,

$$f_k(k,\tau)=e^{-(a+\tau)k} \tag{2.3a}$$

($a>0$), Eq. (2.1b) gives the space–time covariance in $R^{n,1}$,

$$c_X(r,\tau)=c_0A_n(a+\tau)\left[(a+\tau)^2+r^2\right]^{-\frac{n+1}{2}}, \tag{2.3b}$$

where $A_n=2^n\pi^{\frac{n-1}{2}}\Gamma\left(\frac{n+1}{2}\right)$.[1]

The technique of Table 2.1 can also involve generalized PDFs in terms of space–time delta functions that possess a deeper physical meaning (the usefulness and interpretation of this kind of PDFs were discussed in Section 2 of Chapter II). For instance, such a generalized density is given by

$$f_k(\boldsymbol{k},\tau)=f(\boldsymbol{k})\delta\left(\tau-\frac{\lvert\boldsymbol{k}\rvert}{\upsilon(\boldsymbol{k})}\right),$$

where $\upsilon(\boldsymbol{k})$ denotes the velocity of the process, and f is an isotropic function, e.g., $f(k)\propto k^{-\nu}$, $\nu>0$, see Example 2.4 of Chapter II. Another case of a physical interpretable density that can generate an interesting class of space–time covariance models is investigated next.

[1]Eq. (2.3b) is one of the models proposed in Cressie and Huang (1999) and Gneiting (2002). In fact, many of the space–time covariance models listed in the statistics literature can be derived by the general approach of Table 2.1.

Table 2.2 Examples of $f_k(k, \tau)$ Forms ($R^{n,1}$)

Case No.	$f_k(k, \tau)$
1	$e^{-\vartheta(\tau)k^2}$
2	$[k^2 + \lvert\vartheta(\tau)\rvert^2]^{-2\nu}$, $\nu > 0$
3	$\frac{2^n}{(\vartheta(\tau)k)^n}\left[\frac{n}{2}!J_{\frac{n}{2}}(\vartheta(\tau)k)\right]^2$
4	$\left(1 + \frac{(\vartheta(\tau)k)^2}{a^2}\right)^{-\mu}$, $\mu > \frac{n}{2}$, $a > 0$
5	$(\vartheta(\tau)k)^\nu K_\nu(\vartheta(\tau)k)$
6	$\vartheta(\tau)(1 + k^2)^{-(\nu+\mu\lvert\tau\rvert)-1}$, $\mu, \nu > 0$
7	$\vartheta(\tau)e^{-\frac{1}{4}k^2(1+\tau b^{-1})}$, $a \in (0, 1]$, $b > 0$
8	$\Gamma\left(\frac{n+1}{2}\right)\tau\left[\pi\left(k^2 + \tau^2\right)\right]^{-\frac{n+1}{2}}$

Example 2.2

Consider the following generalized PDF (used in physics to define a particle location)

$$f_k(k, \tau) = \frac{1}{S_n}k^{-(n-1)}\delta(k - \tau), \tag{2.4a}$$

where S_n is the surface area of the n-dimensional unit sphere. Then, Eq. (2.1b) gives

$$c_X(r, \tau) = c_0 2^{\frac{n}{2}-1}\Gamma\left(\frac{n}{2}\right)(r\tau)^{1-\frac{n}{2}}J_{\frac{n}{2}-1}(r\tau). \tag{2.4b}$$

Some special cases of Eq. (2.4b) are: $c_X(r, \tau) = c_0 \cos(r\tau)$ (for $n = 1$), $= c_0 J_0(r\tau)$ ($n = 2$), $= \frac{c_0}{r\tau}\sin(r\tau)$ ($n = 3$). The same covariance models can be constructed using more than one technique. So, the last model above can be also obtained using a space transformation technique (see, Example 4.1 later).

Useful classes of valid space–time covariance models in $R^{n,1}$ can be constructed by letting $f_k(k, \tau)$ have one of the forms listed in Table 2.2, where $\vartheta(\tau)$ is a suitable function of τ (Exercise XVI.1).

Example 2.3

For illustration, let $\vartheta(\tau) = a + \tau$ in Case no. 5 of Table 2.2, i.e.,

$$f_k(k, \tau) = ((a + \tau)k)^\nu K_\nu((a + \tau)k) \tag{2.5a}$$

in $R^{3,1}$. Then Eq. (2.1b) gives the following space–time covariance model

$$c_X(r, \tau) = c_0 2^{\nu-1}(\pi)^{-\frac{3}{2}}\Gamma\left(\nu + \frac{3}{2}\right)(a + \tau)^{2\nu}\left[(a + \tau)^2 + r^2\right]^{-\nu-\frac{3}{2}}, \tag{2.5b}$$

for $\nu > 0$ and suitable a. In $R^{2,1}$ one finds

$$c_X(r, \tau) = \frac{c_0}{2^{2\nu}\pi\Gamma(2\nu)}\left(\frac{r}{\lvert a + \tau\rvert}\right)^{2\nu-1}K_{2\nu-1}(\lvert a + \tau\rvert r), \tag{2.5c}$$

which is a nonseparable space–time covariance. When $f_k(k, \tau)$ has the form of Case no. 6 (Table 2.2) with $\vartheta(\tau) = \beta\vartheta^{|\tau|}$, $\beta, \nu > 0$, $\vartheta \in (0, 1)$, and $\tau \in Z$, the space–time covariance in $R^{2,1}$ is

$$c_X(r,\tau) = \frac{\beta\vartheta^{|\tau|} r^{\nu+\nu'|\tau|}}{2^{\nu+\nu'|\tau|-1}\Gamma(\nu+\nu'|\tau|)} K_{\nu+\nu'|\tau|}(r), \tag{2.6}$$

for each $\tau \in Z$, $\nu' > 1$. Lastly, in the Case no. 7 with $\vartheta(\tau) = a^{\tau}$, the space–time covariance in $R^{2,1}$ is

$$c_X(r,\tau) = \frac{4\pi b}{b+\tau} a^{\tau} e^{-br^2(b+\tau)^{-1}}, \tag{2.7}$$

$\tau \geq 0$, $a \in (0, 1]$, and $b > 0$.

The PDF method of Table 2.1 displays some more distinct features in the case of space–time covariance separability (Section 8). Also, another simple yet interesting PDF-based technique of constructing space-time covariance models starting from known univariate PDF and one-to-one functions is discussed in Exercise XVI.16 of the Chapter "Exercises-Comments."

2.2 USING POLYNOMIAL-EXPONENTIAL FUNCTIONS

Noticeably, the idea of involving PDFs in the derivation of valid space–time covariance models, and thus taking advantage of this rich class of functions, was also used by Ma (2002a, 2007) in his derivation of power mixture space–time covariance models. In particular, in place of the functions H_n and $H_{n,1}$ of Eq. (2.2a–b), polynomial-exponential functions were implemented. This limited choice was necessary to take advantage of the Laplace operator in covariance model construction. A few examples are discussed next.

Example 2.4

In place of function $H_{n,1}$ in Eq. (2.1a) one may consider the function

$$H_e = e^{-rk-\tau\omega}. \tag{2.8a}$$

This is a deliberate choice, for which Eq. (2.1a) gives

$$c_X(r,\tau) = \int_0^{\infty}\int_0^{\infty} dk d\omega e^{-rk-\tau\omega} f_{k,\omega}(k,\omega), \tag{2.8b}$$

which is equal to the bivariate Laplace transform $L_T(r, \tau)$. Keeping in mind the properties of the Laplace transform, another convenient choice is

$$H_e = e^{-ark}\left[e^{-b\tau\omega} - e^{-b(t+t')\omega}\right], \tag{2.9a}$$

so that Eq. (2.1a) gives

$$c_X(r,t,t') = \int_0^{\infty}\int_0^{\infty} dk d\omega e^{-ark}\left[e^{-b(t-t')\omega} - e^{-b(t+t')\omega}\right] f_{k,\omega}(k,\omega), \tag{2.9b}$$

where $a, b > 0$, $r = |\boldsymbol{h}|$, and $\tau = |t - t'|$ in $R^{n,1}$.[2] For illustration, Table 2.3 presents some particular cases of bivariate PDFs, $f_{k,\omega}$ (Ma, 2002a, 2007) and the resulting covariance models. If $f_{k,\omega}$ is assumed

[2]Eq. (2.9b) is equal to the bivariate Laplace transform $L_T(ar, b\tau) - L_T(ar, b(t + t'))$.

Table 2.3 Examples of Space–Time Covariance Models in $R^{n,1}$ Generated by H_e and Probability Density Function

H_e and Probability Density Function	Space–Time Covariance Model, $c_X(r, t, t')$	
H_e: Eq. (2.9a) $f_{k,\omega}$: Positive stable $\alpha \in (0, 1)$	$e^{-(ar+b\tau)^{\alpha}} - e^{-(ar+b(t+t'))^{\alpha}}$	(2.10a)
H_e: Eq. (2.9a) $f_{k,\omega}$: CR Gamma $\alpha_1, \alpha_2, \alpha_3 \geq 0$	$(1+ar)^{-\alpha_1}\left[(1+b\tau)^{-\alpha_2}(1+ar+b\tau)^{-\alpha_3} - (1+b\lvert t_1+t_2\rvert)^{-\alpha_2}(1+ar+b\lvert t_1+t_2\rvert)^{-\alpha_3}\right]$	(2.10b)
H_e: Eq. (2.11c) f_k: Gamma $k > 0,\ \beta > n+1$	$(1+r+\tau)^{-\beta-2}\left[1+(2+a\beta)r+(2+b\beta)\tau+(1+a\beta)r^2 + (1+b\beta)\tau^2+(2+a\beta+b\beta+\beta(\beta+1)c)r\tau\right]$, $\tau = \lvert t-t'\rvert$	(2.11g)

to be a positive stable density,[3] then Eq. (2.9b) becomes Eq. (2.10a) of Table 2.3, where α is a constant of the stable PDF. If $f_{k,\omega}$ is a bivariate Cheriyan–Ramabhadran (CR) gamma PDF (Kotz et al., 2000), then Eq. (2.9b) becomes Eq. (2.10b) of Table 2.3, where α_i ($i = 1, 2, 3$) are PDF constants. Both Eqs. (2.10a) and (2.10b) represent spatially isotropic/temporally nonstationary models. Working along similar lines, and using Eq. (2.1b) where in place of H_n one uses

$$H_e = \begin{cases} e^{-(r+\tau)k}, \\ (1+ark+b\tau k)e^{-(r+\tau)k}, \\ \left(1+ark+b\tau k+cr\tau k^2\right)e^{-(r+\tau)k}, \end{cases} \quad (2.11\text{a–c})$$

one obtains, respectively, the covariance models

$$c_X(r,\tau) = \begin{cases} L_T(r+\tau), \\ L_T(r+\tau)-(ar+b\tau)L_T'(r+\tau), \\ L_T(r+\tau)-(ar+b\tau)L_T'(r+\tau)+cr\tau L_T''(r+\tau), \end{cases} \quad (2.11\text{d–f})$$

where L_T' and L_T'' are the first and second derivatives of the Laplace transform. For illustration purposes, Table 2.3 presents a specific case, where a gamma PDF is used, $f_k = \Gamma(\beta)k^{\beta-1}e^{-k}$ ($k > 0$, $\beta > n+1$). Then Eq. (2.11f) leads to Eq. (2.11g), which is a valid space–time covariance in the $R^{n,1}$ domain.

Another technique of space-time covariance model construction using Laplace transforms, also known as the *Montroll-Weiss* technique, is discussed in Exercise XVI.24 of the Chapter "Exercises-Comments".

[3]The readers may recall that no analytical form is available for a general stable PDF (Kotz et al., 2000).

2.3 USING SPECTRAL FUNCTIONS

In many cases, it may be appropriate to work with complete or partial spectra functions $\widetilde{c}_X$ (as discussed in Section 3.3 of Chapter VII). Then, the PDFs can be written in terms of $\widetilde{c}_X$, thus leading to another convenient formulation as regards covariance model construction in which Eq. (2.1a–b) may be replaced by

$$c_X(r,\tau) = \begin{cases} (2\pi)^{\frac{n}{2}} r^{1-\frac{n}{2}} \int_0^\infty \int_0^\infty dk\, d\omega\, k^{\frac{n}{2}} J_{\frac{n}{2}-1}(kr) e^{-i\omega\tau} \widetilde{c}_X(k,\omega), \\ (2\pi)^{\frac{n}{2}} r^{1-\frac{n}{2}} \int_0^\infty dk\, k^{\frac{n}{2}} J_{\frac{n}{2}-1}(kr) \widetilde{c}_X(k,\tau), \\ \int_0^\infty d\omega e^{-i\omega\tau} \widetilde{c}_X(r,\omega) \end{cases} \tag{2.12a–c}$$

(i.e., both complete and partial spectra are involved). Adding to the remarkable flexibility of the spectral technique, partial spectral functions $\widetilde{c}_X$, different than those used in Eqs. (2.12a–c), can be also considered. In this respect, a few additional possibilities are examined in Section 3.

Example 2.5

Some implementation cases of the application of Eqs. (2.12a–c) are presented in Table 2.4. Starting with the wavenumber-frequency spectrum of Eq. (2.13a) of this table, and using Eq. (2.12a), the corresponding space–time model in $R^{2,1}$ is obtained as in Eq. (2.13b). A set of covariance models associated with space–time isostationary (STIS) random fields characterized by the wavenumber-frequency spectrum of Eq. (2.14a) of Table 2.4 were derived by Jones and Zhang (1997). For instance, the space–time covariance model of Eq. (2.14b) is obtained by a direct implementation of Eq. (2.12a). In applications, integrals like that of Eq. (2.14b) are often calculated using numerical techniques. Specifically, Kolovos et al. (2004) used a numerical integration technique to calculate Eq. (2.14b) for $p = n = 2$, and for varying values of α and c, which led to some new covariance models in $R^{2,1}$. Among the notable features of these covariance models is that they decline faster along the time direction than along the space direction, and they are valid for a Euclidean metric (r_E). On the other hand, the covariance model of Eq. (2.15b) is also valid in terms of the spatial Manhattan-time metric (r_M,τ), in which case the conditions on the model coefficients a, b, and c of Eq. (2.15a) must be replaced by $|b - cn| \leq 1 - an$, $|b + cn| \leq 1 + an$ (see, also, Example 3.1 of Chapter XV). The covariance model of Eq. (2.16b) is among the ones derived by Stein (2005), and the corresponding spatiotemporal random fields (S/TRF) is once mean square (m.s.) differentiable along any spatial direction (time origin), and not differentiable in time (space origin), although the model is infinitely differentiable away from the origin. The covariance model Eq. (2.17b) in Table 2.4 is the von Kármán covariance function used in turbulence studies and elsewhere. In statistics, this model is also known as the Matern covariance function.

The approach of Eqs. (2.12a–c) also works for space–time homostationarity (STHS) random fields, i.e., fields that are such that the statistical properties do not change with space–time translation (Chapter VII).

Table 2.4 Examples of Space–Time Covariance Models Generated From Spectral Functions

Spectral Function $\tilde{c}_X(k,\omega)$		Space–Time Covariance Function $c_X(r,\tau)$	
$R^{2,1}$ $\frac{c_0 a^2 I_0\left(\frac{a^2 bk}{2}\right)}{4\pi^2(1+\omega^2)} e^{-\frac{a^2(b^2+k^2)}{4}}$, s.t. $a, b, c_0 > 0$	(2.13a)	$c_0 e^{-\frac{r^2}{a^2}-\tau} J_0(br)$	(2.13b)
$R^{n,1}$ $\left[(k^2+\alpha^2)^{2p}+c^2\omega^2\right]^{-1}$, s.t. $p > \frac{n}{2}$	(2.14a)	$\frac{(2\pi)^{\frac{n}{2}+1} r^{1-\frac{n}{2}}}{2c} \int_0^\infty dk\, k^{\frac{n}{2}} \frac{J_{\frac{n}{2}-1}(kr)}{(k^2+\alpha^2)^p e^{c^{-1}(k^2+\alpha^2)^p \tau}}$	(2.14b)
$R^{n,1}$ $(1+\omega^2)^{-2}(1+k^2)^{-\frac{n+3}{2}}$ $\{(a+c)(n+1)+(1-a+b-c)(1+k^2)$ $+\omega^2[(a-c)(n+1)+(1-a-b+c)(1+k^2)]\}$, s.t. $\lvert b-c\rvert \le 1-a,\ \lvert b+cn\rvert \le 1+an$	(2.15a)	$(1+ar+b\tau+cr\tau)e^{-r-\tau}$	(2.15b)
$R^{3,1}$ $\left[a^2(b^2+k^2)^{-2}+\omega^2\right]^{-2}$, s.t. $a, b > 0$	(2.16a)	$\frac{\pi^2}{16a^6} e^{br} \mathrm{Erfc}\left(ab\lvert\tau\rvert^{0.5}+\frac{r}{2a\lvert\tau\rvert^{0.5}}\right)\left(\frac{1}{b^3}-\frac{r}{b^2}+\frac{4a^4\tau^2}{r}\right)$ $+\frac{\pi^2}{16a^6} e^{-br} \mathrm{Erfc}\left(ab\lvert\tau\rvert^{0.5}-\frac{r}{2a\lvert\tau\rvert^{0.5}}\right)\left(\frac{1}{b^3}+\frac{r}{b^2}-\frac{4a^4\tau^2}{r}\right)$ $+\frac{\pi^{1.5}\lvert\tau\rvert^{0.5}}{4a^5 b^2} e^{-a^2 b^2\lvert\tau\rvert-\frac{r^2}{4a^2\lvert\tau\rvert}}$	(2.16b)
$R^{n,1}$ $\left[a_1(b^2+k^2)+a_2\omega^2\right]^{-\nu}$, s.t. $a_1, a_2 > 0,\ b^2 > 0,\ \nu > \frac{n+1}{2}$	(2.17a)	$b^{\frac{2\nu-n-1}{2}}\left(r^2+\frac{a_1\tau^2}{a_2}\right)^{\frac{2\nu-n-1}{4}} K_{\frac{2\nu-n-1}{2}}\left(b\left(r^2+\frac{a_1\tau^2}{a_2}\right)^{\frac{1}{2}}\right)$	(2.17b)

In this case, e.g., Eqs. (2.12a) and (2.12c) are replaced by, respectively,

$$c_X(\boldsymbol{h}, \tau) = \iint d\boldsymbol{k}d\omega e^{i(\boldsymbol{k}\cdot\boldsymbol{h}-\omega\tau)}\widetilde{c}_X(\boldsymbol{k}, \omega),$$
$$c_X(\boldsymbol{h}, \tau) = \int d\omega e^{-i\omega\tau}\widetilde{c}_X(\boldsymbol{h}, \omega). \tag{2.18a–b}$$

and space−time covariance models can be constructed by assuming several forms for the corresponding spectral functions (full or partial). Depending on the spectral function choice, some particularly simple ways of generating space−time covariance models can be formulated.

Example 2.6

Consider the spectral function $\widetilde{c}_X(\boldsymbol{h}, \omega) = \widetilde{c}_{\widehat{X}}(\omega)e^{ia\omega(\boldsymbol{\varepsilon}\cdot\boldsymbol{h})}$, where $\widetilde{c}_{\widehat{X}}(\omega)$ is the known spectrum of a unidimensional covariance $c_{\widehat{X}}(\widehat{\tau})$, $\widehat{\tau} = \tau - a(\boldsymbol{\varepsilon}\cdot\boldsymbol{h})$, along the direction $\boldsymbol{\varepsilon}$. Then, from Eq. (2.18b) we find that

$$c_X(\boldsymbol{h}, \tau) = c_{\widehat{X}}(\widehat{\tau})\Big|_{\widehat{\tau}=\tau-a(\boldsymbol{\varepsilon}\cdot\boldsymbol{h})}, \tag{2.19}$$

which is of the plane-wave covariance type (Section 3 of Chapter X). It seems that Eq. (2.19) provides an almost trivial way to construct valid space−time covariance models: starting with any of the temporal (or one-dimensional) covariance models $c_{\widehat{X}}(\widehat{\tau})$ available in the relevant literature (see, e.g., list in Table 4.1 of Chapter VIII), and then using the change of arguments $\widehat{\tau} \mapsto \tau - a(\boldsymbol{\varepsilon}\cdot\boldsymbol{h})$, so that Eq. (2.19) can directly generate permissible space−time covariance models. Another interesting covariance model is obtained by assuming in Eq. (2.18a) the spectral function

$$\widetilde{c}_X(\boldsymbol{k}, \omega) = (2\pi)^{-1}\left[(\boldsymbol{k}\cdot(\boldsymbol{a}\boldsymbol{k}) + b)^2 + (\omega + \boldsymbol{v}\cdot\boldsymbol{k})^2\right]^{-1},$$

where $\boldsymbol{a}$ and b are physical vector and coefficient, respectively, and $\widetilde{c}_Y(\boldsymbol{k})$ is a wavenumber spectrum.

A noteworthy observation brought about by the discussion so far, which will be also confirmed by the subsequent analysis, is that many of the covariance construction techniques are interrelated in certain respects.

3. DELTA AND RELATED TECHNIQUES

I will start the discussion of this group of techniques with a straightforward yet fruitful delta (δ)-technique, which can lead to the construction of a rich class of nonseparable, in general, spatiotemporal covariance models in the $R^{n,1}$ domain. An attractive property of this technique is that it has several variants, some of which are also discussed in this section.

3.1 BASIC DECOMPOSITION

The basic steps of the δ-technique are described in Table 3.1 (Christakos, 1992). If $\widetilde{c}_X(\boldsymbol{k})$ is an isotropic density, $\widetilde{c}_X(\boldsymbol{k}) = \widetilde{c}_X(k)$, the wavevector integral in Eq. (3.2a) is simplified to a unidimensional integral over the magnitude of the wavevector, which involves a Bessel function (this one-dimensional integral is evaluated either analytically or by numerical integration).

Table 3.1 The Delta Technique of Space–Time Covariance Model Generation

Step	Description
1	Assume a spectral density in R^n, say $\widetilde{c}_X(\boldsymbol{k})$, and let the corresponding spectral density in $R^{n,1}$ be given by $$\widetilde{c}_X(\boldsymbol{k},\omega)=\widetilde{c}_X(\boldsymbol{k})\widetilde{\rho}_X(\boldsymbol{k},\omega), \quad (3.1)$$ where, say, $\widetilde{\rho}_X(\boldsymbol{k},\omega)=\delta_X(\omega\pm\boldsymbol{k}\cdot\boldsymbol{v})$ is a delta correlation function (nugget effect), and $\boldsymbol{v}$ is a given vector in R^n.
2	Space–time covariance models can be generated by directly inserting Eq. (3.1) into Eq. (2.18a) or by $$c_X(\boldsymbol{h},\tau)=\int d\boldsymbol{k}\, e^{i\boldsymbol{k}\cdot\boldsymbol{h}}\widetilde{c}_X(\boldsymbol{k},\tau), \quad (3.2a)$$ where from Eq. (3.1) it is found that $$\widetilde{c}_X(\boldsymbol{k},\tau)=\widetilde{c}_X(\boldsymbol{k})\widetilde{\rho}_X(\boldsymbol{k},\tau)=\widetilde{c}_X(\boldsymbol{k})e^{\pm i(\boldsymbol{k}\cdot\boldsymbol{v})\tau}, \quad (3.2b)$$ with $\widetilde{c}_X(\boldsymbol{k})>0$ for all $\boldsymbol{k}$, and $\int d\boldsymbol{k}\,\widetilde{c}_X(\boldsymbol{k})<\infty$.

Notice that a link between Steps 1 and 2 of Table 3.1 is that in Eq. (3.2b) the exponential term is the frequency (ω) transform of the delta correlation function. By virtue of this link, if $\widetilde{c}_X(\boldsymbol{k})$ has an inverse transform in R^n, then Eq. (3.2a) may be reduced to the turbulence model in $R^{n,1}$ discussed in Section 2 of Chapter X,[4] i.e.,

$$c_X(\boldsymbol{h},\tau)=c_{\widehat{X}}(\boldsymbol{h}\pm\boldsymbol{v}\tau). \tag{3.3}$$

Similarly to Eq. (2.19), if we start with any of the spatial covariance models $c_{\widehat{X}}(\widehat{\boldsymbol{h}})$ that are available in the literature (Table 4.1 of Chapter VIII again provides a list, by no means exhaustive, of such models), and use the change of coordinates $\widehat{\boldsymbol{h}}\mapsto\boldsymbol{h}\pm\boldsymbol{v}\tau$, then Eq. (3.3) can generate a surprisingly large class of permissible space–time covariance models in an almost trivial manner.

Example 3.1

Particular cases of the spatiotemporal covariance models of the form of Eq. (3.3) are the model nos. 9–10 of Table 5.1 of Chapter VIII. Other spatiotemporal covariance models can be generated in a similar manner. So, starting from the spatial covariance model $c_X(\boldsymbol{h})=\left(1+\frac{h^2}{\beta^2}\right)^{-\frac{\nu}{2}}e^{\frac{h}{\alpha}}$ proposed by Hristopulos (2002), Eq. (3.3) leads to model no. 11 of Table 5.1 of Chapter VIII. Models of this kind can be useful in several physical applications, e.g., in cases in which the "frozen" random field hypothesis applies.

[4]Notice that $c_X(\boldsymbol{h},\tau)=\int d\boldsymbol{k}e^{i\boldsymbol{k}\cdot(\boldsymbol{h}\pm\boldsymbol{v}\tau)}\widetilde{c}_X(\boldsymbol{k})$.

Example 3.2

Certain variants of Eq. (3.3) are possible. For instance, valid covariance models are obtained by combining the frozen and the plane wave models (Section 3 and Eqs. (3.4b) of Chapter X), i.e.,

$$c_X(\boldsymbol{h},\tau) = c_X\left(\boldsymbol{h} - \upsilon\tau, \tau - \frac{\boldsymbol{\eta}\cdot\boldsymbol{h}}{\upsilon}\right), \tag{3.4a}$$

where the vector $\boldsymbol{\eta}$ denotes the propagation direction of the plane wave. Another straightforward extension of Eq. (3.3) is the following expression for generating covariance models in $R^{n,1}$ (Ma, 2002b)

$$c_X(\boldsymbol{h},\tau) = c_{\widehat{X}}(\boldsymbol{h} + \boldsymbol{\upsilon}\tau, \boldsymbol{h} + \boldsymbol{\upsilon}'\tau), \tag{3.4b}$$

where $\boldsymbol{\upsilon}$ and $\boldsymbol{\upsilon}'$ are velocity vectors in R^n. Compared to the turbulence-based models of Eqs. (3.3) and (3.4a), although formally valid, the physical significance of the model of Eq. (3.4b) may not be always clear. A better interpretation of Eq. (3.4b) could be of interest in certain applications.

Another technique of space-time covariance model construction starting from known spatial covariance models is discussed in Exercise XVI.23 of the chapter "Exercises-Comments".

3.2 NORMALIZED ANGULAR SPECTRUM DECOMPOSITION

Motivated by the idea of using a decomposition of the form introduced by Eqs. (3.1) and (3.2b) to construct covariance models, other techniques have been proposed based on different choices of $\widetilde{\rho}_X(\boldsymbol{k},\tau)$. So, if one assumes that in the decomposition of Eq. (3.2b) the $\widetilde{\rho}_X(\boldsymbol{k},\tau)$ is the normalized angular spectrum presented in Section 3 of Chapter VII (a concept used in oceanology, geophysics, optics, acoustics, and radiation sciences),[5] Eq. (3.2a) yields

$$c_X(\boldsymbol{h},\tau) = \int d\boldsymbol{k} e^{i\boldsymbol{k}\cdot\boldsymbol{h}} \widetilde{c}_X(\boldsymbol{k}) \widetilde{\rho}_X(\boldsymbol{k},\tau) \tag{3.5}$$

(a formulation used, e.g., by Cressie and Huang (1999), assuming that the integral exists. In the case of Eq. (3.5), the $\widetilde{\rho}_X(\boldsymbol{k},\tau)$ are real-valued and symmetric functions of τ if the covariance model is fully symmetric (see Section 1.6 of Chapter VII for a definition of fully symmetric space−time covariances). If, on the other hand, the covariance $c_X(\boldsymbol{h},\tau)$ is not fully symmetric, the $\widetilde{\rho}_X(\boldsymbol{k},\tau)$ considered in Eq. (3.5) is complex-valued, in general.

Example 3.3

For an illustration of Eq. (3.5), select the normalized angular spectrum $\widetilde{\rho}_X(k,\tau) = e^{-\frac{1}{4}k^2|\tau|}$, and the wavenumber spectrum $\widetilde{c}_X(k) = e^{-\frac{1}{4}ak^2}$ $(a > 0)$. Then, Eq. (3.5) gives a space−time covariance model of the form

$$c_X(r,\tau) = (|\tau| + a)^{-\frac{1}{2}n} e^{-r^2(|\tau|+\alpha)^{-1}} \tag{3.6}$$

---

[5]For each $\boldsymbol{k}$ the $\widetilde{\rho}_X(\boldsymbol{k},\tau)$ is considered a continuous correlation function with respect to time separation τ, and such that the purely temporal covariance is a combination of the normalized angular spectra weighted by the wavevector spectrum.

in $R^{n,1}$. Also, if one selects the normalized angular spectrum $\widetilde{\rho}_X(k,\tau)=e^{-\left(\frac{1}{4}k^2\tau^2+b\tau^2\right)}$, and the wave-number spectrum $\widetilde{c}_X(k)=e^{-\frac{1}{4}ak^2}$ $(a,b>0)$, Eq. (3.5) gives a space–time covariance model of the form

$$c_X(r,\tau)=\left(\tau^2+a\right)^{-\frac{n}{2}}e^{-\left(r^2\left(\tau^2+a\right)^{-1}+b\tau^2\right)}. \tag{3.7}$$

Both covariance models of Eqs. (3.6) and (3.7) are fully symmetric covariance functions. Several other valid covariance models can be constructed in an obvious way by using the normalized angular spectra that are readily available in the literature.

3.3 NORMALIZED FREQUENCY SPECTRUM (OR COHERENCY FUNCTION) DECOMPOSITION

Apart from the ($\boldsymbol{k}$, τ)-domain decomposition $\widetilde{c}_X(\boldsymbol{k},\tau)=\widetilde{c}_X(\boldsymbol{k})\widetilde{\rho}_X(\boldsymbol{k},\tau)$ used in the preceding Section 3.2 to construct covariance models through Eq. (3.5), it seems natural to also consider the (r, ω)-domain decomposition

$$\widetilde{c}_X(r,\omega)=\widetilde{c}_X(\omega)\widetilde{\rho}_X(r,\omega), \tag{3.8}$$

where $\widetilde{\rho}_X(r,\omega)$ is the normalized frequency spectrum (Section 3 of Chapter VII). In fact, the (r, ω)-approach may have a considerable advantage over the ($\boldsymbol{k}$, τ)-approach, in the sense that it avoids a possible limitation of the ($\boldsymbol{k}$, τ)-approach: the (r, ω)-approach may generate nonfully symmetric covariance models based on normalized frequency spectra that are not necessarily complex-valued.

When selecting a normalized frequency spectrum (or a normalized angular spectrum for that matter) the central issue is the factual accuracy of its interpretation that describes what was observed and experienced. In this context, it is worth-noticing that in earthquake engineering, e.g., the function $\widetilde{\rho}_X(r,\omega)$ is also known as the *coherency function*, and it has a well-understood physical meaning. Several analytical expressions of coherency functions are available. For illustration purposes, Table 3.2 gives a list of nine widely used coherency functions with coefficients, a, c, f, A, α, β, μ, ν, ϑ, λ, $B=(1-A+aA)$, $a(\omega)$, and $\nu(\omega)=\lambda\left[1+\left(\frac{\omega}{2\pi f}\right)^b\right]^{-\frac{1}{2}}$ estimated from physical data. Some of these coherency functions are empirical, whereas some others are semiempirical (i.e., their functional form is derived analytically, but the evaluation of their parameters is based on recorded data). Most coherency functions representing ground motion decrease with spatial lag and frequency. Each coherency function encapsulates specific features of the phenomenon it represents.

Example 3.4

A particular feature of the coherency Function no. 9 (Table 3.2) is that it accounts for differences in coherency behavior at longer separation distances and higher frequencies. In addition to the coherency functions for STIS random fields listed in Table 3.2, there are also direction-dependent coherency functions (e.g., the direction-dependent terms may represent the wave passage effect, motion variability, or that the scattering in the forward direction tends to be in phase with the incident wave *vs.* scattering to the side that tends to loose phase).

Table 3.2 Empirical and Semiempirical Coherency Functions

Function No.	$\widetilde{\rho}_X(r,\omega)$
1	$e^{-ra(\omega)}$
2	$e^{-\frac{ar\omega}{2\pi c}}$
3	$e^{-r\omega^2}$
4	$e^{-a^2r^2\omega^2}$
5	$e^{-\alpha\left(\frac{r\omega}{\beta}\right)^{\nu}}$
6	$e^{-(a+b\omega^2)r}$
7	$e^{-(a\,r^{\mu}+b\omega^{\vartheta})}$
8	$Ae^{-\frac{2Br}{a\nu(\omega)}}+(1-A)e^{-\frac{2Br}{\nu(\omega)}}$
9	$Ae^{-\frac{2r(1-A)}{a\lambda}\left[1+\left(\frac{\omega}{2\pi f}\right)^{b}\right]^{\frac{1}{2}}}+(1-A)$

In sum, the coherency function approach for constructing space–time covariance models consists of two main steps: (1) chose a frequency spectrum, $\widetilde{c}_X(\omega)$, and a coherency function $\widetilde{\rho}_X(r,\omega)$, and (2) in view of Eq. (3.8), generate space–time covariance models in $R^{n,1}$ by using Eq. (2.12c) as

$$c_X(r,\tau)=\int_0^{\infty}d\omega e^{-i\omega\tau}\widetilde{c}_X(\omega)\widetilde{\rho}_X(r,\omega), \tag{3.9}$$

where $r=|\boldsymbol{h}|$, $\widetilde{c}_X(\omega)>0$, $\int d\omega\widetilde{c}_X(\omega)<\infty$.

Example 3.5

By choosing the coherency Function no. 3 of Table 3.2, and letting $\widetilde{c}_X(\omega)=c>0$, Eq. (3.9) gives

$$c_X(r,\tau)=cr^{-\frac{1}{2}}e^{-\frac{\tau^2}{4r}}, \tag{3.10}$$

which is a space–time nonseparable covariance.

Lastly, I reiterate that in many applications it is often preferred or even required to work in terms of spectra, instead of covariances. Among the reasons for this change of perspective is that random field realizations representing important physical attributes in oceanography, seismology, and other applied sciences are generated by means of analytical or computational expressions that explicitly involve the spectral function. In this setting, the coherency function decomposition technique has turned out to be particularly useful in constructing valid spectra. Procedurally, (1) start by choosing a frequency spectrum in R^1, $\widetilde{c}_X(\omega)$, and a coherency function $\widetilde{\rho}_X(r,\omega)$, (2) define $\widetilde{c}_X(r,\omega)$ using Eq. (3.8), as before, and (3) by virtue of the partial spectrum representation

$$\widetilde{c}_X(r,\omega)=\widetilde{c}_X(\omega)\widetilde{\rho}_X(r,\omega)=(2\pi)^{\frac{n}{2}}r^{-\frac{n}{2}+1}\int_0^{\infty}dk\,k^{\frac{n}{2}}J_{\frac{n}{2}-1}(kr)\widetilde{c}_X(k,\omega), \tag{3.11}$$

the $\widetilde{c}_X(k,\omega)$ is obtained by inverting Eq. (3.11). Lets illustrate this approach with the help of an example.

Example 3.6

In $R^{2,1}$, Eq. (3.11) reduces to the equation

$$\widetilde{c}_X(\omega)\widetilde{\rho}_X(r,\omega) = 2\pi \int_0^\infty dk\, kJ_0(kr)\widetilde{c}_X(k,\omega). \tag{3.12}$$

By inverting Eq. (3.12), we get

$$\widetilde{c}_X(k,\omega) = (2\pi)^{-1}\widetilde{c}_X(\omega) \int_0^\infty dr\, rJ_0(kr)\widetilde{\rho}_X(r,\omega). \tag{3.13}$$

As noted earlier, a list of empirical and semiempirical coherency functions exist with coefficients estimated from physical data. So, by choosing the coherency Function no. 4 from Table 3.2, and letting $\widetilde{c}_X(\omega) = \frac{1}{\pi(1+\omega^2)}$, Eq. (3.13) yields

$$\widetilde{c}_X(k,\omega) = \frac{1}{4\pi^2 a^2 \omega^2 (1+\omega^2)} e^{-\frac{k^2}{4a^2\omega^2}}, \tag{3.14}$$

which is a useful wavenumber-frequency spectrum.

4. SPACE TRANSFORMATION TECHNIQUE

Space transformation (ST) operators T_1^n and Ψ_1^n (Chapter V) make it possible to construct new covariance models in higher dimensions ($R^{n,1}$ domain, $n \geq 2$) directly from valid one-dimensional covariance models ($R^{1,1}$ domain). This covariance construction technique, which was initially proposed by Christakos (1984a,b), consists of two basic steps as outlined in Table 4.1. Note that in

Table 4.1 The Space Transformation Technique of Space–Time Covariance Model Construction

Step	Description
1	Select a covariance model $c_{X,1}(h,\tau)$ in $R^{1,1}$.
2	Use a space transformation, T_1^n or Ψ_1^n, defined by (Chapter V) $c_{X,n}(\boldsymbol{h},\tau) = \Psi_1^n[c_{X,1}](\boldsymbol{h},\tau)$ $c_{X,n}(\boldsymbol{h},\tau) = T_1^n[c_{X,1}](\boldsymbol{h},\tau) = \Psi_1^n\Omega[c_{X,1}](\boldsymbol{h},\tau)$, (4.1a–b) to obtain a covariance model $c_{X,n}(\boldsymbol{h},\tau)$ in $R^{n,1}$ $(n = 2, 3)$.

Eqs. (4.1a–b) the form of the covariance in $R^{1,1}$ may vary along different directions $\boldsymbol{\theta}$ (see definition of ST, in Section 3 of Chapter V).[6]

When STIS random fields are considered, Eqs. (4.1a–b) conveniently reduce to the following integral relations between the $R^{n,1}$ and the $R^{1,1}$ domains,

$$\begin{aligned} c_{X,n}(r,\tau) &= \Psi_1^n c_{X,1}(r,\tau) = E_n \int_0^1 du\left(1-u^2\right)^{\frac{n-3}{2}} c_{X,1}(ur,\tau) \\ c_{X,n}(r,\tau) &= T_1^n c_{X,1}(r,\tau) = \Psi_1^n \Omega c_{X,1}(r,\tau), \end{aligned} \qquad (4.2a\text{–}b)$$

where $E_n = \frac{2\Gamma\left(\frac{n}{2}\right)}{\sqrt{\pi}\Gamma\left(\frac{n-1}{2}\right)}$, and Ω is given by Eq. (3.2) of Chapter V. A remarkable property of the ST approach is rather obvious: for each valid covariance model that is available in $R^{1,1}$, two new covariance models can be constructed in $R^{n,1}$ by means of Eqs. (4.1a–b) or (4.2a–b).[7] Of special interest in applications is the case of the $R^{3,1}$ domain, in which the expressions of Eqs. (4.2a–b) reduce to

$$\begin{aligned} c_{X,3}(r,\tau) &= \Psi_1^3\left[c_{X,1}\right](r,\tau) = \frac{1}{r}\int_0^r du\, c_{X,1}(u,\tau), \\ c_{X,3}(r,\tau) &= T_1^3\left[c_{X,1}\right](r,\tau) = -\frac{1}{2\pi r}\frac{\partial c_{X,1}(r,\tau)}{\partial r}. \end{aligned} \qquad (4.3a\text{–}b)$$

These expressions generate STIS models, i.e., models that are independent of rotations and reflections, and which are valid in $R^{n,1}$ ($n \leq 3$). Covariance models for STHS random fields can be also generated by means of STs using the formulas in Table 3.3 of Chapter 5.

Beyond the physical domain, the ST approach can be implemented in the spectral domain, as well. Indeed, ST operators relate the spectrum $\widetilde{c}_{X,1}$ of the covariance function $c_{X,1}$ with the spectrum $\widetilde{c}_{X,n}$ of $c_{X,n}$ by

$$\widetilde{c}_{X,n}(k,\tau) = \begin{cases} k^{-1}\widetilde{c}_{X,1}(k,\tau) & \text{for } n=2, \\ \pi k^{-2}\widetilde{c}_{X,1}(k,\tau) & \text{for } n=3. \end{cases} \qquad (4.4a\text{–}b)$$

In some cases, it may be more convenient to obtain the spectrum from Eqs. (4.4a–b) and then use Eqs. (2.12b) to find the corresponding space–time covariance model. Eqs. (4.2a–b)–(4.4a–b) are instrumental in developing several new classes of space–time nonseparable, in general, covariance models. Since covariance models that are valid in $R^{3,1}$, are also valid in $R^{n,1}$, $n < 3$ (this is the so-called covariance hereditary property discussed in Section 2.1 of Chapter VII, see also Proposition 4.2 of Chapter XV regarding covariance permissibility), a practical approach maybe to use the simpler expression of Eq. (4.3b) to construct covariance functions in $R^{3,1}$, even if the domain of interest is of lower dimensionality.

Example 4.1

Table 4.2 lists space–time covariance models that have been constructed using the ST technique. First, we observe that new covariance models in $R^{n,1}$ ($n > 1$) are derived by applying the ST of Eq. (4.2a) on

[6] For this reason, it might had been more appropriate to use the notation $c_{X,1,\boldsymbol{\theta}}$ in Eqs. (4.1a–b). Yet, for simplicity of presentation, the notation $c_{X,1}$ has been used, instead.

[7] As we will see later, the ST approach includes the construction of geometrically anisotropic covariance models, after an appropriate change of variables.

Table 4.2 Examples of Space Transformation–Generated Space–Time Covariance Models

$c_{X,1}(r,\tau)$	$c_{X,n}(r,\tau)$
Model no. 6, Table 5.1 of Chapter VIII ($n=1$)	$\frac{1}{2}E_n\int_0^1 du\left(1-u^2\right)^{0.5(n-1)}\left[e^{-aur}Erfc\left(a\left(c^{-1}\tau\right)^{0.5}-\frac{1}{2}ur\left(c\tau^{-1}\right)^{0.5}\right)+e^{aur}Erfc\left(a\left(c^{-1}\tau\right)^{0.5}+\frac{1}{2}ur\left(c\tau^{-1}\right)^{0.5}\right)\right]$ $R^{n,1}$ (4.5)
Model no. 2, Table 5.1 of Chapter VIII ($n=1$)	$(4\alpha\pi\tau)^{-0.5}E_n\int_0^1 du\left(1-u^2\right)^{0.5(n-1)}e^{-\frac{ur^2}{4\alpha\tau}}$ $R^{n,1}$ (4.6)
Model no. 3, Table 5.1 of Chapter VIII ($n=1$, $\gamma=0.5$)	$0.25\pi(\beta\tau+1)^{-0.5}E_2\,Kummer\,M\left[0.5,2,-r^2(\beta\tau+1)^{-1}\right]$ $R^{2,1}$ (4.7a) $0.5\pi^{0.5}r^{-1}E_3\left[\left(1-0.5r^{-2}(\beta\tau+1)\right)Erf\left(r(\beta\tau+1)^{-0.5}\right)+\pi r^{-1}(\beta\tau+1)^{\frac{1}{2}}e^{-r^2(\beta\tau+1)^{-1}}\right]$ $R^{3,1}$ (4.7b)
Model no. 22, Table 5.1 of Chapter VIII ($n=1$)	$E_n\int_0^1 du\left(1-u^2\right)^{0.5(n-1)}e^{-\left(\frac{u^2r^2}{a^2}+\frac{\tau^2}{b^2}\right)^{0.5}}$ $R^{n,1}$ (4.8)
$c_0\cos(r\tau)$	$\frac{c_0\tau}{2\pi r}\sin(r\tau)$ $R^{3,1}$ (4.9)
Model of Eq. (4.6) ($n=1$)	$\frac{E_n}{2c(2\pi)^{0.5n}}\int_0^1\int_0^\infty du\,dk\frac{\left(1-u^2\right)^{0.5(n-1)}(urk)^{0.5}}{(k^2+\alpha^2)^p}J_{-\frac{1}{2}}(kur)e^{-c^{-1}(k^2+\alpha^2)^p\tau}$ $R^{n,1}$ (4.10)

any of the covariance models of Table 5.1 of Chapter VII ($n=1$). Indeed, these are the cases with Eqs. (4.5)–(4.8) of Table 4.2. In the model of Eq. (4.5) the values of the coefficients a, c affect the ranges and shapes of the covariance models. Furthermore, the space–time covariance model of Eq. (4.9) was also considered in Example 2.2 earlier. Lastly, starting from Eq. (4.6) with $n=1$, and then transforming it in higher dimensions by means of the ST of Eq. (4.2a), one finds the model of Eq. (4.10). Otherwise said, by letting $n=1$ in any valid covariance model in $R^{n,1}$, and using STs, yet another class of new space–time covariance models can be constructed.

The applicability range of a technique is one of the most common tests of its usefulness. The ST technique of covariance model construction is, perhaps, the best technique in this respect, since it can be applied on any lower-dimensionality function that is a valid covariance—no other restriction limits its applicability. Note that anisotropic covariance models of physical attributes can be derived from the isotropic ones generated by Eqs. (4.2a–b). In the case of geometric anisotropy, e.g., an anisotropic covariance model can be obtained from an isotropic one by means of a coordinate rotation and rescaling of the axes. Alternatively, the anisotropic parameters can be determined directly from the attribute data using the method proposed in Hristopulos (2002, 2004). If the coordinates are then transformed by respective rotation and rescaling transformations the attribute can be modeled by means of an isotropic covariance model.

In sum, ST is a powerful technique, which can be used to construct both isostationary and homostationary covariance models, and is also very easy to implement in the case of $n = 3$. This fact, in combination with the covariance hereditary property, allows the quick and efficient construction of space–time covariance models in lower dimensions.

5. PHYSICAL EQUATION TECHNIQUES

One of the most recurring features in the discussions throughout the book is the view that it is desirable and possible that an attribute's spatiotemporal variation be quantitatively expressed in terms of physical equations codifying lawful regularities of nature that science was able to discover. As far as this chapter is concerned, this means that we need to investigate approaches that derive S/TVFs in the real or in the spectral domain using physical considerations about the attribute. Particularly, in this section we introduce techniques of covariance model construction that are based on physical laws of various mathematical forms. An intriguing feature of these techniques is that they can construct covariance models that not only are valid by construction, but they also have properties that are physically consistent with the space-time variation of the attributes they represent. In this context, the close connection between the developments in the present section and the material discussed in Chapter XIV is obvious.

As was suggested in Chapter XIV, physical laws and empirical relationships in conditions of real-world uncertainty are usually represented in terms of differential equations, and in some cases by means of algebraic equations. It seems then natural to try using these representations to construct valid S/TVFs. I start below with the physical law representation in terms of a stochastic partial differential equation (SPDE), which is the most common and realistic way to represent a physical law mathematically. Later, I will also consider empirical law representations in terms of algebraic equations. While I focus on the construction of space–time covariance models, essentially the same techniques can be used to construct variogram or structure function models, as well.

5.1 COVARIANCE CONSTRUCTION FROM STOCHASTIC PARTIAL DIFFERENTIAL EQUATION REPRESENTATIONS

The process leading to the construction of a covariance model from the SPDE (representing the physical law of an attribute) is based, methodologically, on the covariance permissibility criterion described in Proposition 4.1 of Chapter XV. There are two basic versions of the construction procedure, as outlined in Table 5.1:

(a) in *version A*, one first solves the SPDE governing the attribute of interest, and then constructs the covariance model based on this solution using a statistical averaging process (several cases were examined in Chapter XIV), and

(b) in *version B*, one first obtains the deterministic partial differential equation (PDE) describing the covariance change in space–time from the attribute SPDE, and then solves the covariance PDE to construct the covariance model.

Table 5.1 Physical Stochastic Partial Differential Equation Technique of Space–Time Covariance Model Construction, Versions *A* and *B*

Step	Version *A*	Version *B*
1	Consider the physical stochastic partial differential equation governing the attribute of interest $X(\boldsymbol{p})$, $$\begin{aligned}&\mathcal{L}_X[X(\boldsymbol{p}), Y_l(\boldsymbol{p}), a_{l'}] = 0\\&BIC\left[X_0, Y_{l,0}, a_{l'}\right]\end{aligned} \quad (5.1\text{a–b})$$ where $\mathcal{L}_X$ is a differential operator in $R^{n,1}$; $Y_l(\boldsymbol{p})$ and $a_{l'}(l = 1,\ldots, m, l' = 1,\ldots, k)$ are known attributes and parameters.	Directly from Eqs. (5.1a–b) derive the partial differential equation obeyed by the covariance function of interest $c_X(\boldsymbol{p}, \boldsymbol{p}')$, $$\begin{aligned}&\mathcal{L}_C[c_X(\boldsymbol{p},\boldsymbol{p}'), c_{Y_l}(\boldsymbol{p},\boldsymbol{p}'), c_{l'}] = 0\\&BIC\left[c_{X_0}, c_{Y_{l,0}}, c_{l'}\right]\end{aligned} \quad (5.2\text{a–b})$$ where $\mathcal{L}_C$ is a differential operator in $R^{n,1}$; $c_{Y_l}(\boldsymbol{p},\boldsymbol{p}')$ and $c_{l'}$ $(l = 1,\ldots, m, l' = 1,\ldots, k)$ are known covariances and parameters.
2	Solve Eq. (5.1a–b) for $X(\boldsymbol{p})$.	Solve Eq. (5.2a–b) for $c_X(\boldsymbol{p}, \boldsymbol{p}')$.
3	Construct $c_X(\boldsymbol{p}, \boldsymbol{p}')$ from $X(\boldsymbol{p})$ by means of the stochastic expectation or statistical averaging process.	

In both versions *A* and *B*, the constructed covariance models are permissible by construction, so long as all the inputs to the relevant equations (BIC, sources etc.) are valid. With these approaches, one should first verify the existence and uniqueness of the solutions of the SPDE governing the attribute (this should be done, even when the derivation of the solutions themselves is not a feasible objective). Another important aspect as regards version *B* is the so-called closure problem. More specifically, a closure problem arises when we have a hierarchy of N equations with $N + 1$ statistical moments. It is then necessary to establish a suitable approximation technique of converting the infinite hierarchy of equations into a closed set. There are several good references on the subject of SPDEs, including Syski (1967), Gihman and Skorokhod (1972), Arnold (1974), Friedman (1975, 1976), Da Prato and Tubaro (1987), Sobczyk (1991), and Christakos et al. (1995).

Although the techniques of Table 5.1 are methodologically rather straightforward, the specifics of their implementation vary considerably, depending on the form of the SPDE considered. I start with a large class of space–time nonseparable covariance models associated with natural attributes that obey a physical law represented by the following general SPDE in $R^{n,1}$ (Chapter XIV),

$$\frac{\partial}{\partial t}X(\boldsymbol{p}) = \mathcal{L}_{\boldsymbol{p}}[X(\boldsymbol{p})], \tag{5.3}$$

where $\mathcal{L}_{\boldsymbol{p}}$ is a linear differential operator in $\boldsymbol{p}$. This class includes various families of covariance models, one of which is the family represented by Model no. 14 in Table 5.1 of Chapter VIII; also, covariance models belonging to this family have been studied in Kolovos et al. (2004). These models are space–time heterogeneous, in general, due to a number of reasons, including the boundary and initial condition (BIC) effects. In all these models the covariance functions are expressed as sums over modes, each mode including different functions with separable spatial and temporal components. An example may enlighten certain aspects of our discussion.

Example 5.1

Nonseparable covariance models across space–time are obtained from parabolic SPDE governing the distribution of natural attributes. An example is the covariance Model no. 6 of Table 5.1 of Chapter VIII characterizing the variability of STHS random fields in $R^{1,1}$. A nonseparable covariance model in $R^{n,1}$ (n = 1, 2, 3) obtained from the diffusion SPDE, with several physical applications, is the Model no. 2 of Table 5.1 of Chapter VIII.

SPDEs representing physical laws or empirical models can generate new classes of permissible spatiotemporal covariance models based on other well-known functions. This is the case of physical laws and empirical models represented by an SPDE of the form

$$\mathcal{L}_p[X(\boldsymbol{p})] = Y(\boldsymbol{p}), \tag{5.4}$$

where $Y(\boldsymbol{p})$ is the input attribute with known covariance $c_Y(\boldsymbol{h}, \tau)$.

Example 5.2

Table 5.2 lists physical SPDEs of the form of Eq. (5.4), for various $\mathcal{L}_p$ operators, together with the space−time covariance models constructed on the basis of these SPDEs. Specifically, Christakos (1992) used the SPDE no. 1 of Table 5.2, where $\boldsymbol{p} = (s_1, s_2, t)$, a, b are positive coefficients, and $\widetilde{c}_Y$ is a given SDF model, to derive spatiotemporal covariance models, c_X, via the integral expression shown in Table 5.2, where $\boldsymbol{h} = (h_1, h_2)$, $\boldsymbol{k} = (k_1, k_2)$. For illustration, let $a = b = 1$ and select the SDF

$$\widetilde{c}_X(\boldsymbol{k}, \omega) = 2\pi\delta(\omega - \boldsymbol{k}\cdot\boldsymbol{v})e^{-0.25\alpha^2\boldsymbol{k}^2}, \tag{5.5}$$

where $\boldsymbol{v}$ is a known vector parameter that represents a velocity vector. Then, assuming spatial isotropy, the following nonseparable covariance model is obtained in $R^{2,1}$,

$$c_X(r, \tau) = 0.5\alpha \int_0^\infty dk\, k^{-3}\left(k^2 + v^2\right)^{-2} e^{-0.25\alpha^2 k^2} J_0[k(r + v\tau)], \tag{5.6}$$

where $v = |\boldsymbol{v}|$. Eq. (5.6) has been calculated numerically (see plots in Kolovos et al., 2004). For the SPDE no. 2 of Table 5.2, with $Y(\boldsymbol{p}) = 0$, the initial condition (IC) is $X(\boldsymbol{p})|_{t=0} = X_0(\boldsymbol{s})$, and $H(t)$ is a Heaviside unit step function. The PDE may represent heat conduction equation in isotropic solid and an unbounded domain, in which case the $X(\boldsymbol{p})$ denotes temperature and D is heat diffusivity. This case

Table 5.2 Examples of Stochastic Partial Differential Equations Representing Physical Laws and Associated Covariance Models

No.	Operator $\mathcal{L}_{X,p}$, Eq. (5.4)	Input	Space−Time Covariance $c_X(h, \tau)$, $r = \lvert h \rvert$
1	$\left[\frac{a\partial^2}{\partial t^2} + b\sum_{i=1}^{2}\frac{\partial^4}{\partial s_i^4} + \frac{2b\partial^4}{\partial s_1^2\partial s_2^2}\right]$	$\widetilde{c}_Y$	$\iint d\boldsymbol{k}d\omega \frac{e^{i(\boldsymbol{k}\cdot\boldsymbol{h}-\omega\tau)}}{\left(b\boldsymbol{k}^4 - a\omega^2\right)^2}\widetilde{c}_Y(\boldsymbol{k}, \omega)$ $(R^{2,1})$
2	$\left[\frac{\partial}{\partial t} - D\nabla^2\right]$	Y=0	$\frac{H(\tau)}{D^{1/2}(2\pi\tau)^{3/2}}\int d\boldsymbol{h}' e^{-\frac{\sum_{i=1}^{3}(h_i - h_i')^2}{8D\tau}} c_{X_0}(\boldsymbol{h}')$ $(R^{3,1})$
3	$\left(\sum_{i=1}^{2}\frac{\partial}{\partial s_i} - \alpha^2\right)\left(\frac{\partial}{\partial t} + \beta\right)$	$Y \sim WN$ (0, 1)	$\alpha r K_1(\alpha r)e^{-\beta\tau}$ $(R^{2,1})$
4	$\left[\left(\nabla^2 - a^2\right)^p - b\frac{\partial}{\partial t}\right]$ $p = \frac{2\mu+n}{4}$ $(\mu > 0)$	$Y \sim GWN$ $(0, \sigma^2)$	$\frac{(2\pi)^{\frac{n}{2}+1}\sigma^2 r^{1-\frac{n}{2}}}{2}\int_0^\infty dk\, k^{\frac{n}{2}}\frac{e^{-b^{-1}\tau(k^2+\alpha^2)^p}}{(k^2+\alpha^2)^p}J_{\frac{n}{2}-1}(kr)$ $(R^{n,1})$
5	$\left[\frac{\partial}{\partial t} + \boldsymbol{v}\cdot\nabla - \nabla\cdot\boldsymbol{a}\nabla + b\right]$	$\widetilde{c}_Y$	$c\int d\boldsymbol{k}\, e^{i(\boldsymbol{h}-\tau\boldsymbol{v})\cdot\boldsymbol{k} - (\boldsymbol{k}\cdot(\boldsymbol{ak})+b)\lvert\tau\rvert}[\boldsymbol{k}\cdot(\boldsymbol{ak}) + b]^{-1}\widetilde{c}_Y(\boldsymbol{k})$ $(R^{2,1})$

has been also studied in Example 2.5 of Chapter XII in terms of functional analysis. The corresponding covariance is of the form shown in Table 5.2 (which is the same as Eq. (2.25) of Chapter XII). For further illustration, if $c_{X_0}(\boldsymbol{s}-\boldsymbol{s}')=A_0\delta(\boldsymbol{h})$, $\boldsymbol{h}=\boldsymbol{s}-\boldsymbol{s}'$, then the following space–time covariance model is obtained,

$$c_X(r,\tau)=\frac{A_0H(\tau)}{D^{1/2}(2\pi\tau)^{3/2}}e^{-\frac{r^2}{8D\tau}}, \tag{5.7}$$

$r=|\boldsymbol{h}|$. The IC homogeneity in an unbounded domain implies covariance homogeneity at any later time. In the SPDE no. 3, $\boldsymbol{p}=(s_1,s_2,t)$, α, and β are known coefficients, and $Y(\boldsymbol{p})$ is a white noise (WN) random field. The corresponding space–time covariance model is shown in the table, where K_1 is the modified Bessel function of the second kind. This covariance model has been used in the design of rainfall networks (Rodriguez-Iturbe and Mejia, 1974). The physical SPDEs nos. 4 and 5 have been studied in Examples 3.4 and 3.8 of Chapter XIV, respectively. In the case of no. 4, $Y(\boldsymbol{p})$ is a zero mean Gaussian white noise (GWN) with variance σ^2. Other solutions of these SPDEs exist that yield Gaussian random fields with STIS covariance functions (Exercise XIV.2). In the SPDE 5, $\boldsymbol{p}=(s_1,s_2,t)$, $\boldsymbol{v}$ is a velocity vector, $\boldsymbol{a}$ is a diffusion matrix, $b>0$ is a dumping coefficient, c is a Fourier transformations constant, and $Y(\boldsymbol{p})$ is a Gaussian random field with known wavevector spectrum $\widetilde{c}_Y(\boldsymbol{k})$. For illustration, in the special case that $\boldsymbol{v}=\boldsymbol{a}=\boldsymbol{0}$ (i.e., no advection and diffusion), the space–time covariance model of the SPDE no. 5 is of the form (Sigrist et al., 2015),

$$c_X(\boldsymbol{h},\tau)=cb^{-1}e^{-b|\tau|}c_Y(\boldsymbol{h}),$$

which is permissible in $R^{2,1}$ for any valid model $c_Y(\boldsymbol{h})$ in R^2, where $c_Y(\boldsymbol{h})=\int d\boldsymbol{k}\,e^{i\boldsymbol{h}\cdot\boldsymbol{k}}\widetilde{c}_Y(\boldsymbol{k})$.

Example 5.3

In a similar vein, starting from the general SPDE of Eq. (5.4) in $R^{n,1}$, where, say, $\mathcal{L}_p=\frac{\partial}{\partial t}\mathcal{L}_s$ and $\mathcal{L}_s$ is a linear differential operator in $\boldsymbol{s}$, covariance models can be constructed by means of the equation

$$c_X(\boldsymbol{p},\boldsymbol{p}')=\iint d\boldsymbol{u}\,d\boldsymbol{u}'\;c_Y(\boldsymbol{u},\boldsymbol{u}')G(\boldsymbol{p},\boldsymbol{u})G(\boldsymbol{p}',\boldsymbol{u}'), \tag{5.8}$$

where G is the Green's function that is the solution of $\mathcal{L}_p[G(\boldsymbol{p},\boldsymbol{u})]=\delta(\boldsymbol{p}-\boldsymbol{u})$. This approach produces a versatile class of nonseparable covariance models.

Example 5.4

In other physical applications, the long-range properties of the covariance models may play a key role. There is a family of covariance models with well-defined asymptotic behavior, whereas dependence close to the origin is unspecified. For instance, on the basis of the asymptotic correlation function for the noisy Burgers SPDE (Fogedby, 1998), a nonseparable covariance model for large r and τ values in $R^{1,1}$ can be derived (Christakos, 2000),

$$c_X(r,\tau)\cong 0.25e^{\frac{D}{4}}\left(\pi D^3\right)^{-\frac{1}{2}}\left(r^{z-2\alpha}\tau^{-1}\right)^{\frac{1}{2}}e^{-r^{-z}\tau}, \tag{5.9}$$

where α, z, $D>0$. As noted earlier, the symbol "$\cong$" denotes that the covariance function is of this functional form for finite lags and asymptotically but not close to the origin. The shapes and the correlation ranges of the model of Eq. (5.9) depend on the parameters α and z. The magnitude of the covariance slope increases with the ratio $\frac{\alpha}{z}$.

Another large class of natural laws and scientific models can be represented by the SPDE along the direction of the unit vector $\boldsymbol{\varepsilon}_i$,

$$\mathcal{L}_{X,\boldsymbol{p}}[X(\boldsymbol{s},t)] = Y(\boldsymbol{s}),$$
$$\frac{\partial^k}{\partial s_i^k} X(\boldsymbol{s},t)|_{\boldsymbol{s}=0} = \frac{\partial^k}{\partial s_i^k} X(\boldsymbol{0},t) = 0 \quad (k = 0, 1, \ldots, \nu - 1) \tag{5.10a–b}$$

in $R^{n,1}$, where $\mathcal{L}_{X,\boldsymbol{p}}[\cdot] = \sum_{k=0}^{\nu} a_k(t) \frac{\partial^k}{\partial s_i^k}[\cdot]$, and the attribute of interest is represented by the random field $X(\boldsymbol{s}, t) \in L_2$, i.e., it is m.s. differentiable with respect to the coordinate s_i of the point $(\boldsymbol{s},t) \in R^{n,1}$, $Y(\boldsymbol{s})$ is a known attribute random field, $a_k(t)$ are known coefficients, and Eq. (5.10b) describes BCs. This technique is outlined in Table 5.3. It is worth-noticing that the covariance is defined directly from the physical law of Eqs. (5.10a–b), without the need to know the underlying probability distribution or to involve any data.

Table 5.3 The Physical Equation Technique of Space–Time Covariance Model Generation

Step	Description
1	The mean equations corresponding to the stochastic partial differential equation of Eqs. (5.10a–b) are, $$\sum_{k=0}^{\nu} a_k(t) \frac{\partial^k}{\partial s_i^k} \overline{X(\boldsymbol{s},t)} = \overline{Y(\boldsymbol{s})}, \quad \frac{\partial^k}{\partial s_i^k} \overline{X(\boldsymbol{0},t)} = 0. \tag{5.11a–b}$$ The solution of the deterministic Eqs. (5.11a–b) gives the mean attribute value across space, $\overline{X}$, as a function of the known $a_k(t)$ and $\overline{Y}$.
2	Multiplying both sides of the law (Eq. 5.10a) by $Y(\boldsymbol{s})$ and applying the mean operator yields, $$\sum_{k=0}^{\nu} a_k(t) \frac{\partial^k}{\partial s_i'^k} \overline{Y(\boldsymbol{s})X(\boldsymbol{s}',t')} = \overline{Y(\boldsymbol{s})Y(\boldsymbol{s}')}, \quad \frac{\partial^k}{\partial s_i'^k} \overline{Y(\boldsymbol{s})X(\boldsymbol{0},t')} = 0. \tag{5.12a–b}$$ The solution of the deterministic Eqs. (5.12a–b) provides $\overline{Y(\boldsymbol{s})X(\boldsymbol{s}',t')}$ as a function of the known BC and $\overline{Y(\boldsymbol{s})Y(\boldsymbol{s}')}$.
3	Multiplying both sides of the law of Eq. (5.10a) by $X(\boldsymbol{s}',t')$ and taking the mean yields, $$\sum_{k=0}^{\nu} a_k(t) \frac{\partial^k}{\partial s_i^k} \overline{X(\boldsymbol{s},t)X(\boldsymbol{s}',t')} = \overline{Y(\boldsymbol{s})X(\boldsymbol{s}',t')}, \quad \frac{\partial^k}{\partial s_i^k} \overline{Y(\boldsymbol{0},t)X(\boldsymbol{s}',t')} = 0. \tag{5.13a–b}$$ The solution of Eqs. (5.13a–b) provides the $\overline{X(\boldsymbol{s},t)X(\boldsymbol{s}',t')}$ in terms of $\overline{Y(\boldsymbol{s})X(\boldsymbol{s}',t')}$ derived in Step 2.
4	The required covariance is then obtained as $$c_X(\boldsymbol{s},t,\boldsymbol{s}',t') = \overline{X(\boldsymbol{s},t)X(\boldsymbol{s}',t')} - \overline{X(\boldsymbol{s},t)}\,\overline{X(\boldsymbol{s}',t')}. \tag{5.14}$$

Example 5.5

To gain some insight concerning the physical equation technique of Table 5.3, let us study the following SPDE governing the variation of $X(s, t)$ in $R^{1,1}$,

$$\begin{aligned} &\frac{b}{f(t)}\frac{\partial}{\partial s}X(s,t) + X(s,t) = Y(s), \\ &X(0,t) = 0, \end{aligned} \tag{5.15a–b}$$

where b and $f(t)$ are experimental coefficient and time-dependent parameter, respectively, and $Y(s)$ is a Brownian field with statistics

$$\begin{aligned} &\overline{Y(s)} = 0, \\ &\overline{Y(s)Y(s')} = as. \end{aligned} \tag{5.15c–d}$$

The mean function PDEs that correspond to Eqs. (5.11a–b) of Step 1 (Table 5.3 with $v = 1$) are

$$\frac{b}{f(t)}\frac{\partial}{\partial s}\overline{X(s,t)} + \overline{X(s,t)} = 0,$$

$$\overline{X(0,t)} = 0,$$

with solution $\overline{X(s,t)} = 0$. The covariance PDEs corresponding to Eqs. (5.12a–b) of Step 2 (Table 5.3) are

$$\frac{b}{f(t')}\frac{\partial}{\partial s'}c_{YX}(s,s',t') + c_{YX}(s,s',t') = as,$$

$$c_{YX}(s,0,t') = 0,$$

with solution

$$c_{YX}(s,s',t') = as\left(1 - e^{-\frac{f(t')}{b}s'}\right).$$

Lastly, the covariance PDE corresponding to Eqs. (5.13a–b) of Step 3 is given by

$$\frac{b}{f(t)}\frac{\partial}{\partial s}c_X(s,t,s',t') + c_X(s,t,s',t') = as\left(1 - e^{-\frac{f(t')}{b}s'}\right),$$

$$c_X(0,s',t,t') = 0,$$

with solution

$$c_X(s,s',t,t') = a\left[s - \frac{b}{f(t)}\left(1 - e^{-\frac{f(t)}{b}s}\right)\right]\left(1 - e^{-\frac{f(t')}{b}s'}\right).$$

As was expected, the covariance is defined directly from the physical law (Eq. 5.15a–b), without the need to know the underlying probability distribution or to involve any data.

Example 5.6

Diffusion SPDEs are widely used in applications characterized by dissemination processes (e.g., the distribution of heat or the variation in temperature in a region over time). In this example we start with the diffusion SPDE in $R^{1,1}$,

$$\left[\frac{\partial}{\partial t} - D\frac{\partial^2}{\partial s^2}\right] X(s,t) = 0, \tag{5.16a}$$

where $X(s, t)$ denotes attribute concentration, and D is the diffusion coefficient. The associated BIC is

$$\frac{\partial}{\partial s} X(L,t) = \frac{\partial}{\partial s} X(0,t) = 0, \tag{5.16b}$$

i.e., the boundaries do not permit the concentration to escape (impervious BC), and an initial concentration profile is determined by

$$X(s,0) = c_0\left[\frac{1}{12} + \frac{s^2}{L^2}\left(2 - \frac{s}{3L}\right)\right], \tag{5.16c}$$

where L is the spatial domain of the equation, c_0 is a random variable (RV) with second moment $c_2 > 0$. Then, from Eq. (5.16a) the space–time covariance model of $X(s, t)$ can be derived by (Christakos and Hristopulos, 1998),

$$c_X(s,s',t,t') = c_2 \sum_{l,l'=0}^{\infty} \alpha_l \alpha_{l'} \cos\left(\frac{l\pi s}{L}\right) \cos\left(\frac{l'\pi s'}{L}\right) e^{-D\pi^2(tl^2+t'l'^2)L^{-2}}, \tag{5.17}$$

where $\alpha_0 = \frac{2}{3}$, and $\alpha_l = \frac{6(-1)^l}{l^2\pi^2} - \frac{8\delta_{l,2k+1}}{l^4\pi^4}$ for $l \neq 1$. Next, if we assume that the diffusion coefficient D is a uniformly distributed RV in $[D_1, D_2]$ with $D_2 > D_1$, the covariance model is given by

$$c_X(s,s',t,t') = \frac{c_0^2 L^2}{2\pi^2 \delta D} \sum_{l,l'=0}^{\infty} \alpha_l \alpha_{l'} \cos\left(\frac{l\pi s}{L}\right) \cos\left(\frac{l'\pi s'}{L}\right) e^{-\overline{D}\pi^2(tl^2+t'l'^2)L^{-2}} \frac{\sinh\left(\pi^2 \overline{D}(tl^2 + t'l'^2)L^{-2}\right)}{tl^2 + t'l'^2} \tag{5.18}$$

for all t, $t' > 0$, where $\overline{D} = \frac{D_1+D_2}{2}$ and $\delta D = \frac{D_2-D_1}{2}$. For the purpose of comparing it with the previous case, Eq. (5.17), one can assume that $\overline{D} = D$ and $c_2 = \frac{c_0^2 \overline{D}}{\delta D}$. Since $c_2 \geq c_0^2$, Eq. (5.18) constrains the diffusion coefficient to have a coefficient of variation (CV) less than one. Then, the difference between the two covariance expansions is that Eq. (5.18) involves the mode-dependent factor

$$g_{ll'}(t,t') = \frac{\sinh\left(\lambda(tl^2 + t'l'^2)\right)}{2\lambda(tl^2 + t'l'^2)}, \tag{5.19}$$

where $\lambda = \pi^2 \overline{D} L^{-2}$. Since $g_{ll'}(t,t') \geq 1$, a random diffusion coefficient slows down the homogenization of the initial concentration profile.

Example 5.7

In this case, starting directly from the Langevin SPDE of thermodynamic equilibrium

$$\frac{\partial X(\boldsymbol{s},t)}{\partial t} + D\frac{\delta H[x(\boldsymbol{s})]}{\delta x(\boldsymbol{s})}\bigg|_{x(\boldsymbol{s})=X(\boldsymbol{s},t)} = Y(\boldsymbol{s},t), \tag{5.20}$$

where D is a diffusion coefficient, $\frac{\delta[\cdot]}{\delta x(\boldsymbol{s})}$ is the functional derivative with respect to the field state, and $Y(\boldsymbol{s}, t)$ is the noise field, Hristopulos and Tsantili (2015) obtained the following covariance model in $R^{n,1}$ ($n = 1, 2, 3$),

$$c_X(\boldsymbol{h},\tau) = (2\pi)^{\frac{n}{2}}|\boldsymbol{h}|^{1-\frac{n}{2}}e^{-\widetilde{D}|\tau|}\int_0^\infty dk\frac{\eta_0\xi^n k^{\frac{n}{2}}}{1+\eta_1\xi^2k^2+\mu\xi^4k^4}e^{-\widetilde{D}|\tau|\left(\eta_1\xi^2k^2+\mu\xi^4k^4\right)}J_{\frac{n}{2}-1}(kr), \tag{5.21}$$

where η_0, $\eta_1 > 0$, ξ are coefficients (scale, rigidity, and characteristic length, respectively), $\mu \geq 0$ is a normalization constant, and $\widetilde{D} = \frac{D}{\xi^n\eta_0}$ is the combined diffusion coefficient. If $\mu = 0$, Eq. (5.21) reduces to

$$c_X(\boldsymbol{h},\tau) = \pi^{\frac{n}{2}}e^{-\widetilde{D}|\tau|}\int_0^\infty dk\frac{1}{(\beta_1+\beta_2k)^{\frac{n}{2}}}e^{-\frac{|\boldsymbol{h}|^2}{4(\beta_1+\beta_2k)}-k\beta_0}, \tag{5.22}$$

where $\beta_0 = \frac{1}{\eta_0\xi^n}$, $\beta_1 = \widetilde{D}\eta_1\xi^2|\tau|$, and $\beta_2 = \frac{\eta_1\xi^2}{\eta_0\xi^n}$. In $R^{1,1}$ and $R^{3,1}$, Eq. (5.22) reduces to the models no. 23 and no. 24, respectively, of Table 5.1 of Chapter VIII.

Lastly, many spatiotemporal covariance models are obtained as solutions of covariance PDEs (Chapter XIV) governing the attribute covariance function $c_X(\boldsymbol{p},\boldsymbol{p}')$, i.e.,

$$\mathcal{L}_C[c_X(\boldsymbol{p},\boldsymbol{p}'), c_Y(\boldsymbol{p},\boldsymbol{p}'), c_{XY}(\boldsymbol{p},\boldsymbol{p}'), a_l] = 0, \tag{5.23}$$

where $\mathcal{L}_C$ is a linear differential operator in $R^{n,1}$, and the input covariance $c_Y(\boldsymbol{p},\boldsymbol{p}')$ and the parameters a_l ($l = 1,\ldots, k$) are known. Then, a space–time covariance model $c_X(\boldsymbol{p},\boldsymbol{p}')$ can be constructed as the solution of Eq. (5.23). As in previous cases, the covariance models that are constructed in terms of the PDE solutions are permissible by construction, so long as all the covariance functions that pertain to the PDE inputs (covariance BIC) are valid.

As was noticed at the beginning of this section, a closure problem arises within this approach when we have a hierarchy of N equations with $N+1$ statistical moments (c_X, c_{XY} etc.). In other words, Eq. (5.23) must be supplemented by additional PDEs, like

$$\mathcal{L}'_C[c_Y(\boldsymbol{p},\boldsymbol{p}'), c_{XY}(\boldsymbol{p},\boldsymbol{p}'), a_l] = 0.$$

These additional equations may form an infinite hierarchy. It is then necessary to establish a suitable approximation technique of converting the infinite hierarchy of such equations into a closed set.

Example 5.8

We now revisit the transient single-phase fluid flow governed by the hydrologic equations presented in Example 3.5 of Chapter XIV. The hydraulic head covariance of $X^{(1)}(\boldsymbol{p})$ to the first order in σ_f, i.e., $c_X(\boldsymbol{p},\boldsymbol{p}') = \overline{X^{(1)}(\boldsymbol{p})X^{(1)}(\boldsymbol{p}')}$, is obtained by solving the equation

$$\left[\frac{\partial^2}{\partial s_i^2} + \frac{\partial \overline{f(\boldsymbol{s})}}{\partial s_i}\frac{\partial}{\partial s_i} - \frac{S_s(\boldsymbol{s})}{K_G(\boldsymbol{s})}\frac{\partial}{\partial t}\right] c_X(\boldsymbol{p},\boldsymbol{p}') = \left[J_i(\boldsymbol{p})\frac{\partial}{\partial s_i} + \frac{Y(\boldsymbol{s})}{K_G(\boldsymbol{s})} - \frac{S_s(\boldsymbol{s})}{K_G(\boldsymbol{s})}J_t(\boldsymbol{p})\right] c_{fX}(\boldsymbol{s},\boldsymbol{s}',t'), \tag{5.24}$$

with BIC

$$\begin{aligned} c_X(\boldsymbol{s},0,\boldsymbol{s}',t') &= 0, \quad \boldsymbol{s} \in \Omega, \\ c_X(\boldsymbol{p},\boldsymbol{p}') &= 0, \quad \boldsymbol{s} \in \Gamma_D, \\ \eta_i(\boldsymbol{s})\frac{\partial}{\partial s_i}c_X(\boldsymbol{p},\boldsymbol{p}') &= 0, \quad \boldsymbol{s} \in \Gamma_N, \end{aligned} \tag{5.25a–c}$$

where $J_i(\boldsymbol{p}) = -\frac{\partial}{\partial s_i}X^{(0)}(\boldsymbol{p})$ and $J_t(\boldsymbol{p}) = \frac{\partial}{\partial t}X^{(0)}(\boldsymbol{p})$ are, respectively, the negative of the spatial gradient and the temporal gradient of the zeroth order mean hydraulic head. The cross-covariance function $c_{fX}(\boldsymbol{s}, \boldsymbol{s}', t')$ is the solution of the equation

$$\left[\frac{\partial^2}{\partial s_i^2} + \frac{\partial \overline{f(\boldsymbol{s})}}{\partial s_i}\frac{\partial}{\partial s_i} - \frac{S_s(\boldsymbol{s})}{K_G(\boldsymbol{s})}\frac{\partial}{\partial t'}\right] c_{fX}(\boldsymbol{s},\boldsymbol{s}',t') = \left[J_i(\boldsymbol{p})\frac{\partial}{\partial s_i} + \frac{Y(\boldsymbol{s})}{K_G(\boldsymbol{s})} - \frac{S_s(\boldsymbol{s})}{K_G(\boldsymbol{s})}J_t(\boldsymbol{p})\right] c_{fX}(\boldsymbol{s},\boldsymbol{s}'), \tag{5.26}$$

with BIC

$$\begin{aligned} c_{fX}(\boldsymbol{s},\boldsymbol{s}',0) &= 0, \quad \boldsymbol{s} \in \Omega, \\ c_{fX}(\boldsymbol{s},\boldsymbol{s}',t') &= 0, \quad \boldsymbol{s} \in \Gamma_D, \\ \eta_i(\boldsymbol{s})\frac{\partial}{\partial s_i}c_{fX}(\boldsymbol{s},\boldsymbol{s}',t') &= 0, \quad \boldsymbol{s} \in \Gamma_N. \end{aligned} \tag{5.27a–c}$$

Hence, $c_{fX}(\boldsymbol{s}, \boldsymbol{s}', t')$ is obtained from the solution of Eqs. (5.26), and then substituted into Eqs. (5.24) to find $c_X(\boldsymbol{p}, \boldsymbol{p}')$.

Example 5.9

Next, let us focus on the SPDE of Eq. (3.16) of Chapter XIV satisfied by the S/TRF $X(s, t)$ in $R^{1,1}$. The space–time covariance model of $X(s, t)$ will be

$$c_X(h,t,t') = \frac{1}{4}\int_{|t-t'|}^{t+t'} dt' t'^{-1/2} e^{-\frac{|h|^2}{4t'}} \tag{5.28}$$

where $|h| = |s' - s|$. Working in $R^{2,1}$, the covariance function satisfies the PDE

$$\left[\frac{\partial}{\partial \tau}-\sum_{i=1}^{2}\frac{\partial^2}{\partial h_i^2}\right]c_X(\boldsymbol{h},\tau)=0, \tag{5.29}$$

where $\boldsymbol{h}=(h_1, h_2)$, and with IC

$$c_X(\boldsymbol{h},0)=c_{X,0}(\boldsymbol{h}). \tag{5.30}$$

The solution to Eqs. (5.29) and (5.30) yields the covariance model

$$c_X(\boldsymbol{h},\tau)=\frac{1}{4\pi\tau}\int d\boldsymbol{h}' c_{X,0}(\boldsymbol{h}')e^{-\frac{\left|(h_1-h_1')^2+(h_2-h_2')^2\right|}{4\tau}}. \tag{5.31}$$

For illustration, if $c_{X,0}(\boldsymbol{h})=\delta(h_1-h_0,h_2-h_0')$, i.e., the initial covariance condition is concentrated at a point (h_0,h_0'), Eq. (5.31) leads to the covariance model

$$c_X(\boldsymbol{h},\tau)=\frac{1}{4\pi\tau}e^{-\frac{\left|(h_1-h_0')^2+(h_2-h_0')^2\right|}{4\tau}}, \tag{5.32}$$

with $\tau>0$.

Example 5.10

We now revisit the physical law describing heat transfer phenomena in the Nea Kessani study of Example 3.10 of Chapter XIV. The covariance function of the random field $X'(\boldsymbol{p})$ is derived from Eq. (3.26) of Chapter XIV if one multiplies $X'(\boldsymbol{p})$ by $X'(\boldsymbol{p}')$ and take the stochastic expectation of the product, thus leading to the Laplace covariance equation

$$\nabla^2 c_X(\boldsymbol{p},\boldsymbol{p}')=0, \tag{5.33}$$

with BC

$$\begin{aligned}
c_X(\boldsymbol{p},\boldsymbol{p}')&=\overline{\varepsilon_A(\boldsymbol{p})X'(\boldsymbol{p}')}, && c_X(\boldsymbol{p},\boldsymbol{p}')\in B_D,\\
\frac{\partial}{\partial s_i}c_X(\boldsymbol{p},\boldsymbol{p}')&=\overline{\varepsilon_B(\boldsymbol{p})X'(\boldsymbol{p}')}, \quad i=1,2,3, && c_X(\boldsymbol{p},\boldsymbol{p}')\in B_N,
\end{aligned} \tag{5.34a–b}$$

where the $c_X(\boldsymbol{p},\boldsymbol{p}')$ denotes the simultaneous temperature covariance between points $\boldsymbol{s}$ and $\boldsymbol{s}'$. For each pair of grid nodes $(\boldsymbol{s},\boldsymbol{s}')$, Eqs. (5.33)–(5.34a–b) should be solved with respect to $c_X(\boldsymbol{p},\boldsymbol{p}')$, the latter expressing dependence of the temperature values. Before implementing a numerical solution technique, we need to consider the BC $c_X(\boldsymbol{s}_D,\boldsymbol{s}',t)=\overline{\varepsilon_A(\boldsymbol{s}_D,t)X'(\boldsymbol{s}',t)}$ and $c_X(\boldsymbol{s}_N,\boldsymbol{s}',t)=\overline{\varepsilon_B(\boldsymbol{s}_N,t)X'(\boldsymbol{s}',t)}$, where, as before, the subscripts D and N denote the Dirichlet and Neumann BC, respectively. Specifically, in case that $\boldsymbol{s}=\boldsymbol{s}_D$ or $\boldsymbol{s}_N$, Eqs. (5.33) and (5.34a–b) reduce to

$$\nabla^2 c_X(\boldsymbol{s},\boldsymbol{s}_D,t)=0, \tag{5.35}$$

with BC

$$\begin{aligned}
c_X(\boldsymbol{s},\boldsymbol{s}_D,t)&=\text{cov}\,(\boldsymbol{s},\boldsymbol{s}_D), && c_X(\boldsymbol{s},\boldsymbol{s}_D,t)\in B_D,\\
c_X(\boldsymbol{s},\boldsymbol{s}_N,t)&=0, && c_X(\boldsymbol{s},\boldsymbol{s}_N,t)\in B_N;
\end{aligned} \tag{5.36a–b}$$

or

$$\nabla^2 c_X(\boldsymbol{s}, \boldsymbol{s}_N, t) = 0, \tag{5.37}$$

with

$$\begin{aligned} c_X(\boldsymbol{s}, \boldsymbol{s}_D, t) &= 0, & c_X(\boldsymbol{s}, \boldsymbol{s}_D, t) \in B_D, \\ \frac{\partial}{\partial s_i} c_X(\boldsymbol{s}, \boldsymbol{s}_N, t) &= \text{cov}\ (\boldsymbol{s}, \boldsymbol{s}_N, t), \quad i = 1, 2, 3, & c_X(\boldsymbol{s}, \boldsymbol{s}_N, t) \in B_N. \end{aligned} \tag{5.38a–b}$$

The $c_X(\boldsymbol{s}, \boldsymbol{s}_D, t)$ and $c_X(\boldsymbol{s}, \boldsymbol{s}_N, t)$ denote the covariances between the boundary and the grid points. Such information can be derived from the data or from prior knowledge. For each reference point on the boundary ($\boldsymbol{s}_D$ or $\boldsymbol{s}_N$), the corresponding covariances for the entire region are obtained from the Laplace covariance Eqs. (5.35) or (5.37), depending on the BC assumed. Since $c_X(\boldsymbol{s}, \boldsymbol{s}_D, t) = c_X(\boldsymbol{s}_D, \boldsymbol{s}, t)$, the solutions of Eqs. (5.35) and (5.37) can serve as the BC for Eq. (5.33). Notice that the differential operators are all applied on $\boldsymbol{s}$, whereas the time argument is here suppressed.

Remark 5.1

The assumptions and interpretations made about the attribute governed by the physical equation and the equation parameters can play a critical role in the covariance construction process. For illustration, consider the physical SPDE

$$\frac{\partial^2}{\partial s^2} X(s, t) + e^{-W(t)} Y(s, t) = 0$$

in the $R^{1,1}$ domain, where the RV $W(t) \sim G(0, 1)$ for all t is uncorrelated with $X(s, t)$, and $Y(s, t) = \sin(\pi s)\sin(\pi t)$. Under these conditions, the solution is $X(s, t) = \pi^{-2} e^{-W} \sin(\pi s)\sin(\pi t)$, and the space–time covariance function is

$$c_X(s, t, s', t') = \alpha \sin(\pi s)\sin(\pi s')\sin(\pi t)\sin(\pi t'),$$

where $\alpha = e(e-1)\pi^{-4}$, since $W(t) \sim G(0, 1)$.

In sum, the physical techniques of constructing space–time covariance models discussed above have the significant merit of maintaining close contact between the mathematical description and the physical phenomenon (as a result of the fact that SDPEs have a strong theoretical support such as first principles). Surely, their implementation may be not an easy matter, but will depend on the complexity of the phenomenon of interest. Apart from the differential equations representing physical laws, which are the most common situation in scientific applications, in some cases we also deal with empirical relationships of algebraic forms. These empirical relationships can be also used for covariance construction purposes. This is the topic of the following section.

5.2 COVARIANCE CONSTRUCTION FROM ALGEBRAIC EMPIRICAL RELATIONSHIPS

Space–time covariance models can be also constructed starting from empirical laws linking natural attributes on the basis of purely empirical evidence acquired by observation or experimentation. So, an empirical law may be a phenomenological relationship, that is, a relationship or correlation supported by observation or experimentation but not necessarily supported by theory. Usually, empirical laws are expressed in terms of stochastic or random algebraic equations, i.e., the attributes involved are considered as random fields due to the multisourced uncertainty of the real-world, including measurement error and sparse data. As we saw in Section 2 of Chapter XIV, there are many empirical laws in applied sciences that are expressed in an algebraic form involving elementary mathematical functions (polynomial, logarithms, exponential etc., functions). These empirical laws constitute an important component of scientific explanation.

Also, in Section 2 of Chapter XIV we noticed that in certain real-world situations the study of the phenomenon is aided considerably by empirical equations of the weighted summation form

$$X(\boldsymbol{p}) = \sum_{l=1}^{m} a_l Y_l(\boldsymbol{p}), \tag{5.39}$$

where a_l $(l = 1,\ldots, m)$ is a set of coefficients, usually calculated experimentally. The empirical Eq. (5.39) that is often derived by statistical regression, links the attribute of interest $X(\boldsymbol{p})$, about which we do not have sufficient information or it is physically difficult or expensive to measure, with other attributes $Y_l(\boldsymbol{p})$ about which a significant amount of easily collected information is available. We can apply the expectation or statistical averaging operator on the empirical relationship of Eq. (5.39) to obtain the equations of the attribute mean and covariance functions, respectively,

$$\begin{aligned} \overline{X(\boldsymbol{p})} &= \sum_{l=1}^{m} a_k \overline{Y_l(\boldsymbol{p})}, \\ c_X(\boldsymbol{p},\boldsymbol{p}') &= \sum_{l=1}^{m}\sum_{l'=1}^{m} a_l a_{l'} c_{Y_l Y_{l'}}(\boldsymbol{p},\boldsymbol{p}'). \end{aligned} \tag{5.40a–b}$$

Eq. (5.40a) provides a simple way to construct covariance models of $X(\boldsymbol{p})$, starting from weighted combinations of known models of $Y_l(\boldsymbol{p})$.

Example 5.11

Example 2.1 of Chapter XIV presented a simple empirical relationship, Eq. (2.4) of Chapter XIV, used in soil mechanics to link standard penetration resistance, $Y(s, t)$, and vertical stress, $X(s, t)$, for cohesionless soils, based on in situ measurements. In this case, an elementary implementation of the averaging process of Eqs. (5.40a–b) on the soil mechanics relationship of Eq. (2.4) of Chapter XIV leads to the mean and covariance models

$$\begin{aligned} \overline{X(\boldsymbol{s},t)} &= \overline{a} + \overline{b}\,\overline{Y(\boldsymbol{s},t)}, \\ c_X(\boldsymbol{s},\boldsymbol{s}') &= \sigma_a^2 + \overline{b^2} c_Y(s,s') + \sigma_b^2 \overline{Y}^2, \end{aligned} \tag{5.41a–b}$$

where $\overline{X(\boldsymbol{s},t)} = \overline{X}$ and $\overline{Y(\boldsymbol{s},t)} = \overline{Y}$. Using the empirical statistics of the coefficients stated in Example 2.1 of Chapter XIV, i.e., $\overline{a} = 1.35$, $\sigma_a = 1.05$, $\overline{b} = 7.9$, $\sigma_b = 1.76$, Eqs. (5.41a–b) give $\overline{X(\boldsymbol{s})} = 1.35 + 7.9\overline{Y(\boldsymbol{s})}$ and $c_X(\boldsymbol{s},\boldsymbol{s}') = 1.1 + 65.5\, c_Y(s,s') + 3.1\overline{Y}^2$. In the special case of practically zero model coefficient uncertainty, the covariance equation reduces to $c_X(s, s') = 62.4 c_Y(s, s')$. Hence, the mean and covariance of vertical stress is related to those of standard penetration resistance in a straightforward manner. Eq. (5.41b) is a simple empirical way to generate covariance models of the vertical stress attribute, starting from known models of standard penetration resistance.

In some other applications, the attribute representation of Eq. (5.39) includes separate components that are purely spatial or purely temporal functions.

Example 5.12

Bogaert and Christakos (1997a,b) suggested that in many geoscientific applications an attribute $X(s, t)$ can be expressed empirically in terms of the space–time regressive random field representation

$$X(\boldsymbol{s},t) = Y(\boldsymbol{s},t) + M_1(\boldsymbol{s}) + M_2(t) + \mu(\boldsymbol{s},t), \tag{5.42}$$

where $Y(s, t)$ is an STHS random field, $M_1(s)$ and $M_2(t)$ are a purely spatial and a purely temporal random field, respectively (these random fields have zero mean and are mutually independent), and $\mu(s, t)$ is a deterministic function that represents the mean of $X(s, t)$. $M_1(s)$ and $M_2(t)$ have space homogeneous and time stationary increments, respectively, and the $Y(s, t)$ has a covariance $c_Y(\boldsymbol{h},\tau) = \sigma_Y^2 \rho_{Y,s}(\boldsymbol{h}_{ij})\rho_{Y,t}(\tau_{k\ell})$, where $\rho_{Y,s}(\boldsymbol{h}_{ij})$ and $\rho_{Y,t}(\tau_{k\ell})$ are a spatial and a temporal correlation function, respectively, with $\boldsymbol{h}_{ij} = \boldsymbol{s}_i - \boldsymbol{s}_j$, $\tau_{k\ell} = t_k - t_\ell$.[8] If the $M_1(s)$ is a homogeneous and $M_2(t)$ a stationary random field, the space–time covariance function $c_X(\boldsymbol{h}, \tau)$ is considered; otherwise, the variogram function $\gamma_X(\boldsymbol{h}, \tau)$ can be used. In light of Eq. (5.42), these two functions are given by

$$\begin{aligned} c_X(\boldsymbol{h},\tau) &= c_Y(\boldsymbol{h},\tau) + c_{M_1}(\boldsymbol{h}) + c_{M_2}(\tau), \\ \gamma_X(\boldsymbol{h},\tau) &= \gamma_Y(\boldsymbol{h},\tau) + \gamma_{M_1}(\boldsymbol{h}) + \gamma_{M_2}(\tau). \end{aligned} \tag{5.43a–b}$$

Using Eqs. (5.43a–b), space–time covariance models can be constructed by combining known empirical models. Eqs. (5.42) and (5.43a–b) include, as special cases, certain earlier space–time (additive) separable models, such as that proposed by Bilonick (1985),

$$c_X(\boldsymbol{h},\tau) = c_s(\boldsymbol{h}) + c_t(\tau). \tag{5.44}$$

(see, also, Remark 2.6 of Chapter II). Eq. (5.42) takes into consideration interesting features of physical data, which cannot be accounted for by some earlier models.

There are several other kinds of empirical information that can be taken advantage of by the physical covariance model construction technique. This includes approximations in terms of the first few terms of the Taylor series of the analytical solution of the PDE describing the phenomenon of interest. Yet, we must be aware of empirical relationships that may neglect certain important features of the phenomenon being described, or they may even contradict theory (often the reason these empirical relationships are employed is because they are more mathematically tractable than the theory and may yield what they seem as plausible results).

6. CLOSED-FORM TECHNIQUES

This is a family of techniques that can construct space–time covariance models through closed-form expressions. To orient the readers, nine such techniques are listed in Table 6.1. Operationally, these techniques fall into two groups: (a) techniques that generate new covariance models starting from

[8] While the assumptions underlying Eq. (5.42) may restrict its applicability, they reduce considerably the computational demands of space–time analysis.

Table 6.1 Closed-Form Techniques for Generating Space–Time Covariance Models

No	Input Parameters	Space-Time Covariance Model $c_X(\boldsymbol{h},\tau)$, $r = \lvert\boldsymbol{h}\rvert$	
1	$c_{X,n-2}$: Known covariance in $R^{n-2,1}$.	$-\frac{1}{2\pi r}\frac{\partial}{\partial r}c_{X,n-2}(r,\tau)$, $\left[1+\frac{r}{n-2}\frac{\partial}{\partial r}\right]^{-1}c_{X,n-2}(r,\tau)$.	(6.1a–b)
2	φ: Completely monotone function. ψ: Positive function with completely monotone derivatives.	$\frac{c_0}{\lvert\psi(\tau^2)\rvert^{\frac{n}{2}}}\varphi\left(\frac{\lvert\boldsymbol{h}\rvert^2}{\psi(\tau^2)}\right)$	(6.2)
3	L_T: Bivariate Laplace transform. ψ_i, $i = 1, 2$: Bernstein functions, or variograms (nugget effect), or increasing and concave functions on $(0, \infty)$.	$\frac{c_0}{\psi_1\left(h_1^2\right)^{\frac{1}{2}}\psi_2(\tau^2)^{\frac{1}{2}}}L_T\left(\frac{h_2^2}{\psi_1\left(h_1^2\right)},\frac{h_3^2}{\psi_2(\tau^2)}\right)$ $\boldsymbol{h} = (h_1,h_2,h_3),\ n = 3$	(6.3)
4	$c_{X,i}$: Known covariances in $R^{n,1}$. λ_i, $\nu_i \geq 0$.	$\sum_{i=1}^{N}\lambda_i c_{X,i}(\boldsymbol{h},\tau)$, $\lim_{k\to\infty}\sum_{i\leq k}\nu_i c_{X,i}^{i}(\boldsymbol{h},\tau)$.	(6.4a–b)
5	$L_T\left(\breve{x}_1,\breve{x}_2\right)$: Laplace transform of nonnegative random vector (x_1, x_2). $\gamma(\boldsymbol{h})$, $\gamma(\tau)$: Spatial and temporal variograms.	$L_T(\gamma(\boldsymbol{h}),\gamma(\tau))$	(6.5)
6	γ_k: Stationary variograms. G: Gaussian probability density function. $\boldsymbol{c}$, $\boldsymbol{c}_k$: $n \times n$ covariance matrices. $\boldsymbol{v} \in R^n$: Constant vector.	$G\left(\boldsymbol{h}+\boldsymbol{v}\tau,\boldsymbol{c}+\sum_{k=1}^{m}\gamma_k(\tau)\boldsymbol{c}_k\right)$	(6.6)

Continued

Table 6.1 Closed-Form Techniques for Generating Space–Time Covariance Models—cont'd

No	Input Parameters	Space-Time Covariance Model $c_X(\boldsymbol{h},\tau)$, $r = \lvert\boldsymbol{h}\rvert$
7	ℓ: Continuous and completely monotone function on R^1_+. γ_k: Stationary variograms. $\boldsymbol{v} \in R^n$: Constant vector. γ_I: Intrinsic variogram.	$\left\lvert \boldsymbol{c} + \sum_{k=1}^{m} \gamma_k(\tau)\boldsymbol{c}_k \right\rvert^{-\frac{1}{2}} \ell\left(\frac{(\boldsymbol{h}+\boldsymbol{v}\tau)^T \left(\boldsymbol{c} + \sum_{k=1}^{m}\gamma_k(\tau)\boldsymbol{c}_k\right)^{-1} (\boldsymbol{h}+\boldsymbol{v}\tau)}{2} \right),$ $[1+\gamma_I(\boldsymbol{h})]^{-\frac{1}{2}}\ell\left(\frac{\tau^2}{1+\gamma_2(\boldsymbol{h})}\right).$ (6.7a–b)
8	c_1, c_2: Valid covariances. $\beta_1 = -\delta_{ij}\delta_{kl} + \delta_{ik}\delta_{jl} + \delta_{il}\delta_{jk}$ $+ \frac{3}{r^2}(\delta_{ij}h_k h_l + \delta_{ik}h_j h_l$ $+ \delta_{il}h_j h_k + \delta_{jl}h_i h_k + \delta_{jk}h_i h_l$ $+ \delta_{kl}h_i h_j) - \frac{15}{r^4}h_i h_j h_k h_l,$ $\beta_2 = \delta_{ij}h_k h_l + \delta_{ik}h_j h_l$ $+ \delta_{il}h_j h_k + \delta_{jl}h_i h_k + \delta_{jk}h_i h_l$ $+ \delta_{kl}h_i h_j - \frac{6}{r^2}h_i h_j h_k h_l,$ $\beta_3 = h_i h_j h_k h_l.$	$\left[\frac{h_i h_j}{r^2}\left(\frac{1}{r} - \frac{\partial}{\partial r}\right) - \frac{\delta_{ij}}{r}\right]\frac{\partial c_1(r,\tau)}{\partial r},$ $\left[\frac{h_i h_j}{r^2}\left(\frac{1}{r} - \frac{\partial}{\partial r}\right) - \frac{\delta_{ij}}{r}\right]\frac{\partial c_1(r,\tau)}{\partial r} - \left[\frac{h_i h_j}{r^2}\left(\frac{1}{r} - \frac{\partial}{\partial r}\right) + \delta_{ij}\frac{\partial}{\partial r}\right]\frac{\partial c_2(r,\tau)}{\partial r},$ $\frac{1}{r^2}\left[\beta_1\left(\frac{1}{r}\frac{\partial}{\partial r} - \frac{\partial^2}{\partial r^2}\right) + \frac{\beta_2}{r}\frac{\partial^3}{\partial r^3} + \frac{\beta_3}{r^2}\frac{\partial^4}{\partial r^4}\right]c_1(r,\tau).$ (6.8a–c)
9	r_s: Arc length. b_l, b: Stationary covariances $a \in (0, 1)$, $a_l \geq 0$, $\sum_{l=0}^{\infty} a_l h_{l,n} < \infty$. $E_{l,n} = \frac{h(l,n)}{S_n G_{l,\frac{n-2}{2}}(1)}$. $h(l,n) = (2l+n-2)\frac{(l+n-3)!}{(n-2)!l!}$	$\sum_{l=0}^{\infty} E_{l,n} G_{l,\frac{n-2}{2}}(\cos r_S) b_l(\tau),$ $\sum_{l=0}^{\infty} E_{l,n} a_l G_{l,\frac{n-2}{2}}(\cos r_S) b(\tau),$ (6.9a–d) $S_n^{-1}\left[1 - 2b(\tau)\cos r_S + b^2(\tau)\right]^{-\frac{n}{2}}\left[1 - b^2(\tau)\right],$ $S_n^{-1}\left[1 - 2a\cos r_S + a^2\right]^{-\frac{n}{2}}\left[1 - a^2\right]b(\tau).$

known covariance models, and (b) techniques that generate covariance models starting from properly selected functions. The advantages of group (a) over (b) are rather obvious. Group (a) techniques directly use the already existing families of covariance models, whereas in group (b) before the techniques can be used it is required that some special functions or measures are selected that satisfy the specified conditions (which may limit considerably the number of candidate functions and measures). The class of models that can be potentially produced by the group (a) techniques is larger than that produced by group (b). The techniques nos 1, 4, and 8 are, perhaps, the easiest to implement, since they involve simple or direct operators.

Technique no. 1 of Table 6.1 (Christakos, 1984b, 1992), which is essentially an ST technique with the specific feature of involving the simplest ST operators, starts by selecting any of the known covariance model $c_{X,n-2}(\boldsymbol{h}, \tau)$ that are valid in a lower-dimensionality domain $R^{n-2,1}$, and then constructs new space–time covariance models in the higher-dimensionality domain $R^{n,1}$ using Eqs. (6.1a–b).[9] These new models are also valid in the $R^{n-1,1}$ and $R^{n-2,1}$ domains. In other words, this is a straightforward approach that takes advantage of existing links between valid covariance models in different dimensions. Thus, the gain here lies in easiness of implementation and mathematical treatment.

Example 6.1

Assume that the space–time covariance function of a natural attribute satisfies the following PDE in $R^{1,1}$,

$$\begin{aligned} &\frac{\partial}{\partial \tau} c_X(r,\tau) + k(r) c_X(r,\tau) = 0, \\ &c_X(r,0) = c_0 \lambda(r), \end{aligned} \tag{6.10a–b}$$

where, based on the physics of the situation (say, first-order irreversible reaction), the following conditions apply: $k(r)$ is a known, monotonically increasing function on $[0, \infty)$, with $k(0) = 0$, and $k(r) \to \infty$ as $r \to \infty$, and $\lambda(r)$ is a function such that $\int_0^\infty dr \lambda(r) r = 1$. The solution of Eq. (6.10a–b) is a covariance model in $R^{1,1}$ of the form

$$c_X(r,\tau) = c_0 \lambda(r) e^{-k(r)\tau}. \tag{6.11a}$$

Starting from Eq. (6.11a) and using Eq. (6.1a), a new covariance model can be easily constructed in $R^{3,1}$ as

$$c_X(r,\tau) = \frac{c_0}{2\pi r}\left[\tau \lambda(r) \frac{d}{dr} k(r) - \frac{d}{dr}\lambda(r)\right] e^{-k(r)\tau}, \tag{6.11b}$$

which is also a valid model in $R^{2,1}$. Additional models can be constructed in spaces of different dimensionality starting from Eq. (6.11a) and using other types of ST.

Example 6.2

Given the covariance model $c_{X,1}(r,\tau) = \frac{c_0}{r\tau} \sin(r\tau)$ in $R^{1,1}$, Eq. (6.1a) with $n = 3$ gives the model

$$c_X(r,\tau) = c_{X,3}(r,\tau) = \frac{c_0}{2\pi r^2}\left[\frac{1}{r\tau} \sin(r\tau) - \cos(r\tau)\right] \tag{6.12}$$

[9] The corresponding expressions of Eqs. (6.1a–b) in terms of variogram functions can be found in Christakos (1984b).

in $R^{3,1}$. Also, given the model $c_{X,1}(r,\tau) = c_0 \cos(r\tau)$ in $R^{1,1}$, Eq. (6.1b) with $n = 3$ gives the covariance model

$$c_X(r,\tau) = c_{X,3}(r,\tau) = \frac{c_0}{r\tau}\sin(r\tau) \tag{6.13}$$

in $R^{3,1}$. The above covariance models are also valid in $n = 1$ and two dimensions.

The Technique no. 2, which is sometimes called the monotone function technique (Gneiting, 2002), relies on the adequate choice of the functions φ and ψ in Eq. (6.2) of Table 6.1.[10] Then, Eq. (6.2) is used to construct new covariance models in $R^{n,1}$. Specifically, the technique leads to a class of STHS covariance models, by selecting a completely monotone function, $\varphi(u)$, $u \geq 0$, and a positive function, $\psi(u)$, $u \geq 0$, with completely monotone derivative. A list of φ and ψ functions and the corresponding covariance models can be found in Gneiting (2002), although their interpretation and potential connections with the essentials of the phenomenon the covariance represents are not considered. One must be aware though that the covariance models generated by this technique are fully symmetric, which may be an issue given that this kind of covariance symmetry is not often realized in practice (the readers may also recall that a covariance that belongs to the limited class of separable covariances is also a fully symmetric covariance but the converse is not necessarily valid). Also, Stein (2005) has noticed that the covariance models generated by the monotone function approach cannot be smoother along their axes than at the space–time origin.

Example 6.3

If one chooses the functions $\varphi(u) = e^{-cu^{\gamma}}$ ($c > 0$, $\gamma \in (0, 1]$), and $\psi(u) = (au^{\alpha} + 1)^{\beta}$ ($a, c > 0$, $\alpha, \beta, \gamma \in (0, 1]$), Eq. (6.2) gives the covariance model

$$c_X(\boldsymbol{h},\tau) = c_0\left(a\tau^{2\alpha} + 1\right)^{-\frac{\beta n}{2}} e^{-c|\boldsymbol{h}|^{2\gamma}(a\tau^{2\alpha}+1)^{-\beta\gamma}} \tag{6.14}$$

in $R^{n,1}$ with $\tau \geq 0$. If one chooses $\varphi(u) = (cu^{\alpha} + 1)^{-\beta}$ and the same $\psi(u)$ as above, Eq. (6.2) gives

$$c_X(\boldsymbol{h},\tau) = c_0\left(a\tau^{2\alpha} + 1\right)^{-\beta}\left(1 + \frac{c|\boldsymbol{h}|^{2\alpha}}{(a\tau^{2\alpha} + 1)^{\alpha\beta}}\right)^{-\beta} \tag{6.15}$$

in $R^{2,1}$. Although the functions φ and ψ above were selected in a purely technical way (i.e., to satisfy a set of formal restrictions), it should be worth-investigating ways to select them on the basis of physical considerations. So, the φ and ψ functions of the covariance model no. 17 (Table 5.1 of Chapter VIII) used by Monin and Yaglom (1971) in atmospheric studies are physically determined functions.

A relevant closed-form technique is the so-called Bernstein function technique (e.g., Porcu et al., 2006). This is the Technique no. 3 of Table 6.1, see Eq. (6.3), which can be used to construct a class of nonseparable covariance models in $R^{3,1}$, as follows. The ψ_1 and ψ_2 are selected that are Bernstein functions,[11] variograms (not vanishing at the origin), or increasing and concave functions on $[0, \infty)$. Then, Eq. (6.3) is used to obtain new covariance models in $R^{3,1}$. The method can be extended to $R^{n,1}$.

[10]These models have been criticized, *inter alia*, by Kent et al. (2011) that they do not satisfy the monotonicity property, i.e., for fixed τ the $c_X(r, \tau)$ should decrease with r and for fixed r it should decrease with τ.

[11]That is, positive functions whose first derivative is completely monotonic.

Example 6.4

Using Eq. (6.3), Fernández-Avilés et al. (2011) obtained the space anisotropic time stationary covariance model

$$c_X(\boldsymbol{h},\tau)=\begin{cases}1 & \text{if } \boldsymbol{h}=\boldsymbol{0} \text{ or } \tau=0\\ \left(1-e^{-\frac{|h_2|}{\psi_1\left(h_1^2\right)}-\frac{|h_3|}{\psi_2\left(\tau^2\right)}}\right)\dfrac{\psi_1\left(h_1^2\right)^{\frac{1}{2}}\psi_2\left(\tau^2\right)^{\frac{1}{2}}}{h_2^2\psi_2\left(\tau^2\right)+h_3^2\psi_1\left(h_1^2\right)} & \text{otherwise}\end{cases} \tag{6.16}$$

in $R^{3,1}$, where ψ_1 and ψ_2 are Bernstein functions, variograms (with nugget effect) or increasing concave functions as indicated earlier.

The Technique no. 4 of Table 6.1 relies on a fundamental property of the class C_{n+1} of space–time covariances (Section 2.2 of Chapter IV), namely, that C_{n+1} is closed under addition, multiplication, and passages to the limit. Specifically, the linear superposition technique, Eq. (6.4a), is based on the well-known result that linear combinations of simple covariance models are also valid covariance models. Any known space–time covariance models $c_{X,i}(\boldsymbol{h},\tau)$ or any combination of purely spatial $c_X(\boldsymbol{h})$ and purely temporal models $c_X(\tau)$ are first selected. Then, the covariance models are constructed in $R^{n,1}$ by means of Eqs. (6.4a–b). In particular, the new covariance model $c_X(\boldsymbol{h},\tau)$ in Eq. (6.4a) shares some of the features of the component models $c_{X,i}(\boldsymbol{h},\tau)$ $(i=1,\ldots,N)$. However, the $c_X(\boldsymbol{h},\tau)$ can be a nonseparable covariance model even if the components $c_X(\boldsymbol{h},\tau)$ are space–time separable models. Put differently, the simple but powerful statement of Eq. (6.4a) is that one can combine the models discussed in previous sections to produce new covariance models in space–time. Depending on the weights, λ_i, one can enhance or reduce the effect of the component models on the final covariance model of Eq. (6.4a). Similar comments are valid for the covariance model generated by Eq. (6.4b). Under certain conditions, a linear combination of the same covariance models but with different space–time arguments can yield valid covariance models. And the same is true for the product of a finite number of covariance functions $c_{X,i}(\boldsymbol{h},\tau)$ $(i=1,\ldots,N)$.

Example 6.5

The application of Eq. (6.4a) leads to the power space–time covariance model in $R^{n,1}$,

$$c_X(\boldsymbol{h},\tau)=\left(1+|\boldsymbol{h}|^{\alpha}+\tau^{\beta}\right)^{-0.5(n+1)}, \tag{6.17}$$

where $\alpha,\beta>0$, $\tau\geq 0$. Also, the linear combination of the same covariance model but with different arguments can lead to new space–time covariance models. For illustration, given a valid STHS covariance model $c_X(\boldsymbol{h},\tau)$ in $R^{n,1}$, the functions

$$\begin{aligned}&c_X(\boldsymbol{h}+\boldsymbol{h}',\tau+\tau')+c_X(\boldsymbol{h}-\boldsymbol{h}',\tau-\tau')-2[c_X(\boldsymbol{h},\tau)+c_X(\boldsymbol{h}',\tau')-c_X(\boldsymbol{0},0)],\\&c_X(\boldsymbol{h}-\boldsymbol{h}',\tau-\tau')-c_X(\boldsymbol{h}+\boldsymbol{h}',\tau+\tau'),\end{aligned} \tag{6.18a–b}$$

are also covariance models in $R^{n,1}$ (Ma, 2003b).

Example 6.6

Let $c_X(\boldsymbol{h},\tau)$ be a valid covariance model in $R^{n,1}$. Then, the following are valid covariance models,

$$\begin{aligned}c_X(\boldsymbol{h},\tau)&=\left|c_X(\boldsymbol{h},\tau)\right|^2,\\c_X(\boldsymbol{h},\tau)&=c_X(a\boldsymbol{h},a\tau),\end{aligned} \tag{6.19a–b}$$

where a is a real coefficient.

The techniques nos 5–7 were proposed by Ma (2002b, 2003b). Eq. (6.5) generates STHS covariance models by obtaining the Laplace transform $L_T\left(\breve{x}_1, \breve{x}_2\right)$ of a nonnegative random vector (x_1, x_2), assuming it exists, and then substitute $\breve{x}_1$ and $\breve{x}_2$ with $\gamma_1(\boldsymbol{h})$ and $\gamma_2(\tau)$, respectively. To illustrate this technique, two cases are examined in the example below.

Example 6.7

If $L_T\left(\breve{x}_1, \breve{x}_2\right) = \frac{1-e^{-\left(\breve{x}_1+\breve{x}_2\right)}}{\breve{x}_1+\breve{x}_2}$, then the generated covariance model is

$$c_X(\boldsymbol{h},\tau) = L_T(\gamma_1(\boldsymbol{h}), \gamma_2(\tau)) = \begin{cases} 1 & \text{if } \boldsymbol{h}=0,\ \tau=0 \\ \dfrac{1-e^{-\gamma_1(\boldsymbol{h})-\gamma_2(\tau)}}{\gamma_1(\boldsymbol{h})+\gamma_2(\tau)} & \text{otherwise.} \end{cases} \tag{6.20}$$

If $L_T\left(\breve{x}_1, \breve{x}_2\right) = \frac{e^{-\breve{x}_2}-e^{-\breve{x}_1}}{\breve{x}_1-\breve{x}_2}$, then

$$c_X(\boldsymbol{h},\tau) = L_T(\gamma_1(\boldsymbol{h}), \gamma_2(\tau)) = \begin{cases} e^{-\gamma_1(\boldsymbol{h})} & \text{if } \gamma_1(\boldsymbol{h}) = \gamma_2(\tau) \\ \dfrac{e^{-\gamma_1(\boldsymbol{h})} - e^{-\gamma_2(\tau)}}{\gamma_1(\boldsymbol{h})-\gamma_2(\tau)} & \text{otherwise.} \end{cases} \tag{6.21}$$

Other cases can be considered in the same manner.

The Technique no. 6 constructs space–time covariance models by means of a Gaussian PDF. This technique can be combined with integral techniques, like those to be discussed in the following section, to generate covariance models, although more analytical work needs to be done to obtain specific examples of models that are useful in applications. The Technique no. 7 relies on continuous and completely monotone functions ℓ on R^1_+, i.e., $(-1)^k \frac{d^k\ell(u)}{d^k u} \geq 0$, where $u > 0$, $k \in Z_+$.

The Technique no. 8 is particularly attractive, because it generates a large class of new space–time covariance models by simple differentiations of valid STIS covariance models, see Eqs. (6.8a–c), and the technique's development is based on physical considerations (the generated space–time covariance models may represent the joint space–time morphology of meteorological attributes, like atmospheric pressure and geostrophic winds). For more details concerning the implementation of this technique, the readers may want to revisit Section 2 of Chapter VIII.

Example 6.8

Let $c_1(r,\tau) = c_0 e^{-\frac{r^2}{a^2}-\frac{\tau^2}{b^2}}$ be a covariance model in $R^{2,1}$. Then, for $i=j=1$, and $h_1 = r$, $h_2 = \tau = 0$, Eqs. (6.8a) gives the space–time covariance model

$$c_X(r,\tau) = \frac{2c_0}{a^2}\left(1-\frac{2r^2}{a^2}\right)e^{-\frac{r^2}{a^2}-\frac{\tau^2}{b^2}}. \tag{6.22}$$

Similarly, Eq. (6.8c) leads to

$$c_X(r,\tau) = \frac{16c_0}{a^4}\left[2+\frac{r^4}{a^4}-4\frac{r^2}{a^2}\right]e^{-\frac{r^2}{a^2}-\frac{\tau^2}{b^2}}. \tag{6.23}$$

These are valid STIS covariance models. Although, due to the specific choice of c_1, h_1 and h_2 the above two covariance models happen to be space–time separable, for different initial choices the technique can construct nonseparable models, in general.

Lastly, the Technique no. 9 of Table 6.1 is used for spherical random fields (see, Section 5 of Chapter X). It generates space–time covariance models on $S^{n-1} \times T$, where S^{n-1} denotes a spherical surface (e.g., the surface of the earth) and T is the time domain, as usual. Notice that Eq. (6.9c) is obtained form Eq. (6.9a) by using the Gegenbauer relationship (Mokljacuk and Jadrenko, 1979)

$$\sum_{l=0}^{\infty} \frac{h(l,n)}{G_{l,\frac{n-2}{2}}(1)} G_{l,\frac{n-2}{2}}(u) w^l = \frac{1-w^2}{(1-2wu+w^2)^{\frac{n}{2}}}, \tag{6.24}$$

where $|w| < 1$, $n > 2$, and $w = b(\tau)$.

Example 6.9

By choosing any valid temporal covariance models $b(\tau)$ and $b_l(\tau)$ from Table 4.1 of Chapter VIII and inserting it into Eqs. (6.9a–d), new space–time covariance models on $S^{n-1} \times T$ can be constructed.

An important class of space–time covariance models for heterogeneous random fields, in general, can be constructed by the *generalized decomposition* technique. In particular, one can use the decomposition Eq. (3.11) of Chapter XIII, i.e.,

$$c_X(\boldsymbol{p}, \boldsymbol{p}') = \kappa_X(\Delta \boldsymbol{p}) + \eta_{v,\mu}(\boldsymbol{p}, \boldsymbol{p}'), \tag{6.25a}$$

where $\Delta \boldsymbol{p} = \boldsymbol{p}' - \boldsymbol{p} = (\boldsymbol{h}, \tau)$, κ_X is the generalized spatiotemporal covariance of v/μ-order, and $\eta_{v,\mu}$ are polynomials with variable coefficients of degree v in $\boldsymbol{s}$, $\boldsymbol{s}'$, and degree μ in t, t'. Starting from known κ_X models and standard $\eta_{v,\mu}$ polynomials, new space–time c_X models (nonhomogeneous/nonstationary, in general) can be derived. Moreover, STHS covariances can be obtained from Eq. (3.10) of Chapter XIII, i.e.,

$$c_X(\Delta \boldsymbol{p}) = (-1)^{v+\mu} \nabla^{(2v+2,2\mu+2)} \kappa_X(\Delta \boldsymbol{p}), \tag{6.25b}$$

given the κ_X model. In fact, given the large number of available κ_X models, Eqs. (6.25a–b) offer an efficient and quick way to construct space–time heterogeneous, $c_X(\boldsymbol{p}, \boldsymbol{p}')$, or STHS, $c_X(\Delta \boldsymbol{p})$, covariance models. Yet, the generalized decomposition technique is rather under-appreciated, and the reason is, perhaps, the apparent complexity of the theory of random functionals.

Example 6.10

As an illustration, consider the generalized covariance model of Eq. (3.22) of Chapter XIII, in which case Eq. (6.25a) gives a space–time nonhomogeneous/nonstationary covariance model of the form

$$c_X(\boldsymbol{s}, t; \boldsymbol{s}', t') = \sum_{\rho=0}^{2v+1} \sum_{\zeta=0}^{2\mu+1} (-1)^{s(\rho)+s(\zeta)} a_{\rho\zeta} |\boldsymbol{s} - \boldsymbol{s}'|^{\rho} |t - t'|^{\zeta} + \eta_{v/\mu}(\boldsymbol{s}, t) \eta_{v/\mu}(\boldsymbol{s}', t'), \tag{6.26}$$

where $\eta_{v/\mu}(\boldsymbol{s}, t)$ and $\eta_{v/\mu}(\boldsymbol{s}', t')$ are polynomials of degree v in $\boldsymbol{s}$ (resp. $\boldsymbol{s}'$) and degree μ in t (resp. t'). Also, staring with the generalized covariance model of Eq. (3.21) of Chapter XIII, Eq. (6.25b) gives $c_X(r, \tau) = a\delta(r)\delta(\tau)$, where a is a constant.

I will conclude this section by briefly mentioning two more closed-form techniques of space–time covariance model construction. The first technique, which is discussed in Exercise XVI.25, is the simpler, since all it requires is the availability of valid univariate PDF. The second technique is based on the idea of constructing space–time covariance models by means of physically motivated equations directly relating the PDF with the covariance function of the phenomenon.

Example 6.11
Consider a diffusion process in $R^{1,1}$, where the PDF $f_X(h, \tau)$ of the process and the corresponding covariance function $c_X(h, \tau)$ are related by

$$c_X(h, \tau) = \frac{\upsilon^2 \tau_0}{\lambda - 1} \frac{\partial f_X(h, \tau)}{\partial \tau}, \tag{6.27}$$

where $\lambda > 2$ and υ, $\tau_0 > 0$. By selecting a PDF $f_X(h, \tau)$, Eq. (6.27) can be used to directly derive space–time covariance models $c_X(h, \tau)$.

7. INTEGRAL REPRESENTATION TECHNIQUES

This section presents a group of techniques of space–time covariance model construction based on some kind of integral representation involving properly selected input functions or measures. To gain an understanding about how these techniques work, five of them are listed in Table 7.1. Their implementation will be investigated with the help of specific examples.

Technique no. 1 of Table 7.1 starts by selecting some suitable measures $\varphi(u)$ and $\psi(u)$ (the conditions satisfied by these measures are also shown in Table 7.1) and any valid purely spatial and purely temporal covariance models.

Table 7.1 Integral Representation Techniques for Generating Space–Time Covariance Models

No.	Input Parameters	Space–Time Covariance Model in $R^{n,1}$
1	φ, ψ: Nondecreasing measures. $a, b > 0, A < \infty$. $\int_0^a d\phi(u) < \infty$, $\int_0^b d\psi(v) < \infty$. c_1, c_2: Spatial and temporal covariances.	$A \int_0^a \int_0^b d\phi(u) d\psi(v) e^{uc_1(\lvert\boldsymbol{h}\rvert) + vc_2(\tau)}$. (7.1) $(r = \lvert\boldsymbol{h}\rvert)$
2	$\widetilde{c}_\ell$, $\widetilde{c}_\eta$: Spectra in $R^{1,1}$.	$\int_0^\infty dk \cos(kr) \left[\frac{h_i^2}{r^2} \widetilde{c}_l(k, \tau) + \left(1 - \frac{h_i^2}{r^2}\right) \widetilde{c}_\eta(k, \tau)\right]$. $(i = 1, 2, 3;\ n = 3)$ (7.2)
3	ϕ: Nonnegative finite measure on set $\Phi \neq \varnothing$. c_1, c_2: Spatial and temporal covariances for all $u \in \Phi$. $\int_\Phi du c_1(0, u) c_2(0, u) < \infty$.	$\int_\Phi d\phi(u) c_1(\boldsymbol{h}, u) c_2(\tau, u)$. (7.3)
4	μ: Measure on space U. $c_{\boldsymbol{u}}$: Integrable function on $V \subset U$ for each $(\boldsymbol{p}, \boldsymbol{p}')$.	$\int_V \mu(d\boldsymbol{u}) c_{\boldsymbol{u}}(\boldsymbol{p}, \boldsymbol{p}' \cdot \boldsymbol{u})$ (7.4)
5	ϕ: Measure on sets $(-\infty, +\infty) \times (0, +\infty)$. $\Theta(u) = (2\pi)^{\frac{n}{2}} u^{1-\frac{n}{2}} J_{\frac{n}{2}-1}(u)$	$\int_0^\infty \int_{-\infty}^\infty d\varphi(k, \omega) \Theta(\lvert\boldsymbol{h}\rvert k) e^{-i\omega\tau}$, $\int_0^\infty \int_{-\infty}^\infty dk\, d\omega k^{n-1} \Theta(\lvert\boldsymbol{h}\rvert k) \cos(\omega\tau) \widetilde{c}_X(k, \omega)$. (7.5a–b)

Table 7.1 Integral Representation Techniques for Generating Space–Time Covariance Models—cont'd

No.	Input Parameters	Space–Time Covariance Model in $R^{n,1}$	
6	$\widetilde{c}_\ell, \widetilde{c}_\eta$: Spectra in $R^{1,1}$.	$2\int_0^\infty dk \cos(kr)\widetilde{c}_l(k,\tau)$, $2\int_0^\infty dk \cos(kr)\widetilde{c}_\eta(k,\tau)$, $(n=3)$	(7.6a–b)
7	$\zeta(u,v) \in L_2$, $\iint dudv\lvert\zeta(u,v)\rvert^2 < \infty$, $A = \left(\iint dudv\lvert\zeta(u,v)\rvert^2\right)^{\frac{1}{2}}$.	$A^{-1}\iint dudv\zeta(U_{r,\tau}(u,v))\zeta^*(u,v)$	(7.7a–b)

Example 7.1

Let us choose the functions $\phi(u) = \delta(u - 0.5a)$ and $\psi(v) = \delta(v - 0.5b)$. Then, Eq. (7.1) gives the space–time covariance model

$$c_X(\boldsymbol{h},\tau) = e^{0.5ac_1(|\boldsymbol{h}|)+0.5bc_2(\tau)}, \tag{7.8}$$

which is of the form of Eq. (8.2c) also generated by another technique, see later.

Technique no. 2 of Table 7.1 is a physically motivated technique for constructing STIS covariance models. One starts by selecting wavenumber-time spectra $\widetilde{c}_\ell(k,t)$ and $\widetilde{c}_\eta(k,t)$ in $R^{1,1}$, and then derives covariance models in $R^{3,1}$ by means of Eq. (7.2). Clearly, the covariance $c_X(\boldsymbol{h},\tau)$ is not isostationary, since it depends on the direction of the lag coordinate h_i. An example of the implementation of this technique is discussed in Exercise XVI.9.

Technique no. 3 of Table 7.1 involves the integration of the product of a purely spatial and a purely temporal covariance. By selecting a suitable measure (the conditions the measure must obey are also shown in Table 7.1), and two valid (purely spatial and purely temporal) covariances such that the integral condition shown in Table 7.1 is satisfied, one can derive new covariance models in $R^{n,1}$ using Eq. (7.3).

Example 7.2

Select $c_1(r,u) = e^{-ur^{\lambda_1}a_1^{-2}}$, $r = |\boldsymbol{h}|$, and $c_2(\tau,u) = e^{-u\tau^{\lambda_2}a_2^{-2}}$, and let $\phi(u) = \Gamma(\lambda_3)u^{\lambda_3-1}e^{-u}$ be a gamma function on $\Phi = [0, \infty)$. Then, Eq. (7.3) gives the space–time covariance Model no. 26 of Table 5.1 of Chapter VIII (e.g., De Iaco et al., 2002).

Technique no. 4 of Table 7.1, which has is roots in the work of Matern (1960) and others, directly involves integrable covariance models in an appropriate space. In particular, after selecting a measure μ on a space U and an integrable function c_u satisfying the conditions shown in Table 7.1, one can derive new covariance models in $R^{n,1}$ using Eq. (7.4).

Example 7.3

In Eq. (7.4) let,

$$c_u(\mathbf{p}, \mathbf{p}' \cdot u) = c_u(r - \upsilon\tau, u) = \sigma^2 e^{-\frac{(r-\upsilon\tau)^2}{u^2}}, \tag{7.9a}$$

$\upsilon \in R^1$, and $\mu(du) = \phi(u)du$, so that

$$c_X(r - \upsilon\tau) = \sigma^2 \int_0^\infty du\, e^{-\frac{(r-\upsilon\tau)^2}{u^2}} \phi(u), \tag{7.9b}$$

where the function $\phi(u)$ must be such that $\int_0^\infty du\phi\,(u) = 1$. If we choose $\phi(u) = \frac{1}{\sqrt{\pi a}} e^{-\left(\frac{u}{2a}\right)^2}$ $(a \in R^1)$, Eq. (7.9a) yields the covariance model

$$c_X(r - \upsilon\tau) = \sigma^2 e^{-\frac{r-\upsilon\tau}{a}}; \tag{7.10a}$$

whereas for $\phi(u) = \delta(u - a)$, the same equation gives the covariance model

$$c_X(r - \upsilon\tau) = \sigma^2 e^{-\frac{(r-\upsilon\tau)^2}{a^2}}, \tag{7.10b}$$

which is a nonseparable space–time covariance model.

The theoretical support to Technique no. 5 of Table 7.1 is the spectral theory of random fields discussed in Yadrenko (1983). Furthermore, a comprehensive technique for constructing STIS covariance models is given by Technique no. 6 leading to the space–time covariance models of $c_{\ell\ell}(r, \tau)$ and $c_{\eta\eta}(r, \tau)$. Another way that seems to impose less strict conditions on the selected functions is proposed by Technique no. 7 (the input to this technique is any function $\zeta(u, v) \in L_2$ such that the conditions shown in Table 7.1 are satisfied, and the output is a new covariance model in $R^{n,1}$). Lastly, an integral technique known as the *spatial extension technique* of space–time covariance model construction is discussed in Exercise XVI.23 of the chapter "Exercises-Comments."

As it happens with other covariance model construction techniques, in practice some of the integrals of Table 7.1 will need to be evaluated computationally. In this respect, the generated covariance models should not be solely data-dependent, but they must also conform to the physical constraints concerning the phenomenon of interest.

8. SPACE–TIME SEPARATION TECHNIQUES

A group of covariance model construction techniques have been developed, the inputs to which are space–time separable models. A few of these techniques are listed in Table 8.1. The purely spatial, $c_{X(1)}$, and the purely temporal, $c_{X(2)}$, models that can be used as inputs in the techniques of Table 8.1 can be chosen among the existing classes of such models (see, e.g., Table 4.1 of Chapter VIII). While one must be aware that the introduction of covariance separability may hide some important features of the phenomenon being described, such as the geometrical and topological features, the derived models are, in general, space–time nonseparable.

Table 8.1 Separable Techniques for Generating Space–Time Covariance Models in $R^{n,1}$

Input Parameters	Generated Space–Time Covariance $c_X(h, \tau)$	
$c_{X(1)}$, $c_{X(2)}$: Valid covariance models. $\mu_i \geq 0$, $M = \sum_{i=0}^{\infty} \mu_i < \infty$.	$M^{-2} \sum_{i,j=0}^{\infty} \mu_i \mu_j c_{X(1)}(\boldsymbol{h})^i c_{X(2)}(\tau)^j$	(8.1)
$c_{X(1)}$, $c_{X(2)}$: Valid covariance models. $\alpha_i \geq 0$ $(i = 1, 2, 3)$. $\alpha, \beta, b_1, b_2 > 0$ $a_1, a_2 > 1$; $A, B > 0$.	$\alpha_1 c_{X(1)}(\boldsymbol{h}) c_{X(2)}(\tau) + \alpha_2 c_{x(1)}(\boldsymbol{h}) + \alpha_3 c_{x(2)}(\tau)$, $\dfrac{A}{\left[a_1 a_2 - a_1 c_{X(2)}(\tau) - a_2 c_{X(1)}(\lvert\boldsymbol{h}\rvert) + c_{X(1)}(\lvert\boldsymbol{h}\rvert) c_{X(2)}(\tau)\right]}$, $B e^{\alpha c_{X(1)}(\lvert\boldsymbol{h}\rvert) + \beta c_{X(2)}(\tau)}$.	(8.2a–c)
$c_{X(1)}$, $c_{X(2)}$: Valid covariance models.	$c_{X(2)}(\tau) \prod_{i=1}^{n} c_{X(1)}(h_i)$	(8.3)
$c_{X(1)}$: Homogeneous covariance of $\boldsymbol{h} \in R^n$ for every $\boldsymbol{k} \in R^n_+$, and a measurable function of $\boldsymbol{k}$ for every $\boldsymbol{h}$. $c_{X(2)}$: Stationary covariance of $\tau \in R^1$ for every $\boldsymbol{k} \in R^n_+$, and a measurable function of $\boldsymbol{k}$ for every τ. μ: Nonnegative bounded measure on R^n_+.	$\int_{R^n_+} d\mu(\boldsymbol{k}) c_{X(1)}(\boldsymbol{h}, \boldsymbol{k}) c_{X(2)}(\tau, \boldsymbol{k})$	(8.4)
$\gamma_{X(1)}(\boldsymbol{h})$, $\gamma_{X(2)}(\tau)$ L_T: Laplace transform.	$L_T\left[\gamma_{X(1)}(\boldsymbol{h}), \gamma_{X(2)}(\tau)\right]$	(8.5)

Example 8.1

Using the techniques of Eqs. (8.1) and (8.2a–c) of Table 8.1, or combinations thereof, a variety of nonseparable space–time covariance models in $R^{n,1}$ of the following form can be constructed,

$$
c_X(\boldsymbol{h}, \tau) = \begin{cases}
\left(\dfrac{\theta}{1 - (1 - \theta) c_{X(1)}(\boldsymbol{h}) c_{X(2)}(\tau)} \right)^{\beta}, \\[2ex]
\dfrac{1}{\log \theta} \log \left[1 - (1 - \theta) c_{X(1)}(\boldsymbol{h}) c_{X(2)}(\tau) \right], \\[2ex]
e^{-\alpha - \beta - \gamma + \alpha c_{X(1)}(\boldsymbol{h}) + \beta c_{X(2)}(\tau) + \gamma c_{X(1)}(\boldsymbol{h}) c_{X(2)}(\tau)}, \\[2ex]
\left[1 + \alpha + \beta + \gamma - \alpha c_{X(1)}(\boldsymbol{h}) - \beta c_{X(2)}(\tau) - \gamma c_{X(1)}(\boldsymbol{h}) c_{X(2)}(\tau) \right]^{-\zeta^{-1}}, \\[2ex]
\dfrac{1}{\log \theta} \log \left[1 - \alpha c_{X(1)}(\boldsymbol{h}) - \beta c_{X(2)}(\tau) - \gamma c_{X(1)}(\boldsymbol{h}) c_{X(2)}(\tau) \right],
\end{cases} \tag{8.6a–e}
$$

where $\theta \in (0, 1)$, $\alpha, \beta, \zeta > 0$, $\gamma \in [0, \min(\alpha, \beta)]$, and the coefficients $\mu_{ij} = \frac{\mu_i \mu_j}{M^2}$ of Eq. (8.1) are assumed to be probability distributions of a nonnegative bivariate discrete random vector (Ma, 2002a). Lets choose the input models $c_{X(1)}(\boldsymbol{h}) = e^{-b_1|\boldsymbol{h}|^2}$ in R^n, and $c_{X(2)}(\tau) = e^{-b_2\tau^2}$. Then Eq. (8.6c) with $\alpha = \beta = 1$ and $\gamma = 0.5$ gives the covariance model

$$c_X(\boldsymbol{h}, \tau) = e^{e^{-b_1|\boldsymbol{h}|^2} + e^{-b_2\tau^2} + \frac{1}{2}e^{-b_1|\boldsymbol{h}|^2 - b_2\tau^2} - \frac{5}{2}},$$

in $R^{n,1}$, where $b_1, b_2 > 0$.

Example 8.2

It would be interesting to try in applications space–time covariance models that are obtained in a sequential manner, i.e.,

$$c_X(\boldsymbol{h}, \tau) = e^{e^{e^{-b_1|\boldsymbol{h}|^2}} + e^{e^{-b_2\tau^2}}}, \tag{8.7}$$

where $b_1, b_2 > 0$, and so on. In fact, the following notation may be used $c_X(\boldsymbol{h}, \tau) = e_{[k]}^{(-b_1|\boldsymbol{h}|^2, -b_2\tau^2)}$, where k denotes the number of exponents involved (in the case of Eq. (8.7), obviously $k = 3$).

Example 8.3

Using Eq. (8.3) of Table 8.1, space–time covariance models in $R^{n,1}$ can be constructed, as follows,

$$\begin{aligned} c_X(\boldsymbol{h}, \tau) &= e^{-a_0\tau} \prod_{i=1}^{n} e^{-a_i h_i}, \\ c_X(\boldsymbol{h}, \tau) &= (1 - a_0\tau) \prod_{i=1}^{n} (1 - a_i h_i)_+, \end{aligned} \tag{8.8a–b}$$

where $\alpha_i \geq 0$ $(i = 0, 1, \ldots, n)$. Moreover, valid covariance models are obtained by integrating the above models, such as the cubic model on the unit cube

$$c_X(\boldsymbol{h}, \tau) = \left(1 - a_0\tau^2 + b_0\tau^3\right) \prod_{i=1}^{n} \left(1 - a_i h_i^2 + b_i h_i^3\right), \tag{8.8c}$$

which, in general, are not STIS models.

Example 8.4

The general probability density technique, see Eqs. (2.1a–b) earlier, has a direct application in the case of space–time separability. Consider in $R^{1,1}$, (1) a zero mean Gaussian PDF

$$f_X(k) = \frac{1}{\sqrt{2\pi}\sigma} e^{-\frac{k^2}{2\sigma^2}}; \tag{8.9a}$$

and (2) a uniform density

$$f_X(\omega) = \frac{1}{2b} \quad \text{if } \omega \in (-b, b), = 0 \text{ otherwise.} \tag{8.9b}$$

In this case, letting $f_X(k, \tau) = f_X(k) f_X(\tau)$, and using Eq. (2.1b), the corresponding space–time covariance for $n = 1$ is obtained

$$c_X(r,\tau) = c_0 \begin{cases} \dfrac{\sin(rb)}{rb} e^{-\frac{\sigma^2\tau^2}{2}} & (r,\tau \neq 0) \\ e^{-\frac{\sigma^2\tau^2}{2}} & (r = 0) \end{cases} \tag{8.9c–d}$$

where b, $\sigma > 0$. The corresponding covariance models in $R^{n,1}$ can be constructed by using the ST technique (Section 4).

Example 8.5

Eq. (8.4) of Table 8.1 provides a means for constructing valid space–time covariance models in $R^{n,1}$. Some interesting special cases of Eq. (8.4) are as follows. If we let $c_{X(1)}(\boldsymbol{h}, \boldsymbol{k}) = \cos(\boldsymbol{k} \cdot \boldsymbol{h})$, Eq. (8.4) gives

$$c_X(\boldsymbol{h},\tau) = \int_{R^n_+} d\mu(\boldsymbol{k}) \cos(\boldsymbol{k}\cdot\boldsymbol{h}) c_{X(2)}(\tau,\boldsymbol{k}). \tag{8.10}$$

If $\gamma_{X(2)}(\tau, \boldsymbol{k})$ is a stationary variogram of $\tau \in R^1$ for every $\boldsymbol{k} \in R^n_+$, and a measurable function of $\boldsymbol{k}$ for every τ, then $c_{X(2)}(\boldsymbol{k},\tau) = e^{-\gamma_{X(2)}(\boldsymbol{k},\tau)}$, and

$$c_X(\boldsymbol{h},\tau) = \int_{R^n_+} d\mu(\boldsymbol{k}) \cos\,(\boldsymbol{k}\cdot\boldsymbol{h}) e^{-\gamma_{X(2)}(\tau,\boldsymbol{k})} \tag{8.11}$$

is a valid covariance model in $R^{n,1}$. If $\mu(\omega)$ is a nonnegative bounded measure on R_+, and $c_{X(1)}(\boldsymbol{h}, \omega)$ is a homogeneous covariance of $\boldsymbol{h} \in R^n$ for every $\omega \in R^1_{+,\{0\}}$, and a measurable function of ω for every $\boldsymbol{h}$, then

$$c_X(\boldsymbol{h},\tau) = \int_0^\infty d\mu(\omega) c_{X(1)}(\boldsymbol{h},\omega) \cos\,(\omega\tau) \tag{8.12}$$

is a valid model in $R^{n,1}$. If $\gamma_{X(1)}(\boldsymbol{h}, \omega)$ is a homogeneous variogram of $\boldsymbol{h} \in R^n$ for every $\omega \in R^1_{+,\{0\}}$, and a measurable function of ω for every $\boldsymbol{h}$, then

$$c_X(\boldsymbol{h},\tau) = \int_0^\infty d\mu(\omega) e^{-\gamma_{X(1)}(\boldsymbol{h},\omega)} \cos\,(\omega\tau) \tag{8.13}$$

is a valid model in $R^{n,1}$.

Example 8.6

In the covariance model construction technique based on Eq. (8.5), $\gamma_{X(1)}(\boldsymbol{h})$ and $\gamma_{X(2)}(\tau)$ are purely spatial (R^n) and purely temporal ($T \subseteq R^1$) variograms, respectively. Examples of this kind of models are models nos. 33–35 of Table 5.1 of Chapter VII.

Remark 8.1

As noted earlier, the product as well as the summation of two valid covariances are valid covariances. However, while the product of two correlation functions $\rho_{X(1)}(\boldsymbol{h})\rho_{X(2)}(\tau)$ is a valid correlation, this is not true for the summation of two correlation functions $\rho_{X(1)}(\boldsymbol{h}) + \rho_{X(2)}(\tau)$, since, clearly $\rho_{X(1)}(\boldsymbol{0}) + \rho_{X(2)}(0) = 2$, which cannot be true. The readers may recall that some more examples of space–time covariance models generated as the product or summation of valid ones were discussed in Chapter VII.

I conclude the study of separation techniques by noticing that most techniques of Table 8.1 may be seen as functions of two quantities, $\varphi(\boldsymbol{h})$ and $\psi(\tau)$, selected in a formal manner so that they satisfy certain mathematical conditions (e.g., the choices may be $\varphi(\boldsymbol{h}) = c_{X(1)}(\boldsymbol{h})$, or $= \gamma_{X(1)}(\boldsymbol{h})$; and $\psi(\tau) = c_{X(2)}(\tau)$, or $= \gamma_{X(2)}(\tau)$. Noticeably, covariance models of similar separable forms as those listed in Table 8.1 have been used in physical sciences, where φ and ψ were usually derived on the basis of physical considerations. Let us look at an example.

Example 8.7

In studies of the intensity of radiated sound, covariances of the Lighthill's stress tensor based on a variable separation of the form

$$c_X(\boldsymbol{h}, \tau) = \upsilon^4 \varphi\left(\frac{|\boldsymbol{h}|}{a}\right)\psi(\varpi\tau), \tag{8.14}$$

have been used (e.g., Proudman 1952), where υ, a, and ϖ denote, respectively, reference values of the turbulent velocity, length scale, and frequency. In the case of stationary turbulence they are constants, but in general they can be functions of time during the decay period. Physical guidance as to appropriate functional forms for φ and ψ are sometimes obtained by considering similar functions relating to the two-point velocity covariances. Among the suggested φ and ψ are $\varphi\left(\frac{|\boldsymbol{h}|}{a}\right) = e^{-2\left(\frac{r}{a}\right)^2}$ (which is in agreement with the observation that the dissipation range of wavenumbers and frequencies is finite), and $\psi(\varpi\tau) = e^{-\frac{1}{2}\pi\varpi^2\tau^2}$ (its physical merit is that it captures the characteristics of the lower frequencies and near the peak frequencies in the spectrum).

9. DYNAMIC FORMATION TECHNIQUE

In the dynamic formation techniques, basically a class of random fields is generated from growth and pattern formation processes in which there is a random element (e.g., a porous medium) and the spatiotemporal evolution is governed by a set of dynamic rules instead of a differential equation (for a collection of essays on models of growth and pattern formation, see Stanley and Ostrowsky, 1986).

Example 9.1

A nonseparable covariance of this type that originates from simulations of invasion percolation satisfies the dynamic scaling of Model no. 15 (Table 5.1 of Chapter VIII), where $r = |\boldsymbol{s} - \boldsymbol{s}_0|$and the function $g(x)$ peaks at $x \cong 1$ and scales asymptotically as a power-law $g(x) \sim x^{\alpha}$ $(x << 1)$, and $\sim x^{-b}$ $(x >> 1)$.[12]

[12] At first, one would think that the covariance Model 15 (Table 5.1 of Chapter VIII) is separable, since any power-law in $x = \frac{r^z}{\tau}$ can be expressed as the product of power laws in r and τ. However, separability breaks down because $g(x)$ has a cusp at $x = x_0$ (the point at which $g(x)$ peaks), and two branches with different exponents around the cusp: If τ is held fixed and r is increased so that x crosses over the cusp from the left to the right branch, the characteristic exponent of the spatial power law changes. Hence, Model 15 cannot be expressed as the product of two separable components.

Example 9.2

A special case is the Model no. 16 (Table 5.1 of Chapter VII) in $R^{2,1}$. Fractal models exhibit a power-law asymptotic decay of the correlations with a noninteger exponent. The models can be fractal either in space (in which case the asymptotic behavior refers to $r \to \infty$) or in time (in which case the asymptotic behavior refers to $\tau \to \infty$). The class of Model no. 13 in $R^{n,1}$ (Table 5.1 of Chapter VII) has fractal properties over a corresponding space−time range.

10. ENTROPIC TECHNIQUE

An entropic technique for deriving valid spatiotemporal covariance models by maximizing the relevant entropy function was discussed in Christakos (1992). Let $\{c_X(\boldsymbol{h},\tau), \widetilde{c}_X(\boldsymbol{\kappa},\omega)\}$ be the spatiotemporal covariance-spectral density pair. The entropic technique will go through the following steps:

Step 1: Suppose that the available knowledge about the attribute's spatiotemporal variability is expressed by the constraints

$$\int_{I_k}\int_{I_\omega} d\boldsymbol{k}d\omega e^{i(\boldsymbol{k}\cdot\boldsymbol{r}_\ell-\omega\tau_\ell)}\widetilde{c}_X(\boldsymbol{k},\omega) = c_X(\boldsymbol{h}_\ell,\tau_\ell) \tag{10.1}$$

for $\ell = 1,\ldots, L$, where $\boldsymbol{I}_k$ and I_ω are known wavevector and frequency intervals. The constraints of Eq. (10.1) express mathematically a rather common situation in practise, namely, covariance values are experimentally calculated for a set of space−time intervals ($\boldsymbol{h}_\ell$, τ_ℓ), $\ell = 1,\ldots, L$. Additional constraints may include knowledge regarding the behavior of $c_X(\boldsymbol{h}, \tau)$ at larger space−time lags (e.g., spatial and temporal ranges of influence etc.).

Step 2: Given these constraints, the problem of determining the shape of $c_X(\boldsymbol{h}, \tau)$ is converted into the equivalent one of determining $\widetilde{c}_X(\boldsymbol{k},\omega)$. The latter is obtained by maximizing an entropy function, such as

$$\varepsilon(\widetilde{c}_X) = \int_{I_k}\int_{I_\omega} d\boldsymbol{k}d\omega \, \log\widetilde{c}_X(\boldsymbol{k},\omega), \tag{10.2}$$

with respect to $\widetilde{c}_X(\boldsymbol{k},\omega)$ subject to conditions like those introduced by Eq. (10.1).

Example 10.1

First, consider the unconstrained case, where the only information available is the variance σ^2. Then, the solution of Eq. (10.2) corresponds to a pure nugget-effect covariance,

$$c_X(r,\tau) = c_0\delta(r)\delta(\tau). \tag{10.3}$$

The theory behind the entropic technique suggests that the pure nugget-effect model is the one that most honestly represents the given state of incomplete knowledge regarding spatial variability, without assuming anything else. Any other solution will necessarily take into consideration information not really available. Intuitively this implies that one needs more than just the variance to be able to derive sound conclusions about spatiotemporal correlation. So, assume that in addition to the variance one obtains information about the frequencies $\omega_1 = -\omega_0$ and $\omega_2 = \omega_0$. The entropic approach above will take into consideration the new information, and the corresponding covariance will be of the form

$$c_X(r,\tau) = c_0 \frac{\sin(k_0 r)\sin(\omega_0 \tau)}{k_0 \omega_0 r \tau}, \tag{10.4}$$

which is a space–time hole-effect kind of a model. More details can be found in Christakos (1992).

Other kinds of physical constraints that can be taken into consideration by the entropic technique are discussed in Section 13 below and in Section 5 of Chapter XIV.

11. ATTRIBUTE AND ARGUMENT TRANSFORMATION TECHNIQUES

In principle, there are two kinds of transformation techniques of space–time covariance model construction. In the first kind the transformed quantities are the space–time random field and its covariance function, whereas in the second kind the transformed quantities are the corresponding space and time arguments. Combinations between the two kinds of techniques are also possible, under certain conditions.

11.1 ATTRIBUTE TRANSFORMATION

This technique is based on the assumption that if a transformation exists between two attributes $X(\boldsymbol{p})$ and $Y(\boldsymbol{p})$, i.e.,

$$Y(\boldsymbol{p}) = \mathcal{T}_X[X(\boldsymbol{p})], \tag{11.1}$$

then an associated transformation can be established between the corresponding space–time covariances, i.e.,

$$c_Y(\boldsymbol{p},\boldsymbol{p}') = \mathcal{T}_c[c_X(\boldsymbol{p},\boldsymbol{p}')], \tag{11.2}$$

which can generate new covariance models $c_Y(\boldsymbol{p},\boldsymbol{p}')$ given the existing ones, $c_X(\boldsymbol{p},\boldsymbol{p}')$. The crucial element here is, of course, the selection of a convenient $\mathcal{T}_X$.

Example 11.1

Assume that $\mathcal{T}_X$ is an m-variate log-normal transformation, i.e., $\mathcal{T}_X = L^m$ (Eqs. 4.1a–b of Chapter X). In this case, the corresponding covariances are related by

$$c_Y(\Delta\boldsymbol{p}) = \mathcal{T}_C[c_X(\Delta\boldsymbol{p})] = \overline{Y}^2\left[e^{c_X(\Delta\boldsymbol{p})} - 1\right], \tag{11.3}$$

where $\Delta\boldsymbol{p} = \boldsymbol{p}' - \boldsymbol{p}$. For illustration, consider the space–time extension of the De Wisj covariance model, $c_X(\Delta\boldsymbol{p}) = \sigma_X^2 - \ln|\Delta\boldsymbol{p}|^{\alpha}$. In this case, Eq. (11.3) gives

$$c_Y(\Delta\boldsymbol{p}) = \overline{Y}^2\left[e^{\sigma_X^2 - \ln|\Delta\boldsymbol{p}|^{\alpha}} - 1\right]. \tag{11.4}$$

An interesting expression is obtained if $\sigma_X^2 >> \gamma_X(\Delta\boldsymbol{p}) = \ln|\Delta\boldsymbol{p}|^{\alpha}$. In this case, Eq. (11.4) becomes

$$c_Y(\Delta\boldsymbol{p}) = \sigma_Y^2 e^{-\gamma_X(\Delta\boldsymbol{p})} \tag{11.5}$$

(recall, Eq. 4.4 of Chapter X),[13] or

$$c_Y(\Delta\boldsymbol{p}) = \sigma_Y^2\left|\Delta\boldsymbol{p}\right|^{-\alpha}, \tag{11.6}$$

where $\ln Y(\boldsymbol{p}) \sim G\left(\overline{\ln Y}, \sigma_{\ln Y}^2\right)$. As in many similar cases, the form of the metric $|\Delta\boldsymbol{p}|$ will affect the shape of the covariance model.

Another class of space−time covariance models can be constructed by letting the transformation $\mathcal{T}_X$ of Eq. (11.1) be a space−time differentiation operator of the general form of Eq. (4.2b) of Chapter VI, i.e.,

$$Y(\boldsymbol{p}) = \mathcal{T}_X[X(\boldsymbol{p})] = \frac{\partial^{\nu+\mu}X(\boldsymbol{p})}{\partial s^{\nu}\partial t^{\mu}} = X^{(\nu,\mu)}(\boldsymbol{p}) \tag{11.7}$$

in $R^{n,1}$. It is known that the corresponding covariance models are related by means of

$$c_Y(\boldsymbol{p},\boldsymbol{p}') = \mathcal{T}_c[c_X(\boldsymbol{p},\boldsymbol{p}')] = \frac{\partial^{\nu+\nu'+\mu+\mu'}c_X(\boldsymbol{p},\boldsymbol{p}')}{\partial s^{\nu}\partial s'^{\nu'}\partial s_0^{\mu}\partial s_0'^{\mu'}}, \tag{11.8}$$

which offers another direct way to construct new space−time covariance models $c_Y(\boldsymbol{p}, \boldsymbol{p}')$ by differentiating known covariance models $c_X(\boldsymbol{p}, \boldsymbol{p}')$.

Example 11.2

As a simple illustration, I suggest the following covariance model for $X(s, t)$ in $R^{1,1}$,

$$c_X(s,s',t,t') = a\left[\cos\left(\lambda(s-s')\right) + \frac{b}{\lambda}\sin\left(\lambda|s-s'|\right)\right]e^{-b|s-s'|-c(t-t')^2}, \tag{11.9}$$

where a, b, c, $\lambda > 0$. The covariance model of $Y(s,t) = \mathcal{T}_X X(s,t) = \frac{\partial X(s,t)}{\partial s}$ will be given by

$$c_Y(s,s',t,t') = \frac{\partial^2 c_X(s,s',t,t')}{\partial s\partial s'} = a\left(b^2+\lambda^2\right)\left[\cos\left(\lambda(s-s')\right) - \frac{b}{\lambda}\sin\left(\lambda|s-s'|\right)\right]e^{-b|s-s'|-c(t-t')^2}. \tag{11.10}$$

The covariance model of $\frac{\partial X(s,t)}{\partial t}$, the cross-covariance model between $\frac{\partial X(s,t)}{\partial s}$ and $\frac{\partial X(s,t)}{\partial t}$ etc., can be derived in a similar way.

11.2 ARGUMENT TRANSFORMATION

This is one of the simplest techniques, if not the simplest, for constructing valid space−time covariance models. Indeed, one can choose any valid covariance model in R^1, R^n or $R^{n,1}$, and apply a suitable argument transformation to obtain a new covariance model in $R^{n,1}$.

Useful cases of the argument transformation technique can be found in various parts of this book (see, e.g., the frozen and the plane-wave random fields). For illustration, a list of possible argument

[13]The readers are reminded that if the variogram $\gamma_X(\Delta\boldsymbol{p})$ in Eq. (11.5) is a valid model, then so is the corresponding covariance $c_Y(\Delta\boldsymbol{p})$.

Table 11.1 The Argument Transformation Technique of Space–Time Covariance Model Construction

No.	Original Model	Argument Transformation	$c_X(\boldsymbol{h}, \tau)$ in $R^{n,1}$
1	$c_X(\widehat{\tau})$	$\widehat{\tau} \mapsto \tau - a(\boldsymbol{\varepsilon} \cdot \boldsymbol{h})$	$c_X(\tau - a(\boldsymbol{\varepsilon} \cdot \boldsymbol{h}))$
2	$c_X(\widehat{\boldsymbol{h}})$	$\widehat{\boldsymbol{h}} \mapsto \boldsymbol{h} \pm \boldsymbol{\upsilon}\tau$	$c_X(\boldsymbol{h} \pm \boldsymbol{\upsilon}\tau)$
3	$c_X(\widehat{\boldsymbol{h}}, \widehat{\tau})$	$\widehat{\boldsymbol{h}} \mapsto \boldsymbol{h} - \boldsymbol{\upsilon}\tau$ $\widehat{\tau} \mapsto \tau - \dfrac{\boldsymbol{\eta} \cdot \boldsymbol{h}}{\upsilon}$	$c_X\left(\boldsymbol{h} - \boldsymbol{\upsilon}\tau, \tau - \frac{\boldsymbol{\eta} \cdot \boldsymbol{h}}{\upsilon}\right)$
4	$c_X(\widehat{\boldsymbol{h}}_1, \widehat{\boldsymbol{h}}_2)$	$\widehat{\boldsymbol{h}}_1 \mapsto \boldsymbol{h} + \boldsymbol{\upsilon}\tau$ $\widehat{\boldsymbol{h}}_2 \mapsto \boldsymbol{h} + \boldsymbol{\upsilon}'\tau$	$c_X(\boldsymbol{h} + \boldsymbol{\upsilon}\tau, \boldsymbol{h} + \boldsymbol{\upsilon}'\tau)$
5	$c_X(\widehat{h})$	$\widehat{h} \mapsto \left[(\boldsymbol{h} - \boldsymbol{\upsilon}\tau)^2 + (\boldsymbol{\upsilon}'\tau)^2\right]^{\frac{1}{2}}$	$c_X(\boldsymbol{h}, \tau)$

transformations is given in Table 11.1. Basically, one can start with any of the temporal, spatial or spatiotemporal covariance models that are available in the literature (Tables 4.1 and 5.1 of Chapter VIII provide lists of such models), and use the argument transformations of Table 11.1. Then a large class of permissible space–time covariance models can be constructed in a straightforward manner. Also, the argument transformation is in many cases physically justified (as in the cases of the frozen and the plane-wave random fields, mentioned above).

12. CROSS-COVARIANCE MODEL CONSTRUCTION TECHNIQUES

Techniques are also available for constructing spatiotemporal cross-covariance models of multivariate or vector random fields. Some of these techniques are developed specifically for multivariate or vector random fields, whereas some others are extensions of the ones used above for scalar random fields. Again, the theory presented in previous chapters proves to be valuable for this purpose. So, a technique suggested by Eq. (7.7) of Chapter IX is described in Table 12.1. This technique can generate anisotropic cross-covariance models starting from isotropic ones. A few examples are discussed below.

Example 12.1

Using any known unidimensional spectra $\widetilde{c}_{v,1}$ and $\widetilde{c}_{v,2}$, or any pair of valid STIS covariance models (e.g., selected from Table 5.1 of Chapter VIII) for c_{v_1} and c_{v_2}, and inserting them into Eqs. (12.1a–b) or (12.2a–b), respectively, we obtain the corresponding c_v and c_u models; subsequently, the c_v and c_u models are inserted into Eq. (12.3) to produce new space–time covariance models.

Table 12.1 A Technique of Space–Time Cross-Covariance Model Construction

Step	Description
1	Select **(a)** the unidimensional spectra $\widetilde{c}_{v,1}(k,\tau)$ and $\widetilde{c}_{u,1}(k,\tau)$, or **(b)** the isostationary covariances $c_{v_i}(r,\tau)$, $i = 1, 2$.
2	Calculate the corresponding isotropic covariance models, **(a)** as $$c_v(r,\tau) = 2\int_0^\infty dk\cos(kr)\widetilde{c}_{v,1}(k,\tau), \quad c_u(r,\tau) = 2\int_0^\infty dk\cos(kr)\widetilde{c}_{u,1}(k,\tau); \quad (12.1a–b)$$ or, **(b)** as $$c_v(r,\tau) = -r^{-1}\frac{\partial}{\partial r}c_{v_2}(r,\tau) - \frac{\partial^2}{\partial r^2}c_{v_1}(r,\tau), \quad c_u(r,\tau) = -\frac{\partial^2}{\partial r^2}c_{v_1}(r,\tau) - r^{-1}\frac{\partial}{\partial r}c_{v_2}(r,\tau). \quad (12.2a–b)$$
3	Derive space-time homostationary cross-covariance models in $R^{n,1}$ by $$c_{X_lX_{l'}}(\boldsymbol{h},\tau) = \frac{h_l h_{l'}}{r^2}[c_v(r,\tau) - c_u(r,\tau)] + c_u(r,\tau)\delta_{ll'}, \quad (12.3)$$ $l, l' = 1,\ldots, n$ $(n = 2, 3)$.

Example 12.2

In Step 1b of Table 12.1, select the covariance model $c_{v_i}(r,\tau) = \left(1 + a_i^{-1}r\right)e^{-a_i^{-1}r - b_i^{-1}\tau}$, $i = 1, 2$, and $a_i, b_i > 0$. Then, Eqs. (12.2a–b) of Step 2b give

$$\begin{aligned} c_v(r,\tau) &= a_2^{-1}e^{-a_2^{-1}r-b_2^{-1}\tau}(r+\tau-1)r^{-1} - a_1^{-1}e^{-a_1^{-1}r-b_1^{-1}\tau}\left[a_1^{-1}(r+\tau-1)-1\right], \\ c_u(r,\tau) &= a_1^{-1}e^{-a_1^{-1}r-b_1^{-1}\tau}(r+\tau-1)r^{-1} - a_2^{-1}e^{-a_2^{-1}r-b_2^{-1}\tau}\left[a_2^{-1}(r+\tau-1)-1\right], \end{aligned} \tag{12.4}$$

which are inserted in Eq. (12.3) of Step 3, to find the space–time cross-covariance model

$$\begin{aligned} c_{X_lX_{l'}}(\boldsymbol{h},\tau) = \frac{h_l h_{l'}}{r^2}&\left\{\sum_{i=1}^{2}(-1)^i a_i^{-1}e^{-a_i^{-1}r-b_i^{-1}\tau}\left[\left(r^{-1}+a_i^{-1}\right)(r+\tau-1)-1\right]\right\} \\ &- \delta_{ll'}\left\{-a_2^{-1}e^{-a_2^{-1}r-b_2^{-1}\tau}\left[a_2^{-1}(r+\tau-1)-1\right] + a_1^{-1}r^{-1}e^{-a_1^{-1}r-b_1^{-1}\tau}(r+\tau-1)\right\}. \end{aligned} \tag{12.5}$$

Notice that while c_{v_i}, c_v, and c_u are STIS, the $c_{X_lX_{l'}}$ is STHS, in general.

In many cases, one can start with known STIS covariance functions c_v and c_u, and implement directly Eq. (12.3) of Step 3.

Example 12.3

In $R^{2,1}$, let $l, l' = 1, 2$ with $r^2 = h_1^2 + h_2^2$. Then, Eq. (12.3) gives (for $l = l'=1$, $l = l' = 2$, and $l = 1$, $l = 2$, respectively)

$$\begin{aligned} c_{X_1}(h_1, h_2, \tau) &= \frac{h_1^2}{r^2} c_v(r, \tau) + \left(1 - \frac{h_1^2}{r^2}\right) c_u(r, \tau), \\ c_{X_2}(h_1, h_2, \tau) &= \frac{h_2^2}{r^2} c_v(r, \tau) + \left(1 - \frac{h_2^2}{r^2}\right) c_u(r, \tau), \\ c_{X_1X_2}(h_1, h_2, \tau) &= \frac{h_1 h_2}{r^2} [c_v(r, \tau) - c_u(r, \tau)], \end{aligned} \tag{12.6a–c}$$

where the c_v and c_u are say, of the space–time Gaussian and the combined space-von Kármán time–Gaussian type (see, Table 4.1 of Chapter VIII), respectively, i.e,

$$\begin{aligned} c_v(r, \tau) &= e^{-\frac{r^2}{a^2} - \frac{\tau^2}{b^2}}, \\ c_u(r, \tau) &= \frac{c_0}{2^{\nu-1}\Gamma(\nu)} \left(\frac{r}{\lambda}\right)^{\nu} K_\nu\left(\frac{r}{\lambda}\right) e^{-\frac{\tau^2}{b^2}} \end{aligned} \tag{12.7a–b}$$

$(a, b, \lambda, \nu > 0)$. Then, new space–time covariance and cross-covariance models can be generated from Eqs. (12.6a–c) as

$$\begin{aligned} c_{X_l}(h_1, h_2, \tau) &= c_0 \left[\frac{h_l^2}{r^2} e^{-\frac{r^2}{a^2}} + \left(1 - \frac{h_l^2}{r^2}\right) \frac{1}{2^{\nu-1}\Gamma(\nu)} \left(\frac{r}{\lambda}\right)^{\nu} K_\nu\left(\frac{r}{\lambda}\right)\right] e^{-\frac{\tau^2}{b^2}}, \\ c_{X_lX_{l'}}(h_1, h_2, \tau) &= \frac{c_0 h_l h_{l'}}{r^2} \left[e^{-\frac{r^2}{a^2}} - \frac{1}{2^{\nu-1}\Gamma(\nu)} \left(\frac{r}{\lambda}\right)^{\nu} K_\nu\left(\frac{r}{\lambda}\right)\right] e^{-\frac{\tau^2}{b^2}}, \end{aligned} \tag{12.8a–b}$$

$(l, l' = 1, 2;\ l \neq l')$, which are space–time anisostationary covariance models. Alternatively, one can also assume that the c_{v_1} and c_{v_2} have the forms of Eqs. (12.7a–b), respectively, and use Eqs. (12.2a–b) to define

$$\begin{aligned} c_v(r, \tau) &= c_0 \left[\frac{1}{2^{\nu-1}\Gamma(\nu)\lambda^2} \left(\frac{r}{\lambda}\right)^{\nu-1} K_{\nu-1}\left(\frac{r}{\lambda}\right) + \frac{2}{a^2}\left(1 - \frac{2r^2}{a^2}\right) e^{-\frac{r^2}{a^2}}\right] e^{-\frac{\tau^2}{b^2}}, \\ c_u(r, \tau) &= c_0 \left[\frac{1}{2^{\nu-1}\Gamma(\nu)\lambda^2} \left(\frac{r}{\lambda}\right)^{\nu-1} \left(K_{\nu-1}\left(\frac{r}{\lambda}\right) - \left(\frac{r}{\lambda}\right) K_{\nu-2}\left(\frac{r}{\lambda}\right)\right) + \frac{2}{a^2} e^{-\frac{r^2}{a^2}}\right] e^{-\frac{\tau^2}{b^2}}. \end{aligned} \tag{12.9a–b}$$

Then, Eq. (12.3) gives

$$
\begin{aligned}
c_{X_l X_{l'}}(\boldsymbol{h}, \tau) = {} & \frac{c_0 h_l h_{l'}}{r^2}\left[\frac{1}{2^{\nu-1}\Gamma(\nu)\lambda^2}\left(\frac{r}{\lambda}\right)^{\nu} K_{\nu-2}\left(\frac{r}{\lambda}\right) - \frac{4r^2}{a^4}e^{-\frac{r^2}{a^2}}\right]e^{-\frac{\tau^2}{b^2}} \\
& + \delta_{ll'}c_0\left[\frac{1}{2^{\nu-1}\Gamma(\nu)\lambda^2}\left(\frac{r}{\lambda}\right)^{\nu-1}\left(K_{\nu-1}\left(\frac{r}{\lambda}\right) - \left(\frac{r}{\lambda}\right)K_{\nu-2}\left(\frac{r}{\lambda}\right)\right) + \frac{2}{a^2}e^{-\frac{r^2}{a^2}}\right]e^{-\frac{\tau^2}{b^2}}.
\end{aligned}
\tag{12.10}
$$

Lastly, one can also write $\frac{h_1}{r} = \cos\phi$, where ϕ is the angle between h_1 and r, in which case $\frac{h_2}{r} = \sin\phi$, and the Eq. (12.6a–c) become

$$
\begin{aligned}
c_{X_1}(h_1, h_2, \tau) &= \cos^2\varphi c_\nu(r,\tau) + \sin^2\varphi c_u(r,\tau), \\
c_{X_2}(h_1, h_2, \tau) &= \sin^2\varphi c_\nu(r,\tau) + \cos^2\varphi c_u(r,\tau), \\
c_{X_1X_2}(h_1, h_2, \tau) &= \cos\varphi\sin\varphi[c_\nu(r,\tau) - c_u(r,\tau)].
\end{aligned}
\tag{12.11a–c}
$$

These expressions indicate the anisotropic (ϕ-dependent) features of the generated space–time covariance models.

13. REVISITING THE ROLE OF PHYSICAL CONSTRAINTS

As we saw in Chapter XIV, in applications we often encounter physical constraints imposed by the real-world phenomenon, which, in turn, impose certain conditions on the corresponding covariance models (or any other S/TVF, in general). Some examples are considered below to illustrate some aspects of the situation.

Example 13.1

In Example 2.2 of Chapter XIV the relationship of Eq. (2.5) was assumed that links the attributes of the vector S/TRF $\boldsymbol{X}(\boldsymbol{p}) = [X_1(\boldsymbol{p})...X_k(\boldsymbol{p})]^T$. It seems reasonable to force the cross-covariance matrix $\boldsymbol{c_X} = [c_{X_lX_{l'}}]$, $l, l' = 1,..., k$, constructed by a mathematically valid technique to conform to the physical constraint of Eq. (2.5) of Chapter XIV. In doing so, one obtains the following constraints for the covariance matrix,

$$
\boldsymbol{c_X}(\Delta\boldsymbol{p})\boldsymbol{a}_\rho(\boldsymbol{p}) = 0 \tag{13.1}
$$

for $\rho = 1,..., m$, where $\boldsymbol{a}_\rho^T(\boldsymbol{p}) = [a_1(\boldsymbol{p})...a_m(\boldsymbol{p})]_\rho$ and $b_\rho(\boldsymbol{p})$ are known physical parameters, and the covariance matrix $\boldsymbol{c_X}$ has rank $k - m$.

Example 13.2

Similar is the case of the physical constraint of Eq. (5.1) of Chapter XIV considered in Example 5.1 of Chapter XIV. This physical constraint links the component attributes of the vector S/TRF $\boldsymbol{X}(\boldsymbol{p})$ through an integral relationship. In the frequency domain, Eq. (5.1) of Chapter XIV leads to a set of constraints for the matrix of cross-spectra, i.e.,

$$
\widetilde{\boldsymbol{c}}_{\boldsymbol{X}}(\boldsymbol{w})\widetilde{\boldsymbol{a}}_\rho(\boldsymbol{w}) = 0, \tag{13.2}
$$

where $\widetilde{c}_X = \left[\widetilde{c}_{X_l X_{l'}}\right]$, $l, l' = 1,..., k$, and $\widetilde{a}_\rho(w)$ are the spectral representations of $a_\rho(p)$. These equations show that the calculated cross-spectra are not solely data-dependent but they must also conform to the above constraints. Many other constraints, of varying level of sophistication, can be found in the relevant literature.

The construction of a space−time covariance model can rigorously account for empirical constraints, like those of Eqs. (13.1) and (13.2) above, using the entropic technique of Section 10. I conclude the discussion in this chapter with some comments concerning the appraisal of the covariance model construction techniques discussed in this chapter.

14. CLOSING COMMENTS

In this chapter, a considerable number of techniques for constructing space−time covariance models have been presented. Their pros and cons have been discussed and recommendations have been made. Among the useful properties of an adequate covariance model construction technique in practice is that it produces covariance models that are permissible, physically consistent, interpretable, feasible computationally and flexible in their use.

It is widely acknowledged that initially most practitioners in applied sciences are interested in interpreting their empirical data and results through the existing covariance models, or they tend to simulate results using experimentally realizable covariance models with hypothetical constraints (that may or may not be realizable within current technology) enacted through computational algorithms. However, this approach cannot be continued to be used when they encounter a real-world phenomenon that cannot be reasonably represented by the current covariance models or cannot be explicated within the established frameworks. Then, these practitioners cannot avoid using the covariance model construction techniques discussed in this chapter.

I have suggested that while the formulation of certain of these covariance model construction techniques is appealing, it may also be quite difficult to apply. Otherwise said, it may be challenging to determine the special functions that serve as inputs to some of these techniques, to check the validity of the conditions for their application, to examine the consistency of the constructed model with the physical characteristics of the attribute it describes, or to make sure that the constructed covariance models are interpretable and sufficiently flexible. The decision over which covariance model construction techniques are the most useful for interpretive purposes can influence the kind of models to be constructed. In the end, as is usually the case, the matter essentially comes down to a deeper understanding of the phenomenon and an adequate knowledge of context.

Last but not least, I have also argued that it is not sufficient that covariance model building relies only on the available data. After all, there are several well-known cases in the real-world where relying on data alone can be a very inadequate approach, including studies of extreme events that can turn out to be catastrophic (Coles, 2001), high-throughput screening investigations concerned with the anticipation of new scientific findings (Macarron et al., 2011), and space−time extrapolation beyond the data range (e.g., in the case of regression and forecasting, Brase and Brase, 2010). Similarly, it is not sufficient that the covariance model merely fits reasonably well with the available dataset. It is highly desirable and even necessary that the generated covariance model also satisfies the core and site-specific physical knowledge concerning the attribute variation that the model purports to represent. This fundamental requirement may impose certain physical constraints on the covariance models.

Several kinds of constraints have been discussed throughout the book, which may be categorized as follows:

(a) One kind of constraints concerns the compatibility of the geometrical properties (symmetry, continuity, differentiability) of the generated space–time covariance model with the actual space–time behavior (directionality and smoothness) of the natural attribute. For example, fully symmetric covariance models are often not physically meaningful, and different degrees of smoothness occur along space and with time.

(b) Natural attributes are governed by laws (in the form of stochastic ordinary and partial differential equations, Chapter XIV), empirical associations (in the form of algebraic equations), and physical dispersion equations, which constitute another kind of constraints imposed on the shape of the corresponding covariance models, and, hence, they should be duly taken into account. Furthermore, the covariance equations associated with the natural law of the phenomenon under consideration can impose some definite restrictions on the covariance domain.

Accordingly, in this chapter the physical law-based techniques of covariance model construction focused on physical laws represented as SPDEs that are closely related to the random field models considered in the book and, hence, they can be used to derive the corresponding covariance models.

This chapter concludes by noticing that additional space-time covariance model construction techniques are discussed in the chapter "Exercises-Comments" (see, Exercises XVI.23–VI.25).

EXERCISES

CHAPTER I

Exercise I.1: Section 1 of Chapter I of the book comments on the usefulness of the random field model as a stochastic (nondeterministic) theory with many applications in sciences. Describe natural attributes and associated real-world conditions that are properly represented by the scalar spatiotemporal random field (S/TRF) model.

Exercise I.2: In Section 2 of Chapter I the space–time domains R^{n+1} and $R^{n,1}$, together with the associated coordinate systems, were introduced.

(a) Discuss the main features and differences of the two domains.

(b) Describe applications in which either the R^{n+1} coordinates or the $R^{n,1}$ coordinates are *physical coordinates*, i.e., they have direct metrical significance.

Exercise I.3: A point $\boldsymbol{p}$ in the space–time domain can be determined in terms of space–time composite or separate coordinates, $\boldsymbol{p} = (s_1,\ldots, s_n, s_0) \in R^{n+1}$ or $\boldsymbol{p} = (\boldsymbol{s},t) \in R^{n,1}$, where $\boldsymbol{s} = (s_1,\ldots, s_n)$ is the location vector and $s_0 = t$ denotes the time argument. Write possible mathematical expressions of the location vector defined as

$$\boldsymbol{s} - \boldsymbol{v}s_0, \tag{3.1}$$

where $\boldsymbol{v} = (v_1,\ldots, v_n)$, in terms of coordinates and directional unit vectors.

Exercise I.4: The Cartesian representation of space–time is in terms of the coordinates

$$(\boldsymbol{s}, s_0), \tag{4.1}$$

where $s_0 = t$. An equivalent representation of space–time is in polar coordinates

$$(\rho, \theta). \tag{4.2}$$

In this sense, the representation of Eq. (4.1) measures time along some linear axis, whereas in the representation of Eq. (4.2) time becomes an angle. Consider the domain $R^{1,1}$ and assume that the space–time Cartesian coordinates are linked by the equation $s_1 = \sin(at) + b$. What is the corresponding equation linking the polar coordinates?

Exercise I.5: A major issue in scientific modeling is for a given coordinate system to determine the physical space–time "distance" or metric $m(\boldsymbol{p},\boldsymbol{p}') = |\Delta\boldsymbol{p}|$ between two points $\boldsymbol{p}$ and $\boldsymbol{p}'$ separated by the space–time vector lag $\Delta\boldsymbol{p}$. Show that the Pythagorean space–time metric of Eq. (2.15a) of Chapter I (a) remains unchanged by the translation transformation of Eq. (2.20a–b) of Chapter I, but (b) is not invariant under the Lorentzian transformation of Eq. (2.25) of Chapter I.

Exercise I.6: Consider a disk in $R^{2,1}$ rotating about its center 0 with angular velocity ϖ within a coordinate system with polar coordinates (ρ,θ) centered at 0. Show that the space–time metric is given by Eq. (2.15b) of Chapter I written in polar coordinates, i.e.,

$$\Delta\boldsymbol{p}^2 = c^2\tau^2 - \Delta\rho^2 - (\rho\Delta\theta)^2. \tag{6.1}$$

Next, consider a transformation of the coordinate system to one that rotates with the disk, i.e., $(\rho' = \rho,\ \theta' = \theta + \varpi t)$. Use the general metric Eq. (2.14) of Chapter I to show that the space–time metric can be written as

$$\Delta p^2 = \left[c^2 - (\varpi\rho')^2\right]\tau^2 - \Delta\rho'^2 - (\rho'\Delta\theta')^2 - 2\varpi\rho'^2\tau\Delta\theta', \tag{6.2}$$

where $\tau = \Delta t$. Suggest a comparative interpretation of Eqs. (6.1) and (6.2).

Exercise I.7: Show that the Minkowski (or Einstein) space–time metric of Eq. (2.15b) of Chapter I, which combines space and time into a single whole, is invariant under the Lorentz transformation of coordinates.

Exercise I.8: From a geometry viewpoint, could the special role of time be due to the pseudo-Euclidity of the metric (see Definition 2.5 of Chapter I)?

Exercise I.9: All phenomena in the real-world occur in space and time and, hence, the physical laws that govern space–time connections are the most general. Show that Burgers' Eq. (2.32) of Chapter I in $R^{1,1}$ is a physical law that is invariant under the Galilean transformation.[1]

Exercise I.10: Random field modeling is based on a space–time calculus involving differential operators. For illustration:

(a) Calculate the quantity $\dfrac{\partial^{n+1}\prod_{i=1}^{n,0} s_i}{\prod_{i=1}^{n,0}\partial s_i}$ in R^{n+1}.

(b) Calculate $\dfrac{\partial \boldsymbol{p}}{\partial s_i}$ and $\dfrac{\partial^2 \boldsymbol{p}}{\partial s_i \partial s_j}$ $(i, j = 1, \ldots, n \text{ and } 0)$ in R^{n+1}, where $\boldsymbol{p} = (s_1, \ldots, s_n, s_0)$.

Exercise I.11: Calculate the derivatives $\frac{\partial}{\partial s_i}$ $(i = 1, \ldots, n \text{ and } 0)$ of the point vector $\boldsymbol{s} - \boldsymbol{v}s_0$ in R^{n+1}, and the $\frac{\partial}{\partial s_i}$ of the field $X(\boldsymbol{s} - \boldsymbol{v}s_0)$, assuming that (a) $\boldsymbol{v}$ is a constant vector and (b) $\boldsymbol{v} = \boldsymbol{v}(s_1, \ldots, s_n, s_0)$.

Exercise I.12: Calculate the covariance derivatives with respect to the coordinates s_i $(i = 1, \ldots, n$ and 0; $\nu \geq 1)$,

$$\frac{\partial^\nu}{\partial s_i^\nu} c_X(\Delta\boldsymbol{p}) \text{ and } \frac{\partial^\nu}{\partial s_i'^\nu} c_X(\Delta\boldsymbol{p}),$$

in terms of covariance derivatives with respect to the lags h_i. Let, $\Delta\boldsymbol{p} = \boldsymbol{p} - \boldsymbol{p}' = \left(s_1 - s_1', \ldots, s_n - s_n', s_0 - s_0'\right)$.

Exercise I.13: Derive the formulas of Table 2.4 of Chapter I, assuming that $\Delta\boldsymbol{p} = \boldsymbol{p}' - \boldsymbol{p}$.

Exercise I.14: Do you agree with the view that the different types of space–time coordinate transformations are not equally useful (physically significant) in real-world applications? Why?

Exercise I.15: In some applications one may need to study the changing relationship between space and time at different *timescales*. Specifically, assume that the arguments of a natural attribute X are space $\boldsymbol{s}$, time t at a small timescale (say, week), and time T at a large timescale (say, year), which is denoted as $X(\boldsymbol{s},t,T)$. In addition, assume that during a time period $t_k \in T_k$ the spatial extent of the attribute was $S \subset R^n$ (for $n = 1$, 2, and 3 the S refers to distance, area, and volume, respectively),

[1]Generally, Galilean invariance is used to denote that measured physics must be the same in any nonaccelerating frame of reference.

whereas the same spatial extent S of the attribute occurred during a different time period $t_{k'} \in T_{k'}$. Then, one can define the change in the relationship between space and time as regards $X(s,t,T)$ by

$$R_X = \frac{\Delta t_{kk'}}{\Delta T_{kk'}}, \tag{15.1}$$

where $\Delta t_{kk'} = t_{k'} - t_k$ and $\Delta T_{kk'} = T_{k'} - T_k$. Eq. (15.1) measures space−time convergence if $R_X < 0$, and space−time divergence if $R_X > 0$. For a numerical illustration, consider the spread of flu incidence $X(s,t,T)$ in the Arizona state (United States). In the year $T_k = 1960$ it took for the flu incidence about $t_{1960} = 9.5$ weeks to spread through the area S of Arizona, whereas in the year $T_{k'} = 2000$ it took for the flu about $t_{2000} = 3.4$ weeks to cover the same area S. What is the space−time relationship R_X in this case?

Exercise I.16: The result described by Eqs. (3.7a) and (3.7b) of Chapter I is very useful in applications. Prove the validity of Eqs. (3.7a) and (3.7b) of Chapter I.

Exercise I.17: A classical result of random variable (RV) convergence is that

$$\left.\begin{matrix} x_m \xrightarrow{m.s.} x \\ x_m \xrightarrow{a.s.} x \end{matrix}\right\} \Rightarrow x_m \xrightarrow{P} x \Rightarrow x_m \xrightarrow{F} x. \tag{17.1}$$

Prove the validity of Eq. (17.1).

Exercise I.18: Prove Eq. (3.32a−b) of Chapter I.

Exercise I.19: Show that if x is an RV, it holds that $|\overline{x}|^{\kappa} \leq \overline{|x|}^{\kappa} \leq \overline{|x|^{\kappa}}$ for $\kappa \geq 1$.

CHAPTER II

Exercise II.1: Random field modeling can make especially useful and well-defined contributions. The development of the S/TRF notion in Chapter II assumes that there are considerable interrelations and interactions between space and time and that the spatial variation structure of an attribute represented by an S/TRF changes with time. Show that a function $F_{\boldsymbol{p}_1,\ldots,\boldsymbol{p}_m}$ that satisfies Properties (a)−(e) of Section 2.1 of Chapter II can be considered as a cumulative distribution function (CDF) of the corresponding S/TRF model.

Exercise II.2: Show that if $g(\chi)$ is a Borel-measurable function of the real variable χ and $X(\boldsymbol{p})$ is an S/TRF, then the $Y(\boldsymbol{p}) = g[X(\boldsymbol{p})]$ is an S/TRF too.

Exercise II.3: In light of Eqs. (5.4) and (5.5) of Chapter II, suggest some examples of S/TRFs that may satisfy these equations.

Exercise II.4: Show that if an S/TRF is homostationary in the strict sense (s.s.), it is also wide sense (w.s.) homostationary, but the converse is not necessarily valid.

Exercise II.5: What are the key concepts involved in defining a random field starting from the definition of an RV? Explain.

Exercise II.6: Give examples of characterizing an S/TRF in terms of its generating processes.

Exercise II.7: Are the Definitions 2.1 and 2.2 of Chapter II equivalent and in what way?

Exercise II.8: Calculate the covariance of a random field with a bivariate PDF of the form of Eq. (2.13) of Chapter II with $m = 2$.

Exercise II.9: Prove Eq. (2.19a) of Chapter II that expresses the space−time variogram function in terms of the corresponding covariance function.

Exercise II.10: Assuming that the space–time distribution of a natural attribute modeled as an S/TRF is studied, describe purposes for which the observed interrelations need to be quantified.

Exercise II.11: The sysketogram function and the variogram function both measure space–time variability, although in different ways. In the case of stochastic independence, which one is equal to zero and why?

Exercise II.12: Show that the sysketogram and the contingogram functions are not affected if the random field $X(\boldsymbol{p})$ is replaced by some function $\phi(X(\boldsymbol{p}))$, provided that ϕ is one-to-one.

Exercise II.13: When the natural attribute has random space–time coordinates (e.g., distribution of aerosol particles), is the sysketogram function independent of the coordinate system chosen?

Exercise II.14: Using the formula $f_X(\boldsymbol{p},\boldsymbol{p}') = f_X(\boldsymbol{p})\delta(\chi - \phi(\chi'))$, which defines the bivariate PDF in terms of its univariate PDF, where ϕ is suitable one-to-one function, derive the variance from the space–time covariance of Eq. (2.16a) of Chapter II.

Exercise II.15: Prove Eqs. (2.29a–b) of Chapter II that relate the sysketogram and contingogram functions to the copula function.

Exercise II.16: By using series expansions in Eqs. (2.29a–b) of Chapter II, show that for small copula values the sysketogram and contingogram functions coincide.

Exercise II.17: Separability, when justified, is a useful concept in real-world applications. Yet some important limitations apply.

(a) Show that under certain conditions separability in the physical (real) domain implies separability in the frequency domain. Specifically, starting from Eq. (2.32a) of Chapter II, prove Eq. (2.32b) of Chapter II.

(b) Is it valid that in the case of a separable space–time covariance function there are no essential differences in the regularity properties of the corresponding random field represented by the space–time covariance function compared to the regularity properties associated with the component spatial and temporal covariance functions?

Exercise II.18: Show that Eq. (3.5) of Chapter II is a solution of Eq. (3.4) of Chapter II.

Exercise II.19: If $X(\boldsymbol{p})$ is a normal (Gaussian) random field and $Y(\boldsymbol{p})$ is a lognormal random field, prove the relationships in Table 7.1 of Chapter II.

Exercise II.20: Construct some examples of heterogeneous S/TRFs based on Eq. (7.7) of Chapter II.

CHAPTER III

Exercise III.1: In Section 1 of Chapter III it was pointed out that symmetric space–time metrics are usually assumed in theory, yet there are cases in the real-world that justify the use of an asymmetric metric (i.e., one that satisfies all metric axioms with the exception of symmetry). Is the space–time metric of Eq. (1.13) of Chapter III a symmetric or an asymmetric one?

Exercise III.2: Among the space–time metrics listed in Table 1.1 of Chapter III, which ones are symmetric?

Exercise III.3: Prove Eqs. (1.15a–c) of Chapter III.

Exercise III.4: Derive Eq. (1.16) of Chapter III.

Exercise III.5: Assuming that the real-world situation consists of mountain villages, the metric representing typical walking times between villages should be a symmetric or an asymmetric metric and why?
Exercise III.6: Prove Eqs. (1.17a–d) of Table 1.2 of Chapter III.
Exercise III.7: Given an asymmetric metric $|\Delta \boldsymbol{p}|^{\circ}$, under what conditions it can be linked to a symmetric metric $|\Delta \boldsymbol{p}|$ by $|\Delta \boldsymbol{p}|^{\circ} + |-\Delta \boldsymbol{p}|^{\circ} = 2|\Delta \boldsymbol{p}|$?
Exercise III.8: Prove Eqs. (1.18a–f) of Table 1.3 of Chapter III.
Exercise III.9: Derive Eq. (2.1b) from Eq. (2.1a) of Chapter III.
Exercise III.10: Show that an example of an asymmetric metric is the case of a circle with metric the length of the shortest clockwise path between two points.
Exercise III.11: Consider the metric

$$|\Delta \boldsymbol{p}| = \begin{cases} \boldsymbol{p} - \boldsymbol{p}' \text{ if } \boldsymbol{p} \geq \boldsymbol{p}', \\ 1 + 10^{\boldsymbol{p}' - \boldsymbol{p}} \text{ otherwise.} \end{cases} \tag{11.1}$$

Is it a symmetric or an asymmetric metric?
Exercise III.12: A useful set of equations express the derivatives of covariance functions in terms of the derivatives of the space–time metric. Derive (a) Eqs. (2.3a–b) of Chapter III, (b) Eq. (2.4a–f) of Table 2.1 of Chapter III, and (c) Eq. (2.6a–d) of Table 2.2 of Chapter III.
Exercise III.13: Consider the space–time metric

$$|\Delta \boldsymbol{p}|^{\circ} = \begin{cases} |\Delta \boldsymbol{p}| + \hbar(\boldsymbol{p}') - \hbar(\boldsymbol{p}) \text{ for } \hbar(\boldsymbol{p}') \geq \hbar(\boldsymbol{p}) \\ |\Delta \boldsymbol{p}| \text{ otherwise,} \end{cases} \tag{13.1}$$

where $|\Delta \boldsymbol{p}|$ is a symmetric metric and $\hbar(\boldsymbol{p})$ is a potential field (e.g., $\hbar(\boldsymbol{p})$ may denote the height of point $\boldsymbol{p}$). (a) Is Eq. (13.1) a symmetric or an asymmetric metric and why? (b) What kinds of natural attributes can use the above space–time metric?
Exercise III.14: Consider the space–time metric

$$|\Delta \boldsymbol{p}| = (a - b)H_{\boldsymbol{p}' > \boldsymbol{p}} + b(1 - \delta_{\boldsymbol{p}',\boldsymbol{p}}) \tag{14.1}$$

where $\Delta \boldsymbol{p} = \boldsymbol{p}' - \boldsymbol{p}$, a, $b \in R^1$, $H_{\boldsymbol{p}' > \boldsymbol{p}} = 1$ if $\boldsymbol{p}' > \boldsymbol{p}$, $= 0$, otherwise, and $\delta_{\boldsymbol{p}',\boldsymbol{p}} = 1$ if $\boldsymbol{p}' = \boldsymbol{p}$, $= 0$, otherwise. Is it a symmetric or an asymmetric metric?
Exercise III.15: Derive (a) Eq. (2.7a–b)–(2.9a–b) of Table 2.3 of Chapter III, (b) Eq. (2.10a–d) of Table 2.4 of Chapter III, and (c) Eq. (2.11a–b) of Chapter III.
Exercise III.16: Consider an attribute moving with speed υ (e.g., air pollutant concentration). What is the corresponding space–time metric among those listed in Table 1.1 of Chapter III?
Exercise III.17: Let $c_X(r_E,\tau)$ be a valid covariance model in $R^{n,1}$, where r_E is a Euclidean metric in R^n. Also, let $r_s = 2\rho \sin^{-1}\left(\frac{r_E}{2\rho}\right)$ be the arc length between any two points on the surface of the earth (viewed as a sphere), and ρ is the earth's radius. Is the $c_X(r_s,\tau)$ a valid covariance function on the spherical domain $S^{n-1,1} \subset R^{n,1}$?
Exercise III.18: Among other matters, Chapter III emphasized that if the laws governing a phenomenon are available in the form of differential equations, these equations contain information about

the geometrical structure of space–time and about its physical properties. Discuss the information about the space–time metric form that is provided by:

(a) the transient groundwater flow law of Eqs. (4.32) and (4.33a–c) of Chapter III and
(b) the stochastic wave equation in $R^{3,1}$,

$$\left[\nabla^2 - a^{-2}\frac{\partial^2}{\partial t^2}\right]X(s,t) = Y(s,t),$$

where a is the wave velocity and $Y(s,t)$ is a known random field.

CHAPTER IV

Exercise IV.1: Show that the ρth-order noncentered statistical moments and the ρth-order centered moments are related by Eq. (2.4) of Chapter IV.

Exercise IV.2: In the case of Gaussian (normal) S/TRFs $X(\boldsymbol{p}) = X(s,t)$:

(a) Prove Eqs. (2.6b–c) of Chapter IV.
(b) In some random field modeling cases the conditions of the phenomenon of interest may require that either the space or the time argument is fixed. In these cases, the corresponding space–time statistics are modified accordingly so that one argument is suppressed. For illustration, assume for simplicity that $\overline{X(s,t)} = 0$ in $R^{1,1}$, and consider the Gaussian PDF of Eq. (2.6a) of Chapter IV with zero mean, i.e.,

$$f_X\left(\chi_p, \sigma_X^2\right) = \frac{1}{\sqrt{2\pi}\ \sigma_X} e^{-\frac{1}{2\sigma_X^2}\chi_p^2}, \tag{2.1}$$

with the CDF $F_X\left(\chi_p, \sigma_X^2\right) = \int_{-\infty}^{\chi_p} d\chi_p f_X\left(\chi_p, \sigma_X^2\right)$. Also, define the function

$$G(|s'-s|, \sigma_X^2) = \sigma_X^2 f_X\left(|s'-s|, \sigma_X^2\right) - |s'-s|\left[1 - F_X\left(|s'-s|, \sigma_X^2\right)\right], \tag{2.2}$$

and for the S/TRF $X(s,t)$ assume the covariance function

$$c_X(s,t,s',t') = G(|s'-s|, at) + G(|s'-s|, at') - G(|s'-s|, a|t'-t|). \tag{2.3}$$

Show that for $s = s'$, Eq. (2.3) becomes

$$c_X(s,t,t') = a(2\pi)^{-\frac{1}{2}}\left(t'^{\frac{1}{2}} + t^{\frac{1}{2}} - |t'-t|^{\frac{1}{2}}\right), \tag{2.4}$$

where the space argument has been suppressed.

Exercise IV.3: Show that if $X_{l'}(\boldsymbol{p}) = vX_l(\boldsymbol{p})$, where v is a zero-mean RV independent of $X_l(\boldsymbol{p})$ for any $\boldsymbol{p} \in R^{n+1}$ and $\overline{v^2} = 1$, then $X_l(\boldsymbol{p})$ and $X_{l'}(\boldsymbol{p})$ have the same covariance function although their realizations may be quite different.

Exercise IV.4: Consider the random fields related by $X_{l'}(\boldsymbol{p}') = X_l^2(\boldsymbol{p})$. Assuming that the fields can take the values -1, 0, and 1 with the probabilities shown in Table 4.1, prove that although $f_{X_l,X_{l'}}(\boldsymbol{p},\boldsymbol{p}')$ shows a very strong probabilistic dependency, nevertheless, it is valid that $c_{X_l,X_{l'}}(\boldsymbol{p},\boldsymbol{p}') = 0$.

Table 4.1 Joint Probability Distributions of $X_l(p)$ and $X_{l'}(p')$			
	$X_l(p) = -1$	$= 0$	$= 1$
$X_{l'}(p') = 0$	$f_{X_l,X_{l'}}(p,p') = 0$	$= \frac{1}{3}$	$= 0$
$= 1$	$= \frac{1}{3}$	$= 0$	$= \frac{1}{3}$

Exercise IV.5: Prove Eq. (4.13) of Chapter IV.
Exercise IV.6: Show that if $X(\boldsymbol{p})$ is a normal (Gaussian) random field and $Y(\boldsymbol{p})$ is a lognormal random field, then Eq. (5.7) of Chapter IV is valid.
Exercise IV.7: There are a number of situations (e.g., in spectral analysis of natural attributes) where it is often more convenient to work with complex-valued S/TRFs. In the case of complex S/TRFs, prove Eqs. (6.6a–b) and (6.7) of Chapter IV.
Exercise IV.8: In the case of complex S/TRFs, prove Eqs. (6.9a–d) of Chapter IV.
Exercise IV.9: In the case of complex S/TRFs, consider Eqs. (6.13a–c) of Chapter IV. If $c_X(\Delta\boldsymbol{p})$ is even ($\Delta\boldsymbol{p} \in R^{n+1}$), then it is real too, i.e., $c_X(\Delta\boldsymbol{p}) = \mathrm{Re}\{c_X(\Delta\boldsymbol{p})\}$ and $\mathrm{Im}\{c_X(\Delta\boldsymbol{p})\} = 0$.
Exercise IV.10: Consider a complex random field $Y(\boldsymbol{p})$ with

$$
\begin{aligned}
Y_R(\boldsymbol{p}) &= \frac{1}{2}[\eta(\boldsymbol{p}) + \eta(-\boldsymbol{p})], \\
Y_I(\boldsymbol{s}) &= \frac{1}{2}[\eta(\boldsymbol{p}) - \eta(-\boldsymbol{p})].
\end{aligned}
\tag{10.1}
$$

Show that the $Y(\boldsymbol{p})$ has the same covariances as the random field $X(\boldsymbol{p})$ defined in Eq. (6.18) of Chapter IV, i.e., $c_Y(\boldsymbol{p},\boldsymbol{p}') = c_X(\boldsymbol{p},\boldsymbol{p}')$ and $\mathring{c}_Y(\boldsymbol{p},\boldsymbol{p}') = \mathring{c}_X(\boldsymbol{p},\boldsymbol{p}')$.
Exercise IV.11: Prove Corollary 6.1 of Chapter IV. This proposition presents some useful relationships between the different types of covariance functions for complex S/TRFs.
Exercise IV.12: Theoretical modeling of certain real-world earth and atmospheric phenomena requires the consideration of higher-order spatiotemporal variation functions (S/TVFs, such as covariance, variogram, and structure functions)

$$
\begin{aligned}
c_X^{[2\rho]}(\boldsymbol{p}_i,\boldsymbol{p}_i') &= \overline{\prod_{i=1}^{\rho} X(\boldsymbol{p}_i) \prod_{i=\rho+1}^{2\rho} X(\boldsymbol{p}_i')}, \\
\xi_X^{[2\rho]}(\boldsymbol{p}_i,\boldsymbol{p}_i') &= \overline{\prod_{i=1}^{2\rho-1} \left[X(\boldsymbol{p}_i)X(\boldsymbol{p}_{i+1}) - X(\boldsymbol{p}_i')X(\boldsymbol{p}_{i+1}')\right]}, \\
\xi_X'^{[2\rho]}(\boldsymbol{p}_i,\boldsymbol{p}_i') &= \overline{\prod_{i=1}^{2\rho} \left[X(\boldsymbol{p}_i) - X(\boldsymbol{p}_i')\right]},
\end{aligned}
\tag{12.1a–c}
$$

where ρ is a positive integer. Derive the S/TVF expressions corresponding to Eqs. (12.1a–c) in the case of the jointly Gaussian assumption.

Exercise IV.13: Show that if $X(\boldsymbol{p})$ and $Y(\boldsymbol{p})$ are complex random fields and $Z(\boldsymbol{p}) = aX(\boldsymbol{p})$, $W(\boldsymbol{p}) = bY(\boldsymbol{p})$, where a and b are nonrandom quantities, then $c_{ZW}(\boldsymbol{p},\boldsymbol{p}') = ab^* c_{XY}(\boldsymbol{p},\boldsymbol{p}')$.

Exercise IV.14: Let $\overline{Y(\boldsymbol{p}_i,\boldsymbol{p}_j)} = \overline{X(\boldsymbol{p}_i) - X(\boldsymbol{p}_j)}$ be the incremental mean and $\gamma_X(\boldsymbol{p}_i,\boldsymbol{p}_j,\boldsymbol{p}_k,\boldsymbol{p}_l) = \frac{1}{2}\overline{Y(\boldsymbol{p}_i,\boldsymbol{p}_j)Y(\boldsymbol{p}_k,\boldsymbol{p}_l)}$ the corresponding variogram of the random field $X(\boldsymbol{p})$. Show that

$$(a)\quad \overline{Y(\boldsymbol{p}_i,\boldsymbol{p}_j)} + \overline{Y(\boldsymbol{p}_k,\boldsymbol{p}_i)} = \overline{Y(\boldsymbol{p}_k,\boldsymbol{p}_j)}, \tag{14.1}$$

and

$$(b)\quad \gamma_X(\boldsymbol{p}_i,\boldsymbol{p}_j,\boldsymbol{p}_k,\boldsymbol{p}_l) = \frac{1}{2}\left[c_X(\boldsymbol{p}_i,\boldsymbol{p}_k) - c_X(\boldsymbol{p}_i,\boldsymbol{p}_l) + c_X(\boldsymbol{p}_j,\boldsymbol{p}_l) - c_X(\boldsymbol{p}_j,\boldsymbol{p}_k)\right]. \tag{14.2}$$

Exercise IV.15: Explain why the mean $\overline{X(\boldsymbol{p})}$ and the covariance $c_X(\boldsymbol{p}_i,\boldsymbol{p}_j)$ are more informative than the mean $\overline{Y(\boldsymbol{p}_i,\boldsymbol{p}_j)} = \overline{X(\boldsymbol{p}_i) - X(\boldsymbol{p}_j)}$ and the variogram $\gamma_X(\boldsymbol{p}_i,\boldsymbol{p}_j,\boldsymbol{p}_k,\boldsymbol{p}_l) = \frac{1}{2}\overline{Y(\boldsymbol{p}_i,\boldsymbol{p}_j)Y(\boldsymbol{p}_k,\boldsymbol{p}_l)}$.

Exercise IV.16: Show that

$$\gamma_X(\boldsymbol{p}_i,\boldsymbol{p}_j,\boldsymbol{p}_k,\boldsymbol{p}_l) = \frac{1}{2}\left[\gamma_X(\boldsymbol{p}_i,\boldsymbol{p}_l) - \gamma_X(\boldsymbol{p}_i,\boldsymbol{p}_k) + \gamma_X(\boldsymbol{p}_j,\boldsymbol{p}_k) - \gamma_X(\boldsymbol{p}_j,\boldsymbol{p}_l)\right], \tag{16.1}$$

where

$$\gamma_X(\boldsymbol{p}_i,\boldsymbol{p}_j,\boldsymbol{p}_k,\boldsymbol{p}_l) = \frac{1}{2}\overline{Y(\boldsymbol{p}_i,\boldsymbol{p}_j)Y(\boldsymbol{p}_k,\boldsymbol{p}_l)}, \tag{16.2}$$

and

$$\gamma_X(\boldsymbol{p}_i,\boldsymbol{p}_l) = \frac{1}{2}\overline{[X(\boldsymbol{p}_i) - X(\boldsymbol{p}_l)]^2}, \tag{16.3}$$

Exercise IV.17: Consider a random field $X(\boldsymbol{p})$ such that, by definition, the mean incremental function and the variogram function are given by, respectively,

$$\begin{aligned} \overline{Y(\boldsymbol{p}_i,\boldsymbol{p}_j)} &= \overline{X(\boldsymbol{p}_i) - X(\boldsymbol{p}_i + \Delta\boldsymbol{p}_{ij})} = \overline{Y(\Delta\boldsymbol{p}_{ij})}, \\ \gamma_X(\boldsymbol{p}_i,\boldsymbol{p}_j) &= \gamma_X(\Delta\boldsymbol{p}_{ij}), \end{aligned} \tag{17.1a–b}$$

where $\Delta\boldsymbol{p}_{ij} = \boldsymbol{p}_j - \boldsymbol{p}_i$.

(a) Show that

$$\overline{Y(\Delta\boldsymbol{p}_{ij})} + \overline{Y(\Delta\boldsymbol{p}_{ik})} = \overline{Y(\Delta\boldsymbol{p}_{ij} + \Delta\boldsymbol{p}_{ik})}, \tag{17.2}$$

and

$$\overline{Y(\Delta \boldsymbol{p})} = \sum_{q=1}^{n,0} c_q h_q, \tag{17.3}$$

where $\Delta \boldsymbol{p} = (h_1, \ldots, h_n, h_0)$, and c_q $(q = 1, \ldots, n, 0)$ are constant coefficients; and
(b) if $\overline{X(\mathbf{0})} = c$, then

$$\overline{X(\boldsymbol{p})} = \sum_{q=1}^{n,0} c_q s_q + c, \tag{17.4}$$

i.e., the mean varies linearly with $\boldsymbol{p}$.

Exercise IV.18: Consider the complex S/TRF $X(\boldsymbol{p}) = X_R(\boldsymbol{p}) - iX_I(\boldsymbol{p})$. In many applications (e.g., optical wave propagation in a random inhomogeneous medium) it can be assumed that the space−time variogram function

$$\gamma_X(\boldsymbol{p},\boldsymbol{p}') = \frac{1}{2}\overline{|[X_R(\boldsymbol{p}) - i\,X_I(\boldsymbol{p})] - [X_R(\boldsymbol{p}') - i\,X_I(\boldsymbol{p}')]|^2} \tag{18.1}$$

practically represents all the statistical information about $X(\boldsymbol{p})$, in which case it is possible to also obtain information about the "shape" of $X(\boldsymbol{p})$ from $\gamma_X(\boldsymbol{p},\boldsymbol{p}')$. Show that the variogram function of Eq. (18.1) is given by

$$\gamma_X(\boldsymbol{p},\boldsymbol{p}') = \gamma_{X_R}(\boldsymbol{p},\boldsymbol{p}') + \gamma_{X_I}(\boldsymbol{p},\boldsymbol{p}'). \tag{18.2}$$

CHAPTER V

Exercise V.1: Chapter V discusses domain transformations (in particular, Fourier transformation (FT), Space transformation (ST), and Traveling transformation (TT)) that can simplify considerably the solution of a problem, and they can also provide additional insight that is not available in the original (physical) domain. With this in mind, prove the fundamental Proposition 2.1 of Chapter V, which relates the FT representation of a random field with that of its covariance function.

Exercise V.2: Spectral moments are also very useful in many applications (random vibrations, wave phenomena, harmonic systems, conducting polymers etc.). Show that the spectral moments $\beta_X^{(1_i 1_j)}$ satisfy Eq. (2.19) of Chapter V, and that the sign in the right hand side of the equation depends on the way the FT pair is defined.

Exercise V.3: Consider the S/TRFs $X(\boldsymbol{p})$ and $Y(\boldsymbol{p})$ related by Eq. (2.24) of Chapter V. Show that their spectral density functions (if they exist) are linked in terms of the transfer function as in Eq. (2.29) of Chapter V.

Exercise V.4: The basic idea of ST is to transfer the study of a problem from the original $R^{n,1}$ domain on to the much simpler $R^{1,1}$ domain. Let $Y_3'(-s,t) = Y_3(s,t)$. Show that the ST operators T_3^1 and Ψ_3^1 are such that $\widehat{Y}'_{1,\theta}(-\sigma) = \widehat{Y}_{1,\theta}(\sigma)$ and $Y'_{1,\theta}(-\sigma) = Y_{1,\theta}(\sigma)$, respectively.

Exercise V.5: Using the result of Exercise V.4 above, show that

$$\int dsX_3(s,t)Y_3(s,t) = \Psi_1^3\left[\int d\sigma\widehat{X}_{1,\theta}(\sigma,t)Y_{1,\theta}(\sigma,t)\right], \tag{5.1}$$

where $\widehat{X}_{1,\theta}(\sigma,t) = T_3^1[X_3](\sigma,\boldsymbol{\theta},t)$.

Exercise V.6: Using the result of Exercise V.5 above, show that the following expression is valid for the ST operator T_3^1 of the product of two S/TRFs $X_3(s,t)$ and $Y_3(s,t)$,

$$T_3^1[X_3Y_3](\sigma,\boldsymbol{\theta},t) = \Psi_1^3\left[\int dpX_{1,\theta'}(p,t)\widehat{Y}_1(\sigma,\boldsymbol{\theta},p,\boldsymbol{\theta}',t)\right](\sigma,\boldsymbol{\theta},t), \tag{6.1}$$

where

$$\widehat{Y}_1(\sigma,\boldsymbol{\theta},p,\boldsymbol{\theta}',t) = T_3^1[Y_3(s,t)\delta(\sigma-\boldsymbol{s}\cdot\boldsymbol{\theta})](p,\boldsymbol{\theta}',t). \tag{6.2}$$

Exercise V.7: Prove that the relation

$$T_3^1[X_3Y_3](\sigma,\boldsymbol{\theta},t) = \int dp\widehat{X}_{1,\theta}(\sigma=\boldsymbol{s}\cdot\boldsymbol{\theta},\boldsymbol{\theta},t)\widehat{Y}_{1,\theta'}(p=\boldsymbol{s}\cdot\boldsymbol{\theta}',\boldsymbol{\theta}',t)\delta(\boldsymbol{s}-\boldsymbol{s}') \tag{7.1}$$

follows directly from the definition of the T_3^1 operator.

Exercise V.8: Show that an expression relating products of derivatives in three- and one-dimensional spaces is as follows:

$$T_3^1\left[\sum_{i=1}^{3}\frac{\partial X_3}{\partial s_i}\frac{\partial Y_3}{\partial s_i}\right](\sigma,\boldsymbol{\theta},t) = \Psi_1^3\left[\int dp\frac{\partial X_{1,\theta'}(p,t)}{\partial p}\frac{\partial\widehat{Y}_1(\sigma,\boldsymbol{\theta},p,\boldsymbol{\theta}',t)}{\partial p}\right], \tag{8.1}$$

where the Ψ_1^3 is considered with respect to $\boldsymbol{\theta}'$.

Exercise V.9: Describe the conditionsin order to hold that

$$\begin{aligned} c_X(\tau) &= c_X(\boldsymbol{v}\tau) = c_{\widehat{X}}(\boldsymbol{v}\tau), \\ \gamma_X(\tau) &= \gamma_X(\boldsymbol{v}\tau) = \gamma_{\widehat{X}}(\boldsymbol{v}\tau). \end{aligned} \tag{9.1a–b}$$

Exercise V.10: In the $R^{1,1}$ domain, consider the S/TRF $X(s,t)$, with space–time covariance function

$$c_X(h,\tau) = c_0e^{-\left(\frac{h}{\beta_s}+\frac{\tau}{\beta_t}\right)}. \tag{10.1}$$

Calculate the covariance function of $\widehat{X}(s-vt)$.

Exercise V.11: In the TT case of a random field, select the axis s_j along the direction of velocity $\boldsymbol{v}$ with a constant speed $v = |\boldsymbol{v}|$. Find the differential equations governing the mean and the variance of the corresponding S/TRF.

Exercise V.12: In light of the discussion in Section 4 of Chapter V, show that in the $R^{2,1}$ domain,

$$v_1\left(v_1\frac{\partial^2 c_X}{\partial h_1^2}+v_2\frac{\partial^2 c_X}{\partial h_1\partial h_2}+\frac{\partial^2 c_X}{\partial h_1\partial\tau}\right)+v_2\left(v_1\frac{\partial^2 c_X}{\partial h_1\partial h_2}+v_2\frac{\partial^2 c_X}{\partial h_2^2}+\frac{\partial^2 c_X}{\partial h_2\partial\tau}\right)$$
$$+v_1\frac{\partial^2 c_X}{\partial h_1\partial\tau}+v_2\frac{\partial^2 c_X}{\partial h_2\partial\tau}+\frac{\partial^2 c_X}{\partial\tau^2}=0. \tag{12.1}$$

Exercise V.13: Consider the following TT of coordinates in $R^{3,1}$,

$$\left(\breve{s}_1,\breve{s}_2,\breve{s}_3\right)\mapsto\left(\frac{s_1-v\,t}{\beta},s_2,s_3\right), \tag{13.1}$$

where $\boldsymbol{v}$ is directed along the s_1 axis, and $\beta=\left(1-v^2c^{-2}\right)^{\frac{1}{2}}$. In this case,

$$X(\boldsymbol{s},t)=\widehat{X}\left(\frac{s_1-vt}{\beta},s_2,s_3\right). \tag{13.2}$$

Find the spectra of $X(\boldsymbol{s},t)$ and $\widehat{X}(\boldsymbol{s})$, i.e., $\widetilde{c}_X(\boldsymbol{k},\omega)$, $\widetilde{c}_X(\boldsymbol{k})$, $\widetilde{c}_{\widehat{X}}(\boldsymbol{k})$, and $\widetilde{c}_X(\omega)$.

Exercise V.14: In the case of the Shandong study (Section 4.3 of Chapter V), confirm the TT of Eq. (4.42a–b) of Chapter V.

Exercise V.15: Derive the formulas of Table 3.3 of Chapter V.

Exercise V.16: Prove Eqs. (2.16a–b) and (2.17a–b) of Chapter V.

Exercise V.17: Show that the $\boldsymbol{v}$ solutions of Eqs. (4.33a–b) of Chapter V satisfy Eq. (4.13a–b) of Chapter V for $n=2$.

Exercise V.18: Consider the space–time covariance function $c_X(\boldsymbol{h},\tau)$ that satisfies the partial differential equation (PDE) in $R^{2,1}$,

$$\left(\frac{\partial}{\partial\tau}-\alpha\nabla^2\right)c_X(\boldsymbol{h},\tau)=0, \tag{18.1}$$

with initial condition $c_X(\boldsymbol{h},0)=f(\boldsymbol{h})$, where $\boldsymbol{h}=(h_1,h_2)$, $\boldsymbol{b}=(b_1,b_2)$ is a known vector, and $\tau>0$.

(a) What is the equation obeyed by the associated wavevector–time spectrum, $\widetilde{c}_X(\boldsymbol{k},\tau)$, $\boldsymbol{k}=(k_1,k_2)$?

(b) Given the above initial condition what is the specific form of $\widetilde{c}_X(\boldsymbol{k},\tau)$?

Exercise V.19: Consider the ST in $R^{n,1}$,

$$\mathcal{X}_n(\boldsymbol{s},t)=T^n_{n+1}[\mathcal{X}_{n+1}](\boldsymbol{s},t)=\int ds_{n+1}\mathcal{X}_{n+1}(\boldsymbol{s},s_{n+1},t), \tag{19.1}$$

where $\boldsymbol{s}=(s_1,\ldots,s_n)$.

(a) Show that in polar coordinates, Eq. (19.1) is written as

$$\mathcal{X}_n(s,t)=2\int_0^\infty ds_{n+1}\mathcal{X}_{n+1}\left(\left(s^2+s_{n+1}^2\right)^{\frac{1}{2}},t\right)=2\int_r^\infty du\,u\left(u^2-s^2\right)^{-\frac{1}{2}}\mathcal{X}_{n+1}(u,t), \tag{19.2}$$

where $s=|\boldsymbol{s}|=\left(\sum_{i=1}^n s_i^2\right)^{\frac{1}{2}}$ and $u^2=s^2+s_{n+1}^2$.

(b) Using Eq. (19.1) show that

$$\mathcal{X}_n(\mathbf{s},t) = T^n_{n+2}[\mathcal{X}_{n+2}](\mathbf{s},t) = \iint ds_{n+1}ds_{n+2}\mathcal{X}_{n+2}(\mathbf{s},s_{n+1},s_{n+2},t), \tag{19.3}$$

and in polar coordinates,

$$\mathcal{X}_n(s,t) = \int_0^{2\pi} d\theta \int_0^{\infty} dv\, v\mathcal{X}_{n+2}\left(\left(s^2+v^2\right)^{\frac{1}{2}},t\right) = 2\pi \int_r^{\infty} du\, u\mathcal{X}_{n+2}(u,t). \tag{19.4}$$

(c) Derive Eq. (19.4) from Eq. (19.2).
(d) Show that the inverse of Eq. (19.4) is

$$\mathcal{X}_{n+2}(s,t) = -\frac{1}{2\pi s}\frac{\partial}{\partial s}\mathcal{X}_n(s,t), \tag{19.5}$$

where $s = |\mathbf{s}| = \left(\sum_{i=1}^n s_i^2\right)^{\frac{1}{2}}$.

Exercise V.20: Using the results of Exercise V.19, show that any $\mathcal{X}_n(s,t)$, $s = |\mathbf{s}| = \left(\sum_{i=1}^n s_i^2\right)^{\frac{1}{2}}$, in $R^{n,1}$ ($n = 1, 2,\ldots$) can be expressed in terms of $\mathcal{X}_1(s,t)$ in $R^{1,1}$.

Exercise V.21: STs can be extended in a variety of ways, depending on the application circumstances. Assume that the space−time function $\mathcal{X}$ in $R^{n,1}$ also depends on an additional argument φ, and consider the ST,

$$\mathcal{X}_n(\mathbf{s},t;\varphi) = T^n_{n+1}[\mathcal{X}_{n+1}](\mathbf{s},t;\varphi) = \int ds_{n+1}\cos(s_{n+1}\varphi)\mathcal{X}_{n+1}(\mathbf{s},s_{n+1},t;0), \tag{21.1}$$

where $\mathbf{s} = (s_1,\ldots, s_n)$.
(a) Show that in polar coordinates, Eq. (21.1) becomes

$$\begin{aligned}\mathcal{X}_n(s,t;\varphi) &= 2\int_0^{\infty} ds_{n+1}\cos(s_{n+1}\varphi)\mathcal{X}_{n+1}\left(\left(s^2+s_{n+1}^2\right)^{\frac{1}{2}},t;0\right)\\ &= 2\int_s^{\infty} du\left(u^2-s^2\right)^{-\frac{1}{2}}u\cos\left(\left(u^2-s^2\right)^{\frac{1}{2}}\varphi\right)\mathcal{X}_{n+1}(u,t;0),\end{aligned} \tag{21.2a–b}$$

where $s^2 = \sum_{i=1}^n s_i^2$ and $u^2 = s^2 + s_{n+1}^2$.
(b) Using Eq. (21.1) show that

$$\mathcal{X}_n(\mathbf{s},t;\varphi) = T^n_{n+2}[\mathcal{X}_{n+2}](\mathbf{s},t;\varphi) = \iint ds_{n+1}ds_{n+2}\cos(s_{n+1}\varphi)\mathcal{X}_{n+2}(\mathbf{s},s_{n+1},s_{n+2},t;0), \tag{21.3}$$

and in polar coordinates,

$$\begin{aligned} X_n(s,t;\varphi) &= \int_0^{2\pi} d\theta \int_0^\infty dv\, v \cos(\varphi v \sin\theta) X_{n+2}\left(\left(s^2+v^2\right)^{\frac{1}{2}}, t; 0\right) \\ &= 2\pi \int_s^\infty du\, u\, J_0\left(\left(u^2-s^2\right)^{\frac{1}{2}}\varphi\right) X_{n+2}(u,t;0) \end{aligned} \tag{21.4a–b}$$

(c) Derive Eqs. (21.4a–b) from Eqs. (21.2a–b).
(d) Show that at $\varphi = 0$ Eqs. (21.2b) and (21.4b) give

$$\begin{aligned} X_n(s,t;0) &= 2\int_s^\infty du\, u\left(u^2-s^2\right)^{-\frac{1}{2}} X_{n+1}(u,t;0), \\ X_n(s,t;0) &= 2\pi \int_s^\infty du\, u X_{n+2}(u,t;0). \end{aligned} \tag{21.5a–b}$$

(e) Show that the inverse of Eq. (21.5b) gives

$$X_{n+2}(s,t;0) = -\frac{1}{2\pi s}\frac{\partial}{\partial s} X_n(s,t;0). \tag{21.6}$$

(f) Show that Eqs. (21.5a–b) and (21.6) hold for any φ, i.e.,

$$\begin{aligned} X_n(s,t;\varphi) &= 2\int_s^\infty du\, u\left(u^2-s^2\right)^{-\frac{1}{2}} X_{n+1}(u,t;\varphi), \\ X_n(s,t;\varphi) &= 2\pi \int_s^\infty du\, u X_{n+2}(u,t;\varphi), \\ X_{n+2}(s,t;\varphi) &= -\frac{1}{2\pi s}\frac{\partial}{\partial s} X_n(s,t;\varphi) \end{aligned} \tag{21.7a–c}$$

Exercise V.22: Using the results of Exercise V.21, show that any $X_n(s,t;\varphi)$, $s = |\mathbf{s}| = \left(\sum_{i=1}^n s_i^2\right)^{\frac{1}{2}}$, in $R^{n,1}$ ($n = 1, 2, \ldots$) can be expressed in terms of $X_1(s,t;0)$ in $R^{1,1}$.
Exercise V.23: Consider a TT of the form

$$(\mathbf{s},t) \mapsto \left(\frac{\mathbf{s}_{-n} + \mathbf{v}_{-n}t}{s_n + v_n t}, t\right), \tag{23.1}$$

where $s = (s_1,\ldots, s_n)$, $s_{-n} = (s_1,\ldots, s_{n-1})$, $v = (v_1,\ldots, v_n)$, and $v_{-n} = (v_1,\ldots, v_{n-1})$. Define the S/TRF such that

$$X(s,t) = X\left(\frac{s_{-n} + v_{-n}t}{s_n + v_n t}, t\right). \tag{23.2}$$

Derive the corresponding covariance and variogram functions. What is your interpretation of this random field?

CHAPTER VI

Exercise VI.1: Space–time random field geometrical features (continuity, differentiability, integrability) characterize essential variability aspects (smoothness, roughness, etc.) of a physical attribute. Show that the random field of Eq. (3.10) of Chapter VI is almost surely (a.s.) continuous but not mean square (m.s.) continuous.
Exercise VI.2: Consider the random field $X(s,t)$, $s = (s_1,s_2)$, defined in $R^{2,1}$ as

$$X(s,t) = \begin{cases} \dfrac{s_1 s_2^2}{s_1^2 + s_2^4} v(t) & \text{if } s_1, s_2 \neq 0 \\ 0 & \text{otherwise,} \end{cases} \tag{2.1a–b}$$

where the RV $v(t) \sim N(0,1)$.

(a) Does the directional derivative $D_{\varepsilon_1,\varepsilon_2}(s,t)$ exist in the m.s. sense?
(b) Is the $X(s,t)$ m.s. continuous at $s = \mathbf{0}$?

Exercise VI.3: Let $X(p) = (e^{s_0 s_1} + s_2)v$, $p = (s_1,s_2, s_0)$, denote an S/TRF where the RV $v \sim N(0,1)$. Find:

(a) the directional derivative $D_{\varepsilon_1,\varepsilon_2}X(p)$ in direction $\varepsilon = (\varepsilon_1,\varepsilon_2)$ defined in terms of the partial S/TRF derivatives;
(b) the $D_{\varepsilon_1,\varepsilon_2}X(p)$ at $p = \mathbf{0}$;
(c) the $\frac{\Delta^{1,1}_{\varepsilon_1,\varepsilon_2}X(p)}{h}$; and
(d) the $\frac{\Delta^{1,1}_{\varepsilon_1,\varepsilon_2}X(p)}{h}$ at $p = \mathbf{0}$.

Exercise VI.4: For the random field studied in Example 4.9 of Chapter VI, describe step-by-step the calculations leading to the corresponding limits.
Exercise VI.5: The choice of a space–time covariance function can determine the regularity features of the S/TRF realizations. Outline the main parts of proof of Proposition 4.1 of Chapter VI, which links random field differentiability with mean and covariance function differentiability.
Exercise VI.6: Consider the random field $X(s,t) = v(t)\, s$ in $R^{1,1}$, where $v(t)$ is an RV with zero mean and variance $\sigma_v^2(t)$. The simultaneous two-point covariance of $X(s,t)$ is $c_X(s,s',t) = \sigma_v^2(t)ss'$. Find the partial and the generalized covariance derivatives.
Exercise VI.7: Let $X(s,t)$ be a random field in $R^{1,1}$ that has continuous sample paths with probability 1 and is characterized by a set of finite-dimensional CDFs. Also, consider the $Y(s,t) = X(s,t)$ if $s \neq \nu$, and $Y(\nu,t) = X(\nu,t) + 1$, where ν is an RV independent of $X(s,t)$ with a continuous CDF.

(a) How are the finite-dimensional CDFs of $X(s,t)$ and $Y(s,t)$ related?
(b) Are the $Y(s,t)$ sample paths discontinuous at ν?

Exercise VI.8: Find the expressions of Eqs. (4.6a) and(4.6b) of Chapter VI in the case of space (arbitrary directions)–time partial derivatives.
Exercise VI.9: Describe the steps leading to space–time covariance derivative of Eq. (4.18a) of Chapter VI.
Exercise VI.10: Outline the main parts of the proof of Proposition 4.2 of Chapter VI.
Exercise VI.11: Let $X(s,t)$ in $R^{1,1}$.
(a) If the m.s. derivative

$$X^{(1,0)}(s,t) = \frac{\partial}{\partial s} X(s,t) = \underset{h \to 0}{\text{l.i.m.}} \frac{X(s+h,t) - X(s,t)}{h}$$

exists, derive $\overline{X^{(1,0)}(s,t)}$ in terms of $\overline{X(s,t)}$.
(b) The same for $c_{X^{(1,0)}X}(s,s',t)$, $c_{XX^{(1,0)}}(s,s',t)$ and $c_{X^{(1,0)}}(s,s',t)$ in terms of $c_X(s,s',t)$.

Exercise VI.12: Prove the validity of the analysis in Remark 4.1 of Chapter VI.
Exercise VI.13: Under certain conditions the random field properties can be considered separately in space and time. Example 4.15 of Chapter VI studies links between a random field and its derivatives, as well as between the random field derivatives themselves. Check the validity of the main results using specific random field models of your choice.
Exercise VI.14: Recall that separability may be interpreted as representing purely spatial and purely temporal correlations that involve in a multiplicative way (the main drawback of separability is that it may not incorporate important space–time interactions). Consider the condition of Eq. (3.18) of Chapter VI for a space–time separable covariance function. What is the form of α_t in the separability case?
Exercise VI.15: In Example 4.6 of Chapter VI:
(a) Along which direction the maximum directional derivative is obtained?
(b) Is it valid that $|\nabla X(\boldsymbol{p})| = |D_{\varepsilon_1,\varepsilon_2} X(\boldsymbol{p})|$ in the direction of the vector $\boldsymbol{\varepsilon} = (\varepsilon_1,\varepsilon_2)$?

Exercise VI.16: Let $X(s,t)$ be a zero-mean white noise random field in $R^{1,1}$ with $c_X(s,t,s,t') = W(s,t,s,t')\delta(s-s',t-t')$. Show that Eq. (6.4b) of Chapter VI reduces to Eq. (6.5) of Chapter VI.
Exercise VI.17: Derive Eq. (6.9) in Example 6.1 of Chapter VI.
Exercise VI.18: In $R^{n,1}$, consider the space–time covariance $c_X(\boldsymbol{h}) = e^{-a^{-2}|\boldsymbol{h}|^2 - b^{-2}\tau^2}$, where a, $b > 0$ (the covariance depends on $\boldsymbol{s}$ and $\boldsymbol{s}'$ through the Euclidean metric $|\boldsymbol{h}| = r_E$).
(a) Is the covariance continuous at the origin?
(b) Are the covariance derivatives defined?
(c) Is the associated random field $X(\boldsymbol{s},t)$ m.s. continuous and differentiable?

Exercise VI.19: Given the specific derivatives of the random field $X(\boldsymbol{p})$ in the left column of Table 19.1, write the expressions of the corresponding covariances in the right column of Table 19.1 in terms of the random field covariance function $c_X(\boldsymbol{p},\boldsymbol{p}')$.

Exercise VI.20: Spectra are useful when one seeks to compute the variance of random fields that have been filtered or smoothed by a filtering or smoothing kernel $\alpha(\boldsymbol{s})$. Given the S/TRF $X(\boldsymbol{s},t)$ in $R^{3,1}$, let the filtered random field be given by

$$Z(\boldsymbol{s},t) = \int d\boldsymbol{u}\ \alpha(\boldsymbol{u}) X(\boldsymbol{s}-\boldsymbol{u},t). \tag{19.1}$$

Table 19.1 Spatiotemporal Random Field (S/TRF) and Covariance Derivatives

S/TRF Derivative	Space−time Covariance Derivative
$X_{d,0}^{(\nu,\mu)}(\boldsymbol{p}) = \frac{\partial^{\nu+\mu} X(\boldsymbol{p})}{\partial s_d^{\nu} \partial s_0^{\mu}}$	Covariance expression in general case: $\mathrm{cov}\left(X_{d,0}^{(\nu,\mu)}(\boldsymbol{p}), X_{d,0}^{(\nu',\mu')}(\boldsymbol{p}')\right) = ?$ Covariance expressions in special cases: (a) $\boldsymbol{\nu} = \boldsymbol{\nu}' = \mathbf{1}$: $\mathrm{cov}\left(X_{d,0}^{(\nu,\mu)}(\boldsymbol{p}), X_{d,0}^{(\nu',\mu')}(\boldsymbol{p}')\right) = ?$ (b) $\boldsymbol{\nu} = \boldsymbol{\nu}' = \mathbf{1},\ \rho = \lvert\boldsymbol{\nu}\rvert = \nu = \nu'$: $\mathrm{cov}\left(X_{d,0}^{(\nu,\mu)}(\boldsymbol{p}), X_{d,0}^{(\nu',\mu')}(\boldsymbol{p}')\right) = ?$
$X_{i,0}^{(\nu,\mu)}(\boldsymbol{p}) = \frac{\partial^{\nu+\mu} X(\boldsymbol{p})}{\partial s^{\nu} \partial s_0^{\mu}}$	Covariance expression in general case: $\mathrm{cov}\left(X_{i,0}^{(\nu,\mu)}(\boldsymbol{p}), X_{i,0}^{(\nu',\mu')}(\boldsymbol{p}')\right) = ?$ Covariance expressions in special cases: (a) $\boldsymbol{\nu} = \boldsymbol{\nu}' = \mathbf{1}$: $\mathrm{cov}\left(X_{i,0}^{(\nu,\mu)}(\boldsymbol{p}), X_{i,0}^{(\nu',\mu')}(\boldsymbol{p}')\right) = ?$ (b) $\boldsymbol{\nu} = \boldsymbol{\nu}' = \mathbf{1},\ \lvert\boldsymbol{\nu}\rvert = \nu = \nu' = n$: $\mathrm{cov}\left(X_{i,0}^{(\nu,\mu)}(\boldsymbol{p}), X_{i,0}^{(\nu',\mu')}(\boldsymbol{p}')\right) = ?$

Show that

$$\overline{Z^2(t)} = (2\pi)^{-3} \int d\boldsymbol{k} \lvert \tilde{\alpha}(\boldsymbol{k}) \rvert^2 \tilde{c}_X(\boldsymbol{k}, t). \tag{19.2}$$

CHAPTER VII

Exercise VII.1: Auxiliary random field modeling hypotheses can be insightful and useful to capture and parameterize the essential physics. Space−time homostationarity (STHS) is a key hypothesis that characterizes many random field applications. Show that a homostationary S/TRF is also an ordinary S/TRF-ν/μ for any values of the orders ν and μ, but the converse is not generally true.

Exercise VII.2: Prove the validity of the spectral representations in Eq. (3.6c−d) of Chapter VII.

Exercise VII.3: Prove Propositions 3.1−3.3 of Chapter VII.

Exercise VII.4: Assume that the random field $X(s,t)$ in $R^{1,1}$ is related to the random field $Y(s,t)$ by $Y(s,t) = -aX^{(1,0)}(s,t)$, where a is a physical constant. Determine the covariance of $Y(s,t)$ in terms of the variogram of $X(s,t)$.

Exercise VII.5: For the derivatives of the S/TRF $X(\boldsymbol{p})$ in the left column of Table 5.1, write the expressions of the corresponding covariance functions in the right column of Table 19.1 in terms of the STHS random field covariance function $c_X(\Delta\boldsymbol{p})$.

Exercise VII.6: Prove Proposition 4.1 of Chapter VII.

Table 5.1 Spatiotemporal Random Field (S/TRF) and Covariance and Variance Derivatives ($\Delta p = p' - p$)

S/TRF Derivative	Space–time Covariance Derivative
$X_{d,0}^{(\nu,\mu)}(p) = \frac{\partial^{\nu+\mu} X(p)}{\partial s_d^{\nu} \partial s_0^{\mu}}$	$\mathrm{cov}\left(X_{d,0}^{(\nu,\mu)}(p), X_{d,0}^{(\nu',\mu')}(p')\right) = ?$ $\mathrm{cov}\left(X_{d,0}^{(\nu,\mu)}(p), X_{d,0}^{(\nu,\mu)}(p')\right) = ?$
$X_{d_i,0}^{(\nu,\mu)}(p) = \frac{\partial^{\nu+\mu} X(p)}{\partial s_{d_i}^{\nu} \partial s_0^{\mu}}$	$\mathrm{cov}\left(X_{d_i,0}^{(\nu,\mu)}(p), X_{d_j,0}^{(\nu',\mu')}(p')\right) = ?$
$X_{i,0}^{(\nu,\mu)}(p) = \frac{\partial^{\nu+\mu} X(p)}{\partial s_i^{\nu} \partial s_0^{\mu}}$	$\mathrm{cov}\left(X_{i,0}^{(\nu,\mu)}(p), X_{j,0}^{(\nu',\mu')}(p')\right) = ?$ $\mathrm{var}\left(X_{i,0}^{(\nu,\mu)}(p)\right) = ?$
$X_{n,0}^{(\nu,\mu)}(p) = \frac{\partial^{\nu+\mu} X(p)}{\partial s^{\nu} \partial s_0^{\mu}}$	$\mathrm{cov}\left(X_{n,0}^{(\nu,\mu)}(p), X_{n,0}^{(\nu',\mu')}(p')\right) = ?$ $\mathrm{var}\left(X_{n,0}^{(\nu,\mu)}(p)\right)$, $\mathrm{var}\left(X_{n,0}^{(\nu,\mu)}(0)\right) = ?$

Exercise VII.7: Several interesting expressions are obtained in terms of the linear and homogeneous differential operator $\mathcal{L}_p$.

(a) Let $\Delta p = p - p'$, $\mathcal{L}_p = \frac{\partial^{\nu+\mu}}{\partial s^{\nu} \partial s_0^{\mu}}$, and $\mathcal{L}_{\Delta p} = \frac{\partial^{\nu+\mu}}{\partial h^{\nu} \partial h_0^{\mu}}$, where $\partial s^{\nu} = \prod_{i=1}^{n} \partial s_i^{\nu_i}$, and $\partial h^{\nu} = \prod_{i=1}^{n} \partial h_i^{\nu_i}$. What is the expression for $\mathcal{L}_p c_X(p - p')$ in terms of $\mathcal{L}_{\Delta p} c_X(\Delta p)$?

(b) On the other hand, by letting $\Delta p = p - p'$, $\mathcal{L}_{p'} = \frac{\partial^{\nu+\mu}}{\partial s'^{\nu} \partial s_0'^{\mu}}$, and $\mathcal{L}_{\Delta p} = \frac{\partial^{\nu+\mu}}{\partial h^{\nu} \partial h_0^{\mu}}$, what is the expression of $\mathcal{L}_{p'} c_X(p - p')$ in terms of $\mathcal{L}_{\Delta p} c_X(\Delta p)$?

Exercise VII.8: Under what conditions on the space–time arguments, is it valid that

$$\frac{\partial c_X(p - p')}{\partial s_i} = \frac{\partial c_X(\Delta p)}{\partial h_i} \quad (i = 0, 1, \ldots, n)? \tag{8.1}$$

Exercise VII.9: Answer the following:

(a) For $\nu = \mu = \nu' = \mu' = 1$, what is the expression of $\mathrm{cov}\left(X^{(1,1)}(p), X^{(1,1)}(p')\right)$ in terms of $c_X(\Delta p)$?

(b) For $\nu = \mu = 1, \nu' = 0, \mu' = 2$, what is the expression of $\mathrm{cov}\left(X^{(1,1)}(p), X^{(0,2)}(p')\right)$ in terms of $c_X(\Delta p)$?

Exercise VII.10: Prove that the differentiation formulas in the case of a spatially varying random field (fixed time) $X(s)$ are special cases of Eqs. (4.10a) and (4.10b) of Chapter VII. Specifically,

(a) with $h = s' - s$ express $c_{X^{(\nu)}, X^{(\nu')}}(s, s')$, $c_{X^{(\nu)}, X^{(\nu')}}(s', s)$, and $\sigma^2_{X^{(\nu)}}(s)$ in terms of $c_X(h)$;

(b) also, express $c_{X_i^{(\nu)}, X_j^{(\nu')}}(s, s')$ in terms of $c_{X_{i,j}}(h)$.

Exercise VII.11: Derive the S/TRF derivative expressions shown in Corollary 4.1 of Chapter VII.
Exercise VII.12: Confirm Eq. (4.13) of Chapter VII.

Exercise VII.13: Prove the validity of Eq. (4.29) of Chapter VII.

Exercise VII.14: Let $X(s,t)$, $(s,t) \in R^{1,1}$ be an STHS random field.

(a) Are the random fields $X^{(1,0)}(s,t) = \frac{\partial}{\partial s} X(s,t)$ and $X^{(0,1)} = \frac{\partial}{\partial t} X(s,t)$ (assuming they exist) STHS too?

(b) Can you express the simultaneous covariance $c_X(h,t)$ in terms of $c_X(0,t)$ and $c_{X^{(1,0)}}(0,t)$?

(c) Derive the covariances of the $X(s,t)$ derivatives in the STHS case using the chain rule of composite function differentiation and the corresponding limits.

(d) Show that the same results are obtained in terms of the commuting property of the differentiation and expectation operators.

(e) Are the $c_{X^{(1,0)}X}(h,t)$, $c_{XX^{(1,0)}}(h,t)$ and $c_{X^{(1,0)}X^{(1,0)}}(h,t)$ odd or even functions.

(f) How are the $c_{X^{(1,0)}X}(h,t)$ and $c_{XX^{(1,0)}}(h,t)$ related?

(g) Same question for the pairs $\{c_{X^{(2,0)}X}(h,t),\ c_{X^{(1,0)}}(h,t)\}$, $\{\frac{\partial}{\partial h} c_{X^{(1,0)}}(h,t),\ c_{X^{(1,0)}X^{(2,0)}}(h)\}$, $\left\{\frac{\partial^4}{\partial h^4} c_X(h,t),\ \frac{\partial}{\partial h} c_{X^{(1,0)}X^{(2,0)}}(h,t)\right\}$, and $\left\{\frac{\partial^4}{\partial h^4} c_X(h,t),\ c_{X^{(2,0)}}(h,t)\right\}$.

(h) Lastly, express the spectra $\widetilde{c}_{X^{(1,0)}X}(k,t)$, $\widetilde{c}_{X^{(1,0)}}(k,t)$, $\widetilde{c}_{X^{(2,0)}X^{(1,0)}}(k,t)$, and $\widetilde{c}_{X^{(2,0)}}(k,t)$ in terms of $\widetilde{c}_X(k,t)$.

Exercise VII.15: In $R^{n,1}$, consider the directional space–time derivatives

$$\begin{aligned} &\overline{X(s,t)\frac{\partial^3}{\partial h_i^2 \partial \tau} X(s+h_i\boldsymbol{\varepsilon}_i, t+\tau)} = c_{XX^{(2,1)}}(\boldsymbol{h},\tau), \\ &\overline{\frac{\partial^3}{\partial h_i^2 \partial \tau} X(s-h_i\boldsymbol{\varepsilon}_i, t-\tau) X(s,t)} = -c_{X^{(2,1)}X}(\boldsymbol{h},\tau). \end{aligned} \tag{15.1a–b}$$

When does it hold that $c_{XX^{(2,1)}}(0,0) = -c_{X^{(2,1)}X}(0,0) = 0$?

Exercise VII.16: For a given t in $R^{1,1}$, assume that for $s \in [a,b]$ it holds

$$\overline{|X(s+h,t) - X(s,t)|^{\kappa}} \le \alpha_t |h|^{1+\varepsilon} \tag{16.1}$$

for some α_t, κ, $\varepsilon > 0$ and sufficiently small h. Is the random field a.s. continuous in space?

Exercise VII.17: Answer the following:

(a) If $g(\cdot)$: $L_2 \to R^1$ is a continuous function and $X(\boldsymbol{p})$ is an a.s. continuous random field, is the resulting random field $Y(\boldsymbol{p}) = g(X(\boldsymbol{p}))$ a.s. continuous?

(b) If $g(\cdot)$: $R^1 \to R^1$ is a continuous function that is Lipschitz of order 1 and $X(\boldsymbol{p})$ is an m.s. continuous random field, is the random field $Y(\boldsymbol{p}) = g(X(\boldsymbol{p}))$ m.s. continuous?

Exercise VII.18: For a set of independent STHS random fields $X_i(\boldsymbol{p})$, $i = 1,\ldots, m$, and $u_i(\boldsymbol{p})$ assumed to be continuous deterministic functions, the random field $Y(\boldsymbol{p}) = \sum_{i=1}^{m} u_i(\boldsymbol{p}) X_i(\boldsymbol{p})$ is generally a space–time heterogeneous field.

(a) If the $X_i(\boldsymbol{p})$, $i = 1,\ldots, m$, are a.s. continuous, is the $Y(\boldsymbol{p})$ a.s. continuous too?

(b) Similar question for the case of m.s. continuity.

Exercise VII.19: Answer the following:
(a) In the case of an STHS random field, under what conditions is the covariance function smooth away from the origin?
(b) Prove Corollary 4.2 of Chapter VII.

Exercise VII.20: Prove Eqs. (5.3a–b) of Chapter VII.
Exercise VII.21: In $R^{n,1}$ use Eq. (4.10a) of Chapter VII to derive $\text{cov}\left(X(\boldsymbol{p}), X^{(1,1)}(\boldsymbol{p}')\right)$, $\text{cov}\left(X^{(1,1)}(\boldsymbol{p}), X^{(1,1)}(\boldsymbol{p}')\right)$, and $\text{cov}\left(X^{(1,1)}(\boldsymbol{p}), X^{(0,2)}(\boldsymbol{p}')\right)$ in terms of derivatives of $c_X(\Delta\boldsymbol{p})$.
Exercise VII.22: Is it valid that

$$\begin{aligned} c_X(\tau) &= \int d\boldsymbol{k}\, \tilde{c}_X(\boldsymbol{k})\tilde{\rho}_X(\boldsymbol{k},\tau), \\ c_X(\boldsymbol{h}) &= \int d\omega\, \tilde{c}_X(\omega)\tilde{\rho}_X(\boldsymbol{h},\omega), \end{aligned} \tag{22.1a–b}$$

where $\tilde{\rho}_X(\boldsymbol{k},\tau)$ and $\tilde{\rho}_X(\boldsymbol{h},\omega)$ are the normalized angular spectrum (dynamical correlation function) and the normalized frequency spectrum, respectively?
Exercise VII.23: In the case of a space–time separable (multiplicative) covariance $c_X(\boldsymbol{h},\tau) = c_{X(1)}(\boldsymbol{h})c_{X(2)}(\tau)$,
(a) find the corresponding sample path continuity condition for Eq. (4.37) of Chapter VII; and
(b) find the α_t in Eq. (4.39) of Chapter VII.

Exercise VII.24: In Example 4.7 of Chapter VII prove that Eq. (4.21) of Chapter VII is equivalent to the existence of the variance of Eq. (4.22) of Chapter VII, which, in turn, is equivalent to the existence of $X^{(1,1)}(s,t)$ in the m.s. sense.
Exercise VII.25: What is the necessary and sufficient condition for the existence of $\nabla X(\boldsymbol{p})$ expressed in terms of the corresponding covariance function?
Exercise VII.26: Prove Proposition 3.3 of Chapter VII.
Exercise VII.27: Prove that the random field $X(s,t)$ of Example 4.16 of Chapter VII with the covariance

$$c_X(h,t,t') = \frac{1}{4}\int_{|t-t'|}^{t+t'} du\, u^{-\frac{1}{2}} e^{-\frac{h^2}{4u}} \tag{27.1}$$

in $R^{1,1}$ is a.s. α'-Hölder continuous in space with $\alpha' < \frac{1}{2}$ (i.e., "almost" $\frac{1}{2}$-Hölder continuous).
Exercise VII.28: For fixed space, are the solutions to stochastic ordinary differential equations in general α-Hölder continuous (in time) for $\alpha < \frac{1}{2}$ but not for $\alpha = \frac{1}{2}$?
Exercise VII.29: The space–time covariance expression of Eq. (27.1) of Exercise VII.27 can be extended in $R^{n,1}$ as

$$c_X(h,t,t') = \frac{1}{2^n}\int_{|t-t'|}^{t+t'} du\, u^{-\frac{n}{2}} e^{-\frac{h^2}{4u}}. \tag{29.1}$$

Show that $c_X(h,t,t')$ is finite only in $R^{1,1}$.
Exercise VII.30: For an STHS random field, based on the analysis in Section 5 of Chapter VII, find the spectral moment $\beta_X^{(\nu,\nu',\mu,\mu')}$ in terms of the corresponding spectrum.

CHAPTER VIII

Exercise VIII.1: In the case of space–time isostationarity (STIS), several interesting analytical equations can be derived using spherical coordinates. In this context, derive the covariance function representation pair of Eqs. (1.5a–b) of Chapter VIII.

Exercise VIII.2: Prove that in $R^{n,1}$ ($n = 1, 2, 3$) the corresponding spectral representations of the covariance functions are given by

$$n = 1: \quad c_X(r,\tau) = 2\int_0^\infty\int_0^\infty dk\, d\omega \cos(kr)\cos(\omega\tau)\, \widetilde{c}_X(k,\omega)$$

$$\widetilde{c}_X(k,\omega) = \frac{2}{\pi}\int_0^\infty\int_0^\infty dr\, d\tau \cos(kr)\cos(\omega\tau)\, c_X(r,\tau)$$

$$n = 2: \quad c_X(r,\tau) = 2\pi\int_0^\infty\int_0^\infty dk\, d\omega\, k\, J_0(kr)\cos(\omega\tau)\, \widetilde{c}_X(k,\omega)$$

$$\widetilde{c}_X(k,\omega) = \frac{1}{2\pi}\int_0^\infty\int_0^\infty dr\, d\tau\, r\, J_0(kr)\cos(\omega\tau)\, c_X(r,\tau)$$

$$n = 3: \quad c_X(r,\tau) = 4\pi\int_0^\infty\int_0^\infty dk\, d\omega\, k\frac{\sin(kr)}{r}\cos(\omega\tau)\, \widetilde{c}_X(k,\omega)$$

$$= \int_0^\infty\int_0^\infty dk\, d\omega\, r\left(\frac{2\pi k}{r}\right)^{3/2} J_{\frac{1}{2}}(kr)\cos(\omega\tau)\, \widetilde{c}_X(k,\omega)$$

$$\widetilde{c}_X(k,\omega) = \frac{1}{2\pi^2}\int_0^\infty\int_0^\infty dr\, d\tau\, r\frac{\sin(kr)}{k}\cos(\omega\tau)\, c_X(r,\tau)$$

Exercise VIII.3: If $X(\boldsymbol{p})$ is an STIS random field in $R^{n,1}$ ($n \geq 3$), under what conditions the simultaneous two-point correlation function is $\frac{n-1}{2}$-order space differentiable (this is an extension of a lemma by Yadrenko, 1983)?

Exercise VIII.4: Bessel functions are widely used in analytical calculations of STIS random fields. In this setting, derive Eqs. (2.6a–b) of Chapter VIII.

Exercise VIII.5: Prove Eqs. (2.9a) and (2.9b) of Example 2.5 of Chapter VIII.

Exercise VIII.6: Using Eq. (2.6b) of Chapter VIII, derive analytical expressions for $\left(\frac{1}{r}\frac{\partial}{\partial r}\right)^{\nu} c_X(r,\tau)$ for $\nu = 1$, $n = 1, 2, 3$. Confirm your results starting from Eqs. (1.5a–b) of Chapter VIII.

Exercise VIII.7: In $R^{2,1}$, show that

$$\frac{\partial^2 c_X(r,\tau)}{\partial h_1 \partial h_2} = h_1 h_2\, r^{-2}\left[\frac{\partial^2}{\partial r^2} - r^{-1}\frac{\partial}{\partial r}\right] c_X(r,\tau). \tag{7.1}$$

Exercise VIII.8: Using Eq. (2.1c) of Chapter VIII and given that $Y(\boldsymbol{s},t) = -\frac{\partial^2 X(s,t)}{\partial s_2^2} - \frac{\partial^2 X(s,t)}{\partial s_1^2}$ in $R^{2,1}$, determine $\overline{Y(r,0,0)Y(0,0,0)}$ in terms of $c_X(r, 0)$.

Exercise VIII.9: Given that $c_X(r,\tau) = e^{-\frac{r^2}{a^2}-\frac{\tau^2}{b^2}}$, and using

$$c_Y(r,\tau) = \left[\frac{1}{r^3}\frac{\partial}{\partial r} - \frac{1}{r^2}\frac{\partial^2}{\partial r^2} + \frac{2}{r}\frac{\partial^3}{\partial r^3} + \frac{\partial^4}{\partial r^4}\right] c_X(r,\tau), \tag{9.1}$$

find the exact expression of the space−time covariance $c_Y(r,\tau)$ in terms of a and b.
Exercise VIII.10: Derive the ergodicity Eqs. (7.4a−b) of Chapter VIII.
Exercise VIII.11: Confirm the list of ergodicity conditions in Table 7.1 of Chapter VIII.
Exercise VIII.12: Find the corresponding covariance derivatives for the numerical cases listed in Table 12.1.
Exercise VIII.13: Prove Proposition 2.1 of Chapter VIII.
Exercise VIII.14: Are the following derivatives

$$\frac{\partial^3}{\partial h_1 \partial h_2 \partial h_0} c_X(\boldsymbol{h},\tau), \frac{\partial^2}{\partial h_1 \partial h_2} c_X(\boldsymbol{h},\tau), \frac{\partial^2}{\partial h_1^2} c_X(\boldsymbol{h},\tau), \frac{\partial^4}{\partial h_1^2 \partial h_2 \partial \tau} c_X(\boldsymbol{h},\tau), \frac{\partial^3}{\partial h_1 \partial \tau^2} c_X(\boldsymbol{h},\tau)$$

zero at the origin or not ?
Exercise VIII.15: It is argued that in the case of multiplicative separability a space−time covariance model has the same regularity properties along the spatial axes and the temporal axis as the component spatial and temporal covariances, respectively. Investigate this argument using a specific example. In particular, consider in $R^{n,1}$ the space−time separable covariance function $c_X(\boldsymbol{h},\tau) = e^{-a^{-2}|\boldsymbol{h}|^2 - b^{-2}\tau^2}$, where a, $b > 0$ (the covariance depends on $\boldsymbol{s}$ and $\boldsymbol{s}'$ through the Euclidean metric $|\boldsymbol{h}| = r_E$).
(a) Is the covariance continuous at the origin?
(b) Are the covariance derivatives defined?
(c) Is the associated random field $X(s,t)$ m.s. continuous and differentiable?

Exercise VIII.16: Using Eq. (4.15) of Chapter VII in $R^{1,1}$ what existence condition the spectral density function $\widetilde{c}_X(k,\omega) = ck^{-\ln k}\omega^{-\ln \omega}$ satisfies for any ν, μ of Eq. (4.15) of Chapter VII?
Exercise VIII.17: Answer the following:
(a) Is the covariance $c_X(h,\tau) = e^{-\frac{h}{a}-\frac{\tau^2}{b^2}}\cos(ch)$ in $R^{1,1}$ (a,b,c > 0) continuous at $h = 0$?

Table 12.1 Covariance Derivatives for Different ν and μ

ν_1	ν_1'	ν_2	ν_2'	μ	μ'	$\frac{\partial^{\nu+\nu'+\mu+\mu'}}{\prod_{i=1}^n \partial h_i^{\nu_i+\nu_i'} \partial \tau^{\mu+\mu'}} \widehat{c}_X(r^2,\tau)$
2	0	0	1	1	0	
1	0	0	0	2	0	
2	0	0	0	0	0	
1	0	0	1	0	0	
1	0	0	1	1	0	

(b) Is the associated random field $X(s,t)$, continuous and differentiable in the m.s. sense?

Exercise VIII.18: In $R^{1,1}$ with $\boldsymbol{p} = (s,t)$, for the m.s. differentiable STHS random field $X(s,t)$ find the $\overline{\left(\frac{\partial X(s,t)}{\partial s}\right)^2}$ and $\overline{X(s,t)\frac{\partial^2 X(s,t)}{\partial s^2}}$ in terms of the corresponding spectral moments.

Exercise VIII.19: Consider a random field $X(\boldsymbol{p})$ in $R^{n,1}$ with STIS 0/0 increments $Y(\Delta\boldsymbol{p})$ so that

$$\begin{aligned}&\overline{Y(\boldsymbol{p},\boldsymbol{p}+\Delta\boldsymbol{p})} = \overline{X(\boldsymbol{p}) - X(\boldsymbol{p}+\Delta\boldsymbol{p})} = \overline{Y(\Delta\boldsymbol{p})} = \overline{Y(r,\tau)},\\ &\gamma_X(\boldsymbol{p},\boldsymbol{p}+\Delta\boldsymbol{p}) = \gamma_X(r,\tau).\end{aligned} \tag{19.1a–b}$$

Under what conditions, it holds that

$$\overline{X(\boldsymbol{p})} = ct? \tag{19.2}$$

Exercise VIII.20: Derive the specific form of Eq. (19.2) of Exercise VIII.19 in $R^{2,1}$, assuming the spatial rotation matrix

$$\mathcal{R}_\theta = \begin{bmatrix}\cos\theta & -\sin\theta\\ \sin\theta & \cos\theta\end{bmatrix},$$

so that the space–time matrix is given by

$$\mathcal{R}_{\theta,\alpha} = \begin{bmatrix} \multicolumn{2}{c}{\mathcal{R}_\theta} & \begin{matrix}0\\0\end{matrix}\\ 0\ 0 & & \alpha\end{bmatrix} = \begin{bmatrix}\cos\theta & -\sin\theta & 0\\ \sin\theta & \cos\theta & 0\\ 0 & 0 & \alpha\end{bmatrix}.$$

Exercise VIII.21: Consider the space–time covariance model suggested by Gneiting (2002)

$$c_X(r,\tau) = c_0\left(b\tau^{2\alpha}+1\right)^{-\frac{\beta n}{2}} e^{-\frac{\kappa r^{2\lambda}}{(b\tau^{2\alpha}+1)^{\beta\lambda}}} \tag{21.1}$$

in $R^{n,1}$, where $r = r_E$, and c_0 is the variance. This model has been criticized by Kent et al. (2011) that it does not satisfy the monotonicity property (i.e., for fixed τ the $c_X(r,\tau)$ should decrease with r and for fixed r it should decrease with τ). (a) Do you agree with this criticism and how significant is it in practice? (b) Which model parameters control random field smoothness and which control space–time interaction?

Exercise VIII.22: In many applications we are interested about the spectral distribution of the wavenumber $k = |\boldsymbol{k}|$. In this case, the spectrum $\widetilde{c}_X^{\diamond}(k,\omega)$ is used instead of $\widetilde{c}_X(\boldsymbol{k},\omega)$, which is such that

$$\int_0^\infty dk\, \widetilde{c}_X^{\diamond}(k,\omega) = \int d\boldsymbol{k}\, \widetilde{c}_X(\boldsymbol{k},\omega). \tag{22.1}$$

(a) Show that

$$\widetilde{c}_X^{\diamond}(k,\omega) = S_n k^{n-1}\widetilde{c}_X(k,\omega), \tag{22.2}$$

where $S_n = \dfrac{2\pi^{\frac{n}{2}}}{\Gamma\left(\frac{n}{2}\right)}$.

(b) Which is the corresponding expression for $\widetilde{c}_X^{\diamond}(k,\tau)$?

Exercise VIII.23: Suppose that given the STIS random field $X(\boldsymbol{s},t)$ in $R^{3,1}$, you are interested about its behavior along a certain direction s_1 so that the spectrum of the corresponding random field $X(s_1,t)$ in $R^{1,1}$ is given by,

$$\widetilde{c}_{X,1}(k_1,\omega) = \iint dk_2 dk_3 \widetilde{c}_X(\boldsymbol{k},\omega). \tag{23.1}$$

Show that the spectra in $R^{3,1}$, $\widetilde{c}_X(k,\omega)$, and $R^{1,1}$, $\widetilde{c}_{X,1}(k_1,\omega)$, are related simply by

$$\widetilde{c}_X(k,\omega) = -\frac{1}{2\pi k_1}\frac{\partial \widetilde{c}_{X,1}(k_1,\omega)}{\partial k_1}\Big|_{k_1=k}. \tag{23.2}$$

CHAPTER IX

Exercise IX.1: Let $X_l(\boldsymbol{p})$ be an STHS random field with mean $\overline{X_l(\boldsymbol{p})}$ and constant variance $\sigma^2_{X_l}$, and let us define the random field $X_{l'}(\boldsymbol{p}) = a(\boldsymbol{p})X_l(\boldsymbol{p})$, where $a(\boldsymbol{p})$ is a known deterministic function.

(a) Find the space–time statistics $\overline{X_{l'}(\boldsymbol{p})}$, $\sigma^2_{X_{l'}}(\boldsymbol{p})$, and $c_{X_{l'}}(\boldsymbol{p},\boldsymbol{p}')$ in terms of $a(\boldsymbol{p})$, $\overline{X_l(\boldsymbol{p})}$, $\sigma^2_{X_l}$, and $c_{X_l}(\boldsymbol{p}-\boldsymbol{p}')$.

(b) Find the space–time correlation function $\rho_{X_{l'}}(\boldsymbol{p},\boldsymbol{p}')$ in terms of $\rho_{X_l}(\boldsymbol{p}-\boldsymbol{p}')$. What do you observe?

Exercise IX.2: Two important cases of vector S/TRFs (VS/TRFs) were introduced in Section 4 of Chapter IX: A *solenoidal* VS/TRF, $\boldsymbol{X}^s(\boldsymbol{p})$, was defined in Eq. (4.1a) of Chapter IX as a divergence-free vector field, $\nabla\cdot\boldsymbol{X}^s(\boldsymbol{p}) = 0$. In fact, only solenoidal VS/TRFs are divergenceless, which means that zero divergence is a test for determining if a given VS/TRF is solenoidal. Another characteristic of $\boldsymbol{X}^s(\boldsymbol{p})$ is that it is the only vector field that can be expressed as the curl of some vector field $\boldsymbol{Z}(\boldsymbol{p})$, i.e., $\boldsymbol{X}^s(\boldsymbol{p}) = \nabla\times\boldsymbol{Z}(\boldsymbol{p})$. A *potential* VS/TRF, $\boldsymbol{X}^p(\boldsymbol{p})$, was defined in Eq. (4.2) of Chapter IX as a curl-free vector field, $\nabla\times\boldsymbol{X}^p(\boldsymbol{p}) = 0$. Actually, a necessary and sufficient condition for the existence of a potential VS/TRF is that it is curl free. Another characteristic of $\boldsymbol{X}^p(\boldsymbol{p})$ is that it can be expressed as the gradient of some scalar S/TRF $Y(\boldsymbol{p})$, i.e., $\boldsymbol{X}^p(\boldsymbol{p}) = \nabla Y(\boldsymbol{p})$, see Eq. (4.1b) of Chapter IX. The curl of any VS/TRF always results in a solenoidal VS/TRF. Show that for a potential vector random field $\boldsymbol{X}^p(\boldsymbol{p})$,

(a) $c^p_{\ell\ell}(r,\tau) \le c^p_{\eta\eta}(r,\tau)$ (the equality holding when $r=0$), and

(b) $c^p_{\ell\ell}(0,\tau) = c^p_{\eta\eta}(0,\tau) = -\frac{\partial^2}{\partial r^2}c_Y(0,\tau)$.

Exercise IX.3: Express the relationship in Eq. (7.7) of Chapter IX in terms of cross-variogram functions.

Exercise IX.4: Derive Eq. (7.32b) of Chapter IX under the conditions described in Example 7.9 of Chapter IX.

Exercise IX.5: Do the following:
(a) Check if for an VS/TRF $\boldsymbol{X}(\boldsymbol{p})$ the following vector identity is valid,

$$\nabla \cdot \nabla \times \boldsymbol{X}(\boldsymbol{p}) = \mathbf{0}. \tag{5.1}$$

(b) Check if Eq. (5.1) is analogous to the identity

$$\nabla \times \nabla \cdot X(\boldsymbol{p}) = \mathbf{0} \tag{5.2}$$

which is valid for a scalar S/TRF $X(\boldsymbol{p})$.
Exercise IX.6: Starting from

$$\xi_{X_l X_{l'}}(\boldsymbol{h}, t) = 2c_{X_l X_{l'}}(0, t) - c_{X_l X_{l'}}(\boldsymbol{h}, t) - c_{X_l X_{l'}}(-\boldsymbol{h}, t) \tag{6.1}$$

(Section 7.2 of Chapter IX), show that

$$\xi_{\ell\ell}(r, t) = 2c_{X_l}(0, t) - 2c_{\ell\ell}(r, \tau) \tag{6.2}$$

for any l.
Exercise IX.7: Derive the covariance function approximations of Eqs. (7.16a–b) of Chapter IX.
Exercise IX.8: Derive Eqs. (7.20a–b) of Chapter IX obtained for the longitudinal and the lateral covariances of a solenoidal random field.
Exercise IX.9: Show that in the case of divergence-free random fields the Eqs. (8.5a–b) of Chapter IX hold.
Exercise IX.10: In $R^{2,1}$, let $r^2 = h_1^2 + h_2^2$ and $\frac{h_1}{r} = \cos\phi$, where ϕ is the angle between h_1 and r. Using Eq. (7.7) of Chapter IX with $l, l' = 1, 2$ show that

$$\begin{aligned}
c_{X_1}(h_1, h_2, \tau) &= \cos^2\phi c_{\ell\ell}(r, \tau) + \sin^2\phi c_{\eta\eta}(r, \tau), \\
c_{X_2}(h_1, h_2, \tau) &= \sin^2\phi c_{\ell\ell}(r, \tau) + \cos^2\phi c_{\eta\eta}(r, \tau), \\
c_{X_1 X_2}(h_1, h_2, \tau) &= \cos\phi \sin\phi \left[c_{\ell\ell}(r, \tau) - c_{\eta\eta}(r, \tau)\right].
\end{aligned}$$

Exercise IX.11: Prove Eq. (8.6) of Chapter IX.
Exercise IX.12: In $R^{2,1}$, using Eq. (7.7) of Chapter IX with $l, l' = 1, 2$ find $c_{X_1}(r, 0, \tau)$, $c_{X_1}(0, r, \tau)$, $c_{X_1 X_2}(r, 0, \tau)$, $c_{X_1}(0, r, \tau)$, $c_{X_1 X_2}(0, 0, \tau)$, and $c_{X_l}(0, 0, \tau)$. Which of them are zero and why?
Exercise IX.13: In the Gaussian case find the fourth-order covariance and structure functions considered in Section 6 of Chapter IX, i.e., $c_{X_1 X_2 X_3 X_4}(\boldsymbol{p}, \boldsymbol{p})$, $c'_{X_1 X_2 X_3 X_4}(\boldsymbol{p}, \boldsymbol{p}')$, $\xi_{X_1 X_2 X_3 X_4}(\boldsymbol{p}, \boldsymbol{p}')$, and $\xi'_{X_1 X_2 X_3 X_4}(\boldsymbol{p}, \boldsymbol{p}')$, given $c_{X_l X_{l'}}(\boldsymbol{p}, \boldsymbol{p}') = \frac{1}{2}\left[\overline{X_l(\boldsymbol{p}) X_{l'}(\boldsymbol{p})} + \overline{X_l(\boldsymbol{p}') X_{l'}(\boldsymbol{p}')}\right]$ and $\xi_{X_l X_{l'}}(\boldsymbol{p}, \boldsymbol{p}') = \overline{[X_l(\boldsymbol{p}) - X_l(\boldsymbol{p}')][X_{l'}(\boldsymbol{p}) - X_{l'}(\boldsymbol{p}')]}$.

CHAPTER X

Exercise X.1: There exists a considerable body of evidence (theoretical, numerical, and experimental) regarding attributes with distinctive features that can be best represented by special classes of STRF models. Among these classes, frozen S/TRFs are physically motivated models with a wide range of applicability in sciences. In the frozen S/TRF setting, prove Proposition 2.1 of Chapter X.
Exercise X.2: Prove Corollary 2.1 of Chapter X.
Exercise X.3: Let $X(\boldsymbol{s},t)$ be a frozen random field with its temporal variation determined by a spatial field $\widehat{X}(\widehat{\boldsymbol{s}})$ that travels with velocity $\boldsymbol{\upsilon}$, i.e., $X(\boldsymbol{s},t) = \widehat{X}(\widehat{\boldsymbol{s}})$. Show that in $R^{3,1}$ if $c_{\widehat{X}}(r) = c_0 e^{-\frac{r^2}{a^2}}$, $r = |\boldsymbol{h}|$, then $\widetilde{c}_X(k) = \frac{c_0 a^3}{(2\sqrt{\pi})^3} e^{-\frac{a^2k^2}{4}}$.
Exercise X.4: The frozen S/TRF $X(\boldsymbol{s},t)$ satisfies some interesting stochastic partial differential equations (SPDEs).
(a) Prove Proposition 2.2 of Chapter X.
(b) After making the modification

$$X(\boldsymbol{s},t) = X(\boldsymbol{s} - \boldsymbol{\upsilon}\Delta t, t - \Delta t), \tag{4.1}$$

what is the corresponding SPDE satisfied by $X(\boldsymbol{s},t)$?
(c) In the case of $n = 2$, i.e.,

$$X(s_1, s_2, t) = X(s_1 - \upsilon_1 \Delta t, s_2 - \upsilon_2 \Delta t, t - \Delta t), \tag{4.2}$$

show that (υ_1,υ_2) lie on a straight line.
(d) Again in the case of $n = 2$, show that the movement component in the direction of the random field gradient $(\frac{\partial X(\boldsymbol{p})}{\partial s_1}, \frac{\partial X(\boldsymbol{p})}{\partial s_2})$ is equal to

$$\frac{\frac{\partial X(\boldsymbol{p})}{\partial t}}{\left[(\frac{\partial X(\boldsymbol{p})}{\partial s_1})^2 + (\frac{\partial X(\boldsymbol{p})}{\partial s_2})^2\right]^{\frac{1}{2}}}. \tag{4.3}$$

Exercise X.5: Prove Proposition 2.3 of Chapter X.
Exercise X.6: Consider the covariance $c_X(h,\tau) = e^{-(h^2+\tau^2)^{\frac{1}{2}}}$ in $R^{1,1}$.
(a) Show that

$$\begin{aligned} \upsilon &= \frac{h}{\tau} \pm \left(\frac{h^2}{\tau^2} + 1\right)^{\frac{1}{2}}, \\ \theta &= -\frac{\tau}{h}. \end{aligned} \tag{6.1a–b}$$

(b) Then, show that

$$\left[\frac{\partial}{\partial \tau} - \frac{\tau}{h}\frac{\partial}{\partial h}\right]\gamma_X(h,\tau) = 0,$$
$$\left[\frac{\partial^2}{\partial h \partial \tau} + \frac{\tau}{h^2}\frac{\partial}{\partial h} - \frac{\tau}{h}\frac{\partial^2}{\partial h^2}\right]\gamma_X(h,\tau) = 0. \tag{6.2a–b}$$

Exercise X.7: Derive the important frozen covariance condition described by Eq. (2.22) of Chapter X. The covariance of a frozen S/TRF with space- and time-independent velocity must satisfy this condition.
Exercise X.8: Based on the Taylor expansion, derive Eq. (2.25) of Chapter X.
Exercise X.9: Prove that in the elliptic representation of Eq. (2.28) of Chapter X the υ and $\upsilon\prime$ are given by

$$\upsilon = -\frac{\frac{\partial^2 c_X}{\partial r \partial \tau}(0,0)}{\frac{\partial^2 c_X}{\partial r^2}(0,0)},$$
$$\upsilon' = \left[\frac{\frac{\partial^2 c_X}{\partial \tau^2}(0,0)}{\frac{\partial^2 c_X}{\partial r^2}(0,0)} - \upsilon^2\right]^{\frac{1}{2}}. \tag{9.1a–b}$$

Exercise X.10: Useful information is contained in physical expressions of structure function equations in the physical or the spectral domain. In this context, derive Eqs. (2.39a–c) of Chapter X.
Exercise X.11: Prove Corollary 3.1 of Chapter X.
Exercise X.12: Derive Eq. (4.4) of Chapter X.
Exercise X.13: Show that $s \cdot s' = \rho^2 - \frac{1}{2}r_E^2$, where ρ is the spherical radius of the Earth (considered as a sphere) and r_E is the Euclidean metric.
Exercise X.14: Prove Proposition 3.1 of Chapter X.
Exercise X.15: Prove Corollary 2.2 of Chapter X linking the full and wavevector-time spectra with the traveling spectrum.

CHAPTER XI

Exercise XI.1: A random field is fully characterized stochastically in terms of probability functions, like the multivariate probability density function (M-PDF). Thus, it makes sense to develop techniques of constructing M-PDF. In this setting, prove Eq. (5.24) of Example 5.3 of Chapter XI.
Exercise XI.2: Derive Eq. (5.26) of Chapter XI. For a function g and a characteristic function φ of your choice, construct a PDF model that satisfies Eq. (5.26) of Chapter XI.
Exercise XI.3: Derive Eq. (5.31) of Chapter XI.
Exercise XI.4: In the case of spherical symmetry, derive Eq. (2.7) of Chapter XI.
Exercise XI.5: Derive Eq. (2.12) of Example 2.3 of Chapter XI.
Exercise XI.6: Prove Proposition 3.1. of Chapter XI.

Exercise XI.7: Prove that the factora model satisfies the relationship

$$\int d\chi_{p_1}\left(\chi_{p_1}-\overline{X(p_1)}\right)f_X(p_1,p_2)=c_X(p_1,p_2)\sigma_{X;p_2}^{-2}\left(\chi_{p_2}-\overline{X(p_2)}\right)f_x(p_2), \tag{7.1}$$

for all p_1, p_2.

Exercise XI.8: Prove Eq. (3.10) of Example 3.3 of Chapter XI.

Exercise XI.9: Prove Proposition 4.2 of Chapter XI stating that copulas are scale-invariant, and Proposition 4.3 of Chapter XI linking factoras with copulas.

Exercise XI.10: In Section 2.1 of Chapter II it was noticed that generalized PDFs are very useful in a variety of applications.

(a) Show that if for the random field $X(p)$ it is valid that $X(p')=\phi[X(p)]$, where ϕ is suitable one-to-one function, then its bivariate PDF $f_X(p,p')$ can be written in terms of its univariate PDF $f_X(p)$ as $f_X(p,p')=f_X(p)\delta(\chi'-\phi(\chi))$. This formula can be used to construct new covariance models (see Exercise XVI.16)

(b) Show that in the case of two random fields $X(p)$ and $Y(p')$ such that $Y(p_l')=\phi_l(X(p_1)\ldots X(p_m))$, $l=1,\ldots,m'$, it is valid that

$$f_{XY}(p_1,\ldots,p_m;p_1',\ldots,p_{m'}')=f_X(p_1,\ldots,p_m)\delta(\psi_1'-\phi_1(\chi_1,\ldots,\chi_m))\ldots\delta(\psi_{m'}'-\phi_{m'}(\chi_1,\ldots,\chi_m)).$$

(c) What is the corresponding formulation in the case that $X(p_l')=\phi_l(X(p_l))$, $l=1,\ldots,m$?

CHAPTER XII

Exercise XII.1: The theory of functionals has numerous applications in physical sciences, often making possible the solution of problems that cannot be handled by ordinary functions. Consider the heat conduction equation in isotropic solids

$$\frac{\partial}{\partial t}X(p)=D\Delta X(p)$$

where $p=(s,t)$, and $X(p)$ denotes temperature and D is heat diffusivity, with boundary and initial condition (BIC)

$$\begin{aligned}X(p)&=X(s)\ \ s\in A,\\ X(p)\big|_{t=0}&=X_0(s).\end{aligned}$$

Derive the corresponding PDE governing the covariance function $c_X(s,s',t)$ of $X(p)$.

Exercise XII.2: For the situation described in Exercise XII.1, find the specific expression of $c_X(s,s',t)$ assuming that it is STHS, i.e., $c_X(s,s',t)=c_X(h,t)$, $h=s-s'$, and $c_X(s,s',0)=c_{X_0}(h)$.

Exercise XII.3: For the situation described in Exercise XII.2, find the specific expression of $c_X(s,s',t)$ assuming a delta initial covariance function $c_{X_0}(s,s')=A_0\delta(s-s')$.

Exercise XII.4: Find the characteristic functional (CFL) for the situation described in Exercise XII.1.

Exercise XII.5: Given the CFL

$$\Phi_X(q_1,q_2,q_3,q,t)=e^{\overline{i\left[\sum_{j=1}^{3}q_jX_j+\int ds\,qY\right]}}$$

of the random fields $X_j(\boldsymbol{p})$, $j = 1, 2, 3$, and $Y(\boldsymbol{s})$, find the corresponding $\frac{\delta}{\delta q_j}\Phi_X$ and

$$\frac{\delta^4}{\delta q_1(\boldsymbol{s}_1)\delta q_2(\boldsymbol{s}_2)\delta q_3(\boldsymbol{s}_3)\delta q(\boldsymbol{s}_4)}\Phi_X.$$

Exercise XII.6: For the situation in Exercise XII.5, find the corresponding moments $\overline{X_j(\boldsymbol{s})}$, $j = 1, 2, 3$, and $\overline{X_1(\boldsymbol{s}_1)X_2(\boldsymbol{s}_2)X_3(\boldsymbol{s}_3)Y(\boldsymbol{s}_4)}$.
Exercise XII.7: For the situation in Exercise XII.5, find the time derivative $\frac{\partial}{\partial t}\Phi_X$.
Exercise XII.8: Derive the functional differentiation rules of Eqs. (1.18−1.20) of Chapter XII.
Exercise XII.9: In the case of the string functional of Example 1.10 of Chapter XII, derive Eq. (1.25) of Chapter XII.
Exercise XII.10: In the case of Gaussian CFL, derive Eqs. (2.2a−b) of Chapter XII.
Exercise XII.11: Prove that the fact that for a Gaussian random field all the cumulant functions of order higher than 2 are zero implies Eqs. (2.3b−c) of Chapter XII.
Exercise XII.12: Consider heat conduction in an isotropic medium discussed in Example 2.5 of Chapter XII. Derive the integrodifferential Eq. (2.22) of Chapter XII.
Exercise XII.13: Show that

$$\frac{1}{i^k}\frac{\delta^k}{\delta q^{\nu_1}(\boldsymbol{p}_1)\ldots\delta q^{\nu_k}(\boldsymbol{p}_k)}\Phi_X(q)\Big|_{q=0} = \overline{X^{\nu_1}(\boldsymbol{p}_1)\ldots X^{\nu_k}(\boldsymbol{p}_k)}, \tag{13.1}$$

where $\sum_{i=1}^{k}\nu_k = k$.
Exercise XII.14: Consider the function in $R^{3,1}$ (used in the study of elastic media),

$$S_X(\mathcal{L}) = \int d\boldsymbol{p}\, \mathcal{L}\left(X, \frac{\partial}{\partial t}X, \nabla X\right), \tag{14.1}$$

where $X(\boldsymbol{p}) = X(s_1,s_2,s_3,t)$, and

$$\mathcal{L}\left(X, \frac{\partial}{\partial t}X, \nabla X\right) = \frac{\rho}{2}\left(\frac{\partial}{\partial t}X(\boldsymbol{p})\right)^2 - \frac{T}{2}(\nabla X(\boldsymbol{p}))^2. \tag{14.2}$$

Prove that its functional derivative $\frac{\delta}{\delta X(\boldsymbol{p}_0)}S_X(\mathcal{L})$ leads to the wave equation

$$\nabla^2 X(\boldsymbol{p}) = \frac{\rho}{T}\frac{\partial^2}{\partial t^2}X(\boldsymbol{p}). \tag{14.3}$$

Exercise XII.15: In the case of a Gaussian random field, Prove Eqs. (2.6) of Chapter XII and its vectorial extension Eq. (2.7) of Chapter XII.

CHAPTER XIII

Exercise XIII.1: Construct second-order space−time functionals in the frequency domain corresponding to the covariance $c_X(q_1,q_2)$, the structure $\xi_X(q_1,q_2)$, and the variogram $\gamma_X(q_1,q_2)$ functionals of the physical (real) domain.
Exercise XIII.2: Assuming that Eqs. (1.35a−b) of Chapter XIII are valid, prove Eq. (1.36) of Chapter XIII.

Exercise XIII.3: If $g_{\rho,\zeta}(\boldsymbol{p})$ are space–time expo-polynomials given by Eq. (1.45) of Chapter XIII, show that if

$$\iint dsdt\; q(\boldsymbol{s},t)\boldsymbol{s}^{\rho} t^{\zeta} e^{\alpha\cdot \boldsymbol{s}+\beta t} = 0, \tag{3.1}$$

then

$$\iint dsdt\; q(\boldsymbol{s},t)(\boldsymbol{s}+\boldsymbol{h})^{\rho}(t+\tau)^{\zeta} e^{\alpha\cdot(\boldsymbol{s}+\boldsymbol{h})+\beta(t+\tau)} = 0. \tag{3.2}$$

Exercise XIII.4: Under the conditions of Example 1.8 of Chapter XIII show that

$$\iint dsdt\; g_{\rho,\zeta}(\boldsymbol{s},t)[\delta(\boldsymbol{s},t) - q^{*}(\boldsymbol{s},t)] = (-1)^{\rho+\zeta}[1-\widetilde{q}^{*}(\boldsymbol{k},\omega)]^{(\rho,\zeta)}\Big|_{(\boldsymbol{k},\omega)=(\boldsymbol{0},0)} = 0.$$

Exercise XIII.5: Derive the generalized space–time covariance of Eq. (3.27) in Example 3.6 of Chapter XIII.
Exercise XIII.6: Prove Proposition 4.1 of Chapter XIII concerning the spatiotemporal autoregressive model of order $\nu+1/\mu+1$.
Exercise XIII.7: Prove Proposition 4.2 of Chapter XIII concerning linear space–time estimation.
Exercise XIII.8: Write the full expansion of the space–time variogram function of Eq. (4.29a) of Chapter XIII in terms of the covariance function.
Exercise XIII.9: Derive the space–time variogram function Eq. (4.31) of Chapter XIII.
Exercise XIII.10: Show that the space–time variogram of $\gamma_X(\boldsymbol{p},\Delta\boldsymbol{p})$ of Eq. (4.31) of Chapter XIII can be written in terms of $\gamma_X(\Delta\boldsymbol{p})$.
Exercise XIII.11: Derive the space–time variogram function of Eq. (4.34) of Chapter XIII.
Exercise XIII.12: Consider any set of points with space–time coordinates of your choice and derive the corresponding expressions for $Y_q(\boldsymbol{p})$ and $Y_q(\boldsymbol{s},t)$ of Eqs. (4.3) and (4.4a–b) of Chapter XIII, respectively. Show that they are equivalent.
Exercise XIII.13: As in Exercise XIII.12, chose any set of points with space–time coordinates of your choice and derive expressions for $c_{Y_q}(\boldsymbol{p},\boldsymbol{p}')$, $\xi_X(\boldsymbol{p},\boldsymbol{p}';\Delta\boldsymbol{p},\Delta\boldsymbol{p}')$, and $\gamma_X(\boldsymbol{p},\boldsymbol{p}',\Delta\boldsymbol{p},\Delta\boldsymbol{p}')$. Are they equivalent?
Exercise XIII.14: If $X(\boldsymbol{p})$ is an S/TRF with homostationary 1/1 increments $Y_q(\boldsymbol{p})$, derive expressions for $c_{Y_q}(\boldsymbol{0})$. What happens if $X(\boldsymbol{p})$ is STHS?
Exercise XIII.15: When it is valid that $c_X(\Delta\boldsymbol{p}_{-i,\tau}) = c_X(\Delta\boldsymbol{p}_{i,-\tau})$ for a space–time covariance function, and when it holds that $c_X(\Delta\boldsymbol{p}_{-i,\tau}) = c_X(\Delta\boldsymbol{p}_{i,-\tau}) = c_X(\Delta\boldsymbol{p}_{i,\tau})$?
Exercise XIII.16: Depending on the choice of the space–time arguments, different expressions of the variogram can be derived. What is the definition of $\gamma_X(\Delta\boldsymbol{p},\Delta\boldsymbol{p}')$?
Exercise XIII.17: If $\boldsymbol{p}=\boldsymbol{p}'$ and $\Delta\boldsymbol{p}=\Delta\boldsymbol{p}'$, what is the expression of $\gamma_X(\boldsymbol{p},\Delta\boldsymbol{p})$ of Eq. (4.31) of Chapter XIII in terms of $c_X(\boldsymbol{p},\Delta\boldsymbol{p})$?
Exercise XIII.18: Do the following:
(a) Prove Proposition 4.3 of Chapter XIII.
(b) Under what conditions it is valid that $\gamma_X(\Delta\boldsymbol{p}) \leq a\Delta\boldsymbol{p}^{(2,2)} = ah^2\tau^2$, where a is a constant?
Exercise XIII.19: Prove Proposition 4.4 of Chapter XIII concerning the m.s. continuity of a random field with STHS increments and the continuity of the corresponding variogram function.
Exercise XIII.20: Derive Eq. (4.39) of Chapter XIII.

Exercise XIII.21: As is often the case, the series expansion of a variogram function can produce some interesting results.

(a) Under what conditions can an STIS variogram $\gamma_X(r,\tau)$ be expanded around the origin as

$$\gamma_X(r,\tau) = \gamma_X(0,0) + \sum_k \sum_{k'} \alpha_{2k,2k'} r^{2k} \tau^{2k'} + \sum_l \sum_{l'} b_{l,l'} r^l \tau^{l'} + \log r \log \tau \sum_m \sum_{m'} c_{2m,2m'} r^{2m} \tau^{2m'},$$

where $\alpha_{2k,2k'}$, $b_{l,l'}$, and $c_{2m,2m'}$ are suitable coefficients, $k, k', m, m' = 1, 2, \ldots$, and l, l' are real numbers different from even integers?

(b) Also, under what conditions on the above coefficients the (1,1) derivatives of $X(\boldsymbol{p})$ exist in the m.s. sense?

(c) Generalize your results in the case of the (ν,μ) space–time derivatives of $X(\boldsymbol{p})$.

Exercise XIII.22: Derive the variogram expansion of Eq. (4.41) of Chapter XIII. Under what conditions is it valid?

Exercise XIII.23: Let us rewrite Eq. (4.27) of Chapter XIII as

$$\begin{aligned} \overline{Y_i(\boldsymbol{p})} &= \overline{\Delta^2_{i,0} X(\boldsymbol{p})} = \overline{X(\boldsymbol{p} + \delta_{h_i,\tau}) - X(\boldsymbol{p} + \delta_{0,\tau}) + X(\boldsymbol{p}) - X(\boldsymbol{p} + \delta_{h_i,0})} \\ &= \overline{X(s_1, \ldots, s_i + h_i, \ldots, s_n, t + \tau) - X(s_1, \ldots, s_n, t + \tau) + X(s_1, \ldots, s_n, t) - X(s_1, \ldots, s_i + h_i, \ldots, s_n, t)}, \end{aligned} \tag{23.1a–b}$$

where $\delta_{h_i,\tau} = (h_i\boldsymbol{\varepsilon}_i, \tau)$, $\delta_{0,\tau} = (0,\tau)$, $\delta_{h_i,0} = (h_i\boldsymbol{\varepsilon}_i, 0)$, and the variogram is given by Eq. (4.33) of Chapter XIII, i.e.,

$$\begin{aligned} \gamma_X(\Delta\boldsymbol{p}) &= \frac{1}{2}\overline{Y_i(\boldsymbol{p})^2} = \frac{1}{2}\overline{[X(\boldsymbol{p} + \delta_{h_i,\tau}) - X(\boldsymbol{p} + \delta_{0,\tau}) + X(\boldsymbol{p}) - X(\boldsymbol{p} + \delta_{h_i,0})]^2} \\ &= \frac{1}{2}\overline{[X(s_1, \ldots, s_i + h_i, \ldots, s_n, t + \tau) - X(s_1, \ldots, s_n, t + \tau) + X(s_1, \ldots, s_n, t) - X(s_1, \ldots, s_i + h_i, \ldots, s_n, t)]^2}. \end{aligned} \tag{23.2a–b}$$

This allows us to compare the Ito-Gel'fand 0/0 increment considered in Section 4.2 of Chapter XIII with the omnidirectional 0/0 increment of the form considered in Section 2.4 of Chapter VII, Eq. (2.9a) of Chapter VII, i.e.,

$$\begin{aligned} \overline{Y(\Delta\boldsymbol{p})} &= \overline{X(\boldsymbol{p} + \Delta\boldsymbol{p}) - X(\boldsymbol{p})} \\ &= \overline{X(s_1 + h_1, \ldots, s_n + h_n, t + \tau) - X(s_1, \ldots, s_n, t)}, \end{aligned} \tag{23.3a–b}$$

where $\Delta\boldsymbol{p} = \delta_{\boldsymbol{h},\tau} = (h_1\boldsymbol{\varepsilon}_1, \ldots, h_n\boldsymbol{\varepsilon}_n, \tau)$. In this case the variogram is given by Eq. (2.9a) of Chapter VII, i.e.,

$$\begin{aligned} \gamma_X(\Delta\boldsymbol{p}) &= \frac{1}{2}\overline{Y(\Delta\boldsymbol{p})^2} = \frac{1}{2}\overline{[X(\boldsymbol{p} + \Delta\boldsymbol{p}) - X(\boldsymbol{p})]^2} \\ &= \frac{1}{2}\overline{[X(s_1 + h_1, \ldots, s_n + h_n, t + \tau) - X(s_1, \ldots, s_n, t)]^2}. \end{aligned} \tag{23.4a–b}$$

Show that in the case of Ito-Gel'fand 0/0 increment:
(a) It is valid that

$$\overline{Y_i(\boldsymbol{p})} + \overline{Y_i(\boldsymbol{p}')} = \overline{Y_i(\boldsymbol{p}+\boldsymbol{p}')}, \tag{23.5}$$

where $\overline{Y_i(\boldsymbol{p})} = \overline{\Delta^2_{i,0}X(\boldsymbol{p})}$, $\overline{Y_i(\boldsymbol{p}')} = \overline{\Delta^2_{i,0}X(\boldsymbol{p}')}$, and $\overline{Y_i(\boldsymbol{p}+\boldsymbol{p}')} = \overline{\Delta^2_{i,0}(X(\boldsymbol{p})+X(\boldsymbol{p}'))}$.
(b) Also, show that

$$\overline{Y_i(\boldsymbol{p})} = \frac{\partial^2 \overline{X(\boldsymbol{p})}}{\partial s_i \partial s_0} = \sum_{q=1}^{n,0} c_q s_q = \sum_{q=1}^{n} c_q s_q + c_0 t, \tag{23.6}$$

where $\boldsymbol{p} = (s_1,\ldots, s_n, s_0 = t)$, and c_q $(q = 1,\ldots, n, 0)$ are constant coefficients.
(c) Lastly, show that

$$\overline{X(\boldsymbol{p})} = s_i t\left[\sum_{q=1}^{n} c'_q s_q + c'_0 t\right] = s_i t\left[\sum_{q=1}^{n,0} c'_q s_q\right], \tag{23.7}$$

where $c'_q = c_q \ (q \neq i), \ = \frac{c_i}{2} (q = i), \ c'_0 = \frac{c_0}{2}$.
(d) Compare the results (a)–(c) above with the (a)–(b) results of Exercise 17 of Chapter IV.

CHAPTER XIV

Exercise XIV.1: The material presented in Chapter XIV is based on the perspective that a physical theory is a mathematical model of reality. One such model is the stochastic wave equation in $R^{3,1}$,

$$\left[\nabla^2 - a^{-2}\frac{\partial^2}{\partial t^2}\right] X(\boldsymbol{s},t) = Y(\boldsymbol{s},t),$$

where a is the wave velocity, and $Y(\boldsymbol{s},t)$ is a known random field.
(a) Find the solution $X(\boldsymbol{s},t)$ of the above equation in the physical domain.
(b) What is the corresponding space–frequency solution $\widetilde{X}(\boldsymbol{s},\omega)$ in the spectral domain?

Exercise XIV.2: In the fractional differential equation

$$\left(\nabla^2 - \frac{1}{a(t)^2}\right)^p X(\boldsymbol{p}) - B(\boldsymbol{p}) = 0$$

that governs the attribute $X(\boldsymbol{p}), \boldsymbol{p} \in R^{n,1}$, the $a(t)$ is a known parameter, the $B(\boldsymbol{p})$ is a white noise field, and the $p = \frac{2\mu+n}{4}$ $(\mu > 0)$ is a model parameter. Under what conditions the solution of this stochastic equation is a Gaussian random field with an STIS covariance function?
Exercise XIV.3: Starting from the evolutionary type of Eq. (3.1a) of Chapter XIV, derive Eq. (3.1b) of Chapter XIV subject to the appropriate BIC.

Exercise XIV.4: This exercise illustrates the argument that random field regularity sometimes may depend on improving its differentiability by a fraction of a spatial or temporal derivative. In $R^{n,1}$, Eq. (3.16) of Chapter XIV is written as

$$\left[\frac{\partial}{\partial t} - \nabla^2\right] X(\boldsymbol{s}, t) = Y(\boldsymbol{s}, t), \tag{4.1}$$

where $Y(\boldsymbol{s},t)$ is a known Gaussian white noise random field with $\overline{Y(\boldsymbol{s},t)Y(\boldsymbol{s}',t')} = \delta(t'-t)\delta(\boldsymbol{s}'-\boldsymbol{s})$, and $X(\mathbf{0},0) = X_0$.

(a) Show that the random field solution of Eq. (4.1) is

$$X(\boldsymbol{s}, t) = \int d\boldsymbol{s}' \int_0^t dt' \frac{1}{4\pi|t-t'|^{\frac{n}{2}}} e^{-\frac{|\boldsymbol{s}-\boldsymbol{s}'|^2}{4\pi(t-t')}} Y(\boldsymbol{s}', t'). \tag{4.2}$$

(b) Prove that the $X(\boldsymbol{s},t)$ above represents a centered Gaussian random field with a rather more complicated space–time covariance (discussed earlier in Exercises VII.27–VII.29 and elsewhere).

(c) In $R^{1,1}$, show that the $X(s,t)$ of Eq. (4.2), i.e., Eq. (3.17) of Chapter XIV, is "almost" $\frac{1}{4}$-Hölder continuous in time and "almost" $\frac{1}{2}$-Hölder continuous in space (the matter was also discussed in Exercises VII.27–VII.29). In Hölderian terms, to obtain a continuous $X(s,t)$ seems to somehow require a little more than $\frac{1}{2}$ a derivative. In $R^{n,1}$ $(n > 1)$, the situation is more complex (e.g., $X(\boldsymbol{s},t)$ may not be function valued).

(d) Is the reason for this lower time regularity the fact that the driving white noise random field $Y(\boldsymbol{s},t)$ is not only temporally singular, but also spatially?

(e) In $R^{1,1}$, can the Hölder exponents ($\frac{1}{2}$ for space and $\frac{1}{4}$ for time) be understood as a consequence of the terms $\frac{\partial}{\partial t}$ and $\frac{\partial^2}{\partial s^2}$, meaning that one can exchange spatial field regularity for temporal field regularity at the cost of one time derivative for a double space derivative?

Exercise XIV.5: Consider the physical law of steady-state flow in $R^{2,1}$ governing the random hydraulic head field $X(\boldsymbol{s})$,

$$\sum_{j=1}^{2} \left[\frac{\partial}{\partial s_j} + 2s_j\right] X(\boldsymbol{s}) = 0, \tag{5.1}$$

subject to the boundary condition (BC) $X_2(s = 0) = 1$. Using the ST technique (Section 3 of Chapter V), show that a solution is given by

$$X(\boldsymbol{s}) = e^{-s^2 + \alpha(s_1 - s_2)}, \tag{5.2}$$

where α is an arbitrary real constant.

Exercise XIV.6: The requirement that empirical data and physical insight need to be reflected in a mathematical model of a real-world phenomenon is often satisfied in terms of the model parameters and BICs. For a fixed time t, consider the unidimensional steady-state flow equations based on Darcy's law, i.e.,

$$\frac{dX(h)}{dh}+\frac{dW(h)}{dh}X(s)=0,$$
$$e^{W(s)}X(s)=-q, \tag{6.1a–b}$$

where $X(s)$ is the hydraulic gradient, $W(s)$ is the log-hydraulic conductivity so that $W(s)=\overline{W(s)}+w(s)$ is a normal (Gaussian) random field (with mean value $\overline{W(s)}$ and a homogeneous random fluctuation $w(s)$), and q is a space-independent specific discharge RV. The following BCs are associated with different empirical data and insight:

$BC1$: Deterministic $X(0)$, and random $W(0)$, q.
$BC2$: Deterministic $X(0)$, q, $W(0)$.
$BC3$: Deterministic q, and random $X(0)$, $W(0)$.
$BC4$: Random $X(0)$, $W(0)$, q.

In the above BCs, when random, the $X(0)$ is assumed to be lognormally distributed. On the basis of the unidimensional steady-state flow equations above derive the mean, variance, and covariances of $X(s)$ for the four different BCs (BC1−4).

Exercise XIV.7: Consider the SPDE in $R^{1,1}$,

$$\left(\frac{\partial}{\partial s}-\alpha\frac{\partial}{\partial t}-\beta^2\right)X(s,t)-B(s,t)=0, \tag{7.1}$$

where α and β are known coefficients, and $B(s,t)$ is a white noise random field. Find the corresponding space−time covariance model.

Exercise XIV.8: Consider the SPDE in $R^{2,1}$,

$$\left(\sum_{i=1}^{2}\frac{\partial}{\partial s_i}-\alpha^2\right)X(\boldsymbol{p})-B(\boldsymbol{p})=0, \tag{8.1}$$

where $\boldsymbol{p}=(s_1,s_2,t)$, α and β are known coefficients, and $B(\boldsymbol{p})$ is a white noise random field. Find the corresponding space−time covariance model.

Exercise XIV.9: Let us revisit Example 3.3 of Chapter XIV.

(a) Derive Eq. (3.4) of Chapter XIV specifically in the $R^{2,1}$ domain.

(b) Find the solution of

$$\frac{\partial}{\partial s}\left[a\frac{\partial}{\partial s}X(s,t)\right]=-Y(s,t) \tag{9.1}$$

in the $R^{1,1}$ domain, where $a=e^W$, and $Y(s,t)=\sin(\pi s)\sin(\pi t)$.

Exercise XIV.10: Assume that the random fields $X(\boldsymbol{p})$ and $Y(\boldsymbol{p})$ satisfy the SPDE in $R^{1,1}$,

$$aX^{(1,0)}(s,t) + Y(s,t) = 0, \tag{10.1}$$

where a is a physical constant. Given $\gamma_X(h,\tau)$, express $c_Y(h,\tau)$ in terms of $\gamma_X(h,\tau)$.

Exercise XIV.11: Find the space–time covariance function $c_X(h,\tau)$ that satisfies the PDE in $R^{1,1}$,

$$\left[\frac{\partial^2}{\partial\tau^2} + a\frac{\partial}{\partial\tau} + b^2\right] c_X(h,\tau) = c_Y(h,\tau), \tag{11.1}$$

where $c_Y(h,\tau)$ is a known covariance function, and a and b are known coefficients.

Exercise XIV.12: Consider the system of PDEs satisfied by the covariance function in $R^{1,1}$,

$$\left(\frac{\partial}{\partial\tau} + a\frac{\partial}{\partial h}\right) c_{X_1}(h,\tau) = \frac{1}{2}b[c_{X_2}(h,\tau) - c_{X_1}(h,\tau)],$$
$$\left(\frac{\partial}{\partial\tau} - a\frac{\partial}{\partial h}\right) c_{X_2}(h,\tau) = \frac{1}{2}b[c_{X_1}(h,\tau) - c_{X_2}(h,\tau)], \tag{12.1a–b}$$

where a and b are known coefficients.

(a) Derive a second-order PDE that is satisfied by $c_X(h,\tau) = c_{X_1}(h,\tau) + c_{X_2}(h,\tau)$.
(b) What is the form of this equation when $a \to \infty$, $b \to \infty$, and $b = a^2$?
(c) Interpret the results in terms of random walks.

Exercise XIV.13: Consider the mass conservation law of fluid flow

$$\frac{\partial X(\boldsymbol{p})}{\partial t} + \nabla\cdot[X(\boldsymbol{p})\boldsymbol{v}(\boldsymbol{s})] = 0, \tag{13.1}$$

where $X(\boldsymbol{p})$ denotes fluid flow, and $\boldsymbol{v}(\boldsymbol{s})$ is the fluid velocity. Assuming a constant velocity along trajectories, show that

$$X(\boldsymbol{s} + t\boldsymbol{v}(\boldsymbol{s}), t) = X(\boldsymbol{s}, 0)e^{-t\nabla\cdot\boldsymbol{v}(\boldsymbol{s})}. \tag{13.2}$$

CHAPTER XV

Exercise XV.1: The S/TVFs, in general, and the covariance function, in particular, must satisfy certain permissibility conditions or permissibility criteria (PC). Yet, the underlying assumptions of these PC are often ignored at the modeler's peril. Derive step-by-step the results of the numerical Example 2.3 of Chapter XV, which stress this important point.

Exercise XV.2: Assuming a fixed time t, consider the cosine covariance model $c_Y(h) = \cos(h)$ in $R^{1,1}$. Show that the fact that it is a nonnegative-definite (NND) function implies only that $c_Y(h)$ is the covariance of a normal (Gaussian) random field but not necessarily of a lognormal random field.

Exercise XV.3: In $R^{3,1}$, is the exponential function

$$c_X(r,\tau) = e^{-\frac{r}{a} - \frac{\tau}{b}}, \tag{3.1}$$

where $r = r_E$ and a, $b > 0$, a permissible covariance model?

Exercise XV.4: In light of the analysis in Section 4.4 of Chapter XV prove Eq. (4.27) of Chapter XV.

Exercise XV.5: Consider the random fields $X_1(\boldsymbol{p})$, $X_2(\boldsymbol{p})$, and $X_3(\boldsymbol{p}) = X_1(\boldsymbol{p}) + X_2(\boldsymbol{p})$. Is their covariance matrix NND for the choice of deterministic coefficients $\boldsymbol{\lambda} = [\lambda_1 \;\; \lambda_2 \;\; \lambda_3]^T = [1 \;\; 1 \;\; -1]^T$?

Exercise XV.6: Show that the PC for the generalized space–time covariance function of Eq. (3.23) of Chapter XIII are given by Eqs. (4.20a–b) and (4.21a–b) of Chapter XV.

Exercise XV.7: Assume that the S/TRF $X(\boldsymbol{p})$, $\boldsymbol{p} \in R^{2,1}$, obeys the stochastic wave equation

$$\left(\frac{\partial}{\partial t} + \nabla^2 - a^2\right)X(\boldsymbol{p}) - Y(\boldsymbol{p}) = 0, \tag{7.1}$$

where a is a known coefficient, and $Y(\boldsymbol{p})$ is a white noise random field.

(a) Show that the wavevector–time spectrum of $X(\boldsymbol{p})$ is of the form

$$\widetilde{c}_X(\boldsymbol{k},\tau) = A\left(a^2 + k_1^2 + k_1^2 + \omega^2\right)\Big]^{-2}, \tag{7.2}$$

where A is a constant of proportionality.

(b) Is the corresponding space–time covariance function permissible?

Exercise XV.8: Consider the space–time function

$$c_X(\boldsymbol{p},\boldsymbol{p}') = c_X\left(\alpha_s \boldsymbol{h}^2 + \alpha_t \tau^2\right), \tag{8.1}$$

where $\boldsymbol{h} = \boldsymbol{s} - \boldsymbol{s}'$, $\tau = t - t'$, and α_s and α_t are spatial and temporal coefficients, respectively. Under what conditions on α_s and α_t is the above function a permissible space–time covariance?

Exercise XV.9: In certain applications (e.g., electromagnetic images) the space lag vector $\boldsymbol{h}$ in $R^{3,1}$ is decomposed into its longitudinal coordinate h_1 and its two-dimensional transverse vector $\boldsymbol{\zeta} = (\zeta_1,\zeta_2)$, so that the space–time argument is written as $\boldsymbol{\Delta p} = (h_1,\boldsymbol{\zeta},\tau)$.

(a) Show that the spectral representation of Eq. (3.10a) of Chapter VII at $\tau = 0$ can be written in longitudinal–transverse coordinates as

$$c_X(h_1,\boldsymbol{\zeta}) = \int d\boldsymbol{\lambda}\, e^{i\boldsymbol{\lambda}\cdot\boldsymbol{\zeta}} \widetilde{c}_X(h_1,\boldsymbol{\lambda}), \tag{9.1}$$

where the corresponding wavevector $\boldsymbol{k}$ is decomposed into its longitudinal k_1 and two-dimensional transverse vector $\boldsymbol{\lambda}$ components, and

$$\widetilde{c}_X(h_1,\boldsymbol{\lambda}) = \int dk_1 e^{ik_1 h_1} \widetilde{c}_X(k_1,\boldsymbol{\lambda}). \tag{9.2}$$

(b) What is the PC in the case (a) above?
(c) Assuming a delta correlation along the longitudinal direction so that

$$c_X(h_1 - h_1', \zeta - \zeta') = \delta(h_1 - h_1')g(h_1, \zeta - \zeta'), \tag{9.3}$$

what is the spectral representation of $g(h_1,\zeta)$?
(d) What is the PC in the case (c) above?

Exercise XV.10: Consider a frozen S/TRF with a spectrum of the form of Eqs. (2.7a–b) of Chapter X, i.e.,

$$\begin{aligned} \widetilde{c}_X(\boldsymbol{k}, \omega) &= \widetilde{c}_{\widehat{X}}(\boldsymbol{k})\delta(\omega - \boldsymbol{k}\cdot\boldsymbol{v}), \\ \widetilde{c}_X(\boldsymbol{k}, \tau) &= \widetilde{c}_{\widehat{X}}(\boldsymbol{k})e^{-i\tau(\boldsymbol{k}\cdot\boldsymbol{v})}. \end{aligned} \tag{10.1a–b}$$

How is the Bochnerian PC modified in this case?
Exercise XV.11: Consider a VS/TRF with spectra of the form

$$\begin{aligned} \widetilde{c}_{X_l X_{l'}}(\boldsymbol{k}, \omega) &= \widetilde{c}_{X_l X_{l'}}(\boldsymbol{k})\delta(\omega - k_1 v), \\ \widetilde{c}_{X_l X_{l'}}(\boldsymbol{k}) &= \frac{E(k)}{4\pi k^4}\left(k^2\delta_{ll'} - k_l k_{l'}\right), \end{aligned} \tag{11.1a–b}$$

where $k = |\boldsymbol{k}|$ and $E(k)$ is the energy spectrum. How are the Bochnerian PC for a VS/TRF modified in this case?
Exercise XV.12: Consider the space–time covariance function in $R^{3,1}$ expressed as

$$c_X(\boldsymbol{h}, \tau) = \frac{1}{(2\pi)^{\frac{3}{2}}a^3\tau^3}\int d\boldsymbol{h}'\, c_X(\boldsymbol{h}')\, e^{-\frac{(\boldsymbol{h}-\boldsymbol{h}'-\boldsymbol{v}\tau)^2}{2a^2\tau^2}}, \tag{12.1}$$

where $c_X(\boldsymbol{h})$ is a known purely spatial covariance function, $\boldsymbol{v}$ is a known velocity vector, and $a > 0$ is a known constant.
(a) Does the permissibility of $c_X(\boldsymbol{h})$ imply that of $c_X(\boldsymbol{h},\tau)$?
(b) What is the asymptotic behavior of $c_X(\boldsymbol{h},\tau)$ as $\tau \to \infty$.

Exercise XV.13: Consider the S/TRF $X(\boldsymbol{p})$ and let

$$X(0, s_2, s_3, t) = a_1(t) + a_2(t)s_2 + a_3(t)s_3 + \frac{1}{2}\left[a_4(t)s_2^2 + a_5(t)s_3^2\right] + a_6(t)s_2 s_3, \tag{13.1}$$

where $a_1(t) = X(\boldsymbol{0},t)$, $a_2(t) = \frac{\partial X(\boldsymbol{0},t)}{\partial s_2}$, $a_3(t) = \frac{\partial X(\boldsymbol{0},t)}{\partial s_3}$, $a_4(t) = \frac{\partial^2 X(\boldsymbol{0},t)}{\partial s_2^2}$, $a_5(t) = \frac{\partial^2}{\partial s_3^2}X(0, s_2, s_3, t)|_{\boldsymbol{s}=\boldsymbol{0}}$, and $a_6(t) = \frac{\partial^2 X(\boldsymbol{0},t)}{\partial s_2 \partial s_3}$ (i.e., this is the second-order expansion of $X(\boldsymbol{p})$ about the origin of the coordinate system in the (s_2,s_3) plane).[2] What is the corresponding space–time covariance function and under what conditions is it permissible?

[2]As usual, $\frac{\partial X(\boldsymbol{0},t)}{\partial s_i} = \frac{\partial}{\partial s_i}X(0, s_2, s_3, t)\Big|_{\boldsymbol{s}=\boldsymbol{0}}$ $(i = 2, 3)$ etc.

Exercise XV.14: Consider the following space−time covariance models,

(a) $c_X(r,\tau) = c_0 e^{-a^2r^2-b\tau}$, where a, b, $c_0 > 0$;

(b) $c_X(r,\tau) = c_0 e^{-ar}\begin{cases}(1-\beta\tau) & \text{if } \tau \le \beta^{-1},\\ 0 & \text{if } \tau > \beta^{-1},\end{cases}$ where a, β, $c_0 > 0$;

(c) $c_X(r,\tau) = c_0 e^{-b\tau}\begin{cases}(1-\alpha r) & \text{if } r \le \alpha^{-1},\\ 0 & \text{if } r > \alpha^{-1},\end{cases}$ where α, b, $c_0 > 0$; and

(d) $c_X(r,\tau) = c_0 e^{-ar-b\tau}\cos(k_0 r)$, where a, b, c_0, $k_0 > 0$.

Are the above space−time covariance models permissible in $R^{n,1}$, $n = 1, 2, 3$?

Exercise XV.15: Consider the space−time covariance model in $R^{n,1}$

$$c_X(r,\tau) = c_0\left(\frac{\gamma_{X(1)}(r)+\gamma_{X(2)}(\tau)}{b_0}+1\right)^{-\alpha_0}\left(\frac{\gamma_{X(1)}(r)}{b_1}+1\right)^{-\frac{\alpha_1}{2}}\left(\frac{\gamma_{X(2)}(\tau)}{b_2}+1\right)^{-\alpha_2}$$
$$\frac{K_{\alpha_1}\left(2\left(\left(\alpha_1+\gamma_{X(1)}(r)\right)\kappa\right)^{0.5}\right)}{K_{\alpha_1}\left(2(\alpha_1\kappa)^{0.5}\right)}, \tag{15.1}$$

where $r = r_E$, c_0 is the variance, $\gamma_{X(1)}(r)$ and $\gamma_{X(2)}(\tau)$ are a purely spatial and a purely temporal variogram, respectively, and K_α is the modified Bessel function of the second kind and order α. Derive the conditions on the coefficients α_i, b_i ($i = 0, 1, 2$) and κ for the above covariance to be permissible.

Exercise XV.16: Consider the function

$$c_X(r,\tau) = a^2 b\tau^{-\frac{1}{2}}\left[\delta(r+a\tau)+\delta(r-a\tau)-\frac{1}{2a\tau}H(a\tau-r)\right] \tag{16.1}$$

(r, $\tau > 0$), where H is a Heaviside step function. Is the function of Eq. (16.1) a permissible space−time covariance model in $R^{n,1}$, and under what conditions on the coefficients a, b, and the dimensionality n?

Exercise XV.17: Is the covariance model resulting from a physical law a permissible one and why?

Exercise XV.18: Outline a proof of Proposition 4.8 of Chapter XV.

CHAPTER XVI

Exercise XVI.1: The PDF technique treats space−time covariance models so that both their rigorousness and interpretive values are maintained. Given $f(k,\tau)$ of the forms

$$e^{-(a+\tau)\,k^2},\ \frac{2^n}{(\tau k)^n}\left[\frac{n}{2}!J_{\frac{n}{2}}(\tau k)\right]^2,\ \left(1+\frac{(\tau k)^2}{a^2}\right)^{-\mu},\ \Gamma\left(\frac{n+1}{2}\right)\tau\left[\pi\left(k^2+\tau^2\right)\right]^{-\frac{n+1}{2}},$$
$$\text{and } ((a+\tau)k)^\nu K_\nu((a+\tau)k),$$

in $R^{n,1}$, where a, $\nu > 0$, $\mu > \frac{n}{2}$, and K_ν is a modified Bessel function of the second kind, find the corresponding space−time covariance models using the PDF technique (Table 2.1 of Chapter XVI).

Exercise XVI.2: Show that if $f(k,\tau)$ is a PDF, then the function obtained by the PDF technique is a valid space–time covariance model in $R^{n,1}$.

Exercise XVI.3: Answer the following:

(a) Given the density

$$f(k,\omega) = (2\pi)^{-\frac{1}{2}k^2} e^{-\frac{1}{2}k^2 (1+\omega^2)}, \tag{3.1}$$

use the PDF technique to derive a valid space–time covariance models in $R^{n,1}$.

(b) Do the same with the density in $R^{n,1}$,

$$f(k,\omega) = a\Gamma\left(\frac{n+1}{2}\right)\omega\left[\pi\left(k^2+\omega^2\right)\right]^{-\frac{1}{2}(n+1)}, \tag{3.2}$$

where $a > 0$.

(c) Starting with the densities

$$\begin{aligned} f(k,\omega) &= \frac{1}{a}\lambda\left(1+\frac{\omega}{a}\right)^{-\lambda-1}\delta(|k|-b\omega), \\ f(k,\omega) &= \frac{a}{\pi}\omega\left(k^2+a^2\omega^2\right)^{-1}, \end{aligned} \tag{3.3a–b}$$

use the PDF technique to derive the corresponding space–time covariance models.

Exercise XVI.4: Given the density $f(k,\omega) = f(k)f(\omega)$, where

$$\begin{aligned} f(k) &= ke^{-\frac{1}{2}k^2}, \\ f(\omega) &= 4\Gamma\left(\frac{5}{4}\right)^4 \omega^3 e^{-\Gamma\left(\frac{5}{4}\right)^4 \omega^4}, \end{aligned} \tag{3.1a–b}$$

use the PDF technique to derive a valid space–time covariance model in $R^{n,1}$.

Exercise XVI.5: Assuming that $c_{X,1}(r,\tau) = \frac{c_0}{\tau r}\sin(\tau r)$ and $c_{X,1}(r,\tau) = c_0\cos(\tau r)$ are valid covariance models in $R^{1,1}$, are the

$$c_{X,3}(r,\tau) = \frac{c_0}{2\pi r^2}\left[\frac{1}{\tau r}\sin(\tau r) - \cos(\tau r)\right] \tag{5.1}$$

and

$$c_{X,3}(r,\tau) = \frac{c_0}{\tau r}\sin(\tau r) \tag{5.2}$$

valid covariance models in $R^{3,1}$?

Exercise XVI.6: Consider the SPDE

$$\frac{u}{f(t)}\frac{\partial}{\partial s}X(s,t) + X(s,t) = Y(s), \tag{6.1}$$

where $X(0,t) = 0$, u and $f(t)$ are experimentally obtained parameters, and $Y(s)$ is a Brownian random field with $\overline{Y(s)} = 0$, $\overline{Y(s)Y(s')} = a\, s$. Find the corresponding covariance $c_X(s, t, s', t)$.

Exercise XVI.7: Consider the density

$$f(k, \tau) = \frac{A}{(2\pi)^2 \tau} e^{-\phi(k,\tau)} \left[\cos(\psi(k, \tau)) + \frac{k^2 - \lambda^2}{\nu k} \sin(\psi(k, \nu\tau)) \right], \tag{7.1}$$

for which three cases are to be examined:

Case 1: $\phi(k, \tau) = \frac{\nu k \tau}{2\alpha\lambda (k^2 - \lambda^2)^{0.5}}$, and $\psi(k, \tau) = \frac{(k^2 - \lambda^2)^{0.5} \tau}{\alpha\lambda} \left(k > \lambda,\ k >> \lambda \left(\frac{\lambda}{\nu - \lambda} \right)^{0.5} \right)$;

Case 2: $\phi(k, \tau) = \frac{(\lambda^2 - k^2)^{0.5} \tau}{\alpha\lambda}$, and $\psi(k, \tau) = \frac{\nu k \tau}{2\alpha\lambda (\lambda^2 - k^2)^{0.5}} \left(k < \lambda,\ k << \lambda \left(\frac{\lambda}{\nu - \lambda} \right)^{0.5} \right)$;

Case 3: $\phi(k, \tau) = \psi(k, \tau) = \frac{\tau}{\alpha} \left(\frac{\nu}{2\lambda} \right)^{0.5} \left(k \approx \lambda, \left| 1 - \frac{\lambda^2}{k^2} \right| << \frac{\nu}{\lambda} \right)$.

If A, ν, λ, and α are suitable coefficients, are the corresponding functions obtained by the PDF technique valid space–time covariance models in $R^{n,1}$, and what are their forms?

Exercise XVI.8: Consider the density

$$f(k, \omega) = \frac{1}{S_n k^{n-1}} \delta(k - a) \delta(\omega - b) \quad (a, b > 0). \tag{8.1}$$

(a) Using the PDF technique, derive the corresponding expression for the space–time covariance model in $R^{n,1}$.

(b) What specific forms you get for $n = 1,2,3$? What do you observe?

Exercise XVI.9: You are given the wavenumber–time spectra

$$\begin{aligned} \widetilde{c}_\ell(k, \tau) &= e^{-(a+\tau)\, k}, \\ \widetilde{c}_\eta(k, \tau) &= e^{-(a+\tau)\, k^2}. \end{aligned} \tag{9.1a–b}$$

($a > 0$). Use Eq. (7.2) of Chapter XVI to find the corresponding space–time covariance models $c_X(\boldsymbol{h},\tau)$ in $R^{3,1}$.

Exercise XVI.10: Consider the space–frequency spectra,

$$\begin{aligned} \widetilde{c}_X(\boldsymbol{h}, \omega) &= \widetilde{c}_X(\omega) \frac{\sin(a|\boldsymbol{h}|)}{a|\boldsymbol{h}|}, \\ \widetilde{c}_X(\boldsymbol{h}, \omega) &= \widetilde{c}_X(\omega) J_0(a|\boldsymbol{h}|). \end{aligned} \tag{10.1a–b}$$

Chose suitable expressions for $\widetilde{c}_X(\omega)$, and find the corresponding space–time covariance models in $R^{n,1}$.

Exercise XVI.11: Assume that the covariance function satisfies the following PDEs in $R^{n,1}$,

$$\left(\nabla^2 - \frac{1}{a^2}\frac{\partial^2}{\partial t^2}\right)c_X(s,t,s',t') + \frac{4\pi}{a}\delta(s-s')\delta(t-t') = 0,$$
$$\left(\nabla^2 - a\frac{\partial}{\partial t}\right)c_X(s,t,s',t') + \frac{4\pi}{a}\delta(s-s')\delta(t-t') = 0 \qquad (11.1a\text{–}b)$$

(a is a physical constant). Solve the PDEs to find the explicit forms of the covariance models $c_X(s,t,s',t')$. How do you physically interpret these models?

Exercise XVI.12: Assuming that the wavevector–frequency spectrum is given by

$$\widetilde{c}_X(\boldsymbol{k},\omega) = (2\pi)^{-1}\left[(\boldsymbol{k}\cdot(\boldsymbol{a}\boldsymbol{k}) + b)^2 + (\omega + \boldsymbol{v}\cdot\boldsymbol{k})^2\right]^{-1}, \qquad (12.1)$$

where $\boldsymbol{a}$ and b are physical vector and coefficient, respectively, use Eq. (2.18a) of Chapter XVI to construct the corresponding space–time covariance model $c_X(\boldsymbol{h},\tau)$.

Exercise XVI.13: Assume that the space–time covariance function of a physical attribute satisfies the following PDE in $R^{1,1}$,

$$\frac{\partial}{\partial\tau}c_X(r,\tau) + k(r)c_X(r,\tau) = 0,$$
$$c_X(r,0) = c_0\lambda(r), \qquad (13.1a\text{–}b)$$

where, based on the physics of the situation (say, first-order irreversible reaction), the following conditions apply: $k(r)$ is a known, monotonically increasing function on $[0,\infty)$, with $k(0)=0$, and $k(r)\to\infty$ as $r\to\infty$, and $\lambda(r)$ is a function such that $\int_0^\infty dr\lambda(r)r = 1$. Find the solution of Eqs. (13.1a–b), and then use Eq. (6.1a) of Chapter XVI to construct a covariance model in $R^{3,1}$.

Exercise XVI.14: This exercise also involves the ST technique (Section 4 of Chapter XVI). Recall that among the advantages of this technique is that it can construct valid covariance models from known ones by simple differentiation.

(a) Given $c_{X,1}(r,\tau) = \frac{c_0}{r\tau}\sin(r\tau)$ in $R^{1,1}$, use Eq. (6.1a) of Chapter XVI with $n=3$ to find a covariance model $c_{X,3}(r,\tau)$ in $R^{3,1}$.

(b) Given $c_{X,1}(r,\tau) = c_0\cos(r\tau)$ in $R^{1,1}$, use Eq. (6.1b) of Chapter XVI with $n=3$ to find a covariance $c_{X,3}(r,\tau)$ in $R^{3,1}$.

Exercise XVI.15: Prove the covariance argument transformation Eq. (2.19) of Chapter XVI.

Exercise XVI.16: Another technique of constructing space–time covariance models is based on the results of Exercise XI.10. In particular, use the formulation

$$f_X(\chi_p, \chi_{p'}) = f_X(\chi_p)\delta(\chi_{p'} - \phi(\chi_p)) \qquad (16.1)$$

to derive an integral representation of $c_X(\boldsymbol{p},\boldsymbol{p}')$ that allows you to construct space–time covariance models starting from known univariate PDF $f_X(\chi_p)$ and properly selecting one-to-one functions ϕ.

Exercise XVI.17: In the case of the empirical algebraic Eq. (5.39) of Chapter XVI, find $\gamma_X(\boldsymbol{p},\boldsymbol{p}')$ in terms of $\gamma_{Y_lY_{l'}}(\boldsymbol{p},\boldsymbol{p}')$, $l,l'=1,\ldots,m$, respectively.

Exercise XVI.18: The physical equation techniques (Section 5 of Chapter XVI) provide the means to construct space–time covariance models with properties that are physically consistent with the space–time variation of the attributes they represent. Consider the space–time covariance function $c_X(\boldsymbol{h},\tau)$ that satisfies the PDE in $R^{n,1}$,

$$\left(\frac{\partial}{\partial\tau} - \alpha\nabla^2\right)c_X(\boldsymbol{h},\tau) - c_Y(\boldsymbol{h},\tau) = 0, \tag{18.1}$$

where α is a known coefficient, and $c_Y(\boldsymbol{h},\tau)$ is a known covariance function. Show that the corresponding space–time covariance model is given by

$$c_X(\boldsymbol{h},\tau) = (G * c_Y)(\boldsymbol{h},\tau), \tag{18.2}$$

where $G(\boldsymbol{h},\tau) = (4\pi\alpha\tau)^{-\frac{n}{2}}e^{-\frac{h^2}{4\alpha\tau}}H(\tau)$ is the Green's function, and $H(\tau)$ is the unit step function.

Exercise XVI.19: Consider the space–time covariance function $c_X(\boldsymbol{h},\tau)$ that satisfies the PDE in $R^{2,1}$,

$$\left(\frac{\partial}{\partial\tau} - \alpha\nabla^2\right)c_X(\boldsymbol{h},\tau) = 0 \tag{19.1}$$

($\tau > 0$), with initial condition $c_X(\boldsymbol{h},0) = \delta(\boldsymbol{h} - \boldsymbol{b}) = \delta(h_1 - b_1, h_2 - b_2)$, where $\boldsymbol{b} = (b_1,b_2)$ is a known vector. Solve Eq. (19.1) to obtain a space–time covariance model $c_X(\boldsymbol{h},\tau)$ in $R^{2,1}$.

Exercise XVI.20: Consider the space–time covariance function $c_X(h,\tau)$ that satisfies the unidimensional wave equation in $R^{1,1}$

$$\left(\frac{\partial^2}{\partial\tau^2} - \frac{\partial^2}{\partial h^2}\right)c_X(h,\tau) - \delta(h)g(\tau) = 0, \tag{20.1}$$

with initial condition $\frac{\partial}{\partial h}c_X(h,\tau)\big|_{h=0} = -\frac{1}{2}g(\tau)$, where g is a known function, $h \in (-\infty,\infty)$, and $\tau > 0$.

(a) What is the covariance function $c_X(h,\tau)$ solution of the wave equation above?

(b) What is the form of $c_X(h,\tau)$ if $g(\tau) = \delta(\tau)$?

Exercise XVI.21: The ability to generalize findings to wider groups and circumstances is one of the most important features of the ST technique of constructing space–time covariance models. This and the following exercise use the ST results of Exercises V.19–V.21 to develop useful techniques of constructing space–time covariance functions in $R^{n,1}$, starting from suitable time functions. Let the covariance function satisfy the wave PDE

$$\left(\frac{\partial^2}{\partial\tau^2} - \nabla^2\right)c_X(\boldsymbol{h},\tau) = \delta(\boldsymbol{h})g(\tau), \tag{21.1}$$

where $\delta(\boldsymbol{h}) = \prod_{i=1}^{n}\delta(h_i)$ and $g(\tau)$ is a known time function.

(a) Show that the isotropic covariance solutions in $R^{n,1}$ for $n = 1, 2, 3$ are

$$c_{X,1}(r,\tau) = \frac{1}{2}\int_r^\infty du g(\tau - u),$$
$$c_{X,2}(r,\tau) = \frac{1}{2\pi}\int_r^\infty du\left(u^2 - r^2\right)^{-\frac{1}{2}} g(\tau - u), \qquad (21.2a\text{–}c)$$
$$c_{X,3}(r,\tau) = \frac{1}{4\pi r} g(\tau - r),$$

where $r = |\boldsymbol{h}|$.

(b) Assuming suitable time functions $g(\tau)$ of your choice, use Eqs. (21.2a–c) to construct covariance functions in $R^{n,1}$ ($n = 1, 2, 3$).

Exercise XVI.22: The space–time covariance function satisfies the heat PDE in $R^{2,1}$,

$$\left(\frac{\partial}{\partial \tau} - \nabla^2\right) c_X(\boldsymbol{h},\tau) = \delta(\boldsymbol{h}) g(\tau), \qquad (22.1)$$

where $\delta(\boldsymbol{h}) = \delta(h_1)\delta(h_2)$ and $g(\tau)$ is a known time function.

(a) Show that the isotropic covariance solution of Eq. (22.1) in $R^{2,1}$ is

$$c_{X,2}(r,\tau) = \frac{1}{4\pi}\int_0^\infty du\, u^{-1} e^{-\frac{r^2}{4u}} g(\tau - u), \qquad (22.2)$$

where $r = |\boldsymbol{h}|$.

(b) As in Exercise XVI.21 above, chose suitable time functions $g(\tau)$ and use Eq. (22.2) to construct covariance functions in $R^{2,1}$.

(c) Moreover, starting with $c_{X,2}(r,\tau)$ of Eq. (22.2), use the ST technique to derive covariance functions that are solutions of Eq. (22.1) in $R^{n,1}$ for any dimension n.

Exercise XVI.23: The *spatial extension technique* of space–time covariance model construction. This is a technique for constructing space–time covariance models starting from known purely spatial covariance models.

(a) Using Eq. (12.1) of Exercise XV.12, i.e.,

$$c_X(\boldsymbol{h},\tau) = \frac{1}{(2\pi)^{\frac{3}{2}} a^3 \tau^3}\int d\boldsymbol{h}'\, c_X(\boldsymbol{h}')\, e^{-\frac{(\boldsymbol{h}-\boldsymbol{h}'-\boldsymbol{v}\tau)^2}{2a^2\tau^2}} \qquad (23.1)$$

in $R^{3,1}$, where $\boldsymbol{v}$ is a known velocity vector and $a > 0$ is a known constant, construct new space–time covariance models $c_X(\boldsymbol{h},\tau)$ starting from known spatial covariance models $c_X(\boldsymbol{h})$.

(b) Modifications of the spatial extension technique are possible. What is the form of Eq. (23.1) if we assume that

$$c_X(\boldsymbol{h},\tau) = c_X(\boldsymbol{\varepsilon} v\tau, \tau), \qquad (23.2)$$

where $\boldsymbol{\varepsilon}$ is a unit vector in the s_3 direction?

(c) For illustration, what is the covariance function obtained by the modification of Eq. (23.2), if we assume that

$$c_X(\boldsymbol{h}) = e^{-\frac{\boldsymbol{h}^2}{b^2}}, \quad b > 0? \tag{23.3}$$

In this case, what is the asymptotic behavior of the covariance function when $\tau << b(2a)^{-\frac{1}{2}}$ and when $\tau >> b(2a)^{-\frac{1}{2}}$?

Exercise XVI.24: The *Montroll–Weiss* technique of space–time covariance model construction. This closed-form technique is motivated by the Montroll and Weiss (1965) analysis, as follows. Assume that $w(\boldsymbol{k}) = w(-\boldsymbol{k})$ and $\phi(\omega)$ are PDFs, i.e., $\int d\boldsymbol{k}w(\boldsymbol{k}) = 1$ and $\int_0^\infty d\omega\phi(\omega) = 1$, and let $\Phi(\omega) = \int_0^\omega d\omega\, \phi(\omega)$. Then, the Montroll–Weiss equation can be written as

$$c_X(\boldsymbol{h}, \tau) = \frac{1 - \widehat{\phi}(\tau)}{\tau\left[1 - \widetilde{w}(\boldsymbol{h})\widehat{\phi}(\tau)\right]}, \tag{24.1}$$

where $\widehat{\phi}(\tau)$ is the Laplace transform of $\phi(\omega)$, and $\widetilde{w}(\boldsymbol{h})$ is the Fourier transform of $w(\boldsymbol{k})$. It is often practical to assume that the PDF $w(\boldsymbol{k})$ is rotationally symmetric (with noncorrelated components), i.e., if $k = |\boldsymbol{k}|$ and $\upsilon = \frac{\boldsymbol{k}}{k}$, then $w(\boldsymbol{k})$ can be written as

$$w(\boldsymbol{k}) = u(k)\Omega(\upsilon), \tag{24.2}$$

where for the directions to be uniformly distributed on the unit sphere in R^n, we let $\Omega(\upsilon) = S_n^{-1}$ (S_n is the area of the unit sphere).

(a) Assuming PDFs $w(\boldsymbol{k})$ and $\phi(\omega)$ of your choice, derive space–time covariance models using Eq. (24.1).

(b) A general covariance model in $R^{1,1}$ obtained by the above technique is

$$c_X(h, \tau) = \frac{|h|^{\kappa_1}\tau^{\kappa_2}}{a|h|^{\lambda_1} + b\tau^{\lambda_2}}, \tag{24.3}$$

where $a, b, \kappa_1, \kappa_2, \lambda_1, \lambda_2$ are suitable coefficients. For illustration, Table 24.1 presents some space–time covariance models of the form of Eq. (24.3) in $R^{1,1}$. What are the permissibility conditions (PC) that coefficients β and μ in Table 24.1 must satisfy?

(c) Assuming a Poisson process, i.e., $\phi(\omega) = \lambda^{-1}e^{-\lambda^{-1}\omega}$, $\lambda > 0$, show that

$$c_X(\boldsymbol{h}, \tau) = \frac{\lambda}{1 + \lambda\tau - \widetilde{w}(\boldsymbol{h})}. \tag{24.4}$$

(d) Starting from the space–time covariance models in $R^{1,1}$ of Table 24.1 and/or Eq. (24.4), use the ST of Eq. (6.1a) of Chapter XVI to construct models in $R^{3,1}$.

Exercise XVI.25: The closed-form *univariate PDF* technique of constructing space–time covariance models in $R^{3,1}$ starts from a univariate PDF $f(\tau)$ and uses the equation

$$c_X(r, \tau) = \frac{a}{\tau}\left[\Phi(\tau) - \frac{a^2\tau^2}{\pi(a^2\tau^2 + r^2)}\right] \tag{25.1}$$

Table 24.1 Examples of $c_X(h,\tau)$ Generated by the Montroll–Weiss Technique in $R^{1,1}$

$\dfrac{\tau^{\mu-1}}{a(\beta,\mu)\lvert h\rvert^{\beta}+\tau^{\mu}}$ $a(\beta,\mu)=\dfrac{\pi}{2\sin\left(\frac{\pi\beta}{2}\right)\Gamma(\beta)\Gamma(1-\mu)}$	$\dfrac{\tau^{\mu-1}}{a(\beta,\mu)h^2+\tau^{\mu}}$ $a(\beta,\mu)=\dfrac{1}{(\beta^2-3\beta+2)\Gamma(1-\mu)}$
$\dfrac{1}{a(\beta,\mu)\lvert h\rvert^{\beta}+\tau}$ $a(\beta,\mu)=\dfrac{\pi(\mu-1)}{2\sin\left(\frac{\pi\beta}{2}\right)\Gamma(\beta)}$	$\dfrac{1}{a(\beta,\mu)h^2+\tau}$ $a(\beta,\mu)=\dfrac{\mu-1}{\beta^2-3\beta+2}$

where $\Phi(\tau)=1-\int_0^{\tau}dt\,f(t)$ and $a>0$. Technically, this is obviously a very simple technique that all it requires is the availability of valid PDF. The interpretational usefulness of the constructed covariance models, of course, depends on their physically meaningful features. What is the form of the covariance of Eq. (25.1) if $f(\tau)=\frac{bc^b}{(c+\tau)^{1+b}}$ $(b,c>0)$?

Exercise XVI.26: This exercise is based on the covariance argument transformation technique (Section 11 of Chapter XVI). Among the advantages of this technique are that it constructs valid covariance models from components whose validity is easy to check.

(a) Starting with covariance models $c_{\widehat{X}}(\widehat{\tau})$ available in Table 4.1 of Chapter VIII, use the change of arguments no. 1 in Table 11.2 of Chapter XVI to construct permissible space–time covariance models.

(b) Similarly, starting with covariance models $c_X(\widehat{\boldsymbol{h}})$ available in Table 4.1 of Chapter VIII, use the change of arguments nos. 2 or 5 in Table 11.2 of Chapter XVI to construct permissible space–time covariance models.

Exercise XVI.27: Consider the space–time covariance model in $R^{1,1}$,

$$c_X(r,\tau)=0.5\left[e^{-\beta r}\,Erfc\left(\frac{2\beta\tau-\alpha r}{2\sqrt{\alpha\tau}}\right)+e^{\beta r}\,Erfc\left(\frac{2\beta\tau+\alpha r}{2\sqrt{\alpha\tau}}\right)\right] \tag{27.1}$$

where α and β are known coefficients. Use the ST technique to construct covariance models in $R^{n,1}$ $(n>1)$.

Exercise XVI.28: As was mentioned in Section 2.3 of Chapter XVI, several modifications of the spectral technique of Eqs. (2.12a–c) of Chapter XVI for constructing space–time covariance models are possible. Particularly useful is the representation of the form of Eq. (2.12b) of Chapter XVI in which the starting point is the partial spectrum $\widetilde{c}_X(k,\tau)$. For illustration, consider the spectral representation of the covariance $c_X(r,\tau)$ in $R^{3,1}$,

$$c_X(r,\tau)=2\int_0^{\infty}dk\,(kr)^{-2}\left[\frac{\sin(kr)}{kr}-\cos(kr)\right]\widetilde{c}_X(k,\tau), \tag{28.1}$$

where $a > 0$. Let,

$$\tilde{c}_X(k,\tau) = \tilde{c}_X(k)e^{-ak^2\tau}, \tag{28.2}$$

where $\tilde{c}_X(k)$ is assumed to be of the polynomial form, $\tilde{c}_X(k) = ck^{2(m+2)}$, $c > 0$, $m \in Z_+$. What is the analytical form of $c_X(r,\tau)$ in this case?

REFERENCES

Abrahamsen, P., 1997. A Review of Gaussian Random Fields and Correlation Functions. Norwegian Computing Center, Oslo, Norway.

Adler, R.J., 1981. The Geometry of Random Fields. Jonh Wiley and Sons, New York, NY.

Adler, R.J., Taylor, J., 2007. Random Fields and Geometry. Springer, New York, NY.

Arnold, L., 1974. Stochastic Differential Equations: Theory and Applications. John Wiley, New York, NY.

Bak, P., Chen, K., 1989. The physics of fractals. Physica D 38, 5–12.

Banerjee, S., 2005. On geodetic distance computations in spatial modeling. Biometrics 61 (2), 617–625.

Banerjee, S., Gelfand, A., 2003. On smoothness properties of spatial processes. J. Multivar. Anal. 84, 85–100.

Batchelor, G.K., 1951. Pressure fluctuations in isotropic turbulence. Proc. Camb. Philos. Soc. 47, 359–374.

Baxevani, A., Caires, S., Rychlik, I., 2006. Spatiotemporal Statistical Modelling of Significant Wave Height, Preprint 2006:11. Dept. of Mathematical Sciences, Chalmers Univ. of Technology and Goteborg Univ., Goteborg, Sweden.

Belyaev, YuK., 1972. Point processes and first passage problems. In: Proceedings of the 6th Berkeley Symposium on Mathematical Statistics and Probability, vol. 2. University of California Press, Berkeley, CA, pp. 1–17.

Bell, T.L., 1987. A space-time stochastic model of rainfall for satellite remote-sensing studies. J. Geophys. Res. 92 (D8), 9631–9643.

Beran, M.J., 1968. Statistical Continuum Theories. In: Monographs in Statistical Physics, vol. 9. Interscience Publishers, New York, NY.

Bialynicki-Birula, I., Cieplak, M., Kaminski, J., 1992. Theory of Quanta. Oxford University Press, New York, N.Y.

Billings, S.D., Rick, K., Beatson, R.K., Newsam, G.N., 2002. Interpolation of geophysical data using continuous global surfaces. Geophysics 67 (6), 1810–1822.

Bilonick, R.A., 1985. The space-time distribution of sulfate deposition in the Northeastern United States. Atmos. Environ. 19 (11), 1829–1845.

Bochner, S., 1933. Monotone funktionen Stieltjessche integrale und harmonische analyse. Math. Ann. 108, 378–410.

Bochner, S., 1959. Lectures on Fourier Integrals. Princeton University Press, Princeton, NJ.

Bogaert, P., Christakos, G., 1997a. Spatiotemporal analysis and processing of thermometric data over Belgium. J. Geophys. Res. 102 (D22), 25831–25846.

Bogaert, P., Christakos, G., 1997b. Stochastic analysis of spatiotemporal solute content measurements using a regressive model. Stoch. Hydrol. Hydraul. 11 (4), 267–295.

Bohr, N., 1963. Atomic Theory and Human Knowledge. John Wiley, New York.

Boltzmann, L., 1868. Studien über das Gleichgewicht der lebendigen Kraft zwischen bewegten materiellen Punkten. Wien. Ber. 58, 517–560.

Bongajum, E., Milkereit, B., Huang, J., 2013. Building 3D stochastic exploration models from borehole geophysical and petrophysical data: a case study. Can. J. Explor. Geophys. 38 (1), 40–50.

Borgman, L.E., 1969. Ocean wave simulation for engineering design. J. Waterw. Harb. Div. Proc. ASCE 95 (4), 557–586.

Bouchaud, J.-P., Georges, A., 1990. Anomalous diffusion in disordered media, statistical mechanics, models and physical applications. Phys. Rep. 195, 127–293.

Bouleau, N., 1991. Splendeurs et misères des lois de valeurs extrêmes. Rev. Risques FFSA 4, 85–92.

Bourgine, B., Chilès, J.-P., Watremez, P., 2001. Space-time modeling of sand beach data: a geostatistical approach. In: Geostatistics for Environmental Applications. Kluwer Academic Publishers, Dordrecht, The Netherlands, pp. 101–111.

Bras, R.L., Rodriguez-Iturbe, I., 1985. Random Functions and Hydrology. Addison-Wesley, New York, NY.

Brase, C.H., Brase, C.P., 2010. Understandable Statistics: Concepts and Methods. Brooks/Cole, Belmont, CA.

Carroll, S., 2004. Spacetime and Geometry. Addison-Wesley, San Francisco, CA.
Chandrasekhar, S., 1943. Stochastic problems in physics and astronomy. Rev. Mod. Phys. 15, 1–89.
Christakos, G., 1984a. The space transformations and their applications in systems modeling and simulation. In: Proceedings of the 12th International Conference on Modeling and Simulation (AMSE), vol. 1 (3), pp. 49–68. Athens.
Christakos, G., 1984b. On the problem of permissible covariance and variogram models. Water Resour. Res. 20 (2), 251–265.
Christakos, G., 1986a. Space transformations in the study of multidimensional functions in the hydrologic sciences. Adv. Water Resour. 9 (1), 42–48.
Christakos, G., 1986b. Recursive Estimation of Nonlinear-State Nonlinear-Observation Systems (*Research Rep. OF.86–29*). Kansas Geological Survey, Lawrence, KS.
Christakos, G., 1989. Optimal estimation of nonlinear-state nonlinear-observation systems. J. Optim. Theory Appl. 62, 29–48.
Christakos, G., 1990. A Bayesian/maximum-entropy view to the spatial estimation problem. Math. Geol. 22 (7), 763–776.
Christakos, G., 1991a. Some applications of the Bayesian, maximum-entropy concept in geostatistics. In: Fundamental Theories of Physics. Kluwer, The Netherlands, pp. 215–229.
Christakos, G., 1991b. A theory of spatiotemporal random fields and its application to space-time data processing. IEEE Trans. Syst. Man Cybern. 21 (4), 861–875.
Christakos, G., 1991c. Certain results on spatiotemporal random fields, and their applications in environmental research. In: Proceedings of a NATO Advanced Studies Institute, Lucca, Italy.
Christakos, G., 1992. Random Field Models in Earth Sciences. Academic Press, San Diego, CA (New edition: 2005. Dover Publications, New York, NY).
Christakos, G., 2000. Modern Spatiotemporal Geostatistics. Oxford University Press, New York, NY.
Christakos, G., 2002. On a deductive logic-based spatiotemporal random field theory. Theory Probab. Math. Stat. (Teoriya Imovirnostey ta Matematychna Statystyka) 66, 54–65.
Christakos, G., 2010. Integrative Problem-Solving in a Time of Decadence. Springer, New York, NY.
Christakos, G., Hristopulos, D.T., 1994. Stochastic space transformation techniques in subsurface hydrology-part 2: generalized spectral decompositions and plancherel representations. Stoch. Hydrol. Hydraul. 8 (2), 117–138.
Christakos, G., Hristopulos, D.T., 1996. Characterization of atmospheric pollution by means of stochastic indicator parameters. Atmos. Environ. 30 (22), 3811–3823.
Christakos, G., Hristopulos, D.T., 1997. Stochastic Radon operators in porous media hydrodynamics. Q. Appl. Math. LV (1), 89–112.
Christakos, G., Hristopulos, D.T., 1998. Spatiotemporal Environmental Health Modelling. Kluwer Academic Publishers, Boston, MA.
Christakos, G., Kolovos, A., 1999. A study of the spatiotemporal health impacts of ozone exposure. J. Expos. Anal. Environ. Epidemiol. 9 (4), 322–335.
Christakos, G., Panagopoulos, C., 1992. Space transformation methods in the representation of geophysical random fields. IEEE Trans. Geosci. Remote Sens. 30 (1), 55–70.
Christakos, G., Papanicolaou, V., 2000. Norm-dependent covariance permissibility of weakly homogeneous spatial random fields. Stoch. Environ. Res. Risk Assess. 14 (6), 1–8.
Christakos, G., Hristopulos, D.T., Miller, C.T., 1995. Stochastic diagrammatic analysis of groundwater flow in heterogeneous soils. Water Resour. Res. 31 (7), 1687–1703.
Christakos, G., Hristopulos, D.T., Bogaert, P., 2000. On the physical geometry hypotheses at the basis of spatiotemporal analysis of hydrologic geostatistics. Adv. Water Resour. 23, 799–810.
Christakos, G., Angulo, J.M., Yu, H.-L., 2011. Constructing space-time pdf distributions in geosciences. Bol. Geol. Min. Esp. (BGME) 122 (4), 531–542.

Christakos, G., Zhang, C., He, J., 2017a. Modeling and prediction of space-time disease spread: the traveling epidemic. Stoch. Environ. Res. Risk Assess. 31 (2), 305–314. http://dx.doi.org/10.1007/s00477-016-1298-3.

Christakos, G., Angulo, J.M., Yu, H.-L., Wu, J., 2017b. Space-time metric determination in environmental modeling. J. Environ. Inf. (in press).

Christakos, G., Yang, Y., Wu, J., Zhang, C., He, J., Mei, Y., 2017c. Improved space-time mapping of $PM_{2.5}$ distribution using a domain transformation method. Environ. Pollut. (submitted).

Coles, S., 2001. An Introduction to Statistical Modeling of Extreme Values. Springer-Verlag, London, UK.

Cox, D.R., Isham, V., 1988. A simple spatial-temporal model of rainfall. Proc. R. Soc. Lond. Ser. A 415, 317–328.

Cramer, H., 1946. Mathematical Methods of Statistics. Princeton University Press, Princeton, NJ.

Cramer, H., Leadbetter, M.R., 1967. Stationary and Related Stochastic Processes. John Wiley & Sons, New York, NY.

Cressie, N., Huang, H.C., 1999. Classes of nonseparable, spatio-temporal stationary covariance functions. J. Am. Stat. Assoc. 94, 1330–1340.

Curriero, F.C., 2006. On the use of non-Euclidean distance measures in geostatistics. Math. Geol. 38 (8), 907–926.

Da Prato, G., Tubaro, L., 1987. Stochastic Partial Differential Equations and Applications (Lecture Notes in Mathematics, n. 1236). Springer, New York, NY.

Dagan, G., 1989. Flow and Transport in Porous Formations. Springer, New York, NY.

Daley, R., 1999. Atmospheric Data Analysis. Cambridge University Press, New York, N.Y.

De Iaco, S., Myers, D.E., Posa, D., 2002. Nonseparable space-time covariance models: some parametric families. Math. Geol. 34, 23–42.

De Iaco, S., Palma, M., Posa, D., 2003. Covariance functions and models for complex-valued random fields. Stoch. Environ. Res. Risk Assess. 17, 145–156.

De la Peña, V.H., Ibragimov, R., Sharakhmetov, S., 2006. Characterizations of joint distributions, copulas, information, dependence and decoupling, with applications to time series. In: IMS Lecture Notes–Monograph Series 2nd Lehmann Symposium–Optimality, vol. 49, pp. 183–209.

Dean, R.G., Dalrymple, R.A., 1991. Water Wave Mechanics for Engineers & Scientists, vol. 2. World Scientific Publishing Company, New York, NY.

Dee, D.P., 1991. Simplification of the Kalman filter for meteorological data assimilation. Q. J. R. Meteorol. Soc. 117, 365–384.

Dennis, M.R., 2003. Correlations and screening of topological charges in Gaussian random fields. J. Phys. A Math. Gen. 36, 6611–6628.

Deza, M., Deza, E., 2014. Encyclopedia of Distances, 3rd revised ed. Springer-Verlag, New York, NY.

Dobrovolski, S., 2010. Stochastic Climate Theory: Models and Applications. Springer, New York, NY.

Dudley, R.M., 2014. Uniform Central Limit Theorems. Cambridge University Press, New York, NY.

Einstein, A., 1905. On the motion of small particles suspended in liquids at rest required by the molecular-kinetic theory of heat. Ann. Phys. 17, 549–560 (Also, Investigations on the Theory of Brownian Motion. Dover, New York (1956)).

Etherton, B.J., Bishop, C.H., 2004. Resilience of hybrid ensemble/3DVAR analysis schemes to model error and ensemble covariance error. Mon. Weather Rev. 132, 1065–1080.

Feder, J., 1988. Fractals. Plenum Press, New York, NY.

Feller, W., 1966. An Introduction to Probability Theory and Its Applications, vols. 1 and 2. J. Wiley, New York, NY.

Fernández-Avilés, G., Montero, J.M., Mateu, J., 2011. Mathematical genesis of the spatio-temporal covariance functions. J. Math. Stat. 7 (1), 37–44.

Finke, U., 1999. Space–time correlations of lightning distributions. Mon. Weather Rev. 127 (1), 1850–1861.

Fisher, R., Mayer, M., von der Linden, W., Dose, V., 1997. Enhancement of the energy resolution in ion-beam experiments with the maximum entropy method. Phys. Rev. E 55 (6), 6667–6673.

Fisz, M., 1964. Probability Theory and Mathematical Statistics. Academic Press, New York, NY.

Fogedby, H.C., 1998. Morphology and scaling in the noisy Burgers equation: soliton approach to the strong coupling fixed point. Phys. Rev. Lett. 80 (6), 1126–1129.

Forristall, Z.G., Ewans, K.C., 1998. Worldwide measurements of directional wave spreading. J. Atmos. Oceanic Technol. 15, 440–469.

Frankel, A., Clayton, R.W., 1986. Finite difference simulations of seismic scattering: implications for the propagation of short-period seismic waves in the crust and models of crustal heterogeneity. J. Geophys. Res. 91, 6465–6489.

Frei, C., 2014. Interpolation of temperature in a mountainous region using nonlinear profiles and non-Euclidean distances. Int. J. Climatol. 34, 1585–1605.

Friedman, A., 1975. Stochastic Differential Equations and Applications, vol. 1. Academic Press, New York, NY.

Friedman, A., 1976. Stochastic Differential Equations and Applications, vol. 2. Academic Press, New York, NY.

Fujiwara, D., Morimoto, H., 1977. An L_r theorem of the Helmholtz decomposition of vector fields. J. Fac. Sci. 44, 685–700.

Gaspari, G., Cohn, S.E., 1999. Construction of correlation functions in two and three dimension. Q. J. R. Meteorol. Soc. 125, 723–757.

Gel'fand, I.M., 1955. Generalized random processes. Dokl. Akad. Nauk. SSSR 100, 853–856.

Gel'fand, I.M., Shilov, G.E., 1964. Generalized Functions, vol. 1. Academic Press, New York, NY.

Gel'fand, I.M., Vilenkin, N.Y., 1964. Generalized Functions, vol. 4. Academic Press, New York, NY.

Gelhar, L.W., 1993. Stochastic Subsurface Hydrology. Prentice-Hall, Englewood Cliffs, NJ.

Genest, C., Nešlehová, J., 2007. A primer on copulas for count data. ASTIN Bull. 37 (2), 475–515.

Genest, C., Rivest, L.-P., 1993. Statistical inference procedures for bivariate Archimedean copulas. J. Am. Stat. Assoc. 88, 1034–1043.

Gibbs, J.W., 1902. Elementary Principles in Statistical Mechanics Developed With Especial Reference to the Rational Foundation of Thermodynamics. Charles Scribner's Sons, New York, NY.

Gihman, I.I., Skorokhod, A.V., 1972. Stochastic Differential Equations. Springer-Verlag, Berlin.

Gihman, I.I., Skorokhod, A.V., 1974a. The Theory of Stochastic Processes-1. Springer-Verlag, New York, NY.

Gihman, I.I., Skorokhod, A.V., 1974b. The Theory of Stochastic Processes-2. Springer-Verlag, New York, NY.

Gihman, I.I., Skorokhod, A.V., 1974c. The Theory of Stochastic Processes-3. Springer-Verlag, New York, NY.

Gnedenko, B.V., 1962. The Theory of Probability. Chelsea, New York, NY.

Gneiting, T., 1999. On the derivatives of radial positive definite functions. J. Math. Anal. Appl. 236, 86–93.

Gneiting, T., 2002. Nonseparable, stationary covariance functions for space-time data. J. Am. Stat. Assoc. 97, 590–600.

Gneiting, T., 2013. Strictly and non-strictly positive definite functions on spheres. Bernoulli 19 (4), 1327–1349.

Gradshteyn, T.S., Ryzhik, I.M., 1965. Tables of Integrals, Series, and Products. Academic Press, New York, NY.

Hadsell, F., Hansen, R., 1999. Tensors of geophysics. In: Generalized Functions and Curvilinear Coordinates, vol. 2. Society of Exploration Geophysicists, Tulsa, OK.

Haslett, J., Raftery, A.E., 1989. Space-time modeling with long-memory dependence: assessing Ireland's wind power resource. Appl. Stat. 38 (1), 1–50.

Hauser, W., 1971. Introduction to the Principles of Electromagnetism. Addison-Wesley Publ. Inc., Boston, MA.

He, G.W., Zhang, J.B., 2006. Elliptic model for space–time correlations in turbulent shear flows. Phys. Rev. E 73, 055303.

He, X., Koch, J., Sonnenborg, T.O., Jorgensen, F., Schamper, C., Refsgaard, J.C., 2014. Transition probability-based stochastic geological modeling using airborne geophysical data and borehole data. Water Resour. Res. 50, 3147–3169.

Heisenberg, W., 1930. The Physical Principles of the Quantum Theory. Dover, New York, NY.

Heisenberg, W., 1948. On the theory of statistical and isotropic turbulence. Proc. R. Soc. Lond. A 195, 402–406.

Helgason, S., 1980. The Radon Transform. Birkhauser, Boston.

Helgason, S., 1984. Groups and Geometric Analysis. Academic Press, San Diego, CA.

Heskestad, G., 1965. A generalized Taylor hypothesis with application for high Reynolds number shear flows. Trans. ASME J. Appl. Mech. 87, 735–739.

Hill, R.J., 1996. Pressure–velocity–velocity statistics in isotropic turbulence. Phys. Fluids 8, 3085–3093.

Hill, R.J., Wilczak, J.M., 2001. Fourth-order velocity statistics. Fluid Dyn. Res. 28, 1–22.

Hinze, J.O., 1975. Turbulence. McGraw-Hill, New York, NY.

Hopf, E., 1952. Statistical hydrodynamics and functional calculus. J. Ration. Mech. Anal. 1, 87–123.

Horn, R.A., Johnson, C.R., 1985. Matrix Analysis. Cambridge Univ. Press, Cambridge, UK.

Hristopulos, D.T., 2002. New anisotropic covariance models and estimation of anisotropic parameters based on the covariance tensor identity. Stoch. Environ. Res. Risk Assess. 16, 43–62.

Hristopulos, D.T., 2003. Permissibility of fractal exponents and models of band-limited two-point functions for fGn and fBm random fields. Stoch. Environ. Res. Risk Assess. 17, 191–216.

Hristopulos, D.T., 2004. Anisotropic Spartan random field models for geostatistical analysis. In: Agioutantis, Z., Komnitsas, K. (Eds.), Proceedings of Advances in Mineral Resources Management and Environmental Geotechnology. Heliotopos Conferences, Athens, Greece.

Hristopulos, D.T., Christakos, G., 1997. A variational calculation of the effective fluid permeability of heterogeneous media. Phys. Rev. E 55 (6), 7288–7298.

Hristopulos, D.T., Tsantili, I.C., 2015. Space-time models based on random fields with local interactions. Int. J. Mod. Phys. B 29. http://dx.doi.org/10.1142/S0217979215410076.

Hristopulos, D.T., Christakos, G., Serre, M.L., 1999. Implementation of a space transformation approach for solving the three-dimensional flow equation. SIAM J. Sci. Comput. 20 (2), 619–647.

Huang, C., Zhang, H., Robeson, S.M., 2011. On the validity of commonly used covariance and variogram functions on the sphere. Math. Geosci. 43, 721–733.

Isichenko, M.B., 1992. Percolation, statistical topography, and transport in porous media. Rev. Mod. Phys. 64 (4), 961–1043.

Ito, K., 1954. Stationary random distributions. Univ. Kyoto Memoirs, Series A, XXVIII 3, 209–223.

Joe, H., 2006. Discussion of 'copulas: tales and facts,' by Thomas Mikosch. Extremes 9, 37–41.

John, F., 1955. Plane Waves and Spherical Means. Springer-Verlag, New York, NY.

Jones, R., Zhang, Y., 1997. Models for continuous stationary space-time processes. In: Gregoire, G., Brillinger, D., Diggle, P., Russek-Cohen, E., Warren, W., Wolfinge, R. (Eds.), Modelling Longitudinal and Spatially Correlated Data. Springer, New York, NY, pp. 289–298.

Kent, J.T., 1989. Continuity properties for random fields. Ann. Probab. 17 (4), 1432–1440.

Kent, J.T., Mohammadzadeh, M., Mosammam, A.M., 2011. The dimple in Gneiting's spatial-temporal covariance model. Biometrika 98 (2), 489–494.

Khinchin, A., 1934. Korrelations theorie des stationaren stochastischen prozesse. Math. Anal. 109, 604–615.

Khinchin, A., 1949. Mathematical Foundations of Statistical Mechanics. Dover, New York, NY.

Kinsman, B., 1984. Wind Waves. Dover Inc., New York, NY.

Klafter, J., Sokolov, I.M., 2011. First Steps in Random Walks: From Tools to Applications. Oxford University Press, New York, NY.

Klimes, L., 2002a. Correlation functions of random media. Pure Appl. Geophys. 159, 1811–1831.

Klimes, L., 2002b. Estimating the correlation function of a self-affine random medium. Pure Appl. Geophys. 159, 1833–1853.

Klyatskin, V.I., 2015. Stochastic equations: theory and applications in acoustics. In: Hydrodynamics, Magnetohydrodynamics, and Radiophysics, vol. 1. Springer, New York, NY.

Kobe, D.H., 1984. Helmholtz theorem for antisymmetric second-rank tensor fields and electromagnetism with magnetic monopoles. Am. J. Phys. 52 (4), 354–358.

Kolmogorov, A.N., 1933. Grundbegrjlje der Wahrscheinlichkeitrechnung (Ergebnisse der Mathematik) (English translation: *Foundations of the Theory of Probability*, Chelsea, NewYork. 1950.).

Kolmogorov, A.N., 1941. The distribution of energy in locally isotropic turbulence. Dokl. Akad. Nauk. SSSR 32, 19–21.

Kolovos, A., Christakos, G., Serre, M.L., Miller, C.T., 2002. Computational BME solution of a stochastic advection–reaction equation in the light of site-specific information. Water Resour. Res. 38, 1318–1334.

Kolovos, A., Christakos, G., Hristopulos, D.T., Serre, M.L., 2004. Methods for generating non-separable spatio-temporal covariance models with potential environmental applications. Adv. Water Resour. 27, 815–830.

Kotz, S., Seeger, J.P., 1991. A new approach to dependence in multivariate distributions. In: Dall'Aglio, G., Kotz, S., Salinetti, G. (Eds.), Advances in Probability Distributions. Kluwer, Dordrecht, The Netherlands, pp. 113–127.

Kotz, S., Balakrishnana, N., Johnson, N.L., 2000. Continuous Multivariate Distributions. Wiley, New York, NY.

Kraichnan, R.H., 1958. A theory of turbulence dynamics. In: Second Symposium on Naval Hydrodynamics. Office of Naval Research, Washington DC (Ref. ACR-38).

Kraichnan, R.E., 1974. On Kolmogorov's inertial-range theories. J. Fluid Mech. 62 (2), 305–330.

Lajaunie, C., Béjaoui, R., 1991. Sur le Krigeage des Functions Complexes (Note N-23/91/G). Centre de Geostatistique, Ecole des Mines de Paris, Fontainebleau, France.

Langevin, P., 1908. Sur la theorie du mouvement brownien. C. R. Acad. Sci. Paris 146, 530–533.

Lévy, P., 1948. Processus Stochastiques et Mouvement Brownien. Gauthier-Villars, Paris.

Lewis, R.M., Kraichnan, R.H., 1962. A space-time functional formalism for turbulence. Commun. Pure Appl. Math. XV, 397–411.

Lin, C.C., 1953. On Taylor's hypothesis and the acceleration terms in the Navier–Stokes equations. Q. Appl. Math. 10, 295–306.

Lin, Y.-C., Chang, T.-J., Lu, M.-M., Yu, H.-L., 2015. A space-time typhoon trajectories analysis in the vicinity of Taiwan. Stoch. Environ. Res. Risk Assess. 29, 1857–1866.

Lloyd, C.D., 2010. Local Models for Spatial Analysis. CRC Press, Boca Raton, FL.

Loeve, M., 1953. Probability Theory. Van Nostrand, Princeton, NJ.

Long, D., Krzysztofowicz, R., 1995. A family of bivariate densities constructed from marginals. J. Am. Stat. Assoc. 90 (430), 739–746.

Longair, M.S., 1984. Theoretical Concepts in Physics. Cambridge University Press, Cambridge, UK.

Lurton, X., 2010. An Introduction to Underwater Acoustics Principles and Applications. Springer, New York, NY.

Ma, C., 2002a. Spatio-temporal covariance functions generated by mixtures. Math. Geol. 34, 965–975.

Ma, C., 2002b. Families of spatio-temporal stationary covariance models. J. Stat. Plan. Inference 116, 489–501.

Ma, C., 2003a. Nonstationary covariance functions that model space-time interactions. Stat. Probab. Lett. 62, 411–419.

Ma, C., 2003b. Spatio-temporal stationary covariance models. J. Multivar. Anal. 86, 97–107.

Ma, C., 2007. Stationary random fields in space and time with rational spectral densities. IEEE Trans. Inf. Theory 53 (3), 1019–1029.

Macarron, R., Banks, M.N., Bojanic, D., Burns, D.J., Cirovic, D.A., Garyantes, T., Green, D.V.S., Hertzberg, R.P., Janzen, W.P., Paslay, J.W., Schopfer, U., Sittampalam, G.S., 2011. Impact of high-throughput screening in biomedical research. Nat. Rev. Drug Discov. 10 (3), 188–195.

Mandelbrot, B.B., 1982. The Fractal Geometry of Naure. Freeman & Company, New York, NY.

Mandelbrot, B.B., Van Ness, J.W., 1968. Fractional Brownian motions, fractional noises and applications. SIAM Rev. 10 (4), 422–437.

Marcus, M., Minc, H., 1992. A Survey of Matrix Theory and Matrix Inequalities. Dover Publications, NY.

Matern, B., 1960. Spatial variation. Medd. Fran. Stat. Skogsf. 49, 5 (Stockholm, Sweden).

Matheron, G., 1965. Les Variables Régionalisées et leur Estimation. Masson, Paris.

Matheron, G., 1973. The intrinsic random functions and their applications. Adv. Appl. Prob. 5, 439–468.

Matheron, G., 1976. A simple substitute for conditional expectation. The disjunctive kriging. In: Guarascio, M., et al. (Eds.), Advanced Geostatistics in the Mining Industry. Reidel, Dordrecht, pp. 221–236.

Matheron, G., de Marsily, G., 1980. Is transport in porous media always diffusive? A counterexample. Water Resour. Res. 16 (5), 901–917.

Matthaeus, W.H., Weyland, J.M., Dasso, S., 2016. Ensemble space-time correlation of plasma turbulence in the solar wind. Phys. Rev. Lett. 116, 245101. http://dx.doi.org/10.1103/PhysRevLett.116.245101.

Maxwell, J.C., 1860. Illustrations of the dynamical theory of gases. Philos. Mag. (19, 19–32 and 20, 21–37).

McComb, W.D., 1990. The Physics of Turbulence. Oxford University Press, New York, NY.

McKean, H., 2014. Probability: The Classical Limit Theorems. Cambridge University Press, New York, NY.

Mikosch, T., 2006a. Copulas: tales and facts. Extremes 9, 3–20.

Mikosch, T., 2006b. Copulas: tales and facts—rejoinder. Extremes 9, 55–62.

Mokljacuk, M.P., Jadrenko, M.I., 1979. Linear statistical problems for stationary isotropic random fields on a sphere, I. Theory Prob. Math. Stat. 18, 115–124.

Monin, A.S., Yaglom, A.M., 1971. Statistical Fluid Mechanics. M.I.T. Press, Cambridge, MA.

Montroll, E.W., Weiss, G.H., 1965. Random walks on lattices. II. J. Math. Phys. 6, 167–181.

Müller, T.M., Shapiro, S.A., 2001. Most probable seismic pulses in single realizations of two- and three-dimensional random media. Geophys. J. Int. 144, 83–95.

NAQM, 2014. National Air Quality Monitoring. Shandong Province, China. http://113.108.142.147:20035/emcpublish.

Neeser, F.D., Massey, J.L., 1991. Proper complex random processes with applications to information theory. IEEE Trans. Inf. Theory 39 (4), 1293–1302.

Nelsen, R., 1999. An Introduction to Copulas. Springer, New York, NY.

Olea, R.A., 1999. Geostatistics. Kluwer Academic Publishers, Boston, MA.

Oliver, L.D., Christakos, G., 1996. Boundary condition sensitivity analysis of the stochastic flow equation. Adv. Water Resour. 19 (2), 109–120.

Omatu, S., Seinfeld, J.H., 1981. Filtering and smoothing for linear discrete-time distributed parameter systems based on Wiener-Hopf theory with application to estimation of air pollution. IEEE Trans. Syst. Man Cybern. 11 (12), 785–801.

Orlowski, A., Sobczyk, K., 1989. Solitons and shock waves under random external noise. Rep. Math. Phys. 27 (1), 59–71.

Papoulis, A., 1962. The Fourier Integral and Its Applications. McGraw-Hill, New York, NY.

Pearson, K., 1901. Mathematical contributions to the theory of evolution, VII: On the correlation of characters not quantitatively measurable: Philos. Trans. R. Soc. Lond. Ser. A 195, 1–47.

Phillips, W.R.C., 2000. Eulerian space—time correlations in turbulent shear flows. Phys. Fluids 12, 2056–2064.

Porcu, E., Gregori, P., Mateu, J., 2006. Nonseparable stationary anisotropic space-time covariance functions. Stoch. Environ. Res. Risk Assess. 21, 113–122.

Potvin, G., 1993. Space-Time Correlations and Taylor's Hypothesis for Rainfall (MS thesis). Department of Physics, McGill University, Montreal, Canada.

Praskovsky, A.A., Praskovskaya, E.A., 2003. Structure-function-based approach to analyzing received signals for spaced antenna radars. Radio Sci. 38 (4), 1068. http://dx.doi.org/10.1029/2001RS002544.

Proudman, I., 1952. The generation of noise by isotropic turbulence. Proc. Roy. Soc. A 214, 119.

Pugachev, V.S., Sinitsyn, I.N., 1987. Stochastic Differential Systems: Analysis and Filtering. Wiley, New York, NY.

Purser, R.J., Wu, W.-S., Parrish, D.F., Roberts, N.M., 2003. Numerical aspects of the application of recursive filters to variational statistical analysis. Part II: Spatially inhomogeneous and anisotropic general covariances. Mon. Weather Rev. 131, 1536–1548.

Radon, J., 1917. Uber die Bestimmung von funktionen durch ihre integralwerte langs gewissermannigfaltigkeiten. Berichte Sachs. Aked. Wiss. Leipz. Math. Phys. KL 69, 262–267.

Ramsay, J.O., Silverman, B.W., 2002. Applied Functional Data Analysis: Methods and Case Studies, vol. 77. Springer, New York, NY.

Rao, C.R., 1973. Linear Statistical Inference and Its Applications. John Wiley, New York, NY.

Rasouli, S., Niry, M.D., Rajabi, Y., Panahi, A.A., Niemela, J.J., 2014. Applications of 2-D Moire deflectometry to atmospheric turbulence. J. Appl. Fluid Mech. 7 (4), 651–657.

Renyi, A., 2007. Probability Theory. Dover Publications, New York, NY.

Rhodes, C.J., Anderson, R.M., 1996. Power laws governing epidemics in isolated populations. Nature 381, 600–602.

Robert, A., Roy, A.G., 1990. On the fractal interpretation of the mainstream length-drainage area relationship. Water Resour. Res. 26 (5), 839–842.

Robertson, H.P., 1940. The invariant theory of isotropic turbulence. Camb. Philos. Soc. 36, 209–223.

Rodriguez-Iturbe, I., Mejia, J.M., 1974. The design of rainfall networks in time and space. Water Resour. Res. 10 (4), 713–728.

Rosenblatt, M., 1956. A central limit theorem and the strong mixing conditions. Proc. Natl. Acad. Sci. U.S.A. 42, 43–47.

Scholzel, C., Friederichs, P., 2008. Multivariate non-normally distributed random variables in climate research – introduction to the copula approach. Nonlinear Process. Geophys. 15, 761–772.

Schwartz, 1, 1950–51. Theorie des Distributions 1 and 2. Hermann & Cie, Paris, France.

Shlesinger, M.F., Zaslavsky, G.M., Klafter, J., 1993. Strange kinetics. Nature 363, 31–37.

Shreve, S., 2004. Stochastic Calculus for Finance, I and II. Springer, New York, NY.

Sigrist, F., Künsch, H.R., Stahel, W.A., 2015. Stochastic partial differential equation based modelling of large space-time data sets. J. R. Stat. Soc. Ser. B Stat. Methodol. 77 (1), 3–33.

Sklar, A., 1959. Fonctions de repartition à n dimensions et leurs marges, vol. 8. Publications de l'Institut Statistique de L'Université de Paris, Paris, France, pp. 229–231.

Sobczyk, K., 1991. Stochastic Differential Equations. Kluwer, Dordrecht, The Netherlands.

Sobczyk, K., Kirkner, D.J., 2001. Stochastic Modeling of Microstructures. Springer, New York, NY.

Soong, T.T., 1973. Random Differential Equations in Science and Engineering. Academic Press, London, UK.

Sprössig, W., 2010. On Helmholtz decompositions and their generalizations-An overview. Math. Methods Appl. Sci. 33, 374–383.

Srinivasan, S.K., Vasudevan, R., 1971. Introduction to Random Differential Equations and Their Applications. American Elsevier Publ. Co., Inc., New York, NY.

Stanley, H.E., Ostrowsky, N., 1986. On growth and form. In: NATO ASI Series. Kluwer Academic Publishers, Hingham, MA.

Stauffer, D., Aharony, A., 1992. Introduction to Percolation Theory. Taylor and Francis, London, UK.

Stein, M.L., 2005. Space-time covariance functions. J. Am. Stat. Assoc. 100, 310–321.

Stephani, H., Kramer, D., MacCallum, M., Hoenselaers, C., Herlt, E., 2003. Exact Solutions of Einstein's Field Equations. University of Cambridge Press, Cambridge, UK.

Stewart, A.M., 2003. Vector potential of the Coulomb gauge. Eur. J. Phys. 24, 519–524.

Stewart, A.M., 2011. Longitudinal and transverse components of a vector field. Sri Lankan J. Phys. 12, 33–42.

Stratton, J.A., 1941. Electromagnetic Theory. McGraw Hill, New York, NY.

Strohbehn, J.W., 1978. Modern theories in the propagation of optical waves in a turbulent medium. In: Strohbehn, J.W. (Ed.), Laser Beam Propagation in the Atmosphere. Springer-Verlag, Berlin, Germany.

Sungur, E.A., 1990. Dependence information in parameterized copulas. Commun. Stat. Simul. 19 (4), 1339–1360.

Syski, R., 1967. Stochastic differential equations. In: Saaty, T.L. (Ed.), Modern Nonlinear Equations. McGraw-Hill, New York, pp. 346–456.

Tatarskii, V.I., 1971. The Effects of the Turbulent Atmosphere on Wave Propagation (UDC 551.510). U.S. Department of Commerce, Washington, DC.

Taylor, G.I., 1938. The spectrum of turbulence. Proc. R. Soc. Lond. Ser. A 164, 476–490.

Tennekes, H., 1975. Eulerian and Lagrangian time microscales in isotropic turbulence. J. Fluid Mech. 67, 561–567.

Thiébaux, H.J., 1985. On approximations to geopotential and wind-field correlation structures. Tellus 37A, 126–131.

Tsallis, C., 1998. Generalized entropy-based criterion for consistent testing. Phys. Rev. E 58, 479–487.

Turcotte, D.L., 1997. Fractals and Chaos in Geology and Geophysics. Cambridge University Press, Cambridge, UK.

von Neumann, 1955. Mathematical Foundations of Quantum Mechanics. Princeton University Press, Princeton, NJ.

von Kármán, T., 1948. Progress in the statistical theory of turbulence. J. Mar. Res. 7, 252–264.

Wallace, J.M., 2014. Space-time correlations in turbulent flow. Theory Appl. Mech. Lett. 4, 022003.

Whittle, P., 1954. On stationary processes in the plane. Biometrika 41, 434–449.

Whittle, P., 1963. Prediction and Regulation. English University Press, UK.

Wiener, N., 1930. Generalized harmonic analysis. Acta Math. 55 (2–3), 117–258.

Wikle, C., Berliner, L., Cressie, N., 1998. Hierarchical Bayesian space-time models. Environ. Ecol. Stat. 5 (2), 117–154.

Wikle, C., Milliff, R., Nychka, D., Berliner, L., 2001. Spatiotemporal hierarchical Bayesian modeling tropical ocean surface winds. J. Am. Stat. Assoc. 96 (454), 382–397.

Wilson, W., 1931. On quasi-metric spaces. Am. J. Math. 53 (3), 675–684.

Wu, R.-S., 1982. Attenuation of short period seismic waves due to scattering. Geophys. Res. Lett. 9, 9–12.

Wu, R.-S., Aki, K., 1985. Elastic wave scattering by a random medium and the small-scale inhomogeneities in the lithosphere. J. Geophys. Res. 90B, 10261–10273.

Yadrenko, M.I., 1983. Spectral Theory of Random Fields. Optimization Software, Inc., New York, NY.

Yaglom, A.M., 1962. Stationary Random Functions. Prentice-Hall, Englewood Cliffs, NJ.

Yu, H.-L., Kolovos, A., Christakos, G., Chen, J.-C., Warmerdam, S., Dev, B., 2007. Interactive spatiotemporal modelling of health systems: the SEKS-GUI framework. Stoch. Environ. Res. Risk Assess. 21 (5), 555–572.

Zhang, D., 2002. Stochastic Methods for Flow in Porous Media. Academic Press, San Diego, CA.

Zhou, Y., Rubinstein, R., 1996. Sweeping and straining effects in sound generation by high Reynolds number isotropic turbulence. Phys. Fluids 8, 647–649.

Zhu, Z., Li, T., Hsu, P.-C., He, J., 2015. A spatial–temporal projection model for extended-range forecast in the tropics. Clim. Dyn. 45, 1085–1098.

FURTHER READING

Anderson, R.M., May, R.M., 1991. Infectious Diseases of Humans: Dynamics and Control. Oxford University Press, Oxford, UK.

Angulo, J.M., Yu, H.-L., Langousis, A., Kolovos, A., Wang, J.-F., Madrid, D., Christakos, G., 2013. Spatiotemporal infectious disease modeling: a BME-SIR approach. PLoS One 8 (9), e72168. http://dx.doi.org/10.1371/journal.pone.0072168.

Arkin, P.A., Ardanuy, P.E., 1989. Estimating climatic-scale precipitation from space: a review. J. Clim. 2, 1229–1238.

Boltzmann, L., 1877. Über die Beziehung zwischen dem zweiten Hauptsatz der mechanischen Wärmetheorie und der Wahrscheinlichkeitsrechnung resp. den Sätzen über das Wärmegleichgewicht. Wien. Ber. 76, 373–435.

Boltzmann, L., 1884. Ableitung des stefanschen gesetzes, betreffend die abhängigkeit der wärmestrahlung von der temperatur aus der elektromagnetischen lichttheorie. Wiedemann's Ann. 22, 291–294.

Christakos, G., 1985a. Recursive parameter estimation with applications in earth sciences. Math. Geol. 17 (5), 489–515.
Christakos, G., 1985b. Modern statistical analysis and optimal estimation of geotechnical data. Eng. Geol. 22 (2), 175–200.
Christakos, G., 1987a. Stochastic simulation of spatially correlated geoprocesses. Math. Geol. 19 (8), 803–827.
Christakos, G., 1987b. A stochastic approach in modeling and estimating geotechnical data. Int. J. Numer. Anal. Methods Geomech. 11 (1), 79–102.
Christakos, G., 1987c. The space transformation in the simulation of multidimensional random fields. J. Math. Comput. Simul. 29, 313–319.
Christakos, G., 1988. On-line estimation of nonlinear physical systems. Math. Geol. 20 (2), 111–133.
Conan, R., Borgnino, J., Ziad, A., Martin, F., 2000. Analytical solution for the covariance and for the decorrelation time of the angle of arrival of a wave front corrugated by atmospheric turbulence. J. Opt. Soc. Am. A Opt. Image Sci. Vis. 17 (10), 1807–1818.
David, M., 1977. Geostatistical Ore Reserve Estimation. Elsevier, Amsterdam, The Netherlands.
Gupta, V.K., Waymire, E., 1987. On Taylor's hypothesis and dissipation in rainfall. J. Geophys. Res. 92, 9657–9660.
Hairer, M., 2009. An Introduction to Stochastic PDEs. Lecture Notes. Courant Institute, New York, NY.
Jin, B., Wu, Y., Miao, B., Wang, X.L., Guo, P., 2014. Bayesian spatiotemporal modeling for blending in situ observations with satellite precipitation estimates. J. Geophys. Res. Atmos. 119, 1806–1819.
Johnson, R.S., 1977. A Modern Introduction to the Mathematical Theory of Water Waves. Cambridge University Press, Cambridge, UK.
Manning, R.M., 1993. Stochastic Electromagnetic Image Propagation. McGraw-Hill, New York, NY.
Nieves, V., Llebot, C., Turiel, A., Sole, J., Garcia-Ladona, E., Estrada, M., Blasco, D., 2007. Common turbulent signature in sea surface temperature and chlorophyll maps. Geophys. Res. Lett. 34 (L23602) http://dx.doi.org/10.1029/2007GL030823.
Petersen, D.P., 1973. Static and dynamic constraints on the estimation of space-time covariance and wavenumber-frequency spectral fields. J. Atmos. Sci. 30, 1252–1266.
Porcu, E., Gregori, P., Mateu, J., 2007. La descente et la montée étendues: the spatially d- anisotropic and the spatiotemporal case. Stoch. Environ. Res. Risk Assess. 21 (6), 683–693.
Porcu, E., Mateu, J., Christakos, G., 2010. Quasi-arithmetic means of covariance functions with potential applications to space-time data. J. Multivar. Anal. 100 (8), 1830–1844.
Rand, D.A., Wilson, H.B., 1991. Chaotic stochasticity: a ubiquitous source of unpredictability in epidemics. Proc. R. Soc. Lond. B 246, 179–184.
Sperber, K.R., Kim, D., 2012. Simplified metrics for the identification of the Madden-Julian oscillation in models. Atmos. Sci. Lett. 13, 187–193.
Sun, N.-Z., 1999. Inverse Problems in Groundwater Modeling. Springer, Dordrecht, The Netherlands.
Wiener, N., 1949. Time Series. MIT Press, Cambridge, MA.
Yaglom, A.M., 1955. Correlation theory of processes with stationary random increments of order n. Math. USSR Sb. 37–141 (English translation: *Am. Math. Soc. Trans. Ser.* 2, 8–87, 1958).
Yaglom, A.M., 1957. Some classes of random fields in n-dimensional space, related to stationary random processes. Theory Probab. Appl. (English Transl.) 3, 273–320.
Yaglom, A.M., 1986. Correlation Theory of Stationary and Related Random Functions: Basic Results. Springer-Verlag, New York, NY.
Yaglom, A.M., Pinsker, M.S., 1953. Random processes with stationary increments of order n. Dokl. Acad. Nauk. USSR 90, 731–734.
Yu, H.-L., Christakos, G., 2010. Modeling and estimation of heterogeneous spatiotemporal attributes under conditions of uncertainty. Inst. Electr. Electr. Eng. Trans. Geosci. Remote Sens. 49 (1), 366–376. http://dx.doi.org/10.1109/TGRS.2010.2052624.
Zastavnyi, V.P., Porcu, E., 2011. Compactly supported space-time covariance functions. Bernoulli 17 (1), 456–465.

APPENDIX: USEFUL MATHEMATICAL QUANTITIES, FUNCTIONS, AND FORMULAS

1. FACTORIALS—MULTIFACTORIALS

$n! = n \times (n-1) \times \ldots \times 2 \times 1$ $n!! = \begin{cases} n \times (n-2) \times \ldots \times 5 \times 3 \times 1 & n > 0, \text{ odd} \\ n \times (n-2) \times \ldots \times 6 \times 4 \times 2 & n > 0, \text{ even} \\ 1 & n = -1, 0 \end{cases}$	$0! = 1! = 0!! = -1!! = 1$ $n! = n!!(n-1)!!$	$(2n)!! = 2^n n!$ $(2n-1)!! = \frac{(2n)!}{2^n n!}$

2. GAMMA FUNCTIONS

Definitions $(\chi \in R_+)$:

$\Gamma(\chi) = \int_0^\infty dt\, t^{\chi-1} e^{-\chi}$ $\Gamma(\chi) = 2\int_0^\infty dt\, t^{2\chi-1} e^{-\chi^2}$	$\Gamma^{(k)}(\chi) = \int_0^\infty dt\, (\ln t)^k t^{\chi-1} e^{-\chi}$ $\Gamma(\chi+1) = \chi\Gamma(\chi)$

Basic formulas $(n \in Z_+)$:

$\Gamma(n) = (n-1)!$ $n\Gamma(n) = \Gamma(n+1)$ $\Gamma\left(\frac{n}{2}\right)\Gamma\left(1-\frac{n}{2}\right) = \frac{\pi}{\sin\frac{\pi n}{2}}$ $\Gamma(n+n'+1) = (n+n')!$	$\Gamma\left(n+\frac{1}{2}\right) = \frac{(2n)!\,\pi^{1/2}}{4^n n!}$ $\Gamma\left(\frac{1}{2}-n\right) = \frac{(-4)^n n!\,\pi^{\frac{1}{2}}}{(2n)!}$ $n!! = \begin{cases} 2^{\frac{n}{2}}\Gamma\left(\frac{n+2}{2}\right) & n \text{ even} \\ \pi^{-\frac{1}{2}} 2^{\frac{n+1}{2}}\Gamma\left(\frac{n+2}{2}\right) & n \text{ odd} \end{cases}$ $\Gamma\left(n+\frac{1}{p}\right) = \Gamma\left(\frac{1}{p}\right)\frac{(pn-(p-1))!^{(p)}}{p^n}$	$\Gamma\left(\frac{1}{n}\right)\Gamma\left(\frac{2}{n}\right)\ldots\Gamma\left(\frac{n-1}{n}\right) = \frac{(2\pi)^{(n-1)/2}}{n^{1/2}}$ $\Gamma\left(\frac{n}{2}\right) = \frac{(n-2)!!\,\pi^{1/2}}{2^{(n-1)/2}}$ $\Gamma\left(-\frac{n}{2}\right) = \begin{cases} \frac{(-1)^k}{n!} & n = 2k \\ \frac{(-1)^{k+1} 2^{k+1} \pi^{\frac{1}{2}}}{(2n+1)!!} & n = 2k+1 \end{cases}$

Continued

	*p*th Multifactorial: $(pn-(p-1))!^{(p)}$ $=n(n-p)(n-2p)\ldots(1+p)$ $(pn)!^{(p)}=p^n n!$ $n!^{(p)}=\begin{cases}1, & 0\le n<p\\ n(n-p)!, & n\ge p\end{cases}$	

Particular values:

$\Gamma\left(\frac{1}{2}\right)=\pi^{\frac{1}{2}}$ $\Gamma\left(-\frac{1}{2}\right)=-2\pi^{\frac{1}{2}}$	$\Gamma(0)=-1!=\infty$ $\Gamma(1)=0!=1$ $\Gamma(-1)=-2!=\infty$	$\Gamma\left(\frac{3}{2}\right)=\frac{1}{2}\pi^{\frac{1}{2}}$ $\Gamma\left(-\frac{3}{2}\right)=\frac{4}{3}\pi^{\frac{1}{2}}$	$\Gamma\left(\frac{5}{2}\right)=\frac{3}{4}\pi^{\frac{1}{2}}$ $\Gamma\left(-\frac{5}{2}\right)=-\frac{8}{15}\pi^{\frac{1}{2}}$

3. BESSEL FUNCTIONS

J_ν: Bessel function of the 1st-kind and νth-order.
K_ν: Bessel function of the 2nd-kind and νth-order.

$J_{\frac{1}{2}}(\zeta)=\sqrt{\frac{2}{\pi\zeta}}\sin\zeta$ $J_{-\frac{1}{2}}(\zeta)=\sqrt{\frac{2}{\pi\zeta}}\cos\zeta$	$\frac{\partial}{\partial\zeta}J_n(\zeta)=\begin{cases}\frac{1}{2}(J_{n-1}(\zeta)-J_{n+1}(\zeta)), & n\neq 0\\ -J_1(\zeta), & n=0\end{cases}$ $\frac{\partial}{\partial\zeta}J_\nu(\zeta)=J_{\nu-1}(\zeta)-\frac{\nu}{\zeta}J_\nu(\zeta)=\frac{\nu}{\zeta}J_\nu(\zeta)-J_{\nu+1}(\zeta)$ $\frac{\partial}{\partial\zeta}(\zeta^\nu J_\nu(\zeta))=\zeta^\nu J_{\nu-1}(\zeta)$ $\frac{\partial}{\partial\zeta}(\zeta^{-\nu}J_\nu(\zeta))=-\zeta^{-\nu}J_{\nu+1}(\zeta)$
$J_\nu(\zeta)=\frac{\zeta}{2\nu}[J_{\nu+1}(\zeta)+J_{\nu-1}(\zeta)]$ $J_{-\nu}(\zeta)=J_\nu(-\zeta)=(-1)^\nu J_\nu(\zeta)$	$\left(\frac{1}{\zeta}\frac{\partial}{\partial\zeta}\right)^i\left[\frac{J_\nu(\zeta)}{\zeta^\nu}\right]=(-1)^i\frac{J_{\nu+i}(\zeta)}{\zeta^{\nu+i}}$ $\left(\frac{1}{\zeta}\frac{\partial}{\partial\zeta}\right)^i\left[\frac{J_\nu(k\zeta)}{(k\zeta)^\nu}\right]=(-1)^i k^{2i}\frac{J_{\nu+i}(k\zeta)}{(k\zeta)^{\nu+i}}$
$K_{\frac{1}{2}}(\zeta)=K_{-\frac{1}{2}}(\zeta)=\sqrt{\frac{\pi}{2\zeta}}e^{-\zeta}$ $\frac{\partial}{\partial\zeta}[\zeta^\nu K_\nu(\zeta)]=-\zeta^\nu K_{\nu-1}(\zeta)$	$K_{-n}(\zeta)=(-1)^n K_n(-\zeta)=K_n(\zeta)$ $K_{n+1}(\zeta)-K_{n-1}(\zeta)=\frac{2n}{\zeta}K_n(\zeta)$ $K_{n+1}(\zeta)+K_{n-1}(\zeta)=-2\frac{\partial}{\partial\zeta}K_n(\zeta)$

4. GEGENBAUER POLYNOMIALS

Definition	Compositional Theorem for Spherical Harmonics
$\sum_{l=0}^{\infty} G_{l,q}(u)w^l = \frac{1}{(1-2wu+w^2)^{-q}}$ $G_{l,q}(1) = \frac{(2q+l-1)!}{(2q-1)!\,l!}$ $G_{l,q}(\cos r_s) = \frac{\sin(nr_s)}{\sin r_s},\ r_s \in [0, \rho\theta]$	$\sum_{l=1}^{h(l,n)} S_l^k(s)S_l^k(s') = \frac{h(l,n)}{S_n G_{l,\frac{n-2}{2}}(1)} G_{l,\frac{n-2}{2}}(\cos r_s)$ $S_l^{l'}(s)$: Orthnormal spherical harmonics of degree l on the S^{n-1} domain.

5. THE N-DIMENSIONAL SPHERE AND ITS SURFACE AREA[1]

$S^{n-1} = \{s : s_1^2 + \ldots + s_n^2 = 1\} \subset R^n$	$S_n = \frac{2\pi^{\frac{n}{2}}}{\Gamma\left(\frac{n}{2}\right)} r^{n-1}$ (In the case of a unit sphere, $r = 1$.)

6. COORDINATE SYSTEMS IN N-DIMENSIONS

Main coordinate systems:

Coordinates	Formulas
Cartesian R^n	$s = (s_1, \ldots, s_n),\ \ s = \lvert s \rvert = \left(\sum_{i=1}^{n} s_i^2\right)^{\frac{1}{2}} \geq 0$ $ds = \prod_{i=1}^{n} ds_i$
General R^n	$q = (q_1, \ldots, q_n),\ \ s = q(s)$ $J(q) = J(q_1, \ldots, q_n) = \begin{vmatrix} \frac{\partial s_1}{\partial q_1} & \cdots & \frac{\partial s_1}{\partial q_n} \\ & \vdots & \\ \frac{\partial s_n}{\partial q_1} & \cdots & \frac{\partial s_n}{\partial q_n} \end{vmatrix}$ $ds = \lvert J(q) \rvert \prod_{i=1}^{n} dq_i$

Continued

[1]Here, we use the symbol S^{n-1} to denote the n-dimensional sphere and the symbol S_n to denote its area.

Spherical R^n	$s = (s_1,\ldots,s_n),\ s = \lvert s\rvert \ge 0,\ \theta_i \in [0,\pi]\ (i = 1,\ldots,n-2),\ \theta_{n-1} \in [0,2\pi]$ $s_1 = s\cos\theta_1,\ s_j = s\cos\theta_j \prod_{i=1}^{j-1} \sin\theta_i \quad (j = 2,3,\ldots,n-1),\ s_n = s\prod_{i=1}^{n-1} \sin\theta_i$ $s\cdot s' = ss'\cos\theta_1,$ $ds = \prod_{i=1}^{n} ds_i = s^{n-1}ds\prod_{i=1}^{n-1} (\sin\theta_i)^{n-1-i}d\theta_i$
Polar R^2	$s = (s_1, s_2),\ s = \lvert s\rvert = (s_1^2 + s_2^2)^{\frac{1}{2}} \ge 0,\ \varphi \in [0, 2\pi]$ $s_1 = s\cos\varphi,\quad s_2 = s\sin\varphi,$ $\lvert J(s)\rvert = s,$ $ds = s\ dsd\theta$
Spherical R^3	$s = (s_1, s_2, s_3),\ s = \lvert s\rvert = (s_1^2 + s_2^2 + s_3^2)^{\frac{1}{2}} \ge 0,\quad \theta \in [0,\pi],\ \varphi \in [0,2\pi]$ $s_1 = s\sin\theta\cos\varphi,\quad s_2 = s\sin\theta\sin\varphi,\quad s_3 = s\cos\theta,$ $\lvert J(s,\theta)\rvert = s^2\sin\theta,$ $ds = dsd\theta d\varphi\ s^2\sin\theta$

Alternative spherical coordinate systems (angle-independent):

Coordinates	Formulas
Spherical R^n	$s = (s_1,\ldots,s_n),\ s = \lvert s\rvert = \left(\sum_{i=1}^{n} s_i^2\right)^{\frac{1}{2}} \ge 0$ $\sigma \sim U$: Uniform PDF $d\sigma$ over $S^{n-1} = \{s : s_1^2 + \ldots + s_n^2 = 1\} \subset R^n$ $s = (s_1,\ldots,s_n) \mapsto (s,\sigma),$ $ds = \prod_{j=1}^{n} ds_j = S_n dsd\sigma = \frac{2\pi^{\frac{n}{2}}}{\Gamma(\frac{n}{2})} s^{n-1}dsd\sigma$
Spherical R^2	$s = (s_1, s_2),\ s = \lvert s\rvert = (s_1^2 + s_2^2)^{\frac{1}{2}} \in [0,\infty)$ $\sigma \sim U$: Uniform PDF $d\sigma$ over $S^1 = \{s : s_1^2 + s_2^2 = 1\} \subset R^2,$ $s = (s_1, s_2) \mapsto (s,\sigma),$ $ds = ds_1ds_2 = S_2dsd\sigma = 2\pi sdsd\sigma$ ($2\pi s$ = circumference of the circle)
Spherical R^3	$s = (s_1, s_2, s_3),\ s = \lvert s\rvert = (s_1^2 + s_2^2 + s_3^2)^{\frac{1}{2}} \in [0,\infty)$ $\sigma \sim U$: Uniform PDF $d\sigma$ over $S^2 = \{s : s_1^2 + s_2^2 + s_3^2 = 1\} \subset R^3$ $s = (s_1, s_2, s_3) \mapsto (s,\sigma)$ $ds = ds_1ds_2ds_3 = S_3dsd\sigma = 4\pi s^2dsd\sigma$ ($4\pi s^2$ = area of the sphere).

7. DELTA FUNCTIONS

Coordinate System	Coordinates/Transformations	δ−Formulas
Cartesian $R^{n,1}$	$s = (s_1,\ldots,s_n)$	$\delta(\mathbf{s},t) = \delta(t)\prod_{i=1}^{n}\delta(s_i)$, $\delta(\mathbf{s},t) = 0 \quad (\mathbf{s}\neq 0, t\neq 0)$, $\int d\mathbf{s}\int dt\, \delta(\mathbf{s}-\mathbf{s}^*, t-t^*)f(\mathbf{s},t) = f(\mathbf{s}^*,t^*)$ $\delta(\mathbf{s}' - (\mathbf{A}\mathbf{s}+\mathbf{b}), t' - (at+c))$ $= \dfrac{\delta\left(\mathbf{s}-\mathbf{A}^{-1}(\mathbf{s}'-\mathbf{b})\right)\delta\left(t-a^{-1}(t'-c)\right)}{\lvert a\rvert\lvert\mathbf{A}\rvert}$, $\int d\mathbf{s}\int dt\, \delta(\mathbf{s}-\mathbf{A}^{-1}\mathbf{s}', t-a^{-1}t')f(\mathbf{s},t)$ $= f(\mathbf{A}^{-1}\mathbf{s}', a^{-1}t')$, $\delta(\mathbf{s}-\mathbf{s}', t-t') = \delta(\mathbf{s}-\mathbf{s}')\delta(t-t')$ $= \dfrac{1}{(2\pi)^{n+1}}\int d\mathbf{k}\int d\omega e^{i[\mathbf{s}\cdot(\mathbf{k}-\mathbf{k}')-(\omega-\omega')\tau]}$
Cartesian $R^{3,1}$	$s = (s_1, s_2, s_3)$	$\delta(\mathbf{s},t) = \delta(s_1)\delta(s_2)\delta(s_3)\delta(t)$, $\nabla^2\dfrac{1}{\lvert\mathbf{s}-\mathbf{s}'\rvert} = -4\pi\delta(\mathbf{s}-\mathbf{s}')$
Cartesian $R^{1,1}$	$s = s_1 = s$	$\delta(s,t) = \delta(s)\delta(t)$, $\delta(as-b) = \lvert a\rvert^{-1}\delta(s-a^{-1}b)$, $\delta(-s) = \delta(s)$, $\delta^{(k)}(-s) = (-1)^k\delta^{(k)}(s)$, $s\delta^{(k)}(s) = -k\delta^{(k-1)}(s)$, $\int_{-\infty}^{\infty} ds\delta^{(1)}(s) = 0$, $\int_{-\infty}^{\infty} du\delta(s-u)\delta(s\prime-u) = \delta(s-s\prime)$, $\int_{-\infty}^{\infty} dsf(s)\delta^{(k)}(s) = (-1)^k f^{(k)}(0)$
Curvilinear $R^{n,1}$	$\mathbf{q} = (q_1,\ldots, q_n)$ $s = \mathbf{s}(\mathbf{q}) = \begin{bmatrix} s_1(q_1,\ldots,q_n) \\ \vdots \\ s_n(q_1,\ldots,q_n)\end{bmatrix}$ $J(\mathbf{q}) = J(q_1,\ldots,q_n) = \begin{vmatrix} \dfrac{\partial s_1}{\partial q_1} \cdots \dfrac{\partial s_1}{\partial q_n} \\ \vdots \\ \dfrac{\partial s_n}{\partial q_1} \cdots \dfrac{\partial s_n}{\partial q_n}\end{vmatrix}$	$\delta(\mathbf{s},t) = \frac{\delta(\mathbf{q},t)}{\lvert J(\mathbf{q})\rvert}$, $\int d\mathbf{s}\int dt\delta(\mathbf{s},t) = \int d\mathbf{q}\int dt\lvert J(\mathbf{q})\rvert\frac{\delta(\mathbf{q},t)}{\lvert J(\mathbf{q})\rvert} = 1$, $d\mathbf{s} = d\mathbf{q}\lvert J(\mathbf{q})\rvert$
Polar $R^{2,1}$	$s = (s_1, s_2)$, $s = \lvert\mathbf{s}\rvert = (s_1^2+s_2^2)^{\frac{1}{2}}$ $d\mathbf{s} = s\, dsd\theta$	$\delta(\mathbf{s},t) = \frac{1}{s}\delta(s)\delta(t)$, $\int\int d\mathbf{s}dt\delta(\mathbf{s},t) = \int_0^\infty ds\int_0^{2\pi} d\theta\int_0^\infty dt\delta(s)\delta(t) = 1$
	No dependence on θ: $s = (s_1, s_2)\mapsto(s,\sigma)$ $d\mathbf{s} = 2\pi sdsd\sigma$	$\delta(\mathbf{s},t) = \frac{\delta(s)\delta(\sigma)\delta(t)}{2\pi s}$ $\int d\mathbf{s}\int dt\delta(\mathbf{s},t) = \int_0^\infty ds\int_0^\infty d\sigma\int_0^\infty dt\delta(s)\delta(\sigma)\delta(t) = 1$

Continued

Spherical $R^{3,1}$	$\boldsymbol{q} = (q_1, q_2, q_3) = (s, \theta, \varphi)$ $\boldsymbol{s} = (s_1, s_2, s_3)$ $= (s \sin\theta \cos\varphi,\ s \sin\theta \sin\varphi,\ s \cos\theta)$ $s = \lvert\boldsymbol{s}\rvert = (s_1^2 + s_2^2 + s_3^2)^{\frac{1}{2}}$ $d\boldsymbol{s} = dsd\theta d\varphi s^2 \sin\theta$	$\delta(\boldsymbol{s}, t) = \dfrac{\delta(s)\delta(\theta)\delta(\varphi)\delta(t)}{s^2 \sin\theta}$ $\int d\boldsymbol{s} \int dt \delta(\boldsymbol{s}, t) = \int_0^\infty ds \int_0^{2\pi}$ $d\theta \int_0^\pi d\phi \int_0^\infty dt\ \delta(s)\delta(\theta)\delta(\phi)\delta(t) = 1$
	No dependence on θ, φ: $d\boldsymbol{s} = 4\pi s^2 ds d\sigma$	$\delta(\boldsymbol{s}, t) = \dfrac{\delta(s)\delta(\sigma)\delta(t)}{4\pi s^2}$ $\int d\boldsymbol{s} \int dt \delta(\boldsymbol{s}, t) =$ $\int_0^\infty ds \int_0^\infty d\sigma \int_0^\infty dt\ \delta(s)\delta(\sigma)\delta(t) = 1$

8. SPACE–TIME VECTORS AND MULTIINDEXES (CONVENTIONS AND FORMULAS)

$$\boldsymbol{p} = (s_1, \ldots, s_n, s_0) = \sum_{i=1}^{n,0} s_i \boldsymbol{\varepsilon}_i \in R^{n+1}$$

$\boldsymbol{p}^{\boldsymbol{\lambda}} = \prod_{i=1}^{n,0} s_i^{\lambda_i} = \prod_{i=1}^{n} s_i^{\lambda_i} t^{\lambda_0}$	$\boldsymbol{\lambda} = (\lambda_1, \ldots, \lambda_n, \lambda_0) \in N_0^n$ $\lambda = \lvert\boldsymbol{\lambda}\rvert = \sum_{k=1}^{n,0} \lambda_k$ $\boldsymbol{\lambda}! = \prod_{i=1}^{n,0} \lambda_i! = \lambda_1! \ldots \lambda_n! \lambda_0!$	$\boldsymbol{\kappa} = (\kappa_1, \ldots, \kappa_n, \kappa_0) \in N_0^n$ $\kappa = \lvert\boldsymbol{\kappa}\rvert = \sum_{k=1}^{n,0} \kappa_k$ $\boldsymbol{\kappa}! = \prod_{i=1}^{n,0} \kappa_i! = \kappa_1! \ldots \kappa_n! \kappa_0!$
	$\boldsymbol{\kappa} \le \boldsymbol{\lambda} \Leftrightarrow \kappa_i \le \lambda_i$ $\boldsymbol{\lambda\kappa} = \prod_{i=1}^{n,0} \lambda_i \kappa_i$ $C_{\boldsymbol{\lambda}}^{\boldsymbol{\kappa}} = \binom{\boldsymbol{\lambda}}{\boldsymbol{\kappa}} = \frac{\boldsymbol{\lambda}!}{\boldsymbol{\kappa}!(\boldsymbol{\lambda}-\boldsymbol{\kappa})!},\ \boldsymbol{\kappa} \le \boldsymbol{\lambda}$ $\prod_{i=1}^{n,0} C_{\lambda_i}^{\kappa_i} = \prod_{i=1}^{n,0} \binom{\lambda_i}{\kappa_i} = \dfrac{\prod_{i=1}^{n,0} \lambda_i!}{\prod_{i=1}^{n,0} \kappa_i! \prod_{i=1}^{n,0} (\lambda_i - \kappa_i)!},\ \kappa_i \le \lambda_i$	
$\dfrac{\partial \boldsymbol{p}}{\partial s_i} = \sum_{k=1}^{n,0} \dfrac{\partial s_k}{\partial s_i} \boldsymbol{\varepsilon}_k = \boldsymbol{\varepsilon}_i$ $\dfrac{\partial(\boldsymbol{p} \cdot \boldsymbol{p}')}{\partial s_i} = \dfrac{\partial \boldsymbol{p}}{\partial s_i} \cdot \boldsymbol{p}' + \boldsymbol{p} \cdot \dfrac{\partial \boldsymbol{p}'}{\partial s_i}$ $\dfrac{\partial(\boldsymbol{p} \times \boldsymbol{p}')}{\partial s_i} = \dfrac{\partial \boldsymbol{p}}{\partial s_i} \times \boldsymbol{p}' + \boldsymbol{p} \times \dfrac{\partial \boldsymbol{p}'}{\partial s_i}$	$\nabla \cdot \boldsymbol{p} = n + 1$	$\nabla \times \boldsymbol{p} = 0$
	$\nabla \cdot (\lvert\boldsymbol{p}\rvert^m (\boldsymbol{a} \times \boldsymbol{p})) = 0$ ($\boldsymbol{a}$: *const. vector*) $\nabla \times (\lvert\boldsymbol{p}\rvert^m (\boldsymbol{a} \times \boldsymbol{p})) = (m+2)\lvert\boldsymbol{p}\rvert^m \boldsymbol{a} - m(\boldsymbol{p} \cdot \boldsymbol{a})\lvert\boldsymbol{p}\rvert^{m-2} \boldsymbol{p}$	

$$\Delta \boldsymbol{p} = \boldsymbol{p} - \boldsymbol{p}' = \sum_{i=1}^{n,0} (s_i - s'_i) \boldsymbol{\varepsilon}_i = (s_1 - s'_1, \ldots, s_n - s'_n, s_0 - s'_0) = \sum_{i=1}^{n,0} h_i \boldsymbol{\varepsilon}_i = (h_1, \ldots, h_n, h_0)$$

[2]Which notation is used should be understood from the context.

<table>
<tr>
<td>$\frac{\partial \Delta \boldsymbol{p}}{\partial s_i} = -\frac{\partial \Delta \boldsymbol{p}}{\partial s'_i} = \boldsymbol{\varepsilon}_i \quad (i = 1, \ldots, n, 0)$</td>
<td>$\nabla \cdot \Delta \boldsymbol{p} = n + 1$</td>
<td>$\nabla \times \Delta \boldsymbol{p} = 0$</td>
</tr>
<tr>
<td colspan="3">$$\left|\boldsymbol{p} - \boldsymbol{p}'\right| = \left|\Delta \boldsymbol{p}\right| = \left|\Delta \boldsymbol{p}^2\right|^{\frac{1}{2}} = \Delta p \neq \Delta \boldsymbol{p},$$
$$\partial \boldsymbol{h} = \prod_{i=1}^{n} \partial h_i \text{ (multi-index notation)}, \ \partial \boldsymbol{h}^T = [\partial h_1, \ldots, \partial h_n \ \partial h_0] \text{ (vector notation)}^2$$</td>
</tr>
<tr>
<td>$\frac{\partial \Delta p}{\partial \boldsymbol{h}} = \frac{1}{2} \Delta p^{-1} \frac{\partial \Delta \boldsymbol{p}^2}{\partial \boldsymbol{h}}$
$\frac{\partial \Delta p^{-1}}{\partial \boldsymbol{h}} = -\Delta \boldsymbol{p}^{-2} \frac{\partial \Delta p}{\partial \boldsymbol{h}}$
$\frac{\partial^2 \Delta p}{\partial \boldsymbol{h}^T \partial \boldsymbol{h}} = \frac{\partial}{\partial \boldsymbol{h}} \left(\frac{\partial \Delta p}{\partial \boldsymbol{h}} \right)^T$</td>
<td>$\frac{\partial \Delta p}{\partial h_i} = \frac{1}{2} \Delta p^{-1} \frac{\partial \Delta \boldsymbol{p}^2}{\partial h_i}$
$\frac{\partial \Delta p^{-1}}{\partial h_i} = -\frac{\partial \Delta p}{\partial h_i} \Delta \boldsymbol{p}^{-2}$</td>
<td>$\frac{\partial \Delta p}{\partial \tau} = \frac{1}{2} \Delta p^{-1} \frac{\partial \Delta \boldsymbol{p}^2}{\partial \tau}$
$\frac{\partial \Delta p^{-1}}{\partial \tau} = -\frac{\partial \Delta p}{\partial \tau} \Delta \boldsymbol{p}^{-2}$</td>
</tr>
<tr>
<td></td>
<td colspan="2">$\frac{\partial^2 \Delta p}{\partial h_i \partial h_j} = \frac{1}{2} \frac{\partial}{\partial h_i} \left(\Delta p^{-1} \frac{\partial \Delta \boldsymbol{p}^2}{\partial h_j} \right)$</td>
</tr>
<tr>
<td>$(\boldsymbol{s} + \boldsymbol{h})^{\boldsymbol{\lambda}} = \sum_{\boldsymbol{\kappa} = |\boldsymbol{\kappa}| = 0}^{\lambda = |\boldsymbol{\lambda}|} C_{\boldsymbol{\lambda}!}^{\boldsymbol{\kappa}}! \boldsymbol{s}^{\boldsymbol{\kappa}} \boldsymbol{h}^{\boldsymbol{\lambda} - \boldsymbol{\kappa}}, \ \boldsymbol{\kappa} \leq \boldsymbol{\lambda}$
$\left|\boldsymbol{s}\right|^k = \left| \sum_{i=1}^{n} s_i^2 \right|^{\frac{k}{2}}$
$(t + \tau)^{\lambda} = \sum_{m=0}^{\lambda} C_{\lambda!}^{m}! t^m \tau^{\lambda - m} \quad (\lambda = 1, \ldots)$</td>
<td colspan="2">$\prod_{i=1}^{n,0} (s_i + h_i)^{\lambda_i} = \sum_{\kappa_1=0}^{\lambda_1} \cdots \sum_{\kappa_n=0}^{\lambda_n} \sum_{\kappa_0=0}^{\lambda_0} \prod_{i=1}^{n,0} \binom{\lambda_i}{\kappa_i} s_i^{\kappa_i} h^{\lambda_i - \kappa_i}$
$\frac{\partial |\boldsymbol{s}|}{\partial s_i} = s_i \left|\boldsymbol{s}\right|^{-1} = \zeta_i$
$\sum_{i=1}^{n} \zeta_i \frac{\partial \zeta_k}{\partial s_i} = 0$
$\frac{\partial \zeta_k}{\partial s_i} = (\delta_{ik} - \zeta_i \zeta_k) \left|\boldsymbol{s}\right|^{-1}$
$\sum_{i=1}^{n} \frac{\partial \zeta_i}{\partial s_i} = (n - 1) \left|\boldsymbol{s}\right|^{-1}$
$\nabla \left|\boldsymbol{s}\right|^k = k \left|\boldsymbol{s}\right|^{k-2} \boldsymbol{s}$</td>
</tr>
<tr>
<td colspan="3">$$\breve{\boldsymbol{p}} = \boldsymbol{A} \boldsymbol{p}, \ \boldsymbol{p} = [s_1, \ldots, s_n, s_0]^T, \ \breve{\boldsymbol{p}} = \left[\breve{s}_1, \ldots, \breve{s}_n, \breve{s}_0 \right]^T, \ \boldsymbol{A} = [a_{ij}], \ (i, j = 1, \ldots, n, 0)$$</td>
</tr>
<tr>
<td colspan="2">$\frac{\partial}{\partial s_i} f\left(\breve{\boldsymbol{p}}\right) = \sum_{j=1}^{n,0} \frac{\partial f\left(\breve{\boldsymbol{p}}\right)}{\partial \breve{s}_j} \frac{\partial \breve{s}_j}{\partial s_i} \quad (i, j = 1, \ldots, n, 0)$
$\frac{\partial^2}{\partial s_i \partial s_j} f\left(\breve{\boldsymbol{p}}\right) = \sum_{l=1}^{n,0} \sum_{k=1}^{n,0} \frac{\partial \breve{s}_k}{\partial s_i} \frac{\partial^2 f\left(\breve{\boldsymbol{p}}\right)}{\partial \breve{s}_k \partial \breve{s}_l} \frac{\partial \breve{s}_l}{\partial s_j}$</td>
<td>$\frac{\partial \breve{s}_i}{\partial s_j} = a_{ij} \quad (i, j = 1, \ldots, n, 0)$
$\frac{\partial}{\partial s_i} f\left(\breve{\boldsymbol{p}}\right) = \sum_{j=1}^{n,0} \frac{\partial f\left(\breve{\boldsymbol{p}}\right)}{\partial \breve{s}_j} a_{ji}$
$\frac{\partial^2}{\partial s_i \partial s_j} f\left(\breve{\boldsymbol{p}}\right) = \sum_{l=1}^{n,0} \sum_{k=1}^{n,0} a_{ki} \frac{\partial^2 f\left(\breve{\boldsymbol{p}}\right)}{\partial \breve{s}_k \partial \breve{s}_l} a_{lj}$</td>
</tr>
<tr>
<td colspan="3">$$\Delta p = \left|\Delta \boldsymbol{p}^2\right|^{\frac{1}{2}}, \ \Delta \boldsymbol{p} = \boldsymbol{p} - \boldsymbol{p}'$$</td>
</tr>
<tr>
<td>$\frac{\partial c_X(\Delta p)}{\partial \boldsymbol{h}} = \frac{\partial \Delta p}{\partial \boldsymbol{h}} \frac{\partial c_X}{\partial \Delta p}$
$\frac{\partial^2 c_X(\Delta p)}{\partial \boldsymbol{h}^T \partial \boldsymbol{h}} = \frac{\partial^2 \Delta p}{\partial \boldsymbol{h}^T \partial \boldsymbol{h}} \frac{\partial c_X}{\partial \Delta p} + \left(\frac{\partial \Delta p}{\partial \boldsymbol{h}} \right)^T \frac{\partial \Delta p}{\partial \boldsymbol{h}} \frac{\partial^2 c_X}{\partial \Delta \boldsymbol{p}^2}$</td>
<td colspan="2">$\frac{\partial c_X}{\partial h_i} = \frac{\partial \Delta p}{\partial h_i} \frac{\partial c_X}{\partial \Delta p} \quad (i, j = 1, \ldots, n, 0)$
$\frac{\partial^2 c_X}{\partial h_i^2} = \frac{\partial^2 \Delta p}{\partial h_i^2} \frac{\partial c_X}{\partial \Delta p} + \left(\frac{\partial \Delta p}{\partial h_i} \right)^2 \frac{\partial^2 c_X}{\partial \Delta \boldsymbol{p}^2}$
$\frac{\partial^2 c_X}{\partial h_i \partial h_j} = \frac{\partial^2 \Delta p}{\partial h_i \partial h_j} \frac{\partial c_X}{\partial \Delta p} + \frac{\partial \Delta p}{\partial h_i} \frac{\partial^2 c_X}{\partial \Delta p \partial h_j}$</td>
</tr>
<tr>
<td>$\frac{\partial^{\nu} c_X(\Delta \boldsymbol{p})}{\partial \boldsymbol{s}^{\boldsymbol{\nu}}} = \frac{\partial^{\nu} c_X(\Delta \boldsymbol{p})}{\partial \boldsymbol{h}^{\boldsymbol{\nu}}}$
$\frac{\partial^{\nu} c_X(\Delta \boldsymbol{p})}{\partial \boldsymbol{s}'^{\boldsymbol{\nu}}} = (-1)^{\nu} \frac{\partial^{\nu} c_X(\Delta \boldsymbol{p})}{\partial \boldsymbol{h}^{\boldsymbol{\nu}}}$
$\boldsymbol{\nu} = (\nu_1, \ldots, \nu_n, \nu_0), \ \nu = \left|\boldsymbol{\nu}\right| = \sum_{i=1}^{n,0} \nu_i$
$\partial \boldsymbol{s}^{\boldsymbol{\nu}} = \prod_{i=1}^{n,0} \partial s_i^{\nu_i}, \ \partial \boldsymbol{h}^{\boldsymbol{\nu}} = \prod_{i=1}^{n,0} \partial h_i^{\nu_i}$</td>
<td colspan="2">$\frac{\partial}{\partial h_i} \nabla c_X = \sum_{k=1}^{n,0} \frac{\partial^2 c_X}{\partial h_i \partial h_k} \boldsymbol{\varepsilon}_k$
$\frac{\partial}{\partial s_i} \nabla c_X = \sum_{k=1}^{n,0} \frac{\partial^2 c_X}{\partial s_i \partial h_k} \boldsymbol{\varepsilon}_k = \frac{\partial}{\partial h_i} \nabla c_X$
$= -\frac{\partial}{\partial s'_i} \nabla c_X = -\sum_{k=1}^{n,0} \frac{\partial^2 c_X}{\partial s'_i \partial h_k} \boldsymbol{\varepsilon}_k$</td>
</tr>
</table>

Continued

$\nabla c_X(\Delta \boldsymbol{p}) = \sum_{k=1}^{n,0} \frac{\partial c_X}{\partial h_k}\boldsymbol{\varepsilon}_k$ $\nabla c_X \cdot d\boldsymbol{p} = \sum_{k=1}^{n,0} \frac{\partial c_X(\Delta \boldsymbol{p})}{\partial h_k} dp_k = dc_X$	$\nabla c_X(\Delta \boldsymbol{p}) \cdot \boldsymbol{p} = \sum_{k=1}^{n,0} \frac{\partial c_X}{\partial h_k} s_k$ $\nabla \times \nabla c_X = \nabla \cdot (\nabla \times \nabla c_X) = 0$

9. USEFUL INTEGRALS

$\int_0^\infty du\, u^{\nu+1} J_\nu(ru) e^{-cu}$	$2c(2r)^\nu \Gamma\left(\nu+\frac{3}{2}\right)\pi^{-\frac{1}{2}}\left(c^2+r^2\right)^{-\nu-\frac{3}{2}}$
$\int_0^\infty du\, u^{\nu+1} J_\nu(ru) e^{-cu^2}$	$r^\nu (2c)^{-\nu-1} e^{-\frac{r^2}{4c}}$
$\int_0^\infty du\, u^{\nu-\nu'+1} J_\nu(ru) J_{\nu'}(\lambda u)$	$\begin{cases} \frac{2^{\nu-\nu'+1}}{\Gamma(\nu'-\nu)} r^\nu \lambda^{-\nu'} \left(\lambda^2 - r^2\right)^{-\nu+\nu'-1} & \text{if } r < \lambda \ (\nu' > \nu > -1) \\ 0 & \text{if } r > \lambda \end{cases}$
$\int_0^\infty du\, u^{\nu+\nu'+1} J_\nu(ru) K_{\nu'}(\lambda u)$	$2^{\nu+\nu'} r^\nu \lambda^{\nu'} \Gamma(\nu+\nu'+1)\left(r^2+\lambda^2\right)^{-\nu-\nu'-1}$
$\int_0^\infty d\omega\, e^{-i\omega\tau}\left[\left(k^2+\alpha^2\right)^{2p} + c^2\omega^2\right]^{-1}$	$\pi c^{-1} e^{-c^{-1}\left(k^2+\alpha^2\right)^p \tau}\left(k^2+\alpha^2\right)^{-p}$
$\int_{-\infty}^\infty du\, e^{-\alpha u^2} u \sin(bu)$	$2(\pi)^{\frac{1}{2}} b a^{-\frac{3}{2}} e^{-\frac{b^2}{4a}}$
$\int_{-\infty}^\infty du\, e^{-\alpha u^2} \cos(bu)$	$\left(\pi a^{-1}\right)^{\frac{1}{2}} e^{-\frac{b^2}{4a}}$
$\int_{-\infty}^\infty du\, e^{-\alpha u^2}$	$\left(\pi a^{-1}\right)$
$\int_0^\pi d\theta \sin^\nu\theta$	$2^\nu \Gamma^2\left(\frac{\nu+1}{2}\right)\Gamma^{-1}(\nu+1)$
$\int_0^\pi d\theta\, e^{iu\cos\theta} \sin^{2\nu}\theta$	$\pi^{\frac{1}{2}}\Gamma\left(\frac{2\nu+1}{2}\right)\left(\frac{2}{u}\right)^\nu J_\nu(u)$

10. MISCELLANEOUS

Useful identity	$(a-b)(c-d) \equiv \frac{1}{2}(a-d)^2 - \frac{1}{2}(a-c)^2 + \frac{1}{2}(b-c)^2 - \frac{1}{2}(b-d)^2$
Gaussian random field	$X \sim G\left(\overline{X}, \sigma_X^2\right) : \overline{e^X} = e^{\overline{X}+0.5\sigma_X^2},\ \overline{e^{-X}} = e^{-\overline{X}+0.5\sigma_{-X}^2} = e^{-\overline{X}+0.5\sigma_X^2}$
Combinatorics	$C_n^k = {}_nC_k = \binom{n}{k} = \frac{n!}{k!(n-k)!}$

Arcsine and Arccosine Functions:

$\sin(\text{arc}\sin x) = \sin(\sin^{-1}x) = x \in [-1,1]$	$\cos(\text{arc}\cos x) = \cos(\cos^{-1}x) = x \in [-1,1]$
$\text{arc}\sin(\sin x) = x + 2k\pi \quad (k \text{ is integer})$	$\text{arc}\cos(\cos x) = x + 2k\pi \quad (k \text{ is integer})$
$\text{arc}\sin(-x) = -\text{arc}\sin x$	$\text{arc}\cos(-x) = \pi - \text{arc}\cos x$
$\text{arc}\sin x = \frac{\pi}{2} - \text{arc}\cos x$	$\text{arc}\cos x = \frac{\pi}{2} - \text{arc}\sin x$
$\cos(\text{arc}\sin x) = \sin(\text{arc}\cos x) = (1-x^2)^{\frac{1}{2}}$	$\cos(\text{arc}\sin x) = \sin(\text{arc}\cos x) = (1-x^2)^{\frac{1}{2}}$
$\tan(\text{arc}\sin x) = x(1-x^2)^{-\frac{1}{2}}$	$\tan(\text{arc}\cos x) = \frac{1}{x}(1-x^2)^{\frac{1}{2}}$
$\text{arc}\sin(\cos x) = \frac{\pi}{2} - x \quad \left(-\frac{\pi}{2} < \frac{\pi}{2} - x < \frac{\pi}{2}\right)$	$\text{arc}\cos(\sin x) = -x - \left(2k + \frac{1}{2}\right)\pi$
$\text{arc}\sin x \pm \text{arc}\sin y = \text{arc}\sin(x(1-y^2)^{\frac{1}{2}} \pm y(1-x^2))^{\frac{1}{2}}$	$\text{arc}\cos x \pm \text{arc}\cos y = \text{arc}\cos(xy \mp ((1-x^2)(1-y^2))^{\frac{1}{2}})$
$\frac{d}{dx}(\text{arc}\sin x) = (1-x^2)^{-\frac{1}{2}}$	$\frac{d}{dx}(\text{arc}\cos x) = -(1-x^2)^{-\frac{1}{2}}$
$\int dx\text{arc}\sin x = x\text{arc}\sin x + (1-x^2)^{\frac{1}{2}} + c$	$\int dx\text{arc}\cos x = x\,\text{arc}\cos x - (1-x^2)^{\frac{1}{2}} + c$
$\text{arc}\sin 0 = 0°, \ \text{arc}\sin 1 = \frac{\pi}{2} = 90°$	$\text{arc}\cos 0 = \frac{\pi}{2} = 90°, \ \text{arc}\cos 1 = 0°$
$\lim_{x\to\infty} \text{arc}\sin x = \text{undefined}$	$\lim_{x\to\infty} \text{arc}\cos x = \text{undefined}$

11. SPECIAL CASES—EXAMPLES

Example 1

Let, $\boldsymbol{\lambda} = (\lambda_1, \lambda_2)$. Then, the following standard conventions hold,

$$\lambda = |\boldsymbol{\lambda}| = \lambda_1 + \lambda_2,\ \boldsymbol{\lambda}! = \lambda_1!\lambda_2!,\ \kappa = |\boldsymbol{\kappa}| = \kappa_1 + \kappa_2,\ \boldsymbol{\lambda} - \boldsymbol{\kappa} = (\lambda_1 - \kappa_1, \lambda_2 - \kappa_2),$$
$$\boldsymbol{s}^{\boldsymbol{\lambda}} = s_1^{\lambda_1} s_2^{\lambda_2},\ \boldsymbol{h}^{\boldsymbol{\lambda}-\boldsymbol{\kappa}} = h_1^{\lambda_1-\kappa_1} h_2^{\lambda_2-\kappa_2},$$

$$(\boldsymbol{s}+\boldsymbol{h})^{(\lambda_1,\lambda_2)} = \sum_{\kappa=0}^{\lambda} \frac{\boldsymbol{\lambda}!}{\boldsymbol{\kappa}!(\boldsymbol{\lambda}-\boldsymbol{\kappa})!} \boldsymbol{s}^{\boldsymbol{\kappa}} \boldsymbol{h}^{\boldsymbol{\lambda}-\boldsymbol{\kappa}} = \sum_{\kappa=0}^{\lambda} \frac{\lambda_1!\lambda_2!}{\kappa_1!\kappa_2!(\lambda_1-\kappa_1)!(\lambda_2-\kappa_2)!} s_1^{\kappa_1} s_2^{\kappa_2} h_1^{\lambda_1-\kappa_1} h_2^{\lambda_2-\kappa_2}$$
$$(\lambda_i \geq \kappa_i, i = 1,2).$$

Let, $\lambda_1 = \lambda_2 = 1,\ \lambda = 2,\ \kappa_1,\ \kappa_2 \leq 1$. Then,

$$(\boldsymbol{s}+\boldsymbol{h})^{(1,1)} = \sum_{\kappa_1+\kappa_2=0}^{2} \frac{1!1!}{\kappa_1!\kappa_2!(1-\kappa_1)!(1-\kappa_2)!} s_1^{\kappa_1} s_2^{\kappa_2} h_1^{1-\kappa_1} h_2^{1-\kappa_2}$$
$$= \frac{1}{0!0!1!1!} s_1^0 s_2^0 h_1^{1-0} h_2^{1-0} + \frac{1}{1!0!0!1!} s_1^1 s_2^0 h_1^{1-1} h_2^{1-0} + \frac{1}{0!1!1!0!} s_1^0 s_2^1 h_1^{1-0} h_2^{1-1}$$
$$+ \frac{1}{1!1!0!0!} s_1^1 s_2^1 h_1^{1-1} h_2^{1-1}$$
$$= h_1 h_2 + s_1 h_2 + s_2 h_1 + s_1 s_2 = (s_1 + h_1)(s_2 + h_2).$$

Or,

$$(\boldsymbol{s}+\boldsymbol{h})^{(1,1)}=\prod_{i=1}^{2}(s_i+h_i)^{\lambda_i}=\sum_{\kappa_1=0}^{1}\sum_{\kappa_2=0}^{1}\prod_{i=1}^{2}\binom{1}{\kappa_i}s_i^{\kappa_i}h_i^{1-\kappa_i}$$

$$=\sum_{\kappa_1=0}^{1}\frac{1}{\kappa_1!(\lambda_1-\kappa_1)!}s_1^{\kappa_1}h_1^{1-\kappa_1}\sum_{\kappa_2=0}^{1}\frac{1}{\kappa_2!(\lambda_2-\kappa_2)!}s_2^{\kappa_2}h_2^{1-\kappa_2}$$

$$=\left(\frac{1}{0!(1-0)!}s_1^0h_1^1+\frac{1}{1!(1-1)!}s_1^1h_1^0\right)\left(\frac{1}{0!(1-0)!}s_2^0h_2^1+\frac{1}{1!(1-1)!}s_2^1h_2^0\right)$$

$$=(s_1+h_1)(s_2+h_2).$$

Also, $(\boldsymbol{s}+\boldsymbol{h})^{(\lambda_1,\lambda_2)}=(s_1+h_1,s_2+h_2)^{(\lambda_1,\lambda_2)}=(s_1+h_1)^{\lambda_1}(s_2+h_2)^{\lambda_2}=(s_1+h_1)(s_2+h_2).$

Example 2

Let, $\lambda_1=2,\ \lambda_2=0,\ \kappa_1\le 2,\ \kappa_2=0$. Then,

$$(\boldsymbol{s}+\boldsymbol{h})^{(2,0)}=\sum_{\kappa_1=0}^{2}\frac{2!}{\kappa_1!(2-\kappa_1)!}s_1^{\kappa_1}h_1^{2-\kappa_1}$$

$$=\frac{2}{0!2!}s_1^0h_1^{2-0}+\frac{2}{1!1!}s_1^1h_1^{2-1}+\frac{2}{2!0!}s_1^2h_1^{2-2}=(s_1+h_1)^2.$$

Or,

$$(\boldsymbol{s}+\boldsymbol{h})^{(2,0)}=\prod_{i=1}^{2}(s_i+h_i)^{\lambda_i}=\sum_{\kappa_1=0}^{2}\frac{2!}{\kappa_1!(2-\kappa_1)!}s_1^{\kappa_1}h_1^{2-\kappa_1}$$

$$=\left(\frac{2}{0!(2-0)!}s_1^0h_1^2+\frac{2}{1!(2-1)!}s_1^1h_1^1+\frac{2}{2!(2-2)!}s_1^2h_1^0\right)=(s_1+h_1)^2.$$

Also, $(\boldsymbol{s}+\boldsymbol{h})^{(\lambda_1,\lambda_2)}=(s_1+h_1,s_2+h_2)^{(\lambda_1,\lambda_2)}=(s_1+h_1)^{\lambda_1}(s_2+h_2)^{\lambda_2}=(s_1+h_1)^2.$

Example 3

Let, $\boldsymbol{\Delta p}^2=\boldsymbol{h}^T(\boldsymbol{E}\boldsymbol{h}+\boldsymbol{\varepsilon}),\ \partial\boldsymbol{h}^T=[\partial h_1\ldots\partial h_n\ \partial h_0],\ \Delta p=\left|\boldsymbol{\Delta p}^2\right|^{\frac{1}{2}}$ (see, Eq. (1.13) of Chapter III). Then,

$$\frac{\partial\Delta p}{\partial\boldsymbol{h}}=\frac{\partial(\boldsymbol{\Delta p}^2)^{\frac{1}{2}}}{\partial(\boldsymbol{\Delta p}^2)}\frac{\partial\boldsymbol{\Delta p}^2}{\partial\boldsymbol{h}}=\frac{1}{2}\Delta p^{-1}\frac{\partial\boldsymbol{\Delta p}^2}{\partial\boldsymbol{h}}=\left[\boldsymbol{h}^T\boldsymbol{E}+\frac{1}{2}\boldsymbol{\varepsilon}^T\right]\Delta p^{-1}.$$

$$\frac{\partial \Delta p^{-1}}{\partial \boldsymbol{h}} = \frac{\partial (\Delta \boldsymbol{p}^2)^{-\frac{1}{2}}}{\partial (\Delta \boldsymbol{p}^2)} \frac{\partial \Delta \boldsymbol{p}^2}{\partial \boldsymbol{h}} = -\frac{1}{2} \Delta \boldsymbol{p}^{-3} \frac{\partial \Delta \boldsymbol{p}^2}{\partial \boldsymbol{h}} = -\frac{\partial \Delta p}{\partial \boldsymbol{h}} \Delta \boldsymbol{p}^{-2}$$

$$= -\left[\boldsymbol{h}^T \boldsymbol{E} + \frac{1}{2}\boldsymbol{\varepsilon}^T\right] \Delta \boldsymbol{p}^{-3}.$$

$$\frac{\partial^2 \Delta p}{\partial \boldsymbol{h}^T \partial \boldsymbol{h}} = \frac{\partial}{\partial \boldsymbol{h}^T}\left(\frac{\partial \Delta p}{\partial \boldsymbol{h}}\right) = \frac{\partial}{\partial \boldsymbol{h}^T}\left(\left[\boldsymbol{h}^T \boldsymbol{E} + \frac{1}{2}\boldsymbol{\varepsilon}^T\right] \Delta p^{-1}\right)$$

$$= \left\{\frac{\partial\left[\left(\boldsymbol{Eh} + \frac{1}{2}\boldsymbol{\varepsilon}\right)\Delta p^{-1}\right]}{\partial \boldsymbol{h}}\right\}^T = \left[\left(\boldsymbol{Eh} + \frac{1}{2}\boldsymbol{\varepsilon}\right)\frac{\partial \Delta p^{-1}}{\partial \boldsymbol{h}} + \frac{\partial\left[\left(\boldsymbol{Eh} + \frac{1}{2}\boldsymbol{\varepsilon}\right)\right]}{\partial \boldsymbol{h}} \Delta p^{-1}\right]^T$$

$$= \Delta p^{-1}\boldsymbol{E} - \left(\boldsymbol{Eh} + \frac{1}{2}\boldsymbol{\varepsilon}\right)\left(\boldsymbol{h}^T \boldsymbol{E} + \frac{1}{2}\boldsymbol{\varepsilon}^T\right)\Delta \boldsymbol{p}^{-3},$$

where the transpose is not needed in the right-hand-side of the above equation since the matrix within the bracket is symmetric. Also,

$$\frac{\partial^2 \Delta p}{\partial \boldsymbol{h}^T \partial \boldsymbol{h}} = \left[\boldsymbol{E} - \frac{\partial \Delta p}{\partial \boldsymbol{h}^T}\frac{\partial \Delta p}{\partial \boldsymbol{h}}\right]\Delta p^{-1}.$$

For instance, elements of the above matrices are

$$\frac{\partial \Delta p}{\partial h_i} = \left(\sum_{j=1}^{n,0} \varepsilon_{ij} h_j + \frac{1}{2}\varepsilon_i\right)\Delta p^{-1} = \left(\sum_{j=1}^{n} \varepsilon_{ij} h_j + \varepsilon_{0i}\tau + \frac{1}{2}\varepsilon_i\right)\Delta p^{-1} \quad (i = 1, \ldots, n, 0),$$

$$\frac{\partial \Delta p^{-1}}{\partial h_i} = -\left(\sum_{j=1}^{n,0} \varepsilon_{ij} h_j + \frac{1}{2}\varepsilon_i\right)\Delta \boldsymbol{p}^{-3} = -\left(\sum_{j=1}^{n} \varepsilon_{ij} h_j + \varepsilon_{0i}\tau + \frac{1}{2}\varepsilon_i\right)\Delta \boldsymbol{p}^{-3},$$

$$\frac{\partial \Delta p}{\partial \tau} = \left(\sum_{i=1}^{n,0} \varepsilon_{0i} h_i + \frac{1}{2}\varepsilon_0\right)\Delta p^{-1} = \left(\sum_{i=1}^{n} \varepsilon_{0i} h_i + \varepsilon_{00}\tau + \frac{1}{2}\varepsilon_0\right)\Delta p^{-1},$$

$$\frac{\partial^2 \Delta p}{\partial h_i \partial h_j} = \varepsilon_{ij}\Delta p^{-1} - \left(\sum_{k=1}^{n,0} \varepsilon_{ik} h_k + \frac{1}{2}\varepsilon_i\right)\left(\sum_{l=1}^{n,0} \varepsilon_{lj} h_l + \frac{1}{2}\varepsilon_j\right)\Delta \boldsymbol{p}^{-3},$$

$$\frac{\partial^2 \Delta p}{\partial h_i \partial \tau} = \varepsilon_{0i}\Delta p^{-1} - \frac{\partial \Delta p}{\partial h_i}\frac{\partial \Delta p}{\partial \tau}\Delta p^{-1},$$

$$\frac{\partial^2 \Delta p}{\partial h_i^2} = \varepsilon_{ii}\Delta p^{-1} + \left(\sum_{j=1}^{n,0} \varepsilon_{ji} h_j + \frac{1}{2}\varepsilon_i\right)\frac{\partial \Delta p^{-1}}{\partial h_i}.$$

Example 4

Some specific covariance differentiation formulas are as follows,

$$\frac{\partial c_X(\Delta p)}{\partial \boldsymbol{h}} = \frac{\partial \Delta p}{\partial \boldsymbol{h}} \frac{\partial c_X}{\partial \Delta p}, \quad \partial \boldsymbol{h} = [\partial h_1 \ldots \partial h_n \partial h_0]^T$$

$$\frac{\partial^2 c_X(\Delta p)}{\partial \boldsymbol{h}^T \partial \boldsymbol{h}} = \frac{\partial}{\partial \boldsymbol{h}^T}\left(\frac{\partial c_X}{\partial \boldsymbol{h}}\right) = \frac{\partial}{\partial \boldsymbol{h}^T}\left(\frac{\partial \Delta p}{\partial \boldsymbol{h}} \frac{\partial c_X}{\partial \Delta p}\right) = \left[\frac{\partial}{\partial \boldsymbol{h}}\left(\left(\frac{\partial \Delta p}{\partial \boldsymbol{h}}\right)^T \frac{\partial c_X}{\partial \Delta p}\right)\right]^T$$

$$= \left[\frac{\partial}{\partial \boldsymbol{h}}\left(\frac{\partial \Delta p}{\partial \boldsymbol{h}}\right)^T \frac{\partial c_X}{\partial \Delta p} + \left(\frac{\partial \Delta p}{\partial \boldsymbol{h}}\right)^T \frac{\partial}{\partial \boldsymbol{h}}\left(\frac{\partial c_X}{\partial \Delta p}\right)\right]^T = \left[\frac{\partial^2 \Delta p}{\partial \boldsymbol{h} \partial \boldsymbol{h}^T} \frac{\partial c_X}{\partial \Delta p} + \left(\frac{\partial \Delta p}{\partial \boldsymbol{h}}\right)^T \frac{\partial \Delta p}{\partial \boldsymbol{h}} \frac{\partial^2 c_X}{\partial \Delta p^2}\right]^T$$

$$= \frac{\partial^2 \Delta p}{\partial \boldsymbol{h}^T \partial \boldsymbol{h}} \frac{\partial c_X}{\partial \Delta p} + \left(\frac{\partial \Delta p}{\partial \boldsymbol{h}}\right)^T \frac{\partial \Delta p}{\partial \boldsymbol{h}} \frac{\partial^2 c_X}{\partial \Delta p^2}$$

$$\frac{\partial c_X(\Delta p)}{\partial h_i} = \frac{\partial \Delta p}{\partial h_i} \frac{\partial c_X}{\partial \Delta p} \quad (i = 1, \ldots, n, 0),$$

$$\frac{\partial^2 c_X(\Delta p)}{\partial h_i^2} = \frac{\partial^2 \Delta p}{\partial h_i^2} \frac{\partial c_X}{\partial \Delta p} + \frac{\partial \Delta p}{\partial h_i} \frac{\partial}{\partial h_i}\left(\frac{\partial c_X}{\partial \Delta p}\right) = \frac{\partial^2 \Delta p}{\partial h_i^2} \frac{\partial c_X}{\partial \Delta p} + \left(\frac{\partial \Delta p}{\partial h_i}\right)^2 \left(\frac{\partial^2 c_X}{\partial \Delta p^2}\right),$$

$$\frac{\partial^2 c_X(\Delta p)}{\partial h_i \partial h_j} = \frac{\partial^2 \Delta p}{\partial h_i \partial h_j} \frac{\partial c_X}{\partial \Delta p} + \frac{\partial \Delta p}{\partial h_i} \frac{\partial \Delta p}{\partial h_j} \frac{\partial^2 c_X}{\partial \Delta p^2} = \frac{\partial^2 \Delta p}{\partial h_i \partial h_j} \frac{\partial c_X}{\partial \Delta p} + \frac{\partial \Delta p}{\partial h_i} \frac{\partial^2 c_X}{\partial h_j \partial \Delta p}.$$

Example 5

Let $\breve{\boldsymbol{p}} = \Delta \boldsymbol{p} = (h_1, \ldots, h_n, h_0)$, with $\breve{s}_i = h_i = s_i - s_i'$ $(i = 1, \ldots, n, 0)$, so that $f\left(\breve{\boldsymbol{p}}\right) = c_X(\Delta \boldsymbol{p}) = c_X(h_1, \ldots, h_n, h_0)$. Then, the general expressions in Section 8 above reduce to

$$\frac{\partial}{\partial s_i} c_X(\Delta \boldsymbol{p}) = \sum_{j=1}^{n,0} \frac{\partial c_X(h_1, \ldots, h_n, h_0)}{\partial h_j} \frac{\partial h_j}{\partial s_i} = \frac{\partial c_X(\Delta \boldsymbol{p})}{\partial h_i} \frac{\partial (s_i - s_i')}{\partial s_i} = \frac{\partial c_X(\Delta \boldsymbol{p})}{\partial h_i},$$

$$\frac{\partial^2}{\partial s_i \partial s_j} c_X(\Delta \boldsymbol{p}) = \sum_{l=1}^{n,0} \sum_{k=1}^{n,0} \frac{\partial h_k}{\partial s_i} \frac{\partial^2 c_X(h_1, \ldots, h_n, h_0)}{\partial h_k \partial h_l} \frac{\partial h_l}{\partial s_j} = \frac{\partial^2 c_X(\Delta \boldsymbol{p})}{\partial h_i^2},$$

$$\frac{\partial^2}{\partial s_i \partial s'_j} c_X(\Delta \boldsymbol{p}) = \sum_{l=1}^{n,0} \sum_{k=1}^{n,0} \frac{\partial h_k}{\partial s_i} \frac{\partial^2 c_X(h_1, \ldots, h_n, h_0)}{\partial h_k \partial h_l} \frac{\partial h_l}{\partial s'_j} = -\frac{\partial^2 c_X(\Delta p)}{\partial h_i^2}.$$

Index

'*Note*: Page numbers followed by "f" indicate figures, "t" indicate tables.'

D

E

F

L

M

N

Printed in the United States
By Bookmasters